S0-BKY-220

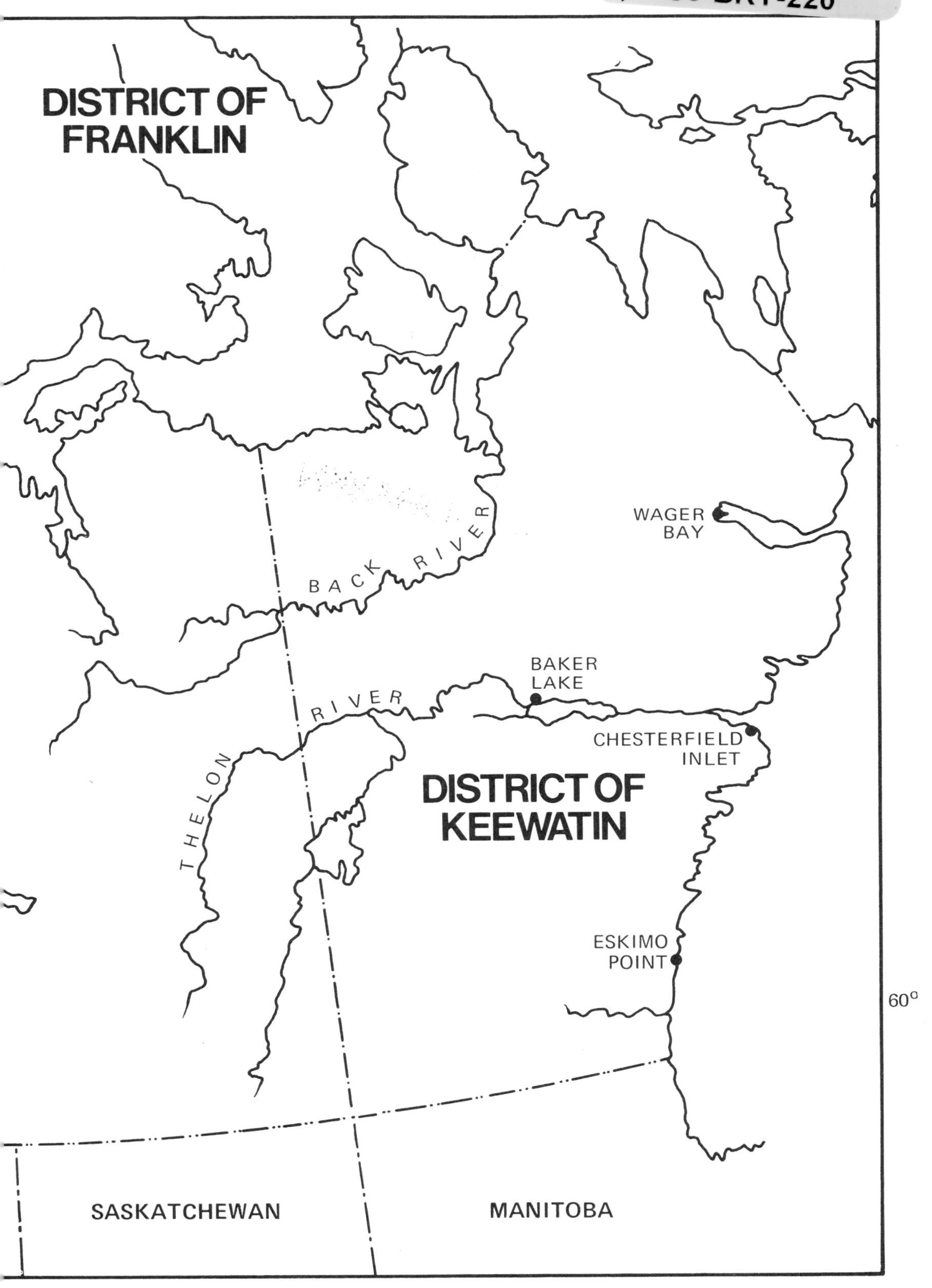
DISTRICT OF FRANKLIN
BACK RIVER
WAGER BAY
BAKER LAKE
RIVER
THELON
CHESTERFIELD INLET
DISTRICT OF KEEWATIN
ESKIMO POINT
60°
SASKATCHEWAN
MANITOBA

Vascular Plants of Continental Northwest Territories, Canada

National Museum of Natural Sciences

Published by the
National Museums of Canada

Musée national des Sciences naturelles

Publié par les
Musées nationaux du Canada

Vascular Plants of Continental Northwest Territories, Canada

A. Erling Porsild†
Curator Emeritus
National Herbarium of Canada
National Museums of Canada

William J. Cody
Curator, Vascular Plant Herbarium
Biosystematics Research Institute
Canada Department of Agriculture

National Museum of Natural Sciences
National Museums of Canada

National Museum of Natural Sciences
National Museums of Canada
Ottawa, Canada

Catalogue No. NM92-71/1979

Musée national des Sciences naturelles
Musées nationaux du Canada
Ottawa, Canada

N° de catalogue NM92-71/1979

Printed in Canada

Imprimé au Canada

ISBN 0-660-00119-5

Contents

Alf Erling Porsild was born in Copenhagen, Denmark, in 1901. He spent his early years in West Greenland, where his father, Morten Pedersen Porsild, was director of the Danish Biological Station at Disko. Later he was sent to boarding school in Denmark. He received the Ph.D. degree from the University of Copenhagen for which his 1955 publication, *The Vascular plants of the Western Canadian Archipelago,* was accepted as the thesis.

During the years 1922 to 1925, Dr. Porsild was assistant botanist at the Danish Biological Station. In 1926 he undertook a study of reindeer-grazing in Arctic Canada and Alaska for the Government of Canada. These studies led to the establishment, under his direction, of the Reindeer Experiment Station on the east branch of the Mackenzie Delta. His reindeer-oriented botanical studies through ten summers and seven winters took him by canoe and dog team from the Richardson Mountains west of the Mackenzie River Delta eastwards to the Coppermine River, about Great Bear Lake, and in 1930 to Central District of Keewatin. His longest survey was a winter trip with his brother Robert Thorbirn Porsild, from Nome, Alaska to Aklavik in the Mackenzie River Delta. In the years that followed these studies, Dr. Porsild carried out field studies in many parts of Canada and elsewhere: Labrador (1937), west Greenland (1940-1943), Alaska and the Yukon Territory (1944), Siberia and Russia (1945), Rocky Mountains in Alberta (1945-1946, 1951, 1955-1960), Mackenzie River (1947), Banks and Victoria Islands (1949), Axel Heiberg Island (1953), western United States (1948, 1961), Hudson Bay lowlands (1956-1957) and Mexico (1963).

Dr. Porsild's publications, which number over one hundred, have not only been on botanical subjects, but also about reindeer, the native people of the arctic regions, birds, mammals, and physiographic features. Major publications of book length, dealt with the floras of Alaska, the Yukon Territory, the Northwest Territories and the Rocky Mountains. More than eighty plant taxa are based on Dr. Porsild's collections, most of which he described and published personally.

Following his return from the north in 1936 he was appointed Acting Chief Botanist of the National Museum of Canada. From 1940 to 1943 he spent most of his time in Greenland as Canadian Vice-Consul to Greenland, and then returned to Canada to continue his full-time duties at the Museum in Ottawa. He became Chief Botanist in the National Herbarium in 1946, which post he held until his retirement in January of 1967. Retirement, however, meant only a release from administrative duties, because he continued his botanical studies actively as Curator Emeritus.

Dr. Porsild received honours from universities and learned societies, not only from Canada and his native country, Denmark, but also from Sweden, Finland, Norway and the United States. For his services as a civilian during the Second World War, he was admitted to Membership of the Order of the British Empire (M.B.E.).

It is most unfortunate that Dr. Porsild did not live to see the fruition of his last work, the *Vascular Plants of Continental Northwest Territories.* He died while on holiday in Vienna, in November 1977.

William James Cody was born in Hamilton, Ontario in 1922. He received his education in that city, gaining his B.A. from McMaster University in 1946.

In 1946, Mr. Cody joined the Botany and Plant Pathology Division of the Canada Department of Agriculture as a botanist and in 1959 he was appointed Curator of the Vascular Plant Herbarium.

Mr. Cody's field work in northern Canada began in 1948 when he was sent to Southampton Island, at the north end of Hudson Bay, to study biting fly habitats, a joint project with the Entomology Division of Canada Department of Agriculture, and in cooperation with the Defence Research Board, Canada Department of National Defence. These studies were continued at Yellowknife (1949), Fort Smith (1950), Big Delta, Alaska (1951), Norman Wells (1953) and Fort Smith (1955). In 1957 and again in 1963 he examined the terrain of the Reindeer Grazing Preserve, east of the Mackenzie River Delta, to ascertain what changes might have taken place in the vegetation following the introduction of reindeer to the area. In 1961, he conducted studies of the vegetation adjacent to the Liard River from the British Columbia border downstream to Fort Simpson, and again in 1965 he engaged in similar studies along the Slave River between Fort Smith and Great Slave Lake. In 1966 he studied the vegetation of that part of the District of Mackenzie east of the Slave River between the 60° parallel and Great Slave Lake. In 1970, 1971 and 1972 he took part in field studies in the Mackenzie Mountains and various other localities in southern District of Mackenzie, to assess sites that had been recommended for preservation in the International Biological Program/Conservation of Terrestrial Habitats (IBP/CT) project. Also in 1971, he was a member of the Northern Roads Task Force that examined the impact of road-building on terrain and vegetation in parts of northern Canada. In addition to these northern activities, studies particularly in relation to Canadian ferns, have taken him at various times to the Maritime Provinces, through much of Ontario and Quebec, and to British Columbia and western Alberta.

Mr. Cody has written extensively on the floristics of the Canadian north. Over forty of his research publications deal with the plants of that most important region.

Abstract

Vascular Plants of Continental Northwest Territories, Canada is a comprehensive treatment of the ferns and flowering plants of the large area lying between Hudson Bay and the Yukon Territory border, and 60° N latitude and the Arctic coast.

The authors have provided keys, descriptions, and habitat and distributional information, not only for the 1112 species found in the region, but also for a number of species which might be expected to occur there. There are distribution maps for all the native species, and 978 line drawings. The work also includes a description of the area and its six phytogeographic zones, a tabulation of families, genera and species, a history of botanical collecting, a selected bibliography, a glossary and an index.

Résumé

L'ouvrage *Vascular Plants of Continental Northwest Territories, Canada* constitue un recensement exhaustif des fougères et des plantes florifères de la partie continentale des Territoires du Nord-Ouest.

Les auteurs y donnent des descriptions, des clefs d'identification, ainsi que des renseignements sur l'habitat et l'aire de distribution, non seulement des 1112 espèces découvertes dans la région mais également d'un certain nombre d'espèces qui pourraient s'y trouver. L'ouvrage comporte une carte de distribution pour chaque espèce indigène, 978 dessins au trait, une description de la région et de ses six zones phytogéographiques, un répertoire des familles, des genres et des espèces, une histoire de l'herborisation, une bibliographie, un lexique et un index.

Introduction

The present work is intended as a guide or manual to the species and major geographical races of flowering plants and ferns of the part of Canada lying between the 60th parallel and the Arctic Ocean, and extending from the Yukon-Mackenzie border to the west coast of Hudson Bay.

As thus defined this area comprises roughly three-quarter million square miles, or 1.6 million square kilometers, approximately the combined areas of Manitoba, Saskatchewan and Alberta.

In Pleistocene time part of northern North America was deeply buried by ice as is most of Greenland today. Today remnants of this ice-sheet are still present on some of the islands of the Arctic Archipelago and in the Mackenzie Mountains.

The Polar tree line, cutting diagonally from the Mackenzie Delta to the west shore of Hudson Bay in latitude 60°, "divides" this area into two congruent triangles of which the western is wooded, and the eastern treeless. Due to climatic fluctuations its position has changed periodically as demonstrated by plant remains preserved in peat deposits, and by the presence of numerous woodland plants now localized in favoured situations well beyond the present limits of forest and, conversely, of isolated pockets of tundra species that have survived well within the forested region on rocky exposures, or in bogs where competition from woodland species has been slight.

Another important but less distinct floristic boundary follows the contact of the mainly calcareous Palaeozoic rocks of the Mackenzie River basin to the West, and the predominantly acidic rocks of the Precambrian Shield to the East. This boundary runs south from Langton Bay on the arctic coast across the eastern arm of Great Bear Lake, and slightly beyond the eastern arm of Great Slave Lake. Finally, the Mackenzie River valley itself is an important phytogeographic boundary across which relatively few plants of Cordilleran or Amphi-Beringian affinity have penetrated.

Throughout the Districts of Mackenzie and Keewatin the climate is continental, with cold winters but relatively warm summers. Everywhere, except in the proximity of rivers, or large bodies of water, the subsoil is permanently frozen, and probably to great depths.

The annual thaw of the surface soil, or the so-called "active layer" varies greatly with the angle of exposure to the sun, with the degree of shading, as in forested areas, and with the texture and water content of the soil; thus, on south-facing slopes, or in well-drained sand or gravel, the seasonal thaw may be relatively deep, whereas in wet, peaty soils the summer thaw penetrates only a short distance. However, the seasonal depth of thaw does not greatly affect plant growth. Thus, in the upper Mackenzie Delta well north of the Arctic Circle, tall spruce, poplar, aspen or larch forest grows on soils that, by the end of the summer, have thawed to a depth of 60 cm or less.

In the District of Mackenzie white spruce *(Picea glauca)* everywhere forms the northern tree-line, whereas in the District of Keewatin the tree-line is commonly formed by the larch *(Larix laricina)* and black spruce *(Picea mariana).*

Thus far few attempts have been made to sub-divide continental Northwest Territories into phytogeographical provinces beyond the broad classification of "tundra" or, "barren grounds" and "woodland". However, with our present knowledge of the flora we are now able to recognize six phytogeographical provinces (See map 1) tentatively defined as follows:

1 The Mackenzie Mountains — a vast mountain fastness extending from the Yukon border east to the Mackenzie Valley, and from the Liard River gap near the British Columbia border north to the Peel River gap southwest of the Mackenzie River delta. This highland is deeply dissected by forested river valleys of an average elevation of from 700 to 1200 m, separated by mountain chains and peaks reaching well above timberline, which is generally found at elevations between 1000 and 1300 m. The flora is comparatively varied and rich in composition and very similar to that of southeastern Yukon (Porsild, 1951 a) and, like it, includes a substantial number of Cordilleran species that, generally, are concentrated in the alpine zone, whereas the valley flora is composed mainly of wide-ranging species of the Boreal Forest region.

2 Our second, and much smaller province, includes the east slope of the Richardson Mountains, and extends north from the Peel River gap to the Arctic Coast. Its mountains are of more gentle relief than those to the south of it, but, owing to its less favourable climate, forests are restricted to the sheltered mountain valleys below 700 m elevation, and are lacking altogether on the Arctic Slope. Species of Cordilleran affinity are few, and in the alpine zone are largely "replaced" by Amphi-Beringian species, of which the majority here find their eastern limits.

3 Our third province includes the Mackenzie River delta and the coastal plain between the lower Mackenzie and Anderson rivers.

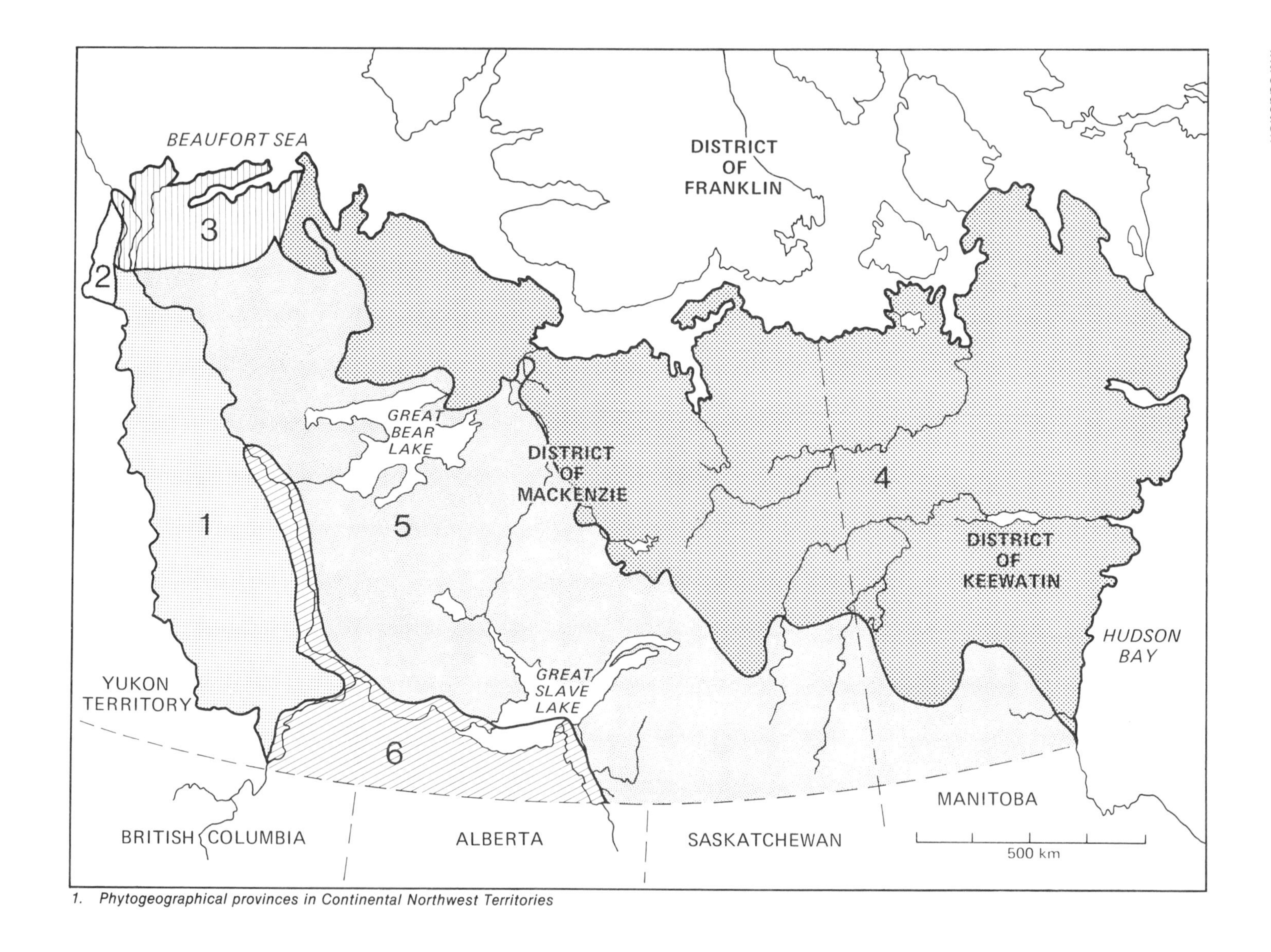

1. *Phytogeographical provinces in Continental Northwest Territories*

Owing to the ameliorating influence of the waters of the great north flowing Mackenzie, the flora of its delta is surprisingly rich in southern species, and may be considered a northern extension of that of the upper Mackenzie Basin; near the Arctic Coast southern species mingle with truly arctic or maritime species. A few Amphi-Beringian species are found on the higher and non-alluvial islands north and east of the delta proper, and on the Caribou Hills which flank the East Branch of the Mackenzie Delta.

4 Our fourth and largest province comprises the treeless northern portion of the District of Mackenzie and all but the southern portion of Keewatin, and thus lies wholly within the Precambrian Shield area. Except for small pockets of woodland species that may persist in protected spots along the river valleys, the flora is composed mainly of wide-ranging, low-arctic tundra species. In eastern Keewatin a small number Amphi-Atlantic species reach their western limits.

5 Like the preceding, our fifth and second largest province is low relief and comprises the wooded western portion of the Precambrian Shield area and the Mackenzie lowlands between the Shield and the Franklin Mountains or peneplain that form the eastern flank of the Mackenzie River valley.

6 The flora of our sixth and last province is essentially similar to that of northern Alberta and Saskatchewan. In it we find the largest concentration of southern and grassland species. Also, it is the only province in which is found timber of commercial size, and land of potential agricultural value. This province embraces an area of low relief extending from the south shore of Great Slave Lake to the Alberta border, and from the Slave River valley to the Liard River, and, as a narrow corridor, extends north along the valley of the Mackenzie River to some distance beyond the 65th parallel.

In the introduction to "Materials for a Flora of Continental Northwest Territories"(Sargentia IV: 1-79, 1943) the senior author briefly outlined the history of botanical exploration of Canada between latitude 60° and the Arctic Coast. The vascular flora, as then known, from the Mackenzie and Keewatin districts comprised 731 taxa of which 345 were dealt with in the main body of that work. Of this 731, no less than one hundred taxa had not then been reported from the area; among these were 14 that were then "new" to the flora of Canada, while a further 17 species and varieties were new to science.

In his "Botany of southwestern Mackenzie" Hugh M. Raup (1947) reported 725 species and varieties of flowering plants and ferns known to him from the southern part of the District of Mackenzie.

In preparation for the present work we published a "Checklist of the Vascular Plants of Continental Northwest Territories, Canada" (Porsild and Cody 1968). Among the 1230 taxa listed we included 76, marked by an asterisk denoting that they had not, as yet, been reported but that we fully expected to occur in that area. A good many of those thus listed have actually been discovered since 1968, and voucher specimens are now deposited in the herbaria of the National Museums of Canada (CAN) or in that of the Department of Agriculture, Ottawa (DAO).

In the present work 978 species are illustrated by line drawings. Among those representing arctic species, users of this work will recognize the excellent drawings by Mrs. Dagny Lid of Oslo, Norway, originally prepared to illustrate the senior author's "Illustrated Flora of the Canadian Arctic Archipelago". The remaining drawings were made especially for the present work mainly by the late Mrs. Lynne Hughes (Lynne Bartosch) but also by Brenda Carter while employed as artists by the National Museums of Canada.

Distribution maps are presented here for the species which are native to the continental Northwest Territories. For the area treated by the flora, these maps are as detailed as could be made from the many specimens examined by the authors. For other parts of Canada, Alaska and Greenland the maps are in no way complete but do show a broad picture of the overall Canadian distribution.

Families, genera and species of vascular plants in the flora of continental Northwest Territories

FAMILIES	GENERA	SPECIES
Ophioglossaceae	1	4
Polypodiaceae	9	18
Equisetaceae	1	8
Lycopodiaceae	1	6
Selaginellaceae	1	2
Isoetaceae	1	1
Pinaceae	5	8
Typhaceae	1	1
Sparganiaceae	1	5
Potamogetonaceae	3	18
Scheuchzeriaceae	2	3
Alismaceae	1	1
Gramineae	38	108
Cyperaceae	6	119
Araceae	2	2
Lemnaceae	1	2
Juncaceae	2	22
Liliaceae	9	12
Iridaceae	1	1
Orchidaceae	10	18
Salicaceae	2	44
Myricaceae	1	1
Betulaceae	2	6
Urticaceae	1	1
Santalaceae	1	1
Polygonaceae	4	19
Chenopodiaceae	8	16
Amaranthaceae	1	1
Portulacaceae	2	3
Caryophyllaceae	13	51
Ceratophyllaceae	1	1
Nymphaceae	2	3
Ranunculaceae	10	40
Papaveraceae	1	5
Fumariaceae	1	3
Cruciferae	23	80
Sarraceniaceae	1	1
Droseraceae	1	3
Crassulaceae	2	2
Saxifragaceae	8	35
Rosaceae	14	52
Leguminosae	9	41
Geraniaceae	1	2
Linaceae	1	2
Callitrichaceae	1	3
Empetraceae	1	1
Aceraceae	1	1
Balsaminaceae	1	1
Elatinaceae	1	1
Cistaceae	1	1
Violaceae	1	8
Elaeagnaceae	2	2
Onagraceae	2	11
Haloragaceae	2	6
Araliaceae	1	1
Umbelliferae	8	10
Cornaceae	1	3
Pyrolaceae	2	6
Ericaceae	11	22
Diapensiaceae	1	2
Primulaceae	7	14
Plumbaginaceae	1	1
Gentianaceae	2	11
Menyanthaceae	1	1
Apocynaceae	1	2
Polemoniaceae	3	6
Hydrophyllaceae	1	1
Boraginaceae	5	8
Labiatae	9	9
Scrophulariaceae	9	30
Orobanchaceae	1	1
Lentibulariaceae	2	6
Plantaginaceae	1	4
Rubiaceae	1	7
Caprifoliaceae	4	5
Adoxaceae	1	1
Valerianaceae	1	3
Campanulaceae	1	4
Lobeliaceae	1	2
Compositae	37	146
	332	1113

Many people have gathered specimens of vascular plants throughout the vast region covered by this flora, and as can be readily understood, for some parts there is a much better coverage than others. The early travellers brought back examples of the flora from a new and unexplored country for scientific study. Their expeditions by canoe, and on foot were most arduous, and it is a wonder that any specimens were gathered at all. Thus Hooker's Flora Boreali-Americana (1829-40) was based in part on the collections of John Richardson. These collections and those of many others, such as the early geologists made, were incidental to some other work which was their main reason for travelling in the territories. More recently however, specimens have been gathered to document detailed ecological, environmental or wildlife studies, or solely to gain information on the distribution of the various plant species within the territory. Some specimens have been gathered just so that a visitor might supplement memories of a never-to-be-forgotten trip.

As can readily be gathered, the tempo of botanical exploration in the Continental Northwest Territories has been on the upswing, particularly in the last twenty years, and more especially in the last ten years. The Mackenzie Valley Pipeline surveys have employed teams of investigators throughout the length of the river. Our northern regions too have now opened up for detailed studies by students working on masters and doctoral theses, primarily in ecology. The advent of the bush plane and more recently the helicopter has opened up even the most remote areas to the ready access of the scientific investigator. It is now no longer possible to keep track of every botanical collector in the region. There are still many blank spots in our knowledge of plant distribution in the Continental Northwest Territories as can readily be observed when almost every collecting expedition brings back interesting and important new records and extensions of range, but much has been learned since Hearne, Mackenzie, Franklin and Richardson toiled through the wilderness on their journeys of exploration.

Following is a list of collections which have been made in the Continental Northwest Territories. The list is chronological, and thus for ease in searching out individual collectors, their names have been italicized. Herbarium abbreviations of herbaria where specimens are located are those given in Holmgrem and Keuken (1974), but duplicate specimens may, in addition, be found in several other herbaria. Not all of these specimens have been examined in the preparation of this flora, but certainly the more important have been seen. Additional information on some collectors can be found in Polunin (1940), Porsild (1943) and Raup (1935, 1936, 1947), for the adjacent Yukon Territory and Alaska, in Hultén (1940).

The bibliography which follows is intended as a guide to show the use to which various collections have been put, and is not intended to be all-inclusive. Other more extensive bibliographies related to the region can be found in Porsild and Cody (1968), Polunin (1940), and Raup (1935, 1946 and 1947).

In 1819 the British Government dispatched an expedition under *John Franklin* (1823) to trace the arctic coastline of North America east of the Coppermine River. He was accompanied, among others, by *John Richardson*.

In the spring of 1821 they proceeded northward to Point Lake from their winter camp at Fort Enterprise on Winter Lake, and then down the Coppermine River to the Arctic Coast which they finally reached on 18 July. Specimens which had been collected up to this time were then sent south by the same route with Mr. Wentzel, and some of the voyageurs, while the rest of the party went eastward along the Arctic Coast a distance of more than 500 miles to Cape Turnagain, in the process exploring Bathurst Inlet; they then moved up the Hood River and proceeded across country towards Point Lake, and it was near here on 21 September that "Dr. Richardson was obliged to deposit his specimens of plants and minerals, collected on the seacoast, being unable to carry them any farther". The expedition had an absolutely awful time on their way south from the coast: cold, wet, near drowning and freezing, starvation, cannibalism, murder, execution.

At the end of the first edition of Franklin's Journal (1823) there are several appendixes. No. 7 paginated 729-768, is entitled "Botanical Appendix" and was written by *John Richardson* (K). It was also paginated separately, hence page references for descriptions of new taxa described by Richardson and Robert Brown may be given differently by subsequent authors.

The following quotation is from Richardson's introduction:

"The following list of plants is not offered, as containing anything like a full catalogue of the Flora of the country through which we travelled. During our summer journeys, only a small portion of time could be allotted to Botanical researches, and the constant and more important duties of the other officers prevented them from aiding me in collecting objects of Natural History, which they were otherwise anxious to do. Under such circumstances, a large proportion of plants must have escaped our notice, and the disasters attending our return across the Barren Grounds from the sea-coast, caused us to leave behind the whole collection made during the summer of 1821, with the exception of a few plants collected during the descent of the Copper-Mine River, which were intrusted to Mr. Wentzel's care when he left us. The part of the collection, which is lost, contained some plants, which I deemed to be new or curious."

"In drawing up the list, imperfect as it is, I have received much assistance from able botanists. To Mr. [Robert] Brown, I am under the greatest obligations, not only for the liberal use of the Herbarium and Library, which, so happily for science, have been placed in his possession; but, also for the friendly manner in which he aided my researches, and condescended to solve the doubts so frequently presenting themselves to one little versant in these pursuits. In addition to this general assistance, he kindly superintended the botanical drawings, and has enriched my catalogue with the lists of the Cyperoideae (including the Carices) the Gramineae, Junci and Filices, and, with the accounts of the genera Eutoca [Phacelia], Heuchera, and Crytogramma."

In the catalogue which follows, Richardson used abbreviations to save space. The abbreviations which concern us are:

(W.) Denotes the wooded country from latitude 54° to 64° north.
(B.) Denotes the Barren Grounds from Point Lake to the Arctic Sea.
(A.) Denotes the Arctic Sea-Coast.

These designations were certainly adequate for describing a country which was otherwise botanically unknown, but for those of us interested in pin-pointing actual collections to delimit distributions, they now pose problems. The (W.) thus should not be interpreted as meaning a plant occurred as far north as 64° north, but only that Richardson found it somewhere in the woody country between 54° and 64°. Indeed, these designations have been the cause of some unwarranted ranges given by monographers, which must now be discounted.

In 1825 the British Government again sent out an expedition under *John Franklin* (1828). This expedition was directed to trace the Arctic Coast between the Coppermine River and the Mackenzie River and also the shoreline west of the Mackenzie River to the northwest extremity of America. The officers for this trip were *John Franklin, George Back, John Richardson, Ernest Kendall* and *Thomas Drummond*. Drummond, the Assistant Naturalist, was left at Cumberland House on the Saskatchewan River in order that he might make collections "in the vicinity of the Rocky Mountains", while the rest of the expedition was working in the north.

On 21 June 1826 the bulk of the expedition left Fort Franklin, proceeded down the Bear River and then the Mackenzie River to Point Separation, the upper end of the Delta. They then separated into two groups, *Franklin* and *Back* going down the west side of the delta to the Arctic Coast and thence west along the shore past Herschel Island to about longitude 149°W and then retraced their way back to Fort Franklin, where they arrived on 21 September, 1826.

Richardson who was in charge of the Eastern Detachment went down the eastern side of the Delta, worked his way eastward along the Arctic Coast to the Coppermine River and returned to the east end of Great Bear Lake overland, where they were met by canoes which had been sent out for them, and arrived back at Fort Franklin on 1 September.

Franklin in May 1845, left England with the ill-fated "Erebus" and "Terror" in search of a Northwest Passage. Because of the concern in England for his expedition, several relief expeditions were sent out. *John Richardson* led an overland expedition in the years 1847 to 1849.

This party entered what is now the District of Mackenzie on 15 July 1848, when they completed the Portage of the Drowned on the Slave River, continued down the river following the route of the second Franklin expedition to Fort Norman where they arrived on 26 July, then on down the Mackenzie River and reached Cape Bathurst on the Arctic Coast east of the Mackenzie Delta on 4 August. Coppermine was reached about 3 September. The party then walked up the route of the Coppermine River. On 11 September Richardson wrote "I deposited my packet of dried plants and some books in a tree, intending to send for them in the winter". This was along the Kendall River near its juncture with the Coppermine River. They then went overland to the Dease River and descended to its mouth on Great Bear Lake and thence a few miles along this line to Fort Confidence, where they arrived on 15 September. Buildings were being prepared for them by others sent from the west to meet them, and it was here the party spent the winter. On 7 May, 1849, Richardson left Fort Confidence for Fort Franklin where he arrived 12 May, then to Fort

Simpson arriving 25 May, thence to Fort Resolution (11 July), Fort Chipewyan (19 July) and finally arrived in England on 6 November.

Unfortunately Richardson's search party found no trace of the ships "Erebus" and "Terror" or their crews. His journal (1851) does however contain considerable information on the country through which he travelled, and there is an appendix "on the geographical distribution of plants in the country north of the 49th parallel of latitude". It would appear that Richardson did send back for the specimens he left in a tree near Kendall River because Francis Boott in a letter giving identifications of *Carex* species collected on this "excursion", lists a number of species collected as "Arctic Sea Coast".

Of the specimens collected by *Richardson*, those from his first two expeditions were by far the most important, because they were used by W.J. Hooker (1829-1840) in writing Flora Boreali — Americana. The first set of Richardson's specimens (including types of taxa described by himself, Hooker and Robert Brown) is preserved in the herbarium at Kew (K). Some "duplicates" are at Gray Herbarium (GH), New York Botanical Garden (NY) and National Museums of Canada (CAN).

Also in 1838, *Peter Dease* and *Thomas Simpson* of the Hudson's Bay Company made a trip along the Arctic Coast from Coppermine east to near Boothia Peninsula (K). An appendix listing 150 plants collected is given in Simpson's Narrative (1843).

Captain W.J.S. Pullen of the British navy collected a few plants along the Arctic Coast in 1850 (K) as did *Captain McClure*. *Pullen* also collected a few plants along the Mackenzie River in 1852, following his unsuccessful search for the Franklin Expedition (Selman 1852-7).

Between the years 1853 and 1860 a trader by the name of *McTavish* made a number of collections along the Mackenzie River (CAN [Herb Lawson]).

James Anderson another trader, reputedly made collections along the Mackenzie River in 1855 (MTMG). The accuracy of his localities is discussed by Boivin (1972). According to B. Boivin (personal communication) there may have been two individuals by the name of *James Anderson*, one who collected along the Mackenzie and went down the Great Fish (Back) River in 1852-55 and another who collected on Great Slave Lake in 1876 (MTMG, MT).

In 1859-60 *Robert Kennicot*, an American naturalist worked in the western part of Great Slave Lake, along the Liard River, and then descended the Mackenzie River to the Peel River which he ascended on his way to Alaska. Unfortunately his death in 1866 precluded his publishing any account of his specimens (NY, F, US — The first set of his specimens was destroyed in the great Chicago fire of 1871).

In 1860 *B.R. Ross*, a trader, collected specimens at Fort Simpson and Fort Norman on the Mackenzie River (US) (Ross 1862).

In 1861 both *J.S. Onion* and *W.L. Hardisty* gathered specimens along the Mackenzie River (US).

In 1888 and 1889, *R.G. McConnell* gathered some specimens at Fort Good Hope on the lower Mackenzie River (CAN).

In 1892 *Miss Elizabeth Taylor* collected plants along the Mackenzie River. Her locality data and dates of collection are excellent. She did not publish on her collections and Raup (1947) was the first to systematically record them (CAN, GH, NY).

In 1893 *James W. Tyrrell* accompanied his brother Joseph Burr Tyrrell on an exploration of the hitherto unknown terrain from the east end of Lake Athabaska north through the barren grounds to Baker Lake and Chesterfield Inlet. *James Tyrrell* gathered many plants along the route of their journeys, and an annotated list of his collections (CAN) was published by J.B. Tyrrell (1898) as an appendix to a geological report.

In 1893-4 *G. Comer* made some collections of plants at Whale Point and Depot Island and vicinity, on Hudson Bay (G, K). Also in 1893-94 *Frank Russell*, a zoologist from the University of Iowa, gathered specimens about Great Slave Lake and along the Mackenzie River (IA, GH, MT) (Russell 1898).

From 1893 to 1900 *Rev. I.O. Stringer* made collections along the Arctic Coast of the Yukon and Mackenzie Territories from Herschel Island to Warren Point (CAN, TRT).

In 1899 *Robert Bell*, Geological Survey of Canada made a few collections about Great Slave Lake while conducting the first comprehensive geological study of that region (CAN).

In 1900 *J.W. Tyrrell* conducted an exploratory survey of the country between Great Slave Lake and Hudson Bay (CAN) and *J.M. Bell* gathered specimens in the vicinity of Great Bear Lake (CAN).

In 1900 *E.A.* and *A.E. Preble* conducted a biological investigation in eastern District of Keewatin (US) (E.A. Preble 1902).

In 1903 *A.E. Preble* and *Merritt Cary* collected along the south shore of Great Slave Lake and down the Mackenzie River as far north as Wrigley. At the same time *E.A. Preble* crossed the terrain from the north arm of Great Slave Lake to Great Bear Lake, down the Bear River and then up the Mackenzie to Fort Simpson, where he spent the winter. The following summer he worked along the Mackenzie River as far north as the Peel River. *E.A. Preble* (1908) published an extensive biological study of the area which they surveyed. This included an annotated list of trees and shrubs (US).

In 1904 *L.E. Borden* made collections at Fullerton on the Hudson Bay shore (CAN).

In 1910 *J.M. Macoun*, son of the pioneer Canadian botanist John Macoun, made extensive collections on the Hudson Bay shore at Wager Bay, Fullerton, Rankin Inlet and Cape Eskimo (CAN).

In 1911 *W. Jones* collected a few specimens at Arctic Red River on the lower Mackenzie River (CAN).

During the years 1913-1918 the Canadian Arctic Expedition worked along the Arctic Coast between the Alaska border and Bathurst Inlet. Members of this expedition, chiefly *Frits Johansen*, but also *R.M. Anderson, J.R. Cox* and *J.J. O'Neil*, collected plants (CAN). Their specimens were identified in large part by J.M. Macoun and are reported upon by Macoun and Holm (1921). Also included in this report were records of collections made in this area by *Rev. Isaac O. Stringer, Rev. H. Girling* and *Captain Joseph F. Bernard*.

In 1914 *Francis Harper* (1931), who accompanied *Charles Camsell* on the Geological Survey of Canada expedition of that year, made plant collections in the Tazin and Taltson (Rocher) river region south of Great Slave Lake, east of the Slave River (CAN).

In 1916, *C.F. Howe* collected some specimens in the vicinity of Great Slave Lake (ALTA). Also in 1916, *Rev. H. Girling* made some collections at Clifton Point on the south side of Amundsen Gulf (CAN).

In 1919 *E.M. Kindle*, Geological Survey of Canada, made collections in the delta of the Slave River (CAN) (Kindle 1928).

In 1922 *C.L. Crickmay* who was an assistant on a Geological Survey of Canada expedition which worked along the lower Liard and upper Mackenzie rivers made collections in that area (CAN).

Other geologists who collected about this time in the Great Slave Lake and Mackenzie River region were *C.F. Howe* (1916), *R.A. Brooke* (1919), *G.S. Hume* (1920) and *R.F. Bedford* (1925) (CAN).

In 1922-3 *K. Birket-Smith* made collections at Wager Inlet, Depot Island, Chesterfield, Eskimo Point and Tikeraijualas, south of Eskimo Point. He was one of the members of the "Fifth Thule" Expedition (C) (Grøntved 1936; Polunin 1940).

In 1922, 1923 *Peter Freuchen*, also a member of the "Fifth Thule" Expedition made collections at Wager Bay, Roes Welcome south of Wager Bay, Whale Point, Walrus Island near Cape Fullerton and Chesterfield (C, K) (Grøntved 1936, Polunin 1940).

In 1923 *J. Russell* made some collections at Old Fort Rae on the northern arm of Great Slave Lake (CAN).

From 1924 to 1926 *L.T. Burwash* made collections at Chesterfield (CAN).

In 1926 *John Russell*, while working for the Topographical Survey of Canada, made a few collections in that part of Wood Buffalo Park which lies north of the 60° parallel (CAN).

In 1927 *A.E. Porsild*, together with his brother *R.T. Porsild*, made a grazing survey in the northwestern part of the District of Mackenzie between the Mackenzie Delta and the Anderson River (CAN). In 1928 these grazing surveys were continued about Great Bear Lake (CAN). In 1931 *A.E. Porsild* made a botanical survey of central District of Keewatin from the Yathkejed Basin down the Kazan River to Baker Lake (CAN). On the basis of recommendations made following these surveys the Canadian Government made the decision to purchase a herd of reindeer in Alaska in order to establish a reindeer industry for the natives in the Mackenzie Delta region.

In 1931 *O. Bryant* gathered specimens on the east slope of the Richardson Mountains, west of the Mackenzie River Delta (CAN).

During the years 1932 to 1935 *A.E. Porsild* was in charge of the Reindeer Project in the Mackenzie Delta. His collections from that area were added to considerably during this period.

With the exception of Richardson's collections, Porsild's collections are probably the most important for our region up to this date. His paper (1943) "Materials for a Flora of the Continental Northwest Territories of Canada" contains only the most interesting of his finds.

In 1927, 1928, and 1933 *M.O. Malte*, who succeeded John Macoun at the Geological Survey of Canada Herbarium (now National Herbarium of Canada), as a member of the Canadian Eastern Arctic Expedition, made extensive collections at Chesterfield (CAN).

Also in 1927 *Hugh M. Raup* made collections along the lower Slave River and about Great Slave Lake. Again in 1929 Raup crossed a part of Wood Buffalo Park lying north of the District of Mackenzie border while en route to the southern parts of the Park (A, GH, CAN). The results of his studies were: "Botanical Investigations in Wood Buffalo Park" which was published by the National Museum of Canada (1935) and "Phytogeographic Studies in the Athabaska–Great Slave Lake Region I" (1936) and "II" (1946), two most important papers for our area.

In 1930 *G.M. Sutton* made a few collections at Chesterfield (GH).

In 1932 *W.S. Güssow*, (1933), son of the then Dominion Botanist, made collections in southeastern Keewatin District along the Maguse River, at Maguse Lake and at Padlei, while in the employ of the Geological Survey of Canada (DAO, CAN).

In 1933 *G. Gardner* made collections at Chesterfield and Cape Eskimo (QFA, GH) (Gardner 1973).

In 1934 *T.M. Cope* made collections at Cape Eskimo (CU, PH).

In 1934 and again in 1940, *Father Arthème Dutilly* from the Catholic University of America at Washington, D.C., travelled down the Mackenzie River system (Louis-Maire, 1961). His collections are mainly from areas adjacent to the various settlements along the river (QFA, DAO).

In 1936, *Father Dutilly* (Louis-Maire 1961) visited Chesterfield and Fairway Island and made collections there (QFA).

In 1936 and 1937 a study of the Thelon Game Sanctuary and Artillery Lake, east of Great Slave Lake, was conducted by *C.H.D. Clarke*

In 1939 *Hugh Raup,* Arnold Arboretum of Harvard University, accompanied by *James H. Soper* (now Chief Botanist at the National Museums of Canada) made extensive collections at Fort Simpson and at Brintnell Lake in the Mackenzie Mountains. This mountain site is almost inaccessible except by air. The specimens (A, GH, CAN) which were collected formed a large part of the basis for Raup's monumental study (1947) entitled "The Botany of Southwestern Mackenzie". In this volume Raup made an effort to account for all the specimens which had ever been collected in the southwest District of Mackenzie, a formidable task but a most useful contribution, but more than that, he attempted to explain the Mackenzie Mountain flora in the light of Eric Hultén's (1937) "Outline of the History of Arctic and Boreal Biota during the Quaternary Period". The work also included Canadian distribution maps for all of the Brintnell Lake species, a most important contribution.

At Brintnell Lake, *Raup* was also accompanied by his wife *Lucy*, who made extensive collections of lichens.

In 1940, *J. Robertson* made some collections at Kittigazuit on the Arctic Coast west of the Mackenzie River Delta (ISC).

In 1944 *V.C. Wynne-Edwards* made a number of collections along the Mackenzie River, at Nahanni and Lone Mountains near the mouth of the North Nahanni River on the Mackenzie, and in late July made an excursion up the Canol Road into the Mackenzie Mountains as far as the Bolstead Creek Pump Station (CAN).

Also early in September of 1944, *A.E. Porsild* made a reconnaissance along the Canol Road between Macmillan Pass and the Mackenzie River, a side trip from his detailed examination of the terrain adjacent to the Canol Road in the Yukon Territory (CAN). *A.E. Porsild* (1945) wrote up the results of this survey in "The Alpine Flora of the East Slope of the Mackenzie Mountains, N.W.T.". He recorded the collections of *Wynne-Edwards* mentioned above in an addendum to this publication.

Also, in 1944 *Frank S. Nowosad* of the Canada Dept. of Agriculture made a small collection of plants along the Liard River between the British Columbia border and Fort Simpson (DAO, CAN); and *C.O. Hage* made a small collection of plants at Trout Lake east of the Liard River (GH).

In 1945 *C.L. Crickmay* made a small collection of plants in the Franklin Mountains which lie east of and parallel to the Mackenzie River (CAN).

F. Harper gathered specimens in central Keewatin District as a member of the Nueltin Lake Expedition of 1947 (CAN).

In 1947, *T.N. Freeman*, entomologist with the Canada Department of Agriculture collected specimens at Baker Lake (DAO) and *I. McT. Cowan* gathered some specimens in the Mackenzie River Delta (UBC, DAO).

In 1948 *William Steere* lead the Port Radium Expedition of the Botanical Gardens, University of Michigan, for the detection of hereditary mutations in plants. His party, which included *H.T. Schacklette* (MICH, CAN) worked about Port Radium, Sawmill Bay and Dease Arm of Great Bear Lake, and as well made some observations at Coppermine and Great Slave Lake. *Steere* made extensive collections of mosses during this expedition.

In 1948 *W.J. Cody*, Canada Department of Agriculture, began his work in the north as a part of a team of botanists and entomologists studying biting flies and their habitats for the Defence Research Board (Canada). His first visit to the Continental Northwest Territories was in 1949 when he was based at Yellowknife. He was accompanied here by *J.B. McCanse*. From there he was fortunate to obtain flights out to the Snare River Power Dam, Eastern Great Bear Lake and Norman Wells (DAO).

In 1949 *A.E. Porsild*, while en route to Banks Island, made a side trip to the Sweet Grass Hills on Great Bear Lake (CAN); *I. McT. Cowan* gathered a few specimens in the Mackenzie River Delta region (CAN); *P. Lawson* gathered some specimens along the lower Thelon River (CAN); *A.L. Wilk* gathered a few specimens in the vicinity of Fort Smith (ALTA); and *B.J. Woodruff* while working with a Geodetic Survey of Canada party gathered a few specimens at Bathurst Inlet and Kigyik Lake (DAO).

In 1950 *W.J. Cody*, accompanied by *C.C. Loan*, was based at Fort Smith on the 60° parallel. From here he was able to study the Salt Plains which lie a few miles west of Fort Smith (DAO). This same year *D.B.O. Savile*, a Canada Department of Agriculture mycologist and botanist, worked out of Chesterfield Inlet on Hudson Bay (DAO); *A.L. Morrison* made some collections from south of Yellowknife (ALTA); *J.P. Kelsall* and *E.H. McEwen*, Canadian Wildlife Service, gathered a few specimens at Bathurst Inlet (CAN); and *W.K.W. Baldwin* and *H.J. Scoggan* of the National Museum of Canada, gathered specimens and studied the vegetation at Baralzon Lake in southern Keewatin District in relation to the latter's work on the flora of Manitoba (CAN) (Baldwin 1953).

In 1951 *Alton A. Lindsey* (1952) served as botanist on the Canadian Arctic Permafrost Expedition sent to the Mackenzie Basin by Purdue University in cooperation with the National Research Council of Canada. They worked about Great Slave Lake, along the Mackenzie River to its mouth and about Great Bear Lake. *Lindsey* had a particular interest in the vegetation of ancient beaches (CAN); two Canada Department of Agriculture student assistants, *W.H. Lewis* (now at the Missouri Botanical Garden) (Cody 1956) and *W.I. Findlay* (Cody 1954) were based at Hay River on the southwest shore of Great Slave Lake and Coppermine on the Arctic Coast respectively (DAO). *J.R. Mackay*, a geographer from the University of British Columbia, made a few collections from Darnley Bay and lower Hornaday River (DAO); and *E. McEwen* made some collections at Bathurst Inlet (CAN).

In 1952 *D.K. Brown (Mrs. D.K. Beckel)* who was based at the Defence Research Board (Canada) establishment at Fort Churchill, Man., visited Ennadai Lake in central Keewatin District (CAN); and *J.S. Tener*, Canadian Wildlife Service gathered specimens in the Thelon Game Sanctuary in Keewatin District (CAN).

In 1953, *W.J. Cody* (1960), accompanied by *R. Gutteridge*, was based at Norman Wells on the Mackenzie River. From here he made one trip across the river and out the now abandoned Canol Road into the eastern slope of the Mackenzie Mountains, and a four day tour late in the season to Aklavik in the Mackenzie River Delta (DAO); *J.G. Chillcott*, an entomologist with the Canada Department of Agriculture gathered specimens at Matthews and Muskox lakes which lie northeast of the east end of Great Slave Lake (DAO) (Cody & Chillcott 1955).

In 1955 *W.J. Cody*, accompanied by *J.M. Matte*, (1961) made intensive collections in the area about Fort Simpson (DAO).

In 1957, *W.J. Cody* (1965a, 1965b) examined the range on the Reindeer Grazing Preserve which embraces Richards Island at the northwest part of the Mackenzie Delta and a large area lying between the East Branch of the Mackenzie River east to the Anderson River (DAO). He was accompanied by *D.H. Ferguson*.

During parts of summers of 1958, 1959, 1961 *John W. Thieret* (1961b, 1962, 1963a, 1963b) then of the Field Museum of Natural History, Chicago, worked along the Yellowknife and Mackenzie highways (F, DAO, CAN).

Thieret in 1959 was also able to make a brief visit by air to the Horn Plateau, an elevated area rising out of the Slave lowland about 32 miles east-northeast of Fort Simpson (F, DAO) (Thieret 1961a).

Also in 1959 *W.W. Jeffrey* (1961) of the Canada Dept. of Forestry made a detailed study of the flora of the region lying adjacent to the Liard River between the British Columbia border and Nahanni Butte (CAN). He undoubtedly would have carried out other similar studies but unfortunately he died in an aeroplane accident shortly after.

In 1959, 1960, 1961, and 1963, *James Larsen* of the University of Wisconsin made botanical collections and rather detailed ecological studies in the Ennadai Lake region of Keewatin District (1965, 1967) (WISC), in connection with his meteorological studies and their relationships to vegetation. He continued these studies in the Fort Reliance — Artillery Lake — Aylmer Lake area at the east end of Great Slave Lake in 1962 and 1964, (1971). *Dr. John W. Thomson*, lichenologist, worked with Larsen in 1962 (1969).

In 1959 *H.J. Scoggan* of the National Museum of Canada gathered specimens at Ennadai

Lake in central Keewatin District, in studies related to his work on the flora of Manitoba (CAN).

In 1960 *Edward Arnold* visited a small glacial lake in the Mackenzie Mountains known as Hole-in-the-Wall Lake. He collected a surprising number of species which were "new" to the flora of Mackenzie District and again surprisingly, 28 taxa of the 105 which he found were not known from Brintnell Lake, the nearby locality which *H.M. Raup* had studied so intensively (CAN) (Porsild 1961).

In 1961, *E. Kuyt*, Canadian Wildlife Service made collections along the Thelon River (CAN).

In 1961, *J.S. Maini* and *J.M.A. Swan* of the University of Saskatchewan visited Small Tree Lake in southwestern Mackenzie District in connection with their forest ecological studies (SASK, DAO).

John Packer (1964) of the University of Alberta, visited Summit Lake in the Richardson Mountains in 1961. On the basis of his work there and in the Rocky Mountains he wrote a paper entitled "Chromosome Numbers and Taxonomic Notes on Western Canadian and Arctic Plants" (ALTA, DAO).

In 1961 *W.J. Cody*, accompanied by *K.W. Spicer*, (1963a) was attached to a Canada Department of Agriculture soil survey party which worked along the Liard River from the British Columbia border to its juncture with the Mackenzie River (DAO).

In 1961 *George W. Scotter* (1966), Canada Wildlife Service conducted caribou range studies in an area northeast of Yellowknife. He collected a large number of specimens in this region (CWS-Edmonton, DAO).

In 1962 *George Scotter* (1966) continued his studies of the winter range of the barren ground caribou in an area lying east of the Slave River and south of Great Slave Lake. (CWS-Edmonton, DAO). On both of these expeditions, *Scotter* collected long series of bryophytes and lichens as well as vascular plants.

In 1962 *P.M. Youngman* and *G. Tessier* of the National Museum of Canada gathered some specimens in the Mackenzie River Delta (CAN); *James A. Calder*, Canada Department of Agriculture made collections along the Yukon-Mackenzie border in the Richardson Mountains (DAO). He was supported here and in the northern Yukon by a Geological Survey of Canada field party. Also in 1962 two prospectors, *E.W. Johnson* and *D. Munro* made collections in the Redstone River region in the Mackenzie Mountains (DAO). They returned to this general area in 1963 (DAO) and in addition, two other prospectors, *Einar Kvale* and *Keith Haggard* worked nearby the second year (DAO) (Cody 1963b).

Tom Barry, Canadian Wildlife Service maintained a cabin for several years at the mouth of the Anderson River in order to study breeding bird populations in the area. Around 1962 he made a representative collection of the flora (CAN, DAO) of that area.

In 1963 *W.J. Cody* (1965a, 1965b) again visited the Reindeer Grazing Preserve at the mouth of the Mackenzie River, to ascertain if there had been any changes in the range since his visit there in 1957 (DAO). For comparative studies, with *S. Johansson*, he made a survey of the vegetation about Canoe Lake in the Richardson Mountains, about 25 miles west of Aklavik (DAO). While he was at Canoe Lake, *Vladimir Krajina* and some of his students from the University of British Columbia arrived to begin ecological studies in the area (UBC).

Also in 1963, *J.A. Parmelee*, a Canada Department of Agriculture mycologist visited DEW Line sites between Coronation Gulf and the Alaska border. His particular interest was in parasitic fungi, but he made full collections of the vascular plants at each stop he made (DAO, DAOM); *I.R. Rowlands* gathered a few specimens in the southern Mackenzie Mountains (DAO) — she was a member of a University of Ottawa Astronomical Research expedition; *Michael Alesiuk* collected some specimens in the Perry River region in northeastern District of Mackenzie (ALTA, TRT); and *J.P. Ryder* also collected in the same area in 1963 (ALTA, TRT).

In 1964 *Kaye MacInnes* accompanied her husband who was studying geese at the mouth of the McConnell River on Hudson Bay. She made extensive collections of the flora of this area. During the summers of 1965, 6, 7, 8 and 1970, she made some additional collections here while she was specifically concerned with ecological studies related to the genus *Pedicularis* (UWO, DAO).

In 1964 *George Scotter*, Canadian Wildlife Service, made some collections in the Thelon region, again in connection with his caribou range studies (CWS-Edmonton, CAN).

In 1964 *John Lambert* made collections from near the mouth of the Anderson River on the Arctic Coast (UBC, DAO). During 1965 and 66 he conducted research in the Richardson Mts. and adjacent British Mts. in northern Yukon in connection with his doctoral thesis (UBC, DAO).

In 1965 *W.J. Cody*, in a Department of Agriculture field party, worked along the Slave River from Fort Smith at the 60° parallel north to its mouth on Great Slave Lake (DAO).

In 1965 *George Rossbach*, West Virginia Wesleyan College flew in to the Thelon River with two companions and then travelled down the river by canoe to Chesterfield Inlet, collecting plants along the way. He spent the winter of 1967-68 working in the Herbarium of the National Museum of Canada going over his material with the intention of publishing the information (CAN).

In 1965 and 1966 *George Scotter* conducted range studies in the Reindeer Grazing Preserve between the Mackenzie Delta and the Anderson River (CWS-Edmonton, DAO).

In 1966 *W.J. Cody* was able to examine the vegetation of a large part of the southeastern District of Mackenzie lying east of the Slave River and from the 60° parallel north to the latitude of Great Slave Lake (DAO); *J. Carrol* gathered a few specimens in the Yellowknife Preserve (CAN); and *S. Sverre* gathered some specimens at Kangowan Lake in northwestern Keewatin District (CAN).

In 1967 *W.J. Cody* joined a Geological Survey of Canada field party working in the southwestern parts of the Mackenzie Mountains. This party worked out of camps at June Lake and O'Grady Lake by helicopter, thus affording excellent opportunity to gather specimens from inaccessible situations. *K.W. Spicer,* an assistant, was stationed at the Canada Tungsten Mine just over the border from the Yukon (DAO); *N.W. Simmons* wildlife biologist with the Canadian Wildlife Service gathered a few specimens at "Sterile" Lake in the Mackenzie Mountains (DAO).

In 1967, *E.O. Höhn*, University of Alberta, gathered a few specimens at Chesterfield Inlet while he was studying birds in that region (ALTA).

In 1968 *John Lambert*, Carleton University, worked for a time in the vicinity of Rankin Inlet and later about the Reindeer Grazing Preserve at the mouth of the Mackenzie River (CCO).

In 1969 *Lambert* again worked on the Reindeer Grazing Preserve (CCO).

In 1970, *George Scotter* (Scotter & Cody 1974) made extensive collections in an area of the Mackenzie Mountains adjacent to the South Nahanni River which is now Nahanni National Park (DAO, CWS-Edmonton).

Also in 1970, *W.J. Cody* was invited by the Canadian Committee of the International Biological Programme, Conservation of Terrestrial Biological Communities Subcommittee, Region 10 Panel (IBP/CT), to examine sites in the District of Mackenzie which had been proposed for preservation. He was thus able to examine the terrain and collect specimens at Porter Lake, Horn Plateau, Ebbutt Hills, "Cartridge" Lakes, Blackwater River and finally the Plains of Abraham in the Mackenzie Mountains (DAO); *M.G. Dumais*, University of Alberta travelled along the Mackenzie and Yellowknife highways in southern Mackenzie District (ALTA); *R.J. Clark (Mrs. R.J. Maltais)* gathered a few specimens in the vicinity of Fort Providence (ALTA); and *S.L. Welsh* visited Inuvik and the Mackenzie River Delta (ISC). He was accompanied by *J.K. Rigby* (BRY, ALTA).

In 1970 and 1971 *John Lambert*, Carleton University, worked in the Mackenzie River Delta on project ALUR (Arctic Land Use Research Program) (CCO).

From 1970 to 1973, *H. Hernandez* collected at Norman Wells and in the Mackenzie River Delta Region (ALTA, TRT) (Hernandez 1973a, 1973b).

In 1971 *L. Dahlke* made some collections at Hay River in southwestern District of Mackenzie (WAT, CAN).

In 1971, *K. Taylor* gathered a few specimens in the vicinity of the Discovery Mine, north of Yellowknife (ALTA, DAO).

In 1971, *W.E. Younkin* collected some specimens in the vicinity of Tuktoyaktuk, in connection with his autecological studies (ALTA) (Younkin 1973).

In 1971 *I.G.W. Corns* studied the ecology of plant communities east of the Mackenzie River Delta (ALTA) (Corns 1974).

In 1971 *G.H. La Roi*, University of Alberta, gathered specimens in the Mackenzie Delta Region in connection with his ecological studies (ALTA); *P. Ducruc*, Laval University collected specimens along the Hay River–Fort Smith Highway in the southwestern District of Mackenzie (QFA); *C.D. Bird* and *A.H. Marsh*, University of Calgary, studied the vegetation of Mount Clark and the Plains of Abraham. Their particular interests centered around mosses and lichens but vascular plants were also collected (UAC).

In 1971 and 1972 *Ross W. Wein*, University of New Brunswick, together with *L.R. Hettinger* and *A.J. Janz*, Northern Engineering Service, Calgary, made collections in the northern Yukon Territory and the Peel Plateau in northwestern District of Mackenzie, while investigating soils — vegetation relationships (DAO) (Wein, *et al.* 1974).

In 1971 *George Scotter* and *W.J. Cody* examined sites in the Mackenzie Mountains which had been proposed for conservation by IBP/CT Panel 10.

In the years 1971 to 1973 *S. Zoltai* and *R.M. Strang*, Canadian Forestry Service, conducted environmental studies related to the Mackenzie Valley Pipeline Survey (CFS-Edmonton).

During the years 1971 to 1973, *S. Talbot* conducted detailed ecological studies in the vicinity of the University of Alberta Biological Station at Heart Lake in southwestern District of Mackenzie (DAO, ALTA). This was a part of his Ph.D. program.

Again in 1972, *C.D. Bird*, with *G. Benson*, worked in the Peel River region of the Yukon Territory, and then in the vicinity of Wrigley on the Mackenzie River. In both years their particular interests centered around bryophytes and lichens, but vascular plants were also collected as part of site analyses (UAC).

In 1972 *A.H. Marsh* worked on the Mackenzie Valley Pipeline Survey with *R.M. Strang*. His particular interest was in bryophytes and lichens, but some vascular plants were collected (UAC); also in 1972, *W.J. Cody* continued his examination of IBP sites in the Mackenzie Mountains.

About 1972 *L.C. Bliss* collected some specimens in the Mackenzie River Delta in relation to his ecological studies (ALTA) (Wein and Bliss 1973a, 1973b; Bliss and Wein 1972).

During the years 1972 to 1976 *G.W. Scotter* spent short periods in the southern Mackenzie Mountains assessing areas for a new National Park, making impact assessments, preparing checksheets for the IBP/CT Panel 10 program, and working on a sheep survey (DAO; CWS-Edmonton) (Scotter and Cody 1974). In addition to vascular plants Scotter also collected long series of bryophytes and lichens.

In 1973, *S.L. Welsh* and *J.K. Rigby* gathered some specimens from along the Rat River east of the Slave River (ALTA) and *J. Nolan* and *B. Goski*, Canadian Wildlife Service, made collections in the Richardson Mountains of northwestern District of Mackenzie, in wildlife-related studies (CWS-Edmonton).

In 1974 *D. Gubble* and *D. Burr* gathered some specimens in the vicinity of Chick Lake in northwestern District of Mackenzie (CAN). The same year a Canadian Wildlife Service party made up of *A.M. Pearson, C.B. Larsen* and *W.H. Owen* gathered specimens of plants on the Tuktoyaktuk Peninsula west of the Mackenzie River Delta, while studying animal life in that area (DAO, CWS-Edmonton); and *G.C. Trottier* in another Canadian Wildlife Service party, collected specimens at Mills Lake, an expansion of the Mackenzie River below Fort Providence, in connection with investigations of potential waterfowl–agricultural conflicts (CWS-Edmonton, DAO).

During the years 1974 to 1976 *A.H. Marsh* studied the vegetation of several parts of Nahanni National Park in the southern Mackenzie Mountains while working on contract with the Canadian Wildlife Service (DAO, CWS-Edmonton).

In 1973-5 *S.M. Lamont* made extensive collections at Fisherman Lake and at "Grass" Lake in southwestern District of Mackenzie (SASK).

In the summers of 1975 and 1976 *S. Talbot* and *B.J. Meuleman* conducted ecological studies in Nahanni National Park while under contract to the Forest Management Institute of Environment Canada (DAO).

In the years 1975 and 1976 *Sam Miller*, Northwest Territories Fish and Wildlife Service, made collections in the Godlin Lakes region in the central Mackenzie Mountains in relation to his wildlife studies (DAO).

In 1976 *M. Kershaw* conducted an ecological study of the Prairie Creek area of Deadman Valley, Nahanni National Park as a part of an M.Sc. program (ALTA, DAO).

In 1976 *S. Zoltai*, Canadian Forestry Service, Environment Canada, conducted environmental studies in the Baker Lake — Kazan River region of Central Keewatin District in relation to a possible polar gas pipeline; and *Wayne Neily*, Parks Canada, made a few collections in the southern Mackenzie Mountains (DAO).

KEYS TO THE MAJOR GROUPS AND FAMILIES

a. Plants without flowers or seeds; reproducing by spores; mostly fern-like or rush-like . . . . Division Pteridophyta, (p. 15)
a. Plants producing seeds . . . . Division Spermatophyta, (p. 15)
  b. Ovules not enclosed in an ovary; seeds in a dry cone (rarely in a berry-like fruit); pollen sacs on scales arranged in cones; trees or shrubs; leaves needle-like or scale-like, mostly evergreen . . . . Subdivision Gymnospermae, (p. 15)
  b. Ovules enclosed in an ovary, which at maturity becomes the fruit; pollen sacs terminating stamens borne in flowers; trees, shrubs or herbs; leaves of various forms, mostly deciduous . . . . Subdivision Angiospermae, (p. 15)
    c. Leaves usually parallel-veined; vascular bundles of stem irregularly arranged; cambium absent; parts of flowers usually in 3's or 6's, never in 5's; embryo with one cotyledon; herbs (rarely shrubby) . . . . Class Monocotyledoneae, (p. 15)
    c. Leaves usually net-veined; vascular bundles of stem commonly in a single ring; cambium usually present; parts of flowers usually in 5's or 4's; embryo with two cotyledons; herbs, shrubs or trees . . . . Class Dicotyledoneae, (p. 16)

Division PTERIDOPHYTA
(Ferns and Fern Allies)

a. Leaves broad, usually more than 2 cm long, often quite large, variously incised or dissected
  b. Spore-cases relatively large, borne in a terminal (grape-like) cluster, the sterile blade appearing lateral on a common stalk with it . . . . Ophioglossaceae, (p. 23)
  b. Spore-cases minute, borne in clusters (sori) on the back or near the margins of green blades, or on separate modified fronds . . . . Polypodiaceae, (p. 26)
a. Leaves slender, often scale-like, simple, sessile, mostly small
  c. Stems conspicuously jointed and hollow; leaves scale-like, in sheath-like whorls at the nodes; spore-cases on the scales of terminal cone-like spikes . . . . Equisetaceae, (p. 37)
  c. Stems not as above; leaves mostly imbricated; spore-cases in cone-like spikes or otherwise
    d. Leaves small, less than 1.5 cm long; stems slender, elongated, branched, the main ones usually horizontal and creeping; spore-cases in club-like spikes; plants resembling coarse mosses
      e. Leafy shoots mostly 7-15 mm wide; homosporous . . . . . . Lycopodiaceae, (p. 41)
      e. Leafy shoots less than 5 mm wide; heterosporous . . . . . . Selaginellaceae, (p. 44)
    d. Leaves slender, grass-like, 5-15 cm long, dilated and sheathing at base; stem short, corm-like, lobed; plant rooting in mud, commonly covered by water . . . . Isoetaceae, (p. 46)

Division SPERMATOPHYTA (Seed Plants)
Subdivision Gymnospermae
Class CONIFERAE (Conifers), (p. 46)
Subdivision Angiospermae (Flowering Plants)
Class MONOCOTYLEDONEAE (Monocotyledons)

a. Plants not differentiated into stem and leaf, small, thallus-like, usually ellipsoid or oblong; free-floating or immersed aquatics . . . . Lemnaceae, (p. 192)
a. Plant with stem and leaves, not thallus-like
  b. Perianth lacking, or inconspicuous, often of bristles or scales, not petal-like (see also Juncaceae and Juncaginaceae)

c. Flowers enclosed or subtended by scales (glumes); plants grass-like, with jointed stems, sheathing leaves and 1-seeded fruit
  d. Stems usually hollow, terete or flattened; leaves 2-ranked; leaf-sheaths usually split (open); anthers attached at the middle . . . . Gramineae, (p. 68)
  d. Stems solid, usually more or less 3-sided; leaves usually 3-ranked; leaf-sheaths not split (closed); anthers attached at the base . . . . . Cyperaceae, (p. 129)
c. Flowers not enclosed in scales (though sometimes in involucrate heads); plants not as above
  e. Aquatic plants, the stems and leaves flaccid, either immersed or floating
    f. Leaves alternate or sub-opposite . . . . Potamogetonaceae, (p. 55)
    f. Leaves opposite or whorled . . . . Najadaceae, (p. 64)
  e. Terrestrial plants, or bases in water and the stems rigid enough to support shoots above water level
    g. Flowers in globose heads; perianth of flat scales . . . . . Sparganiaceae, (p. 52)
    g. Flowers mostly in spikes
      h. Flowers unisexual, the pistillate in a thick spike, the staminate above; perianth of slender bristles . . . . Typhaceae, (p. 52)
      h. Flowers perfect, in a thick dense fleshy spike, this subtended by a large bract (spathe) . . . . Araceae, (p. 191)

b. Perianth of 2 distinct whorls, the inner often petal-like and conspicuous
  i. Perianth relatively inconspicuous, green or brownish; plants rush-like
    j. Perianth dry, often scarious; flowers commonly in panicles or heads; carpels 3, united, forming a small capsule . . . . Juncaceae, (p. 193)
    j. Perianth herbaceous; flowers in racemes or spikes; carpels 3, 6, almost distinct, separating as follicles when ripe . . . . Scheuchzeriaceae, (p. 65)
  i. Perianth conspicuous, at least the inner whorl brightly coloured
    k. Carpels numerous, in a ring or cluster, becoming achenes; plants of marshes and bogs . . . . Alismaceae, (p. 67)
    k. Carpels 3, their ovaries united; fruit a capsule or berry
      l. Ovary superior (rarely partly inferior) . . . . Liliaceae, (p. 205)
      l. Ovary inferior
        m. Flowers regular (actinomorphic); stamens 3 . . . . Iridaceae, (p. 211)
        m. Flowers very irregular; stamens 1 or 2 . . . . Orchidaceae, (p. 212)

Class DICOTYLEDONEAE (Dicotyledons)

Key to the Groups

a. Corolla none; calyx present or lacking
  b. Plants monoecious or dioecious; staminate and pistillate flowers, one or both kinds in catkins or dense heads . . . . Group I, (p. 17)
  b. Plants mostly with perfect flowers and these not in catkins (often in dense clusters). . . . . Group II, (p. 17)
a. Corolla and calyx both present
  c. Corolla of separate petals
    d. Stamens usually numerous, at least more than 10, and more than twice as many as the sepals or calyx-lobes . . . . Group III, (p. 18)
    d. Stamens not more than twice as many as the petals . . . . Group IV, (p. 18)
  c. Corolla with petals more or less united
    e. Stamens more numerous than the corolla-lobes . . . . Group V, (p. 20)
    e. Stamens not more numerous than the corolla-lobes . . . . Group VI, (p. 20)

GROUP I

a. Staminate and pistillate flowers both in catkins or catkin-like heads; trees or shrubs
  b. Ovary becoming a many-seeded capsule; seeds hairy-tufted . . . . . . Salicaceae, (p. 222)
  b. Ovary becoming a 1-seeded nut or winged nutlet
    c. Pistillate flowers 2 or 3 at each bract of the catkin; fruit not resin-dotted. . . . . Betulaceae, (p. 254)
    c. Pistillate flowers single at each bract; fruit resin-dotted. . . . . Myricaceae, (p. 254)
a. Staminate or pistillate flowers (not both) in catkins or catkin-like heads; herbs. . . . . Urticaceae, (p. 259)

GROUP II

a. Ovary or its locules with many ovules
  b. Ovary and fruit partly inferior . . . . Saxifragaceae, (p. 383)
  b. Ovary or ovaries superior
    c. Ovaries 2 or more, separate . . . . Ranunculaceae, (p. 313)
    c. Ovary single
      d. Leaves compound . . . . Ranunculaceae, (p. 313)
      d. Leaves simple
        e. Sepals separate. . . . . Caryophyllaceae, (p. 279)
        e. Sepals more or less united. . . . . Primulaceae, (p. 498)
a. Ovary or its locules with only 1 or 2 ovules
  f. Pistil of more than one carpel; carpels separate or nearly so . . . Ranunculaceae, (p. 313)
  f. Pistil solitary, simple or compound
    g. Ovary superior; flower hypogynous
      h. Stipules (ocreae) sheathing the stem at the nodes . . . . Polygonaceae, (p. 261)
      h. Stipules, if present, not sheathing the stem
        i. Herbs
          j. Plants usually aquatic, immersed or nearly so
            k. Leaves whorled, finely dissected; style 1 . . . . . Ceratophyllaceae, (p. 310)
            k. Leaves opposite, entire; styles 2 . . . . Callitrichaceae, (p. 457)
          j. Plants terrestrial, but sometimes growing in marshy places (see also *Callitriche*)
            l. Style (if any) 1 and stigma 1
              m. Flowers unisexual; ovary of fertile flowers 1-loculed. . . . . Urticaceae, (p. 259)
              m. Flowers perfect; ovary 2-loculed and 2-seeded . . . . Cruciferae, (p. 339)
            l. Styles 2-3 or of 2-3 branches
              n. Flowers subtended by spiny scarious bracts; sepals scarious . . . . Amaranthaceae, (p. 276)
              n. Flowers not subtended by scarious bracts; sepals herbaceous or fleshy . . . . Chenopodiaceae, (p. 269)
        i. Trees or shrubs . . . . Aceraceae, (p. 459)
    g. Ovary inferior or partly so, or so closely enclosed by the calyx as to appear so
      o. Aquatic herbs; leaves whorled . . . . Haloragaceae, (p. 472)
      o. Terrestrial plants; leaves opposite or alternate
        p. Shrubs or small trees; leaves scurfy; calyx not corolla-like . . . . Elaeagnaceae, (p. 465)
        p. Herbs; calyx corolla-like. . . . . Santalaceae, (p. 259)

GROUP III

a. Ovary superior; flowers hypogynous or perigynous
  b. Carpels few-numerous, entirely separate or united only at the base
    c. Stamens inserted below the ovaries; leaves without stipules .... Ranunculaceae, (p. 313)
    c. Stamens inserted on a hypanthium or on a disk; leaves commonly stipulate .... Rosaceae, (p. 404)
  b. Carpels solitary, or few-several with their ovaries completely united
    d. Ovary simple, 1-loculed
      e. Ovules 2; seed solitary; fruit a drupe .... Rosaceae, (p. 404)
      e. Ovules numerous; leaves compound .... Ranunculaceae, (p. 313)
    d. Ovary compound as shown by the number of its locules, placentae, styles or stigmas
      f. Leaves hollow, hooded; stigma dilated, umbrella-shaped . Sarraceniaceae, (p. 379)
      f. Leaves and stigma not as above
        g. Ovary 1-loculed; placentae parietal .... Papaveraceae, (p. 334)
        g. Ovary several-loculed .... Nymphaeaceae, (p. 311)
a. Ovary inferior; flowers epigynous
  h. Plants aquatic; leaves floating; blades peltate .... Nymphaeaceae, (p. 311)
  h. Terrestrial shrubs .... Rosaceae, (p. 404)

GROUP IV

a. Stamens of the same number as the petals and opposite them
  b. Style and stigma 1; calyx usually 5-parted .... Primulaceae, (p. 498)
  b. Styles and stigmas 2, 3 or more; sepals 2 .... Portulacaceae, (p. 277)
a. Stamens not of the same number as the petals, or if of the same number alternate with them
  c. Ovary superior; flowers hypogynous or perigynous
    d. Ovaries 2 or more, wholly separate or somewhat united
      e. Stamens inserted at the base of the ovary on the receptacle
        f. Carpels quite separate or united only at the base, each several-seeded
          g. Leaves fleshy, entire .... Crassulaceae, (p. 381)
          g. Leaves not fleshy, usually lobed or compound ..... Ranunculaceae, (p. 313)
        f. Carpels more or less united, with a common style; ovary 5-lobed .... Geraniaceae, (p. 454)
      e. Stamens inserted on the perianth or on a hypanthium
        h. Plants fleshy; stamens just twice as many as the carpels .... Crassulaceae, (p. 381)
        h. Plants not fleshy; stamens not just twice as many as the carpels
          i. Stipules present .... Rosaceae, (p. 404)
          i. Stipules absent .... Saxifragaceae, (p. 383)
    d. Ovary 1 (simple or compound)
      j. Ovary simple, with 1 parietal placenta and 1 style........ Leguminosae, (p. 434)
      j. Ovary compound, as shown by the number of its locules, placentae, styles or stigmas
        k. Ovary 1-loculed
          l. Corolla quite irregular
            m. Petals 4; stamens 6 .... Fumariaceae, (p. 338)
            m. Petals 5; stamens 5 .... Violaceae, (p. 461)
          l. Corolla regular or nearly so
            n. Fruit 1-seeded .... Cruciferae, (p. 339)
            n. Fruit with more than 1 seed

o. Placentation central .... Caryophyllaceae, (p. 279)
o. Placentation parietal
p. Leaves with gland-tipped hairs (tentacles); small insectivorous plants of bogs .... Droseraceae, (p. 379)
p. Leaves without gland-tipped tentacles
q. Low heath-like shrub with small appressed grey-pubescent leaves and numerous small flowers .... Cistaceae, (p. 461)
q. Plants not as above
r. Stamens 4, 5, 8 or 10; petals 4 or 5... Saxifragaceae, (p. 383)
r. Stamens 6; petals 4................. Cruciferae, (p. 339)
k. Ovary 2-several-loculed
s. Flowers irregular (zygomorphic)................ Balsaminaceae, (p. 459)
s. Flowers regular or nearly so
t. Stamens 6; petals 4; ovary 2-loculed.............. Cruciferae, (p. 339)
t. Plants not as above
u. Ovules and seeds 1 or 2 in each locule
v. Depressed evergreen shrubs with small flowers; fruit berry-like...... .... Empetraceae, (p. 458)
v. Herbs with showy flowers; fruit a capsule
w. Capsule rounded, splitting from the top, 10-seeded .... Linaceae, (p. 456)
w. Capsule long-beaked, splitting from below upward into 5 parts, each 1-seeded..................... Geraniaceae, (p. 454)
u. Ovules and usually the seeds several-many in each locule
x. Stipules present between opposite leaves; small semi-aquatic plants .... Elatinaceae, (p. 460)
x. Stipules none when the leaves are opposite
y. Styles 2-5; leaves opposite.......... Caryophyllaceae, (p. 279)
y. Style 1
z. Plants with scale-like, non-green leaves; saprophytes........ .... Pyrolaceae, (p. 482)
z. Plants with mainly green leaves
aa. Herbs, or somewhat woody at base; less than 5 dm tall .... Pyrolaceae, (p. 482)
aa. Shrubs; over 5 dm tall .... Ericaceae, (p. 486)
c. Ovary inferior; flowers epigynous
bb. Ovules and seeds 1 in each locule
cc. Stamens 5 or 10
dd. Fruit berry-like; styles 5......................... Araliaceae, (p. 475)
dd. Fruit dry, splitting at maturity; styles 2............ Umbelliferae, (p. 475)
cc. Stamens 4 or 8
ee. Fruit a drupe; style and stigma 1; large shrub or small herb .... Cornaceae, (p. 481)
ee. Fruit a capsule or nut-like; herbs
ff. Style 1; stigma 2-4-lobed; terrestrial plants..... Onagraceae, (p. 467)
ff. Styles or sessile stigmas 2-4; aquatic plants with dissected leaves..... .... Haloragaceae, (p. 472)
bb. Ovules and seeds more than 1 in each locule
gg. Ovary 1-loculed, fruit a berry.................... Saxifragaceae, (p. 383)
gg. Ovary 2-many-loculed
hh. Style 1; stamens 4 or 8 .... Onagraceae, (p. 467)
hh. Styles 2-3; stamens 5 or 10.................. Saxifragaceae, (p. 383)

GROUP V

a. Ovary 1-loculed; corolla irregular; placentae parietal
  b. Placenta 1; sepals usually 5 or wholly united; petals 5, of which 2 are usually united along one edge. . . . . Leguminosae, (p. 434)
  b. Placentae 2; sepals 2, small; petals 4, of which one is spurred at the base. . . . . Fumariaceae, (p. 338)
a. Ovary 3-several-loculed
  c. Corolla irregular, with a large pouch (sac) terminating in a spur; anthers united . . . . Balsaminaceae, (p. 459)
  c. Corolla regular
    d. Saprophytic herbs, without green foliage . . . . Pyrolaceae, (p. 482)
    d. Shrubs or herbs, not saprophytic; foliage green
      e. Shrubs; leaves all simple . . . . Ericaceae, (p. 486)
      e. Small herb; basal leaves ternate. . . . . Adoxaceae, (p. 555)

GROUP VI

a. Stamens opposite the corolla-lobes and of the same number. . . . . . . Primulaceae, (p. 498)
a. Stamens alternate with the corolla-lobes and of the same number, or fewer
  b. Ovary superior
    c. Corolla irregular (zygomorphic)
      d. Ovules solitary in the locules; ovary separating at maturity into 4 nutlets. . . . . Labiatae, (p. 524)
      d. Ovules 2-many in each locule; ovary not separating into nutlets
        e. Plants without green foliage; parasites on the roots of other plants. . . . . Orobanchaceae, (p. 544)
        e. Plants with green foliage
          f. Ovary 1-loculed; stamens 2; small carnivorous plants, in water or on moist ground. . . . . Lentibulariaceae, (p. 545)
          f. Ovary 2-loculed; stamens 4, or sometimes 5 or 2. Scrophulariaceae, (p. 528)
    c. Corolla regular
      g. Stamens fewer than the lobes of the corolla
        h. Ovary 4-lobed . . . . Labiatae, (p. 524)
        h. Ovary 2-loculed, not 4-lobed
          i. Leaves mostly basal; corolla scarious; capsule circumscissile. . . . . Plantaginaceae, (p. 548)
          i. Leaves mostly cauline and opposite; corolla not scarious; dehiscence not circumscissile . . . . Scrophulariaceae, (p. 528)
      g. Stamens as many as the corolla-lobes
        j. Carpels 2, separate except sometimes at the apex; herbs with milky juice . . . . Apocynaceae, (p. 513)
        j. Carpels united; plants without milky juice
          k. Ovary deeply 4-lobed, at maturity separating into 2 or 4 nutlets
            l. Leaves alternate; stems usually rounded; flowers mostly regular . . . . Boraginaceae, (p. 519)
            l. Leaves opposite; stems usually 4-sided; flowers mostly irregular . . . . Labiatae, (p. 524)
          k. Ovary not deeply 4-lobed
            m. Ovary 1-loculed
              n. Leaves entire, opposite. . . . . Gentianaceae, (p. 507)
              n. Leaves toothed, lobed or compound

o. Leaves with 3 entire leaflets; corolla white-bearded on upper surface . . . . Menyanthaceae, (p. 513)
o. Leaves, if compound, with toothed leaflets; corolla not as above . . . . Hydrophyllaceae, (p. 518)
m. Ovary 2-10-loculed
p. Stamens free from the corolla or nearly so . . . . Ericaceae, (p. 486)
p. Stamens inserted on the corolla-tube
q. Stamens 4
r. Leaves mostly basal, strong-nerved; corolla scarious . . . . Plantaginaceae, (p. 548)
r. Leaves mostly cauline; corolla not scarious
s. Fruit a capsule, many-seeded . . . . Scrophulariaceae, (p. 528)
s. Fruit of 2 or 4 nutlets. . . . . Labiatae, (p. 524)
q. Stamens 5
t. Fruit of 2 or 4 nutlets; ovary somewhat 4-lobed . . . . Boraginaceae, (p. 519)
t. Fruit a capsule or berry; ovary not 4-lobed
u. Styles 2, or style 1 and 2-cleft; ovary 2-loculed; inflorescence more or less scorpioid . . . . Hydrophyllaceae, (p. 518)
u. Style single, 3-cleft; ovary 3-loculed; inflorescence not scorpioid
v. Dwarf evergreen shrubs with entire leathery leaves . . . . Diapensiaceae, (p. 497)
v. Herbs with entire or pinnately divided leaves . . . . Polemoniaceae, (p. 515)
b. Ovary inferior; flowers epigynous
w. Stamens free from the corolla
x. Corolla regular; anthers separate . . . . Campanulaceae, (p. 558)
x. Corolla irregular; anthers united . . . . Lobeliaceae, (p. 560)
w. Stamens inserted on the corolla
y. Ovary with 2-many fertile locules and 2-many ovules; calyx not modified as a pappus or other special structure
z. Shrubs (erect or climbing), or trailing plants; stems rounded; leaves opposite, not stipulate . . . . Caprifoliaceae, (p. 553)
z. Herbs; stems mostly 4-sided; leaves whorled . . . . Rubiaceae, (p. 550)
y. Ovary with 1 fertile locule; calyx often modified as a pappus
aa. Flowers in panicled cymes, not in heads . . . . Valerianaceae, (p. 556)
aa. Flowers in involucrate heads . . . . Compositae, (p. 561)

OPHIOGLOSSACEAE Adder's-tongue Family

Botrychium Sw. Moonwort

Mostly low, perennial, essentially glabrous herbs from a short, erect rhizome and clustered, fleshy roots. Fronds with a single, pinnately or ternately divided green, sterile blade above which appear the fertile segments bearing double rows of mostly sessile, naked sporangia.

a. Sterile blade thick and fleshy, once or twice pinnate
  b. Blade oblong, the pinnae of equal size, orbicular or reniform
    c. Pinnae sub-imbricate, their margins entire . . . . *B. Lunaria* ssp. *Lunaria*
    c. Pinnae somewhat distant, their margins always somewhat erose or shallowly lobed . . . . *B. Lunaria* ssp. *minganense*
  b. Blade deltoid, the lowermost pinnae longest, all more or less pointed
    d. Fertile blade almost completely sessile; pinnae narrowly lanceolate, acute. . . . . *B. lanceolatum*
    d. Fertile blade distinctly short-petioled, pinnae rhomboid, obtuse. . . . . *B. boreale* ssp. *obtusilobum*
a. Sterile blade 3-pinnate, broadly triangular
  e. Blade thin, sessile; frond 15-30 cm tall . . . . *B. virginianum* ssp. *europaeum*
  e. Blade leathery, wintergreen, its long petiole attached at the base of the stipe; frond 5-20 cm tall . . . . *B. multifidum*

Botrychium boreale Milde
ssp. **obtusilobum** (Rupr.) Clausen
Stems 6 to 15 cm tall, the oblong-deltoid sterile frond short petioled, attached at or above the middle of the plant, the oblong pinnae pinnatifid, obtuse.

Grassy tundra slopes, rare, Richardson Mountains.

General distribution: Amphi-Beringian, arctic-alpine.

Fig. 1 Map 2

***Botrychium lanceolatum** (Gmel.) Ångstr.
ssp. **lanceolatum**
Stems 10 to 20 cm tall, the deltoid sterile frond sessile, attached near the summit, pinnate, the pinnae mostly pinnatifid, lanceolate.

Alpine meadows and grassy places.

General distribution: Circumpolar.

Fig. 2 Map 3

Botrychium Lunaria (L.) Sw.
ssp. **Lunaria**
Stems 7 to 15 cm tall, the sterile frond usually attached above the middle, pinnate, the pinnae flabellate, imbricated.

Rare and local in the western part of our area in meadows and usually open grassy situations.

In the ssp. *minganense* (Vict.) Calder & Taylor the frond is yellowish-green and the pinnae remote, obovate and cuneate at the base, their margins frequently incised.

Of similar habitats and distribution as the ssp. *Lunaria*, but less frequent.

General distribution: ssp. *Lunaria*, Circumpolar; ssp. *minganense*, boreal American.

Fig. 3 Map 4 (ssp. *Lunaria*); Map 5 (ssp. *minganese*)

Botrychium multifidum (Gmel.) Rupr.
Stems 5 to 20 cm tall, the deltoid, sterile and somewhat leathery frond long petioled and attached near the base of the stem, tri-pinnatifid, the ultimate segments ovate to suborbicular, blunt, usually imbricated.

Rare and local in prairie clearings, north to Fort Simpson.

General distribution: Circumpolar, widespread but local.

Fig. 4 Map 6

Botrychium virginianum (L.) Sw.
ssp. **europaeum** (Ångstr.) Clausen
Rattlesnake Fern.
Stems 15 to 30 cm tall, the sterile frond ternate, sessile attached above the middle of the stem, pinnately decompound, the ultimate segments oblong-lanceolate.

Rare and local in rich woodland.

General distribution: Circumpolar.

Fig. 5 Map 7

**Species not yet recorded from the Continental Northwest Territories, but to be expected.*

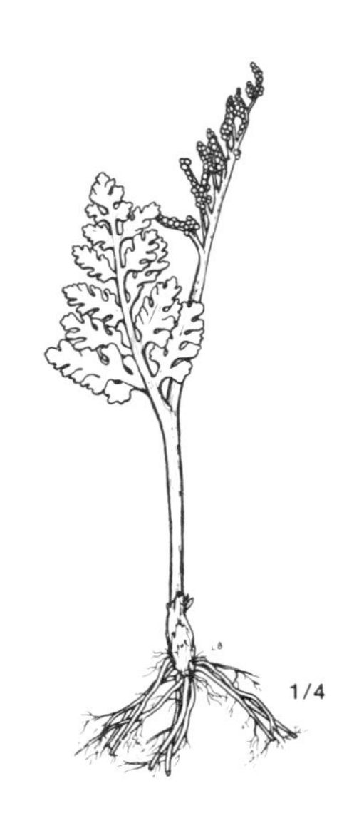

1. *Botrychium boreale* ssp. *obtusilobum*

2. *Botrychium lanceolatum* ssp. *lanceolatum*

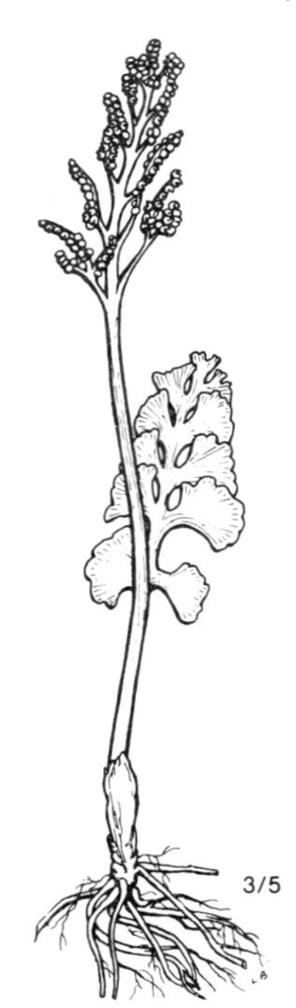

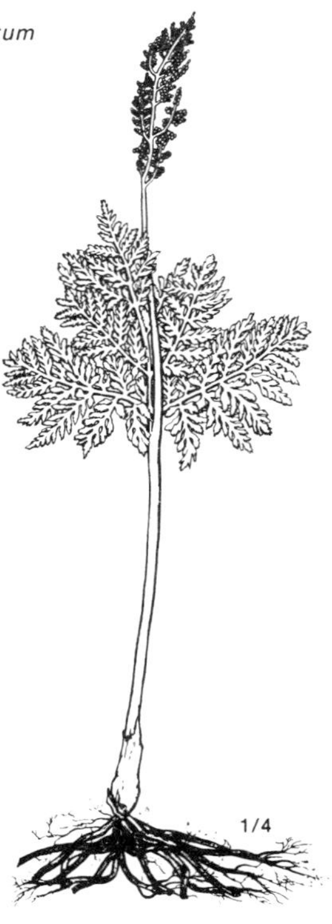

3. *Botrychium Lunaria* ssp. *Lunaria*

4. *Botrychium multifidum*

5. *Botrychium virginianum* ssp. *europaeum*

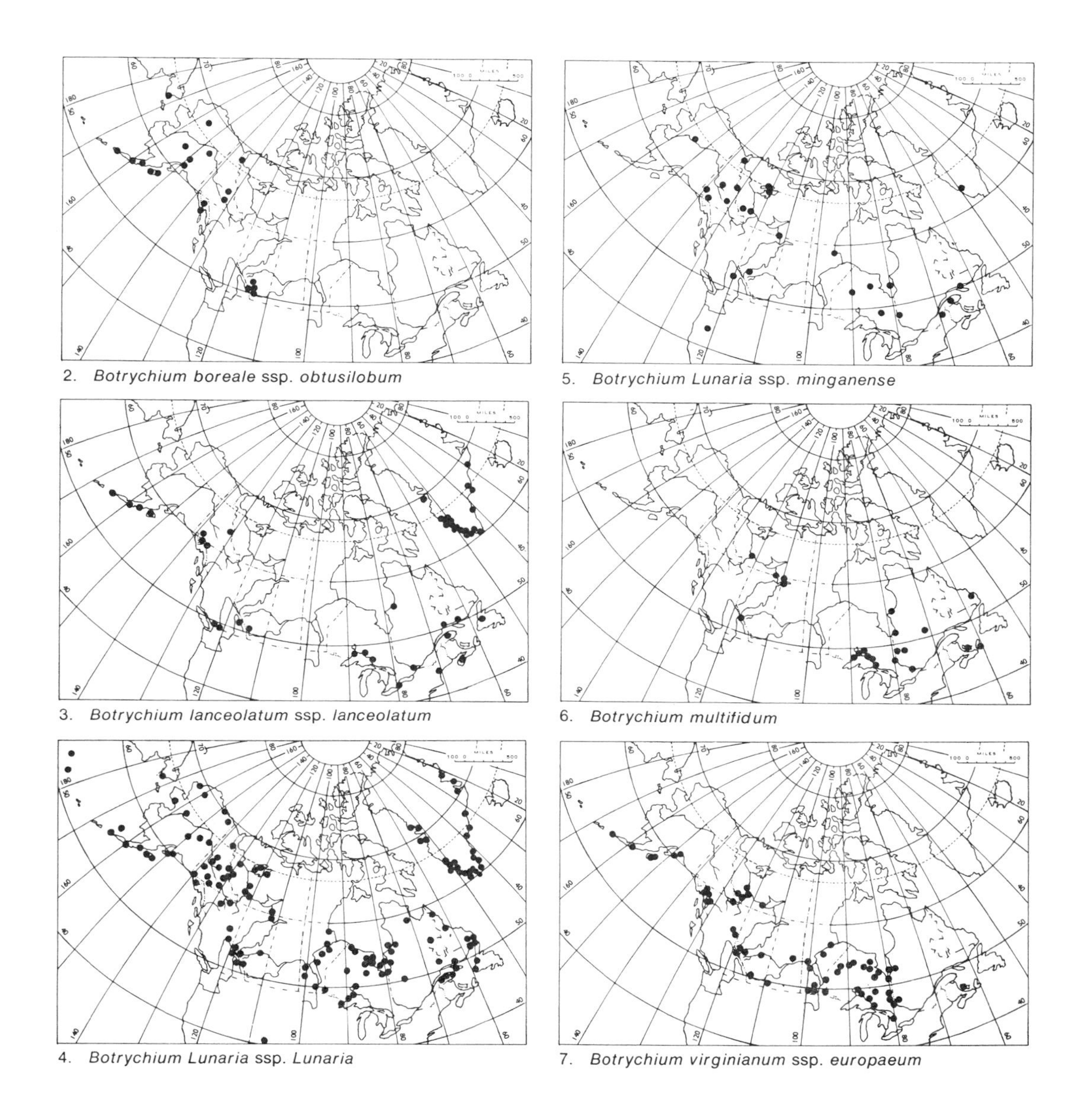

2. *Botrychium boreale* ssp. *obtusilobum*

3. *Botrychium lanceolatum* ssp. *lanceolatum*

4. *Botrychium Lunaria* ssp. *Lunaria*

5. *Botrychium Lunaria* ssp. *minganense*

6. *Botrychium multifidum*

7. *Botrychium virginianum* ssp. *europaeum*

POLYPODIACEAE Fern Family

Perennial spore-bearing herbs (ours), with creeping or ascending rhizomes from which rise the commonly feather-shaped, mostly once- to thrice-divided fronds; sporangia (sori) in round, oblong or linear clusters on the underside or margins of the frond or its divisions, and covered (at least when young) by a variously shaped, membranaceous shield called the indusium, or by the revolute frond-margins, or naked.

a. Sori partly covered by the revolute margins of the frond
  b. Sterile and fertile fronds dissimilar
    c. Sori marginal; low, delicate or subcoriaceous rock ferns . . . . *Cryptogramma*
    c. Sori intermarginal; coarse, tall ferns of damp woodland, sterile frond broad, pinnate-pinnatifid, green; fertile frond shorter, contracted, becoming dark brown in age . . . . *Matteuccia struthiopteris*
  b. Sterile and fertile fronds alike; small, brittle cliff ferns with tufted fronds from a short, hairy rhizome . . . . *Pellaea glabella*
a. Sori free; margins of frond not revolute
  d. Sori roundish
    e. Frond simple, once pinnate, or the pinnae merely toothed
      f. Fronds densely tufted, pinnae toothed
        g. Pinnae teeth spinulose-tipped; indusium round; fronds evergreen, subcoreaceous . . . . *Polystichum lonchitis*
        g. Pinnae teeth blunt; indusium lacerate or hair-like; fronds not evergreen . . . . *Woodsia*
      f. Fronds from a slender, creeping spongy rhizome covered by felt-like scales; pinnae entire-margined . . . . *Polypodium vulgare* ssp. *virginianum*
    e. Fronds 2- or 3-pinnatifid
      h. Sori mostly lacking indusia; frond deltoid, with a long stipe. . . . . *Dryopteris*
      h. Sori with indusia
        i. Indusia inferior or lateral, attached at the base or side of the sori
          j. Indusia inferior, stellate or divided into spreading, linear lobes . . . . . *Woodsia*
          j. Indusia lateral, soon falling . . . . *Cystopteris*
        i. Indusia superior, orbicular or reniform, attached at the sinus . . . . . . *Dryopteris*
  d. Sori not roundish
    k. Sori oblong; small plants with delicate, pinnate fronds . . . . *Asplenium viride*
    k. Sori crescent-shaped; tall plants with plumose 2- or 3-pinnate fronds . . . . *Athyrium Filix-femina*

Asplenium L. Spleenwort

Asplenium viride Huds.
Dwarf plants with delicate linear fronds 4 to 14 cm long, rachis green, pinnate, the pinnules round- or rhombic-crenate; sori elongate; indusia straight or slightly curved, attached to the upper side of the fertile vein.

Rare on wet rocks; Nahanni Range, Mackenzie Mountains.

General distribution: Circumpolar.

Fig. 6 Map 8

Athyrium Roth

Athyrium Filix-femina (L.) Roth
ssp. **cyclosorum** (Rupr.) C. Chr.
Lady-Fern.
Fronds tall tufted from a stout rhizome, broadly lanceolate and tapering to the tip, twice pinnate-pinnatifid; sori oblong to horse-shoe shaped; indusia attached to the inner side of the fertile veinlet.

Rare and local; Liard River Valley and by hot springs in Flat River Valley.

General distribution: Aleutian Islands to California, east to S. W. District of Mackenzie.

Fig. 7 Map 9

Cryptogramma R. Br. Rock-brake

Fronds dimorphous, the fertile longer and with narrower divisions than the sterile; margins of fertile fronds revolute to form a false indusium.

a. Fronds tufted from a stout ascending rhizome; sterile frond subcoriaceous
  b. Sterile fronds narrowly deltoid, twice pinnate to twice pinnate-pinnatifid, the ultimate segments oblong to ovate, crenate .... *C. crispa* var. *acrostichoides*
  b. Sterile fronds broadly deltoid, thrice pinnate, the ultimate segments small, obovate.. *C. crispa* var. *sitchensis*
a. Fronds scattered from a slender creeping rhizome; sterile frond delicate .... *C. Stelleri*

Cryptogramma crispa (L.) R. Br.
var. **acrostichoides** (R. Br.) C. B. Clarke
Mountain-Parsley
Sterile fronds shorter than the fertile.

Frequent to common in shallow organic soil and crevices throughout the Precambrian shield area north to Great Bear Lake.

General distribution: Circumpolar, with large gaps.

Fig. 8 Map 10

Cryptogramma crispa (L.) R. Br.
var. **sitchensis** (Rupr.) C.Chr.
Similar to var. *acrostichoides* but with sterile fronds broadly triangular, finely dissected, the ultimate segments obovate; the fertile segments shorter.

Rare on calcareous rocks in the Mackenzie Mountains.

General distribution: Alaska eastward to the Mackenzie Mts.

Map 11

Cryptogramma Stelleri (Gmel.) Prantl
Fragile Rock-brake
Small delicate ferns with fronds rising singly from the slender horizontal rhizome; in the sterile frond the blade is ovate to ovate-deltoid with oblong to obovate flabelliform segments; the fertile is slightly longer and stiffer, with linear, lanceolate or narrowly oblong segments.

Rare on moist shale slopes, Richardson and Mackenzie Mountains.

General distribution: Nearly circumpolar with a broad gap from Lab. to the Ural Mountains; in America, Alaska to Lab. but very local and restricted to limestone outcrops.

Fig. 9 Map 12

Cystopteris Bernh.

Delicate rhizomatose ferns; sori dorsal, round, attached to the veins, separate, the indusium hood-like, attached on the side towards the mid-rib.

a. Blade lanceolate; stipe shorter than blade .... *C. fragilis*
a. Blade deltoid; stipe twice as long as the blade........................... *C. montana*

Cystopteris fragilis (L.) Bernh.
Fragile Fern
Fronds up to 30 cm tall, the blade lanceolate, twice to thrice-pinnate, the pinnules shiny and somewhat translucent, ovate to oblong-ovate; stipes brittle.

Damp rocky slopes

General distribution: Circumpolar, wide-ranging arctic-alpine.

Fig. 10 Map 13

Cystopteris montana (Lam.) Bernh.
Fronds 10 to 30 cm long, arising singly from a cord-like rhizome, the blade deltoid on a long stipe, thrice-pinnate to thrice-pinnate pinnatifid, the ultimate segments small, oblong to ovate-oblong.

Damp calcareous woodlands; rare in the southern Mackenzie Mountains and southwest Mackenzie lowlands.

General distribution: Circumpolar but always rare and local, mainly growing in damp calcareous places.

Fig. 11 Map 14

Dryopteris Adans. Shield-Fern

Small to large ferns, the fronds either in clusters from stout rhizomes or singly from a slender rhizome; fronds pinnate or ternate, the stipes densely covered or with but few scales; sori round, situated on the veins of the lower surface of the pinnae; indusium, when present, roundish-reniform, attached at its sinus, persistent.

a. Rootstock thick, covered by persistent remains of old stipes
  b. Fronds leathery, stiff, persisting and green for several seasons, densely clustered; stipe and blade chaffy, indusia present; dwarf species up to 25 cm tall . . . . . *D. fragrans*
  b. Fronds plumose deciduous, taller species up to 1 m tall
    c. Fronds twice pinnate pinnatifid; basal pinnule of the lowest pinna only slightly larger than its opposite. . . . . *D. spinulosa*
    c. Fronds thrice pinnate pinnatifid; basal pinnule of the lowest pinna much larger than its opposite . . . . *D. dilatata*
a. Rootstock slender, creeping; low, delicate ferns with smooth stipe and triangular blade; indusia lacking
  d. Fronds ternate
    e. Rachis and blade glandless . . . . *D. disjuncta*
    e. Rachis and blade densely glandular . . . . *D. Robertiana*
  d. Fronds pinnate . . . . *D. Phegopteris*

Dryopteris dilatata (Hoffm.) Gray *s.l.*
D. austriaca (Jacq.) Woynar
Spinulose Wood Fern.
Fronds tufted from a stout rhizome, to 1 m long, the blade delicate, oblong to ovate, acuminate, tri-pinnate, the segments toothed, the teeth spinulose-tipped; the stipes scaly.

Rare, in our area known only from rich soil by hot springs, Flat River, Mackenzie Mountains.

General distribution: Circumpolar, but with wide gaps. Because of taxonomic problems, only the western population is mapped.

Map 15

Dryopteris disjuncta (Ledeb.) Morton
D. Linnaeana C. Chr.
Gymnocarpium dryopteris (L.) Newm.
Oak Fern
Fronds up to 3 dm long singly or few together from a brittle, black and shiny rhizome, the blade pale green, ternate and broadly deltoid, on a slender stipe; the sori small, lacking indusia.

Rich and mainly deciduous woods; rare.

General distribution: Circumpolar, widespread.

Fig. 12 Map 16

Dryopteris fragrans (L.) Schott. *s. lat.*
Fragrant Shield Fern
Fronds coarse and densely tufted from a stout rootstock which is covered by the persistent remains of shrivelled and curled old fronds. Fronds 15 to 20 cm high, stout, wintergreen and somewhat leathery, linear-lanceolate, the pinnae pinnatifid or crenate, dark green above, rust-coloured beneath, as well as on the rachis and stipe, from coarse chaff. Fronds spicy-fragrant in drying.

On cliffs and rock screes, and always on non-calcareous rocks.

General distribution: Circumpolar, arctic-alpine.

Fig. 13 Map 17

Dryopteris Phegopteris (L.) C. Chr.
Thelypteris phegopteris (L.) Slosson
Phegopteris connectilis (Michx.) Watt
Beech-Fern
Fronds scattered or tufted from slender forking rhizomes; stipe brittle, chaffy at first, the blade herbaceous, triangular, long-acuminate.

On rich humus in shade; in our area known only from the vicinity of hot springs along Flat River in the Mackenzie Mountains.

General distribution: Circumpolar, widespread.

Fig. 14 Map 18

Dryopteris Robertiana (Hoffm.) C. Chr.
Gymnocarpium Robertianum (Hoffm.) Newm.
Similar to *D. disjuncta*, but the frond more narrowly deltoid; rachis and frond dull green and minutely glandular-puberulent.

Shaded calcareous ledges; rare in the southern Mackenzie Mountains and southwest Mackenzie lowlands.

General distribution: Circumpolar, but always local because of its special habitat preferences.

Map 19

Dryopteris spinulosa (O.F. Müll.) Watt
Similar to *D. dilatata* but shorter (up to 5.5 dm), and the blades less divided.

Within our area known only from rich woods north of the height of land between Great Slave and Great Bear Lakes and by a hotspring in the southern Mackenzie Mountains.

General distribution: Boreal American, widespread and common, particularly in eastern N. America.

Fig. 15 Map 20

Matteuccia Todaro Ostrich Fern

Matteuccia Struthiopteris (L.) Todaro
var. **pensylvanica** (Willd.) Morton;
Pteretis pensylvanica (Willd.) Fern.
Fronds dimorphous, the sterile green, 1 m or more long, pinnate-pinnatifid, forming a crown at the top of a stout caudex, the fertile few, dark brown in age and much shorter, persistent during winter; the sori borne on the margins of the shallowly lobed, tightly inrolled, pod-like pinnae.

In the District of Mackenzie known from alluvial terraces of the Liard River between Nahanni Butte and the British Columbia border, and about a hotspring on the South Nahanni River.

General distribution: Particularly common in eastern N. America.

Fig. 16 Map 21

Pellaea Link Cliff-brake

Pellaea glabella Mett. ex Kuhn
var. **nana** (Richards.) Cody
Fronds brittle, up to 1.2 dm long, pinnate, the pinnae simple, oblong, the basal ones sometimes lobed, greyish-green; stipes lustrous dark brown with numerous brown scales at the base; sori continuous, covered by the inrolled margins of the fertile pinnae.

Rare; in our area known only from two collections: West ridge of the Franklin Mountains and Nahanni Range of the Mackenzie Mts.; crevices of sunny limestone rock faces.

General distribution: Western North America; very local and apparently restricted to limestone exposures.

Fig. 17 Map 22

Polypodium L. Polypody

Polypodium vulgare L.
ssp. **virginianum** (L.) Hultén
Fronds evergreen, occasionally up to 2 dm tall but usually less, singly along the rope-like and scaly rhizome; blade lanceolate, deeply pinnatifid, the segments oblong-linear; sori large, rounded, lacking an indusium.

Rare or occasional on mossy ledges and cliffs in wooded parts of the Precambrian Shield area, north to Great Bear Lake; in the Mackenzie Mts. known only from Mt. Coty opposite Ford Liard.

General distribution: *P. vulgare*, circumpolar; the ssp. *virginianum* wide-ranging from Lab. to B. C. and S. W. District of Mackenzie.

Fig. 18 Map 23

Polystichum Roth

Polystichum Lonchitis (L.) Roth
Holly Fern
Rhizome stout and scaly; fronds tufted, up to 4 dm tall, narrowly lanceolate, coriaceous and evergreen, the pinnae very numerous, spiny; the sori mostly in two rows.

An alpine species of moist and often somewhat shaded and rocky slopes north in the Rocky Mountains, with a single disjunct station in the Mackenzie Mountains on the Yukon-Mackenzie divide.

General distribution: Circumpolar with large gaps.

Map 24

Woodsia R. Br. Woodsia

Low, tufted ferns with short, freely branched rhizome. Fronds pinnate. Sori round, borne on the back of the veins; indusium attached under the sori, divided into numerous hair-like or lacerate segments.

a. Stipe articulate near the base; dwarf species
 b. Stipe and rachis pale or straw-coloured; fronds linear, glabrous, yellowish-green, thin and delicate . . . . *W. glabella*
 b. Stipe brown, firm, wiry, and more or less chaffy
  c. Fronds oblong-lanceolate, glabrous or promptly glabrous; stipe pale brown, glabrous or with a few caducous scales . . . . *W. alpina*
  c. Fronds lanceolate and stiff, permanently hairy and chaffy on the underside; stipes dark brown, covered with brownish chaff . . . . *W. ilvensis*
a. Stipe not articulate at base; larger species, fronds glabrous or the blade glandular, narrowly lance-oblong . . . . *W. oregana*

Woodsia alpina (Bolton) S. F. Gray
Northern Woodsia
Fronds to 8 cm long, oblong-lanceolate, glabrous or promptly glabrate, chaffless; sori near the margin.

Rock crevices and rock screes; rare and local, and perhaps always on calcareous or non-acid rocks.

General distribution: Circumpolar, wide-ranging, arctic-alpine.

Fig. 19 Map 25

Woodsia glabella R. Br.
Smooth Woodsia
Fronds glabrous, bright green, linear, to 15 cm long, but usually shorter; pinnae sub-orbicular to ovate, variously cleft or lobed.

Rare in moist, calcareous and often shaded rock crevices or talus slopes.

General distribution: Circumpolar, wide-ranging, arctic-alpine.

Fig. 20 Map 26

Woodsia ilvensis (L.) R. Br.
Rusty Woodsia
Similar to but coarser than *W. alpina*; forming dense tufts from a stout rhizome; fronds to 20 cm long, the pinnae oblong, crenate to pinnatifid, often with revolute margins; rachis and blade usually quite rusty chaffy.

Dry rocky places; usually on Precambrian or acid rocks; occasional northward to Great Bear Lake and Coppermine; rare in the Mackenzie Mountains and apparently absent in the mainly alluvial Mackenzie lowlands.

General distribution: Circumpolar, wide-ranging, low-arctic.

Fig. 21 Map 27

***Woodsia oregana** D. C. Eaton
Oregon Woodsia
Densely tufted, the fronds bright green, lance-oblong, glabrous to somewhat glandular, pinnate, to 25 cm long; pinnae triangular-oblong, deeply pinnatifid, the lobes crenate and often revolute.

Known from a number of places in the upper Mackenzie drainage just south of the 60th parallel in B.C., Alta. and Sask. and to be expected in the southern part of the Mackenzie Mts.

Mainly on calcareous rocks.

General distribution: Cordilleran foothill species; local and rare eastward to the Great Lakes and Gaspé Peninsula.

Fig. 22 Map 28

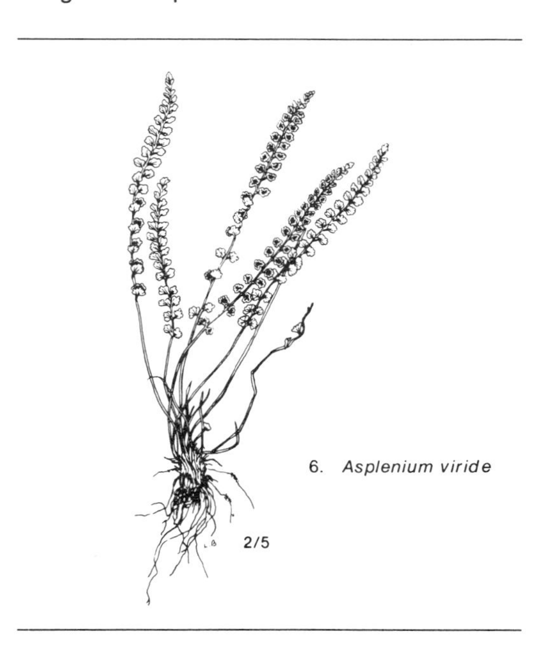

6. *Asplenium viride*

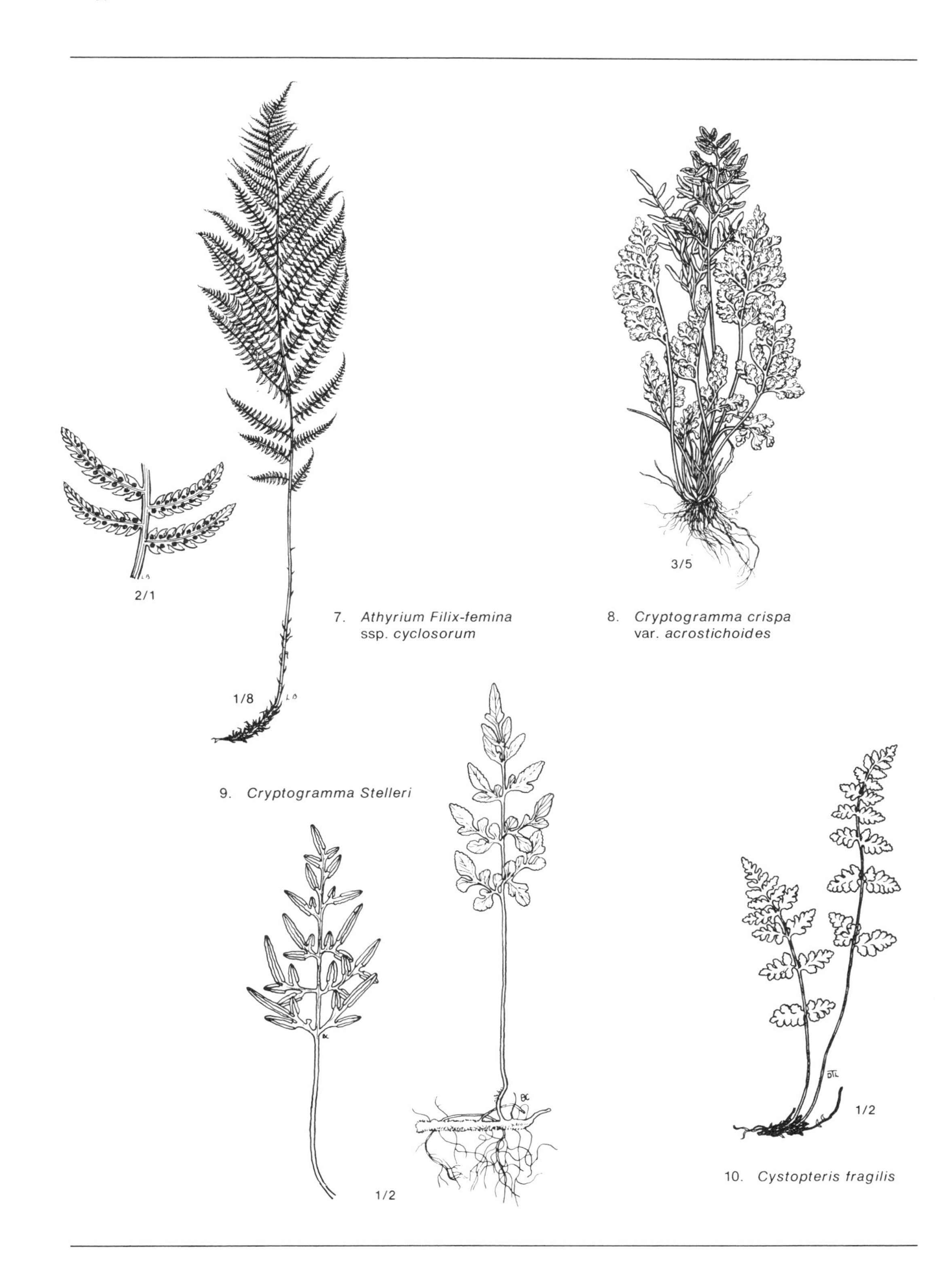

7. *Athyrium Filix-femina* ssp. *cyclosorum*

8. *Cryptogramma crispa* var. *acrostichoides*

9. *Cryptogramma Stelleri*

10. *Cystopteris fragilis*

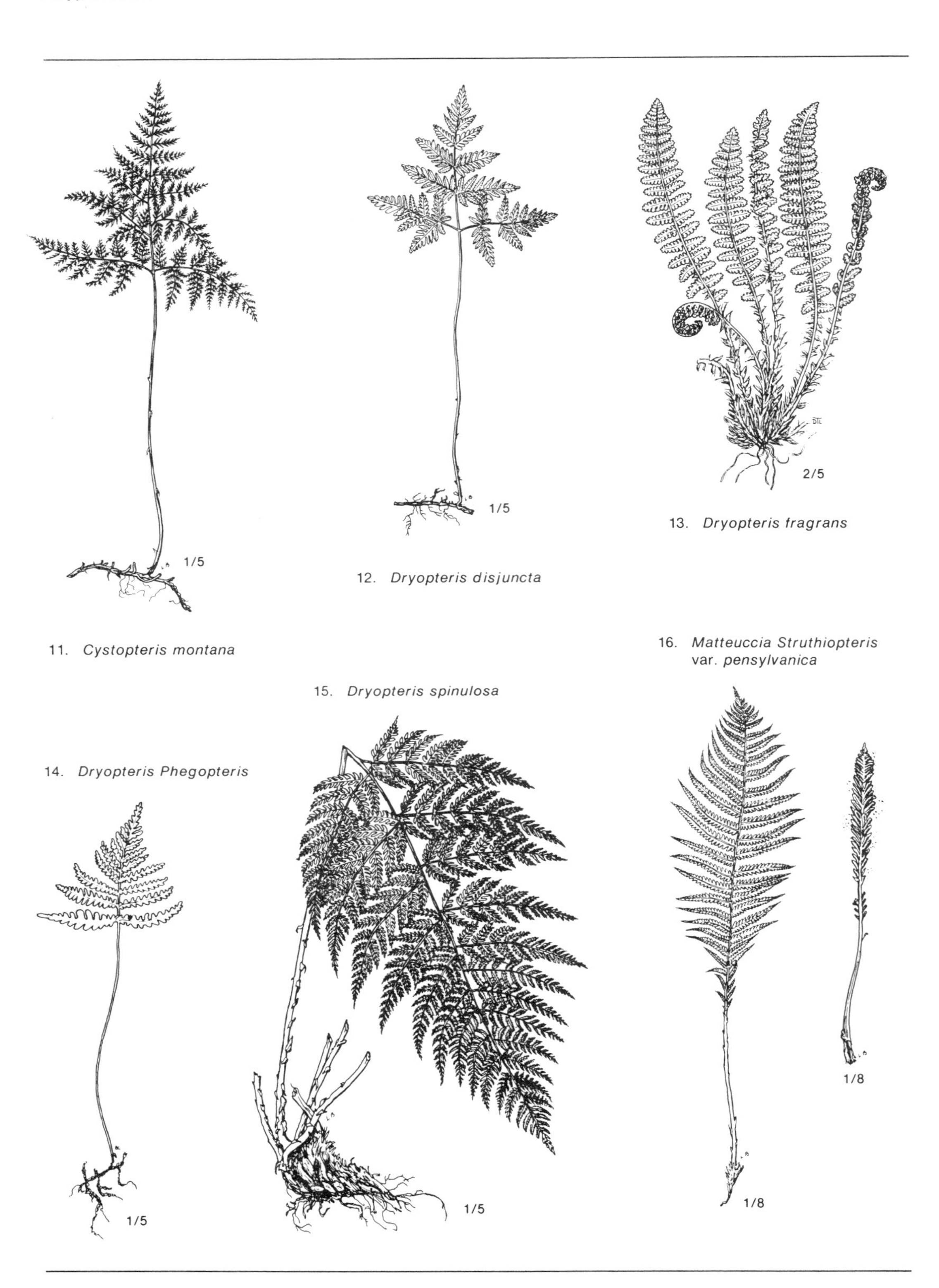

11. *Cystopteris montana*

12. *Dryopteris disjuncta*

13. *Dryopteris fragrans*

14. *Dryopteris Phegopteris*

15. *Dryopteris spinulosa*

16. *Matteuccia Struthiopteris* var. *pensylvanica*

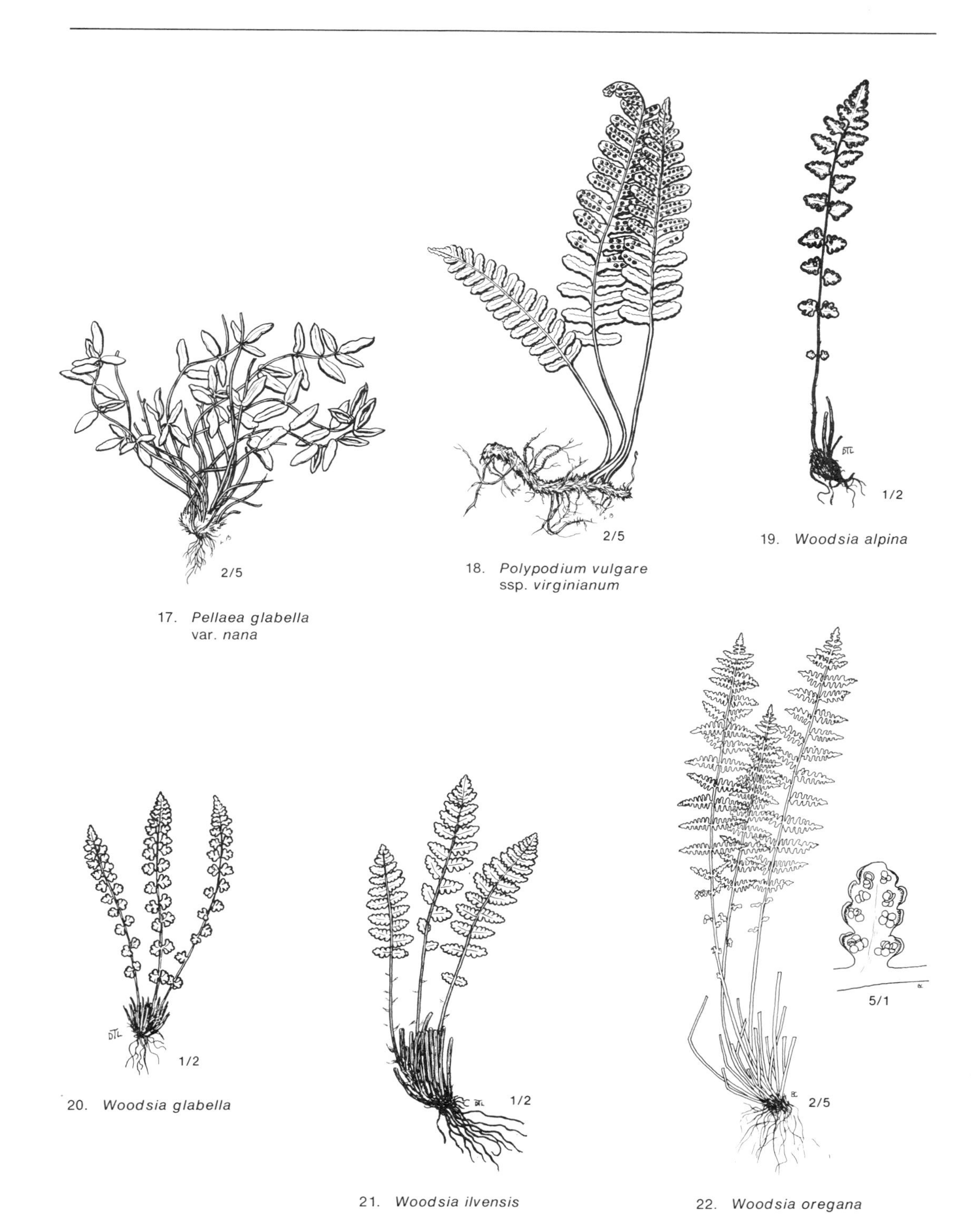

17. *Pellaea glabella* var. *nana*

18. *Polypodium vulgare* ssp. *virginianum*

19. *Woodsia alpina*

20. *Woodsia glabella*

21. *Woodsia ilvensis*

22. *Woodsia oregana*

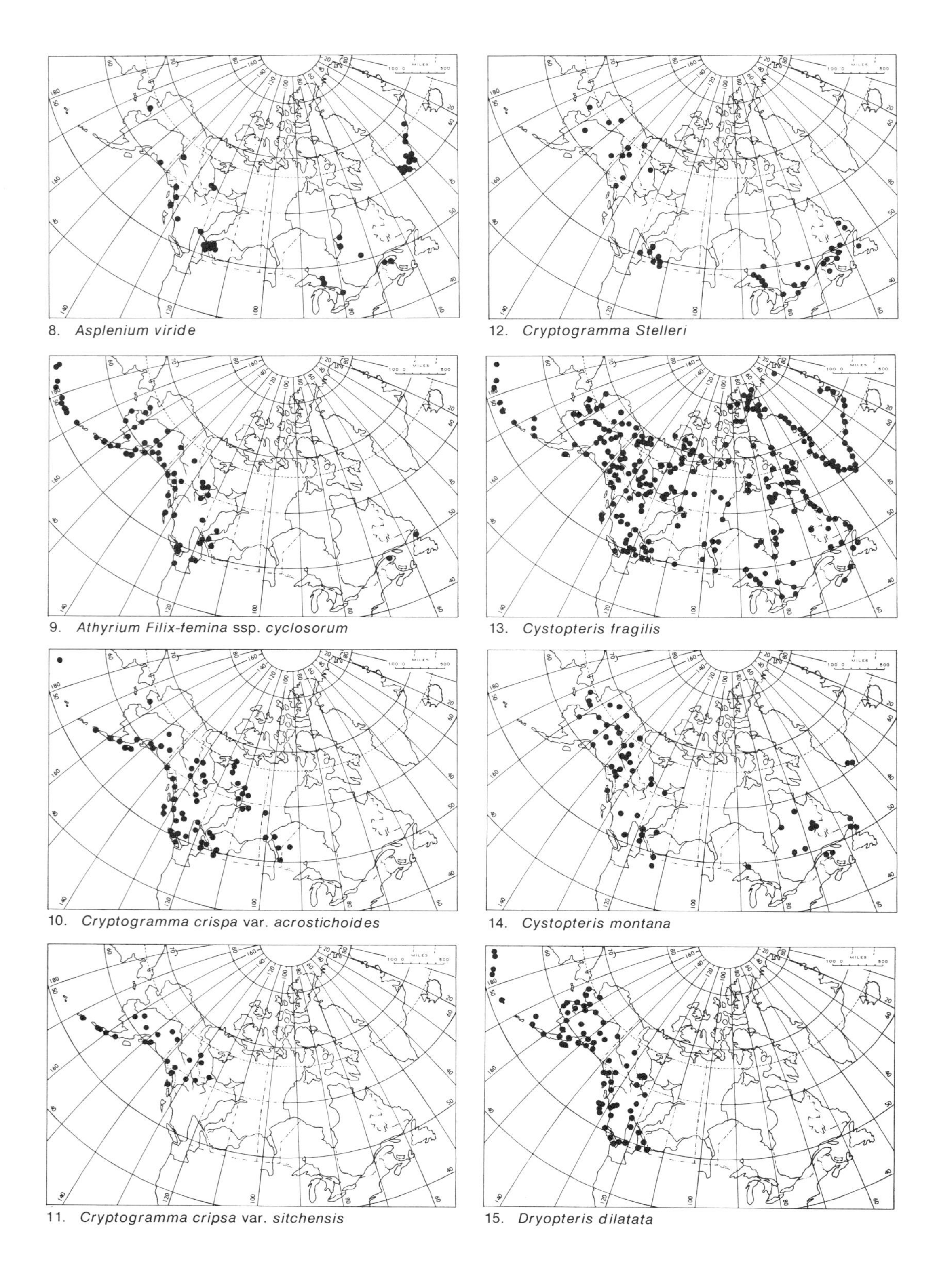

8. *Asplenium viride*

9. *Athyrium Filix-femina* ssp. *cyclosorum*

10. *Cryptogramma crispa* var. *acrostichoides*

11. *Cryptogramma cripsa* var. *sitchensis*

12. *Cryptogramma Stelleri*

13. *Cystopteris fragilis*

14. *Cystopteris montana*

15. *Dryopteris dilatata*

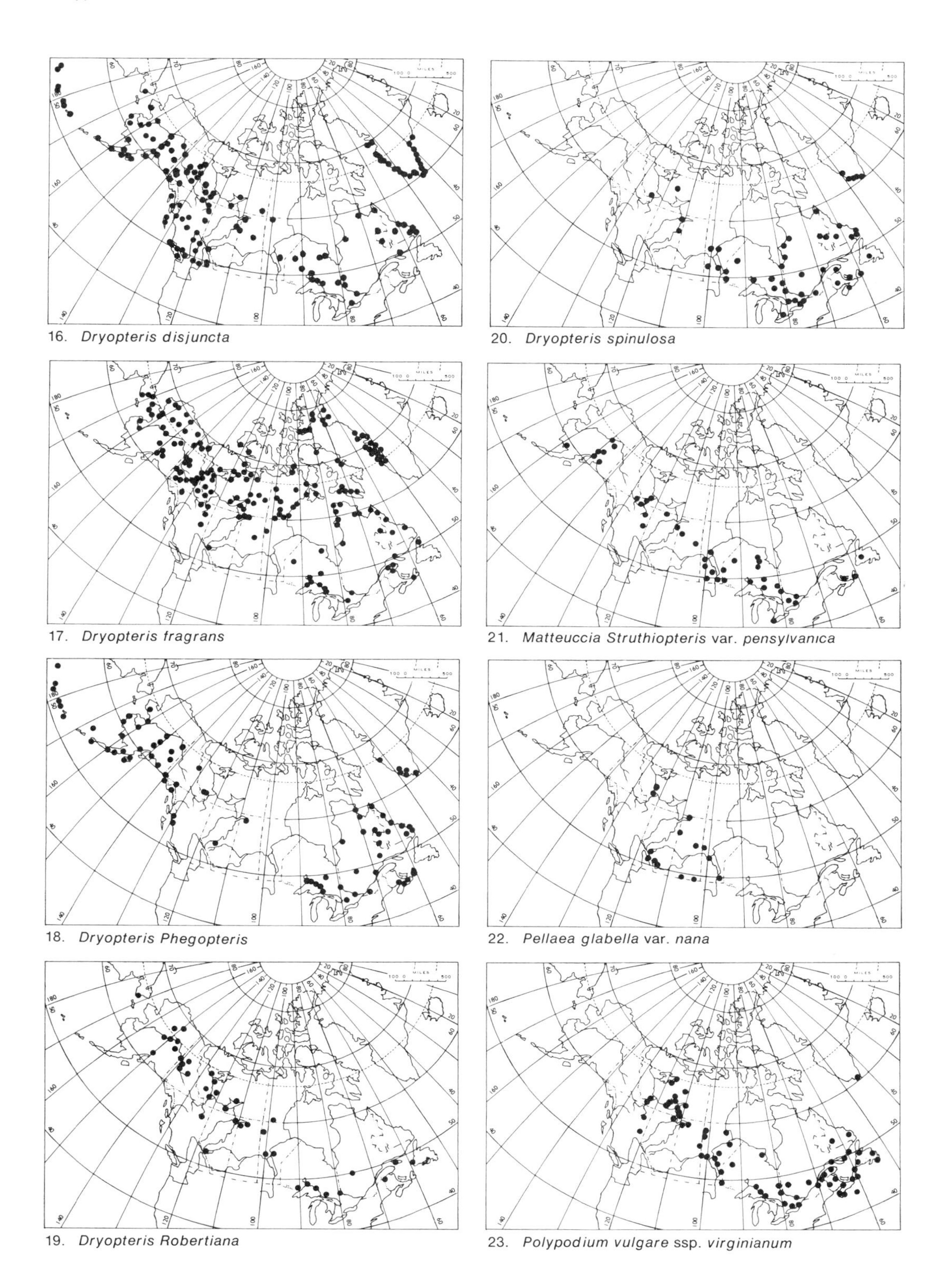

16. *Dryopteris disjuncta*

20. *Dryopteris spinulosa*

17. *Dryopteris fragrans*

21. *Matteuccia Struthiopteris* var. *pensylvanica*

18. *Dryopteris Phegopteris*

22. *Pellaea glabella* var. *nana*

19. *Dryopteris Robertiana*

23. *Polypodium vulgare* ssp. *virginianum*

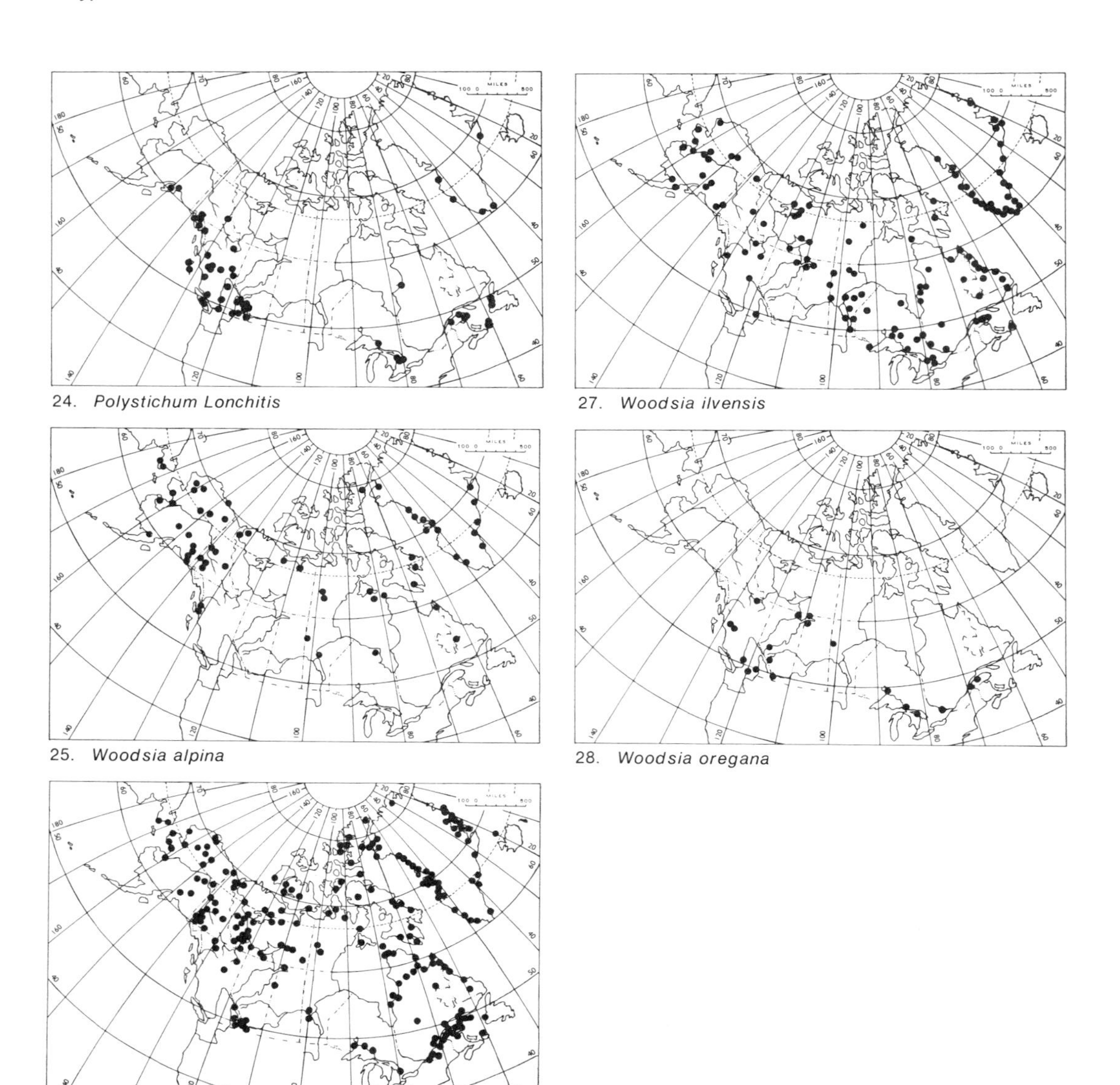

24. *Polystichum Lonchitis*

25. *Woodsia alpina*

26. *Woodsia glabella*

27. *Woodsia ilvensis*

28. *Woodsia oregana*

EQUISETACEAE Horsetail Family

Equisetum L. Horsetail

Perennial spore-bearing plants with creeping, freely branched, shiny, black or reddish-brown rhizomes; stems annual or perennial, jointed, hollow and cylindrical, some simple, reed-like, some with whorls of green branches from the nodes; leaves reduced to toothed sheaths which encase the solid joints. Sporophylls aggregated in terminal, cone-like strobils.

a. Stems annual
  b. Stems dimorphous the vernal fertile, followed by sterile, green stems bearing whorls of branches
    c. Fertile stems short lived, lacking chlorophyll; branches of sterile stems mostly simple, ascending
      d. Teeth of primary sheaths dark brown throughout . . . . *E. arvense*
      d. Teeth of primary sheaths with conspicuous white margins. . . . . *E. pratense*
    c. Fertile stems fleshy, soon forming green branches; branches of sterile stems compound recurving; teeth of sheaths light brown . . . . *E. sylvaticum*
  b. Sterile and fertile stems similar, green
    e. Central cavity of main stems about 1/6 of the diameter . . . . *E. palustre*
    e. Central cavity of main stems 1/2 or more of the diameter . . . . *E. fluviatile*
a. Stems evergreen, simple or forking, lacking whorled branches from the nodes
  f. Central cavity of stem large; stems stout, stiff, harsh to the touch . . . . *E. hiemale* var. *affine*
  f. Central cavity small or almost lacking
    g. Stems filiform, flexuous; teeth of the sheaths 3 to 4, long-acuminate . . . *E. scirpoides*
    g. Stems stiff and straight; teeth of the sheaths 6 to 8, broadly lanceolate. . . . . *E. variegatum*

Equisetum arvense L.
Common Horsetail
Stems dimorphic, the fertile appearing early in the spring, simple, erect, shortlived and lacking chlorophyll; the sterile stems green and branched, erect or decumbent, deciduous; teeth of primary sheaths dark throughout; teeth of sheaths of branches lance-attenuate.

Wide-ranging arctic-boreal species ubiquitous in not too dry habitats throughout the area.

General distribution: Circumpolar, wide-ranging from the high Arctic south to temperate regions.

Fig. 23 Map 29

Equisetum fluviatile L.
E. limosum L.
Water Horsetail
Stems green, deciduous, hollow, usually unbranched, but occasionally with whorls of short branches at the nodes; teeth of sheaths 15 to 20, fine, dark.

Boreal forest species of sloughs and wet river meadows throughout the Mackenzie lowlands north to the Mackenzie River Delta and eastwards into the Eskimo Lake Basin and Great Bear Lake.

General distribution: Circumpolar, wide-ranging.

Fig. 24 Map 30

Equisetum hyemale L.
var. **affine** (Engelm.) A. A. Eaton
E. prealtum Raf.
Scouring-Rush
Stems up to 1 m tall, grey-green and rough to the touch, hollow, mostly single; sheaths usually with blackish rings near the base and summit, the teeth narrow, deciduous.

River bars, sandy lakeshores and upland prairies north along the Mackenzie River to lat. 65°N.

General distribution: Boreal North American; *E. hyemale s. lat.* is circumpolar; wide-spread.

Fig. 25 Map 31

Equisetum palustre L.
Marsh-Horsetail
Stems erect deciduous to 2 dm or more tall, simple or branched, deeply 5 to 10 ridged, the central cavity small; sheaths of main stems expanding upwards, the teeth 10 or less, blackish, usually with hyaline margins.

Occasional to common on wet river meadows

and mud flats where it is often a pioneer species; north to the limit of trees.

General distribution: Circumpolar, wide-ranging.

Fig. 26 Map 32

Equisetum pratense Ehrh.
Meadow-Horsetail
Similar to *E. arvense* but not dimorphous; the fertile stems eventually develop simple, ascending branches paler than those of *E. arvense;* also the nodes not swollen, and the teeth rarely over 3 mm long, with pale white margins.

Forms dense stands in woodland on river terraces north to Fort Simpson, rare in moist wooded upland habitats; rarer northward to Norman Wells and the Mackenzie Mts.

General distribution: Circumpolar, wide-ranging.

Fig. 27 Map 33

Equisetum scirpoides Michx.
Dwarf Scouring-Rush
Low and often densely matted, the slender stems often interwoven among other plants; sheaths dark, the 3 or 4 teeth deltoid with broadly scarious margins and brown subulate more or less deciduous tips.

Common in dry, mossy woods throughout the region, becoming less frequent north beyond the limit of forest to the Arctic Coast and islands.

General distribution: Circumpolar, low-arctic.

Fig. 28 Map 34

Equisetum sylvaticum L.
var. **pauciramosum** Milde
Wood-Horsetail
Stems 1.5 to 3.5 dm high; sheaths of the main stem inflated, brownish, its membranous teeth fused to form 3 or 4 lobes; teeth of sheaths of branches usually divergent.

In muskeg forest and occasionally in sandy situations northward to near the limit of trees.

General distribution: *E. sylvaticum s. lat.* circumpolar; var. *pauciramosum* in boreal North America and southern Greenland.

Fig. 29 Map 35

Equisetum variegatum Schleich.
Variegated Horsetail
Similar to *E. scirpoides* but more robust and with upright straighter stems; stems hollow at the centre; lower sheaths dark, the upper ashy, the 5 to 10 teeth lanceolate, broadly scarious margined, with or without filiform tips.

Occasional throughout most of our area in mostly wet alluvial sand and clay and sometimes in moss by brooks; more frequent in alpine and arctic situations.

General distribution: Circumpolar, wide-ranging, arctic-alpine.

Fig. 30 Map 36

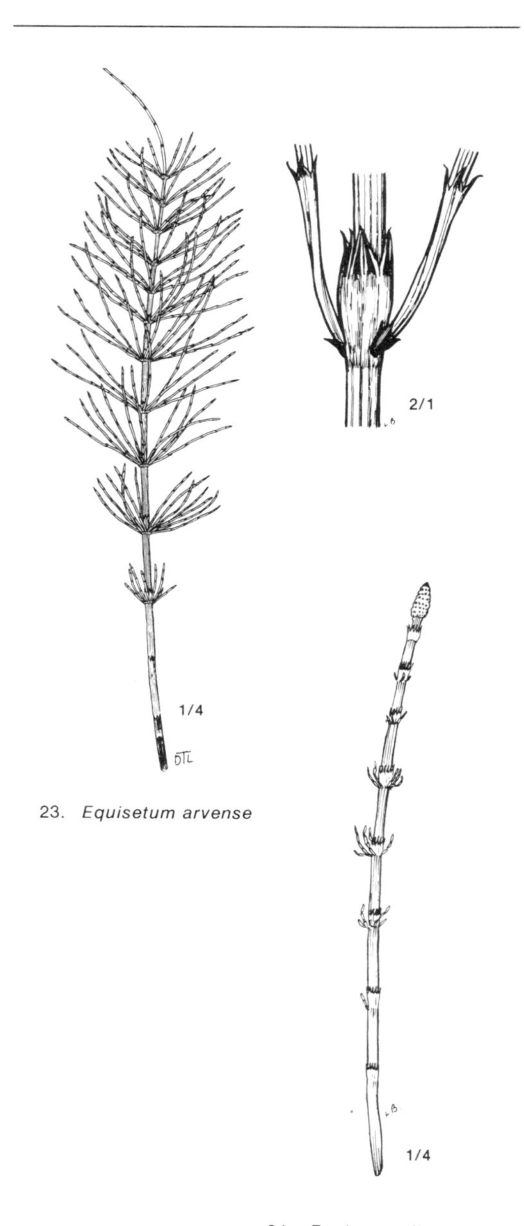

23. *Equisetum arvense*

24. *Equisetum fluviatile*

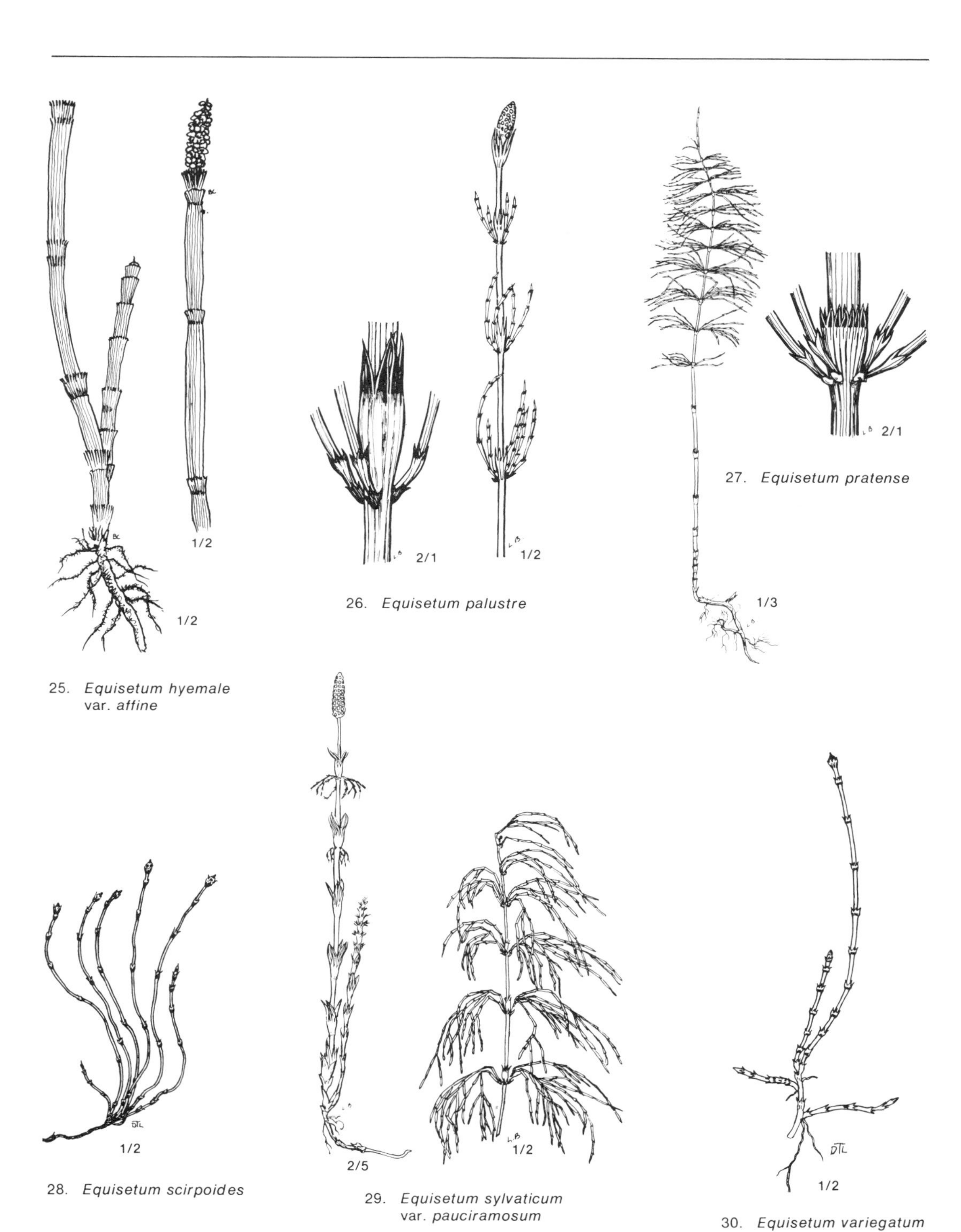

25. *Equisetum hyemale* var. *affine*

26. *Equisetum palustre*

27. *Equisetum pratense*

28. *Equisetum scirpoides*

29. *Equisetum sylvaticum* var. *pauciramosum*

30. *Equisetum variegatum*

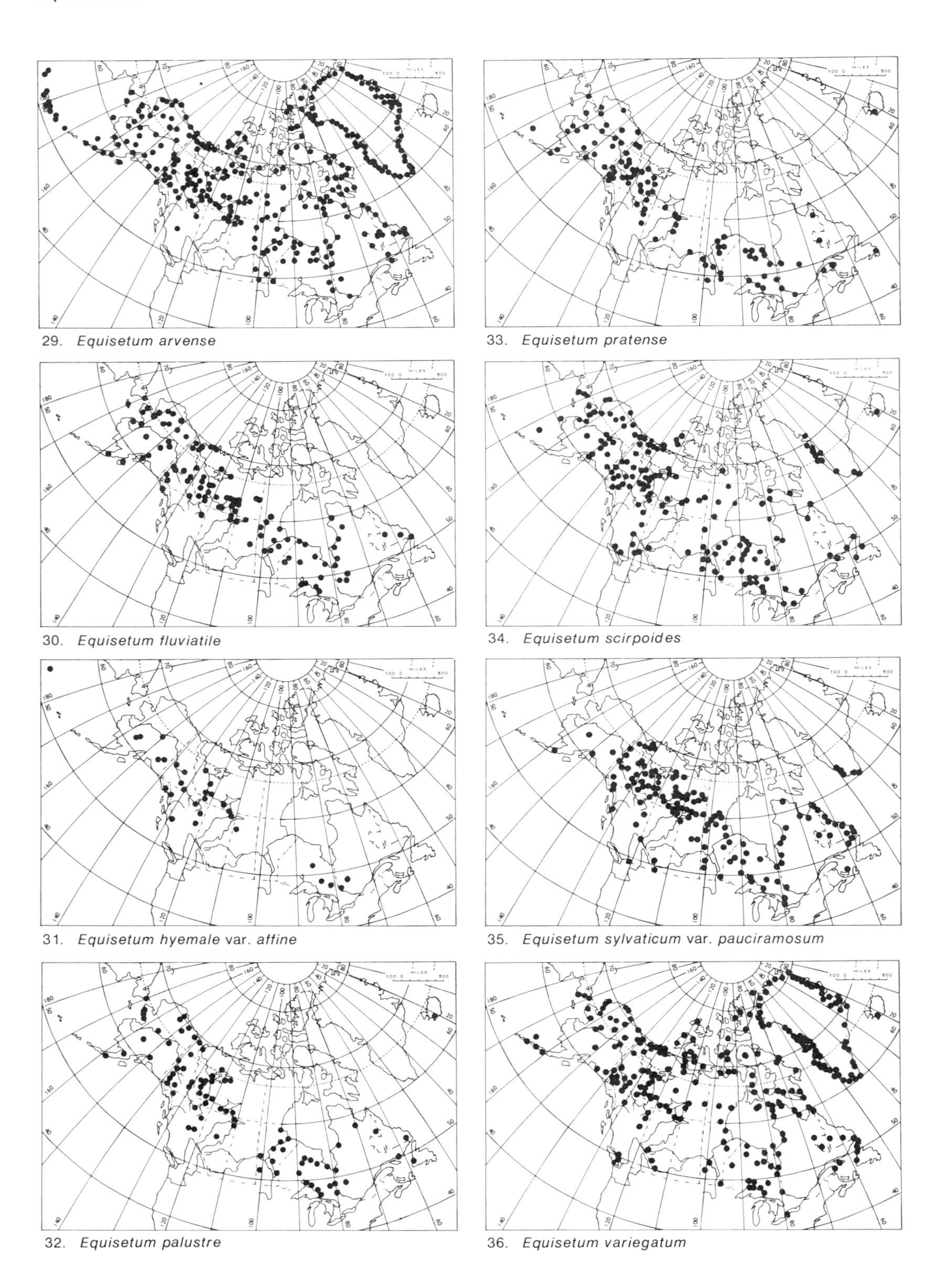

29. *Equisetum arvense*

30. *Equisetum fluviatile*

31. *Equisetum hyemale* var. *affine*

32. *Equisetum palustre*

33. *Equisetum pratense*

34. *Equisetum scirpoides*

35. *Equisetum sylvaticum* var. *pauciramosum*

36. *Equisetum variegatum*

LYCOPODIACEAE Club-Moss Family
Lycopodium L. Club-Moss

Evergreen, with creeping or tufted, freely branching stems covered by rows of imbricated or spreading, scale-like leaves. Sporophylls in terminal sessile or peduncled cone-like strobils, except in *L. Selago* where the sporangia are axillary and reniform. Spores very numerous, powdery, yellow and, being highly inflammable, formerly used in the manufacturing of fireworks.

a. Sporangia in terminal cone-like, leafy-branched strobiles; plant rhizomatose
  b. Leaves of aerial branches in 6 to 10 ranks
    c. Creeping stems deeply buried; strobiles sessile. *L. obscurum*
    c. Creeping stems on or slightly beneath the surface
      d. Leaves with sharp spine-like tips or bluntish; strobiles sessile
        e. Leaves thick, the lower part adnate to the stem, bluntish . . . . *L. sabinaefolium* var. *sitchense*
        e. Leaves thinner, decurrent but not adnate, long attenuate, cuspidate, serrated or entire . . . . *L. annotinum*
      d. Leaves with long, hair-like tips; strobiles long-peduncled . . . . *L. clavatum*
  b. Leaves of aerial branches mostly 4-ranked
    f. Strobiles sessile . . . . *L. alpinum*
    f. Strobiles pedunculate. . . . . *L. complanatum*
a. Sporangia axillary, not cone-like; plant tufted . . . . *L. Selago*

Lycopodium alpinum L.
Alpine Club-Moss
Stem pale green, creeping slightly beneath the surface, bearing tufts of glaucous branches; sterile branches flattened, with 4-ranked leaves, these acuminate, entire, the dorsal and marginal adnate and decurrent, the ventral trowel-shaped; sporangia in a sessile strobile.
Alpine snowbeds in the Mackenzie and Richardson Mts. and summit of Scented Grass Hills, Great Bear Lake.
General distribution: Circumpolar, low-arctic, alpine.
Fig. 31 Map 37

Lycopodium annotinum L., *s. lat.*
Bristly Club-Moss
Stems creeping mostly above the ground, up to one metre or more long, the curved branches in fascicles along the stem; leaves 8-ranked, stiff, long attenuate with sharp spine-like tips, appressed to more or less reflexed; sporangia in a single sessile strobile.
Dry sunny places such as open heath or in herbmats.
General distribution: Circumpolar, low-arctic.
Fig. 32 Map 38

Lycopodium clavatum L.
var. **monostachyon** Grev. & Hook.
Running Club-Moss
Stems prostrate, branched, the aerial occuring singly or in fascicles along the stem; leaves many-ranked, linear-subulate, spreading, tipped with a long hair-like bristle; strobile pedunculate.
Dry open mossy woods.
General distribution: *L. clavatum s. lat.* circumpolar, wide-spread; var. *monostachyon*, South Greenland and boreal North America.
Fig. 33 Map 39

Lycopodium complanatum L. *s. lat.*
Ground-Cedar
Similar in appearance to *L. alpinum*, but larger, the branches up to 2 dm tall, strongly flattened; leaves somewhat glaucous, 4-ranked; the single strobile pedunculate.
Common in open woodland on sands and rocks north to the limit of trees.
General distribution: Circumpolar, wide-spread.
Fig. 34 Map 40

Lycopodium obscurum L.
var. **dendroideum** (Milde) D.C.Eat.
Ground-Pine
Bushy-branched erect stems to 2 dm tall from a deeply buried rhizome; strobiles sessile, singly at the ends of the branches.
Rare or occasional in moist woods and mainly in the southern parts of our area.
General distribution: Wide-spread across boreal N. America to E. Asia
Fig. 35 Map 41

***Lycopodium sabinaefolium** Willd.
ssp. **sitchense** (Rupr.) Calder & Taylor
Ground-Fir.
Upright, densely compact branched stems arising at intervals along a creeping stem which lies at or near the surface; branches terete, the sterile 2 to 10 cm long, usually overtopped by the elongate fertile branches; leaves uniform, bluntish, the upper free part longer than the adnate portion; strobile sessile at the end of the branch.

Dry sub-alpine coniferous woods and barren knolls.

General distribution: Frequent in northwestern United States and adjacent Canada, thence near the coast through Alaska to Kamchatka.

Map 42

Lycopodium Selago L.
Mountain Club-Moss
Stems tufted 2 to 15 cm high with erect ascending simple or dichotomously forked branches; leaves lanceolate, thickened, entire or nearly so, densely clothing the branches; reproductive buds frequently present; sporangia in the upper leaf axils.

Locally common in turfy and mossy tundra.

General distribution: Circumpolar, wide-ranging, high arctic-alpine, reaching lat. 84° in Greenland.

Fig. 36 Map 43

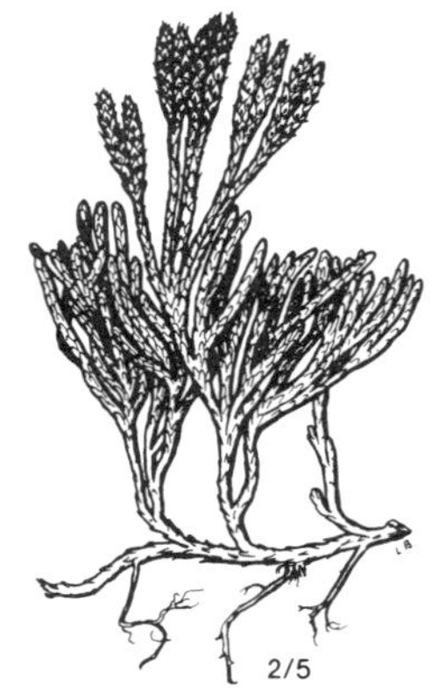

31. *Lycopodium alpinum*

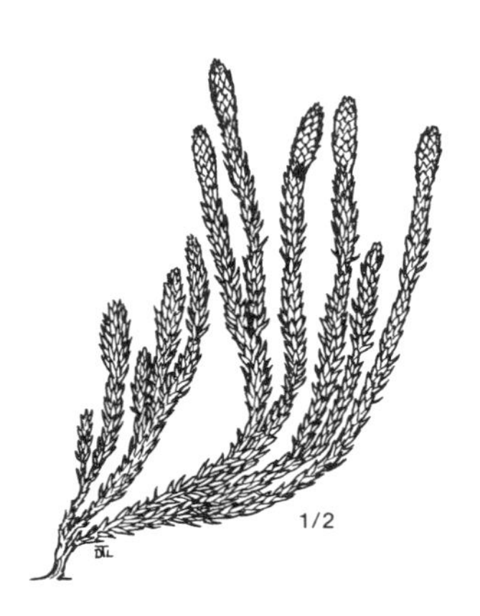

32. *Lycopodium annotinum*

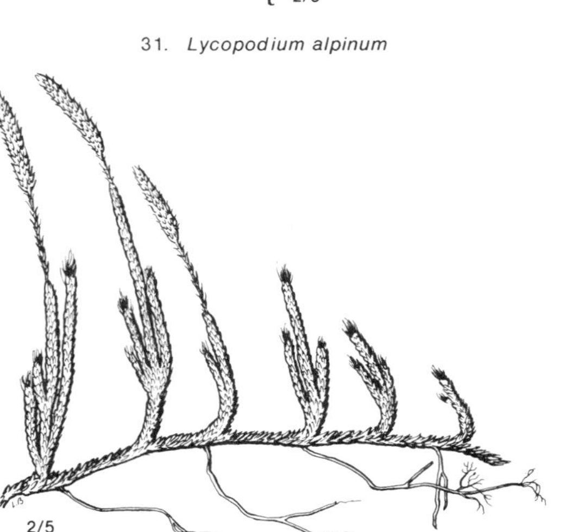

33. *Lycopodium clavatum* var. *monostachyon*

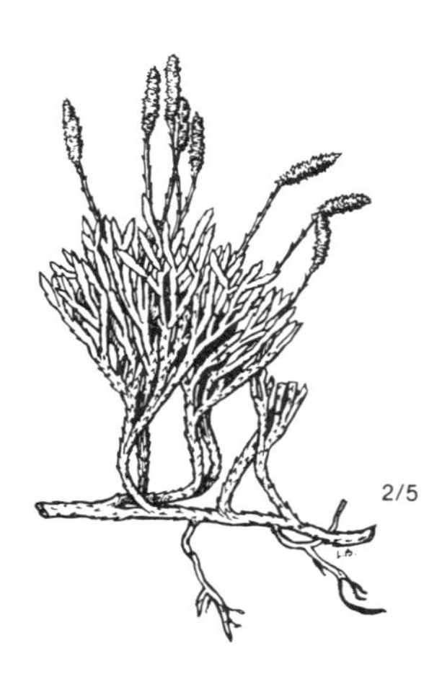

34. *Lycopodium complanatum*

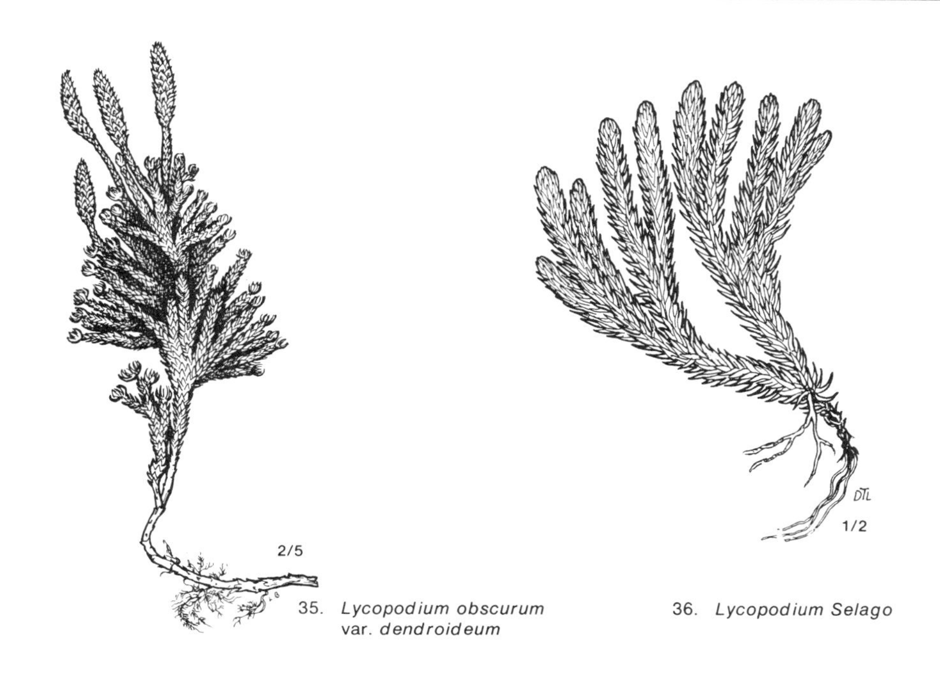

35. *Lycopodium obscurum* var. *dendroideum*

36. *Lycopodium Selago*

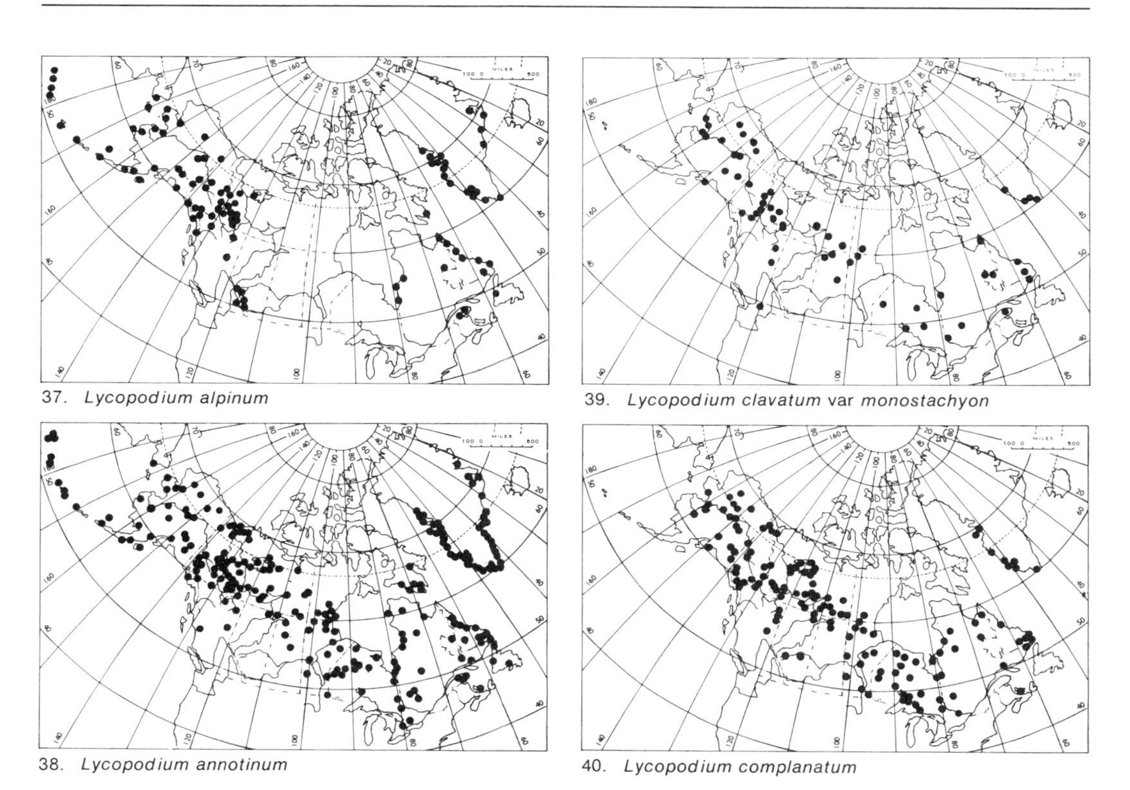

37. *Lycopodium alpinum*

38. *Lycopodium annotinum*

39. *Lycopodium clavatum* var *monostachyon*

40. *Lycopodium complanatum*

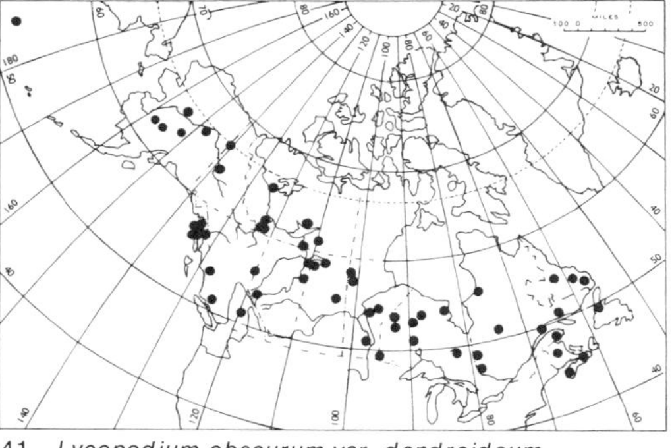
41. *Lycopodium obscurum* var. *dendroideum*

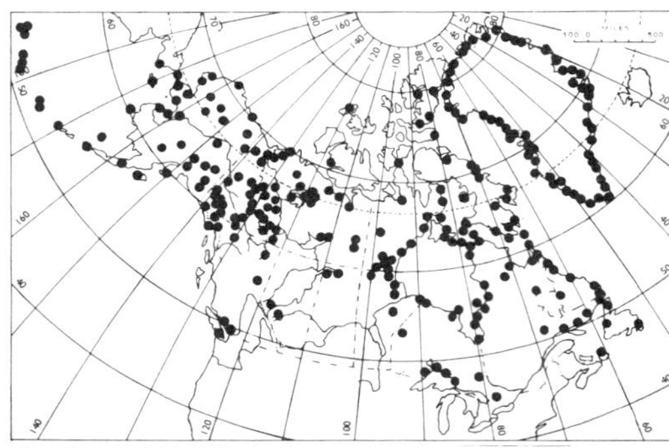
43. *Lycopodium Selago*

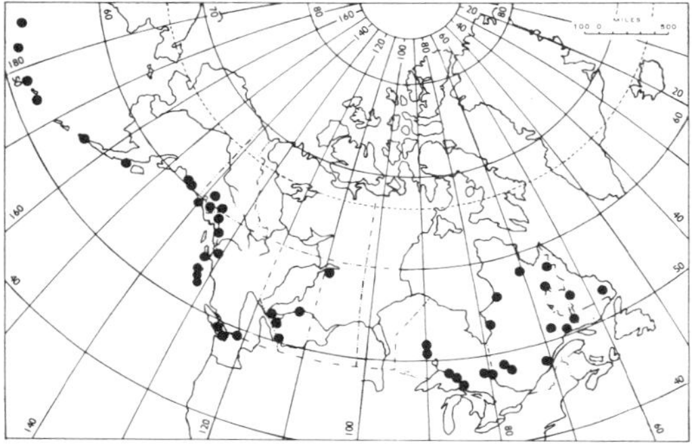
42. *Lycopodium sabinaefolium* ssp. *sitchense*

SELAGINELLACEAE Spikemoss Family

Selaginella Beauv. Spikemoss

Ours dwarf, terrestrial, perennial moss-like plants with branched creeping stems. Leaves small, scale-like, densely imbricated. Sporangia in the axils of leafy bracts of a terminal spike, some containing macrospores, some microspores.

a. Plant delicate herbaceous, yellowish-green . . . . *S. selaginoides*
a. Plant stiff, evergreen
  b. Apex of leaf acute; sporophylls narrowly triangular . . . . *S. rupestris*
  b. Apex of leaf bluntish; sporophylls broadly triangular . . . . *S. sibirica*

Selaginella rupestris (L.) Spring
Forming small mats, the stiff fertile branches 2 to 3 cm long; leaves densely imbricated, tapering to the acute apex, bristle tipped, ciliate; strobiles erect, the sporophylls narrowly triangular.

Usually in exposed rocky and sandy habitats.

General distribution: Boreal eastern America, north to Fort Fitzgerald, Alta. on the Slave River and to be looked for on Precambrian rocks south of Great Slave Lake.

Map 44

Selaginella selaginoides (L.) Link
Forming small mats, the sterile branches prostrate and filiform, with weak yellow-green, broadly lanceolate, acute, remotely dentate leaves; fertile branches upright to 5 cm in height

with a terminal strobile; sporophylls longer than the leaves, spreading.

Often partly buried in mosses in muskeg and similar moist situations north nearly to the limit of trees.

General distribution: Circumpolar, widespread, but probably often overlooked.

Fig. 37 Map 45

Selaginella sibirica (Milde) Hieron.

Similar to *S. rupestris*, but the leaves shorter and bluntish, and the sporophylls broadly triangular.

Rare and local in sandy gravel and rocky places; on the Caribou Hills west of Mackenzie Delta, Richards Island and in the Richardson Mountains.

General distribution: Amphi-Beringian.

Fig. 38 Map 46

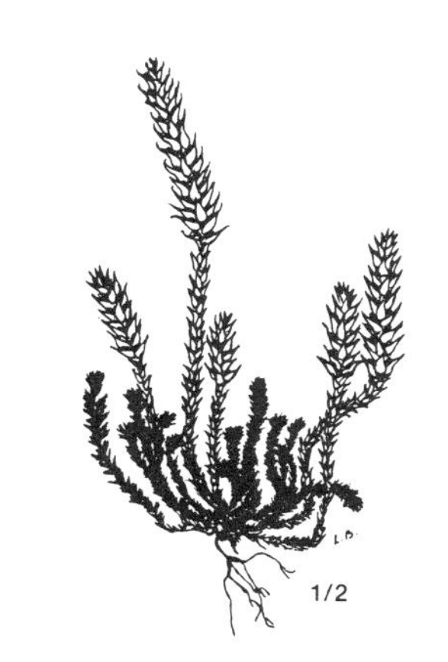

37. *Selaginella selaginoides*

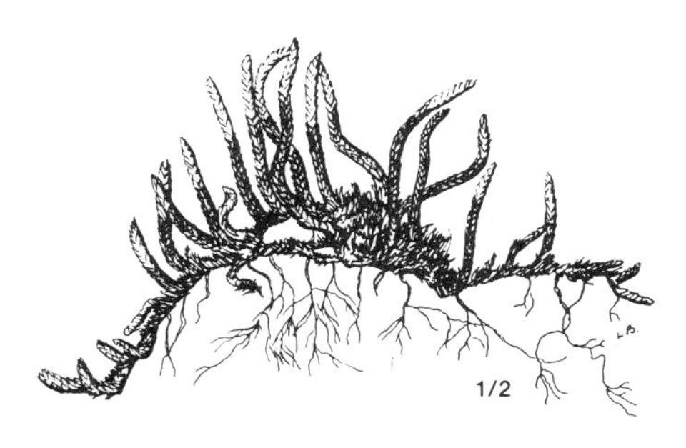

38. *Selaginella sibirica*

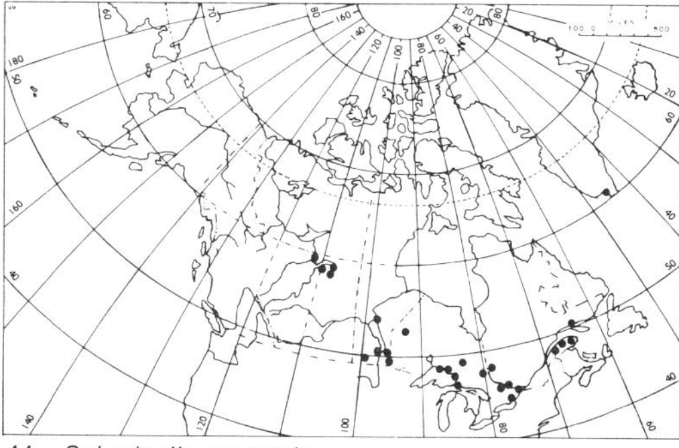

44. *Selaginella rupestris*

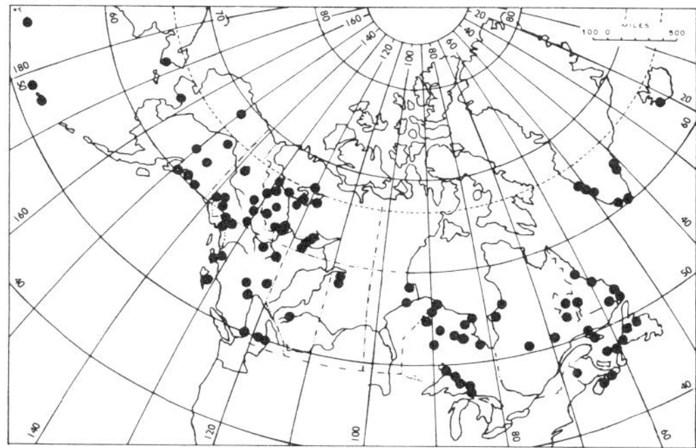

45. *Selaginella selaginoides*

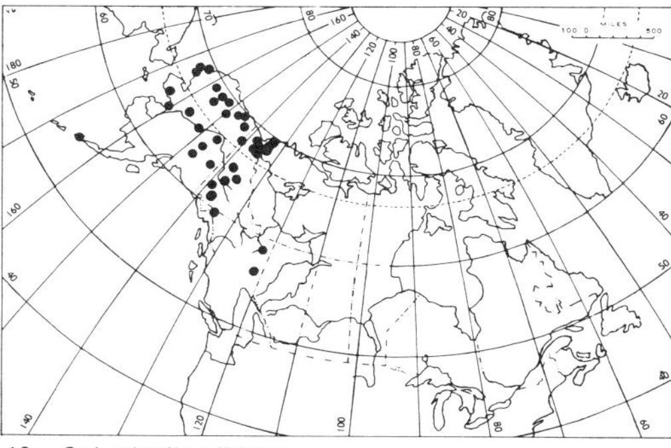

46. *Selaginella sibirica*

ISOETACEAE Quillwort Family

Isoetes L. Quillwort

Isoetes muricata Dur.
var. **Braunii** (Dur.) Reed
Small tufted aquatics with arching tufted leaves; the leaves to 12 cm in length, hyaline margined towards the bulbous base which contains the sporangia; megaspores covered with numerous spinules.

Rooted in silty bottom of shallow bays and small lakes or ponds in the southern part of the Precambrian region north to Great Bear Lake; apparently rare and local although perhaps often overlooked.

General distribution: Boreal America, wide-ranging.

Fig. 39 Map 47

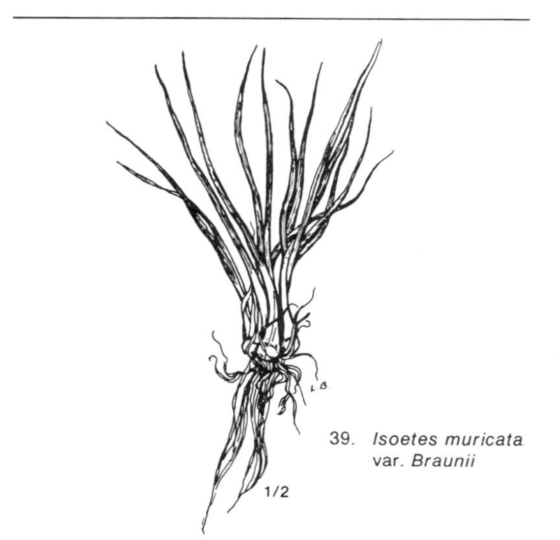

39. *Isoetes muricata* var. *Braunii*

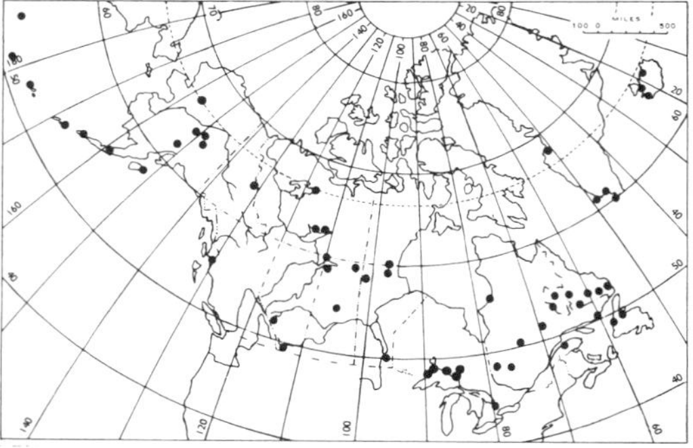

47. *Isoetes muricata* var. *Braunii*

PINACEAE Pine Family

Resinous trees or shrubs with needle-shaped or scale-like, mainly evergreen leaves and chiefly monoecious flowers borne in scaly aments of which the pistillate become dry, scaly cones, or berry-like in *Juniperus*.

a. Leaves linear, needle-shaped, more than 1 cm long; seeds in a dry cone; normally tree-like
  b. Leaves in fascicles of 2 to many
    c. Leaves evergreen, 2 together . . . . *Pinus*
    c. Leaves deciduous, many. . . . . *Larix*
  b. Leaves evergreen, attached singly to the branches
    d. Leaves 4-angled, pointed and prickly . . . . *Picea*
    d. Leaves flat, blunt . . . . *Abies*
a. Leaves scale-like or subulate, less than 1 cm long; fruits drupe-like with a bluish bloom; ours low and usually prostrate shrubs . . . . *Juniperus*

Abies Mill. Fir

Abies lasiocarpa (Hook.) Nutt.
Alpine Fir
Tree-like but often shrub-like at timberline; bark smooth, greyish and usually covered by resin blisters. Leaves linear, flattened in cross-section, leathery, shiny green above, somewhat whitish beneath, blunt and not prickly as are those of the spruces. Cones, unlike those of the spruces, erect and mature at the end of the first season and then promptly disintegrate, except for the slender, spike-like axes that remain on the branches long after the broadly rounded cone scales have fallen.

The alpine fir is Cordilleran and barely enters our area in the Mackenzie Mountains where it is restricted to soils derived from acid crystalline rocks, and where it commonly ascends above the upper limit of spruces. It is probably always sterile at timberline where it reproduces entirely by layering when a low scrub is formed around the mother tree.

The taxonomically closely related eastern balsam fir (*A. balsamea*) is essentially a lowland species of the eastern boreal forest element reaching eastern Alberta and north to Lake Athabasca. It was reported by John Richardson as having been seen by him in latitude of 62°, which, presumably would have been somewhere near the western end of Slave Lake. The report has not been confirmed by subsequent travellers and may have been based on the observation of the resin blisters on the bark of *Picea glauca* var. *Porsildii* that, on young trees, to a remarkable degree resemble those of *Abies balsamea*, especially when only the lower part of the tree trunk is seen, as would be the case when the observer is carrying a heavy back load, as Richardson may well have done habitually when crossing portages.

General distribution: Cordilleran, from New Mexico north to southeastern central Y.T. and S. W. District of Mackenzie.

Fig. 40 Map 48

Juniperus L. Juniper

Ours low erect or prostrate shrubs with scale or needle-like leaves and blue berry-like cones.

a. Mature leaves acicular, spreading or imbricated . . . . *J. communis*
a. Mature leaves scale-like, imbricated . . . . *J. horizontalis*

Juniperus communis L.
Ground Juniper
Prostrate or spreading shrub commonly forming low, open thickets or, in alpine situations, matforming. The awl-shaped 5 to 10 mm long leaves crowded in whorls of 3. Fruit sessile, ripening tardily, with a whitish-blue bloom.

Variable according to habitat. In exposed or alpine situations the branches are depressed or decumbent, the leaves short and relatively broad and often imbricated (var. *saxatilis* Pall.). In lowland or sheltered situations often bushy and up to 1 m high with slender spreading leaves (var. *depressa* Pursh).

General distribution: Circumpolar, north to and slightly beyond the northern limit of trees.

Fig. 41 Map 49

Juniperus horizontalis Moench
Creeping Juniper
Prostrate and matforming with long, freely rooting main branches. Leaves blue-green, in very young plants acicular as in *J. communis* but soon becoming scale-like and imbricated. Fruit recurved on a short stalk.

Rocky and gravelly places and apparently restricted to soils derived from calcareous rocks.

General distribution: North America. From Nfld. and southern Lab. northwestward to northern District of Mackenzie, and barely reaching S. W. Alaska.

Fig. 42 Map 50

Larix Mill. Larch

Larix laricina (DuRoi) Koch
L. alaskensis Wight
A small tree, with us rarely more than 5 to 6 m high, with thin, scaly bark and long, slender and pliable branches. Leaves deciduous, soft, pale green, in small lateral clusters of 12 to 20, turning bright yellow in autumn. Male and female aments globular or egg-shaped, terminating very short and peg-like shoots, the female erect and cone-like, their scales dark red at flowering

time, becoming leathery and brown in age. The wood is dense and tough but of very limited use because of its small size.

The distribution of the larch is always spotty in the Northwest Territories where it is generally confined to soils derived from rocks rich in lime, such as well-drained screes in the mountains, or in minerotrophic bogs in the lowland. In Keewatin the larch, together with black spruce (*P. mariana*) forms the polar tree line, whereas in the District of Mackenzie neither attain the northern limit of white spruce.

General distribution: Boreal North America from Nfld. and Lab. to central Alaska, northward to the polar limit of trees.

Fig. 43 Map 51

Picea Dietr. Spruce

Large evergreen, monoecious trees with rough and flaky bark; leaves needle-shaped, from persisting bases (sterigmata); cones ovoid, drooping, maturing the first season.

a. Branchlets glabrous, cones soon falling, 3-6 cm long
  b. Bark scaly, becoming furrowed in age; crown spire-like . . . . *P. glauca* var. *albertiana*
  b. Bark smooth, resin-blistered, scaly and furrowed only in age; crown ample . . . . *P. glauca* var. *Porsildii*
a. Branchlets pubescent; cones short-ovoid, 1.5-3 cm long, persisting on the tree for many years . . . . *P. mariana*

Picea glauca (Moench) Voss
White spruce

Two races of white spruce are common in the Mackenzie drainage where the var. *albertiana* (S. Brown) Sarg. commonly grows in alpine or exposed situations and sometimes also in bogs and muskegs; in the District of Mackenzie it always forms the polar and alpine tree line. The western race, var. *Porsildii* Raup, is dominant on rich alluvial soils of the lowland where it is at once distinguished from var. *albertiana* by its comparatively smooth bark which generally, and always in young trees, shows distinct resin blisters not unlike those of *Abies*. In outline the crown of var. *Porsildii* is pyramidal or somewhat bulb-shaped, never narrow and spire-like as in var. *albertiana*. Both are important local sources of timber logs and firewood, but the best stands of timber along the Mackenzie and its tributaries are composed of the var. *Porsildii* that in such situations may attain trunks over 30 m high, and 1.5 m DBH. The difference between the two races is well recognized by wood cutters and sawmill operators in the area who commonly identify the var. *albertiana* as "black spruce".

Near the tree-line var. *albertiana* commonly forms a low scrub and may even become prostrate and mat-forming.

General distribution: Wide-ranging from Lab. to Alaska.

Fig. 44 Map 52

Picea mariana (Mill.) B.S.P.
Black spruce

A small, slender tree, rarely exceeding 10 m in height, with rough and scaly bark and pubescent twigs. Black spruce is the dominant tree in lowland muskegs, on poorly drained clays and glacial tills of alpine valleys and slopes throughout the Mackenzie drainage. Everywhere in the District of Mackenzie its northern limit falls slightly short of that of white spruce. In the District of Keewatin the situation is reversed. Here the white spruce is restricted to more favoured habitats in sheltered river valleys and black spruce and larch form the polar tree line.

Owing to its slow rate of growth the wood of black spruce is dense and hard but is used mainly for firewood.

General distribution: Boreal North America from Lab. to Alaska north to or near the polar tree line.

Fig. 45 Map 53

Pinus L. Pine

Ours low to medium-sized, monoecious and evergreen trees with needle-shaped, 3 to 5 cm long leaves in 2's from a peg-like sheathed base; cones ovoid, with thick persisting and somewhat prickly scales, maturing tardily.

a. Cone scales smooth . . . . *P. Banksiana*
a. Cone scales with a single often minute but sharp and prickly dorsal spine . . . . *P. contorta* var. *latifolia*

Pinus Banksiana Lamb.
Banksian or Jack Pine
Medium-sized tree at once distinguished as a pine by its crooked branches, paired leaves and conical, divergent and often oblique cones that remain closed on the tree for many years.

Common only locally and forming pure stands on well-leached, sandy, non-calcareous soils such as river terraces and flood plains. In the Mackenzie valley and between Great Slave Lake and Great Bear Lake along the western edge of the Precambrian Shield, it extends north almost to the 65th parallel. Although well-grown trees may attain a diameter of 18 inches, the wood of the pine has been put to a limited use in the district because of the more abundant and ready supply of spruce logs.

General distribution: North America from N.S. and central Que. northwest across northern Man., Sask., and Alta. to upper Mackenzie drainage.

Fig. 46 Map 54

Pinus contorta Loud.
var. **latifolia** Engelm.
Lodgepole Pine
Similar and closely related to Banksian pine from which it can usually be distinguished by its commonly contorted and prickly cones. Like the Banksian pine it forms pure stands on sandy soil but is not intolerant of lime. Its cones persist for many years and open only by the heat from a forest fire.

In our area known only from the lower slopes of the southwestern Mackenzie Mountains. Hybrids between lodgepole and Banksian pine have been reported from the Liard River.

General distribution: Western N. America from N. California north to southern and central Y.T., eastward to foothills of Rocky Mts.

Fig. 47 Map 55

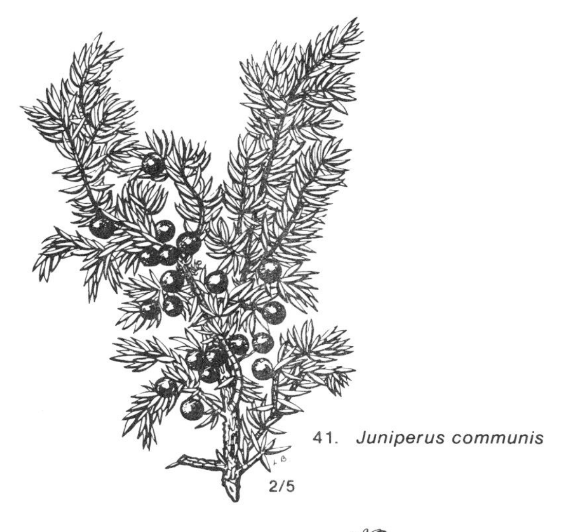

41. *Juniperus communis*

42. *Juniperus horizontalis*

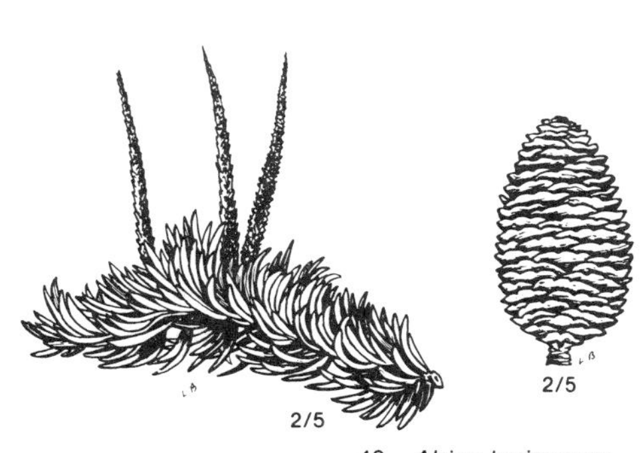

40. *Abies lasiocarpa*

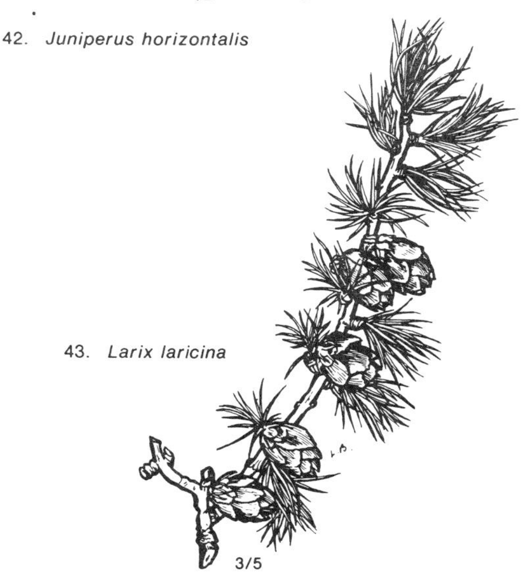

43. *Larix laricina*

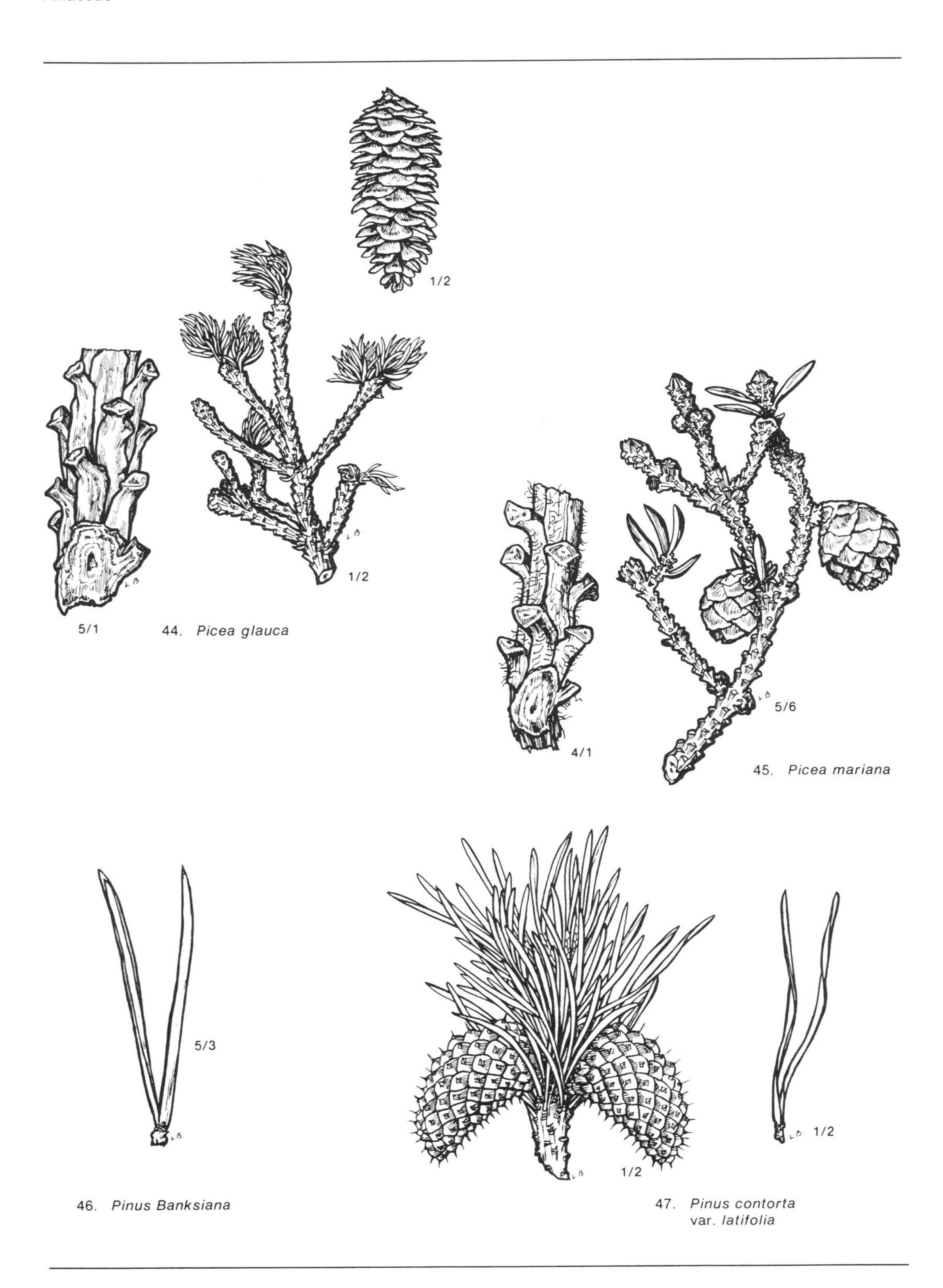

44. *Picea glauca*

45. *Picea mariana*

46. *Pinus Banksiana*

47. *Pinus contorta* var. *latifolia*

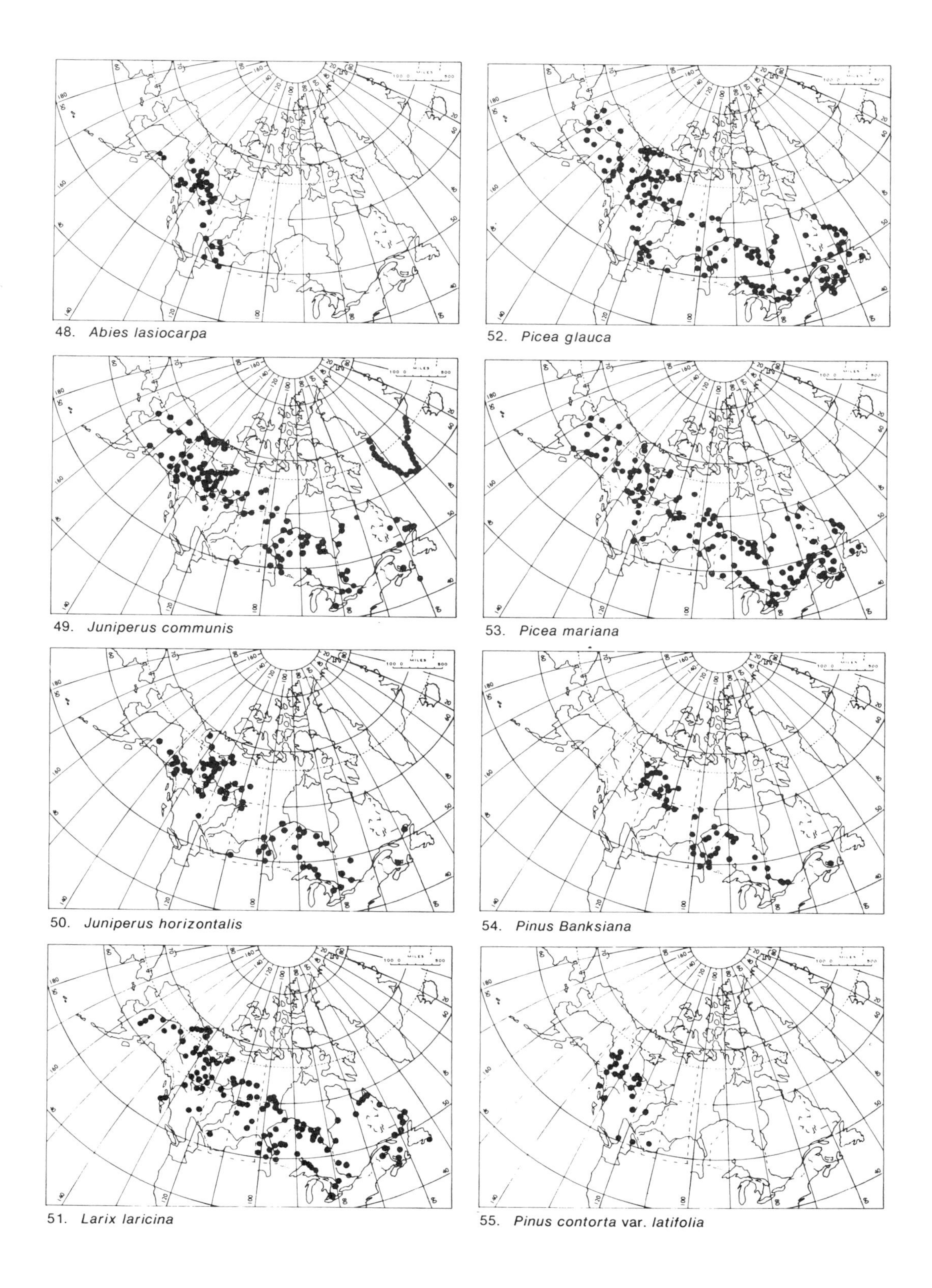

48. *Abies lasiocarpa*

49. *Juniperus communis*

50. *Juniperus horizontalis*

51. *Larix laricina*

52. *Picea glauca*

53. *Picea mariana*

54. *Pinus Banksiana*

55. *Pinus contorta* var. *latifolia*

TYPHACEAE Cat-tail Family
Typha L. Cat-tail

Typha latifolia L.
Stout, perennial 1 to 2 m tall marsh plants, with simple, glabrous and terete stems from a thick, fleshy rhizome. Leaves linear, flat, 1 to 2.5 cm wide, often overtopping the stems. Flowers monoecious, crowned in a dense, terminal contiguous spike, of which the lower half is pistillate, dark brown and cylindrical, 7 to 15 cm long and up to 3 cm thick.

Common locally along the Mackenzie River where it is now known from half a dozen stations, north to lat. 65° 17′, and also occasional in the lower South Nahanni River valley.

General distribution: Non-arctic, circumpolar.

Fig. 48 Map 56

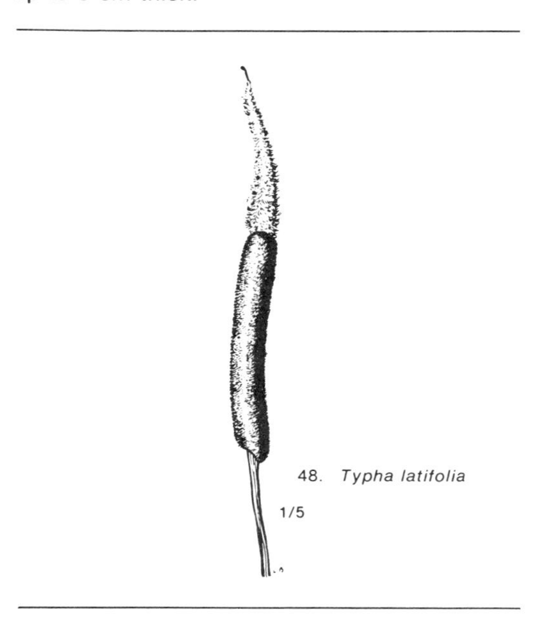
48. *Typha latifolia*
1/5

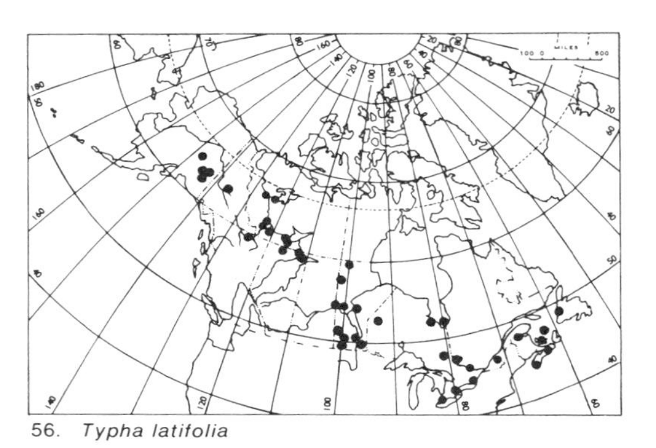
56. *Typha latifolia*

SPARGANIACEAE Bur-reed Family
Sparganium L. Bur-Reed

Perennial, grass-like, rooted aquatics with slender, leafy, jointed stems. Leaves linear, sheathing at the base, erect or floating. Flowers small and inconspicuous, unisexual, in globular sessile or pedunculate heads, the uppermost containing the male, the lower the female flowers. Fruit nut-like, obovoid to ellipsoid.

a. Stigmas 2; mature heads 2 to 3 cm in diameter; achenes broadly obpyramidal. . . . . *S. eurycarpum*
a. Stigmas 1; mature heads 0.5 to 2.0 cm in diameter; achenes ellipsoid, tapering at each end
   b. Staminate heads 2-several; achenes prominently beaked

c. Leaves 2 to 5 mm wide, rounded on the back . . . . *S. angustifolium*
c. Leaves 5 to 10 mm wide, flat . . . . *S. multipedunculatum*
b. Staminate head one; beak of the achene short or none
d. Achene short-beaked; pistillate heads axillary, separate from staminate head . . . . *S. minimum*
d. Achene beak-less; staminate head nearly contiguous with upper pistillate head . . . . *S. hyperboreum*

Sparganium angustifolium Michx.
Stems submersed or floating with elongated 2 to 5 mm broad leaves, flat or convex on the back; fruiting heads 1 to 3, 1.0 to 1.5 cm in diameter, the lowermost often pedunculate; staminate heads 2-several, mostly close together.

Common in suitable places in the lowland along the Mackenzie River and its tributaries northward to near the limit of trees in the Mackenzie Delta.

General distribution: Circumpolar with large gaps.

Fig. 49 Map 57

Sparganium eurycarpum Engelm.
Giant Bur-reed
By its stiffish ascending up to 12 mm broad leaves and sessile, angular and very firm pistillate heads 2.0 to 2.5 cm in diameter, readily distinguished from other bur-reeds in our area where, thus far, it has been collected but once, near Fort Norman.

Muddy bottom of ponds or sloughs.

General distribution: N. American endemic. Que. to B.C.

Fig. 50 Map 58

Sparganium hyperboreum Laest.
Stems 2 to 3 dm long, leaves 3 to 6 dm long, very slender but rather firm and shiny, 2 to 4 mm wide. Pistillate heads 1 to 3 and 0.5 to 1.0 cm in diameter, when more than one, the lowermost is stalked; staminate head solitary and contiguous with the uppermost pistillate one. Fruits ellipsoid 3.5 to 4.5 mm long, beakless.

In shallow bog pools to slightly beyond the tree limit.

General distribution: Circumpolar, subarctic.

Fig. 51 Map 59

Sparganium minimum (Hartm.) Fries
Similar but usually smaller than *S. hyperboreum* from which fertile specimens may at once be distinguished by the single staminate head being stalked well above the sessile pistillate heads. Fruit distinctly beaked.

Apparently rare or perhaps often overlooked in our area where it has been reported from a few widely scattered stations within the wooded zone north to the Mackenzie Delta.

In shallow bog pools.

General distribution: Circumpolar (with large gaps), subarctic.

Fig. 52 Map 60

Sparganium multipedunculatum (Morong) Rydb.
Similar to *S. angustifolium* but coarser, the leaves ribbon-like, commonly floating, up to 10 mm wide. Fruiting heads up to 2.5 cm in diameter, the lowermost often long-pedunculate.

Thus far reported from half a dozen stations in the Mackenzie Valley north to the delta and about Great Slave Lake.

In quiet lakes and ponds.

General distribution: N. America from Lab. to Alaska.

Fig. 53 Map 61

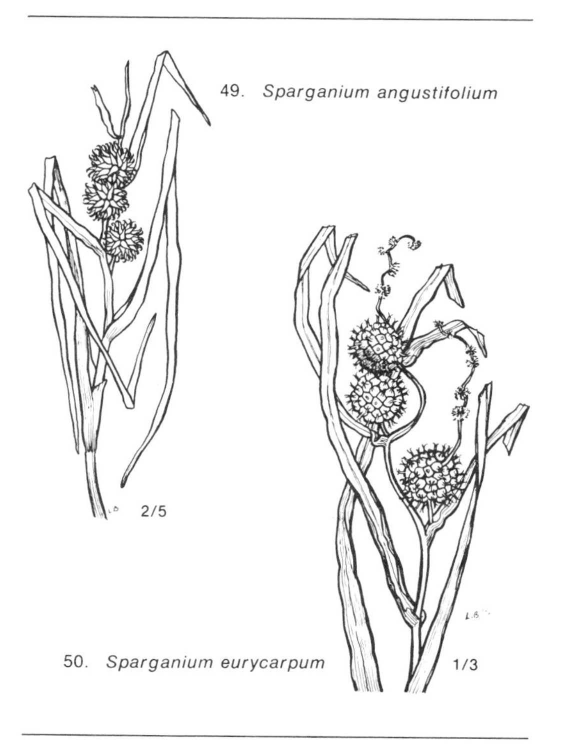

49. *Sparganium angustifolium*

50. *Sparganium eurycarpum*

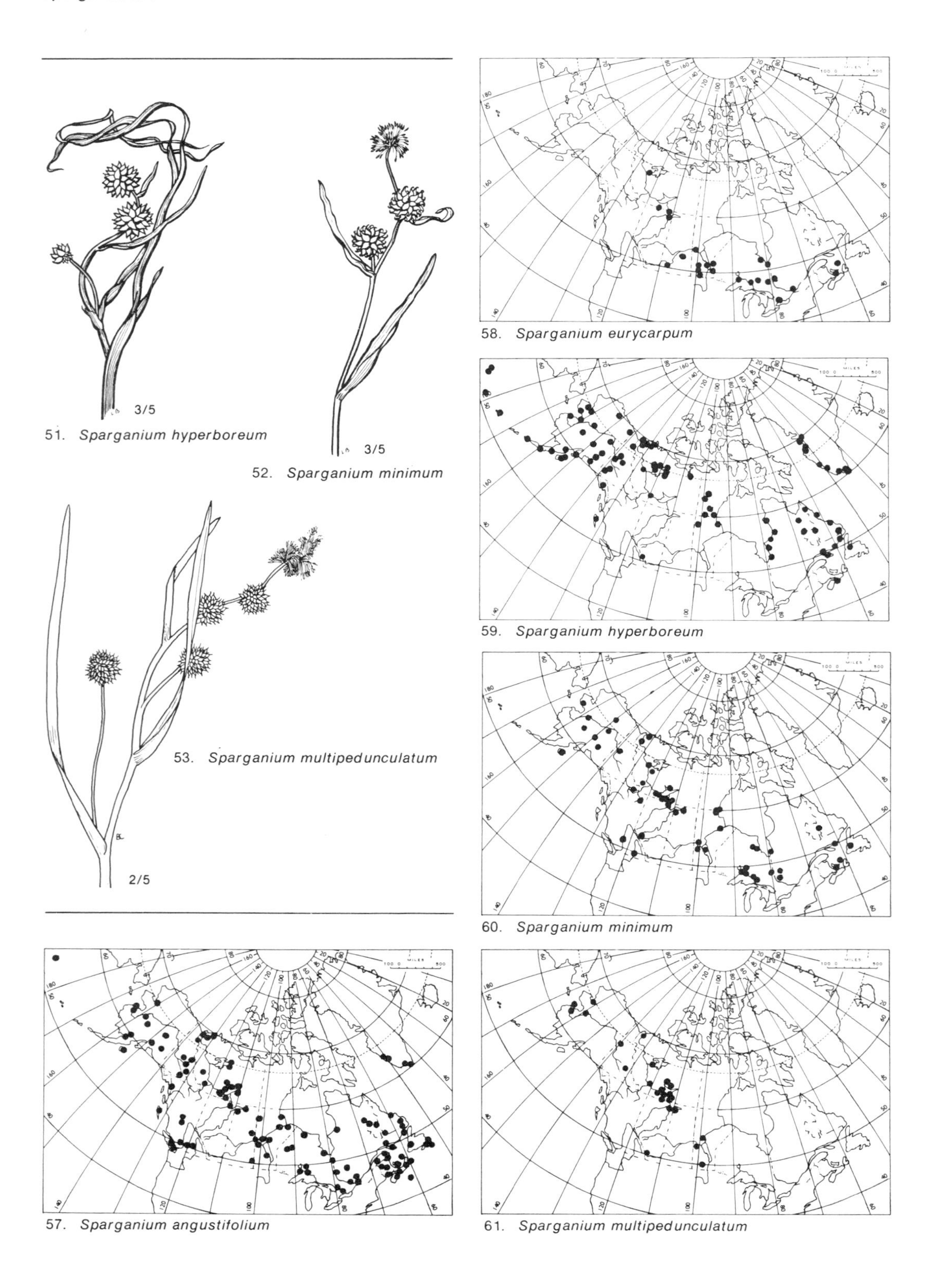

51. *Sparganium hyperboreum*

52. *Sparganium minimum*

53. *Sparganium multipedunculatum*

57. *Sparganium angustifolium*

58. *Sparganium eurycarpum*

59. *Sparganium hyperboreum*

60. *Sparganium minimum*

61. *Sparganium multipedunculatum*

POTAMOGETONACEAE Pondweed Family

Rooted aquatic herbs with jointed stems and small, perfect, monoecious or dioecious flowers lacking perianth. The fruit an achene.

a. Leaves mostly alternate
  b. Plants with submersed or floating leaves, of fresh or rarely brackish water; flowers and fruits in spikes or heads . . . . *Potamogeton*
  b. Plants wholly submersed
    c. Coarse marine plants with ribbon-like, 3-10 mm wide bright green leaves; flowers and fruits wholly concealed in leaf-like sheaths . . . . *Zostera*
    c. Delicate plants of alkaline waters; leaves linear-capillary; fruits long-pedicellate, umbellate on an elongated spiraling peduncle. . . . . *Ruppia*
a. Leaves opposite, 3-10 cm long, capillary; fruits, 2 to 4, sessile, umbellate . . . . *Zannichellia*

Potamogeton L. Pondweed

Mostly flaccid, perennial rooted aquatic plants with leafy, jointed stems, some wholly submersed, some with floating and submersed leaves and emergent spike-like inflorescences. The leaves alternate or opposite, the submersed mostly thin and linear, oblong or lanceolate, the floating ones, of firmer texture and shiny, always different in shape from the submersed. Flowers small and inconspicuous, bisexual, lacking perianth, mostly 4-parted, in spikes or axillary clusters. Fruit an ovoid to subglobose achene commonly covered with a thin and spongy pericarp. Several species, especially those reaching subarctic regions, rarely produce mature fruits and thus regularly propagate vegetatively, some by axillary winter-buds, others mainly by rooting stolons or root-tubers. The fruits and root tubers are an important source of food for waterfowl, and in winter the root tubers are eaten by muskrats.

a. Floating leaves usually present
  b. Stipules firm, 4-10 cm long
    c. Submersed leaves linear and very narrow, soon disintegrating . . . . *P. natans*
    c. Submersed leaves lanceolate-oblanceolate, 8-20 cm long and 2-5 cm wide; floating leaves not always present. . . . . *P. illinoensis*
  b. Stipules soft, 1-3 cm long
    d. Submersed leaves 5-20 cm long; whole plant with a reddish tinge; floating leaves not always present . . . . *P. alpinus*
    d. Submersed leaves 3-8 cm long; plant green, floating leaves normally present . . . . *P. gramineus*
a. Leaves all submersed and alike
  e. Spike contiguous, oblong or capitate; stipules free, not united to the leaf-base
    f. Leaves not linear, 4 mm wide or wider, many-nerved
      g. Stipules 3-5 cm long, white; leaves half-clasping, more than 10 cm long . . . . *P. praelongus*
      g. Stipules inconspicuous, soon disintegrating; leaves cordate-clasping, less than 10 cm long. . . . . *P. Richardsonii*
    f. Leaves all linear, less than 4 mm wide
      h. Leaves 3.0-8.0 mm wide, the margins minutely serrulate. . . . . *P. Robbinsii*
      h. Leaves 0.3-5.0 mm wide, the margins entire
        i. Leaves 9-35-nerved (often apparent only under magnification)
          j. Plant bright green; leaves 2-4 mm wide, 15-35-nerved . . . . . *P. zosteriformis*
          j. Plant dull green or fulvous, leaves 1.5-2.0 mm wide, 9-17-nerved. . . . . *P. Porsildiorum*
        i. Leaves 3-7-nerved
          k. Stipules connate, the margins united two thirds their length
            l. Stipules strongly fibrous, in age breaking into shreds
              m. Leaves thin and translucent, 5-7-nerved, flat, abruptly narrowed to a blunt tip; peduncles flattened. . . . . *P. Friesii*

m. Leaves firm, 3- or rarely 5-nerved, often revolute, tapered to a slender tip; peduncles filiform . . . . *P. strictifolius* var. *rutiloides*
l. Stipules membranaceous or scarious; leaves 3-5-nerved
n. Gland present at the base of leaf; peduncles 1-5 cm long . . *P. pusillus*
n. Gland none; peduncles less than 1 cm long. . . . . *P. foliosus*
k. Stipules not connate . . . . *P. obtusifolius*
e. Spike moniliform; stipules united with the leaf-base or petiole; leaves 1-3-nerved
p. Stipular sheaths prominent, 2-5 cm long; leaves blunt. . . . . *P. vaginatus*
p. Stipular sheaths inconspicuous, less than 2 cm long
q. Leaves with long-tapering, acute tips; fruits 3-4 mm long . . . . *P. pectinatus*
q. Leaves mostly obtuse-tipped; fruits 2-3 mm long . . . . *P. filiformis*

Potamogeton alpinus Balbis
ssp. **tenuifolius** (Raf.) Hult.
Entire plant commonly reddish tinged. Stems elongate, simple or branched, from a slender, creeping rhizome. Submersed leaves linear-lanceolate to oblong, up to 25 cm long and usually eight times as long as broad, rounded at the apex. Floating leaves, when present, mostly opposite, slender-petioled, the blade narrowly obovate to oblanceolate. Spikes dense, cylindric, on 8 to 15 cm long peduncles. Achenes flattened-ellipsoid, 3 mm long.

Still waters, northward to the limit of trees.

General distribution: Circumpolar, non arctic.

Fig. 54 Map 62

***Potamogeton epihydrus** Raf.
var. **Nuttallii** (Cham. & Schlecht.) Fern.
Well preserved fruits and leaves have been reported from peat deposits on the east branch of the Mackenzie Delta, about 1500 km north of the nearest present day Canadian station for this species (Prince George, B.C.) and over 1000 km from its nearest Alaskan station in the Kuskokwim basin.

Potamogeton filiformis Pers.
Wholly submersed with widely creeping filiform stolons terminating in slender, white tubers. Stems branched, 10 to 15 cm long, bearing clusters of flaccid leaves 0.2 to 0.5 mm wide. Peduncles 5 to 10 cm long, filiform. Spike moniliform. In our area represented by the northern race var. *borealis* (Raf.) St. John.

In shallow lakes and ponds.

General distribution: Circumpolar, low-arctic.

Fig. 55 Map 63

Potamogeton foliosus Raf.
var. **macellus** Fern.
Stems thread-like, flattened, simple or branching; leaves bright green, narrowly linear, obscurely 3-nerved, lacking translucent glands at the base; spikes sub-capitate, commonly of 2 to 3 whorls, on short peduncles rising from the upper nodes.

Forming dense colonies on muddy bottom of shallow, still waters. Rare or perhaps often overlooked; along the Mackenzie River to Fort Norman.

General distribution: North America: Nfld. to Alaska.

Fig. 56 Map 64

Potamogeton Friesii Rupr.
Stems compressed, freely branching, 5 to 15 dm long. Leaves bright green and translucent, blade linear, 4 to 7 cm long and 2 to 3 mm wide, obtuse or mucronate at the tip, usually with two small, pellucid glands at the base. Spikes short-peduncled, short-cylindric, commonly of 3 whorls; fruit about 2 mm long. Freely propagating by its prominent and abundant winter-buds 1.5 to 2.5 cm long, covered by the persisting, fibrous stipules.

Calcareous or brackish waters. Northward to the limit of trees.

General distribution: Circumpolar (with large gaps in Siberia).

Fig. 57 Map 65

Potamogeton gramineus L.
Stems slender up to 40 cm long, usually freely branched below. Submersed leaves linear to linear-lanceolate, oblong-elliptic to oblanceolate, sessile, 3- to 11-nerved, in youth finely denticulate. Floating leaves firm, shiny, ovate to elliptic, mostly shorter than the petioles. Spikes dense, 1.0 to 2.5 cm long, the fruits obovate 1.7 to 2.8 mm long.

A very variable species common in still water 0.5 to 3.0 m deep; northward to the limit of trees.

General distribution: Circumpolar (with large gaps).

Fig. 58 Map 66

Potamogeton illinoensis Morong
Stems stout, simple or branching. Floating leaves (when present) opposite, oval with a rounded or narrowed base, short-petioled; submersed leaves oblong, tapering at both ends, 10 to 20 cm long, the stipules prominent, strongly keeled, 5 to 7 cm long; spikes 4 to 5 cm long, clustered near the summit; fruits 3.0 to 4.5 cm long, prominently keeled dorsally.
In quiet water.
General distribution: N. America; S. W. Que. to B. C. In our area known only from near Fort Providence.
Fig. 59 Map 67

Potamogeton natans L.
Stems simple or slightly branched, up to 2 m long; floating leaves long-petioled, their blades coriaceous, ovate, cordate or rounded; submersed leaves, narrowly linear, commonly disintegrating before the end of the summer leaving the fibrous and very durable, peg-like stipules. Spike compact-cylindrical, 3 to 5 cm long; fruits obovoid, green.
Lakes and quiet streams. In our area thus far known only from a single collection on Rabbitkettle Lake, South Nahanni River drainage.
General distribution: Circumpolar with some gaps, non-arctic.
Fig. 60 Map 68

***Potamogeton obtusifolius** M. & K.
Somewhat similar to *P. Friesii* but stem more branched, with shorter and less compressed internodes; leaves narrower, 3-nerved, and darker green and the stipules enclosing the winter-buds but not so stiff and fibrous.
Small shallow lakes and ponds; north to Lake Athabasca and to be expected in the southern parts of the Precambrian Shield area.
General distribution: Amphi-Atlantic.
Fig. 61 Map 69

Potamogeton pectinatus L.
Stems slender, freely branched and repeatedly forking, commonly 6 to 10 dm long from a slender, tuber-bearing rhizome. Leaves 0.2 to 1.0 mm wide, tapering to a sharply pointed tip in flattened comb-like fascicles. Peduncles filiform, 3 to 10 cm long; spike moniliform, of from 2 to 6 whorls; flowering under water.
In fresh, saline or alkaline shallow lakes. Thus far reported only from a few stations north to the Mackenzie Delta.
General distribution: Circumpolar, with some gaps.
Fig. 62 Map 70

Potamogeton Porsildiorum Fern.
Similar to *P. obtusifolius*, but stems not flattened and the leaves with a prominent median and 8 to 16 lateral nerves.
Shallow ponds and lakes.
General distribution: Alaska to James Bay, north to Mackenzie Delta.
Fig. 63 Map 71

Potamogeton praelongus Wulfén
Stems pale, 1 m long or more, branching and somewhat flexuous, leaves all submersed narrowly oblanceolate, bright green, half clasping at the base, the white stipules prominent and persisting. Spikes cylindric, 3 to 5 cm long, on up to 5 dm long, slender peduncles.
In still waters 1 to 2 m deep.
General distribution: Circumpolar.
Fig. 64 Map 72

Potamogeton pusillus L.
P. panormitanus Biv. var. *major* G. Fisch.
Wholly submersed with slender and much branched stems 2 to 4 dm long; leaves fresh green, linear, 2 to 5 cm long and 0.5 to 1.5 mm wide, the central nerve prominent, the lateral ones obscure, with translucent glands at the base; spikes capitate on short curved peduncles from the upper nodes.
Upper Mackenzie valley in quiet waters up to 2 m deep where it forms dense colonies; perhaps always sterile near its northern limit.
General distribution: Circumpolar.
Fig. 65 Map 73

Potamogeton Richardsonii (Benn.) Rydb.
Stems very leafy, wholly submersed, up to 1 m long, from a creeping rhizome. Leaves thin, lanceolate, 5 to 10 cm long and 8 to 15 mm wide near the broad and somewhat clasping base; stipules whitish, soon disintegrating into white shreds. Peduncles from the upper leaf-axils, 3 to 4 cm long; the spikes cylindric, about 1 cm long; fruits obovoid, about 4 mm long.
Common in quiet lakes and sluggish streams, often growing in water up to 3 m deep. North to the limit of trees.
Potamogeton Richardsonii and several other species of pond-weeds are important sources of food for muskrats and waterfowl and, with *Myriophyllum* spp., supply oxygen to the waters of icebound lakes during the long winter. Some large lakes in the Mackenzie Delta region, especially such as are situated at a level high enough so as not to be subject to flooding and consequent sedimentation during the spring breakup of the Mackenzie R., are often surprisingly well-

stocked with whitefish (*Coregonus* spp.) that here may attain a very large size. These large whitefish presumably are not anadromous, and spend the winter in these lakes. Despite the much reduced intensity of light under the thick, snow-covered winter ice, the dense growth of *Potamogeton Richardsonii* on the bottom of such lakes apparently is able to maintain an adequate supply of oxygen to keep the water "sweet". The Mackenzie Delta Eskimo, who formerly depended on these lakes for winter fishing under the ice, knew well that only lakes "with much grass on the bottom" provided good fishing in winter.

General distribution: North America: wide-ranging from Ungava to Alaska.

Fig. 66 Map 74

Potamogeton Robbinsii Oakes
Stems 5 to 10 dm tall from creeping rhizome; leaves two-ranked, always totally submersed, 6 to 10 cm long and about 1 cm wide, acute; spikes interrupted, flowering under water; in the northern parts of its range it is propagated entirely by the freely rooting detached branchlets.

In the District of Mackenzie known only from drift collected on the upper Thelon River, the sources of which are within the District of Mackenzie.

General distribution: N. America; Que. to B. C.

Map 75

Potamogeton strictifolius A. Benn.
var. **rutiloides** Fern.
Stems slender and wiry, freely branching above, 3 to 5 dm long; leaves rather stiff, 2 to 3 cm long and about 1 mm wide; stipules as long as the upper internodes; the short-cylindric and somewhat interrupted spikes on short, stiff peduncles. Resembling *P. pusillus* from which it differs by its more rigid leaves and larger and more abundant winter-buds.

Along the Mackenzie River valley north to its delta.

General distribution: N. America: Que. to the District of Mackenzie and S. E. Yukon.

Map 76

Potamogeton vaginatus Turcz.
Similar in habit but much coarser than *P. filiformis*, with stems commonly up to 1.5 m long from creeping 2 to 4 mm thick rhizomes from which issue 4 to 5 mm long, white tubers; leaves 4 to 5 dm long, 1 to 2 mm wide with short, blunt tips; spike long-peduncled, interrupted or moniliform of 3 to 6 whorls of flowers.

In fresh or somewhat brackish water up to 2 m deep. North to, or slightly beyond, the limit of trees.

General distribution: Circumpolar (with large gaps).

Fig. 67 Map 77

Potamogeton zosteriformis Fern.
Stems distinctly compressed and winged, freely branching; the fresh green leaves flat and grass-like, 2 to 4 mm wide and as long as the internodes, abruptly pointed, with 3 prominent and many fine nerves. Spikes cylindrical on rather stout and stiff peduncles much longer than the spike. Perhaps mainly propagated by winter-buds.

Thus far reported from a few stations along the Mackenzie River north to its delta.

General distribution: Circumpolar, N.B. to B.C. north to Alaska and the Mackenzie Delta.

Fig. 68 Map 78

Ruppia L. Ditch-grass

Ruppia spiralis L.
R. occidentalis S. Wats.
Submersed freely branching aquatic; stems slender, bearing alternate, linear-capillary, 5 to 15 cm long leaves with membranous sheaths. Flowers usually 2, on a long spiraling peduncle arising from a leaf axil; fruit on elongated stalks, pear-shaped, about 2 mm long, tipped by the blunt stigma.

Shallow alkaline waters; in our area known only from near Yohin Lake in S. W. District of Mackenzie.

General distribution: Circumpolar, with gaps.

Map 79

Zanichellia L. Horned pondweed

***Zanichellia palustris** L.
Fully submersed freely branched aquatic; stems slender and fragile, bearing whorls of filiform, 3 to 8 cm long leaves from sheathing stipules. Flowers unisexual, 3 to 5 in the leaf-axils. Fruit an oblong, narrow achene terminated by the beak-like persistent style.

Zanichellia has been reported from the Wood Buffalo Park just south of Fort Smith and should be looked for in shallow, fresh or mildly saline ponds or sluggish streams in S. W. District of Mackenzie.

General distribution: Circumpolar, but in the northern part of its range perhaps often overlooked.

Fig. 69 Map 80

Zostera L. Eelgrass

Zostera marina L.
Grasslike wholly submersed marine herb of sheltered tidal flats. Leaves fresh green, ribbon-like and up to one metre long, and 3 to 5 mm wide, in small fascicles from the widely creeping fleshy rhizome well buried in the silty sea-bottom. Flowers and fruits rarely observed, being hidden from view within the closed leaf-sheaths.

An important source of food for geese and other waterfowl.

Within our area known only from near Eskimo Point on the west coast of Hudson Bay where it was first observed in 1930 in a sheltered bay, but in Alaska known from the Pacific Coast north to Norton Sound and Cape Prince of Wales.

General distribution: Circumpolar, with large gaps along the arctic shores of N. America and Eurasia.

Fig. 70 Map 81

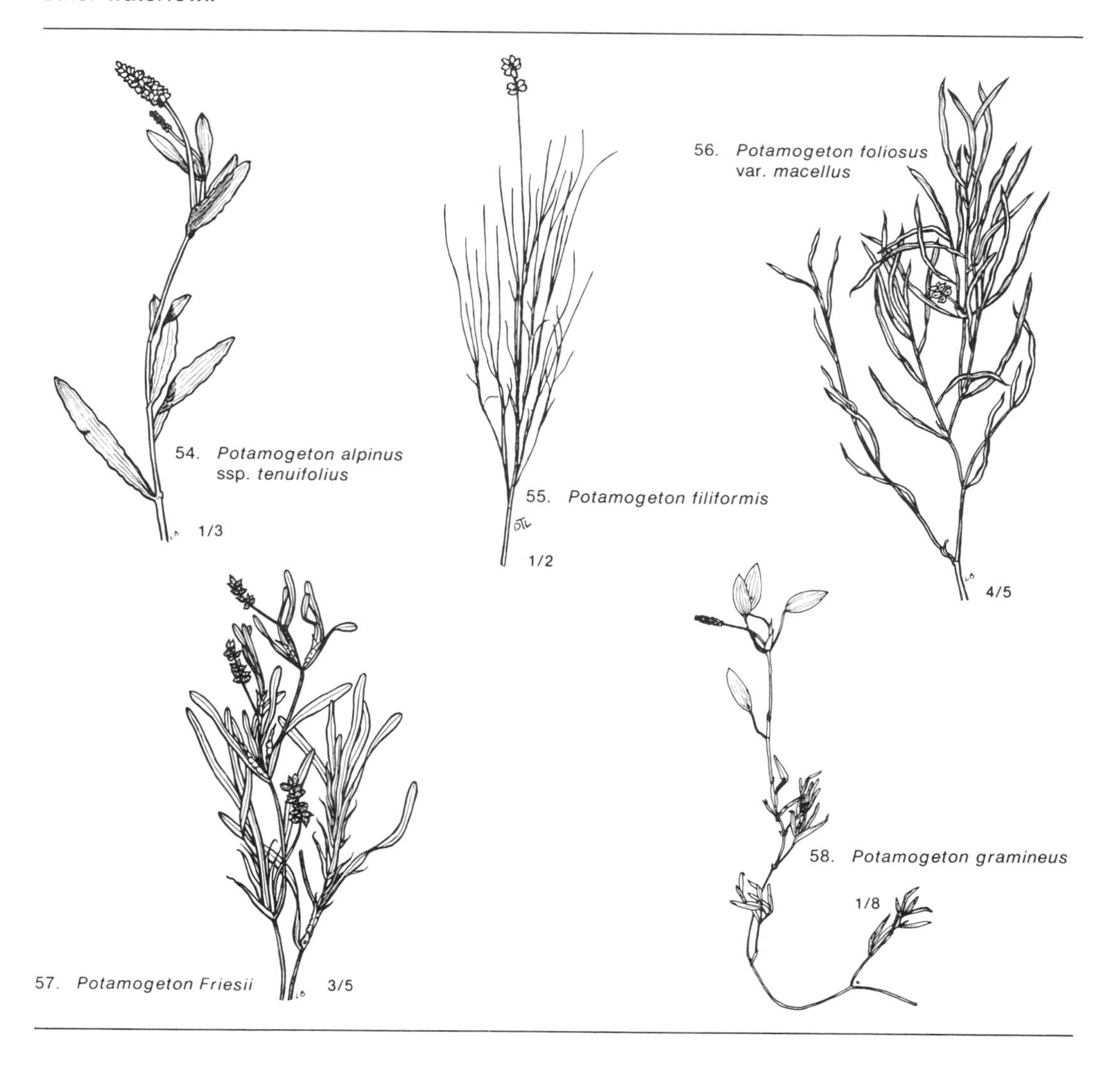

54. *Potamogeton alpinus* ssp. *tenuifolius*

55. *Potamogeton filiformis*

56. *Potamogeton foliosus* var. *macellus*

57. *Potamogeton Friesii*

58. *Potamogeton gramineus*

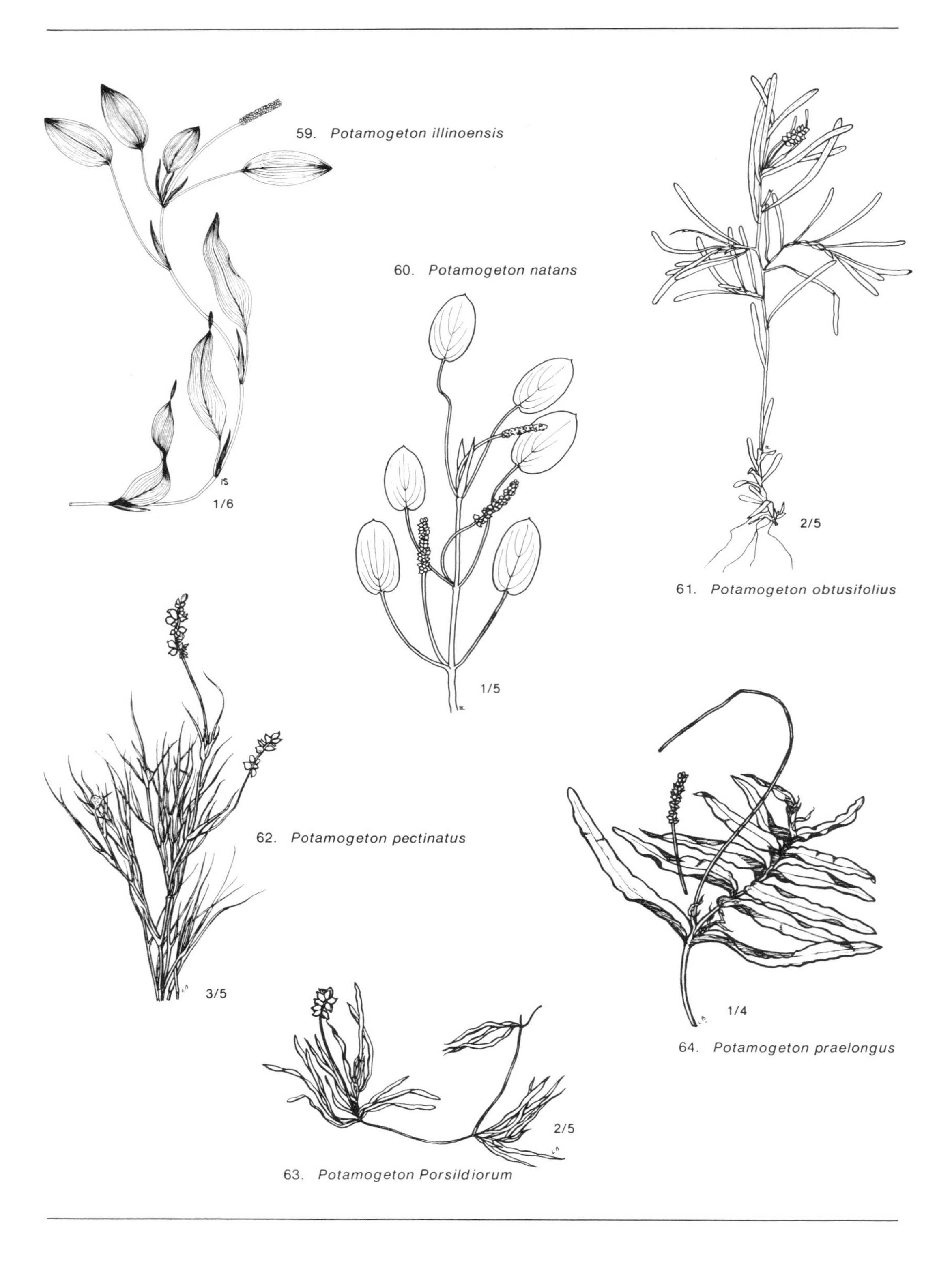

59. *Potamogeton illinoensis*

60. *Potamogeton natans*

61. *Potamogeton obtusifolius*

62. *Potamogeton pectinatus*

63. *Potamogeton Porsildiorum*

64. *Potamogeton praelongus*

65. *Potamogeton pusillus*

66. *Potamogeton Richardsonii*

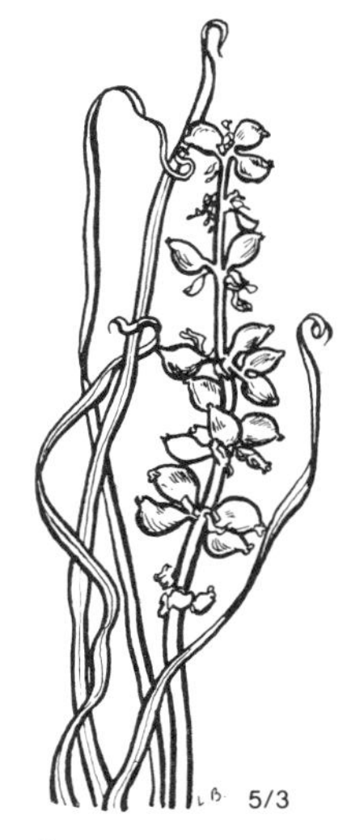

67. *Potamogeton vaginatus*

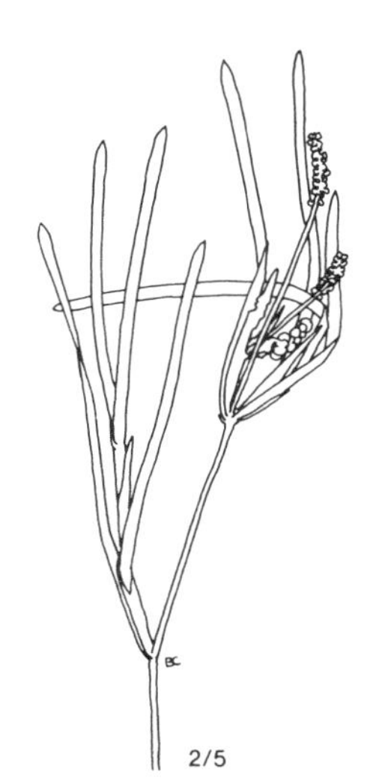

68. *Potamogeton zosteriformis*

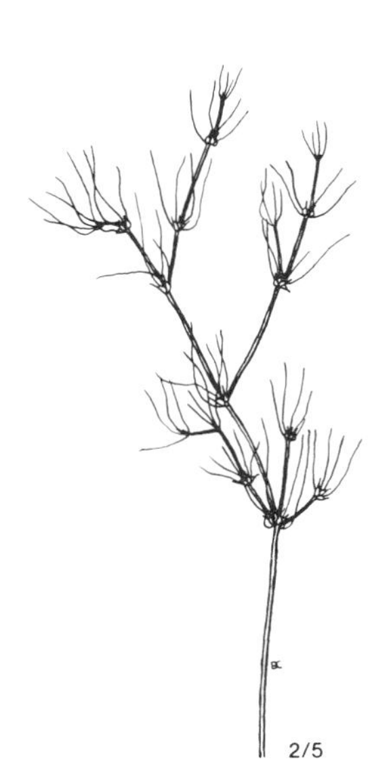

69. *Zanichellia palustris*

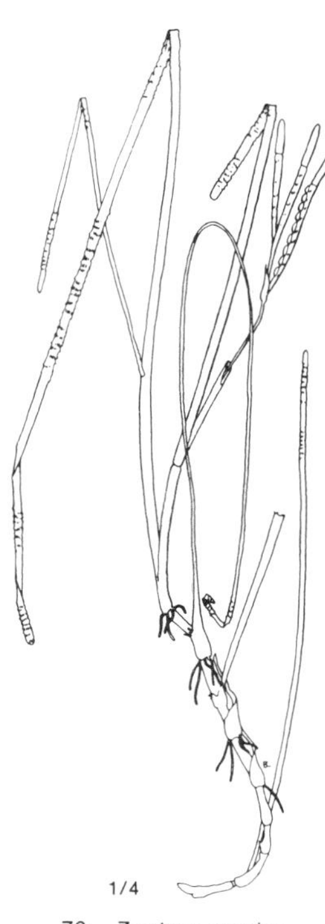

70. *Zostera marina*

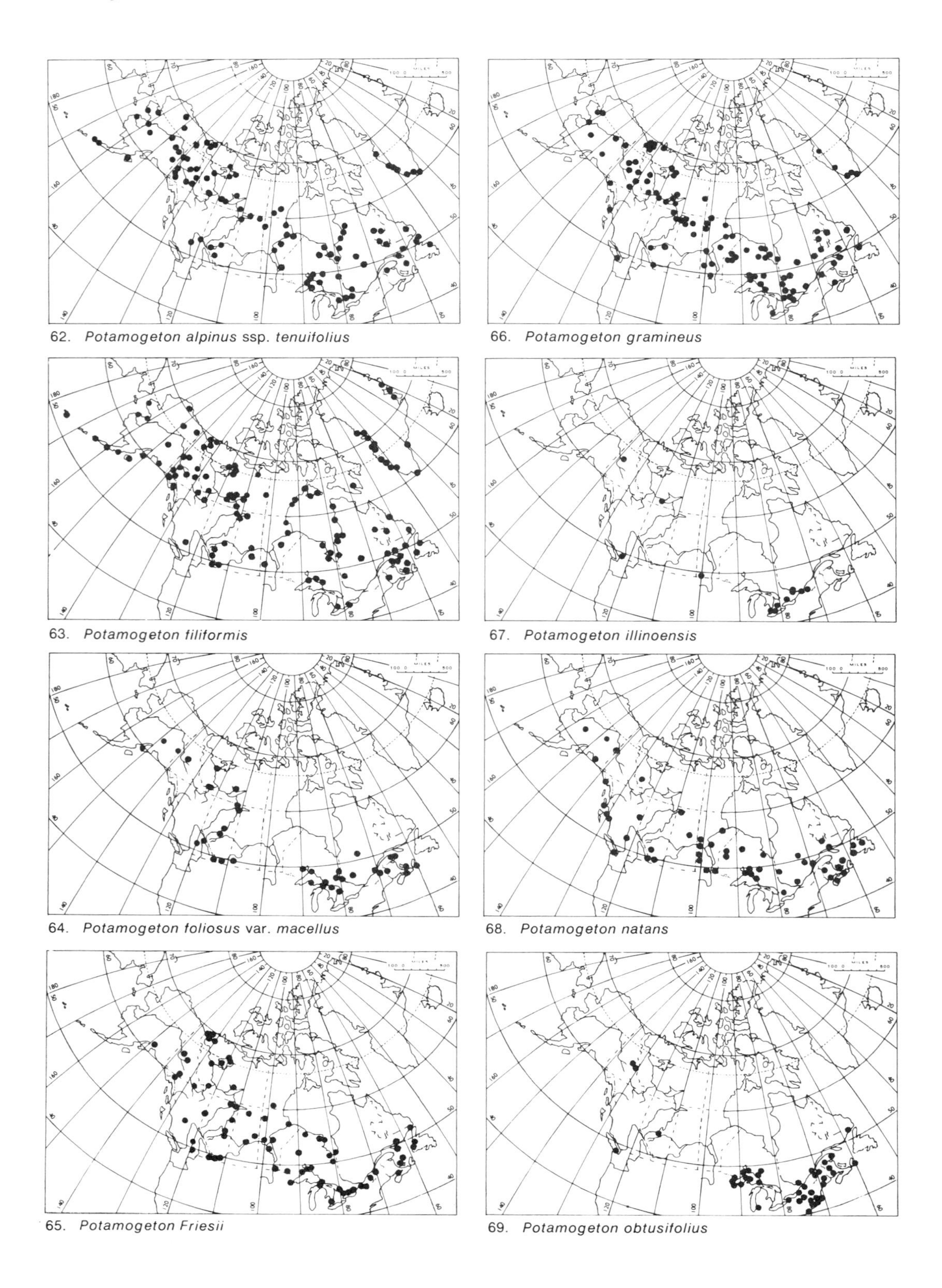

62. *Potamogeton alpinus* ssp. *tenuifolius*

63. *Potamogeton filiformis*

64. *Potamogeton foliosus* var. *macellus*

65. *Potamogeton Friesii*

66. *Potamogeton gramineus*

67. *Potamogeton illinoensis*

68. *Potamogeton natans*

69. *Potamogeton obtusifolius*

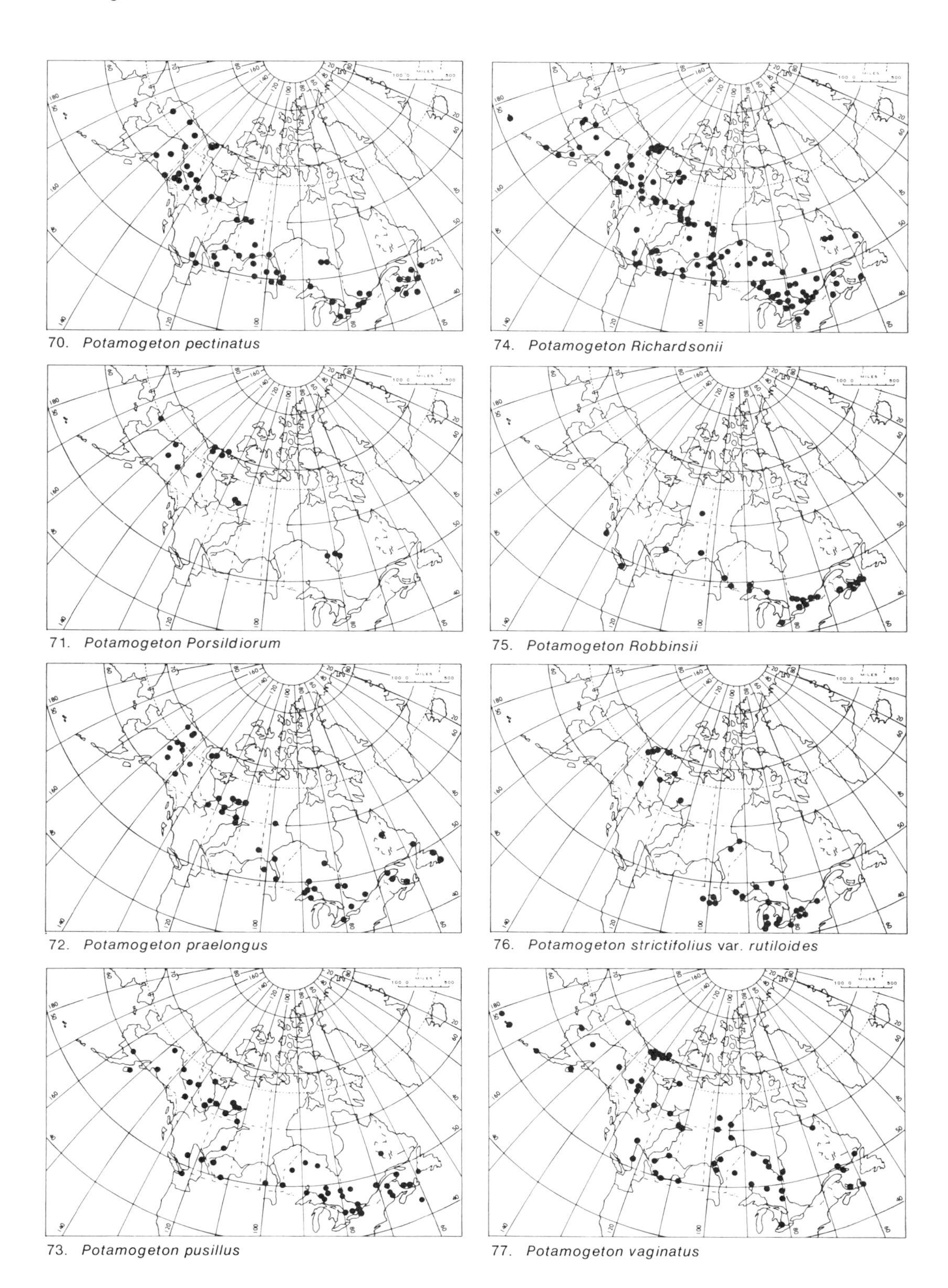

70. *Potamogeton pectinatus*

71. *Potamogeton Porsildiorum*

72. *Potamogeton praelongus*

73. *Potamogeton pusillus*

74. *Potamogeton Richardsonii*

75. *Potamogeton Robbinsii*

76. *Potamogeton strictifolius* var. *rutiloides*

77. *Potamogeton vaginatus*

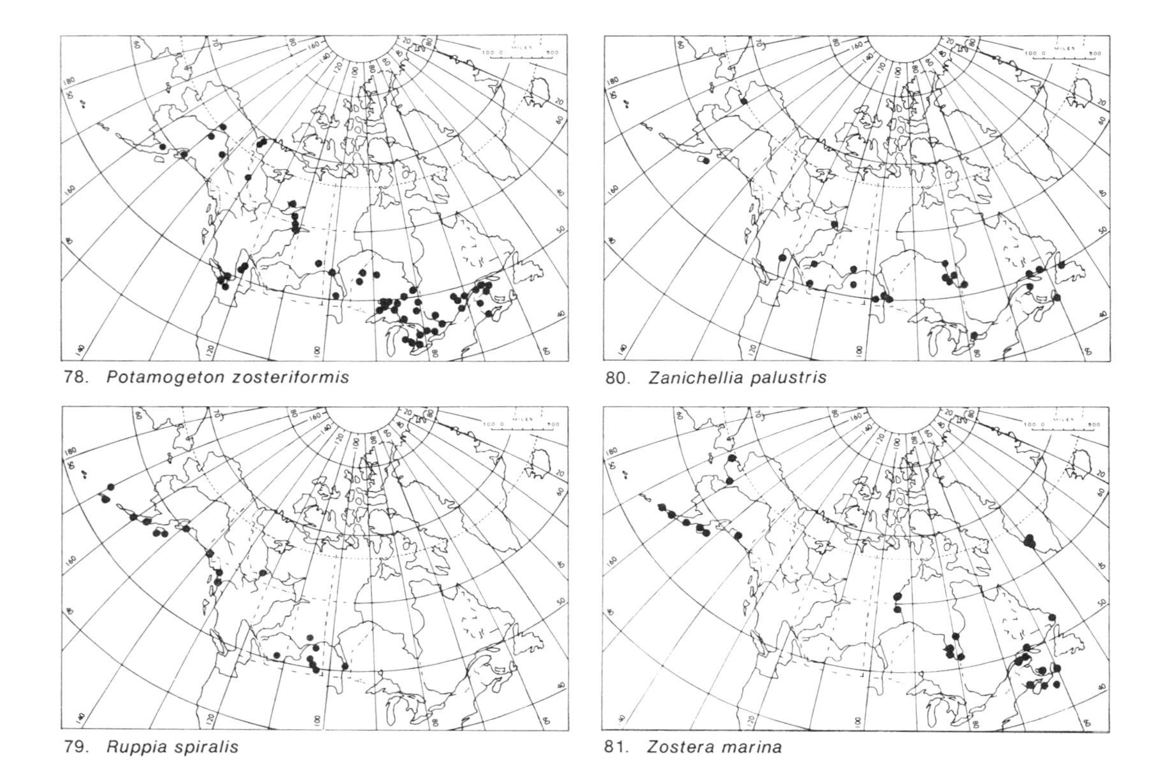
78. *Potamogeton zosteriformis*
80. *Zanichellia palustris*
79. *Ruppia spiralis*
81. *Zostera marina*

NAJADACEAE Naiad Family
Najas L. Naiad

***Najas flexilis** (Willd.) Rostk. & Schmidt
Delicate branched aquatic annual with narrow, finely serrated, opposite and commonly whorled leaves, mostly clustered toward the end of the branches. Flowers unisexual, borne in the axils of the branches and from the sheaths of the leaf-bases.

In shallow fresh or brackish water.

Najas flexilis has been reported from the Wood Buffalo Park just south of Fort Smith and may thus be expected to occur also in S. W. District of Mackenzie.

General distribution: Amphi-Atlantic, Nfld. to Alta.

Fig. 71 Map 82

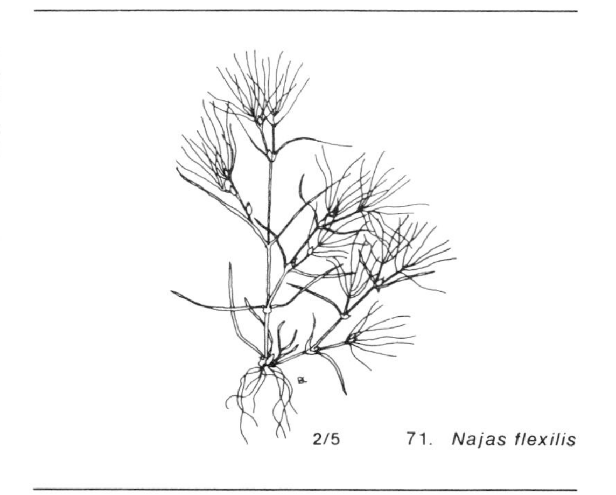
2/5 71. *Najas flexilis*

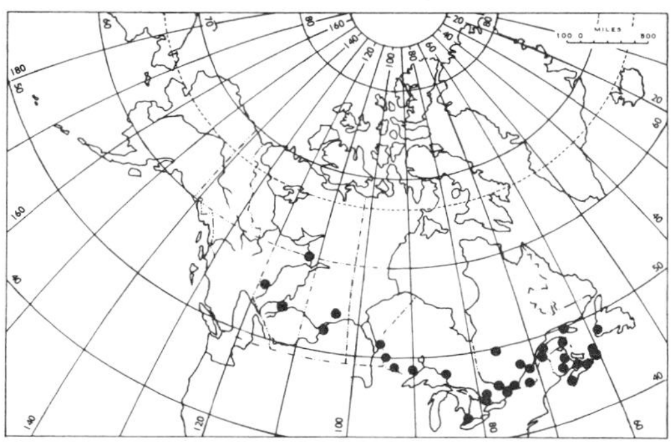

82. *Najas flexilis*

SCHEUCHZERIACEAE Arrow-grass Family

Smooth, perennial marsh plants with terete, bladeless leaves, and terminal perfect flowers in spikes or racemes; perianth of 3 green sepals and 3 similar petals. Fruit dry, indehiscent, 1- or 2-seeded.

a. Stem scapose from tuft of basal leaves. Flowers small in a narrow, spike-like raceme . . . . *Triglochin*
a. Stems leafy. Flowers in short, few-flowered receme . . . . *Scheuchzeria*

Scheuchzeria L.

Scheuchzeria palustris L.
var. **americana** Fern.
Herbaceous, yellowish-green bog plant with a creeping, freely branched rhizome from which rise the simple, characteristically flexuous stems, 1 to 2 dm tall and sheathed at the base by a few rather stiff and tightly rolled leaves. Inflorescence racemose, few-flowered, each flower supported by a sheathing bract. Fruit an inflated, 3-valved pod containing 1 or 2 rather large seeds.

Pools on quaking bogs along the southern edge of the Precambrian Shield and north in the Mackenzie River Valley to about lat. 64° N.

General distribution: *S. palustris s. lat.* circumpolar with large gaps.

Fig. 72 Map 83

Triglochin L. Arrow-grass

Glabrous herbs with linear somewhat fleshy basal leaves and small, regular 3-parted bisexual flowers arranged in an elongated spike-like raceme.

a. Fruit ovate, splitting into 6 segments; coarse, tufted plants up to 30-40 cm tall from an obliquely ascending rhizome . . . . *T. maritimum*
a. Fruit linear-clavate, splitting into 3 segments; delicate plants from a short, erect rhizome from which issue short, filiform bulb-bearing stolons . . . . *T. palustre*

Triglochin maritimum L.
Coarse, tufted, with smooth stiff scapes 3 to 5 dm tall and short, linear somewhat fleshy leaves from an obliquely ascending thick rhizome. Inflorescence oblong, 4 to 5 cm long, much elongated in fruit. The flowers small, with greenish sepals and petals.

Saline or alkaline flats or river meadows, north to, or slightly beyond, the limit of trees.
General distribution: Circumpolar.
Fig. 73 Map 84

Triglochin palustre L.
Delicate, with slender terete stems and leaves from a short, bulb-like rhizome producing filiform, bulb-bearing stolons. Scape 1.5 to 3.0 dm high, terminating in a few-flowered spike-like raceme.
Wet, brackish places, north to the Arctic Coast.
General distribution: Circumpolar.
Fig. 74 Map 85

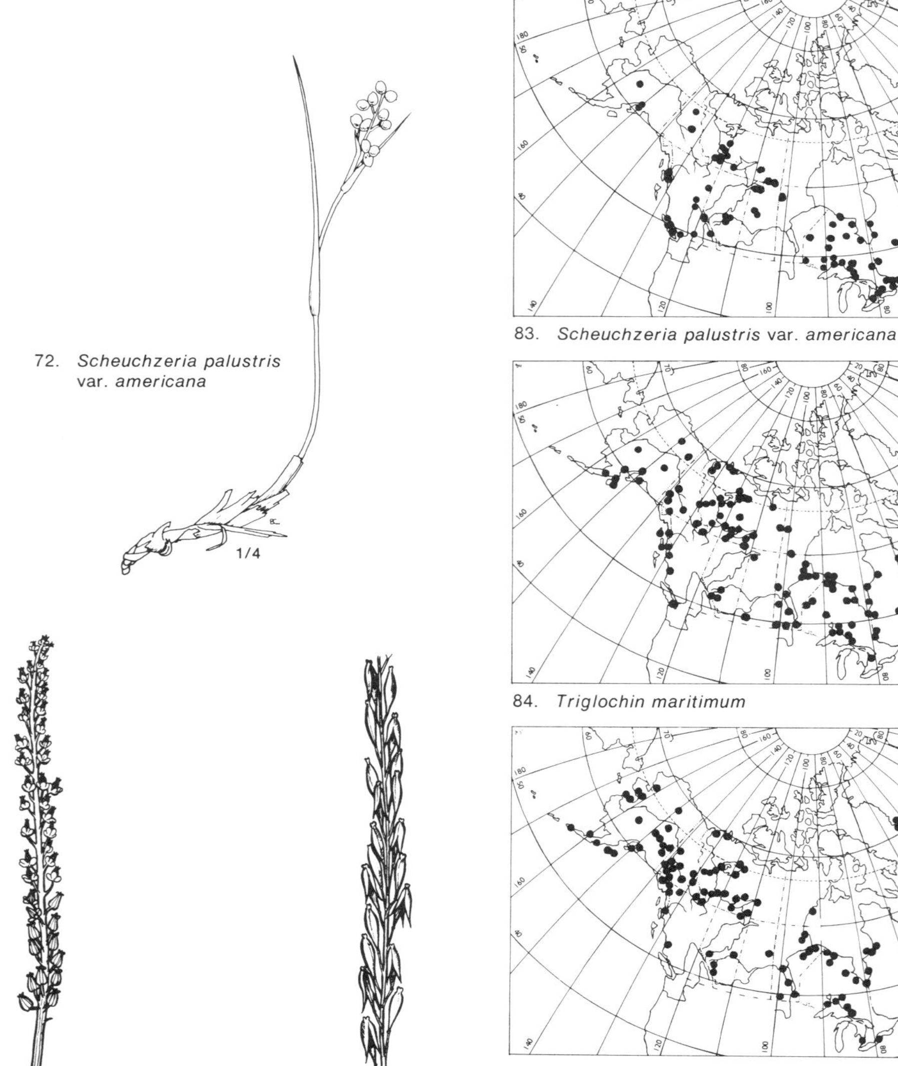

72. *Scheuchzeria palustris* var. *americana*

73. *Triglochin maritimum*

74. *Triglochin palustre*

83. *Scheuchzeria palustris* var. *americana*

84. *Triglochin maritimum*

85. *Triglochin palustre*

ALISMACEAE Water Plantain Family

Glabrous, perennial marsh plants with scape-like stems and sheathing leaves; flowers perfect or monoecious; perianth of 3 persistent sepals and an equal number of deciduous petals.

a. Flowers all perfect; scape with whorled panicle branches and long-petioled, narrowly ovate to heart-shaped leaves. . . . . *Alisma triviale*
a. Flowers monoecious, the lower pistillate, the upper and later developed staminate; leaf-blades commonly shaped like an arrowhead . . . . *Sagittaria cuneata*

Alisma L. Water-plantain

***Alisma triviale** Pursh
Emergent fibrous-rooted aquatic with erect, long-petioled and lanceolate-elliptic blades 4 to 8 cm long. Inflorescence paniculate with numerous small white flowers in small umbellate clusters at the end of the branches.

May be expected in marshy places and edges of sloughs in S. W. District of Mackenzie.

General distribution: N. S. to B. C., north to Fort Fitzgerald in N. Alta.

Fig. 75 Map 86

Sagittaria L. Arrowhead

Sagittaria cuneata Sheld.
Emergent aquatic with milky juice and tuber-bearing, nodose rhizome and long-petioled leaves with sagittate or sometimes lanceolate blade, 6 to 18 cm long, the narrow lobes somewhat spreading. Flowers verticillate, petals large and white. Fruiting heads 10 to 15 mm in diameter; the winged achenes 2 mm long.

In shallow water along calcareous, muddy shores. Southern Mackenzie basin with a widely disjunct station in the Mackenzie Delta.

General distribution: North America: From Que. to central Alaska.

Fig. 76 Map 87

75. *Alisma triviale*

76. *Sagittaria cuneata*

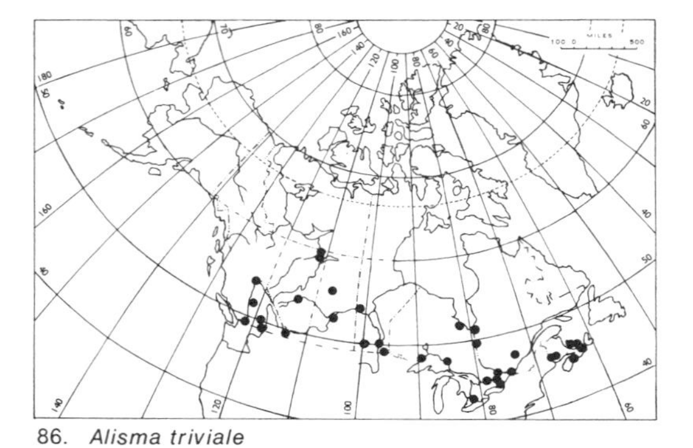

86. *Alisma triviale*

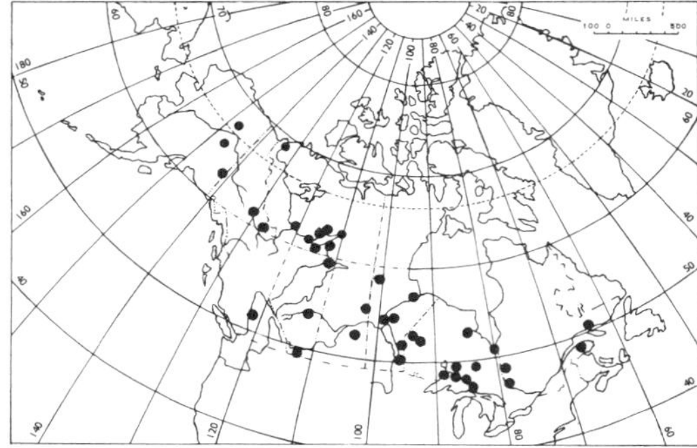

87. *Sagittaria cuneata*

GRAMINEAE Grass Family

Annual, or ours mainly perennial, herbaceous plants with jointed, usually terete and hollow culms, and linear, alternate, 2-ranked, flat, canaliculate or convolute leaves, each composed of a sheath, ligule, and blade. The much reduced, usually bisexual flowers in spikelets of two-ranked, scale-like leaves. In the spikelet the lower scales (glumes) are sterile; each flower is supported by an outer scale (lemma) and usually enclosed by a smaller inner scale (palea); perianth lacking, stamens usually 3, stigmas plumose, 2, and the ovary superior; the fruit a grain. The spikelets variously arranged in panicles, racemes, or spikes.

Fig. 77

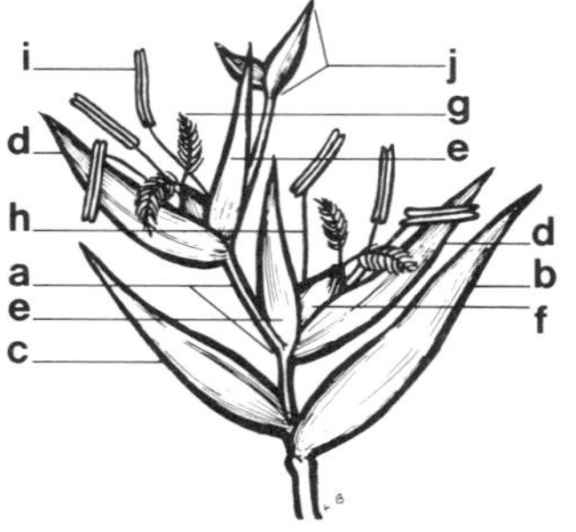

Fig. 77. Stylized grass spikelet

a. rachilla
b. first glume
c. second glume
d. lemma
e. palea
f. ovary
g. stigma
h. filament
i. anther
j. sterile floret

Key to the sections of grasses

a. Spikelets more or less laterally compressed or terete; rachilla articulate above the glumes (except in *Cinna*) leaving the persistent glumes after the rest of the spikelet has dropped
  b. Spikelets pedicellate forming a panicle or raceme
    c. Spikelets 1-flowered . . . . I. *Agrostideae* (Fig. 78)
    c. Spikelets 2-many-flowered
      d. Spikelets with only one terminal functional flower and two lateral staminate or sterile florets . . . . II. *Phalarideae* (Fig. 79)
      d. Spikelets normally with only bisexual flowers
        e. Lemmas (except *Sphenopholis*) with a bent awn arising from the back; glumes usually as long as the spikelet . . . . III. *Aveneae* (Fig. 80)

e. Lemmas awnless or with an apical awn; glumes shorter than the first floret (except in *Dupontia*) . . . . IV. *Festuceae* (Fig. 81)

b. Spikelets sessile or subsessile in spikes or spike-like racemes

f. Spikelets on one side of the rachis; spikes usually more than one, digitate or racemose . . . . V. *Chlorideae* (Fig. 82)

f. Spikelets on opposite sides of rachis; spike mostly solitary and terminal . . . . VI. *Hordeae* (Fig. 83)

a. Spikelet dorsally compressed; rachilla articulate below the glumes so that the entire spikelet is dropped when mature . . . . VII. *Paniceae*

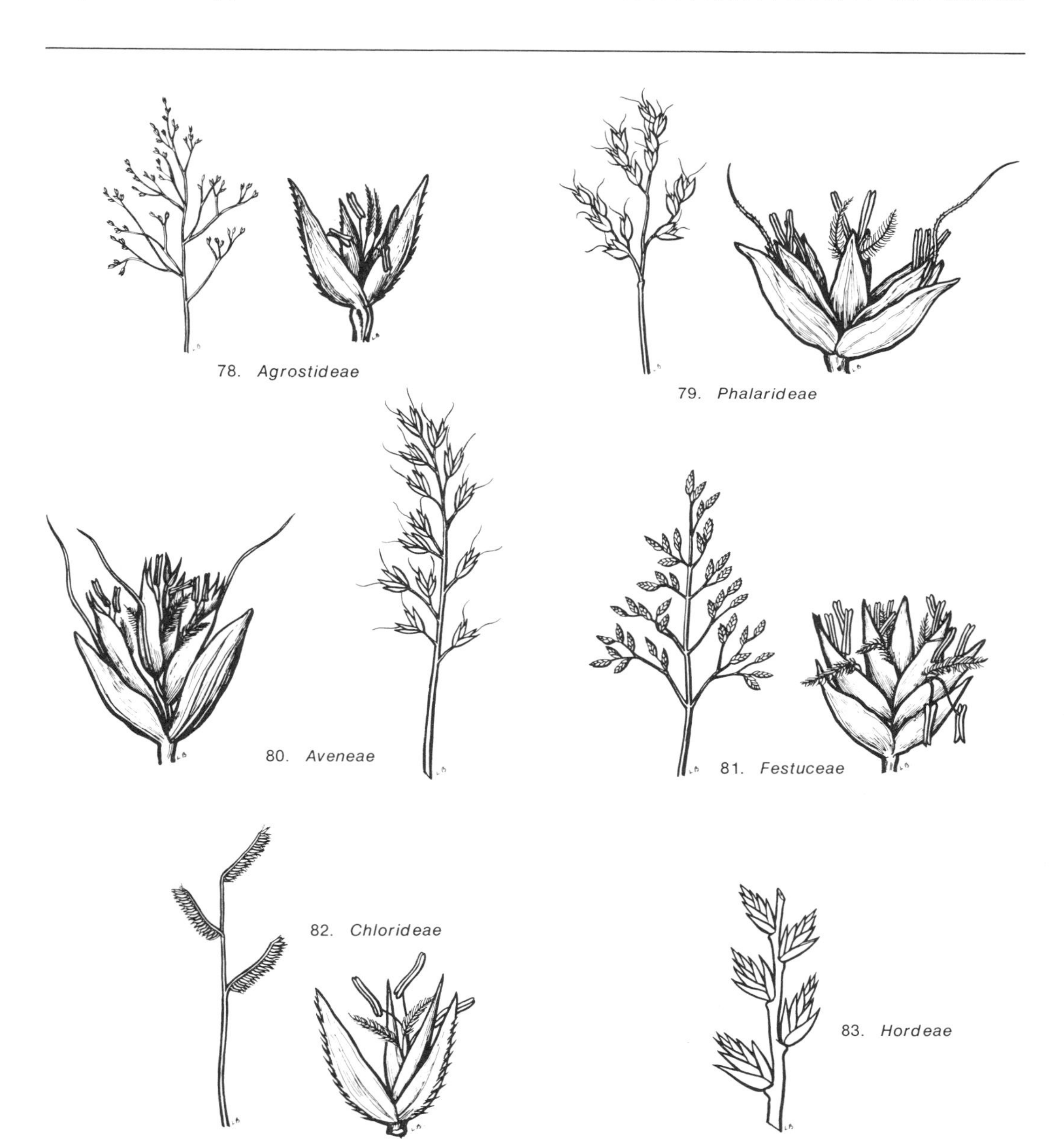

78. *Agrostideae*

79. *Phalarideae*

80. *Aveneae*

81. *Festuceae*

82. *Chlorideae*

83. *Hordeae*

I. **Agrostideae**

a. Inflorescence cylindrical or ovoid, spike-like
  b. Glumes awnless . . . . *Alopecurus*
  b. Glumes awned . . . . *Phleum*
a. Inflorescence not spike-like
  c. Inflorescence narrow and short; glumes minute, much shorter than the lemma; fruit at maturity loosely enclosed in the lemma; dwarf arctic-alpine species . . . . *Phippsia*
  c. Inflorescence paniculate or racemose; fruit at maturity adherent to the lemma
    d. Lemmas membranaceous
      e. Spikelet disarticulating below the glumes; broad-leaved forest species . . . . . *Cinna*
      e. Spikelet disarticulating above the glumes
        f. Lemmas surrounded at base by a tuft of stiff hairs, in some species reaching to the tip of lemma; lemmas awned dorsally . . . . *Calamagrostis*
        f. Lemmas lacking tuft of hairs at base
          g. Glumes longer than the lemma, the latter with or without dorsal awn . . . . *Agrostis*
          g. Glumes shorter than lemma
            h. Lemmas awnless
              i. Coarse, broad-leaved grasses with 10-20 cm long panicle . . . . *Arctagrostis*
              i. Medium to small wiry grasses, panicle rarely over 12 cm long . . . . *Muhlenbergia*
            h. Lemmas awned from the tip . . . . *Muhlenbergia*
    d. Lemmas hard, firmly encasing the fruit, with a long terminal awn; callus bearded; prairie or woodland species
      j. Awn not twisted, less than three times as long as lemma . . . . *Oryzopsis*
      j. Awn twisted and bent, 4-10 times as long as lemma . . . . *Stipa*

II. **Phalarideae**

a. Inflorescence evidently paniculate; spikelets 3-flowered, the central perfect, the lateral staminate; medium to small grasses . . . . *Hierochloe*
a. Inflorescence dense, spike-like; spikelets appearing 1-flowered, the central floret perfect, the lateral reduced to sterile scales; tall, coarse, broad-leaved grasses . . . *Phalaris*

III. **Aveneae**

a. Articulation below the glumes, the spikelet falling entire; glumes dissimilar, the second obovate; lemma awnless . . . . *Sphenopholis*
a. Articulation above the glumes
  b. Lemmas bifid at the apex, with a flat, bent awn attached between the lobes . . . . *Danthonia*
  b. Lemmas merely toothed at apex, awned from the back of the body
    c. Glumes 2.0-3.5 cm long, spikelets pendulous, 2-flowered; annual . . . . *Avena*
    c. Glumes not over 1 cm long; spikelets not pendulous; perennials
      d. Spikelet 3-6-flowered, 1.0-1.5 cm long . . . . *Helictotrichon*
      d. Spikelet 2-4-flowered, less than 1 cm long
        e. Lemmas keeled, awned from above the middle
          f. Awn straight . . . . *Koeleria*
          f. Awn bent and exserted . . . . *Trisetum*
        e. Lemmas convex, awned from below the middle
          g. Glumes shorter than spikelet . . . . *Deschampsia*
          g. Glumes equalling or exceeding the uppermost floret . . . . *Vahlodea*

IV. **Festuceae**

a. Coarse and reed-like plants, 1-3 m tall, with large plumose panicle .... *Phragmites*
a. Plants of small or medium stature
  b. Plant dioecious; culms erect, very leafy, from a creeping rhizome. Prairie grass of saline or alkaline soil .... *Distichlis*
  b. Plant not normally dioecious
    c. Lemmas densely villous on the callus, 7-nerved, tall, coarse prairie grasses of wet places .... *Scolochloa*
    c. Lemmas not villous on the callus, 3-5-nerved
      d. Glumes distinctly longer than the spikelet; arctic plants of wet places ... *Dupontia*
      d. Glumes shorter than spikelet
        e. Palea with two conspicuous bristles on each side of its base; high-arctic emergent aquatic .... *Pleuropogon*
        e. Palea without such bristles
          f. Lemmas bifid at the apex, awned from the notch or just below
            g. Callus of floret bearded .... *Schizachne*
            g. Callus of floret not bearded .... *Bromus*
          f. Lemmas not bifid at apex, if awned the awn is apical
            h. Glumes nearly as long as lowermost floret; species of arctic range
              i. Tall marsh plants from a creeping rhizome; leaves broad and flat .... *Arctophila*
              i. Dwarf terrestrial tufted plants with yellowish-green foliage and crinkly roots .... *Colpodium*
            h. Glumes shorter than the lowermost floret
              j. Lateral nerves of lemma arched or converging; lemmas awnless, acuminate or awned
                k. Lemmas distinctly rounded on the back
                  l. Lemmas awned or with acuminate tip .... *Festuca*
                  l. Lemmas awnless, or the tip blunt or erose .... *Poa*
                k. Lemmas distinctly keeled, the tip blunt or erose .... *Poa*
              j. Lateral nerves of lemma parallel; lemmas awnless, with blunt or rounded tip
                m. Lemmas prominently nerved .... *Glyceria*
                m. Lemmas obscurely nerved; glumes minute, much shorter than the lowermost lemma .... *Puccinellia*

V. **Chlorideae**

a. Spikelets 1-2-flowered, in two rows along one side of a slender continuous rachis. Coarse annual slough-grass .... *Beckmannia*
a. Spikelets 1-flowered, closely imbricated on one side of rachis. Coarse, perennial marsh grass from a stout, cord-like rhizome .... *Spartina*

VI. **Hordeae**

a. Spikelets solitary at each joint of the rachis .... *Agropyron*
a. Spikelets 2 or 3 at each joint of the rachis
  b. Spikelets 2 at each joint, not readily disarticulating .... *Elymus*
  b. Spikelets 3 at each joint, readily disarticulating .... *Hordeum*

VII. **Paniceae**

Spikelets surrounded by more or less connate bristles forming an involucre; the bristles remain after spikelet drops. Represented with us by only one genus, *Setaria*.

Agropyron Gaerntn. Wheat Grass

Tufted or stoloniferous coarse grasses, ours mostly with elongated and narrow spikes from 7 to 20 cm long. Spikelets solitary, alternating at the continuous joints or notches of the rachis, 3- to 5-flowered, readily disarticulating above the glumes. Glumes 3- to several-nerved, mucronate; lemmas mucronate or awned, the awn scabrous, sometimes as long or even several times longer than body of lemma.

The following natural crosses between species of *Agropyron* with *Elymus* and *Hordeum* have been reported from our area: *Agropyron sericeum* crosses with *Elymus sibiricus* to form X *Agroelymus palmerensis* Lepage. *Agropyron trachycaulum* crosses with *Hordeum jubatum* to form X *Agrohordeum Macounii* (Vasey) Lepage (*Elymus Macounii* Vasey). *Agropyron sericeum* crosses with *Hordeum jubatum* to form X *Agrohordeum pilosilemma* Mitchell & Hodgson.

a. Non-native, introduced and weedy species
  b. Plants with long, creeping yellowish and wiry rhizomes . . . . *A. repens*
  b. Plants tufted
    c. Spike oval and somewhat flattened, the very numerous spikelets laterally compressed, divergent on the rachis, comb-like . . . . *A. pectiniforme*
    c. Similar, but rather smaller; spikelets less numerous and ascending rather than divergent . . . . *A. sibiricum*
a. Native species, ours tufted, creeping rhizomes lacking or very short; spike elongated, narrow
  d. Glumes thin, with conspicuous hyaline margins; lemmas glabrous or more often somewhat pubescent, awnless or with awn not longer than body of lemma; spike commonly purplish, rarely over 7 cm long
    e. Glumes nearly as long as spikelet, 3- to 4-nerved, broadest above the middle, with broad, somewhat unsymmetrical hyaline margins
      f. Glumes always glabrous . . . . *A. violaceum*
      f. Glumes and lemmas densely pubescent . . . . *A. violaceum* var. *hyperarcticum*
    e. Glumes rarely more than half as long as spikelet, 3-5-nerved, tapering evenly towards the tip, with narrow hyaline margins; lemmas distinctly soft-pubescent except near the tip . . . . *A. sericeum*
  d. Glumes firm, lacking hyaline margins, commonly 5-7-nerved; lemmas glabrous; spike commonly green, usually more than 7 cm and often up to 15 or 20 cm long
    g. Lemmas awnless or very short-awned
      h. Spikelets large, well spaced on the rachis, their tips barely, if at all, reaching the base of the spikelet above; spike commonly 15 to 20 cm long or more . . . . *A. trachycaulum* var. *trachycaulum*
      h. Spikelets smaller, closely imbricated; spike short and relatively broad . . . . *A. trachycaulum* var. *novae-angliae*
    g. Lemmas awned; spikelets imbricated
      i. Spikes very slender, commonly 6-7 cm long; awn thin, usually less than 1 cm long . . . . *A. trachycaulum* var. *glaucum*
      i. Spikes coarse, commonly 10 cm long or longer; awn stout and very scabrous, 2-3 cm long . . . . *A. trachycaulum* var. *unilaterale*

Agropyron pectiniforme R. & S.
Tufted, the culms up to 7 dm or more tall, the leaves up to 8 mm broad but usually much narrower; spikes ovate-oblong, 1.5 to 6.5 cm long, flattened, the spikelets laterally compressed and divergent on the rachis.

Occasional in waste places, near settlements in southern District of Mackenzie.

General distribution: Introduced from Eurasia.

Agropyron repens (L.) Beauv.
Culms up to 8 dm tall, with thin flat, sparsely soft pilose, glaucous green blades 5 to 8 mm wide, their sheaths often pubescent; spikes variable, thin and elongated, 5 to 15 cm long. Glumes and lemma short-awned, or that of the lemma sometimes as long as the body.

Waste places, southern District of Mackenzie.

General distribution: Introduced from Eurasia.

Agropyron sericeum Hitchc.
Loosely tufted, sometimes short rhizomatose; culms erect-ascending, 4 to 8 dm high; leaf-blades short, flat, 4 to 8 mm wide. Spike slender and narrow, 7 to 14 cm long, green, or less often purplish-tinged, the lowermost spikelets commonly somewhat remote; glumes glaucous, awnless or merely apiculate; lemmas awnless, or with an awn shorter than the body, evenly soft-appressed pubescent. Anthers 1.0 to 1.5 mm long.

Alluvial flats and river banks, from Alaska, Yukon and western District of Mackenzie, barely reaching the polar tree line.

General distribution: Northwest subarctic America and (?) E. Siberia.

Map 88

Agropyron sibiricum (L.) P.B.
Similar to *A. pectiniforme* from which it may be distinguished by the usually longer (6 to 10 cm long) and unflattened spikes and less divergent spikelets.

Occasional in waste places about settlements in southern District of Mackenzie.

General distribution: Introduced from Eurasia.

Agropyron trachycaulum (Link) Malte
A. pauciflorum (Schwein.) Hitchc.
Loosely tufted, commonly with numerous sterile shoots; culms erect 6 to 10 dm tall, green or glaucous; leaf-blades long and narrow, scabrous. Spike narrow and green, the spikelets usually well spaced on the rachis. Glumes glabrous, nearly as long as spikelet, firm, glabrous, broadest at or below the middle, awnless or merely apiculate. Lemmas glabrous, awnless or awned.

In our area *A. trachycaulum* is represented by four rather distinct varieties, by some authors considered subspecies or even full species:

var. *trachycaulum*. Plant commonly forming large, leafy and loose tussocks from which rise the up to 1 m tall culms; spike slender, 20 cm long, the lowermost spikelets usually somewhat remote.

Claybanks, not too dry river banks and meadows, readily invading roadside clearings.

General distribution: Low-arctic, from southern S.W. Greenland and Lab. to Alaska, northward to or near the polar tree line.

Fig. 84.

var. *novae-angliae* (Scribn.) Fern. Similar, but culms shorter and commonly fewer together; the spikes shorter and denser, the spikelets overlapping and commonly in two distinct rows.

General distribution: N. America, Nfld. to Alaska, north to southwestern District of Mackenzie.

var. *glaucum* (Pease & Moore) Malte. Plant glaucous throughout; culms few together, slender; lower leaf-sheaths pubescent. Awn of lemma very thin, usually as long as the body.

Range similar to that of preceding and, like it, reaching southwestern District of Mackenzie.

var. *unilaterale* (Cassidy) Malte, *A. subsecundum* (Link) Hitchc. Forming small, loose and leafy tussocks; spikes dense, erect, slightly curved or nodding and slightly one-sided.

Prairie parkland and open aspen woods in the southwestern parts of the District of Mackenzie.

General distribution: Que. and Ungava west to B.C. southern District of Mackenzie, Y.T. and Alaska.

Map. 89

Agropyron violaceum (Hornem.) Lange ssp. **violaceum**
A. latiglume (Scribn. & Sm.) Rydb.
A. alaskanum Scrib. & Merr.
A. boreale auctt.
Tufted, culms geniculate-ascending; leaf-blades 3 to 5 mm wide; scabrous or sparsely short-haired on the ribs; spike dense, averaging from 7 to 9 cm in length; spikelets overlapping, 9 to 15 mm long, 3- to 5-flowered, purple; glumes 7 to 11 mm long, broadly oblanceolate or narrowly obovate, acuminate toward the tip, 3- to 5-ribbed, with broad hyaline margins unilaterally broadest toward the tip; lemmas short-pubescent or glabrous, awnless, long-acuminate or with an awn shorter than the body; anthers 1.0 to 1.5 mm long.

Well drained calcareous sand or clay of river banks, floodplains or lake shores; low- arctic-alpine.

General distribution: ? Circumpolar, low-arctic. In N. America from central and southern W. Greenland to Alaska.

Fig. 85 Map 90

Agropyron violaceum (Hornem.) Lange var. **hyperarcticum Polunin**
A. boreale ssp. *hyperarcticum* (Polunin) Melderis
Similar to typical *A. violaceum* in habit and differing mainly in having the glumes and lemmas densely hirsute, and the nodes of the culms short-pubescent (scarcely visible except under a strong lens). In specimens from Greenland

and from the eastern islands of the Canadian Arctic Archipelago the glumes are always pubescent whereas on the western Canadian islands and on the mainland, plants with either pubescent or glabrous glumes frequently occur together.

Sandy or gravelly calcareous floodplains or river banks.

General distribution: N. American, arctic and high-arctic, from Alaska to N.E. Greenland.

Agrostis L. Bentgrass

Perennials mostly tufted; the panicle open, with numerous 1-flowered spikelets, disarticulating above the glumes; lemma awned or awnless.

a. Caespitose, native species
  b. Culms stout up to 1 m tall; panicle narrowly contracted, almost spike-like . . . *A. exarata*
  b. Culms rarely up to 50 cm tall, panicle open
    c. Lemma prominently awned (awn rarely absent), anthers 0.6-0.9 mm long . . . . *A. borealis*
    c. Lemma awnless (or minute if present), anthers 0.2-0.4 mm long . . . . *A. scabra*
a. Stoloniferous, not native, with up to 1 m tall culm and large panicle . . . . *A. gigantea*

Agrostis borealis Hartm.
Low, densely tufted with pyramidal open panicle of small spikelets on long capillary branches.

In dry rocky and turfy places on acid rocks, mainly in the Precambrian Shield area.

General distribution: Circumpolar, low-arctic.

Fig. 86 Map 91

Agrostis exarata Trin.
Tufted, with culms 2 to 10 dm tall; leaf-blades scabrous, flat, 2 to 10 mm wide, with a prominent up to 10 mm long ligule. Panicle green, very dense and narrow, almost spike-like, with numerous short branches; glumes 2.0 to 2.5 mm long, longer than the lemma; palea minute or lacking.

Moist, springy places.

General distribution: A Pacific coast species. From our area known only from the S.W. District of Mackenzie.

Map 92

Agrostis gigantea Roth
Culms up to 1 m tall, sometimes decumbent or geniculate at the base, from a creeping or stoloniferous rhizome; panicle 5 to 20 cm long, pyramidal but contracted after flowering.

In waste places; introduced.

General distribution: Circumpolar.

Agrostis scabra Willd.
Tufted, with culms 3 to 6 dm tall, very slender, erect or ascending. Leaves flat, very scabrous, 1 to 3 mm wide. Panicle large and diffuse, 5 to 25 cm long, with erect-spreading, hairlike, and scabrous branches 5 to 20 cm long; the spikelets crowded near the tips. Glumes commonly purple, about 2 mm long, acuminate, scabrous on the keel, two-thirds as long as the awnless lemma; palea wanting.

Occasional to common on floodplains, river banks, and occasionally in bogs; commonly pioneering on disturbed soil by roadsides.

General distribution: N. America: Wide ranging from Nfld. to Alaska.

Fig. 87 Map 93

Alopecurus L. Foxtail

Rhizomatose perennials with soft and dense spike-like panicles.

a. Panicle ovoid-short cylindrical; culms erect . . . . *A. alpinus*
a. Panicle narrowly cylindrical, several times longer than broad . . . . *A. aequalis*

Alopecurus aequalis Sobol
Marsh grass with decumbent creeping stems, and flaccid floating leaves. Panicle spike-like, pale green.

In shallow, often summer-dry ponds, north to the limit of forest.

General distribution: Circumpolar, wide ranging.
Fig. 88 Map 94

Alopecurus alpinus J.E. Smith
Culm smooth 15 to 30 cm high from a short rhizome; panicle contracted into a short, spike-like head; leaves short, flat, and rather succulent, the upper sheaths somewhat inflated.

Common and widespread arctic grass growing in wet tundra and by brooks and lake shores. Although not littoral, the species is generally restricted to the sea coast. It is strongly nitrophilous and often forms a dense turf on manured soil, near human habitations or below bird cliffs.

General distribution: Circumpolar, wide ranging, arctic.
Fig. 89 Map 95

Arctagrostis Griseb.

Coarse, stoloniferous mostly broad-leaved perennials with large panicles; the glumes shorter than the awnless lemmas.

a. Anthers 2 mm long or longer, dark purple; spikelets 4 mm long or longer
 b. Panicle stiff, narrowly lanceolate, dense, often somewhat interrupted below; anthers rarely over 2.0 mm long . . . . *A. latifolia* ssp. *latifolia*
 b. Panicle open, branches spreading, 2-6 cm long; anthers 2.7-3.0 mm long . . . . *A. latifolia* ssp. *nahanniensis*
a. Anthers less than 2 mm long; spikelets small, less than 4 mm long
 c. Leaves broad, 5-7 mm broad; panicle open and large, 15-25 cm long, pyramidal . . . . *A. arundinacea*
 c. Leaves narrower, averaging about 3 mm broad; panicle narrower, often somewhat nodding . . . . *A. angustifolia*

Arctagrostis angustifolia Nash
Differs strikingly from *A. arundinacea* by its much narrower leaf-blade, averaging about 3 mm in width (against 5 to 7 mm in *A. arundinacea*), by its 5 to 6 mm long ligules, more slender and about 1 m tall culms, and by its narrower, commonly pale and somewhat nodding panicle.

General distribution: Described from near Dawson, Y.T. Otherwise known from a few stations in central Alaska and Y.T., and in our area from near Virginia Falls on the South Nahanni R.
Map 96

Arctagrostis arundinacea (Trin.) Beal
var. **arundinacea**
A tall and coarse grass somewhat similar to *A. latifolia* from which it can usually be distinguished by its larger and more open panicle, smaller spikelets and somewhat shorter anthers.

River flats, alpine meadows and open forest.

The var. *crassispica* Bowden is said to differ from *A. arundinacea* by having "coarser spikes and culms". Also it is reported to be octoploid (2n=56), as is also *A. latifolia*, whereas *A. arundinacea* is tetraploid. Thus far the var. *crassispica* is known only from the type locality at Norman Wells on the middle Mackenzie River.

General distribution: An Amphi-Beringian species which from E. Siberia extends through Alaska and Y.T. a short distance east of the Mackenzie Delta and in the mountains south into northernmost Alta. and B.C.
Fig. 90 Map 97

Arctagrostis latifolia (R.Br.) Griseb.
ssp. **latifolia**
Stoloniferous; culms smooth, 10 to 50 cm high; leaves short, broad and flat, scabrous on both sides. Panicle narrowly lanceolate or, in southern and more favoured situations narrowly pyramidal, dark purple, often somewhat interrupted below.

A common or even ubiquitous species rarely absent in moderately damp and turfy tundra.

General distribution: Circumpolar, wide spread, arctic-alpine.
Fig. 91 Map 98

Arctagrostis latifolia (R.Br.) Griseb.
ssp. **nahanniensis Porsild**
A rather well marked race with 40 to 60 cm tall culms and 6 to 18 cm long panicle with spreading branches of which the lower and middle ones are 4 to 6 cm long. The spikelets are large, 4.0 to 4.5 mm long, and the dark purple anthers from 2.7 to 3.0 mm long.

General distribution: Thus far known only from the type locality on the upper South Nahanni R., and from Bolstead Creek, a tributary to Keele R., both in the Mackenzie Mts.

Arctophila Rupr.

Arctophila fulva (Trin.) Rupr.
Coarse marsh grass with a stout, creeping rhizome and smooth, leafy culms 30 to 50 cm high; leaves broad, flat, and smooth. Panicle pyramidal with long, drooping, capillary branches bearing from 2- to 7-flowered spikelets.

Growing in shallow water by the edge of tundra ponds and sloughs. Near its northern limit often sterile, propagating vegetatively.

General distribution: Circumpolar, low-arctic.

Fig. 92 Map 99

Avena L. Oats

Annual, non-native and introduced grasses with coarse and rough leaves and open panicles of large drooping spikelets.

a. Awn 3-4 cm long . . . . *A. fatua*
a. Awn short or lacking . . . . *A. sativa*

Avena fatua L.
Wild Oats
Lemma usually hairy on the back, with a 3 to 4 cm long twisted and sharply bent awn; rachilla disarticulating and the grain promptly shedding on maturity.

Occasional weed in waste places.

General distribution: Cosmopolitan

Avena sativa L.
Oats
Lemma glabrous, awnless or with a short awn; rachilla not disarticulating and the grain, therefore, not promptly shed.

In waste places.

General distribution: cultivated widely.

Beckmannia Host

Beckmannia Syzigachne (Steud.) Fern.
Sloughgrass
Tall, pale blue-green, smooth annual with culms up to 1 m tall and broad, flat leaves. Panicle large, simple or compound, 10 to 15 cm long; the spikelets 1- to 2-flowered, laterally compressed, subcircular, nearly sessile, and closely imbricated in two rows along the slender, continuous rachis, disarticulating below the glume and falling entire. Glumes membranaceous, inflated; lemma acute.

Wet meadows and sloughs subject to spring flooding; along the Mackenzie River in settlements where it was likely introduced with shipments of native hay from upriver points. However the small but very numerous seeds are eaten by waterfowl and, no doubt, distributed by them, as well as by flood water.

General distribution: Circumpolar.

Fig. 93 Map 100

Bromus L. Brome-grass

Ours coarse perennial grasses with broad and flat leaves and open or contracted panicle of large, many-flowered ascending or pendulous spikelets on slender, flexuous branches.

a. Plant tufted, lacking creeping rhizomes; panicle branches spreading or drooping; lemmas glabrous except for a row of long hairs on the marginal nerve . . . . *B. ciliatus*
a. Plant with creeping rhizomes; panicle branches ascending
  b. Lemmas glabrous, introduced species . . . . *B. inermis*
  b. Lemmas pubescent, native species . . . . *B. Pumpellianus*

Bromus ciliatus L.
Culms slender, singly or few together, 7 to 12 dm tall; nodes pubescent, sheaths short pilose, ligule short, and the blades soft, up to 10 mm wide, commonly with scattered long, soft hairs on both surfaces. Panicle 15 to 25 cm long,

open, the branches drooping, the lowermost up to 15 cm long. Spikelets 1.5 to 2.5 cm long, 4- to 9-flowered; glumes glabrous, the first 1-nerved, the second 3-nerved; lemmas 10 to 15 mm long, long-ciliate, glabrous across the back, awn 3 to 5 mm long. Anthers variable, 1.0 to 1.8 mm long.

Moist, open woods and rich meadows; thus far reported only from a few stations in the southern parts of Mackenzie River drainage.

General distribution: N. America and E. Asia.

Map 101

Bromus inermis Leyss.

Similar to *B. Pumpellianus*, differing chiefly by its totally glabrous or somewhat scabrous, and usually awnless lemmas.

General distribution: A native of Eurasia widely cultivated, in our area occasional in waste places and by roadsides.

Bromus Pumpellianus Scribn.
var. **Pumpellianus**

Culms erect, 5 to 10 dm tall, mostly solitary from a creeping rhizome. Leaves flat, glabrous, the ligule 1.5 to 2.0 mm long. Panicle 1 to 2 dm long, ellipsoid, the spikelets green, bronze or purple-tinged, 2- to 11-flowered; lemmas membranaceous, glabrous on the back, villous along the marginal nerves, 10 to 12 mm long, exclusive of the 2 to 3 mm long, straight awn. Anthers 4 to 5 mm long, exserted at anthesis.

Common on sandy or gravelly river banks and islands along the Mackenzie River and some of its tributaries north to the Delta.

The var. *arcticus* (Shear) Porsild of Amphi-Beringian range is distinguished by lower growth, hirsute to densely villous glumes and lemmas, and by its villous sheaths, and is of more arctic range, occasional to common on dunes eastward beyond the Mackenzie River delta, and south into the treeless country east of Great Bear Lake.

General distribution: W. and N.W. America and E. and central Asia.

Fig. 94 Map 102, 103

Calamagrostis Adans. Reed-Bentgrass

Ours mostly tall, narrow-leaved perennials from a creeping rhizome. Panicle terminal, lax or contracted, of numerous 1-flowered spikelets that disarticulate above the glumes; lemma equalling or shorter than the glumes, awned below the middle of the back, and surrounded at the base with a tuft of white hairs.

a. Awn exserted, geniculate
  b. Panicle compact, its branches appressed; awn much exserted; leaves stiff and harsh . . . . *C. purpurascens*
  b. Panicle lax, more or less pyramidal at time of anthesis, the lower branches spreading or even reflexed; awn straight, not greatly exserted
    c. Leaves and culm scabrous . . . . *C. montanensis*
    c. Leaves and culm smooth . . . . *C. deschampsioides*
a. Awn straight, included or barely exserted
  d. Panicle large, lax and open in anthesis, nodding in youth; callus hairs copious, all about the same length, equalling or not much shorter than the lemma
    e. Culm with 2-3 nodes . . . . *C. lapponica* var. *nearctica*
    e. Culm with 5-6 nodes
      f. Glumes 3 mm long, lanceolate . . . . *C. canadensis* var. *canadensis*
      f. Glumes 5 mm long, long-acuminate . . . . *C. canadensis* var. *Langsdorfii*
  d. Panicle more or less contracted
    g. Blades flat, scabrous . . . . *C. chordorrhiza*
    g. Blades involute, or if flat, rigid and becoming involute
      h. Leaves, sheath and culm stiff and harsh; panicle pale; glumes firm . . . *C. inexpansa*
      h. Leaves, sheath and culm smooth; panicle dark purplish-tinged; glumes thin . . . . *C. neglecta*

Calamagrostis canadensis (Michx.) Beauv.
var. **canadensis**

Culms leafy 5 to 15 dm tall from a creeping rhizome; leaves long and flat, soft, often involute in drying. Panicle large, purple or green, open and spreading during anthesis, 6 to 30 cm long and 2 to 10 cm broad. Spikelets numerous and small, the glumes about 3 mm long, scabrous, about as long as the lemma; callus hairs copious, as long as the lemma.

Although actual collections are few, the var. *canadensis* has been noted as common along the Mackenzie River to its delta. Wide ranging species of the Boreal Forest zone.

General distribution: N. America: Nfld. to Alaska.

Calamagrostis canadensis (Michx.) Beauv. var. **Langsdorfii** (Link) Inman

Similar to the var. *canadensis* but of more alpine and subarctic range. At once distinguished by its commonly nodding panicle, its 5 to 6 mm long, narrow glumes distinctly tapering towards the tip, and by its broader, up to 8 mm broad leaves. The whole plant is often suffused with purple.

The var. *Langsdorfii* appears to be of consistently more northern range than the var. *canadensis*. In our area it is mainly found near or slightly beyond the limit of trees reaching the arctic coast in the Mackenzie Delta and at the head of Bathurst Inlet.

General distribution: Circumpolar, low arctic.

Fig. 95 Map 104

Calamagrostis chordorrhiza Porsild

Tufted, with long, cord-like widely branching, shiny rhizomes from which rise fascicles of flat, glaucous leaves much shorter than the mostly solitary, geniculate-erect, 30 to 35 cm tall, scabrous culms. Panicle pale reddish-purple, lax and narrow, 5 to 6 cm long, 1.0 to 1.5 cm wide, more or less interrupted in the lower part.

Well drained calcareous sandy or gravelly places. Described from the Caribou Hills along the east branch of the Mackenzie River delta where it is common locally.

General distribution: N.W. District of Mackenzie.

Fig. 96 Map 105

Calamagrostis deschampsioides Trin.

Culms geniculate-erect, rarely more than 25 cm high, solitary or few together from a thin, creeping rhizome, leaves smooth, much shorter than the culm, those of the culm 3 to 5 cm long; panicle lax, pyramidal, the small spikelets at the end of the smooth branches; glumes 3 to 4 mm long, equalling or slightly longer than the lemma.

Littoral species of damp tundra.

General distribution: Circumpolar, arctic, of very spotty and disrupted range.

Fig. 97 Map 106

Calamagrostis inexpansa A. Gray

Tufted and stoloniferous; culms stiff, commonly from 3 to 10 dm high, very scabrous, at least in the upper part; leaves scabrous and harsh, flat, becoming strongly involute. Panicle pale, usually dense and spike-like, 5 to 20 cm long, its stiff and ascending branches very scabrous; glumes firm and opaque, acute, 3 to 4 mm long; the lemma slightly shorter; callus hairs shorter than the lemma.

A very variable species, common and often dominant on floodplain meadows.

General distribution: N. America: Nfld. to Alaska.

Fig. 98 Map 107

Calamagrostis lapponica (Wahlenb.) Hartm. var. **nearctica** Porsild

Loosely caespitose with a short, thin rhizome; culms solitary or few together, slender, erect, 30 to 60 cm high, glabrous except just below the panicle, with 2 or 3 nodes; the cauline leaves strongly involute; panicle lax, narrow, 4 to 8 cm long and 1.5 or rarely 2.0 cm wide. Glumes 4.0 to 4.5 mm long, slightly lustrous and strongly suffused with purple, turning bronze, abruptly acuminate, hispidulous only along the keel; callus hairs copious, unequal, the longest as long as the lemma. The var. *nearctica* differs from the species by its shorter stature, shorter panicle and shorter and narrower glumes.

Stable dunes, or in peaty soil of open heath or muskeg.

General distribution: Alaska to Lab.

Fig. 99 Map 108

***Calamagrostis montanensis** Scrib.

Stoloniferous; culms stiffly erect, 2 to 4 dm high, scabrous mainly below the panicle. Leaves short, involute and scabrous. Panicle oblong-lanceolate, 4 to 8 cm long and 6 to 12 mm wide, purplish, occasionally interrupted below. Spikelets 4 to 5 mm long; glumes acuminate, scabrous on the keel, slightly longer than the lemma.

Dry, sandy prairie.

General distribution: N. America: Plains and foothills of Alta. to western Man., north almost to lat. 60°N. on the Slave R.

Map 109

Calamagrostis neglecta (Ehrh.) Gaertn., Mey. and Schreb.

Stoloniferous, often turf-forming; culms slender, smooth, 2 to 4 dm long. Leaves soft, smooth or slightly scabrous at the tips and along the margins, commonly involute. Panicle dense and spike-like, more rarely narrowly pyramidal; spikelets purplish, bronze or greenish; glumes ovate-lanceolate, hyaline and somewhat translucent, about 3 mm long; lemma firm with a short awn, the callus hairs unequal, shorter than the lemma.

A very variable species, occasional to common in wet meadows northward to the Arctic Coast, but in the southern part of the District of Mackenzie perhaps mainly in alpine situations.

General distribution: Circumpolar.

Fig. 100 Map 110

Calamagrostis purpurascens R. Br.
Coarse and tufted with stiff and usually scabrous culms 2 to 7 dm high. Leaves glaucous, much shorter than the culms, very scabrous especially on the margins, flat, becoming strongly involute in age. Panicle 6 to 8 cm long, narrowly pyramidal in youth, dense or interrupted below, purplish or bronze, the branches short and floriferous to the base. Glumes firm, about 6 mm long, scabrous, especially on the prominent keel; lemma thin, two-thirds as long as the glumes.

Common in well drained, calcareous, sandy or stony places northward and beyond the Arctic Coast.

General distribution: N. America. Arctic-alpine, from E. Greenland to E. Asia.

Fig. 101 Map 111

Cinna L. Wood Reedgrass

Cinna latifolia (Trev.) Griseb.
Tall, tufted woodland grass with short and broad leaf-blades and conspicuous hyaline ligules. Panicle 15 to 25 cm long, open, commonly nodding. Spikelets small and numerous, 1-flowered, articulate below the glumes. Glumes awnless, persistent; lemmas similar, short-awned.

Rich, moist and somewhat shady, deciduous woods; occasional in the southernmost parts of the District of Mackenzie.

General distribution: Circumpolar.

Fig. 102 Map 112

Colpodium Trin.

Colpodium Vahlianum (Liebm.) Nevski
Puccinellia Vahliana (Liebm.) Scribn. & Merr.
Densely caespitose with pale, fibrous, crinkly roots; foliage yellowish-green with a peculiar metallic lustre. Culms stout, erect, 15 to 20 cm tall, bearing 2 to 3 lax and flat leaves; withered basal leaf-sheaths very soft, early decaying. Panicle, 3 to 5 cm long, contracted; branches in pairs, glabrous, ascending, bearing only a few spikelets. Spikelets lanceolate, 6.0 to 7.5 mm long, dark purple, shining, 2- to 4-flowered. Glumes and lemmas thin, translucent, subacute, their tips erose in age. Glumes subequal, 3.5 to 4.0 mm long; lemmas 4.0 to 4.5 mm long, copiously hairy in their lower part. Palea keels long-hairy. Anthers 1.0 to 1.2 mm long. Grains about 2 mm long.

A non-littoral species, often growing in moist clay by brooks and on snowbeds. Within our area mainly along the Arctic Coast.

General distribution: Amphi-Atlantic, high-arctic.

Fig. 103 Map 113

Danthonia DC. Wild Oat-grass

Tufted perennials with narrow leaves and few-flowered open or spike-like panicles. Ligule represented by a collar of long soft hairs. Spikelets several-flowered, the pilose rachilla readily disarticulating above the glumes. Glumes persistent, broad and papery, acute or acuminate, mostly exceeding the uppermost floret; lemmas obscurely nerved, the bifid apex with a stout, twisted and geniculate awn attached between the lobes.

a. Lemma glabrous on the back, pilose on the margins, lobes of the lemma aristate . . . . *D. intermedia*
a. Lemma pilose on the back, lobes of the lemma acuminate . . . . *D. spicata*

Danthonia intermedia Vasey
Culms 3 to 5 dm high; sheaths glabrous except at the throat. Panicle narrow, 3 to 6 cm long, with from 5 to 10 spikelets; glumes acuminate,

about 15 mm long, purplish in youth, turning papery white.

Open grassy places. Thus far known only from the Nahanni Mountains but to be expected also east of the Mackenzie valley along the southern rim of the Precambrian Shield.

General distribution: N. America: Lab. to Alaska, non-arctic, mainly alpine.

Fig. 104 Map 114

***Danthonia spicata** (L.) Beauv.
Culms tufted, commonly up to 5 dm tall, slender, with numerous filiform, curled and somewhat hirsute leaves around their bases. Panicle lax, 3 to 5 cm long, few-flowered, the branches short, erect or ascending.

Sterile and rocky places. As yet not reported from the Northwest Territories where it is likely to occur along the southern rim of the Precambrian Shield, east of the Slave River.

General distribution: N. America: Nfld. to B.C., non-arctic.

Fig. 105 Map 115

Deschampsia Beauv. Hairgrass

Tall, medium to low, perennial grasses with somewhat glossy, purplish, 2-flowered spikelets in terminal panicles; lemmas with a slender, barely exserted awn.

a. Tufted, lacking stolons
  b. Panicle open
    c. Culms 3-7 dm high, panicle more or less pyramidal . . . . *D. caespitosa*
    c. Culms 7-10 dm high, panicle elongated rather than pyramidal . . . . . *D. mackenzieana*
  b. Panicle narrow; culms rarely up to 3 dm high . . . . *D. brevifolia*
a. Mat-forming; culms 15-25 cm high, panicle open . . . . *D. pumila*

Deschampsia brevifolia R.Br.
Densely tufted, the culms usually 5 to 15 cm, but occasionally up to 30 cm tall, equalling or somewhat longer than the leaves. Panicle narrow and contracted, strongly suffused with purple or bronze.

Described from Melville Island in the Canadian Arctic Archipelago and long known as a high-arctic circumpolar species, *D. brevifolia* is now known also in high-alpine situations in the Mackenzie Mountains and in mountains of central Yukon and Alaska, where it is more highgrown, with culms commonly 30 cm high, or even taller. Hummocky and frost-heaved rather wet soils, and occasionally in turfy places in tundra.

General distribution: Circumpolar, arcticalpine.

Fig. 106 Map 116

Deschampsia caespitosa (L.) Beauv.
A coarse but smooth and densely tufted species with culms 3 to 7 dm high, and large pyramidal panicle.

The var. *glauca* (Hartm.) Lindm., distinguished by its lower stature, narrower leaves and smaller and generally less open panicle, apparently is less common and has been collected mainly east of the Paleozoic boundary. A viviparous form of *D. caespitosa* has been reported from the vicinity of Fort Simpson on the Mackenzie River.

Wet meadows and on stony or gravelly lakeshore and river banks, northward to and slightly beyond the limit of trees.

General distribution: Circumpolar and lowarctic.

Fig. 107 Map 117

***Deschampsia mackenzieana** Raup
Similar to *D. caespitosa* but taller and stouter, with up to 1 m tall culms and an open and more elongated panicle. Thus far known only from the type locality in damp hollows among dunes along the south shore of Athabasca Lake, but may be expected to occur elsewhere in the southern parts of the District of Mackenzie. Perhaps not distinct from the very similar *D. beringensis* Hultén of the N.W. Pacific coast and S.W. Alaska.

General distribution: Northern Sask.

Map 118

Deschampsia pumila (Trin.) Ostenf.
Low, mat-forming and often sterile. Culms, when present, solitary or few together and characteristically geniculate at the base, the panicle open with few and small spikelets on capillary branches.

A seashore species usually growing in wet sand or clay among rocks.

General distribution: Circumpolar, high-arctic.

Fig. 108 Map 119

Distichlis Raf. Alkali-grass

Distichlis spicata (L.) Greene
var. **stricta** (Torr.) Beetle
Stiffly erect, pale or glaucous, dioecious grass with culms 1 to 4 dm tall from a stout, scaly and much branched, creeping rhizome; the rigid, involute leaves ascending, their blades in two vertical ranks; panicle small and spicate of large, smooth and compressed spikelets.

Salt plains west of Fort Smith.

General distribution: Salt or alkaline plains of central interior N. America.

Fig. 109 Map 120

Dupontia R. Br.

Stoloniferous, arctic, turf-forming, smooth marsh grasses with culms 10 to 25 cm high, and panicle of somewhat shiny, bronze coloured spikelets.

a. Panicle usually narrow and contracted with short, ascending branches bearing 2-4-flowered spikelets . . . . *D. Fisheri*
a. Panicle usually pyramidal, open and interrupted, with few long, spreading branches bearing 1-2-flowered spikelets . . . . *D. Fisheri* ssp. *psilosantha*

Dupontia Fisheri R. Br.
ssp. **Fisheri**
Spikelets 2- to 4-flowered. Culms from 10 to 25 cm high, rather stout. Glumes firm, broadly lanceolate, 3-nerved, often obtuse, and shorter than the spikelet. Lemmas always appressed hirsute along the keel and margins, occasionally aristate (var. *aristata* Malte); anthers 2 to 2.5 mm long.

The high-arctic, polyploid race of the more southern and more widely distributed ssp. *psilosantha*. See following.

Wet meadows on alluvial soils, such as in estuaries and on brackish lagoon shores, where occasionally flooded by high tides. Perhaps always littoral.

General distribution: Circumpolar, high-arctic.

Fig. 110 Map 121

Dupontia Fisheri R. Br.
ssp. **psilosantha** (Rupr.) Hultén
D. psilosantha Rupr.
Similar, but as a rule lower, with narrower, somewhat involute leaves, and spreading, interrupted, and more or less pyramidal panicle. Spikelets 1- to 2-flowered, the glumes thin, narrow, and long-acuminate, equalling the golden yellow, purple-tipped lemmas, which are perfectly glabrous on the back and have a tuft of hairs at their base. Anthers 1.5 to 2.0 mm long, distinctly shorter than in ssp. *Fisheri*.

In similar places as preceding but not strictly littoral and of less pronounced arctic range.

General distribution: Circumpolar, low-arctic.

Fig. 111 Map 122

Elymus Wild Rye or Lyme-Grass

Coarse tufted or rhizomatose perennials with flat or sometimes involute leaves and slender elongated spikes. Spikelets commonly in pairs at each node, 2- to 6-flowered. Glumes narrow to subulate, awned or awnless, often spreading from the base of the spikelet. Lemmas merely acute, or awned from the tip.

Elymus sibiricus crosses with *Agropyron sericeum* to form X *Agroelymus palmerensis* Lepage.

a. Lemmas awnless
  b. Glumes lanceolate; glaucous dune species of sea-shore and shores of large northern lakes . . . . *E. arenarius* ssp. *mollis*
  b. Glumes narrowly subulate; non-glaucous plant of dry, open forest and grassland . . . . *E. innovatus*
a. Lemmas with a long divergent awn
  c. Tufted, spike relatively short and thick . . . . *E. canadensis*
  c. Tufted with short runners, spike long and thin, flexuous and nodding . . . . *E. sibiricus*

Elymus arenarius L.
ssp. **mollis** (Trin.) Hultén
Coarse, stoloniferous beach-grass with glaucous, broad, flat leaves, and stout culms, 15 to 100 cm high, which are soft villous in the upper part. Spike 8 to 15 cm long; glumes and lemmas soft villous.

In some parts of the Arctic the leaves of lyme-grass are used by the Eskimo for the weaving of baskets, and in Greenland also to provide an insole or insulating layer between the outer and inner native boot *(kamik)*. Inadvertently the seeds of lyme-grass may thus be spread by Eskimo when boot "hay" is discarded near campsites. The recent discovery of *Elymus* near Eureka, in Ellesmere Island, six hundred miles north of its previously known northern limit, may thus be a recent introduction.

Dunes and sandy places near the sea or sometimes by lake shores.

General distribution: Circumpolar, littoral, low-arctic.

Fig. 111 Map 123

Elymus canadensis L.
Culms loosely tufted, erect or ascending at base, smooth, glaucous, up to 1.5 m high; cauline leaves coarse, 5 to 15 mm broad, glabrous or somewhat scabrous above; spikes dense, up to 20 cm long, sometimes slightly nodding; glumes glabrous, strongly nerved, with a straight awn as long as the body, the lemmas villous-hirsute or merely scabrous, with a long curved or arching awn more than twice as long as the body.

Sandy and gravelly places, barely entering southernmost District of Mackenzie.

General distribution: N. American: N.B. to Alaska.

Map 124

Elymus innovatus Beal
Culms smooth, 3 to 10 dm high, rising in tufts from long, creeping rhizomes; sheaths glabrous, blades firm, glabrous beneath, 2 to 5 mm wide, flat or involute in drying. Spike rather short and dense, 5 to 12 cm long, the rachis villous; the narrow glumes and lemmas densely purplish or greyish villous, the latter with a short awn 2 to 4 mm long. Anthers purple.

Common on calcareous, sandy and shaly slopes and river banks along the Mackenzie valley from the lower Liard to the Mackenzie Delta.

General distribution: N. America: N. Ont. to central and northern Alaska.

Fig. 113 Map 125

Elymus sibiricus L.
Similar to *E. canadensis* from which it is readily distinguished by its much longer, thin, flexuous and usually pendulous spike, and by its shorter glumes. Also, although of tufted habit, *E. sibiricus* produces short, but distinct rhizomes.

Eroding river banks and cabin clearings along the Liard and lower South Nahanni rivers. In Alaska as well as in the District of Mackenzie the sporadic occurrence of *E. sibiricus* suggests that it is a recent introduction.

General distribution: East and central Siberia.

Fig. 114 Map 126

Festuca L. Fescue

Perennial (ours) caespitose or stoloniferous grasses, mostly with involute leaves and with variously shaped panicle; glumes somewhat unequal; the lemmas rounded on the back, terminating in a short awn.

a. Stoloniferous; culms decumbent at base
  b. Culms low, rarely over 2 dm high, always viviparous .... *F. prolifera* var. *lasiolepis*
  b. Culms 2-6 dm high
    c. Lemmas glabrous .... *F. rubra* ssp. *rubra*
    c. Lemmas villous .... *F. rubra* ssp. *Richardsonii*
a. Caespitose
  d. Tall and coarse, culms 5-10 dm tall; spikelets large, lustrous, body of lemmas 8-12 mm long .... *F. altaica*
  d. Low to medium; culms rarely over 3 dm high; body of lemmas 4-6 mm long
    e. Anthers 1.0 mm or less
      f. Leaves smooth

g. Culms scabrous below the panicle with minute curved hairs; panicle commonly 1-sided . . . . *F. baffinensis*
g. Culms smooth; panicle not 1-sided . . . . *F. brachyphylla*
f. Leaves very scabrous . . . . *F. saximontana*
e. Anthers 2 mm or longer . . . . *F. ovina* ssp. *alaskana*

Festuca altaica Trin.
Densely tufted; the old sheaths and leaf-bases persist for many years causing the formation of large, firm tussocks. Culms erect, smooth, 3 to 10 dm high. Leaf-blades coarse, involute, 1.5 to 3.0 dm long, scabrous along the margins. Panicle lax and open, 1 to 2 dm long, the 3 to 8 cm long branches spreading, drooping in age. Spikelets 3- to 5-flowered, 12 to 15 mm long, green or purplish-bronze. Glumes glabrous; lemmas ovate, long-acuminate but not awned, prominently nerved and finely hispid.

Alpine grassland and tundra.

General distribution: Amphi-Beringian. Common in the Mackenzie Mts. but with only a few scattered stations east of the Mackenzie valley. Similar to *F. scabrella* Torr., a dominant species of foothill-prairies of Alta.

Fig. 115 Map 127

Festuca baffinensis Polunin
Similar to *F. brachyphylla* but with somewhat longer and softer leaves that in length almost equal the culm, its upper half with short, stiff, and often curved hairs, visible only under a lens. Panicle narrowly ovoid, often one-sided and usually densely suffused with purple; anthers, 0.4 to 0.6 mm long.

On calcareous soil in sandy and gravelly places.

General distribution: N. American, high-arctic-alpine, south in the Rocky Mts. to Wyoming and Colorado.

Fig. 116 Map 128

Festuca brachyphylla Schultes
Densely tufted; the smooth, wiry culms 10 to 40 cm high, usually much longer than the stiff, bristle-like, smooth leaves. Panicle narrow and spike-like, sometimes viviparous (f. *vivipara*); anthers about 0.75 mm long.

Ubiquitous in exposed sandy, gravelly, or rocky places.

General distribution: Circumpolar, wide ranging, high-arctic-alpine.

Fig. 117 Map 129

Festuca ovina L.
ssp. **alaskana** Holmen
Somewhat similar to *F. brachyphylla*, but culms very slender and not so stiff, and panicle shorter and broader, purplish (or yellowish in f. *pallida* Holmen). Anthers about 2.0 mm long.

Alpine slopes.

General distribution: Described recently from near Old John L. on the north slope of Brooks Range, Alaska, and now known also from the Richardson Mts. west of Mackenzie Delta.

Map 130

Festuca prolifera (Piper) Fern.
var. **lasiolepis** Fern.
Mat-forming; culms smooth, capillary, geniculate at base 1 to 3 dm high; leaves soft, shorter than the culms; panicle slender, of commonly viviparous, glabrous spikelets. Glumes and lemmas merely pointed, in most spikelets modified into leaves.

Damp sandy river banks and lake shores.

General distribution: Subarctic Eastern N. America.

Fig. 118 Map 131

Festuca rubra L.
ssp. **rubra**
Of loosely tufted or matted growth with freely branching rhizomes and extravaginal offshoots emerging from the base of the culms. Culms erect or geniculate from the base, smooth and commonly from 2 to 6 dm high; basal leaves narrow and setaceous. Panicle 4 to 10 cm long with ascending or slightly spreading branches; spikelets green or purplish, 4- to 6-flowered; glumes and lemmas glabrous, the latter awn-tipped and distinctly nerved.

A cultivated variety has been introduced in the upper Mackenzie River area.

General distribution: Circumpolar, at least when considered in a broad sense.

Festuca rubra L.
ssp. **Richardsonii** (Hook) Hultén
F. rubra
ssp. *arenaria* of authors, not Osbeck.
Similar to preceding from which it can usually be distinguished by its glaucous leaves and panicle and, always, by its densely villous lemmas.

An arctic or sub-arctic race of the polymorphic *F. rubra* found on moist sandy river banks and lake shores or on well-established dunes.
Fig. 119 Map 132

Festuca saximontana Rydb.
Similar to *F. brachyphylla* from which it differs by its taller culms, often 5 dm tall or more, by its narrower, linear-cylindrical, and often somewhat interrupted panicle, by its pale glaucous leaf-blades that usually are very scabrous on the margins, and by its anthers, about 1.0 mm long against 0.75 mm in *F. brachyphylla* and 0.4 to 0.6 mm in *F. baffinensis*.

Dry grassy and rocky places of the southern parts of the District of Mackenzie.

General distribution: Central Alaska to L. Superior with widely disjunct stations in Ungava and W. Greenland.
Fig. 120 Map 133

Glyceria R. Br. Manna-grass

Perennial, tufted or rhizomatose, glabrous aquatic or marsh grasses with flat leaves and medium to tall culms. Panicle contracted or open, often quite large, erect or nodding. Spikelets compressed; the lemmas prominently nerved and their tips membranaceous.

a. Spikelets linear, nearly terete, 10 mm long or over; panicle narrow, erect ...... *G. borealis*
a. Spikelets ovate or oblong, rarely more than 5 mm long; panicle lax and commonly nodding
 b. First glume less than 1 mm long, lemmas small, prominently nerved .... *G. striata* var. *stricta*
 b. First glume 1.0-1.5 mm long
  c. Plant stout, often more than 1 m high, with a very large and compound panicle .... *G. grandis*
  c. Plant slender, rarely more than 5 dm high; panicle less than 2 dm long, few-flowered .... *G. pulchella*

Glyceria borealis (Nash) Batchelder
Culms erect or ascending from a creeping and decumbent base, slender, 6 to 10 dm high, the leaf-blades flat or folded, commonly 2 to 4 mm wide, elongate and often floating. Panicle 2 to 4 dm long, simple or with few stiffly erect or spreading branches. Spikelets numerous, narrowly cylindrical, 7- to 13-flowered.

In shallow water or very wet meadows. Apparently rare and restricted to the southernmost parts of the District of Mackenzie.

General distribution: Boreal N. America.
Map 134

Glyceria grandis S. Wats.
G. maxima (Hartm.) Holmb.
ssp. *grandis* (S. Wats.) Hultén
Culms tufted; leaf-blades flat, 6 to 12 mm wide; panicle nodding at the summit, its spikelets 4- to 7-flowered, 5 to 6 mm long; glumes pale or whitish; lemmas purplish, 1.5 to 2.0 mm long; anthers 0.6 mm long.

Wet river meadows, lake shores and ditches where it often forms pure stands. Southern District of Mackenzie north to Fort Simpson.

General distribution: N. America from Nfld. to Alaska.
Fig. 121 Map 135

Glyceria pulchella (Nash) K. Schum.
Culms erect, smooth, from a creeping rhizome; leaf-blades flat, 2 to 4 mm wide with somewhat involute margins, the lowermost sheaths rough. Panicle lax, its capillary branches smooth, somewhat flexuous, erect-ascending, the lowermost 8 to 10 cm long; spikelets 5 to 6 mm long and 4- to 6-flowered; glumes obovate, thin, pale purplish-brown or whitish, the lemmas firm, with broad, scarious margins.

Wet meadows, in the Mackenzie valley north to Norman Wells.

General distribution: Boreal N. W. America.
Fig. 122 Map 136

Glyceria striata (Lam.) Hitchc.
var. **stricta** (Scribn.) Fern.
Culms loosely to densely tufted, slender, 3 to 10 dm high; leaf-blades erect or ascending, flat or folded, 2 to 6 mm wide. Panicle ovoid, lax and open, the capillary branches in age becoming divergent or even reflexed; spikelets greenish or purple, 3- to 7-flowered, 2 to 4 mm long; glumes and lemmas minute, the latter falling soon after maturity.

Wet meadows or along shallow pond-margins; southwestern District of Mackenzie north to lat. 63°.

General distribution: Boreal N. America.

Fig. 123 Map 137

Helictotrichon Besser Wild Oats

Helictotrichon Hookeri (Scribn.) Henry
Avena Hookeri Scribn.
Coarse, dense and tufted perennial with culms 2 to 4 dm tall; leaf-blades glabrous, flat and firm, 1.5 to 3.0 mm wide. Panicle 8 to 12 cm long, narrow, the branches erect or ascending; spikelets 3- to 6-flowered, about 1.5 cm long; glumes thin and greenish, somewhat shiny, about as long as the spikelet; lemmas with a 1.0 to 1.5 cm long geniculate and twisted awn.

Dry foothill prairies.

General distribution: N. America, western foothills and plains barely entering our area on the lower Hay R.

Fig. 124 Map 138

Hierochloë R. Br. Holy Grass

Glabrous perennial grasses with flat leaves and terminal panicles of somewhat shiny, yellowish-brown spikelets. The entire plant is fragrant in drying.

a. Culms densely tufted; lemma of sterile florets with a long, geniculate and twisted awn . . . . *H. alpina*
a. Culms solitary from a creeping rhizome; awn straight or lacking
  b. Panicle narrow, composed of a few spikelets; dwarf arctic species . . . . *H. pauciflora*
  b. Panicle pyramidal; tall, subarctic species . . . . *H. odorata*

Hierochloë alpina (Sw.) R. & S.
Culms 10 to 30 cm high, caespitose, basal leaves narrow, their sheaths conspicuously purplish. Panicle contracted.

Dry tundra and rocky places, chiefly on acidic rocks.

General distribution: Circumpolar, widespread, arctic-alpine.

Fig. 125 Map 139

Hierochloë odorata (L.) Beauv.
Sweet Grass
Culms solitary, from a thin creeping rhizome; basal leaves short and broad, their sheaths not conspicuously purplish.

Sandy meadows and lake shores, in our area mainly within the Paleozoic formation and in the lowland north or slightly beyond the limit of forest.

General distribution: Circumpolar, wide ranging, low-arctic-alpine.

Fig. 126 Map 140

Hierochloë pauciflora R.Br.
Low and delicate grass with very slender culms, 10 to 15 cm high from a thin, creeping rhizome; panicle narrow, one-sided.

Generally growing in sphagnum in wet tundra; although not littoral, a distinctly coastal and lowland species.

General distribution: A wide ranging arctic species.

Fig. 127 Map 141

Hordeum L. Wild barley

Hordeum jubatum L.
Squirrel-tail Grass
Tufted perennial with erect-ascending culms 3 to 6 dm high; leaf-blades 2 to 5 mm wide, scabrous. Spikes 5 to 10 cm long, nodding; glumes awn-like, spreading, 2 to 6 cm long; lemmas with equally long, finely scabrous and often purplish-tinged awns.

Hordeum jubatum crosses with *Agropyron trachycaulum* to form X *Agrohordeum Macounii* (Vasey) Lepage (*Elymus Macounii* Vasey). *Hordeum jubatum* crosses with *Agropyron sericeum* to form X *Agrohordeum pilosilemma* Mitchell & Hodgson.

Often a troublesome weed common in waste places near townsites north to the Mackenzie Delta.

General distribution: Native of N. America and eastern Asia.

Fig. 128 Map 142

Koeleria Pers.

Ours small tufted slender and rather low short-leaved perennials, with 2- to 4-flowered, compressed smooth spikelets in a spike-like panicle.

a. Loosely tufted, with slender rhizomes; culms finely pubescent; lemma awnless . . *K. asiatica*
a. Densely tufted, culms glabrous and smooth; lemma awned. . . . . *K. cristata*

Koeleria asiatica Domin
K. Cairnesiana Hultén
Culms slender and finely pubescent, commonly 10 to 25 cm high; basal leaves numerous, flat, or involute, those of the culm short. Panicle dense and contracted, commonly purplish or bronze-coloured; glumes glabrous, the lemmas finely pilose, awnless. Superficially resembling a slender *Trisetum spicatum* from which, however, it is readily distinguished by its entire and awnless lemmas.

Dry tundra and shale screes.

General distribution: An Amphi-Beringian species until recently known from Ural Mts. to Anadyr but now known also from widely disjunct N. American stations in the western parts of the arctic slope of northern Alaska, S.W. Y.T., and from mountains west of the Mackenzie Delta, N.W.T.

Fig. 129 Map 143

Koeleria cristata (L.) Pers.
K. yukonensis Hultén
Densely tufted with slender, smooth, erect-ascending culms 2 to 5 dm high, short, narrow basal leaves and a dense-flowered 4 to 15 cm long cylindrical, spike-like panicle. Spikelets 2- to 4-flowered, compressed, 4 to 5 mm long, somewhat shiny; glumes and lemmas smooth or slightly scabrous, the lemmas awned.

Species of dry grassland, open woods and roadsides, barely entering our area in the southwestern part where almost certainly it is a recent introduction.

General distribution: Circumpolar.

Fig. 130 Map 144

Muhlenbergia Schreb.

Panicle of 1-flowered spikelets, in ours, narrow and contracted. Glumes membranaceous, persistent, acute to short-awned. Lemma awned or merely acute, firm and indurate, enclosing the palea and the perfect flower.

a. Leaves broad and flat; glumes awl-shaped . . . . *M. glomerata*
a. Leaves involute; glumes ovate, obtuse . . . . *M. Richardsonis*

Muhlenbergia glomerata (Willd.) Trin.
var. **cinnoides** (Link) F.J. Herm.
Culms leafy, commonly 3 to 5 dm high, rarely up to 1 m, with scabrous internodes, simple, erect, rarely branching from the base, from an elongated, slender and scaly rhizome; leaves erect firm and scabrous, 2 to 6 mm wide; panicle greenish-purple, narrow and spike-like, the spikelets 4 to 7 mm long; lemma half as long as the glumes, villous in the lower part; anthers 1.0 to 1.5 mm long.

Wet rocky or peaty places along sheltered lake shores, barely entering our area in the southernmost parts of the District of Mackenzie.

General distribution: Boreal N. America: Nfld. to S.E. Y.T.

Fig. 131 Map 145

Muhlenbergia Richardsonis (Trin.) Rydb.
Tufted or matted from a short, freely branched rhizome. Culms wiry, 2 to 4 dm high, erect-ascending. Leaves usually involute. Panicle very slender, interrupted and spike-like, commonly 5 to 7 cm long and 2 to 4 mm wide, spikelets lead-coloured, lance-subulate; glumes awnless, much shorter than the strongly involute, glabrous, slender-pointed but awnless lemma.

Wet, calcareous and gravelly lake shores and river banks north along the Mackenzie River to Great Bear River.

General distribution: Boreal N. America: N.B. to S.W. Y.T.

Fig. 132 Map 146

Oryzopsis Michx. Mountain-rice

Tufted perennials with 1-flowered spikelets in narrow to somewhat diffuse panicles. Glumes subequal, acute, lemma broad, indurate, enclosing the palea and the perfect flower, and terminating in an early deciduous, straight awn.

a. Leaves flat, spikelets (exclusive of awn) 6 to 8 mm long . . . . *O. asperifolia*
a. Leaves involute, spikelets (exclusive of awn) 3 to 4 mm long . . . . *O. pungens*

Oryzopsis asperifolia Michx.
Culms 3 to 5 dm high; basal leaves erect, scabrous, 4 to 10 mm wide, flat or with involute margins, not much shorter than the culms, those of the culms with a blade 1 cm long or less. Panicle contracted, spike-like, 5 to 12 cm long, the branches simple and erect. Glumes glabrous, 6 to 8 mm long, short ciliate at the apiculate summit, many-nerved; lemma whitish, sparingly pubescent, with a tuft of hairs at the base; awn 7 to 10 mm long.

Dry, open pinewoods and thickets; barely entering our area in the southwest part.

General distribution: Boreal N. America: Nfld. to B.C.

Fig. 133 Map 147

Oryzopsis pungens (Torr.) Hitchc.
Culms 1.5 to 3.0 dm high, forming dense tussocks; basal leaves 1 to 2 dm long, those of the culm with a minute setaceous blade. Panicle 3 to 6 cm long, with erect-ascending branches; spikelets 3 to 4 mm long; glumes subequal, obtuse, glabrous; the lemma appressed pubescent, as long as the glumes, the awn 0.5 to 2.0 mm long, or none.

Dry peaty or turfy slopes. Southern parts of the District of Mackenzie north to Great Bear Lake.

General distribution: Boreal N. America: Que. to B.C. and barely reaching S.E. Y.T.

Fig. 134 Map 148

Phalaris L. Canary-grass

Glabrous, annual or perennial grasses with 1-flowered spikelets in a narrow spike-like panicle. Glumes equal, much longer than the awnless lemmas of which the sterile is scale-like, and the fertile firm and often glossy.

a. Coarse, rhizomatose perennial with elongated panicle . . . . *P. arundinacea*
a. Slender, non-native annual with ovoid, spike-like inflorescence . . . . *P. canariensis*

Phalaris arundinacea L.
Reed-Grass
Smooth native species with erect 6 to 15 dm high, leafy culms from a stout, horizontally creeping rhizome. Leaf-blades 1.0 to 2.5 dm long and 6 to 15 mm wide. Panicle 8 to 20 cm long, open in anthesis, tightly contracted in fruit, the spikelets lanceolate, laterally flattened, 5 to 6 mm long, with one perfect terminal flower, and two scale-like, sterile lemmas at its base.

Moist, sandy lake and river banks. Upper Mackenzie drainage north to Fort Simpson.

General distribution: Circumpolar (with gaps), non-arctic.

Fig. 135 Map 149

Phalaris canariensis L.
Canary Grass
Annual to 6 dm high with short blades 4 to 7 mm wide; panicle dense, ovate to oblong-ovate, 1.5 to 4.0 cm long; spikelets broad, 5 to 6 mm long, imbricated, pale with green stripes.

Accidentally introduced in settled areas where probably always ephemeral.

General distribution: Cosmopolitan

Phippsia R. Br.

Phippsia algida (Sol.) R. Br.
Caespitose, diminutive plant, 2 to 15 cm high, with flat, yellowish-green, smooth leaves. Culms ascending, barely overtopping the leaves, panicle narrow; spikelets very small and caducous.

Occasional to rare on wet clay or among rocks by lake shores and alpine brooks, often near perennating snowbanks. Strongly nitrophilous and frequently found near human habitations.

General distribution: Circumpolar, widespread, high-arctic.

Fig. 136 Map 150

Phleum L. Timothy

Tufted perennials with flat leaves, slender culms that are leafy below, and dense, spike-like panicles. Glumes longer than the lemma and awned.

a. Sheath of the upper cauline leaf inflated; panicle spike-like, short-cylindrical or ovoid; native and alpine . . . . *P. commutatum*

a. Sheath of upper cauline leaf not inflated; panicle cylindrical, many times longer than broad; introduced but becoming naturalized . . . . *P. pratense*

Phleum commutatum Gaud.
P. alpinum L., in part
Culms 2 to 5 dm high, singly or tufted from prolonged, creeping rhizome; leaves smooth; the blade short and flat. Spikelets 3.5 mm long, 1-flowered, the glumes compressed, ciliate on the keels, awned; anthers 1 mm long, much exserted during anthesis.

Moist alpine slopes and in moss by alpine brooks. In our area thus far collected a few times in the Mackenzie Mountains.

General distribution: Circumpolar, with large gaps.

Fig. 137 Map 151

Phleum pratense L.
Culms slender, solitary or in clumps, up to 1.5 m tall, from a bulbous base. Anthers 1.5 to 2.5 mm long.

Introduced and becoming naturalized near settlements of the upper Mackenzie valley.

Phragmites Trin. Reed

Phragmites australis (Cav.) Trin. ex Steud.
P. communis Trin. var. *Berlandieri* (Fourn.) Fern.
Stout perennial with leafy culms 2 to 3 m high from a stout, creeping rhizome. Leaves broad and flat. Panicle plumose, 1 to 4 dm long; spikelets very numerous, 10 to 15 mm long, 3- to several-flowered, rachilla with long, spreading, silky hairs.

Fresh to alkaline marshes commonly forming pure stands in shallow water along the shores of sheltered bays. In our area reported from a single station by a hot spring near the juncture of the South Nahanni and Liard Rivers.

General distribution: Circumpolar and non-arctic, with large gaps in Siberia and N. W. America.

Fig. 138 Map 152

Pleuropogon R. Br. Semaphore Grass

***Pleuropogon Sabinei** R.Br.
Small, stoloniferous, aquatic grass with smooth, leafy culms 10 to 20 cm high, and soft, flat, short leaves. Panicle racemose, one-sided, with from 5 to 7 linear, dark purplish, 5- to 8-flowered, drooping spiklets.

Occasional in shallow water growing on soft muddy shores of small sheltered ponds and tarns, where it may sometimes form pure stands.

General distribution: Although fairly common on islands of the Canadian Arctic Archipelago and northernmost E. and W. Greenland, this nearly circumpolar, high-arctic dwarf species is known from the N. American mainland only from Melville Pen. and northernmost Ungava, both outside the area covered by this manual. It should, however, be looked for in suitable places along the Arctic Coast east of Bathurst Inlet and in northern Keewatin.

Fig. 139 Map 153

Poa L. Blue Grass

Ours perennial (except the introduced *P. annua*), tufted or turf-forming grasses, mostly with rather narrow and commonly involute leaf-blades ending in a boat-shaped tip. Inflorescence paniculate; spikelets more or less laterally compressed to terete, 2- to several-flowered; the flowers perfect (except in the dioecious *P. Porsildii*), the uppermost floret sterile or rudimentary; glumes boat-shaped, 1- to 3-nerved, somewhat unequal, distinctly keeled (except in *P. ammophila* and *P. scabrella*); lemmas awnless, glabrous or hairy, commonly somewhat membranaceous, scarious-tipped, keeled and with 1 or 2 pairs of lateral veins that converge near the apex; in some species with a tuft of long, cobwebby hairs at the base (the web); stigmas plumose; stamens commonly 3 but often 2 or 1, or lacking, the anthers from 0.5 to 3.2 mm long, their length usually constant within the species and, therefore of some diagnostic value.

A large and taxonomically difficult genus in which asexual (vegetative or parthenogenetic) reproduction is widespread causing the fixation of small morphological variations. When geographically isolated, such "minor" variations are sometimes treated as distinct taxa.

a. Spikelets distinctly compressed, the glumes and lemmas distinctly keeled
  b. Plants with creeping rhizome or at least with leafy innovations
    c. Culms conspicuously flattened; non-native . . . . *P. compressa*
    c. Culms terete, or at most very slightly flattened; native species
      d. Culms slender, rarely exceeding 30 cm in height, with 1-2 nodes, lowermost panicle branches usually in pairs
        e. Panicle mostly contracted, narrowly lanceolate, the branches approximately of equal length, each bearing 3-7 spikelets . . . . *P. alpigena*
        e. Panicle usually pyramidal, the lowermost branches spreading, spikelets mostly 2-3, at the tip of the very slender branches
          f. Lemmas lacking a web at the base; spikelets 3-4 mm long, leaves narrow . . . . *P. arctica*
          f. Lemmas webbed at base; spikelets 6-8 mm long, leaves flat and broad, gradually tapering to a narrow point . . . . *P. Williamsii*
      d. Culms stout, mostly more than 30 cm high, panicle more or less pyramidal
        g. Lemmas glabrous between the keel and marginal nerves; lowermost branches of panicle usually in whorls of 5 . . . . *P. pratensis*
        g. Lemmas lanate also between the keel and marginal nerves; lowermost branches of panicle usually in whorls of 2-4 . . . . *P. lanata*
  b. Plant tufted, lacking creeping rhizomes
    h. Lemmas webbed at base; panicle open, usually pyramidal
      i. Panicle branches capillary, smooth, the lower usually in pairs . . . . . *P. paucispicula*
      i. Panicle branches not capillary, the lowermost usually in whorls of 4 or 5; culm decumbent-ascending from the usually purplish base; panicle 10-20 cm long . . . . *P. palustris*
    h. Lemmas not webbed at base
      j. Panicle open, about as long as broad
        k. Leaves short, densely crowded at the base of the culm, the blade 2-5 mm wide with a blunt tip . . . . *P. alpina*
        k. Leaves narrow, the blade 1-2 mm wide
          l. Anthers 2.5-3.2 mm long; plant dioecious . . . . *P. Porsildii*
          l. Anthers 1.5 mm long; plant monoecious . . . . *P. arctica* ssp. *caespitans*
      j. Panicle not open, spike-like
        m. Culms rarely more than 12 cm tall
          n. Culms barely longer than leaves; lemmas distinctly pubescent . . . . *P. abbreviata*
          n. Culms much longer than leaves; lemmas essentially glabrous . . . . . *P. Jordalii*
        m. Culms usually 10-30 cm tall
          o. Plant glaucous, pruinose; the culms usually scabrous, panicle narrow, stiff . . . . *P. glauca*
          o. Plant not glaucous
            p. Culms stiffly erect 20-30 cm tall, culms and leaves yellowish-green . . . . *P. ammophila*
            p. Culms ascending flexuous, rarely over 15 cm tall, culm and leaves fresh green . . . . *P. flexuosa*

a. Spikelets little compressed, more or less terete in cross-section, much longer than broad; plant densely tufted
  q. Sheaths somewhat scabrous, lemmas puberulent in the lower half .... *P. scabrella*
  q. Sheaths smooth; lemmas glabrous or merely scabrous, anthers 1.5-2.0 mm long
    r. Culms 2.0-3.5 dm tall; panicle 4 cm long .... *P. ammophila*
    r. Culms 5-8 dm tall; panicle 6-20 cm long .... *P. juncifolia*

Poa abbreviata R. Br.
Caespitose dwarf species, often forming small, flat or hemispherical tussocks. Culms 5 to 15 cm high, smooth, somewhat arching, barely longer than the convolute, densely crowded and somewhat curved leaves; ligules short, the sheaths papery, white and inflated. Panicle short, oval, usually from 10 to 15 mm long and from 5 to 8 mm wide; spikelets 2- to 4-flowered, glumes purplish with broad, membranaceous margins; lemma distinctly short-pubescent; anthers 1 mm long.

Common in dry tundra and turfy places, frequently growing on the sides of owl perches. Apparently a pronounced calciphile. Wide ranging in the Arctic Archipelago but thus far reported from our area only from Cape Parry and an area south of Spence Bay.

General distribution: Amphi-Atlantic, high-arctic.

Fig. 140 Map 154

Poa alpigena (Fr.) Lindm.
P. pratensis s. lat. of some authors on arctic flora
Turf-forming, with rather stout, widely creeping rhizomes. Culms mostly solitary, 20 to 30 cm high, smooth, somewhat longer than the folded or rarely flat leaves. Spikelets 2- to 3-flowered; glumes equal; the lemmas obscurely 5-nerved, long, soft-pilose on the median and lateral nerves, with a copious tuft of cobwebby hairs at the base.

Occasional to common in sandy tundra and in meadows by streams and lake shores, where it often forms a firm turf. It is strongly nitrophilous and greatly favours refuse and manure near human habitation or below bird cliffs and bird nesting sites. Although rather short for mowing, the species provides valuable pasture in Greenland, especially for goats and sheep.

In wet or heavily manured places; especially in the Far North represented by the mostly viviparous, narrow-panicled var. *colpodea* (Fr.) Schol., which barely enters our area.

General distribution: Circumpolar, wide ranging arctic-alpine tundra species.

Fig. 141 (a. var. *alpigena*, b. var. *colpodea*)
Map 155 (var. *alpigena*), 156 (var. *colpodea*)

Poa alpina L.
Densely tufted with stout, crowded bases covered by the persistent white sheaths. Basal leaves short and flat, fresh green. Culms wiry, 15 to 30 cm high; panicle green or purplish, oval to broadly pyramidal. Often viviparous, especially in the Eastern Arctic.

Alpine slopes, usually on snowbeds.

General distribution: Circumpolar, low arctic-alpine.

Fig. 142 Map 157

Poa ammophila Porsild
Densely tufted with short, wiry rhizomes; culms to 20 or rarely up to 35 cm tall, glabrous and wiry, erect or somewhat ascending, twice as long as the glabrous, involute or conduplicate, grey-green leaves; the persistent sheaths somewhat inflated, shiny, the ligule 1.5 to 2.0 mm long. Panicle narrow, about 4 cm long and less than 1 cm wide; spikelets about 6 mm long; glumes equal, 4 mm long, glabrous and lustrous, pale purplish with thin margins; the lemma similar, short pubescent along the keel and near the base; anthers about 1.5 mm long.

General distribution: *P. ammophila* is thus far known only from the Arctic Coast from about long. 142° eastward to Darnley Bay and south to Great Bear L. Within this area it is common locally on old, well established dunes and hilltops.

Fig. 143 Map 158

***Poa annua** L.
Tufted, annual with 5 to 20 cm long ascending culms, narrow panicle and soft, flat leaves.

A weedy introduced species common in lawns and by road sides. As yet not actually reported from our area where it is likely to occur around settlements.

General distribution: Circumpolar.

Poa arctica R.Br. ssp. **arctica**
P. brintnellii Raup
Similar to *P. alpigena* but with short, ascending rhizomes and not turf-forming. Culms erect-ascending, often somewhat geniculate, usually several together, 15 to 30 cm high, and mostly longer than the short and usually involute

leaves. As a rule easily distinguished from *P. alpigena* by its pyramidal panicle. Spikelets mostly 2-flowered, 3 to 4 mm long; glumes equal; the lemmas copiously soft pilose, especially in the lower part. Anthers average 1.6 mm in length.

Common or even ubiquitous in not too moist tundra, by lake shores and brooks. In the Eastern Arctic often represented by the well-marked ssp. *caespitans* (Simm.) Nannf., which is densely tufted, only occasionally rhizomatose, and has flat, soft leaves. The ssp. *caespitans* is probably always apomictic, and although anthers are present, they rarely contain good pollen. Whereas var. *vivipara* Hook. is strictly eastern arctic, the ssp. *caespitans* has been reported westward to the Mackenzie Delta.

General distribution: *P. arctica* s. str. Circumpolar, wide ranging, arctic-alpine.

Fig. 144 (a. ssp. *arctica*, b. ssp. *caespitans*)

Map 159 (ssp. *arctica*), 160 (ssp. *caespitans), 161 (var. vivipara)*

Poa compressa L.

Bluish-green perennial with slender rhizomes and geniculate-ascending wiry and strongly flattened culms, 3 to 4 dm high; leaf-blades rather short, 1 to 4 mm broad. Panicle 3 to 10 cm long, commonly with short, paired branches; the spikelets crowded, 3- to 6-flowered.

General distribution: Cosmopolitan weedy species, introduced and seemingly established at Ft. Simpson and perhaps elsewhere in southern District of Mackenzie.

Poa flexuosa Sm.

P. laxa Haenke

Densely caespitose with soft, fresh green leaves; ligules 3 to 4 mm long. Culms smooth, 10 to 20 cm high, usually procumbent. Panicle narrow, contracted, and flexuous, often somewhat interrupted, with few-flowered, greyish-violet spikelets. Anthers very short.

In gravelly, not too dry places, often pioneering on fresh moraines.

General distribution: Amphi-Atlantic, low-arctic-alpine, barely entering the District of Keewatin.

Fig. 145 Map 162

Poa glauca M. Vahl

P. interior sensu Cody

Densely tufted. Culms stiff, 15 to 25 cm high, slightly longer than the leaves. Culm and leaves glaucous, usually faintly scabrous, at least below the panicle. Ligule very short and truncate. Panicle narrowly lanceolate.

Notoriously variable species which is common throughout our area in open, sandy or gravelly placesanfrom the Arctic Coast southward not only in the mountains, but apparently common also in the lowlands wherever suitable habitats are available.

General distribution: Circumpolar wide ranging arctic-alpine

Fig 146 Map 163

Poa Jordalii Porsild

P. pseudoabbreviata sensu Hultén, *non* Roshv.

Caespitose dwarf species with glabrous and striate culms 8 to 12 cm long, and much longer than the narrowly duplicate leaves; ligules truncate and very short; panicle short and dense, rarely over 2 cm long, the lemmas essentially glabrous and the anthers 0.5 to 0.75 mm long.

In habit somewhat similar to the high-arctic Amphi-Atlantic, *P. abbreviata* R.Br. from which it is at once distinguished by its short leaves emerging from thin, cylindric crowns formed by the papery-white and marescent sheaths, and by its essentially glabrous lemmas.

From *P. pseudoabbreviata* Roshv. (*P. brachyanthera* Hultén), an Amphi-Beringian dwarf species, originally described from Sayan Mountains, but wide ranging across arctic Siberia and in mountains of Alaska east to the St. Elias Range of Yukon, *P. Jordalii* is at once distinguished by its always dense and contracted panicle contrasting strongly with the open panicle of spreading, long-pedicelled spikelets of *P. pseudoabbreviata*.

In turfy, alpine situations, thus far known only on soils derived from calcareous rocks.

General distribution: Endemic, Brooks Range, Alaska and the Mackenzie Mts., N.W.T.

Fig. 147 Map 164

Poa juncifolia Scribn.

Densely tufted; culms 5 to 8 dm tall, smooth, two or more times as long as the glabrous, involute leaves; ligule rounded, about 1 mm long; panicle narrow 6 to 20 cm long; spikelets 3- to 6-flowered, 4 to 5 mm long; glumes subequal, 3-nerved; lemmas about 4 mm long, glabrous and lustrous, rounded on the back, not webbed; anthers about 1.5 mm long.

Rare in alkaline meadows adjacent to the Slave R.

General distribution: western prairies and foothills to 1000 m.
Map 165

Poa lanata Scribn. & Merr.
Similar to *P. arctica* ssp. *caespitans* but a coarser and taller plant with rather stout culms about 30 cm high or over, and leaves 2 to 4 mm broad, short, flat or folded. Panicle pyramidal, erect, open; the branches faintly scabrous, somewhat flexuous, the lowermost often in whorls of 4, and frequently somewhat reflexed, bearing 2 to 4 spikelets near the tip. Lemmas obtuse with broad, hyaline margins, 5-nerved, with a dense web at the base, and villous on the lower half of the keel and on the marginal nerves. The anthers are 2 mm long, but often lacking.

Occasional or common locally in moist alpine floodplain meadows and on moist gravel below snow beds of the Mackenzie and Richardson Mountains, southwards in alpine situations to Banff National Park, Alberta.

General distribution: Cordilleran species which over Y.T. and Alaska reaches easternmost Siberia.
Fig. 148 Map 166

Poa palustris L.
P. nemoralis sensu Raup, *non* L.
Culms tufted, firm 30 to 70 cm tall, with 4 to 6 nodes, from a somewhat geniculate and commonly purplish base; ligules 3 to 5 mm long, leaf-blades 1 to 2 mm wide. Panicle narrowly pyramidal to ellipsoid, the filiform, spreading branches in rather distant fascicles of 3 to 10; spikelets 2- to 4-flowered, glumes 2 to 3 mm long, lanceolate; lemma 2.0 to 3.0 mm long, usually bronzed at the tip, copiously webbed at the base.

Floodplain meadows and lake shores north along the Mackenzie valley to Norman Wells.

General distribution: Circumpolar, non arctic.
Fig. 149 Map 167

Poa paucispicula Scribn. & Merr.
P. leptocoma auct. non Trin.
Densely tufted; culms slender, smooth, 20 to 30 cm high, with numerous thin and narrow basal leaves. Panicle open, its paired branches capillary, smooth, the lowermost somewhat remote and often drooping, each branch with 1 to 3 spikelets near the apex. Spikelets strongly compressed, 2- to 3-flowered, dark purple, turning bronze in age. Lemma 3 mm long, 5-nerved, broadly obtuse with hyaline margins, pubescent on the keel and marginal nerves, webbed at the base. Anthers short, rarely over 0.7 mm long.

P. leptocoma Trin., erroneously reported from our area, is related to *P. paucispicula* from which it can usually be separated by its longer, narrower and green spikelets and more widely spaced florets between which the rachilla joints are commonly visible.

P. paucispicula is probably common throughout non-calcareous parts on the Mackenzie Mountains, growing in damp places by alpine brooks and near snow beds.

General distribution: North Cordilleran with disjunct areas in boreal E. Asia.
Fig. 150 Map 168

Poa Porsildii Gjaerevoll
P. Vaseyochloa sensu Hultén, *non* Scribn.
Strictly dioecious, loosely tufted and often forming large and matlike tussocks; culms 35 to 40 cm high, smooth, with one or two nodes; basal leaves very numerous and crowded around the base of the culm; basal sheaths pale, persisting; leaf-blades about 2 mm wide but often so tightly rolled as to appear terete when less than 1 mm in diameter. The ligule, described as being "1.5 to 2.0 mm long, ovate-transversely cut and laciniate" in the more abundant material now available averages 0.5 to 1.0 mm in length and is hyaline and very fragile. Panicle 3 to 4 cm long, its branches single or in pairs, the lowermost often slightly deflexed; spikelets mostly crowded near the end of the branches, 6.0 to 7.5 mm long, 3- to 5-flowered; glumes prominently nerved, suffused with bronze-purple; lemma 3.5 mm long, totally glabrous; anthers 2.5 to 3.2 mm long.

The type of *P. Porsildii* came from MacMillan Pass on the Yukon—Mackenzie watershed in the Mackenzie Mountains and had originally been reported as *Colpodium Wrightii* (Porsild, 1951). In addition *P. Porsildii* is now known also from the White Mountains of central Alaska, from rich collections from the Ogilvie Mountains and the Stewart Plateau of central Yukon, and besides, from additional stations in the Richardson and Mackenzie Mountains east of the Mackenzie—Yukon divide. Although the dioecious nature of the species was not observed in the diagnosis, the abundant material now available consists entirely of male and female plants. In the several hundreds of panicles examined no bisexual florets were found.

In moist alpine heath near the edge of snow beds well above timberline. *P. Vaseyochloa*

Scribn., to which Hultén (1967 and 1968) referred *P. Porsildii*, is a dwarf species endemic to dry and hot interior valleys of the states of Washington and Oregon. It is not dioecious and in size and general appearance does not even remotely resemble *P. Porsildii.*

General distribution: Endemic of unglaciated mountains of central Alaska, Y. T. and the east slopes of Richardson and Mackenzie Mts.

Fig. 151 Map 169

Poa pratensis L.
? *P. agassizensis* Boivin & Lovë
Kentucky Bluegrass
Stoloniferous, the culms tufted, smooth, 30 to 100 cm hi e; leaves soft, flat or folded, 2 to 4 mm wide, the basal ones often very numerous. Panicle open, pyramidal or oblong, 10 to 15 cm long, the lowermost branches commonly in whorls of 4 to 5. Spikelets crowded, 3- to 5-flowered, 3 to 6 mm long. Lemmas copio aly webbed at the base, silky pubescent on the lower half of the keel and marginal nerves, otherwise glabrous.

A lowland species of moist river-flat meadows and open woods, in the District of Mackenzie north to Norman Wells and mainly restricted to settlements and roadsides and, probably not indigenous. In the northern parts of our area and in the mountains the closely related and clearly indigenous *P. alpigena.*

General distribution: Circumpolar, non arctic; several races of *P. pratensis* are grown as l grasses and hay crops.

Fig. 152 Map 170

Poa scabrella (Thurb.) Benth.
Poa Buckleyana Nash
Densely tufted and glaucous; culms smooth, 20 to 50 cm high, more than twice as long as tne flat or involute leaves; panicle 3 to 10 cm long, narrow and dense, the spikelets 2- to 5-flowered, dark green, tinged wth purple, not compressed.

General distribution: Cordilleran species barely entering our area in the southern extension of the Mackenzie Mts.

Fig. 153 Map 171

***Poa Williamsii** Nash
P. arctica R. Br.
ssp. *Williamsii* (Nash) Hultén
Loosely tufted; culms ascending from a geniculate base, 20 to 40 cm high; leaves flat, 2 to 3 mm wide, gradually tapering to a narrow point. Panicle pyramidal, open, commonly 5 to 6 cm long, the spikelets 6 to 8 mm long, 3- to 5-flowered, prominently suffused with purple; lemmas copiously hirsute on the back and at the base; anthers averaging about 3 mm in length or nearly twice as long as in *P. arctica.*

A rather handsome alpine species probably related through *P. arctica* ssp. *caespitans* to the stoloniferous circumpolar *P. arctica* R. Br.

In mountains of interior Yukon where *P. Williamsii* is commonly growing together with *P. arctica*, no intermediate forms have been detected; for this reason and because of its consistently longer anthers it seems best retained as distinct.

General distribution: *P. Williamsii* thus far is known only from Alaska and Y. T. but undoubtedly will be found also in alpine situations in the Mackenzie Mts.

Map 172

Puccinellia Parl.[1] Goose Grass

Caespitose or stoloniferous, yellowish-green, smooth grasses, usually growing near the seashore, although several are non-littoral, growing in alkaline or saline soil far from the sea. Spikelets 2- to many-flowered in open or narrow panicles; glumes very unequal in length, and much shorter than the spikelet.

a. Arctic species with erect-ascending culms commonly less than 20 cm tall, inhabiting arctic tundra or sea-shores
  b. Glumes and lemmas entire, not evidently erose-ciliolate even in age
    c. Anthers 1.2-2.0 mm long
      d. Stoloniferous; anthers without pollen; plant propagating vegetatively . . . . *P. phryganodes*
      d. Caespitose and without stolons, plant fertile

[1] *Key and descriptions adapted from Th. Sørensen (1953) and Sørensen in Porsild (1957)*

e. Lemmas distinctly nerved, often with a few spines on the back; keels of palea densely ciliate . . . . *P. arctica*
e. Lemmas obscurely nerved, not spiny on back; keels of palea sparsely ciliate to almost glabrous . . . . *P. agrostidea*
c. Anthers 0.5 to 1.0 mm long
f. Lemmas herbaceous throughout, prominently veined, commonly with hairs at their base; anthers about 0.5 mm long . . . . *P. Langeana* ssp. *Langeana*
f. Lemmas firm throughout, or at most scarious-translucent in the upper third; veins not prominent; anthers 0.6-1.0 mm long
g. Pedicels distinctly lustrous from tumid cells of the epidermis; spikelets pale greenish . . . . *P. pumila*
g. Pedicels not lustrous; spikelets purplish tinged
h. Lemmas glabrous or only sparsely hairy on the nerves near the base; keels of palea glabrous or sparsely spinulose near the summit; glumes and lemmas commonly coarsely toothed . . . . *P. Andersonii*
h. Lemmas conspicuously hairy, especially in lower part
i. Glumes firm or opaque; lemmas not hairy on inter-nerves . . . *P. angustata*
i. Glumes thin and translucent; lemmas conspicuously hairy also on the inter-nerves . . . . *P. contracta*
b. Glumes and lemmas erose-ciliolate, at least in age
j. Glumes always ciliolate; panicle comparatively large, one third to one half the length of culm . . . . *P. vaginata*
j. Glumes and lemmas entire, becoming erose or toothed but not ciliolate in age
k. Panicle contracted and dense; dwarf species with ascending or prostrate culms 5-10 cm long, anthers 0.6-0.8 mm long. . . . . *P. Bruggemannii*
k. Panicle open, one fifth to one fourth the length of the culm; culms erect, 20-40 cm tall; anthers 1.2-1.5 mm long
l. Lemmas distinctly nerved; panicle branches spreading. . . . . *P. arctica*
l. Lemmas obscurely nerved; panicle branches ascending . . . . *P. agrostidea*
a. Non-arctic, non-littoral species with culms 20-55 cm tall, inhabiting flood plains, river banks or waste places (*P. distans*)
m. Glumes keeled, pruinose; lemmas acute . . . . *P. deschampsioides*
m. Glumes not keeled; lemmas obtuse or truncate
n. Spikelets loose; lemmas commonly 2.2-2.4 mm long, distinctly hairy on veins of their lower fourth . . . . *P. borealis*
n. Spikelets rather tight; lemmas commonly 1.8-2.2 mm long, sparsely hairy at base, or almost glabrous
o. Anthers 0.8-1.0 mm long; the leaves 2-5 mm broad . . . . *P. distans*
o. Anthers 0.5-0.6 mm long; the leaves 1-2 mm broad . . . . *P. interior*

Puccinellia agrostidea Th. Sør.
Caespitose, slender plant. Culms erect, 20 to 30 cm tall with about 3 long-sheathing leaves. Ligules very thin, about 2 mm long, truncate, often lacerate. Panicle intensely purple, glossy, contracted, linear-lanceolate, the branches fascicled, capillary, subglabrous, and the pedicels scaberulous. Spikelets 4 to 6 mm long, 2- to 4-flowered; glumes and lemmas thin, translucent, obscurely nerved, entire; glumes acutish, the first 0.8 to 1.0 mm, the second 1.8 to 2.0 mm long; lemmas obtuse or abruptly pointed, 2.3 to 3.0 mm long, faintly pilose at the very base. Anthers 1.2 to 1.5 mm long.

A non-littoral species of turfy places in tundra.

General distribution: Endemic of western N. American Arctic.

Fig. 154 Map 173

Puccinellia Andersonii Swallen
Caespitose, glaucescent. Culms stout, geniculate, 10 to 20 cm tall. Leaves rather rigid, abruptly pointed; ligules 1.5 to 2.0 mm long, acute, decurrent. Panicle up to 8 cm long; branches glabrous, commonly in pairs, stiffly ascending or spreading. Spikelets rather- ew, comparatively large, 6 to 9 mm long, 4- to 7-flowered, pale purple or reddish, the upper ones subsessile, appressed. Glumes and lemmas usually coarsely toothed. Glumes firm, the first 1.5 to 2.0, the second 2.5 to 3.0 mm long, acute, evi-

dently alternately inserted on the more or less thickened pedicels; lemmas rather firm, thinner, and with a peculiar fatty lustre toward the margin, 3 to 4 mm long, sparsely pilose below. Anthers 0.8 to 1.0 mm long. Grains 1.8 to 2.0 mm long.

Wet, grassy places by the sea-shore.

General distribution: N. America, high-arctic, from E. Greenland across the Canadian Arctic Archipelago to Banks Island, and widely disjunct at Richards Island in the Mackenzie Delta and Pt. Lay, N.W. Alaska (type locality), and to be expected elsewhere along the arctic mainland coast.

Fig. 155 Map 174

***Puccinellia angustata** (R. Br.) Rand & Redf.
Densely caespitose. Culms stout, 15 to 30 cm tall, erect or prostrate, bearing only 1 to 2 leaves. Leaf-blades, especially those of the innovations, comparatively wide and flat. Ligules thin, 2 to 4 mm long, tapering. Panicle purple, 4 to 10 cm long, commonly contracted; branches stout, more or less scabrous, ascending, in pairs or with accessory weaker branchlets from the lower node. Spikelets lanceolate, 6 to 8 mm long, 3- to 5-flowered. Glumes usually acute, variable in shape, the tips rarely erose; lemmas about 3 to 4 mm long, tapering, commonly with hyaline tips, more or less hairy on the nerves in their lower part. Anthers 0.6 to 0.8 mm long. Grains 1.9 to 2.1 mm long.

A non-littoral species commonly growing in moist clay spots in otherwise dry barrens.

General distribution: Circumpolar, high-arctic, in N. America from northern E. Greenland westward across the Canadian Arctic Archipelago. To be expected also along the arctic mainland coast.

Fig. 156 Map 175

Puccinellia arctica (Hook.) Fern. & Weath.
Caespitose, culms stout, erect, often geniculate at base, 20 to 40 cm tall, mostly 2-leaved; leaves deeply furrowed or involute when dry; ligule lacerate, 1.5 to 2.0 mm long. Panicle 5 to 11 cm long, lanceolate-oblong, the branches ascending, smooth or slightly scabrous, slender, 2 to 3 from each node, each bearing 1 to 3 spikelets on slender pedicels. Spikelets purplish tinged, 6 to 11 mm long, 5- to 9-flowered; glumes and lemmas very thin and translucent, distinctly nerved, first glume about 1.5 mm and second about 2.5 mm long, obtuse; lemmas 3.0 mm or slightly over, obtuse, strongly arched, beautifully variegated, the lower part purple, the tips bright golden yellow and slightly erose, nerves of lemma slightly pilose; callus hairs short; keels of palea strongly ciliate above; anthers 1.8 to 2.0 mm.

Beaches and strand flats subject to flood. Of weedy habit and often taller and more lush near nesting sites of sea-birds and near human habitations.

Thus far known only from the arctic coast of Canada between Herschel Island and Cape Bathurst, and from southern Banks Island and Cambridge Bay, Victoria Island. The type is a Richardson collection labelled "Arctic Sea-Coast". In view of its present known range this was probably somewhere between the Mackenzie delta and Cape Bathurst.

According to Th. Sørensen, *P. arctica* belongs in the *P. Nuttalliana* group.

General distribution: Endemic of arctic N. W. Canada.

Fig. 157 Map 176

Puccinellia borealis Swallen
Caespitose, culms somewhat geniculate at base, 25 to 40 cm tall; basal sheaths reddish in the lower part, mostly a little longer than the internodes, blades flat, 1 to 2 mm wide, ligule about 2 mm long. Panicle 10 to 16 cm long, open and pyramidal in age, the slender branches spreading, the lower commonly in pairs 4 to 8 cm long; spikelets 4- to 6-flowered, green or purplish-tinged in age, short-pedicelled. First glume 1.0 to 1.5 mm, the second 1.5 to 2.0 mm long; the lemmas 2.0 to 2.3 mm long, the anthers 0.6 to 0.7 mm long.

A non-littoral species of weedy habit, and readily spread by man and animals. Described from Teller on Seward Peninsula, Alaska. *P. borealis* is common along rivers and around trading posts and settlements of interior Alaska and Yukon; along the Mackenzie River north to its delta, where it first became established about 1931.

General distribution: Amphi-Beringian.

Fig. 158 Map 177

***Puccinellia Bruggemannii** Th. Sør.
Densely caespitose. Culms stout, erect or prostrate, 5 to 10 cm tall, with mostly withered leaves at the base; basal sheaths wide, scarious, glossy. Ligules stout, 1 to 2 mm long, acutish, decurrent. Panicle dark purple, about 2 cm long, dense, contracted; branches glabrous, in pairs or singly, bearing 1 or 2 spikelets. Spikelets ovate, 4.5 to 6.0 mm long, 3- to 4-flowered; glumes acute, sometimes sinuate-dentate at the tip, commonly herbaceous and greenish at the base; lemmas 3.0 to 3.5 mm long, obtusish,

incurved, pilose on the nerves in their lower third. Anthers 0.6 to 0.8 mm long. Grains about 2 mm long.

General distribution: Apparently a non-littoral endemic of the Canadian Arctic Archipelago which should be looked for in the extreme northeastern parts of our area.

Fig. 159 Map 178

Puccinellia contracta (Lge.) Th. Sør.
Densely caespitose, culms decumbent at base, up to 30 cm tall, 1- to 3-leaved; blades flat 1 to 2 mm wide, soft; ligule 2.0 mm long. Panicle 3 to 8 cm long, branches ascending, usually 3 from the lower node, with from 1 to 5 spikelets, each on slender pedicels. Spikelets pale purple-tinged, 5.0 to 6.5 mm long, 3- to 4-flowered. Glumes thin, erose-ciliolate, the first 1.5 to 1.8 mm, the second 2.2 to 2.8 mm long, obscurely 3-nerved. Lemmas 3.3 to 3.8 mm long, obscurely 5-nerved, the nerves hairy in the lower third, the internerves pubescent. Palea about as long as the lemma and crisp-hairy towards the base.

A littoral species in habit similar to *P. angustata*. Widely distributed along the arctic shores of Russia and Siberia and recently identified in material from between the Mackenzie Delta and Cape Bathurst. To be looked for also along the arctic coast of Yukon and Alaska.

General distribution: Amphi-Beringian, arctic.

Fig. 160 Map 179

Puccinellia deschampsioides Th. Sør.
Densely caespitose, glaucous. Culms 30 to 50 cm tall, decumbent at base, stout and stiffly erect, 2-leaved, naked in their upper part. Leaves stiff, strongly involute, scabrous on the margins their sheaths short, those of the base marcescent. Ligule 2 mm long, truncate. Panicle stiff, 10 to 15 cm long, the branches commonly paired, elongated, scabrous, stiffly ascending, spreading in age, bearing 5 to 15 spikelets. Spikelets compact, 4 to 7 mm long, 3- to 5-flowered, variegated, reddish-purple and golden yellow, shiny or pruinose. First glume 1.0 to 1.5 mm, second 1.8 to 2.1 mm long. Lemmas 2.4 to 2.8 mm long. Anthers 0.7 to 0.9 mm long.

P. deschampsioides is a species of dry, mildly alkaline or saline flats. It was described from continental parts of West Greenland but is known also from isolated and disjunct stations in Canada from Ungava, S.E. Y.T., north to Great Bear L. and Liverpool Bay on the Arctic Coast.

General distribution: N. America, arctic.

Fig. 161 Map 180

Puccinellia distans (L.) Parl.
Caespitose, culms erect or decumbent at base, 20 to 50 cm tall; leaf-blades flat or involute, 2 to 3 mm wide. Panicle pyramidal in age, 5 to 15 cm long, the branches in fascicles of 3 or 4, spreading, or the lower reflexed; spikelets 4- to 6-flowered, 4 to 5 mm long.

General distribution: A weedy species native of Eurasia, in our area collected a few times near townsites along the upper Mackenzie R.

Puccinellia interior Th. Sør.
Densely caespitose, glaucous; culm erect 30 to 55 cm tall, 2- to 3-leaved. Upper sheaths much longer than the lower; leaf-blades 1.5 to 2.0 mm wide, involute, the ligule 1.8 to 2.3 mm long; blades of sterile innovations comparatively short and bristle-like when dry. Panicle pyramidal, 15 to 20 cm long, the branches fascicled and more or less distant, long and capillary, naked in their lower part, bearing numerous, clustered and approximate spikelets on short somewhat thickened pedicels. Spikelets pale yellow or variegated, more or less shiny.

In habit rather similar to *P. borealis* and, like it, of weedy habit and readily spread by man.

General distribution: Described from Alaska, its known range has now been extended across Y.T. to the upper Mackenzie Basin where it may be a recent but already established introduction, mainly near human habitations.

Fig. 162 Map 181

Puccinellia Langeana (Berl.) Th. Sør.
P. paupercula (Holm) Fern. & Weath.
Plant loosely caespitose, often reddish tinged. Culms stout, erect, somewhat geniculate or even procumbent, usually 5 to 10 cm tall, rarely taller. Leaves folded, abruptly pointed; ligules 1.5 to 2.0 mm long, acutish. Panicle 3 to 5 cm long, contracted or sometimes spreading; branches commonly in pairs, glabrous. Spikelets purple or greenish-variegated, 4 to 6 mm long, florets 3 to 7, loosely imbricated, appressed, the upper ones often sessile; glumes and lemmas firm, herbaceous, entire, distinctly nerved. Glumes 1 to 2 mm long, acutish, subcarinate; lemmas 1.5 to 2.5 mm long, incurved, obtuse or bluntly acute. Lemma and palea keels

glabrous; callus hairs absent. Anthers about 0.5 mm long. Grains 1.2 to 1.4 mm long.

An obligate littoral species, in our area thus far known only from the west coast of Hudson Bay, but to be looked for along the Arctic Coast.

General distribution: N. American, low-arctic, from W. Greenland, Alaska and E. Asia, but with a large gap between Hudson Bay and Alaska.

Fig. 163 Map 182

Puccinellia phryganodes (Trin.) Scribn. & Merr.
Turf-forming, matted with widely trailing stolons, from the leaf-axils of which develop short, rooting, easily detached shootlets. Flowering plant often caespitose with sessile innovations. Culms geniculate at the base or procumbent, the upper leaf-sheath elongated, often reaching the panicle. Leaves of flowering plant rigid, folded, abruptly pointed, glaucous; those of the stolons commonly flat, bright green. Panicle narrow and thin, of a few comparatively large spikelets; branches glabrous, stiffly ascending or reflexed. Spikelets variable up to 1 cm long or more, 3- to 6-flowered, pale purple or whitish. Glumes more or less flattened, obtuse; lemmas 3.5 to 4.5 mm long, obtuse or even emarginate, entire, glabrous throughout; callus hairs absent. Anthers commonly 1.5 to 2.0 mm long, thin, lacking pollen. Grains do not develop.

An obligate littoral species, common along the shores of the mainland and the arctic islands. Near the Mackenzie Delta it forms a dense turf along sheltered bays and lagoon shores flooded by high tide.

General distribution: Circumpolar, arctic.

Fig. 164 Map 183

Puccinellia pumila (Vasey) Hitchc.
Tufted, the up to 30 cm tall culms erect or decumbent at base; leaves flat, scaberulous; panicle with branches stiffly ascending to reflexed; spikelets 4- to 6- flowered, pale greenish, callus hairs present; anthers 0.8 to 1.0 mm long.

Our only justification for including this taxon is a single collection from Rankin Inlet on the west coast of Hudson Bay, (J. M. Macoun, August 30, 1910, CAN. No. 79,116) originally named *Glyceria angustata* by the collector, by Fernald and Weatherby changed to *Puccinellia paupercula* (Holm) Fern. & Weath. and, finally, by Th. Sørensen (in herb., 1953) to *P. pumila* (Vasey) Hitchc. The four specimens on the sheet are overmature; of most spikelets only the empty glumes remain and the determinations, at best, can be conjectural.

In view of the recent discovery (Porsild, 1969) in the Hudson Bay region of *P. ambigua* Th. Sør., otherwise known from only the Gulf of St. Lawrence, the Rankin Inlet plant may indeed belong to the latter species rather than to *P. pumila* which, otherwise, is known only from the shores of the Northwest Pacific.

Map 184

Puccinellia vaginata (Lge.) Fern. & Weath.
Caespitose. Culms 10 to 20 cm tall, slender, ascending; leaves soft, involute in age, the uppermost leaf-sheath usually subinflated. Panicle comparatively large, barely exserted; branches commonly in pairs, slender, somewhat scabrous, ascending or drooping. Spikelets pedicelled, more or less purple, glossy, oblong, 5 to 10 mm long, 4- to 6-flowered; glumes and lemmas thin, translucent, erose-ciliolate; glumes rounded, very faintly nerved; lemmas obtuse, 2.5 to 3.0 mm long, sparsely pilose at the very base. Anthers 0.6 to 0.8 mm long. Grains 1.7 to 1.9 mm long.

On clay by the sea shore; often forming large tussocks near bird cliffs and human habitations.

General distribution: N. American, arctic from E. Greenland to Bering Strait.

Fig. 165 Map 185

Schizachne Hack.

Schizachne purpurascens (Torr.) Swallen
Loosely tufted perennial with slender, 3 to 6 dm long, culms, exceeding the flat leaves. Panicle about 10 cm long, somewhat flexuous, the branches filiform, single or in pairs, erect or drooping in age, each bearing one or rarely two spikelets, 2.0 to 2.3 cm long. Glumes unequal, purplish with pale, hyaline margins, much shorter than the 8 to 10 mm long lemmas, the latter with a 1 cm long, straight or somewhat spreading awn.

A handsome woodland species, in our area collected only a few times north to Great Bear Lake, and mainly within the Precambrian Shield area.

General distribution: *S. purpurascens s. lat.*, circumpolar with large gaps.

Fig. 166 Map 186

Scolochloa Beauv.

Scolochloa festucacea (Willd.) Link
Fluminea festucacea (Willd.) Hitchc.
Culms leafy, erect, stout, 1.0 to 1.5 m tall, somewhat spongy at the base, from extensively creeping, whip-like rhizomes. Leaf-blades flat, commonly up to 10 mm wide, from large, somewhat papery and inflated sheaths. Panicle open, 20 cm long or more, the branches in distant fascicles, the lowermost 10 to 15 cm long. Spikelets about 8 mm long, 3- to 4-flowered.

In shallow water or wet marshes. A plains species barely entering the Mackenzie basin.

General distribution: Wide ranging from W. Europe to western Siberia with large gaps in N. America.

Fig. 167 Map 187

Setaria Beauv. Bristlegrass or Bristly foxtail

Ours annual, weedy plants, usually much branched and geniculate, spreading at base, with flat and broad leaf-blades and hairy ligules; the panicles dense and spike-like and the small spikelets subtended by 1 to several bristles.

a. Bristles retrorsely scabrous, as long or slightly longer than the spikelets; panicle erect, thin . . . . *S. verticillata*
a. Bristles antrorsely scabrous, 3 to 4 times as long as the spikelets, panicle nodding, bushy . . . . *S. viridis*

Setaria verticillata (L.) Beauv.
For differences between this and following see key.
Ephemeral and accidental introduction, in waste places and gardens.

General distribution: Introduced from Europe.

Setaria viridis (L.) Beauv.
For differences between this and preceding see key.
Ephemeral and accidental introduction, in waste places and gardens.

General distribution: Introduced from Europe.

Spartina Schreb. Cord-Grass

Coarse and tall perennial grasses of saline or alkaline wet meadows or marshes, ours with stout, creeping rhizomes, usually solitary culms, and tough, wiry leaf-blades. Inflorescence racemose, 10 to 30 cm long of 1-sided appressed spikes 4 to 5 cm long, each composed of numerous 1-flowered laterally flattened spikelets.

a. Plant robust, 1-2 m tall, leaf-blades 5 mm wide, scabrous on the margins; inflorescence of 10 or more spikes; second glume prolonged into a very scabrous awn as long as the glume . . . . *S. pectinata*
a. Plant slender, rarely up to 1 m tall; leaf-blades narrower, scabrous above; inflorescence of 4-8 spikes; second glume awnless . . . . *S. gracilis*

Spartina gracilis Trin.
Alkali Cord-Grass
Culms rarely to 1 m tall; leaf blades narrow, to 2 mm wide, becoming involute, very scabrous above; spikes 4 to 8, closely appressed, 2 to 4 cm long.

Common on salt plains west of Slave River and lately reported from opposite Simpson on the upper Mackenzie.

General distribution: N. America, central western plains and foothills.

Fig. 168 Map 188

Spartina pectinata Link
Prairie Cord-Grass
Culms robust, 1 to 2 m tall; leaf blades to 5 mm wide, becoming involute when dry, scabrous on

the margins; spikes 10 or more, ascending, 4 to 8 cm long.

Brackish marshes along coasts, and on wet salt plains; our only authority for the inclusion of this species is an Onion, Kennicott and Hardisty specimen gathered in 1861-62 at Fort Resolution.

General distribution: Eastern and central N. America.

Fig. 169 Map 189

Sphenopholis Scribn. Wedge-Grass

Sphenopholis intermedia (Rydb.) Rydb.
Tufted perennial; culms slender, smooth and shiny, 3 to 7 dm high; leaf-blades 5 to 15 cm long, flat, 2 to 5 mm wide. Panicle 5 to 15 cm long, fairly dense but not spike-like, 1.0 to 2.5 cm thick, with the spikelets densely crowded on the short, ascending branches; spikelets 2.5 to 3.0 mm long, glumes subequal, scabrous on the keel, the first linear-subulate, the second obovate; lemmas awnless.

Wet prairies and flood-plain meadows, north in the Mackenzie valley to Fort Simpson.

General distribution: N. America: Nfld. to B.C., north to southern District of Mackenzie.

Fig. 170 Map 190

Stipa L. Feathergrass

Tall and tufted, perennial prairie grasses with 1-flowered spikelets in terminal panicles. Glumes somewhat papery, narrow, keeled, acute or bristle-tipped; lemma indurate, more or less terete and strongly involute so that its margins overlap and tightly envelop both palea and grain, usually with a sharp-pointed retrorse-hairy callus at the base, and terminating in a long persistent sharply bent awn which is twisted spirally with numerous tight turns below the bend.

a. Glumes 20 mm long or more
  b. Lemma 20-25 mm long; awn short and straight above the bend, 12-20 cm long. . . . . *S. spartea*
  b. Lemma 8-12 mm long; awn slender and curled above the bend, 10-15 cm long . . . . *S. comata*
a. Glumes 7-10 mm long
  c. Panicle loose and open. *S. Richardsonii*
  c. Panicle dense and more or less spike-like . . . . *S. viridula*

***Stipa comata** Trin. & Rupr.
Culms 3 to 6 dm high, smooth; the uppermost and conspicuously inflated sheaths usually enclosing the base of the panicle. Leaf-blades somewhat scabrous, involute, the basal ones filiform, 1 to 3 dm long. Panicle 15 to 20 cm long, open. Glumes 18 to 25 mm long, glabrous; lemma about 10 mm long.

Thus far not recorded from the Mackenzie River basin north of lat. 60°, but to be expected along the south shore of Slave Lake and along the lower Liard River.

Prairie grassland.

General distribution: Man., Alta., and southern Y.T., south to Texas and California.

Fig. 171 Map 191

***Stipa Richardsonii** Link
Culms slender, 5 to 10 dm high. Leaf-blades involute, filiform, 5 to 10 cm long, scabrous. Panicle open, 10 to 20 cm long, the slender branches distant and spreading, bearing 3 to 4 spikelets near their tips; glumes 8 to 9 mm long, lemma about 5 mm, terminating in a 2.5 to 3.0 cm long awn.

Thus far not reported from the District of Mackenzie where it may be expected in dry grassy places of the upper Mackenzie Basin.

General distribution: Similar to that of *S. comata*.

Fig. 172 Map 192

Stipa spartea Trin. var. **curtiseta** Hitchc.
Culms densely tufted, erect, smooth, 0.6 to 1.5 m high; basal leaves 2 to 5 dm long, strongly involute, scabrous above; panicle narrow, 10 to 25 cm long, with erect branches, each bearing 1 to 2 spikelets; glumes 3 to 4 cm long, pale and papery; lemma subcylindric, 1.5 to 2.5 cm long, with a stout and twice geniculate awn.

In the District of Mackenzie thus far reported only from the vicinity of Fort Simpson.

General distribution: Ont., Man. to Alta., south to Montana, S. Dakota and Wyoming.

Fig. 173 Map 193

Stipa viridula Trin.
Culms slender, 4 to 8 dm tall; leaves 2 to 3 mm wide involute; panicle 15 to 20 cm long, narrow with appressed-erect branches, greenish or straw-coloured; awns 2 to 3 cm long, spreading. At once distinguished from the otherwise somewhat similar *S. columbiana* Macoun (and from other *Stipa* in the area) by having a conspicuous collar of hairs at the throat of the sheaths.

Plains and dry foothills. Thus far collected a few times along the upper Mackenzie River.

General distribution: Sask. to Alta., north to upper Mackenzie valley, south to Montana and Arizona.

Map 194

Trisetum L.

Ours perennial, with flat leaf-blades and spike-like or open panicles. Spikelets 2-flowered; glumes membranaceous, unequal, acute and awnless; lemma 2-toothed at the apex, its twisted awn inserted below the apex and arising between the teeth.

a. Panicle spike-like; plants tufted
  b. Panicle dense except below, commonly bronze-purple. . . . . . . *T. spicatum* var. *Maidenii*
  b. Panicle less dense, larger, pale or silvery green
    c. Culms thinly hirsute, at least below the panicle; sheaths soft pilose . . . . *T. spicatum* var. *molle*
    c. Culms, sheaths and leaves glabrous or nearly so . . . . *T. spicatum* var. *majus*
a. Panicle open and lax, strongly bronze-tinged; plants from slender rhizomes . . . *T. sibiricum*

***Trisetum sibiricum** Rupr.
Culms 15 to 30 cm tall, glabrous, singly or several together from a slender rhizome; leaf-blades flat, up to 6 mm broad, totally glabrous; panicle open, ovate or elongate, 5 to 8 cm long, yellowish-brown and somewhat shiny.

Moist grassy slopes and tundra, and in willow and alder thickets.

General distribution: From eastern arctic Europe across Siberia to Alaska and Y.T. To be expected in the Richardson Mts. west of the Mackenzie Delta.

Fig. 174 Map 195

Trisetum spicatum (L.) Richt.
Caespitose, with erect culms 10 to 50 cm high, and flat, more or less pubescent leaves and sheaths. Panicle lanceolate-oblong, dense and spike-like, somewhat shiny, suffused with purple or green and with a "fuzzy" appearance because of the spreading prominently bent and twisted awns.

Common in rocky and gravelly places in dry tundra.

Within our area represented by three not too well differentiated races or varieties (see key):

var. *Maidenii* Fern. of arctic-alpine range in the upper Mackenzie drainage, is known chiefly from the treeless country east and west of the Mackenzie valley, and from the Precambrian Shield area to the east. To the north it extends across the arctic islands.

var. *molle* (Michx.) Beal is, perhaps, the most common variety in the central parts of the Mackenzie basin, where it is mainly found in the lowland.

var. *majus* Vasey is Cordilleran and differs from the preceding by being more loosely tufted, by its totally glabrous and shiny, up to 7 dm high culms, glabrous sheaths, flat leaf-blades, and pale green, lax, erect-ascending or open and somewhat spreading panicle. In our area thus far recognized only from a few collections in the Mackenzie Mountains.

General distribution: *T. spicatum s. lat.*: Circumpolar, arctic-alpine.

Fig. 175 (a. var. *Maidenii*, b. var. *spicatum*)
Map 196 (*T. spicatum s. lat.*)

Vahlodea Fries

Vahlodea atropurpurea (Wahlenb.) Fries
Deschampsia atropurpurea (Wahlenb.) Scheele
Loosely tufted, essentially glabrous perennial with culms from 2 to 6 dm high. Leaves flat, short acuminate, the basal ones with blades up to 30 cm long and 6 mm wide, those of the culm shorter, their sheaths commonly purplish. Panicle open, few-flowered, from 5 to 15 cm long, commonly somewhat flexuous or nodding, in tall plants the lowermost branches up to 5 cm long. Spikelets 2-flowered, dark purple, turning bronze; glumes 5 mm long, lanceolate, enclosing the florets; lemma firm, truncate, 3 mm long, bearing a dorsal awn attached above the mid-

dle of the body and protruding from the side of the spikelet.

East of long. 110° represented by the eastern ssp. *atropurpurea* of damp tundra and mossy lake shores, which is separated by a broad corridor of grassland and forest from the western ssp. *latifolia* (Hook.) Porsild of alpine and subalpine meadows and is distinguished by its taller growth, broader and longer leaves, and by its shorter callus hairs.

General distribution: *V. atropurpurea s. lat.*: Circumpolar with large gaps.

Fig. 176 (a. ssp. *atropurpurea*, b. ssp. *latifolia*)

Map 197 (ssp. *atropurpurea*) 198 (ssp. *latifolia*)

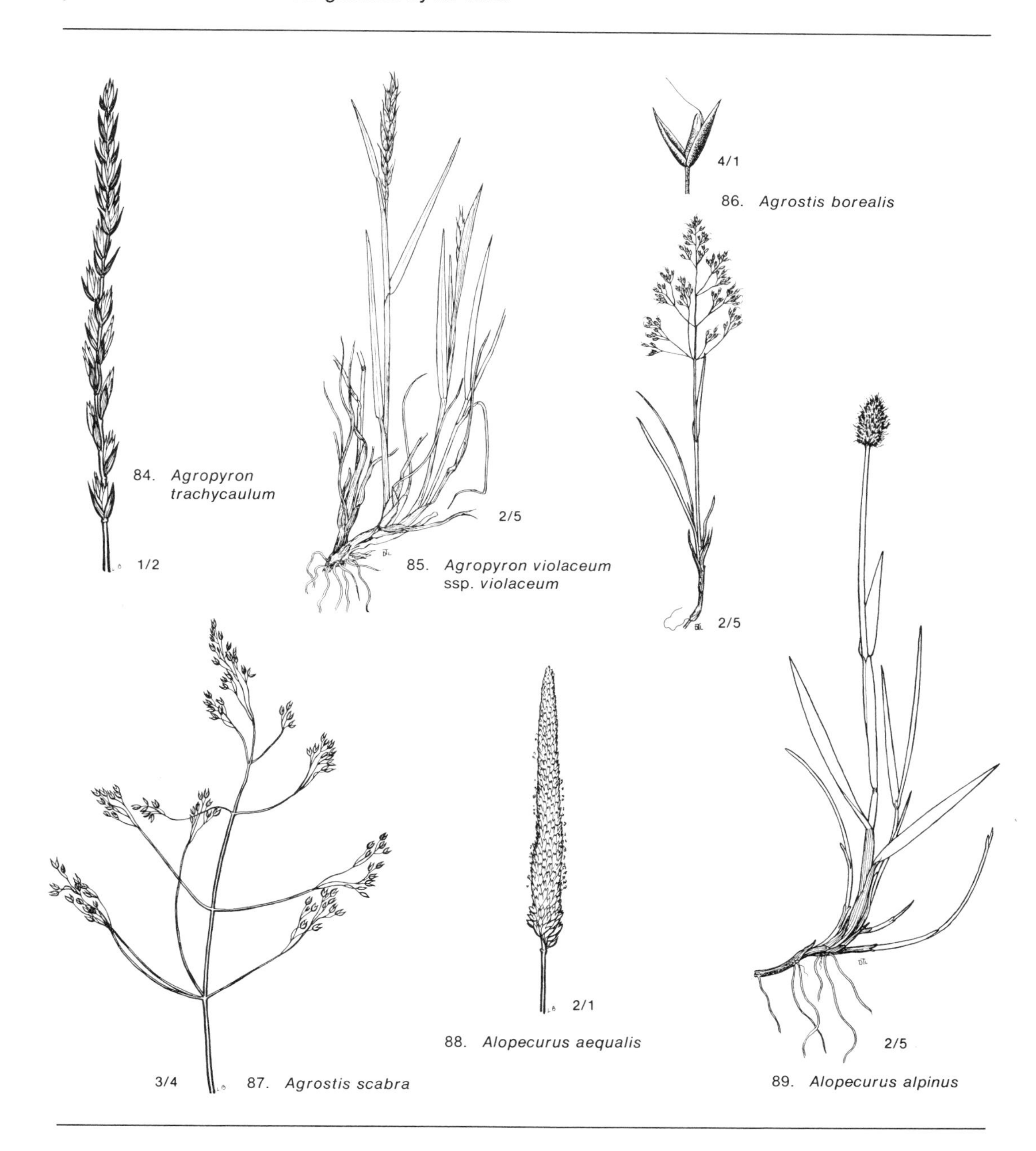

84. *Agropyron trachycaulum*

85. *Agropyron violaceum* ssp. *violaceum*

86. *Agrostis borealis*

87. *Agrostis scabra*

88. *Alopecurus aequalis*

89. *Alopecurus alpinus*

90. *Arctagrostis arundinacea* var. *arundinacea*

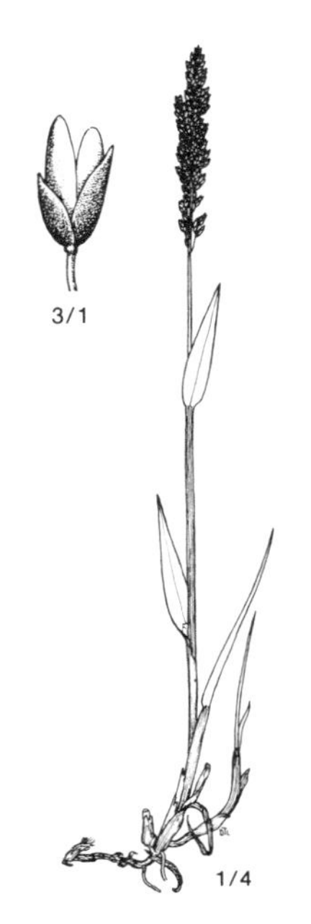

91. *Arctagrostis latifolia*

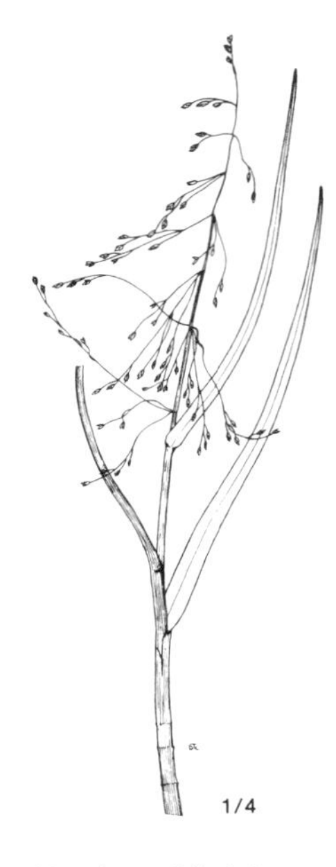

92. *Arctophila fulva*

93. *Beckmannia Syzigachne*

94. *Bromus Pumpellianus* var. *Pumpellianus*

95. *Calamagrostis canadensis* var. *Langsdorffii*

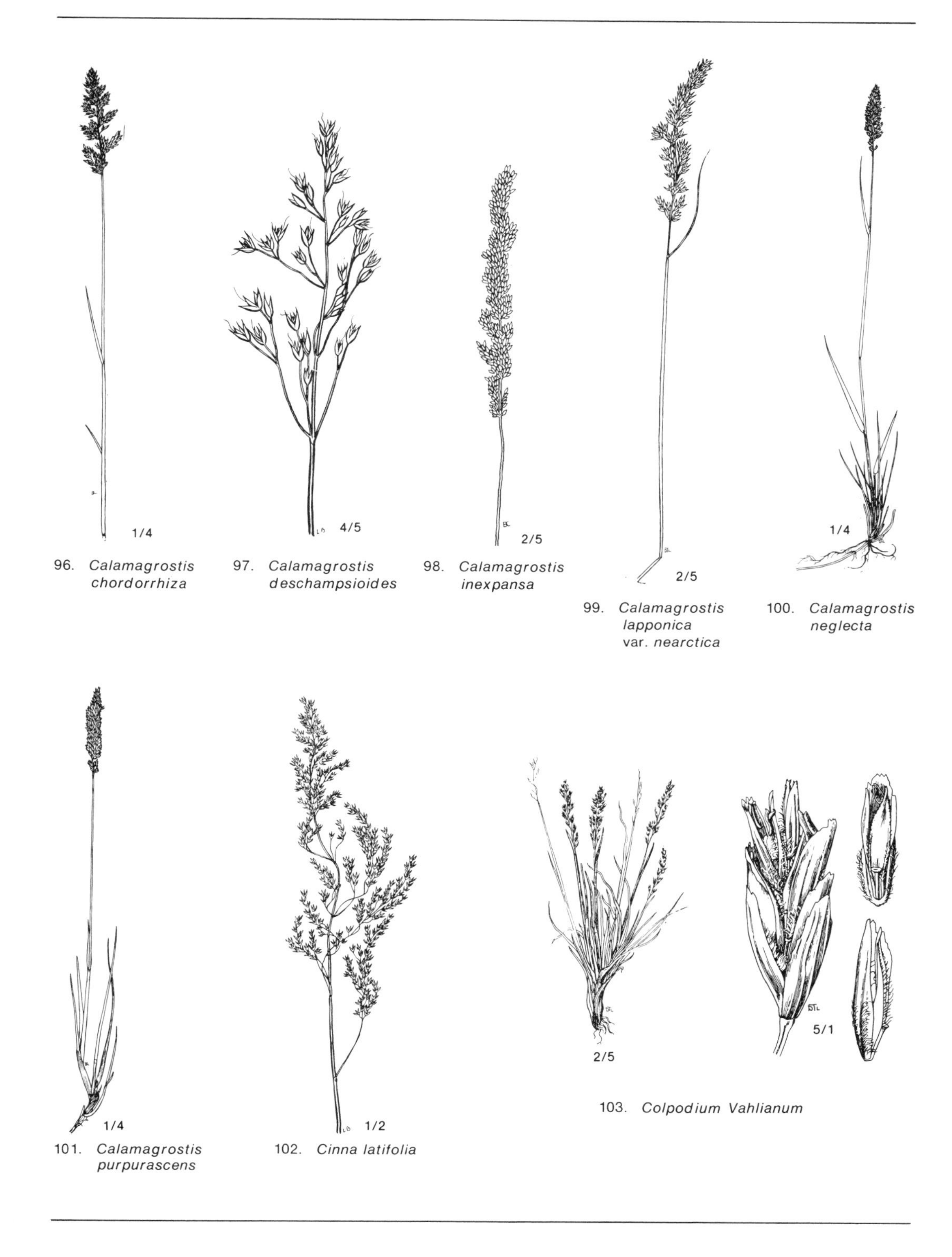

96. *Calamagrostis chordorrhiza*

97. *Calamagrostis deschampsioides*

98. *Calamagrostis inexpansa*

99. *Calamagrostis lapponica* var. *nearctica*

100. *Calamagrostis neglecta*

101. *Calamagrostis purpurascens*

102. *Cinna latifolia*

103. *Colpodium Vahlianum*

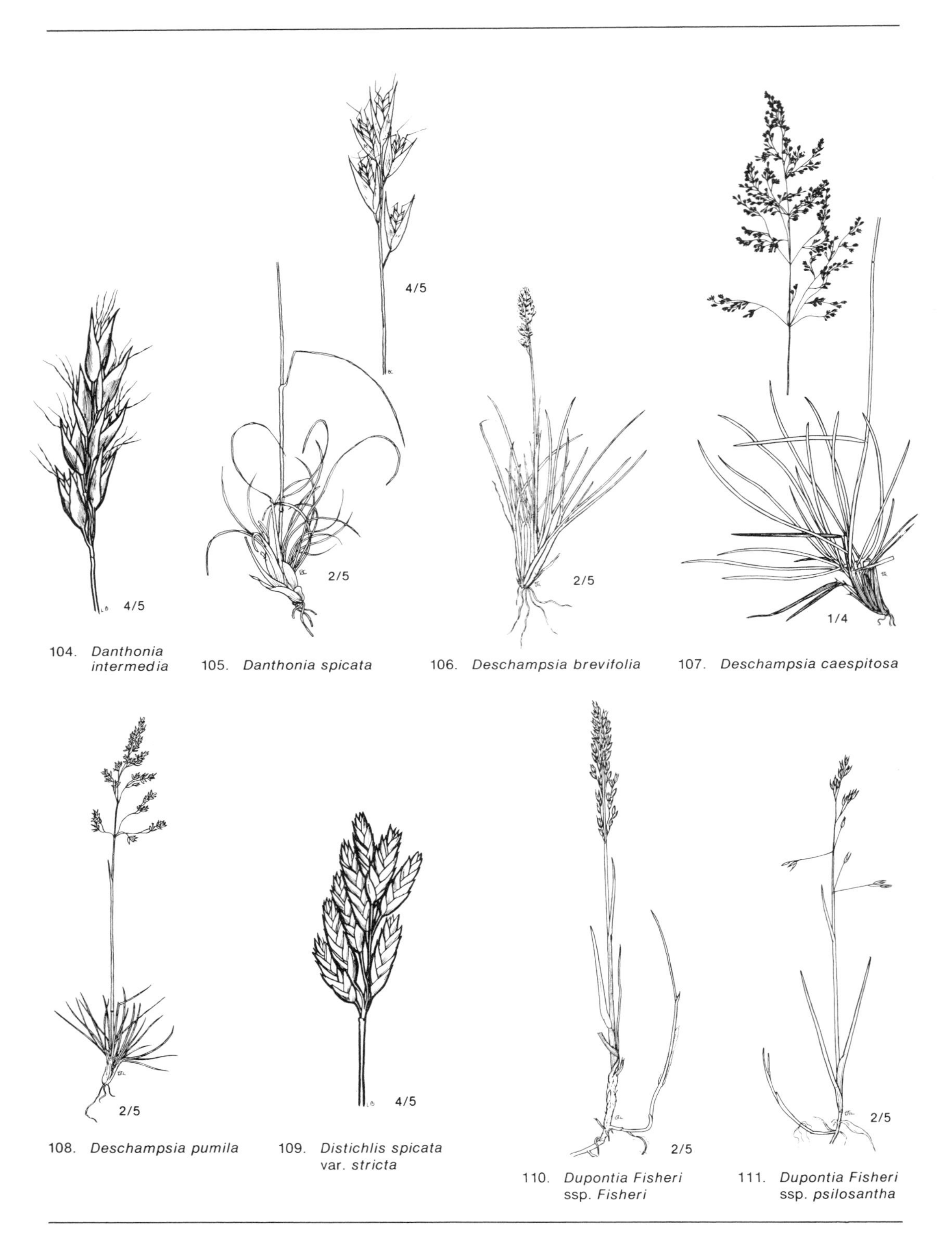

104. *Danthonia intermedia*

105. *Danthonia spicata*

106. *Deschampsia brevifolia*

107. *Deschampsia caespitosa*

108. *Deschampsia pumila*

109. *Distichlis spicata* var. *stricta*

110. *Dupontia Fisheri* ssp. *Fisheri*

111. *Dupontia Fisheri* ssp. *psilosantha*

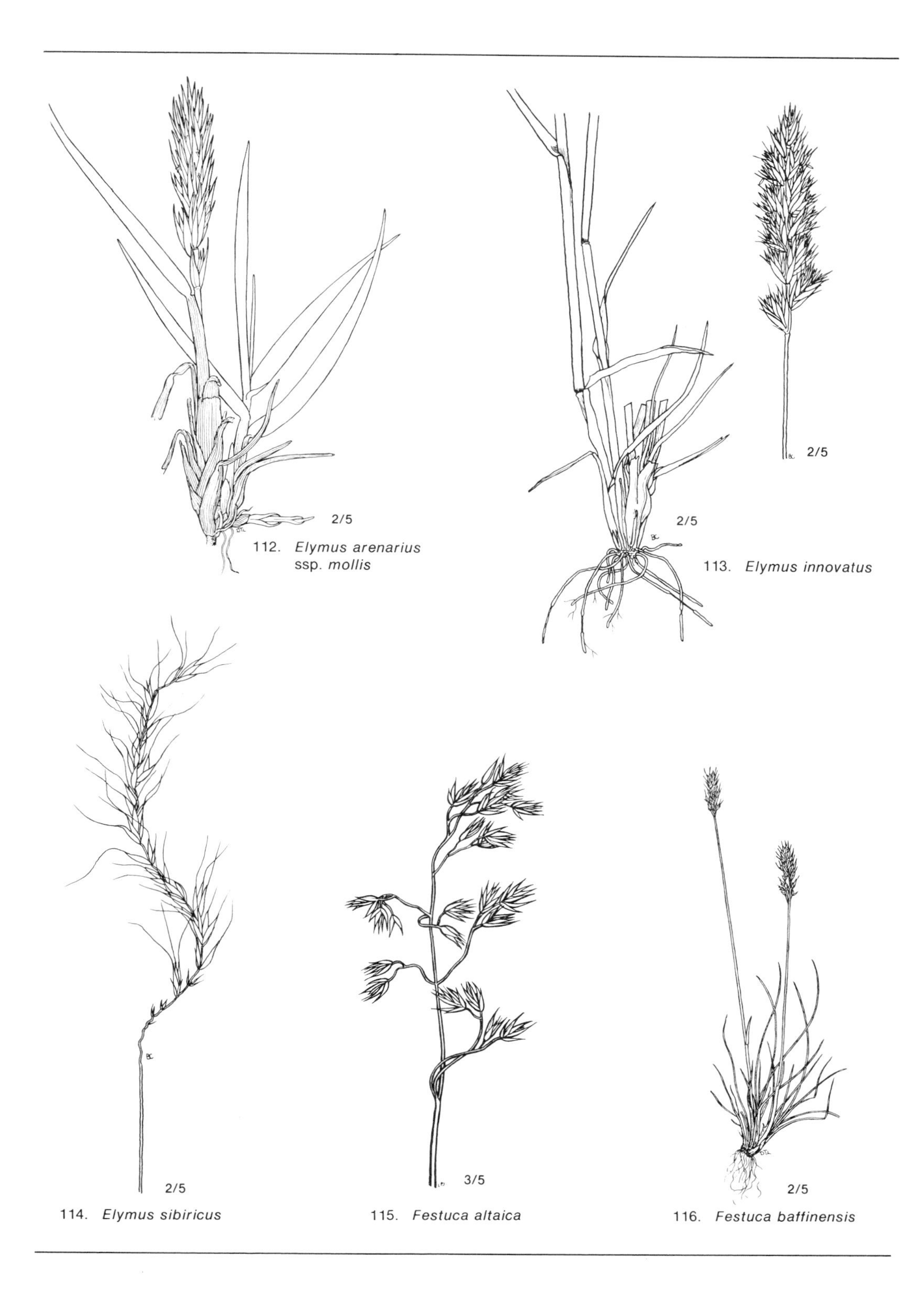

112. *Elymus arenarius* ssp. *mollis*

113. *Elymus innovatus*

114. *Elymus sibiricus*

115. *Festuca altaica*

116. *Festuca baffinensis*

117. *Festuca brachyphylla*

118. *Festuca prolifera* var. *lasiolepis*

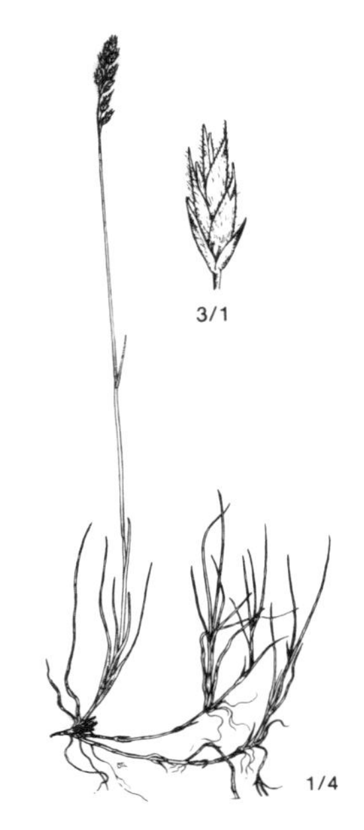

119. *Festuca rubra*

120. *Festuca saximontana*

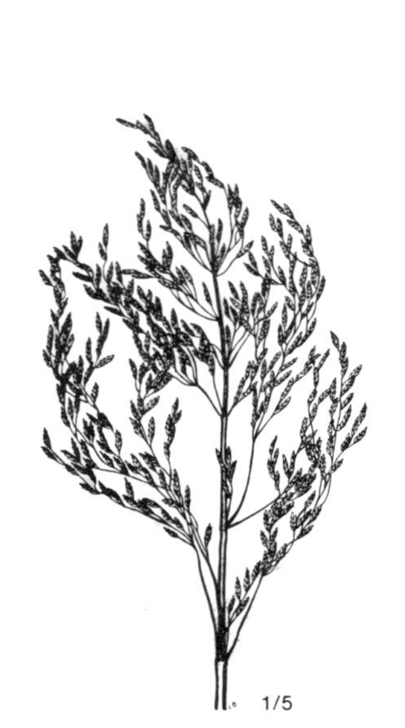

121. *Glyceria grandis*

122. *Glyceria pulchella*

123. *Glyceria striata* var. *stricta*

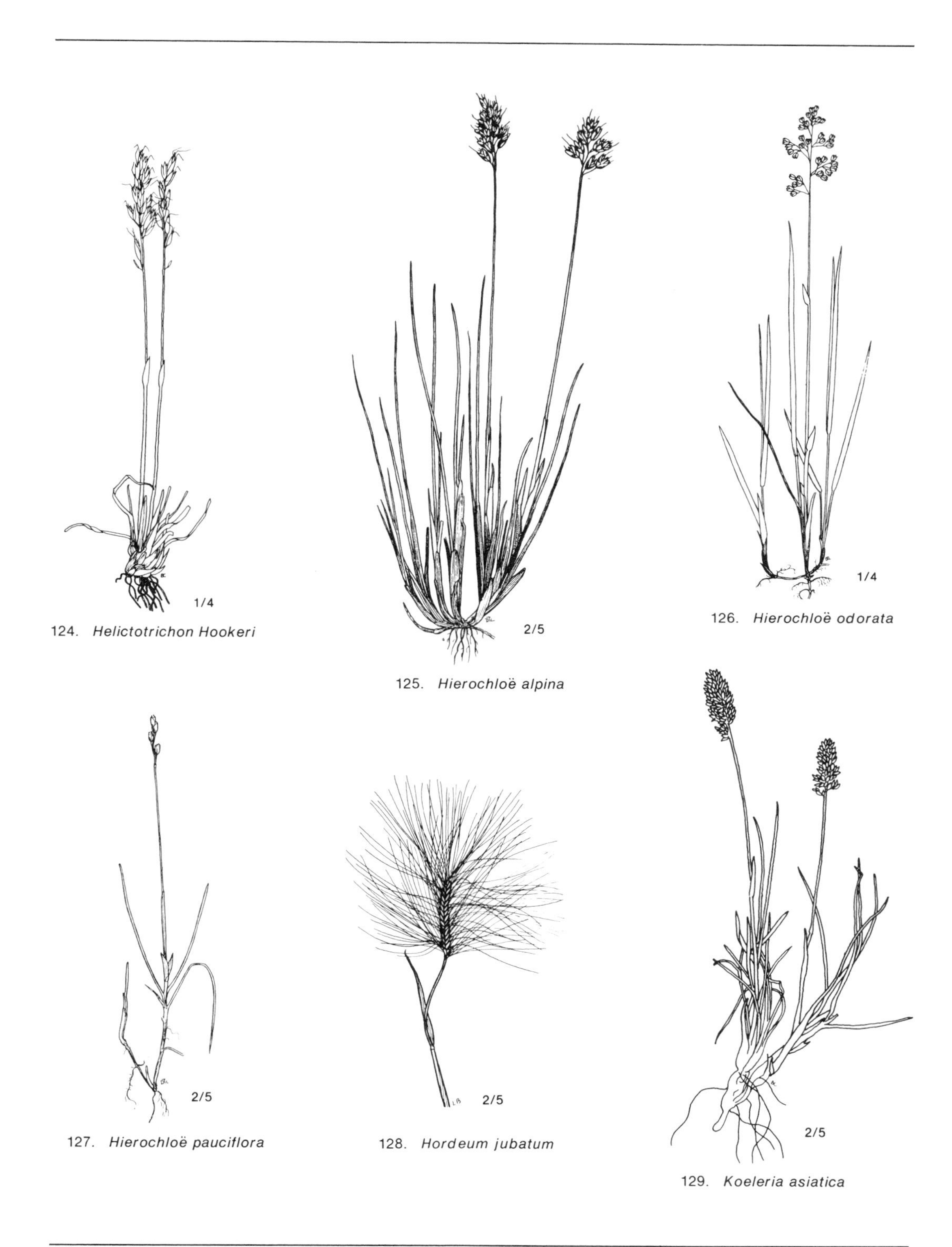

124. *Helictotrichon Hookeri*

125. *Hierochloë alpina*

126. *Hierochloë odorata*

127. *Hierochloë pauciflora*

128. *Hordeum jubatum*

129. *Koeleria asiatica*

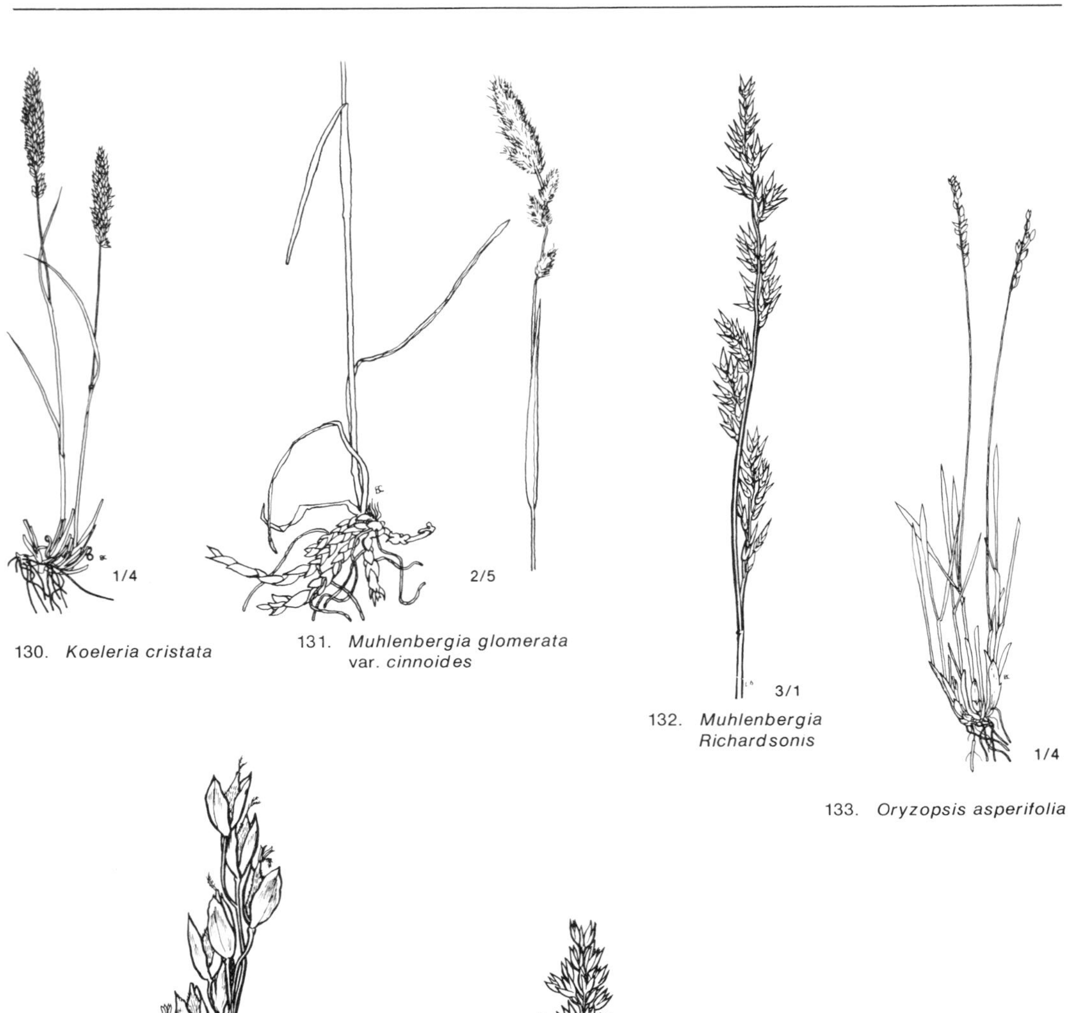

130. *Koeleria cristata*

131. *Muhlenbergia glomerata* var. *cinnoides*

132. *Muhlenbergia Richardsonis*

133. *Oryzopsis asperifolia*

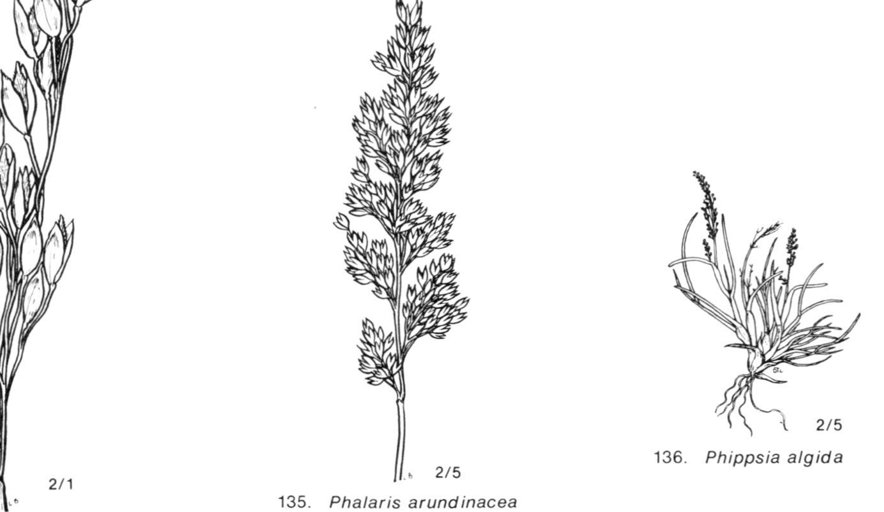

134. *Oryzopsis pungens*

135. *Phalaris arundinacea*

136. *Phippsia algida*

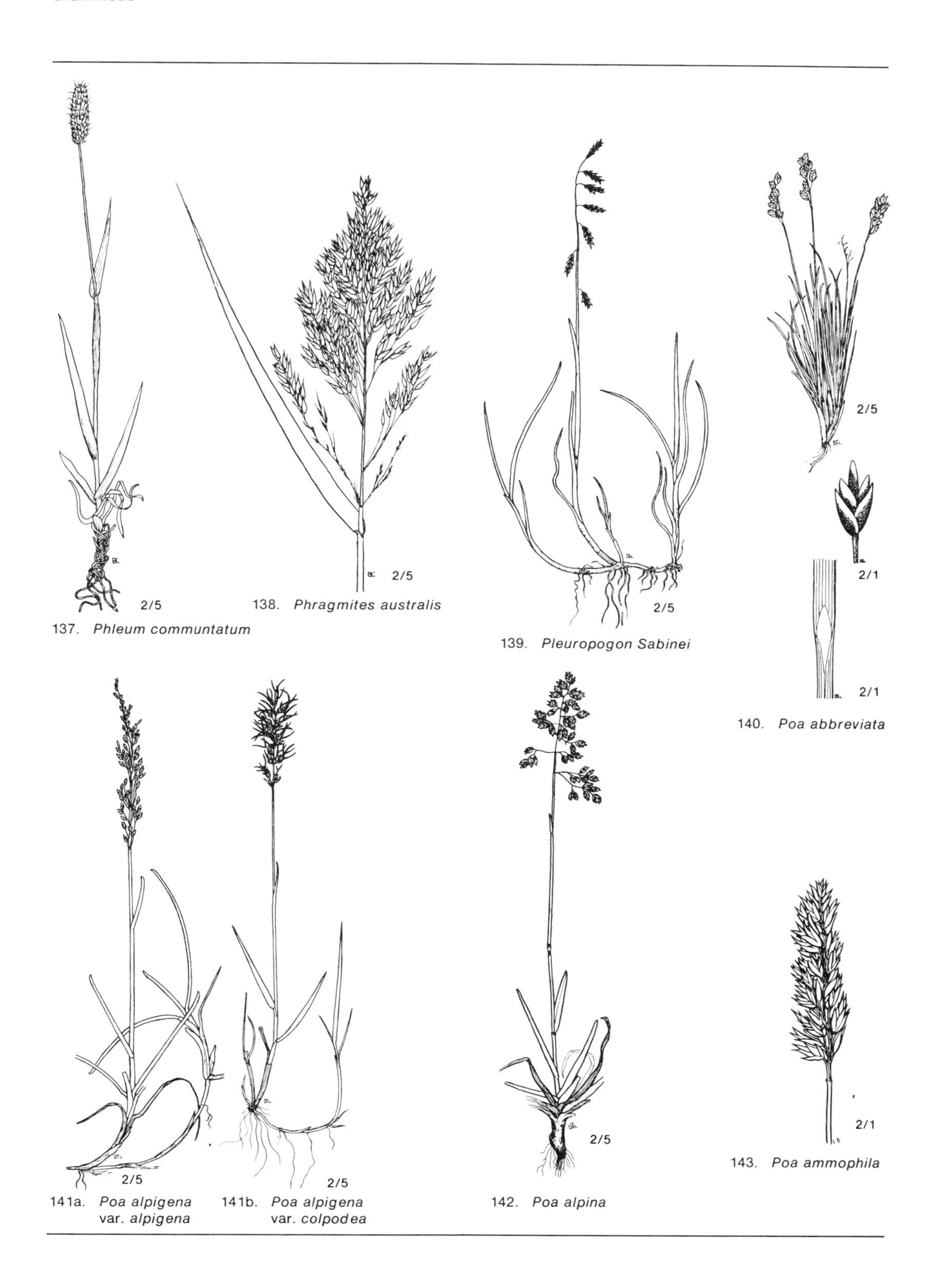

137. *Phleum communtatum*

138. *Phragmites australis*

139. *Pleuropogon Sabinei*

140. *Poa abbreviata*

141a. *Poa alpigena* var. *alpigena*

141b. *Poa alpigena* var. *colpodea*

142. *Poa alpina*

143. *Poa ammophila*

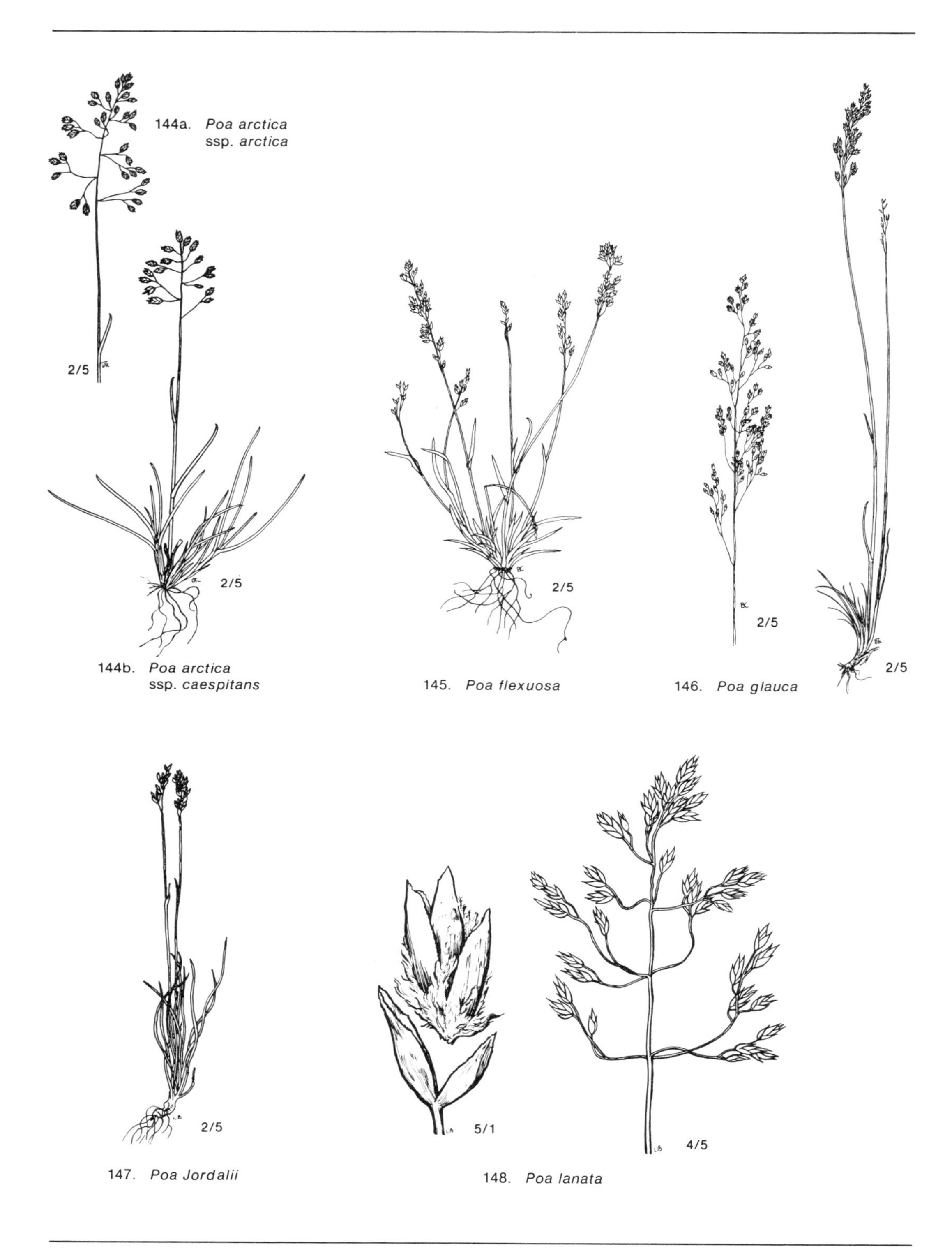

144a. *Poa arctica* ssp. *arctica*

144b. *Poa arctica* ssp. *caespitans*

145. *Poa flexuosa*

146. *Poa glauca*

147. *Poa Jordalii*

148. *Poa lanata*

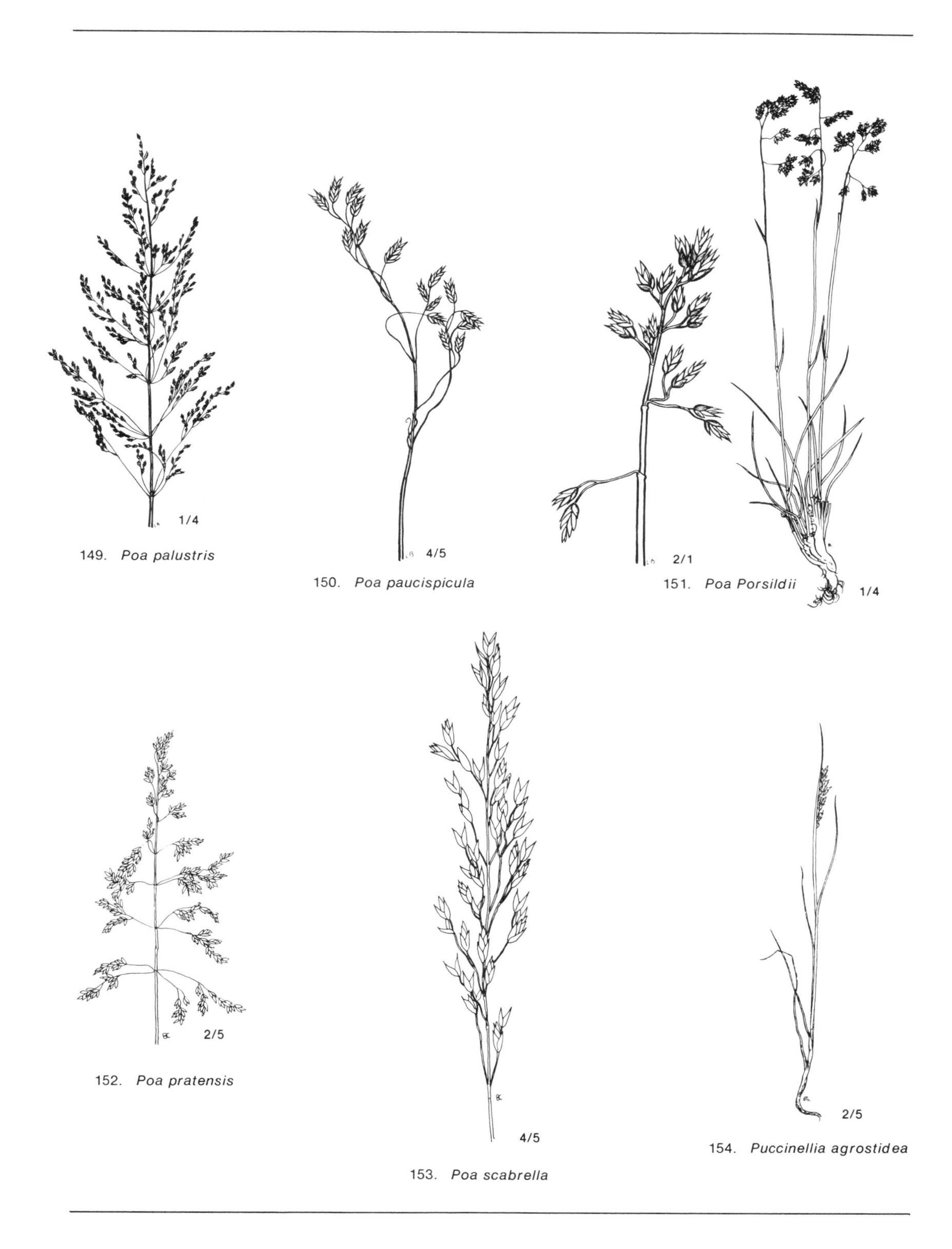

149. *Poa palustris*

150. *Poa paucispicula*

151. *Poa Porsildii*

152. *Poa pratensis*

153. *Poa scabrella*

154. *Puccinellia agrostidea*

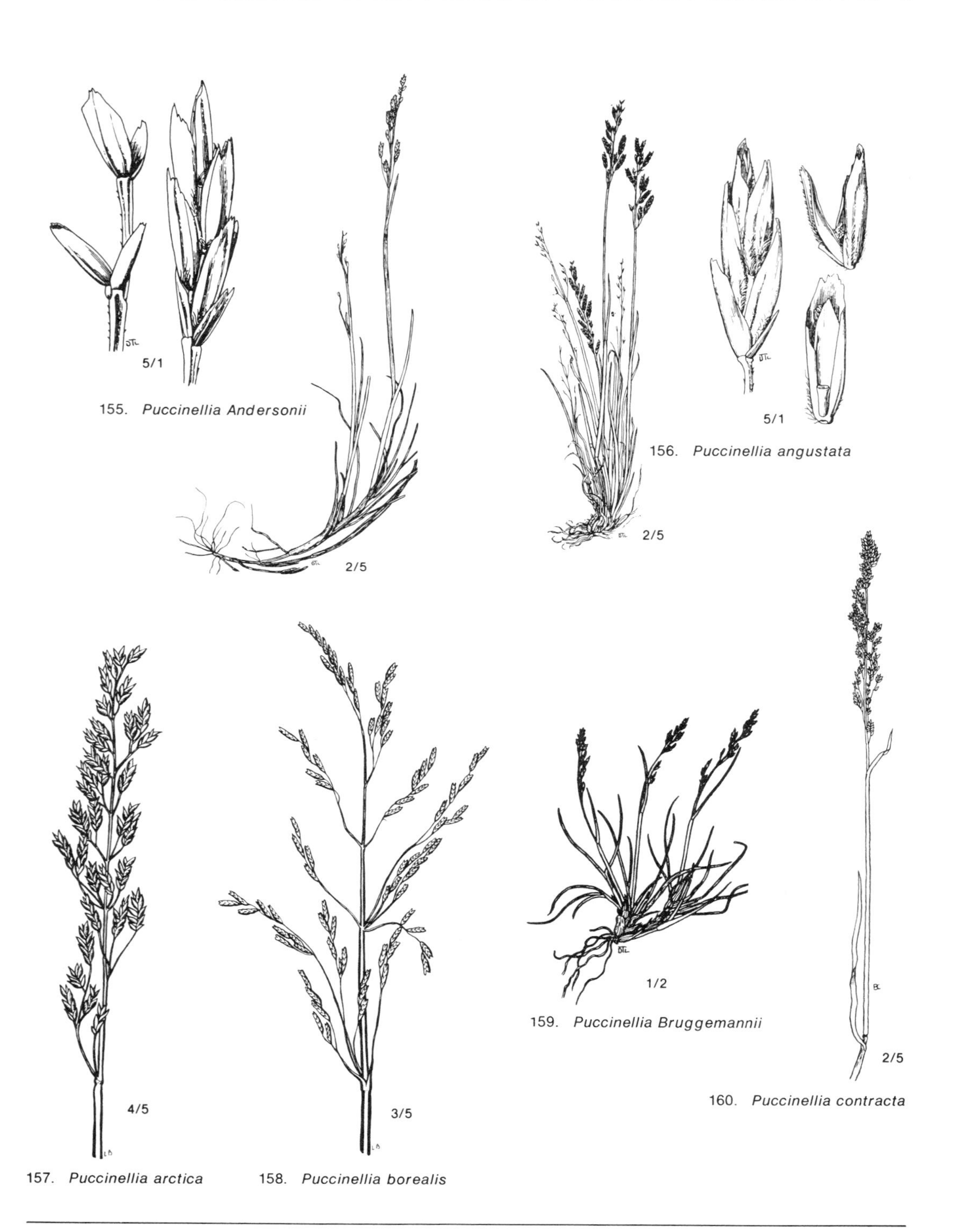

155. *Puccinellia Andersonii*

156. *Puccinellia angustata*

157. *Puccinellia arctica*

158. *Puccinellia borealis*

159. *Puccinellia Bruggemannii*

160. *Puccinellia contracta*

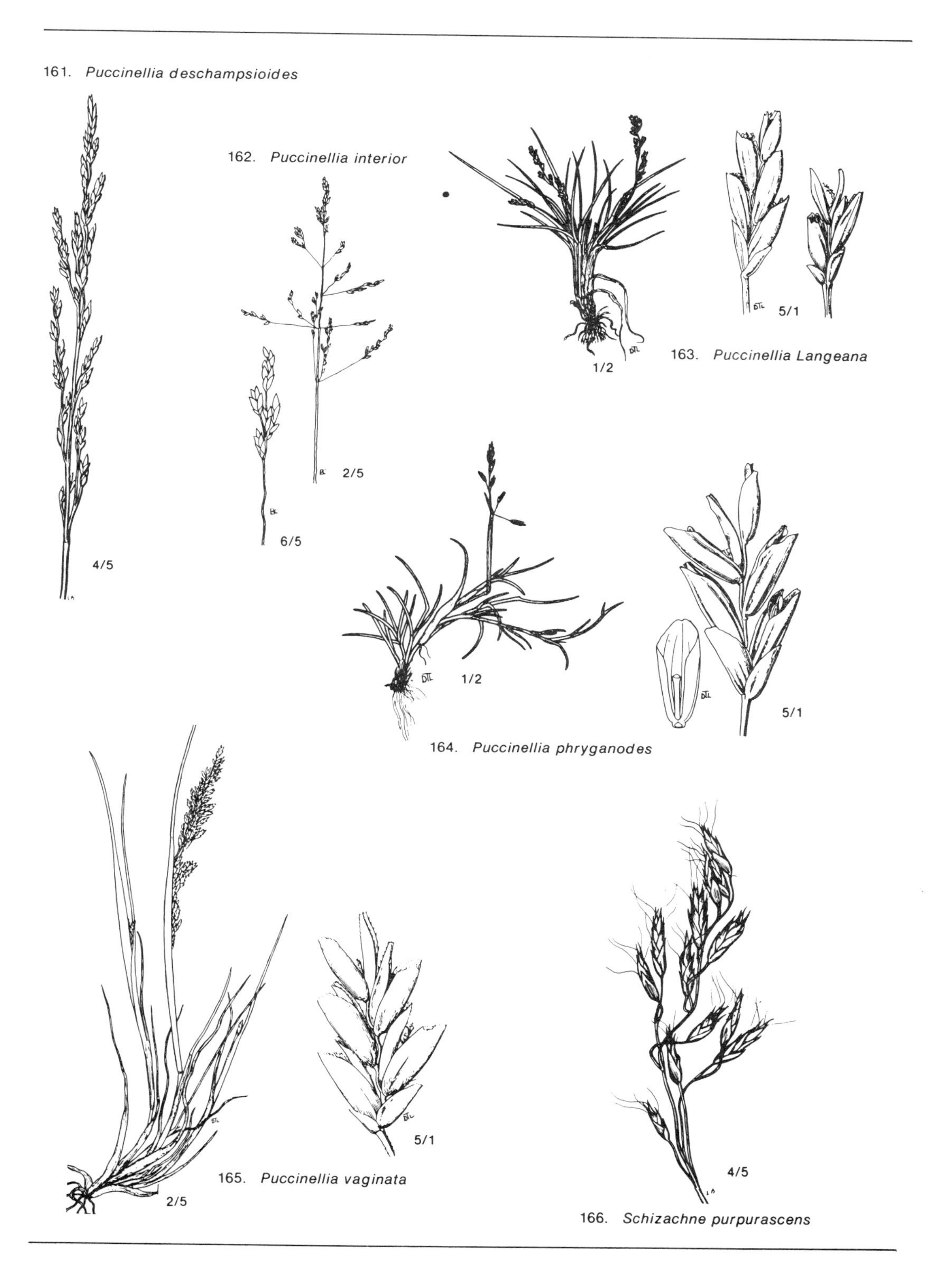

161. *Puccinellia deschampsioides*

162. *Puccinellia interior*

163. *Puccinellia Langeana*

164. *Puccinellia phryganodes*

165. *Puccinellia vaginata*

166. *Schizachne purpurascens*

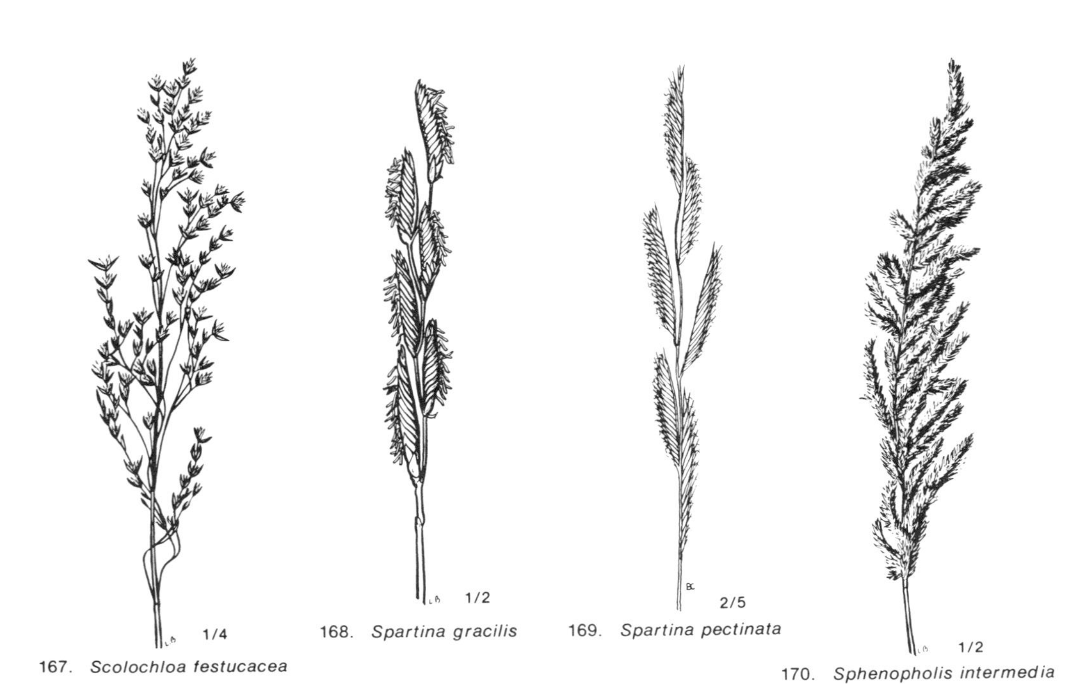

167. *Scolochloa festucacea*

168. *Spartina gracilis*

169. *Spartina pectinata*

170. *Sphenopholis intermedia*

171. *Stipa comata*

172. *Stipa Richardsonii*

173. *Stipa spartea* var. *curtiseta*

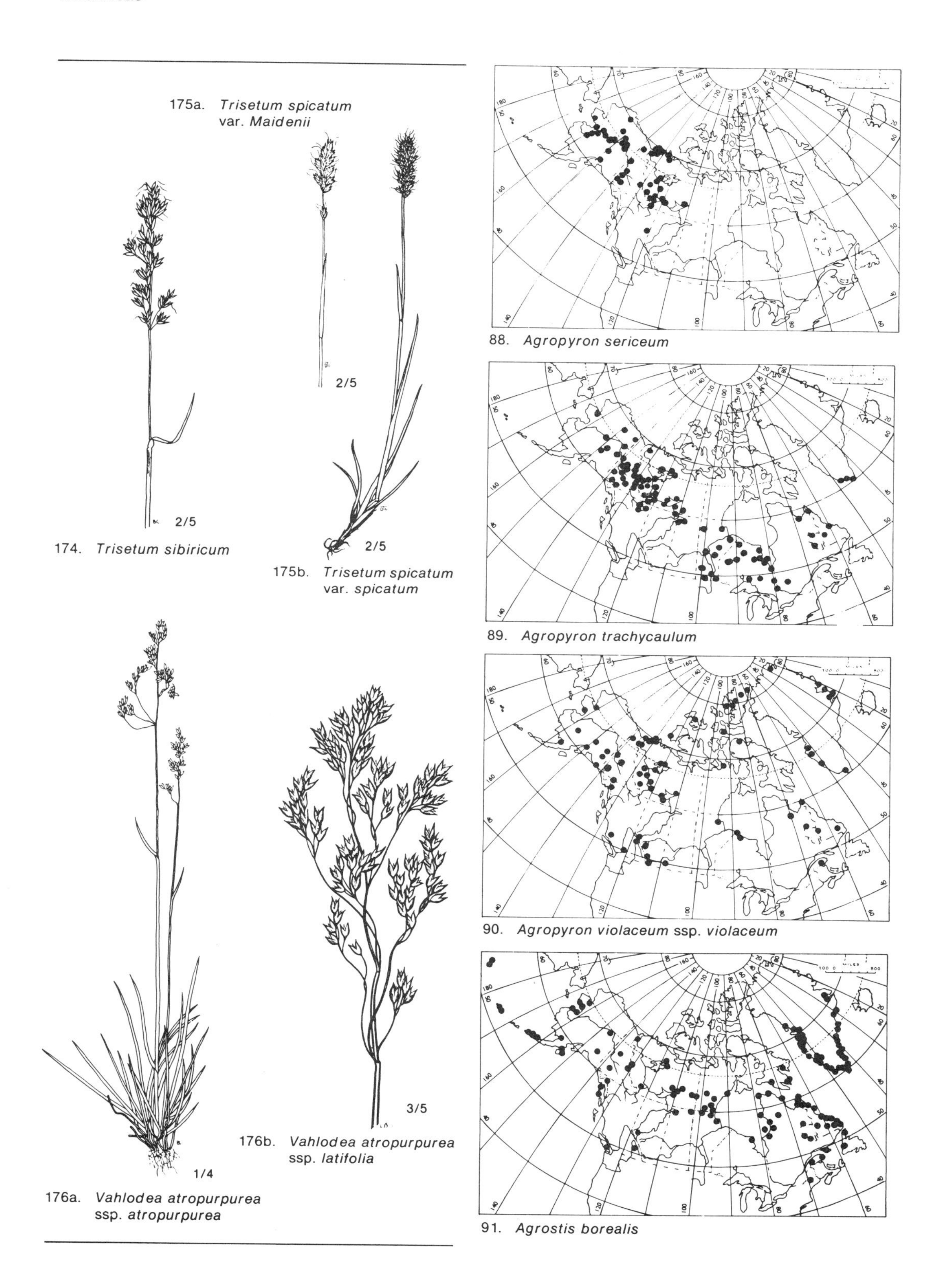

174. *Trisetum sibiricum*

175a. *Trisetum spicatum* var. *Maidenii*

175b. *Trisetum spicatum* var. *spicatum*

176a. *Vahlodea atropurpurea* ssp. *atropurpurea*

176b. *Vahlodea atropurpurea* ssp. *latifolia*

88. *Agropyron sericeum*

89. *Agropyron trachycaulum*

90. *Agropyron violaceum* ssp. *violaceum*

91. *Agrostis borealis*

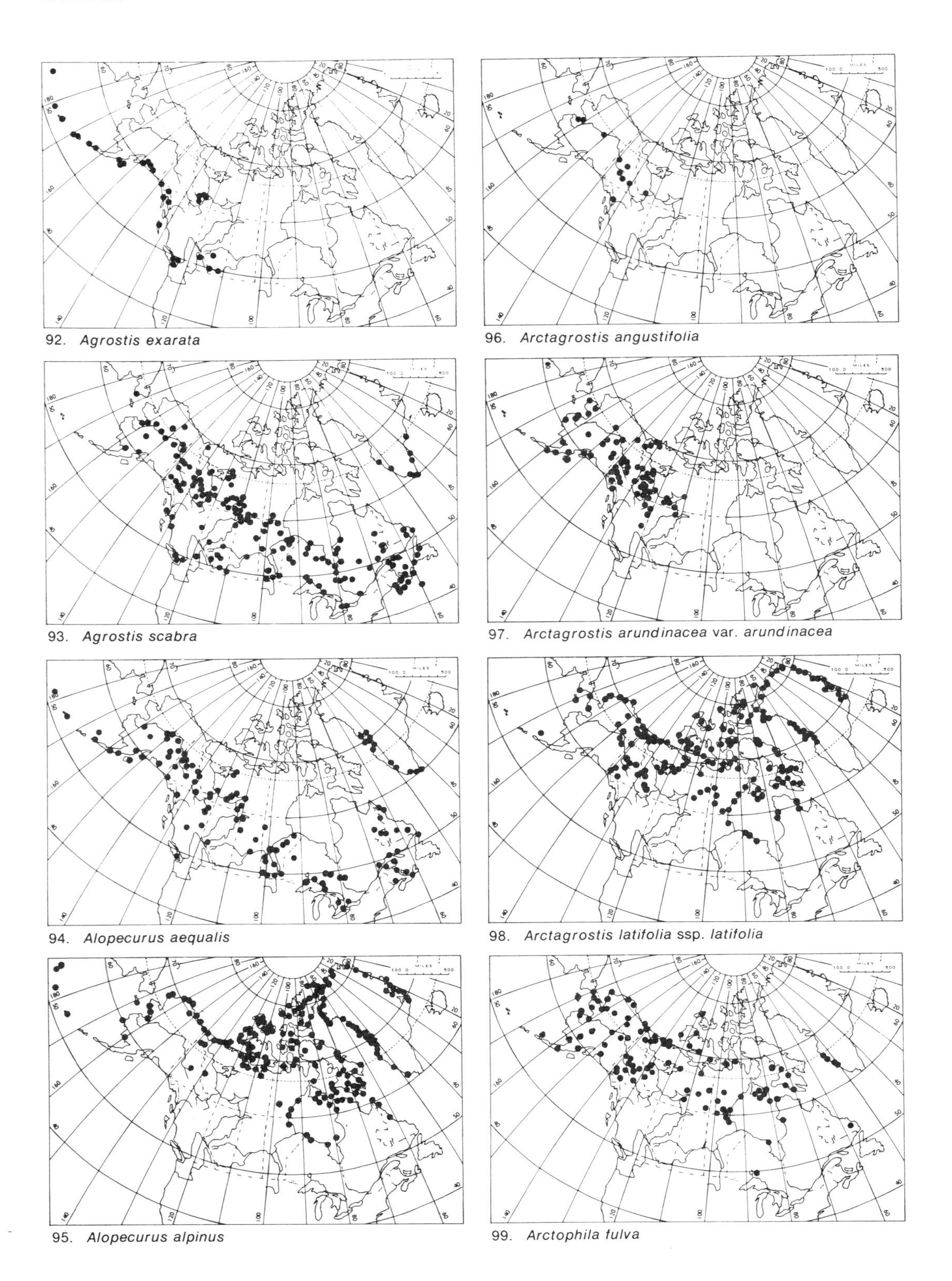

92. *Agrostis exarata*

96. *Arctagrostis angustifolia*

93. *Agrostis scabra*

97. *Arctagrostis arundinacea* var. *arundinacea*

94. *Alopecurus aequalis*

98. *Arctagrostis latifolia* ssp. *latifolia*

95. *Alopecurus alpinus*

99. *Arctophila fulva*

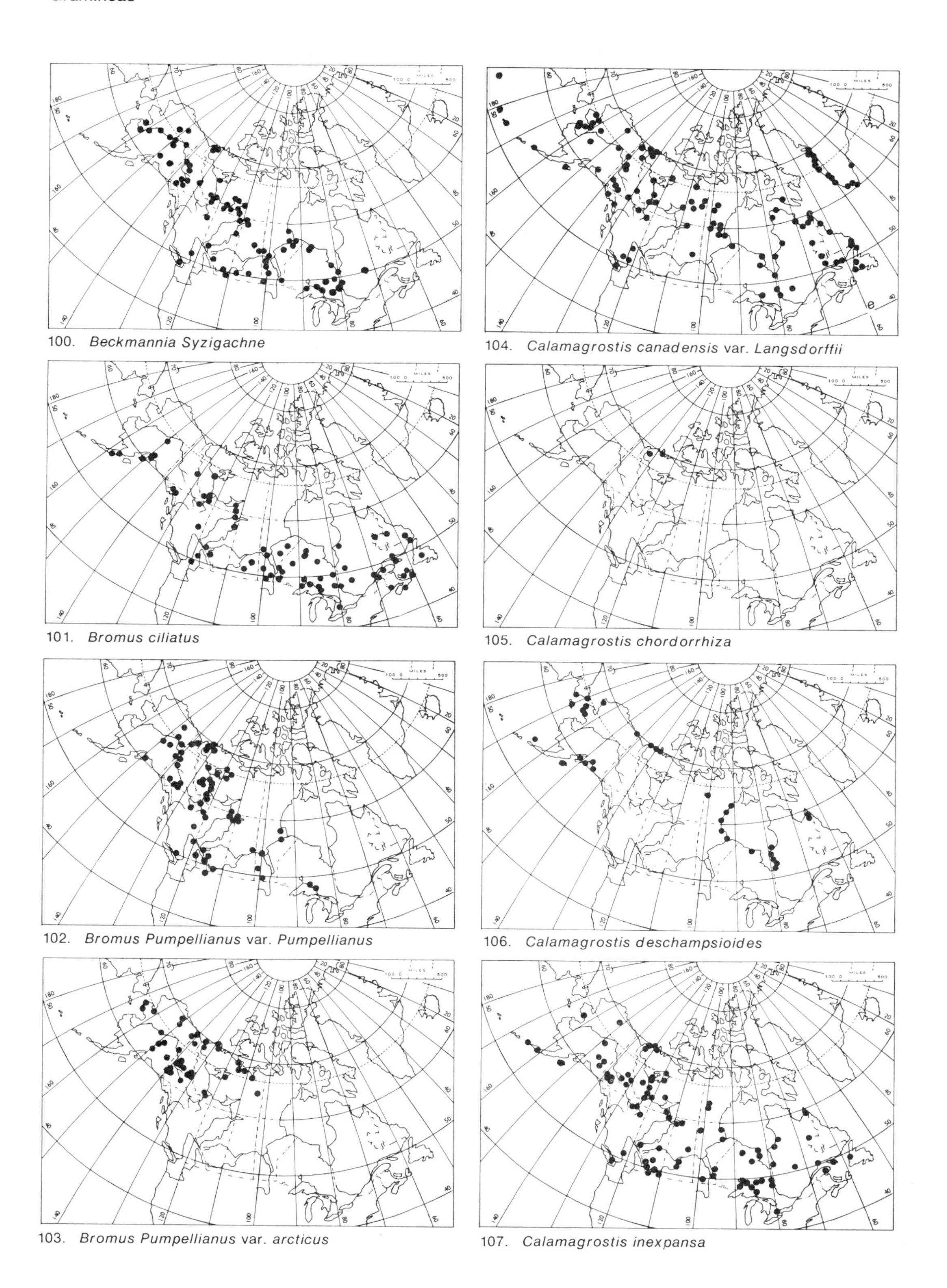

100. *Beckmannia Syzigachne*

101. *Bromus ciliatus*

102. *Bromus Pumpellianus* var. *Pumpellianus*

103. *Bromus Pumpellianus* var. *arcticus*

104. *Calamagrostis canadensis* var. *Langsdorffii*

105. *Calamagrostis chordorrhiza*

106. *Calamagrostis deschampsioides*

107. *Calamagrostis inexpansa*

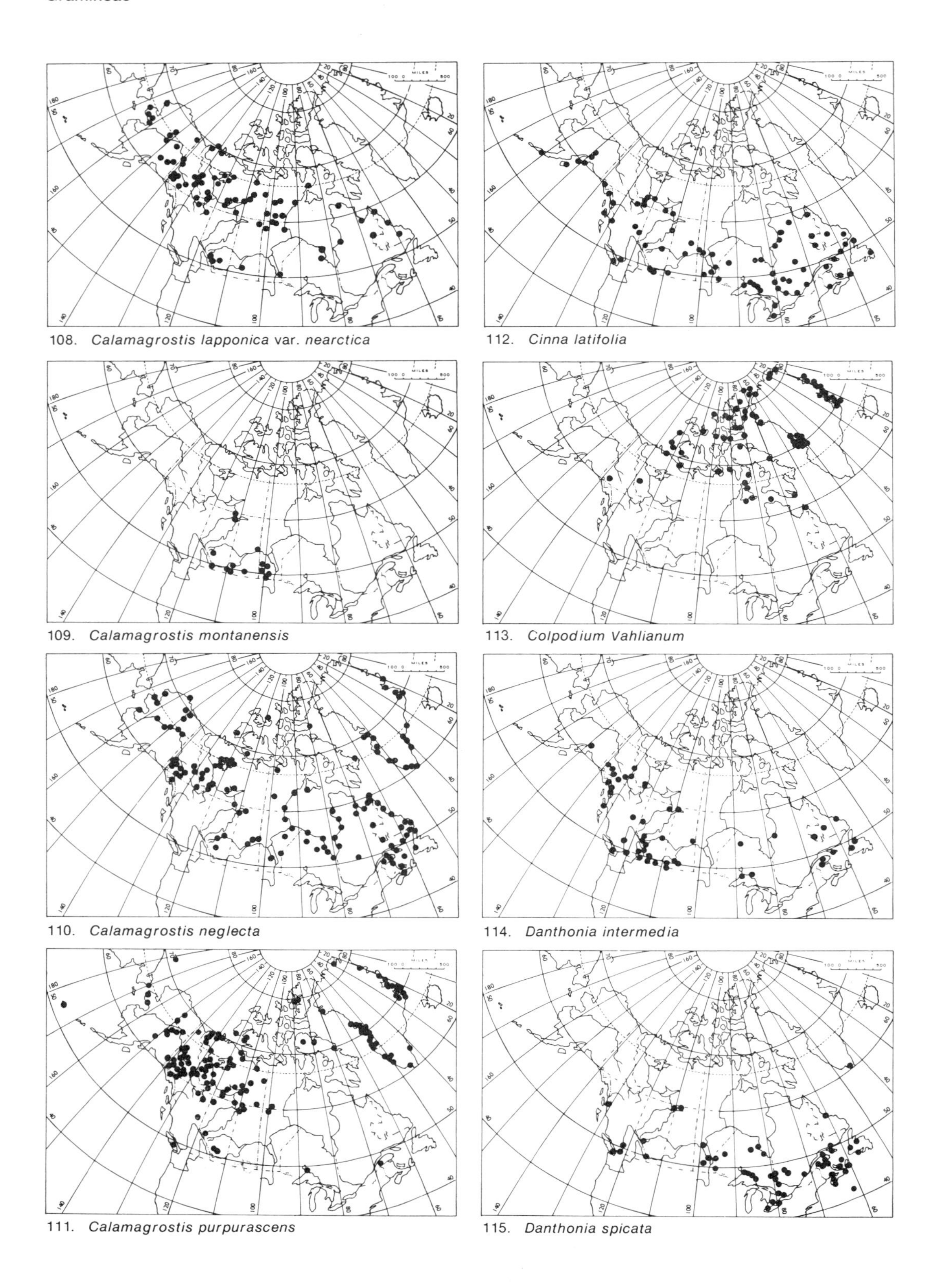

108. *Calamagrostis lapponica* var. *nearctica*

109. *Calamagrostis montanensis*

110. *Calamagrostis neglecta*

111. *Calamagrostis purpurascens*

112. *Cinna latifolia*

113. *Colpodium Vahlianum*

114. *Danthonia intermedia*

115. *Danthonia spicata*

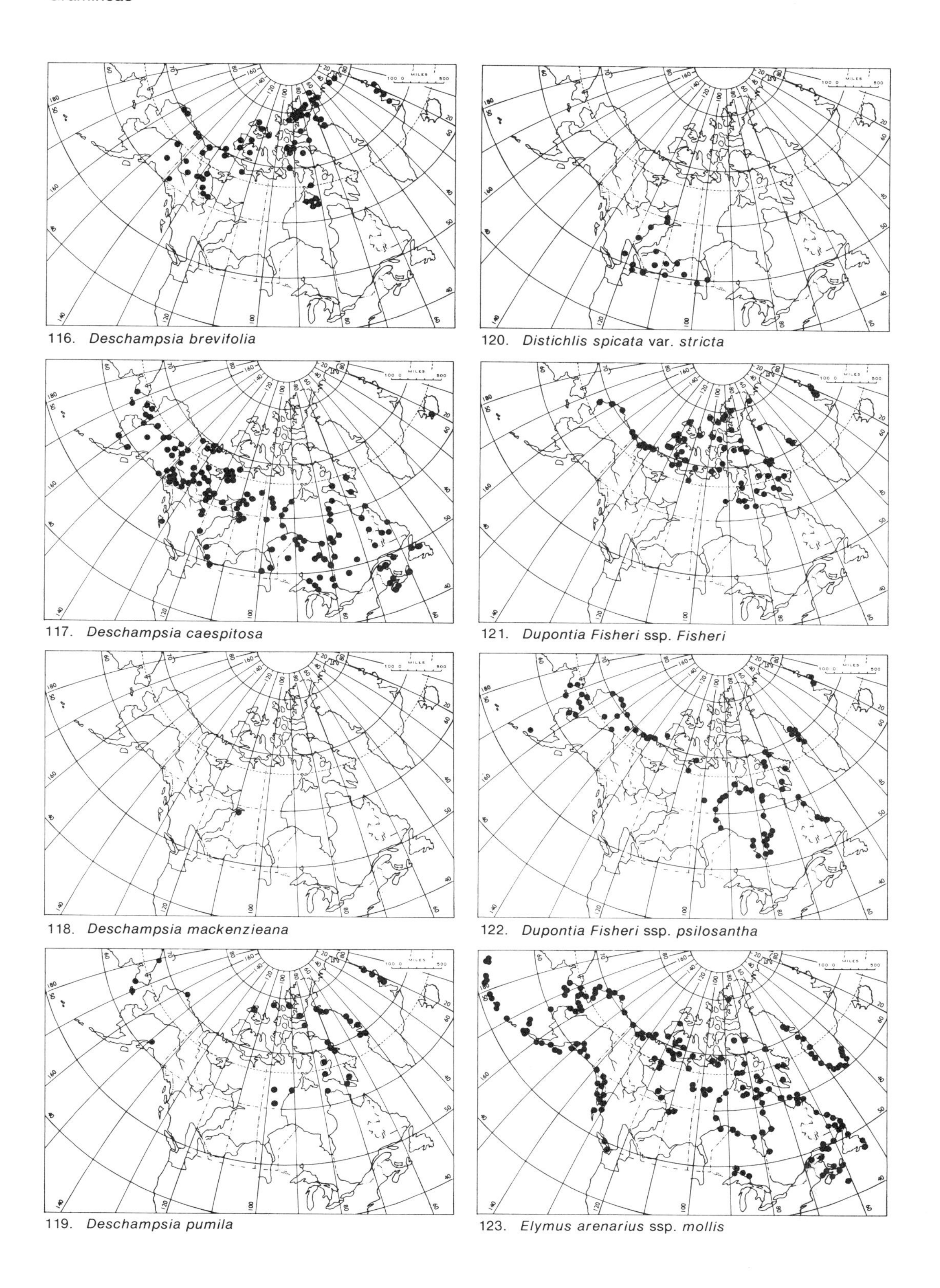

116. *Deschampsia brevifolia*

117. *Deschampsia caespitosa*

118. *Deschampsia mackenzieana*

119. *Deschampsia pumila*

120. *Distichlis spicata* var. *stricta*

121. *Dupontia Fisheri* ssp. *Fisheri*

122. *Dupontia Fisheri* ssp. *psilosantha*

123. *Elymus arenarius* ssp. *mollis*

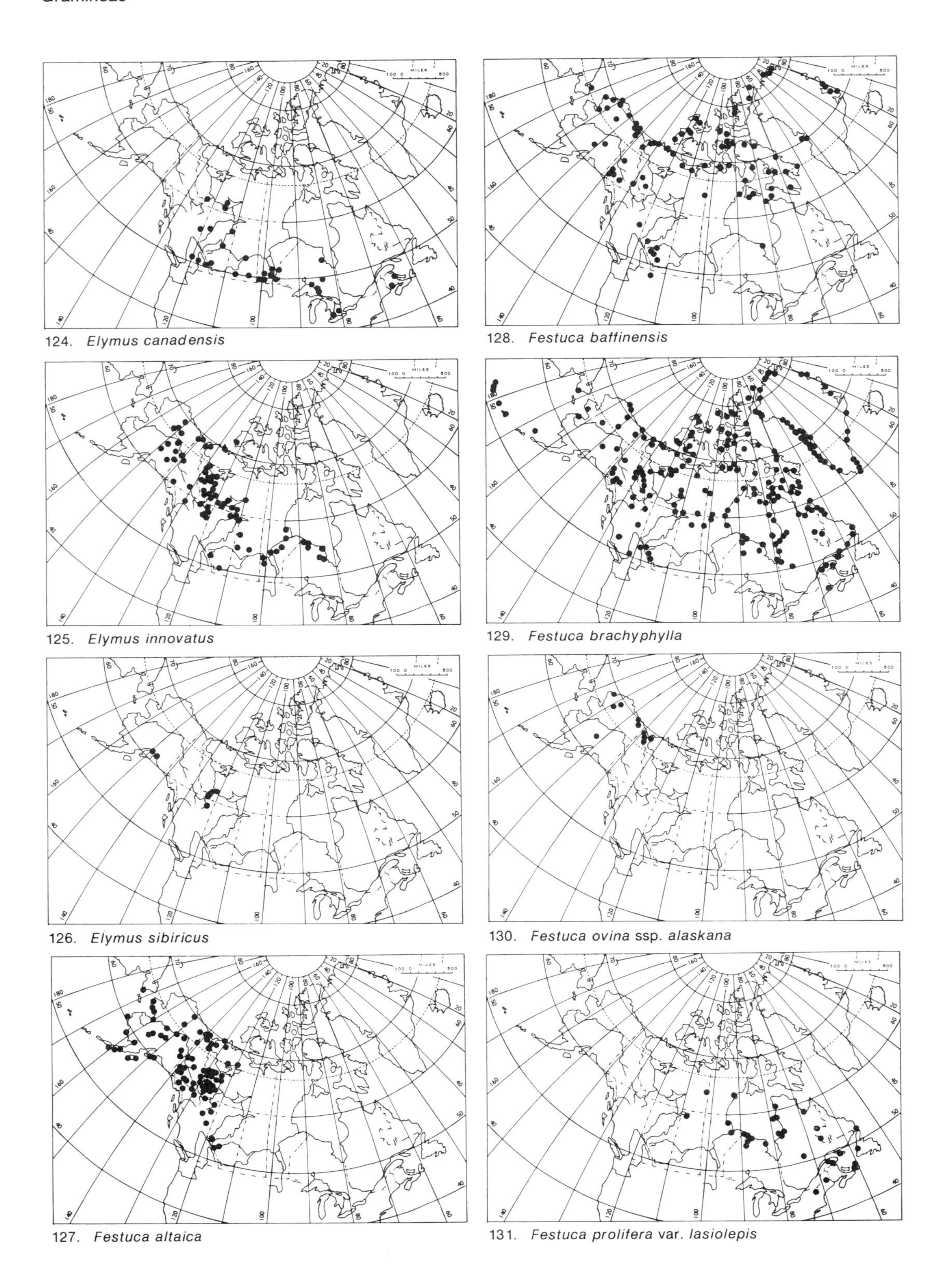

124. *Elymus canadensis*

125. *Elymus innovatus*

126. *Elymus sibiricus*

127. *Festuca altaica*

128. *Festuca baffinensis*

129. *Festuca brachyphylla*

130. *Festuca ovina* ssp. *alaskana*

131. *Festuca prolifera* var. *lasiolepis*

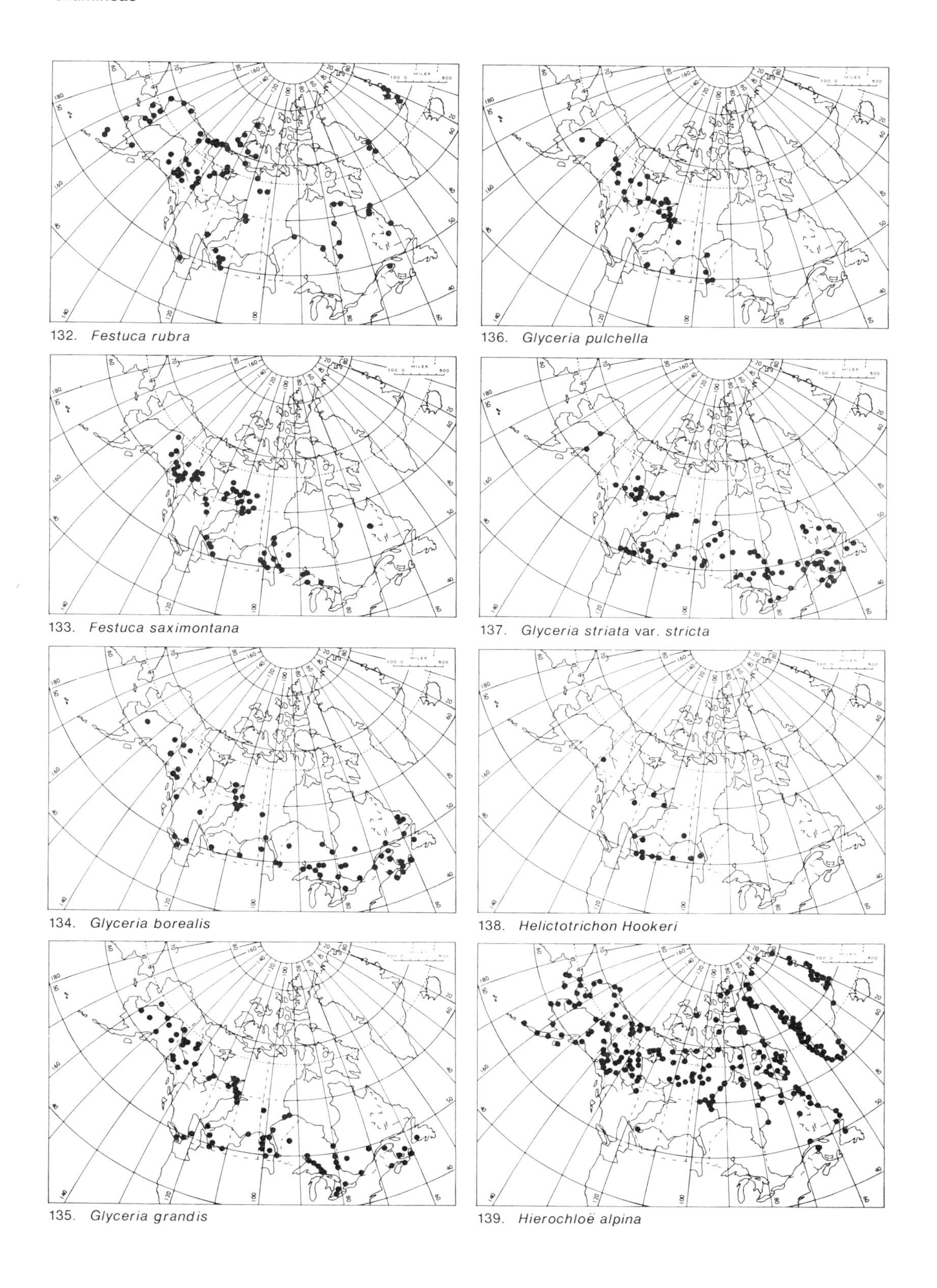

132. *Festuca rubra*

133. *Festuca saximontana*

134. *Glyceria borealis*

135. *Glyceria grandis*

136. *Glyceria pulchella*

137. *Glyceria striata* var. *stricta*

138. *Helictotrichon Hookeri*

139. *Hierochloë alpina*

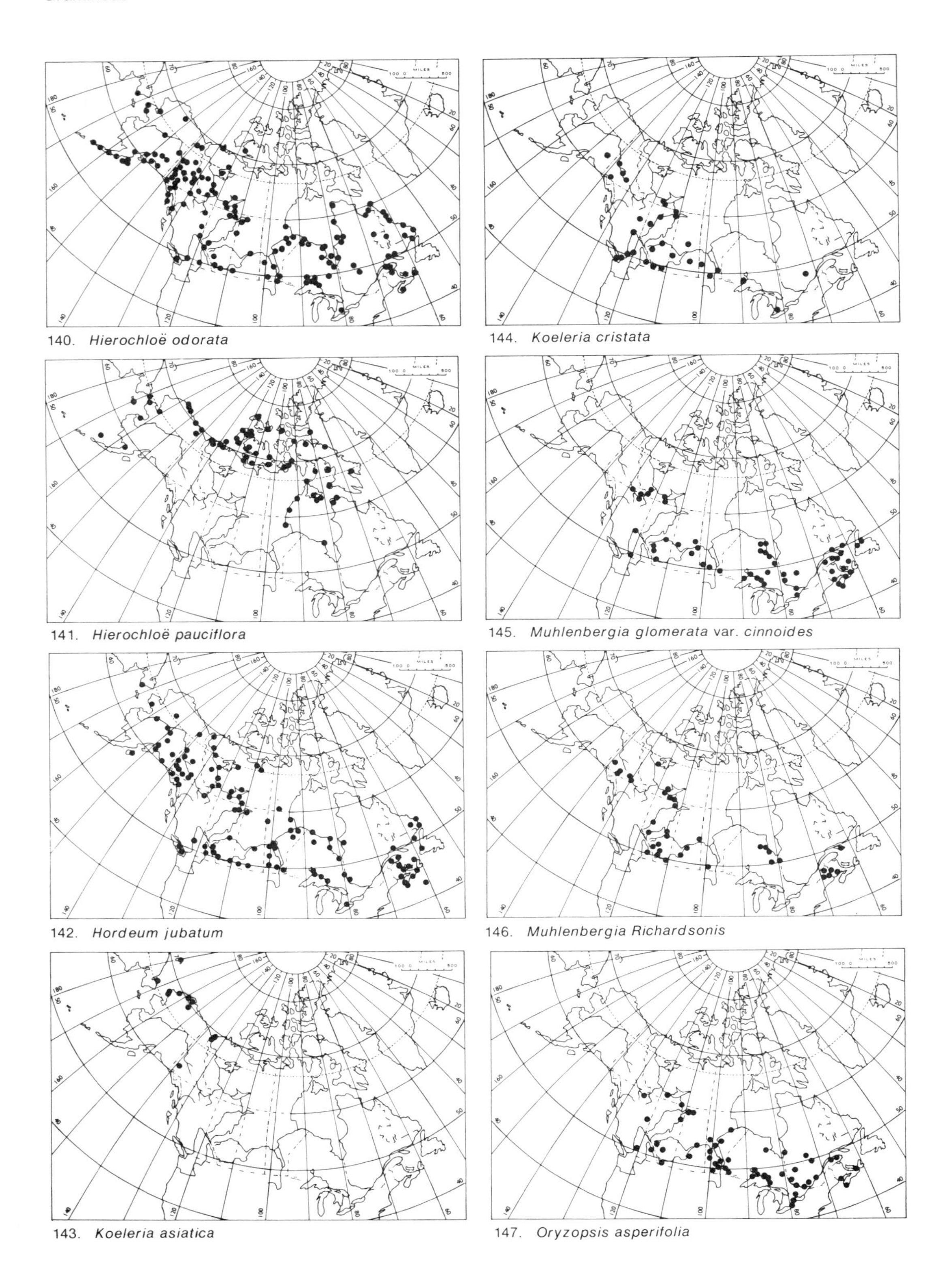

140. *Hierochloë odorata*

144. *Koeleria cristata*

141. *Hierochloë pauciflora*

145. *Muhlenbergia glomerata* var. *cinnoides*

142. *Hordeum jubatum*

146. *Muhlenbergia Richardsonis*

143. *Koeleria asiatica*

147. *Oryzopsis asperifolia*

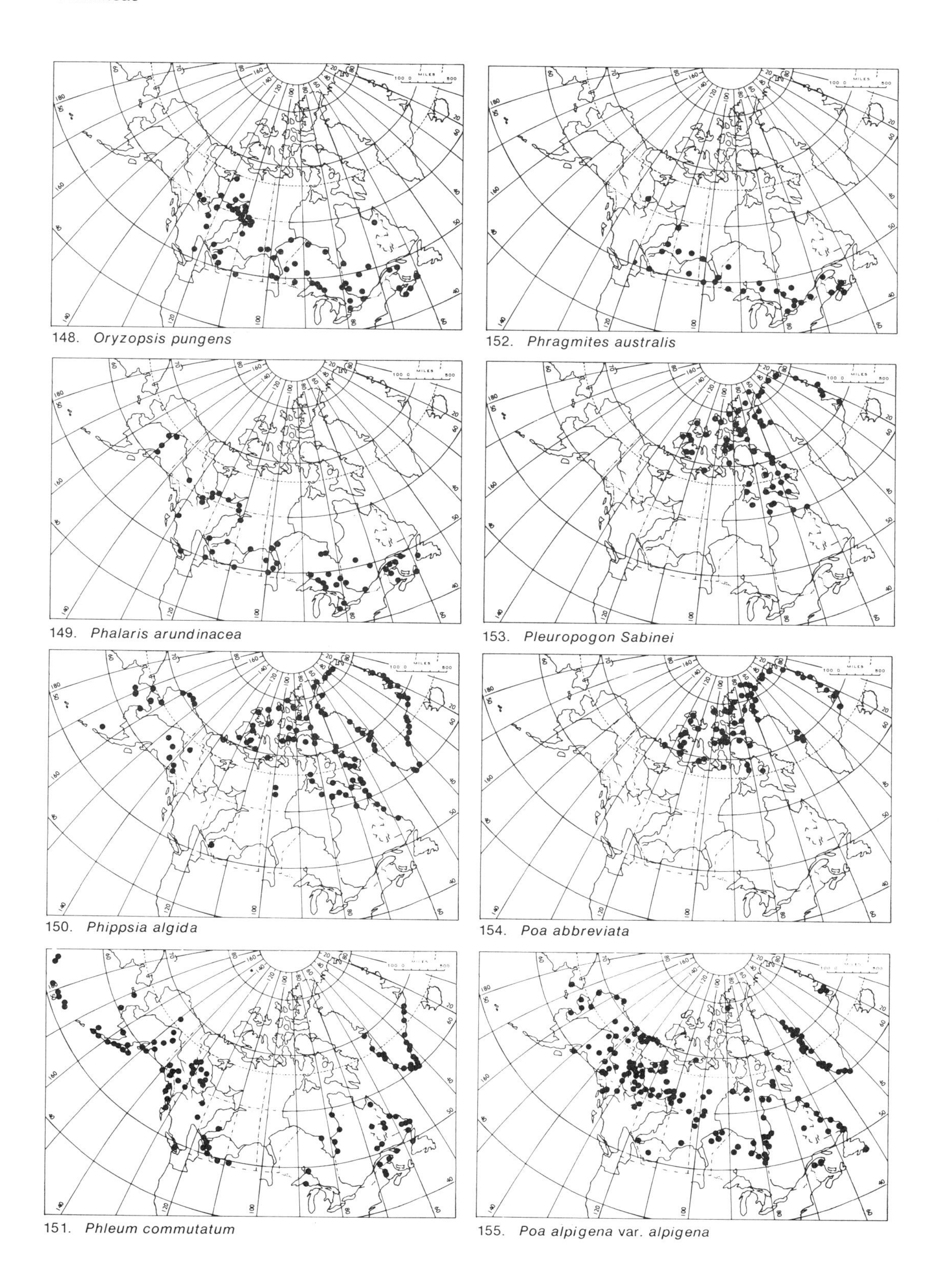

148. *Oryzopsis pungens*

149. *Phalaris arundinacea*

150. *Phippsia algida*

151. *Phleum commutatum*

152. *Phragmites australis*

153. *Pleuropogon Sabinei*

154. *Poa abbreviata*

155. *Poa alpigena* var. *alpigena*

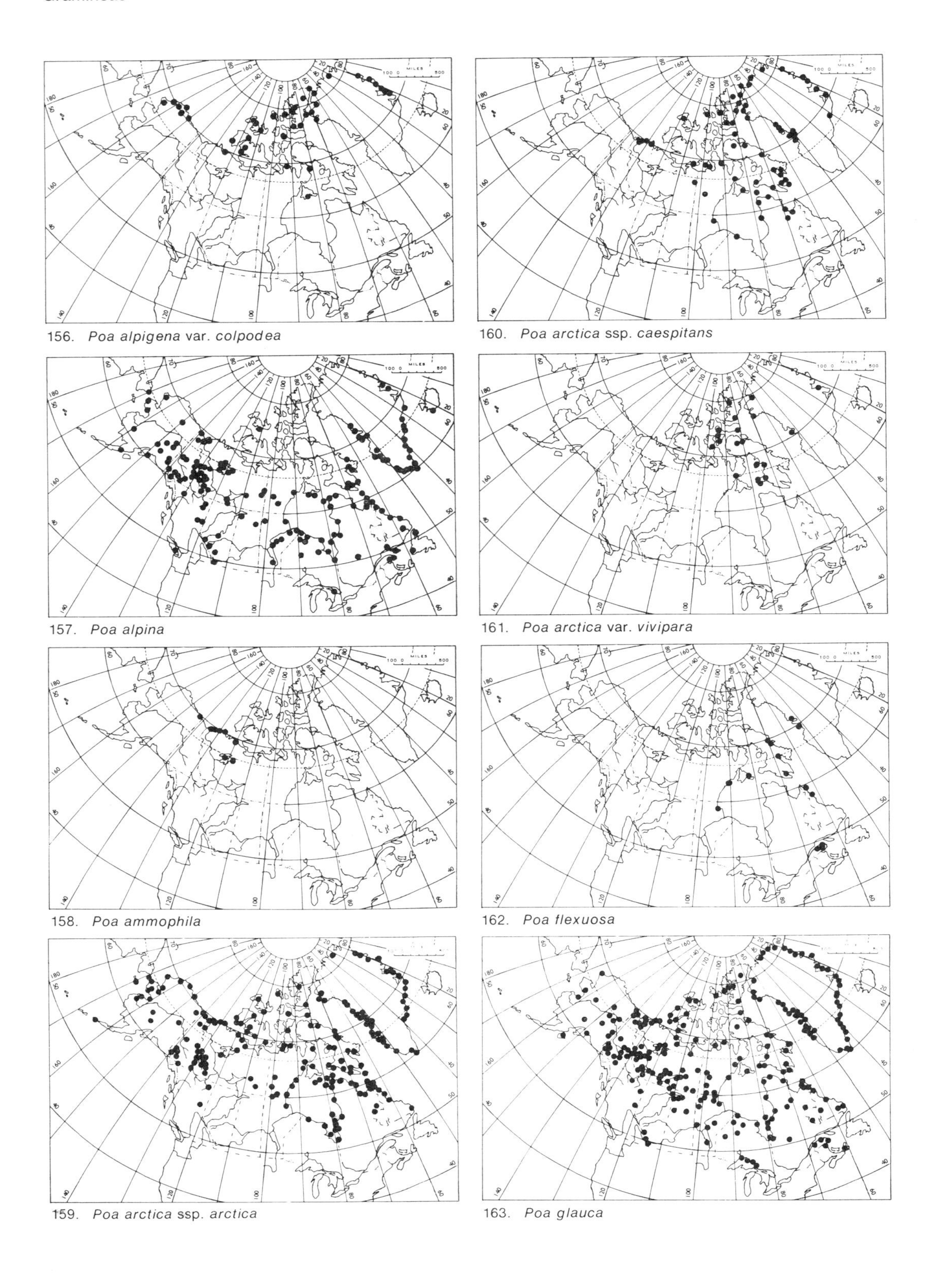

156. *Poa alpigena* var. *colpodea*

157. *Poa alpina*

158. *Poa ammophila*

159. *Poa arctica* ssp. *arctica*

160. *Poa arctica* ssp. *caespitans*

161. *Poa arctica* var. *vivipara*

162. *Poa flexuosa*

163. *Poa glauca*

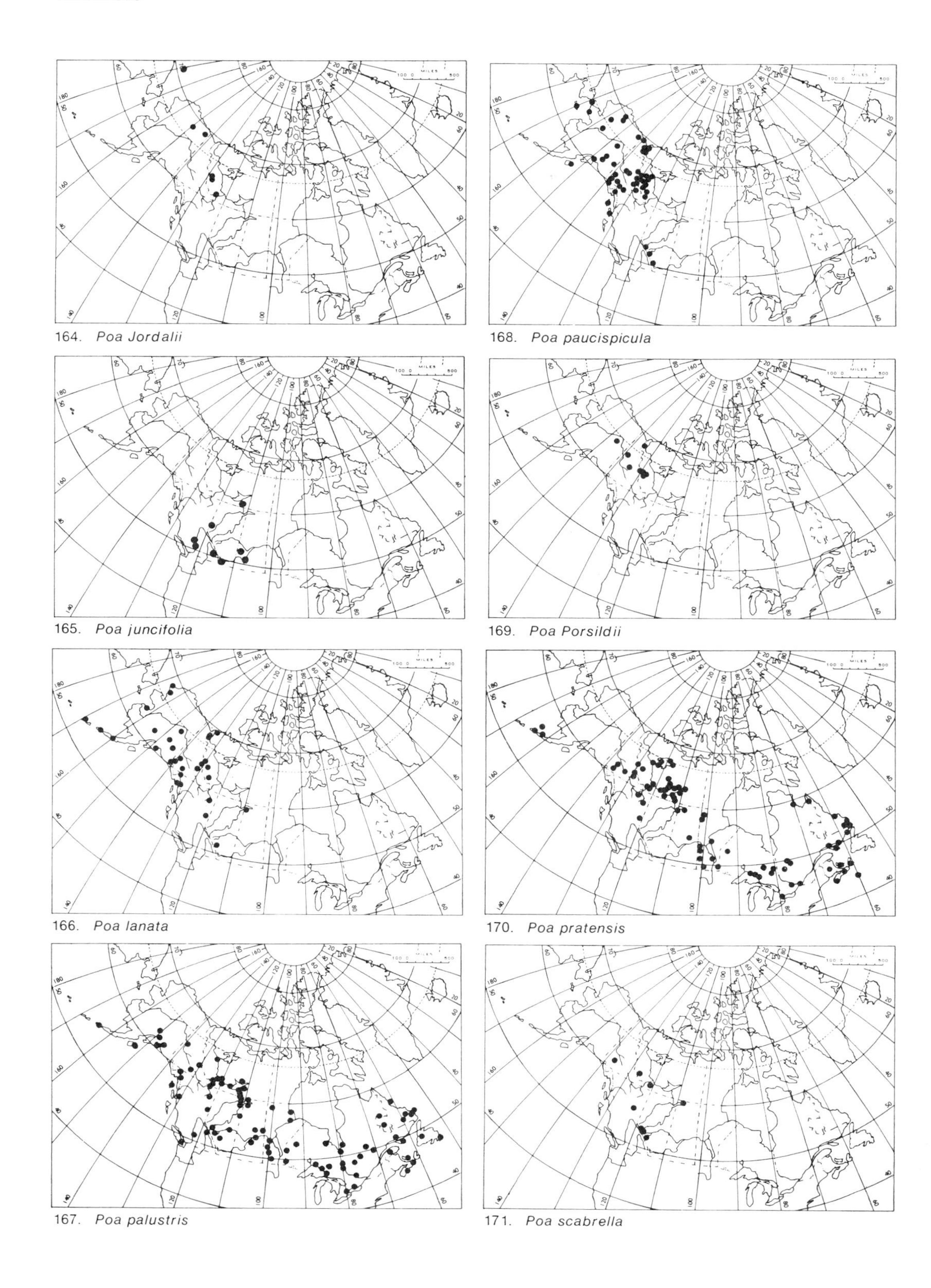

164. *Poa Jordalii*

165. *Poa juncifolia*

166. *Poa lanata*

167. *Poa palustris*

168. *Poa paucispicula*

169. *Poa Porsildii*

170. *Poa pratensis*

171. *Poa scabrella*

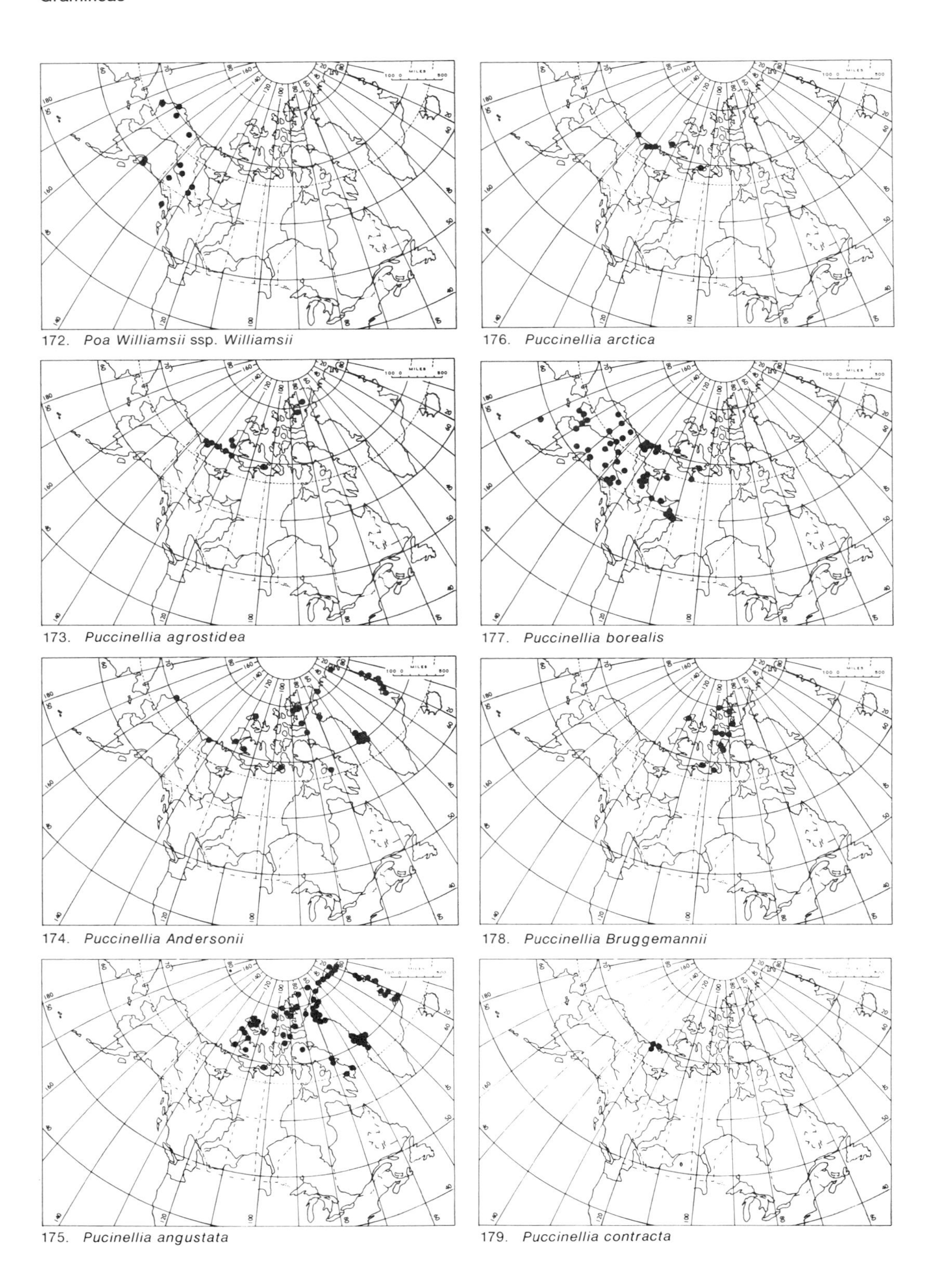

172. *Poa Williamsii* ssp. *Williamsii*

173. *Puccinellia agrostidea*

174. *Puccinellia Andersonii*

175. *Pucinellia angustata*

176. *Puccinellia arctica*

177. *Puccinellia borealis*

178. *Puccinellia Bruggemannii*

179. *Puccinellia contracta*

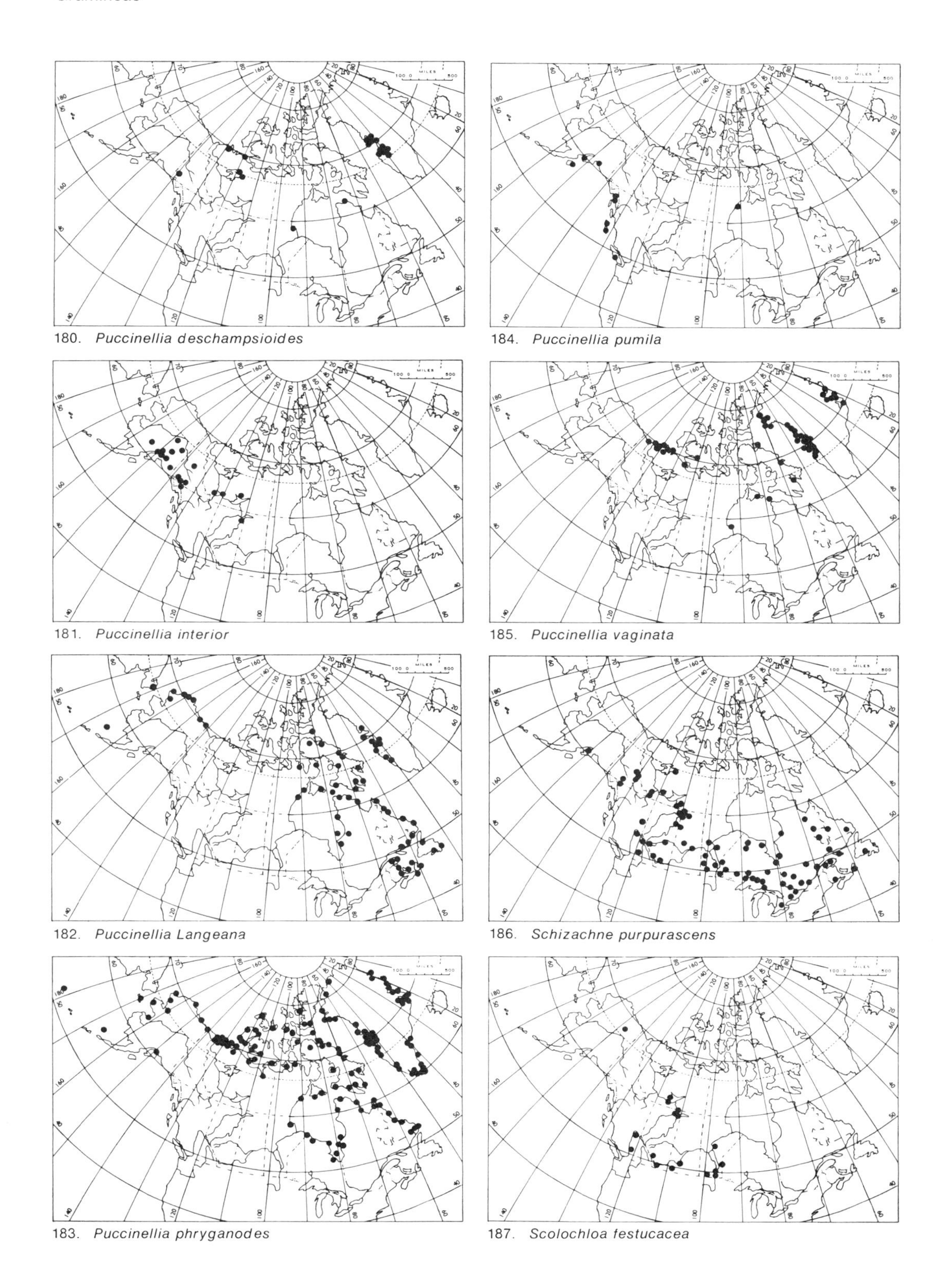

180. *Puccinellia deschampsioides*

181. *Puccinellia interior*

182. *Puccinellia Langeana*

183. *Puccinellia phryganodes*

184. *Puccinellia pumila*

185. *Puccinellia vaginata*

186. *Schizachne purpurascens*

187. *Scolochloa festucacea*

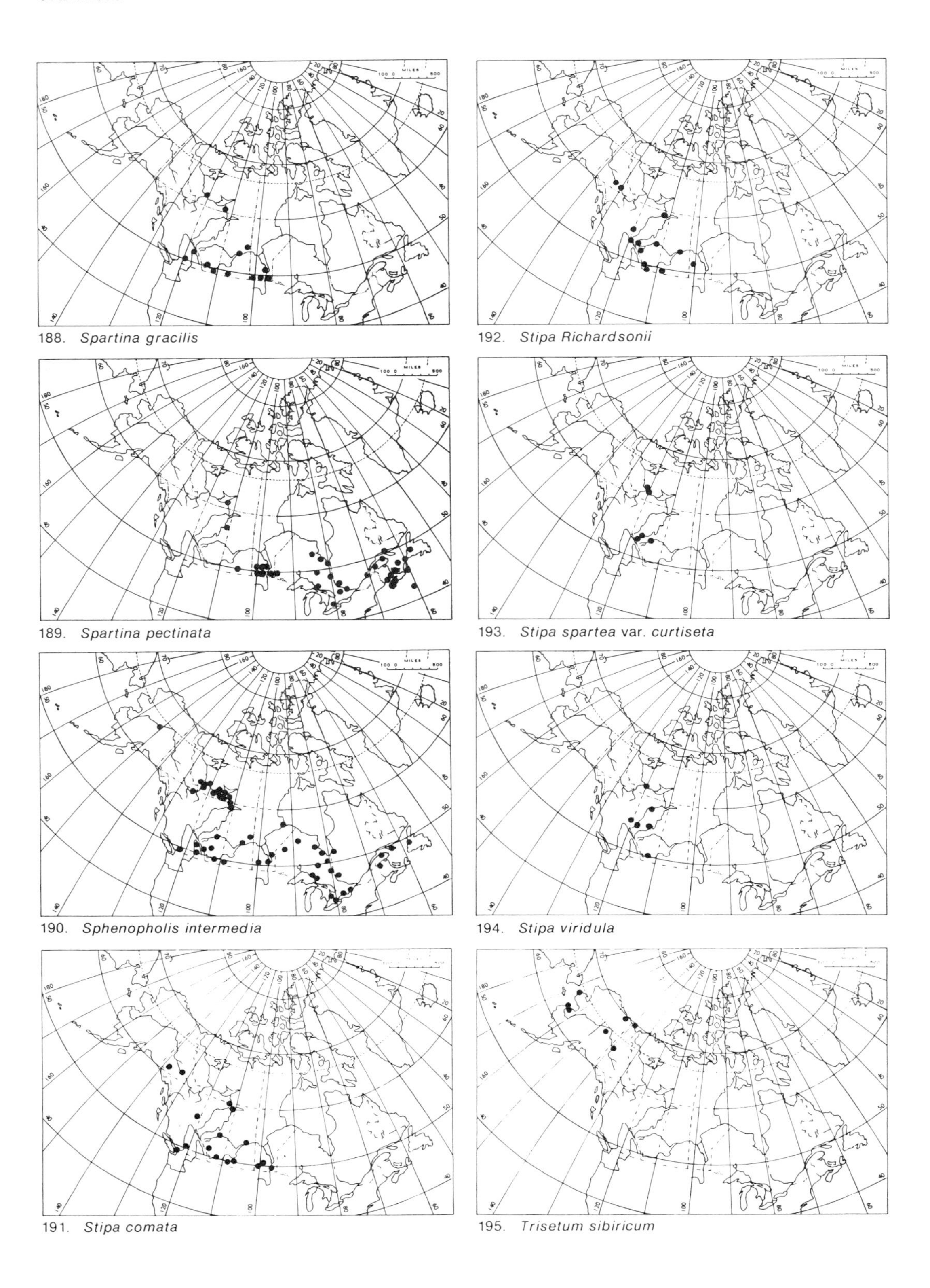

188. *Spartina gracilis*

192. *Stipa Richardsonii*

189. *Spartina pectinata*

193. *Stipa spartea* var. *curtiseta*

190. *Sphenopholis intermedia*

194. *Stipa viridula*

191. *Stipa comata*

195. *Trisetum sibiricum*

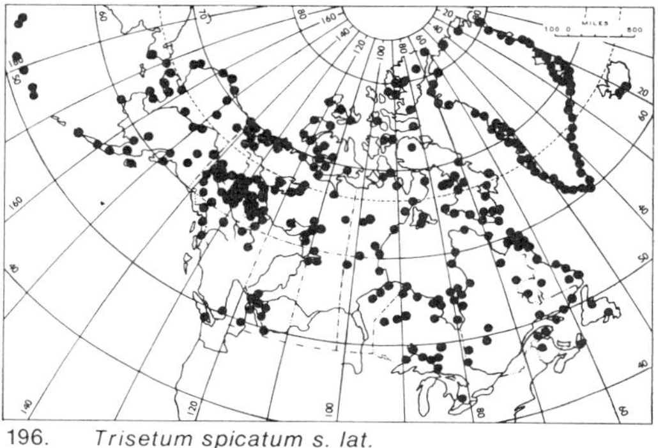

196. *Trisetum spicatum s. lat.*

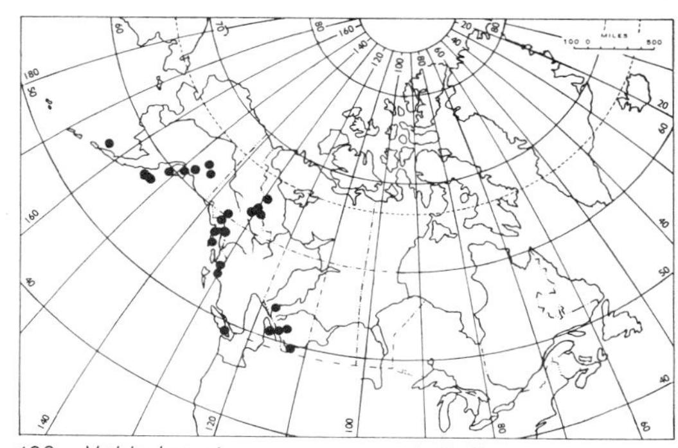

198. *Vahlodea atropurpurea* ssp. *latifolia*

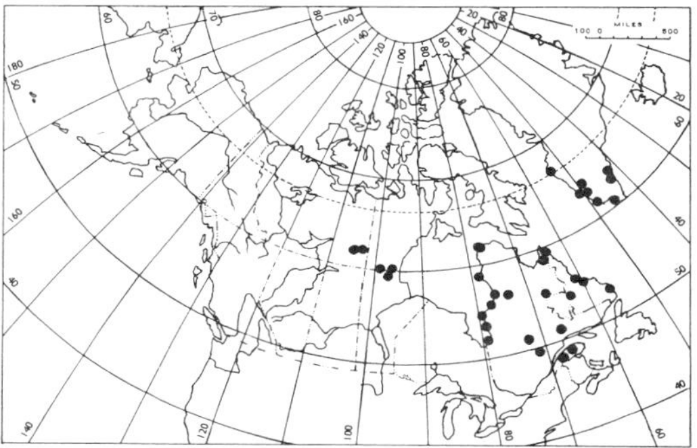

197. *Vahlodea atropurpurea* ssp. *atropurpurea*

CYPERACEAE Sedge Family

Perennial grass- or rush-like plants with solid, 3-angled or terete culms; leaves 3-ranked with closed sheaths and narrow, linear blades. Flowers perfect or unisexual, arranged in one or more spikes; perianth sack-like, composed of bristles, or lacking; stamens mostly 3, their filaments slender, anther 2-celled; ovary superior, 1-celled, stigmas 2 or 3; fruit an achene.

a. Flowers mostly perfect; perianth represented by bristles
  b. Spikes mostly many-flowered
    c. Bristles much elongated in fruit, silky . . . . *Eriophorum*
    c. Bristles short, inconspicuous or lacking
      d. Base of style enlarged, bulbiform, separated from the summit of the achene . . . . *Eleocharis*
      d. Base of style not enlarged, continuous with the achene . . . . *Scirpus*
  b. Spikes 1-2-flowered . . . . *Rhynchospora*
a. Flowers unisexual, the staminate and pistillate in the same or in different spikes
  e. Achene enclosed in a spathe which is open on one side . . . . *Kobresia*
  e. Achene completely enclosed in a bottle-shaped sack (perigynium) . . . . *Carex*

Carex

Perennial monoecious or rarely dioecious grass-like plants with 3-ranked long and narrow leaves and triangular, solid culms bearing an inflorescence composed of one or several spikes which may be wholly staminate, pistillate, or composed of both staminate and pistillate flowers; in the latter the staminate flowers may be placed either at the summit of the spike, and the pistillate below (androgynous), or the pistillate above the staminate (gynaecandrous). The position of staminate and pistillate flowers within the spike is constant for each species.

The flowers lack floral envelopes, and are borne in the axil of a scale-like bract; the staminate flower consists of three stamens, and the pistillate of a single bifid or trifid style emerging from the neck of a bottle-shaped sack (perigynium) which envelops the 3-angled or lens-shaped achene.

Mature fruits are essential for the determination of some species of *Carex*.

In the flora of Continental Northwest Territories *Carex*, represented by well over one hundred species is the largest single genus.

Key to subdivisions.

a. Spike solitary . . . . *Primocarex* p. 130
a. Spikes compound
  b. Spikes all bisexual and sessile; stigmas mostly 2 . . . . *Vignea* p. 131
  b. Spikes mostly unisexual, the terminal commonly staminate, less commonly with pistillate flowers at the base (androgynous), or with pistillate flowers at the summit (gynaecandrous); spikes pedunculate (although peduncle sometimes so short as to appear sessile)
    c. Stigmas 2 . . . . *Eucarices distigmatae* p. 132
    c. Stigmas 3 . . . . *Eucarices tristigmatae* p. 133

Primocarex

a. Pistillate scales persistent
  b. Stigmas 2
    c. Densely caespitose; spike monoecious
      d. Spike gynaecandrous (pistillate flowers at summit, staminate flowers at the base) . . . . *C. ursina*
      d. Spike androgynous (staminate at summit, pistillate at the base)
        e. Spike globose; perigynia broader than scales
          f. Spike commonly 5-6 mm long, the male part prominent . . . . *C. capitata*
          f. Spike smaller usually less than 5 mm long, the male part less prominent . . . . *C. arctogena*
        e. Spike ovate; perigynia narrower than scales . . . . *C. nardina* var. *atriceps*
    c. Stoloniferous; spike usually dioecious but not infrequently with a few staminate flowers at the summit of the pistillate spike . . . . *C. gynocrates*
  b. Stigmas 3
    g. Plants with creeping rhizomes
      h. Plant strictly dioecious; perigynia hairy . . . . *C. scirpoidea*
      h. Plant monoecious; spike staminate at summit; perigynia not hairy
        i. Perigynia beakless, membranaceous, flat . . . . *C. leptalea*
        i. Perigynia with short beak, trigonous
          j. Perigynia obovate; rhizomes brown . . . . *C. rupestris*
          j. Perigynia ovate; rhizomes thin and black . . . . *C. obtusata*
    g. Plant densely tufted; leaves filiform, convolute . . . . *C. filifolia*
a. Pistillate scales soon falling; perigynia spreading
  k. Spikes dense-flowered; perigynia long-stipitate, only the lower spreading in age . . . . *C. pyrenaica*
  k. Spike few-flowered; perigynia 4–5 mm long, spreading in age . . . . *C. microglochin*

Vignea

a. Some or all the spikes androgynous
  b. Rhizomes slender and cord-like; culms mostly solitary, not tufted
    c. Perigynia wingless or essentially so
      d. Culms from subterranean rhizomes
        e. Spikes crowded, head-like; perigynia membranaceous, inflated . . . *C. maritima*
        e. Spikes distinct in elongated head; prairie species
          f. Culms rarely more than 1 dm tall; mat-forming . . . . *C. stenophylla* ssp. *Eleocharis*
          f. Culms 2-3 dm tall, appearing serially from widely creeping rhizome . . . . *C. praegracilis*
      d. Culms from the leaf-axils of last year's stolons . . . . *C. chordorrhiza*
    c. Perigynia with distinct but narrow wing or margin; spikes often dissimilar, some androgynous, some staminate, pistillate or the sexes mixed
      g. Spikes 8-20; scales reddish-brown . . . . *C. Sartwellii*
      g. Spikes 2-8; scales yellowish . . . . *C. foenea*
  b. Rhizome short and stout, tussock forming . . . . *C. diandra*
a. Some (especially the terminal), or all spikes gynaecandrous, (except in *C. disperma*)
  h. Perigynia not wing-margined, minutely white-puncticulate
    i. Loosely caespitose, with slender rhizomes; spikes small, few-flowered
      j. Spikes androgynous; head moniliform on slender filiform culms . . . . . *C. disperma*
      j. Spikes gynaecandrous (at least the uppermost)
        k. Head subglobose of 2-4 silvery-green, loose-flowered spikes . . . . . *C. tenuiflora*
        k. Head not subglobose
          l. Spikes 3-4, in moniliform head . . . . *C. loliacea*
          l. Spikes 2-3, the uppermost long-peduncled, the lowermost subtended by a long bract . . . . *C. trisperma*
    i. Caespitose, from short and stout rhizome
      m. Perigynia beakless or with short, entire beak
        n. Spikes 2-4, more or less congested
          o. Leaves 2 mm broad, flat; culm smooth, stiff
            p. Perigynia nerveless, lanceolate, abruptly beaked . . . . *C. Lachenalii*
            p. Perigynia many-nerved, elliptic-ovate, beakless . . . . *C. amblyorhyncha*
          o. Leaves narrow, canaliculate
            q. Culms smooth, weak, ascending, somewhat flexuous . . . . *C. glareosa* var. *amphigena*
            q. Culms stiff, very scabrous above . . . . *C. Heleonastes*
        n. Spikes 4-8, the lowermost more or less remote
          r. Beak of perigynia scabrous on the margins
            s. Perigynia loosely spreading, the beak bidentate; leaves not glaucous . . . . *C. brunnescens*
            s. Perigynia appressed-ascending, the short beak entire or merely emarginate; leaves glaucous
              t. Perigynia 1.50-1.75 mm long; spikes small, subglobose; scales dark brown . . . . *C. bonanzensis*
              t. Perigynia 1.8-3.0 mm long; spikes larger; scales hyaline, yellowish with a green centre . . . . *C. canescens*
          r. Beak of perigynia smooth
            u. Scales yellowish-brown, concealing the ca. 3 mm long perigynia . . . . *C. Mackenzii*
            u. Scales yellowish-green, shorter than the ca. 1.7 mm long perigynia . . . . *C. lapponica*
      m. Perigynia with a long, bidentate beak
        v. Leaves 2-3 mm broad, soft, pale green; spikes pale, ovate, about 1 cm long . . . . *C. Deweyana*

v. Leaves 1-2 mm broad, dark green; spikes small, subglobose
w. Perigynia short-stipitate . . . . *C. laeviculmis*
w. Perigynia not stipitate, broadest at the base
x. Spikes congested into an oblong-triangular head . . . . *C. arcta*
x. Spikes distinct, in a moniliform head . . . . *C. interior*
h. Perigynia prominently wing-margined, not minutely white-puncticulate
y. Bract subtending inflorescence none, or not leaf-like, and shorter than the inflorescence
z. Spikes densely congested into a globular head
aa. Scales and perigynia brownish copper-coloured . . . . *C. macloviana*
aa. Scales and perigynia dark chestnut-brown . . . . *C. festivella*[1]
z. Spikes not densely congested
bb. Beak of perigynia flattened and serrulate to the tip
cc. Head moniliform
dd. Head flexuous nodding, spikes 8-10 mm long . . . . *C. aenea*
dd. Head stiff; spikes smaller . . . . *C. tenera*
cc. Head oblong-ovate, spikes approximate
ee. Perigynia narrowly lanceolate, 3-4 times as long as broad . . . . *C. Crawfordii*
ee. Perigynia ovate, 2 times as long as broad . . . . *C. Bebbii*
bb. Beak of perigynia terete, smooth
ff. Culms stout and stiff forming a large, firm tussock; spikes 3-4
gg. Perigynia 6-8 mm long, scales light reddish-brown . . . . . *C. petasata*
gg. Perigynia 4-6 mm long, scales brownish-black . . . . *C. phaeocephala*
ff. Culms slender, forming a small tussock, from a short, black rhizome; spikes 4-7 in a slender, flexuous moniliform head . . . . *C. praticola*
y. Bracts subtending inflorescence many times exceeding the globular-ovate head
hh. Bracts leaf-like, green . . . . *C. sychnocephala*
hh. Bracts brown, not leaf-like . . . . *C. athrostachya*

Eucarices distigmatae

a. Perigynia membranaceous, more or less inflated, shiny, obscurely nerved dorsally; beak distinct, bidentate; marsh species with stout, short rhizomes; stigmas 2 or rarely 3
b. Perigynia with distinct beak; culms 3-5 dm tall; staminate spikes commonly 2 (Western species) . . . . *C. physocarpa*
b. Perigynia with short beak; culms lower; staminate spike usually solitary (Eastern species) . . . . *C. saxatilis* var. *rhomalea*
a. Perigynia neither membranaceous, inflated or shiny; nerveless or faintly nerved
c. Culms slender; leaves soft and flat, commonly equalling the culms; terminal spike gynaecandrous (except in *C. aurea*), the lower pistillate; perigynia falling when ripe
d. Caespitose; culms more or less decumbent
e. Lowermost bract short, barely equalling inflorescence, with broad basal auricle; culms 5-15 cm tall; spikes usually 3, ovate, 6-8 mm long . . . . . . *C. bicolor*
e. Lowermost bract foliaceous, equalling or longer than inflorescence; culms 2-3 dm long; spikes 3-5, cylindrical 1-2 cm long . . . . *C. eleusinoides*
d. Loosely caespitose or matted
f. Terminal spike gynaecandrous
g. Dwarf species with culms less than 10 cm long rarely overtopping the involute leaves; perigynia blotched with black . . . . *C. rufina*
g. Culms slender 2.5-4.0 dm tall, much longer than the flat leaves; perigynia greenish-white . . . . *C. Garberi*
f. Terminal spike mostly staminate; perigynia spongy, orange . . . . *C. aurea*
c. Culms stiff; terminal spike staminate, the lateral pistillate, short-pedunculate or sub-sessile, erect; perigynia persistent

- g. Lowermost bract shorter than inflorescence; pistillate spikes rarely over 2 cm long
  - h. Densely caespitose, tussock-forming; culms 2-5 dm tall, slender; leaves about 2 mm wide, somewhat revolute; stigmas 2, rarely 3; lowermost bract setaceous . . . . *C. lugens*
  - h. With stolons; lowermost bract foliaceous
    - i. Loosely caespitose; culms 2-3 dm tall, with reddish-brown basal sheaths; head dense and pistillate spikes short-peduncled (Western species) . . . . *C. consimilis*
    - i. Stoloniferous; culms 1-3 dm long, with dark brown basal sheaths (Eastern species)
      - j. Pistillate spikes short-peduncled . . . . *C. Bigelowii*
      - j. Pistillate spikes with slender peduncles as long or longer than inflorescence . . . . *C. anguillata*[1]
- g. Lowermost bract foliaceous, equalling or longer than inflorescence; pistillate spikes many-flowered, 2-6 cm long
  - k. Culms phyllopodic
    - l. Caespitose; perigynia finely nerved, falling when ripe . . . . *C. lenticularis*
    - l. Stoloniferous; perigynia nerveless, tardily falling
      - m. Tall, non-littoral species
        - n. Culms 4.5-6.0 dm tall, leaves green with involute margins; staminate spikes cylindrical, 1-4, pistillate spikes 2-6 . . . . *C. aquatilis*
        - n. Culms shorter; leaves flat, yellowish-green, shorter than culms; staminate spike clavate, solitary, pistillate spikes 2-4 . . . . *C. aquatilis* var. *stans*
      - m. Low to medium littoral species
        - o. Culms 2-10 cm tall, curved or ascending, turf-forming; pistillate spikes few-flowered . . . . *C. subspathacea*
        - o. Culms 1.0-1.5 dm tall, erect; pistillate spikes many-flowered (Amphi-Beringian species) . . . . *C. Ramenskii*
  - k. Culms aphyllopodic; pistillate spikes elongate, pedunculate, erect or nodding (Pacific Coast species) . . . . *C. sitchensis*

Eucarices tristigmatae

- a. Terminal spike staminate (rarely with one or two pistillate flowers at the base)
  - b. Inflorescence capitate, the lateral spikes short-peduncled, erect or spreading but never drooping.
    - c. Staminate spike inconspicuous, more or less concealed between the pistillate spikes
      - d. Culms 15-60 cm tall
        - e. Perigynia hairy . . . . *C. Peckii*
        - e. Perigynia glabrous
          - f. Pistillate spikes mostly 2, cylindrical, commonly opposite, perigynia concealed by the black scales; leaves shorter than the 15-20 cm tall culms . . . . *C. holostoma*
          - f. Pistillate spikes mostly 3, ovoid; leaves flat 2-3 mm wide, nearly equalling the culm
            - g. Perigynia pale, greenish-white . . . . *C. media*
            - g. Perigynia pale brown . . . . *C. norvegica*
      - d. Culms 5-15 cm long, weak, more or less deflexed; perigynia hairy

---

[1]*See under C. Bigelowii*

[2]*In some members of this group the number of stigmas may be variable; thus whereas C. petricosa and some other species normally have three stigmas, spikes may occasionally be found in which two stigmas predominate.*

h. Basal sheaths reddish-brown shiny . . . . *C. deflexa*
h. Basal sheaths dull, pale brown . . . . *C. concinna*
c. Staminate spike prominent; dwarf species with culm 5-15 cm tall
i. Stoloniferous, with reddish-brown, cord-like stolons; base of leaves reddish; scales fulvous; perigynia coriaceous, chestnut-brown, shiny, persistent . . . . *C. supina* ssp. *spaniocarpa*
i. Tufted
j. Leaf bases drab or pale brown; scales purplish-brown, hyaline-margined, shorter than the caducous perigynia . . . . *C. glacialis*
j. Leaves and culm yellowish-green; scales pale brown, membranaceous, shorter than the persistent strongly nerved, green perigynia . . . . *C. viridula*
b. Inflorescence not capitate
k. Pistillate spikes remote, sessile or nearly so
l. Beak of perigynia entire or at most emarginate; terminal spike staminate, pistillate spikes mostly 2
m. Perignynia not crowded, pubescent; scales with broad, silvery-hyaline margins; lowermost bract short and scale-like . . . . *C. Richardsonii*
m. Perigynia densely crowded, inflated, shiny; lowermost bract foliaceous but shorter than inflorescence
n. Pistillate spikes oblong-cylindrical; scales and perigynia almost black; leaves 4-5 mm wide, flat . . . . *C. membranacea*
n. Pistillate spikes globular-ovoid; perigynia and scales chestnut-brown; leaves narrow canaliculate . . . . *C. rotundata*
l. Beak of perigynia deeply cleft or bidentate; staminate spikes mostly more than one; pistillate spikes 3 or more; lowermost bract foliaceous, equalling or exceeding inflorescence
o. Pistillate spikes cylindric to ± clavate, 3-5 cm long; perigynia prominently beaked; leaves and sheaths septate-nodulose
p. Pistillate spikes congested; perigynia densley crowded, retrorse . . . . *C. retrorsa*
p. Pistillate spikes remote; perigynia ascending
q. Sheaths and lower parts of leaves pubescent . . . . *C. atherodes*
q. Sheaths and blades glabrous; strongly septate . . . . *C. rostrata*
o. Pistillate spikes short-cylindric-ovate, 1.5-2.0 cm long; leaves narrow, canaliculate, not septate-nodulose
r. Perigynia coriaceous, strongly nerved, shiny . . . . *C. oligosperma*
r. Perigynia hairy . . . . *C. lasiocarpa* var. *americana*
k. Pistillate spikes on slender peduncles as long or longer than the spikes, these erect spreading or at least the lowermost nodding or drooping; if short-peduncled or erect, inflorescences not head-like or moniliform
s. Plants tall, with flat leaves 3-4 mm broad commonly 2-6 dm high
t. Culms aphyllopodic (i.e. with few much reduced leaves at the base)
u. Tip of pistillate scale awn-like, as long as body . . . . *C. macrochaeta*
u. Tip of pistillate scale merely pointed . . . . *C. podocarpa*
t. Culms enclosed in cluster of normal leaves
v. Pistillate spikes ovate; plant tufted from ascending rhizome . . . . *C. atrofusca*
v. Pistillate spikes linear, at least twice as long as broad; loosely tufted with short horizontal rhizome . . . . *C. microchaeta*
s. Plants low (except *C. capillaris* ssp. *chlorostachys*) with narrow leaves about 2 mm wide
w. Plant tussock-forming
x. Spikes arising near summit of culm (except in *C. capillaris* ssp. *chlorostachys* in which lowermost pistillate spike is distant); perigynia glabrous, lanceolate, beakless or merely tapering into short entire beak
y. Pistillate spikes 2-3, ovate, dark brown or black; perigynia flat with scabrous margins; leaves flat 3-5 mm wide, shorter than culms . . . . *C. atrofusca*

y. Pistillate spikes 2-6, chestnut-brown, on capillary peduncles; perigynia fusiform, smooth
  z. Culms 2-20 cm tall; pistillate spikes all from summit of culm
    aa. Basal leaves short, flat, yellowish-green; perigynia distinctly beaked; lateral spikes drooping .... *C. capillaris*
    aa. Basal leaves canaliculate, less than 0.75 mm wide, their margins scabrous; perigynia beakless; lateral spikes erect .... *C. Williamsii*
  z. Culms 15-60 cm tall; lateral spikes loose-flowered, the lowermost distant and long-peduncled; basal leaves long, fresh green .... *C. capillaris* ssp. *chlorostachys*

x. At least some of the pistillate spikes arising from the base of the culm; perigynia fusiform, thinly pubescent and with a conical, bidentate beak
  bb. Culm strict, equalling or longer than the pale green stiff leaves .... *C. Rossii*
  bb. Culm short or none; spikes all axillary and largely concealed among the short, stiff, persistent leaves .... *C. tonsa*

w. Plant stoloniferous
  cc. Lateral spikes erect
    dd. Leaves pale green, filiform, 0.2-1.0 mm wide, canaliculate-involute, in fascicles; culms 10-20 cm tall, filiform; pistillate spikes few-flowered, minute, often exceeding the tiny staminate spike; perigynia chestnut-brown, shiny .... *C. eburnea*
    dd. Leaves 3 mm wide or wider, flat or canaliculate
      ee. Leaves glaucous, canaliculate; perigynia beakless .... *C. livida* var. *Grayana*
      ee. Leaves green, short and flat, perigynia prominently beaked .... *C. vaginata*
  cc. Lateral spikes, or at least the lowermost drooping
    ff. Perigynia more or less membranaceous, elliptic, flat, 4.5-5.0 mm long with scabrous beak and margins; leaves canaliculate; plants loosely tufted with long stolons
      gg. Culms slender, 40-90 cm tall; spikes 4-6; leaves with long attenutate curly tips, almost as long as culm .... *C. Franklinii*
      gg. Culms slender, 30-40 cm tall; spikes 3-5, leaves somewhat curly, shorter than culms .... *C. petricosa*
    ff. Perigynia leathery, smooth, about 3 mm long, not thin or flat, with smooth margins
      hh. Perigynia plano-convex, pale greenish-grey
        ii. Lowermost bract short-sheathing
          jj. Culm stiff, obtuse-angled; smooth; leaves flat .... *C. rariflora*
          jj. Culms slender, sharp-angled, scabrous above
            kk. Stolons long; scales broad with a short point .... *C. limosa*
            kk. Loosely tufted; scales narrow with long slender point, soon falling .... *C. paupercula*
        ii. Lowermost bract long-sheathing; scales blunt; loosely tufted .... *C. laxa*
      hh. Perigynia plump, terete, about 4 mm long, dark brown, shiny and densely crowded on the rachis
        ll. Perigynia 4 mm long, distinctly beaked; western species .... *C. physocarpa*[1]
        ll. Perigynia 3 mm long, with short beak; eastern species .... *C. saxatilis var. rhomalea*[1]

---

[1] *See also under Eucarices distigmatae*

a. Terminal spike gynaecandrous (or in *C. Morrisseyi* commonly staminate or androgynous)
  mm. Spikes sessile or with very short, stiff peduncles
    nn. Culms caespitose; inflorescence more or less capitate
      oo. Scales purplish-black with conspicuous pale hyaline margins; culms stiff . . . . . . *C. albo-nigra*
      oo. Scales uniformly purplish-black; culms slender, more or less nodding at summit . . . . . *C. atrosquama*
    nn. Stoloniferous; inflorescence not capitate, the lower spikes more or less remote
      pp. Terminal spike oblong-cylindric, twice as long as the lateral spikes . . . . . *C. albo-nigra*
      pp. Terminal spike clavate, similar to lateral spikes
        qq. Pistillate scales awned . . . . . *C. Buxbaumii*
        qq. Pistillate scales merely acute . . . . . *C. Morrisseyi*
  mm. Spikes on capillary peduncles, the lateral more or less nodding
    rr. Densely caespitose; culms slender, 10-25 cm tall; leaves short and flat
      ss. Perigynia flat, black-tipped with serrulate margins; spikes dark brown or black; long narrow tips of basal leaves characteristically curled . . . . . *C. misandra*
      ss. Perigynia fusiform, light brown . . . . . *C. capillaris* ssp. *robustior*
    rr. Loosely tufted; culms 40-60 cm tall; leaves long flat, fresh green; perigynia ovoid, smooth . . . . . *C. atratiformis* ssp. *Raymondii*

Carex aenea Fern.
Densely caespitose from a short, ascending rhizome. Culms up to 1 m long or more, smooth and somewhat flexous near the tip, much longer than the soft, flat but rather thick leaves. Spikes 5 to 6, oblong, 6 to 8 mm long, pale green turning light bronze, in an open moniliform 5 to 6 cm long, slightly nodding head. Perigynia firm, 4 to 5 mm long, nerveless, gradually tapered into a winged, serrulate beak.
Sandy woodland meadows and clearings.
General distribution: N. America: Nfld. to eastern Alaska.
Fig. 177 Map 199

Carex albo-nigra Mack.
Loosely tufted from a stout, ascending rhizome; culms 2 to 4 dm tall, naked, very stiff, smooth and obtuse-angled, much longer than the short, 3 to 4 mm broad and smooth basal leaves; spikes 3, in a 2 to 3 cm long head subtended by a short foliaceous bract, the terminal staminate at the base, the lateral and pistillate short-peduncled; pistillate scales dark brown with a narrow and paler margin; perigynia obovate, dark yellowish-brown, abruptly beaked, the beak short and entire; stigmas 3.
Turfy places in dry, alpine tundra, north to Great Bear Lake and the Mackenzie Delta area.
General distribution: Cordilleran.
Fig. 178 Map 200

Carex amblyorhyncha Krecz.
Loosely tufted; culms slender but wiry, 15 to 30 cm tall and longer than the narrow and flat leaves. In general habit similar to *C. Lachenalii* but at once distinguished by its elliptic-ovate, yellowish-brown, densely punctate, many-nerved, conical, and very short-beaked, perigynia. Stigmas 2.
A truly arctic species growing in mossy places in wet bogs.
General distribution: Circumpolar.
Fig. 179 Map 201

Carex aquatilis Wahlenb.
var. **aquatilis**
Coarse marsh species forming small dense tufts, each composed of half a dozen sterile, leafy shoots around one or several fertile culms, from the base of which issue stout, scaly yellowish or brown horizontal stolons. The culms smooth, obtuse-angled, up to 1 m tall, very leafy, their terminal spikes barely overtopping the sterile, flat or channelled, 2 to 5 mm broad glaucous leaves; their sheaths and marcescent leaf-bases purplish-black. Staminate spikes terminal, 1 or 2, the upper usually 2 cm long; pistillate spikes 2 to 5, cylindric, erect, 2 to 5 cm long, each subtended by a foliaceous bract that overtops the inflorescence; pistillate scales narrowly ovate, black, with a pale midvein, shorter than the elliptic, obovate 2.5 to 3.0 mm long, pale green, nerveless perigynia; beak very short and entire; stigmas 2.
In shallow water by margins of ponds and sloughs, sheltered river flats etc.; north along

the Mackenzie River to its delta, forming extensive and often nearly pure stands.

General distribution: Circumpolar, non-arctic.

Fig. 180 Map 202

Carex aquatilis Wahlenb.
var. **stans** (Drej.) Boott
Similar but smaller, with fertile culms usually 2 to 4 dm long, the leaves commonly with strongly revolute margins, yellowish-green rather than glaucous. Sometimes confused with tall specimens of *C. Bigelowii* from which it is distinguished by its longer leafy bracts of the inflorescence and by its nodulose-septate leaf-veins clearly observed through a lens.

The var. *stans* is ubiquitous on rather wet arctic tundra and by shallow ponds, mainly beyond the limit of trees or above timberline. Although by some authors considered a minor variation of *C. aquatilis* the distinct arctic range of var. *stans* supports the view that it is, at least, a well-defined geographic race. This is well illustrated in Greenland where typical *C. aquatilis* does not occur, and where the var. *stans* is wide ranging in the northern half of that island, but totally lacking south of lat. 70° on the east coast and lat. 68° on the less arctic west coast.

General distribution: Circumpolar, arctic.

Fig. 181 Map 203

Carex arcta Boott
Densely tufted marsh species with slender and soft culms commonly 40 to 60 cm tall, equalling or slightly longer than the glaucous, flat, 2 to 3 mm wide leaves. Spikes usually 10 to 12, in a dense, oblong and pale green head; perigynia oblong-ovate with a distinct beak, faintly nerved dorsally, and slightly longer than its narrow and pale brown scale.

Wet woodland bogs and marshes where it often forms large tussocks. In the Northwest Territories known from a single collection in lat. 61°01′N., long. 109°15′W.

General distribution: Boreal N. America.

Fig. 182 Map 204

***Carex arctogena** H. Smith
Densely caespitose with persistent, firm, greyish-brown sheaths. Culms 10 to 20 cm long, longer than the narrow, trigonous channelled leaves. Spike ovate-globose, 6 to 8 mm long and 4 to 5 mm wide, the staminate part conspicuous, forming a small projecting point at the summit. Perigynia globose averaging 1.9 mm in length and 1.5 mm in width, with a short beak. Stigmas 2.

Moist, peaty soil. An arctic-alpine species which is closely related to the more widely distributed *C. capitata* Sol., from which it differs by its consistently smaller and darker brown spike and smaller perigynia.

Thus far, *C. arctogena* has not been reported from the Districts of Keewatin and Mackenzie; it is, however, known from North West Manitoba and should be looked for on the Precambrian Shield east of Great Slave Lake.

General distribution: Amphi-Atlantic, low arctic.

Fig. 183 Map 205

Carex atherodes Spreng.
Loosely tufted from a long scaly, stolon-like rhizome; culms very leafy, up to 1 m tall, their bases covered by dark-purplish leaf sheaths, blades 4 to 10 mm wide, septate-nodulose, their underside, and the upper sheaths hairy; inflorescence 1.5 to 2.0 dm long, overtopped by a foliaceous bract; terminal spikes staminate, 1 or 2; pistillate spikes 2 to 3, commonly 5 to 7 cm long, sessile or nearly so; perigynia ascending, stongly ribbed, lanceolate, 8 to 10 mm long, including the slender beak; the latter with prominent spreading teeth as long, or longer, than the beak; stigmas 3.

Wet calcareous lake shore meadows; local in southernmost District of Mackenzie.

General distribution: Circumpolar, non-arctic.

Fig. 184 Map 206

***Carex athrostachya** Olney
Tufted from a short rhizome; culms up to 7 dm tall with smooth and rounded angles, overtopping the flat, 5 mm broad leaves. Spikes numerous, densely aggregated in an ovoid, 2 cm long light brown head, subtended by 2 to 3 leafy bracts. Perigynia lanceolate, 3.5 mm long, pale, smooth with a narrow, scabrous margin.

In wet meadows; to be looked for in southwestern District of Mackenzie.

General distribution: Cordilleran-Pacific reaching Yukon T. and S. E. Alaska.

Map 207

Carex atratiformis Britt.
ssp. **Raymondii** (Calder) Porsild
Loosely tufted from a stout, fibrillose, horizontal rhizome; culms slender, smooth, obtuse-angled, 5 to 7 dm high, leafy in the lower part, and longer than the soft and glaucous 3 to 4 mm broad, flat leaves; inflorescence of 4 to 6, slender, short-cylindric, pedunculate spikes, the lowermost subtended by a foliaceous bract

overtopping the inflorescence; terminal spike gynaecandrous, the lateral pistillate and somewhat drooping, on short, slender peduncles; pistillate scales narrowly lanceolate, black; perigynia golden-yellow, ovate-oblong, tapering into a bidentate, dark brown beak; stigmas 3.

Open coniferous woods and meadows.

General distribution: A western subspecies closely related to the eastern N. American *C. atratiformis s. str.* but in most cases easily distinguished by its two-coloured, reddish-brown spikes.

Fig. 185 Map 208

Carex atrofusca Schk.
Loosely caespitose, with erect, slender, smooth culms, 15 to 30 cm high, and flat revolute-margined, glaucous leaves; spikes 3 to 4, the terminal staminate, erect, club-shaped, the pistillate ovoid, 1 to 2 cm long, drooping, on capillary peduncles subtended by sheathing, bristle-pointed bracts. Pistillate scales narrowly lanceolate, purplish black, as long as the purplish-black ovate-lanceolate, nerveless, beaked perigynia; stigmas 3.

Wet places in tundra.

General distribution: Circumpolar, high-arctic.

Fig. 186 Map 209

Carex atrosquama Mack.
Densely tufted with a short, horizontal rhizome; culms 3 to 5 dm tall, slender, sharp-angled, and rough above; basal leaves 2.0 to 2.5 mm wide, flat or with revolute margins, their sheaths dark reddish-purple; spikes commonly 3, oblong-elliptic and short-peduncled, aggregated into an oval, 2.5 to 3.0 cm long head, subtended by a short, herbaceous bract, commonly drooping in age; terminal spike gynaecandrous; pistillate scales broadly lanceolate, ebony black, shorter than the yellowish-green or purplish spotted and finely papillate, elliptic-ovoid perigynia, the latter abruptly contracted into a short minutely bidentate beak; stigmas 3.

Calciphilous species of well drained alpine herbmat slopes.

General distribution: Cordilleran.

Fig. 187 Map 210

Carex aurea Nutt.
Resembling *C. Garberi* but of lower stature and more densely tufted; the bracts subtending the inflorescence much longer, and the pistillate spikes more lax-flowered; the terminal spike commonly staminate throughout, the pistillate scales orbicular and short-mucronate, and the perigynia almost globular, beautifully golden-yellow in life, drying brown, falling when ripe and thus readily dispersed by water standing between tussocks of the wet meadows that are its natural habitat.

General distribution: Non-arctic, from Nfld. to S. W. Alaska, north to near the limit of trees.

Fig. 188 Map 211

Carex Bebbii Olney
Densely tufted from a very short ascending rhizome; culms smooth and very slender up to 1 m long, overtopping the flat, soft leaves; spikes 5 to 6, ovoid, in a compact, yellowish-brown, oblong and nearly 2 cm long head; perigynia 3 mm long, narrowly lanceolate.

Wet meadows in southern District of Mackenzie.

General distribution: Boreal N. America from Nfld. to S. E. Alaska.

Fig. 189 Map 212

Carex bicolor All.
Loosely caespitose. The culms weak and slender, often recurved or spreading, 5 to 15 cm long, slightly longer than the flat, pale glaucous leaves. Spikes usually 3, oval or short cylindrical, the terminal staminate at the base, the lower short-peduncled; the bract scarcely foliaceous, equalling or slightly longer than the inflorescence. Pistillate scales reddish-brown, with a pale midvein, shorter than the elliptic, white or pale green, nerveless or very finely nerved, papillose and almost beakless perigynia. Stigmas 3.

In wet sand by brooks and lake shores.

General distribution: Circumpolar, low-arctic.

Fig. 190 Map 213

Carex Bigelowii Torr.
Coarse, stoloniferous with stout, scaly stolons terminating in leafy shoots. Culms stiff, smooth, and sharply trigonous, 10 to 25 cm high, slightly longer than the flat, revolute-margined, strongly ribbed but not septate, glaucous and usually somewhat curved leaves. Terminal spike staminate, linear, the lower 2, or less commonly 3, pistillate, occasionally with a few staminate flowers at the summit, short cylindrical, erect, sessile or short-peduncled; pistillate scales short, obtuse, purplish-black with a pale margin and greenish midvein, equalling or shorter than the broadly elliptic-lanceolate, greenish or pale brown blotched, smooth, nerveless short-beaked perigynia. Stigmas 2.

A very variable species. The typical plant normally is found in rather dry, turfy places; in wet places it becomes taller, occasionally with longer and more lax-flowered pistillate spikes. Extreme or possibly distinct is *C. anguillata* Drej. [*C. Bigelowii* f. *anguillata* (Drej.) Fern.], with linear, long-peduncled pistillate spikes, of which the lowermost may even rise from the base of the culm.

General distribution: Although considered circumpolar by some authors, *Carex Bigelowii* is an arctic Amphi-Atlantic species, and west of the Mackenzie Valley is replaced by the related but amply distinct Amphi-Beringian *C. consimilis* Holm and *C. lugens* Holm.

Fig. 191 Map 214

Carex bonanzensis Britt.

Forming small, compact tussocks from a short ascending rhizome. Culms slender, stiffly erect, sharply trigonous, smooth except near the tip, overtopping the flat, 2 mm broad, rough-margined leaves. Spikes 7 to 10, ovate, 4 to 5 mm long, the uppermost somewhat crowded, the lower 2 or 3 remote; each spike subtended by one or more hyaline, deciduous bracts, the lowermost of these commonly with a filiform, 5 to 6 mm long awn; the scales ovate with a pale, hyaline margin, shorter than the perigynia; these about 2 mm long, plano-convex, prominently nerved and finely punctate on both sides, terminating in a minutely bidentate, short and conical beak.

River banks and alluvial river flats. Long known only from the type locality (Bonanza Creek near Dawson, Yukon Territory), *C. bonanzensis* is now known from scattered and disjunct stations from central Alaska eastward over the Mackenzie Delta to Great Bear Lake, south along the upper Mackenzie River and to the upper tributaries of the Yukon River.

General distribution: N. W. sub-arctic N. America.

Map 215

Carex brunnescens Poir.

Densely to loosely tufted; culms slender but stiff, scabrous above, 20 to 40 cm long (or more), much longer than the flat, fresh-green leaves; spikes small 3 to 4 mm long, commonly 5 to 7 in an elongated head, the lowermost subtended by a short linear bract. The terminal spike often somewhat clavate and conspicuously staminate at the base, while the rest are ovoid, usually with only one or two male flowers at the base. Scales oval with pale hyaline margins and a narrow darker centre; perigynia about 2 mm long including the short but distinct beak, brown and spreading when mature.

A highly variable circumpolar species represented with us by ssp. *alaskana* Kalela of subarctic-alpine range but separated only with difficulty from ssp. *brunnescens* from eastern subarctic North America and Greenland, and distinguished mainly by its lower growth, rather dense caespitose habit, and shorter and rather compact heads; ssp. *alaskana* was described from S. E. Yukon Territory, and is thought to be endemic in central Alaska and Yukon Territory reaching eastward into central and southern District of Mackenzie where its range overlaps that of ssp. *sphaerostachya* (Tuckerm.) Kalela, which latter, is an eastern woodland race, more high-grown, with slender and rather weak culms 30 to 80 cm long, and fewer but well spaced spikes.

A lowland species of damp, turfy places.

General distribution: *C. brunnescens s. lat.*, circumpolar.

Fig. 192 Map 216

Carex Buxbaumii Wahlenb.

Zoosely tufted, with slender rhizomes; culms 6 to 7 dm tall, longer than the flat or somewhat revolute leaves; inflorescence 3 to 5 cm long, of 3 to 5 spikes, the terminal largest, clavate and always gynaecandrous, the lower pistillate, small and ovate, the lowermost remote and subtended by a leafy bract about as long as the inflorescence; perigynia 3 mm long, pale green and distinctly nerved, beakless, and with a very short-protruding style; pistillate scales lanceolate, purplish black, prominently awned. Stigmas 3.

Woodland bogs of the boreal forest, from Newfoundland to Alaska.

General distribution: Circumpolar.

Fig. 193 Map 217

Carex canescens L.

Loosely tufted, characteristically grey-green with soft, smooth culms commonly 25 to 30 cm long and soft flat leaves. Heads commonly somewhat bent, 3 to 4 cm long, composed of from 3 to 7 ovoid, pale yellowish-green, 5 to 6 mm long and well separated spikes. The terminal spike, but often all, with a few male flowers at the base. Scales ovoid, narrower than the ovoid 2.0 to 2.5 mm long, pale green perigynia.

River meadows and lake shores north to the limit of forest.

General distribution: Circumpolar.

Fig. 194 Map 218

Carex capillaris L.

ssp. **capillaris**

Culms 5 to 20 cm high, longer than the yellowish-green 0.75 to 2.0 mm wide, flat or somewhat canaliculate leaves; spikes usually 4, of which the terminal is staminate; pistillate spikes commonly 4 to 10 flowered, short-peduncled and aggregated near the summit, overtopping and often concealing the staminate spike; pistillate scales not readily deciduous; perigynia distinctly beaked, lustrous and nerveless. Stigmas 3.

Moist mineral spots in tundra, pond margins etc.

General distribution: Arctic-alpine, ? circumpolar.

Fig. 195 Map 219

Carex capillaris L.
ssp. **chlorostachys** (Steven) Löve *et al.*
C. capillaris L. var. *elongata* Olney
Similar to ssp. *capillaris*, but culms 20 to 30 cm tall, fresh green like the commonly up to 3 mm broad soft and flat leaves; pistillate spikes loose-flowered, the lowermost often very distant, on long, capillary peduncles; pistillate scales readily deciduous. Stigmas 3.

Moist, open woods and bogs north to the limit of trees.

General distribution: N. America from Nfld. to central Alaska but lacking in Greenland, and not reaching the shores of either Bering Sea or the Pacific coast.

Fig. 196 Map 220

Carex capillaris L.
ssp. **robustior** (Drej. ex Lange) Böcher
C. capillaris L. var. *major* Drej.
Resemblng a tall ssp. *capillaris* but the terminal spike always gynaecandrous; the pistillate spikes 6 to 7, each 10- to 20-flowered, the lowermost often remote but not droopng as in ssp. *chlorostachys*; in Yukon and Alaska the lowermost spikes sometimes with short, lateral and ascending branches from their base (fastigate); perigynia fusiform, not lustrous, and the pistillate scales persistent. Stigmas 3.

Dry, calcareous or alkaline soils, mainly restricted to dry, continental parts.

General distribution: Arctic and alpine; from Iceland over E. and W. Greenland, across arctic Canada to Alaska, south in the Rocky Mts. to Colorado and Utah.

Fig. 197 Map 221

Carex capitata L.
Similar to *C. arctogena* but perigynia consistently larger (2.3 to 3.0 mm long and 1.3 to 2.1 mm broad as against 1.7 to 2.1 mm long and 1.2 to 1.9 mm broad in *C. arctogena*).

Pronounced calciphile and, unlike the Amphi-Atlantic *C. arctogena*, boreal-alpine rather than arctic.

General distribution: Circumpolar.

Fig. 198 Map 222

Carex chordorrhiza L. f.
Tufted bog species with widely trailing and ascending, leafy stems, up to 1 m long. Culms smooth, 10 to 20 cm long, equalling or slightly longer than the rather broad, strongly keeled leaves. Spikes 3 to 8, staminate at the summit and aggregated into an ovoid head; pistillate scales broader than the ovoid, short-beaked, and strongly nerved perigynia. Stigmas 2.

Occasional to rare in wet sedge bogs.

General distribution: Circumpolar, low-arctic.

Fig. 199 Map 223

Carex concinna R. Br.
Loosely tufted from a widely creeping slender and dark brown, subterranean rhizome, from which rise fascicles of flat, somewhat curved leaves, 2 to 3 mm wide, and very slender and arching, naked culms 10 to 15 cm long; inflorescence of 3 to 4 densely congested, 4 to 5 mm long spikes, of which the terminal is staminate and exserted above, or sometimes, concealed among the sessile, pistillate spikes; pistillate scales obovate, brown, with a pale centre and margins, shorter than the grey-hairy, trigonous, perigynia; stigmas 3, rarely 2.

Common in calcareous, gravelly places.

General distribution: Wide ranging N. America, from Nfld. to central Alaska, north to the limit of trees, and reaching the Arctic Coast at the Mackenzie Delta.

Fig. 200 Map 224

Carex consimilis Holm
Loosely tufted from stout, ascending stolons; culms stiff, sharply trigonous, 2.5 to 4.0 dm long, leafy only below; leaves dark green, 3 to 5 mm wide, flat with revolute margins, their sheaths and marcescent bases purplish-brown and somewhat shiny. Spikes 3 to 5, subsessile, the terminal staminate, the lateral short cylindric 0.8 to 1.4 cm long, the uppermost commonly with a few staminate flowers at their summits. Perigynia about 2 mm long-elliptic-obovate, flat, purplish-tipped, with a very short, entire beak, their scales with sooty-black centre and a pale edging along the sides and tip.

General distribution: *C. consimilis*, and to a lesser degree *C. lugens*, resemble the eastern *C. Bigelowii*. Both are of Amphi-Beringian range and appear to be restricted to mature peaty tundra and woodland bogs of central and

northern Alaska and Y.T., reaching the Arctic Coast east and west of the Mackenzie Delta, with disjunct stations on Great Bear L.
Fig. 201 Map 225

Carex Crawfordii Fern.
Densely tufted with up to 6 to 8 dm long slender, sharp-angled culms as long or longer than the soft, flat, 2 to 3 mm broad leaves. Spikes ovate, 4 to 8, loosely aggregated into a slender oblong, yellowish-green and up to 2 cm long head. Scales yellowish-brown; perigynia narrowly lanceolate with a narrow but distinct wing.
Dampish, well drained lake and river meadows.
General distribution: Boreal N. America. Nfld. to Alaska.
Map 226

Carex deflexa Hornem.
Loosely tufted from a short, ascending rhizome from which rise numerous leafy shoots, their bases and sheaths characteristically reddish-purple, the blades pale green, 1.5 to 2.0 mm broad, flat or somewhat involute, shorter than the naked, slender and arching, 1.5 to 2.5 dm long culms; inflorescence short and congested, of 3 to 4 small spikes, the lower subtended by short, foliaceous bracts; terminal spike staminate, commonly concealed among the few-flowered pistillate spikes; pistillate scales broadly lanceolate, dark brown with pale centre and margins, barely concealing the short-pubescent 2.5 to 3.0 mm long perigynia; stigmas 3.
Gravelly or sandy places, or in lichen mats of open forest, chiefly on soils derived from acid, crystalline rocks.
General distribution: N. American subarctic-alpine species, wide ranging from S. Greenland to mountains of Alta. and B.C. to central E. Alaska.
Fig. 202 Map 227

Carex Deweyana Schw.
Loosely tufted from a short rhizome; culms soft, commonly 5 to 8 dm long, much longer than the soft, flat and pale blue-green leaves. The 2 to 7 pale-green spikes in a flexuous and commonly nodding inflorescence, the uppermost sub-approximate, the lower remote and subtended by a long, filiform bract. Perigynia pale-green, membranaceous, lanceolate-ovate, 4 mm long, faintly nerved, with a long and wing-margined beak.
Open woods and river bank thickets.
General distribution: N. American from S. Lab. to B.C., central Yukon T. and Alaska barely reaching our area on the lower Liard R.
Fig. 203 Map 228

Carex diandra Schrank.
Densely tufted with a short, ascending rhizome; culms to 1 m tall, obtuse-angled, smooth below but very scabrous near the tip, as long or longer than the brownish-green soft, flat or channeled leaves, their sheaths characteristically brown-black. Heads cylindrical, 2 to 3 cm long composed of from 5 to 7 dark brown and almost contiguous spikes. Perigynia onion-shaped, black and lustrous, spreading at maturity.
Wide ranging marsh species commonly growing in water on the floating margin of woodland bogs.
General distribution: Circumpolar, non-arctic.
Fig. 204 Map 229

Carex disperma Dew.
Delicate bog species with filiform 30 to 40 cm long culms rising singly from a fascicle of narrow, flat leaves that are often partly buried in the sphagnum carpet through which the slender, horizontal rhizome grows. Head moniliform, of from 3 to 4 spikes, the lower composed of 1 to 3, the terminal of 4 to 6 perigynia and a single staminate flower. Perigynia plump, ellipsoid, about 3 mm long, finely nerved, pale grey-green.
Wide ranging in woodland sphagnum bogs.
General distribution: Circumpolar.
Fig. 205 Map 230

Carex eburnea Boott
Of similar habit as *C. concinna* but with a light, yellowish-brown rhizome, filiform leaves less than 1 mm wide, and capillary, naked culms commonly 2 to 3 dm long; inflorescence of 3 to 4 very small spikes of which the terminal is staminate and sessile, but the pistillate on stiffly erect peduncles; pistillate scales papery white, shorter than the distinctly beaked, dark brown or black, shiny perigynia; stigmas 3.
Woodland species confined mainly to calcareous soils.
General distribution: N. America, wide ranging from Nfld. to south central Alaska and along the Mackenzie Valley north to Pt. Separation.
Fig. 206 Map 231

Carex eleusinoides Turcz.
C. kokrinensis Porsild
Loosely caespitose; culms smooth, 2 to 3 dm long, arching or reclining, barely longer than the

yellowish-green, 2 to 3 mm broad, flat leaves rising from purplish-brown, bladeless sheaths. Spikes 4 to 5, cylindrical, 1.2 to 2.0 cm long, the terminal sessile, gynaecandrous, the lateral pedunculate in a fascicle almost level with the terminal spike; uppermost bract equalling the lower, much exceeding the inflorescence. Perigynia falling when ripe, 2 mm long, obscurely nerved, pale grey, smooth, the achene filling the perigynium, beak very short, entire; stigmas 2.

In moist, alpine river bank and floodplain meadows of central Alaska and Yukon Territory, and southern Mackenzie Mountains. To be looked for in the Richardson and northern Mackenzie Mountains of Northwest Territories.

General distribution: Amphi-Beringian.

Fig. 207 Map 232

Carex filifolia Nutt.
C. elynaeformis Porsild
Tussock forming with short and firm rhizomes; basal sheaths conspicuous and durable; culms slender and wiry, grey-green, 20 to 30 cm tall, somewhat arching, equalling or somewhat longer than the leaves; spike 15 to 20 mm long, the male part linear and prominent; pistillate scales ovate with brown centres and broad hyaline margins; perigynia 6 to 8, obovoid, somewhat inflated, hirsute on the beak and margins; achene triangular, stipitate.

Apparently rare and local within our area where, thus far, it has been reported from Great Bear Lake, the east bank of Mackenzie River near Wrigley and from Nahanni Butte.

General distribution: N. American prairie and foothill species of calcareous, sandy or gravelly cut banks.

Fig. 208 Map 233

Carex foenea Willd.
C. siccata Dew.
Rhizome cord-like, 2 to 3 mm thick, light brown and fibrillose, widely creeping just below the surface; culms erect, slender, somewhat flexuous, 25 to 40 cm long, arising singly at intervals, overtopping the flat, pale green leaves; spikes 4 to 8, aggregated into a 2 to 3 cm long ovate-oblong light brown head, the terminal gynaecandrous and commonly clavate, the lower androgynous or with mixed male and female flowers.

In habit somewhat similar to *C. praegracilis* from which it may at once be distinguished by its pale brown, more slender and less deeply buried rhizome.

General distribution: Dry sandy prairie from S. W. Que. to S. W. Yukon T. north to southern District of Mackenzie.

Fig. 209 Map 234

Carex Franklinii Boott
Loosely tufted, with long cord-like and fibrous rhizomes from which rise well spaced fascicles of sterile or fertile shoots; culms 3 to 8 dm high, slender, obtuse-angled and smooth, not much longer than the flat or somewhat revolute-margined, 1.5 to 2.0 mm broad leaves that taper into long, curly and whiplash points, their bases pale brown and fibrous; inflorescence of from 4 to 6 spikes, nodding on 2 to 3 cm long peduncles, the terminal clavate and staminate, the next 2 pistillate but with a few staminate flowers at the summit, and the lower entirely pistillate; scales broadly lanceolate, chestnut-brown with paler midvein and margins; perigynia lanceolate, 5 to 6 mm long, with a short hyaline beak; stigmas 3 (2) long-exserted.

Calcareous, sandy river banks and alluvial flats.

General distribution: Northern Cordilleran foothill species, north to Mackenzie Mts., Y.T. and Alaska.

Fig. 210 Map 235

Carex Garberi Fern.
Loosely tufted with short, ascending stolons; culms weak, somewhat arching, 2 to 3 dm tall and longer or equalling the flat, 2 mm wide, blue-green leaves; spikes commonly 3 to 4, the terminal gynaecandrous and clavate, the remainder pistillate and short-peduncled, 1.0 to 1.5 cm long, the lowermost subtended by a foliaceous bract usually overtopping the inflorescence; pistillate scales ovate-oblong, brown with a broad, pale green midvein, equalling or shorter than the obovate, pale green, prominently nerved and beakless 2.0 to 2.5 mm long perigynia; stigmas 2.

Damp, calcareous mud in wet lake shore and river bank meadows.

In our area typical *C. Garberi* appears to be confined to the wooded parts of the Precambrian Shield whereas the var. *bifaria* Fern., distinguished by its more lax habit and slightly larger and distinctly stipitate perigynia, is known from the Mackenzie Delta and from Brintnell Lake in the southern Mackenzie Mountains.

General distribution: E. Que. to B.C. and north to central Alaska and reaching the Arctic Ocean at the Mackenzie Delta.

Fig. 211 Map 236

Carex glacialis Mack.
Densely caespitose with smooth culms, 8 to 15 cm long, slightly longer than the narrow, flat, strongly keeled, somewhat curved leaves; the terminal spike staminate, the lateral 2 to 3 pistillate, erect, short-peduncled, and few-flowered; the lowermost supported by a sheathing bract terminating in a bristle-like blade, the pistillate scale obtuse, broadly ovate, with a pale midvein and broad translucent margins, shorter than the smooth, ovoid, brown, distinctly beaked perigynia; stigmas 3.
Calcareous sandy and gravelly places.
General distribution: Circumpolar, arctic-alpine.
Fig. 212 Map 237

Carex glareosa Wahlenb.
var. **amphigena** Fern.
C. marina Dew.
Densely caespitose with slender, ascending, and generally somewhat flexuous culms, slightly longer than the narrow, glaucous leaves. Spikes usually 3, close together, ovoid, the terminal club-shaped, staminate at the base. Scales brown with a paler midvein about as long as the ovate, greyish-brown, short-beaked, many-nerved perigynia. Stigmas 2.
A seashore plant which forms dense, compact, flat tussocks on sand and clay beaches subject to spring floods. Often a dominant species around the nesting grounds of sea birds.
General distribution: Circumpolar, arctic.
Fig. 213 Map 238

Carex gynocrates Wormskj.
Stoloniferous, culms filiform, 5 to 15 cm high, longer than the narrow, bristle-like leaves. Spike either staminate and linear, pistillate and cylindric, or staminate above and pistillate in the lower part. Perigynia at first ascending, spreading when mature, plump, dark brown, and somewhat shiny. Stigmas 2.
In springy, wet places and fens, generally on calcareous soil.
General distribution: Circumpolar, low-arctic.
Fig. 214 Map 239

Carex Heleonastes Ehrh.
Tufted from a short ascending rhizome; culms stiff, the upper half with sharp and scabrous angles, 20 to 40 cm high, solitary or few together; leaves flat, grey-green, slightly shorter than the culms. Spikes pale brown, 3 to 4, forming a short dense head. Perigynia 3 mm long, ash-coloured and faintly nerved dorsally.
Northern peat bog species of rather spotty distribution from Ungava to Alaska; in our area known from a single collection in Nahanni National Park.
General distribution: Circumpolar, with large gaps.
Fig. 215 Map 240

Carex holostoma Drej.
Stoloniferous; culms rather stout, smooth, 15 to 20 (30) cm high, longer than the flat, light green leaves; the basal sheaths dark purplish; spikes 3, the terminal staminate and minute, often concealed between the cylindrical and mostly opposite, stiffly erect, short-peduncled, 8 to 15 mm long pistillate spikes; bract leafy, shorter or equalling the inflorescence; pistillate scales ovate, blackish-brown, with a faint midvein, equalling the ovate-obovate, smooth, nearly beakless, pale green or brownish blotched perigynia; stigmas 3.
In turfy places in tundra and by the edge of small ponds. A pronounced acidophyte.
General distribution: Arctic. Circumpolar, with large gaps.
Fig. 216 Map 241

Carex interior Bailey
Caespitose, from ascending, dark-coloured rhizome; culms slender and wiry, smooth, except near the tip, 15 to 30 cm long and usually longer than the 2 mm broad flat or somewhat canaliculate fresh green leaves. Head 2 to 3 cm long of 2 to 5 subglobose, sessile and commonly somewhat spreading, pale greenish-brown spikes, the terminal somewhat clavate with staminate base; perigynia spreading, pale green becoming light brown, prominently beaked, nerveless.
Damp calcareous meadows, barely entering southern District of Mackenzie and the Yukon Territory.
General distribution: N. America: from Nfld. to B.C. south to Virginia and New Mexico.
Map 242

Carex Lachenalii Schk.
Loosely caespitose, with stiff culms, 15 to 25 cm high, and flat, rather short leaves. Spikes gynaecandrous, 3 to 4, in an oblong head 1 to 2 cm long, the terminal tapering at the base, the lowermost often somewhat distant. Pistillate scales brown with a pale midvein, shorter than the smooth or faintly scabrous margined, lanceolate, nerveless, rather abruptly beaked perigynia that turn a rich, golden brown when mature. Stigmas 2.

Wet sand and turfy places by brooks and lake shores.
General distribution: Circumpolar, arctic-alpine.
Fig. 217 Map 243

***Carex laeviculmis** Meinsh.
Loosely tufted with a slender, ascending rhizome; culms very slender, smooth, 3 to 7 dm long, much longer than the 2 mm broad, flat leaves. Head 3 to 6 cm long composed of from 3 to 8 spikes, the lowermost widely remote and subtended by a 1 to 5 cm long leafy bract, the upper 2 to 3 commonly aggregated, the terminal gynaecandrous, and distinctly clavate; perigynia faintly nerved dorsally, greenish-brown, ascending or spreading.
Damp, woodland meadows.
General distribution: Pacific Coast with some alpine inland stations, and to be expected in the upper Liard River drainage.
Map 244

Carex lapponica O.F. Lang
C. canescens L. var. *subloliacea* Laest.
Similar to *C. canescens* but culms usually solitary or a few together rather than caespitose, and the leaves narrower. Spikes short-ovoid to subglobose, 4 to 7 mm long, usually well separated, the perigynia about 2 mm long and much smaller than those of *C. canescens*, smooth or very faintly nerved.
Wet woodland meadows and bogs.
General distribution: Circumpolar, with a large gap in Central Siberia.
Fig. 218 Map 245

Carex lasiocarpa Ehrh.
var. **americana** Fern.
Tufted, from a horizontal rhizome; culms slender, smooth, obtuse-angled, up to 1 m tall, with smooth, narrow, involute, or folded leaves, shorter than culms; inflorescence 10 to 12 cm long, usually of two linear and contiguous staminate spikes placed well above the sessile and widely separated, 1.5 to 2.0 cm long ovate-oblong pistillate spikes, the latter subtended by long, foliaceous bracts; pistillate scales narrowly triangular, with a broad, pale midvein; perigynia about 4 mm long, densely short-hirsute, ovate, tapering into a short but sharply bidentate beak; stigmas 3.
Wet margins of peat-bog ponds.
General distribution: non-arctic, Nfld. to S. E. Alaska, north in the Mackenzie R. valley to the Arctic Circle.
Map 246

Carex laxa Wahlenb.
Similar to, and of similar habitats, as *C. limosa* from which it may be distinguished by its smaller stature, more slender culms, smaller and shorter pistillate spikes on much shorter peduncles; also its roots lack the dense, felt-like cover so characteristic of *C. limosa*. However, the most reliable difference is in the sheath of the lowermost bract of the inflorescence which in *C. limosa* consists of a 2 mm long black tube, whereas in *C. Laxa* the sheath is green and more than 1 cm long. Stigmas 3.
Wet tundra bogs.
General distribution: A Eurasian species, in N. America thus far known only from two stations in Alaska, one in central Y.T. and one in the Mackenzie Delta.
Map 247

Carex lenticularis Michx.
Densely tufted riparian species lacking stolons or creeping rhizomes; culms slender, sharply trigonous, usually overtopped by the very numerous pale grey-green, 1 to 2 mm broad leaves; spikes 4 to 5 (7), rather closely aggregated, narrowly cylindric about 2 to 3 cm long, the terminal staminate or at least androgynous, the lowermost subtended by a leafy bract rarely overtopping the inflorescence; pistillate scales narrowly ovate, reddish-brown, with a paler midvein, barely concealing the obscurely veined pale green perigynia, its beak very short and entire; stigmas 2.
Forming pure stands along muddy shores of sheltered ponds, lakes or river flats.
General distribution: Non-arctic, from Nfld. to B.C.; in our area thus far collected only a few times north to Great Bear L.
Fig. 219 Map 248

Carex leptalea Wahlenb.
Loosely tufted from a mass of filiform elongated rhizomes; culms very slender, with us commonly 1.5 to 3.0 dm tall, not much longer than the soft and narrow leaves; spike rarely more than 5 mm long consisting of 3 to 5 pale, green and beakless perigynia and one or rarely two male flowers at the summit. Entire plant, including the spike, characteristically pale green.
Wet, calcareous fens, north to Great Bear Lake.
General distribution: N. American boreal forest species.
Fig. 220 Map 249

Carex limosa L.
Widely spreading by slender, branching rhi-

zomes; the roots cord-like, densely covered by yellow, velvety felt; sterile leaves fascicled, and the culms mostly solitary, 1.5 to 2.5 dm long; the terminal spike staminate, linear and long-peduncled, the pistillate 1, 2 or rarely 3, the uppermost erect, the lower long-peduncled and drooping; perigynia ovate, 2.5 to 4.0 mm long, prominently nerved and well concealed by persisting reddish-brown scales; stigmas 3.

In shallow water or very wet places in peat-bogs or quagmires, north to the limit of forest.

General distribution: Circumpolar.

Fig. 221 Map 250

Carex livida Willd.
var. **Grayana** (Dew.) Fern.
Pale, blue-green and loosely tufted, with slender, scaly stolons from which rise fascicles of sterile leaves, or fascicles of leaves containing mostly one, commonly 2 to 4 dm tall, fertile culm. Terminal spike staminate, linear, 1 to 2 cm long, projecting well above the pistillate spikes, these mostly 2, about 1 cm long, sessile, subsessile, or the lowermost sometimes subradical and long-peduncled, subtended by a foliaceous bract that equals or overtops the inflorescence; perigynia pale, glaucous, fusiform or oblong-ovoid, 2-keeled and very faintly nerved, beakless and about 4 mm long; pistillate scale pale brown with a broad, pale centre; stigmas 3.

Wet, calcareous lake-shore meadows, within our area thus far known only from the Mackenzie Delta region and the west end of Great Slave Lake.

General distribution: *C. livida s. lat.* circumpolar. Low-arctic, with large gaps.

Fig. 222 Map 251

Carex loliacea L.
Loosely caespitose with slender stolons that are well buried in the sphagnum mat in which the species commonly grows. Culms slender, 1.5 to 4.0 dm long, scabrous above and about twice as long as the 2 mm wide, flat and attenuate leaves; spikes 3 to 5 (8), gynaecandrous, subglobose and few-flowered, the lower well spaced and the lowermost subtended by a short, setaceous bract; scales ovate, hyaline-margined, about half as long as the 2.5 to 3.0 mm long, light-green and conspicuously many-nerved and beakless perigynia.

Uncommon but probably often overlooked in sphagnum bogs in the southern parts of our area.

General distribution: Circumpolar, with large gaps; N. America from northern Ont. to central Alaska.

Fig. 223 Map 252

Carex lugens Holm
Caespitose, often forming very large and rather dense tussocks. Culms slender, 2 to 5 dm long, but stiff and wiry, sharply trigonous and scabrous on the edges, much longer than the 2 mm wide blue-green, flat or somewhat revolute leaves. Spikes 3 to 4, the terminal staminate and linear, the lower pistillate or with a few male flowers at the summit, short-cylindrical or ovate, commonly about 1 cm long, subsessile, in a 2 to 3 cm long head. Perigynia 2 mm long, plump, obovate with a short entire beak, purplish-mottled near the tip, their scales narrowly ovate, black with pale margins. Stigmas 2.

General distribution: Amphi-Beringian, subarctic-alpine species of mature tundra which, over central and northern Alaska and Y.T., reaches N.W. District of Mackenzie, with disjunct stations on Banks and Victoria Islands, and in central District of Keewatin.

Fig. 224 Map 253

Carex Mackenziei Krecz.
C. norvegica Willd. *non* Retz.
Loosely caespitose with widely creeping, fibrillose rhizomes from which rise fertile culms or fascicles of glaucous or yellowish-green leaves; culms smooth 10 to 30 cm long, as long or somewhat longer than the soft, flat leaves; spikes 4 to 6 in an elongated head, the terminal clavate and gynaecandrous; perigynia about 3 mm long, grey-green, faintly nerved.

A littoral species of brackish marshes; in our area known only from two collections from the Mackenzie Delta area and a third from the west coast of Hudson Bay.

General distribution: Circumpolar, with large gaps.

Map 254

Carex macloviana d'Urv.
Densely caespitose, forming large, tough tussocks from the firmly interwoven brownish and fibrillose rhizomes; culms stout, 20 to 35 cm long, sharply trigonous and rough above, leafy below, their bases densely clothed by the remains of last year's leaves; leaf-blades flat, pale green, 2.4 to 4.0 mm wide. Spikes 3 to 8, densely aggregated, although the two lowermost are usually distinct, into a triangular-ovoid, copper-coloured head up to 14 to 20 mm long; perigynia about 4 mm long, greenish, turning copper-

brown, with a broad wing gradually tapering into a slender beak, faintly nerved dorsally.

Well watered grassy alpine or subalpine snowbed slopes; on Great Bear Lake also in hollows of stabilized dunes.

General distribution: The world distribution of *C. macloviana* is bi-polar. It was described from the Falkland Islands and is known also from southern S. America. Its main area, however, is in the northern hemisphere: from N. Scandinavian mountains (lat. 64° to 70°) over Iceland and southern half of E. and W. Greenland to northern Lab.–Ungava, with an isolated station in Gaspé, Que. This eastern and rather uniform population is separated by a 6,000 km gap from a western and less uniform population represented by *C. macloviana* ssp. *pachystachya* (Cham.) Hult., besides other and perhaps less well understood segregates (*C. festivella* Mack. and *C. Soperi* Raup)

Fig. 225 Map 255

***Carex macrochaeta** C. A. Mey.

Somewhat similar in appearance and habit to *C. podocarpa* from which, however, it is at once distinguished by the long awn-like tips of its pistillate scales. Although *C. macrochaeta* is essentially maritime, it is known from a few inland and alpine stations in Alaska and southern Y.T., the most easternly in the Selwyn Mountains. It should be looked for also in the southern extension of the Mackenzie Mountains.

General distribution: N. W. Pacific.

Map 256

Carex maritima Gunn.

C. incurva Lightf.

Dwarf species with a creeping rhizome and low, 5 to 15 cm high, usually curved culms that barely overtop the leaves. Head globose or ovoid, 8 to 14 mm in diameter, composed of 3 to 5 densely aggregated androgynous spikes. Pistillate scales brown, ovate and obtuse, thin, with a broad, pale margin. The mature perigynia membranaceous, chestnut-brown, somewhat inflated, and at length divergent or spreading. Stigmas 2.

A mostly littoral species generally confined to sandy, gravelly, or turfy places along the seacoast, but in the Western Arctic (Alaska, Yukon Territory, District of Mackenzie, and the Western Archipelago) not infrequently growing at some distance inland and at considerable elevation above sea-level.

General distribution: Circumpolar, arctic.

Fig. 226 Map 257

Carex media R. Br.

Loosely tufted with culms commonly 3 to 4 dm tall and much overtopping the fresh green, flat 2 to 3 mm broad leaves of which the basal sheaths are characteristically bright reddish-purple; spikes usually 3, sub-sessile in a head-like cluster subtended by a short foliaceous bract; the terminal spike staminate at the base; pistillate scales broadly lanceolate, dark brown, contrasting strongly with the pale, greenish-white perigynia. In habit similar to the eastern and truly arctic *C. norvegica* Retz. from which it may usually be distinguished by its taller growth and pale perigynia, 2.5 to 3.0 mm long and distinctly longer than those of *C. norvegica*.

Common in moist spruce woods and fens north to the limit of continuous forest.

General distribution: Circumpolar, but lacking in Greenland.

Fig. 227 Map 258

Carex membranacea Hook.

Caespitose and stoloniferous, of the habit of *C. Bigelowii*. Culms stout and stiff, 15 to 40 cm high, obtusely triangular, smooth, the base covered by marcescent leaves and their reddish-purple sheaths. Leaves flat, somewhat curled, with revolute margins, 3 to 5 mm wide, septate-nodulose (visible under a lens), equalling or shorter than the culms; spikes 3 to 5, somewhat crowded, the terminal staminate, the lower all pistillate, sessile or short-peduncled, erect or somewhat spreading, dense-flowered, cylindrical, 1 to 3 cm long, the lowermost subtended by long, leafy, non-sheathing bracts; pistillate scales obtuse to acute, dark brown or purplish-black, shorter than the inflated and tightly crowded, sub-orbicular, membranaceous, purplish-black, lustrous and abruptly beaked perigynia; stigmas 3, rarely 2.

In turfy places in dry tundra.

General distribution: Arctic N. America reaching E. Asia but not Greenland.

Fig. 228 Map 259

Carex microchaeta Holm

Loosely tufted from a horizontal, stout and scaly rhizome from which rise fascicles of short, 2 to 5 mm broad leaves, some sterile, some enclosing a solitary culm 2 to 3 dm tall, equalling or slightly longer than the leaves; spikes 2 to 4, the terminal staminate and clavate, the lateral drooping, on slender peduncles; pistillate scales black with a pale midvein, pointed, but not awned, and concealing the dark brown and beakless perigynia; stigmas 3. In habit somewhat similar to *C. podocarpa* which, however, forms firm tussocks.

Alpine tundra.

General distribution: Arctic-alpine Alaska and Yukon, reaching the east slope of the Mackenzie and Richardson Mts.

Fig. 229 Map 260

Carex microglochin Wahlenb.
Loosely tufted with short stolons and short, strongly involute leaves; culms singly or few together, filiform, smooth, 4 to 15 cm high. Spike few-flowered, staminate at the summit. At the time of flowering, the perigynia are erect but soon become reflexed. Pistillate scales early deciduous; the 6 mm long, straw-coloured perigynia subulate-lanceolate, tapering to a long beak from which the stiff, bristle-like rachilla projects beyond the 3 stigmas.

In wet, springy places on calcareous soil.

General distribution: Circumpolar, with large gaps, low-arctic.

Fig. 230 Map 261

Carex misandra R. Br.
Densely caespitose; culms slender, smooth, somewhat flexuous, 15 to 25 cm high, much longer than the short, flat, yellowish-green and somewhat curled leaves; spikes 3 to 4; the terminal club-shaped, staminate at the base, the rest pistillate throughout, ovoid, 10 to 15 mm long, erect or somewhat nodding, on long, capillary peduncles, the lowermost subtended by a long-sheathing bract terminating in a bristle-like blade; pistillate scales oblanceolate, dark brown with pale, membranaceous margins, slightly shorter than the lanceolate, dark brown, scabrous margined, beaked perigynia; stigmas 3.

Dry turfy places in tundra and rocky places.

General distribution: Circumpolar, wide ranging high-arctic, and alpine.

Fig. 231 Map 262

Carex Morrisseyi Porsild
C. adelostoma auctt. non Krecz.
Similar to *C. Buxbaumii* from which it is easily distinguished by its smaller size, shorter and darker inflorescence of which the terminal spike is nearly always staminate and the pistillate scales darker, always awnless and, therefore, shorter than the perigynia. From the otherwise somewhat similar *C. adelostoma* Krecz. of alpine, northern Eurasia, *C. Morrisseyi* differs by its nearly always staminate terminal spike; in *C. adelostoma* the terminal spike is mostly gynaecandrous. Stigmas 3 (2).

Minerotrophic *Larix* fens.

General distribution: N. America; subarctic and thus far known only from three widely disjunct areas from northern Lab. to northwestern District of Mackenzie.

Fig. 232 Map 263

Carex nardina Fr.
Densely caespitose, with persistent brown sheaths, forming small, firm and compact tufts. Culms 10 to 20 cm high, equalling or slightly longer than the narrow, trigonous, often somewhat curved leaves. Spike ovoid, about 10 mm long and 5 to 6 mm wide. Staminate part inconspicuous. Pistillate scales dark brown with an inconspicuous pale midvein, broader than the lanceolate, distinctly stipitate perigynia. Stigmas 2.

In the Canadian Arctic represented by the eastern race, var. *atriceps* Kük., distinguished by its taller growth and less curved culms and leaves.

Calcareous sand and gravel, dry, grassy or rocky slopes.

General distribution: Amphi-Atlantic, arctic.

Fig. 233 Map 264

Carex norvegica Retz.
C. Vahlii Schkuhr
Caespitose with stiff, sharply trigonous, scabrous culms which are much longer than the soft, flat, 1.5 to 2.5 mm broad, scabrous leaves; spikes usually 3, in a dense head-like inflorescence supported by a leaf-like bract, which equals or exceeds the inflorescence; the terminal spike club-shaped, staminate at the base, the lateral pistillate 5 to 7 mm long, sub-sessile or short-peduncled; pistillate scales ovoid, dark brown, monochrome, or with a very faint midvein, equalling the obovate, pale brown, papillose, and faintly nerved, abruptly beaked, 2.0 to 2.5 mm long perigynia; stigmas 3.

Turfy places in dry tundra and on grassy slopes, always on Precambrian acid rocks.

General distribution: *C. norvegica s. lat.* is Amphi-Atlantic; in our area represented by ssp. *inserrulata* Kalela which is low arctic and reaches only a short distance west of Hudson Bay.

Fig. 234 Map 265

Carex obtusata Liljeb.
Culms 5 to 15 cm tall, longer than the leaves, usually well-spaced along the purplish-black,

cord-like rhizome. Spike androgynous and the perigynia shiny and dark brown.

Calcareous, gravelly places.

General distribution: Western arctic-alpine. From Alaska, Yukon T., and western District of Mackenzie southward in the Rocky Mts.; also Eurasia.

Fig. 235 Map 266

Carex oligosperma Michx.
Culms 7 to 10 dm tall, wiry and smooth, singly or few from horizontally spreading pale and scaly stolons; leaves involute, wiry, about as long as the culms; spikes 2 to 4, the staminate terminal and linear, 2 to 3 cm long, on a stiffly ascending peduncle, the pistillate oval, few-flowered, and about 1 cm long, 1 to 3 but often solitary, sessile and subtended by a wiry bract 5 to 10 cm long; perigynia 4 to 5 mm long, coriaceous, shiny and strongly nerved, with a prominently bidentate beak; stigmas 3.

Wet, sandy lake shores.

General distribution: Boreal eastern N. America west to L. Athabasca; from our area thus far known only from north of the east arm of Great Slave L.

Fig. 236 Map 267

Carex paupercula Michx.
Loosely tufted, with 2 to 3 mm broad, flat blue-green leaves shorter than the slender, smooth and stiffly erect 1.5 to 3.5 dm long culms; terminal spike staminate, linear, about 1.0 to 1.5 cm long, and barely overtopping the 2 to 3, approximate and pendulous or spreading, short-cylindric, 1 to 2 cm long pistillate spikes; perigynia ovate or obovate, blue-green, finely papillate, beakless and few-nerved, the scales chestnut-brown, long-acuminate, longer than the perigynia, soon falling; stigmas 3.

Minerotrophic woodland fens or bogs, north to the limit of trees.

General distribution: Circumpolar.

Fig. 237 Map 268

Carex Peckii Howe
Loosely tufted from short ascending rhizome; culms 1 to 5 dm long, reddish purple at base, much longer than the soft, pale 1.5 to 3.0 mm broad leaves; inflorescence of 3 to 4 spikes, the lower subtended by foliaceous bracts often much longer than the inflorescence, the terminal spike staminate, often exceeded by the closely approximate pistillate spikes; pistillate scales light reddish-brown with broad white-hyaline margins, shorter than the prominently stipitate, narrowly obovoid, grey-pubescent, 3 to 4 mm long, perigynia; stigmas 3.

Open woodland.

General distribution: North America: N.B. to Alta, with disjunct stations in W. District of Mackenzie, Y.T. and Alaska.

Map 269

***Carex petasata** Dew.
Caespitose from a short ascending rhizome; culms 20 to 30 cm long, very stiffly erect, smooth except below the head, much longer than the 3 mm broad leaves; their bases conspicuously clothed by the persisting remains of previous years' leaf-bases; spikes gynaecandrous, ovoid, 2 to 6, aggregated into a 2 to 4 cm long head; scales lanceolate-ovate with chestnut-coloured centre and broad, pale margins, the perigynia boat-shaped, 6 to 7 mm long, membranaceous.

General distribution: N. Cordilleran species reaching dry interior prairies of S. E. Yukon T. and northern B.C. It should be looked for in S. W. District of Mackenzie.

Map 270

Carex petricosa Dew.
C. petricosa var. *Edwardsii* Boivin
C. Franklinii var. *nicholsonis* Boivin
C. magnursina Raymond
Stoloniferous with long scaly rhizomes; culms slender, 15 to 35 cm high, obtusely angled, smooth, slightly longer than the 2 to 3 mm broad, revolute-margined, greyish-green, somewhat curly leaves; spikes 3 to 5, oblong-oblanceolate, 10 to 15 mm long, the uppermost 1 to 2 staminate or sometimes pistillate at the base, the lower entirely pistillate, or with a few staminate flowers at their summit, erect-ascending, long-peduncled, the lowermost usually remote; bracts long-sheathing, their blade long and leaf-like; pistillate scales oblong-ovate, obtuse or acute, dark brown with pale membranaceous margins and a pale midvein, slightly shorter than the ovate-elliptic, 4 to 5 mm long, dark brown, scabrous-beaked perigynia; stigmas 3, or rarely 2.

Stony, calcareous slopes.

General distribution: Canadian Rocky Mts., north to the Arctic Coast and western islands, Y.T. and Alaska.

Fig. 238 Map 271

Carex phaeocephala Piper
Loosely tufted with a short, horizontal rhizome; culms slender, somewhat flexuous, commonly

15 to 20 cm long, but occasionally taller, overtopping the flat 2 mm broad leaves; spikes 3 to 4 (rarely 5), gynaecandrous, aggregated into a 2 to 3 cm long head; perigynia oblong-ovate, 4 to 5 mm long, somewhat membranaceous, their scales ovate, reddish-brown with a lighter centre and broad, hyaline margins, completely covering the perigynia.

Alpine meadows and sliderock slopes.

General distribution: N. American Cordillera north to S. W. Y. T. and with us reported only from Glacier L. in the southern Mackenzie Mts.

Fig. 239 Map 272

Carex physocarpa Presl

Coarse, with a stout, creeping rhizome. Culms 3 to 5 dm high, purplish at the base, smooth or slightly scabrous above, not much longer than the 3 to 4 mm broad, revolute-margined, somewhat keeled, pale-green leaves. Spikes 3 to 4, the uppermost 2 (or 1) staminate and linear, the lower pistillate, cylindrical, 2 to 3 cm long and 8 to 10 mm wide, the upper subsessile and erect, the lowermost drooping, on a long capillary peduncle and subtended by a short-sheathing foliaceous bract as long as the inflorescence. Pistillate scales ovate, obtuse, narrower and shorter than the ovate, distinctly beaked, dark brown and lustrous perigynia. Stigmas 2, rarely 3.

In shallow water along the sheltered margins of ponds or in springy calcareous places in wet tundra.

General distribution: E. Asia and subarctic-alpine N. W. America.

Fig. 240 Map 273

Carex podocarpa R. Br.

Densely tufted from a stout, ascending rhizome covered by the shredded and fibrillose remains of dead leaves; culms slender, smooth, 3 to 6 dm tall, bearing 3 to 4 short and broad leaves abruptly tapered to a short point, and shorter basal leaves; spikes 3 to 4, the terminal staminate and clavate, the lateral 2, or rarely 3, pistillate and drooping, on slender peduncles; pistillate scales sooty black, lanceolate, tapering to a short, obtuse point, and almost as long as the equally black, 4 mm long, beakless perigynia; stigmas 3.

Common in not too dry, subalpine herbmat meadows where it forms firm tussocks and where it may be recognized from a distance by its graceful culms and nodding black spikes.

General distribution: Amphi-Beringian. Common locally in mountains of Alaska and Y.T. east to Mackenzie Valley, with a few disjunct stations along the Arctic Coast east to Bathurst Inlet (103° W.), and in mountains of eastern B.C. and Alta.

Fig. 241 Map 274

***Carex praegracilis** Boott

Rhizome stout, black and fibrillose, widely creeping 4 to 5 cm below the surface; culms slender, erect, somewhat flexuous, 25 to 40 cm long, mostly solitary from a cluster of leaves nearly as long as the culm; spikes mostly 2- to 3-flowered, androgynous, 7 to 15, densely aggregated into an oblong 2.0 to 2.5 cm long head; each spike subtended by a short hyaline-margined and strongly keeled bract of which only the lowermost is longer than its spike; scales ovate, chestnut-brown with brown hyaline margins and a prominent, pale mid-rib and tightly covering the ovate-lanceolate, 4 to 5 mm long, dark brown and long-beaked perigynia of which there are only 2 or 3 in each spike; stigmas 2.

General distribution: N. American. Prairie and foothill species ranging from Man. to central Alaska. To be looked for in S. W. District of Mackenzie.

Fig. 242 Map 275

Carex praticola Rydb.

In habit rather similar to *C. aenea* but culms commonly less than 50 cm long, more slender and gracile, with a flexuous, nodding and moniliform head; the spikes pale brown and commonly with a whitish or silvery sheen.

Rather dry, open grassland. The only record from our area is from a cultivated field at Simpson where it may have been introduced with imported hay. It should be looked for elsewhere in the southern half of the District of Mackenzie.

General distribution: N. America; from W. Greenland to central Alaska.

Fig. 243 Map 276

***Carex Preslii** Steud.

Densely tufted with a short, thickened, branching and rather brittle rhizome; culms 5 to 8 cm long, slender, longer than the flat, 2 to 3 mm broad leaves; spikes gynaecandrous, 3 to 5, loosely aggregated into a 1.5 to 2.0 cm long ovate-oblong, reddish-brown head; perigynia 10 or less in each spike, 3.5 to 4.0 mm long, winged, faintly nerved and abruptly tapering into a long beak; stigmas 2.

To be looked for in alpine meadows in the southern Mackenzie Mountains.

General distribution: Alta. and B.C. to S. E. Alaska, south to California.
Map 277

Carex pyrenaica Wahlenb.
ssp. **micropoda** (C.A. Mey.) Hultén
Very densely tufted with wiry culms from 10 to 20 cm tall, at time of flowering shorter than the short and linear and flat leaves but rapidly elongating as the fruits ripen; the spike androgynous, 1 to 2 cm long, oblong; pistillate scales chestnut-brown; perigynia small, about 3 mm long, spindle-shaped, shiny and nerveless, falling as soon as the achenes ripen. Stigmas 3.

A high-alpine snowbed species inhabiting moist slopes, ravines, and the frost-heaved rock-paved floors of hanging valleys, where it is often a dominant species and forms large, firm tussocks.

In the District of Mackenzie known only from a few stations in the Mackenzie Mountains north to Macmillan Pass.

General distribution: Amphi-Beringian southward in the Rocky Mts.

Fig. 244 Map 278

Carex Ramenskii Kom.
Several collections of a plant intermediate between *C. subspathacea* Wormskj. and *C. salina* Wahlenb. were reported as *C. salina*, from brackish meadows of the Mackenzie Delta region (Porsild, 1943). By their tufted growth, taller stature and blunt, ovate one-nerved scales and nerveless perigynia they match a plant reported as *C. Ramenskii* from the shores of Bering Sea, Bering Strait and the north shore of Alaska (Hultén, 1968).
Map 279

Carex rariflora (Wahlenb.) Sm.
Stoloniferous dwarf sedge with erect, smooth culms, 8 to 20 cm high, and short, flat, glaucous leaves; terminal spike staminate, erect, the lower 1 to 3 pistillate, short-cylindrical, at length drooping, on capillary peduncles; bract sheathing, terminating in a short, subulate blade; pistillate scales broadly ovate, often somewhat mucronate, purplish-black with a conspicuous pale midvein, broader than the pale green, faintly nerved, nearly beakless perigynia; stigmas 3.

The var. *androgyna* Porsild, thus far known only from the Arctic Coast just east of the Mackenzie Delta, differs by its terminal spike being mainly pistillate, with a few staminate flowers near the summit, and by its pistillate flowers having 2 instead of 3 stigmas.

Wet, peaty places in tundra or in wet moss by the edge of brooks and ponds.

General distribution: Circumpolar, widespread, low-arctic.

Fig. 245 Map 280 (var. *rariflora*), 281 (var. *androgyna*)

Carex retrorsa Schw.
Tussock-forming with stout culms up to 1 m tall, and commonly shorter than the soft, broad and septate-nodulose leaves; staminate spikes 1 to several, sessile and linear, the pistillate 3 to 4, cylindric, 3 to 6 cm long, sessile and densely crowded, or the lowermost somewhat remote and subtended by an up to 6 dm long leafy bract; perigynia 5 to 8 mm long, densely crowded, inflated, horizontally spreading or even reflexed at maturity, with a slender, 2-toothed beak as long as the body, and much longer than the almost completely concealed pistillate scales; stigmas 3.

Woodland marshes.

General distribution: An eastern N. American species which barely enters our area on the lower Liard R.

Fig. 246 Map 282

Carex Richardsonii R. Br.
Loosely tufted from an ascending, dark brown and fibrillose rhizome from which rise fascicles of flat or slightly involute 2 to 4 mm broad leaves; culms slender, erect or somewhat arching, 1.5 to 2.5 dm long; inflorescence of 3 to 4 spikes, each subtended by a prominent purplish-black sheathing bract; the terminal spike pedunculate, staminate, 1.5 cm long, the pistillate spikes clustered, or the lowermost remote, on short, erect peduncles; pistillate scales dark brown with a prominent white margin, longer than the elliptic-ovoid, grey-pubescent perigynia; stigmas 3.

Calcareous meadows.

General distribution: N. American species from Great Lakes region to interior B.C., barely entering our area in the upper Mackenzie drainage.

Map 283

Carex Rossii Boott
Densely tufted from short, ascending rhizome; culms 2 to 3 dm long equalling or shorter than the flat, 2 to 3 mm broad leaves; inflorescence of 4 to 5 spikes, the lower subtended by foliaceous bracts 2 to 5 cm long, the terminal spike linear and staminate, the pistillate few- and lax-flowered, commonly approximate and short-peduncled, at the summit of the culm; long-

peduncled pistillate spikes may occasionally arise from the base of the culm where they remain concealed among the leaves; pistillate scales pale, membranaceous, broadly lanceolate and prominently keeled, as long as the oval, pale, membranaceous, prominently stipitate and beaked perigynia; stigmas 3.

Dry, rocky slopes.

General distribution: N. American western species extending north into the upper Mackenzie, southern Y.T. and southwestern Alaska.

Fig. 247 Map 284

Carex rostrata Stokes

Culms up to 1 m long, singly or few together in rooted, leafy fascicles from stout, widely spreading scaly stolons; leaves septate-nodulose, as long as the culms, 4 to 5 mm wide, flat or with revolute margins; terminal spikes staminate, 1 to 3, linear and 4 to 6 cm long, projecting well above the commonly 2, cylindric, 4 to 6 cm long, sessile or short-peduncled pistillate spikes which are subtended by leafy bracts, the lowermost overtopping the inflorescence; perigynia 3 to 4 mm long, light brown and lustrous, the ovoid body abruptly tapering to a slender, minutely 2-toothed beak; pistillate scales dark brown, narrow and not concealing the perigynia; stigmas 3.

Wide-ranging woodland swamp species commonly growing in water along sheltered slough and lake margins, north to the limit of continuous forest.

General distribution: Circumpolar.

Fig. 248 Map 285

Carex rotundata Wahlenb.

Widely spreading from pale, cord-like subterranean stolons from which rise single culms or clusters of leaves and culms, the latter slender and smooth from 2.5 to 5.0 dm tall, equalling or slightly longer than the involute, curly-tipped leaves, their bases tightly enclosed in the pale, grey-green leaf-sheaths; spikes 3 to 4, the terminal staminate and linear, 2.0 to 2.5 cm long, the lateral 2 (or rarely 3) pistillate, well-separated, sessile or nearly so, oblong or short-cylindric, about 1 cm long, the lowermost subtended by a foliaceous bract about as long as the inflorescence; perigynia chestnut-brown, crowded, shiny and inflated, the body orbicular-ovate, abruptly contracted into a beak and barely covered by their brown membranaceous scales.

Moist turfy tundra.

General distribution: Amphi-Beringian, extending from Alaska, Y.T. over northern District of Mackenzie to the District of Keewatin and northwestern Man. Lacking in the eastern and northern Canadian Arctic, and in Greenland.

Fig. 249 Map 286

Carex rufina Drej.

Forming rather dense, flat tussocks of soft dark-green leaves, among which the culms are commonly hidden. Spikes 3 to 4, densely clustered, and subtended by long, leafy bracts, the terminal androgynous.

Wet, stony places, often by the edge of ponds or on snowbeds.

General distribution: Amphi-Atlantic; Greenland, in N. America thus far known from a few stations in the District of Keewatin and northeastern Man., but to be expected also in Baffin Island and Lab.

Fig. 250 Map 287

Carex rupestris All.

Tufted, with a slender, cord-like brown rhizome. Culms rather stout, scabrous, 8 to 15 cm high, barely longer than the scabrous, flat, and somewhat curly leaves. Spike cylindrical, 6 to 20 mm long, mostly staminate, with a few pistillate flowers near the base. Pistillate scale obtuse, dark brown, broadly hyaline-margined, broader and longer than the oblanceolate, obtusely trigonous, and faintly nerved, dull, castaneous, 3.0 to 3.5 mm long perigynia. Stigmas 3.

In dry, turfy places in tundra, on gravelly ridges and rocky ledges. A pronounced calcicole, often associated with *Dryas integrifolia*.

General distribution: Circumpolar, arctic-alpine.

Fig. 251 Map 288

Carex Sartwellii Dew.

Similar to *C. praegracilis* but much stouter, with culms 4 to 8 dm long and much larger spikes, that are loosely aggregated in a 5 cm long head, each containing 10 to 15 perigynia only 2.5 to 3.0 mm long.

Wet prairie or lake shore meadows.

General distribution: N. America: in Canada from the St. Lawrence valley to interior B.C., north to southwestern District of Mackenzie.

Fig. 252 Map 289

Carex saxatilis L.
var. **rhomalea** Fern.

Similar to *C. physocarpa* but with a shorter and denser rhizome, narrower and involute leaves, and one terminal staminate spike, 1 to 2 short peduncled, erect, rarely drooping pistillate spikes, usually with dull, blackish-brown perigynia. Stigmas 2, rarely 3.

In shallow water of acid tundra ponds and by the edge of small lakes.

General distribution: *C. saxatilis*, circumpolar; the var. *rhomalea* arctic and subarctic E. American.

Fig. 253 Map 290

Carex scirpoidea Michx.

Loosely caespitose from a stout, black, scaly rhizome. Culms stiff, sharply trigonous, 15 to 25 cm high, much longer than the flat, 3 to 4 mm broad, scabrous leaves. Spike of the male plant obovate, that of the female linear-cylindrical, 20 to 30 mm long and 3 to 5 mm wide, the scales black with pale hyaline margins; perigynia hairy, with a short beak. Stigmas 3.

Dry, turfy places; usually on calcareous soil.

General distribution: Wide ranging N. American radiant.

Fig. 254 Map 291

***Carex sitchensis** Presl

Similar in habit to *C. aquatilis* from which, however, it is readily distinguished by its more numerous staminate spikes, its long-peduncled and commonly drooping lowermost pistillate spike, and by its linear-lanceolate pistillate scales.

In as much as several other Pacific Coast species have reached S.W. District of Mackenzie through the Liard gap, *C. sitchensis* should be looked for on the lower Liard River, and near hot springs in the southern Mackenzie Mountains.

General distribution: A Pacific coastal species of river meadows and estuaries north along the B.C. coast to Cook Inlet of S. Alaska.

Fig. 255 Map 292

Carex stenophylla Wahlenb.
ssp. **Eleocharis** (Bailey) Hult.

Matted with a much branched, slender, cord-like rhizome and black, fibrillose rootlets; culms 5 to 20 cm long (near its northern limit generally low and dwarfed) slender but stiffly erect, slightly longer than the filiform involute or canaliculate leaves; spikes 3 to 5, densely aggregated into a compact, ovoid 1.0 to 1.5 cm long head; spikes androgynous, the staminate part prominent during anthesis; perigynia 2.5 to 3.0 mm long, obscurely nerved; stigmas 2.

Prairie species of summer-dry, often somewhat alkaline hollows and edges of sloughs.

General distribution: N. America from S.W. Man. to S. W. Y. T. and S. central Alaska. In the District of Mackenzie thus far known from a single station near Ft. Simpson.

Map 293

Carex subspathacea Wormskj.

A turf-forming dwarf species with a creeping rhizome and very short, erect-ascending culms, which are frequently hidden among the short, flat, involute-margined, yellowish-green leaves. The bract of the inflorescence leafy. Terminal spike staminate, the pistillate short-peduncled, somewhat remote, and few-flowered. Pistillate scales light brown with a pale midvein, longer than the nerveless perigynia. Stigmas 2.

A widespread arctic species, which is confined to the seacoast where it grows in wet clay and sheltered beaches flooded by high tide.

General distribution: Circumpolar, high-arctic.

Fig. 256 Map 294

Carex supina Wahlenb.
ssp. **spaniocarpa** (Steud.) Hult.

Tufted dwarf species with slender, brown stolons; culms 8 to 15 cm long, erect, sharply trigonous and scabrous, longer than the narrow, crowded leaves; spikes 3, the terminal staminate and linear, the lateral pistillate, few-flowered, subglobose, sessile; the lowermost supported by short scale-like bract; pistillate scales broadly ovate, chestnut-brown with broad translucent and pale margins, slightly shorter than the ovate-globose brown, shiny, prominently beaked perigynia; stigmas 3.

In non-calcareous, dry, rocky and sandy places.

General distribution: N. American, low-arctic.

Fig. 257 Map 295

Carex sychnocephala Carey

Densely tufted; culms smooth, leafy, 1 to 6 dm long but near its northern limit rarely more than 20 cm long; the compact, ovate, 1.5 to 2.0 cm long head subtended by 2 to 3, leaf-like bracts, the lowermost longest and commonly 10 to 15 cm long; spikes 4 to 10, gynaecandrous, yellowish-green perigynia numerous, very narrow, and with a long, serrulate beak.

Wet places in prairie and open woodland meadows, by cold springs or moose-licks.

General distribution: S. W. Que. to interior B.C. north to S. District of Mackenzie, S. E. Y. T. and Alaska.

Fig. 258 Map 296

***Carex tenera** Dewey

Densely tufted; culms slender, 3 to 7 dm long,

with 3 or more well-developed, pale-green, flat, 2 to 3 mm broad leaves; spikes gynaecandrous, 3 to 5, tawny-greenish and broadly ovate, 7 to 8 mm long; perigynia narrowly ovate, 3.5 to 4.0 mm long and 2 mm wide, broad-winged and nerveless, with a prominent beak.

In habit somewhat similar to *C. aenea* and *C. praticola*, but in all parts distinctly smaller, but as densely tufted as the former and taller than the less densely tufted *C. praticola* in which the spikes are of a uniform light brown with a characteristic silvery whitish sheen.

General distribution: Damp woodland meadows from Que. to interior B.C. north through Alta. to L. Athabasca. To be looked for in southern District of Mackenzie.

Map 297

Carex tenuiflora Wahlenb.
Loosely tufted with slender stolons; culms very slender, 5 to 6 dm long, overtopping the flat or canaliculate 1.0 to 1.5 mm broad leaves; spikes 2 to 3, gynaecandrous, few-flowered, pale or whitish-green, subglobose, aggregated into a nearly spherical head 5 to 7 mm in diameter; perigynia elliptic-oblong 3.0 to 3.5 mm long, faintly nerved and beakless, partly covered by the papery, ovate scale.

Sphagnum bogs of rich woodland fens, from Newfoundland to Alaska, north to the limit of trees.

General distribution: Circumpolar.

Fig. 259 Map 298

***Carex tonsa** (Fern.) Bickn.
In habit somewhat similar to *C. Rossii* but lower and of more densely caespitose or even matted growth; culms so short that the inflorescence normally remains completely concealed among the sheaths of the stiff and very scabrous leaves; pistillate scales membranaceous and keeled, narrowly lanceolate, the perigynia obovate, finely pubescent, 3.5 to 4.0 mm long, prominently stipitate and with a prominent serrulate beak; stigmas 3.

Sandy and rocky places.

General distribution: N. American non-arctic species reaching north to L. Athabasca and to be looked for in the upper Mackenzie Valley.

Fig. 260 Map 299

Carex trisperma Dew.
Loosely tufted, and stoloniferous; culms very slender and weak, 2 to 7 dm long, overtopping the soft, flat, 1 to 2 mm broad, light green leaves; spikes gynaecandrous, 2 to 3, each with 2 to 5 ascending perigynia, the lowermost subtended by a 5 to 7 cm long herbaceous bract, the terminal 1 or 2 on a commonly 4 cm long, slender, peduncle-like continuation of the culm; perigynia 3 to 4 mm long, elliptic-oblong, finely nerved, short-beaked, longer than the ovate, hyaline and prominently green-keeled scale.

Woodland bogs; in our area thus far known from a single depauperate specimen from south east District of Mackenzie.

General distribution: Lab. to E. Alta. north to L. Athabaska.

Map 300

Carex ursina Dew.
Densely caespitose dwarf species; the somewhat ascending culms 4 to 6 cm long, barely emerging above the tufted, involute leaves. Spike globose, 4 to 5 mm long and 4 mm wide, staminate at the base; pistillate scales shorter than the broadly ovate, faintly nerved and nearly beakless perigynia. Stigmas 2.

A pronounced littoral species of sheltered sea and lagoon shores, subject to flood by high tide.

General distribution: Circumpolar, high-arctic.

Fig. 261 Map 301

Carex vaginata Tausch
C. sparsiflora (Wahlenb.) Steud.
Stoloniferous with stiff, smooth culms, 15 to 20 cm high, and short, flat, broad leaves; terminal spike staminate, linear, on a stout, often somewhat divergent peduncle; pistillate spikes, 1 to 2, lax-flowered, cylindrical, 1 to 2 cm long, usually somewhat distant, on erect, short and rather stout peduncles, each subtended by a sheathing, leafy bract; pistillate scales lanceolate-ovate, light brown, with a broad, green midvein shorter than the yellowish-green, ovoid, nerveless, smooth, somewhat obliquely beaked perigynia; stigmas 3.

Moist, calcareous rocky and turfy places.

General distribution: Circumpolar.

Fig. 262 Map 302

Carex viridula Michx.
C. Oederi Retz.
var. *pumila* (Cosson & Germ.) Fern.
Caespitose, dwarf species with characteristically yellowish-green culms and leaves, forming small tufts that with us are rarely more than 2 dm tall; spikes 4 to 5, the terminal staminate, linear and commonly less than 1 cm long, the pistillate, short-cylindric or globular, often less than 1 cm long, densely crowded, or the lowermost somewhat remote and subtended by a

long foliaceous bract; often also a long-peduncled pistillate spike may be hidden among the basal leaves; pistillate scales membranaceous, pale brown; perigynia 2 to 3 mm long, orbicular-ovate, densely crowded and spreading, green, strongly nerved, abruptly tapered into a prominent beak, half as long as the body of the perigynium.

Damp calcareous lake shores.

General distribution: Nfld. to S. W. Y. T., and eastern Alaska north to Great Bear L.

Fig. 263 Map 303

Carex Williamsii Britt.
Resembling a small *C. capillaris* from which it differs by its tightly canaliculate or even convolute, bristle-like leaves, and few-flowered, and usually approximate and erect-ascending spikes. Also, in *C. Williamsii*, the mature perigynium is dull, reddish-brown with a few but prominent nerves, and its tip is characteristically bent outwards.

Calcareous sandy soil, often on hummocks by the edge of small ponds.

General distribution: N. America, from Lab. to Alaska.

Fig. 264 Map 304

Eleocharis R. Br. Spike-Rush

Scapose herbs of marshy habitats. Culms simple, mostly tufted, from a horizontal rhizome. The leaves reduced to mere basal sheaths. Spike solitary and terminal.

a. Bulbiform base of style nearly continuous with summit of achene . . . . *E. pauciflora*
a. Bulbiform base of style articulate with achene
  b. Culm and rhizome capillary; achene with prominent longtitudinal ridges; plant often completley submersed and sterile . . . . *E. acicularis*
  b. Culm and rhizome not capillary; achene with smooth or variously sculptured surface
    c. Achene lens-shaped; style 2-cleft
      d. Basal scales of spike 2 to 3; culm stout . . . . *E. palustris*
      d. Basal scale of spike solitary, sub-orbicular, usually completely encircling base of spike . . . . *E. uniglumis*
    c. Achene 3-sided; style 3-cleft . . . . *E. compressa*

Eleocharis acicularis (L.) R. & S.
A tiny aquatic plant with bristle-like, 3 to 6 cm long, filiform culms, from a thin, creeping rhizome. Spike 3 to 4 mm long.

Near its northern limit nearly always sterile and submerged (f. *submersa* Hj. Nilss.), forming soft, green carpets on muddy bottom of shallow ponds.

General distribution: Circumpolar, subarctic, wide ranging.

Fig. 265 Map 305

Eleocharis compressa Sulliv.
Culms compressed, 2 or more dm tall, from a stout darkened rhizome; spikes ovate to oblong, 5 to 12 mm long; achene obovoid, bluntly three-sided, 1.0 to 1.3 mm long, granular roughened, the tubercule short, depressed-pyramidal.

Calcareous sandy and muddy shorelines.

General distribution: North America: Quebec to S. W. District of Mackenzie.

Map 306

Eleocharis palustris (L.) R. & S.
Culms stout but not wiry, 2 to 3 dm tall and 2 to 3 mm in diameter, from a stout, freely branching rhizome; sheath reddish-brown with a distinctly oblique summit; spike oblong-linear, 5 to 15 mm long and 2 to 3 mm wide, the achenes narrowly obovoid, 1.5 to 1.7 mm long.

Commonly forming dense colonies, rooted in soft mud along sheltered margins of lakes and ponds, north to limit of trees.

General distribution: Circumpolar, non-arctic.

Fig. 266 Map 307

Eleocharis pauciflora (Lightf.) Link
var. **Fernaldii** Svens.
Forming small, neat tufts from a slender, elongated rhizome that may often produce tiny, bulb-like tubers which, however, are rarely seen on herbarium specimens. Culms capillary, commonly no more than 15 cm high. Spike few-flowered, lanceolate, somewhat flattened, 4 to 8 mm long.

Calcareous seepages along river banks and lake shores, or by mineral springs, north to Great Bear Lake and central Yukon Territory.
General distribution: *E. pauciflora s. lat.*: Circumpolar, non-arctic.
Fig. 267 Map 308

Eleocharis uniglumis (Link) Schult.
E. macrostachya sensu Cody 1956
Similar to *E. palustris* but culms generally shorter and more wiry, commonly only 15 to 25 cm tall; the summit or orifice of the sheaths square or barely oblique.
Calcareous or saline seepages along river banks or lake shores, but unlike *E. palustris*, rarely growing in water; in the Mackenzie drainage north to Norman Wells.
General distribution: Circumpolar, non-arctic.
Fig. 268 Map 309

Eriophorum L. Cotton-Grass

Caespitose or stoloniferous bog or marsh plants with smooth, obtusely triangular or terete culms terminating in from one to several spikes. Perianth bristles long and silky in fruit.

a. Spikes several, drooping or clustered, subtended by one or two foliaceous bracts
  b. Leaves linear, trigonous, 1.0-1.5 mm broad; bract one . . . . *E. gracile*
  b. Leaves broader, flat, conduplicate or involute; bracts 2 or 3
    c. Midrib of scales prominent, extending to the very tip . . . . *E. viridi-carinatum*
    c. Midrib of scale slender, not reaching the tip
      d. Peduncles glabrous . . . . *E. angustifolium*
      d. Peduncles scabrous . . . . *E. triste*
a. Spikes solitary, not subtended by leafy bracts
  e. Plant stoloniferous
    f. Anthers minute (1 mm long, or less); fruiting head globose . . . . *E. Scheuchzeri*
    f. Anthers 1.5 mm, or longer; fruiting head oblong . . . . *E. russeolum* var. *albidum*
  e. Plant caespitose
    g. Scales monochrome, without pale margins, not spreading or reflexed at maturity; fruiting heads obovate; anthers 1-2 mm long
      h. Culm slender, commonly 3-6 dm long; uppermost sheath mostly above middle of culm; anthers more than 1 mm long; mature bristles dull white . . . . *E. brachyantherum*
      h. Culm stout and stiff, commonly 2-3 dm tall; the uppermost sheath commonly below middle of culm; anthers not more than 1 mm long; mature bristles with a sheen, mostly pure white . . . . *E. callitrix*
    g. Scales pale-margined, divergent or spreading at maturity; fruiting spikes globose; bristles dull white; anthers 2-3 mm . . . . *E. vaginatum*

Eriophorum angustifolium Honck.
Culms 20 to 40 cm high. Leaves flat, keeled. Spikes several, nodding.
In wet bogs or in water by the edge of ponds where it often forms pure stands.
General distribution: Circumpolar, low-arctic.
Fig. 269 Map 310

Eriophorum brachyantherum Trautv.
E. opacum (Björnstr.) Fern.
Loosely tufted, with slender, terete culms 20 to 30 cm long. Sheaths remote, the uppermost above the middle of the culm. Spike small with lead-coloured scales; the bristles dull white or yellowish.
Lowland muskeg and tundra species, which is rare and restricted to favourable habitats beyond the limit of trees.
General distribution: Circumpolar, low-arctic.
Fig. 270 Map 311

Eriophorum callitrix Cham.
Forming small, solitary and compact tufts. Culms slender, 10 to 20 cm high, the uppermost sheath usually below the middle of the culm. Bristles pure white with silky sheen. Anthers about 1 mm long.
In calcareous, turfy places.
General distribution: Circumpolar, with large gaps; arctic-alpine.
Fig. 271 Map 312

Eriophorum gracile Koch
Culms smooth and slender up to 5 dm tall, loosely tufted or even turf-forming, from slender creeping and much branched rhizomes; leaves trigonous, shorter than the culms and soon wilting. Involucral bracts, 1 to 2, short and green; the inflorescence of 2 to 5 small spikes on short, scabrous peduncles of inequal length. Anthers 1.5 mm long; bristles short, bright, white.

Wet peaty bogs and lake shores.

General distribution: Circumpolar, non arctic. Apparently rare in the Northwest Territories where thus far collected only a few times in the southern Precambrian Shield area and in Nahanni National Park.

Fig. 272 Map 313

Eriophorum russeolum Fr.
var. **albidum** Nyl.
E. russeolum Fr. var.
leucothrix (Blomgr.) Hult.
Similar to *E. Scheuchzeri* but at once distinguished by its oblong fruiting head and much longer anthers. Bristles dull, dirty-white or yellowish.

In mud by shallow ponds.

General distribution: Circumpolar, low-arctic.

Fig. 273 Map 314

Eriophorum Scheuchzeri Hoppe
Culms 15 to 30 cm high, mostly solitary, soft and sparingly leafy at the base, from a thin, creeping rootstock. Leaves strongly involute.

Common in wet meadows and in shallow water by the edge of ponds, where it often forms pure stands.

General distribution: Circumpolar, wide ranging arctic-alpine.

Fig. 274 Map 315

Eriophorum triste (Th. Fr.) Hadac & Löve
Similar to *E. angustifolium* of which it may be considered a high-arctic race (*E. angustifolium* ssp. *triste* (Fries) Hult.), but, besides being of lower stature, at once distinguished by its distinctly scabrous peduncles, narrower leaves and black scales and spathes.

In mineral soil, often in open clay spots in tundra and by the edge of snowbeds, never in water by the edge of ponds.

General distribution: Circumpolar, arctic.

Map 316

Eriophorum vaginatum L.
Forming large and compact tussocks. Culms stout, 20 to 40 cm high, the uppermost sheath usually below the middle of the culm. Anthers 2.5 to 3.0 mm long. East of the 100th meridian gradually passing into ssp. *spissum* (Fern.) Hult., distinguished chiefly by its depressed, globose fruiting head, and slightly shorter, about 2 mm long anthers.

Common in peaty soils.

General distribution: *E. vaginatum s. lat.*: Circumpolar.

Fig. 275 (ssp. *spissum*) Map 317 (ssp. *vaginatum*), 318 (ssp. *spissum*)

Eriophorum viridi-carinatum (Engelm.) Fern.
Plant bright green, forming small tufts; culms slender 0.3 to 1.0 m tall; leaves flat, mainly basal; involucral bracts green with a light brown base and usually overtopping the inflorescence; spikes 4 to 10, drooping, on slender, scabrous peduncles; scales greenish, turning lead-grey.

Peaty meadows and rich bogs.

General distribution: Subarctic N. America from Nfld. to S. W. Alaska; in the Mackenzie Mts. north to lat. 65°.

Fig. 276 Map 319

Kobresia Willd.

Tufted, sedge-like perennials with a terminal spike composed of few to several spikelets; of the general habit and appearance of certain species of *Carex* of the Section *Vigneae*, but readily distinguished by the achene being enclosed in a membraneous scale open in one side.

a. Spike linear, composed of numerous small, 2-flowered spikelets . . . . *K. myosuroides*
a. Spike oval-oblong, composed of several spikelets
  b. Spikelets bi-sexual with 1 staminate and 2-3 pistillate flowers . . . . *K. hyperborea*
  b. Spikelets unisexual, 1-2-flowered, the upper staminate, the lower pistillate . . . . *K. simpliciuscula*

Kobresia hyperborea Porsild
Densely caespitose with a short, stout, ascending rhizome. Culms stout, 12 to 35 cm high, at maturity much longer than the narrow leaves. Spike 10 to 18 mm long, 5 to 8 mm wide.

Common locally in rather dry, peaty tundra.

General distribution: Endemic of arctic and subarctic N. W. America, and most closely related to *K. macrocarpa* Clokey of alpine Colorado and to *K. sibirica* Turcz. of central and eastern arctic and subarctic Siberia.
Fig. 277 Map 320

Kobresia myosuroides (Vill.) Fiori & Paol.
K. Bellardii (All.) Degl.
Densely tufted, with slender, more or less terete culms, 10 to 40 cm high; old sheaths long-persistent, densely packed, fibrillose, brown. Leaves numerous, somewhat flexuous, narrow and convolute, shorter than the culms. Spike linear, 10 to 15 mm long and 2.0 to 2.5 mm wide.

Although not confined to the area of Palaeozoic rocks, the species is a pronounced calciphile, preferring dry, calcareous, sandy heath and windswept ridges.

General distribution: Circumpolar, wide ranging.
Fig. 278 Map 321

Kobresia simpliciuscula (Wahlenb.) Mack.
Densely caespitose. Culms 10 to 30 cm high, at maturity much longer than the strongly convolute, somewhat flexuous leaves. Spike, oblong-ovoid, 15 to 25 mm long, 5 to 8 mm wide; in tall specimens the lowermost spikelets somewhat remote.

In damp calcareous, gravelly places in open tundra.

General distribution: Circumpolar, arctic-alpine.
Fig. 279 Map 322

Rhynchospora Vahl

Rhynchospora alba (L.) Vahl
Perennial delicate and loosely tufted, pale grey-green bog plant from a short ascending rhizome. Culm leafy, slender, 15 to 20 cm tall and trigonous, its leaves setaceous and the involucral bract barely exceeding the inflorescence which consists of two or more fascicles of mostly 2-flowered spikelets; these have whitish scales that turn light brown in age. Flowers with 10 to 12 retrorsely scabrous bristles that overtop the pale, lustrous and prominently tubercled achene.

Inconspicuous and easily overlooked, sedge-like plant of wet hollows and muddy margins of bog pools. Rare in S. W. District of Mackenzie.

General distribution: Circumpolar with large gaps, non-arctic.
Fig. 280 Map 323

Scirpus L. Bulrush

Stoloniferous or tufted perennials (ours) with terete or triangular culms sheathed at the base, and grass-like leaves, the latter in some species reduced to a basal sheath; inflorescence capitate or umbellate, subtended by one or more involucral, scaly or leaf-like bracts. Spikes 3 to several flowered, the flowers perfect, stigmas and stamens 2 to 3, the perianth consisting of 1 to several bristles; achene triangular or lenticular, mostly obovate.

a. Inflorescence subtended by the outermost and deciduous scale of the terminal spike or spikelet.
  b. Spike solitary
    c. Culms wiry, terete, smooth; perianth bristles not elongating
      d. Coarse plant forming dense, firm tussocks . . . . *S. caespitosus* ssp. *austriacus*
      d. Delicate, stoloniferous plant forming small tussocks . . . . *S. Rollandii*
    c. Culms soft, 3-angled, scabrous; perianth bristles elongating, white, simulating those of *Eriophorum* . . . . *S. hudsonianus*
  b. Spikes several in compound, terminal head . . . . *S. rufus*
a. Inflorescence subtended by persistent foliaceous bracts
  e. Bracts single, terete, appearing as a continuation of the culm; culm naked; spikelets numerous in glomerules. Plant with stout horizontal rhizome . . . . *S. validus*
  e. Bracts leaf-like; culm leafy
    f. Culm solitary, sharply 3-angled . . . . *S. maritimus* var. *paludosus*
    f. Culms several, obtuse-angled, from leafy bases

g. Sheaths conspicuously tinged with red; styles bifid; achenes lenticular . . . . *S. microcarpus*
g. Sheaths green, drying brown or black; styles 3-fid; achenes triangular . . . . *S. atrocinctus*

***Scirpus atrocinctus** Fern.
Forming small to medium-sized tussocks firmly anchored by a mass of fibrous roots. Culms up to 1 m high, slender and smooth. Leaves mainly from the base of the culm, very scabrous and harsh, 3 to 5 mm broad, brownish black at the base, and as long as the culms. Inflorescence elongated, compound-umbelliform with numerous ovate 2 to 3 mm long, blackish spikes on slender, scabrous branches; scales 1.5 mm long, papery, greenish-black. Bristles light brown, persisting.

Should be looked for in wet lake-shore meadows and marshes of southern District of Mackenzie.

General distribution: Eastern N. America, non-arctic, from Nfld. to Alta., reaching north to L. Athabasca.

Map 324

Scirpus caespitosus L.
ssp. **austriacus** (Pallas) Asch. & Graeb.
Trichophorum caespitosum (L.) Hartm.
Low and very densely caespitose, with wiry, terete, smooth, erect-ascending culms, 10 to 15 cm high; the bases of culms firmly encased in sheaths which persist for many years; the uppermost sheaths bearing a short, callous-tipped, bract-like blade. Spike terminal, 3 to 5 mm long.

Forming firm tussocks in peaty soil, in springy tundra and along brooks; in the North perhaps always in calcareous bogs.

General distribution: Circumplolar, low-arctic.

Fig. 281 Map 325

Scirpus hudsonianus (Michx.) Fern.
Trichophorum alpinum (L.) Pers.
Loosely to densely tufted from a slender horizontally creeping rhizome. Culms 15 to 30 cm tall, rather weak, with scabrous, acute angles. Spike narrowly ovoid, 4 to 7 mm long, much surpassed by the tuft of persistent, slender white bristles. Achenes 1.5 mm long, brownish green.

Common locally from south of Great Slave Lake, north to about lat. 65°, mainly in springy and somewhat calcareous bogs.

General distribution: Circumpolar, with several large gaps.

Fig. 282 Map 326

Scirpus maritimus L.
var. **paludosus** (A. Nels.) Kük.
Culms up to 1 m tall, smooth, 3-angled, from an elongated, nodulose rhizome; leaves flat, pale green, from tight, cylindrical sheaths of the lower half of the culm. Inflorescence glomerulate, of 3 or more sessile spikes, each subtended by one or more leaf-like bracts.

In the District of Mackenzie known only from saline flats in Wood Buffalo Park west of Fort Smith.

General distribution: N. America, non-arctic.

Fig. 283 Map 327

Scirpus microcarpus Presl
Culms stout and leafy, up to 1 m tall, from a stout rhizome; the leaves flat, 10 to 15 mm wide, scabrous beneath. Inflorescence large, compound-umbellate; the spikes 6 to 7 mm long, in small glomerules, their scales and involucres greenish-black. Achenes whitish, distinctly flattened and thin.

Wet marshes and lake shores.

General distribution: Western N. America, north to the upper Mackenzie drainage.

Fig. 284 Map 328

Scirpus Rollandii Fern.
Trichophorum pumilum (M. Vahl) Schinz & Thell. var. *Rollandii* (Fern.) Hult.
Culms 5 to 12 cm long, fresh green, slender and delicate, in small tufts from a deeply buried slender rhizome. Spikes ovoid, few-flowered, 3 to 4 mm long, the flowers lacking perianth; achenes plano-convex, dull black.

In our area thus far known from a single station on the Yellowknife Highway, and from two sites in the Mackenzie Mountains. It should be looked for elsewhere on damp, marly lake shores, and by alkaline seepages.

General distribution: Cordilleran species, with a few widely disjunct stations in Que. and southern Lab.

Map 329

Scirpus rufus (Huds.) Schrad.
Blysmus rufus (Huds.) Link
Culms 10 to 30 cm tall, leafy below, commonly in a straight row from an extensively creeping and freely branched rhizome; leaves narrow and stiff, shorter than the culms. The solitary spike compound, 1 to 2 cm long, formed of two

rows of flattened and spreading, dark brown spikelets.

A rare species, with us known only from two stations; one on saline, wet river banks near Wrigley in the Mackenzie Valley, and the other in the Mackenzie Mountains in a saline meadow adjacent to the Keele River; its nearest known stations are Churchill, Manitoba and Anchorage, Alaska.

General distribution: Circumpolar, with very large gaps.

Fig. 285 Map 330

Scirpus validus Vahl
Culms smooth, spongy and easily compressed, 0.5 to 2.0 m high, from a stout fibrous-rooted and creeping rhizome. Inflorescence umbelliform, of short-peduncled glomerules. Spikes ovoid, 5 to 10 mm long, dark olivaceous-brown when mature; achenes obovate, plano-convex, 1.5 to 2.5 mm long.

Forming dense colonies in water up to 1 m deep along sheltered lake shores, north to Norman Wells on the Mackenzie River.

General distribution: N. America from Nfld. to southeast and central Alaska.

Fig. 286 Map 331

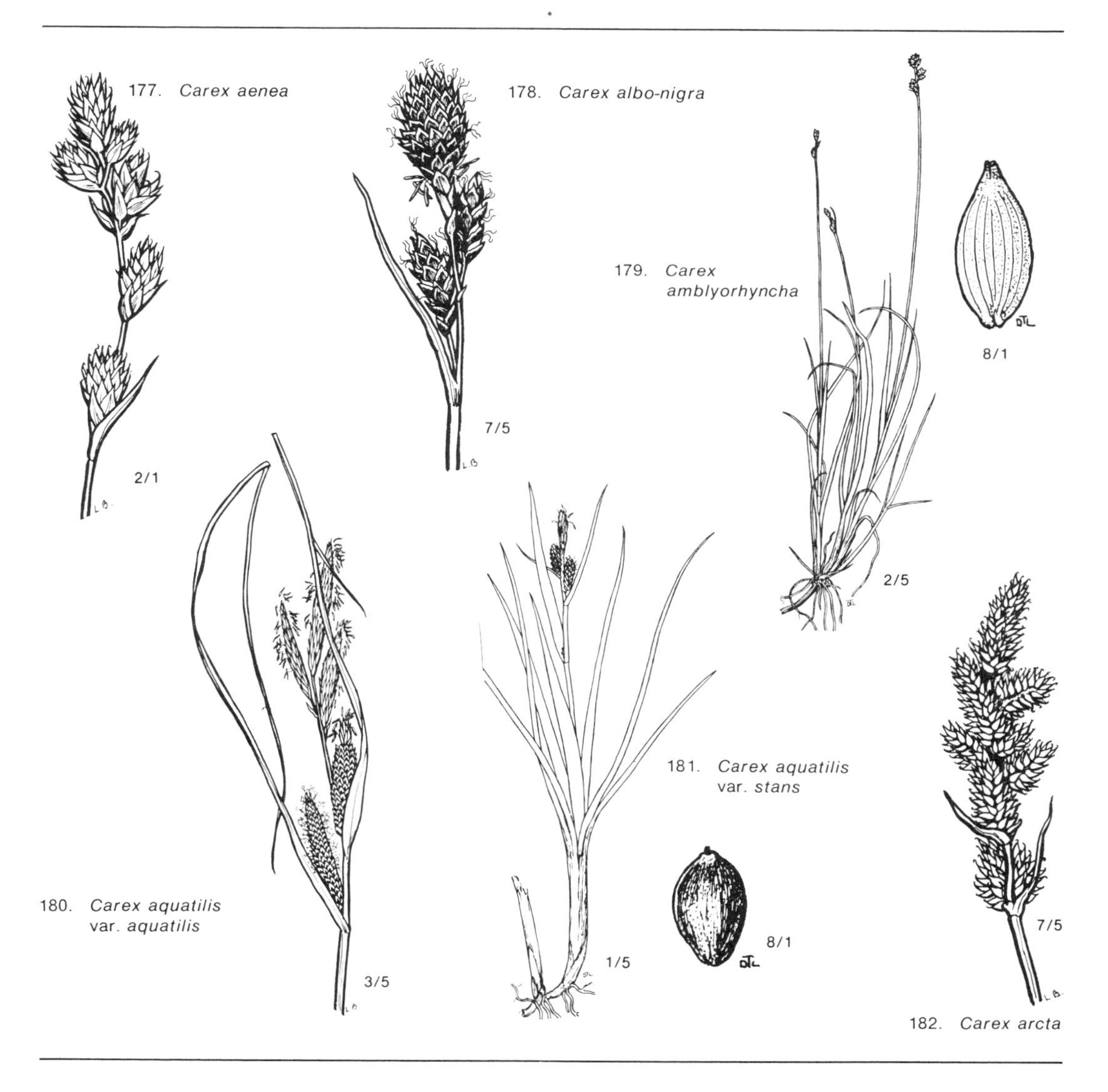

177. *Carex aenea*

178. *Carex albo-nigra*

179. *Carex amblyorhyncha*

180. *Carex aquatilis* var. *aquatilis*

181. *Carex aquatilis* var. *stans*

182. *Carex arcta*

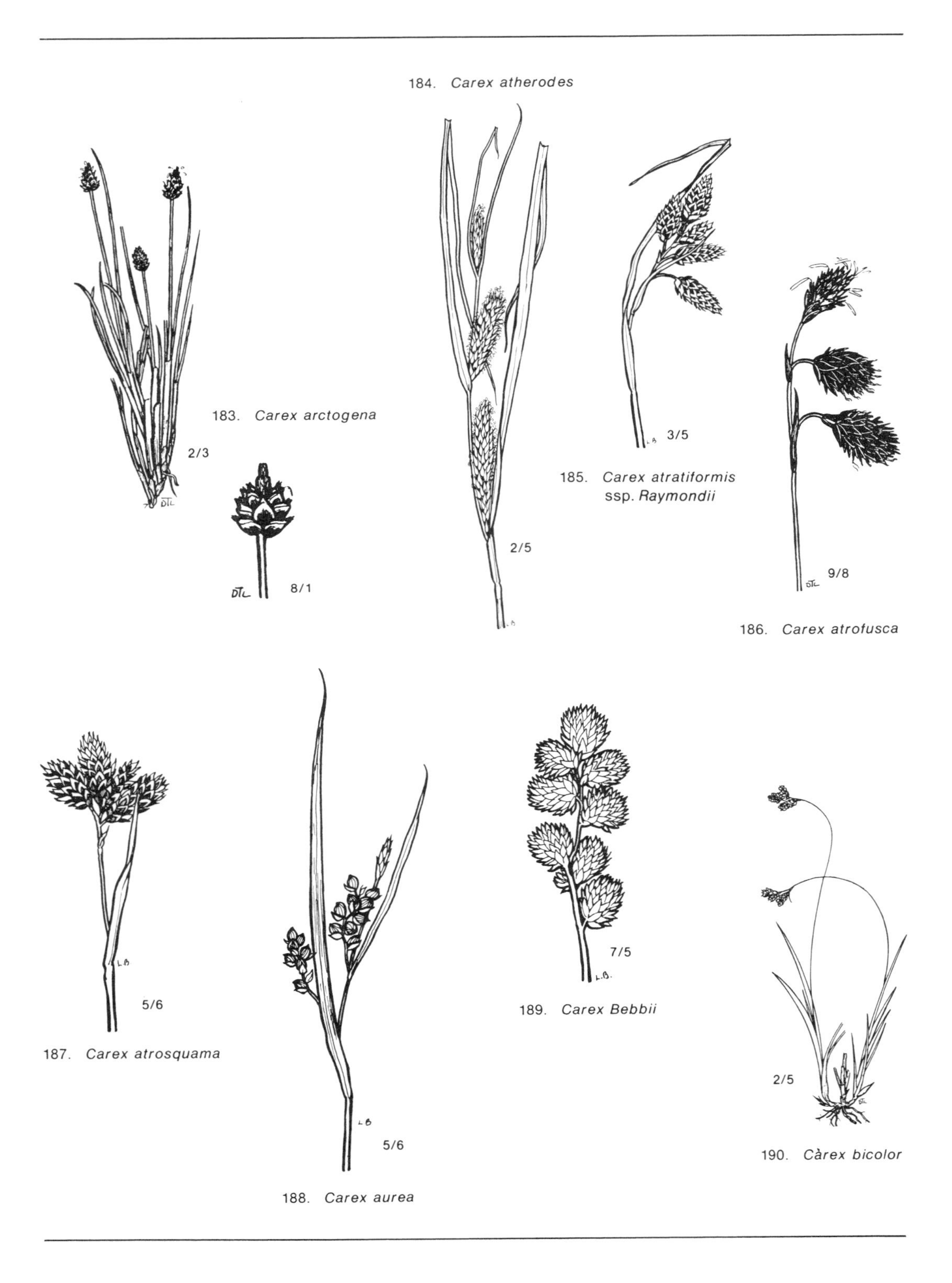

183. *Carex arctogena*

184. *Carex atherodes*

185. *Carex atratiformis* ssp. *Raymondii*

186. *Carex atrofusca*

187. *Carex atrosquama*

188. *Carex aurea*

189. *Carex Bebbii*

190. *Càrex bicolor*

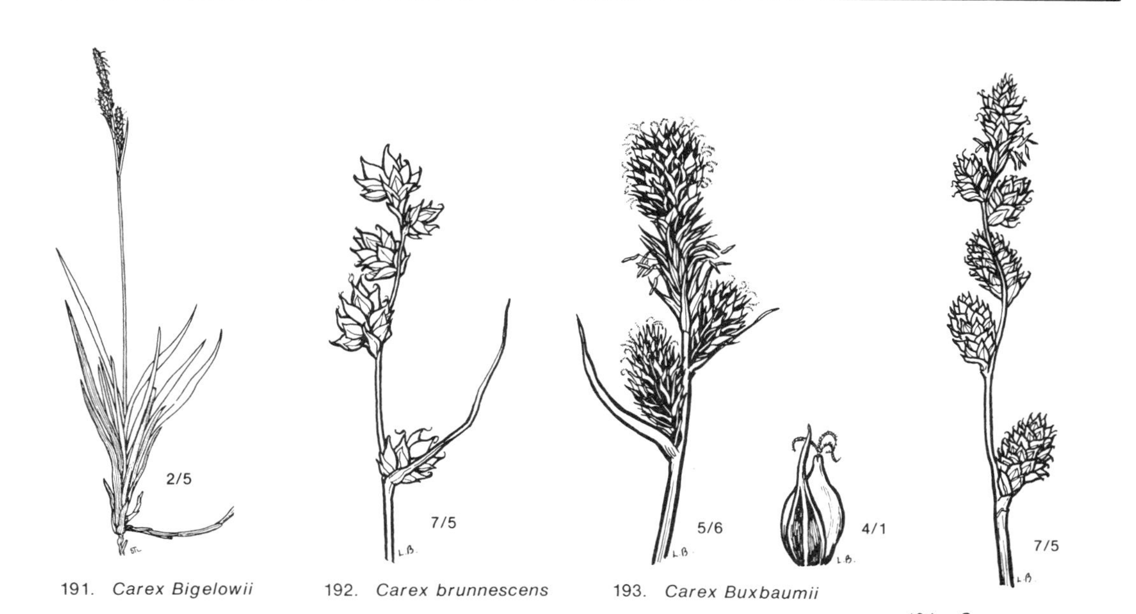

191. *Carex Bigelowii*

192. *Carex brunnescens*

193. *Carex Buxbaumii*

194. *Carex canescens*

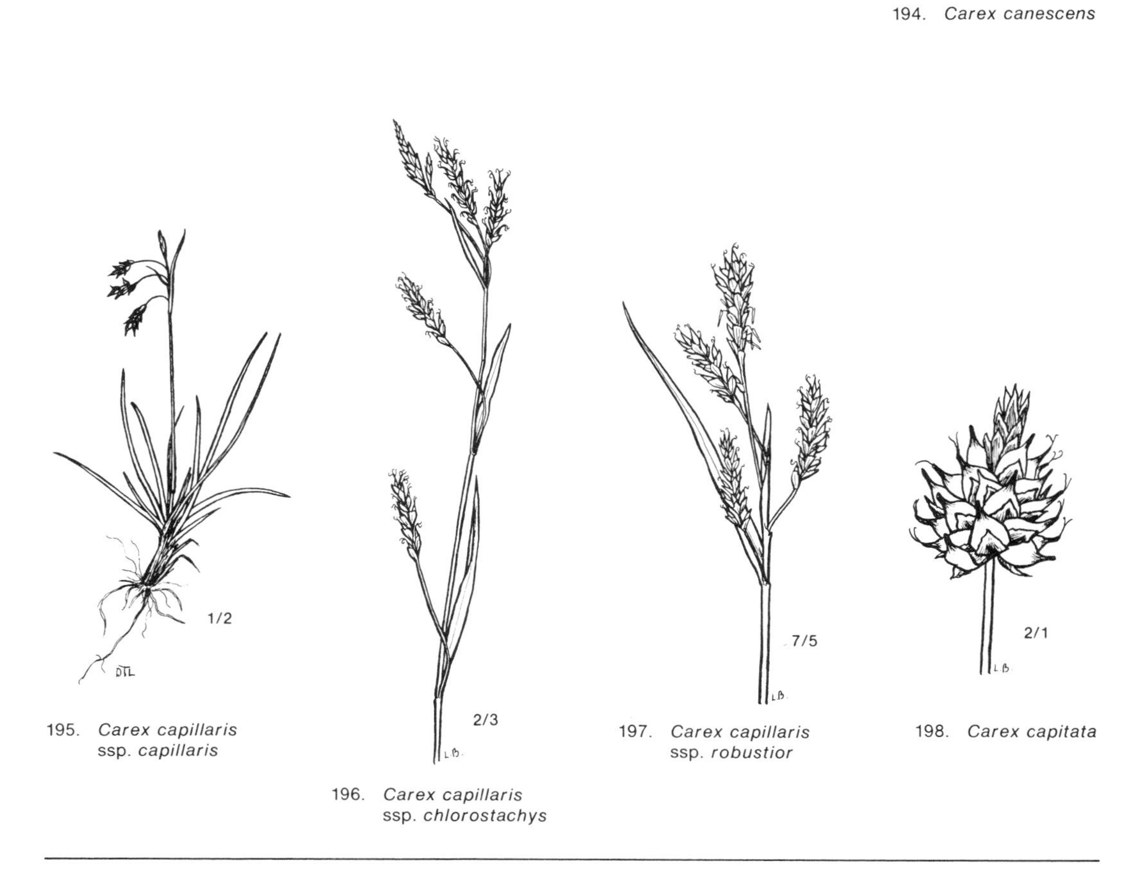

195. *Carex capillaris* ssp. *capillaris*

196. *Carex capillaris* ssp. *chlorostachys*

197. *Carex capillaris* ssp. *robustior*

198. *Carex capitata*

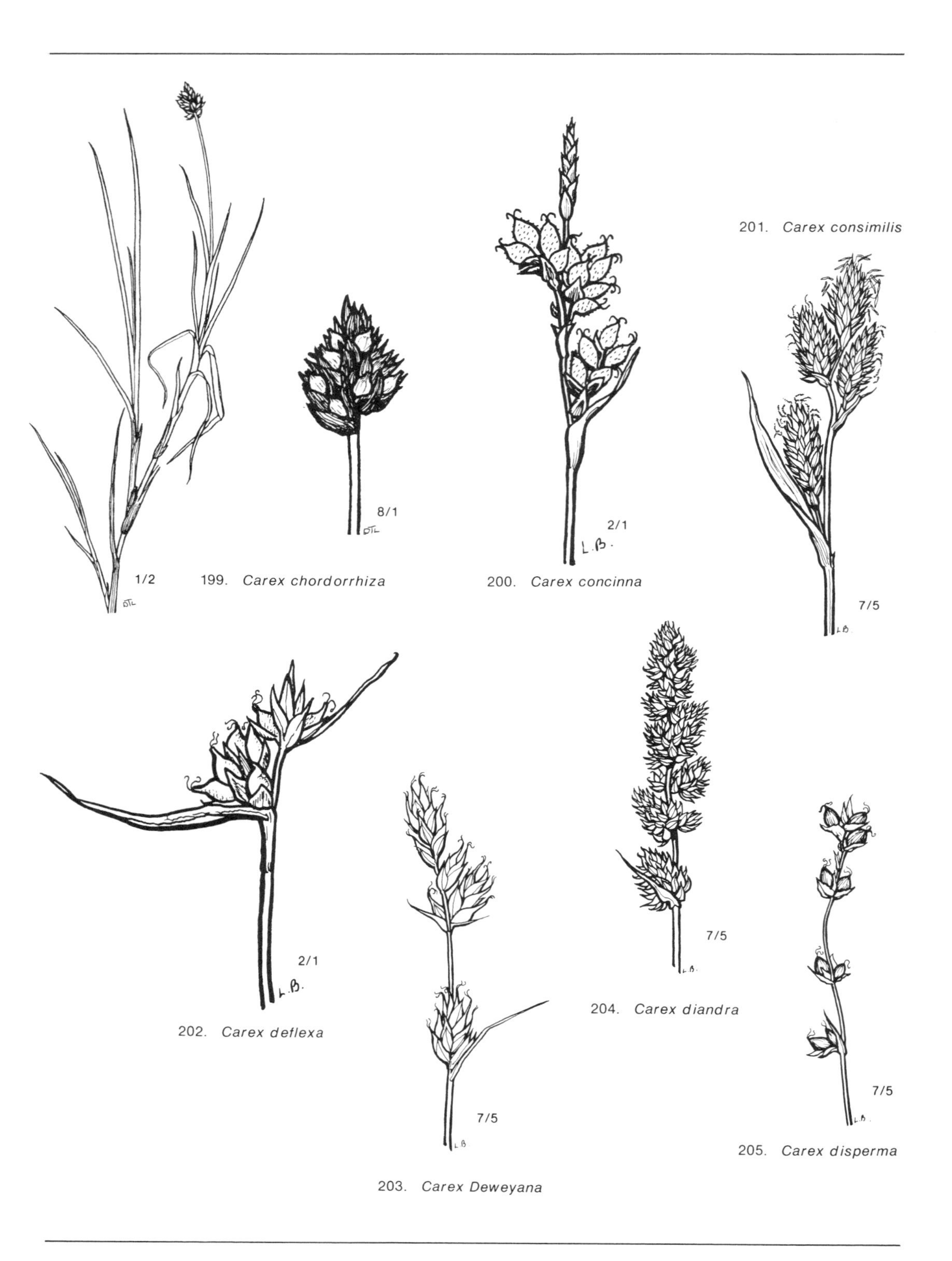

199. *Carex chordorrhiza*

200. *Carex concinna*

201. *Carex consimilis*

202. *Carex deflexa*

203. *Carex Deweyana*

204. *Carex diandra*

205. *Carex disperma*

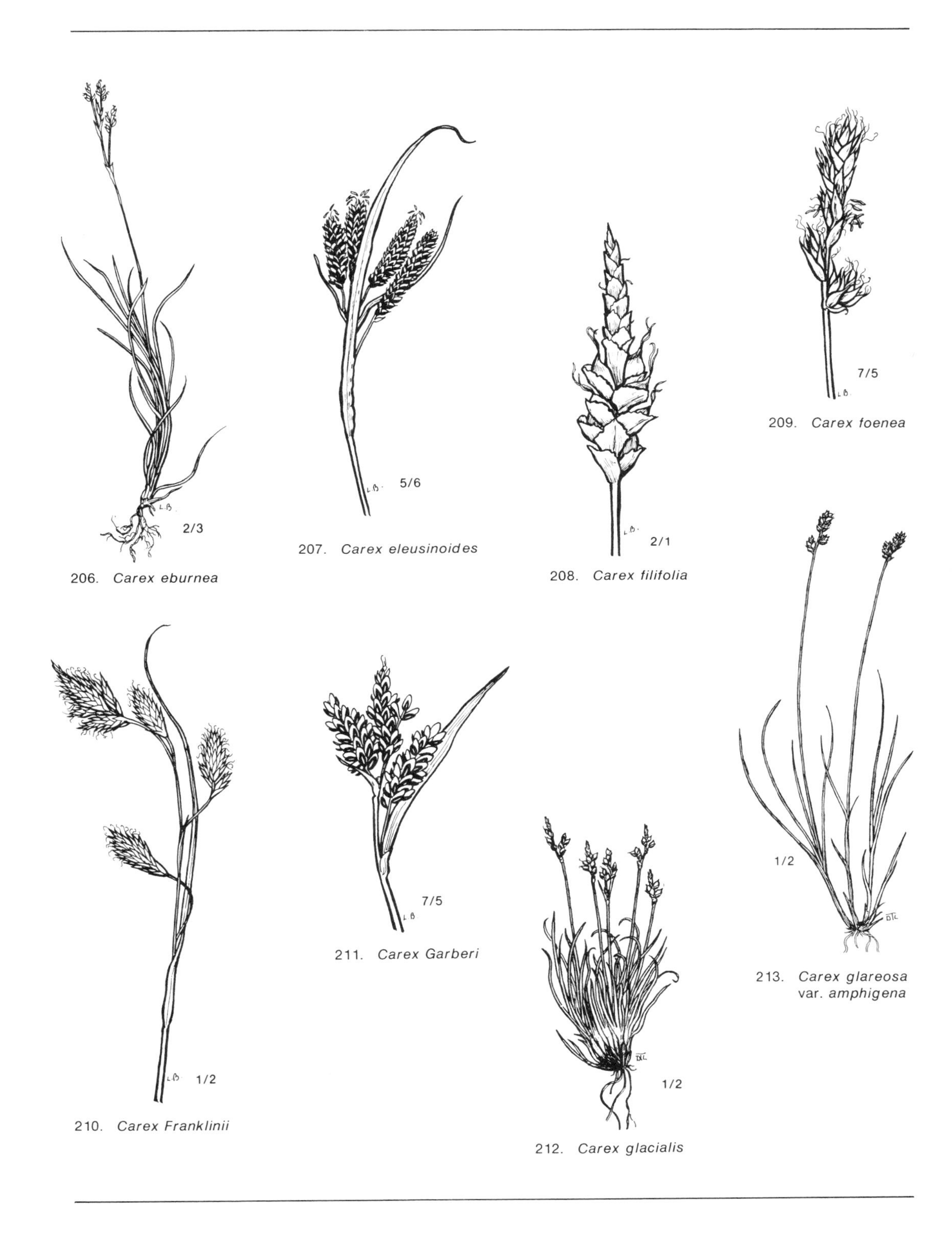

206. *Carex eburnea*

207. *Carex eleusinoides*

208. *Carex filifolia*

209. *Carex foenea*

210. *Carex Franklinii*

211. *Carex Garberi*

212. *Carex glacialis*

213. *Carex glareosa* var. *amphigena*

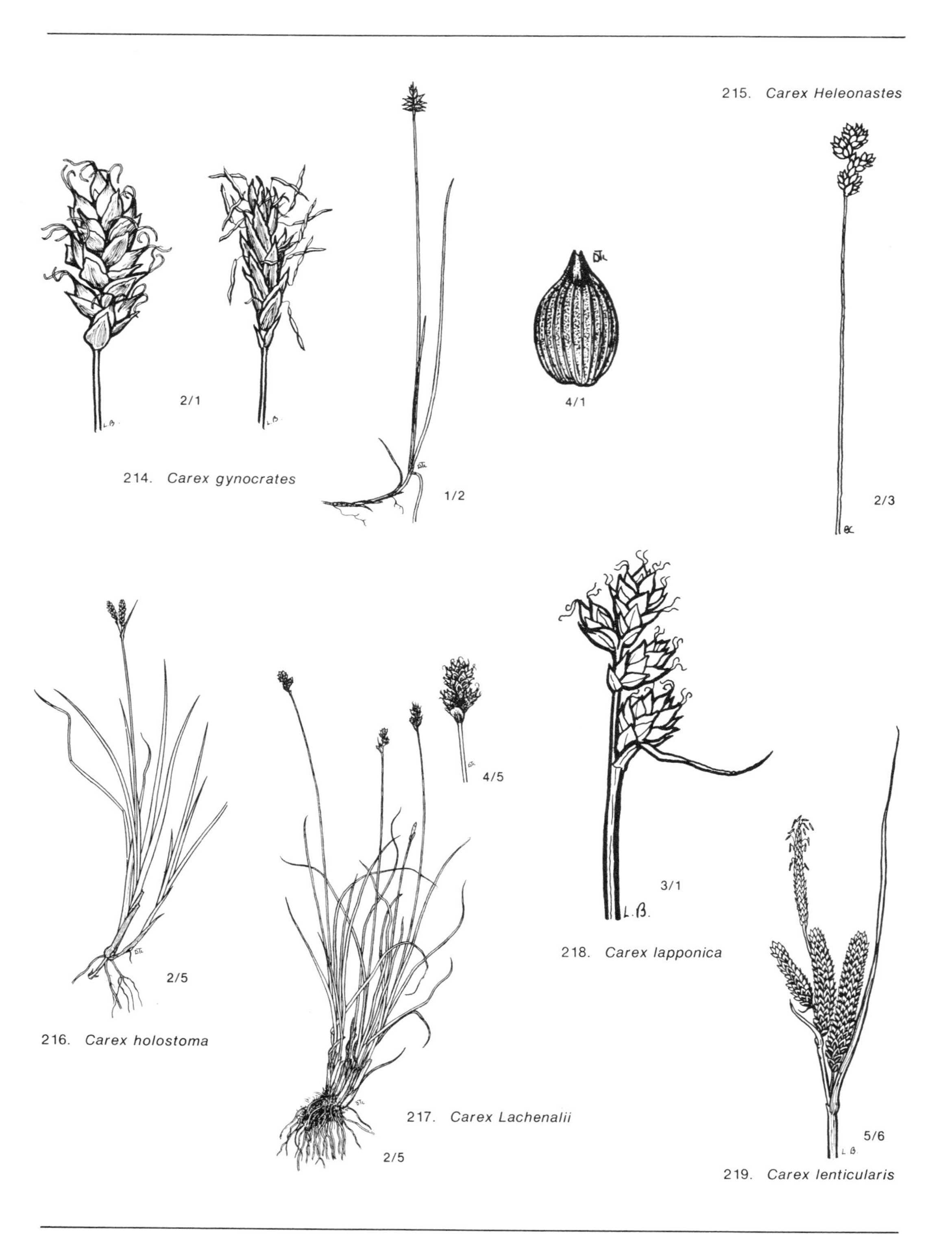

214. *Carex gynocrates*

215. *Carex Heleonastes*

216. *Carex holostoma*

217. *Carex Lachenalii*

218. *Carex lapponica*

219. *Carex lenticularis*

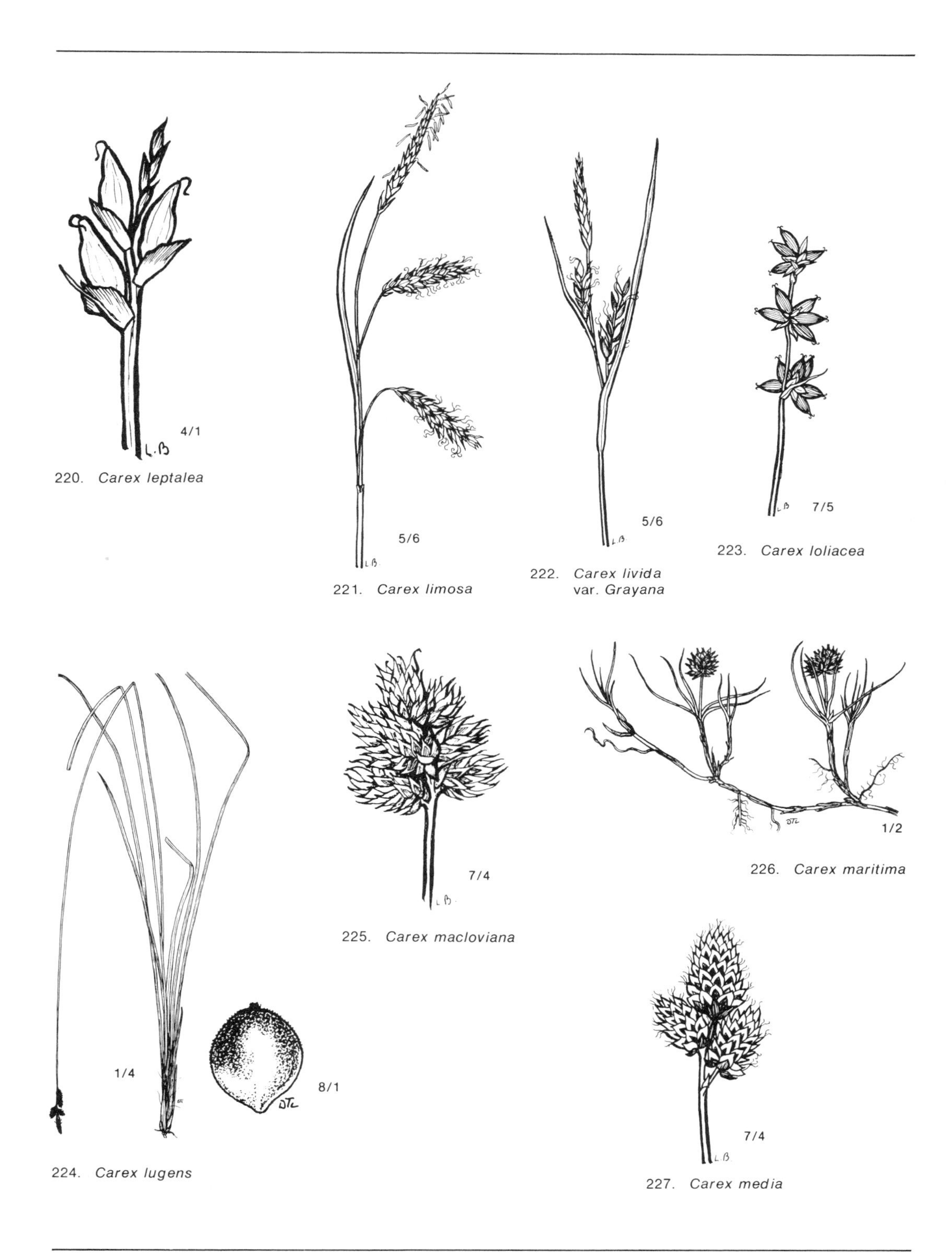

220. *Carex leptalea*

221. *Carex limosa*

222. *Carex livida* var. *Grayana*

223. *Carex loliacea*

224. *Carex lugens*

225. *Carex macloviana*

226. *Carex maritima*

227. *Carex media*

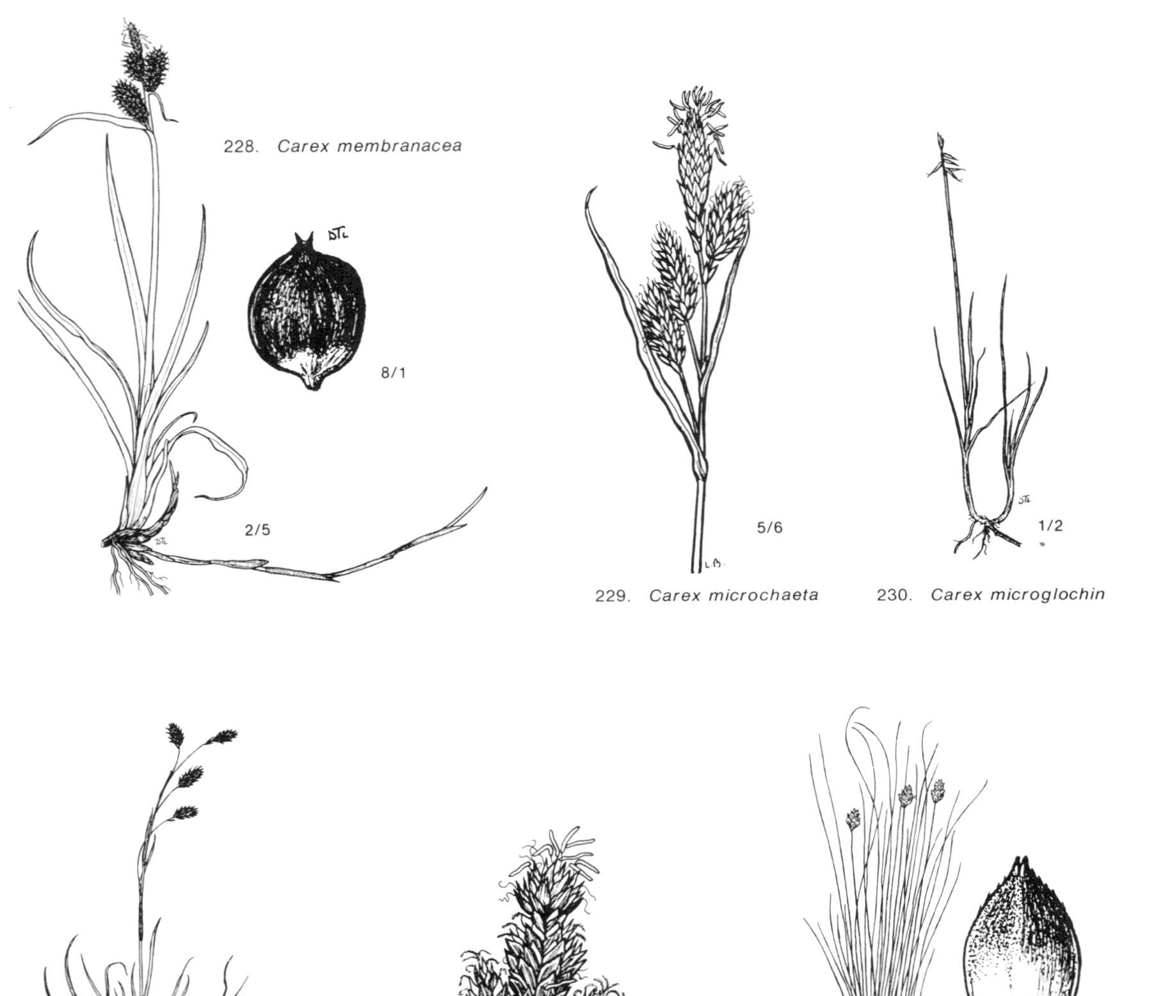

228. *Carex membranacea*

229. *Carex microchaeta*

230. *Carex microglochin*

231. *Carex misandra*

232. *Carex Morrisseyi*

233. *Carex nardina*

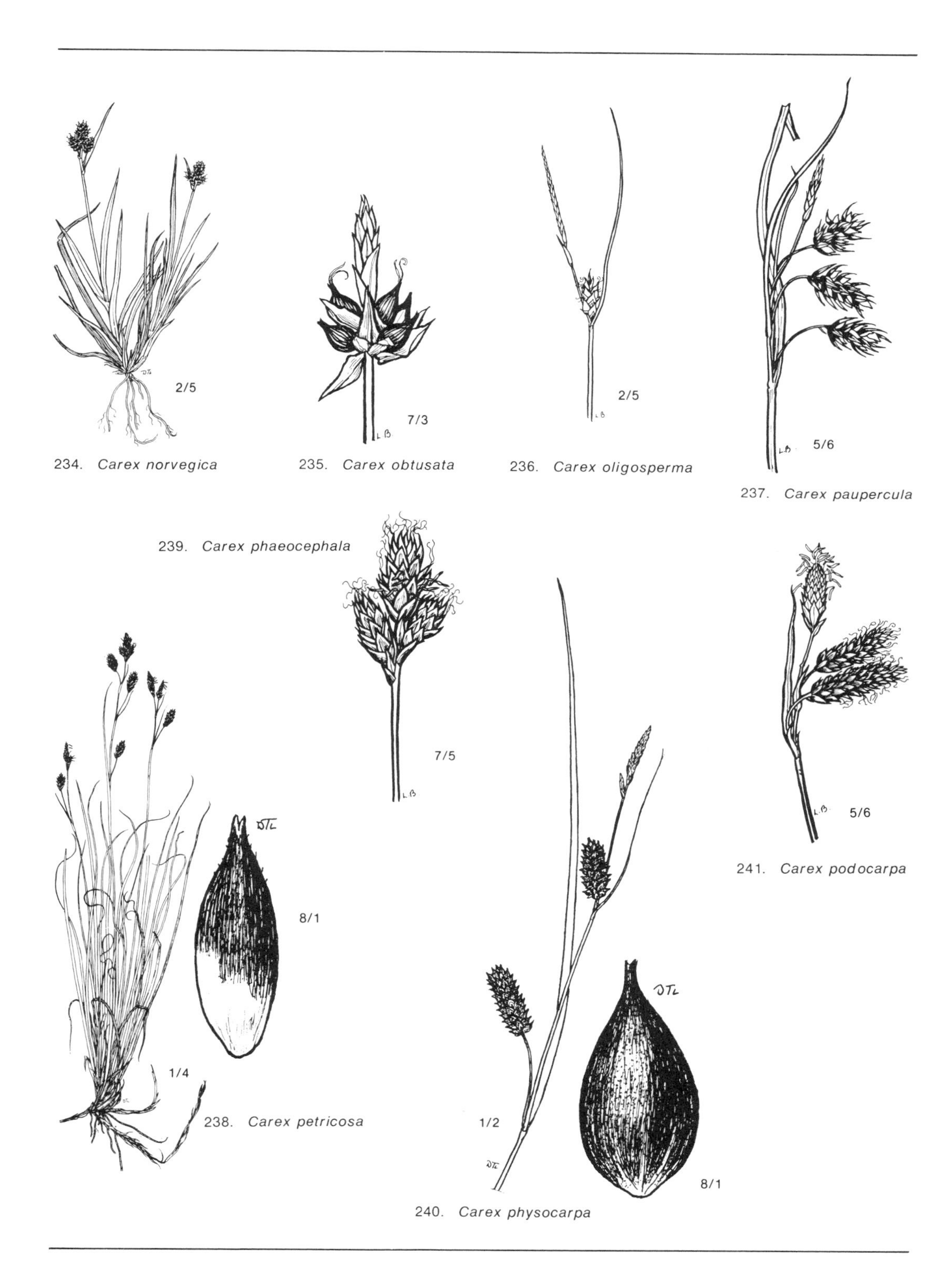

234. *Carex norvegica*

235. *Carex obtusata*

236. *Carex oligosperma*

237. *Carex paupercula*

238. *Carex petricosa*

239. *Carex phaeocephala*

240. *Carex physocarpa*

241. *Carex podocarpa*

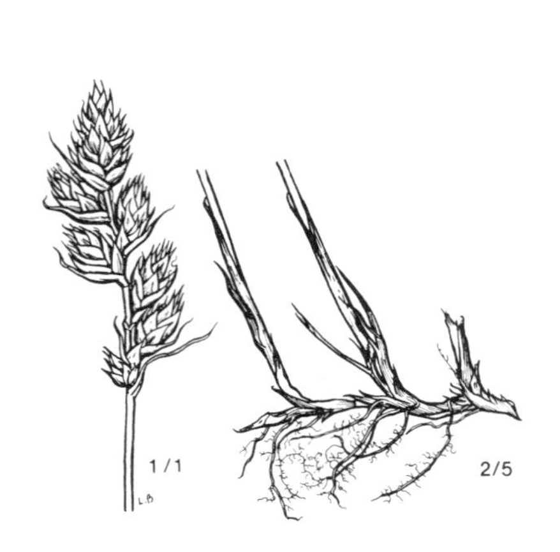

242. *Carex praegracilis*

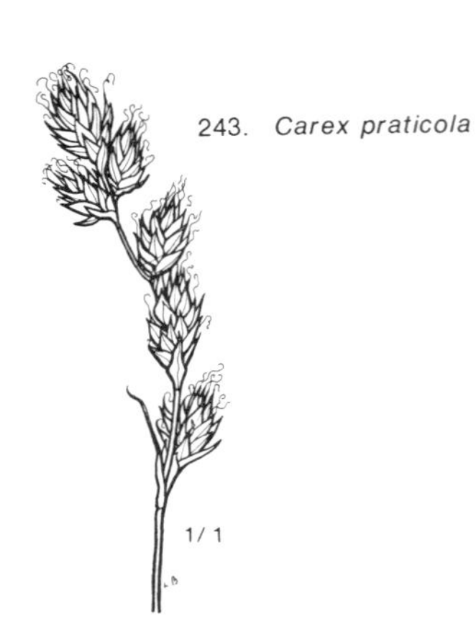

243. *Carex praticola*

245. *Carex rariflora*

244. *Carex pyrenaica* ssp. *micropoda*

246. *Carex retrorsa*

247. *Carex Rossii*

248. *Carex rostrata*

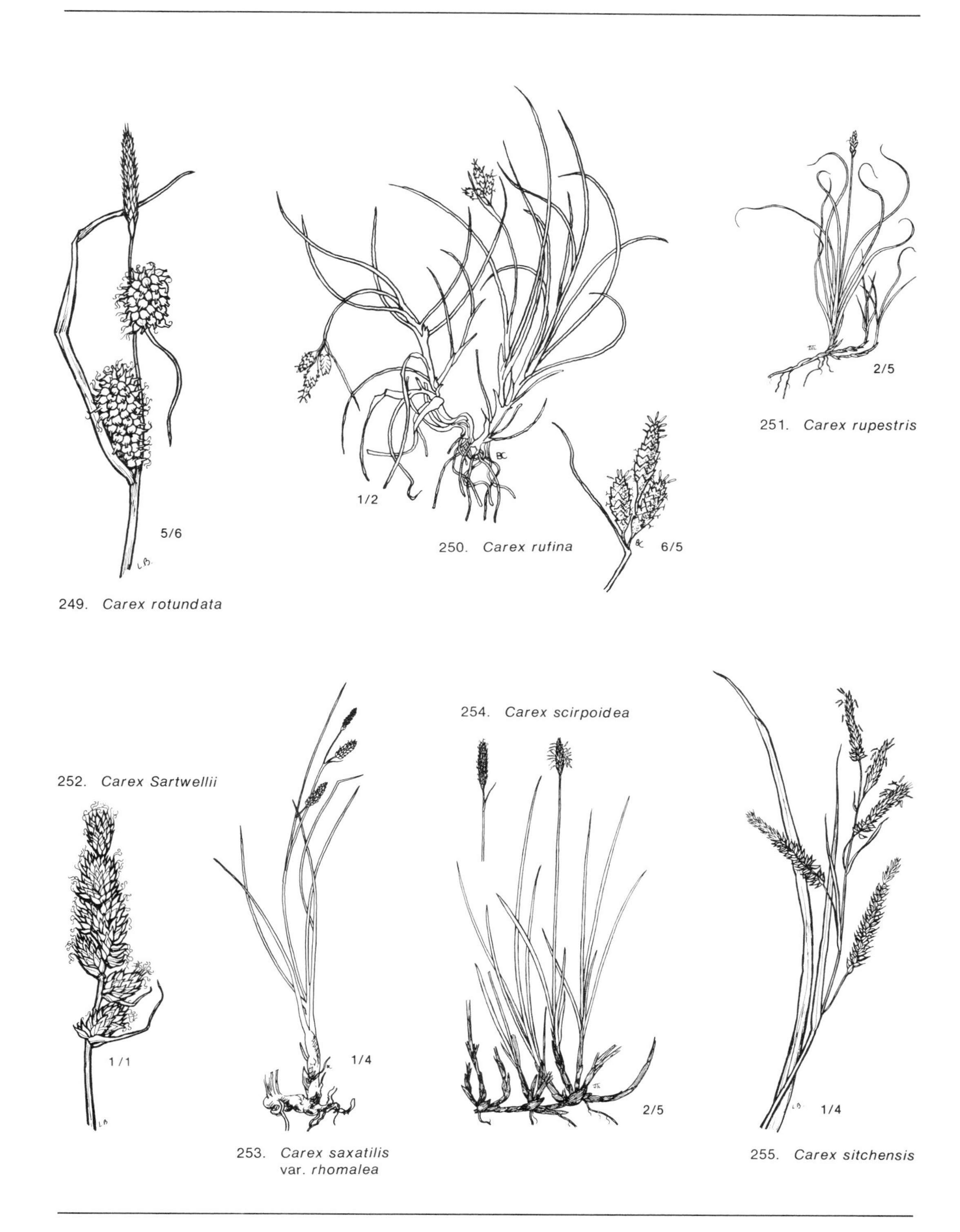

249. *Carex rotundata*

250. *Carex rufina*

251. *Carex rupestris*

252. *Carex Sartwellii*

253. *Carex saxatilis* var. *rhomalea*

254. *Carex scirpoidea*

255. *Carex sitchensis*

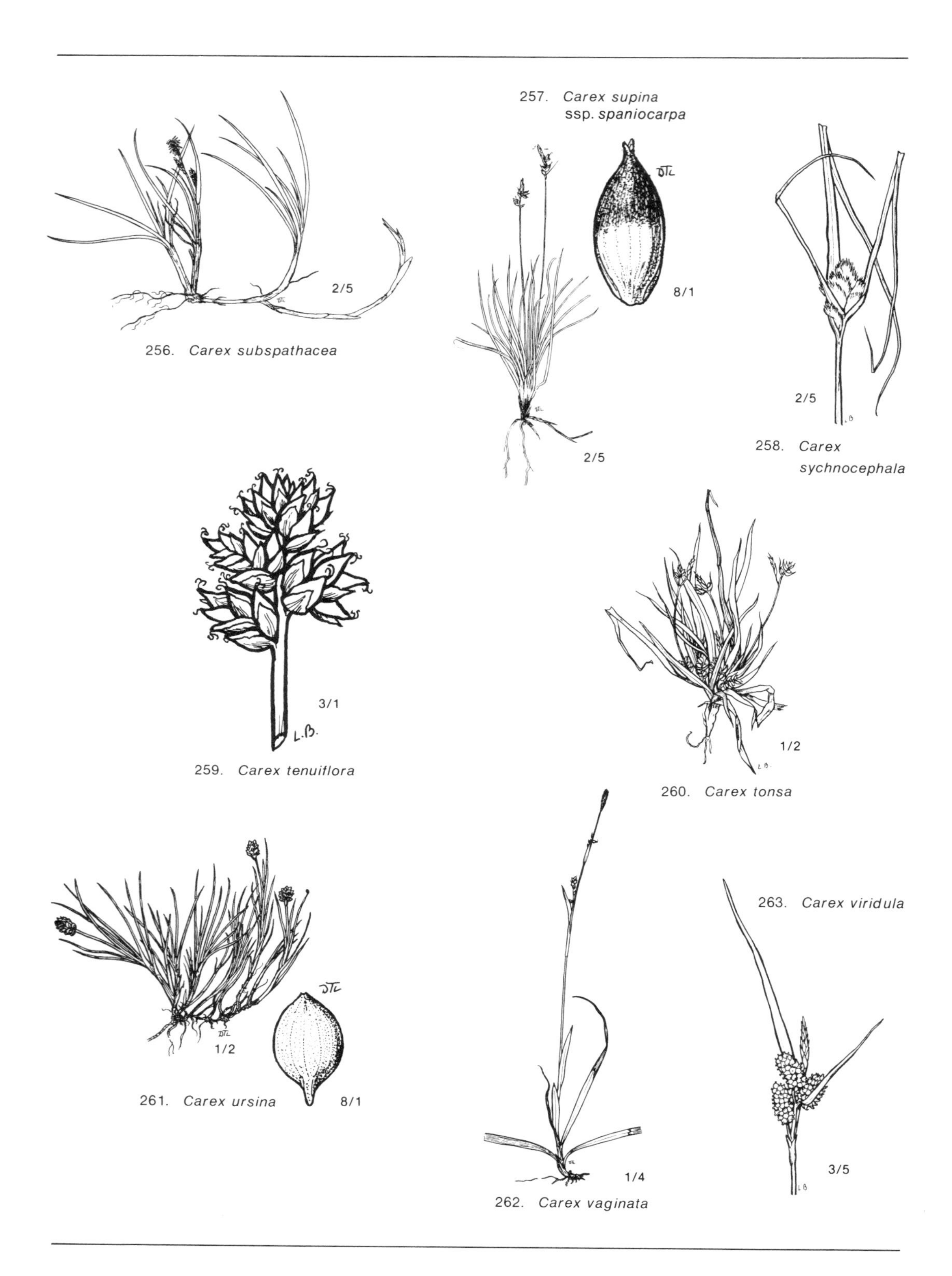

256. *Carex subspathacea*

257. *Carex supina* ssp. *spaniocarpa*

258. *Carex sychnocephala*

259. *Carex tenuiflora*

260. *Carex tonsa*

261. *Carex ursina*

262. *Carex vaginata*

263. *Carex viridula*

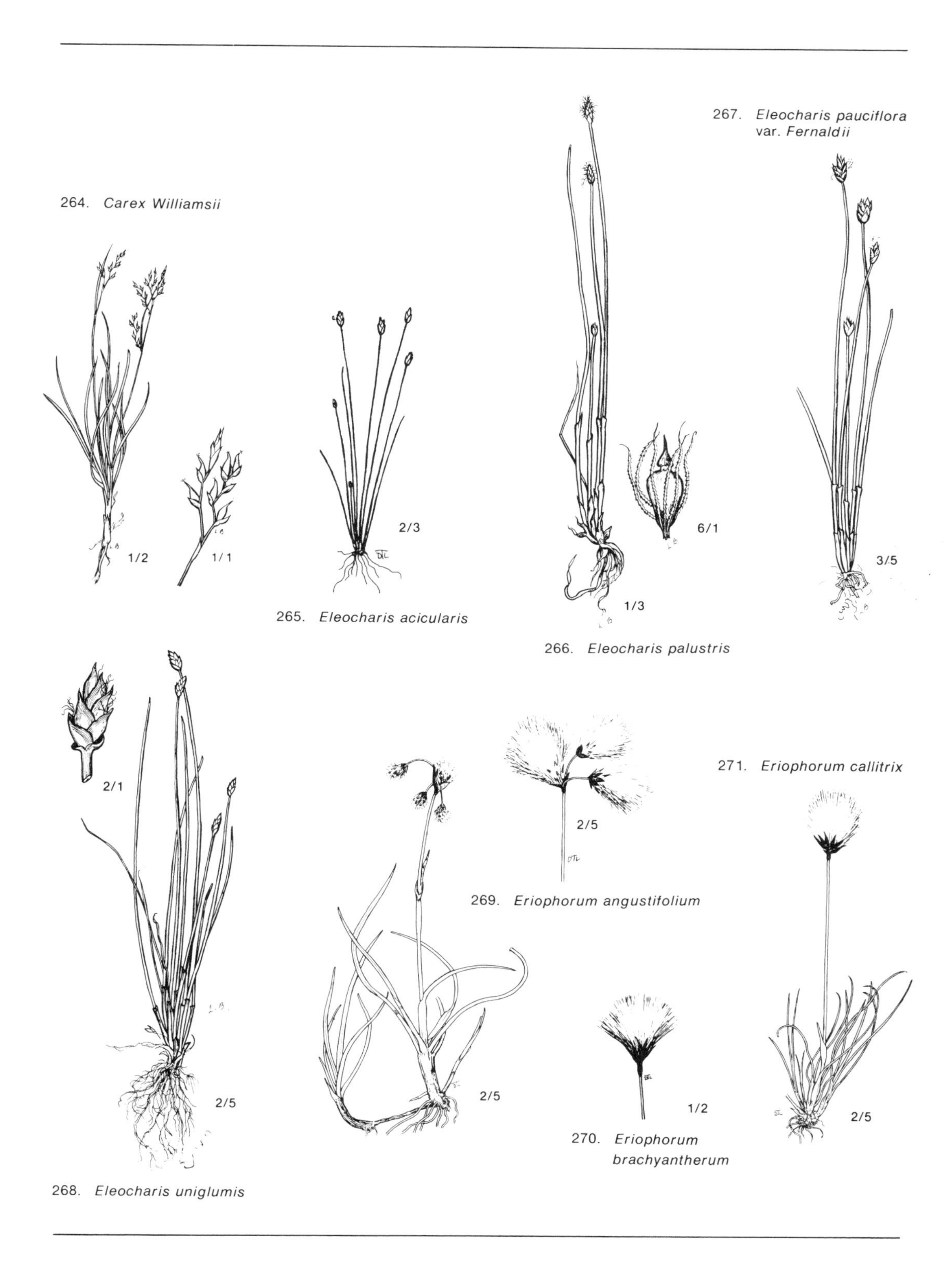

264. *Carex Williamsii*

265. *Eleocharis acicularis*

266. *Eleocharis palustris*

267. *Eleocharis pauciflora* var. *Fernaldii*

268. *Eleocharis uniglumis*

269. *Eriophorum angustifolium*

270. *Eriophorum brachyantherum*

271. *Eriophorum callitrix*

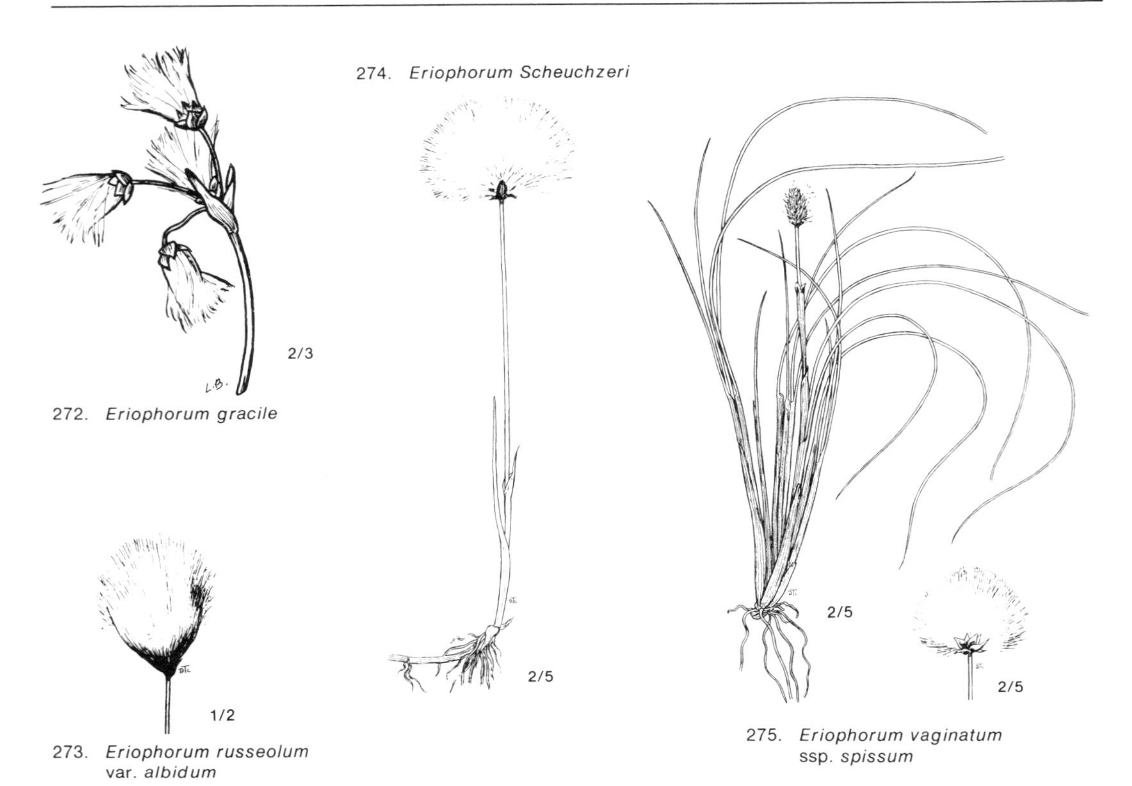

272. *Eriophorum gracile*

273. *Eriophorum russeolum* var. *albidum*

274. *Eriophorum Scheuchzeri*

275. *Eriophorum vaginatum* ssp. *spissum*

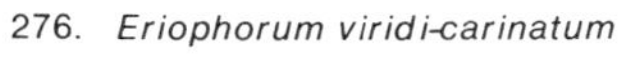

276. *Eriophorum viridi-carinatum*

277. *Kobresia hyperborea*

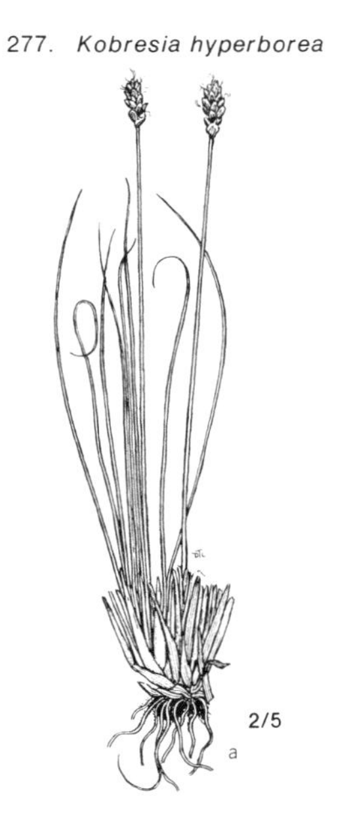

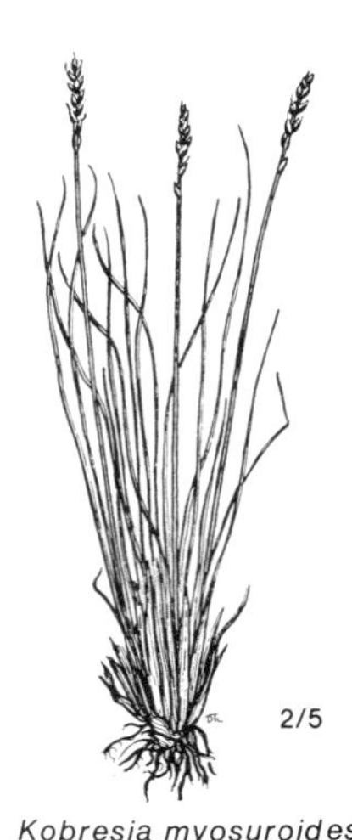

278. *Kobresia myosuroides*

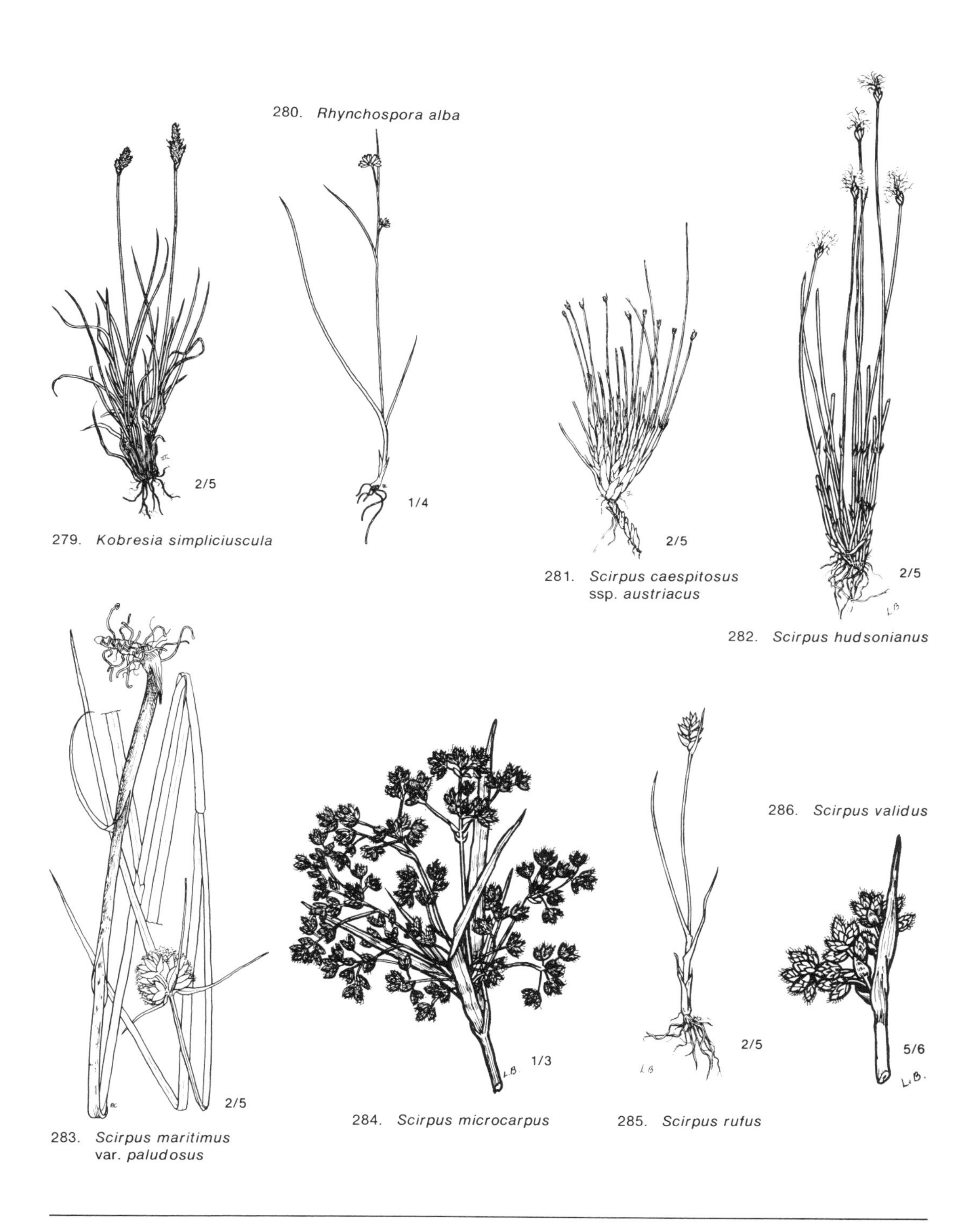

279. *Kobresia simpliciuscula*

280. *Rhynchospora alba*

281. *Scirpus caespitosus* ssp. *austriacus*

282. *Scirpus hudsonianus*

283. *Scirpus maritimus* var. *paludosus*

284. *Scirpus microcarpus*

285. *Scirpus rufus*

286. *Scirpus validus*

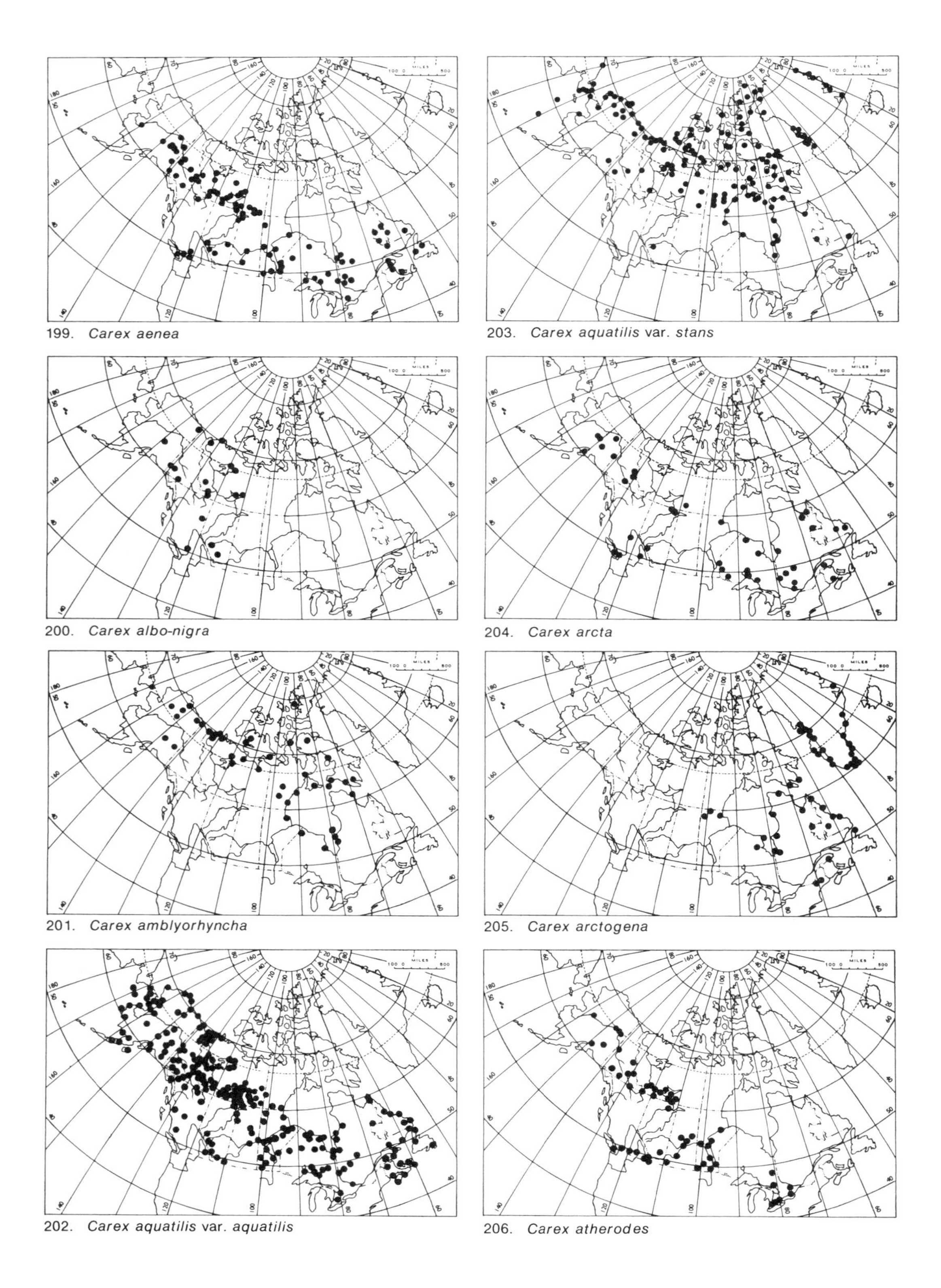

199. *Carex aenea*

200. *Carex albo-nigra*

201. *Carex amblyorhyncha*

202. *Carex aquatilis* var. *aquatilis*

203. *Carex aquatilis* var. *stans*

204. *Carex arcta*

205. *Carex arctogena*

206. *Carex atherodes*

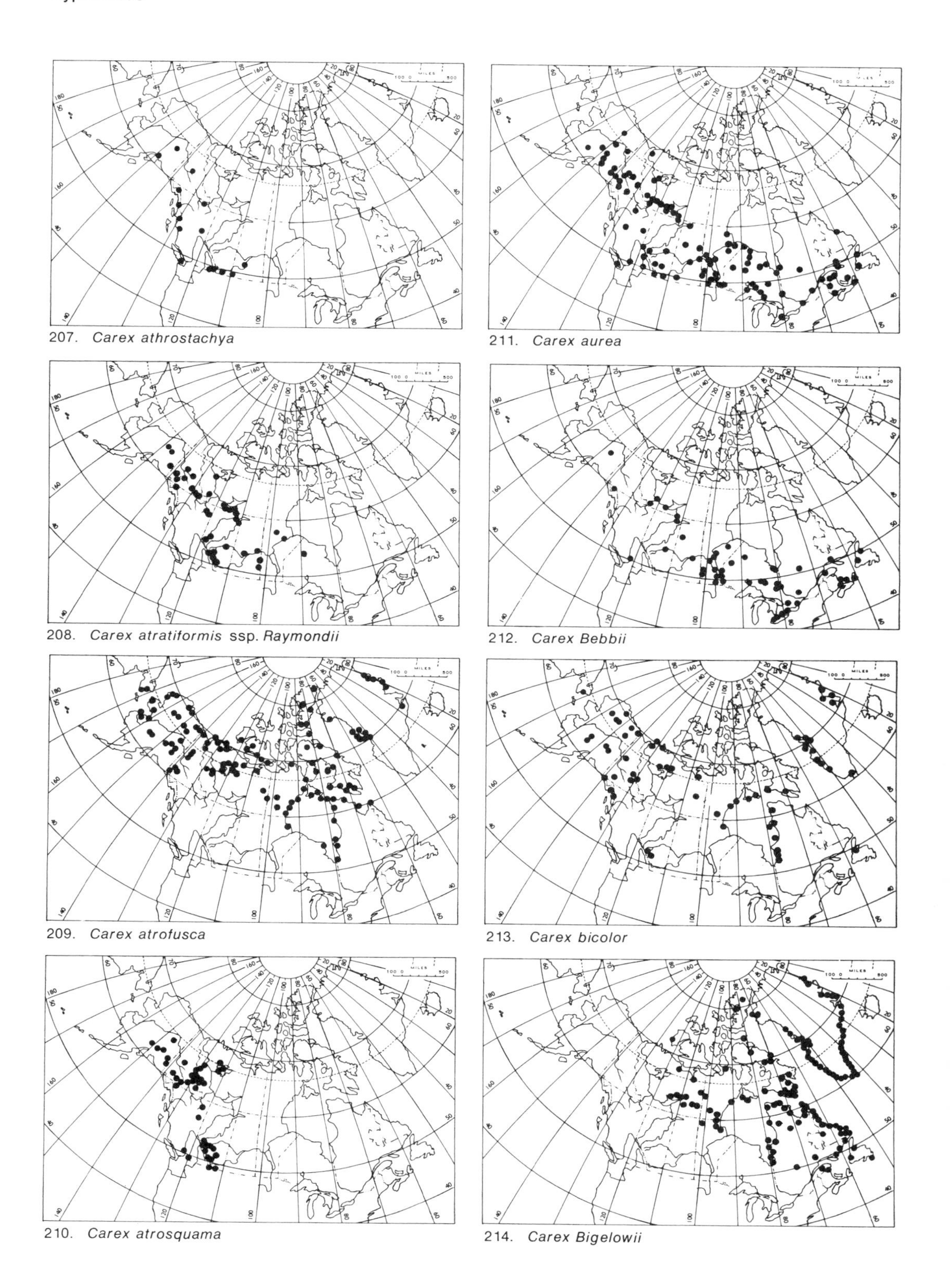

207. *Carex athrostachya*

208. *Carex atratiformis* ssp. *Raymondii*

209. *Carex atrofusca*

210. *Carex atrosquama*

211. *Carex aurea*

212. *Carex Bebbii*

213. *Carex bicolor*

214. *Carex Bigelowii*

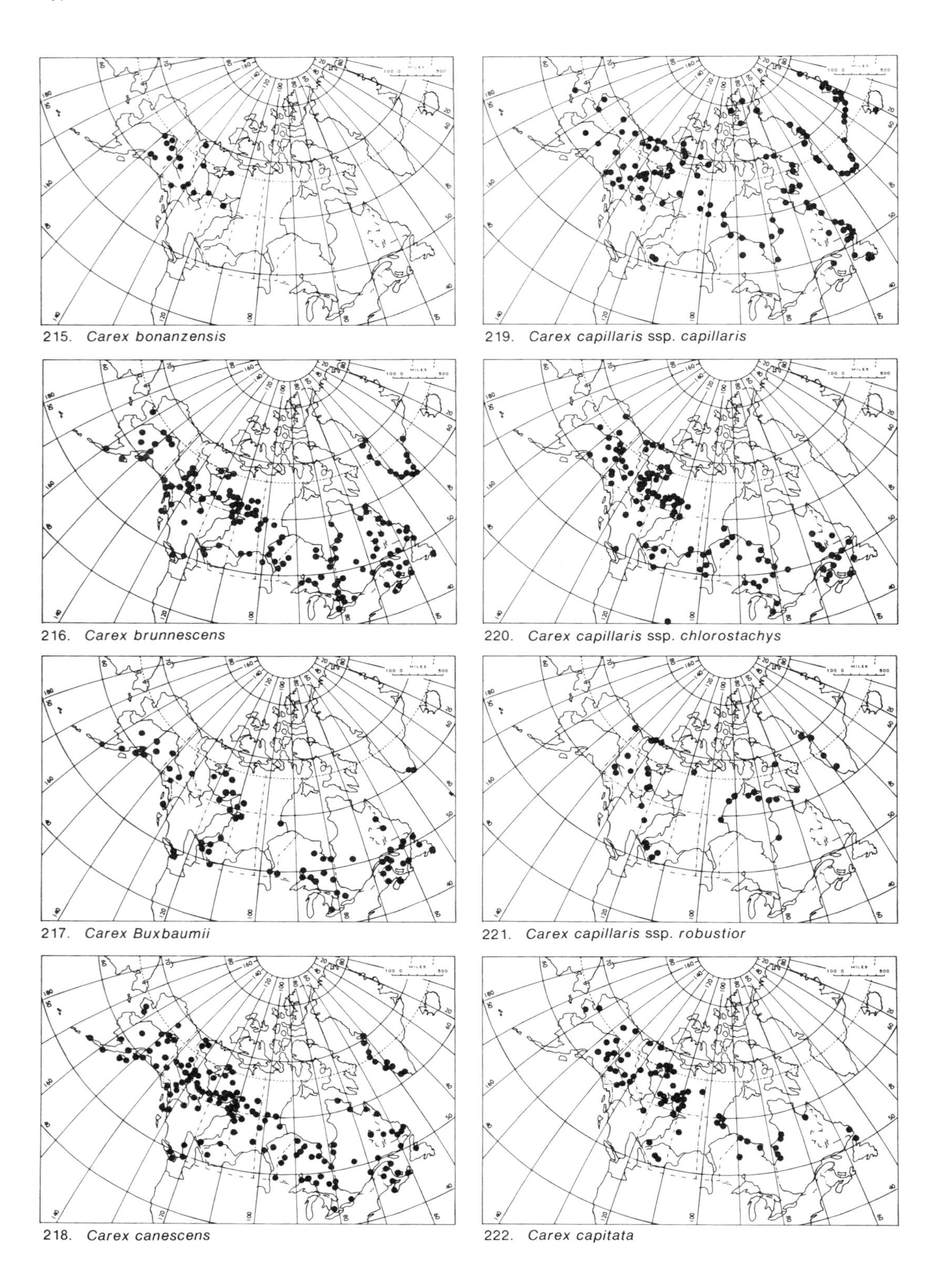

215. *Carex bonanzensis*

216. *Carex brunnescens*

217. *Carex Buxbaumii*

218. *Carex canescens*

219. *Carex capillaris* ssp. *capillaris*

220. *Carex capillaris* ssp. *chlorostachys*

221. *Carex capillaris* ssp. *robustior*

222. *Carex capitata*

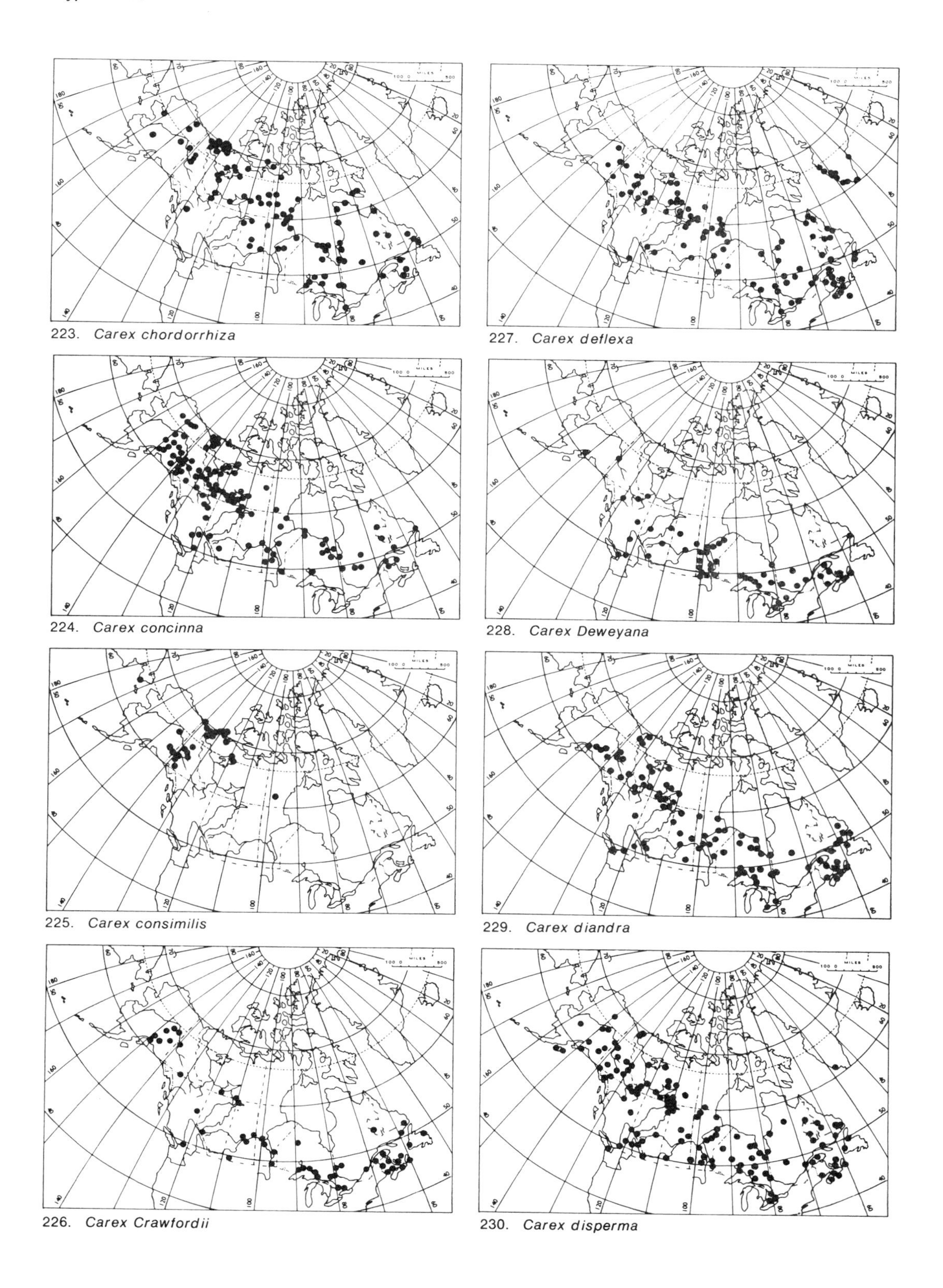

223. *Carex chordorrhiza*

224. *Carex concinna*

225. *Carex consimilis*

226. *Carex Crawfordii*

227. *Carex deflexa*

228. *Carex Deweyana*

229. *Carex diandra*

230. *Carex disperma*

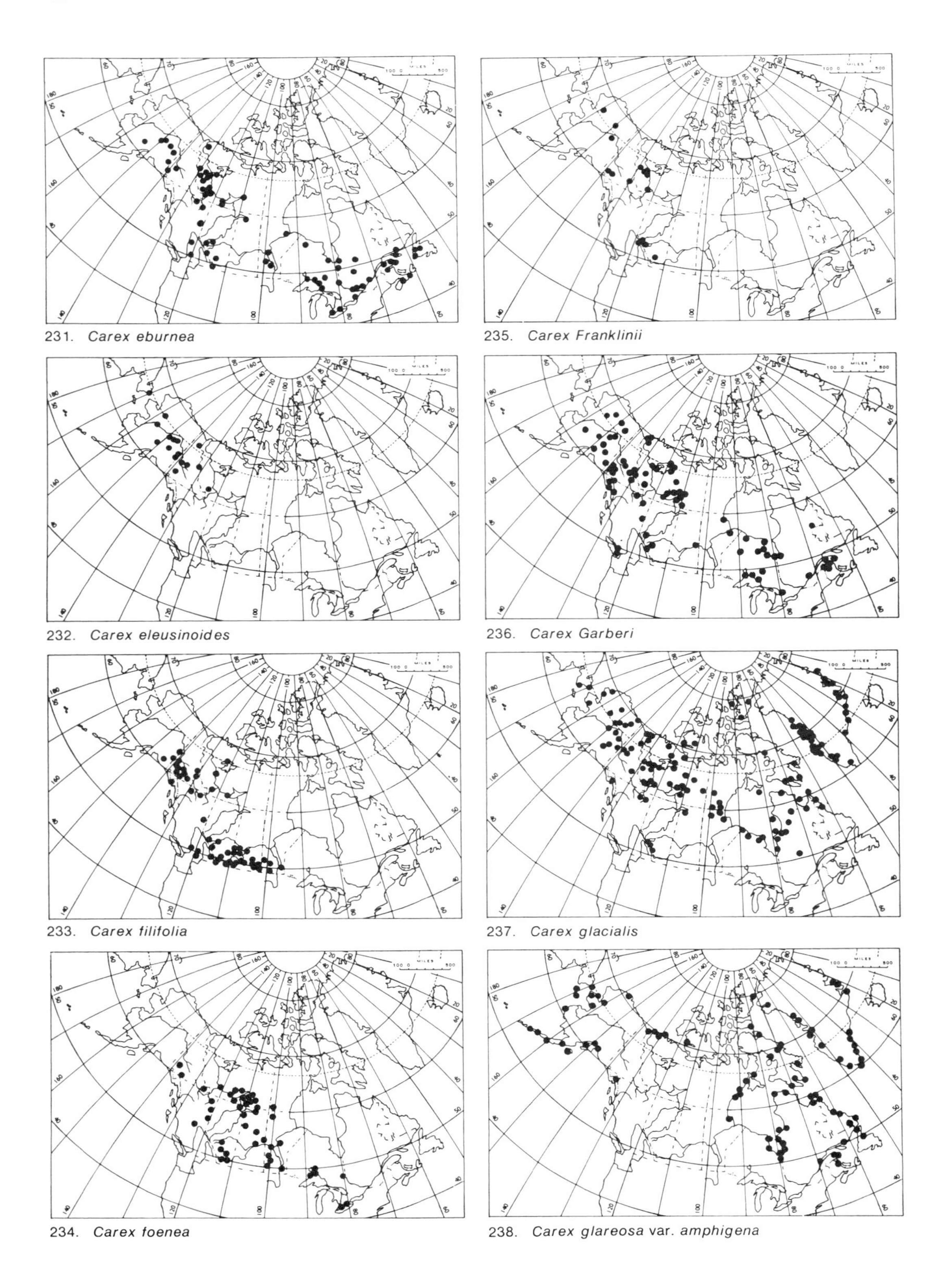

231. *Carex eburnea*

232. *Carex eleusinoides*

233. *Carex filifolia*

234. *Carex foenea*

235. *Carex Franklinii*

236. *Carex Garberi*

237. *Carex glacialis*

238. *Carex glareosa* var. *amphigena*

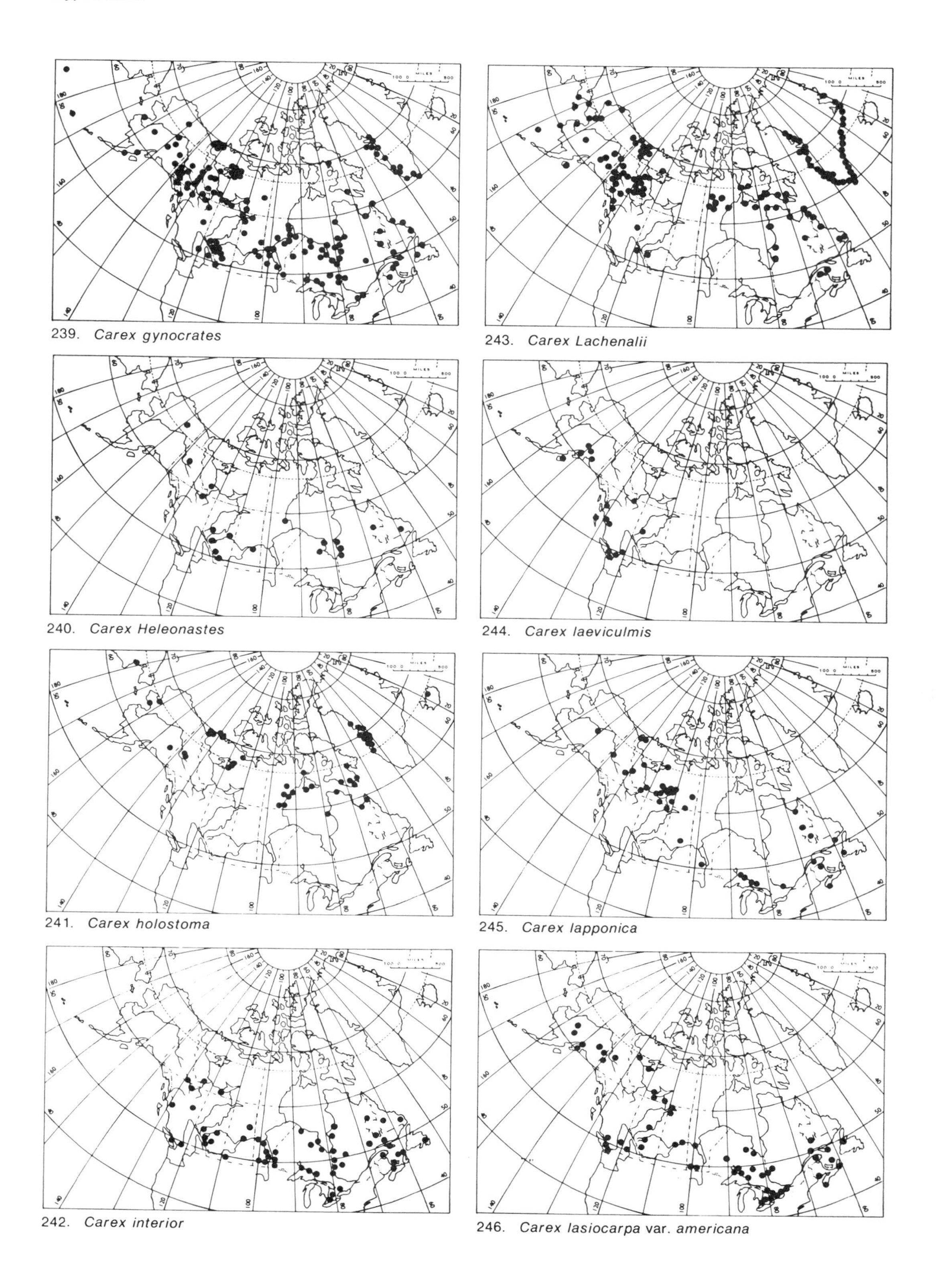

239. *Carex gynocrates*

240. *Carex Heleonastes*

241. *Carex holostoma*

242. *Carex interior*

243. *Carex Lachenalii*

244. *Carex laeviculmis*

245. *Carex lapponica*

246. *Carex lasiocarpa* var. *americana*

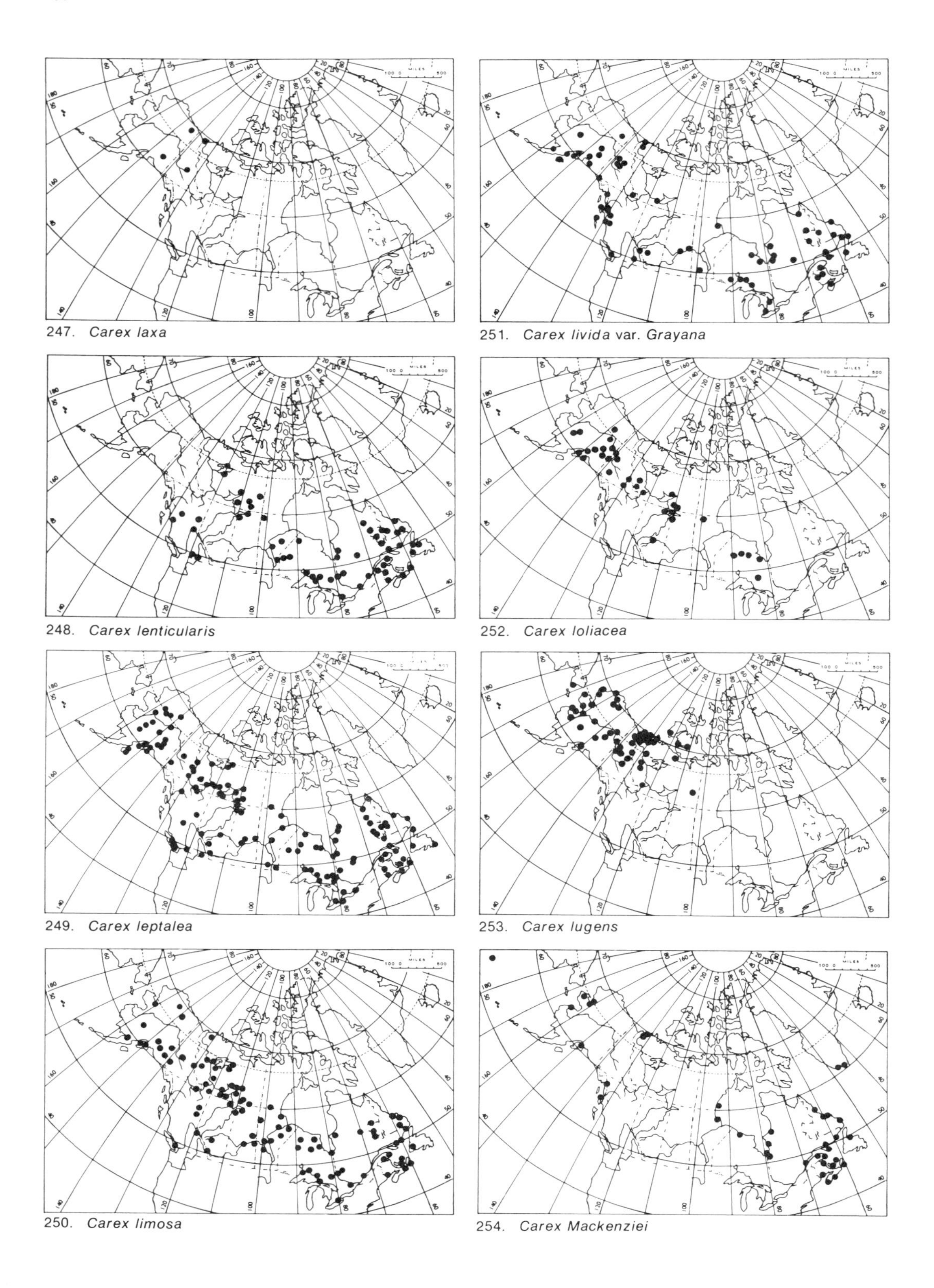

247. *Carex laxa*

248. *Carex lenticularis*

249. *Carex leptalea*

250. *Carex limosa*

251. *Carex livida* var. *Grayana*

252. *Carex loliacea*

253. *Carex lugens*

254. *Carex Mackenziei*

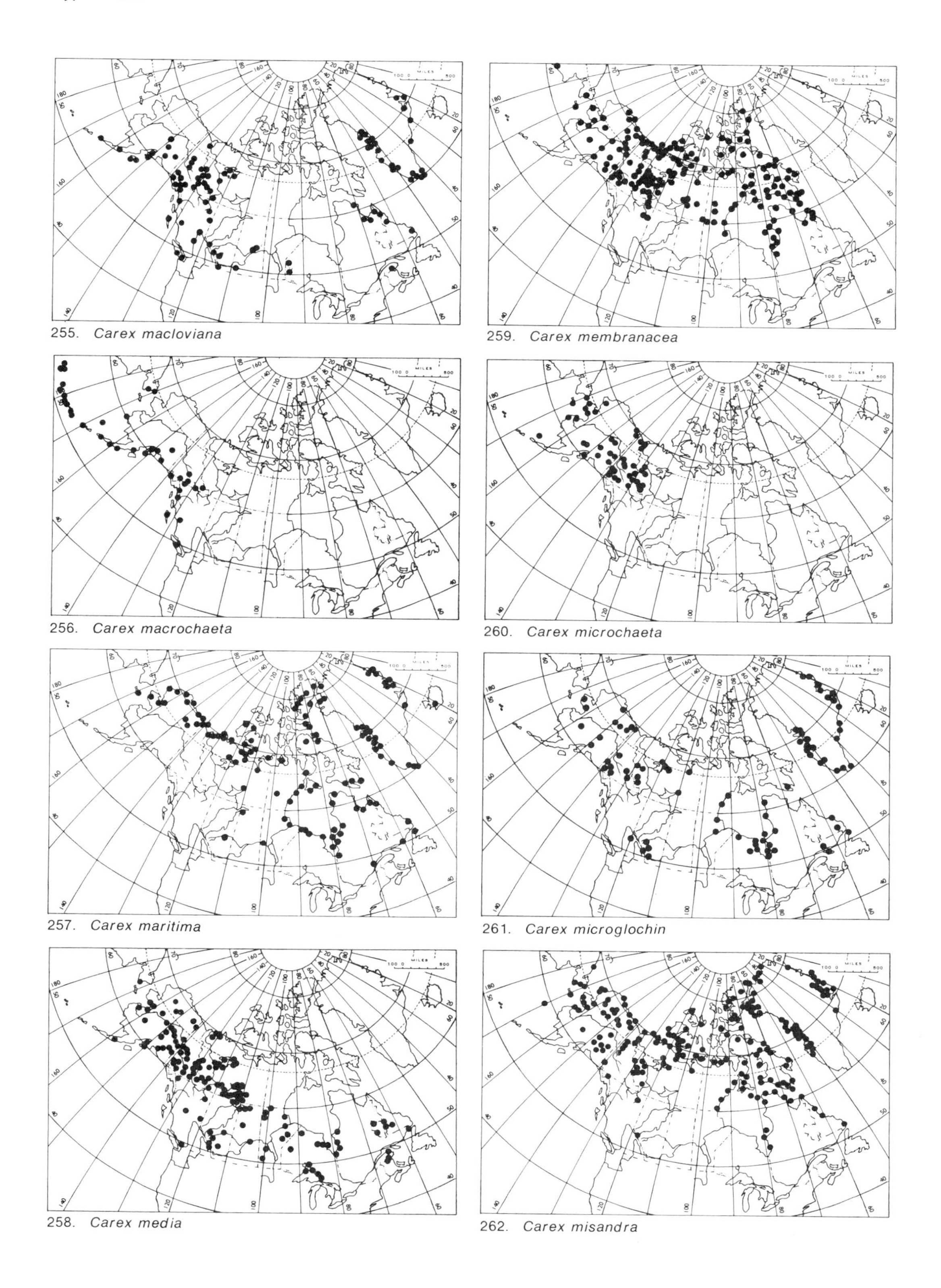

255. *Carex macloviana*

256. *Carex macrochaeta*

257. *Carex maritima*

258. *Carex media*

259. *Carex membranacea*

260. *Carex microchaeta*

261. *Carex microglochin*

262. *Carex misandra*

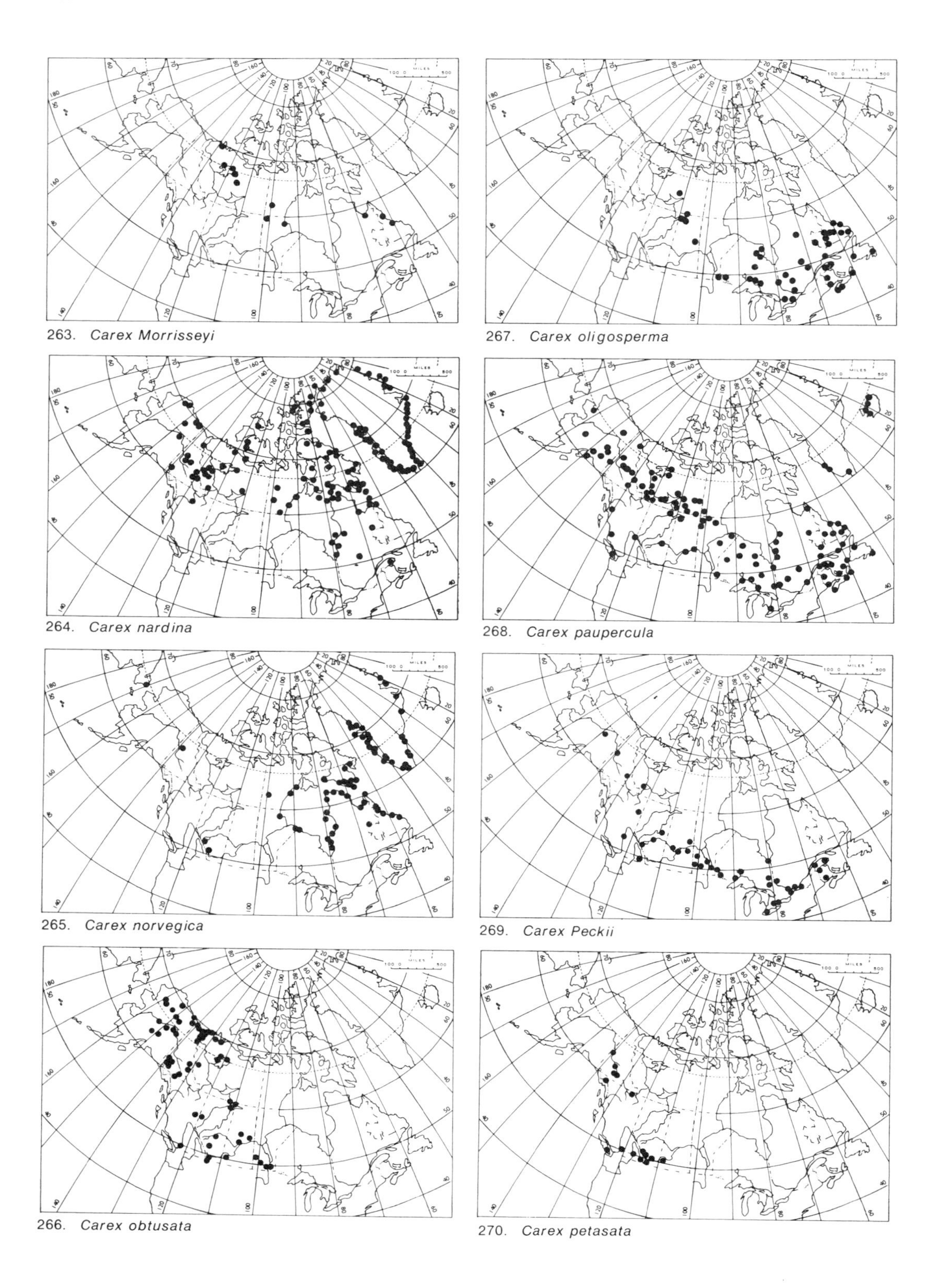

263. *Carex Morrisseyi*

264. *Carex nardina*

265. *Carex norvegica*

266. *Carex obtusata*

267. *Carex oligosperma*

268. *Carex paupercula*

269. *Carex Peckii*

270. *Carex petasata*

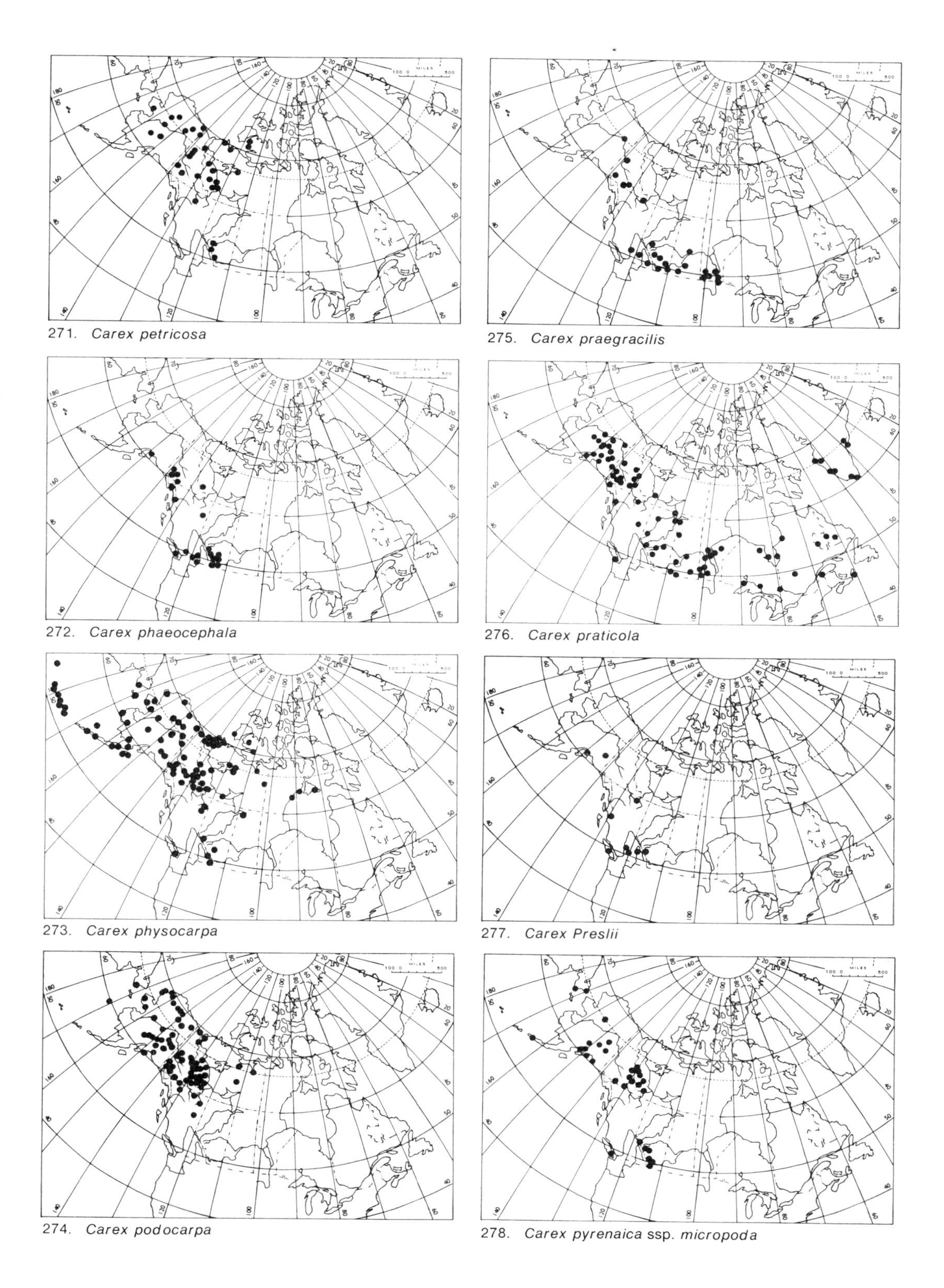

271. *Carex petricosa*

272. *Carex phaeocephala*

273. *Carex physocarpa*

274. *Carex podocarpa*

275. *Carex praegracilis*

276. *Carex praticola*

277. *Carex Preslii*

278. *Carex pyrenaica* ssp. *micropoda*

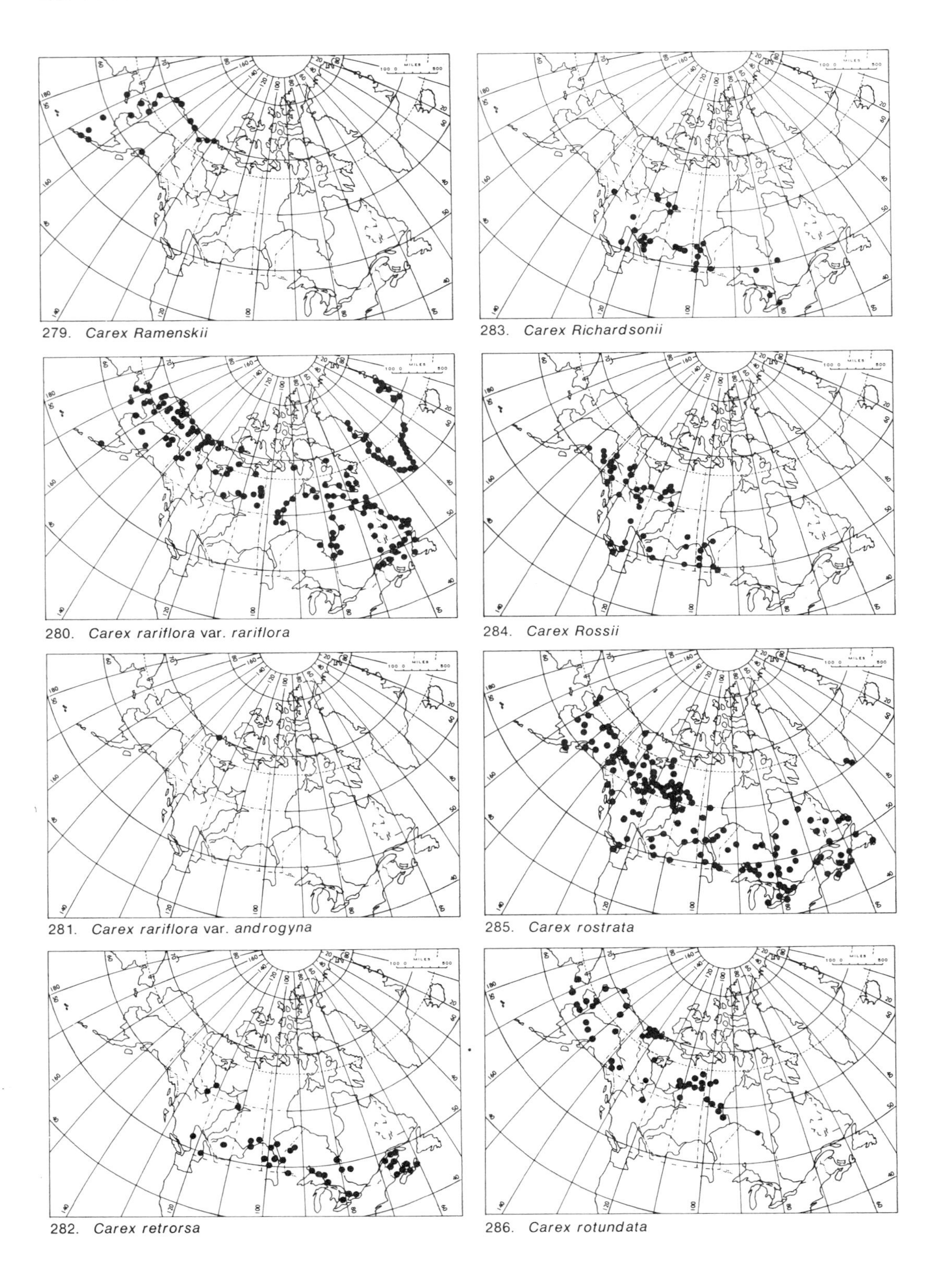

279. *Carex Ramenskii*

283. *Carex Richardsonii*

280. *Carex rariflora* var. *rariflora*

284. *Carex Rossii*

281. *Carex rariflora* var. *androgyna*

285. *Carex rostrata*

282. *Carex retrorsa*

286. *Carex rotundata*

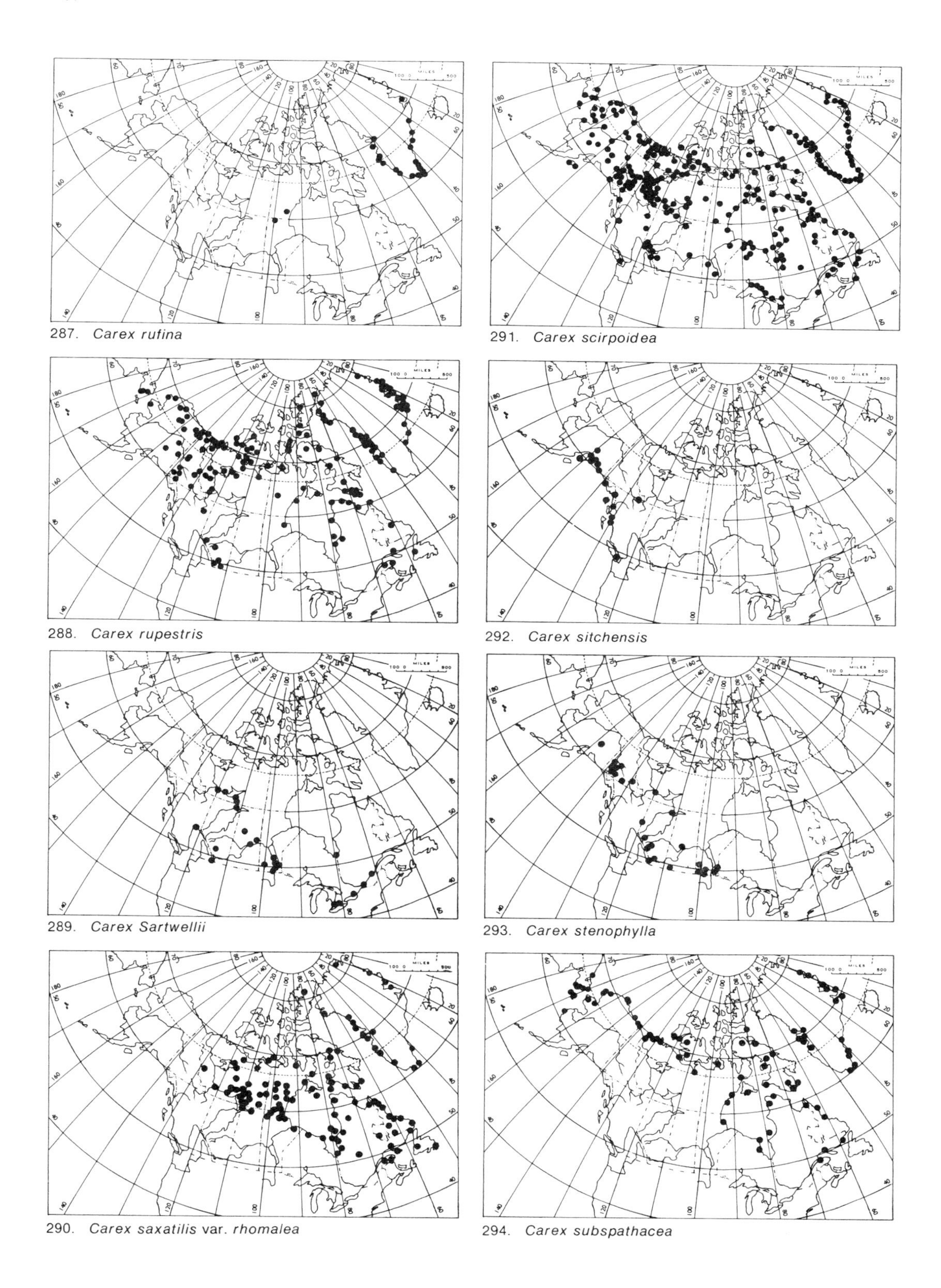

287. *Carex rufina*

288. *Carex rupestris*

289. *Carex Sartwellii*

290. *Carex saxatilis* var. *rhomalea*

291. *Carex scirpoidea*

292. *Carex sitchensis*

293. *Carex stenophylla*

294. *Carex subspathacea*

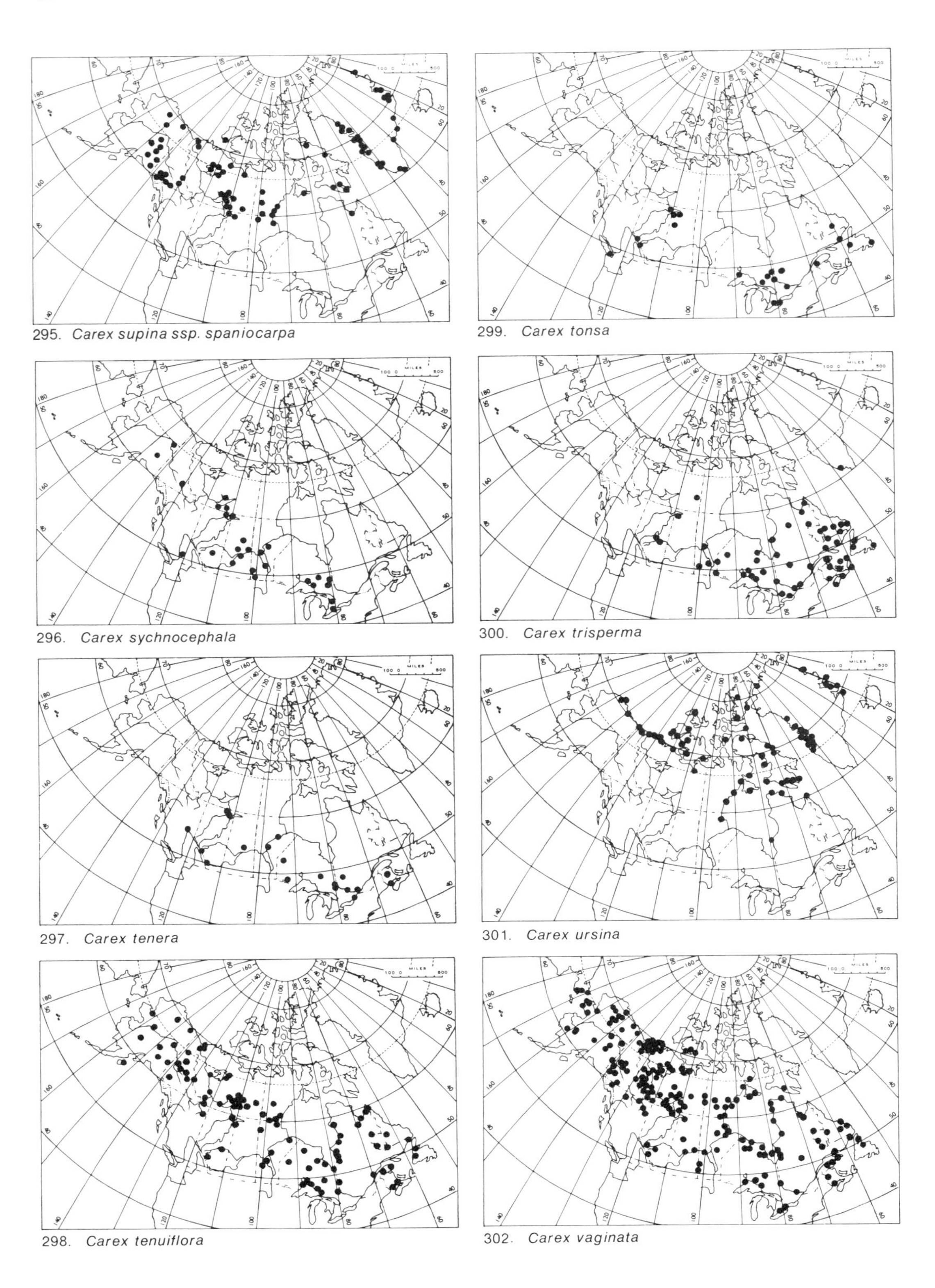

295. *Carex supina ssp. spaniocarpa*

296. *Carex sychnocephala*

297. *Carex tenera*

298. *Carex tenuiflora*

299. *Carex tonsa*

300. *Carex trisperma*

301. *Carex ursina*

302. *Carex vaginata*

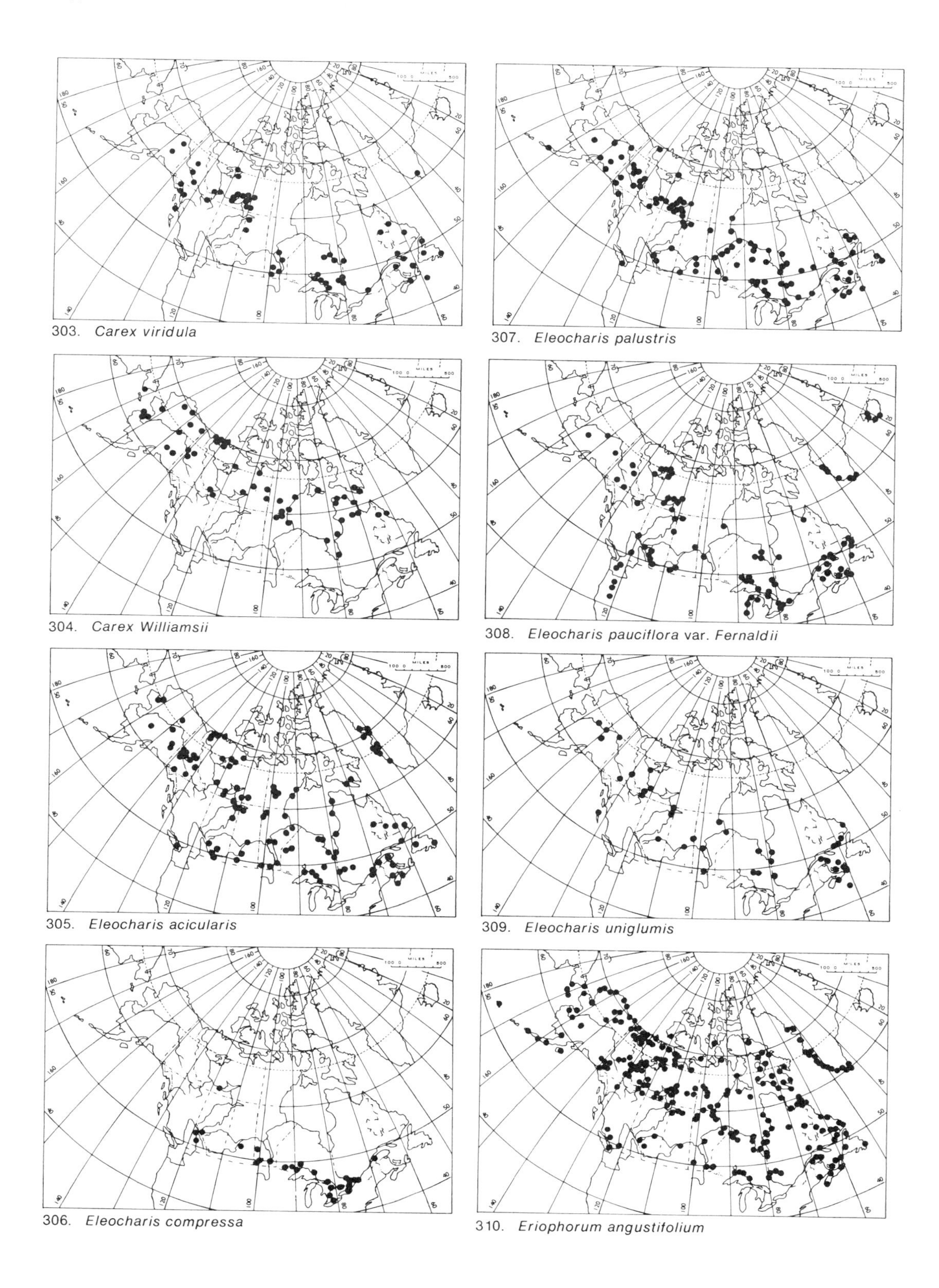

303. *Carex viridula*

304. *Carex Williamsii*

305. *Eleocharis acicularis*

306. *Eleocharis compressa*

307. *Eleocharis palustris*

308. *Eleocharis pauciflora* var. *Fernaldii*

309. *Eleocharis uniglumis*

310. *Eriophorum angustifolium*

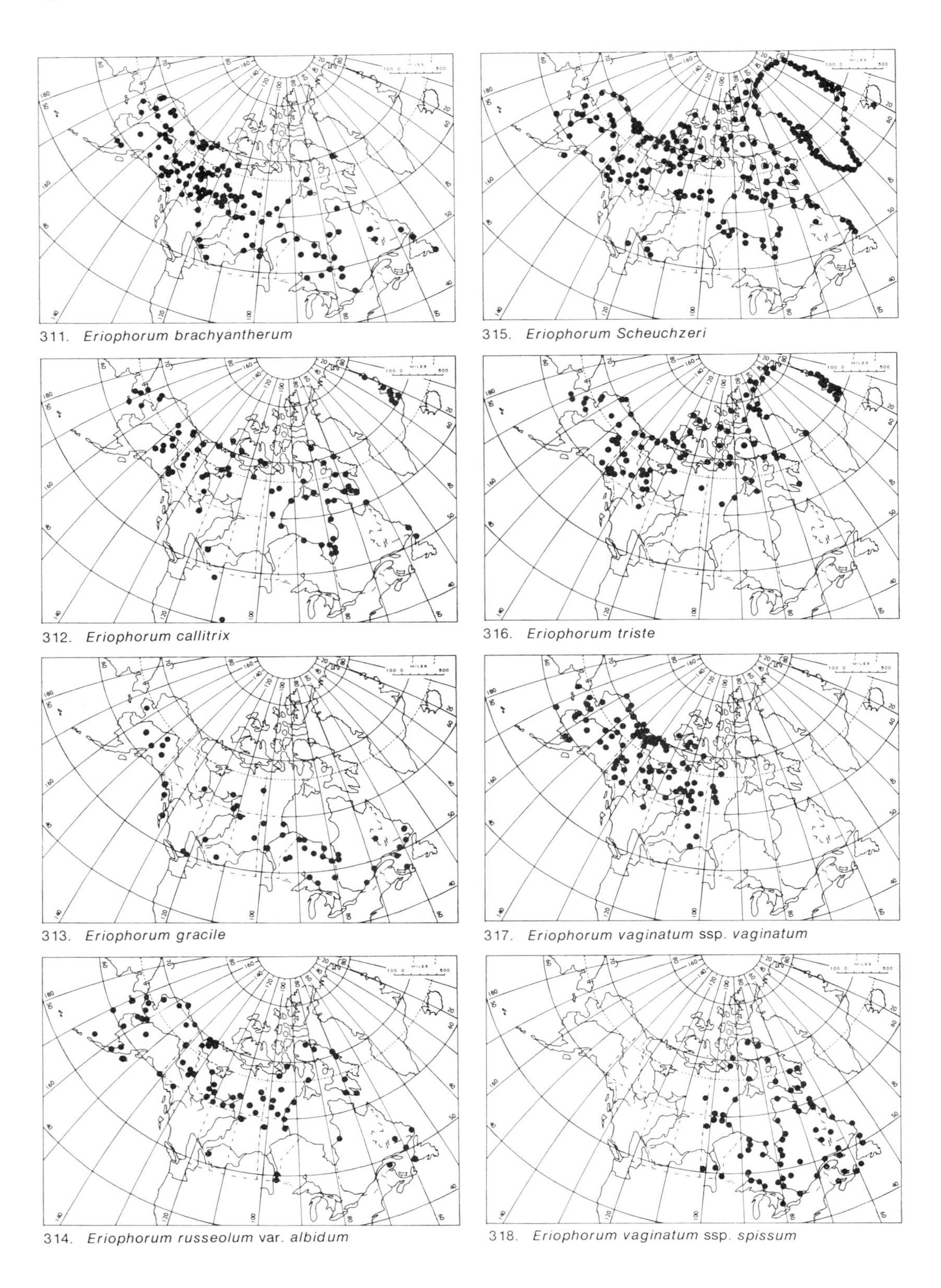

311. *Eriophorum brachyantherum*

312. *Eriophorum callitrix*

313. *Eriophorum gracile*

314. *Eriophorum russeolum* var. *albidum*

315. *Eriophorum Scheuchzeri*

316. *Eriophorum triste*

317. *Eriophorum vaginatum* ssp. *vaginatum*

318. *Eriophorum vaginatum* ssp. *spissum*

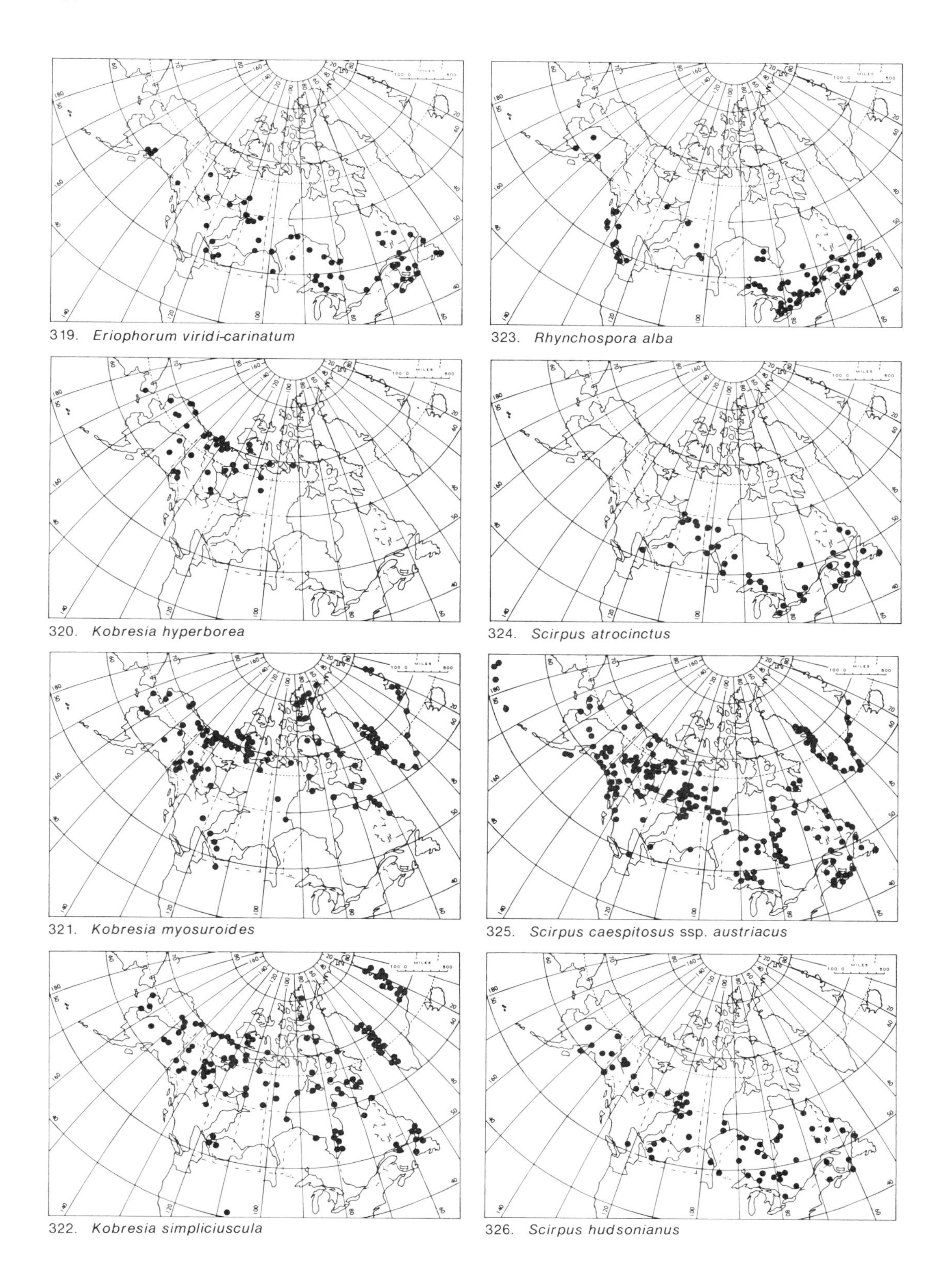

319. *Eriophorum viridi-carinatum*

323. *Rhynchospora alba*

320. *Kobresia hyperborea*

324. *Scirpus atrocinctus*

321. *Kobresia myosuroides*

325. *Scirpus caespitosus* ssp. *austriacus*

322. *Kobresia simpliciuscula*

326. *Scirpus hudsonianus*

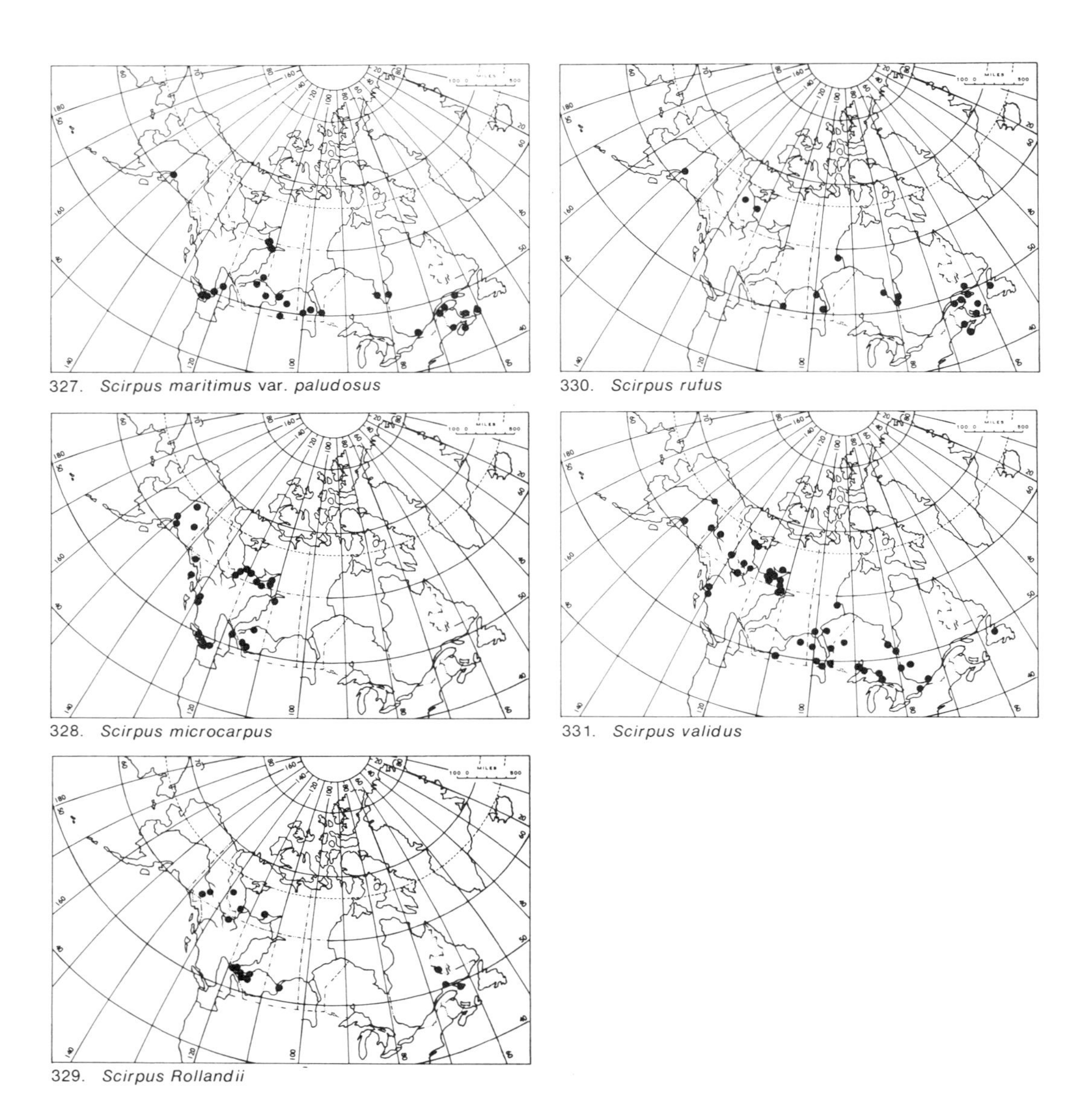

327. *Scirpus maritimus* var. *paludosus*

330. *Scirpus rufus*

328. *Scirpus microcarpus*

331. *Scirpus validus*

329. *Scirpus Rollandii*

ARACEAE Arum Family

Herbaceous and perennial bog or fen plants with acrid, watery juice and simple (ours), veiny leaves, from a stout and fleshy rhizome. Flowers small, crowded in a dense, fleshy spike or spadix subtended by, or enclosed in, a leaf-like spathe. Perianth none, or consisting of sepals.

a. Spathe leaf-like, green; leaf-blades linear . . . . *Acorus*
a. Spathe prominent, petaloid, white; leaf-blades ovate with cordate base . . . . *Calla*

Acorus L. Sweetflag

Acorus Calamus L.
Rhizome thick and creeping; leaves sword-shaped, crowded at the base, 4 to 8 dm long and 1.0 to 2.5 cm wide; scape similar to the leaves and prolonged beyond the spadix into an erect, green spathe; spadix cylindric, 5 to 10 cm long, composed of small brownish flowers.

All parts of the plant, but especially the rhizome, with a sweet but somewhat nauseating odor and taste. The candied root was formerly used as an aromatic confection.

Wet places or quiet borders of streams; the single station known for the Northwest Territories, along the newly constructed Yellowknife Highway near Fort Rae, may be a recent and accidental introduction, but it is interesting to note that the Rev. Father Petitot in his memoirs published in 1891, reported *Acorus Calamus* growing along stream banks near Lac la Martre some 100 km northwest of Fort Rae.

General distribution: Native of Eurasia, widely introduced and established, particularly in eastern N. America.

Fig. 287 Map 332

Calla L. Water-Arum

Calla palustris L.
Low, glabrous herb from a creeping, fleshy rhizome rooting at the nodes, and bearing long-petioled ovoid-cordate leaves and solitary scapes; spadix short-cylindric, subtended by the showy, pure white and petaloid spathe; berries bright red, non edible, in a tight, ovoid cluster.

The roots of *Calla palustris*, properly boiled and dried to remove the acrid and poisonous properties, were formerly used to mix with flour in the making of bread.

Common in fens or wet bogs, north along the Mackenzie River valley to its delta.

General distribution: Circumpolar; wide ranging.

Fig. 288 Map 333

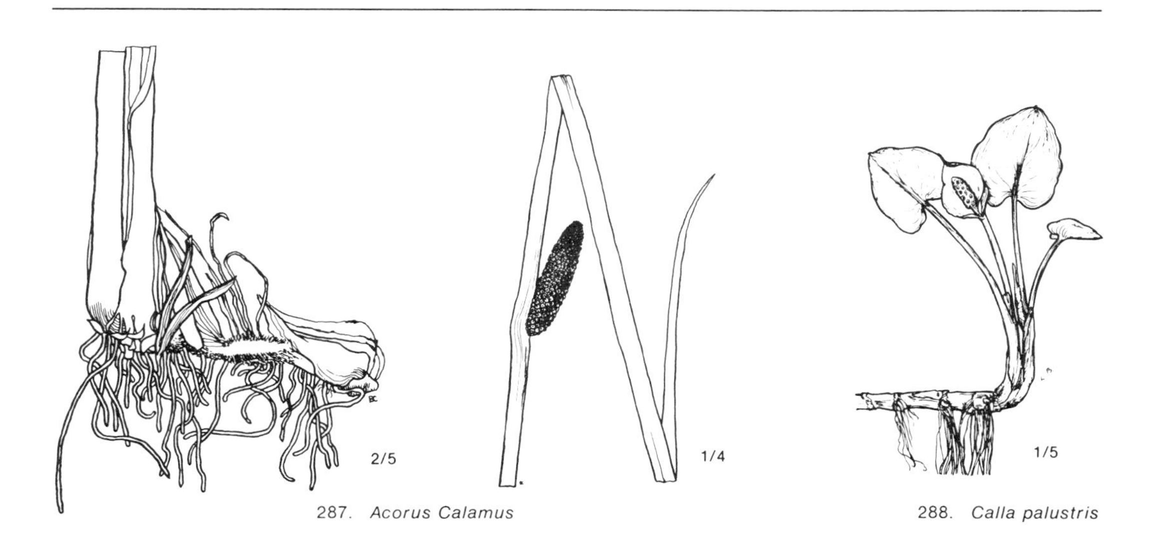

287. *Acorus Calamus*

288. *Calla palustris*

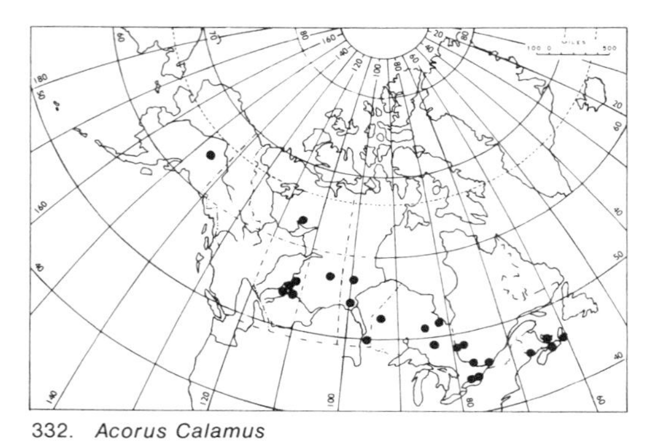

332. *Acorus Calamus*

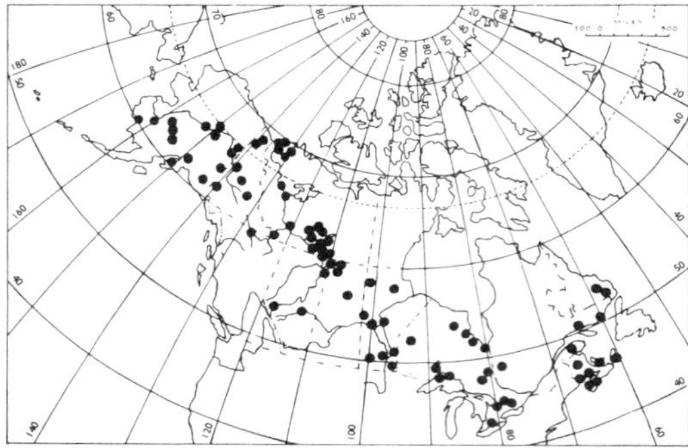

333. *Calla palustris*

LEMNACEAE Duckweed Family

Minute, non-rooted, stemless aquatic plants commonly floating or submersed in stagnant pools; each plant consists of a simple or branched frond on which the tiny monoecious flowers, when present, appear along the edge or the upper surface. However, the plants multiply mainly from tiny buds which form along the edge of the parent frond, or on the upper surface. Represented with us by two species of *Lemna*.

Lemna L. Duckweed

a. Fronds 2-5 mm long, round or elliptic, floating on the surface . . . . *L. minor*
a. Fronds 6-10 mm long, 3-lobed and stalked, often with a few short rootlets, forming tangled and mostly submerged mats . . . . *L. trisulca*

Lemna minor L.
Commonly forming a green scum on shallow, stagnant pools; thus far known mainly from near settlements in western parts of the District of Mackenzie.
General distribution: Circumpolar.
Map 334

Lemna trisulca L.
Forming submerged tangled mats, often among the stems of sedges and other plants in quiet streams, beaver ponds, etc.
Of more northern range than *L. minor*; in the Mackenzie Delta north to the limit of trees, and near Great Bear Lake ascending to 350 m.
General distribution: Circumpolar.
Fig. 289 Map 335

289. *Lemna trisulca*

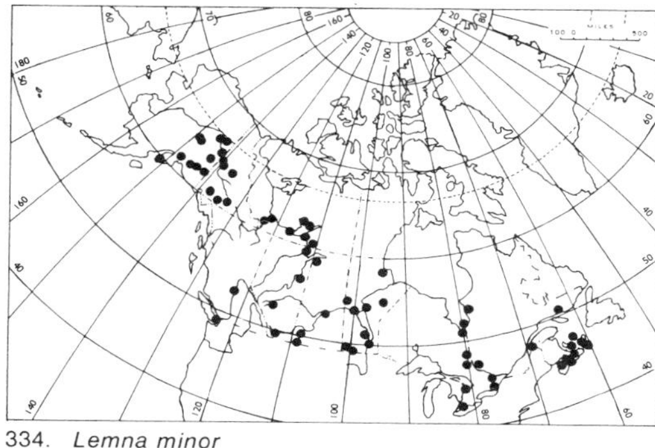
334. *Lemna minor*

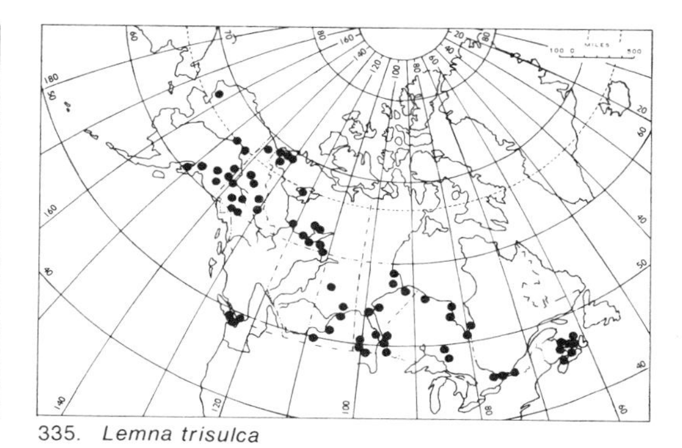
335. *Lemna trisulca*

JUNCACEAE Rush Family

Mostly perennial, essentially glabrous and grass-like herbs with linear, flat or canaliculate leaves (in *Juncus* often reduced to scale-like bracts). Flowers cymose or capitate, small and inconspicuous, with 3 scale-like sepals and petals, 3 to 6 stamens, a single short style, and 3 stigmas. The fruit a 3-valved capsule.

a. Capsule many-seeded; plant totally glabrous . . . . *Juncus*
a. Capsule 3-seeded; leaves and sheaths, at least in youth, with a few, soft and white hairs . . . . *Luzula*

Juncus L. Bog-Rush

Annual or perennial, glabrous, grass-like herbs with channelled more or less terete, or flattened leaves; inflorescence a tight or loose cyme; seeds minute, often with tail-like appendages at the ends.

a. Perennials
  b. Inflorescence appearing lateral, at the base of a terete bract which looks like the continuation of the culm; culm leafless but with firm, persistent sheaths at the base. Coarse plants mostly with stout, subligneous horizontal rhizomes
    c. Flowers few, 2-5; plant densely tufted . . . . *J. Drummondii*
    c. Flowers several; plants with creeping rhizomes
      d. Bract slender, as long or longer than the stem . . . . *J. filiformis*
      d. Bract stout, much shorter than the stem
        e. Inflorescence open; anthers longer than their thick and short filaments . . . . *J. balticus*
        e. Inflorescence dense; anthers distinctly shorter than their slender filaments . . . . *J. arcticus*
  b. Inflorescence appearing terminal, or if lateral the bract flat, not like a continuation of the culm; culm with well-developed green leaves from the base; rhizomes various but not subligneous
    f. Inflorescence capitate (in *J. Mertensianus* sometimes with several heads)
      g. Inflorescence few-flowered; sepals and mature capsules pale, straw-coloured or pale reddish-brown

h. Bract leaf-like, exceeding the mostly 2-flowered inflorescence; capsule retuse . . . . *J. biglumis*
h. Bract scale-like, equalling or slightly longer than the inflorescence; capsule acuminate
i. Flowering stems mostly solitary or few together, filiform, 10-30 cm tall, with 1-3 leaves in the lower third; seeds 3 mm long . . . . *J. stygius* ssp. *americanus*
i. Flowering stems tufted, 5-15 cm tall, rather stout; leaves all basal . . . . *J. albescens*
g. Inflorescence many-flowered, spherical, dark brown or black; leaves septate or nodulose (tall, densely tufted Cordilleran species) . . . . *J. Mertensianus*
f. Inflorescence compound
j. Leaves distinctly septate-nodulose, terete
k. Stems from slender rhizome, heads 1-3, spherical, few-flowered . . . . *J. nodosus*
k. Stems tufted, heads not spherical
l. Seeds 0.5 mm long, tail-less . . . . *J. alpinus* ssp. *nodulosus*
l. Seeds 1.0 mm long, with tails . . . . *J. brevicaudatus*
j. Leaves not septate-nodulose, flat or terete
m. Densely tufted; leaves narrow; heads numerous
n. Seeds minute, 0.3-0.4 mm long, tail-less . . . . *J. Dudleyi*
n. Seeds larger, about 1.3 mm long, with white, tail-like appendages . . . . *J. Vaseyi*
m. Stoloniferous; leaves broad; heads 1-3 (4); seeds spindle-shaped, about 3 mm long, with prominent white tails . . . . *J. castaneus*
a. Annual, much branched dwarf species . . . . *J. bufonius*

Juncus albescens (Lange) Fern.
J. triglumis L. ssp. *albescens* (Lange) Hultén
Caespitose, forming small rather compact tufts of slender, but stiffly erect culms 5 to 20 cm, or more, high; inflorescence terminal in a 3- to 5-flowered head, subtended by a very short, scaly bract; sepals pale reddish-brown or straw-coloured; capsules elliptic, obtuse, never retuse, pale yellowish-brown.

Calcareous clay or sand, or by the edge of ponds.

General distribution: N. American arctic-alpine; closely related to *J. triglumis* L. of Eurasia.

Fig. 290 Map 336

Juncus alpinus Vill.
ssp. **nodulosus** (Wahlenb.) Lindm.
J. alpinus var. *rariflorus* (Hartm.) Hartm.
Loosely tufted from a rather stout, horizontal rhizome; culms 15 to 45 cm tall bearing 2 to 3 short and stiff leaves; the basal sheaths pinkish; heads few to several, on stiffly erect branches; glomerules commonly 4 to 5 and rarely more than 12, 3- to 5-flowered; seeds spindle-shaped, 0.5 mm long and tail-less.

Wet calcareous, gravelly beaches and river banks, north to the polar tree-line.

General distribution: Circumpolar.

Fig. 291 Map 337

Juncus arcticus Willd.
Similar to *J. balticus* var. *alaskanus* from which it differs by its more densely capitate inflorescence of fewer and more or less sessile flowers, blackish-brown, obtuse or even retuse capsules, shorter sepals and shorter anthers, on longer and more slender filaments.

Wet sand and clay shores of lakes and rivers, and not infrequently associating with obligate halophytes on strand flats.

General distribution: Amphi-Atlantic; low-arctic.

Fig. 292 Map 338

Juncus balticus L.
Culms 20 to 50 cm high, terete, stiff and wiry, from a stout, scaly rhizome; the bladeless sheaths at the base of the culms reddish brown, firm and somewhat lustrous; inflorescence of several clusters of short-pedunculate flowers; capsule light brown, elliptic, with a short, conical point.

In the Mackenzie Basin *J. balticus* is represented by two rather distinct and easily separated races: var. *alaskanus* (Hult.) Porsild and var. *littoralis* Engelm. The first, which is distinguished from other North American races of *J. balticus* by its more slender growth, rather short involucral bract, rather dense inflorescence, and anthers that in length about equal the length of their filament, is arctic-alpine. It is common along the Arctic Coast and in the mountains of Alaska and Yukon, and extends east along the Arctic Coast at least to Bathurst Inlet and is common also about Great Bear Lake and elsewhere on the Precambrian Shield east of the Mackenzie River. It reaches the Arctic Archipelago in Banks and Victoria Islands.

The var. *littoralis (J. ater* Rydb.; *J. arcticus* ssp. *ater*(Rydb.) Hult.) differs by its shorter and often bright yellow rhizome, longer involucral bracts and more open inflorescence with numerous flowers on dichotomous branches up to 5 to 6 cm long. The anthers are 1.5 to 2.0 mm long, and always much longer than the short and stout filaments. Its name notwithstanding, it is a non-arctic and essentially inland race which ranges from the mountains of central Alaska to Newfoundland. It is probably always calciphilous, and in the Mackenzie Basin is common on alluvial river flats along the Mackenzie River and its tributaries, extending north to Good Hope, and to the eastern limit of the Paleozoic formation.

General distribution: *J. balticus s. lat.*, circumpolar.

Fig. 293 (var. *alaskanus*), 294 (var. *littoralis*) Map 339 (var. *alaskanus*), 340 (var. *littoralis*)

Juncus biglumis L.

Similar to *J. albescens*, but inflorescence 1- to 3-flowered, subtended by a dark brown, leafy bract 1 to 3 cm long; sepals dark brown; capsules narrowly club-shaped, dark brown, and conspicuously retuse.

Common in wet sand or clay throughout the treeless parts of our region.

General distribution: Circumpolar; wide ranging high-arctic-alpine.

Fig. 295 Map 341

***Juncus brevicaudatus** (Engelm.) Fern.

Culms leafy, 5 to 50 cm tall, in small dense tufts; the leaves nodulose, their pinkish sheaths auriculate; inflorescence elongate, 5 to 10 cm long, the numerous glomerules turbinate, 3- to 6-flowered; seeds, including their tails, about 1 mm long.

Wet meadows and river banks, north to Lake Athabaska, and to be expected in S. W. District of Mackenzie.

General distribution: Lab. to Alta.

Map 342

Juncus bufonius L. *s. lat.*

Tufted annual, commonly less than 10 cm tall, with spreading dichotomously branched leafy stems; flowers remote in an open cyme; occasionally viviparous. Seeds ovoid, 0.3 to 0.5 mm long.

In our area thus far recorded only from damp roadsides in or near settlements of the wooded parts, north to the Mackenzie Delta.

General distribution: Cosmopolitan, non-arctic weedy species.

Fig. 296 Map 343

Juncus castaneus Smith

Culms 10 to 30 (50) cm high, with well-developed leaves at their base from a stolon-like rhizome; inflorescence of 1 to 3 somewhat distant, chestnut-brown heads, each composed of from 4 to 10 flowers; the lowermost bract flat, leafy, and green, longer than the inflorescence; capsules about 0.6 mm long, castaneous to purplish-black; seeds about 3 mm long with white tails longer than the body of the seed.

In wet sand or clay on lake shores or by brooks.

General distribution: Circumpolar, wide ranging, arctic-alpine.

Fig. 297 Map 344

Juncus Drummondii E. Mey.

Densely tufted, from a short, ascending rhizome. Culms slender and wiry, 20 to 50 cm tall, bladeless; inflorescence compact, few-flowered, its bracts rarely exceeding the inflorescence; seeds 1.5 to 2.0 mm long, including the tail-like appendage.

Alpine snowbeds and damp slopes; barely entering our area along the east slope of the Mackenzie Mountains.

General distribution: Cordilleran.

Fig. 298 Map 345

Juncus Dudleyi Wieg.

J. tenuis Willd. var. *Dudleyi* (Wieg.) Hermann

Loosely caespitose from a slender, ascending rhizome; culms 20 to 60 cm tall, wiry; leaves flat, becoming involute in drying, half as long as the culms; sheaths auricled; inflorescence cymose, compact or with unequal branches, each bearing small clusters of from 2 to 6 flowers; perianths pale green or straw-coloured; seeds 0.3 to 0.4 mm long, tail-less.

A lowland species of wet, calcareous meadows and river banks, in the N.W.T. thus far collected but once along the Slave River north of Fort Smith.

General distribution: Nfld. to Alaska and southward.

Fig. 299 Map 346

Juncus filiformis L.
Stems grey-green, very slender and finely fluted, commonly in a single row from a thin creeping and branched rhizome; leaves basal, reduced to blade-less sheaths, or the blade merely a short bristle; the involucral bract almost as long as the stem proper; the inflorescence few-flowered; the capsules spherical and the seeds 0.5 mm long, tail-less.

In the District of Mackenzie apparently mainly on moist, non-calcareous woodland meadows and sandy lake shores along the southern rim of the Precambrian Shield.

General distribution: Circumpolar.

Fig. 300 Map 347

***Juncus Mertensianus** Bong.
Culms leafy, 20 to 30 cm high, loosely tufted, from a much branched rhizome; leaves flat, green; inflorescence capitate, solitary or sometimes with a second and smaller pedunculate head; perianths coal black and shiny; seeds tail-less.

An alpine species of gravelly stream beds and meadows; to be expected along the east slope of the central and southern Mackenzie Mountains.

General distribution: Cordilleran, Pacific Coast.

Fig. 301 Map 348

Juncus nodosus L.
Culms 20 to 30 cm high rising singly or a few together from a creeping, thin, white and tuber-bearing rhizome; leaves prominently septate; inflorescence of from 2 to 3 spherical, bur-like, reddish-brown heads; seeds 0.4 mm long, tail-less.

Common locally in damp calcareous river meadows north along the Mackenzie River to the Good Hope Ramparts.

General distribution: Nfld. to Alaska.

Fig. 302 Map 349

Juncus stygius L.
ssp. **americanus** (Buch.) Hult.
Culms loosely tufted, or single, filiform, leafy; inflorescence terminal, few-flowered, subtended by a short, scarious awl-like bract; seeds about 3 mm long including their broad tails.

Wet margins of woodland bog pools or seepages; in the District of Mackenzie known only from west of Great Slave Lake.

General distribution: var. *stygius* Eurasian; var. *americanus* Nfld. and Lab. to Alaska and southward, wide ranging but apparently in a series of disjunct populations.

Fig. 303 Map 350

Juncus Vaseyi Engelm.
Culms slender and wiry, 30 to 60 cm high, from a short ascending rhizome; leaves all basal, half to nearly as long as the culms, their sheaths not auricled; inflorescence compact, few-flowered; seeds about 1.3 mm long, including the thin white tails.

Lowland slough-margins, damp river margins and moist sandy lake shores.

General distribution: Boreal N. America, in the N.W.T. north to Great Slave L. and Ft. Simpson.

Fig. 304 Map 351

Luzula DC. Wood Rush

Perennials with hollow stems and flat or longitudinally grooved leaves, usually with soft, white hairs along the leaf-margins and in the inflorescence. Fruit a capsule containing three smooth and tail-less seeds.

a. Flowers solitary at the end of the ultimate branches of the paniculate or umbellate, somewhat drooping inflorescence; branches capillary.
  b. Inflorescence umbellate . . . . . *L. rufescens*
  b. Inflorescence a loose compound cyme
    c. Bracts of the inflorescence with much lacerated, long-ciliate tips; basal leaves 3-8 mm wide . . . . . *L. Wahlenbergii*
    c. Bracts of inflorescence sub-entire; basal leaves 6-10 mm broad . . . . . *L. parviflora*
a. Flowers crowded in few- to many-flowered glomerules or spikes
  d. Leaves involute
    e. Inflorescence spike-like, drooping . . . . . *L. spicata*
    e. Inflorescence commonly composed of several glomerules on erect or spreading stiff branches . . . . . *L. confusa*

d. Leaves flat
 f. Bracts of the inflorescence short and inconspicuous
  g. Leaves 3-5 mm wide, with blunt, callous tips
   h. Inflorescence capitate . . . . *L. nivalis*
   h. Inflorescence of several erect-ascending stalked glomerules . . . . *L. nivalis* var. *latifolia*
  g. Leaves 2-3 mm wide, with thin, attenuate tips; stems slender; inflorescence of several few-flowered drooping glomerules on capillary branches . . . . . . *L. arcuata*
 f. Bracts usually conspicuous, exceeding the inflorescence, this commonly dense and glomerate, occasionally with smaller lateral and stalked glomerules
  i. Perianth leaves 2.5-3.5 mm long; seeds 1.1-1.4 mm long . . . . *L. multiflora* ssp. *frigida* var. *contracta*
  i. Perianth leaves 2.0-2.5 mm long; seeds 0.8-1.1 mm long . . . . *L. groenlandica*

Luzula arcuata (Wahlenb.) Sw.
Loosely caespitose, with short underground stolons; culms 10 to 25 cm high, slender and somewhat arched, bearing 1 to 3 leaves; basal leaves narrow, more or less involute, shorter than the culm; inflorescence open with arching, filiform branches, each with one or more few-flowered glomerules; bracts linear, scarious, the upper ones tipped with white hairs; bractlets with membranaceous hairy margins; sepals narrowly lanceolate and acuminate, as long as the dark brown, ovate capsules.

Alpine slopes, gravelly or sandy beaches and moraines.

General distribution: Northern Europe and Iceland; Siberia, Alaska, Y.T. and western District of Mackenzie, south through B.C. to Washington.

Fig. 305 Map 352

Luzula confusa Lindebl.
Loosely caespitose; culms 10 to 30 cm high, stiffly erect and leafy; leaves narrow, channelled or involute and often somewhat curled; the basal sheaths dark brown, sometimes purplish and persisting for many years; inflorescence capitate or with lateral glomerules on stalked, erect, or arching branches; the subtending bract short and scale-like; bractlets with membranaceous, hairy margins; sepals narrowly lanceolate and acuminate, as long as the dark brown, ovate capsules.

Common in dry, turfy places in tundra or on rocky slopes and ledges.

General distribution: Circumpolar, wide ranging arctic-alpine.

Fig. 306 Map 353

Luzula groenlandica Böcher
Densely tufted; culms 10 to 30 cm high, leafy; basal leaves flat, often reddish-brown, much shorter than the culm; inflorescence glomerate, or occasionally with an additional pair of small short-stalked glomerules, subtended, by an obliquely projecting leafy bract; sepals 2.0 to 2.5 mm long, chestnut-brown with a broad membranaceous margin; seeds 0.8 to 1.1 mm long.

Turfy tundra, often by lakes and ponds, and on gravelly alpine flats.

General distribution: Subarctic barrens; Western Greenland, Lab. to Alaska.

Map 354

Luzula multiflora (Retz.) Lej.
ssp. **frigida** (Buch.) Krecz. var. **contracta** Sam.
L. campestris var. *alpina sensu* Raup
Similar to *L. groenlandica*, but inflorescence mostly sub-umbellate with one large and several smaller heads on stiffly erect branches; also the sepals are longer (2.5 to 3.5 mm long) and blackish-brown, and the seeds larger (1.1 to 1.4 mm long).

Turfy places, often by the edge of ponds.

General distribution: Subarctic barrens from Nfld. and Greenland to the Bering Sea.

Fig. 307 Map 355

Luzula nivalis (Laest.) Beurl.
L. arctica Blytt
Densely caespitose; culms stiffly erect, 10 to 20 cm high, often dark brown, leafy; basal leaves flat, 3 to 5 mm wide and marcescent; inflorescence subtended by a leafy bract, subcapitate, occasionally with 1 or 2 lateral, stalked glomerules; bractlets with white, membranaceous, lacerated margins; sepals ovate or lanceolate, dark brown, shorter than the ovate and mucronate capsules.

The var. *latifolia* (Kjellm.) Sam. (*L. tundricola* Gorodk.) is readily distinguished from var. *nivalis* by its strikingly open and branched inflorescence simulating that of *L. arcuata*, and by its distinctly paler leaves.

Common in heath and not too dry tundra; on

snowbeds frequently associated with *Cassiope tetragona*.

General distribution: var. *nivalis*, circumpolar, high-arctic; var. *latifolia*, Amphi-Beringian.

Fig. 308 (var. *nivalis*), 309 (var. *latifolia*) Map 356 (var. *nivalis*), 357 (var. *latifolia*)

Luzula parviflora (Ehrh.) Desv.
Tufted; the culms leafy, 25 to 50 cm high; basal leaves flat, 6 to 10 mm broad, linear, abruptly acuminate, glabrous; inflorescence cymose, the branches few-flowered, arching, or sometimes drooping; bracts of the inflorescence sub-entire.

Open woods and willow thickets; in sheltered ravines and on herbmat slopes ascending to 1300 m.

General distribution: Circumpolar; wide ranging low-arctic-alpine.

Fig. 310 Map 358

Luzula rufescens Fisch. & Mey.
Loosely tufted; culms leafy, 10 to 30 cm high; basal leaves flat narrow, reddish at the base; inflorescence umbellate, the branches filiform, each with a single flower; bracts much shorter than the inflorescence; bractlets erose, scarious; sepals broadly lanceolate, chestnut-brown, scarious-margined, shorter than the pale acuminate capsule; seeds plump, 1.1 mm long, with a short, thin, yellowish tail.

Borders of bogs and marshes and on moist sand and gravel bars; in our area known only from the Richardson Mountains.

General distribution: Amphi-Beringian.

Fig. 311 Map 359

Luzula spicata (L.) DC.
Densely caespitose, with slender, leafy culms 20 to 35 cm high; leaves narrow and channelled; the interrupted, spike-like inflorescence nodding, and subtended by a narrow, leafy bract; bractlets lanceolate, with narrow, white, membranaceous, ciliated margins; sepals lanceolate, bristle-pointed, equalling or slightly longer than the broadly ovate dark brown capsules.

Dry, sunny slopes or rocky ledges; apparently rare in our area where it is as yet known only from a few stations on the Precambrian Shield and in the Mackenzie Mountains.

General distribution: Circumpolar, low-arctic, with several large gaps.

Fig. 312 Map 360

Luzula Wahlenbergii Rupr.
Similar to *L. parviflora*, but culms shorter, with fewer and narrower leaves, narrower and more tapered basal leaves, and the bracts of the inflorescence much lacerated.

Usually growing in sphagnum bogs in tundra, or less commonly in moss by brooks or on lake shores.

General distribution: Circumpolar; low-arctic.

Fig. 313 Map 361

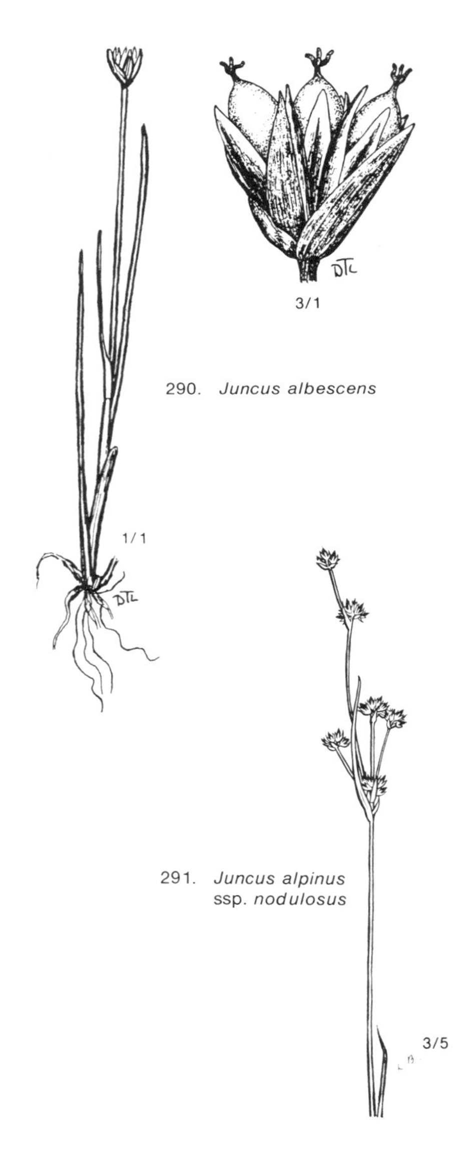

290. *Juncus albescens*

291. *Juncus alpinus* ssp. *nodulosus*

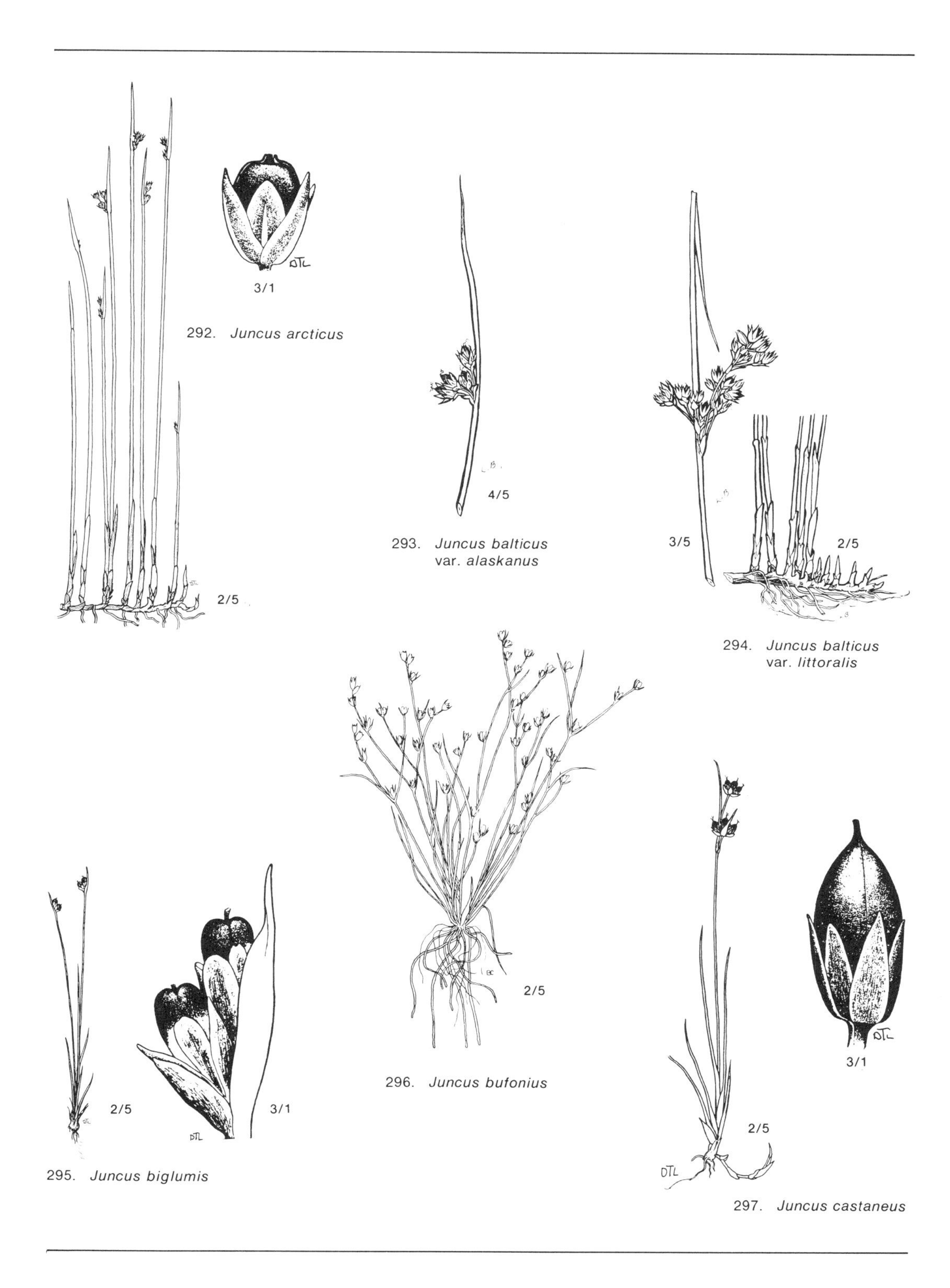

292. *Juncus arcticus*

293. *Juncus balticus* var. *alaskanus*

294. *Juncus balticus* var. *littoralis*

295. *Juncus biglumis*

296. *Juncus bufonius*

297. *Juncus castaneus*

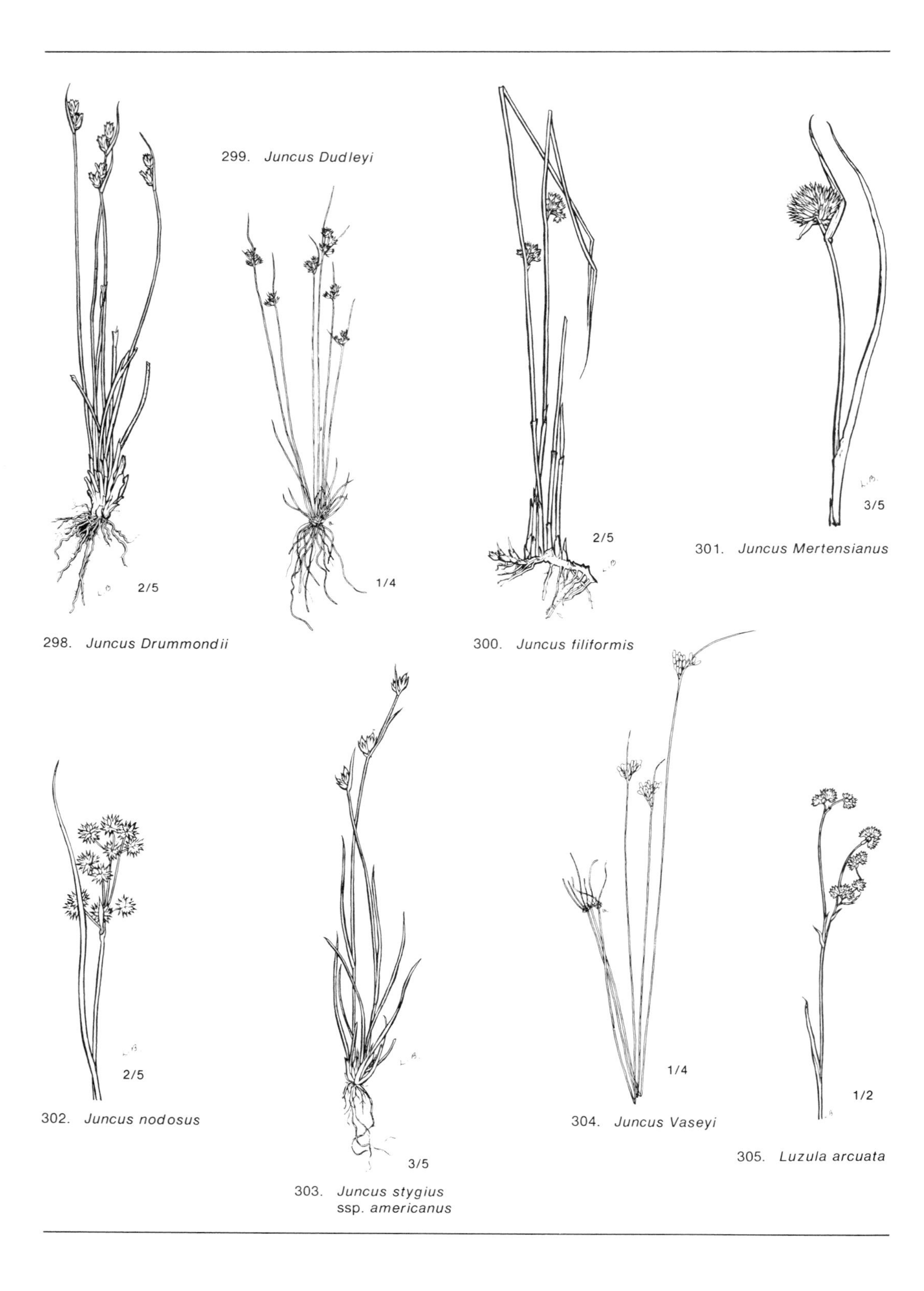

298. *Juncus Drummondii*

299. *Juncus Dudleyi*

300. *Juncus filiformis*

301. *Juncus Mertensianus*

302. *Juncus nodosus*

303. *Juncus stygius* ssp. *americanus*

304. *Juncus Vaseyi*

305. *Luzula arcuata*

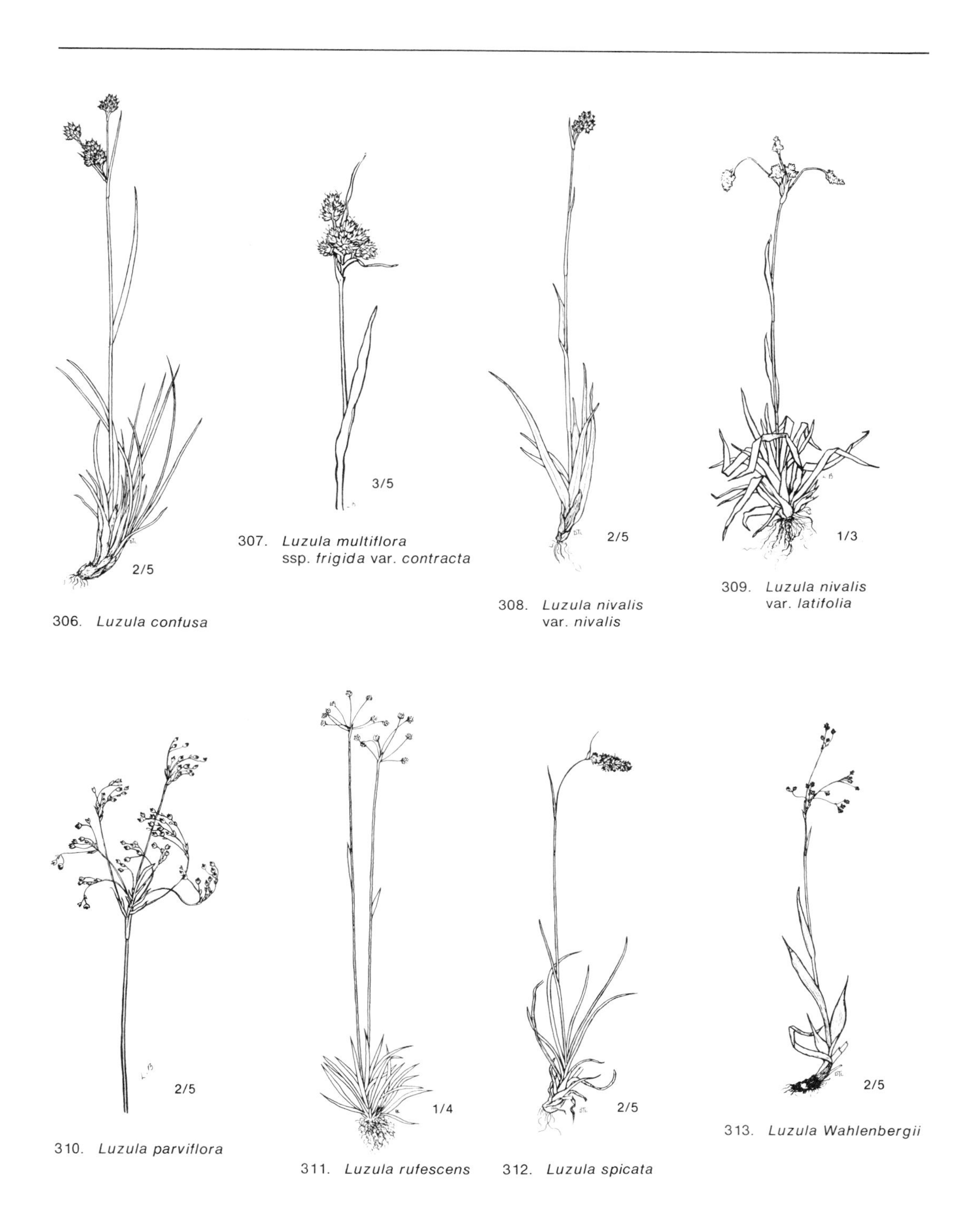

306. *Luzula confusa*

307. *Luzula multiflora* ssp. *frigida* var. *contracta*

308. *Luzula nivalis* var. *nivalis*

309. *Luzula nivalis* var. *latifolia*

310. *Luzula parviflora*

311. *Luzula rufescens*

312. *Luzula spicata*

313. *Luzula Wahlenbergii*

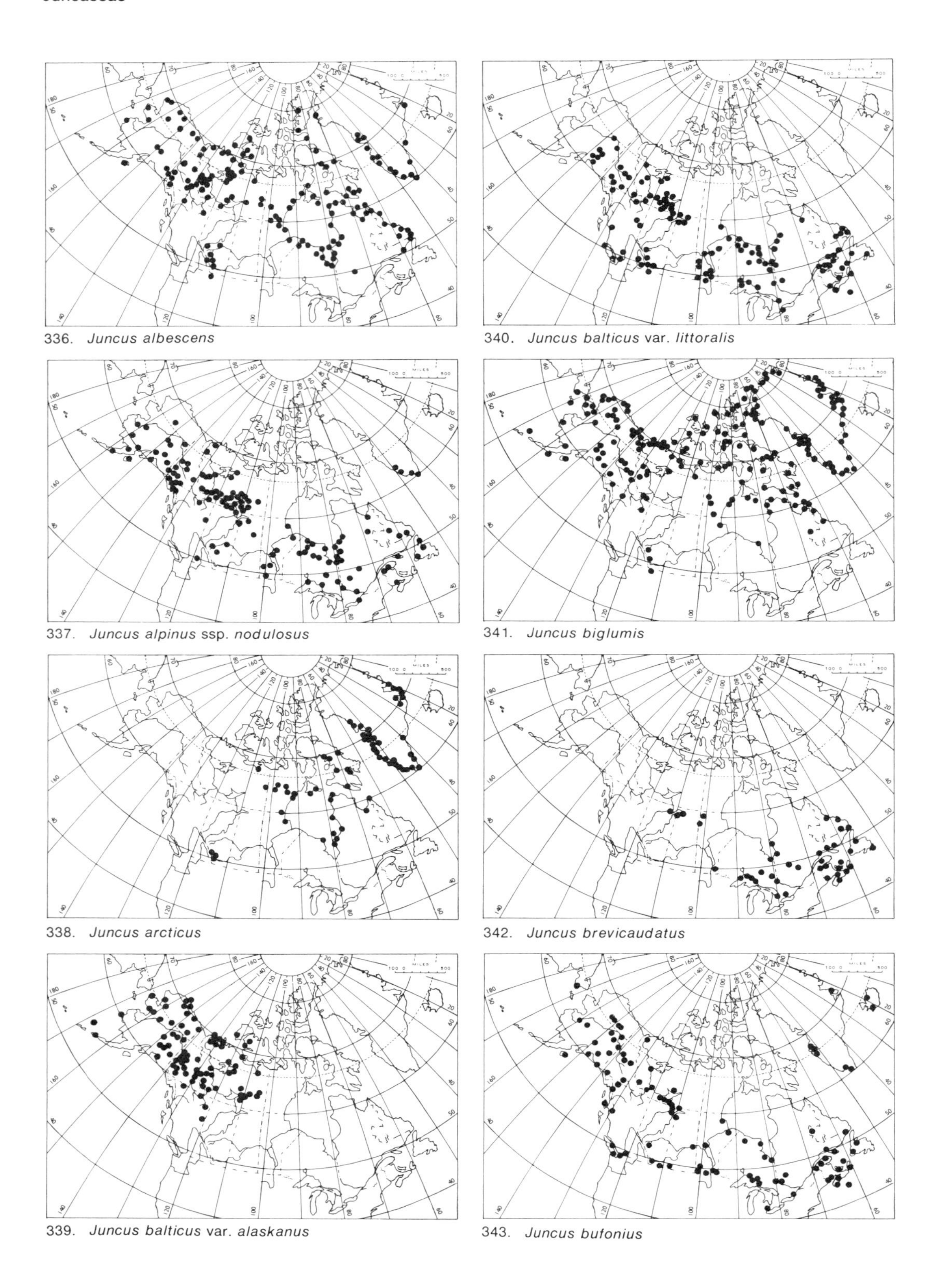

336. *Juncus albescens*

337. *Juncus alpinus* ssp. *nodulosus*

338. *Juncus arcticus*

339. *Juncus balticus* var. *alaskanus*

340. *Juncus balticus* var. *littoralis*

341. *Juncus biglumis*

342. *Juncus brevicaudatus*

343. *Juncus bufonius*

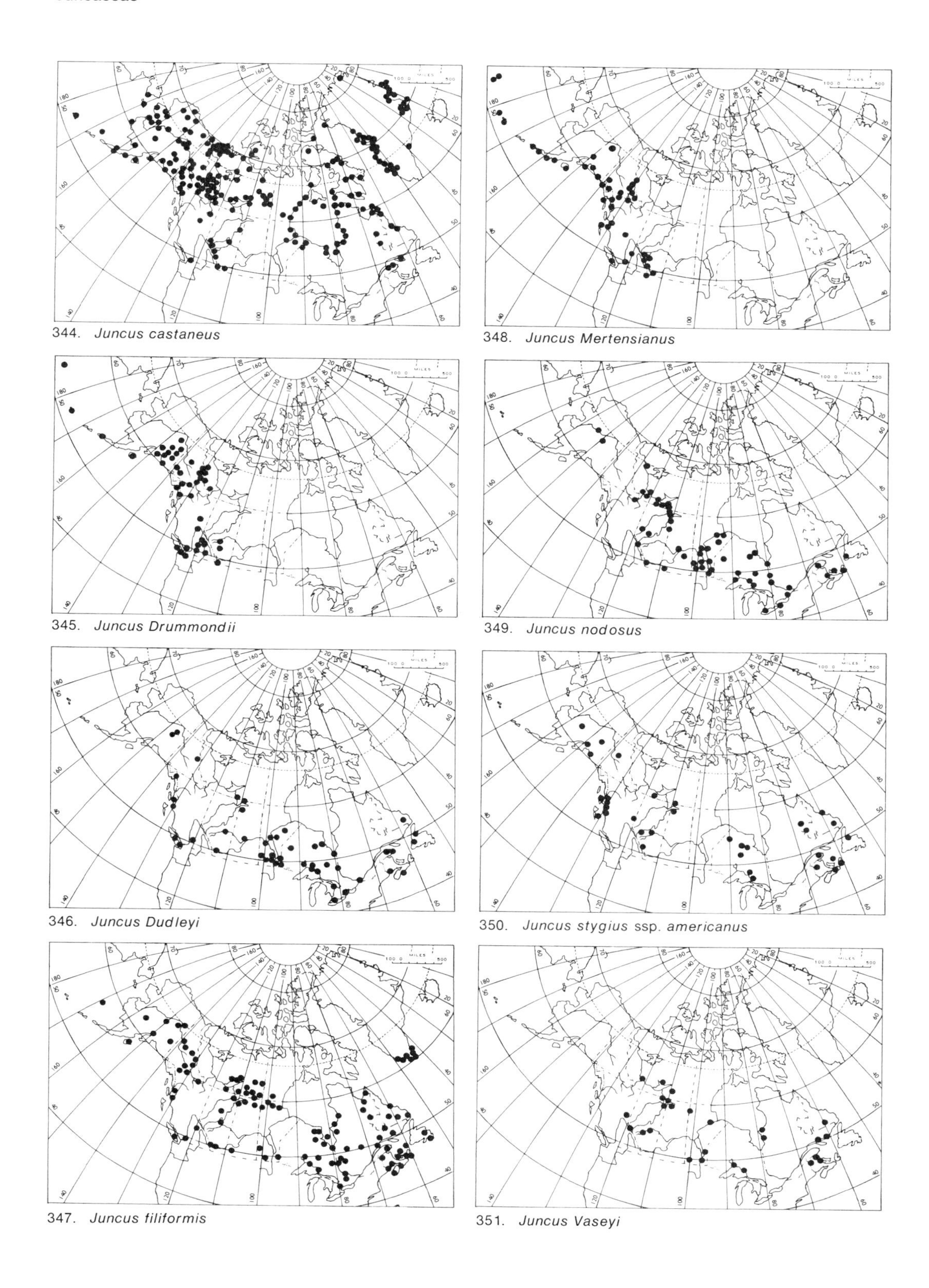

344. *Juncus castaneus*

345. *Juncus Drummondii*

346. *Juncus Dudleyi*

347. *Juncus filiformis*

348. *Juncus Mertensianus*

349. *Juncus nodosus*

350. *Juncus stygius* ssp. *americanus*

351. *Juncus Vaseyi*

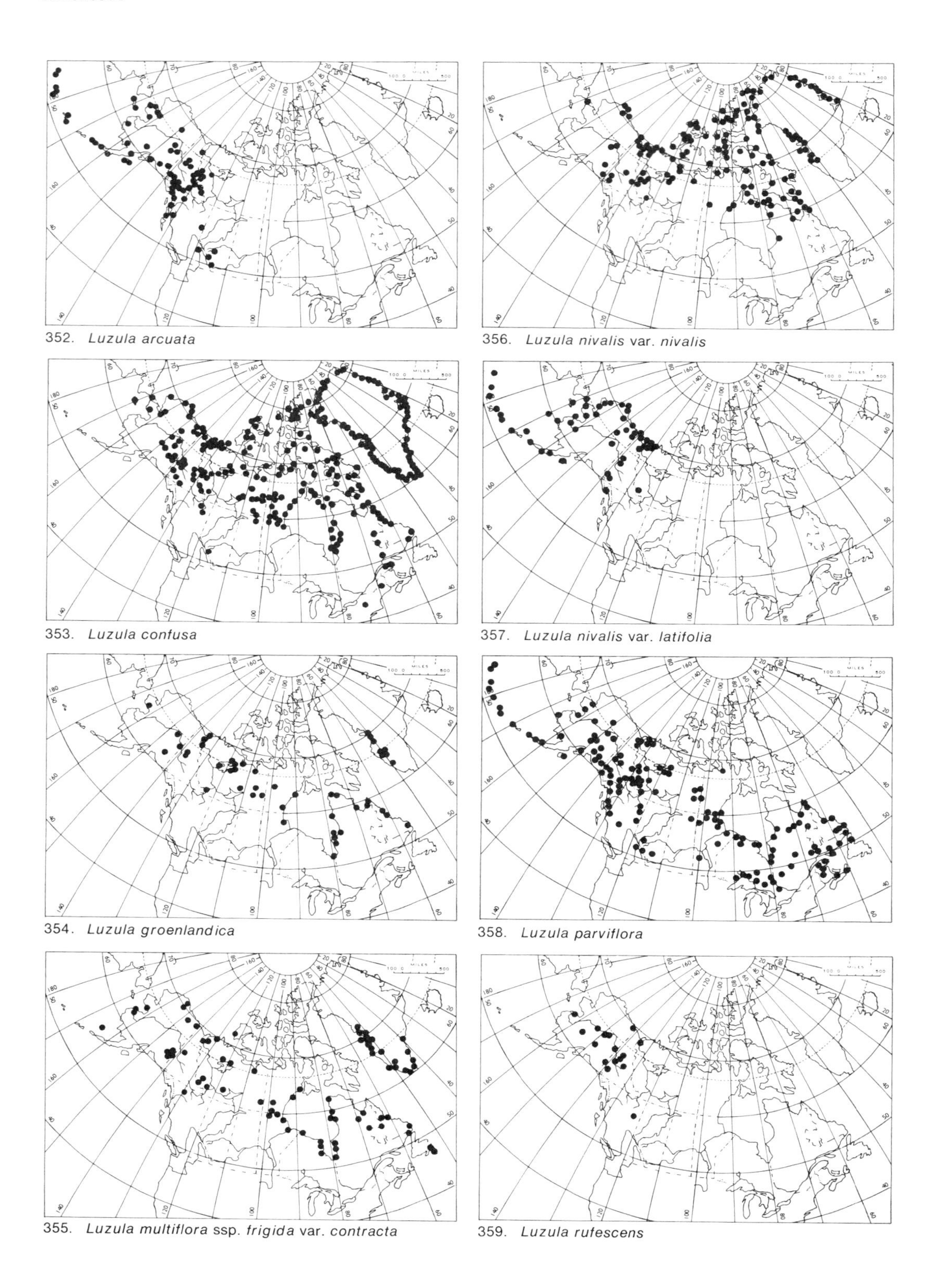

352. *Luzula arcuata*

353. *Luzula confusa*

354. *Luzula groenlandica*

355. *Luzula multiflora* ssp. *frigida* var. *contracta*

356. *Luzula nivalis* var. *nivalis*

357. *Luzula nivalis* var. *latifolia*

358. *Luzula parviflora*

359. *Luzula rufescens*

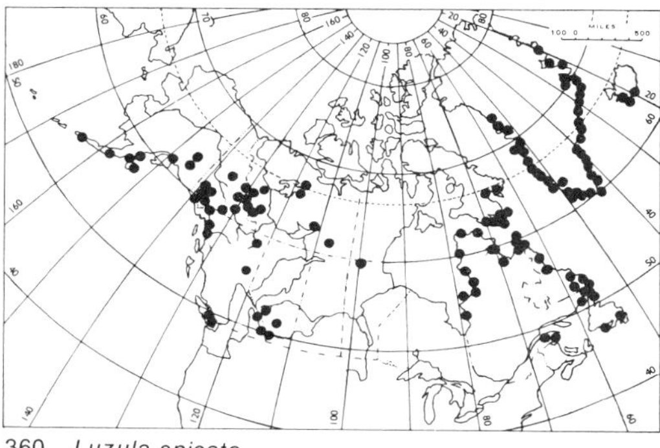

360. *Luzula spicata*

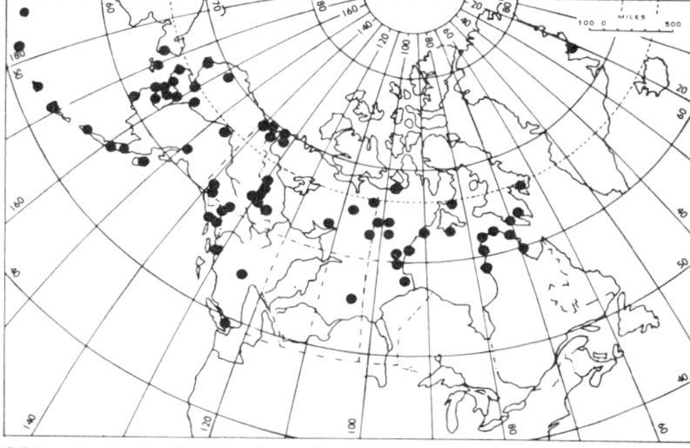

361. *Luzula Wahlenbergii*

LILIACEAE Lily Family

Perennial herbaceous (ours) and mostly somewhat poisonous plants of very diverse habit, with regular and symmetrical, mainly 6-parted and usually perfect, mostly yellowish-green or white flowers or purple in *Allium* (ours) ovary superior, 3-parted; fruit a few to many-seeded capsule, or a berry.

a. Styles 3, distinct; fruit a capsule, or a berry in *Trillium*
  b. Plants with rhizomes
    c. Leaves linear; inflorescence racemose . . . . *Tofieldia*
    c. Leaves broad; stems leafy
      d. Flowers numerous, small, in a terminal raceme. *Veratrum*
      d. Flower solitary, large . . . . *Trillium*
  b. Plants with bulbs; leaves linear and elongate. . . . . *Zygadenus*
a. Style 1; fruit a capsule or berry
  e. Plants with bulbs; leaves linear, terete, fruit a capsule
    f. Flower mostly solitary or few together. . . . . *Lloydia*
    f. Flowers numerous; umbellate . . . . *Allium*
  e. Plants with rhizomes and leafy stems; fruit a berry
    g. Perianth 4-parted; leaves cordate; inflorescence racemose . . . . *Maianthemum*
    g. Perianth 6-parted
      h. Flowers small in a terminal raceme. . . . . *Smilacina*
      h. Flowers axillary, 1-2 together . . . . *Streptopus*

Allium L. Onion

Allium Schoenoprasum L.
var. **sibiricum** (L.) Hartm.
Wild Onion
Onion-smelling plants 1.5 to 4.5 dm tall with hollow leaves shorter than the scape, arising from clusters of membranous-coated oblong-ovoid bulbs. Flowers umbellate; the 6-parted perianths dark pink to purple, the sepals lanceolate-attenuate; fruit a capsule.

Moist to turfy places along river and lake shores, mainly on calcareous or basic rocks, in the western parts of the Precambrian region about Great Slave Lake and northward through the Mackenzie lowlands to the delta of the Anderson River on the Arctic Coast; rare on the east slope of the Mackenzie Mountains ascending at least to 400 m altitude.

The tender young leaves and bulbs of the wild onion are palatable and make a very acceptable substitute for the garden variety.

General distribution: Circumpolar, wide ranging.
Fig. 314 Map 362

Lloydia Salisb.

Lloydia serotina (L.) Rchb.
Dwarf caulescent herb from a small, oblong, fibrous-coated bulb, with basal narrow and grass-like leaves about half as long as the 1- to 2-bracted flowering stem. Flowers mostly solitary, the six perianth segments ovate, about 1 cm long, creamy coloured with purplish veins. Fruit a capsule.

Alpine tundra.

General distribution: Wide ranging Eurasian alpine or sub-arctic species which over Alaska—Y.T. extends along the Richardson and Mackenzie Mts., southward to Nevada and Colorada.
Fig. 315 Map 363

Maianthemum Weber Wild Lily-of-the-Valley

Maianthemum canadense Desf.
var **interius** Fern.
Low herb from a slender, freely branching rhizome; stems 5 to 25 cm tall, solitary, each with from 1 to 3 somewhat pubescent alternate, cordate leaves, and a short, terminal raceme of small, fragrant white flowers; fruit a reddish berry.

Sandy pine woods and rich woodland.

General distribution: *M. canadense s. lat.*; Nfld. to northern B.C. and District of Mackenzie; in eastern Canada and U.S.A. represented by the essentially glabrous var. *canadense*.
Fig. 316 Map 364

Smilacina Desf. False Solomon's Seal

Stems erect with alternate, ovate or broadly lanceolate clasping leaves, from a branching rhizome; flowers in racemes, the perianth 6-parted; fruit a 1 to 2-seeded berry.

a. Stems with numerous leaves, 1.5-5.5 dm tall . . . . *S. stellata*
a. Stems with 2-3 leaves, 0.4-2.0 dm tall . . . . *S. trifolia*

Smilacina stellata (L.) Desf.
Stems slightly flexuous from a stout rhizome; leaves sessile, lanceolate, partly clasping, finely pubescent below; raceme short-peduncled, the flowers creamy white, anthers yellow. Berries green with black stripes, turning black when ripe.

Common locally in dry open woodlands, on calcareous river banks or lake shores, north along the Mackenzie Valley to Bear River, and central Mackenzie Mountains.

General distribution: Boreal American; Nfld. to Alaska.
Fig. 317 Map 365

Smilacina trifolia (L.) Desf.
Rhizome slender; leaves 2 or 3, elliptic, glabrous; raceme peduncled, flowers creamy-white and the anthers purple or dotted with purple. Berries red when ripe.

A peat bog species restricted to the lowland and boreal parts of our region as far north as Norman Wells on the Mackenzie River.

General distribution: Eastern Siberia; boreal America from S.E. Y.T. to Nfld., but apparently lacking in Alaska and most of Y.T.
Fig. 318 Map 366

Streptopus Michx. Twisted-Stalk

Streptopus amplexifolius (L.) DC.
var. **americanus** Schultes
Rhizome short and thick; stems flexuous, simple or forking, up to 9 dm tall; leaves alternate, cordate, clasping, membranaceous and glabrous; flowers axillary, singly or in pairs on slender jointed peduncles; perianth 6-parted, whitish-green, the sepals recurved; anthers longer than the filament; ovary 3-loculed, fruit a red ellipsoid to globose berry about 1.5 cm long.

In our area known only from the Upper Liard River valley and from hot springs in the South Nahanni drainage in the Mackenzie Mountains.
General distribution: S. Greenland, Nfld. to Alaska; boreal with large gaps between L. Superior and B.C.
Fig. 319 Map 367

Tofieldia Huds. False Asphodel

Tufted, from a short rhizome; leaves 2-ranked, linear, basal, or nearly so; flowers racemose on scape-like stems, perianth 6-parted, persistent; styles 3; fruit a many-seed capsule.

a. Stems glabrous
  b. Stems scapose, pale green; flowers yellowish-green .... *T. pusilla*
  b. Stems with one or more leafy bracts, dark coloured; flowers purplish ..... *T. coccinea*
a. Stems glandular pubescent, especially above .... *T. glutinosa*

Tofieldia coccinea Richards.
Stems low, up to 15 cm tall, commonly purplish-tinged, inflorescence dark purplish-brown; capsules reflexed when mature, about 2.5 mm long; seeds not appendaged.
Occasional to common, mainly alpine or beyond the limit of trees, and perhaps always on calcareous soils.
General distribution: E. Greenland to central E. Asia, with large gaps, arctic-alpine.
Fig. 320 Map 368

Tofieldia glutinosa (Michx.) Pers.
T. occidentalis S. Wats.
Stems stout, 1.2 to 3.5 dm tall, glutinous, occasionally with a single bract; leaves broadly linear, usually more than half the length of the stem; perianth segments creamy-white; capsules erect about 6 mm long, yellow to red; seeds with a twisted, tail-like appendage at each end.
Moist calcareous seepages and lake shores, usually at low elevations but known also from the north peak, Nahanni Mountain.
General distribution: N. America, Nfld. to B.C. and Alaska; boreal, wide ranging.
Fig. 321 Map 369

Tofieldia pusilla (Michx.) Pers.
T. palustris Huds.
Stems green, 0.5 to 2.5 dm tall, bractless or occasionally bracted near the base; leaves narrower than in *T. coccinea*; perianth segments greenish white; capsules erect, about 2.5 mm long, green or straw-coloured; seeds not appendaged.
Common locally in moist calcareous turfy places, north beyond the limit of trees.
General distribution: Circumpolar, arctic-alpine.
Fig. 322 Map 370

Trillium L. Wakerobin

***Trillium cernuum** L.
var. **macranthum** Eames & Wieg.
Stems simple, to 3 dm tall, 1 to several from an ascending thick, brown and fleshy rhizome, bearing at the summit a whorl of 3 rhombic-ovate, short-petioled leaves, and a single, nodding and recurved flower which is almost hidden under the leaves. Sepals long-acuminate, petals rhombic-ovate, white, creamy or roseate, 1.5 to 2.0 cm long.
Raup (1947) reported a Richardson collection in the Gray Herbarium merely labelled "Mackenzie River", noting that "it is the only evidence for the occurrence of any species of *Trillium* in the entire Mackenzie basin".
Fig. 323 Map 371

Veratrum L. False Hellebore

Veratrum Eschscholtzii A. Gray
Stems stout and leafy up to 2 m tall, from a thick rhizome; leaves large, veiny, pubescent on the under surface, clasping, the lower oval, flat and blunt, the upper elliptic to lance-elliptic and acute, and often folded; flowers small but very numerous, in an open, drooping and branched panicle; perianth 6-parted, the segments yellowish-green, about 1 cm long; ovary 3-lobed, each lobe with a short recurved style; fruit a many-seeded capsule.
Rare to occasional in wet meadows and about hot springs near treeline in the southwestern Mackenzie Mountains.

Veratrum Eschscholtzii is closely related to the eastern N. American *V. viride* of which some authors consider it a geographic race. The latter is known to contain poisonous alkaloides.

General distribution: Cordilleran-Pacific Coast timber line, snowbed species.

Fig. 324 Map 372

Zygadenus Michx. Death-Camass

Zygadenus elegans Pursh
Somewhat glaucous herb with erect stems up to 6 dm tall, with elongate linear leaves near the base, arising from a bulb; inflorescence a bracted raceme; perianth 6-parted, the sepals similar, oblong-ovate, 6 to 10 mm long, greenish-yellowish-white; ovary 3-lobed; fruit an ovate-lanceolate many seeded capsule.

Common on open sunny slopes, river banks and gravelly places in the upper Mackenzie River valley and mountains, mainly in calcareous soils, eastward to the Palaeozoic boundary and north to the Mackenzie and Anderson River deltas.

All parts of the plant contain the poisonous alkaloid zygadenine.

General distribution: Plains and mountains. Man. to B.C. and Alaska and southward.

Fig. 325 Map 373

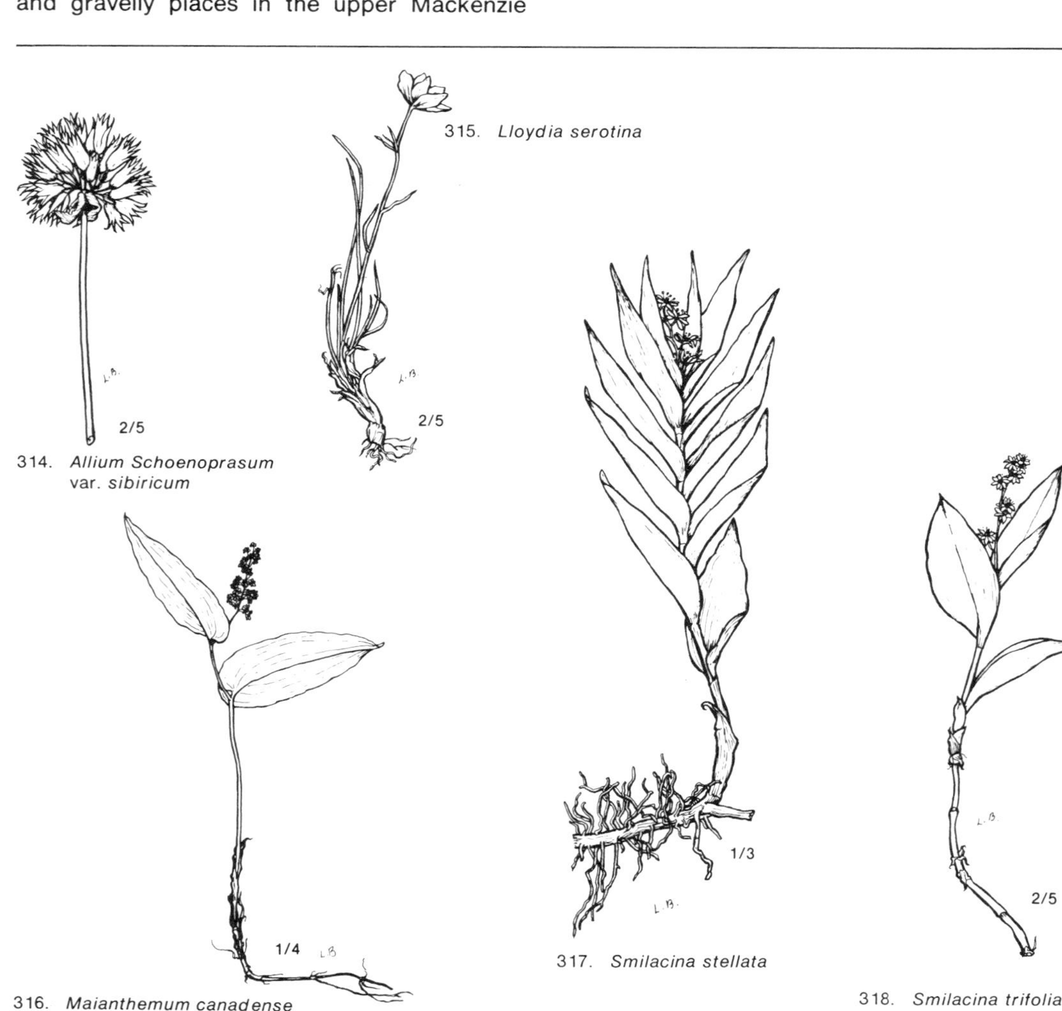

314. *Allium Schoenoprasum* var. *sibiricum*

315. *Lloydia serotina*

316. *Maianthemum canadense* var. *interius*

317. *Smilacina stellata*

318. *Smilacina trifolia*

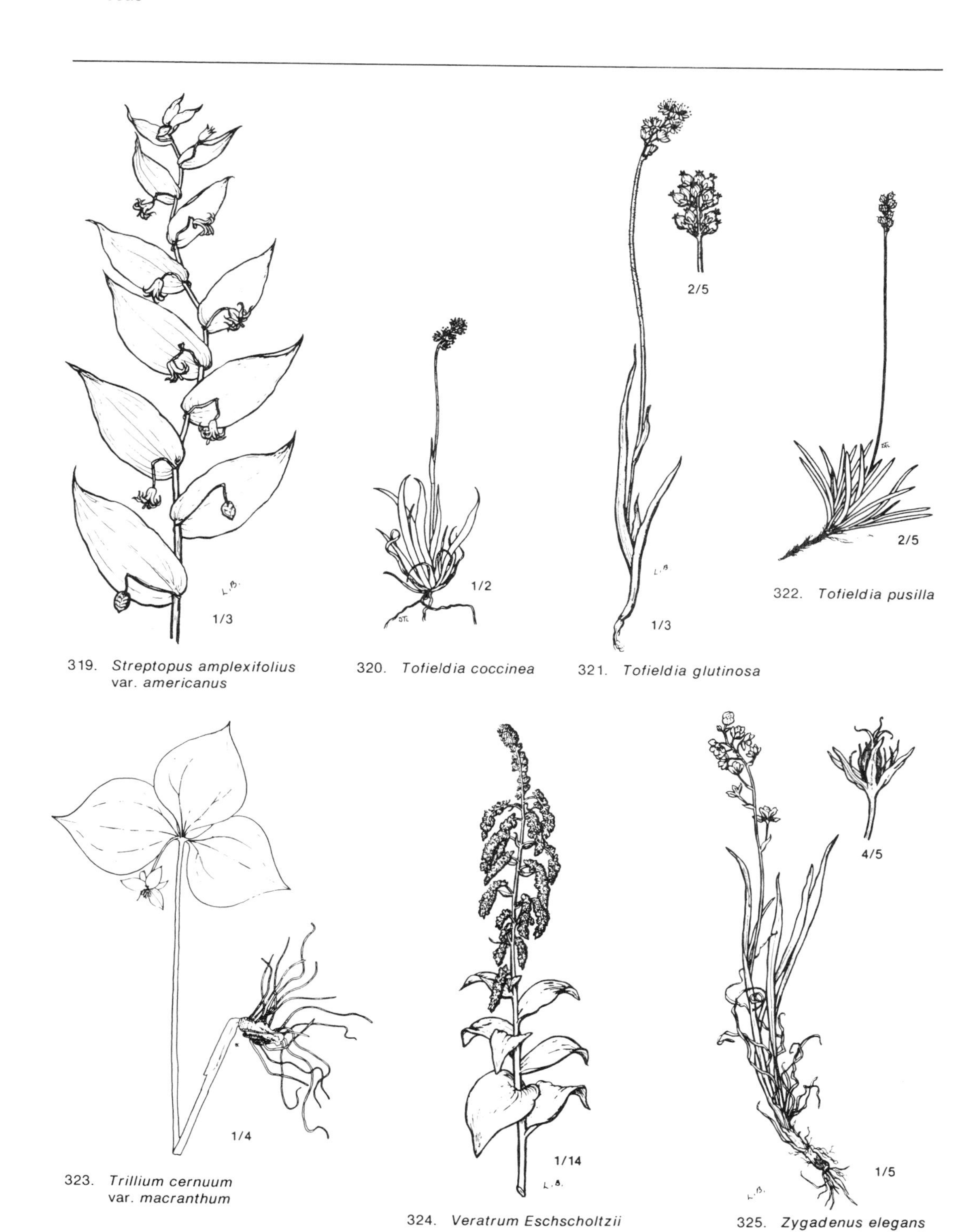

319. *Streptopus amplexifolius* var. *americanus*

320. *Tofieldia coccinea*

321. *Tofieldia glutinosa*

322. *Tofieldia pusilla*

323. *Trillium cernuum* var. *macranthum*

324. *Veratrum Eschscholtzii*

325. *Zygadenus elegans*

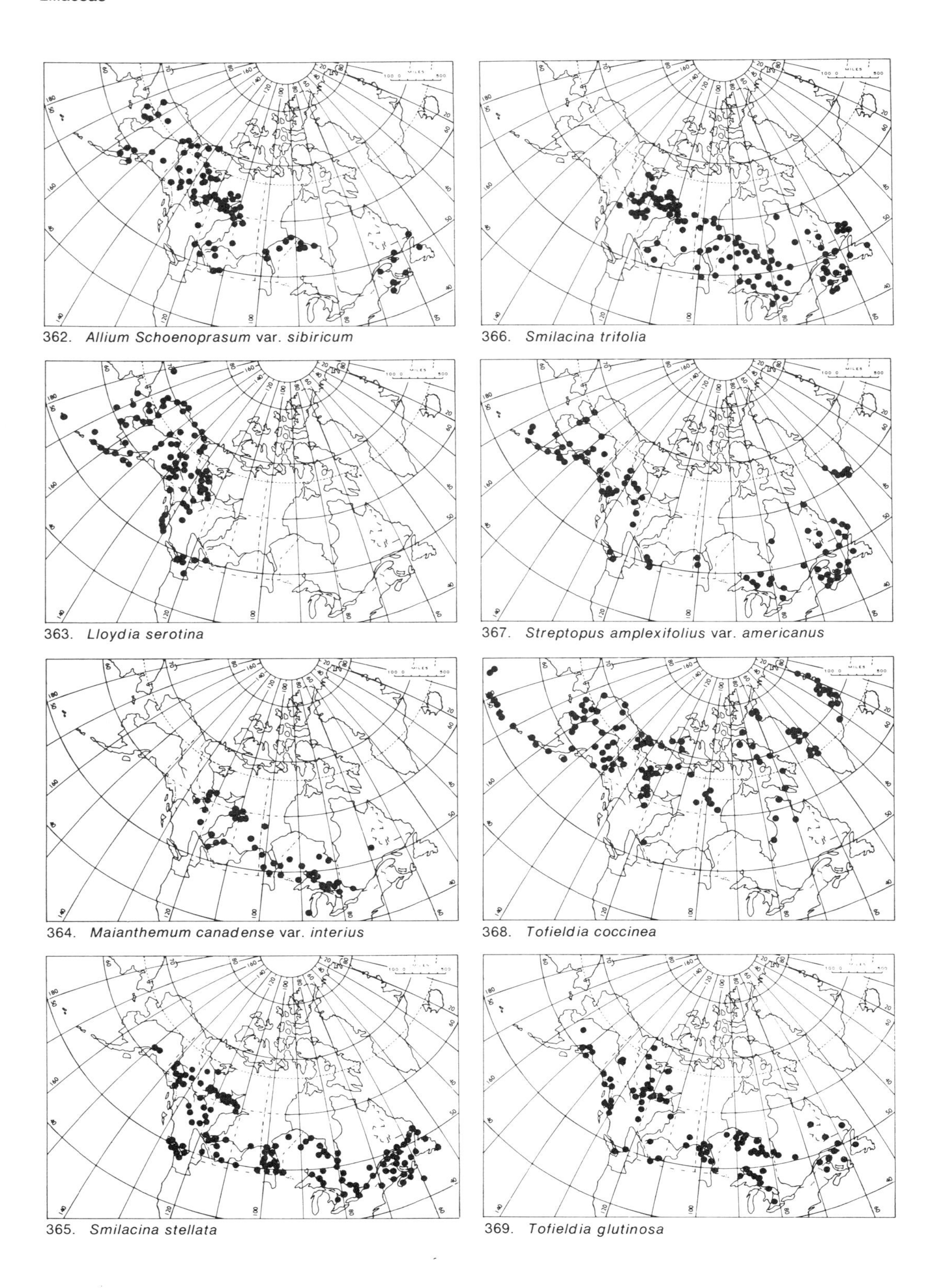

362. *Allium Schoenoprasum* var. *sibiricum*

363. *Lloydia serotina*

364. *Maianthemum canadense* var. *interius*

365. *Smilacina stellata*

366. *Smilacina trifolia*

367. *Streptopus amplexifolius* var. *americanus*

368. *Tofieldia coccinea*

369. *Tofieldia glutinosa*

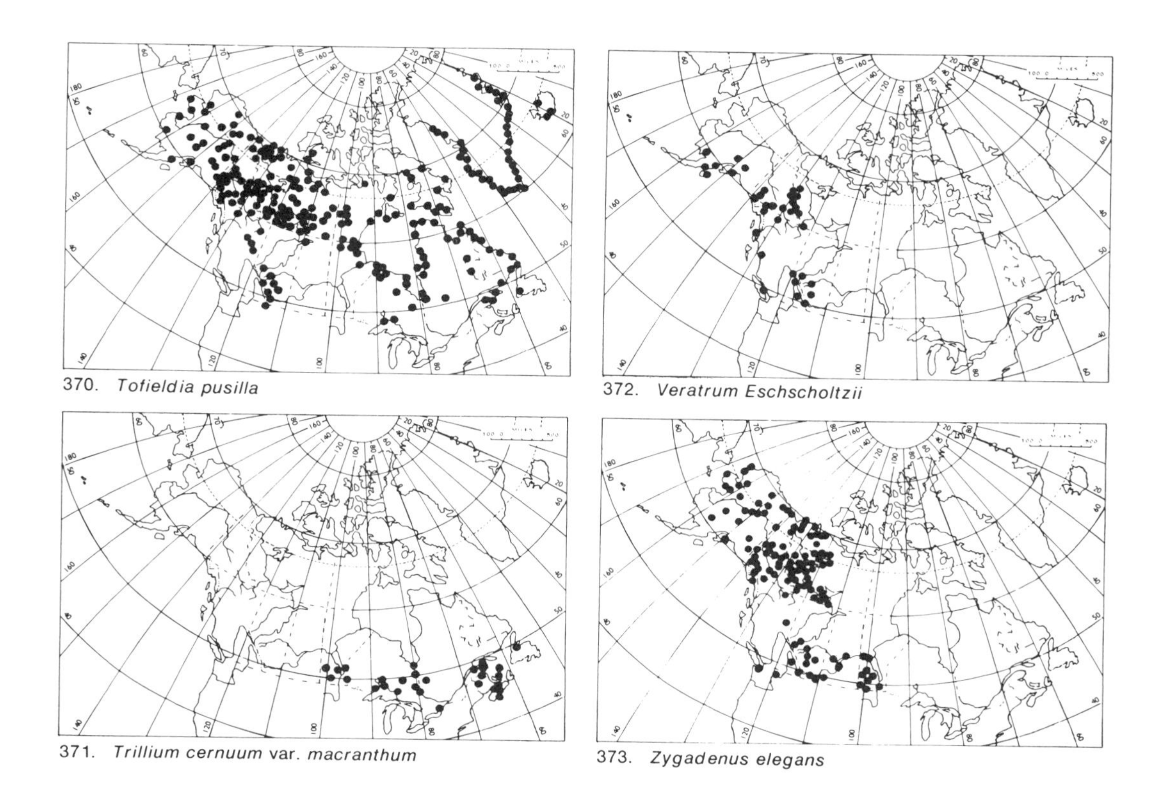

370. *Tofieldia pusilla*

371. *Trillium cernuum* var. *macranthum*

372. *Veratrum Eschscholtzii*

373. *Zygadenus elegans*

IRIDACEAE Iris Family

Sisyrinchium L. Blue-eyed Grass

Sisyrinchium montanum Greene
Tufted grass-like herbs from a short rhizome and fibrous roots; stems flattened, wing-margined, 2 to 3 dm tall; leaves basal, about half the length of the stem; flowers umbellate, between two greenish bracts; perianth 6-parted, the segments blue-violet, oblong-ovate, apiculate, about 1 cm long; fruit a globose capsule opening at the top; seeds ovoid, black.

Occasional to frequent in meadows, on open hillsides and river banks in the southwestern Mackenzie lowland region, extending as far north as Norman Wells, and barely reaching into the Precambrian Shield area east of the Slave River.

General distribution: S.W. Greenland; boreal America, Nfld. to the District of Mackenzie and northeastern B.C., and apparently local and disjunct in S.W.Y.T.

Fig. 326 Map 374

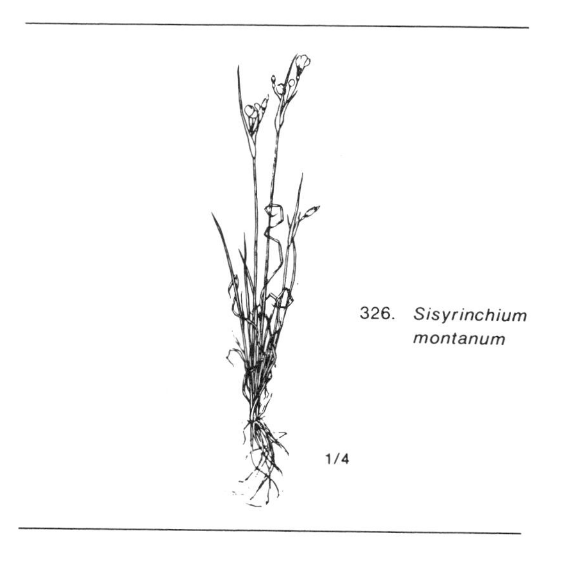

326. *Sisyrinchium montanum*

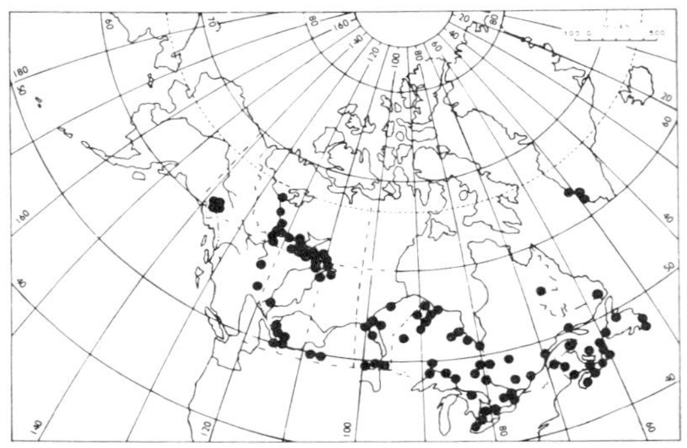

374. *Sisyrinchium montanum*

ORCHIDACEAE Orchid Family

Ours terrestrial, perennial herbs with rhizomes, fleshy roots, root-tubers, corms or bulbs, and simple, entire leaves, or mere sheathing scales in the saprophytic *Corallorhiza*. Flowers perfect, zygomorphic, 3-merous, often showy; sepals mostly 3, green or coloured; petals 3 of which the two lateral are alike, but different from the middle and lower, called the lip, which moreover, in some genera is spurred at the base. Ovary inferior, 1-celled; fruit a 3-valved capsule containing inumerable, minute seeds.

a. Plants entirely saprophytic, lacking green leaves . . . . *Corallorhiza*
a. Plants with normal green leaves
  b. Lip inflated, pouch- or boat-shaped, or saccate, at least at the base
    c. Flowers 1-3, large and brightly coloured
      d. Lip bearded; flower solitary, variegated in purple, yellow and pink; leaf solitary, at the base of the stem . . . . *Calypso*
      d. Lip smooth, the margins inrolled; flowers 1-3; leaves 2 or more . . . . . *Cypripedium*
    c. Flowers numerous, small, in a spike or spike-like raceme; leaves fleshy, wintergreen, forming a rosette at the base of the stem . . . . *Goodyera*
  b. Lip not inflated or boat-shaped
    e. Spur conspicuous
      f. Petals spotted with pink; leaf solitary, basal . . . . *Orchis*
      f. Petals white, or yellowish-green; leaves more than one, except in *H. obtusata*. . . . . *Habenaria*
    e. Spur lacking
      g. Leaves 2, opposite
        h. Leaves near the middle of the stem . . . . *Listera*
        h. Leaves basal . . . . *Liparis*
      g. Leaves one, two or several, alternate on the stem
        i. Stem arising from elongated fleshy roots . . . . *Spiranthes*
        i. Stem arising from solid, tuber-like corm or bulb. . . . . *Malaxis*

Calypso Salisb.

Calypso bulbosa (L.) Oakes
Stems about 1.5 dm tall with a single very showy flower, and a solitary basal oval and petiolate leaf, from a solid bulb; capsule cylindrical, erect.

Occasional to rare in damp moss, usually in

Picea glauca forest about Great Slave Lake and in the river drainages of the Slave, Liard and Mackenzie, north to Great Bear Lake and Norman Wells.

General distribution: Circumpolar with large gaps.

Fig. 327 Map 375

Corallorhiza Chat. Coral-root

Corallorhiza trifida Chat.
Low, yellowish-brown herbs lacking green leaves, from a coral-like rhizome. The inflorescence a few-flowered raceme of small flowers; sepals and lateral petals similar, 4 to 5 mm long, greenish-yellow, the lip broadly ovate, notched at the tip, whitish and often spotted with purple; capsule reflexed.

Occasional to common locally in turfy, open places, mainly on calcareous soils, and often growing in the centre of well-established *Dryas integrifolia* mats, extending northward beyond the treeline to the arctic coast near the Mackenzie Delta, the mouth of the Coppermine River and the head of Bathurst Inlet; in the Mackenzie Mountains ascending to above the treeline in Macmillan Pass.

General distribution: Circumpolar; wide ranging.

Fig. 328 Map 376

Cypripedium L. Lady's-slipper

Stems leafy, or with two large basal leaves from a short or elongate rhizome and coarse fibrous roots; flowers showy, leafy-bracted, the sepals three, or two of them united under the sac-like lip (lower petal), the lateral petals oblong or linear; capsules large, ovate-cylindrical.

a. Plant scapose, with 2 basal leaves; flower solitary, lip pink, cleft in front .... *C. acaule*
a. Plant not scapose; flowers 1-3; lip not cleft
  b. Leaves 2, sub-opposite on the lower third of stem; flower solitary, lip white, spotted with purple; plant turns black in drying .... *C. guttatum*
  b. Leaves 3 or more
    c. Flower large, 1 or 2, petals purplish brown, spirally twisted; lip yellow .... *C. Calceolus* var. *parviflorum*
    c. Flowers small, mostly one, petals white, not twisted, lip white or pale magenta .... *C. passerinum*

Cypripedium acaule Ait.
Stemless Lady's-slipper
At once distinguished from the following three species by its large and always solitary flower with greenish-brown sepals and petals and deeply fissured pink lip. The peduncle is characteristically bent just below the flower to form a "goose neck".

In moist moss or sandy woodland and bogs; as noted by Raup (1947) a Richardson specimen of *C. acaule*, in the Gray Herbarium labelled "Ft. Franklin" is likely the basis for the range given in Hooker's Fl. Bor.-Am. "from Canada to Fort Franklin on the Mackenzie River". No later collectors have reported *C. acaule* in the Mackenzie drainage from north of lat. 60°.

General distribution: Boreal America, Nfld. to N.E. Alta. and the District of Mackenzie and southward.

Fig. 329 Map 377

Cypripedium Calceolus L.
var. **parviflorum** (Salisb.) Fern.
c. parviflorum Salisb.
Yellow Lady's-slipper
Stems leafy, 1.5 to 3 dm tall; flowers 1 to 3, fragrant, the lateral sepals united, lateral petals longer than the lip, purplish-brown, the sac-like lip 2 to 3 cm long, yellow, often purple-spotted on the inner surface, especially near the mouth; capsule upright.

Moist calcareous woodlands along the Liard River and upper Mackenzie River north to Norman Wells, and western Great Bear Lake and the Mackenzie Mountains.

General distribution: The species is circumpolar with gaps, the var. *parviflorum* in America from Nfld. to B.C., Y.T. and S.E. Alaska and southward.

Fig. 330 Map 378

Cypripedium guttatum Sw.
Stems slender, 1 to 2 dm tall; lateral sepals united almost to the tip, lateral petals ovate, tapering to a blunt tip, white with purple blotches, similar to the sac-like and about 2 cm long lip; capsule drooping.

Rare and localized in open, mossy woods and perhaps always restricted to calcareous soils.

General distribution: An Amphi-Beringian species which extends through Alaska and Y.T. into the District of Mackenzie.

Fig. 331 Map 379

Cypripedium passerinum Richards.
Stems 1.0 to 3.5 dm tall, leafy; flowers 1 to 3, small and fragrant, the lateral sepals free almost to the base, lateral petals white, linear-elliptic to oblong, about as long as the lip which is egg-shaped and about 1.5 cm long, pure white but so translucent that the purple spots on the inner surface are faintly visible; in life the lip is reminiscent of a sparrow's egg, hence the specific name bestowed by its discoverer, John Richardson.

Occasional or common locally throughout the Mackenzie drainage north to the delta, always growing in mineral soils derived from basic or calcareous rocks.

General distribution: Boreal N. America from Alaska to James Bay, with disjunct stations on the northeast shore of L. Superior, and on Mingan Islands on the north shore of the St. Lawrence R.

Fig. 332 Map 380

Goodyera R.Br. Rattlesnake-plantain

Goodyera repens (L.) R.Br.
Plant rhizomatose, with a cluster of basal ovate-lanceolate, petiolate, 2 to 3 cm long reticulate-veined leaves; scape 1.5 to 2.5 dm tall, glandular-puberulent; flowers greenish-white in a bracted, one-sided and often twisted raceme; capsules ovoid, 5 to 9 mm long, ascending to divergent, tipped by the marcescent flowers.

In the field, plants in which the leaves are green throughout, or white-reticulate (var. *ophioides* Fern.), are frequently found growing side by side.

Occasional in damp, mossy spruce woods in S.W. District of Mackenzie, extending as far north as the Bear River and Sans Sault Rapids on the Mackenzie River and to the Mackenzie Mountains.

General distribution: Boreal America: Southern Lab. to Alaska

Fig. 333 Map 381

Habenaria Willd. Bog orchid

Small-flowered glabrous plants with leafy or scapose stems from fleshy or tuber-like roots; flowers in a terminal bracted raceme or spike; sepals and lateral petals approximately the same length, the lip entire or notched at the apex, spurred; capsule oblong to ovoid, ascending or spreading.

a. Stem scapose
  b. Leaves 2, orbicular-oval, shiny above, 10-20 cm in diameter, commonly flat on the ground . . . . *H. orbiculata*
  b. Leaves mostly one, oblanceolate, erect-ascending. . . . . *H. obtusata*
a. Stem leafy
  c. Lip notched or 2-3 toothed at the apex; spur pouch-like; flowers scentless . . . . *H. viridis* var. *bracteata*
  c. Lip entire; spur linear; flowers fragrant
    d. Flowers greenish, lip not conspicuously dilated at the base . . . . *H. hyperborea*
    d. Flowers pure white; lip conspicuously dilated at the base . . . . *H. dilatata*

Habenaria dilatata (Pursh) Hook.
White Orchid
Stems leafy, 1.5 to 4.0 dm tall; raceme dense, bracts appressed and thus appearing shorter than the flowers; flowers very fragrant, white, the lip dilated at the base, the filiform spur usually longer than the lip.

In life readily distinguished from the somewhat similar *H. hyperborea* by its white flowers.

In the District of Mackenzie thus far known only in the South Nahanni River drainage in the southern Mackenzie Mountains, from two sites where it grew in meadows by hot springs.

General distribution: A wide-ranging N. American species of wet meadows, fens and open moist woods.

Fig. 334 Map 382

Habenaria hyperborea (L.) R. Br.
Northern Green Orchid
Stems leafy, 1.0 to 5.5 dm tall; raceme dense, bracts appressed and thus appearing shorter than the flowers; flowers greenish, smaller than in *H. dilatata*; lip lanceolate, tapering to the blunt apex, the spur about as long as the lip.

Occasional to common on moist grassy river banks and lake shores, in mossy spruce woods and muskegs north to near the Arctic Circle on Great Bear Lake, and east to the end of Great Slave Lake. In the valleys of the Mackenzie Mountains ascending at least 1000 m.

General distribution: Wide ranging, boreal N. America, from Greenland to B.C. and Alaska.

Fig. 335 Map 383

Habenaria obtusata (Pursh) Richards.
Northern Bog-Orchid
Leaf usually solitary, oblanceolate, bluntish, ascending; scape 0.6 to 2.0 dm tall; raceme usually open, few to many-flowered, the flowers greenish, small; lip linear with a somewhat dilated base, spur tapering from a broad base, shorter than the lip.

The flowers of the northern bog-orchid are regularly cross-pollinated by mosquitoes in a rather interesting manner. The mosquito enters the flower in search of honey, which the orchid produces and stores in the slender, downward-projected spur. On departing, the insect often can be seen carrying on its head what looks like one or two tiny, yellow "horns". Under a lens these prove to be the club-shaped pollen masses, or pollinia, of the orchid flower. When the insect visits a second flower, the pollinia are unavoidably brought into contact with the orchid's stigma, causing a transfer of pollen and subsequent fertilization. A simple experiment in which the tip of a lead pencil is substituted for the head of the mosquito, will demonstrate what happens. At the slightest touch the elastic pollinia spring forward and attach themselves by their sticky, discoid base to the head or eyes of the mosquito, in this case, the tip of the lead pencil.

Common or even ubiquitous throughout the wooded parts of the district and locally extending far beyond the treeline to the Thelon River, the arctic coast near Coppermine, Bathurst Inlet and the shores of Hudson Bay.

General distribution: Wide spread; boreal N. America, Nfld. to B.C. and Alaska.

Fig. 336 Map 384

Habenaria orbiculata (Pursh) Torr.
Round-leaved Orchid
Leaves two, basal, orbicular-oval, 10 to 20 cm in diameter, usually flat on the ground; scape with 1 to several bracts; raceme open, the flowers pedicellate, greenish-white, the lip linear-oblong, blunt, about 10 mm long, the spur club-shaped, twice as long as the lip.

Known in our area from spruce and tamarack woodland in the southern Mackenzie Mountains.

General distribution: Boreal N. America, Nfld. to B.C. and southern-most Alaska, north to central Y.T. and S.W. District of Mackenzie.

Fig. 337 Map 385

Habenaria viridis (L.) R.Br.
var. **bracteata** (Muhl.) Fern.
Bracted Green Orchid
Stems leafy 1.5 to 3.0 dm tall; raceme usually dense, bracts, particularly the lower ones, much longer than the flowers, and spreading; flowers small and green, the lip narrowly oblong-spatulate, the apex 2-toothed (rarely also with a smaller median tooth); spur short, saccate, less than half as long as the lip.

Moist, sandy places in mixed woods; in the District of Mackenzie thus far known only from the vicinity of Fort Smith, and from the east flanks of the Mackenzie Mountains adjacent to the Liard and Mackenzie Rivers, north to near Norman Wells.

General distribution: *H. viridis s. lat.*, circumpolar.

Fig. 338 Map 386

Liparis Rich. Tway-blade

Liparis Loeselii (L.) Rich.
Low herbs from a bulbous base; leaves lustrous, elliptical, in pairs at the base of the up to 1.5 dm tall scape; flowers in a raceme, yellowish-green on short ascending pedicels.

Wet organic soil in fens; thus far collected only once in our area at Yohin Lake in SW District of Mackenzie.

General distribution: Que. to Sask. and disjunct to SW District of Mackenzie.

Map 387

Listera R.Br. Twayblade

Small, slender plants with fibrous roots and two opposite and sessile leaves near or slightly above the middle of the stem, and an open raceme of small bracted flowers.

a. Flowers purplish or less commonly yellowish-green; leaves cordate; lip deeply cleft into 2 narrow lobes . . . . *L. cordata*
a. Flowers green; lip oblong, narrowest in the middle, notched at the apex, and with 2 lobes near the base . . . . *L. borealis*

Listera borealis Morong
Stems 0.5 to 2.0 dm tall, glandular pubescent above; leaves ovate-elliptic.

Occasional to common in rich open spruce forest north through the Mackenzie Valley to the Delta. In the Mackenzie Mountains known from Brintnell Lake and Nahanni National Park.

General distribution: Boreal N. America.

Fig. 339 Map 388

Listera cordata (L.) R.Br.
Stems 0.8 to 2.5 dm tall, glabrous; leaves ovate-cordate or deltoid.

Rare and always local in our area where, thus far, it is known only from two collections, both from southern Mackenzie Mountains: One in forest by edge of hot springs, 1300 m, Hole-in-the-Wall Lake; the other in lodgepole pine forest and at slightly lower elevation in the Liard Range.

General distribution: Circumpolar with large gaps.

Fig. 340 Map 389

Malaxis Sw. Adder's-mouth

Small and delicate plants from a solid, ovoid corm; stems erect, bearing one or several leaves at or near the base; inflorescence an elongated raceme of very small, greenish or yellowish flowers.

a. Leaf 1, elliptic oval to ovate, sheathing; lip reflexed, ovate with an abrupt long point . . . . *M. brachypoda*
a. Leaves 2-several, small and scale-like, alternate, obovate, subtending small new tubers; lip ovate, erect . . . . *M. paludosa*

***Malaxis brachypoda** (Gray) Fern.
Stem 0.8 to 2.0 dm tall; leaf single, usually near the base, elliptic-oval to ovate, to 6 or more cm long, clasping; flowers minute, on short divergent pedicels.

Damp calcareous fens.

General distribution: Nfld. to northern B.C. and Alaska; to be looked for in the southwestern parts of our area.

Fig. 341 Map 390

Malaxis paludosa (L.) Sw.
Stems 0.6 to 1.5 dm tall; new small tubers formed in axils of the leaves; flowers larger than in *M. brachypoda*, the lip ovate, erect.

Thus far collected but once in our area in open black spruce muskeg in the Liard River lowland near the British Columbia border.

General distribution: Circumpolar, apparently rare and local with very large gaps.

Fig. 342 Map 391

Orchis L. Orchis

Orchis rotundifolia Banks
Rhizome short, with fleshy-fibrous roots; leaf solitary, basal or nearly so, orbicular to elliptic, 3 to 7 cm long; flowers few, showy, in a short bracted raceme; sepals and lateral petals whitish to pink, the 3-lobed lip white with purple spots, and a basal spur.

Frequent in the lowland drainage area of the Mackenzie River basin north to its delta; commonly growing in moist, well-drained calcareous soils of open woods, and by seepages.

General distribution: S.W. Greenland to Alaska.

Fig. 343 Map 392

Spiranthes Richard Ladies'-tresses

Spiranthes Romanzoffiana Cham. & Schlecht. Stems 0.8 to 2.0 dm tall, leafy in the lower half; flowers white and very fragrant, in a bracted, dense and spirally twisted spike.

Occasional to frequent in rich fens and wet meadows in the western lowland of our area, north to Great Bear Lake.

General distribution: Boreal N. America.

Fig. 344 Map 393

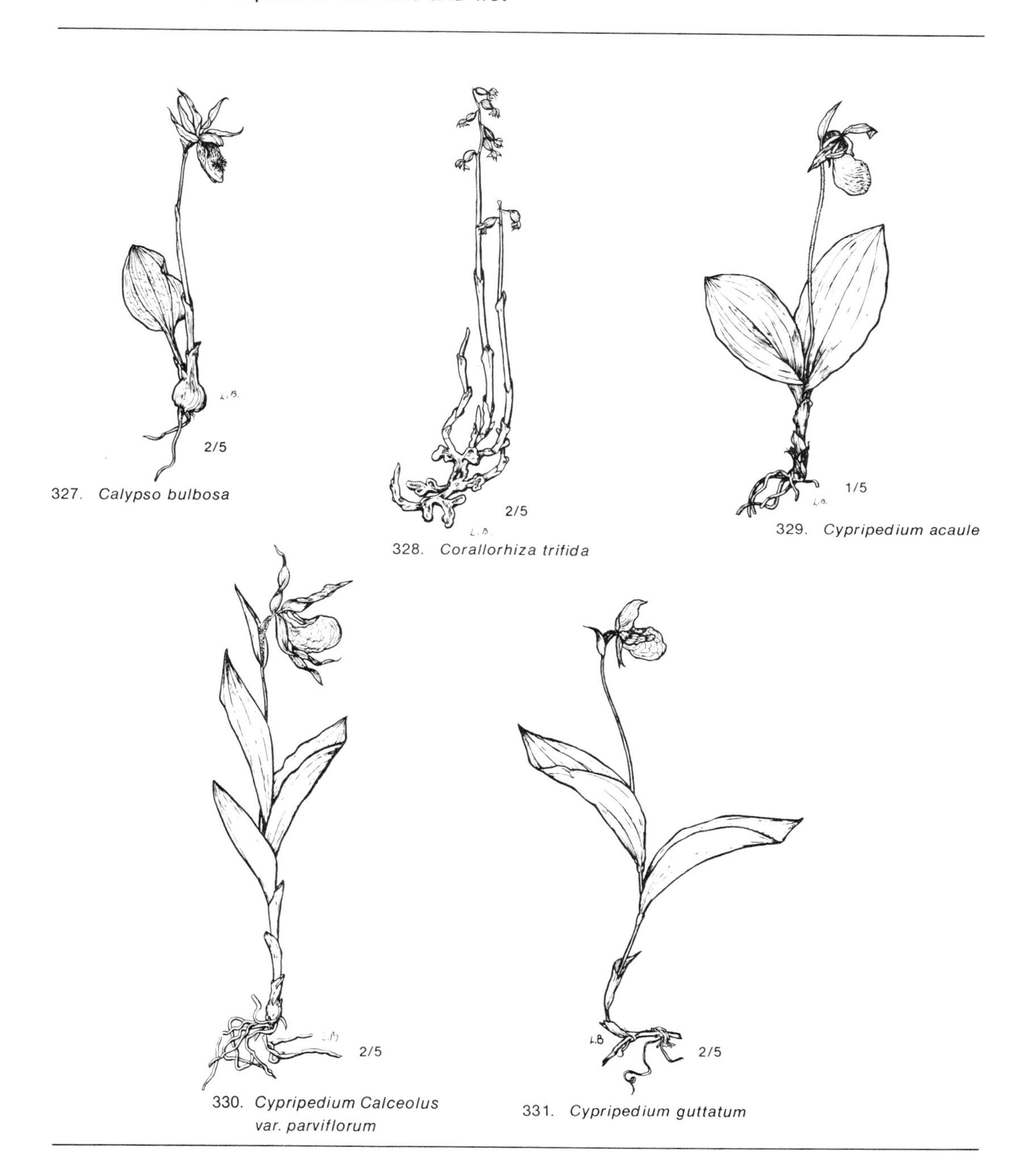

327. *Calypso bulbosa*

328. *Corallorhiza trifida*

329. *Cypripedium acaule*

330. *Cypripedium Calceolus* var. *parviflorum*

331. *Cypripedium guttatum*

332. *Cypripedium passerinum*

333. *Goodyera repens*

334. *Habenaria dilatata*

335. *Habenaria hyperborea*

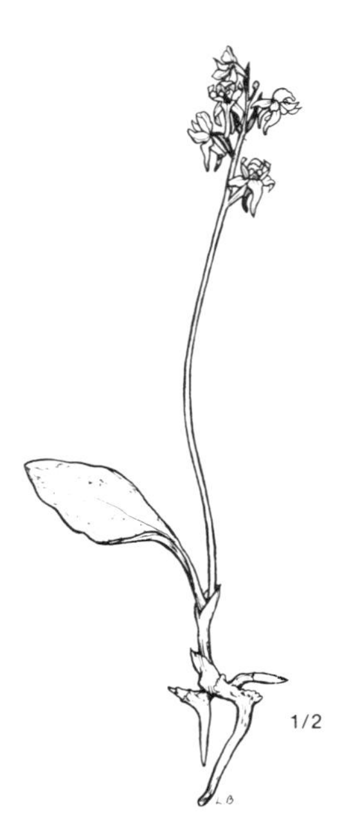

336. *Habenaria obtusata*

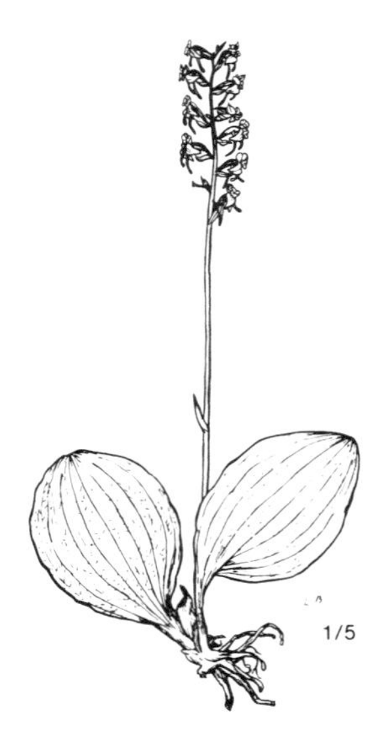

337. *Habenaria orbiculata*

338. *Habenaria viridis* var. *bracteata*

339. *Listera borealis*

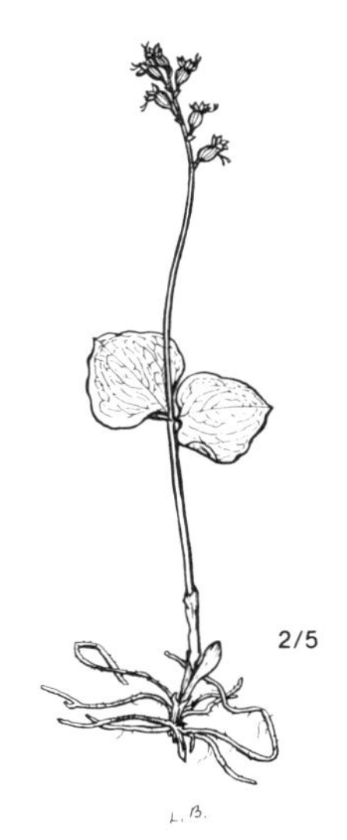

340. *Listera cordata*

341. *Malaxis brachypoda*

342. *Malaxis paludosa*

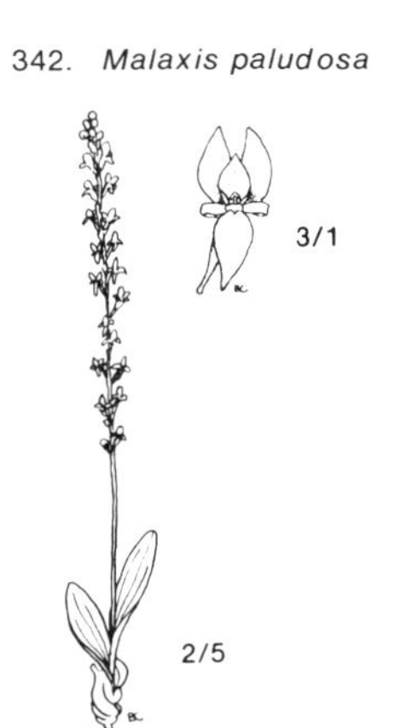

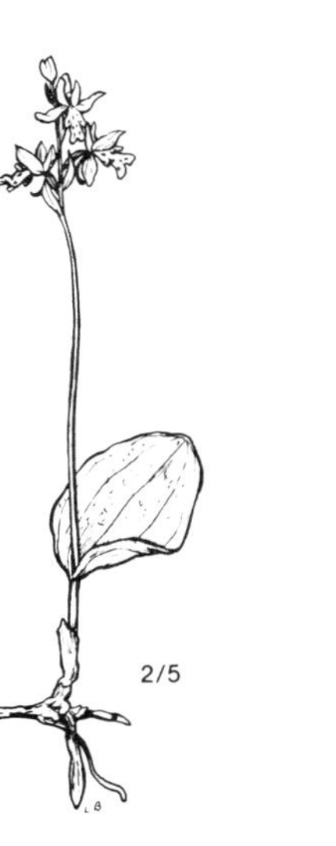

343. *Orchis rotundifolia*

344. *Spiranthes Romanzoffiana*

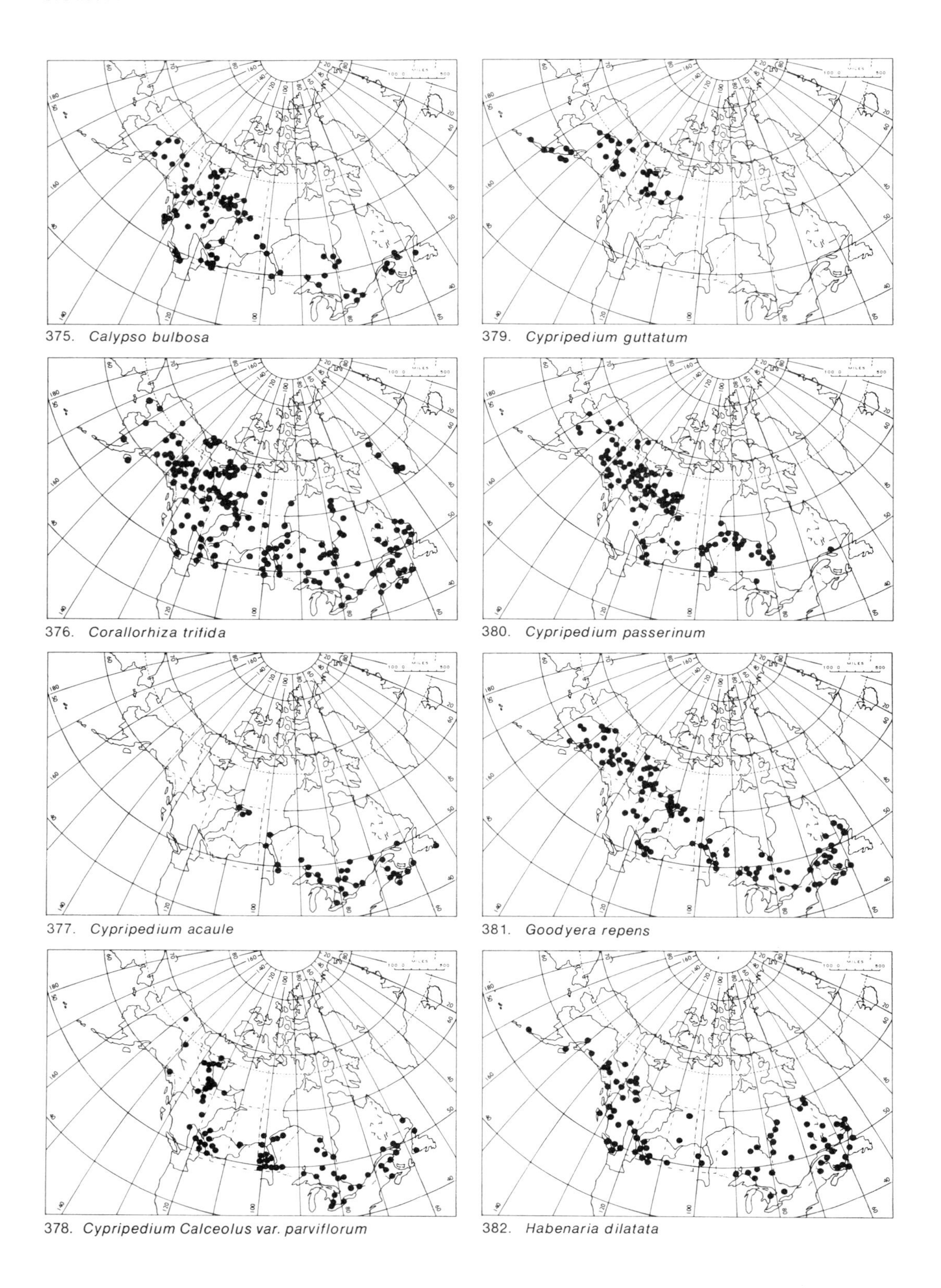

375. *Calypso bulbosa*

379. *Cypripedium guttatum*

376. *Corallorhiza trifida*

380. *Cypripedium passerinum*

377. *Cypripedium acaule*

381. *Goodyera repens*

378. *Cypripedium Calceolus var. parviflorum*

382. *Habenaria dilatata*

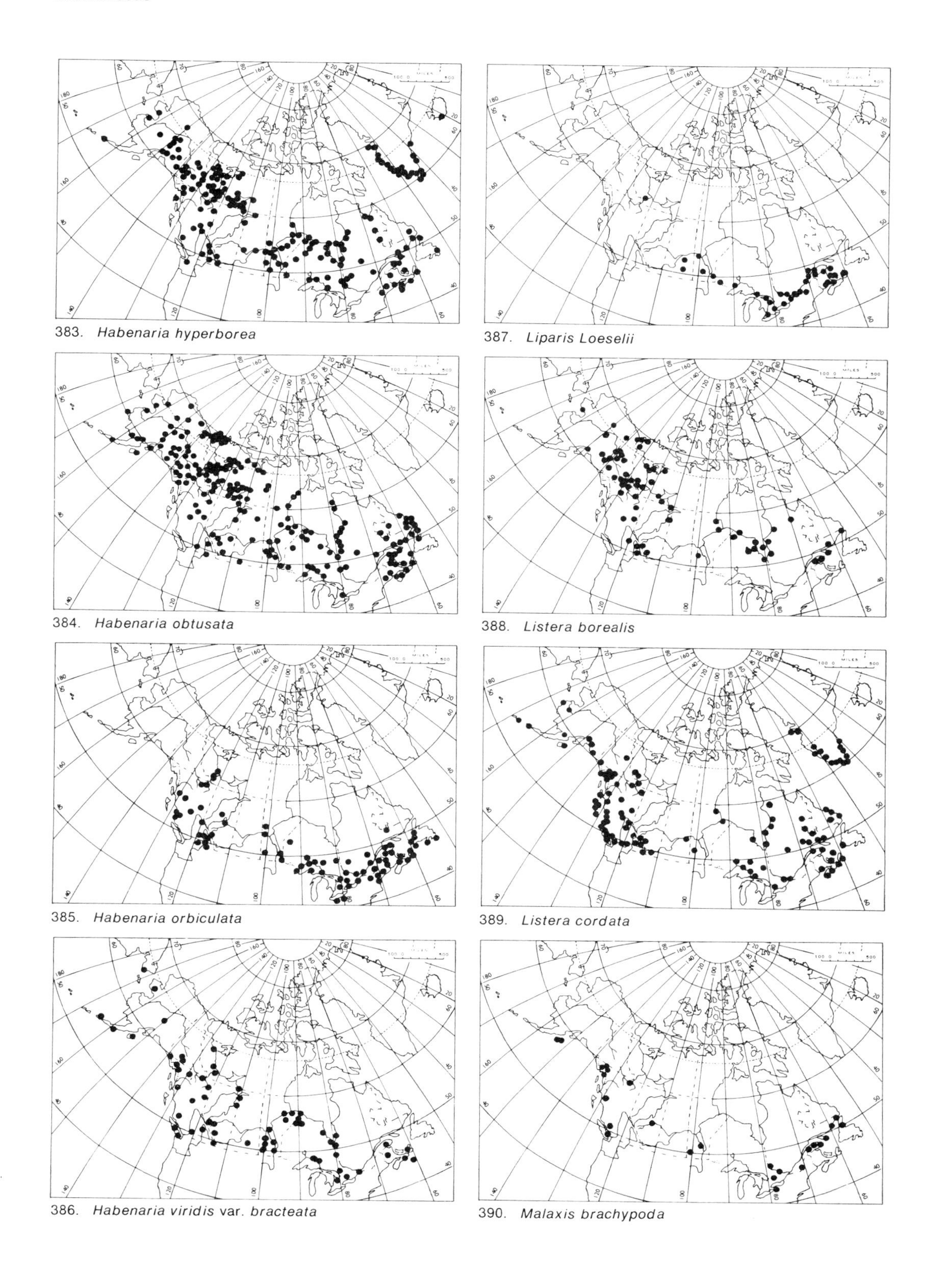

383. *Habenaria hyperborea*

384. *Habenaria obtusata*

385. *Habenaria orbiculata*

386. *Habenaria viridis* var. *bracteata*

387. *Liparis Loeselii*

388. *Listera borealis*

389. *Listera cordata*

390. *Malaxis brachypoda*

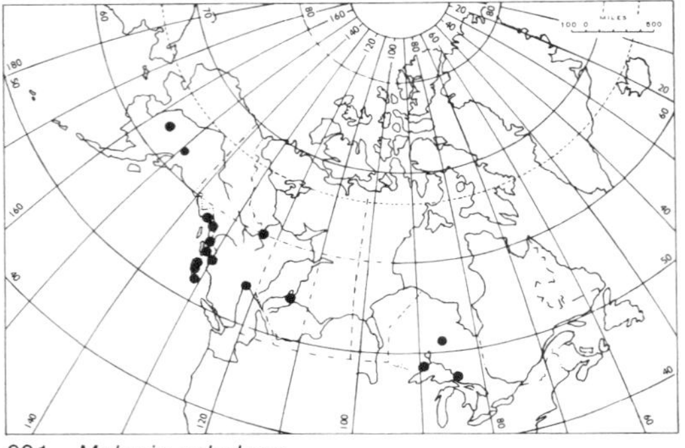

391. *Malaxis paludosa*

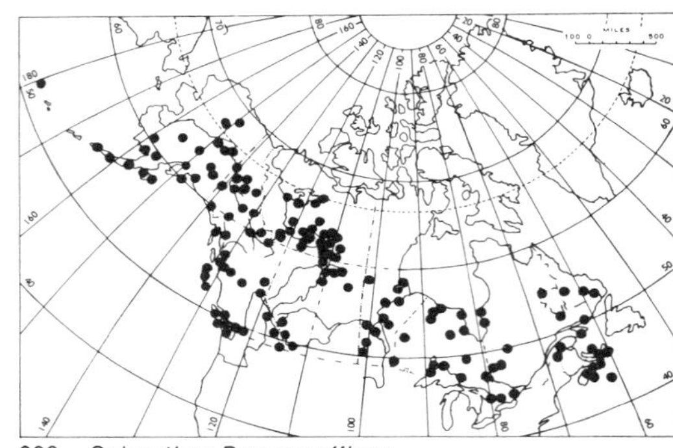

393. *Spiranthes Romanzoffiana*

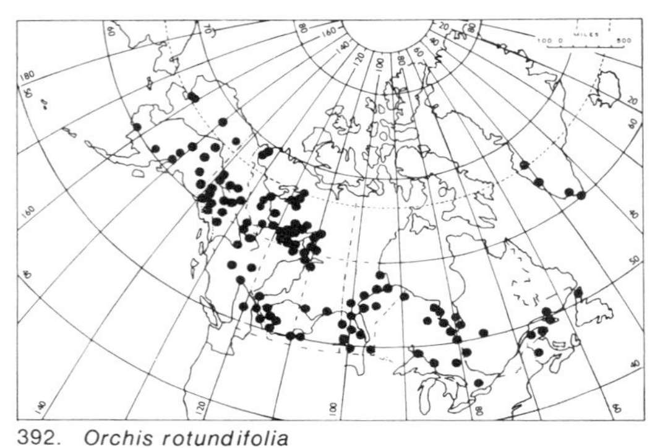

392. *Orchis rotundifolia*

SALICACEAE Willow Family

Dioecious trees or shrubs with deciduous, alternate, entire or at most serrate-margined, stipulate leaves; flowers in erect or drooping, cylindrical male or female catkins (aments). Flowers without perianth, but each subtended by a tiny bract; stamens one to several; ovary solitary; the fruit a dehiscent capsule containing numerous small seeds, each bearing a tuft or long, silky down; stigmas 2 to 4, often 2-lobed.

a. Catkins erect-ascending, each flower with 1-4 basal glands; winter buds covered by a single scale; ours mainly shrubs or small trees. . . . . *Salix*
a. Catkins soon pendulous; each flower arising from a basal cup-shaped disc; winter buds covered by several scales; tree-like. . . . . *Populus*

Populus L. Poplar

Dioecious trees with alternate, simple, and deciduous and petioled leaves and caducous stipules. Flowers with 10 or more stamens, always in drooping aments.

a. Petioles round in cross-section; leaves lanceolate-cordate, shiny above, with strong balsamic fragrance, especially when young; large tree with dark deeply furrowed bark . . . . *P. balsamifera*

a. Petioles flattened laterally; leaf-blade small, suborbicular, crenulate; a slender tree up to 20 m high with smooth greenish-grey or whitish bark. . . . . *P. tremuloides*

Populus balsamifera L.
Balsam Poplar
A common and wide ranging flood-plain species in the District of Mackenzie, with approximately similar boreal and alpine limits as the white spruce, and with that species commonly found far beyond the limit of continuous forest along most of the north-flowing rivers. Its northern limit is on Eskimo Lakes, east of the Mackenzie Delta where a small isolated grove of trees 1.5 to 4.5 m high was observed in 1931-34 near the narrows between 3rd and 4th lake, lat. 68°45′. On the upper Liard River, balsam poplars up to 30 m tall, 1.5 to 2.0 m in diameter, may be seen.

Large drift logs of balsam poplar sometimes up to 1.5 m in diameter, presumably from the Liard River, reach the Mackenzie Delta and the arctic coast to the eastward where, at least formerly, they were in considerable demand for firewood.

General distribution: Boreal N. America.
Fig. 345 Map 394

Populus tremuloides Michx.
Trembling Aspen
Common to abundant throughout the Mackenzie basin forests where it is found mainly on dry, well drained ridges along the rivers north to the Mackenzie Delta where, at its northern limit, in lat. 68°40′, it is a low and often prostrate shrub. In September, when its leaves turn a deep, golden yellow, aspen-covered ridges and slopes strongly contrast with the sombre green of the spruce forest.

The species is of little economic importance, being used only for firewood.

General distribution: Boreal N. America.
Fig. 346 Map 395

Salix L. Willow

Catkins appearing with or before the leaves. Male flowers with 2 to 8 stamens, the female with 2 entire or divided stigmas.

A taxonomically difficult genus in the District of Mackenzie, represented by 40-odd species, a few tree-like, but mostly low, erect or prostrate shrubs.

The willows are normally insect-pollinated and hybridization is wide spread in many groups.

The larger species of willows provide an important source of food for a number of arctic herbivorous animals. Thus, in winter, the twigs and young branches are eaten by caribou and muskoxen, and the bark and small twigs are eaten by hares and lemmings. Willow buds are one of the principal winter foods of ptarmigan, and in spring and early summer the young leaves and flowering catkins are eagerly eaten by herbivorous mammals and birds. The young leaves and buds of several species of willow are collected regularly by Eskimos and Chukchi for food and have been found to be rich in vitamin C. Willows, especially the dry, wind-killed stems of *S. arctica*, provide a ready fuel for outdoor cooking in some parts of the Arctic.

The following key to the willows of Continental Northwest Territories is largely adapted from H. M. Raup (1959): The Willows of boreal Western America, Contr. Gray Herb., Harvard University 185:1-95.

a. Bracts of the flowers not persisting in fruit, straw-coloured; catkins borne on leafy peduncles; usually tall shrubs of alluvial soils, or occasionally in muskegs . . . . Series I.
a. Bracts of the flowers persisting in fruit, brownish, yellowish, or black, rarely straw-coloured; catkins sessile or on leafy peduncles.
  b. Low, prostrate shrubs, some species with rooting branches. . . . . Series II.
  b. Erect shrubs or small trees, with branches never prostrate or rooting though many are of low and spreading form . . . . Series III.

Series I.

a. Leaves distinctly petioled; catkins borne singly on leafy peduncles; leaves finely serrate

b. Leaves mostly acute at the apex, scarcely acuminate, the width not much less than ⅓ of the length, late-flowering, in mid to late summer . . . . *S. serissima*
b. Leaves usually acuminate at the apex, width commonly ¼ of the length, or less; flowering early . . . . *S. lasiandra*
a. Leaves nearly sessile or on very short petioles, linear, the width commonly 1/10 of the length, usually with sharp rather widely spaced teeth; catkins usually appear with or before the leaves . . . . *S. interior* var. *pedicellata*

Series II

a. Pistillate flowers with 2 glands, on the inner and outer side of the base of the pedicels; catkins appearing to be terminal on the branchlets; leaves prominently net-veined. . . . . *S. reticulata*
a. Pistillate flowers with only 1 gland, on the inner side of the base of the pedicel; catkins usually on leafy lateral branchlets; leaves not prominently veined
b. Ovaries and pedicels glabrous, even when young
c. Leaves more or less distinctly serrate on the margins
d. Leaves regularly serrate around the whole of the margin, green on both sides, not glaucous beneath
e. Tiny arctic plants with slender subterranean branches, the aerial twigs usually with only 2-4 rounded, thin, shiny leaves commonly about 1 cm long . . . . *S. herbacea*
e. Larger plants with copiously leafy twigs, the leaves oblong to ovate or oblong-lanceolate . . . . (See *S. Barclayi* group) *S. myrtillifolia*
d. Leaves irregularly serrate or entire, or with minute more or less regular teeth around the lower half of the blades; green or glaucous beneath . . . . (See *S. arbutifolia* group) *S. fuscescens*
c. Margins of the leaves entire, or at most with only occasional small teeth
f. Densely matted plants of arctic and alpine tundra, formed by frequent branching of the stems; leaves 2 cm long or less
g. Leaves 1-2 cm long, those of former years marcescent, often skeletonized, crowded on the stems and forming part of the mat . . . . *S. phlebophylla*
g. Leaves very small, rounded or ovate, 4-10 mm long, some persisting for one year but not forming conspicuous part of mat . . . . *S. Dodgeana*
f. Arctic-alpine plants with elongated prostrate branches; leaves 1-4 cm long. . . . . *S. ovalifolia*
b. Ovaries and pedicel more or less densely pubescent (sometimes only on the lower parts of the ovaries)
h. Leaves finely though distinctly serrate all around the margin, obovate, up to 4-5 cm long and half as wide or more. . . . . *S. Chamissonis*
h. Leaves entire or nearly so, or serrate only around the lower half
i. Bracts of the flowers 2-3 times longer than wide, uniformly straw-coloured, yellowish, or pale brown, or sometimes darkened a little toward the apex, rather short-haired, with the hairs much shorter than the bracts . . . . *S. fullertonensis*
i. Bracts of the flowers broader in proportion to their length, the width commonly more than half the length, variously coloured but usually with long hairs or merely ciliate at the apex
j. Prostrate and matted shrubs with leaves 1-2 cm long; catkins usually 3 cm long or less
k. Leaves marcescent, those of previous years becoming skeletonized and forming part of the mat . . . . *S. phlebophylla*
k. Leaves not marcescent and not forming a conspicuous part of the mats; stems commonly very thin and often buried in the vegetation mat . . . . *S. polaris* ssp. *pseudopolaris*
j. Prostrate shrubs with larger leaves and longer catkins

l. Leaves 1-3 cm long, usually more than half that in width, finely serrate around approximately the lower half; mature catkins about 2 cm long, the capsules glabrous in age . . . . *S. fuscescens*

l. Leaves up to 8 cm long (commonly 2-6 cm), entire or obscurely and irregularly toothed; catkins upright, sometimes as much as 10 cm long

m. Leaves usually dark green and lustrous above; the lateral veins numerous and extending from the midribs at wide, sometimes nearly right angles; glands short and broad, scarcely twice as long as thick, broadly truncate at the apex, half the length of the pedicels of the capsules . . . . *S. arctophila*

m. Leaves usually pale green above, opaque or slightly lustrous, the lateral veins few and extending out from the midribs at much smaller angles, when young rather copiously villous but glabrate at maturity; glands oblong, mostly 2 ½-4 times as long as thick, as long or longer than the pedicels of the capsules . . . . *S. arctica*

Series III

a. Ovaries and pedicels glabrous, even when young

b. Pedicels distinct, 2-6 times as long as the glands

c. Leaves with entire margins, or serrate only around the lower half, smooth, green above and glaucous beneath . . . . *S. fuscescens* group

c. Margins of the leaves distinctly serrate

d. Leaves green on both surfaces, the lower surface sometimes of a slightly lighter shade, rounded or obtuse at the apex; shrubs 2 dm to 2 m high with dark-coloured branches . . . . *S. Barclayi* group

d. Leaves green above and glaucous beneath

e. Leaves fragrant (even long after they are dried), broadly ovate, elliptic or elliptic-oblong; pedicels of the capsules 2-4 mm long . . . . *S. pyrifolia*

e. Leaves not fragrant

f. Leaves usually broadly lanceolate, ovate or obovate, the width commonly ⅓ to ⅗ the length; pedicels of the capsules 2-4 mm long . . . . *S. padophylla*

f. Leaves usually lanceolate, oblanceolate or narrowly obovate, the width commonly ⅙ to ⅓ the length

g. Twigs yellowish, though sometimes wth reddish blotches or streaks; pedicels 1-3 mm long . . . . *S. lutea*

g. Twigs reddish to brown, or yellowish only when very young; pedicels of the fruiting capsules 3-4 mm long; leaves glabrous from the first, or nearly so . . . . *S. mackenzieana*

b. Pedicels none or very short, at most and even in fruit not more than twice the length of the gland

h. Stipules of preceding year persistent on the twigs, conspicuous, linear- to half-cordate-lanceolate or reniform, glandular serrate on the margins or entire; catkins sessile on the twigs of the preceding year, not on leafy peduncles . . . . *S. lanata*

h. Stipules not persisting after the growing season; catkins borne on leafy peduncles

i. Margins of the leaves distinctly and regularly serrate, at least on the lower part and sometimes all around

j. Pedicels 2-3 mm long; shrubs usually less than 1 m high and commonly 2-3 dm, with branches spreading in the moss of muskegs; leaves about 2-2.5 times as long as wide, oblong, oblong-ovate or oblong-oblanceolate . . . . *S. myrtillifolia*

j. Pedicels about 0.5 mm long; shrubs commonly 1-2 m tall; leaves 2-4 times longer than wide, ovate or obovate, acute at the apex; commonly drying black . . . . *S. Barclayi*

i. Margins of the leaves entire, or with occasional teeth at irregular intervals or toward the base; styles 2 mm long or less
   k. Pedicels about 0.5 mm long; leaves usually serrate, but sometimes entire . . . . *S. Barclayi* and *S. myrtillifolia*
   k. Pedicels 1-1.8 mm long
      l. Leaves glaucous beneath, all entire or occasionally with minute teeth, the blades completely glabrous . . . . *S. Farrae*
      l. Leaves green beneath, serrate or entire, pubescent on both sides at least on the veins . . . . *S. commutata*

a. Ovaries pubescent (rarely only on the base)
   m. Leaves green on both sides, though sometimes of a lighter shade beneath, not glaucous, glabrous except when very young, 3-7 cm long and ¼ to ⅖ as wide, glandular-serrate on the margins. . . . . *S. MacCalliana*
   m. Leaves glaucous with a bluish bloom on the lower surfaces
      n. Pedicels distinct, 2-6 times as long as the glands
         o. Pedicels 0.5-2.4 mm long; catkins sessile and precocious or on very short leafy peduncles, and appearing with the early leaves . . . . *S. Scouleriana*
         o. Pedicels conspicuously longer, 2.5-5.0 mm long; catkins borne on well-developed, leafy peduncles, and appearing with the leaves
            p. Leaves lanceolate, acuminate, 4-5 times as long as wide, entire or finely toothed, pubescence if present giving a somewhat silvery appearance to the under surfaces; bracts 1-2 mm long . . . . *S. gracilis*
            p. Leaves ovate, oblong, obovate, or ovate-lanceolate, the width ⅓ to ½ the length, entire or dentate, pubescence if present not silky in appearance; bracts commonly 2.5-3.5 mm long
               q. Pistillate catkins on distinct peduncles appearing with the leaves . . . . *S. Bebbiana*
               q. Pistillate catkins sessile, appearing before the leaves . . . . *S. discolor*
      n. Pedicels none or very short, even in fruit scarcely more than twice the length of the glands
         r. Flowering twigs more or less distinctly pruinose with a bluish bloom; catkins appearing before the leaves
            s. Lower surfaces of the leaves with a permanent covering of whitish felt-like, opaque pubescence; styles 2 mm long or more
               t. Twigs, even when several years old, densely white woolly. . . . . . *S. alaxensis*
               t. Twigs soon glabrate, usually covered with a bluish, waxy bloom (pruinose) . . . . *S. longistylis*
            s. Leaves glabrous beneath except on the midribs and main veins, entire or with a few small, irregularly spaced teeth; catkins sessile or with a few small leaves at the base
               u. Stipules persistent for 3 or more years as dried, linear or linear-lanceolate appendages on old wood; last year's leaves persist until new leaves full-grown . . . . *S. pulchra*
               u. Stipules not persistent after one growing season; leaves deciduous in autumn . . . . *S. planifolia*
         r. Flowering twigs not pruinose; catkins appearing with or after the leaves
            v. Bracts of the flowers more or less uniformly coloured, yellowish-straw-coloured or yellowish-brown, pubescent with short hairs that are much shorter than the bracts themselves; mostly upright (thought often low) shrubs
               w. Petioles well developed, more than 2 mm long
                  x. Leaves usually much longer than broad, those of the peduncles about 2.5-4.0 times longer than wide, mostly upright (though often low) shrubs . . . . *S. glauca*
                  x. Leaves usually short-oblong or short-ovate to rounded, those of the peduncles usually less than 2.5 times longer than wide; very low, spreading or somewhat matted shrubs . . . . . *S. cordifolia* var. *callicarpea*

w. Petioles of the leaves short, none, or at most 1-2 mm long
  y. Catkins short-oblong or spherical; leaves averaging 1.5-3.0 cm long . . . . *S. brachycarpa*
  y. Catkins cylindrical, usually more than twice as long as thick; leaves averaging 2-5 cm long . . . . *S. niphoclada*

v. Bracts of the flowers of 2 colours, greyish-brown to black towards the apex, or if nearly all of one colour, greyish-brown to black, with hairs often as long or longer than the bracts themselves
  z. Lower surfaces of the leaves, even in age, densely white or light grey-tomentose or silky
    aa. Leaves linear-lanceolate or narrowly oblong, commonly 5-7 times longer than wide, with entire or undulate margins; young twigs, lower leaf-surfaces, and capsules with a dense, whitish wool; upper leaf-surfaces commonly greyish-pubescent, giving the whole shrub a hoary appearance . . . . *S. candida*
    aa. Leaves broader in proportion to their length, usually only 3-4 times longer than broad; pubescence of the lower surfaces of the leaves dull white or dull greyish
      bb. Leaves thickly felted beneath or on both sides with a white tomentum, the midveins on the lower surfaces, where visible through the tomentum, yellowish; stipules rather distantly glandular on the margins . . . . *S. alaxensis*
      bb. Leaves whitish or light greyish hairy-pubescent on both sides, somewhat silky in appearance; stipules densely glandular on the margins; erect, thicket-forming shrub; stipules, young leaves and twigs glutinous to the feel, and when pressed exude an oily substance that stains and penetrates the pressing papers . . . . *S. Barrattiana*
  z. Lower surfaces of the leaves glabrous or somewhat pubescent, but not with a dense whitish or greyish pubescence or tomentum
    cc. Leaves distinctly, regularly and often conspicuously glandular-serrate on the margins, delicately silky-pubescent beneath with short, appressed hairs; small tree . . . . *S. arbusculoides*
    cc. Leaves entire, or the margins with occasional small glandular teeth at irregular intervals, or the serrations only around the lower half
      dd. Catkins coetaneous, borne on leafy peduncles; bracts short, 1.3 mm long or less; styles 0.3-0.8 mm long
        ee. Leaves glabrous, the margin entire or with a few teeth near the base; pedicels 1.3 mm long; ovaries glabrous at maturity. . . . . *S. fuscescens*
        ee. Leaves thinly appressed-hairy, entire or minutely toothed around the whole margin; pedicels 1.0-1.5 mm long; ovaries thinly hairy, or glabrate in age . . . . . . *S. athabascensis*
      dd. Catkins precocious, appearing before the leaf-buds break, sessile on the branchlets; bracts 1.5-2.0 mm long, and the styles 0.8-2.0 mm long
        ff. Stipules persistent for 3 or more years as dried, linear or linear-lanceolate appendages on old wood. . . . . *S. pulchra*
        ff. Stipules not persistent after one growing season . . *S. planifolia*

Salix alaxensis (Anderss.) Cov.
Erect shrub or small tree with gnarled trunk and branches; young twigs and branches permanently and conspicuously velvety-tomentose; mature leaves oblanceolate-elliptic, velvety-tomentose beneath, dull green above, with a short petiole and persistent, linear stipules, the blade 4 to 8 cm long and 2.0 to 2.5 cm wide; catkins sessile appearing before the leaves, the pistillate 4 and frequently up to 12 cm long. The young leaves are very fragrant.

Thicket-forming on sandy and gravelly stream banks and lake shores.

General distribution: Amphi-Beringian, east

to Hudson Bay and northern tip of Ungava; subarctic-alpine (See also *S. longistylis* Rydb. here considered distinct).
Fig. 347 Map 396

Salix arbusculoides Anderss.
Erect shrub or slender tree up to 5 to 7 m tall with smooth, reddish and shiny twigs; leaves narrowly elliptic-lanceolate, 2 to 6 cm long, with finely serrate or sometimes entire margins, the underside silky with short, appressed hairs; catkins appearing with the leaves, sessile or on short, leafy peduncles, the pistillate up to 5 cm long.

One of the commonest willows of the wooded parts of the District of Mackenzie, extending north along the Mackenzie River to its delta where, near its northern limit, it forms thickets 3 to 5 m tall, and is often a pioneer species on alluvial flats. Between the Mackenzie River and Hudson Bay this willow may be seen along north-flowing rivers a considerable distance beyond the true tree line. In the Mackenzie Mountains ascending to 1200 m.

General distribution: Western N. America from Hudson Bay to Alaska.
Fig. 348 Map 397

Salix arctica Pall. *s. lat.*
S. anglorum Cham.
S. crassijulis Trautv.
S. hudsonensis Schneider
Decumbent, often matted but never erect shrub with trailing or ascending, glabrous branches and variously shaped, entire-margined leaves; the mature leaves are elliptic-obovate to broadly rounded-ovate, 2.5 to 7.0 cm long and 1.5 to 2.5 cm wide, glabrate or glabrous and dull green above, scarcely paler beneath, with sparse but long-villous pubescence; petioles about one third the length of blade, without stipules. Catkins on leafy peduncles, appearing with the leaves, the pistillate not stiffly erect as in *S. arctophila*, 3 to 7 cm long, their bracts long-bearded, with a dark brown or blackish tip and a pale base; ovaries and capsules permanently short-pubescent.

The most arctic of our willows, and in North America the only one ranging north beyond the 80th parallel, and perhaps also the most adaptable as to habitat, although in the far North commonly restricted to snow patch habitats; common also in alpine tundra west of the Mackenzie Valley.

General distribution: Circumpolar, arctic-alpine.
Fig. 349 Map 398

Salix arctophila Cockerell
Always prostrate shrub with trailing, brownish-green branches that tend to turn black in drying; leaves short-petioled, elliptic to obovate or oblanceolate, with sub-entire margins, and acutish or rounded tips, 2 to 4, or rarely 6 cm long and 1 to 2, rarely 2.5 cm wide, totally glabrous, dark green and lustrous above, distinctly paler beneath; catkins large, on leafy peduncles, usually stiffly erect, appearing with the leaves; the pistillate from 3 to 10 cm long at maturity; the bracts brown, bearded at the apex; ovaries and capsules thinly pubescent, becoming glabrescent or totally glabrous in f. *lejocarpa* (Anderss.) Fern.

An eastern arctic species of wet, mossy tundra, which, in the District of Mackenzie, extends a short distance into the forest tundra. Although still common in the Mackenzie Delta, *S. arctophila* has only recently been collected in northern Yukon Territory and northeastern Alaska.

General distribution: Eastern N. American Arctic.
Fig. 350 Map 399

Salix athabascensis Raup
Erect shrub 3 to 12 dm high, with finely grey-hairy twigs; leaves lanceolate 1 to 3 cm long and about half as wide, green above, paler beneath, both surfaces thinly silky pubescent, glabrate in age, the margins entire and slightly involute, on petioles up to 6 mm long; catkins appearing with the leaves, the pistillate 1.5 to 2.0 cm long, on leafy peduncles; capsules short pubescent, on 1.0 to 1.5 mm long pedicels; styles about 3 mm long; staminate catkins 5 to 10 mm long.

A woodland species, in the District of Mackenzie thus far known only from the upper Mackenzie drainage, north to Lat. 65° N.

General distribution: S. central Alaska, southern Y.T. and upper Mackenzie drainage, east to Hudson Bay.
Fig. 351 Map 400

Salix Barclayi Anderss.
Thicket-forming shrub with stems commonly 2 to 3 m tall, but much lower in alpine situations; young twigs black, thinly pubescent, turning dark brown in age; leaves green above, glaucous beneath, drying black, 3 to 8 cm long and about half as wide, elliptic-obovate with acute tips, tapering or rounded bases, and finely serrate or sometimes entire margins; petioles short, those of vigorous, sterile shoots commonly with leafy stipules up to 1 cm long; catkins on leafy peduncles, appearing with the

leaves, the pistillate up to 5 cm long; capsules up to 1 cm long, glabrous or thinly silky, on very short pedicels; styles 1.0 to 1.5 mm long.

General distribution: *Salix Barclayi* is a rather variable N. Cordilleran-Pacific Coast species, related to, but usually much taller than *S. myrtillifolia;* it ranges north along the east slope of the Mackenzie and Richardson Mts. to the Mackenzie R. delta and west across central Y.T. to central and southern Alaska.

Fig. 352 Map 401

Salix Barrattiana Hook.

Upright, much branched and gnarled shrub commonly about 1 m tall, forming dense and often pure thickets that, when in leaf, may be recognized from a distance by their characteristic blue-green colour; twigs and branches black, sparsely but permanently pubescent and ragged from the black remains of stipules of former years; leaves narrowly oblanceolate, 3 to 5 cm long, appressed-tomentose above, and grey-silky beneath, subsessile or short-petioled; stipules oblong, glandular-margined; catkins appearing before the leaves, sessile on the twigs, the pistillate upright, commonly 4 to 7 cm long, grey-shaggy from the densely grey-silky capsules and their black, long-pilose bracts; styles 1.5 to 2.5 mm long. The twigs and stipules exude an oily, yellowish substance which stains and penetrates the drying papers.

Moist but well-drained alpine meadows and river flats north along the Mackenzie Mountains to the Arctic Circle, and probably also in the Richardson Mountains, east to the Anderson River.

General distribution: N. Cordilleran.

Fig. 353 Map 402

Salix Bebbiana Sarg.

Long-beaked willow

Shrub or small tree, up to 6 m tall; leaves oblanceolate, 3 to 5 cm long and about half as wide, tomentose when young, glabrescent or glabrous in age, their petioles and the young twigs grey-velvety pubescent; stipules deciduous; catkins on leafy peduncles, appearing with the leaves, the pistillate 3 to 4 cm long; capsules not crowded, grey-pubescent 6 to 8 mm long including the slender beak, on pedicels 3 to 4 mm long; styles short, almost obsolete.

One of the commonest willows of the wooded parts of the District of Mackenzie where, with *S. arbusculoides*, it commonly forms an understory in open *Picea glauca* forest on dry, well-drained river banks north to the Mackenzie Delta.

General distribution: N. American boreal forest, from Nfld.—Lab. to Alaska, north to the limit of trees.

Fig. 354 Map 403

Salix brachycarpa Nutt.

Low, upright and freely branching shrub, commonly less than 1 m tall; leaves oblong, 2 to 3 cm long and 0.7 to 1.0 cm broad, entire, densely grey-hirsute beneath, less so above, glabrescent in age, sub-sessile on the twigs; catkins appearing with the leaves, on short, leafy peduncles, the pistillate commonly 1 to 2 cm long, the capsules crowded, 5 to 6 mm long, grey-woolly, sessile; styles short or nearly obsolete.

Calciphilous boreal forest species which has been collected a number of times along the Mackenzie River and its tributaries, north to Great Bear River; commonly forming low thickets near the top of limestone screes.

General distribution: Gaspé and Anticosti Island, Que. to Alaska.

Fig. 355 Map 404

Salix candida Flügg

Hoary willow

Upright shrub, in the northern parts of its range rarely exceeding 1 m in height, the twigs smooth, characteristically "frosted" in youth from short, woolly pubescence; leaves 2 to 10 cm long and 0.5 to 1.5 cm wide, the smooth margins somewhat inrolled, dull, velvety, white-tomentose beneath, dark green-glabrescent above; catkins appearing before the leaves, the pistillate 2 to 5 cm long, on short, leafy peduncles; styles about 1 mm long; capsules densely white-tomentose on 1 mm long pedicels.

A woodland species of damp swales and muskeg bogs.

General distribution: N. America, from Nfld. to east central Alaska, north along the Mackenzie R. drainage to Great Bear R.

Fig. 356 Map 405

Salix Chamissonis Anderss.

Always prostrate, with trailing reddish-brown, smooth branches; leaves petioled, the blade 3 to 4 cm long and 2 to 3 cm wide, obovate, with obtuse, rounded or even slightly retuse tips, and finely glandular-serrate margins, deep green and lustrous above, somewhat paler beneath; catkins appearing with the leaves, the pistillate 3 to 6 cm long, on stout, few-leaved branchlets; capsules about 5 mm long, sparsely pubescent, glabrous in age, on very short pedicels; bracts about 2 mm long, oval, long-bearded; styles 2 mm long.

In habit somewhat similar to the eastern *S. arctophila* whose western range just overlaps that of *S. Chamissonis*, but at once distinguished by its finely serrate and broadly obovate leaves.

General distribution: Amphi-Beringian species which barely enters our area where it has been collected a few times in damp and grassy snowbed habitats in the Richardson Mts., west of the Mackenzie Delta.

Fig. 357 Map 406

Salix commutata Bebb

Thicket-forming, rarely more than 1 m tall; twigs dark brown under a loose, floccose pubescence; leaves 3 to 7 cm long and about half as wide, short-petioled, elliptic or slightly obovate, tapering at both ends, the margins finely glandular-serrate (sometimes visible only under a lens), fresh green on both sides under a loose, tardily lost, long-woolly pubescence; stipules glandular-serrulate, early deciduous; catkins on leafy peduncles, appearing with the leaves, the pistillate 3 to 4 cm long; capsules glabrous, with a reddish tinge, on 1 mm long pedicels; styles 1.0 to 1.5 mm long.

Alpine meadows near and above timberline.

General distribution: N. Cordilleran species barely entering our area in the Mackenzie Mts.

Fig. 358 Map 407

Salix cordifolia Pursh

var. **callicarpea** (Trautv.) Fern.

Branches rather stout, erect-ascending, but in the Arctic often prostrate, although never creeping; twigs of the year usually pubescent; mature leaves elliptic to ovate, obovate, 1 to 9 cm long and 0.5 to 5.0 cm broad, greyish pilose or villous on both sides and on the slender petioles, glabrate in maturing or merely silky on the nerves beneath; catkins appearing with the leaves, the pistillate 1 to 7 cm long, their bracts yellowish-brown; capsules densely grey, hirsute; styles long.

General distribution: Endemic of Greenland and the eastern N. American Arctic, becoming progressively scarcer westward.

Fig. 359 Map 408

Salix discolor Muhl.

Tall shrub or small tree; branches spreading, yellowish-brown, hairy when young but soon glabrous; leaves bright green above, glaucous beneath, their margins finely crenate. Catkins sessile, 3 to 5 cm long, appearing before the leaves; capsules 8 to 10 mm long, with a long beak, finely short-pubescent, on a slender pedicel.

Damp thickets in meadows or on river banks.

General distribution: Nfld. to Alta., and southward; disjunct to near the west end of Great Slave Lake and Ft. Simpson on the upper Mackenzie R., about 800 km beyond its nearest known station in Alta.

Map 409

Salix Dodgeana Rydb.

S. phlebophylla sensu Porsild (1945) and Raup (1947), *non* Anderss.

Caespitose, mat-forming dwarf shrub from a central tap-root, with branches commonly 1 mm in diameter; leaves marcescent, 3 to 5 mm long, obovate, with retuse tips, glabrous except for the soft-ciliate margins, dark green above, paler beneath, with prominent arching lateral nerves; catkins sessile, appearing with the leaves, each composed of only 3 to 4 flowers, the pistillate catkins less than 1 cm long; capsules glabrous, reddish-brown; styles obsolete.

Rare and local in alpine, turfy places on calcareous plateaus and scree-slopes.

Salix Dodgeana is one of the smallest known willows and, perhaps, is most closely related to the Amphi-Beringian *S. rotundifolia* Trautv. and *S. phlebophylla* Anderss.

General distribution: Originally described from Yellowstone Park, Wyoming, *S. Dodgeana* is now known also from the east slopes of Richardson Mts. west of the Mackenzie Delta, and Mackenzie Mts., disjunct by about 3000 km from its nearest station in S. W. Montana.

Fig. 360 Map 410

Salix Farriae Ball

S. hastata L.

var. *Farriae* (Ball) Hult.

Thicket-forming, freely branching shrub up to 1.5 m tall with smooth, black bark; leaves short-peduncled, elliptic-lanceolate, 4 to 7 cm long and about half as wide, with acute tips and narrowly rounded bases, entire margins, glabrous, green above and paler beneath; stipules oval, soon deciduous; catkins appearing with the leaves, the pistillate 2 to 5 cm long, on leafy peduncles; capsules glabrous their pedicels very short.

Within our area known only from the upper Mackenzie Delta area where, together with *S. pulchra* var. *yukonensis*, it forms open thickets on alluvial river banks. District of Mackenzie specimens in the National Herbarium, Ottawa, by B. Floderus were named *S. Walpolei* Cov. &

Ball. Hultén (1968) considered it *S. hastata* L.; to us the Mackenzie Delta plant is indistinguishable from *S. Farriae* of the Canadian Rocky Mountains.

General distribution: Cordilleran; if taken in the broader sense of Hultén (1968) as *S. hastata* it is nearly circumpolar.

Fig. 361 Map 411

Salix fullertonensis Schneid.

Prostrate shrub with slender rooting stems, and entire, lanceolate leaves, 1.5 to 2.9 cm long, pale green and glabrate above, paler beneath; catkins 2 to 3 cm long on short, leafy peduncles; capsules pubescent.

Moist, stony lake shores and barrens.

General distribution: Apparently endemic to Hudson Bay area and barely enters the District of Mackenzie.

Map 412

Salix fuscescens Anderss.

S. arbutifolia sensu Porsild (1943).

Creeping and freely rooting; twigs and branches reddish-brown, smooth and shiny; leaves short-petioled and without stipules, obovate, up to 2 cm long and half as wide, obtuse or rounded at the tip, dark green and lustrous above, paler and glaucous beneath, the veins prominent, margins entire or, at most with a few teeth near the base; catkins on leafy peduncles, appearing with the leaves, the pistillate 3 to 5 cm long with dark reddish, short-pedicellate capsules; styles short or nearly obsolete.

Damp and mossy tundra.

General distribution: Amphi-Beringian: Arctic-alpine from Alaska and northern District of Mackenzie to the west shore of Hudson Bay.

Fig. 362 Map 413

Salix glauca L. *s. lat.*

Blue-green willow

Erect or spreading shrubs occasionally up to 2 m tall, but in unfavourable situations much lower or even depressed; young twigs greyish-pubescent, glabrate and dark brown in age; leaves petioled mostly oblanceolate, 3 to 5 cm long and about half as wide, with entire margins, obtuse or acute at the base, obtuse or somewhat rounded to acute at the apex, dull green above, glaucous beneath, pubescent or even densely hairy on both surfaces when young, but commonly glabrescent above; catkins appearing with or soon after the leaves, on leafy peduncles; the pistillate commonly 3 to 5 cm long; capsules densely and permanently grey-pubescent, and the styles usually less than 1 mm long; pedicels none or very short; bracts oblong, half as long at the capsule, sparsely pubescent, pale yellowish-brown.

The above description will probably accomodate all but the most extreme variations of *S. glauca s. lat.* as represented by the ample District of Mackenzie material in the Ottawa herbaria.

General distribution: *S. glauca s. lat.* Circumpolar.

Following Raup (1959) it is now possible to recognize four varieties within the large number that answer the above description of *S. glauca s. lat.* These four may be separated as follows:

a. Stipules persisting on the twigs for many years, appearing as narrowly lanceolate, dried appendages, often 1 cm long or longer, their margins glandular . . . . var. *stenolepis*

a. Stipules not persisting, falling within the first year

  b. Leaves densely greyish-hairy or tomentose on both surfaces, even in age, averaging shorter and broader than in the other varieties . . . . var. *Aliceae*

  b. Leaves greyish-pubescent on both surfaces when young, but lacking the long-hairy appearance of var. *Aliceae*, and usually glabrate in age, at least on the upper surface

    c. Most of the leaves with obtuse or almost rounded apex . var *glauca*

    c. Most of the leaves with acute apex . . . . var. *acutifolia*

Salix glauca L.

var. **glauca**

In the District of Mackenzie late-flowering and often thicket-forming, common throughout the wooded parts, reaching the Mackenzie Delta and Great Bear Lake.

Salix glauca L.

var **acutifolia** (Hook.) Schneid.

A tundra willow apparently of more subarctic-alpine range than the var. *glauca*. It is a northwestern type ranging from Alaska and Yukon Territory eastward to the shores of southern Hudson Bay, but not reaching the arctic islands.

Salix glauca L.

var **Aliceae** Ball

By its densely tomentose, characteristically light yellowish-grey or sometimes pure white capsules, its shaggy white indument of the young twigs and the underside of the leaves, and by its up to 5 mm broad oval stipules, the var. *Aliceae* is easily the most distinctive of the races of *S. glauca* known to occur in the District of Mackenzie where it is confined to lowland situations along the Mackenzie River and its tributaries, north to the Mackenzie River delta.

Salix glauca L.
var. **stenolepis** (Flod.) Polunin,
S. glauca L.
var. *perstipula* Raup
Apparently a subarctic-alpine race which, although first detected in eastern Canadian Arctic (Lake Harbour, Malte, CAN 118812), and reported from a number of stations in northern Ungava and the Hudson Bay region, the var. *stenolepis* may yet prove to have its centre of distribution in the Northwest, where it has been reported from Great Bear Lake, Mackenzie Delta, Richardson Mountains, S. E. Yukon Territory and central Alaska.

Fig. 363 (var. *glauca*) Map 414 (var. *glauca*), 415 (var. *acutifolia*), 416 (var. *Aliceae*), 417 (var. *stenolepis*)

Salix gracilis Anderss.
S. petiolaris of auth. not Sm.
Slender shrub 1 to 3 m high with dark brown bark, ascending branches and glabrous or soon glabrate twigs; leaves linear-lanceolate, those of the fertile twigs commonly 3 to 4 cm long and about one fifth as wide, their margins entire or minutely denticulate, appressed silky or glabrate above, yellowish-silky beneath; catkins on leafy peduncles, appearing with the leaves, the pistillate 2 to 3 cm long; capsules pale yellowish-green, finely short pubescent, on slender pedicels 2.5 to 3.0 mm long, or as long as the bearded bracts; styles obsolete or nearly so.

An eastern woodland species within our area known only from a few stations along the Slave and upper Mackenzie Rivers.

General distribution: Que. to Alta., north to the upper Mackenzie R. drainage.

Fig. 364 Map 418

Salix herbacea L.
Least Willow
A tiny, creeping shrub with thin, slender, freely rooting stems and branches, which are often completely buried in moss; leaves orbicular, usually retuse, crenulate, glabrous, and dark green on both sides; catkins small and few-flowered, appearing with the leaves and often hidden among them; capsules reddish, glabrous.

In herbmats, generally in places where the snow remains late.

General distribution: Amphi-Atlantic, low-arctic species common in the eastern Canadian Arctic, becoming progressively less common west of Hudson Bay, and thus far collected only in a few places in the District of Mackenzie west to Great Bear L.

Fig. 365 Map 419

Salix interior Rowlee
var. **pedicellata** (Anderss.) Ball
Sandbar Willow
Upright shrub with us rarely over 2 m tall, with smooth brown or greyish bark; leaves linear, about 10 times longer than wide, those of the fertile shoots 8 to 11 cm long, but often much longer on the sterile shoots, glabrate or thinly appressed pubescent, their margins entire or more commonly with minute, widely spaced teeth; catkins sessile or very short petioled, on leafy branches, appearing with the leaves, the pistillate spreading or even somewhat pendulous, 5 to 8 cm long; capsules yellowish-brown, glabrous, on short pedicels, their bracts promptly deciduous; styles obsolete.

A common and important pioneering species on sandbars and mudflats along the Mackenzie River and its tributaries north almost to the head of the Mackenzie Delta. Its slender, smooth and horizontally spreading roots are used by Indians for the making of baskets; one such single and unbranched root was traced across a sandbar for more than 10 meters; for the entire distance its diameter held a constant 5 mm.

General distribution: Que. to Alaska, north to the limit of forest.

Fig. 366 Map 420

Salix lanata L.
ssp. **calcicola** (Fern. & Weath.) Hult.
S. calcicola Fern. & Wieg.
Low, erect ascending shrub with stout, gnarly stems and branches; those of the year lanate-pubescent; leaves oblong-ovate to suborbicular or even subcordate, glabrescent, entire-margined, green above, glaucous beneath, subtended by large, persistent, glandular-dentate stipules; catkins large, the pistillate 3 to 7 cm long, appearing before the leaves.

Calcareous rocky and gravelly places.

General distribution: Eastern Canadian

subarctic, with disjunct stations in Gaspé Pen., Que., and Banff Nat. Park, Alta.
Fig. 367 Map 421

Salix lanata L.
ssp. **Richardsonii** (Hook.) Skvortsov
Richardson's Willow
Similar to ssp. *calcicola* from which it differs by its normally narrower, oblanceolate leaves that are not glaucous beneath, and by its narrower, linear-lanceolate stipules; catkins appear before the leaves. In the southern part of its range *S. lanata* ssp. *Richardsonii* becomes taller and may attain a height of from 2 to 3 m.

Well watered sandy and gravelly places such as lake shores and river banks.

General distribution: Amphi-Beringian tundra species, in N. America common and wide spread in Alaska—Y.T., ranging eastward across the Barren Grounds to the west shore of Hudson Bay and Baffin Island, lacking in Greenland.
Fig. 368 Map 422

Salix lasiandra Benth.
Tall shrub, up to 8 m tall, but with us rarely tree-like, with light brown bark on the branches, and reddish-brown shiny twigs that in youth are densely white-pubescent; leaves lanceolate or sometimes long-acuminate, 7 to 12 cm long, although those on older wood and on flowering branchlets are much smaller, thick and of leathery texture, bright green, glabrous, paler beneath, the margins finely serrate; the petiole short, with two small glands at its base; pistillate catkins 5 to 10 cm long on leafy peduncles, appearing with the leaves; capsules glabrous 5 to 7 mm long on short, slender pedicels; stamens 5.

Common on sand bars and with us mainly along streams of the upper Mackenzie Valley.

General distribution: Cordilleran, lowland species reaching Alaska and the upper Mackenzie R. valley.
Fig. 369 Map 423

Salix longistylis Rydb.
Erect shrub up to 7 m tall, with straight or ascending branches; the young twigs smooth with a dark bluish-black and waxy bloom; mature leaves of fertile shoots oblanceolate, 4 to 5 cm long and 1.5 cm wide, but those of the sterile shoots commonly three times as large, glabrous above, and the underside covered by a firm, white felt; stipules linear, 1 to 2 cm long; catkins sessile on the twigs appearing with the leaves, but fully developed and functional while the leaves are still quite small, the pistillate 13 to 16 cm long, at first pendulous, later stiffly erect; capsules sessile or nearly so, pale yellow, and contrasting strongly with their much shorter, ovate, black and bearded bracts; styles 2.0 to 2.5 mm long.

Thicket-forming species of gravelly floodplains, by some authors considered a variety of *S. alaxensis* from which it is readily distinguished by its growth habit, and by its naked, blue-black twigs; also its young leaves are not noticeably fragrant as are those of *S. alaxensis.*

General distribution: Endemic of Alaska, Y.T., western District of Mackenzie and northern B.C.
Fig. 370 Map 424

Salix lutea Nutt.
Yellow Willow
Upright shrub or small tree 3 to 5 m tall with yellowish-green twigs; leaves oblanceolate, up to 7 cm long and 1.5 cm broad, with shallowly crenate or sub-entire margins, dark green and smooth, but not shiny above, paler beneath, reddish purple when young, on petioles 5 to 6 mm long; stipules lacking or soon deciduous; catkins on leafy but very short branchlets, appearing before the leaves, the pistillate 4 to 5 cm long; capsules about 5 mm long, glabrous, pale brown, on 2 to 3 mm long pedicels; styles 0.5 mm long.

Well watered stream banks.

General distribution: N. American plains or prairie species extending north along the Slave and Mackenzie rivers to at least Lat. 65°N.
Fig. 371 Map 425

Salix MacCalliana Rowlee
Upright much branched shrub, rarely over 1 m tall, with shiny, dark brown or even black twigs and branches; mature leaves narrowly elliptic, 5 cm long or less, with finely glandular-serrate margins, glabrous and fresh green above, slightly paler beneath on very short petioles; catkins appearing with the leaves, on short, leafy branchlets, the pistillate 2 to 3 cm long, densely silky-grey pubescent, in age thinly white-tomentose; bracts oblong, papery, light brown three-fourths the length of the capsule; styles very short but thick.

A woodland peat bog species, in habit somewhat similar to *S. glauca* var. *acutifolia*, from which it is readily distinguished by its bright green and finely serrate leaves.

General distribution: N. American foothills

and plains, reaching the southern District of Mackenzie and southeastern Y.T.

Fig. 372 Map 426

Salix mackenzieana Barratt
Tall shrub with spreading branches; the twigs rich brown, smooth; leaves with well-formed petioles, those of the fertile twigs oblanceolate, 2 to 7 cm long and 0.5 to 2.0 cm wide, smooth and dull green above, paler beneath with finely glandular-denticulate, rarely smooth, margins, those of sterile shoots 10 to 15 cm long, each supported by a pair of small, green stipules; young leaves often dark reddish-brown; catkins appearing with the leaves, on short, leafy branchlets, the pistillate 3 to 5 cm long; capsules pale green and glabrous, on 2 to 3 mm long pedicels, their bracts 1 mm long; styles 0.5 mm long.

Somewhat similar to *S. lutea* from which it is readily distinguished by its reddish-brown twigs.

Thicket-forming on damp, sandy river banks.

General distribution: A Western floodplain species found along the Mackenzie R. and its tributaries north to approximately Lat. 65° N.

Map 427

Salix myrtillifolia Anderss.
S. novae-angliae Anderss.
Low shrub with smooth, spreading, reddish-brown branches, sometimes 1 m high or more, but generally much lower; mature leaves of the fertile branches lanceolate with finely crenate-dentate margins, green and lustrous above, slightly paler beneath, those of sterile shoots much larger and commonly with well-formed stipules; catkins appearing with the leaves, on leafy branchlets, the pistillate 2 to 3 cm long; capsules glabrous, pale yellowish-green, on 2 mm long pedicels; styles less than 1 mm long; staminate catkins very sweetly scented.

A very variable species of the North American boreal forest, common or ubiquitous in damp, mossy woodland fens, and along small streams.

General distribution: Nfld. to Alaska, north almost to the limit of trees.

Fig. 373 Map 428

Salix niphoclada Rydb.
Low, much branched, ascending or erect shrub with stems up to 5 dm tall but often much lower, the young twigs grey pubescent becoming glabrous and brown in age; leaves sub-sessile, narrowly lanceolate, with acute or rounded tips, rarely over 2 cm long, silky pubescent beneath when young, glabrescent, green and smooth above; catkins appearing with the leaves, on leafy branchlets, the pistillate loosely flowered, 2 cm long or often less; capsules thinly short-pubescent, on very short pedicels, their bracts short-pubescent and pale brown; styles obsolete.

Subarctic-alpine, mainly of calcareous tundra and talus slopes; resembling *S. brachycarpa*, but with longer catkins.

General distribution: Alaska eastward to the District of Keewatin.

Fig. 374 Map 429

Salix ovalifolia Trautv.
var. **arctolitoralis** (Hult.) Argus
S. arctolitoralis Hult.
Trailing and freely branched shrub with slender, whip-like smooth and reddish-brown twigs; leaves broadly lanceolate, 2 to 4 cm long and about half as wide, yellowish-green above, paler and prominently veined beneath, with a slender petiole up to 4 to 5 mm long; pistillate catkins on leafy peduncles, the capsules glabrous, greenish-purplish, with a very short style.

General distribution: Coastal tundra from Kotzebue Sd., Alaska to the Mackenzie Delta.

Fig. 375 Map 430

Salix padophylla Rydb.
S. pseudomonticola Ball
S. monticola Bebb *p.pte.*
Upright shrub commonly 2 to 3 m tall with black, lustrous twigs; leaves petioled, thin, narrowly ovate-elliptic coming to a sharp tip, rounded at the base, those of the fertile twigs 4 to 6 cm long and almost half as wide, but on the sterile shoots 6 to 8 cm long and with up to 1 cm long stipules, glabrous and green above, glaucous and prominently veined beneath, the young commonly reddish-purple, their margins finely glandular serrate; catkins appearing with the leaves, sessile on the twigs, or on very short branchlets, the pistillate 3 to 4 cm long; capsules glabrous, on short pedicels.

Thicket-forming, and with *S. Bebbiana* perhaps the most common willow on alluvial river banks along the Mackenzie River, extending north almost to the Arctic Circle.

General distribution: Boreal forest, from James Bay to central Alaska, south to Colorado.

Fig. 376 Map 431

Salix pedicellaris Pursh
var. **hypoglauca** Fern.
Erect shrub seldom more than 1 m high and mostly lower, with glabrous, reddish-brown twigs and dark grey bark; leaves almost sessile or very short-petioled, firm and somewhat leathery, oblanceolate or oblong, with obtuse or rounded tips, entire-margined, 2 to 3 cm long and about one third as wide, glabrous, green above, glaucous beneath; catkins sessile, appearing with the leaves, on leafy twigs, the pistillate about 2 cm long; capsules glabrous, dark reddish-purple when young, on slender, 2 to 4 mm long pedicels; styles obsolete.

An eastern woodland species, commonly inhabiting open *Larix* fens.

General distribution: Nfld. to S. E. Y.T.

Fig. 377 Map 432

Salix phlebophylla Anderss.
Skeleton Willow
Small and mat-forming; the leaves firm and leathery, elliptic, rarely more than 1 cm long, glabrous and green on both sides, with entire margins, in time skeletonized, and marcescent on the twigs for many years; pistillate catkins 1.5 to 2.5 cm long, appearing with the leaves; capsules glabrous or thinly pubescent, dark purplish-black, on very short pedicels, their bracts black, with long, white hairs; styles short and thick.

Gravelly tundra ridges.

General distribution: Amphi-Beringian, arctic-alpine species barely entering our area in the Richardson Mts., and reaching the Arctic Coast east of the Mackenzie Delta.

Fig. 378 Map 433

Salix planifolia Pursh
Much branched, erect or spreading shrub, which in the southern part of our area may become 2 to 3 m high, with glabrous, black sometimes pruinose twigs; leaves glabrous, elliptic-lanceolate to oblong, with entire or remotely dentate margins, dark green and lustrous above, glaucous beneath; petioles short; stipules minute, soon falling; catkins sessile, appearing before the leaves, those of the female plant 3.0 to 4.5 cm long, bracts black, villous; capsules whitish-silky.

Snowpatches and sheltered slopes.

General distribution: An eastern subarctic forest-tundra species reaching the upper and central Mackenzie Basin, southern Y.T. and southwestern Alaska, but thus far not collected on the lower Mackenzie beyond Norman Wells and on Great Bear L., although ranging beyond the general limit of forest in northern Y.T. and along the rivers across the Barren Grounds.

Fig. 379 Map 434

Salix polaris Wahlenb.
ssp. **pseudopolaris** (Flod.) Hult.
Snow-bed Willow
A tiny creeping shrub with slender, thin, freely rooting, pale yellowish stems which, as a rule, are deeply buried in moss; the mature leaves obovate-elliptic, usually 1.0 to 1.5 cm (rarely 2.0 to 2.5 cm) long, thin, fresh green on both sides, somewhat lustrous above and prominently veined beneath; catkins appearing with the leaves, the pistillate large for the size of the plant, varying from 2 to 4 cm in length; capsules thinly pubescent, glabrous in age, dark purplish-brown, their obovate-orbicular, often retuse and white-villous bracts reddish-purple, drying black.

In not too dry, mossy tundra in places where the snow remains late.

General distribution: An Amphi-Beringian, arctic-alpine species, in the District of Mackenzie known only from the Mackenzie and Richardson Mts., and from the Caribou Hills east of the Mackenzie Delta.

Fig. 380 Map 435

Salix pulchra Cham.
In habit and general appearance similar to *S. planifolia* from which however, it is readily distinguished by its marcescent leaves and stipules that remain on the twigs at least until next year's catkins mature.

Common in the Mackenzie and Richardson Mountains, and ranging east across the barrens beyond Great Bear Lake, at least to Coronation Gulf. A common species of inland and alpine tundra where it is characteristic of sheltered but well watered slopes and hollows and of the banks of lakes and streams in places where snow accumulates in winter. The flowers appear very early, often when the snow still covers the lower parts of the bushes. On the gently sloping tundra the reddish-brown bands formed by *Salix pulchra* thickets beautifully outline the drainage systems; seen from the air in early winter, when the willow thickets are not yet completely covered by snow drifts, the effect of the brown bands meandering across the white plains is most striking.

The var. *yukonensis* Schneid. differs consistently from typical *S. pulchra* by its taller growth, diamond-shaped and distinctly broader and larger leaves, and puberulent young twigs. It is a dominant component of the Mackenzie Delta willow flats, but its range does not extend far beyond the edge of the forest.

General distribution: Amphi-Beringian, arctic-alpine.

Fig. 381 Map 436

Salix pyrifolia Anderss.
Balsam Willow
Upright shrub, in the Mackenzie District never tree-like and rarely over 2 m high, with shiny reddish-purple twigs; leaves on slender petioles about one sixth as long as the thin, oval or obovate, and finely serrate blade which, when young, is delicate and almost translucent, fresh green above, pale and prominently veined beneath; stipules small or often lacking; catkins appearing with the leaves, on leafy peduncles, the pistillate 4 to 8 cm long; capsules glabrous, dark green in youth, turning dark purplish, their bracts linear-oblong, 2 to 3 mm long, light brown under the persisting woolly pubescence; styles less than 1 mm long. In drying, the leaves give off an agreeable spicy odor, that, in untreated herbarium specimens may persist for many years (hence the vernacular name).

Moist woodland clearings.

General distribution: Nfld. to southern District of Mackenzie with a single widely disjunct station in mountains of northern Y.T.

Fig. 382 Map 437

Salix reticulata L.
Net-veined Willow
Prostrate, with rather stout, freely rooting, few-leaved stems; leaves orbicular-elliptic, leathery, glabrous, dark green and rugose above, glaucous and strongly reticulate beneath, in youth the underside of the leaves is silky pubescent; catkins small, appearing with the leaves, borne on subterminal, slender, naked, and usually villous peduncles.

Calcareous and not too dry, sandy, gravelly, or turfy places; near its northern limit usually found only in protected places where the snow remains late.

General distribution: Circumpolar, wide spread arctic-alpine; lacking in Greenland.

Fig. 383 Map 438

Salix Scouleriana Barratt
Shrub or small tree up to 10 m tall; twigs stout, commonly yellow and densely pubescent when young, glabrescent and first grey, becoming black and shiny in age, the bark of older branches dark grey and smooth; leaves rather thick and firm with entire margins; the blade oblanceolate with obtuse or rounded apex and wedge-shaped base, 5 to 8 cm long, dark green and shiny above, in youth velvety and light grey, in age characteristically rust-coloured beneath due to the emergence from beneath the velvety pubescence of variously branched, thick, flattened and shiny reddish-brown trichomes; petioles permanently hoary, stipules minute or lacking; the catkins sessile appearing before the leaves, the pistillate 4 to 5 cm long; capsules permanently hoary, their bracts oval, black under the long, silky pubescence; styles about 1 mm long.

Well drained not too densely wooded slopes or river banks.

General distribution: A western plains or foothill species from Man. to B.C., north to central Y.T., Alaska and S.W. District of Mackenzie.

Fig. 384 Map 439

Salix serissima (Bailey) Fern.
Autumn Willow
Erect shrub up to 4 m tall, with shiny, greenish-brown twigs; leaves oblong-lanceolate, short-acuminate, about three times longer than broad, glandular-serrate, fresh green above, glaucous beneath; catkins appearing with the leaves, on leafy twigs; capsules glabrous, very crowded, their pedicels stout and white-pubescent; styles short or nearly obsolete; stamens 5.

In habit rather similar to *S. lasiandra* which, however, flowers nearly one month earlier, and has larger leaves that are long-attenuate at the apex and green on both sides.

Lake shores and stream banks.

General distribution: N. America from Nfld. west to Alta. and upper Mackenzie drainage, southward in the Rocky Mts.

Fig. 385 Map 440

***Salix Tyrrellii** Raup
Slender, erect shrub up to 2 m tall with reddish and glabrous twigs; leaves narrowly lanceolate, 2.0 to 2.5 cm long and 4 to 8 mm wide, green and shiny on both sides, of leathery texture, with minutely but conspicuously glandular-serrate margins, on petioles 2 to 6 mm long; catkins appearing with the leaves, on short, leafy branchlets; capsules 3 to 4 mm long, densely white-hairy, sessile or on short pedicels.

S. Tyrrellii by Raup placed in the section *Glaucae*, and thought to be most closely related to *S. MacCalliana* Rowlee, is known only from the type locality among shifting sand dunes along the south shore of Lake Athabasca; it should be looked for in similar habitats elsewhere in the Mackenzie River drainage.

Map 441

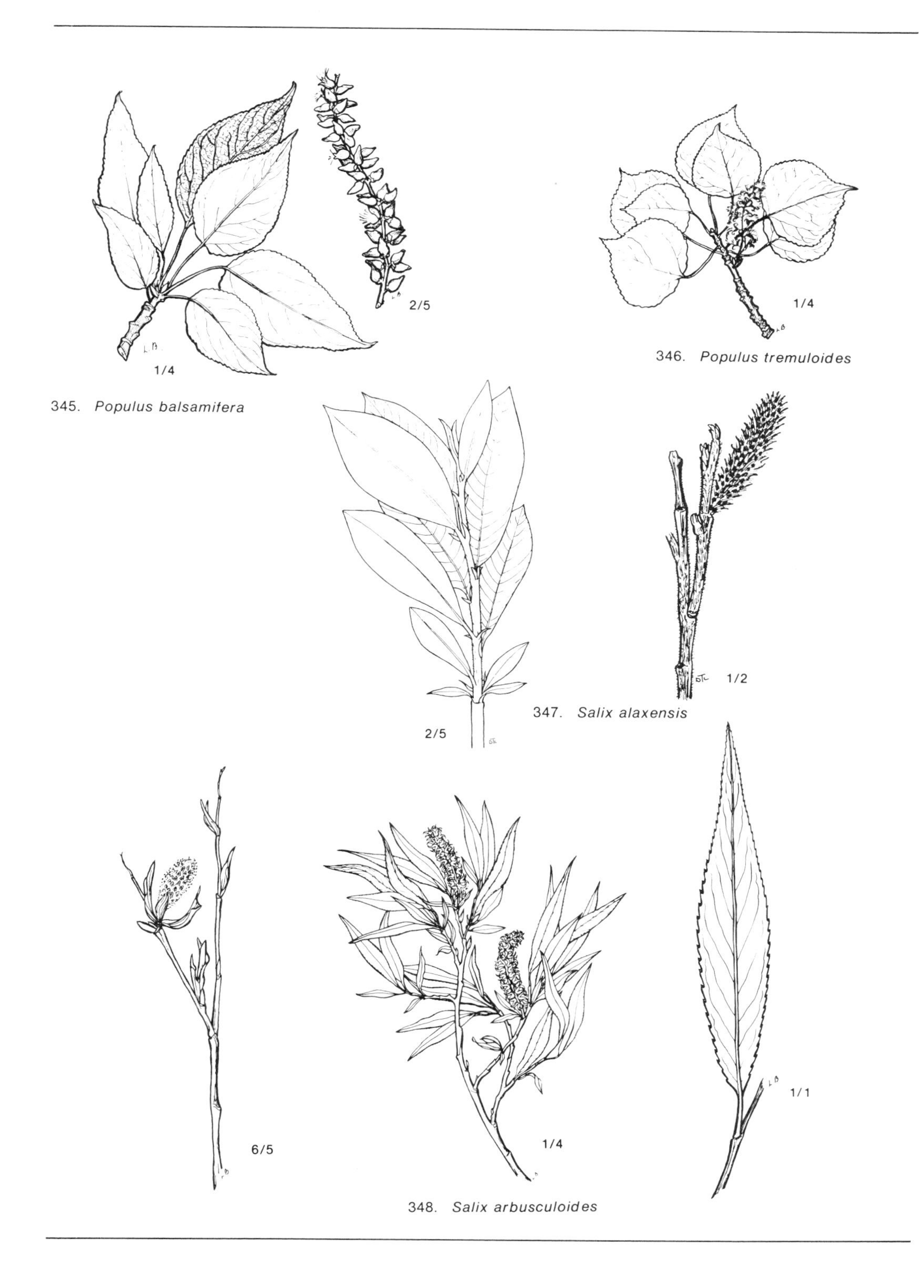

345. *Populus balsamifera*

346. *Populus tremuloides*

347. *Salix alaxensis*

348. *Salix arbusculoides*

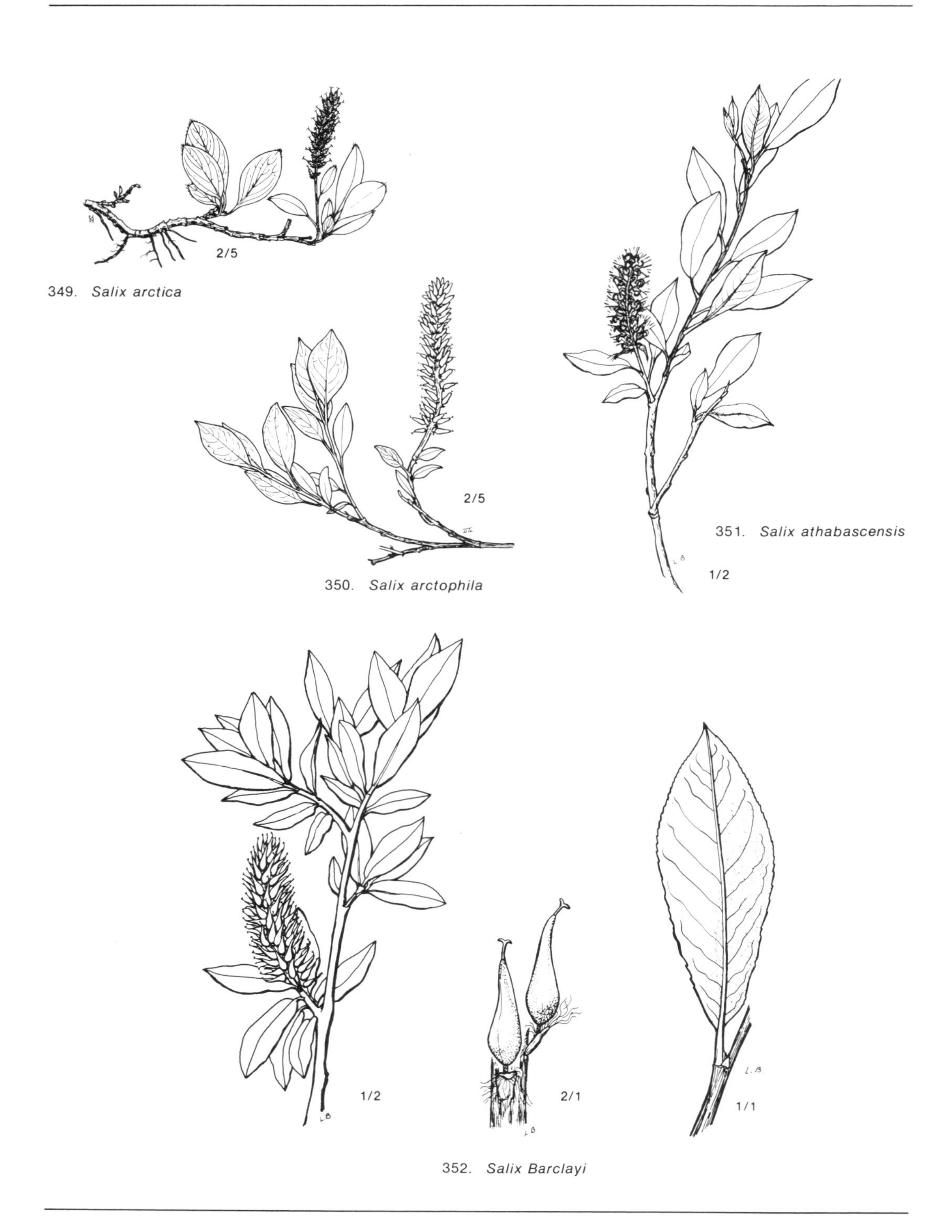

349. *Salix arctica*

350. *Salix arctophila*

351. *Salix athabascensis*

352. *Salix Barclayi*

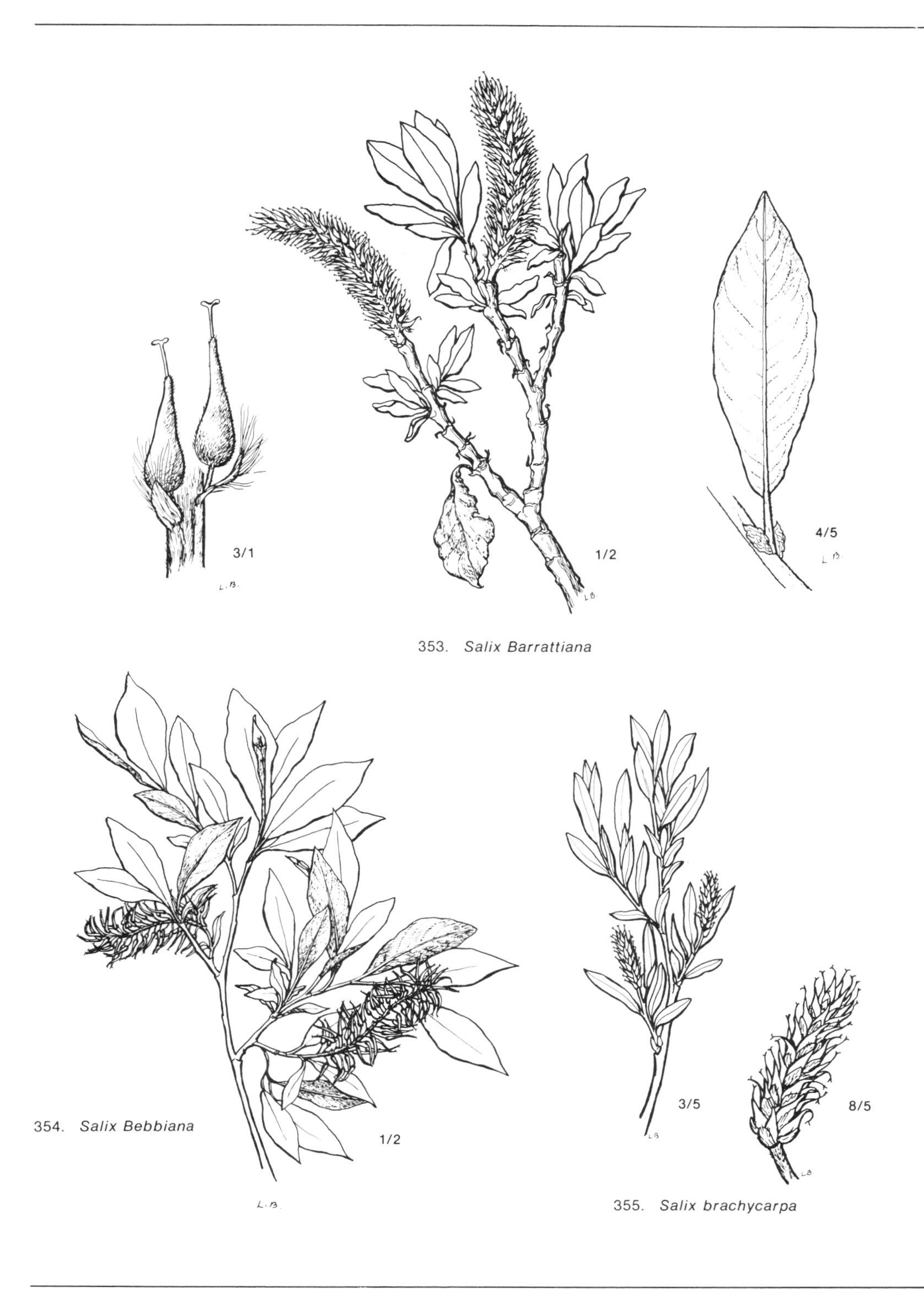

353. *Salix Barrattiana*

354. *Salix Bebbiana*

355. *Salix brachycarpa*

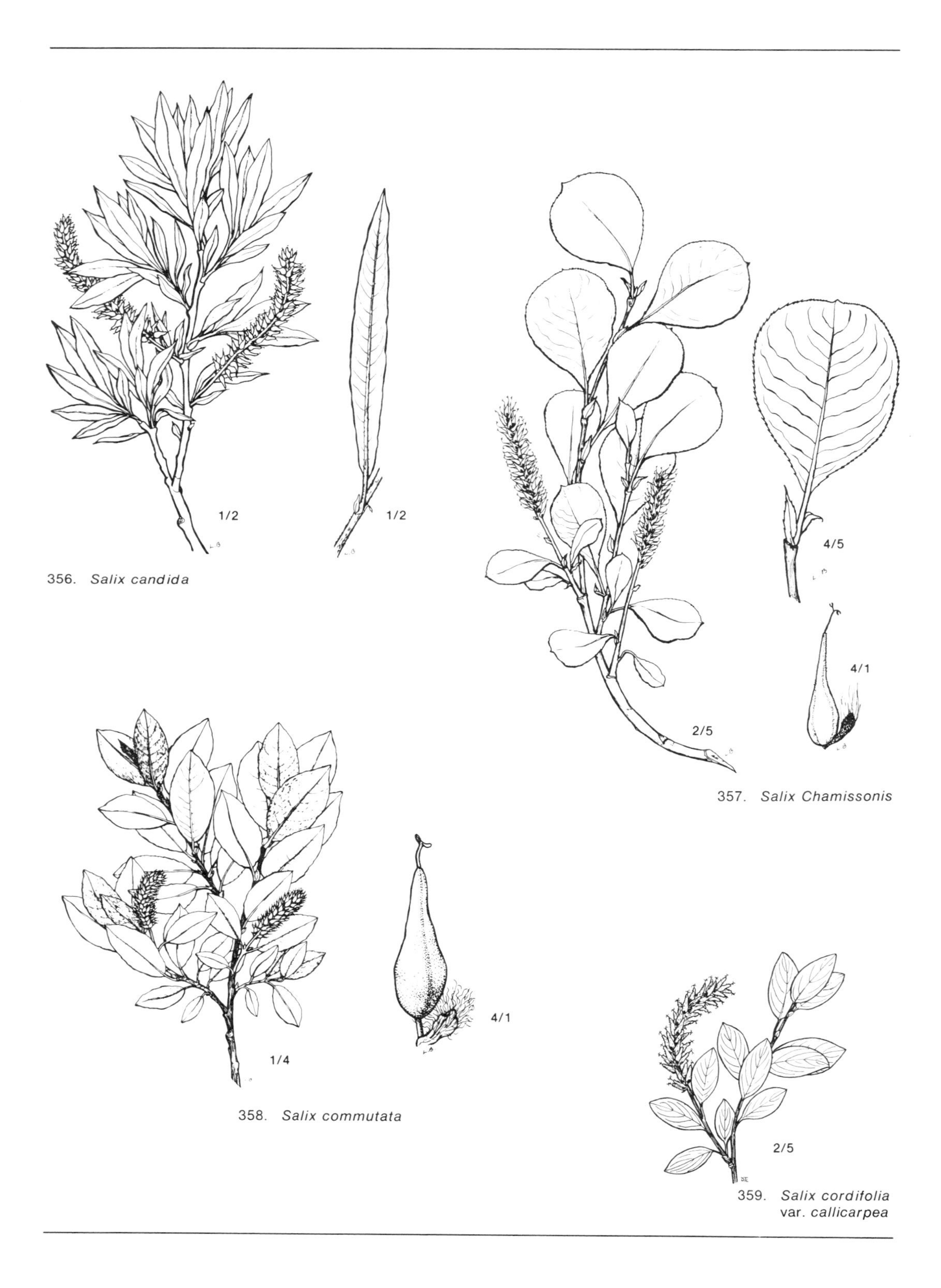

356. *Salix candida*

357. *Salix Chamissonis*

358. *Salix commutata*

359. *Salix cordifolia* var. *callicarpea*

360. *Salix Dodgeana*

361. *Salix Farriae*

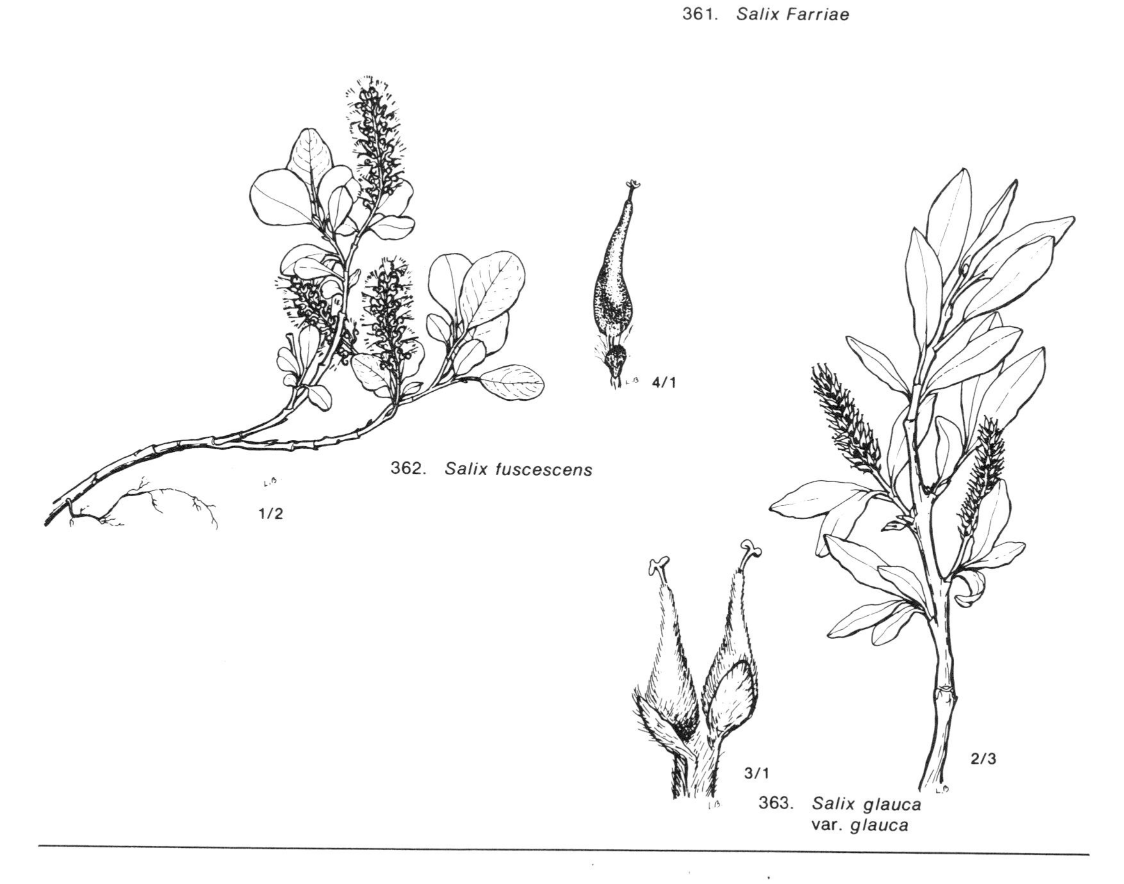

362. *Salix fuscescens*

363. *Salix glauca* var. *glauca*

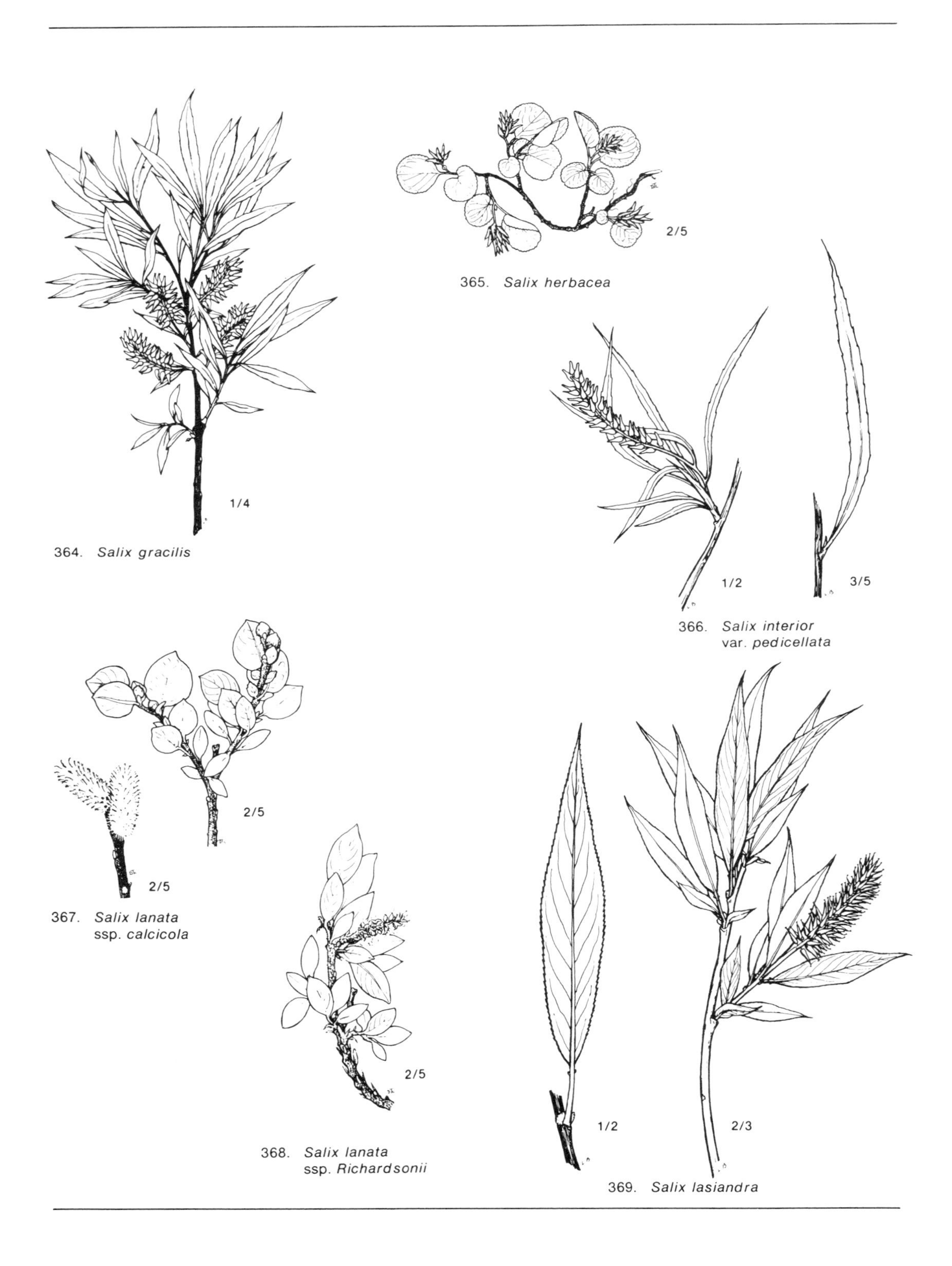

364. *Salix gracilis*

365. *Salix herbacea*

366. *Salix interior* var. *pedicellata*

367. *Salix lanata* ssp. *calcicola*

368. *Salix lanata* ssp. *Richardsonii*

369. *Salix lasiandra*

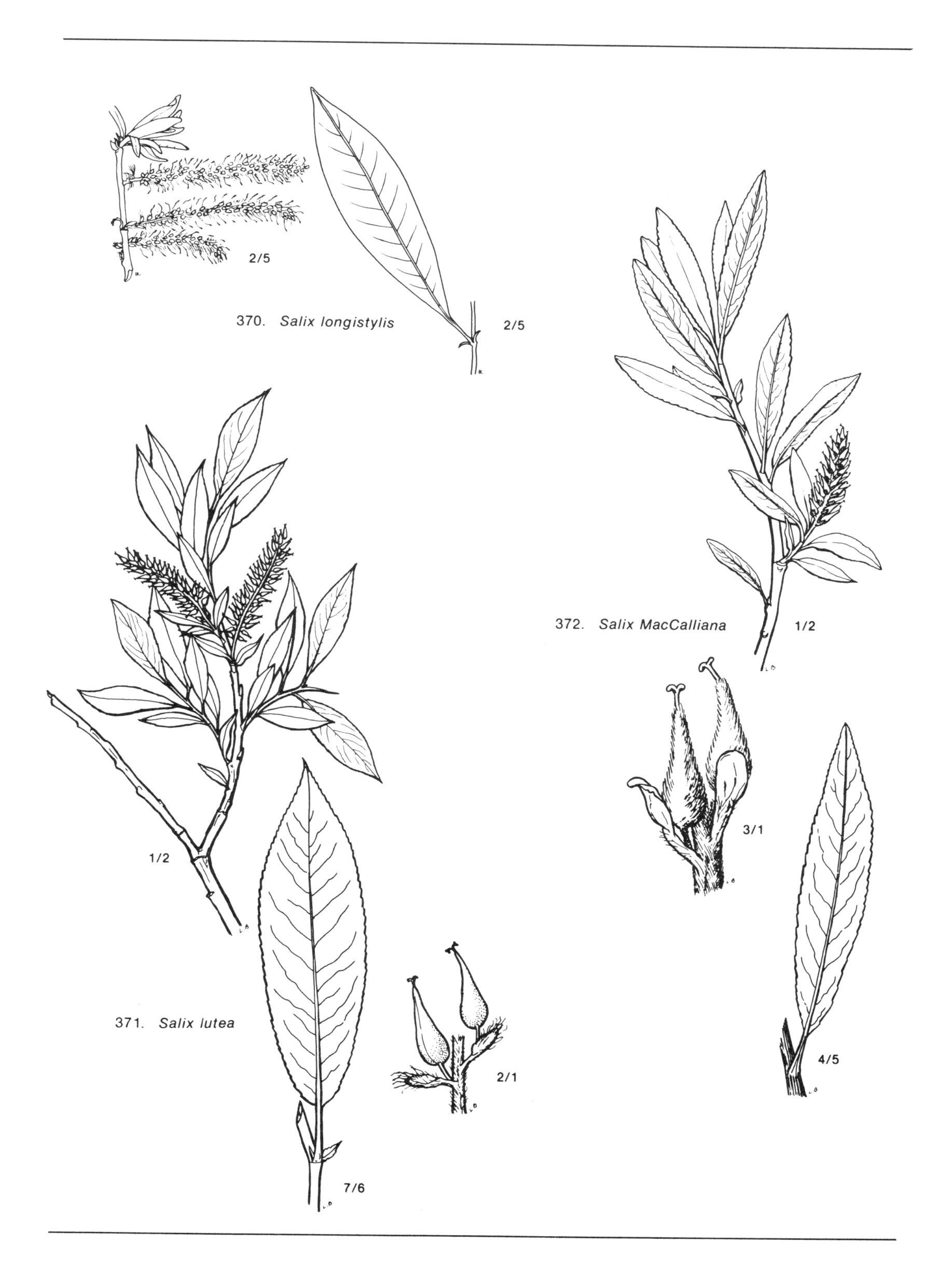

370. *Salix longistylis*

371. *Salix lutea*

372. *Salix MacCalliana*

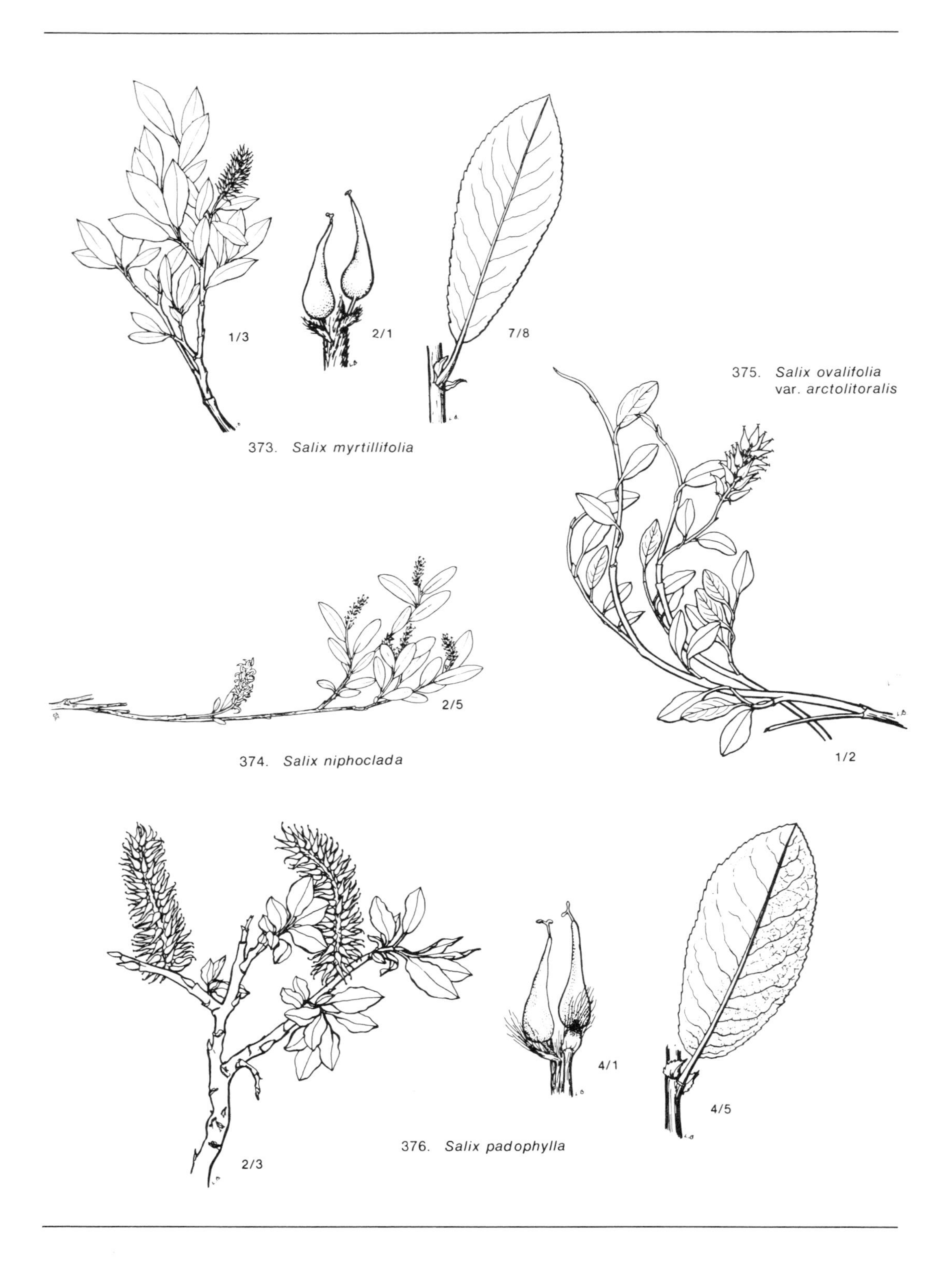

373. *Salix myrtillifolia*

374. *Salix niphoclada*

375. *Salix ovalifolia* var. *arctolitoralis*

376. *Salix padophylla*

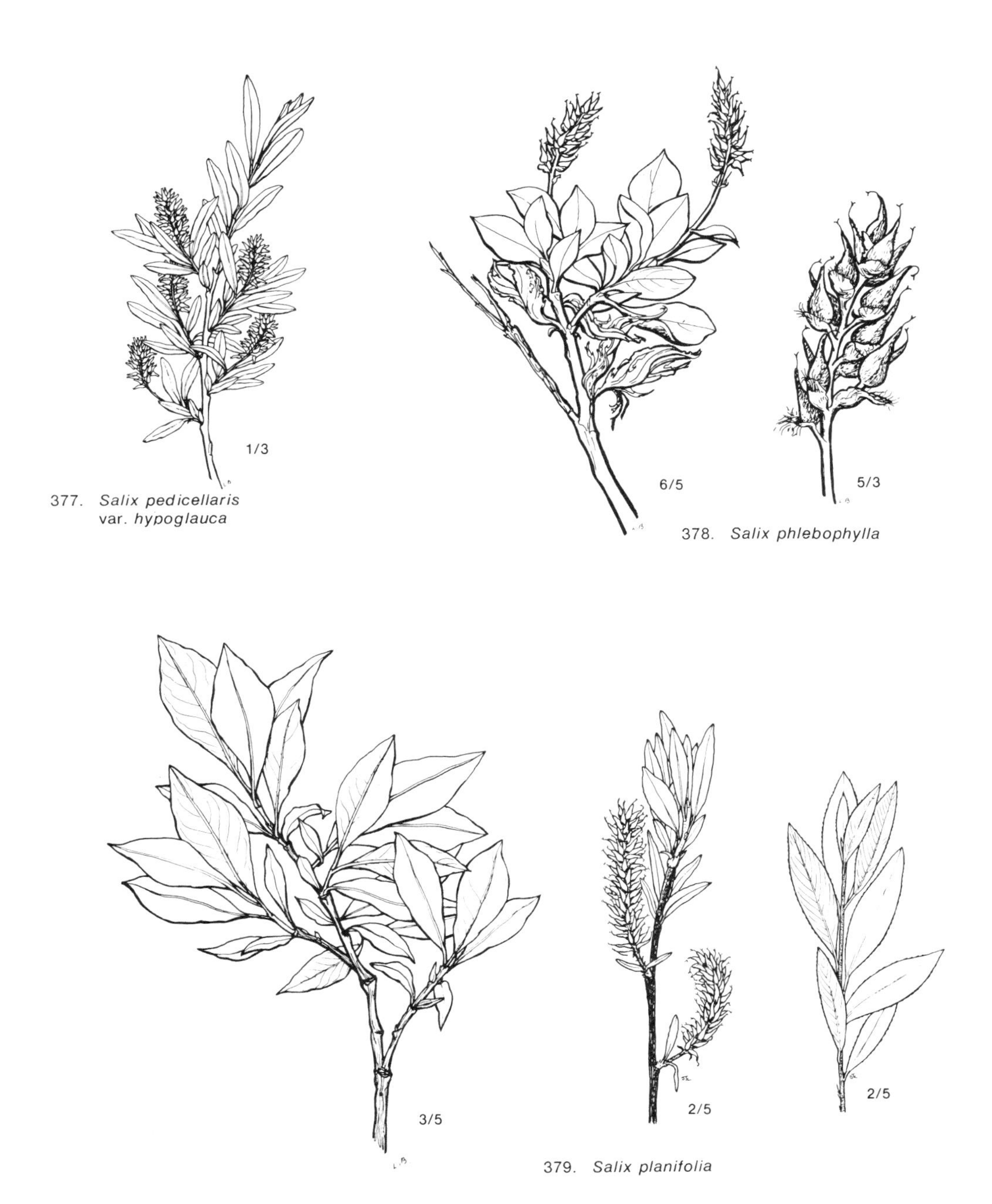

377. *Salix pedicellaris* var. *hypoglauca*

378. *Salix phlebophylla*

379. *Salix planifolia*

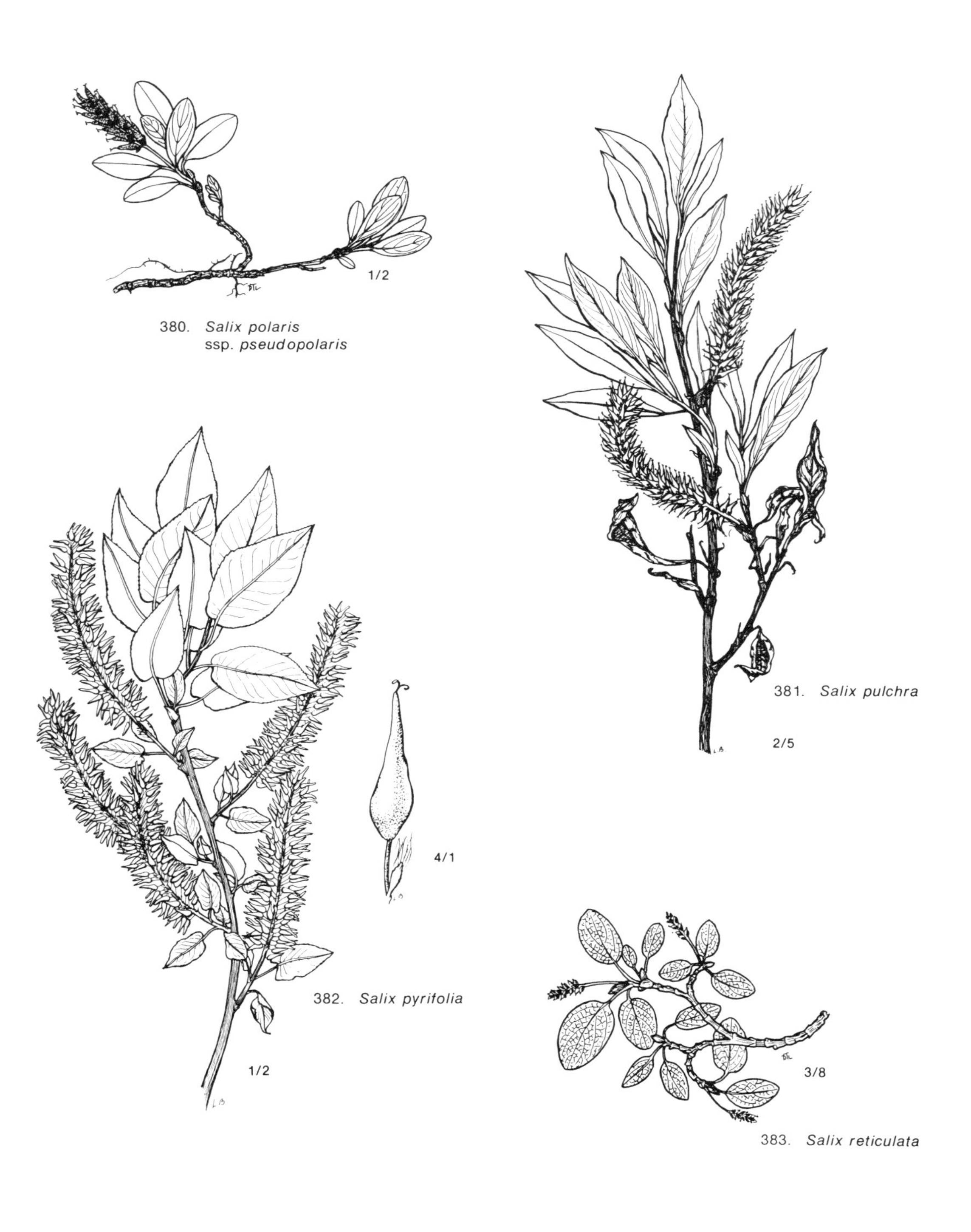

380. *Salix polaris* ssp. *pseudopolaris*

381. *Salix pulchra*

382. *Salix pyrifolia*

383. *Salix reticulata*

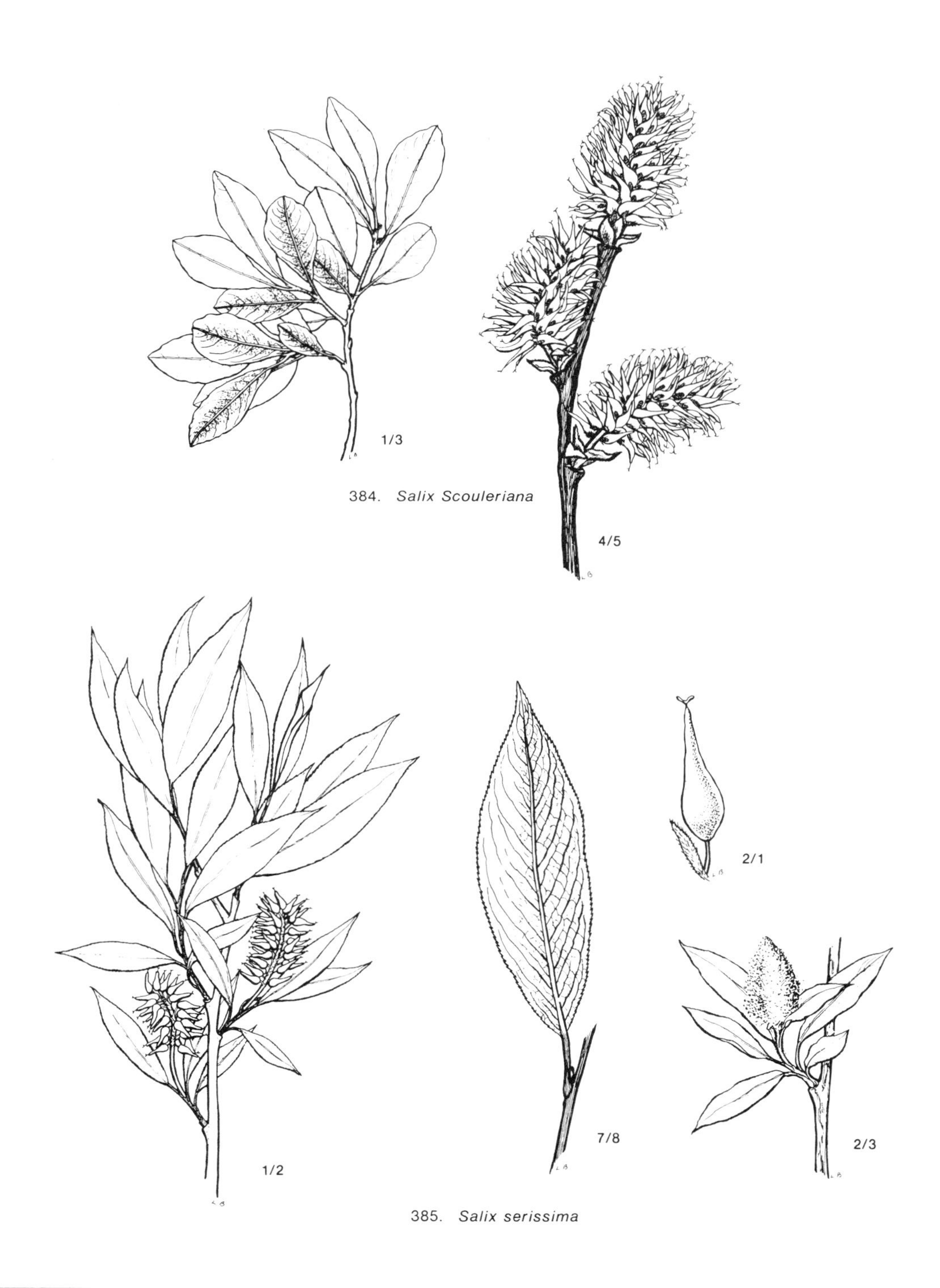

384. *Salix Scouleriana*

385. *Salix serissima*

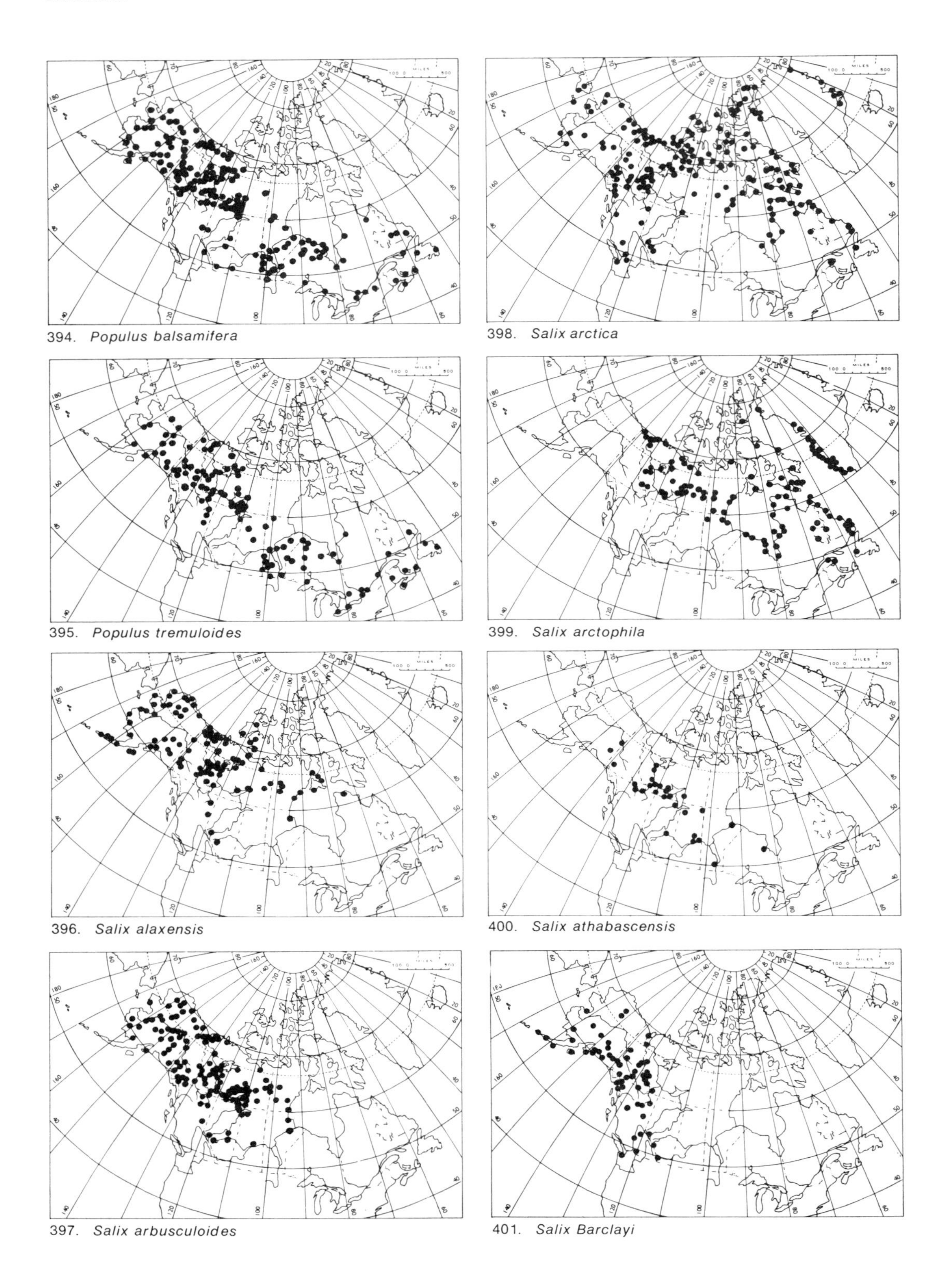

394. *Populus balsamifera*

395. *Populus tremuloides*

396. *Salix alaxensis*

397. *Salix arbusculoides*

398. *Salix arctica*

399. *Salix arctophila*

400. *Salix athabascensis*

401. *Salix Barclayi*

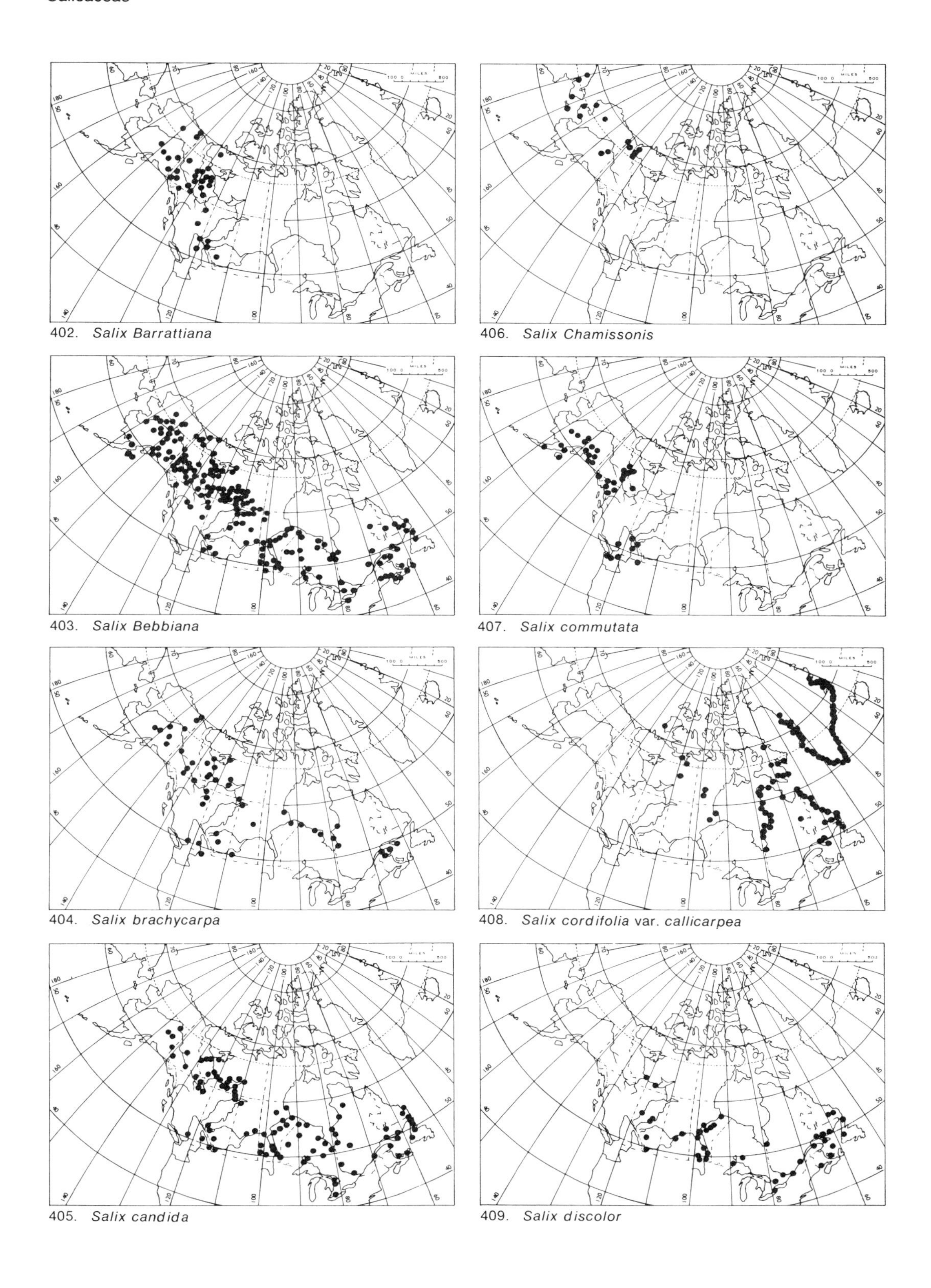

402. *Salix Barrattiana*

406. *Salix Chamissonis*

403. *Salix Bebbiana*

407. *Salix commutata*

404. *Salix brachycarpa*

408. *Salix cordifolia* var. *callicarpea*

405. *Salix candida*

409. *Salix discolor*

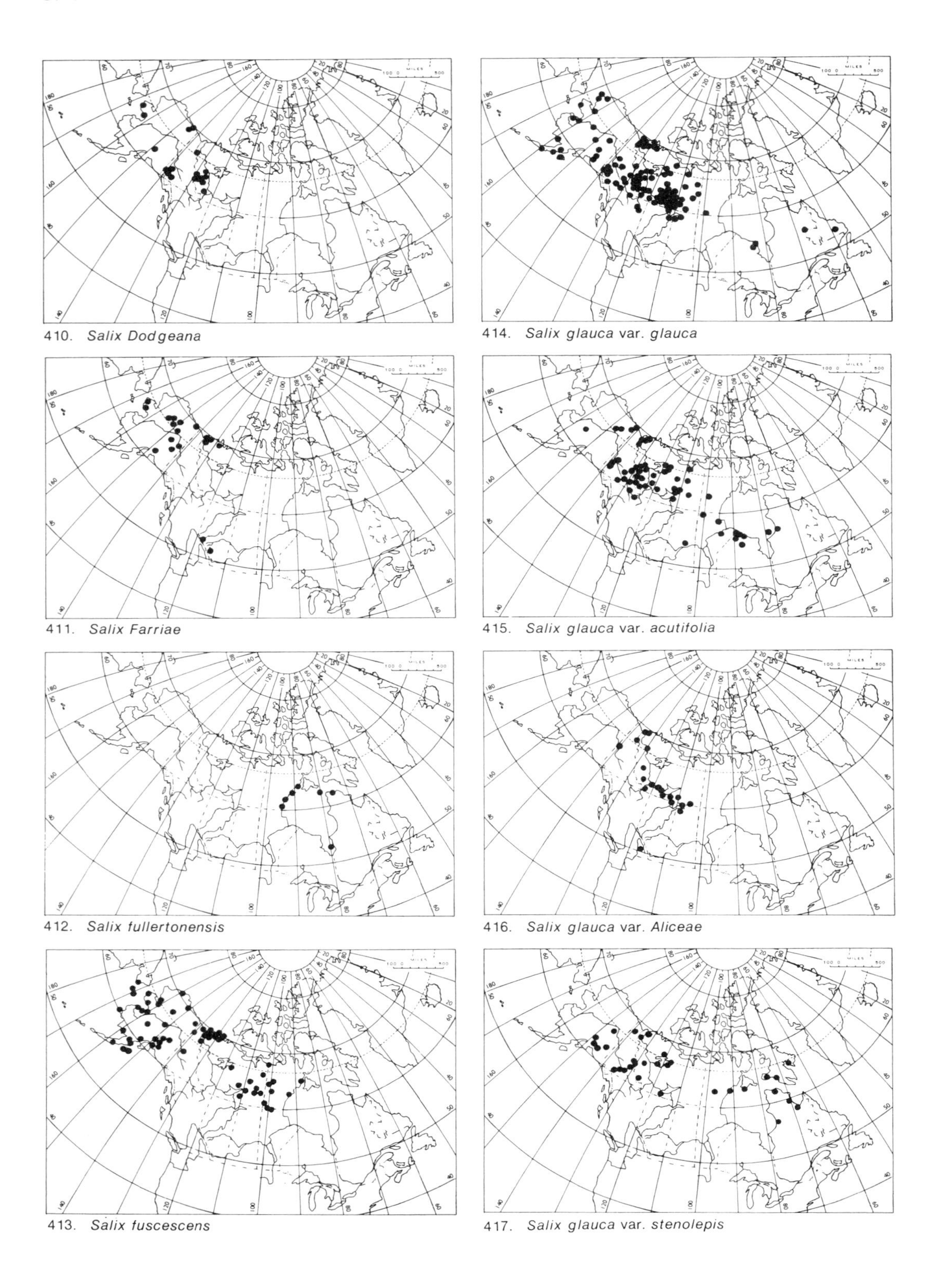

410. *Salix Dodgeana*

411. *Salix Farriae*

412. *Salix fullertonensis*

413. *Salix fuscescens*

414. *Salix glauca* var. *glauca*

415. *Salix glauca* var. *acutifolia*

416. *Salix glauca* var. *Aliceae*

417. *Salix glauca* var. *stenolepis*

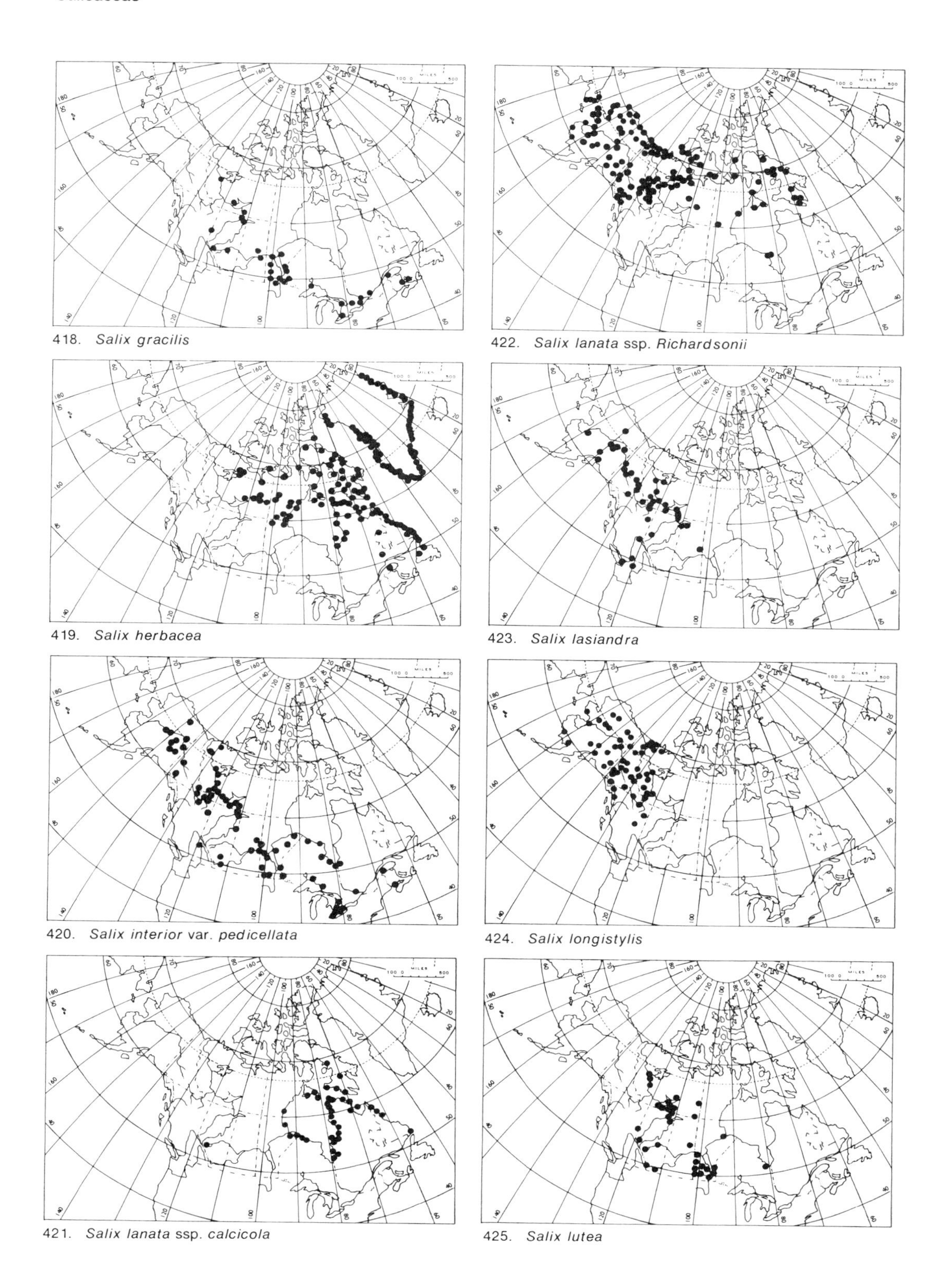

418. *Salix gracilis*

419. *Salix herbacea*

420. *Salix interior* var. *pedicellata*

421. *Salix lanata* ssp. *calcicola*

422. *Salix lanata* ssp. *Richardsonii*

423. *Salix lasiandra*

424. *Salix longistylis*

425. *Salix lutea*

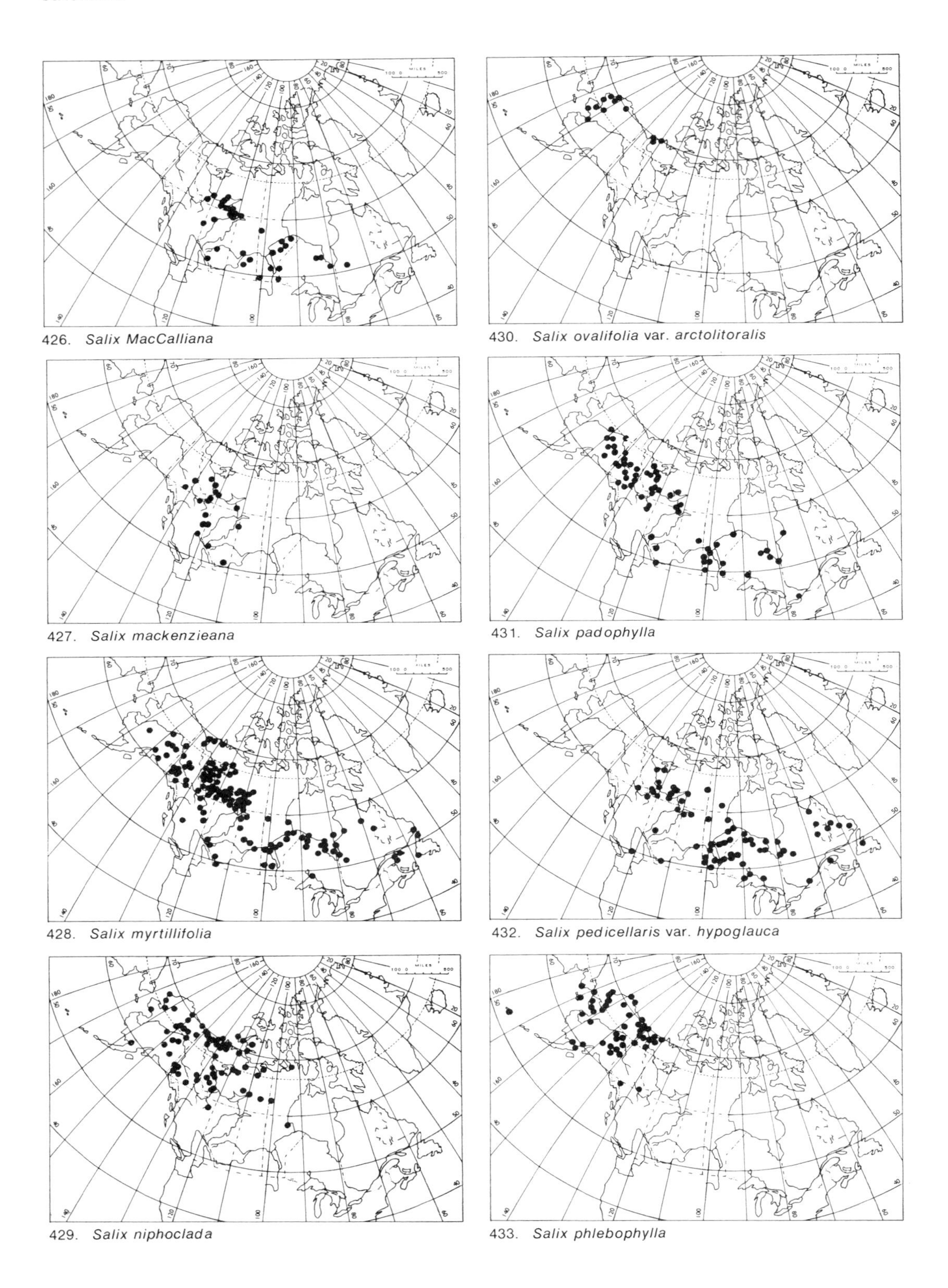

426. *Salix MacCalliana*

427. *Salix mackenzieana*

428. *Salix myrtillifolia*

429. *Salix niphoclada*

430. *Salix ovalifolia* var. *arctolitoralis*

431. *Salix padophylla*

432. *Salix pedicellaris* var. *hypoglauca*

433. *Salix phlebophylla*

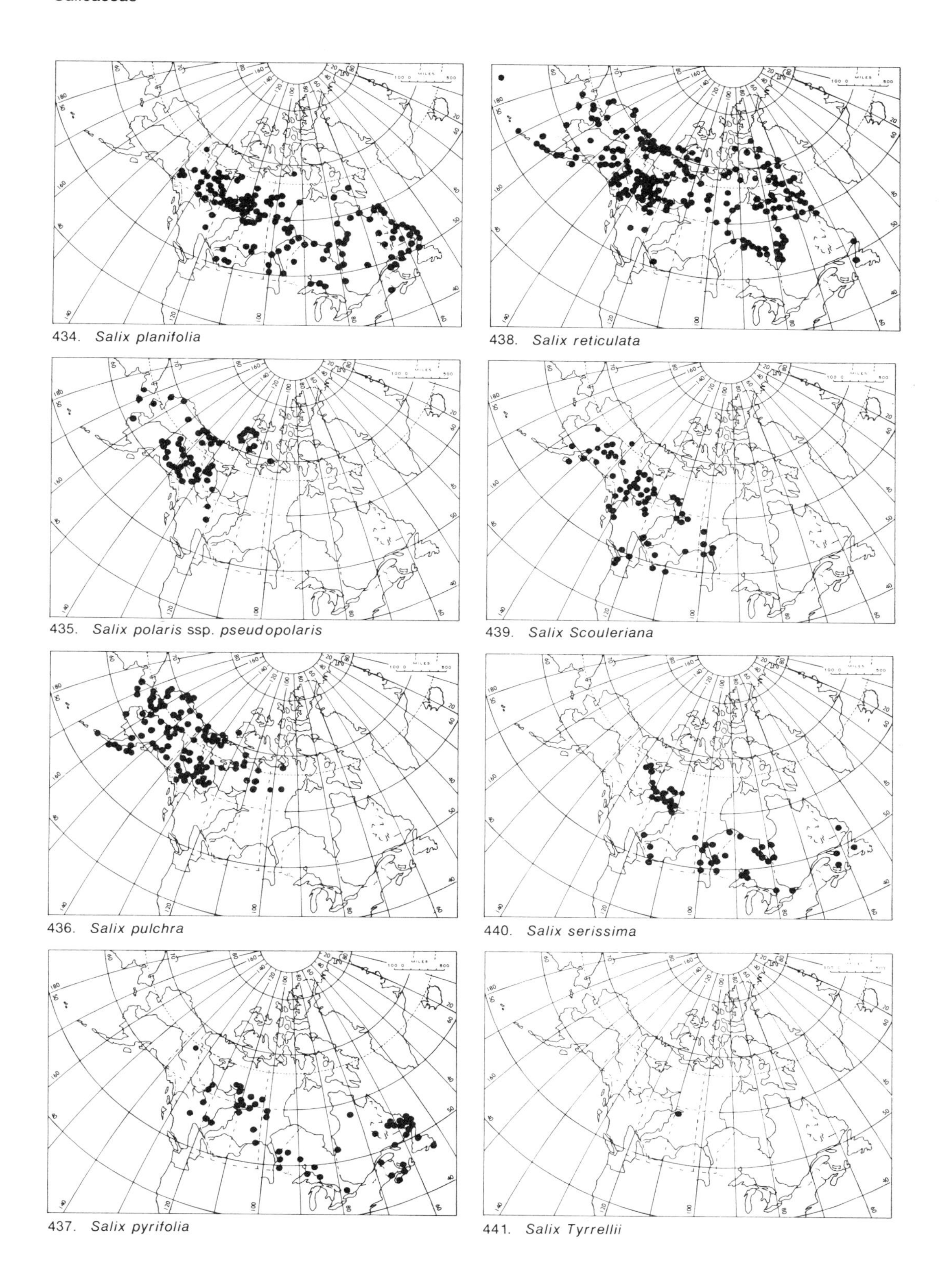

434. *Salix planifolia*

435. *Salix polaris* ssp. *pseudopolaris*

436. *Salix pulchra*

437. *Salix pyrifolia*

438. *Salix reticulata*

439. *Salix Scouleriana*

440. *Salix serissima*

441. *Salix Tyrrellii*

MYRICACEAE Wax-myrtle Family

Myrica L. Sweet Gale

Myrica Gale L.
Low mostly dioecious, shrub with strongly ascending, brown branches, with us rarely more than 1 m high, with strongly aromatic, cuneate-oblanceolate greyish, deciduous leaves that are sparingly toothed towards the tip. The pistillate aments are cone-like and borne near the tip of last year's wood.

Probably common throughout the wooded parts of the Precambrian Shield area growing in shallow water by the edge of small lakes, poor in plant nutrients, or occasionally in wet muskeg bogs northward along the Mackenzie River and its tributaries. Its known northern limit in the District of Mackenzie is in black spruce bogs, on the upper east branch of the Mackenzie R. Delta.

General distribution: Lab. to Alaska; E. Asia and northern Europe; north nearly to the limit of trees.

Fig. 386 Map 442

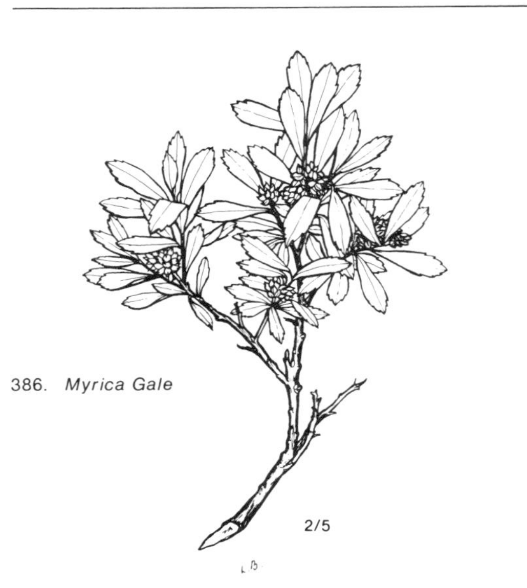

386. *Myrica Gale*

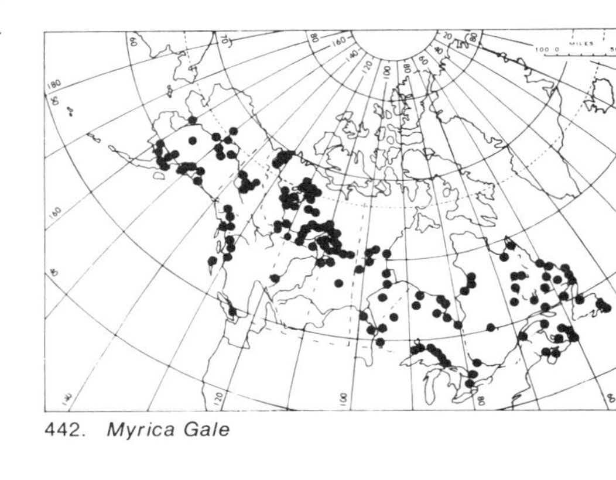

442. *Myrica Gale*

BETULACEAE Birch Family

Monoecious trees or shrubs with deciduous, alternate, simple and usually prominently serrated leaves; flowers in catkins or aments appearing before the leaves, the staminate mostly drooping, the pistillate spike-like, solitary and sessile or peduncled, or in pairs; fruit (in ours) a small 1-seeded, winged or narrow-margined nutlet.

a. Bracts of pistillate catkins thin, 3-lobed, deciduous . . . . *Betula*
a. Bracts of pistillate catkins firm and woody, persistent . . . . *Alnus*

Alnus B. Ehrh. Alder

Shrubs or small trees with reddish or greyish-brown bark; stamens 4; fruiting catkins woody, persistent, the nutlets with or without a membranaceous wing.

a. Leaves serrulate or biserrulate with sharp teeth; nutlet broadly wing-margined. . . *A. crispa*
a. Leaves doubly serrate or serrate-dentate, undulate or shallowly lobed; nutlet wingless . . . . *A. incana* ssp. *tenuifolia*

Alnus crispa (Ait.) Pursh
Green Alder
Ascending and bushy shrubs to 3 m in height; branches more or less glutinous when young, becoming glabrous; leaf-buds sessile; leaves ovate to broadly elliptic, irregularly serrulate or biserrulate with sharp teeth, rounded or cuneate to the base; flowers developed with the leaves on the twigs of the season; nutlets broadly wing-margined.

This is the common alder of the wooded parts of the District of Mackenzie where its northern limit coincides with that of the white spruce, although isolated stands of both may be found in the treeless barrens along some of the north-flowing rivers, almost to the shores of the Arctic Ocean; in the Mackenzie Delta it reaches lat. 68°46′ and in Bathurst Inlet lat. 66°50′. In the Richardson Mountains west of the Mackenzie Delta green alder forms dense thickets on moist slopes above timberline at elevations up to 650 m and in lat. 62° it reaches timberline approximately at an elevation of 1400 m. Although on the upper Mackenzie River *Alnus crispa* is largely replaced in the flood plains by *A. incana,* the former is by no means uncommon on high river banks or along cold-water creeks and seepages, and was noted to form tall thickets on river banks above Fort Simpson.

The Pacific coast race, ssp. *sinuata* (Regel) Hult., distinguished by larger and thinner, somewhat cordate, undulate or double-serrate leaves, is known from S. W. Yukon Territory but apparently does not enter the District of Mackenzie.

General distribution: S. W. Greenland to E. Asia.

Fig. 387 Map 443

Alnus incana (L.) Moench
ssp. **tenuifolia** (Nutt.) Breitung
Grey or Hoary Alder
Large shrub or small tree to 5 m in height; leaf-buds stipitate; leaves oval to broadly ovate, doubly serrate or serrate-dentate, undulate or shallowly lobed; flowers developed before the leaves in early spring; nutlets wingless, merely margined.

The hoary alder is common on river banks and on lake shores northward in the Mackenzie Valley to the Arctic Circle, but beyond is known from isolated stands in the Mackenzie Delta.

General distribution: N. America from Man. to Alaska.

Fig. 388 Map 444

Betula L. Birch

Trees or shrubs with white, reddish or grey-brown bark, the branchlets often dotted with wart-like glands; leaves serrate or dentate; stamens 2; fruiting catkins deciduous, the nutlets winged.

a. Low depressed or upright shrubs rarely more than 1 m tall; mature leaves 1-2 cm long; young twigs not downy or pubescent; staminate catkins 1-2 cm long, erect or spreading; fruiting catkins commonly 1.0-1.5 cm long, erect, sessile or short-peduncled; wing-margin narrower than nutlets, or wanting
  b. Leaves green on both sides or at most slightly paler in drying, glandular-glutinous on the underside; young twigs densely covered with resinous, wart-like glands; nutlet with wing narrower than body . . . . *B. glandulosa*
  b. Leaves notably paler and scarcely glutinous beneath; young twigs less glandular-warty, and mainly near the tip; nutlets with wing as broad as body . . . . *B. pumila* var. *glandulifera*
a. Trees or tall shrubs (although *B. occidentalis* is a low shrub near its northern limit); mature leaves over 2.5 cm long; young twigs downy or pubescent; staminate catkins 3-8 cm long, drooping, wing-margin of nutlets as broad or broader than the body

c. Pistillate catkins erect, short peduncled; wing of nutlets about as broad as body . . . . *B. occidentalis*
c. Pistillate catkins drooping; wing of nutlets broader than body
  d. Mature leaves 3-5 cm long, rhombic-ovate to triangular-ovate in outline, simple-serrate, commonly cuneate at the base. . . . . *B. papyrifera* var. *neoalaskana*
  d. Mature leaves 4-8 cm long, ovate, rounded or truncate at the base, commonly double-serrate . . . . *B. papyrifera* var. *commutata*

Betula glandulosa Michx.
Ground or Dwarf Birch
Wide ranging North American low-arctic tundra species, growing mainly on acidic rocks or in woodland muskegs or peat-bogs, where it commonly forms thickets up to 1.5 m, but usually lower, generally associated with willows and other shrubs.

General distribution: Boreal and sub-arctic N. America from S.W.Greenland to Alaska.

Fig. 389 Map 445

Betula occidentalis Hook.
B. microphylla Am. *auct. non* Bunge
B. fontinalis Sarg.
An upland species of dry ridges and slopes, north to the Mackenzie Delta and the Mackenzie and Richardson Mountains.

General distribution: N. America from James Bay to southern and central Alaska.

Fig. 390 Map 446

Betula papyrifera Marsh
var. **commutata** (Regel) Fern.
B. papyrifera Marsh
var. *occidentalis* Sarg., *non B. occidentalis* Hook.
Western species of rich woodlands of the upper Mackenzie drainage. Similar to the more common and wide ranging Paper Birch and rare and local in the District of Mackenzie where it is, perhaps, always restricted to richer, non-peaty soils such as occur in the vicinity of hot springs. The wood is harder and darker than that of the Paper Birch and was formerly used in the making of canoes.

General distribution: Cordilleran; in the District of Mackenzie north to Ft. Norman and barely entering southeastern Y.T.

Fig. 391 Map 447

Betula papyrifera Marsh
var. **neoalaskana** (Sarg.) Raup;
B. papyrifera
var. *humilis* (Regel) Fernald and Raup
B. resinifera Britt.
Paper Birch
This is the common paper birch of the western boreal forest, in the Mackenzie District growing mainly on Precambrian rocks and in acid, peaty not too wet places in the Mackenzie Valley and in the mountains to the west. The best stands known to us are on a low peat-covered island in the west end of Great Slave Lake where a number of stems measured 30.5 cm or better at the butt.

General distribution: N. America from western Ont. to Alaska, north to the limit of trees and in the Mackenzie Delta reaching lat. 68°40′ on the East Branch.

Fig. 392 Map 448

Betula pumila L.
var. **glandulifera** Regel
An eastern woodland bog species with a western limit in the Great Slave Lake area and the upper Mackenzie Valley. Of similar habit but taller than *B. glandulosa* with which it is sometimes confused, and with which is probably hybridizes. However, in most cases the characters given in the key will suffice to separate mature and fruiting specimens.

General distribution: N. America; Nfld. to S. W. District of Mackenzie.

Fig. 393 Map 449

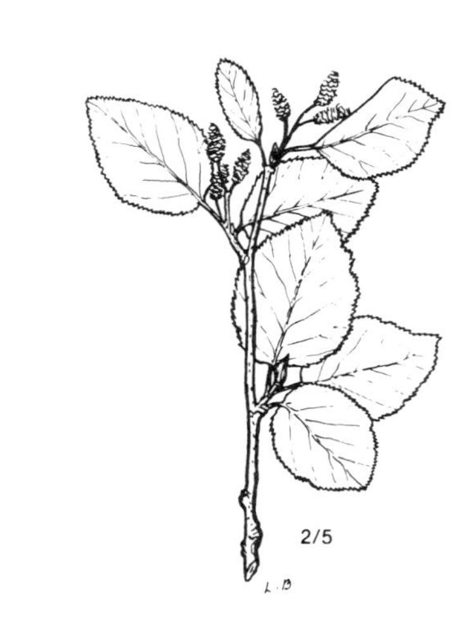

387. *Alnus crispa*

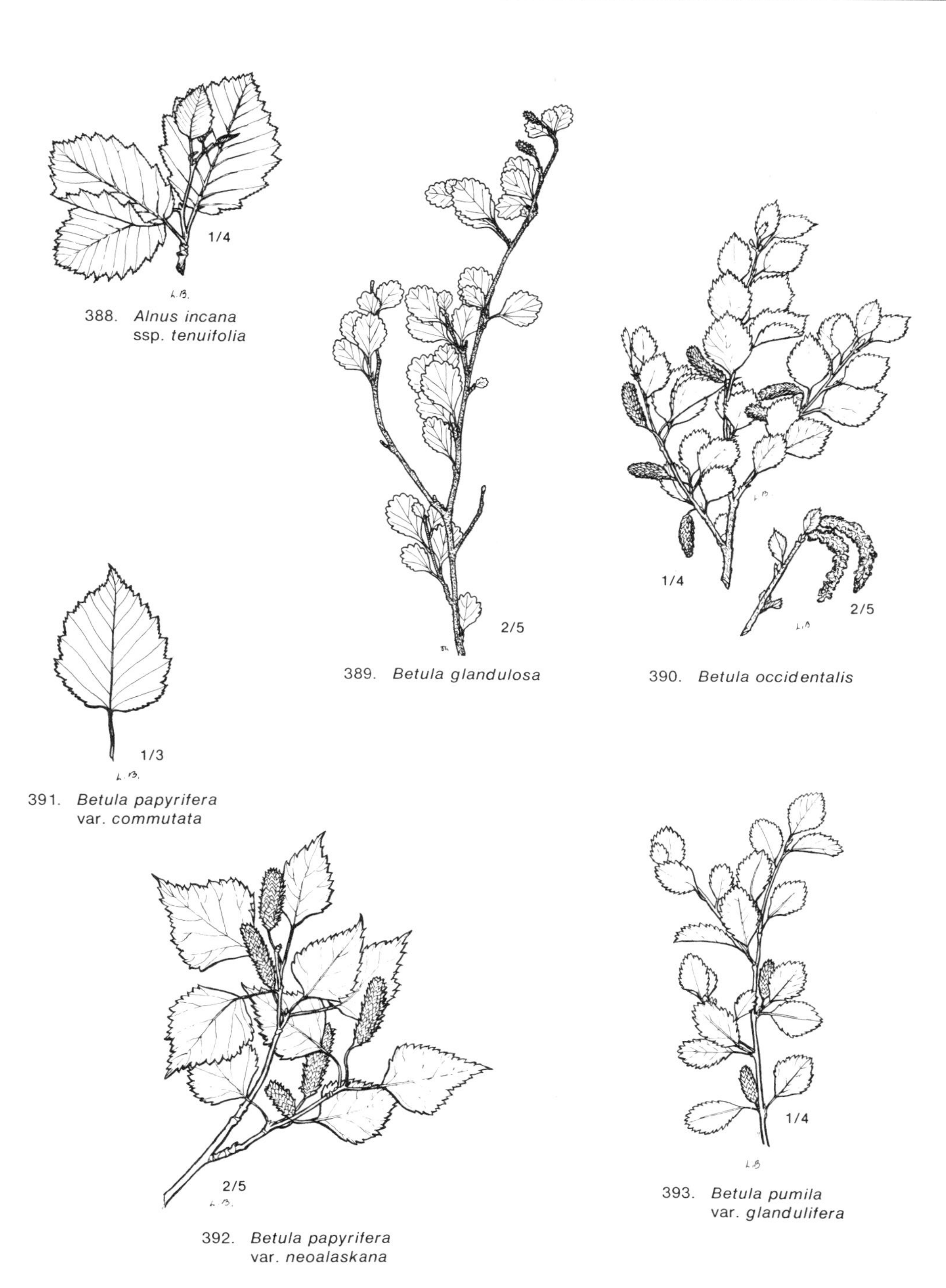

388. *Alnus incana* ssp. *tenuifolia*

389. *Betula glandulosa*

390. *Betula occidentalis*

391. *Betula papyrifera* var. *commutata*

392. *Betula papyrifera* var. *neoalaskana*

393. *Betula pumila* var. *glandulifera*

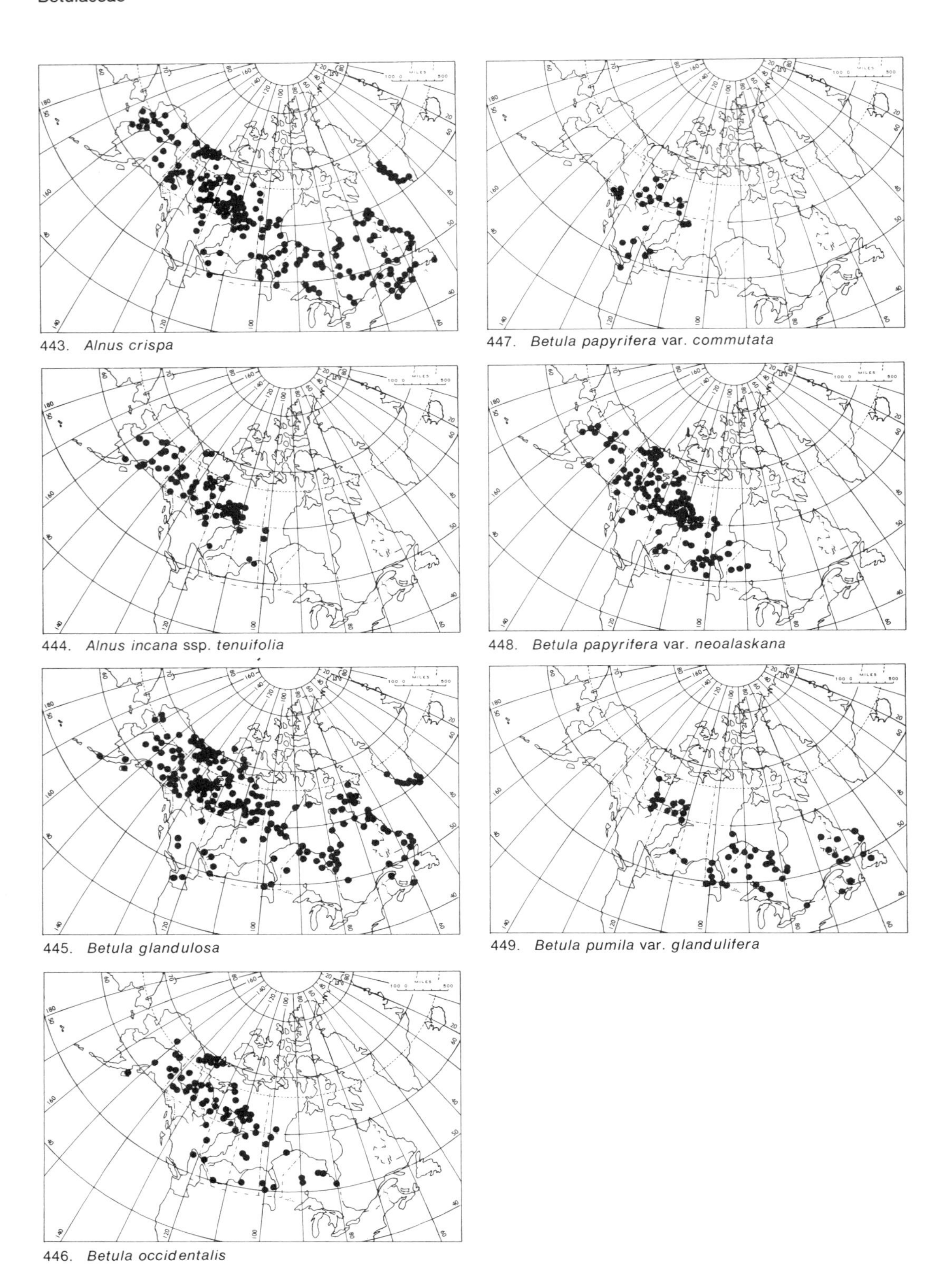

443. *Alnus crispa*

444. *Alnus incana* ssp. *tenuifolia*

445. *Betula glandulosa*

446. *Betula occidentalis*

447. *Betula papyrifera* var. *commutata*

448. *Betula papyrifera* var. *neoalaskana*

449. *Betula pumila* var. *glandulifera*

URTICACEAE Nettle Family

Urtica L. Nettle

Urtica gracilis Ait.
Perennial herb with tough, simple or branched stems, and stinging hairs, the latter mainly on the leaves which are narrowly lanceolate, coarsely toothed, commonly two or more, clustered at the nodes, the lowermost with blades up to 15 cm long; flowers small and green, the staminate and pistillate commonly in separate, elongated and drooping racemes from the upper nodes.

Thickets and rich damp soil; although anthropochorus, and in the District of Mackenzie reported mainly from places where live-stock has been regularly kept for more than a century, some collections, at least from the upper Mackenzie Valley, undoubtedly represent indigenous plants.

General distribution: Nfld. to Alaska.

Fig. 394 Map 450

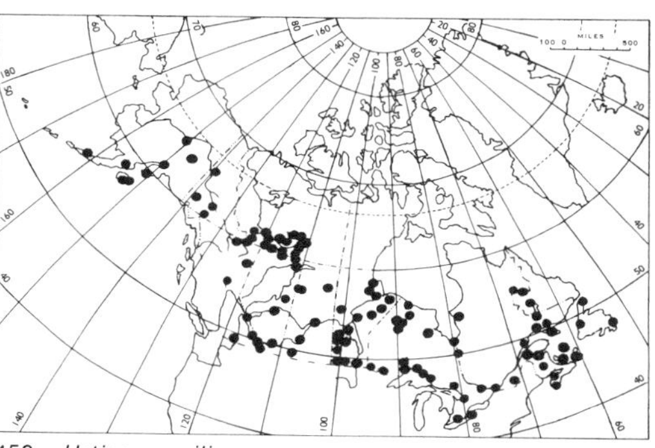

450. *Urtica gracilis*

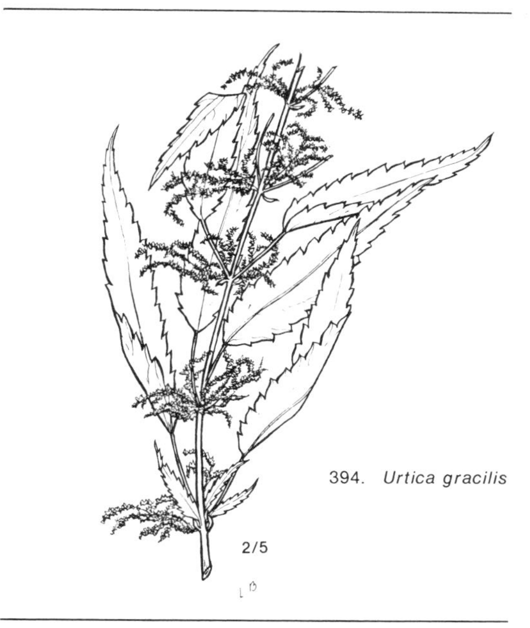

394. *Urtica gracilis*

2/5

SANTALACEAE Sandalwood Family

Ours herbs with extensive creeping subterranean rhizomes and simple, erect leafy stems; leaves alternate, entire, lanceolate; flowers small, mostly perfect. Fruit drupaceous, juicy or dry.

a. Flowers in terminal cymes or panicles; fruit dry, nut-like . . . . *Comandra*
a. Flowers axillary; fruit drupaceous, red and juicy . . . . *Geocaulon*

Comandra Nutt. Bastard-toadflax

***Comandra pallida** A. DC.
The only evidence of this species occurring within our area is an undated specimen now in the National Herbarium (CAN), by Geo Lawson

labelled: "Comandra umbellata Nutt. On the Anderson River and Fort Hope, McT. [McTavish]".

A prairie and foothill species, in the Northwest otherwise known from a single station in central Yukon Territory, and from widely disjunct stations on the upper Nelson and Peace Rivers in British Columbia and from Lake Athabaska and Wood Buffalo Park, Alberta.

General distribution: Western N. America.

Fig. 395 Map 451

Geocaulon Fern. Northern Comandra

Geocaulon lividum (Richards.) Fern.
A woodland species of mossy flood-plain spruce and poplar forest common northward in the Mackenzie drainage to about Lat. 65° N. and again locally at the head of the Mackenzie Delta, in Lat. 68° 20′.

General distribution: Boreal America, from Nfld. to Alaska.

Fig. 396 Map 452

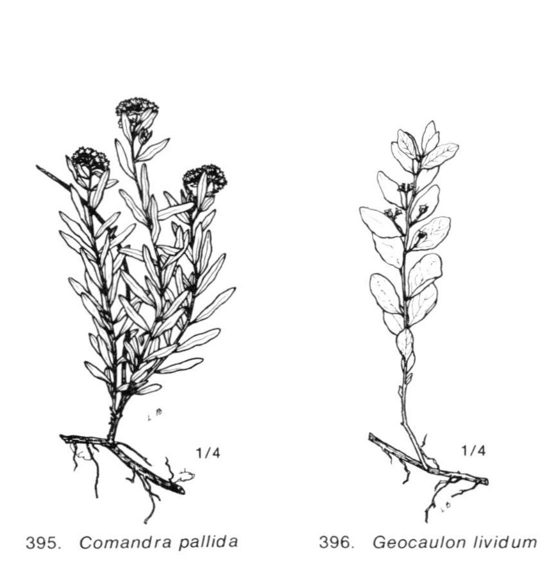

395. *Comandra pallida* 396. *Geocaulon lividum*

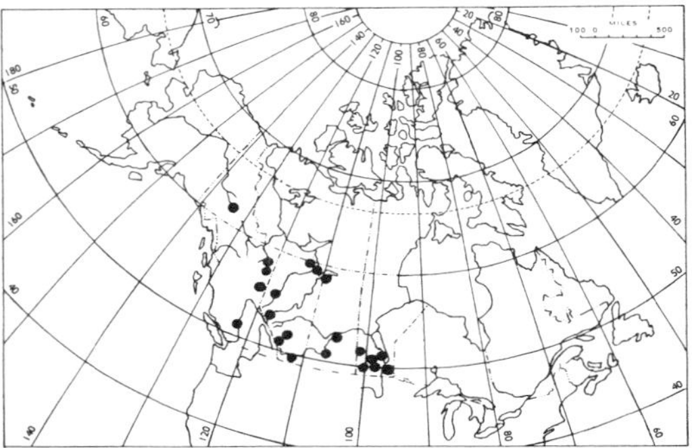
451. *Comandra pallida*

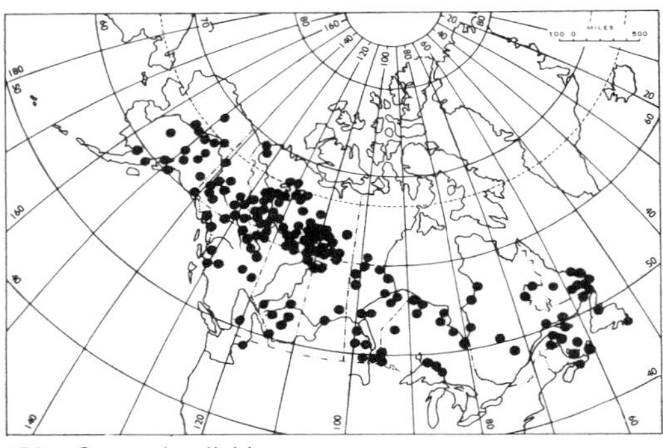
452. *Geocaulon lividum*

POLYGONACEAE Buckwheat Family

Ours herbs with alternate, simple leaves, and membranaceous, sheathing stipules above the swollen joints of the stems; flowers small, perfect and regular, in spike-like racemes, panicles or axillary clusters; petals lacking, sepals 3 to 6, sometimes petaloid; stamens 4 to 12; the ovary superior, 1-locular; styles 1 to 3 and the fruit an achene.

a. Sepals 3; diminutive arctic or high-alpine annual, often only a few centimetres high, with slender, simple or branched stems . . . . *Koenigia*
a. Sepals more than 3; larger plants
  b. Stigmas tufted, sepals 4 or 6
    c. Leaves reniform; sepals 4 . . . . *Oxyria*
    c. Leaves not reniform; sepals 6 . . . . *Rumex*
  b. Stigmas capitate; sepals 5, often petaloid . . . . *Polygonum*

Koenigia L.

Koenigia islandica L.
Dwarf annual, often only a few centimetres high, with erect or ascending, filiform, simple or branched stems bearing a few obovate, often reddish-purple, sessile leaves, those near the summit usually arranged in a whorl around the tiny, 3-parted flowers.

In wet moss or springy places, or in wet sand or silt by the edge of small lakes or tarns. One of the few annual members of the arctic flora.

General distribution: Circumpolar, arctic-alpine, with a few widely disjunct, high-alpine stations in the Rocky Mountains.

Fig. 397 Map 453

Oxyria Hill Mountain Sorrel or Scurvy-Grass

Oxyria digyna (L.) Hill
Perennial herb with a short, fleshy rhizome bearing clusters of long-petioled, reniform, fresh green somewhat succulent leaves. The scapose or few-leaved flowering stems 10 to 30 cm high, terminating in an open, panicled raceme of small reddish-green or crimson flowers on slender pedicels. The achene lenticular, thin and flat, much larger than the calyx and surrounded by a broad, translucent wing.

Common in cool, moist ravines where the snow remains late. The species responds readily to fertilizer as shown by its luxurient growth on moist ledges below bird cliffs, and near human habitations.

The succulent, mildly acid leaves and young stems of the mountain sorrel are edible and may be eaten raw as a salad, cooked as a green or pot-herb, or stewed as rhubarb. Owing to its wide distribution and local abundance, mountain sorrel is among the most important edible plants of the Arctic, where it is gathered regularly by the Eskimo who eat the leaves fresh, or preserved with seal oil. The fresh leaves and stems of the mountain sorrel are eagerly sought by caribou, muskoxen, and geese, whereas arctic hares and lemmings prefer the fleshy rhizomes.

General distribution: Circumpolar, widely distributed arctic-alpine.

Fig. 398 Map 454

Polygonum L. Knotweed

Perennial or annual, mostly terrestrial but some aquatic herbs, commonly with swollen joints; leaves alternate and entire, the flowers small, mostly perfect, spicate-racemose or axillary; calyx mostly 5-parted, stamens 3 to 9; stigmas 2 to 3 and the achenes lenticular or triangular.

a. Perennials
  b. Stems simple, unbranched; terrestrial plants from a corm-like, short rhizome; inflorescence spike-like
    c. Spike narrow of normal (but often sterile) white or pinkish flowers in the upper part, and bulbils in the lower . . . . *P. viviparum*
    c. Spike usually over 1 cm thick, short-cylindric, always lacking bulbils; flowers commonly deep pink . . . . *P. Bistorta* ssp. *plumosum*

- b. Stems freely branching; terrestrial or marsh plants
  - d. Aquatic, amphibious or marsh plants with spike-like racemose inflorescence
    - e. Aquatic or amphibious plants commonly with floating, elliptic-oblong to lanceolate, blunt-tipped leaves on glabrous peduncles . . . . *P. amphibium* var. *stipulaceum*
    - e. Marsh plants; leaves lanceolate-acuminate, never floating; peduncles pubescent . . . . *P. coccineum*
  - d. Terrestrial, coarse plants with up to 2 m tall simple or branched stems; leaves lanceolate-acuminate, 5-10 cm long; inflorescence in open, many-flowered panicles from the upper leaf-axils . . . . *P. alaskanum*
- a. Annuals, often of weedy habit
  - f. Stems erect-ascending; leaves lanceolate
    - g. Racemes often drooping; flowers pale pink; achenes ca. 2 mm long . . . . *P. lapathifolium*
    - g. Racemes strictly erect; flowers mostly greenish; achenes ca. 3 mm long . . . . *P. scabrum*
  - f. Stems mostly low, prostrate, erect-ascending or twining; inflorescence few-flowered, in axillary clusters
    - h. Stems twining; leaves ovate-sagittate . . . . *P. Convolvulus*
    - h. Stems prostrate or ascending, not twining; leaves elliptic-oblong
      - i. Leaves 2-4 mm wide, short-petioled; stems filiform; native species . . . . *P. caurianum*
      - i. Leaves 3-8 mm wide, sessile; stems coarse; introduced, weedy species
        - j. Stems smooth; leaves oblong . . . . *P. aviculare*
        - j. Stems striate; leaves oval . . . . *P. achoreum*

Polygonum achoreum Blake
Striate Knotweed
Similar in habit to *P. aviculare*, but a coarser plant with conspicuously fluted or striate stems, and larger, firm, oval or obovate leaves that commonly are rather densely crowded on the branches.

Common by roadsides and in waste places near settlements, along the Slave and upper Mackenzie Rivers where probably a recent introduction.

General distribution: N. America, from Lab. to Alaska; non arctic.

Polygonum alaskanum (Small) Wight
P. alpinum All. var. *lapathifolium* Cham. & Schlecht.
Essentially glabrous perennial from a much branched crown and woody rhizome; stems simple or branching, occasionally up to 2 m tall; leaves all cauline, short-petioled or sub-sessile, the blade lanceolate or lanceolate-oval, 5 to 20 cm long including the attenuate tip, strongly crisped along the margins, dark green above, paler beneath; stipules sparingly hirsute, papery and pale brown, 1.5 to 2.0 cm long; flowers small, numerous, with greenish-white perianth, in terminal or axillary panicles.

A pioneer species on freshly exposed clay of recent landslides where it may form pure stands. The succulent young stems and leaves are edible and provide an acceptable substitute for fresh rhubarb.

General distribution: Sub-arctic-alpine, Alaska, barely reaching northwest District of Mackenzie.

Fig. 399 Map 455

Polygonum amphibium L.
var **stipulaceum** (Coleman) Fern.
Water Smartweed
Perennial marsh or aquatic species with a thick sub-woody base from which rise leafy flowering stems; the leaves oblong-ellpitic, long-petioled, 8 to 14 cm long, somewhat leathery; flowers commonly pink in 1- to several panicles of cylindric spikes, rarely over 3 cm long.

Not uncommon by the edge of shallow ponds and on muddy lakeshores, northward in the upper Mackenzie Valley and Great Bear Lake Region, nearly to the Arctic Circle, with disjunct stations in the Mackenzie Delta.

General distribution: N. America; Nfld. to Alaska.

Fig. 400 Map 456

Polygonum aviculare L. *s. lat.*
Knot-weed
Much branched prostrate or ascending annual with elliptic-oblong, alternate and sessile blue-

green 1 to 2 cm long leaves; flowers sessile and small, solitary or few together; achene about 2.5 mm long enclosed in the persistent calyx.

This, and probably other introduced and weedy members of the Section *Avicularia*, is common near settlements and towns along the MacKenzie River northward to the Delta.

General distribution: Circumpolar—Cosmopolitan.

Polygonum Bistorta L.
ssp. **plumosum** (Small) Hult.
Similar to the circumpolar and wide-ranging *P. viviparum* but a much stouter plant with a thicker spike composed entirely of fertile and nearly always deep pink flowers. As in *P. viviparum* the starchy rhizome is edible.

Moist, peaty tundra.

General distribution: Amphi-Beringian: Over Alaska and unglaciated central and northern Y.T., eastward slightly beyond the Mackenzie Delta.

Fig. 401 Map 457

Polygonum caurianum Robins.
Similar in habit and appearance to *P. aviculare* but stems more slender, the leaves smaller and narrower, and usually with a short but distinct petiole; the flowers smaller and the calyx commonly pink.

P. caurianum is probably the only member of the Sect. *Avicularia* which is indigenous to the District of Mackenzie where it inhabits wet, gravelly pond margins and lake shores.

General distribution: N. America, subarctic-alpine.

Fig. 402 Map 458

Polygonum coccineum Muhl.
Similar to *P. amphibium* var. *stipulaceum* but readily distinguished by its up to 12 cm long spikes on pubescent and glandular peduncles, and by its longer acuminate rather than blunt-tipped leaves.

Stream banks and wet meadows.

General distribution: N. American, non arctic species which barely reaches the District of Mackenzie on the south shore of Great Slave L.

Fig. 403 Map 459

Polygonum Convolvulus L.
Bind-weed
Weedy annual with simple or branched twining stems; leaves petioled, the blade sagittate, 2 to 6 cm long and 1 to 4 cm broad, alternate at the nodes; flowers small, in axillary clusters or open racemes; achene 3-angled, 3 to 4 mm long.

Occasional in gardens and on cultivated land near settlements along the Mackenzie, north at least to Fort Simpson.

General distribution: Cosmopolitan.

Polygonum lapathifolium L. *s. lat.*
Branched weedy annual with ascending or sometimes prostrate stems; leaves petioled, the blade narrowly lanceolate and usually 4 times longer than broad, light green on both sides; flowers small, pale pink, in slender cylindrical, erect or spreading axillary spikes.

Waste places and roadsides of the upper Mackenzie drainage.

General distribution: Circumpolar.

Map 460

Polygonum scabrum Moench
Branching weedy annual resembling *P. lapathifolium* but of more stiffly erect habit with stems 2 to 5 dm high; leaves short-petioled, the blade lanceolate 10 to 20 cm long and about one third as wide, dark green above, paler beneath; flowers small, with greenish-white perianth, in densely cylindrical, 2 to 3 cm long, axillary and erect spikes.

Introduced roadside weed.

General distribution: Circumpolar .

Polygonum viviparum L.
Bistort
Perennial herb with a short, often contorted or twisted, starchy rhizome terminating in a cluster of slender-petioled oblong-lanceolate, dark green, somewhat shiny leaves. Stem solitary, simple, from a few to 30 cm high, usually bearing one or two short-petioled or sessile leaves, and terminating in a 2 to 10 cm long spike in which the lowermost flowers are replaced by bulbils, that often sprout while still on the mother plant. The white or pale pink flowers of the upper half of the spike appear normal, but are mostly sterile, but may occasionally produce mature and viable achenes.

Common or even ubiquitous, in turfy places, rocky barrens, as well as in moist grassy herbmats. It strongly favours manured places and attains luxuriant growth near human habitations, animal dens, or below bird cliffs.

The pecan-sized, starchy and slightly astringent rhizome of the bistort may be eaten raw, but is more palatable when cooked. It was formerly a choice delicacy of the Eskimo who preserved the rhizomes by freezing, or in seal oil. The bulbils are eaten by ptarmigan and by lemmings.

General distribution: Circumpolar, widely distributed arctic-alpine.

Fig. 404 Map 461

Rumex L. Dock

Ours coarse and mostly perennial, glabrous herbs with alternate, entire and often large leaves; the flowers small, axillary or spicate; the achenes 3-angled, in some with a grain-like tubercle.

a. Flowers dioecious; basal leaves hastate or sagittate, somewhat succulent . . . . *R. Acetosa* ssp. *alpestris*
a. Flowers perfect or monoeciously polygamous; leaves not hastate or sagittate, nor succulent
  b. Annual or biennial with soft hollow stems; the inner sepals (valves) with marginal spinules . . . . *R. maritimus* var. *fueginus*
  b. Perennial, valves with entire or merely denticulate margins
    c. Stems erect from a stout branching root
      d. Leaves narrowly oblong, mostly with an obtuse tip and a cuneate base
        e. Plant strongly tinged with purple; alpine or subarctic . . . . *R. arcticus*
        e. Plant dark green, non arctic . . . . *R. orbiculatus*
      d. Leaves oblong, usually with a truncate base and acute tip . . . . *R. occidentalis*
    c. Stems ascending to depressed-branching, from a horizontally spreading rhizome . . . . *R. triangulivalvis*

Rumex Acetosa L.
ssp. **alpestris** (Scop.) Löve
Glabrous perennial commonly 3 to 6 dm tall from a short, ascending rhizome; cauline leaves subsessile, the lower long-penduncled; the blade hastate, broadly lanceolate-ovate, with a broad sinus; inflorescence narrowly paniculate, 1 to 2 dm long; flowers small, the staminate soon falling; in the pistillate flower the valves enclose the maturing achene of which the wing is broad, reticulate-veined and wine-coloured.

Indigenous and commonly growing in moist, alpine or subalpine meadows but within our area known only from a single station in the Richardson Mountains west of the Mackenzie Delta.

R. Acetosa L. ssp. *Acetosa* (Garden Sorrel) is similar to ssp. *alpestris* but the leaf-blade with a narrower sinus; it is a naturalized European potherb, and is wide-ranging across Canada north to the limits of agriculture.

General distribution: Amphi-Beringian.

Fig. 405 Map 462

Rumex arcticus Trautv.
Arctic Dock
Glabrous perennial with erect, simple stems from a stout, fleshy rhizome; leaves mostly basal, the blade dark green and somewhat fleshy, oblong-oval to narrowly lanceolate, 7 to 30 cm long and 2 to 5 cm wide. Flowering stems 25 to 100 cm high, terminating in a simple or short-branched panicle of small, reddish flowers.

Common locally in damp, turfy places.

General distribution: Amphi-Beringian, arctic-alpine, common eastward to Coronation Gulf, and with a widely disjunct station on the headwaters of Athabasca R. in Jasper Park, Alta.

Fig. 406 Map 463

Rumex maritimus L.
var **fueginus** (Phil.) Dusén
Annual, or with us probably always biennial; stems erect, thick and hollow, simple or branching, reddish-purple, rarely over 6 dm tall but often much less; leaves linear-oblong with a truncate base; the flowers small, 2 to 3 mm long, in tight, reddish-brown glomerules.

Apparently rare and local by saline or alkaline seepages north to lat. 62° on the Mackenzie River.

General distribution: N. America; Que. to B.C.

Fig. 407 Map 464

Rumex occidentalis S. Wats.
Stout, essentially glabrous perennial, with yellowish-green stems 1 m tall, or taller; leaves mainly basal, long-petioled, the blade oblong-lanceolate, up to 30 cm long, cordate at base, crisp-margined; inflorescence paniculate, very large, with erect branches; flowers verticillate,

very numerous, nodding or reflexed, on 5 to 7 mm long, filiform pedicels; the mature valves membranaceous, reddish-brown, reticulate-veiny.

In marshy places, north along the Mackenzie River to lat. 65°.

General distribution: N. America; Que. to Alaska.

Fig. 408 Map 465

Rumex orbiculatus Gray

Stems stout, dark green, and erect, 1 to 2 m tall; leaves oblong-lanceolate, with entire or finely crenulate margins, the lower long-petioled; valves orbicular, all grain-bearing.

In swamps and wet meadows.

General distribution: N. America, from Nfld. to North Dakota, barely entering the District of Mackenzie between Ft. Smith and Hay River.

Map 466

Rumex triangulivalvis (Danser) Rech. f.

Decumbent, ascending or even erect, glabrous perennial from a branching tap-root; stems leafy below, 3 to 6 dm tall; the leaves linear-lanceolate, 5 to 15 cm long and 1.5 to 3.0 cm wide; inflorescence up to 3 dm long, with ascending branches; the small, reddish-purple flowers in whorls; valves deltoid-cordate and strongly reticulate, each with a large, lanceolate grain.

Common locally on moist river banks and lake shores along the Mackenzie River to its delta, and along some of its tributaries.

In the District of Mackenzie this plant has been dealt with, by different authors as *R. salicifolius* (Hooker, 1838), *R. mexicanus* (Raup, 1936 and 1947), *R. pallidus* (Porsild, 1943), *R. sibiricus* (Hultén, 1944 and 1947) and as *R. triangulivalvis* (Cody, 1960).

General distribution: Nfld. to Alaska, southward to New Mexico and California.

Fig. 409 Map 467

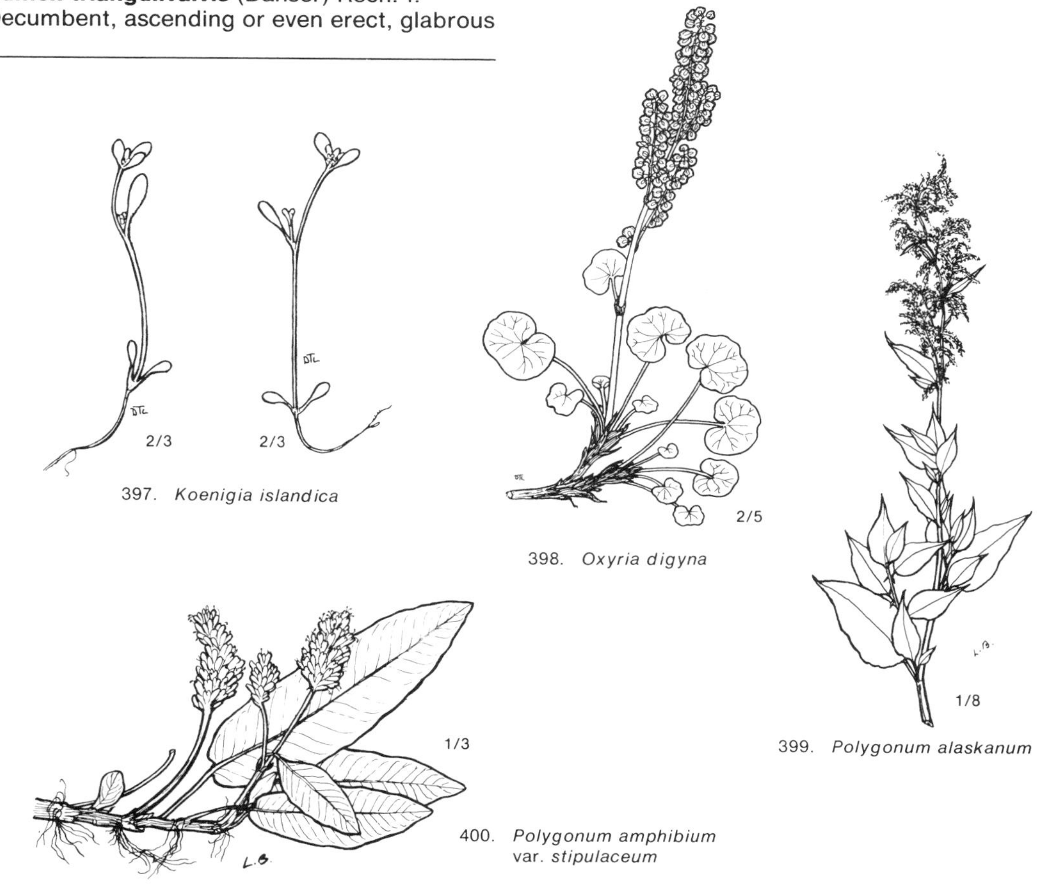

397. *Koenigia islandica*

398. *Oxyria digyna*

399. *Polygonum alaskanum*

400. *Polygonum amphibium* var. *stipulaceum*

401. *Polygonum Bistorta* ssp. *plumosum*

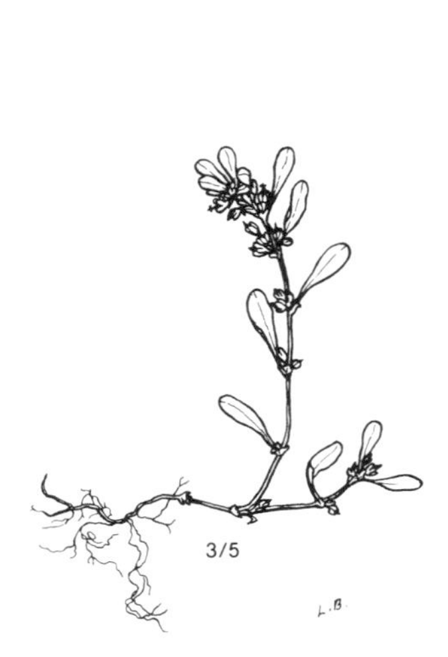

402. *Polygonum caurianum*

403. *Polygonum coccineum*

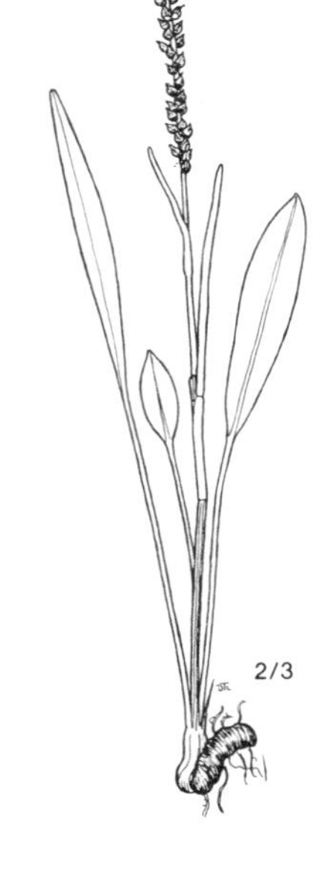

404. *Polygonum viviparum*

405. *Rumex Acetosa* ssp. *alpestris*

406. *Rumex arcticus*

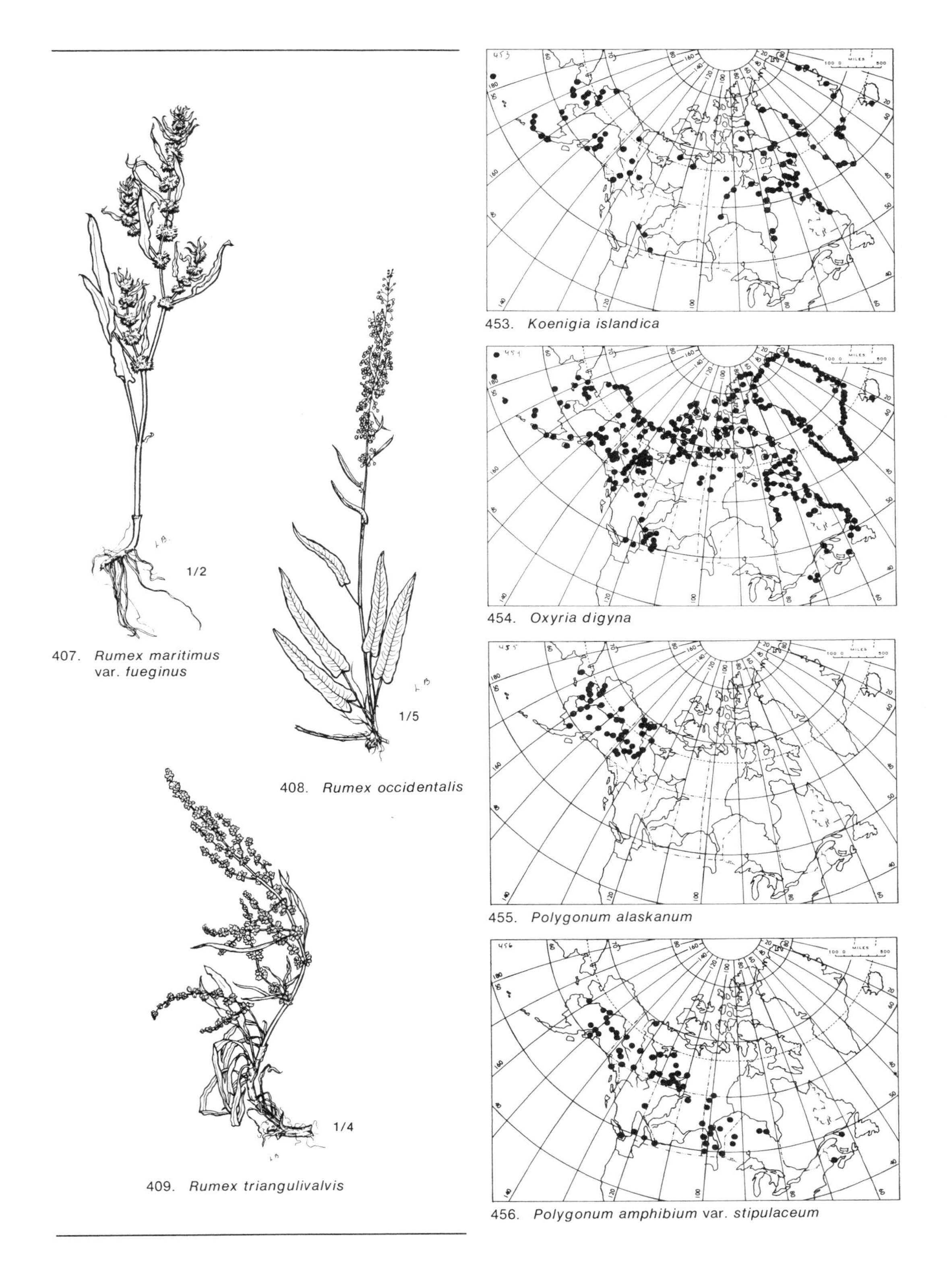

407. *Rumex maritimus* var. *fueginus*

408. *Rumex occidentalis*

409. *Rumex triangulivalvis*

453. *Koenigia islandica*

454. *Oxyria digyna*

455. *Polygonum alaskanum*

456. *Polygonum amphibium* var. *stipulaceum*

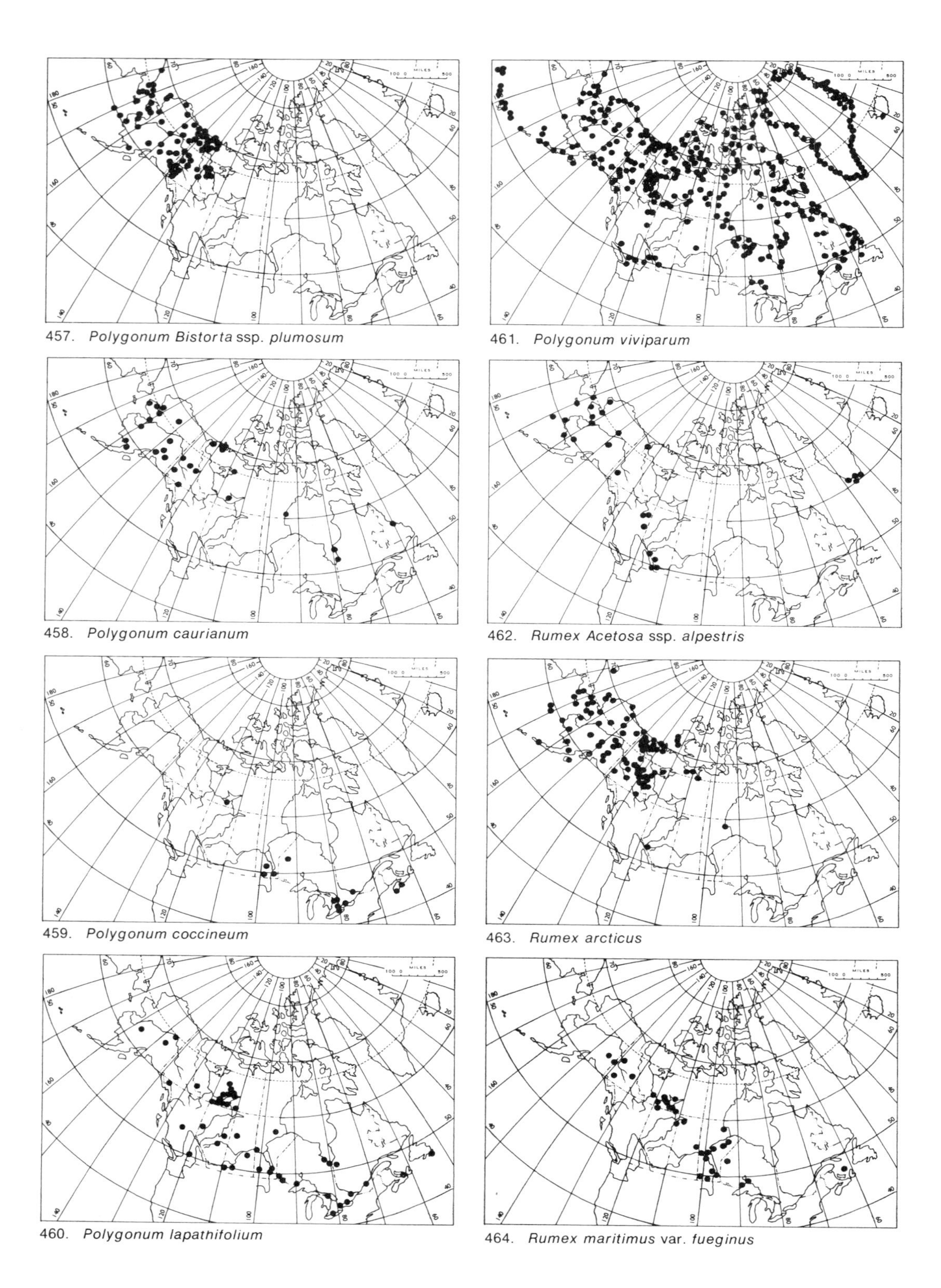

457. *Polygonum Bistorta* ssp. *plumosum*

458. *Polygonum caurianum*

459. *Polygonum coccineum*

460. *Polygonum lapathifolium*

461. *Polygonum viviparum*

462. *Rumex Acetosa* ssp. *alpestris*

463. *Rumex arcticus*

464. *Rumex maritimus* var. *fueginus*

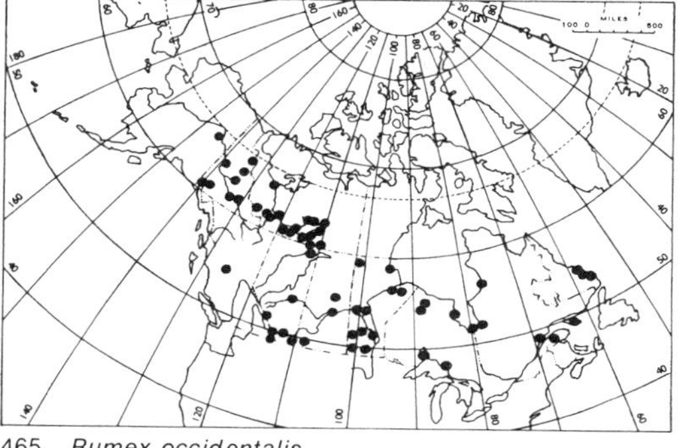
465. *Rumex occidentalis*

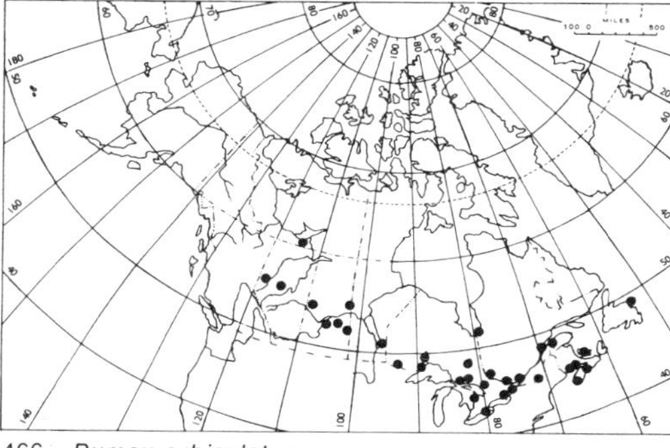
466. *Rumex orbiculatus*

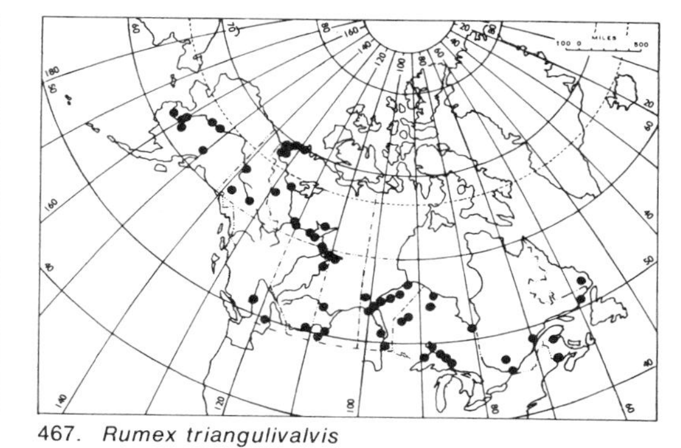
467. *Rumex triangulivalvis*

CHENOPODIACEAE Goosefoot Family

Ours mostly annual herbs, many of them of weedy habit; leaves simple, lacking stipules, mostly alternate; flowers small and inconspicuous, greenish, commonly arranged in clusters or in compound inflorescences; perianth of 5 (or fewer) sepals; stamens 1 to 5; pistil 1; fruit small, one-seeded, commonly enclosed by the persistent calyx, or by bracts.

a. Stems jointed; leaves succulent, scale-like; flowers axillary, sunk into the fleshy stem . . . . *Salicornia*
a. Stems not jointed
  b. Flowers monoecious
    c. Pistillate flowers lacking calyx; fruit enclosed by bract
      d. Stigmas 2 . . . . *Atriplex*
      d. Stigmas 4 or 5 . . . . *Spinacia*
    c. Pistillate and staminate flowers with calyx; fruit lacking bract . . . . *Axyris*
  b. Flowers perfect
    e. Fruit mostly free, not enclosed by the calyx lobes
      f. Flowers appearing in the axils of the upper and mostly hastate leaves . . . . *Monolepis*
      f. Flowers with non-leafy bracts; leaves linear, entire . . . . *Corispermum*

e. Fruit mostly enclosed in the persisting calyx
  g. Leaves fleshy, glabrous, narrowly linear or oblong; dwarf, non-weedy annual of damp, saline places . . . . *Suaeda*
  g. Leaves not fleshy, entire or toothed, glabrous or mealy-puberulent annuals, mostly of weedy habit . . . . *Chenopodium*

Atriplex L. Orach

Ours annual, with triangular or oblong to lanceolate leaves and small monoecious flowers in dense, axillary clusters, the pistillate with bracteoles which enlarge and become hardened in fruit, and are often rough from sharp projecting points (tubercules).

a. Plants small (to 7 cm tall with us); leaves oblong to lanceolate . . . . *A. Gmelinii*
a. Plants much taller; leaves lanceolate or ovate to triangular, frequently hastate
  b. Pistillate flowers of two kinds, some with lobed calyx, lacking bracteoles, and some without calyx but enclosed between two flat, round or oval bracteoles . . . . . *A. hortensis*
  b. Pistillate flowers all alike; with bracteoles but lacking calyx
    c. Bracteoles broadly triangular to ovate triangular bearing one or more dorsal tubercules . . . . *A. subspicata*
    c. Bracteoles triangular to triangular-hastate, lacking tubercules . . . . *A. patula*

Atriplex Gmelinii C.A.Meyer
A. glabriuscula sensu Porsild and Cody (1968).
With us a small, 3 to 7 cm tall, branched semi-erect annual with oblong to lanceolate mostly alternate grey-green leaves. Flowers in small glomerules, supported by leafy bracts. Bracteoles oblong to ovate or orbicular, the lower entire or with a few slender teeth at the base, the dorsal surface smooth or with a few minute tubercules.

An obligate littoral species in our area known only from gravelly beaches at the mouth of the Anderson River on the Arctic Coast. Its nearest known station is on the Bering Sea coast of Alaska, disjunct by more than 1000 km.

General distribution: Amphi-Beringian.

Map 468

Atriplex hortensis L.
Garden Orach
Coarse, weedy annual 5 to 7 dm tall; leaves yellowish-green to reddish, petioled, the blade 5 to 20 cm long, deltoid, with cuneate, truncate or cordate base and entire or somewhat sinuate margins; inflorescence of broad spikes in an open panicle; bracteoles sub-rotund, flat, in maturity 1.0 to 1.5 cm broad.

Occasional escape from cultivation.

General distribution: Native of Asia.

Atriplex patula L.
Coarse glabrous (or commonly somewhat mealy when young) weedy annual, erect or prostrate with branching stem up to 1 m long, but with us usually much lower, with green, triangular or halberd-shaped leaves, 5 to 10 cm long, on petioles about half as long as the blade; flowers small and scaly, clustered in slender axillary spikes; bracteoles triangular to triangular-hastate, lacking tubercules.

Somewhat saline or disturbed peaty situations in southern District of Mackenzie where it is presumably introduced. When boiled and seasoned the succulent young leaves of *A. Gmelinii* and *A. patula* are edible and very tasty.

General distribution: Circumpolar; wide spread.

Atriplex subspicata (Nutt.) Rydb.
Erect or semi-erect branching annual to 3 dm or more tall (taller southwards) with grey-green (mealy when young) triangular leaves; bracteoles broadly triangular to ovate-triangular with entire margins or occasionally short sharp teeth, and bearing one or more tubercules on the dorsal surface.

Saline or alkaline soils; in our area known only from the Salt Plain west of Fort Smith.

General distribution: N. America; Nfld. to B.C.

Map 469

Axyris

Axyris amaranthoides L.
Russian Pigweed
Simple to much branched smooth annual 4 to 6 dm high with green, striped stems and alternate, short-petioled, lanceolate-elliptic leaves with entire or sinuate-dentate margins; flowers small, the staminate in terminal spikes, the pistillate axillary or, sometimes, mixed with the staminate.
Introduced weed reported from Enterprise on the Hay River Highway where, obviously, a recent and, likely, ephemeral introduction.
General distribution: Eurasian.

Chenopodium L. Goosefoot

Ours annuals of weedy habit, some glabrous and some mealy-coated; flowers small, sessile, green or reddish, perfect, in small glomerules aggregated in interrupted axillary spikes or panicles; calyx 5-lobed, persisting, enveloping the single, lenticular seed.

a. Plants glabrous, green or reddish
  b. Calyx fleshy, at maturity bright red and juicy . . . . *C. capitatum*
  b. Calyx not fleshy or red
    c. Stamens 1-2; leaves rhombic-ovate, coarsely sinuate-dentate . . . . *C. rubrum*
    c. Stamens 5; leaves ovate with rounded or subcordate base and triangular lobes . . . . *C. gigantospermum*
a. Plants grey or glaucous, mealy at least on the underside of the leaves; leaves linear to ovate, with broadly triangular-dentate or entire margins
  d. Plants prostrate . . . . *C. glaucum* var. *salinum*
  d. Plants upright
    e. Leaves pale grey, linear to narrowly oblong-lanceolate, entire, or with a pair of small teeth near the base, thick and firm, densely farinose . . . . *C. leptophyllum*
    e. Leaves glaucous, farinose beneath, rhombic or broadly ovate
      f. Pericarp reticulate, adhering to the 0.8-1.0 mm broad seed . . . . *C. Berlandieri* var. *Zschakei*
      f. Pericarp smooth, not adhering to the 1.0-1.5 mm broad seed . . . . *C. album*

Chenopodium album L.
Lamb's-quarters
Much branched annual, with us rarely over 6 dm tall; leaves with rhombic or lanceolate blade, usually twice as long as the petiole, green above and paler and mealy beneath; the small flowers clustered in spicate terminal and axillary panicles; pericarp smooth; seeds 1.0 to 1.5 mm in diameter, black and shiny.
Introduced weed now well established near settlements north along the Mackenzie River to its delta. Cooked young plants of *C. album* make an acceptable substitute for garden spinach.
General distribution: Circumpolar.
Fig. 410

Chenopodium Berlandieri Moq.
var. **Zschackei** (Murr) Murr
Similar to *C. album* from which it may be distinguished by the reticulate pericarp which adheres to the smaller 0.8 to 1.0 mm broad seed.
Waste situations, clearings and townsites north to latitude 67°22′N.
General distribution: Ontario to B.C. and southward.
Map 470

Chenopodium capitatum (L.) Asch.
Strawberry Blite
Simple or branched annual, in the north rarely over 7 dm high; leaves with triangular or halberd-shaped blade equalling or shorter than the petiole; flowers small, in globose, axillary clusters, the calyx becoming fleshy, bright red and juicy in fruit.
Indigenous, but of weedy habit; common in clearings and often spontaneous in burned-over areas.
The purple juice extracted from the "fruits" of Strawberry Blite was used formerly by Indian tribes of the North as war paint and as a dye, also the young plants are edible when cooked. Through these uses its tiny seeds were readily distributed by man.
General distribution: Amphi-Atlantic; Que. to Alaska.
Fig. 411 Map 471

Chenopodium gigantospermum Aellen
Simple or branched glabrous, fresh green and weedy annual; leaves oval, thin, deeply sinuate-dentate, on long and slender petioles; the flowers mostly solitary in panicles of slender, interrupted spikes; seeds black, lustrous, about 2 mm in diameter.

A recent introduction near settlements in southwest District of Mackenzie.

General distribution: Que. to Y.T.

Fig. 412 Map 472

Chenopodium glaucum L.
ssp. **salinum** (Standl.) Aellen
Low, indigenous and non-weedy annual with glabrous, erect or spreading stems. Leaves alternate, the blade lanceolate, shallowly sinuate-dentate, gradually tapering into the short petiole, glabrous and glaucous above, pale and white-mealy beneath; flowers singly, or in few-flowered axillary glomerules; seeds dark brown and lustrous, about 1 mm in diameter.

Rare or local on moist, saline seepages along river banks north to the Arctic Ocean where fruiting specimens measured only 4 cm in length, with only one or two pairs of leaves.

General distribution: Que. to central Alaska.

Fig. 413 Map 473

***Chenopodium leptophyllum** Nutt.
Slender, erect, simple or branched grey-green annual, usually mealy, especially in the inflorescence and on the underside of the leaves, these sessile or short-petioled, the blade linear-lanceolate or oblong with entire or rarely few-toothed margins; glomerules of flowers in interrupted and erect paniculate spikes.

In the Northwest Territories thus far known only from a Richardson specimen collected a Century and a half ago, at Fort Franklin on Great Bear Lake, where it was surely an ephemeral introduction.

General distribution: A native of saline places in Western Central N. America; in Alta.—Sask. reported north to L. Athabasca.

Fig. 414 Map 474

Chenopodium rubrum L.
Erect, glabrous, often reddish-tinged and somewhat fleshy annual with simple or branched stems, with us rarely more than 3 to 4 dm high; leaves with slender petioles, the blade ovate to deltoid, with deeply sinuate-dentate margins; flowers in small axillary glomerules.

In moist, saline places, and occasionally on disturbed soil, north to Great Slave Lake.

General distribution: Circumpolar.

Map 475

Corispermum L. Bugseed

Corispermum hyssopifolium L.
2 to 3 dm tall annual branched from the base, grey-green villous from simple or variously branched hairs; leaves linear, 2 to 5 cm long and about 2 mm wide; flowers small, perfect, in the axils of short leafy bracts arranged in a short and dense, terminal spike.

Common locally on sand dunes along the upper Mackenzie drainage but thus far collected only a few times north to about lat. 65° N.

General distribution: Circumpolar; Que. to Alaska (Eurasia).

Fig. 415 Map 476

Monolepis Schrad.

Monolepis Nuttalliana (R. & S.) Greene
Glabrous or somewhat fleshy winter annual, in the second year branching from the base; leaves slender-petioled or nearly sessile in the upper half of the branches, the blade triangular-hastate; flowers in small glomerules in the upper leaf axils.

Weedy species of saline areas; with us known from a single collection near Yellowknife where it appeared to be a recent introduction.

General distribution: Native of saline or alkaline soils of N. American prairies and plains.

Map 477

Salicornia L. Glasswort

Salicornia rubra A. Nels.
A tiny succulent annual, with us rarely more than a few centimeters high, with jointed, freely branching stems, commonly turning reddish

when mature; leaves scale-like; flowers mostly perfect, minute and inconspicuous, in spikes at end of the branches.

Common and dominant locally in the Wood Buffalo Park on saline flats where commonly demarcating the upper level of spring inundation. Closely related to, and by some authors considered a western race of the Old World and East American *S. europaea*.

General distribution: Man. to Alaska.

Fig. 416 Map 478

Spinacia L. Spinach

Spinacia oleracea L.
Garden Spinach
Annual or biennial and normally dioecious with somewhat succulent, triangular-hastate leaves; the male flowers in short, moniliform, axillary spikes, the pistillate and sometimes perfect flowers in small globular clusters.

Garden escape.

General distribution: Cosmopolitan.

Suaeda Forsk. Sea-Blite

Suaeda calceoliformis (Hook.) Moq
S. depressa (Pursh) S. Wats.
S. maritima sensu Porsild and Cody (1968).
Freely branched erect or spreading annual with linear, fleshy, blue-green leaves and small perfect flowers 2 or 3 together in the leaf-axils.

Seashores and saline praire depressions.

General distribution: Nfld. to B.C. and north in the District of Mackenzie to the Arctic Coast.

Fig. 417 Map 479

410. *Chenopodium album*

411. *Chenopodium capitatum*

412. *Chenopodium gigantospermum*

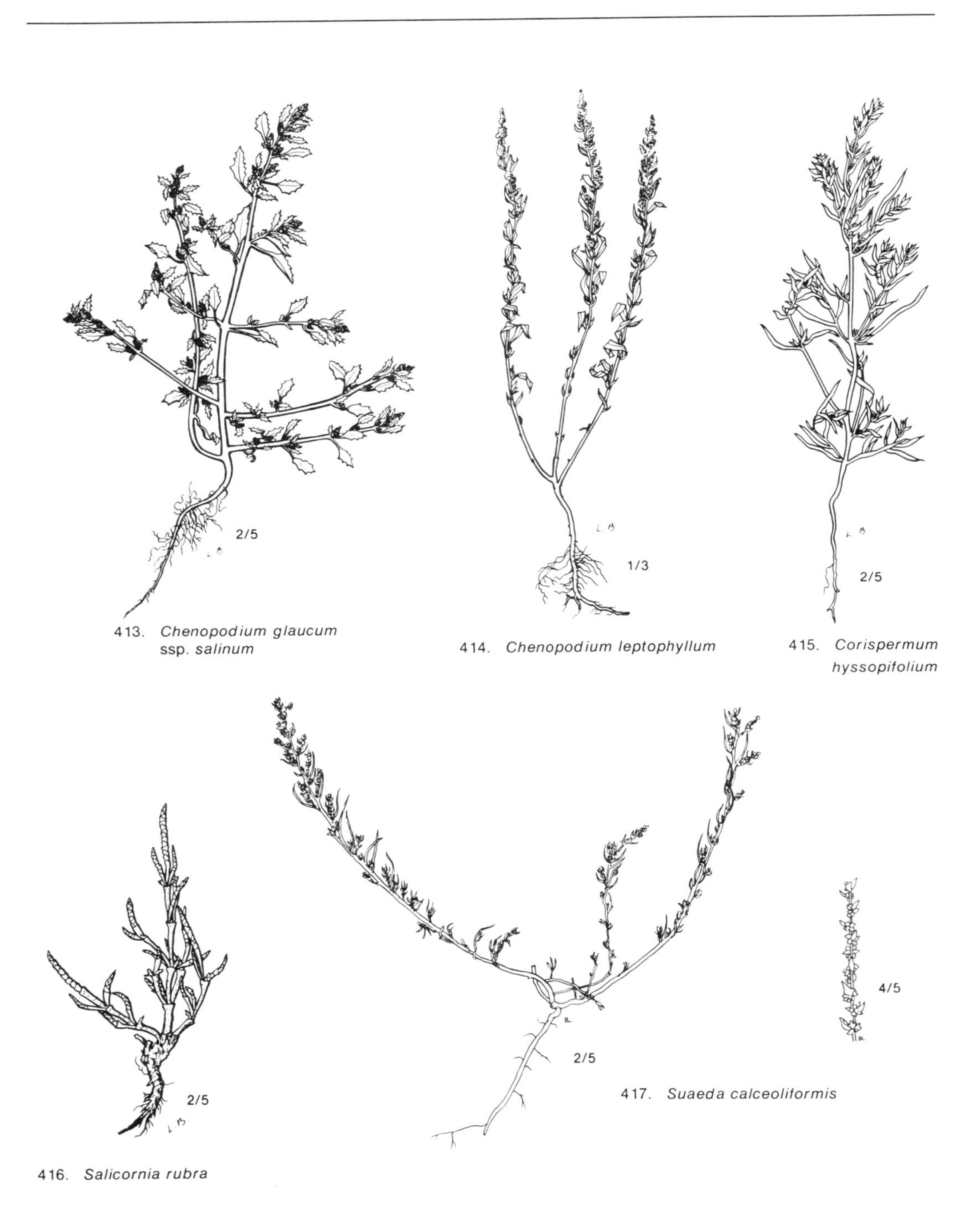

413. *Chenopodium glaucum* ssp. *salinum*

414. *Chenopodium leptophyllum*

415. *Corispermum hyssopifolium*

416. *Salicornia rubra*

417. *Suaeda calceoliformis*

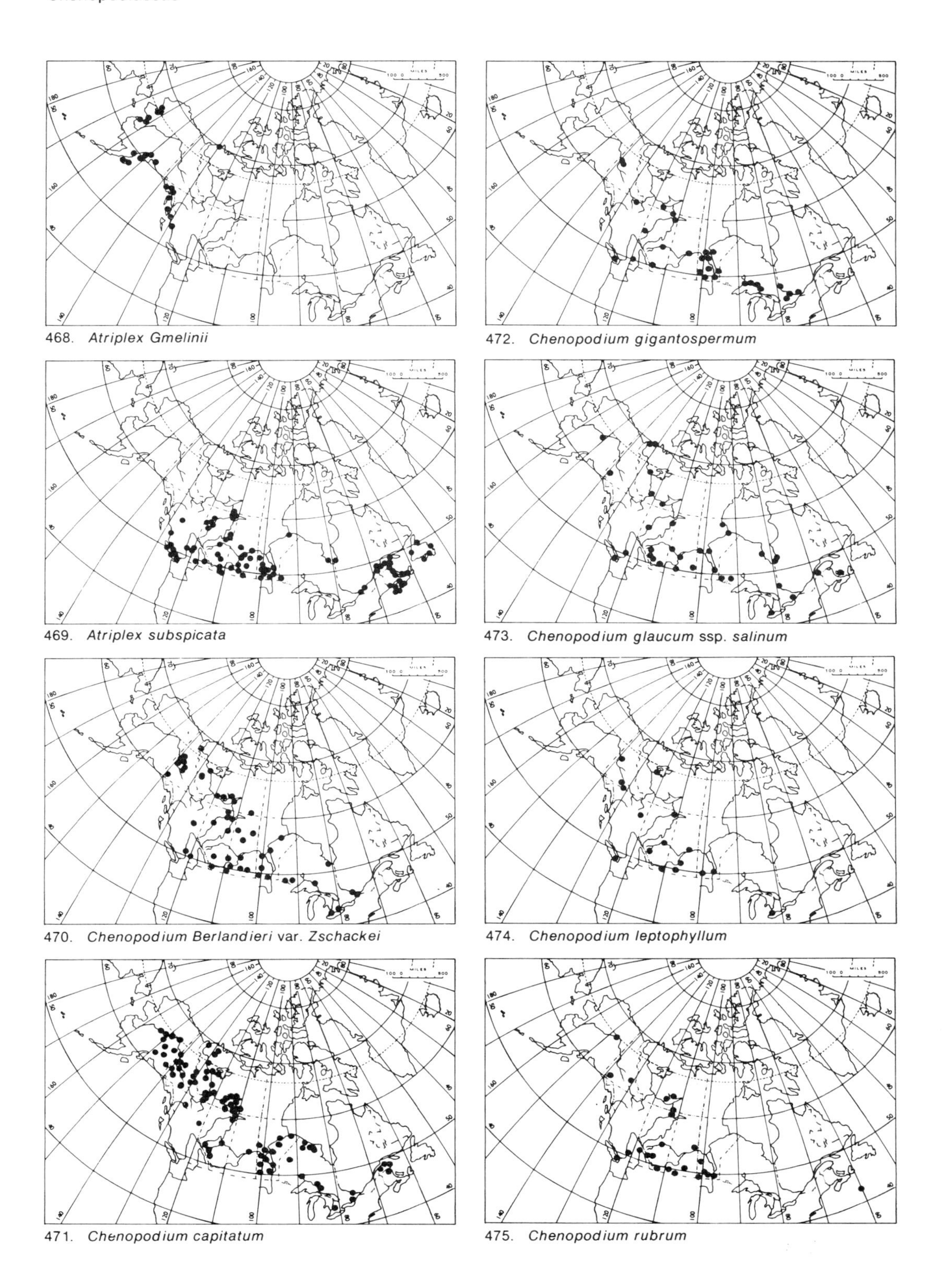

468. *Atriplex Gmelinii*

472. *Chenopodium gigantospermum*

469. *Atriplex subspicata*

473. *Chenopodium glaucum* ssp. *salinum*

470. *Chenopodium Berlandieri* var. *Zschackei*

474. *Chenopodium leptophyllum*

471. *Chenopodium capitatum*

475. *Chenopodium rubrum*

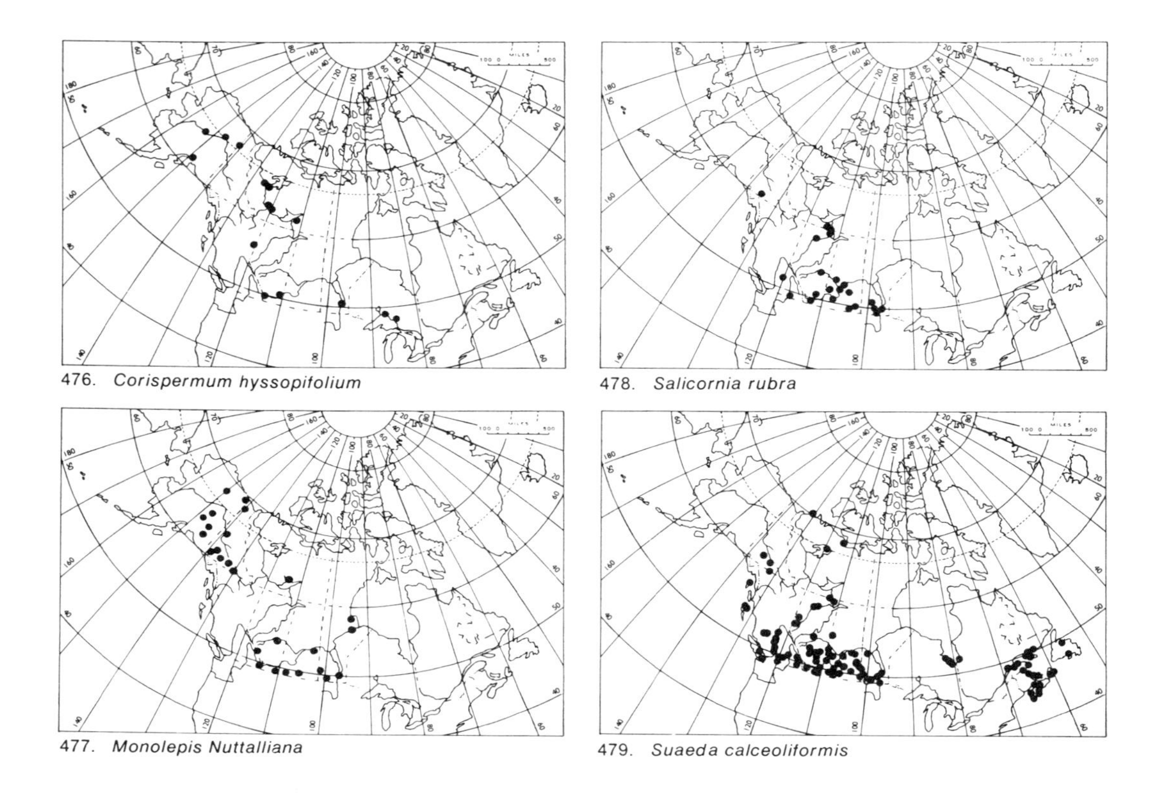

476. *Corispermum hyssopifolium*

478. *Salicornia rubra*

477. *Monolepis Nuttalliana*

479. *Suaeda calceoliformis*

AMARANTHACEAE Amaranth Family

Amaranthus L. Amaranth

Amaranthus retroflexus L.
Red-root Pigweed
Dioecious weedy annual with simple to freely branching and lanate stems; leaves triangular-lanceolate, petioled; the greenish or white, fragrant flowers in terminal clusters.

A recent and no doubt ephemeral adventive.

General distribution: Cosmopolitan weed.

PORTULACACEAE Purslane Family

Perennial or annual, mostly somewhat fleshy herbs with simple leaves; flowers perfect and regular; sepals 2, free or joined at the base; petals and stamens 3 to 5 (in ours), pistil 1, composed of 3 carpels; fruit a capsule dehiscing by 3 valves, containing from 1 to several black and shiny seeds. In the District of Mackenzie represented by the annual *Montia lamprosperma*, and by three perennial species of *Claytonia*.

a. Tiny, weak stemmed annual with several pairs of opposite leaves and minute pedicellate axillary or terminal flowers . . . . *Montia*
a. Perennials, larger, with or without a single pair of opposite leaves and with or without basal leaves; flowers showy, in a terminal raceme . . . . *Claytonia*

Claytonia L.

Perennial herbs acaulescent or with opposite cauline leaves, with or without basal leaves; flowers in a terminal raceme; sepals 2; petals 5, showy, pink or white; stamens 5, attached to the petal claw; fruit a 1-locular, 3-valved, 3-6-seeded capsule.

a. Acaulescent, from a thick, fleshy tap-root terminating in a dense rosette of fleshy leaves . . . . *C. megarrhiza*
a. Caulescent
  b. Slender, delicate stem from a starchy corm . . . . *C. tuberosa*
  b. Matted, from slender horizontal rhizome and filiform runners (the latter often missing in herbarium specimens) . . . . *C. sarmentosa*

Claytonia megarrhiza (A. Gray) Parry
Plant with stout, fleshy tap-root terminating in a dense rosette of short-spatulate, fleshy leaves that commonly over-top the few-flowered racemes; petals white or pale pink, with a yellow claw.

High-alpine, on rather moist scree-slopes; in the District of Mackenzie known from a few collections in the Mackenzie Mountains, widely disjunct from its main range, from Colorado to British Columbia and Alberta.

General distribution: Cordilleran.

Fig. 418 Map 480

***Claytonia sarmentosa** C.A.Mey.
Loosely matted and spreading by slender, subterranean rhizomes as well as by rooting, filiform runners (often lacking in herbarium specimens); basal leaves long-petioled, the blade elliptic-lanceolate, 1 to 2 cm long, those of the flowering stems sessile and opposite; flowers in few-flowered racemes, the petals rose-pink, with a yellow claw.

Alpine herbmats and on snowbeds.

General distribution: Amphi-Beringian, reaching mountains of northern B. C. and southern Y.T.; to be looked for along the east slope of the Richardson and Mackenzie Mts.

Map 481

Claytonia tuberosa Pall.
Stems very slender and brittle, 10 to 15 cm high, from a spherical corm 1.0 to 1.5 cm in diameter normally deeply buried below the surface; basal leaves (often wanting in herbarium specimens) lanceolate, tapering into a narrow petiole; the two stem leaves sessile and opposite well below the racemose inflorescence of 3 to 5 flowers; petals white or more often pink; seeds black and shiny.

The corms are edible and, at least formerly, were preserved in seal oil by the Alaskan Eskimo.

Rather similar to, and perhaps closely related to, the Cordilleran *C. lanceolata* Pursh.

General distribution: An alpine Amphi-Beringian species extending from Alaska and Y.T. to the Richardson and Mackenzie Mts.

Fig. 419 Map 482

Montia L.

Montia lamprosperma Cham.
Blinks
Small, pale-green or yellowish, somewhat succulent annual with simple or forked, weak stems, and opposite, narrow, and ligulate leaves; flowers small, solitary or in pairs, from

the leaf-axils or from the tip of the stem. Sepals 2; petals and stamens 3 to 5; capsule 1-locular, containing a few black and highly lustrous seeds.

Rare or no doubt often overlooked in wet, springy places among mosses, along brooks, or among tall sedges or grasses by the edge of stagnant and muddy ponds; less often in open places in wet sand or mud, when the stems become prostrate and somewhat tufted.

General distribution: Circumpolar, low-arctic.

Fig. 420 Map 483

418. *Claytonia megarrhiza*

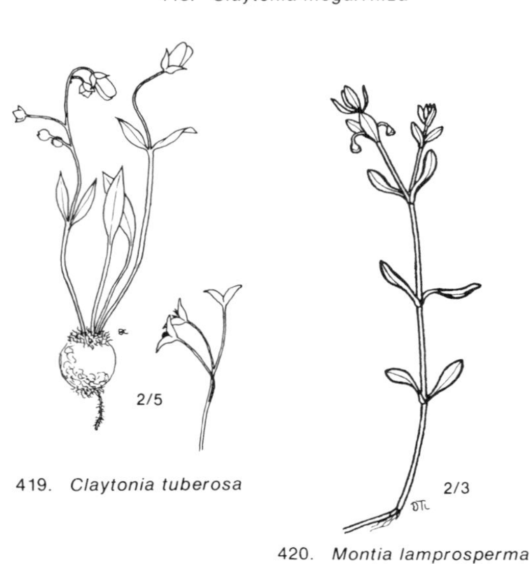

419. *Claytonia tuberosa*

420. *Montia lamprosperma*

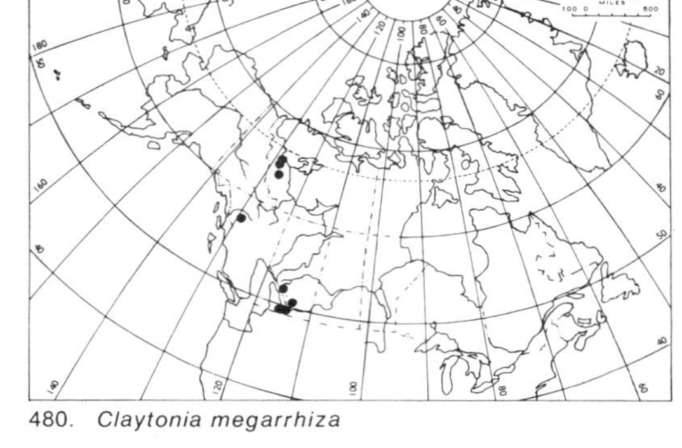

480. *Claytonia megarrhiza*

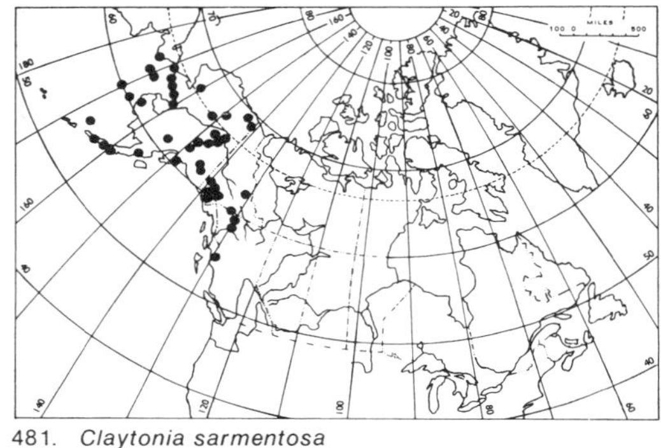

481. *Claytonia sarmentosa*

482. *Claytonia tuberosa*

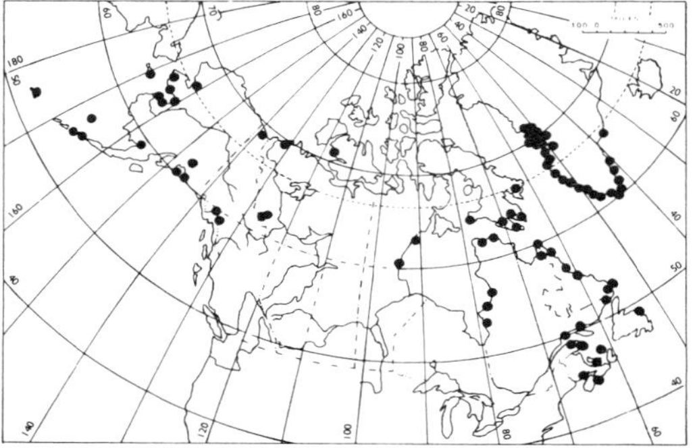

483. *Montia lamprosperma*

CARYOPHYLLACEAE Pink Family

Herbs with opposite, entire leaves, often united at the base; flowers regular, bisexual, commonly 5-merous, with distinct calyx and corolla; styles 2 to 5; fruit a 1-locular (except *Wilhelmsia*) capsule, dehiscing at the apex by teeth or valves of equal or twice the number of the styles; seeds few to many, commonly reniform.

a. Sepals and petals free, not united below; petals white
  b. Leaves without stipules
    c. Petals deeply cleft, notched, or none
      d. Styles usually 3; capsule ovoid or ellipsoid . . . . *Stellaria*
      d. Styles usually 5; capsule cylindrical . . . . *Cerastium*
    c. Petals entire or shallowly notched
      e. Leaves linear, acute; styles 3, 4 or 5
        f. Styles 4-5; flower-buds spherical . . . . *Sagina*
        f. Styles normally 3; flower-buds oblong
          g. Capsules dehiscing by 3 teeth . . . . *Minuartia*
          g. Capsules dehiscing by 6 teeth . . . . *Arenaria*
      e. Leaves broad, oval or oblong; styles 3
        h. Capsules inflated, 3-4-locular . . . . *Wilhelmsia*
        h. Capsules 1-locular
          i. Capsule opening with 3 valves; leaves fleshy . . . . *Honckenya*
          i. Capsule opening with 6 valves; leaves not fleshy
            j. Seeds shiny, with a pale, spongy appendage; leaves elliptic or lanceolate . . . . *Mohringia*
            j. Seeds dull, without a spongy appendage; leaves lanceolate-oblanceolate . . . . *Arenaria humifusa*
  b. Leaves with scarious stipules
    k. Leaves opposite; styles 3 . . . . *Spergularia*
    k. Leaves whorled; styles 5 . . . . *Spergula*
a. Sepals united into a tube; petals free, pink or white
  l. Styles 2
    m. Petals pink (native species) . . . . *Dianthus*
    m. Petals white (adventive) . . . . *Gypsophila*
  l. Styles more than 2
    n. Styles 3; petals pink or white . . . . *Silene*
    n. Styles 5; petals white or purple . . . . *Melandrium*

Arenària L.

Low perennial herbs with small flowers terminal, 1 to 3; petals 5, entire; styles 3; capsule opening with 6 valves.

a. Plant 10-20 cm tall; leaves filiform 5-8 cm long . . . . *A. capillaris* var. *nardifolia*
a. Plant low, tufted or matted; leaves imbricated, lanceolate or oblanceolate, 3-5 mm long . . . . *A. humifusa*

Arenaria capillaris Poir
var. **nardifolia** (Ledeb.) Regel
Densely tufted from a stout, somewhat woody base; leaves filiform in dense fascicles; flowering stem smooth, erect or ascending, 12 to 20 cm tall, terminating in a cluster of small flowers.

A. capillaris in Alaska, Yukon Territory and the District of Mackenzie, as in East Siberia, is totally glabrous, whereas in the Rocky Mountains it is glandular in the inflorescence and has smaller and usually more congested flowers (*A. capillaris* ssp. *americana* Maguire). In the District of Mackenzie plant the calyx lobes are oval with obtuse or rounded tips whereas in Alaska and Yukon Territory some populations have acute calyx lobes while in others they are obtuse.

Calcareous, warm, south-facing slopes.
General distribution: Amphi-Beringian.
Fig. 421 Map 484

Arenaria humifusa Wahlenb.
A. longipedunculata Hult.
Low, tufted or loosely matted perennial with filiform subterranean creeping stems; leaves generally imbricated, lanceolate or oblanceolate, glabrous and somewhat fleshy, 2 to 10 mm long. Flowering stems decumbent or ascending, leafy and usually only a few centimetres high, puberulent above, one- or rarely few-flowered. Sepals oblong, obtuse, nerveless, about as long as the white petals. Capsule cylindric, longer than the calyx; seeds lustrous and faintly rugulose.

Moist, calcareous or serpentine gravels or moist rock crevices.

In sheltered and shaded situations throughout its range *A. humifusa* specimens of taller growth may occur in which the flowering and fruiting peduncle is elongated so that the capsule is raised well above the plant mat (*A. longipedunculata* Hult.).

General distribution: Amphi-Atlantic, wide ranging arctic-alpine, from Alaska to Greenland and N. W. Europe.

Fig. 422 Map 485

Cerastium L. Mouse-ear chickweed

Matted or tufted, pubescent or glabrate perennials, or weedy short-lived annual herbs; flowers in cymes or umbels; petals and sepals 5, rarely 4, the petals white, shallowly notched, or rarely entire; styles (in ours) 5, opposite the petals; capsule 1-locular, many-seeded; seeds rough.

a. Perennial, native species; petals much longer than sepals
  b. Plant matted, rooting at the nodes
    c. Stems, leaves and peduncles pubescent
      d. Leaves oblong or oblanceolate
        e. Flowers commonly 1 or 2 at the end of leafy shoots . . . . *C. alpinum s. lat.*
        e. Flowers commonly 3 or more at the end of the elongated flowering stems . . . . *C. Beeringianum*
      d. Leaves linear and narrowly lanceolate, short pubescent, not glandular . . . . *C. arvense*
    c. Stems and leaves glabrous, flowering peduncle (when present) short-pubescent or glabrate . . . . *C. Regelii*
  b. Plant tufted, 3-6 dm tall, from ascending, subligneous rhizome; leaves linear-attenuate, 3-12 cm long; flowers 3-12 in a terminal umbel . . . . *C. maximum*
a. Annual, or short-lived perennial, weedy species; petals equalling or shorter than the sepals
  f. Plant strictly annual; small flowers in terminal umbels . . . . *C. nutans*
  f. Plants with overwintering basal shoots and decumbent or reclining, much branched flowering stems . . . . *C. vulgatum*

Cerastium alpinum L. *s. lat.*
A variable species which, as a rule, may be distinguished by its matted growth and densely lanate-villous and glandular-viscid or even clammy basal leaves and flowering stems, the latter weak, bearing one or several large flowers in an open corymb; the sepals green, with broad, scarious margins.

In rocky, sandy, or gravelly places, but frequently also on manured soil such as on bird cliffs or near human habitations.

General distribution: Amphi-Atlantic, arctic, thus far not collected west of long. 126°.

Fig. 423 Map 486

Cerastium arvense L. *s.lat.* (incl. *C. campestre* Greene)
Matted or loosely caespitose, non-glandular, short crisply pubescent. Leaves narrowly lanceolate; flowering stems erect-ascending, 2 to 4 dm tall, simple or branching, commonly with conspicuous axillary fascicles of short, sterile shoots that root when detached; inflorescence few- to many-flowered, on slender, elongated peduncles; petals white, twice as long as the somewhat spreading sepals, commonly turning yellowish-brown in drying.

The plant described above is the Cordilleran *C. campestre* Greene of dry, subalpine limestone ledges and screes, the range of which extends north to Yukon Territory and southeastern Alaska, and along the Mackenzie Valley and foothills, to the Mackenzie Delta. The circumpolar and rather similar *C. arvense* L. is always weedy and distinguished by its more caespitose habit and lower stature; it has been collected in the southern District of Mackenzie, but always in disturbed areas by roadsides and near townsites where, obviously, it is a recent introduction.

General distribution: *C. campestre* Greene, Cordilleran; *C. arvense* L., circumpolar.

Fig. 424 Map 487

Cerastium Beeringianum Cham. & Schlecht.
Similar to *C. alpinum* but less densely matted and lacking the long, lanate and glandular-viscid pubescence so characteristic of that species. The flowering stems are stiffer and often repeatedly forking, bearing from a few to a dozen flowers; these are smaller than in *C. alpinum*, and the sepals are usually brownish-purple, and only slightly shorter than the petals.

Rocky, gravelly or sandy places.

General distribution: Arctic-alpine, from northeastern Asia, extending across Alaska eastward through Y.T., the District of Mackenzie and southward in the Rocky Mts. Less common eastward across northern Canada to Nfld.; lacking in Greenland but reaching lat. 80°N. in Ellesmere Island.

Fig. 425 Map 488

Cerastium maximum L.
By its tall stature, viscid-pubescent stems, attenuate leaves and large flowers *C. maximum* is at once distinguished from other northern members of the genus.

Grassy slopes and open willow thickets.

General distribution: Amphi-Beringian, and in N. America known from unglaciated mountain plateaux of central Alaska and Y.T., barely entering our area in the Richardson Mts.

Fig. 426 Map 489

Cerastium nutans Raf.
Weak annual of weedy habit, with erect, glandular-hirsute stems and linear-lanceolate leaves. The open cymose inflorescence terminal. The specific name alludes to its nodding, mature capsules.

A recent introduction in S.W. District of Mackenzie.

General distribution: N. America. Wide ranging from N. S. to B. C.

Map 490

Cerastium Regelii Ostf.
Essentially glabrous, loosely caespitose or matted perennial, usually with trailing, long-internoded, freely branching and rooting stems, but in exposed situations a dense, pulvinate and always sterile form occurs, with extremely short internodes and closely imbricated leaves. Leaves oblanceolate-ovate to sub-orbicular, 5 to 12 mm long, somewhat fleshy, glabrous on both sides, or sparingly and softly ciliated along the margins, often subtending axillary fascicles or elongate shoots terminating in wintering buds. Marcescent and bleached leaves persist on the stems for several years, and often only the growing tip of each shoot is fresh and green. The species apparently reproduces chiefly by adventitious buds and flowers are rarely, if ever seen, except near the southern edge of its range. When present, the 1- to 4-flowered stems are prostrate, and may be 15 to 20 cm long, with internodes 5 to 8 cm long. Sepals soft pubescent, about 5 mm long, greenish purple, two-thirds as long as the shallowly notched petals. Mature capsules 10 mm long, opening with 10 teeth.

Common or even ubiquitous throughout the Canadian Arctic Archipelago in wet swales of low, calcareous tundra, on lake shores and in low places in general where the snow remains late; on the Canadian mainland collected only on Melville and Tuktoyaktuk Peninsulas. The caespitose or pulvinate and mostly sterile form is generally restricted to wet clay on frost-activated limestone barrens.

General distribution: Circumpolar, high-arctic.

Fig. 427 Map 491

Cerastium vulgatum L.
Short-lived, hirsute and glandular-viscid perennial with ascending stems; inflorescence a few-flowered cyme; the petals white, equalling or shorter than the sepals.

Cosmopolitan weed, with us thus far reported only from Fort Smith, but likely to turn up elsewhere in southern District of Mackenzie by roadsides and in disturbed areas.

General distribution: Circumpolar; non arctic.

Dianthus L.

***Dianthus repens** Willd.
Tufted glaucous and glabrous perennial from a stout tap-root, with 10 to 20 cm tall, leafy, erect-ascending stems. Leaves linear-lanceolate, 3 to 4 cm long, the flowers showy, terminal and mostly solitary, with a cylindrical greenish-purple, 1.0 to 1.4 cm long calyx, and deep pink or purple 2 to 3 cm long petals.

Gravelly river banks, terraces and slopes of unglaciated central and northern Alaska, barely entering the north east corner of Yukon Territory. To be looked for in the foothills of Richardson Mountains west of the Mackenzie Delta.

General distribution: Amphi-Beringian.

Map 492

Gypsophila L.

Gypsophila paniculata L.
Diffusely branched up to 1 m tall perennial with lanceolate leaves and a large, paniculate inflorescence of small, white flowers.

Reported from Fort Smith where it was noted as persisting and spreading from gardens.

General distribution: Cosmopolitan.

Honckenya Ehrh.

Honckenya peploides (L.) Ehrh.
var. **diffusa** (Hornem.) Mattf.
Arenaria peploides L. var. *diffusa* Hornem.
Seabeach-Sandwort
Glabrous, fleshy perennial with much branched and freely rooting stems deeply buried in sand. Leaves elliptic or oblong-lanceolate. Flowers in terminal, few-flowered, leafy cymes, or from the upper leaf axils. Petals white, inserted on a conspicuous 10-lobed disk; capsule globular, leathery, 6 to 12 mm in diameter, containing 5 or 6 large, smooth seeds.

Common and ubiquitous along sandy seabeaches.

General distribution: Circumpolar, low-arctic; rare on inland lake shores in northern District of Keewatin.

Fig. 428 Map 493

Melandrium Roehl. Bladder-Campion

Perennial herbs with a strong tap-root, opposite, oblanceolate, mostly basal leaves, and erect, slender, one- to several-flowered stems. Calyx large, more or less inflated, prominently striped; petals clawed, often included, 2-cleft, white, pale pink, or purple. Styles 5; stamens 10, alternating with the petals. Ovary one-celled; capsule opening with 5 or 10 reflexed teeth, containing numerous seeds. (*Wahlbergella* Fries; *Lychnis* Am. auth., in part).

a. Seeds large, 1.5-2.4 mm in diameter, with a distinct wing surrounding the body
  b. Petals barely exserted; seeds over 2.0 mm in diameter, with a broad wing
    c. Calyx much inflated, almost globular, petals pink or pale purple . . . . *M. apetalum* ssp. *arcticum*
    c. Calyx less inflated, ellipsoid never globular; petals dark purple . . . . *M. apetalum* ssp. *attenuatum*
  b. Petals much exserted, 1.5-2 times as long as calyx; flowers commonly 3, rarely solitary; seeds less than 1.8 mm in diameter, with a narrow wing
    d. Long-lived perennial from a stout tap-root; flowering stems stiffly erect, 1.5-2.5 dm high; cauline leaves linear, smaller than the basal ones; petals milky white . . . . *M. affine*
    d. Short-lived perennial from a weak tap-root; flowering stems slender, 3-6 dm tall, somewhat flexuous; cauline leaves as large as the basal; petals roseate . . . . *M. Taylorae*
a. Seeds smaller, 1 mm or less in diameter, wingless, angular, the testa tuberculate or merely granular; calyx never inflated; petals exserted
  e. Flowers on slender peduncles; calyx tubular
    f. Flowers on stout erect peduncles; the lowermost subtended by green, leafy bracts; seeds dull, strongly tuberculate . . . . *M. Drummondii*
    f. Flowers on slender, spreading peduncles; seeds granular, lustrous . . . . *M. Ostenfeldii*
  e. Flowers congested, sessile or very short-peduncled; fruiting calyx narrowly campanulate, seeds angular-prismatic, 0.6-0.8 mm in diameter, the testa smooth and dull . . . . *M. taimyrense*

Melandrium affine J. Vahl
M. furcatum (Raf.) Hult.
Lychnis brachycalyx Raup
L. Gillettii Boivin
Flowering stems stiffly erect, stout, 5 to 30 cm high, somewhat glandular-viscid, and usually with 2 or 3 pairs of prominent, lanceolate leaves. Inflorescence in few-flowered cymes, usually of 3 erect flowers on short, erect peduncles that rapidly elongate toward maturity. Calyx ovoid, somewhat glandular-viscid, barely inflated; the petals milky white and much exserted; seeds small, reniform, less than 2 mm in diameter, with a narrow wing.

Common in not too dry, stony, gravelly, or sandy places, but readily established near animal dens.

General distribution: Circumpolar, wide ranging arctic, alpine.

Fig. 429 Map 494

Melandrium apetalum (L.) Fenzl
ssp. **arcticum** (Fr.) Hult.
Basal leaves linear-oblong to oblanceolate, obtuse, crowded around the base of the stems. Flowering stems simple, somewhat flexuous, purplish and short pubescent above, usually from 5 to 12 cm high, with one or two pairs of sessile linear leaves. The usually solitary, large flower is nodding during anthesis, but as the capsule matures becomes erect. The inflated, bladder-like, purple-striped, calyx resembles a miniature Japanese lantern. Petals short-exserted, purple; seeds reniform or circular in outline, 1.5 to 2.4 mm in diameter, including the somewhat swollen and prominent wing.

Common in wet tundra, in moist meadows along lagoons and lake shores, in moss by alpine brooks, or in wet clay on gravelly barrens.

General distribution: *M. apetalum s. lat.*, Circumpolar, high-arctic.

Fig. 430 Map 495

Melandrium apetalum (L.) Fenzl
ssp. **attenuatum** (Farr) Hara
Lychnis attenuata Farr
L. macrospermum sensu Raup *non* Porsild
Similar to ssp. *arcticum* but usually of more compact and matted growth, often with a dozen or more flowering stems from a multicipital or sometimes rhizomatose base. Stems rarely over 15 cm high, with 2 to 3 pairs of rather well-developed leaves. Petals usually darker purple and slightly more exserted than in ssp. *arcticum*; whereas the latter is circumpolar and commonly grows in wet arctic tundra, ssp. *attenuatum* is a Cordilleran endemic, and is almost exclusively found on well drained, calcareous slide-rock.

General distribution: Cordilleran, extending through northeastern B.C. to the central Mackenzie Mts., and west through Y.T. to the south slope of Endicott Mts., Alaska.

Map 496

***Melandrium Drummondii** (Wats.) Porsild
Stems viscid-puberulent, mostly several, 2 to 5 dm tall from a stout, branching base. Inflorescence several-flowered, the flowers on long conspicuously appressed-erect pedicels; petals white or purplish, barely exserted; capsule cylindric; seeds small and wingless.

General distribution: Prairies and foothills from Man. to B.C. and southward, north to L. Athabasca and to be expected in S.W. District of Mackenzie.

Map 497

Melandrium Ostenfeldii Porsild
M. taimyrense Hult. *non* Tolm.
Densely caespitose, often with a stout tap-root and much branched crown. Basal leaves linear-spatulate, and the flowering stems somewhat flexuous, 10 to 30 cm tall with 2 to 3 pairs of linear leaves, glandular-viscid above; flowers 1 to 8, in small, dense clusters; calyx sub-cylindrical, with broad, greenish stripes, glandular viscid; petals narrow, much exserted and deeply notched, white or pale pink; seeds about 0.8 by 0.6 mm, angular, wingless, dark brown and granulate.

Sunny, non-calcareous cliffs, ledges or talus slopes.

General distribution: Endemic of N.W. North America, mainly on Precambrian rocks from S.E. Y.T. to Great Slave and Great Bear L., east to Bathurst Inlet, and north to Victoria Island.

Fig. 431 Map 498

Melandrium taimyrense Tolm.
Stems strict, few to many from a stout tap-root, 2 to 5 dm tall, in vigorous specimens with freely branching inflorescences of up to a dozen or more flowers; basal leaves oblong, 3 to 5 cm long, like those of the flowering stems short, hoary-canescent. Calyx tubular, 8 to 12 mm long with dark, purple lines. Seeds wingless, angular, rugose, somewhat granulate and light brown.

Dry gravelly river terraces

General distribution: Amphi-Beringian; over Alaska and Y.T. east to the Mackenzie Delta and beyond.

Map 499

Melandrium Taylorae (Robins.) Tolm.
Short-lived perennial from a weak tap-root. Flowering stems 1 to 5, very slender, 3 to 6 dm tall, nearly glabrous below, increasingly puberulent and glandular-viscid towards the inflorescence, with 3 or very rarely 4 pairs of linear-lanceolate, 3 to 5 cm long leaves; inflorescence few- to 9-flowered, in vigorous species with flowering peduncles 5 to 10 cm long, rising from the second and third internodes. Calyx ovate, scarcely inflated, viscid-puberulent, 7 to 8 mm long and 6 to 7 mm broad, with greenish-purple stripes. Petals roseate, much exserted. Seeds small wedge-shaped, with a narrow wing, wrinkled but not granulate.

Locally abundant on calcareous river terraces and rocky ledges of the Mackenzie and Yukon river basins.

General distribution: Alaska, Y.T., east to Mackenzie Delta and Great Bear L. area.

Map 500

Minuartia Loefl. Sandwort

Small, loosely tufted or mat-forming perennials with linear and mostly sessile leaves lacking stipules; flowers small, solitary, or in few-flowered cymes; styles 3, sepals 5, and 5 entire or barely notched white petals, and 10 stamens; the capsule few- to many-seeded, opening with 3 valves.

a. Leaves stiff, prominently 3-nerved (best seen in last year's marcescent leaves) . . . *M. rubella*
a. Leaves 1-nerved, or appearing nerveless
  b. Stems and leaves glabrous
    c. Matted or loosely caespitose; stems always 1-flowered . . . . *M. Rossii*
    c. Tufted, at least some stems 2- to several-flowered
      d. Long-lived perennial from a much branched tap-root; sepals obtuse; seeds brown . . . . *M. stricta*
      d. Short-lived perennial from a weak tap-root; sepals acute; seeds black . . . . *M. dawsonensis*
  b. Stems and leaves pubescent, or the latter at least ciliate-margined
    e. Leaves very narrow, 7-8 times longer than broad, appearing glabrous, but distinctly ciliate-margined . . . . *M. yukonensis*
    e. Leaves obtuse, 4-5 times longer than broad
      f. Petals narrowly oblong, rarely much longer than the sepals; seeds smooth . . . . *M. biflora*
      f. Petals spatulate or clawed, much longer than sepals; seeds tuberculate
        g. Leaves narrowly linear, stiff . . . . *M. obtusiloba*
        g. Leaves soft
          h. Leaves obscurely 3-nerved, ciliate; flowers short-peduncled; capsules more than three times as long as calyx . . . . *M. macrocarpa*
          h. Leaves 1-nerved, flowers long-peduncled; capsule not much longer than calyx . . . . *M. arctica*

Minuartia arctica (Stev.) Aschers. & Graebn.
Arenaria arctica Stev.
Loosely caespitose from much branched base. Leaves linear, glabrous or sparsely ciliate. Flowers solitary, well raised above the plant cushion, on slender, leafy and usually about 5 cm long, glandular-pubescent branches; sepals glandular-puberulent, obtuse, usually dark purple, the petals white or often pink or purple-tinged, not much longer than the sepals. Capsule, blunt, not much longer than the calyx; seeds chestnut brown, the testa finely sculptured, but otherwise smooth.

Alpine snowbed slopes or stony tundra.

General distribution: Amphi-Beringian, reaching east to Great Bear L. Disjunct by over 3000 km as *M. marcescens* (Fern.) House in Gaspé, Que. and Nfld.

Fig. 432 Map 501

Minuartia biflora (L.) Schinzl. & Thell.
Arenaria sajanensis Willd.
Densely matted with widely creeping, firm, and almost sub-ligneous stems with crowded, 5 to 10 mm long, somewhat fleshy, flat, linear and obtuse fresh-green leaves. Flowering stems minutely glandular-hirtellous above, 1- to 3-flowered. Sepals linear-oblong, obtuse and prominently 3-nerved, distinctly shorter than the oblanceolate, pure white petals.

On favourably exposed, calcareous, grassy slopes and herbmats having abundant snow-cover in winter.

General distribution: Circumpolar, low arctic.

Fig. 433 Map 502

Minuartia dawsonensis (Britt.) Mattf.
Arenaria dawsonensis Britt.
Glabrous, short-lived, loosely tufted perennial from a weak, fibrous root. Leaves linear, fresh green, 1-nerved, in fascicles from the nodes. Flowering stems 1 to 2 dm long, slender, green, spreading, erect or declining, branching above; inflorescence a few-flowered cyme. Petals lacking, or shorter than the strongly 3-nerved, acuminate sepals. Seeds globose or slightly ovate, about 0.7 mm in diameter; the testa coal-black, dull, finely sculptured.

Moist sandy places.

General distribution: N. America: Nfld. to central Alaska, north to the limit of forest.

Fig. 434 Map 503

Minuartia macrocarpa (Pursh) Ostenf.
Arenaria macrocarpa Pursh
Loosely caespitose with decumbent, trailing branches. Leaves soft, flat and linear, ciliate along the margins. Flowers solitary on glandular-pubescent up to 2 cm long peduncles; sepals linear, obtuse, green and softly glandular-pubescent; petals white, much longer than the sepals. The mature capsule 15 to 18 mm long, and more than twice as long as the calyx.

Stony, well drained alpine tundra.

General distribution: Amphi-Beringian: Reaching east across the mountains of Alaska and Y.T. into the Mackenzie Mts.

Fig. 435 Map 504

Minuartia obtusiloba (Rydb.) House
Arenaria obtusiloba (Rydb.) Fern.
Similar to *M. arctica* but more densely caespitose from a sub-ligneous tap-root and more glandular-pubescent throughout; leaves firmer and narrower, and the obtuse sepals green, rarely with a dark tip, and distinctly 3-nerved.

Very variable as to size according to habitat and exposure; thus in specimens from sunny river terraces the flowers may be raised 8 to 10 cm above the mat, whereas in arctic or high-alpine situations the peduncle is often very short.

General distribution: Amphi-Beringian east to the Mackenzie Delta area, and south in the Rocky Mts.

Fig. 436 Map 505

Minuartia Rossii (R.Br.) House
Arenaria Rossii R.Br.
Low, glabrous and loosely caespitose, with crowded, 3 to 5 mm long, linear, obtuse, somewhat fleshy and prominently keeled leaves. Flowering stems spreading and always 1-flowered, the peduncles filiform, usually about 1.0 to 1.5 cm long; sepals lanceolate, obscurely 3-nerved, dark purple, spreading in anthesis, slightly shorter than the linear-oblong white petals. Capsules slightly longer than the sepals. In the Far North the reproduction may be largely vegetative, by axillary or terminal and easily detached leaf buds.

Common in not too dry, turfy, gravelly or sandy calcareous barrens throughout the Arctic Archipelago and along the mainland coast.

In the Richardson and Mackenzie Mountains and in mountains of central and northern Yukon Territory and Alaska to East Asia largely replaced by the ssp. *elegans* (Cham. & Schlecht.) Rebr. of more loosely matted habit, and with flowering peduncles 2 to 3 cm long.

General distribution: N. American, high arctic-alpine.

Fig. 437 Map 506

Minuartia rubella (Wahlenb.) Hiern.
Arenaria verna L.
var. *pubescens* (Cham. & Schlecht.) Fern.
A. rubella (Wahlenb.) Sm.
Low, glandular-hirtellous, tufted or matted, but in exposed situations commonly pulvinate, when the flowers barely emerge above the surface of the plant cushion. Leaves linear-subulate, crowded on the basal shoots, prominently 3-nerved, and from 3 to 5 mm long. Flowering stems decumbent at the base, erect-ascending, filiform, usually 3 to 6 cm long and one- to several-flowered; sepals lanceolate-acuminate, 3-nerved, greenish-purple, slightly longer than the entire, white petals, and slightly shorter than the capsule.

Common in dry, gravelly, sandy or rocky places.

General distribution: Circumpolar, arctic-alpine.

Fig. 438 Map 507

Minuartia stricta (Sw.) Hiern. *non Arenaria stricta* Michx.
Arenaria uliginosa Schleich.
Glabrous, loosely tufted perennial from a much branched taproot. Flowering stems very slender, erect or ascending, commonly 5 to 10 cm high. Flowers one or rarely two, on slender and usually purplish peduncles. Sepals oval, green-

ish-purple, faintly 1- to 3-nerved; petals shorter than sepals, or lacking. Seeds globular, 0.5 mm in diameter, chestnut brown, the testa very finely sculptured, and faintly lustrous.

Wet places in alpine or arctic tundra.

General distribution: Circumpolar (with many and large gaps) but in N. America with an Amphi-Atlantic type of range.

Fig. 439 Map 508

Minuartia yukonensis Hult.
Arenaria laricifolia sensu Hultén *non* Schinz & Thell.
Caespitose from a much branched subligneous taproot from which issue trailing, repeatedly branched leafy shoots, 5 to 10 cm long and densely clothed by fascicles of linear, fresh green, glabrous or short-ciliate leaves, 7 to 8 mm long and 0.5 mm wide. Flowering peduncles 8 to 15 cm long, with several pairs of bracts, and terminating in a 1- to 4-flowered cyme. Sepals glandular pubescent, 3-nerved, linear, obtuse. Petals 1 to 1 ½ times as long as sepals; seeds kidney-shaped, about 1 mm long.

In alpine or less protected situations *M. yukonensis* becomes more compact and more hirsute when it may resemble condensed forms of *M. obtusiloba*.

General distribution: Amphi-Beringian, alpine, over Alaska and Y.T. reaching the east slope of the Richardson Mts. west of the Mackenzie Delta.

Fig. 440 Map 509

Moehringia L.

Low perennial woodland plants with slender horizontal rhizomes, erect, simple or branching stems, and sessile, elliptic or narrowly lanceolate leaves. Flowers in small terminal cymes.

a. Stems terete; leaves elliptic, puberulent and white punctate beneath . . . . *M. lateriflora*
a. Stems angled; leaves narrowly lanceolate, glabrous and not punctate beneath . . . . *M. macrophylla*

Moehringia lateriflora (L.) Fenzl
Arenaria lateriflora L.
Stems simple or branching, crisply puberulent, 1.0 to 2.5 dm tall, bearing 2 to 5 pairs of sessile, opposite, narrowly elliptic and puberulent leaves. Flowers solitary, or several, in few-flowered cymes, on very slender 2-bracted peduncles. Petals white, longer than sepals.
Willow thickets and open woodland.

General distribution: Circumpolar; N. America from Nfld. to Alaska, north to the limit of trees.

Fig. 441 Map 510

Moehringia macrophylla (Hook.) Torr.
Arenaria macrophylla Hook.
Similar to preceding but stems stouter and stiffer. The leaves glabrous, 2 to 5 cm long and 3 to 5 mm wide. Flowers 3-several in a terminal or lateral cyme; petals white, shorter than the sepals.

Dry wooded slopes and always on soils derived from magnesian or ultrabasic rocks and hence of spotty and disjunct range.

General distribution: Lab., L. Superior, Athabasca and Great Slave L., southern B.C.

Fig. 442 Map 511

Sagina L. Pearlwort

Small, tufted herbs with filiform, subulate leaves, and small, inconspicuous terminal or axillary flowers. Petals 4 or 5, or none, white, entire or notched. Styles as many as the sepals. Capsule many seeded, 4- or 5-valved.

a. Flowers normally 5-merous
  b. Petals inconspicuous; upper leaves of flowering branches slender, not subtending leafy bulbils in their axils . . . . *S. caespitosa*
  b. Petals large, twice as long as the sepals; upper leaves of flowering branches subtending small, firm bulbils . . . . *S. nodosa*
a. Flowers normally 4-merous
  c. Tuft compact, branches ascending, non-rooting . . . . *S. intermedia*
  c. Tuft loose, branches declining often rooting at nodes . . . . *S. Linnaei*

Sagina caespitosa (J. Vahl) Lge.
Glabrous, dwarf, perennial forming small, reddish-green, rather compact, hemispherical tufts from a few to 6 or 8 cm in diameter. The solitary, erect, short peduncled, globular flowers 2 to 3 mm in diameter; petals white, slightly longer than the sepals. Stamens 10.

Occasional or rare, but perhaps often overlooked; in moist gravelly places. In our area thus far not reported west of long. 100°W.

General distribution: Amphi-Atlantic, arctic.

Fig. 443 Map 512

Sagina intermedia Fenzl
Glabrous, dwarf perennial, forming small, flat or depressed, and usually reddish-green tufts, from few to several centimeters in diameter, with spreading or ascending, mostly 1-flowered leafy stems radiating from the central, leafy rosette. Petals 4, white, barely longer than the sepals. Stamens 4 or 5.

Occasional to common in moist sand or gravel by lake shore or in wet clay in open tundra.

General distribution: Circumpolar, arctic-alpine.

Fig. 444 Map 513

Sagina Linnaei Presl
S. saginoides (L.) Karst.
Caespitose, with procumbent, leafy and yellowish-green stems rooting at the nodes. Flowers solitary on slender peduncles, nodding when young; petals 5, rarely 4.

Edge of snowbeds, in moist sand or moss by brooks.

General distribution: Circumpolar, low-arctic.

Fig. 445 Map 514

Sagina nodosa (L.) Fenzl
Tufted glabrous or sparingly glandular-hirtellous perennial with simple or freely branched, capillary stems up to 20 cm long, bearing numerous bulb-like fascicles of axillary, minute leaves, and sometimes entirely lacking flowers, or more commonly with each branch terminating in a single flower. Petals 5, white, about twice as long as the sepals; seeds black.

Moist, gravelly or peaty soil.

General distribution: Amphi-Atlantic, low-arctic.

Fig. 446 Map 515

Silene L. Campion

Perennial herbs (ours), with opposite, entire leaves. The flowers with tubular or ovoid, 5-toothed calyx, petals 5, with a narrow claw, and bilobed or variously cleft blade, styles 3 and the capsule 3-locular at the base.

a. Plant densely pulvinate; flowers solitary, pink
  b. Leaves flat and short; flowers sub-sessile . . . . *S. acaulis* ssp. *acaulis*
  b. Leaves narrow, 1.0-3.5 cm long; flowers peduncled . . . . . *S. acaulis* ssp. *subacaulescens*
a. Plant not pulvinate; flowers white, in terminal cymes
  c. Calyx purple, tubular . . . . *S. repens* ssp. *purpuratus*
  c. Calyx green, campanulate . . . . *S. Menziesii*

Silene acaulis L.
ssp. **acaulis**
Moss-Campion
Caespitose, glabrous perennial, forming compact, hemispherical, or flat cushions up to 50 cm or more in diameter. The densely crowded branches are clothed with marcescent linear leaves and terminate in a single, short-peduncled, monoecious or perfect flower. Calyx tubular, glabrous, Petals pink, lilac or pale purple, rarely pure white (f. *albiflora* Hartz.). Capsule twice as long as the calyx.

The moss-campion is common throughout the Arctic, mainly on well-drained gravelly or turfy barrens, or on cliffs or rocky ledges.

General distribution: Circumpolar, arctic-alpine.

Fig. 447 Map 516

Silene acaulis L.
ssp. **subacaulescens** (F.N.Williams) Hult.
Differs from ssp. *acaulis* by its narrower ciliate and usually much longer leaves, and by its flowering peduncle which elongates so that the maturing capsules are raised well above the plant cushion. The calyx and capsule are cylindrical and much longer than in ssp. *acaulis*.

General distribution: Cordilleran race, with us

apparently restricted to the Mackenzie and Richardson Mts.
Map 517

Silene Menziesii Hook.
Low, loosely tufted or matted, minutely pubescent or glabrate perennial from a slender, much branched rhizome from which rise fascicles of freely branched and leafy stems, commonly 10 to 20 cm tall. Leaves mainly cauline, oblanceolate, sessile, 3 to 4 cm long and 3 to 10 mm wide. Flowers solitary, or more often in simple and terminal few-flowered and leafy cymes. Calyx campanulate, 6 to 8 mm long, several-nerved, corolla white, exserted. Seeds 0.8 mm in diameter, shiny and black.
Sunny slopes and open aspen or spruce woods, often pioneering and weedy on disturbed soils.
General distribution: Cordilleran species barely entering S. W. District of Mackenzie where it appears to be rapidly spreading along roads.
Fig. 448 Map 518

Silene repens Patrin
ssp. **purpurata** (Greene) Hitchc. & Maguire
Perennial with solitary or fascicled, simple or rarely branched stems, 2 to 5 dm high, from a slender, freely branching rhizome. Leaves cauline, 5 to 6 pairs, sessile, linear-lanceolate, 3 to 6 cm long and 0.3 to 0.6 cm wide, soft ciliate; stems crisply pubescent, increasingly so toward the inflorescence which is an elongating raceme composed of several axillary and bracted cymes. Calyx tubular or narrowly campanulate, dark purple and finely crisp-pubescent, 10 to 15 mm long; corolla white or more often roseate.
Non-arctic, locally abundant on sandy river terraces, gravelly ridges or occasionally in open willow thickets.
General distribution: Endemic of central Alaska, and Y.T. to northwestern District of Mackenzie.
Fig. 449 Map 519

Spergula L. Spurrey

Spergula arvensis L.
Glandular-pubescent annual with simple or branching stems 2 to 5 dm tall. Leaves narrowly linear in well-spaced whorls of 8 to 10 leaves. The small white flowers on slender pedicels in an open terminal raceme.
General distribution: Cosmopolitan weed, in the N.W.T. known from a single collection near the town of Yellowknife on Great Slave L.

Spergularia J. & C. Presl Sand-Spurrey

Spergularia marina (L.) Griseb.
Glandular-pubescent annual with simple or freely branched and spreading stems, with us rarely more than 5 to 7 cm long. Leaves opposite, linear, 2 to 4 cm long. Flowers small in a few-flowered, leafy cyme, the petals white or pink.
Abundant locally by the edge of salt springs on Salt Plains west of Slave River just north of the 60th parallel.
General distribution: Circumpolar, non arctic, with disjunct stations from N.S. to S.E. Alaska.
Fig. 450 Map 520

Stellaria L. Chickweed

Essentially glabrous perennial herbs (except the annual and weedy *S. media*) with ascending, prostrate or creeping and freely rooting stems bearing rather small, in ours sessile, opposite leaves. The rather small white flowers solitary or cymose, terminal or appearing lateral, by the prolongation of the stem, from the upper leaf-axils. Petals 5 (or sometimes lacking), deeply cleft, white; styles 3; capsule 1-locular, opening by 6 valves.
In the Arctic, all native species, in addition to normal sexual reproduction, reproduce vegetatively by axillary leaf buds.

a. Annual, weedy; the basal leaves at least distinctly petioled . . . . *S. media*
a. Perennial, native species, all with sessile leaves
b. Leaves fresh green, soft and not prominently keeled

c. Flowers in the axils of small, scarious bracts or scarious-margined leaves
d. Petals as long or longer than the sepals . . . . *S. longifolia*
d. Petals shorter than the sepals, or lacking . . . . *S. umbellata*
c. Flowers in the axils of green leaves, if cymose, with leafy bracts
e. Leaves somewhat fleshy, plants of matted habit
f. Leaves oval-elliptic; seeds smooth; sea-shore plants . . . . *S. humifusa*
f. Leaves linear-oblong, thin; seeds rugose; plants of wet meadows . . . . *S. crassifolia*
e. Leaves not fleshy, commonly ciliate-margined; plant not matted . . . . *S. calycantha*
b. Leaves firm, keeled, green or often glaucous . . . . *S. longipes s. lat.*
g. Sepals entirely glabrous
h. Flowers terminal in few- to several-flowered cymes, their peduncles supported by small, scarious bracts; leaves commonly green and more or less lustrous
i. Plants mostly high-grown, 1-3 dm tall
j. Stems and leaves glabrous or essentially so
k. Inflorescence with spreading branches, commonly rather few-flowered; flowers large, the petals much longer than the calyx
l. Capsule black and shiny, its valves not reflexed at maturity . . . . *S. longipes s. str.*
l. Capsule straw-coloured, its valves reflexed and rolled at maturity . . . . *S. arenicola*
k. Inflorescence many-flowered, with characteristically stiff and strongly ascending branches, the terminal flowers small, commonly approximate; petals scarcely longer than the calyx . . . . *S. stricta*
j. Stems strongly pubescent, especially on the internodes; leaves commonly villous, especially on the midrib dorsally, and on the margins, rarely glabrate . . . . *S. subvestita*
i. Plants mostly of low and matted habit; flowers mostly solitary, pedicelled, in the axils of small, scarious bracts or scarious-margined leaves; capsules (rarely present) pale with reflexed teeth (high-arctic) . . . . *S. crassipes*
h. Flowers pedicelled, mostly solitary, in the axils of normal green leaves, or if cymose with leaf-like bracts; leaves commonly bluish-green . . . . *S. monantha*
g. Sepals ciliated, glabrous or pubescent on the back
m. Flowers pedicelled, in few-flowered cymes, or solitary in the axils of scarious and mostly ciliated bracts; sepals ciliated; internodes glabrous or sparingly pubescent . . . . *S. Edwardsii*
m. Flowers pedicelled, in the axils of normal green leaves; sepals normally pubescent on the back especially towards the tip, less commonly short-ciliate as well; plant fresh green and lustrous . . . . *S. laeta*

***Stellaria arenicola** Raup
Similar to *S. longipes* and *S. stricta* but easily distinguished when in fruit (see key) but perhaps not at other times.

General distribution: Thus far known only from the type locality on the south shore of L. Athabasca (Sask.) where it is said to be common in damp places among sand dunes. It should be looked for in similar places north of the N.W.T. border.

Map 521

Stellaria calycantha (Ledeb.) Bong. *s. lat.*
S. borealis Bigel.
Perennial from a slender rhizome. Stems glabrous, weak, erect or ascending from 2 to 5 dm tall, commonly freely branched upwards from the middle nodes. Leaves soft 1 to 3 cm long, their margins commonly, but not always, soft ciliate at the base. The inflorescence an open leafy cyme, the individual flowers on slender often arching peduncles. Petals shorter than the sepals, or absent.

Common locally in moist, shaded willow thickets throughout the wooded parts of the area.

General distribution: Circumpolar, non-arctic.

Fig. 451 Map 522

Stellaria crassifolia Ehrh.
Similar to *S. humifusa* but smaller and more

delicate in all parts, with flaccid, freely branching and densely matted, slender stems. Leaves narrower, linear-oblong.

Near its northern limit *S. crassifolia* probably does not normally produce viable seeds, depending entirely on vegetative reproduction for which it is well equipped by the formation of terminal as well as axillary wintering buds that develop at the tip of aerial stems or branches, and also at the tips of filiform and very fragile runners issuing from the lower leaf-axils, or from the subaerial stems, in the same manner as in *S. umbellata*.

Occasional in wet meadows, where it often forms mats around the bases of tall grasses and sedges.

General distribution: Circumpolar, low-arctic.

Fig. 452 Map 523

Stellaria crassipes Hult.
The compressed or reduced arctic or high-arctic extreme of *S. longipes*, of habit similar to *S. monantha* and *S. laeta*, differing from the first by having scarious bracts or at least a pair of reduced scarious-margined green leaves supporting the flowering pedicel, and from the latter by its totally glabrous sepals. Its capsule, like that of *S. arenicola*, is said to be pale straw-coloured, and the tip of the valves reflexed at maturity. This character, however, is of small diagnostic value because mature capsules thus far have not been reported from North America.

Wet tundra, herbmats and rocky slopes.

General distribution: Amphi-Atlantic, arctic.

Fig. 453 Map 524

Stellaria Edwardsii R.Br.
S. ciliatosepala Trautv.
Similar in habit to *S. monantha* from which it can be distinguished by the presence of scarious and mostly ciliated bracts in the few flowered cyme, and the ciliated sepals.

Grassy or gravelly rather moist tundra.

General distribution: Circumpolar, arctic; on the Canadian mainland extending south across the treeless country to the edge of the forest, but in central Alaska and Yukon Territory found also in sub-alpine forest. In the east somewhat disjunct along the coast of Labrador.

Fig. 454 Map 525

Stellaria humifusa Rottb.
Densely matted perennial with long trailing or ascending, freely rooting stems, bearing numerous somewhat crowded oval-elliptic, somewhat fleshy, flat leaves. Flowers solitary in the axils or in few-flowered, leafy cymes.

Wide ranging, sea shore species ubiquitous along sheltered beaches and in brackish and occasionally flooded meadows where it is often associated with turf-forming halophytic sedges and grasses such as *Puccinellia phryganodes* and *Carex glareosa* var. *amphigena*.

General distribution: Circumpolar.

Fig. 455 Map 526

Stellaria laeta Richards.
Similar in habit to *S. monantha*, but differing in fresh green and shiny aspect, and the sepals which are normally pubescent on the back.

Wet tundra, herbmats and rocky slopes.

General distribution: N. America, arctic: distinctly of North American range, from West Greenland to Alaska, north in the Canadian Arctic Archipelago to northern Ellesmere Island, but with a southern limit generally short of the tree-line. It is a very rare species in the Rocky Mountains where it has been collected only a few times south to lat. 51°24′ (Lake Louise).

Fig. 456 Map 527

Stellaria longifolia Muhl.
S. atrata (J.W.Moore) Boivin
Stems weak, 2 to 4 dm tall. In mature fruiting plants the secondary branches of the inflorescence (and also the somewhat curved petioles) are usually inserted at a blunt angle. Petals only slightly longer than the sepals.

River meadows.

General distribution: Circumpolar, non arctic.

Fig. 457 Map 528

Stellaria longipes Goldie *s. str.*
Upright to densely tufted, essentially glabrous perennials with stiff, commonly green and more or less shiny leaves; inflorescence a few-flowered terminal cyme, the peduncles supported by small scarious bracts; capsules black and shiny.

Considered in a strict sense, *S. longipes* is a North American species of the boreal forest region and, like so many others in that group, with a distinctly eastern type of range from Labrador, south to northern and western New York, west through Quebec and Ontario, south to Indiana and Minnesota, and in the West to Montana, Idaho and Colorado, north through Alberta and interior British Columbia, but not reaching the Pacific coast, southern District of Mackenzie and the Yukon Territory; in interior parts of Alaska where it has been collected mainly in disturbed soil along recently constructed highways, airports, etc., where it appears to be weedy and probably is a recent introduction.

In the northern parts of its range dwarfed and reduced forms cannot always be separated from *S. crassipes* and *S. monantha*.

General distribution: N. America, non-arctic.

Fig. 458 Map 529

Stellaria media (L.) Cyr.

Glabrous branching, trailing or often matted annual with ovate leaves; flowers solitary or in few-flowered leafy cymes.

An introduced, but now a widely established weed in gardens and cultivated fields, north along the Mackenzie River to its delta.

General distribution: Cosmopolitan weed.

Fig. 459

Stellaria monantha Hult.

In its typical form *S. monantha* is densely matted or tufted, in general habit often resembling *S. laeta*. It is totally glabrous and normally glaucous, so much so that even from a distance a mat of it appears distinctly blue. Like the other arctic members of the *S. longipes* complex it is well equipped for vegetative reproduction by the formation of over-wintering, green, axillary buds, and by its subterranean stems. The flowers are pedicellate in the axils of normal stem leaves or leaf-like bracts.

North American tundra species with a similar range as *S. laeta* although in the East reaching south to Newfoundland and Gaspé, and in the Rocky Mountains fairly common in high alpine situations south to Montana, Wyoming, Colorado and Utah.

General distribution: N. America, arctic-alpine.

Fig. 460 Map 530

Stellaria stricta Richards.

High-grown and perfectly glabrous like *S. longipes s. str.* from which it differs by its narrower, linear-lanceolate leaves, narrower and more rich-flowered and mainly terminal inflorescence, with characteristically stiff and strongly ascending branches on which the terminal flowers are often approximate; the flowers are smaller than in *S. longipes*, with petals not much longer than the sepals, and the black and shiny capsule about one third longer than the calyx.

Essentially a prairie and foothill species commonly growing in damp meadows and on river banks, from James Bay and western Ontario west to Alberta and interior British Columbia north to southwestern Yukon Territory and the upper Mackenzie (with widely disjunct stations in the Mackenzie Delta and on Great Bear Lake, south to Washington, Colorado and California.

General distribution: N. America from James Bay to S. W. Y. T.

Fig. 461 Map 531

Stellaria subvestita Greene

In general habit resembling *S. longipes s. str.* more than *S. stricta*, but rarely as high-grown. It differs from the former by its green, non-glaucous, pilose to glabrate, narrower and strongly attenuate leaves and shorter capsules, and from both by its always densely pubescent inter-nodes. As in *S. stricta*, the petals are but slightly longer than the glabrous calyx.

A prairie and foothill species commonly growing in well-drained, grassy and rather dry places, with a general range rather similar to *S. stricta* but not known to us from south of the United States border.

General distribution: N. America, from James Bay to southern Y.T.

Fig. 462 Map 532

Stellaria umbellata Turcz.

Slender, erect-spreading, glabrous, dwarf perennial from a weak spindly root; leaves lanceolate; flowers small, commonly lacking petals, in small, axillary or terminal cymes.

Moist, alpine slopes.

General distribution: Amphi-Beringian across Alaska east to Richardson Mts. west of the Mackenzie Delta, south, with a large gap, to Alta., Nevada and Oregon.

Map 533

Wilhelmsia Rchb.

Wilhelmsia physodes (Fisch.) McNeill

Arenaria physodes Fisch. in DC.

Merckia physodes (Ser.) Cham. & Schlecht.

Perennial, of matted habit with freely branching and rooting stems. Leaves elliptic-lanceolate, about 1 cm long, sessile, opposite or whorled, punctate beneath, and with soft-ciliate margins. Flowers terminal and solitary; petals linear-oblong, white, longer than the sepals. Capsule spherical and inflated, 3 to 4-locular, its valves leathery, reddish-brown and shiny; the 2 mm large, brown seeds remain in the capsules that stay unopened during winter to be dispersed by spring floods.

Wet, sheltered river banks, lake shores and shores of brackish lagoons.

General distribution: Amphi-Beringian; Alaska and Y.T. to Franklin Bay east of the Mackenzie Delta.

Fig. 463 Map 534

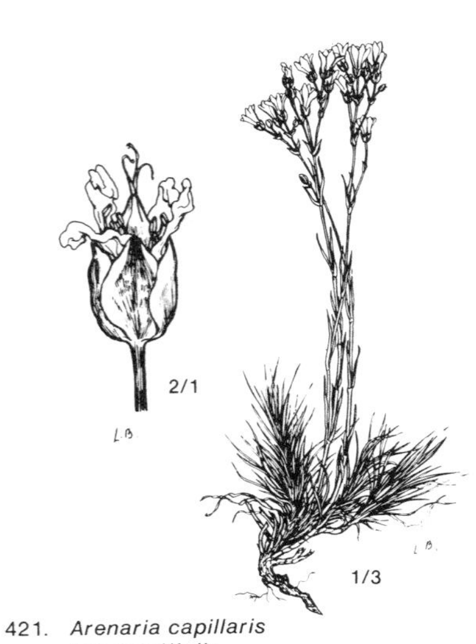

421. *Arenaria capillaris* var. *nardifolia*

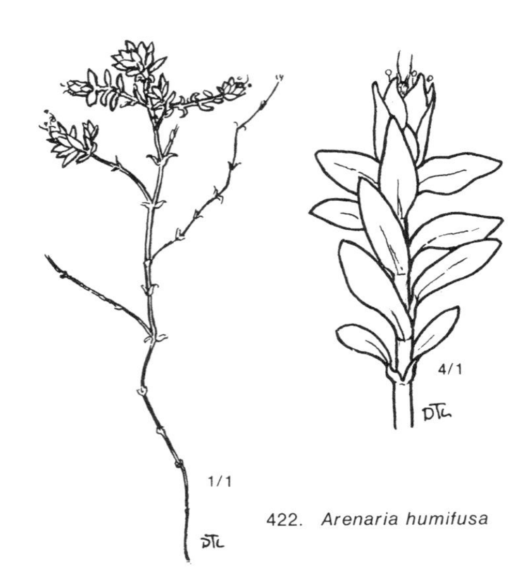

422. *Arenaria humifusa*

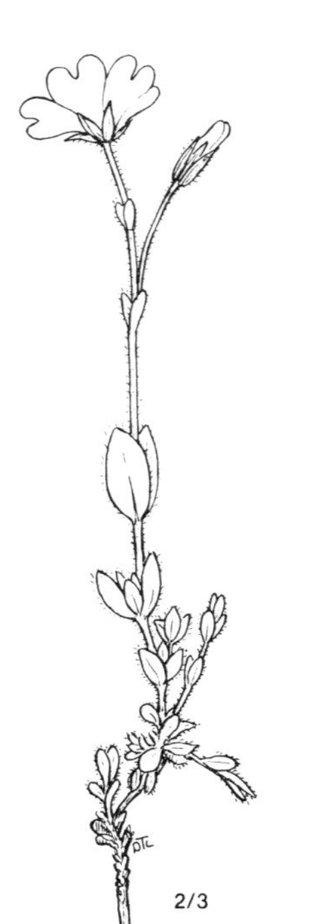

423. *Cerastium alpinum*

424. *Cerastium arvense*

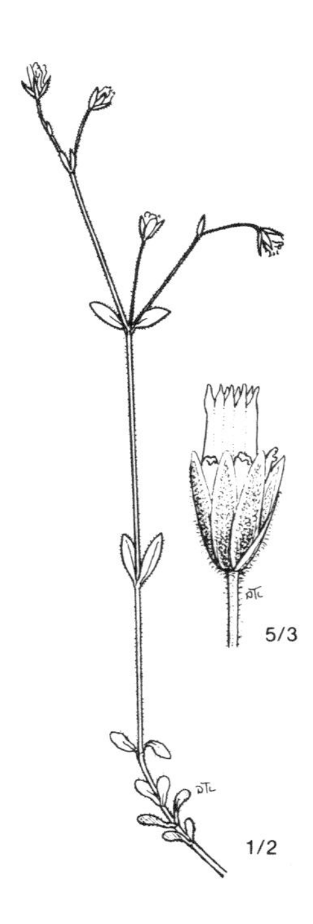

425. *Cerastium Beeringianum*

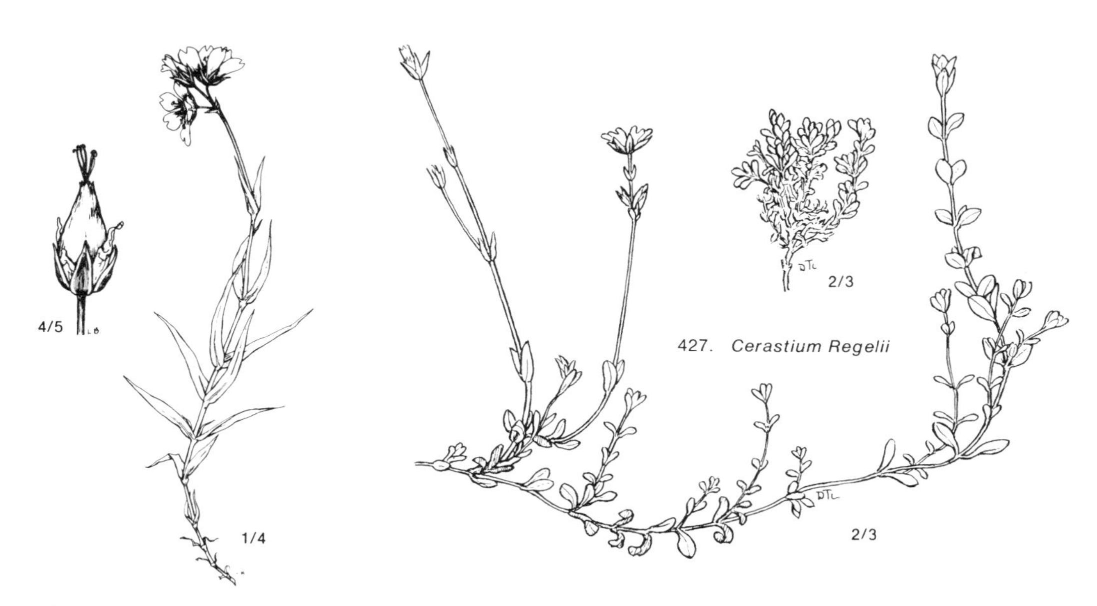

427. *Cerastium Regelii*

426. *Cerastium maximum*

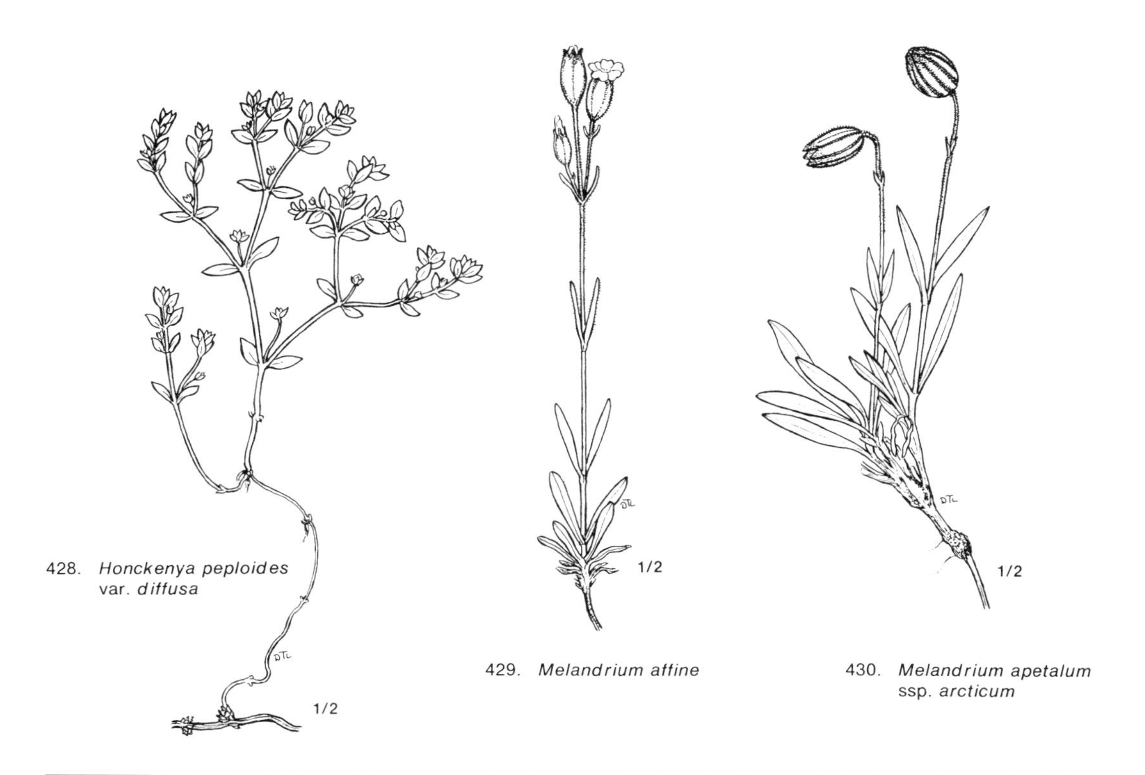

428. *Honckenya peploides* var. *diffusa*

429. *Melandrium affine*

430. *Melandrium apetalum* ssp. *arcticum*

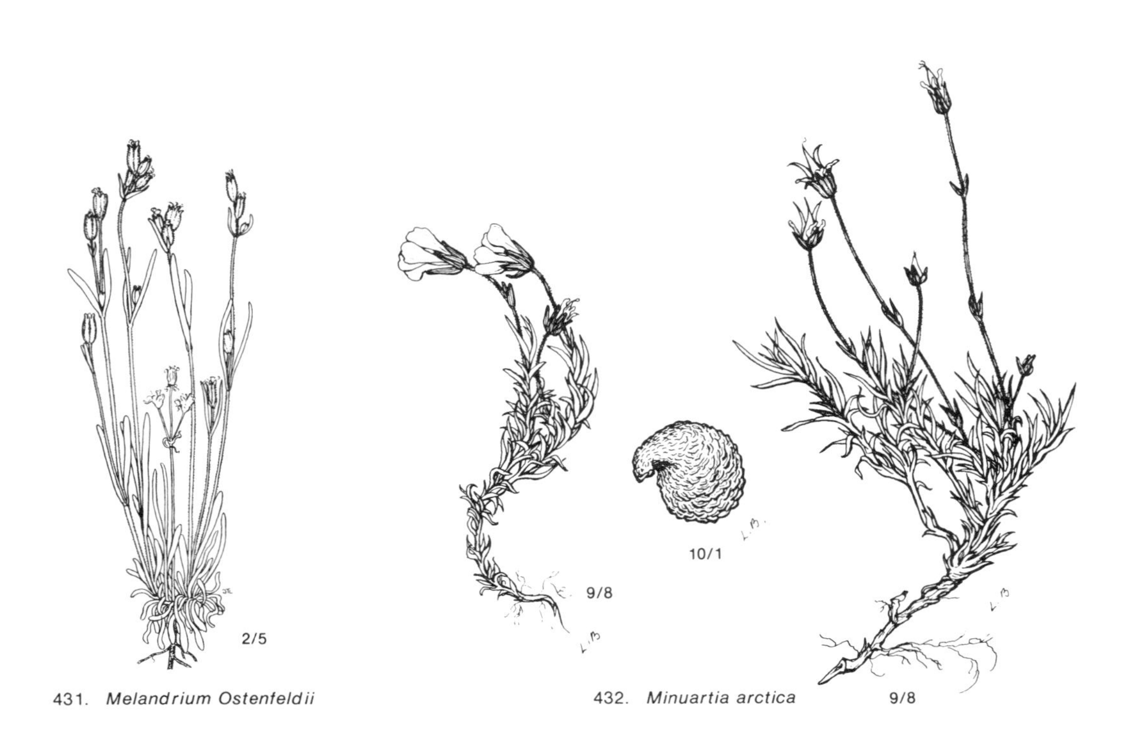

431. *Melandrium Ostenfeldii*

432. *Minuartia arctica*

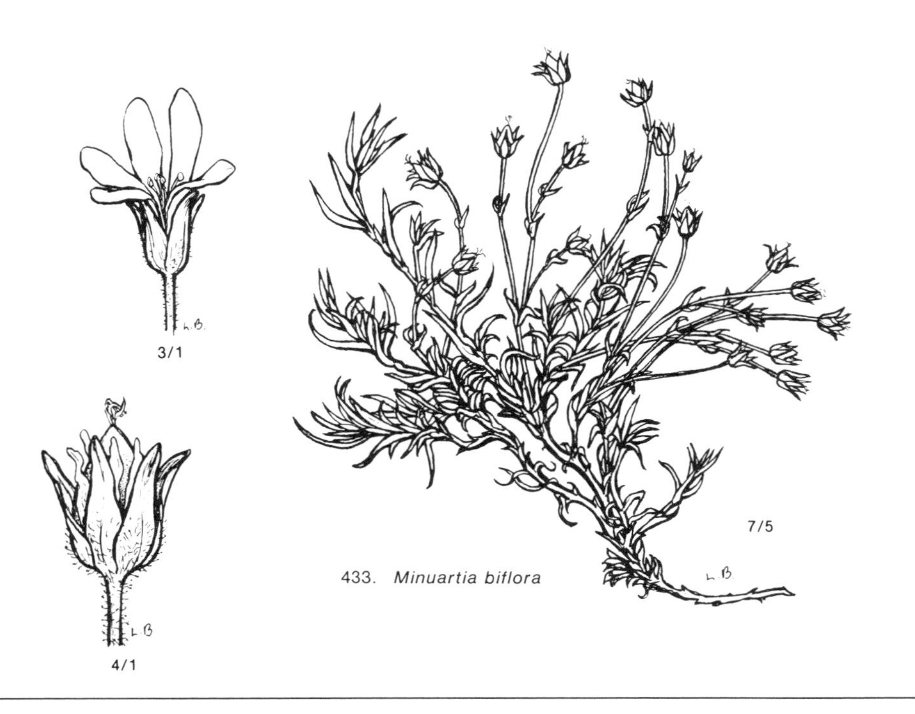

433. *Minuartia biflora*

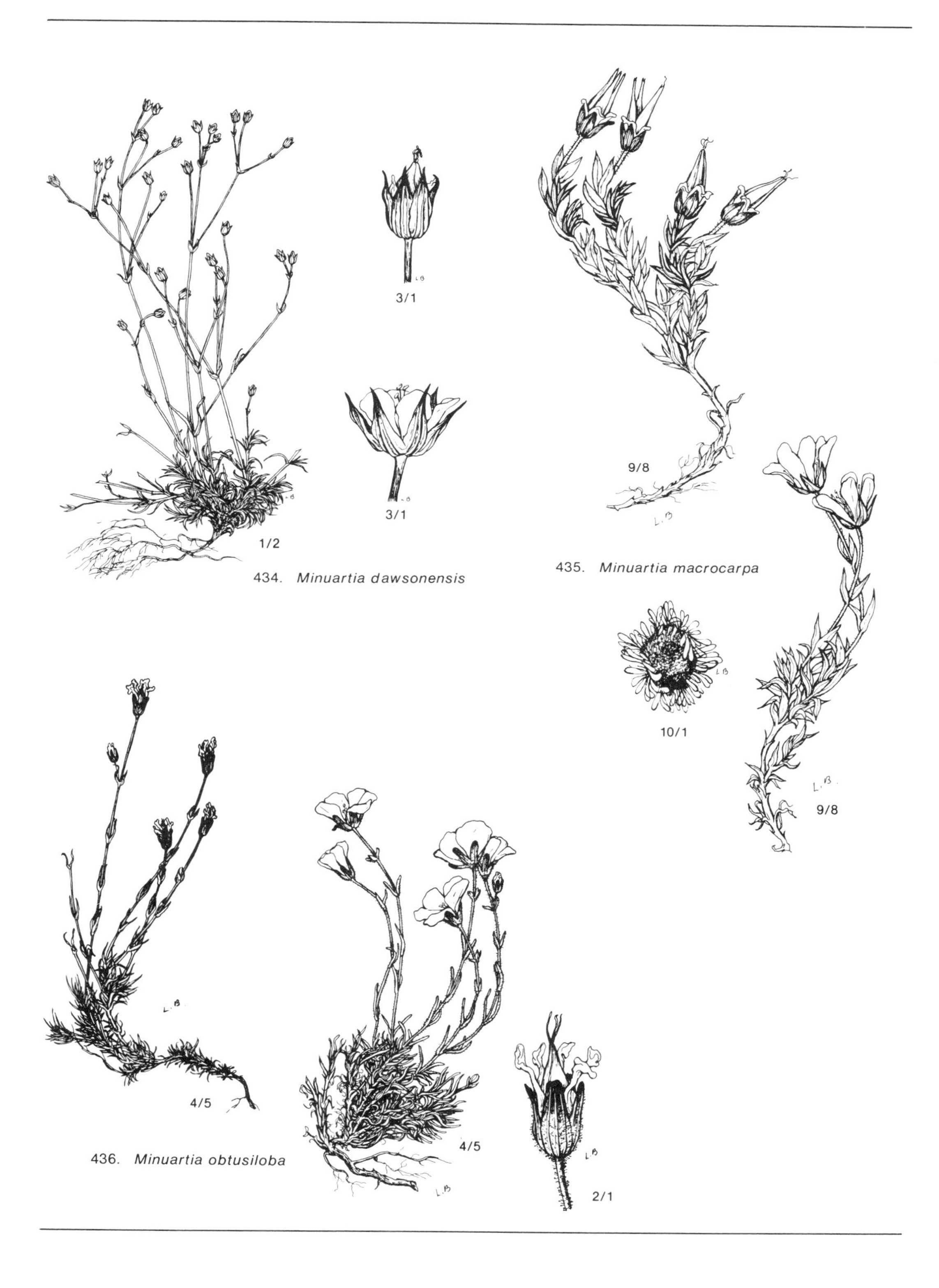

434. *Minuartia dawsonensis*

435. *Minuartia macrocarpa*

436. *Minuartia obtusiloba*

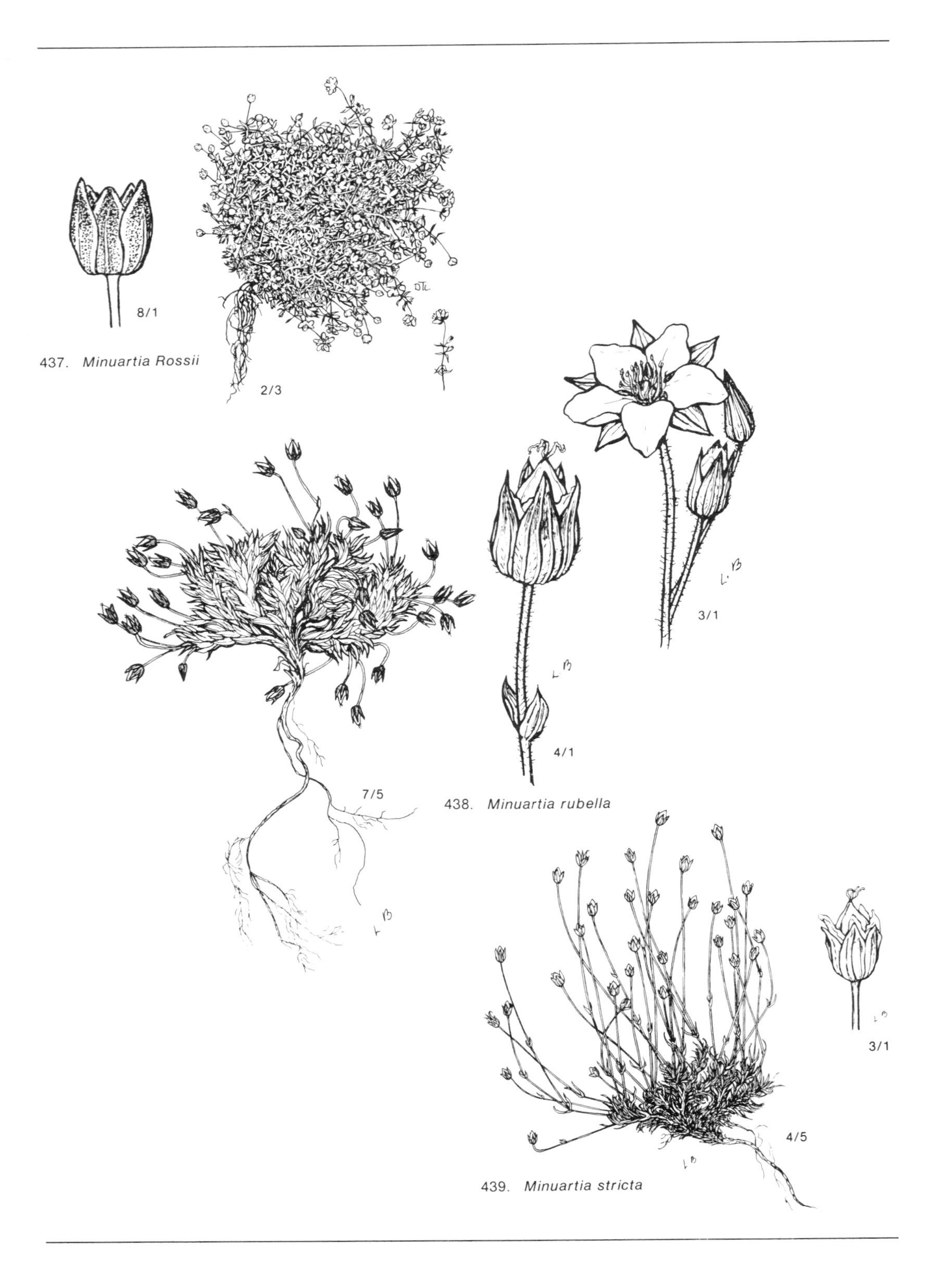

437. *Minuartia Rossii*

438. *Minuartia rubella*

439. *Minuartia stricta*

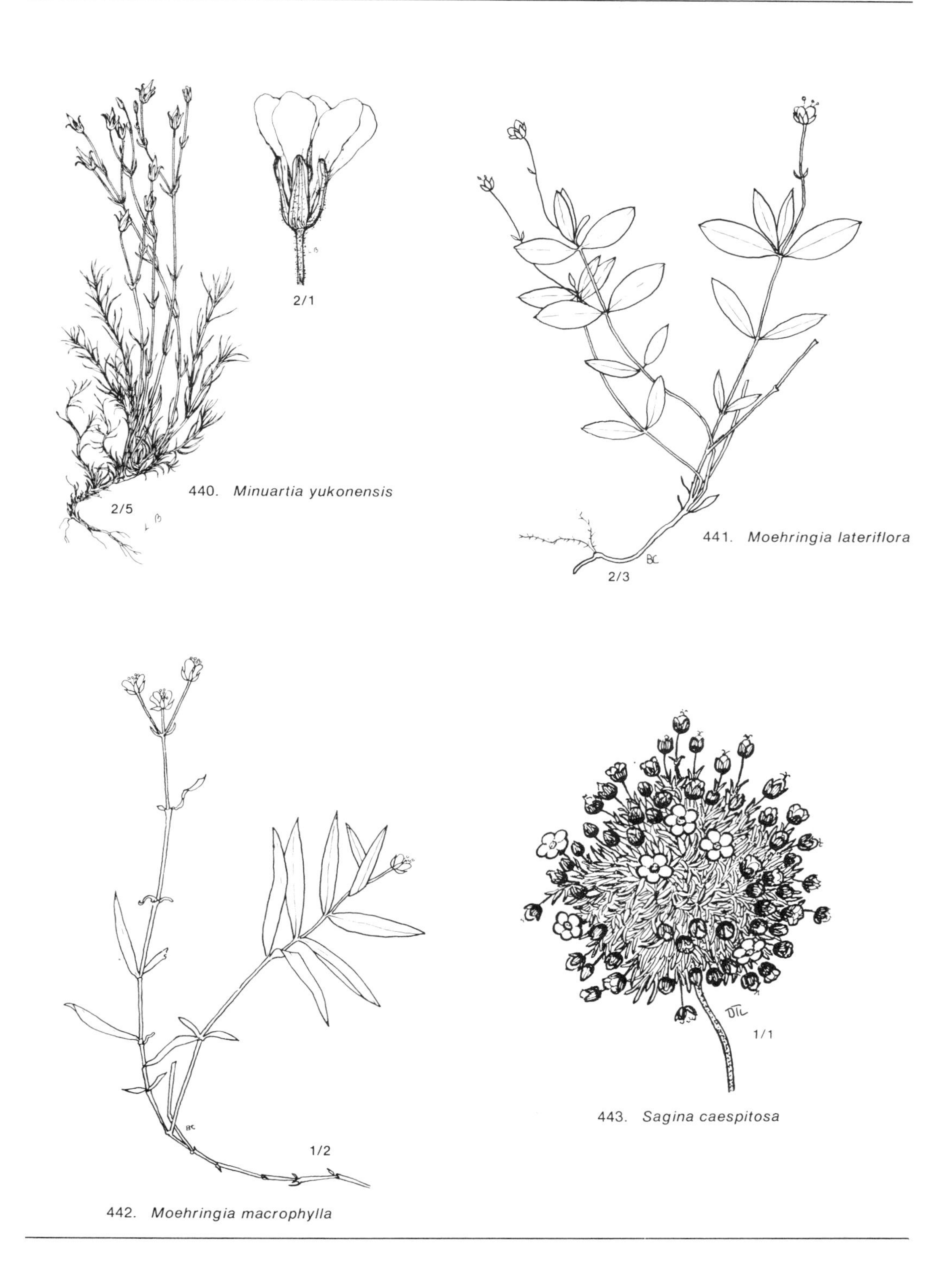

440. *Minuartia yukonensis*

441. *Moehringia lateriflora*

442. *Moehringia macrophylla*

443. *Sagina caespitosa*

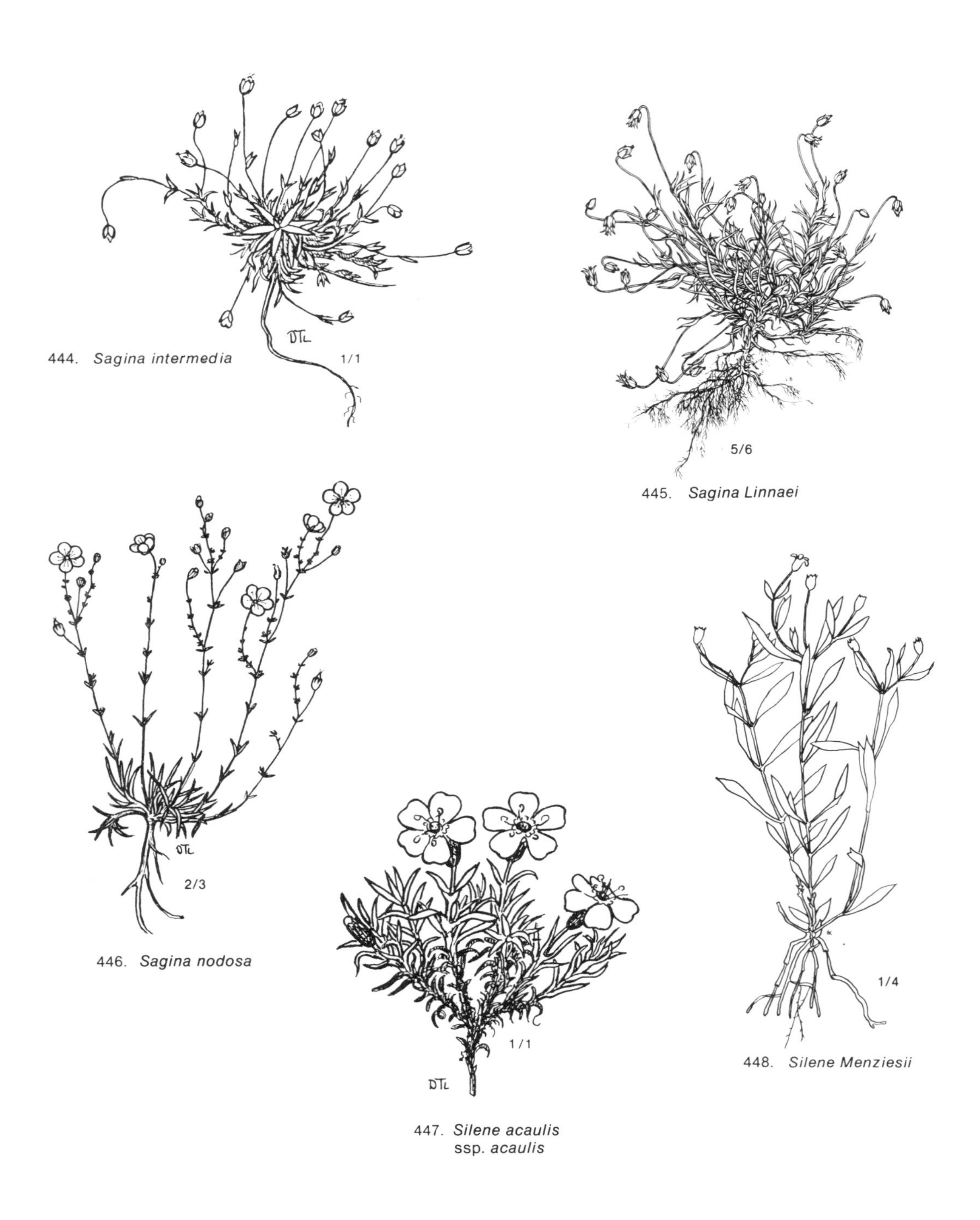

444. *Sagina intermedia*

445. *Sagina Linnaei*

446. *Sagina nodosa*

447. *Silene acaulis* ssp. *acaulis*

448. *Silene Menziesii*

449. *Silene repens* ssp. *purpurata*

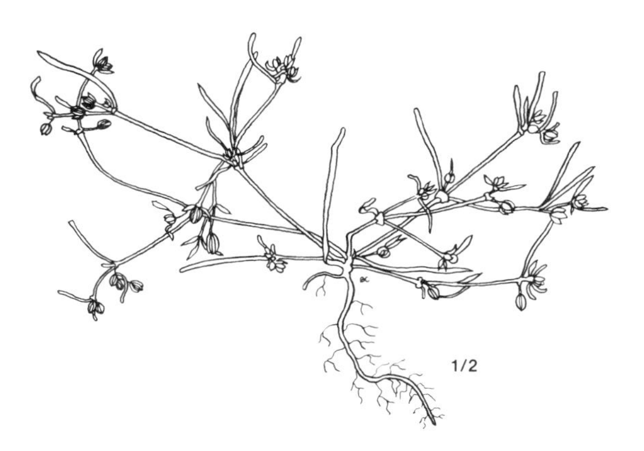

450. *Spergularia marina*

451. *Stellaria calycantha*

452. *Stellaria crassifolia*

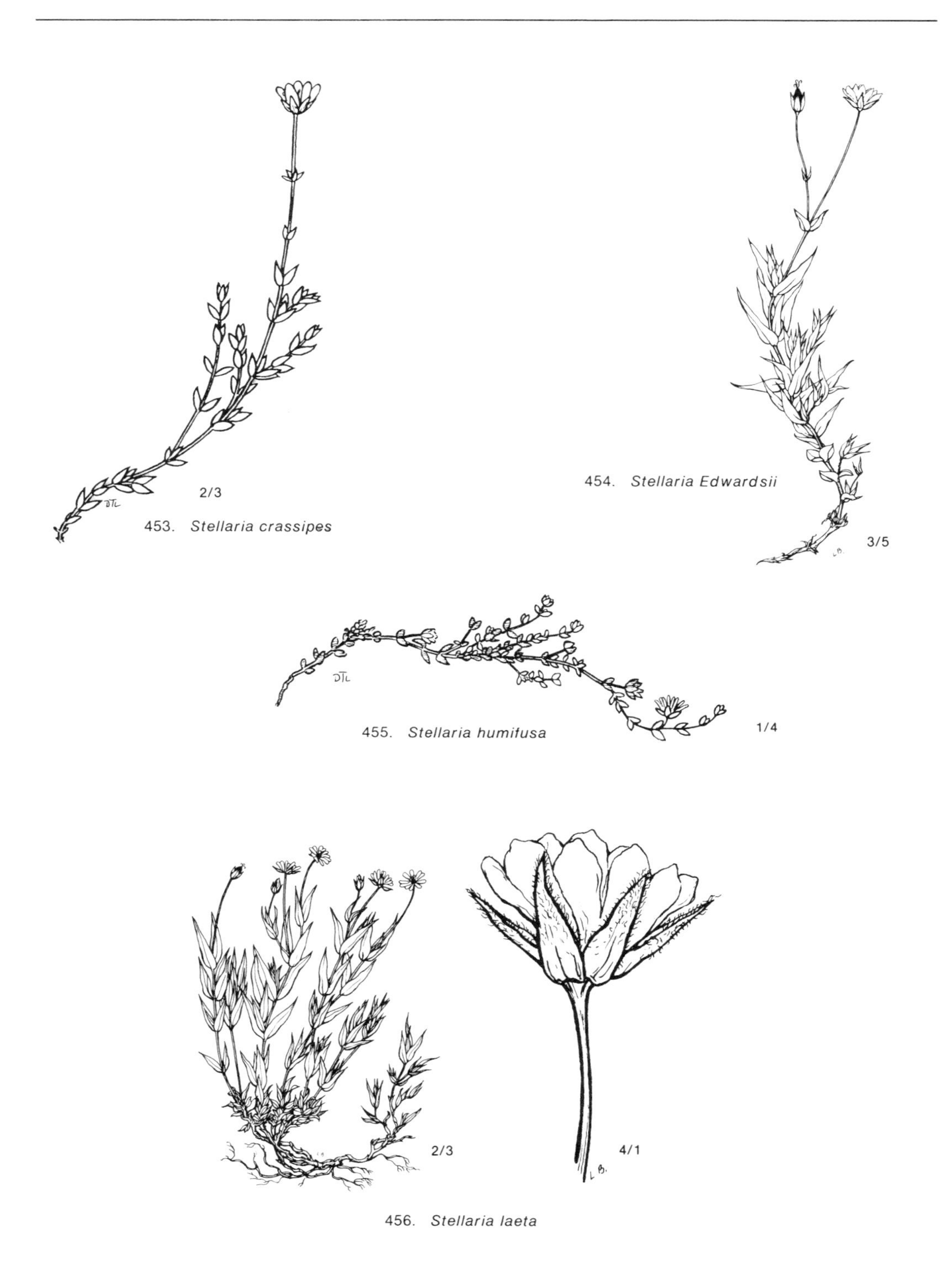

453. *Stellaria crassipes*

454. *Stellaria Edwardsii*

455. *Stellaria humifusa*

456. *Stellaria laeta*

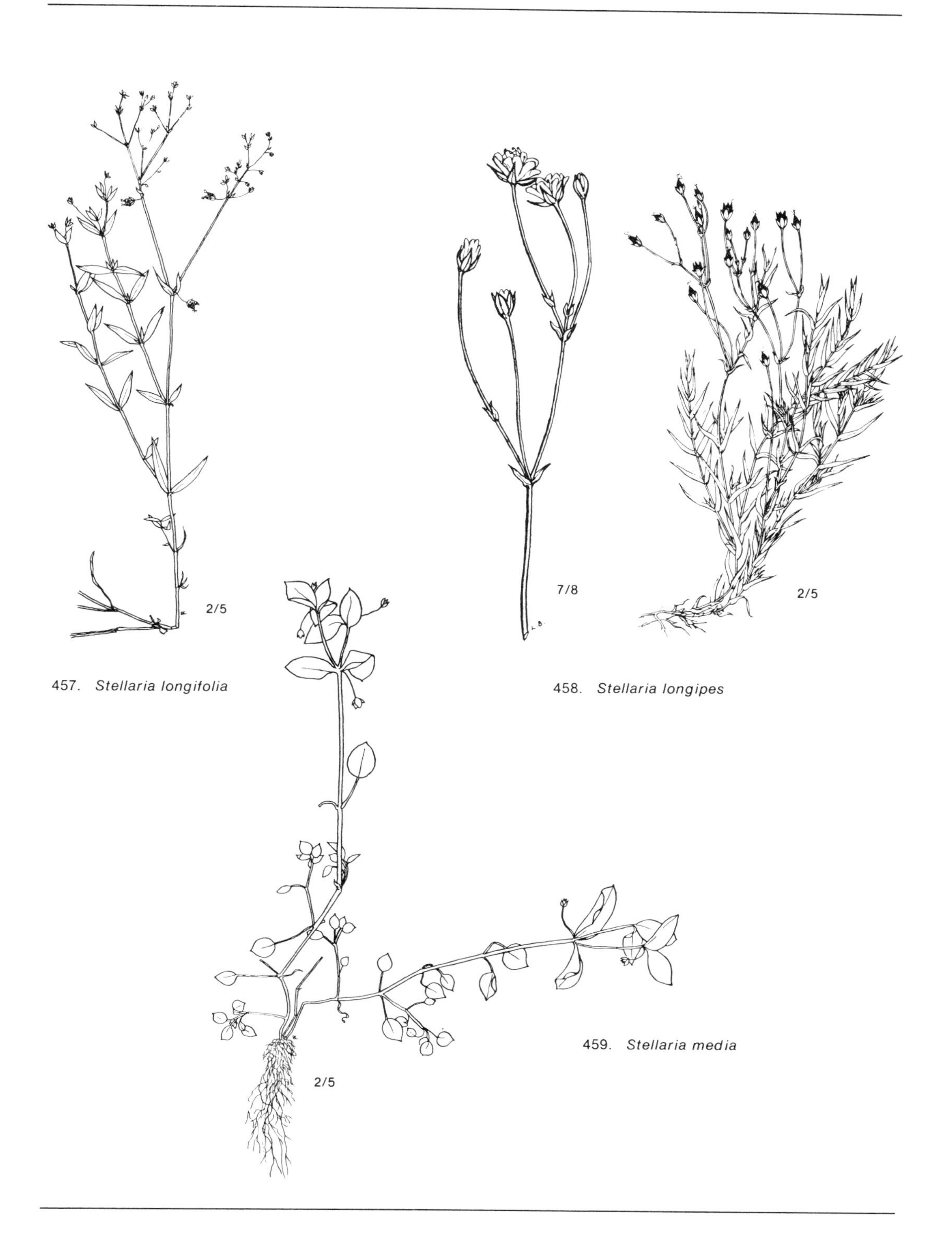

457. *Stellaria longifolia*

458. *Stellaria longipes*

459. *Stellaria media*

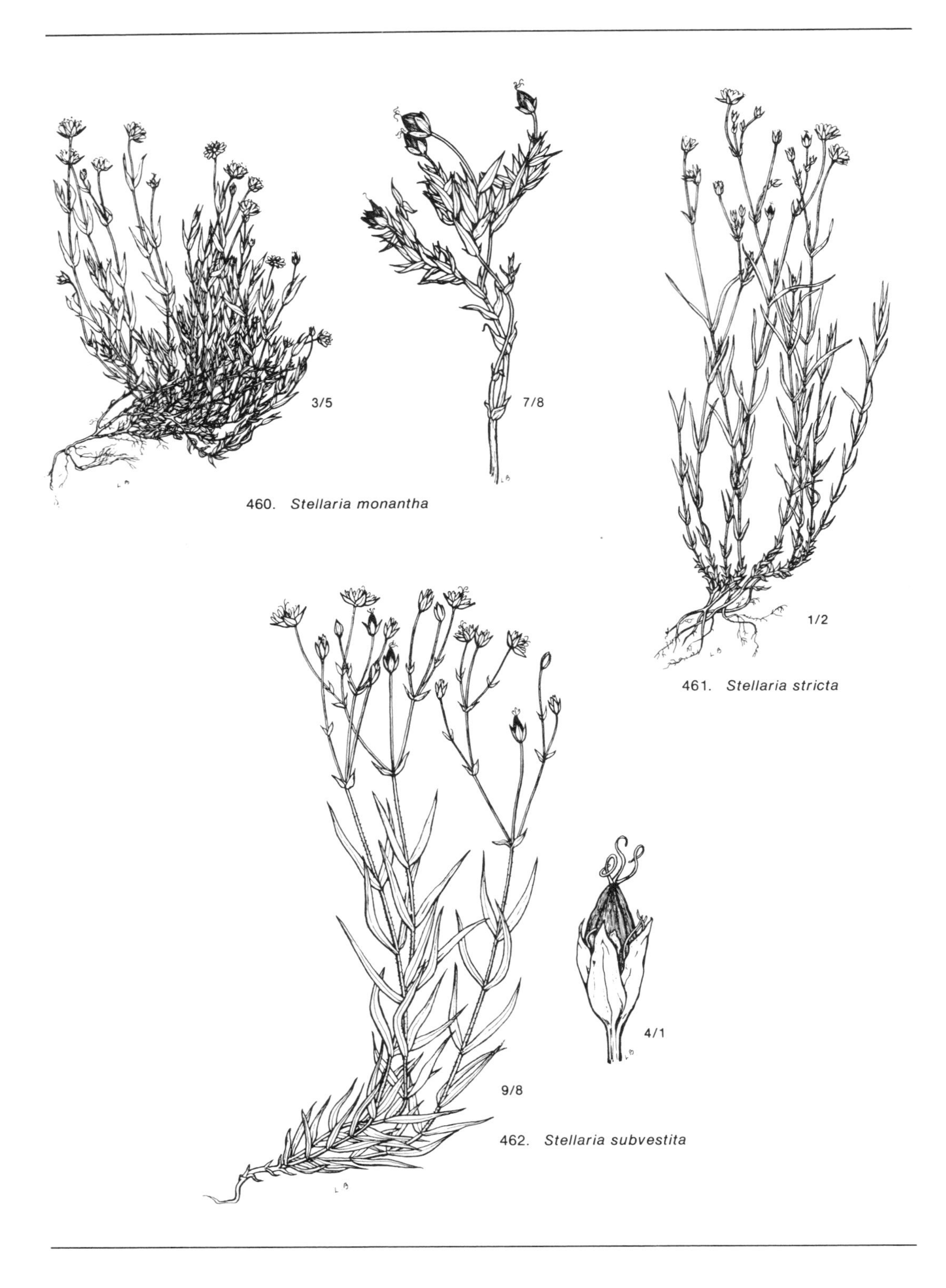

460. *Stellaria monantha*

461. *Stellaria stricta*

462. *Stellaria subvestita*

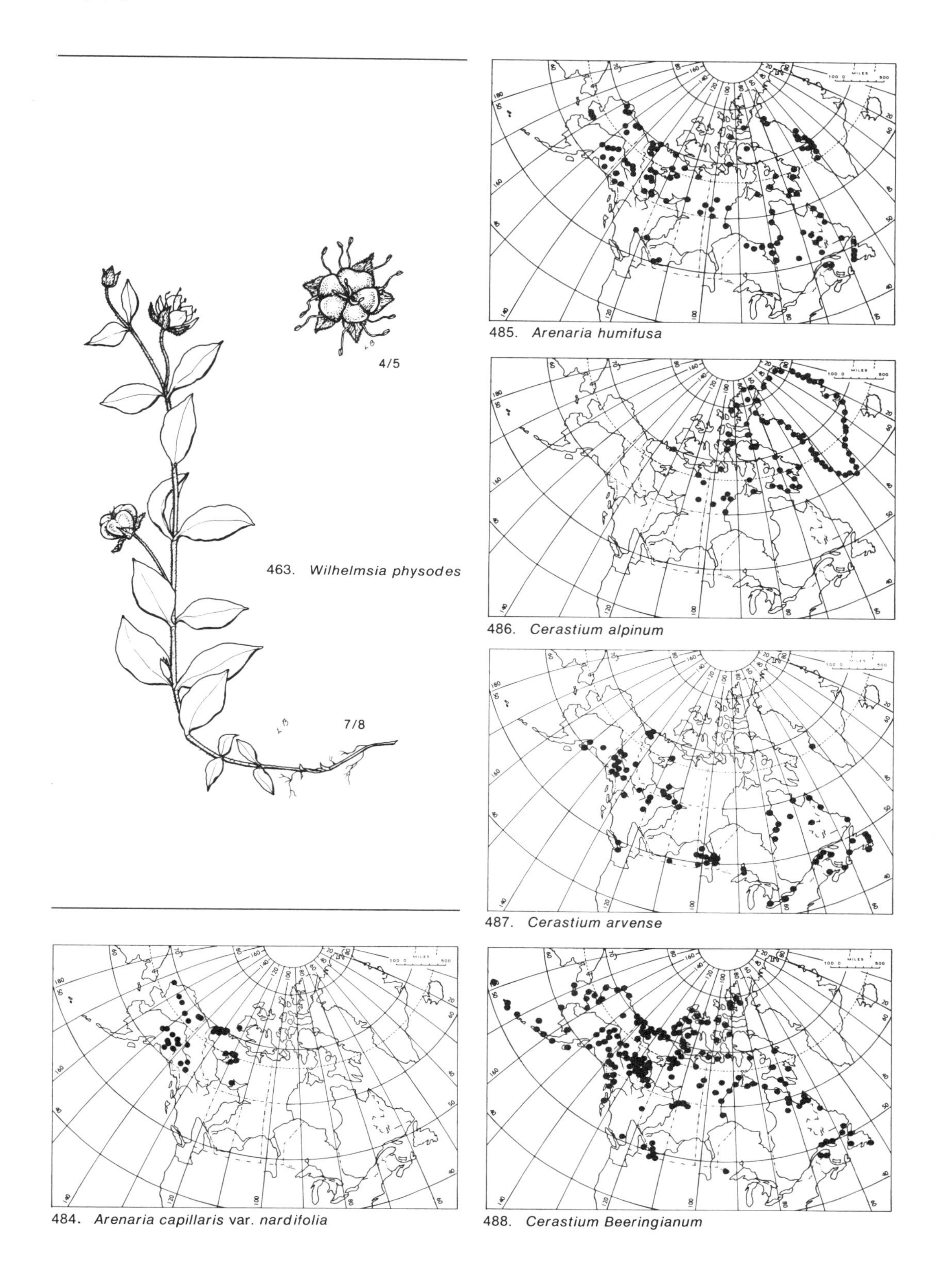

463. *Wilhelmsia physodes*

485. *Arenaria humifusa*

486. *Cerastium alpinum*

487. *Cerastium arvense*

484. *Arenaria capillaris* var. *nardifolia*

488. *Cerastium Beeringianum*

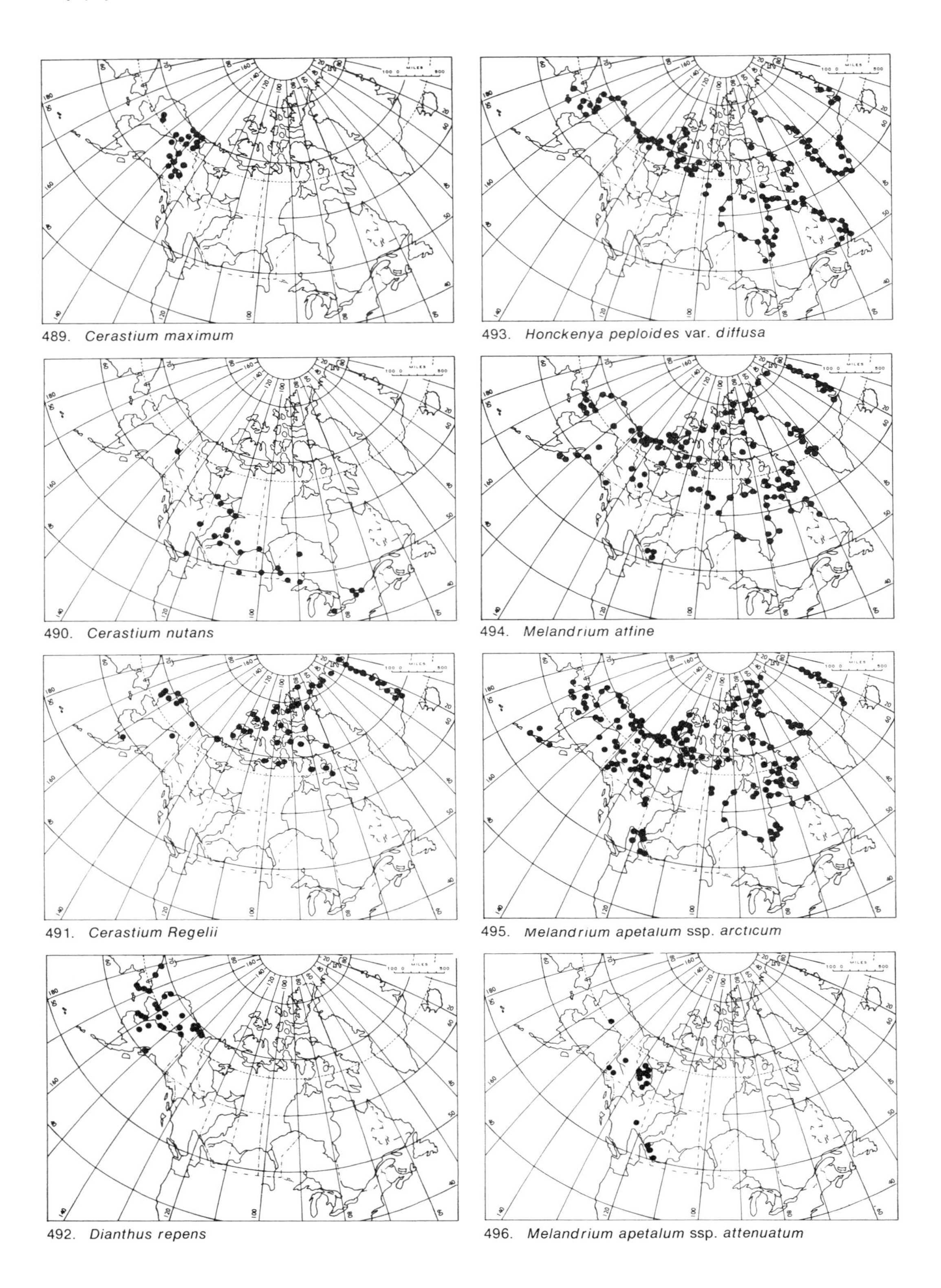

489. *Cerastium maximum*

490. *Cerastium nutans*

491. *Cerastium Regelii*

492. *Dianthus repens*

493. *Honckenya peploides* var. *diffusa*

494. *Melandrium affine*

495. *Melandrium apetalum* ssp. *arcticum*

496. *Melandrium apetalum* ssp. *attenuatum*

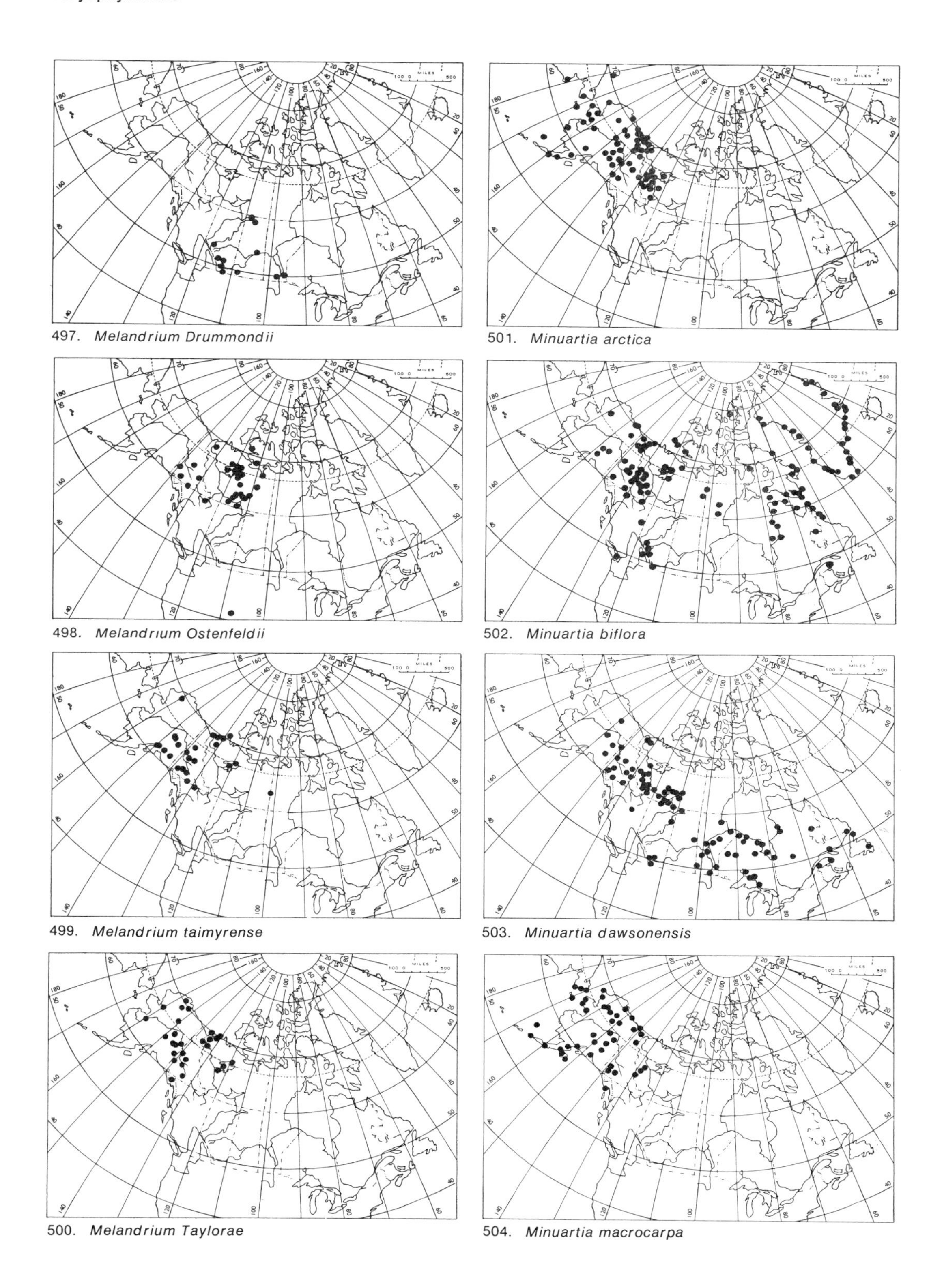

497. *Melandrium Drummondii*

498. *Melandrium Ostenfeldii*

499. *Melandrium taimyrense*

500. *Melandrium Taylorae*

501. *Minuartia arctica*

502. *Minuartia biflora*

503. *Minuartia dawsonensis*

504. *Minuartia macrocarpa*

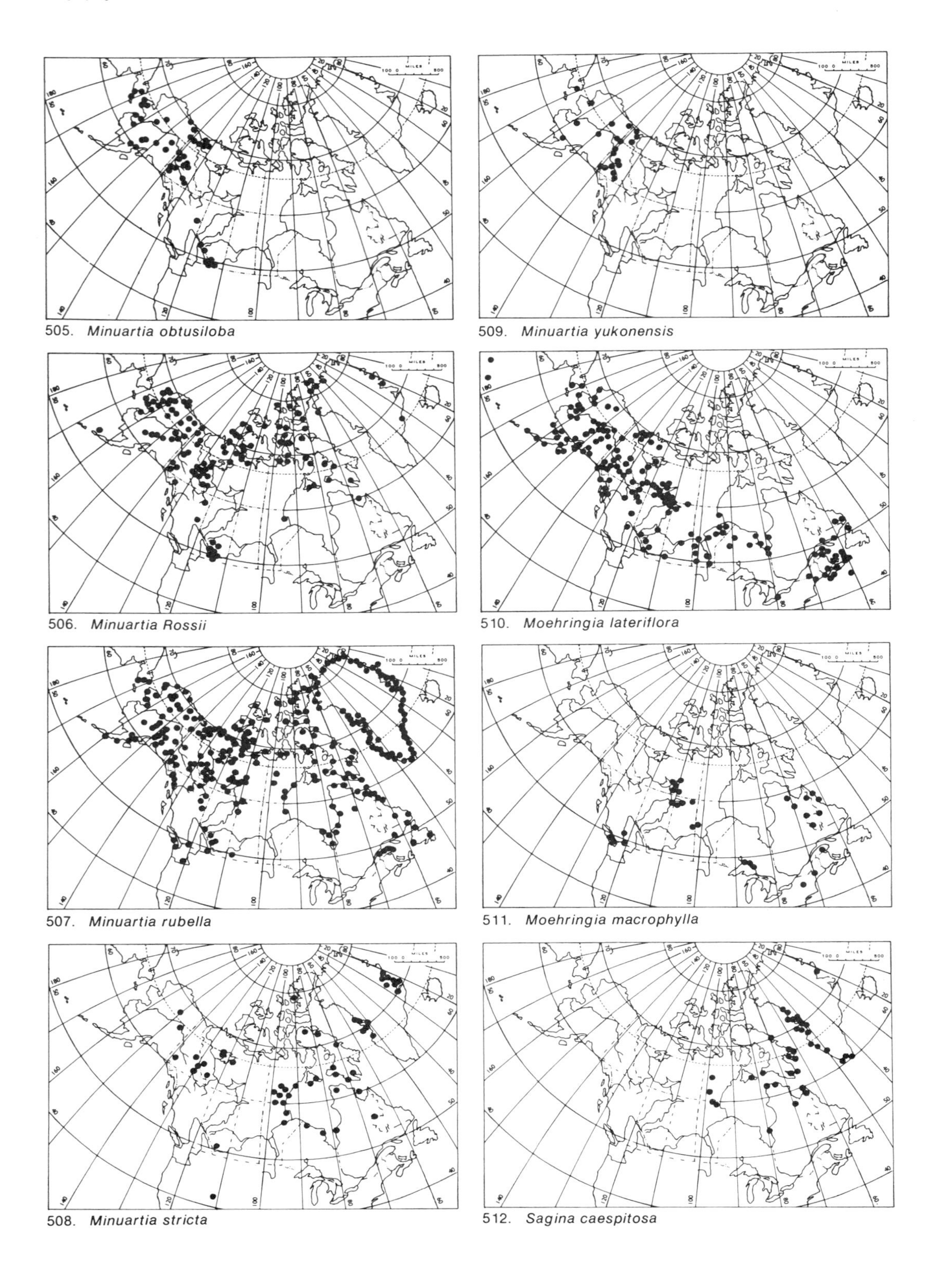

505. *Minuartia obtusiloba*

506. *Minuartia Rossii*

507. *Minuartia rubella*

508. *Minuartia stricta*

509. *Minuartia yukonensis*

510. *Moehringia lateriflora*

511. *Moehringia macrophylla*

512. *Sagina caespitosa*

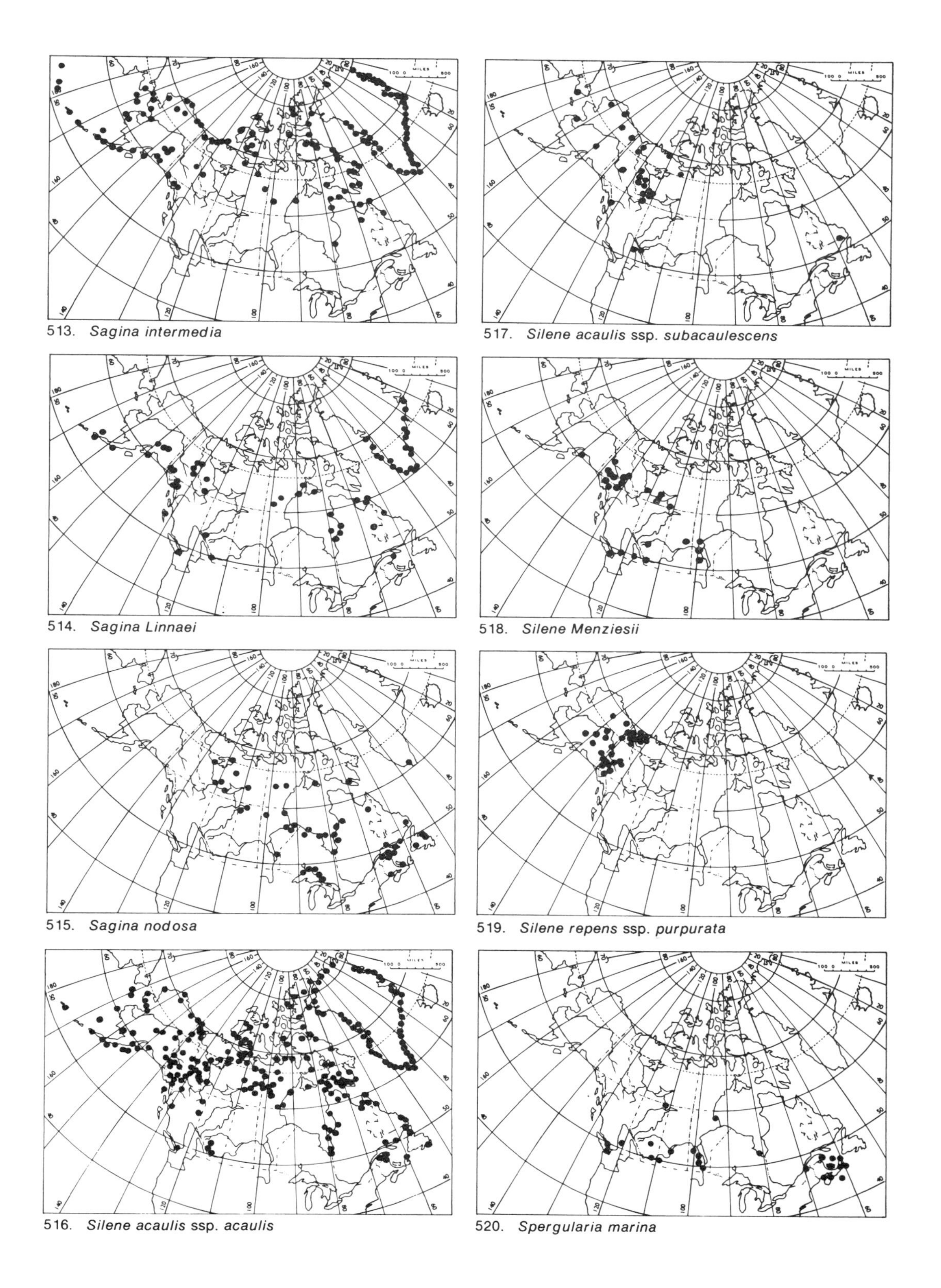

513. *Sagina intermedia*

514. *Sagina Linnaei*

515. *Sagina nodosa*

516. *Silene acaulis* ssp. *acaulis*

517. *Silene acaulis* ssp. *subacaulescens*

518. *Silene Menziesii*

519. *Silene repens* ssp. *purpurata*

520. *Spergularia marina*

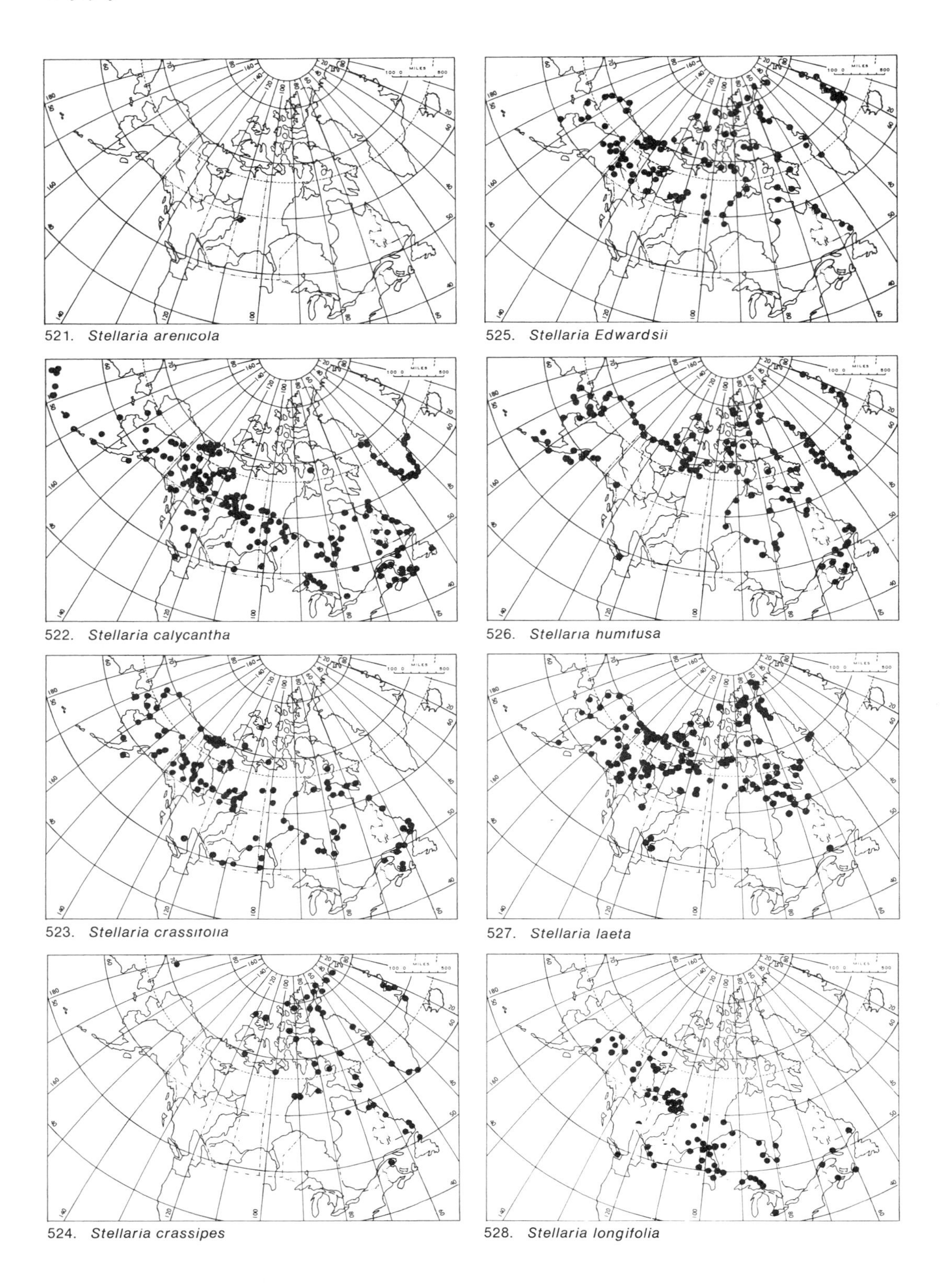

521. *Stellaria arenicola*

522. *Stellaria calycantha*

523. *Stellaria crassifolia*

524. *Stellaria crassipes*

525. *Stellaria Edwardsii*

526. *Stellaria humifusa*

527. *Stellaria laeta*

528. *Stellaria longifolia*

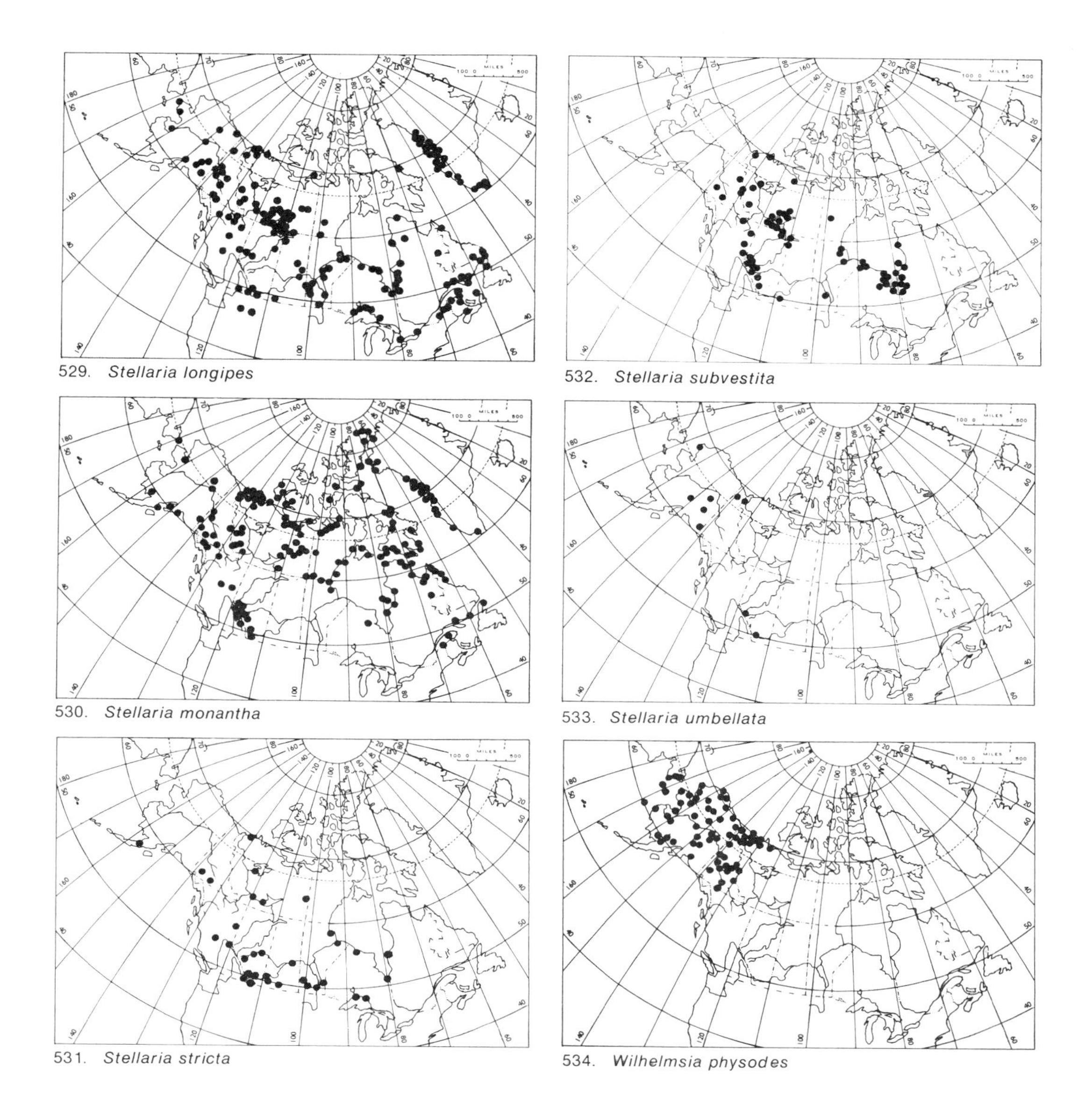

529. *Stellaria longipes*

530. *Stellaria monantha*

531. *Stellaria stricta*

532. *Stellaria subvestita*

533. *Stellaria umbellata*

534. *Wilhelmsia physodes*

CERATOPHYLLACEAE
Ceratophyllum L.

Ceratophyllum demersum L.
Hornwort
Submersed, non-rooted aquatic with slender and freely branched stems with whorled and finely dissected leaves, and sessile, axillary and monoecious flowers lacking a true perianth. Fruit a 4 to 6 mm long narrowly elliptic achene with two prominent basal spines.

The hornwort superficially resembles a sterile *Myriophyllum* from which, however, it is readily distinguished by the submerged leaves that, in hornwort, are dichotomously divided but pinnate in *Myriophyllum*.

Although in our area only recently reported from Great Slave Lake, and southern Nahanni National Park, where, apparently, it is rare, well preserved *Ceratophyllum* fruits, radiocarbon dated at 5500 ± 250 years BP, are known from peat deposits in a pingo in the upper Thelon River valley, Northwest Territories.

General distribution: Circumpolar, non-arctic.

Fig. 464 Map 535

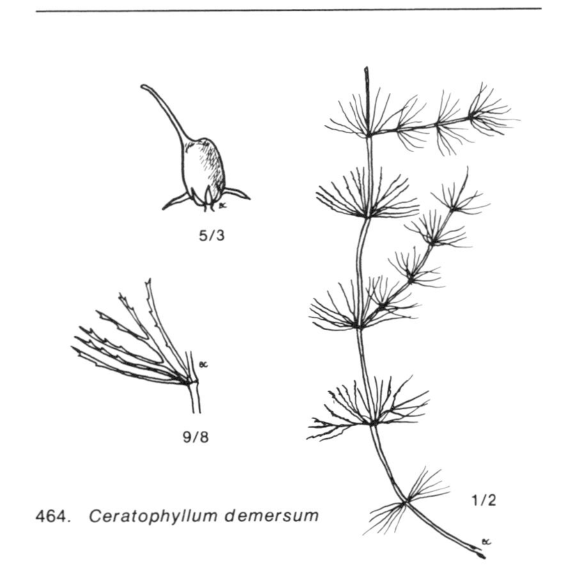

464. *Ceratophyllum demersum*

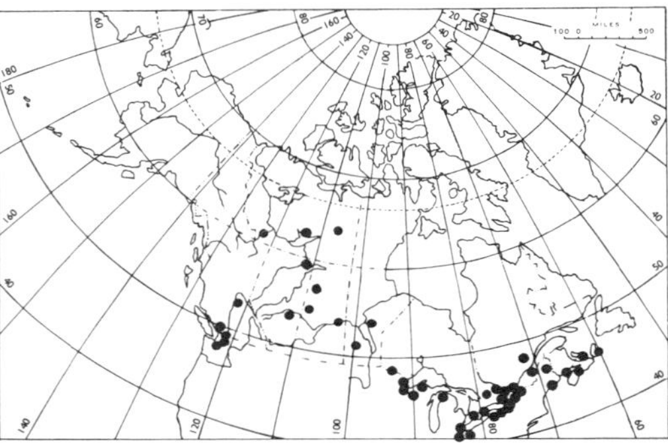
535. *Ceratophyllum demersum*

NYMPHAEACEAE Water-lily Family

Perennial aquatics with fleshy rhizomes from which rise long-petioled floating, oblong and deeply cordate leaves and equally long-peduncled emergent flowers.

a. Rhizome horizontal and long; leaf-blades 10-30 cm long, their lobes rounded; sepals petal-like, yellow, the petals small and scale-like . . . . *Nuphar*
a. Rhizome short erect-ascending; leaf-blade 6-7 cm long, their lobes acute; sepals green, petals white . . . . *Nymphaea*

Nuphar Sm. Yellow Pond-lily

Rhizome thick and cylindrical, with numerous leaf scars; leaves floating, large, with a deep sinus at the base; flowers single, held above the water; sepals yellow, leathery, but petal-like, the petals numerous and scale- or stamen-like; fruit ovoid.

a. Sepals 7-9, yellow; anthers reddish; petioles terete . . . . *N. polysepalum*
a. Sepals 6 or fewer, the inner red at base inside; anthers yellow; petioles flattened . . . . *N. variegatum*

Nuphar polysepalum Engelm.
Floating leaves commonly 2 to 3 dm long; the expanded flowers 8 to 13 cm in diameter; the number of sepals 7 to 9, and the anthers reddish-purple, drying black, the mature fruit 4 to 8 cm long.

In quiet waters up to 3 m deep; common throughout the forested parts of Alaska and Yukon Territory, east to the Mackenzie Delta and the Eskimo Lakes where it is common.

General distribution: Endemic of northwestern N. America.

Map 536

Nuphar variegatum Engelm.
Similar to *N. polysepalum* but somewhat smaller in all parts; thus the floating leaves are commonly 10 to 20 cm long and very rarely exceed 25 cm; the diameter of the expanded flower rarely exceeds 4 cm, and the length of the mature fruit is usually between 2.5 to 4.0 cm. However, the most reliable distinguishing character is the number of sepals which does not exceed 6, and the yellow stamens that dry yellow in *N. variegatum* but black in *N. polysepalum*.

The rhizomes of yellow water-lilies provide an important year long source of food for beaver, muskrat, and in summer also for the moose.

Common throughout the wooded lowlands west to the upper Mackenzie Basin and north to Great Bear Lake.

General distribution: N. America.

Fig. 465 Map 537

Nymphaea L. Water-lily

Nymphaea tetragona Georgi
ssp. **Leibergii** (Morong) Porsild
Much smaller and more delicate in all parts than the yellow water-lilies, and always growing in shallower water, mostly less than one meter deep. Flower and leaf peduncles very slender, rarely more than 3 mm in diameter. Flowers white, unscented and from 4 to 6 cm in diameter.

Boreal forest species, with us known from a single collection made 60 years ago on an island in Great Slave Lake, 40 miles northeast of Fort Resolution.

General distribution: Circumpolar; in N. America from Que. to Alaska.

Fig. 466 Map 538

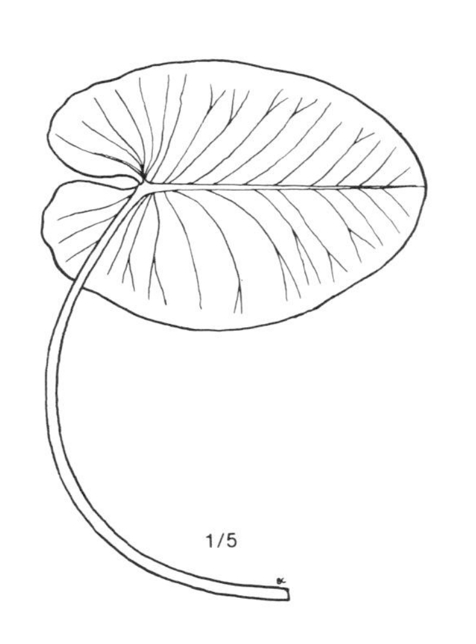

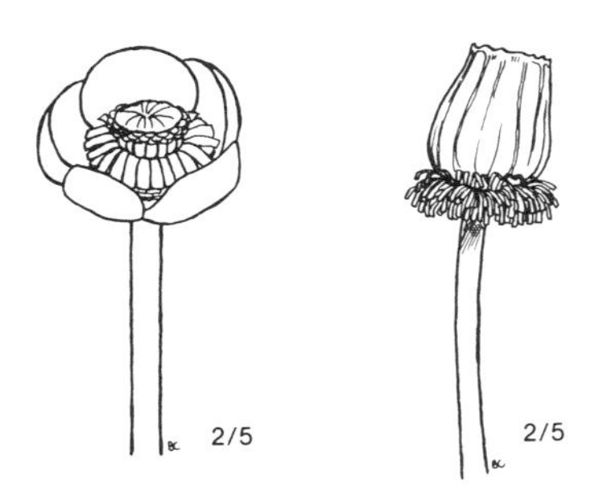

465. *Nuphar variegatum*

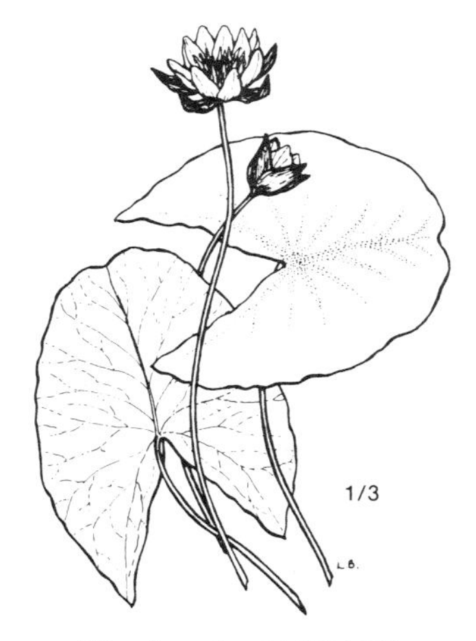

466. *Nymphaea tetragona* ssp. *Leibergii*

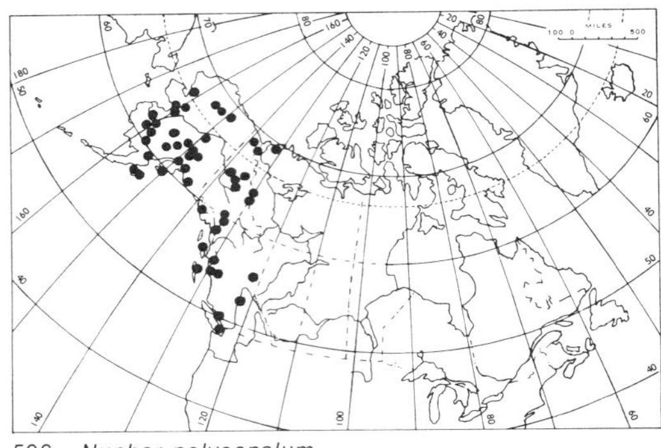

536. *Nuphar polysepalum*

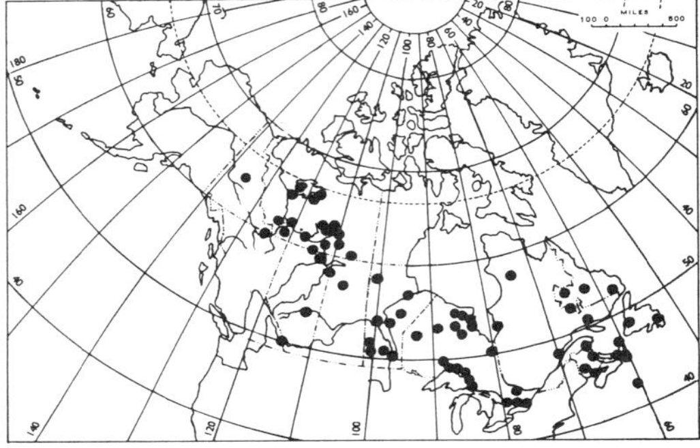

537. *Nuphar variegatum*

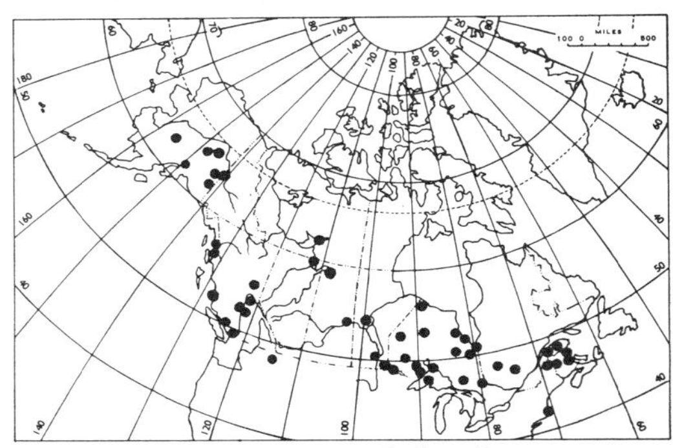

538. *Nymphaea tetragona* ssp. *Leibergii*

RANUNCULACEAE Crowfoot Family

Ours perennial herbs with colourless, acrid juice and alternate simple or compound leaves without stipules, but often with the petioles dilated at the base. Flowers symmetrical or zygomorphic, polypetalous or apetalous with petaloid calyx. Fruit an achene, a many-seeded pod or follicle, opening along one side, or a berry in *Actaea*.

a. Fruit berry-like, flowers small, white or greenish in a dense, cylindrical raceme . . . . . *Actaea*
a. Fruit not berry-like
  b. Fruit a follicle, opening along one side
    c. Flowers regular
      d. Flowers yellow or white, petals without spurs
        e. Follicles sessile, leaves simple or merely cleft . . . . *Caltha*
        e. Follicles stipitate, leaves ternately cleft . . . . *Coptis*
      d. Flowers purple and white, petals spurred . . . . *Aquilegia*
    c. Flowers irregular, blue
      f. Upper sepal hooded, flowers few in an open raceme . . . . *Aconitum*
      f. Upper sepal with long spur; flowers numerous in a spike-like raceme . . *Delphinium*
  b. Fruit a dry achene borne in a head, or in a dense short-cylindric spike
    g. Petals wanting, sepals petaloid
      h. Flowers large and showy, 1-several
        i. Styles long and plumose, flowers purple . . . . *Pulsatilla*
        i. Styles short, not plumose; flowers yellow, white or pale blue . . . . *Anemone*
      h. Flowers small and numerous . . . . *Thalictrum*
    g. Petals normally present; sepals 3 or 5 . . . . *Ranunculus*

Aconitum L. Monkshood

Aconitum delphinifolium DC.
Glabrous, stems simple, 2 to 5 dm tall from a short, corm-like rhizome. Leaves long-peduncled and mainly cauline, palmately lobed, the divisions deeply dissected. Flowers 3 to 5, deep blue, 2 to 4 cm in diameter, in open racemes; follicles erect, 1.0 to 1.5 cm long.

Moist subalpine meadows, Mackenzie and Richardson Mountains, north to the Arctic Coast.

General distribution: Amphi-Beringian.

Fig. 467 Map 539

Actaea L. Baneberry

Actaea rubra (Ait.) Willd.
Red Baneberry
Glabrous; stems one to several, 5 to 10 dm tall, from a thick fleshy rhizome; leaves thin and very large, 2 to 3-ternately divided, the segments sharply toothed. The numerous small and white flowers in a stout, terminal raceme. The fruit berry-like, bright red or rarely pink, and poisonous.

General distribution: Wide ranging boreal forest species; from Nfld. to Alaska.

Map 540

Anemone L. Anemone

Perennial herbs with radical, petioled leaves, those of the flowering stem whorled and forming a leafy involucrum below the solitary or umbellate flowers; petals lacking but the sepals coloured and petaloid; fruit a dry achene.

a. Plants with slender, creeping rhizome; sepals yellow . . . . *A. Richardsonii*
a. Plant tufted; sepals white, bluish or red
  b. Achenes small, densely woolly, the style not hooked
    c. Plant rarely more than 2 dm high; flowers solitary
      d. Sepals white or pale blue on the outside

e. Leaf-segments cuneate; sepals glabrous . . . . *A. parviflora*
e. Leaf-segments oblong or linear, sepals silky-haired . . . . *A. Drummondii*
d. Sepals deep purplish blue on both surfaces . . . . *A. multiceps*
c. Plants 2-4 dm tall, flowers 1-3 . . . . *A. multifida*
b. Achenes elongated, glabrate or merely hirsute
f. Achenes sessile, appressed pubescent; styles straight . . . . *A. canadensis*
f. Achenes stipitate and glabrous; styles short and hooked . . . . *A. narcissiflora*

Anemone canadensis L.
Tufted, villous herbs 2 to 5 dm tall from a short but slender rhizome. Basal leaves long-petioled, the blade deeply 3 to 5-cleft; involucral leaves sessile but otherwise similar. Flowers creamy white, in vigorous specimens 2 or 3, the terminal largest, up to 4 cm in diameter. Fruiting head globose.

Common on calcareous, gravelly beaches and river banks of the upper Mackenzie River north to Fort Simpson.

General distribution: N. America; Que. to B.C. and southward.

Fig. 468 Map 541

Anemone Drummondii Wats.
Tufted, 2 to 3 dm tall, from a stout, branching taproot, the leaves and flowering peduncle soft silky-pubescent; basal leaves long-peduncled, the blade 3 to 4-cleft into linear segments; involucral leaves similar, sessile. Flowers normally solitary, 2 to 3 cm in diameter, the sepals appressed pubescent, white or commonly pale blue on the outer side. Fruiting head globose or ovate; the achenes densely woolly, and the persisting, long and straight style creamy-white.

Somewhat variable species with us restricted to non-calcareous rocks along the Mackenzie and Richardson Mountains north to the Mackenzie Delta.

General distribution: Cordilleran.

Fig. 469 Map 542

***Anemone multiceps** (Greene) Standl.
Pulsatilla multiceps Greene
Dwarf species from a branching sub-ligneous rhizome; radical leaves ternate, the segments cleft into narrow lobes, densely silky-villous, glabrate in age. Flowering stems 7 to 10 cm tall when the solitary flower is barely raised above its involucrum, the peduncle elongating as the fruit matures; sepals broadly elliptic, 1.5 to 2.0 cm long, deep purplish-blue on both sides, and silky on the outside; filaments dark purple, and the anthers almost black. Fruiting head globose. Achenes woolly and the wine-red styles persistent.

Rare or local, commonly growing in *Dryas* mats in calcareous, alpine tundra.

General distribution: Endemic to northern Alaska and northwest Y.T. To be looked for in the Richardson Mts.

Fig. 470 Map 543

Anemone multifida Poir. *s. lat.*
Tufted, 2 to 5 dm tall, from a strong, many-headed tap-root. Flowering stems and leaves silky-villous; basal leaves long-peduncled, the blade 2 to 4 times ternately divided; involucral leaves similar, sessile or nearly so. Flowers solitary or rarely 2 or 3, on long, erect peduncles, sepals 5 to 10 mm long, creamy white or sometimes red or purplish (forma *sanguinea* (Pursh) Fern.). Fruiting heads globose, the achenes silky-villous.

A variable species common on gravelly, calcareous slopes, river banks or lake shores north to the limit of forest or slightly beyond.

General distribution: Nfld. to southeast and central Alaska; S. America.

Fig. 471 Map 544

Anemone narcissiflora L.
Tufted, from a thick fibrous, erect-ascending rhizome. Flowering stems 3 to 6 dm tall, sparingly villous, glabrescent in age, the basal leaves long-petioled, their blades ternately cleft, and the cuneate divisions deeply incised into linear segments; those of the involucre sessile but smaller and similar. Sepals creamy white, 1.0 to 1.5 cm long. Fruiting head globose, the achenes 5 to 6 mm in diameter, glabrous, blackish-green and flat, with a narrow wing.

In lowland situations, even north to near the Arctic Coast, plants may grow 6 dm tall, usually with three flowers (ssp. *alaskana* Hult.), whereas close by, at higher elevations, are always 1-flowered and of much lower stature (ssp. *interior* Hult.), but not otherwise different.

General distribution: Amphi-Beringian with a disjunct population in high mountains of Colorado and Wyoming, where it has been separated as *A. zephyra* A. Nels.

Fig. 472 Map 545

Anemone parviflora Michx.
Rhizomes slender, dark fibrillose, terminating in rosettes of dark green, ternate and lustrous

leaves and solitary flowering stems 10 to 20 cm high. Flowers 2 to 3 cm in diameter, the sepals pure white above, silky and tinged with blue beneath. Fruiting head spherical; achenes woolly.

Open spruce woods and calcareous river flats.

General distribution: Wide ranging subarctic, chiefly on calcareous soil, from Nfld. to E. Asia.

Fig. 473 Map 546

Anemone Richardsonii Hook.
Slender and delicate, sparingly hirsute with filiform rhizomes and deeply 5-cleft basal leaves; the involucral leaves similar but 3-cleft. Scapes 5 to 20 cm high, solitary, flowers 1.5 to 2.5 cm in diameter; sepals yellow, fruiting heads subglobose, and the achenes few and glabrous, with a slender, hooked beak.

Moist herbmats and willow thickets.

General distribution: Subarctic, E. Asia, N. America, and W. Greenland.

Fig. 474 Map 547

Aquilegia L. Columbine

Aquilegia brevistyla Hook.
Stems slender, 3 to 8 dm tall, from a stout, fibrous tap-root. Leaves glabrous, the radical long-petioled, trifoliate, the cauline much smaller, short-petioled or sessile. Flowers commonly 1 to 4, nodding, about 1.5 cm long, sepals blue, petals white, with a pale bluish spur. Mature follicles stiffly erect, 2.0 to 2.5 cm long.

Sunny, rocky slopes, chiefly on acid rocks, north to the limit of trees.

General distribution: N. America from James Bay to central Alaska.

Fig. 475 Map 548

Caltha L. Marsh-marigold

Low soft-stemmed glabrous herbs with simple petioled cordate or reniform leaves; sepals petaloid, petals none; fruit composed of several-many carpels; follicles dehiscent on the ventral suture.

a. Stems ascending or decumbent; flowers yellow; carpels 6-12 . . . . *C. palustris*
a. Stems floating or creeping; flowers white; carpels about 30 . . . . *C. natans*

Caltha natans Pall.
Creeping or floating marsh plants rooting at the nodes. Leaves petioled, blades cordate, 3 to 5 cm broad, with a narrow sinus. Sepals petaloid, white, 3 to 7 mm long. Fruiting head globular, composed of 4 mm long follicles.

General distribution: Amphi-Beringian, boreal.

Fig. 476 Map 549

Caltha palustris L.
var. **palustris**
Glabrous perennial with long-petioled, cordate or reniform leaves. Flowers solitary in the axils of the subsessile cauline leaves, 1.5 to 2.5 cm in diameter. Sepals petaloid, pale yellow, often purplish on the back, petals wanting. Fruiting heads turbinate, composed of 10 to 12 compressed and ascending follicles.

The Amphi-Beringian var. *arctica* (R.Br.) Huth. is a smaller and more delicate plant usually creeping and rooting at the nodes.

In shallow water or in wet marshy places, often among tall grasses or sedges.

General distribution: var. *palustris*, wide ranging from Nfld. to B.C. north to Hudson Bay with disjunct stations on Great Slave L. and in the Mackenzie Delta; var. *arctica*, E. Asia and arctic N.W. America.

Fig. 477 (var. *palustris*), 478 (var. *arctica*)
Map. 550 (var. *palustris*), 551 (var. *arctica*)

Coptis Salisb. Goldthread

Coptis trifolia (L.) Salisb.
Dwarf perennial with thread-like bright yellow rhizomes, evergreen, lustrous, and slender-petioled 3-nate leaves and solitary white flowers on filiform scapes 3 to 10 cm high. Flowers 1.0 to 1.5 cm in diameter; sepals petaloid, narrowly oblanceolate, white; petals inconspicuous and yellowish-green, about one-third as long as the sepals. Follicles spreading

on slender stalks, each containing a few large and glossy black seeds.

In sheltered, turfy places, or in leaf mould under willows.

General distribution: E. Asia to E. Greenland. Thus far known in our area only from a single station in southern District of Keewatin.

Fig. 479 Map 552

Delphinium L. Larkspur

Erect herbs with palmately lobed or cleft leaves and a terminal raceme of blue or purple flowers; sepals 5, petal-like, the upper prolonged backward into a spur; petals 4, irregular; carpels few, ripening into follicles; seeds black, angled or winged.

a. Flowers about 2 cm long, few; follicles densely pubescent; pedicels villous . . . . *D. brachycentrum*

a. Flowers 1.0 to 1.5 cm long, many; follicles glabrous or nearly so; pedicels pilose . . . . *D. glaucum*

***Delphinium brachycentrum** Ledeb.
Similar to *D. glaucum* but smaller and with pubescent stems and follicles; also, the flowers are fewer and larger.

Alpine meadows.

General distribution: Amphi-Beringian, subarctic-alpine, reaching northern Y.T., and to be expected in the Richardson Mts. west of the Mackenzie Delta.

Map 553

Delphinium glaucum Wats.
D. brownii Rydb.
Stems stout, glabrous and commonly purplish tinged, 1 to 2 m tall from a thick, subligneous and fibrous much branched ascending rhizome. Leaves mainly cauline, palmately lobed, the sections deeply incised, glabrous and glaucous above, glabrous or short-pubescent beneath. Flowers very numerous, 2.0 to 2.5 cm long, in an elongating, up to 4 dm long raceme; follicles 2 to 3 together, about 1 cm long.

Moist woodland meadows and clearings.

General distribution: Cordilleran foothills north through western District of Mackenzie to the Arctic Coast, west to Alaska.

Fig. 480 Map 554

Pulsatilla Pers. Prairie Crocus or Pasque-flower

Pulsatilla Ludoviciana (Nutt.) Heller
Anemone patens L.
var. *Wolfgangiana* (Bess.) Koch
Tufted perennial with a stout, branched, ascending rhizome. Basal leaves long-petioled, in a rosette, the blade bipinnate, the leaflets pinnately divided into long, linear-segments; the stem-leaves sessile, deeply cleft into linear lobes. All leaves, including the slender petioles of the basal leaves long silky-villous. Flowers solitary, terminal; the sepals 5 to 7, purple, 3 to 5 cm long and spreading; the styles plumose, rapidly elongating and becoming 2 to 3 cm long in fruit, resembling those of *Dryas*.

Sandy and well-drained places where its large and showy flowers appear soon after the snow disappears. The peduncles elongate rapidly and in fruit may be up to 50 cm long.

General distribution: Man. to B.C. and northward along the Mackenzie R. drainage to the Arctic Coast and Western Arctic Islands west to central and northern Alaska.

Fig. 481 Map 555

Ranunculus L. Buttercup; Crowfoot

Mostly perennial herbs with acrid juice. Leaves mainly radical, petioled, the blade variously dissected and usually different in shape and cut from the mostly sessile or short-petioled cauline leaves. Flowers yellow or less commonly white, one to several, on long pedicels; sepals and petals mostly 5, the latter often showy; stamens mostly numerous; carpels from 5 to many, each with a single ovule. Fruit a globose or oblong head of variously beaked achenes.

a. Stoloniferous, rooting at the nodes; flowers yellow or white
  b. Aquatic or amphibious plants with creeping or floating stems
    c. Aquatic plants with white flowers and finely dissected leaves

d. Leaves flaccid; flowers rarely emergent .... *R. aquatilis* var. *eradicatus*
d. Leaves rigid, not collapsing when withdrawn from the water; flowers normally emergent .... *R. aquatilis* var. *subrigidus*
c. Amphibious plants; flowers yellow or white
e. Primary divisions of leaves entire
f. Flowers white; coarse, fleshy plant .... *R. Pallasii*
f. Flowers yellow; low, creeping and often matted; emergent or submersed .... *R. hyperboreus*
e. Primary divisions of leaves again divided; the floating leaves firm, the submersed flaccid
g. Plant delicate; diameter of leaves rarely over 1 cm .... *R. Gmelinii*
g. Plant more robust, diameter of leaves 2-3 cm .... *R. Purshii*
b. Terrestrial plants with yellow flowers
h. Plants scapose; sepals 3 .... *R. lapponicus*
h. Plants decumbent with filiform runners; sepals 5
i. Leaves reniform-ovate, 0.5-2.5 cm long .... *R. Cymbalaria*
i. Leaves linear or linear-lanceolate
j. Leaves filiform-setaceous .... *R. flammula* var. *filiformis*
j. Leaves broader, with distinct blade and petiole .... *R. flammula* var. *ovalis*
a. Tufted, terrestrial or paludal plants with erect or ascending stems
k. Radical leaves with entire or at most crenulate blade
l. Petals 3 mm long or less, radical leaves cordate .... *R. abortivus*
l. Petals 3-10 mm long; radical leaves oval .... *R. rhomboideus*
k. Radical leaves always lobed or incised
m. Plants glabrous or very nearly so
n. Stems fistulose; non-arctic paludal species .... *R. sceleratus* var. *multifidus*
n. Stems firm; arctic or alpine species
o. Petals not longer than sepals; dwarf species
p. Radical leaves deeply lobed or cleft
q. Segments merely 3-lobed .... *R. pygmaeus*
q. Segments again ternately divided .... *R. gelidus*
p. Radical leaves cuneate, merely 3-lobed .... *R. Sabinei*
o. Petals longer than sepals
r. Radical leaves merely 3-lobed
s. Sepals densely brown or black hairy
t. Receptacle glabrous .... *R. nivalis*
t. Receptacle with reddish-brown hairs .... *R. sulphureus*
s. Sepals sparsely pale yellowish pubescent .... *R. Eschscholtzii*
r. Radical leaves deeply incised; flowering peduncle hirsute
u. Sepals elliptic-oblong with a conspicuous black tip; petals pale yellow .... *R. pedatifidus* var. *leiocarpus*
u. Sepals linear-lanceolate; petals shiny, deep golden yellow ..... *R. Turneri*
m. Plant distinctly pubescent or hirsute, especially on the leaf-petioles; non-arctic, lowland species
v. Terminal or middle segment of the radical leaves not stalked
w. Petals 3 mm long or less, as long as the sepals; achenes flat, with a prominently recurved beak .... *R. Macounii*
w. Petals much longer and twice as long as the sepals; achenes flat, the beak minute .... *R. acris*
v. Terminal or middle segment of radical leaves stalked
x. Flowers small, petals usually less than 5 mm long; plant not glaucous-green
y. Petals shorter than the sepals; fruiting head cylindric-oblong .... *R. pensylvanicus*
y. Petals as long or longer than the sepals; fruiting head globose . . *R. Macounii*
x. Flowers larger, petals usually more than 5 mm long; plant glaucous-green sparsely pubescent .... *R. septentrionalis*

Ranunculus abortivus L.
Biennial or short-lived perennial with slender fibrous roots. Stems 1 to several, 2 to 5 dm tall, simple or in tall specimens repeatedly forked and glabrous or sparingly pubescent above; radical leaves long petioled, the blade cordate or reniform, shallowly crenulate, the cauline short-petioled or sessile, mostly ternately divided into linear-lanceolate lobes. Petals pale yellow, shorter than the 2 to 3 mm long reflexed sepals. Fruiting head globose or ovate, the achenes orbicular, with a minute beak.

Damp grassy places, pastures and roadsides.

General distribution: Boreal N. America.

Fig. 482 Map 556

Ranunculus acris L.
Tall or Common Buttercup
Coarsely hirsute weedy perennial from a fleshy fibrous root. Stems 3 to 8 dm tall, few to several, branching above, commonly somewhat hairy; radical leaves long-petioled, the blade palmately lobed, the blade and petiole commonly coarsely hirsute. Flowers 1.5 to 2.0 cm in diameter, on long, slender peduncles, the petals deep yellow and shiny. Fruiting heads globular, the achenes smooth, 2 to 3 mm long, with a minute beak.

General distribution: Native of Europe, widely naturalized by roadsides and farms across N. America, barely entering our area at Ft. Smith.

Ranunculus aquatilis L.
var. **eradicatus** Laest.
R. trichophyllus Chaix
var. *eradicatus* (Laest.) W.B. Drew
R. confervoides Fr.
White Water-buttercup.
Totally submersed, much-branched aquatic, with finely divided flaccid leaves, which collapse when withdrawn from water. The mostly submerged but sometimes emergent flowers solitary on peduncles 1 to 6 cm long; the petals pure white, 2 to 3 times as long as the sepals. The achenes in this and the following variety characteristically transverse wrinkled.

In shallow ponds or sometimes in running water, less commonly in brackish lagoons.

General distribution: Circumpolar, boreal-arctic and one of the few truly aquatic species whose northern range extends far beyond the northern limit of trees.

Fig. 483 Map 557

Ranunculus aquatilis L.
var. **subrigidus** (W.B. Drew) Breitung
R. subrigidus W.B. Drew
R. circinatus Sibth.
var. *subrigidus* (W.B. Drew) L. Benson
Similar to *R. trichophyllus* var. *eradicatus* but the submersed leaves are stiffer and hold their shape when removed from the water.

In shallow, calcareous ponds.

General distribution: N. America, boreal-subarctic.

Fig. 484 Map 558

Ranunculus Cymbalaria Pursh
Northern Seaside Buttercup
Low, caespitose, glabrous, and somewhat fleshy, with widely trailing runners rooting freely at the nodes. Leaves long-petioled, the blade reniform-ovate, from 5 to 25 mm long, crenately toothed or shallowly lobed. Scapes one- to several-flowered, from a few to 20 cm long. Flowers 5 to 9 mm in diameter with bright yellow, narrowly oblanceolate petals. Fruiting head short-cylindrical, the achenes small, obovate, longitudinally ribbed, and with a short, erect beak.

Beyond the limit of trees and in alpine situations *R. Cymbalaria* becomes much reduced in size, when the leaf-blade is merely 3-toothed and the flowers solitary (var. *alpinus* Hook.).

Saline meadows or brackish shores, in moist clay or sand.

General distribution: Circumpolar, with large gaps.

Fig. 485 Map 559

Ranunculus Eschscholtzii Schlecht.
A rather variable species resembling the circumpolar arctic-alpine *R. nivalis*. Where the ranges of *R. Eschscholtzii* and *R. nivalis* overlap flowering specimens can nearly always be distinguished by the pubescence of the sepals that in *R. nivalis* is copious and dark brown, but sparse and pale yellowish in *R. Eschscholtzii;* also, in the latter 2-flowered stems are not uncommon, but very rare in *R. nivalis*.

Moist alpine meadows and herbmats.

General distribution: Cordilleran-Pacific coast species, reaching north to the Mackenzie Delta and across S.W. Alaska to E. Asia.

Fig. 486 Map 560

Ranunculus flammula L.
Low, essentially glabrous perennial forming flat, leafy rosettes from which issue prostrate, slender stems that root freely at the nodes; radical leaves 5 to 7 cm long, entire-margined. Flowers solitary, ascending on short and bracted peduncles; sepals glabrous, petals pale yellow, about 6 mm long. Achenes subglobose with a short, stout beak; var. *filiformis* (Michx.) Hook. has filiform or linear leaves, var. *ovalis* (Bigel.) L. Benson (*R. reptans* L. var. *ovalis* (Bigel.) T. & G.) has the blade narrowly lanceolate tapering into a filiform petiole.

Moist clayey and sandy shores and wet meadows.

General distribution: var. *filiformis*; circumpolar, wide ranging, north beyond the limit of trees; var. *ovalis*: N. America, non-arctic, north along the Mackenzie R. to Ft. Simpson.

Fig. 487 Map 561

Ranunculus gelidus Karel. & Kiril.
Dwarf, grey-green, essentially glabrous perennial forming small firmly rooted tussocks, from a branched base, each branch composed of numerous radical leaves and one flowering stem. Leaf-petioles 3 to 5 cm long, the blade cordate, 1.0 to 1.5 cm in diameter, bi-ternately lobed. Flowers mostly solitary, on a 3 to 7 cm long peduncle; petals 3 to 5 mm long, pale yellow, often purplish tinged, longer than the reflexed sepals. Achene lenticular, smooth, 2.6 to 3.0 mm in diameter, with a slender, curved beak.

Moist, gravelly talus slopes, chiefly alpine.

General distribution: Amphi-Beringian, from the arctic coast of Y.T., south through Richardson and Mackenzie Mts., with disjunct stations south to Colorado.

Fig. 488 Map 562

Ranunculus Gmelinii DC.
(including var. *yukonensis* (Britt.) Benson)
Submerged or emergent aquatic resembling *R. Purshii* but smaller and more delicate in all parts, the diameter of its leaves rarely exceeding 10 mm, and that of its flowers rarely over 8 mm. From submerged specimens of the circumpolar *R. hyperboreus* it can always be distinguished by its more finely divided leaves.

In shallow fresh-water ponds, often among the bases of sedges and grasses.

General distribution: Eurasia and N.W. America, east to Hudson Bay.

Fig. 489 Map 563

Ranunculus hyperboreus Rottb.
Low, creeping, densely matted amphibious plant with small, deeply 3-lobed leaves and tiny solitary yellow flowers from the leaf axils. Fruiting heads globular, about 4 mm in diameter.

Common in shallow fresh or brackish water, but frequently growing among tall sedges and grasses by the edge of ponds.

General distribution: Circumpolar, arctic-alpine, reaching the northern tip of Ellesmere Island.

Fig. 490 Map 564

Ranunculus lapponicus L.
Lapland Buttercup
Terrestrial, with filiform creeping stems, deeply buried in moss, freely rooting from the nodes, and bearing slender, simple or forking, scapose or 1-leaved flowering stems, from a few to 25 cm high; basal leaves long-petioled, solitary or a few together, the blade reniform and deeply 3-cleft, its lobes broadly obovate and coarsely crenate. Flowers solitary, sweet-scented, about 1 cm in diameter, with 3 sepals and from 6 to 10 pale yellow or whitish petals. Fruiting head globose, of from 3 to 15 achenes, each slightly longer than the slender and hooked style.

In damp moss, often under willows or in sphagnum bogs.

General distribution: Nearly circumpolar, wide ranging, low-arctic.

Fig. 491 Map 565

Ranunculus Macounii Britt.
Stems erect to ascending, 3 to 7 dm tall, sometimes rooting from the nodes, freely branching and forked above. Leaves long-petioled, the blade divided into 3 stalked and deeply toothed segments, usually coarsely yellowish hirsute. Flowers small, rarely more than 1.0 cm in diameter, the sepals soon deciduous, shorter than the pale yellow petals; achenes plump, ovate, 2.5 to 3.0 mm long with a straight beak, in a globose head.

Wet meadows, roadsides etc. Of weedy habit but undoubtedly native along the upper Mackenzie and its tributaries.

General distribution: E. Que. and Lab. to Alaska and southward.

Fig. 492 Map 566

Ranunculus nivalis L.
Snow-Buttercup
Tufted, essentially glabrous, perennial from a short fibrous base, with from a few to several

slender, erect, 1- to 3-leaved flowering stems. Basal leaves petioled, their blades reniform, deeply 3-lobed, each lobe cleft half-way to the base. Flowering stems 5 to 10 cm high, much elongated in fruit; the flowers 1.5 to 2.5 cm in diameter, sepals 5, densely brownish-black villous on the back, petals 5, dark yellow, twice as long as the sepals. Fruiting head ovoid to short-cylindrical; the receptable naked; achenes 1.5 to 2.0 mm long, with a short beak.

In wet moss by brooks and in herbmats, often near the edge of melting snowbanks.

General distribution: Circumpolar, arctic.

Fig. 493 Map 567

Ranunculus Pallasii Schlecht.
Pallas' Buttercup
Glabrous and somewhat fleshy marsh- or emergent aquatic plant with a creeping or floating, and freely branched stem, rooting at the nodes. Basal leaves with 10 to 30 cm long petioles, and deeply cleft, ternate or sometimes entire blades. Flowering stems erect-ascending, bearing one or two reduced leaves, and one or rarely two long-pedunculate flowers from the leaf axils. Flowers strongly perfumed, 1.5 to 2.5 cm in diameter, with 6 to 10 elliptic-oblong white petals. Fruiting heads globose; the plump achenes 4.5 to 5.5 mm long, including the short beak.

Wet, brackish meadows and sloughs and mainly along seacoasts and estuaries.

General distribution: Circumpolar, low-arctic, with several large gaps.

Fig. 494 Map 568

Ranunculus pedatifidus Sm.
var. **leiocarpus** (Trautv.) Fern.
R. affinis R.Br.
Tufted, essentially glabrous perennial from a short, erect and fibrous base, bearing from a few to several erect, slender, commonly short-pubescent, simple or slightly forking flowering stems, 10 to 30 cm high. Basal leaves long-petioled, cordate to reniform in outline, pedately cleft nearly to the centre, into 5 to 9 entire, linear or lobed segments. Cauline leaves sessile or short-petioled, the lowermost similar to the basal leaves, the upper merely 3-cleft into linear segments. Sepals thinly grey-villous on the back; petals obovate or almost orbicular, from pale to bright yellow, 0.5 to 1.0 cm long; fruiting heads ovoid or short-cylindrical; achenes plump, glabrous, about 2.0 mm long, with a curved, subulate, readily deciduous beak.

Calcareous gravelly, sandy or grassy places.

General distribution: Circumpolar, arctic-alpine.

Fig. 495 Map 569

Ranunculus pensylvanicus L. f.
Coarse and strongly hirsute annual or short-lived perennial, not rooting from the nodes, resembling *R. Macounii* from which flowering specimens are readily distinguished by their much smaller petals, usually 2 to 3 mm long, or only half the length of the sepals. In the fruiting state *R. pensylvanicus* is readily distinguished from *R. Macounii* by its cylindric-oblong fruiting head.

Wet ditches and marshy places; with us perhaps a recent introduction. Thus far reported only from a few stations along the upper Mackenzie River.

General distribution: Boreal N. America, Lab. to Alaska.

Fig. 496 Map 570

Ranunculus Purshii Richards.
Aquatic; stems freely branched, creeping or floating, freely rooting at the nodes. Leaves submersed and emergent, the blade 2 to 3 cm in diameter, circular or reniform in outline, divided into 3 or more segments, the floating leaves commonly smaller and not so finely divided. Flowers emergent, yellow and terminal or axillary, on slender, naked pedicels; achenes obovate 1.0 to 1.5 mm long, with a straight beak. An eastern North American relative of the Old World *R. Gmelinii* but a much coarser plant, and belonging in the boreal forest rather than the tundra.

Shallow ponds and wet bogs, mainly within the boreal forest zone, from east Quebec to central Alaska.

General distribution: N.S. to B.C. and north to Alaska.

Map 571

Ranunculus pygmaeus Wahlenb.
Dwarf Buttercup
Tufted, glabrous, dwarf perennial from a simple or branched base, bearing from a few to many slender, weak, ascending or spreading one-flowered stems 10 to 15 cm long. Basal leaves numerous, slender-petioled, the blade flabelliform to reniform, deeply palmately cleft into 3 to 5 oblanceolate-cuneate lobes. Cauline leaves 1 to 2, or wanting. Flowers 5 to 8 mm in diameter; petals pale yellow, readily deciduous. Fruiting head globular to sub-cylindrical, 3 to 5 mm long.

Moist, grassy, and turfy places; often on snowbeds by the edge of melting snowbanks.

General distribution: Circumpolar, arctic-alpine.

Fig. 497 Map 572

Ranunculus rhomboideus Goldie
Erect, soft-pubescent perennial, 2 to 3 dm tall, from slender and fibrous roots; radical leaves long-petioled, the blade oval with crenulate margins, or very rarely with an apical notch; cauline leaves sessile and ternately divided into linear segments. Flowers 3 to 5, small and mostly solitary, on slender peduncles from the axils of the upper leaves; petals reflexed, narrow, pale yellow and much longer than the sepals; the small, plump achenes in a globular head.

Salt marshes and damp hollows of prairies and plains barely entering our area west of Fort Smith.

General distribution: N. America, Ont. to B.C. and southward.

Fig. 498 Map 573

Ranunculus Sabinei R. Br.
R. pygmaeus Wahlenb.
ssp. *Sabinei* (R.Br.) Hult.
Tufted, glabrous dwarf species from a simple non-fibrous caudex, bearing from one to a few rather stout, stiffly erect, 1- to 2-leaved flowering stems 3 to 10 cm high. Basal leaves somewhat fleshy, petioled, the blade flabellate-reniform in outline, shallowly 3- to 5-lobed; cauline leaves more deeply 3-cleft. Flowers solitary, 8 to 10 mm in diameter; sepals purplish tinged, grey villous on the back, about as long as the oblanceolate, pale yellow petals. Fruiting heads globular to short cylindrical; the achenes 1.5 mm long, exclusive of the slender, straight, or somewhat curved beak.

Moist, gravelly or turfy places. With us known from a few stations along the Arctic Coast.

General distribution: Amphi-Beringian; high arctic reaching northwest Greenland.

Fig. 499 Map 574

Ranunculus sceleratus L.
ssp. **multifidus** (Nutt.) Hult.
Annual or short-lived perennial with glabrous, fistulose, erect-ascending 1.5 to 6.0 dm tall stems from a fibrous root; radical and lower cauline leaves petioled, the reniform blade ternately divided, the divisions cuneate and again variously toothed or dissected. Flowers small, their sepals and petals reflexed, soon deciduous, the latter pale yellow, linear-oblong, 6 to 7 mm long; achenes small and very numerous, in a cylindric to globose head.

Wet, peaty places by pools or in springy places.

General distribution: Amphi-Beringian; Que. to Alaska, north along the Mackenzie R. drainage nearly to the Arctic Coast.

Fig. 500 Map 575

Ranunculus septentrionalis Poir.
Glaucous perennial with erect, ascending, declining or trailing, sparsely hirsute to glabrous stems from a fibrous root; radical and lower cauline leaves long-petioled, with ternately divided blade, the divisions distinctly stipitate and coarsely toothed, their petioles and stipules brown and coriaceous. Flowers small, on slender peduncles, the sepals strongly reflexed, soon deciduous; petals pale yellow, broadly ovate, 6 to 7 mm long, longer than the sepals; achenes wingless, 2.5 to 3.0 mm long with a straight beak, in a small globose head.

Damp willow thickets and slough margins.

General distribution: N. America; N.B. to central Y.T., non-arctic.

Map 576

Ranunculus sulphureus Sol.
Basal leaves fleshy, broadly wedge-shaped, truncate, toothed or shallowly lobed. Flowering stem stiff and stout. Resembling *R. nivalis* from which it is distinguished by its coarser habit, the pale brown pubescence of the sepals and, always, by the presence of rusty-brown hairs on the receptacle, which in *R. nivalis* is always naked.

Moist, gravelly arctic and alpine tundra.

General distribution: Circumpolar, arctic-alpine.

Fig. 501 Map 577

Ranunculus Turneri Greene
Glabrous or sparsely hirsute perennial. Stems few to several, 2 to 3 dm tall, from a branching, fibrous root; radical leaves long-peduncled, the blade ternate, its lobes merely toothed; cauline leaves sessile, commonly divided into linear segments. Flowers large, 1 to 3, the petals deep yellow and shiny, 1.5 to 1.8 cm long, the sepals much smaller, sparingly pale hairy, soon reflexed. Fruiting head globose, the achenes 2.0 to 2.5 mm long, the beak prominently curved into a broad hook.

A very handsome species of sub-alpine

meadows, superficially resembling a low and very large-flowered *R. acris*.

General distribution: Amphi-Beringian, barely entering our area along the east flank of Richardson Mts. west of Mackenzie Delta.

Fig. 502 Map 578

Thalictrum L. Meadow-Rue

Erect herbs with 2-3-ternately compound basal and cauline leaves; the small flowers in a panicle or raceme, perfect or dioecious; carpels several, becoming grooved or ribbed achenes.

a. Alpine-arctic dwarf species with scapose stem rarely over 2 dm tall . . . . *T. alpinum*
a. Woodland or prairie species with leafy stems 3 to 10 dm tall
  b. Flowers perfect . . . . *T. sparsiflorum* var. *Richardsonii*
  b. Flowers dioecious . . . . *T. venulosum*

Thalictrum alpinum L.
Tufted; basal leaves dark green and glossy as if varnished. Stems capillary, scapose, 8 to 20 cm high, terminating in a simple, few-flowered raceme of small, apetalous, bisexual flowers; sepals 4 to 5, petaloid, promptly deciduous. Anthers bright yellow, nodding on the slender filaments.

Alpine herbmats.

General distribution: Circumpolar but of alpine and disrupted range. Alaska through Y.T. and western District of Mackenzie to northern B.C., then isolated in the southern Cordillera, the Gulf of St. Lawrence, northern Lab., southern E. and W. Greenland; Eurasia.

Fig. 503 Map 579

Thalictrum sparsiflorum Turcz.
var. **Richardsonii** (Gray) Boivin
Stems tall and slender, the inflorescence open, few-flowered, the bisexual flowers in small, nodding clusters.

Rich subalpine thickets. In our area thus far known only from along the Liard River.

General distribution: Amphi-Beringian, east to Hudson Bay.

Fig. 504 Map 580

Thalictrum venulosum Trel.
T. Turneri Boivin
Coarser than *T. sparsiflorum*; basal as well as cauline leaves petioled. Flowers unisexual, on separate plants.

Prairie and woodland species. With us north along the Mackenzie Valley to its delta.

General distribution: Boreal N. America.

Fig. 505 Map 581

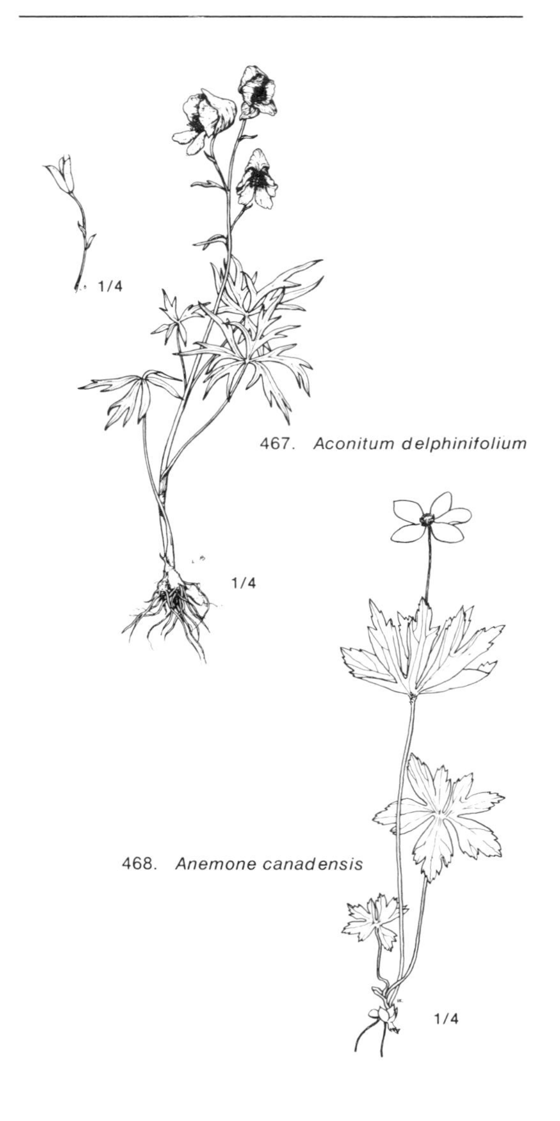

467. *Aconitum delphinifolium*

468. *Anemone canadensis*

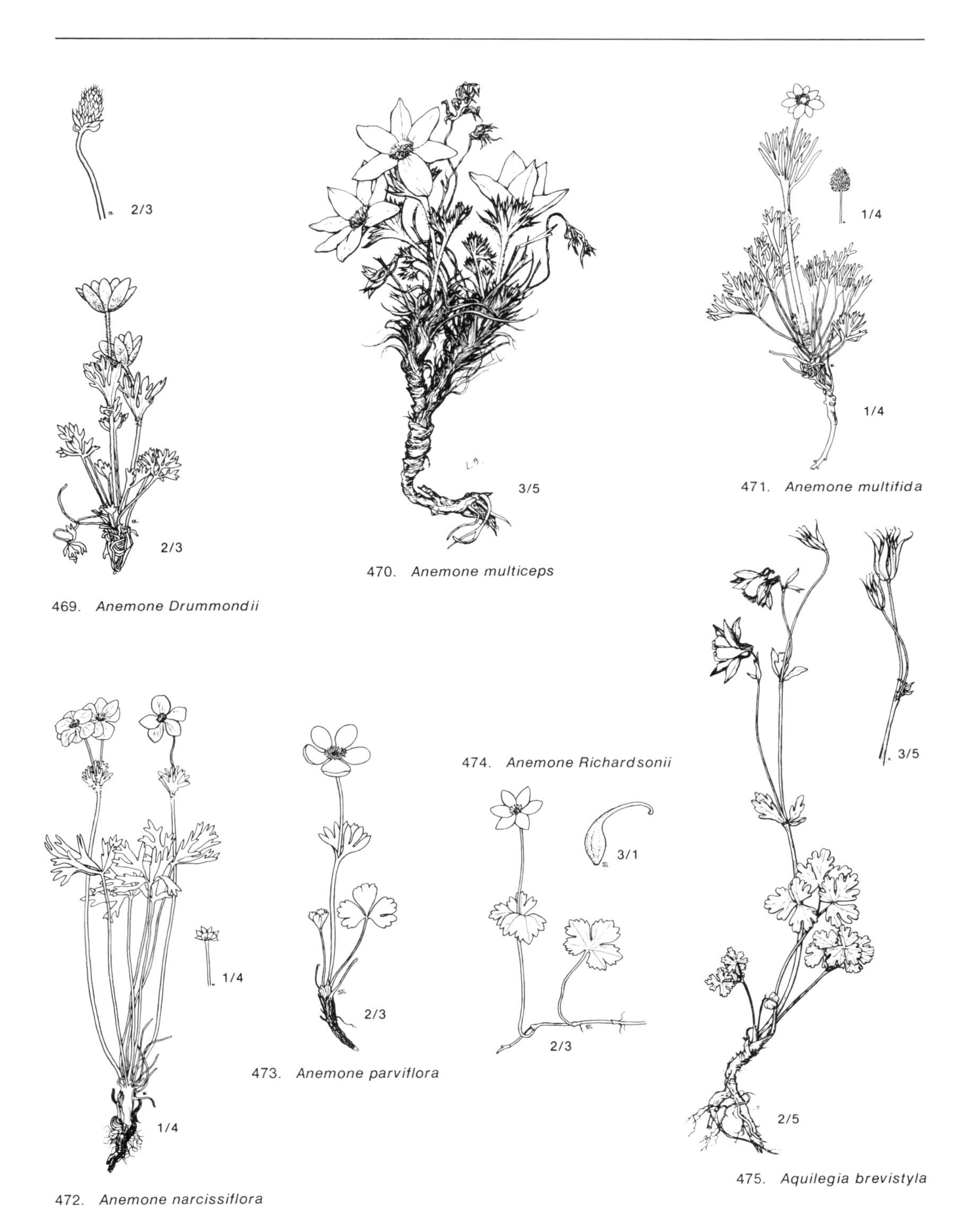

469. *Anemone Drummondii*

470. *Anemone multiceps*

471. *Anemone multifida*

472. *Anemone narcissiflora*

473. *Anemone parviflora*

474. *Anemone Richardsonii*

475. *Aquilegia brevistyla*

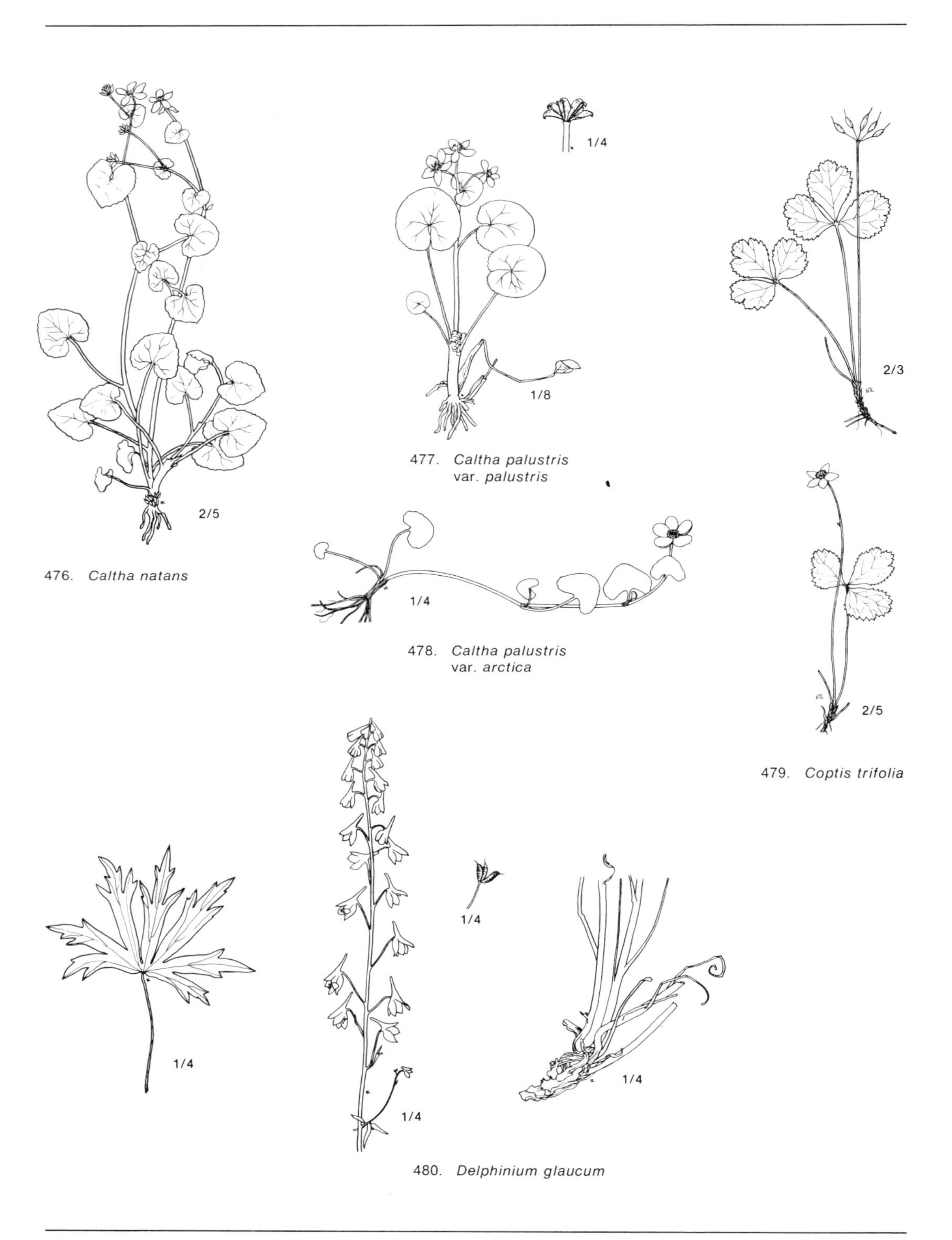

476. *Caltha natans*

477. *Caltha palustris* var. *palustris*

478. *Caltha palustris* var. *arctica*

479. *Coptis trifolia*

480. *Delphinium glaucum*

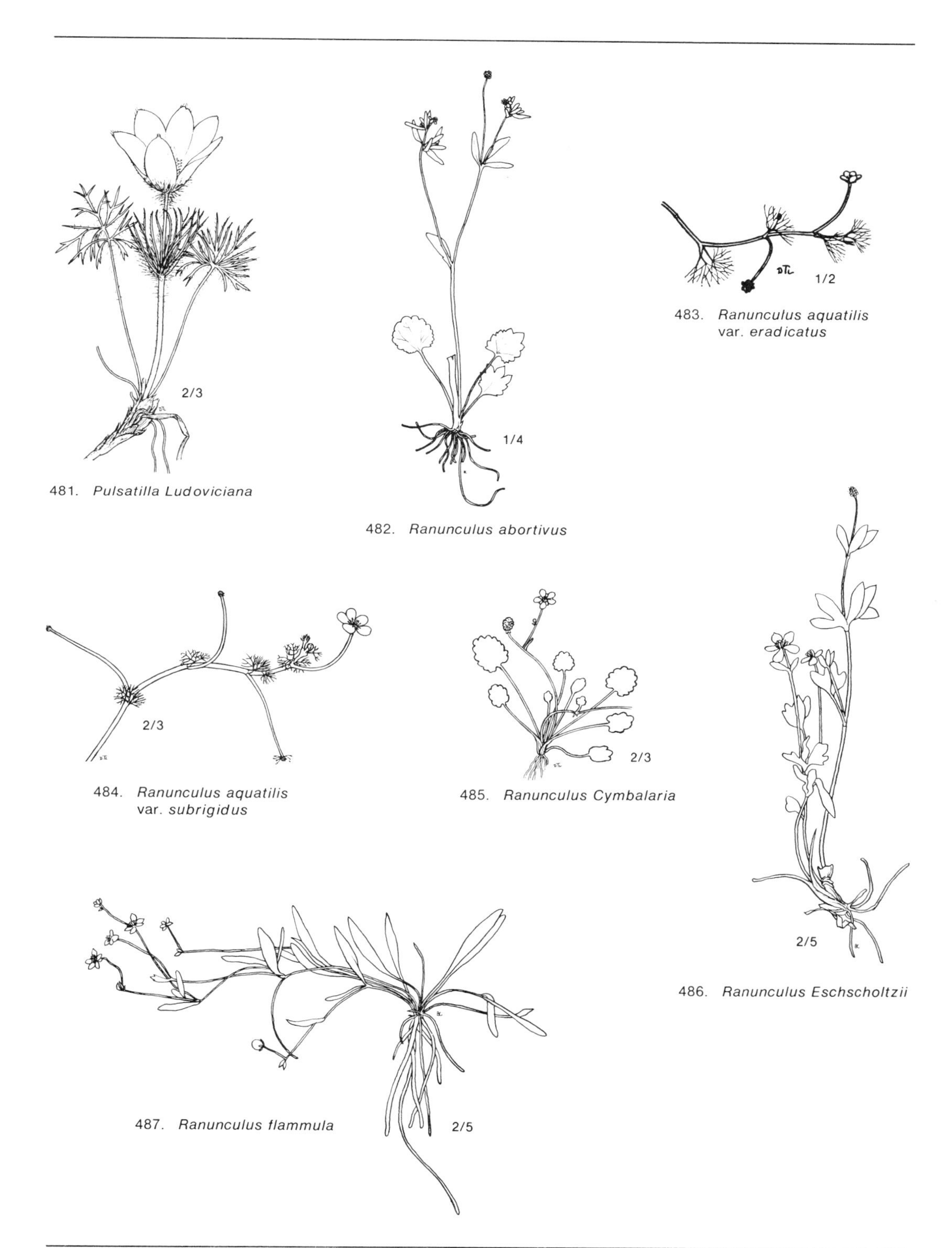

481. *Pulsatilla Ludoviciana*

482. *Ranunculus abortivus*

483. *Ranunculus aquatilis* var. *eradicatus*

484. *Ranunculus aquatilis* var. *subrigidus*

485. *Ranunculus Cymbalaria*

486. *Ranunculus Eschscholtzii*

487. *Ranunculus flammula*

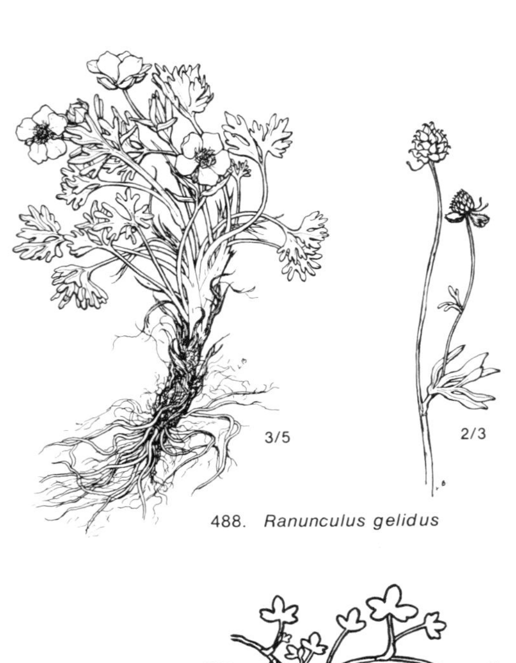

488. *Ranunculus gelidus*

489. *Ranunculus Gmelinii*

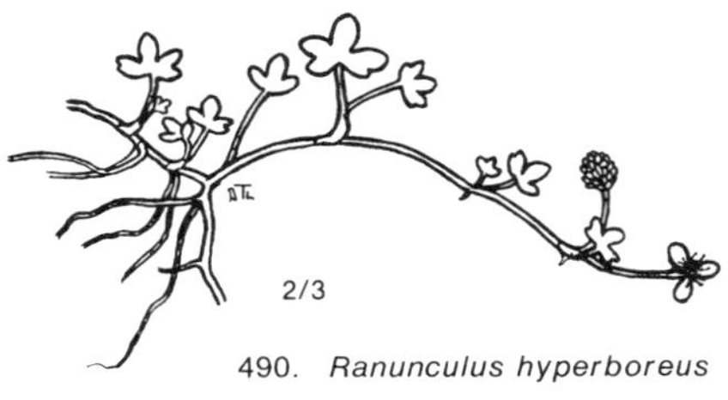

490. *Ranunculus hyperboreus*

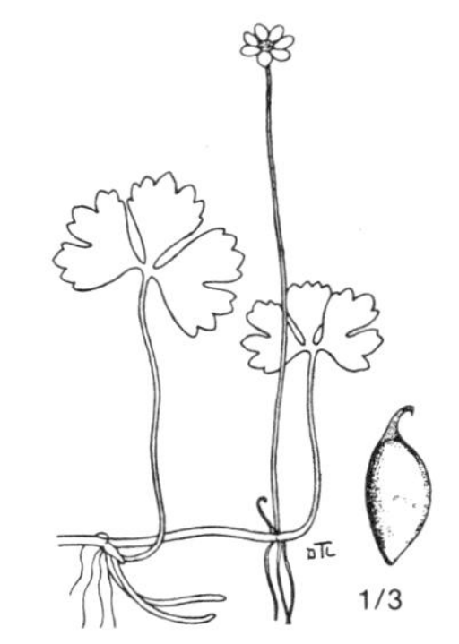

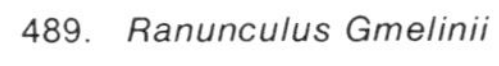

491. *Ranunculus lapponicus*

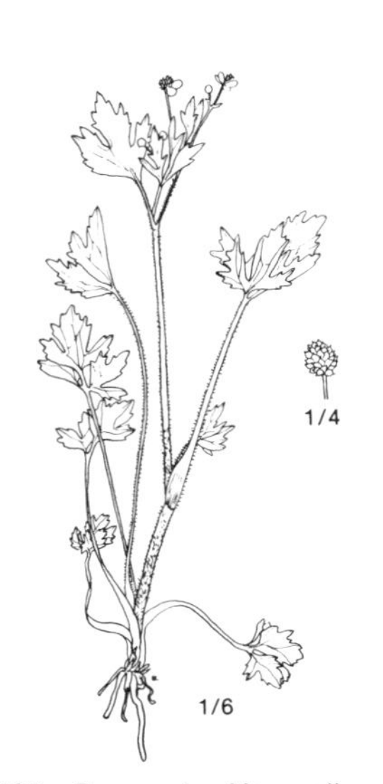

492. *Ranunculus Macounii*

493. *Ranunculus nivalis*

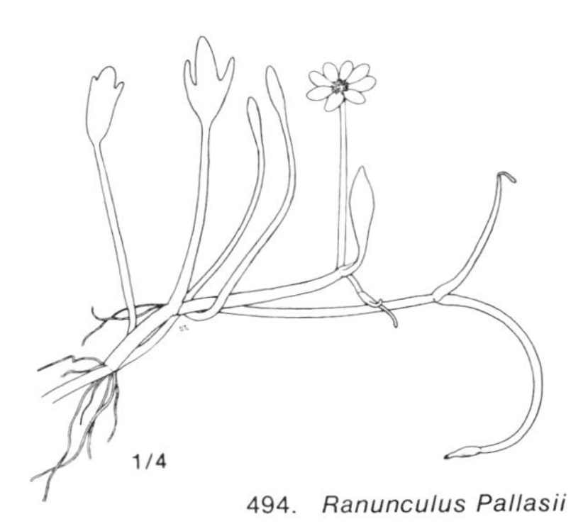

494. *Ranunculus Pallasii*

495. *Ranunculus pedatifidus* var. *leiocarpus*

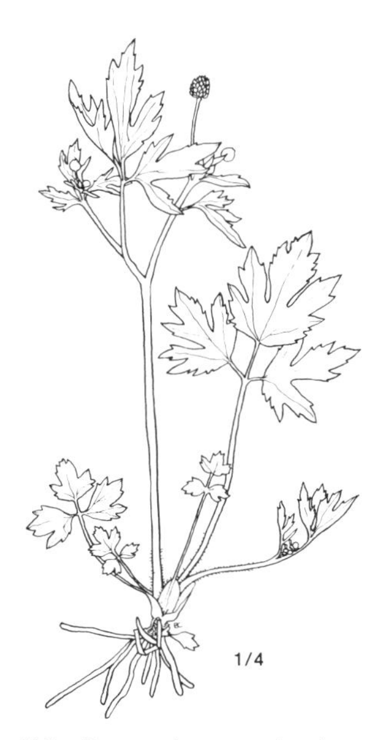

496. *Ranunculus pensylvanicus*

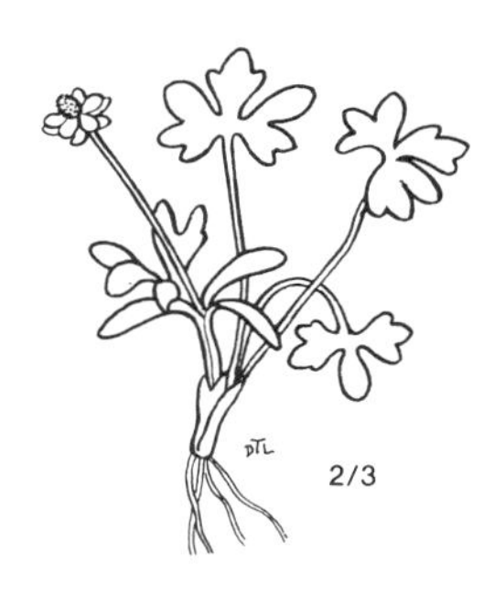

497. *Ranunculus pygmaeus*

498. *Ranunculus rhomboideus*

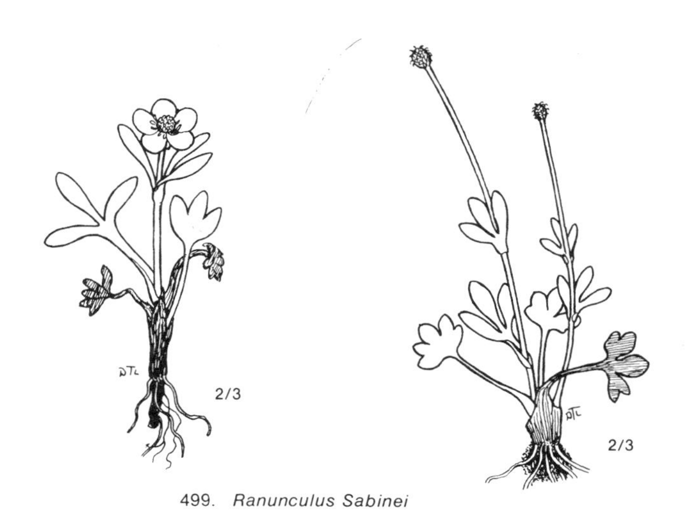

499. *Ranunculus Sabinei*

500. *Ranunculus sceleratus* ssp. *multifidus*

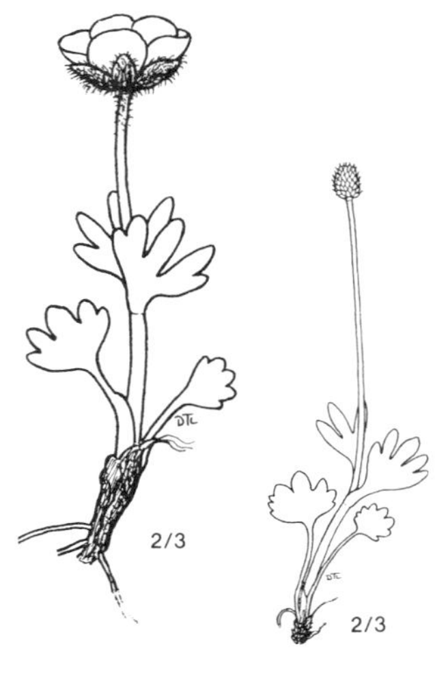

501. *Ranunculus sulphureus*

502. *Ranunculus Turneri*

503. *Thalictrum alpinum*

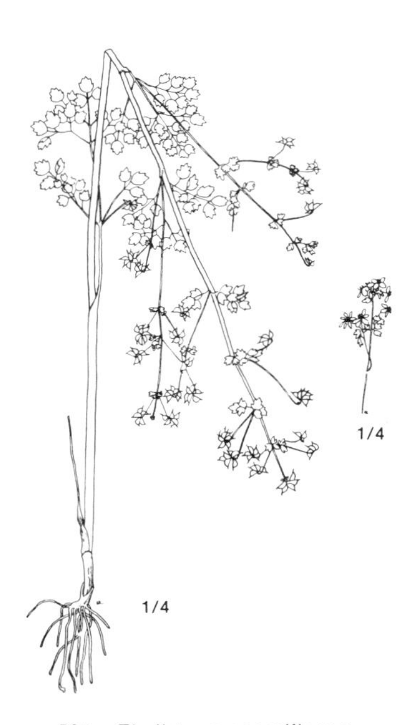

504. *Thalictrum sparsiflorum* var. *Richardsonii*

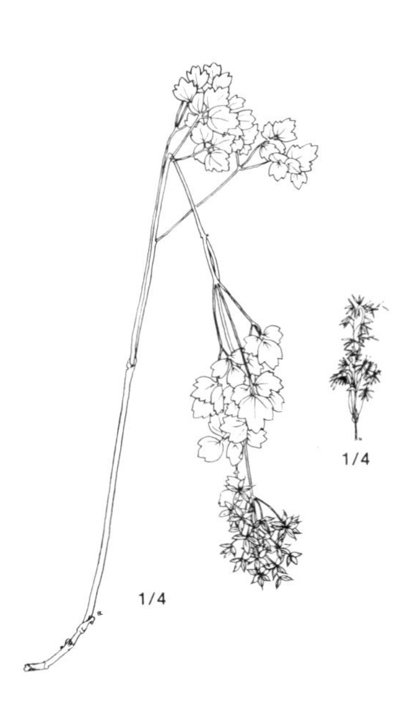

505. *Thalictrum venulosum*

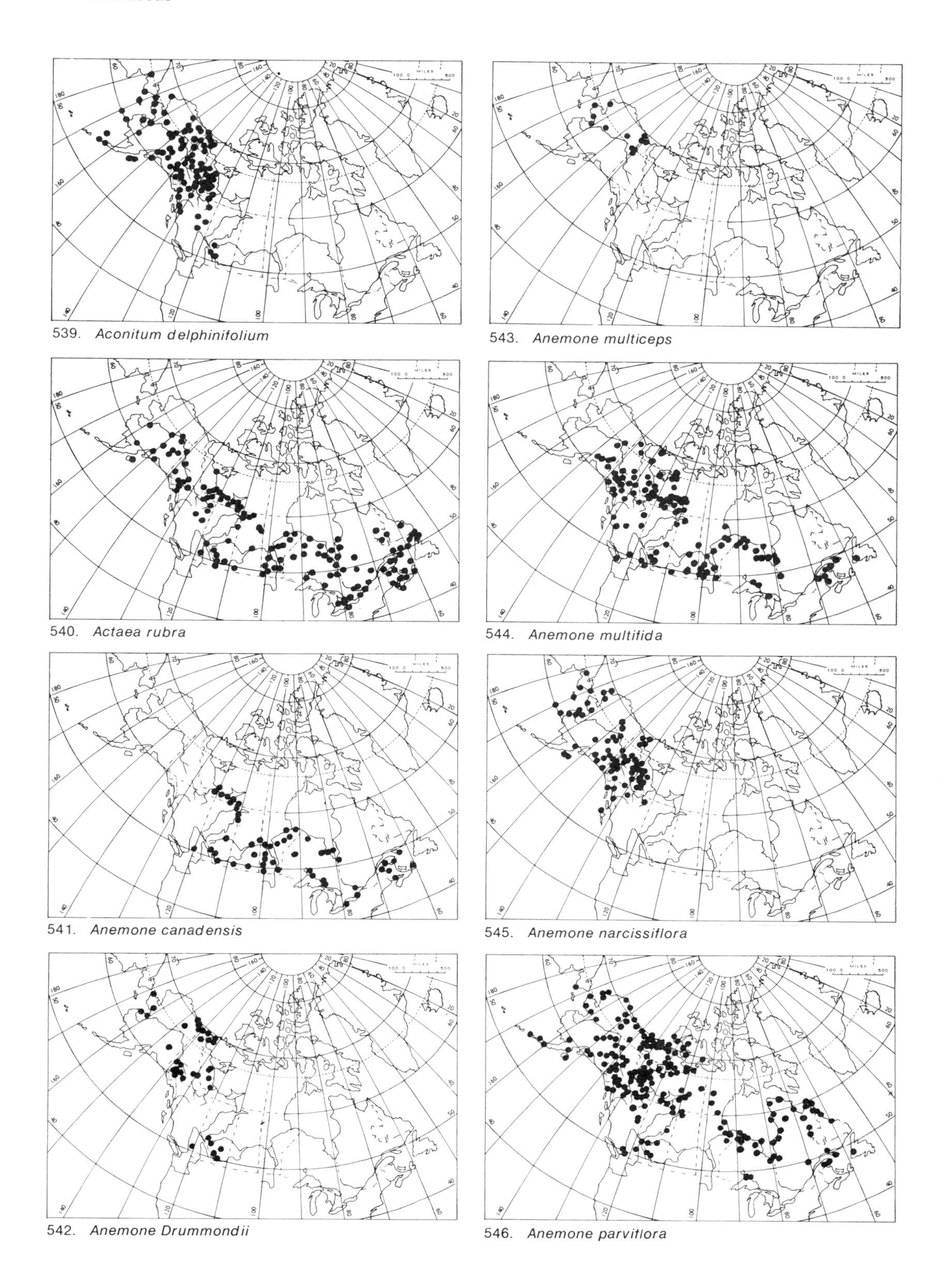

539. *Aconitum delphinifolium*

540. *Actaea rubra*

541. *Anemone canadensis*

542. *Anemone Drummondii*

543. *Anemone multiceps*

544. *Anemone multifida*

545. *Anemone narcissiflora*

546. *Anemone parviflora*

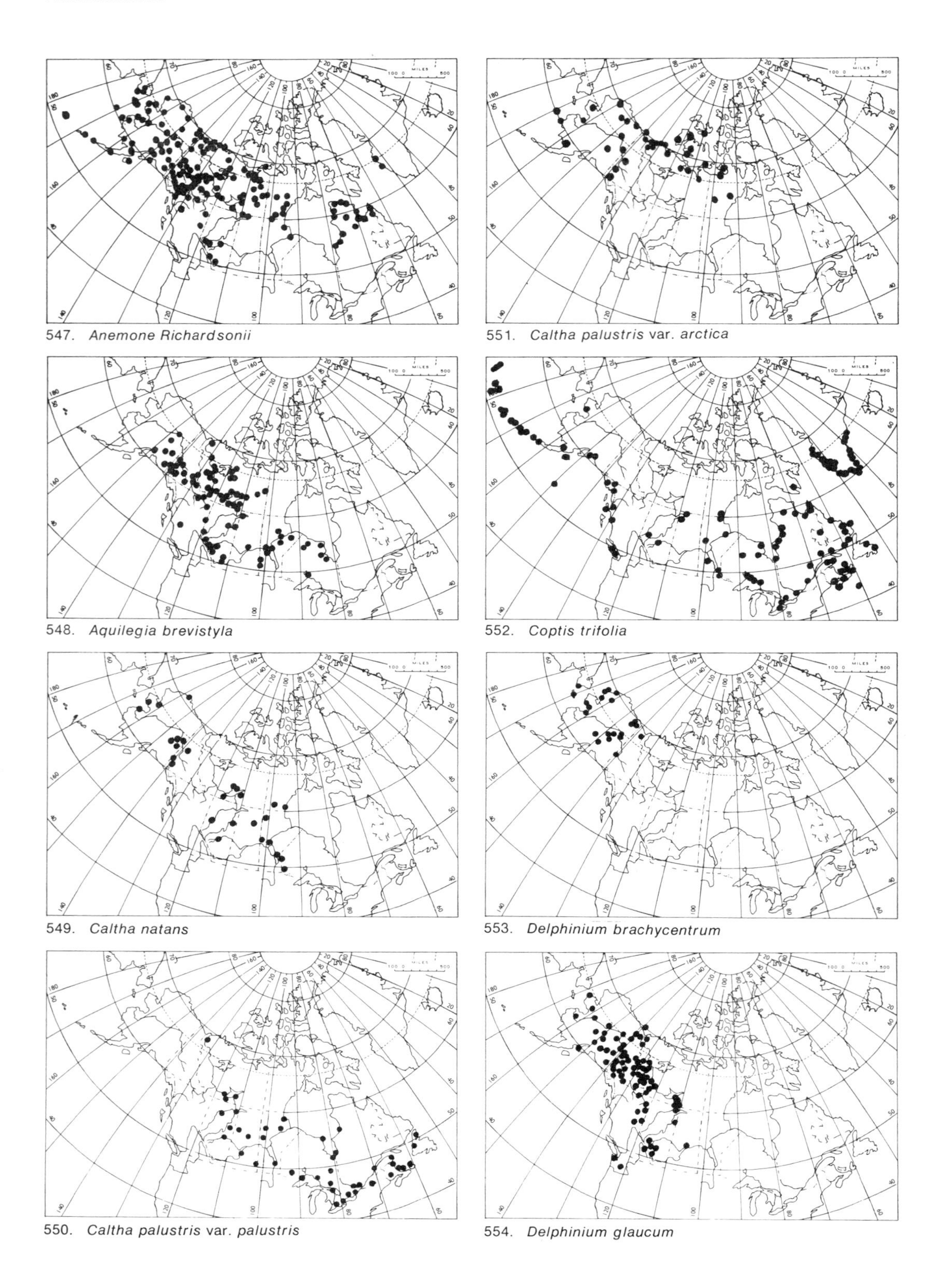

547. *Anemone Richardsonii*

548. *Aquilegia brevistyla*

549. *Caltha natans*

550. *Caltha palustris* var. *palustris*

551. *Caltha palustris* var. *arctica*

552. *Coptis trifolia*

553. *Delphinium brachycentrum*

554. *Delphinium glaucum*

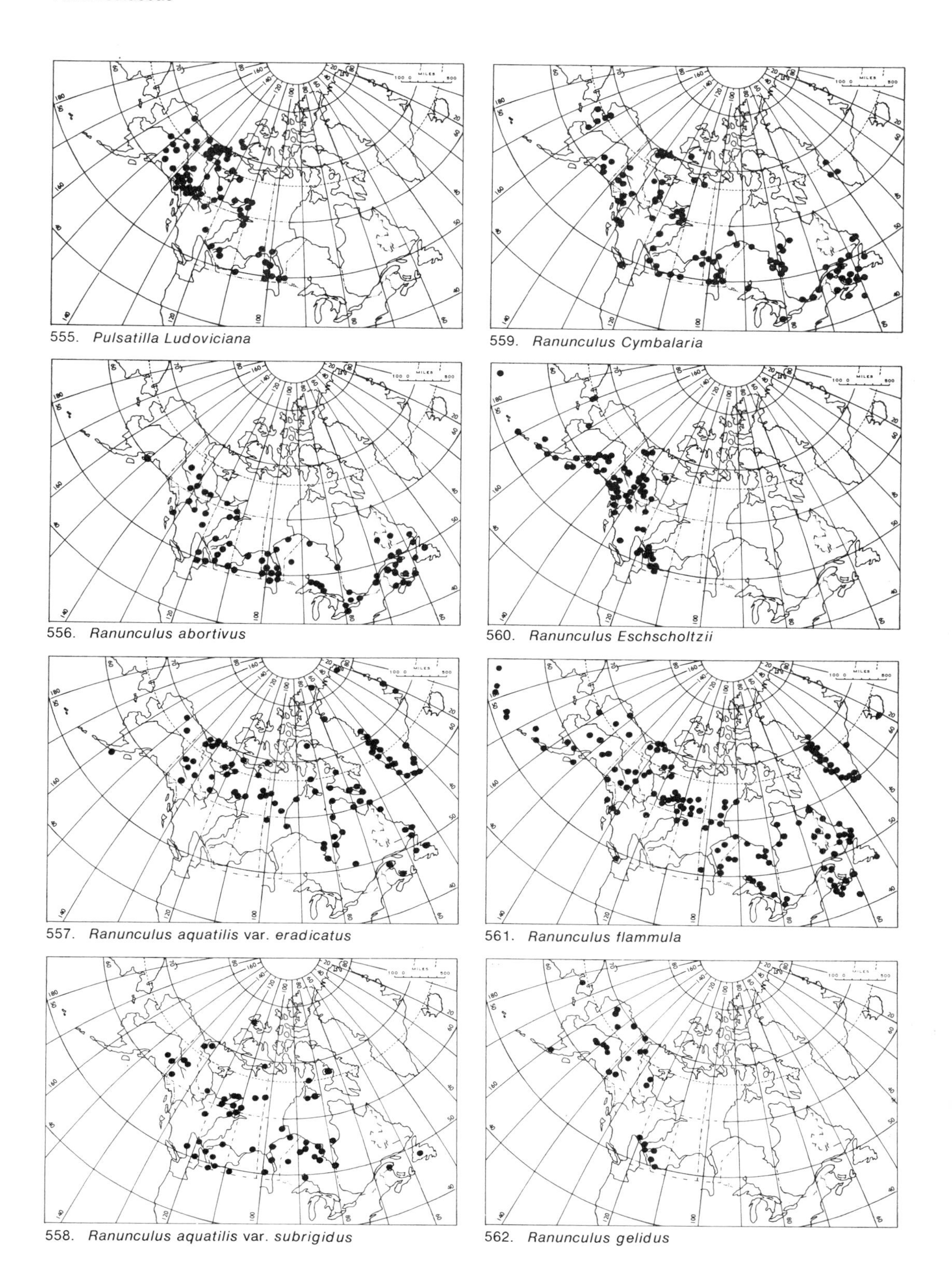

555. *Pulsatilla Ludoviciana*

556. *Ranunculus abortivus*

557. *Ranunculus aquatilis* var. *eradicatus*

558. *Ranunculus aquatilis* var. *subrigidus*

559. *Ranunculus Cymbalaria*

560. *Ranunculus Eschscholtzii*

561. *Ranunculus flammula*

562. *Ranunculus gelidus*

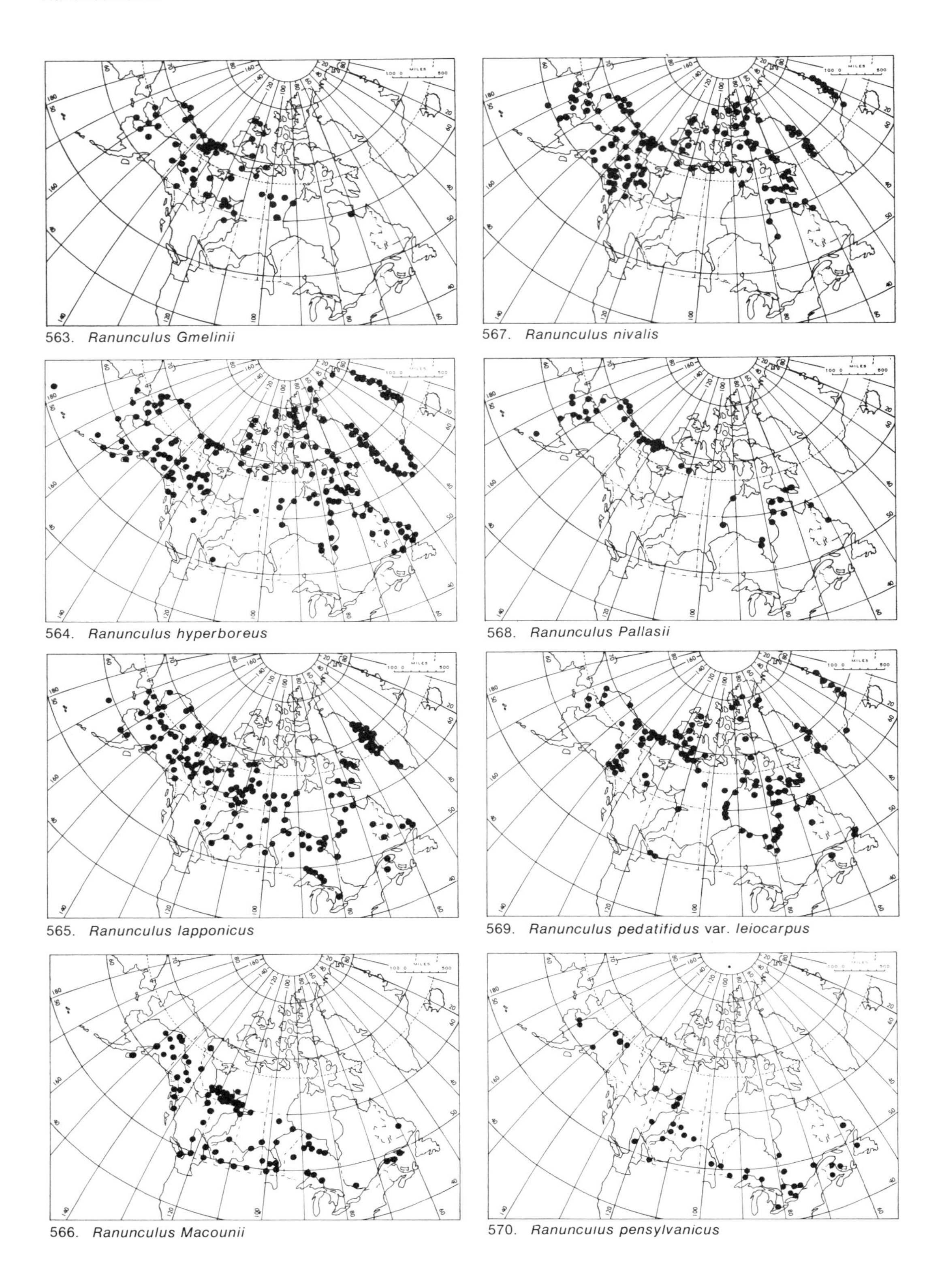

563. *Ranunculus Gmelinii*

564. *Ranunculus hyperboreus*

565. *Ranunculus lapponicus*

566. *Ranunculus Macounii*

567. *Ranunculus nivalis*

568. *Ranunculus Pallasii*

569. *Ranunculus pedatifidus* var. *leiocarpus*

570. *Ranunculus pensylvanicus*

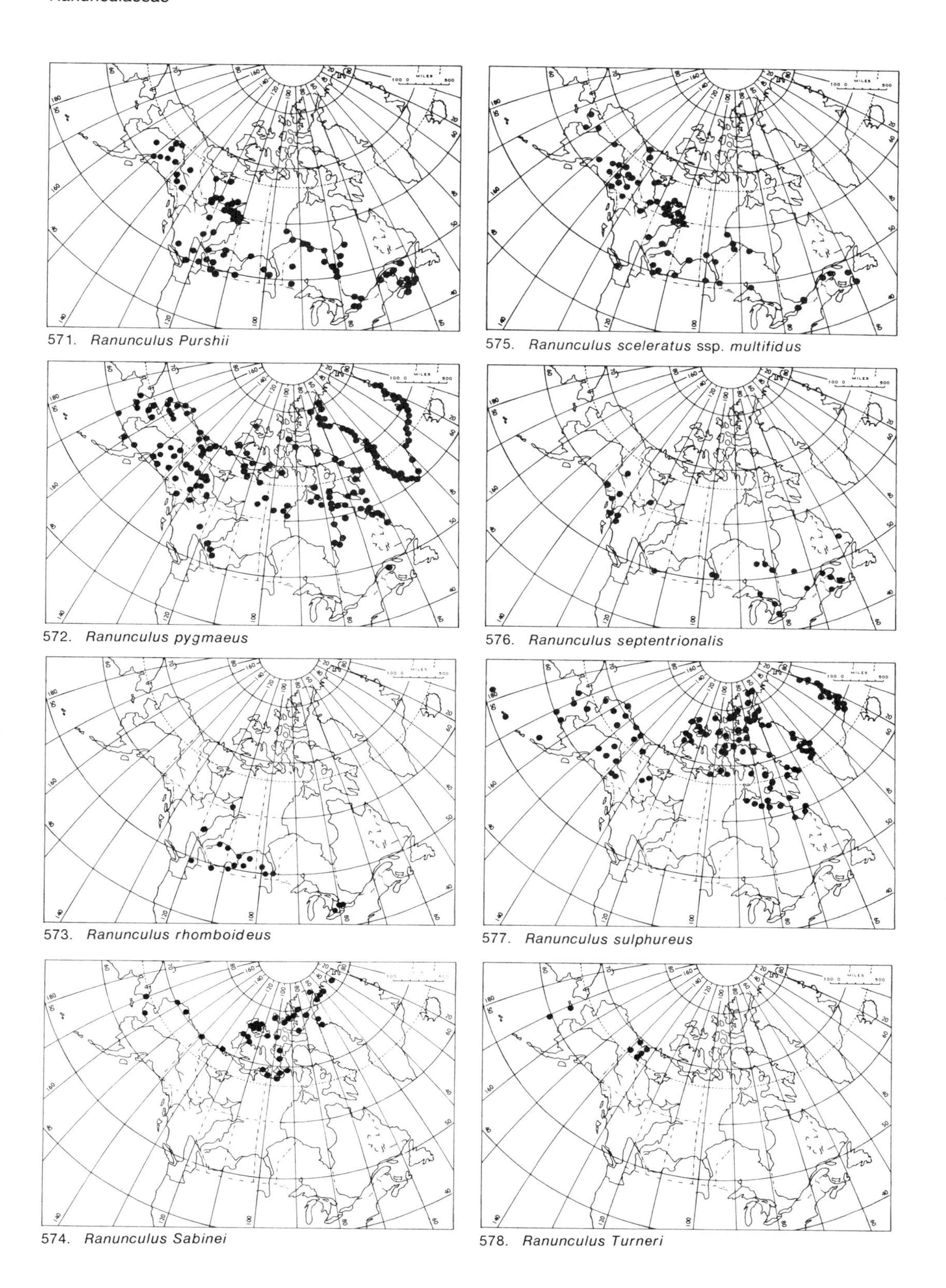

571. *Ranunculus Purshii*

572. *Ranunculus pygmaeus*

573. *Ranunculus rhomboideus*

574. *Ranunculus Sabinei*

575. *Ranunculus sceleratus* ssp. *multifidus*

576. *Ranunculus septentrionalis*

577. *Ranunculus sulphureus*

578. *Ranunculus Turneri*

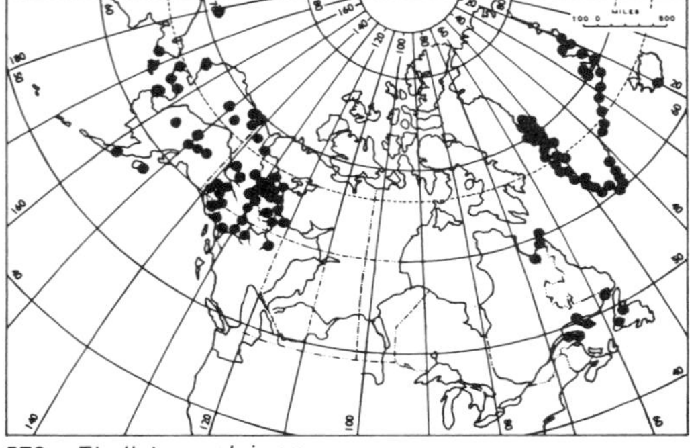

579. *Thalictrum alpinum*

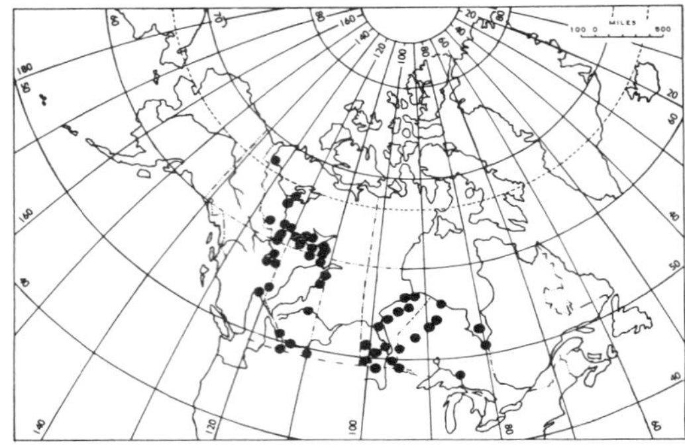
581. *Thalictrum venulosum*

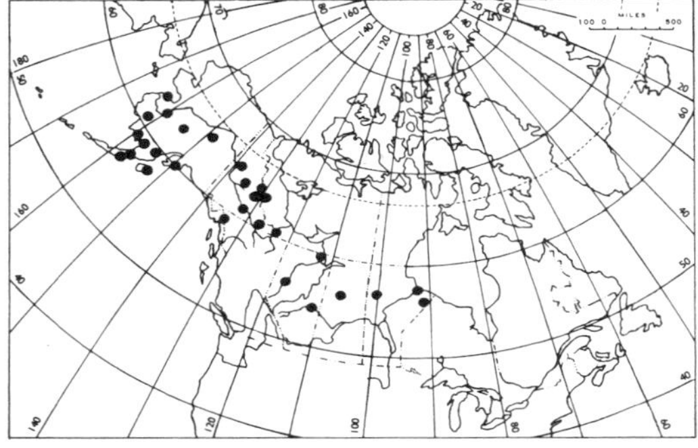
580. *Thalictrum sparsiflorum* var. *Richardsonii*

PAPAVERACEAE Poppy Family

Papaver L. Poppy

Caespitose, perennial, scapose (ours) arctic or high-alpine herbs with yellow or white milky latex and basal rosettes of hairy, pinnately lobed or divided leaves. The large and showy flowers hypogynous; sepals 2, soon deciduous, the petals thin, normally 4. Fruit a barrel-shaped capsule, containing numerous small seeds.

The colour of the latex apparently is constant and diagnostic when obtained from living material; it should be observed on cuts made either through the lower part of an immature capsule, through the scape just below it, or through a leaf-petiole.

a. Capsule narrowly ovoid, about 2 cm long and 3 times longer than broad
  b. Stigmatic disc flat, rays 6-7 . . . . *P. Hultenii*
  b. Stigmatic disc conical, rays usually 5 (4), converging to a central small but prominent point . . . . *P. Keelei*
a. Capsule obovoid usually less than twice as long as broad
  c. Stigmatic disc flat, not broader than the capsule
    d. Scapes relatively short, rarely over 12 cm long, densely villous from spreading, soft, dark brown or black hairs (high-arctic); rays 5 . . . . *P. cornwallisensis*
    d. Scapes commonly over 12 cm long, soft pubescent, but hairs much sparser, and ascending rather than spreading; rays 8-9 . . . . *P. radicatum*

c. Stigmatic disc prominently vaulted and broader than the capsule, the latter turbinate, its 5-6 rays converging to a small but prominent central point
   e. Leaves bipinnate; flowers 2-3 cm in diameter .... *P. McConnellii*
   e. Leaves entire or 3-lobed, leathery .... *P. Walpolei*

Papaver cornwallisensis A. Löve
Dwarf species of densely caespitose habit; leaves short, the blade rarely more than 1 cm long and commonly 3-lobed, although the terminal lobe is often shallowly 3-cleft. Scapes from a few to half a dozen or more, very rarely over 12 cm long, stout but commonly somewhat arching, densely villous from spreading brown or black hairs, nodding in bud; petals pale yellow or almost as often white or pinkish-yellow, 1.5 to 2.0 cm long, tardily deciduous; capsule less than 1 cm long, and almost as broad just below the stigma.

Local and always in rather wet, calcareous, gravelly places; high-arctic and high-alpine.

General distribution: Northernmost Greenland across the Canadian Arctic islands, with a disjunct alpine population in the Richardson and Mackenzie Mountains.

Map 582

Papaver Hultenii Knaben
Differs from other members of the *P. radicatum* group by its slender petiole five times as long as the oblong, pinnate blade; the scapes slender, with spreading brown hairs, in anthesis often 25 to 30 cm tall. Petals about 2 cm long, bright yellow in life, and the capsule narrowly cylindrical, commonly about 2 cm long.

Apparently common on sandy and gravelly beaches and tundra ridges.

General distribution: Endemic of arctic-alpine N.W. America.

Map 583

Papaver Keelei Porsild
P. Macounii auct. non Greene
Loosely tufted from an ascending and branching rhizome, the caudices characteristically covered by the persisting brown remains of leaf-bases. Leaves dark green and lustrous in life, glabrous above, sparsely pubescent beneath, about 5 cm long, the blade oblong 1.0 to 1.5 cm long, usually 5-lobed, the lateral lobes flat, narrowly lanceolate and entire, the terminal often shallowly 3-lobed or cleft. Scapes slender, mostly solitary and rarely more than 25 cm tall, sparsely pubescent from soft brown spreading or ascending hairs. Petals spreading in anthesis, bright yellow and drying yellow, soon deciduous. Capsule narrowly obovoid or clavate, commonly about 1 cm long and half as broad, the stigmatic disc conical, the 4 (5) rays converging into a small but prominent point.

The somewhat similar *R. Macounii* of southwest Alaska is a much larger and coarser plant with stout and pubescent scapes 20 to 25 cm tall, and much larger flowers. In the type (from St. Paul Island in Bering Sea) the petals are 2.5 cm long and the nearly ripe capsules 2 cm long and 0.5 cm in diameter, with an obtuse vaulted stigmatic disc lacking the projecting central point of *P. Keelei.*

Moist alpine herbmats or tundra.

General distribution: Endemic of arctic-alpine N.W. America.

Fig. 506 Map 584

Papaver McConnellii Hult.
Loosely tufted from a branching caudex; leaves glaucous, sparsely pubescent, 5 to 8 cm long, the petiole about as long as the rather short, bipinnate blade, its divisions linear and obtuse-pointed. Scapes 5 to 15 cm long, slender, usually somewhat arching, sparsely pubescent with soft, pale-brown spreading hairs. Flowers relatively large, the petals sulphur yellow, 2 to 3 cm long, drying yellow. Capsule obovate, 9 to 10 mm long, becoming distinctly turbinate in age; the stigmatic disc prominently vaulted, broader than the capsule, its 5 to 6 rays meeting in a small but distinct central point.

Sparsely vegetated, shaly slopes at elevations of about 900 m.

General distribution: Alpine. Endemic of northern Y.T. and N.W.T.

Fig. 507 Map 585

Papaver radicatum Rottb. *s. lat.*
(including ssp. *radicatum*
ssp. *labradoricum* (Fedde) Fedde and
ssp. *Porsildii* (Knaben) Löve
P. alaskanum auctt. non Hultén
Densely tufted, leaves short and crowded, blue-green, the blade lanceolate, commonly 3- to 5-lobed, the lobes entire or occasionally notched, sparsely pubescent on both sides; scapes usually several, commonly 10 to 12 cm tall, with sparse and soft, spreading pubescence. The buds globose, nodding; flowers relatively large, the petals sulphur yellow, rarely white, becoming green when bruised, tardily deciduous, 1.0 to 1.5 cm long; capsule obovoid 1.0 cm long and 0.6 cm in diameter, turbinate

in drying, the stigma flat, rays averaging 8 to 9. Latex yellow.

Open gravelly places; distinctly arctic and alpine.

General distribution: Circumpolar.

Fig. 508 Map 586

***Papaver Walpolei** Porsild
Dwarf species at once distinguished by its densely crowded, short-petioled leaves in which the blade is sub-entire or 3-lobed, entirely glabrous, dark green, shiny and somewhat leathery; the scapes rarely over 15 cm tall, the pale yellow or white flower mostly 2 cm in diameter, and the obovoid-pyriform capsule strongly tapering from the very broad and strongly arching stigma.

General distribution: Long known only from mountains of Seward Pen., Alaska, *P. Walpolei* is now known also from easternmost Siberia, and was recently reported from mountains of northern Y.T. (Porsild, 1972). It should be looked for also along the east slopes of Richardson and Mackenzie Mts.

Map 587

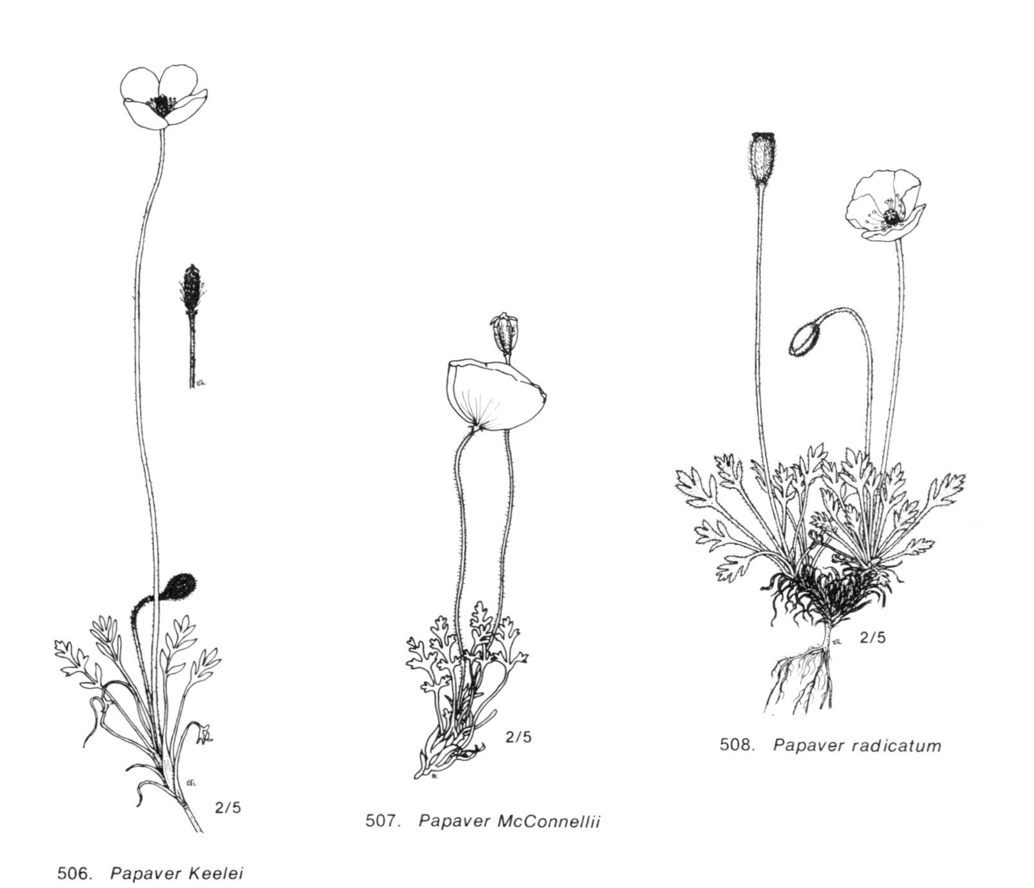

506. *Papaver Keelei*

507. *Papaver McConnellii*

508. *Papaver radicatum*

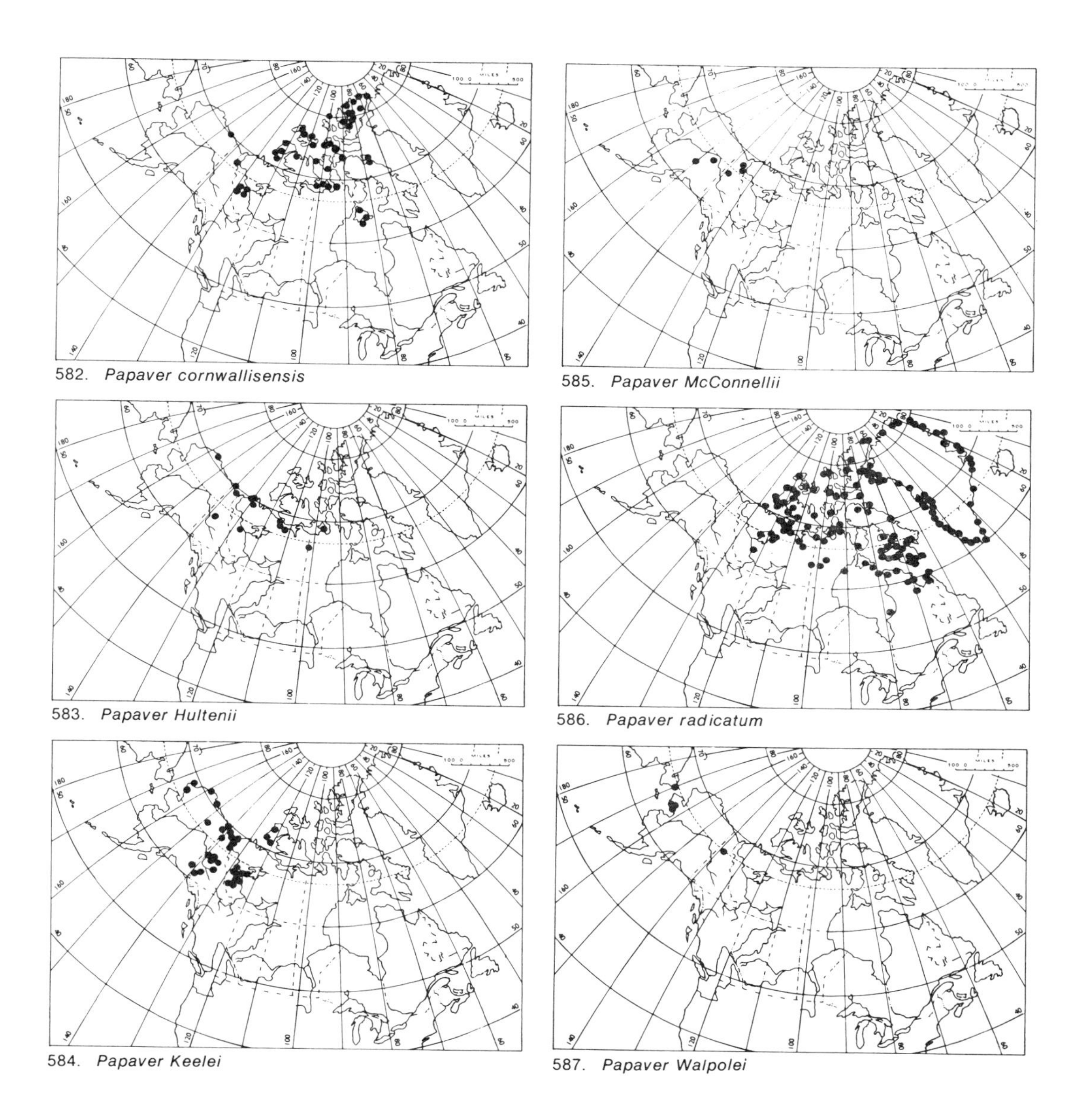

582. *Papaver cornwallisensis*

583. *Papaver Hultenii*

584. *Papaver Keelei*

585. *Papaver McConnellii*

586. *Papaver radicatum*

587. *Papaver Walpolei*

FUMARIACEAE Fumitory Family

Corydalis Medic

Tender, glabrous herbs with watery juice, dissected leaves and hypogynous, irregular flowers. Fruit a slender, pod-like, 2-valved capsule, containing small black, shiny seeds.

a. Perennial, from a small spherical, and deeply buried tuber; flowers blue or purple, in a few-flowered raceme . . . . *C. pauciflora*
a. Annual or biennial, from a weak tap-root; flowers in elongating racemes
  b. Stems much branched, spreading; corolla yellow . . . . *C. aurea*
  b. Stems simple, or branched above, erect, corolla pink with yellow tip . . . *C. sempervirens*

Corydalis aurea Willd.
Glaucous; raceme many-flowered, sessile from the upper leaf-axils; capsules spreading or somewhat pendulous, torrulose.
Gravelly slopes, roadsides, often weedy and pioneering on disturbed soil.
General distribution: Boreal Forest region from Que. to central Alaska and southwards.
Fig. 509 Map 588

Corydalis pauciflora (Steph.) Pers.
Stems 1 to 2 dm tall; leaves few from the lower part of stem, peduncled, their blade cordate, 3-parted, the divisions 3-lobed. Flowers 2 to 6, in a terminal cluster; corolla pale sky-blue or rarely white (var. *albiflora* Porsild); capsules pendant, short, 3-mm broad.
Damp herbmats, near and above timberline.
General distribution: Amphi-Beringian; from Alaska and Y.T. east to Mackenzie Mts., south to northern B.C.
Fig. 510 Map 589

Corydalis sempervirens (L.) Pers.
Stems up to 6 dm tall, mainly branched above; racemes stalked, commonly 3- to 6-flowered, one or more from the uppermost leaf-axils. Capsules erect-ascending, smooth.
Rocky clearings in coniferous northern forests.
General distribution: Boreal Forest region from Nfld. to Alaska and southwards.
Fig. 511 Map 590

509. *Corydalis aurea*

510. *Corydalis pauciflora*

511. *Corydalis sempervirens*

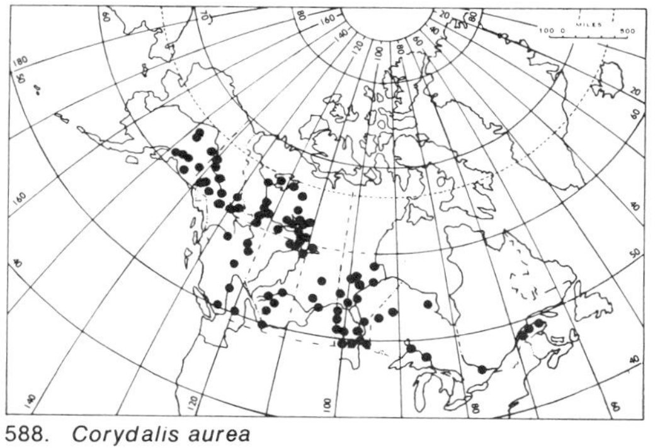
588. *Corydalis aurea*

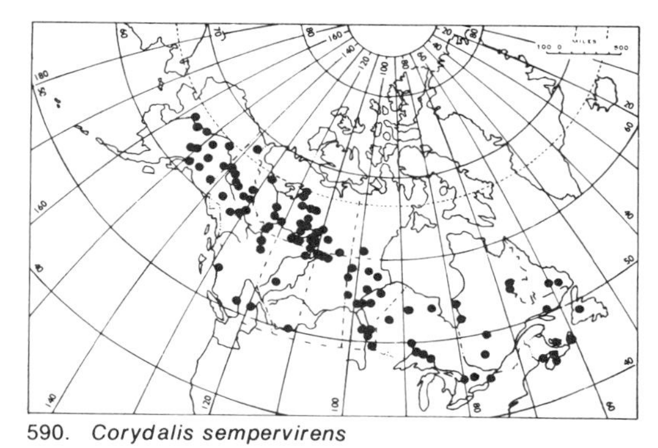
590. *Corydalis sempervirens*

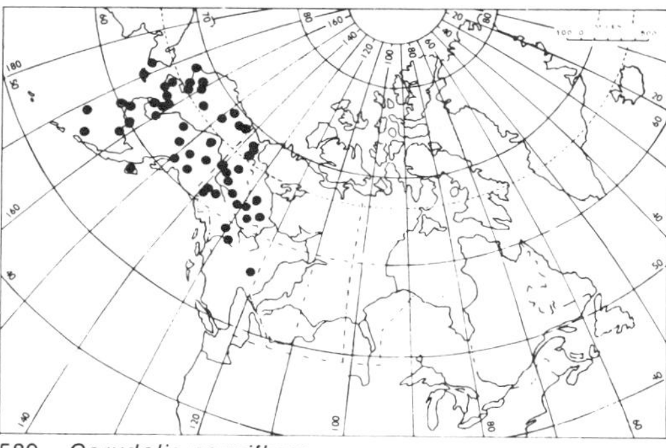
589. *Corydalis pauciflora*

CRUCIFERAE Mustard Family

Herbs with acrid, watery juice, alternate leaves lacking stipules, and regular, cruciform flowers with 4 deciduous sepals, 4 petals, and 6 stamens. Fruit a 2-valved pod or silique, 2-locular by a thin, membranaceous partition or septum stretched between the marginal placenta. Flowers in terminal racemes or corymbs.

a. Fruit short, not more than three times as long as broad
  b. Plant essentially glabrous
    c. Petals yellow . . . . *Draba*
    c. Petals white
      d. Siliques wingless
        e. Leaves subulate; small, submerged aquatic plant . . . . *Subularia*
        e. Leaves elliptic-oblong; terrestrial plants
          f. Basal leaves roundish, somewhat fleshy; sea shores . . . . *Cochlearia*
          f. Basal leaves lanceolate, not fleshy . . . . *Draba*
      d. Siliques broadly winged . . . . *Thlaspi*
  b. Plant pubescent
    g. Hairs simple
      h. Petals purple . . . . *Parrya*
      h. Petals white or yellow

i. Basal leaves entire or toothed . . . . . *Draba*
i. Basal leaves pinnatifid or pinnate
j. Valves of silique winged; seeds solitary in each locule . . . . . *Lepidium*
j. Valves not winged; seeds several in each locule . . . . . *Rorippa*
g. Hairs forked, branched or stellate, sometimes mixed with simple hairs
k. Petals deep purple; siliques ovoid oblong or terete . . . . . *Smelowskia borealis*
k. Petals white or yellow
l. Petals white
m. Siliques obcordate-triangular . . . . . *Capsella*
m. Siliques not obcordate-triangular
n. Plant densely pubescent . . . . . *Smelowskia*
n. Plant not densely pubescent
o. Siliques torulose . . . . . *Braya*
o. Siliques not torulose . . . . . *Draba*
l. Petals yellow (or creamy white in *Smelowskia calycina*)
p. Plant short-grey-pubescent . . . . . *Smelowskia*
p. Plant not densely grey-pubescent
q. Siliques flat
r. Siliques obovate-orbicular, 1-2-seeded . . . . . *Alyssum*
r. Siliques oblong, several-seeded . . . . . *Draba*
q. Siliques globular
s. Siliques 1-seeded (non-native) . . . . . *Neslia*
s. Siliques several-seeded (native) . . . . . *Lesquerella*
a. Fruit more than three times as long as broad
t. Plants with glabrous stem and leaves
u. Petals purplish
v. Leaves pinnate; valves of silique veinless . . . . . *Cardamine*
v. Leaves not pinnate; valves veined
w. Siliques terete; style long . . . . . *Parrya*
w. Siliques linear, flat; style short . . . . . *Arabis*
u. Petals white, pale lavender or yellowish
x. Leaves pinnate, digitate or 3-parted . . . . . *Cardamine*
x. Leaves simple
y. Basal leaves sessile or nearly so . . . . . *Arabis*
y. Basal leaves long-petioled
z. Low, densely tufted alpine arctic plant . . . . . *Cardamine bellidifolia*
z. Tall, not densely tufted plants
aa. Stem leaves clasping . . . . . *Thellungiella*
aa. Stem leaves not clasping . . . . . *Eutrema*
t. Plant pubescent, with simple, forked, branched or stellate hairs
bb. Hairs all simple
cc. Petals purplish
dd. Basal leaves lyrate, pinnate or entire
ee. Valves of siliques veined; leaves entire . . . . . *Cardamine bellidifolia*
ee. Valves of siliques veinless; leaves pinnate . . . . . *Arabis*
dd. Basal leaves simple, entire, toothed or sinuate
ff. Siliques 4-5 mm broad . . . . . *Parrya*
ff. Siliques narrower
gg. Siliques torulose . . . . . *Braya*
gg. Siliques not torulose . . . . . *Arabis*
cc. Petals white or yellow
hh. Petals white . . . . . *Arabis*
hh. Petals yellow or yellowish
ii. Lower leaves bi- or tri-pinnate . . . . . *Descurainia sophioides*
ii. Leaves not bi- or tri-pinnate
jj. Flowers leafy-bracted . . . . . *Erucastrum*
jj. Flowers not leafy-bracted

kk. Siliques with stout beak; seeds globular
 ll. Beak of siliques flat . . . . *Sinapis*
 ll. Beak of siliques terete . . . . *Brassica*
kk. Silique beakless, with a short, thick style; seeds oblong
 mm. Stem angular; lower leaves lyrate . . . . *Barbarea*
 mm. Stem terete
  nn. Silique linear, up to 10 cm long . . . . . . *Sisymbrium*
  nn. Silique oblong-elliptic, terete to orbicular, 2-4-valved, rarely up to 1.0 cm long . . . . . . *Rorippa*

bb. Hairs forked, branched or stellate, sometimes mixed with simple hairs
 oo. Petals purplish
  pp. Stem scapose or 1-3-leaved; siliques torulose . . . . *Braya*
  pp. Stem with several leaves; siliques not torulose . . . . *Arabis*
 oo. Petals white or yellow
  qq. Petals white
   rr. Stem leaves clasping
    ss. Stems erect, simple . . . . *Arabis*
    ss. Stems much branched, spreading . . . . *Halimolobos*
   rr. Stem leaves not clasping
    tt. Plant densely pubescent with short, greyish hairs . . . . *Smelowskia*
    tt. Plant not densely pubescent with short, greyish hairs
     uu. Siliques torulose . . . . *Braya*
     uu. Siliques not torulose, flat
      vv. Siliques less than 15 mm long . . . . *Draba stenoloba*
      vv. Siliques longer . . . . *Arabis*
  qq. Petals yellow; lower leaves pinnate
   ww. Stems with appressed, forked hairs attached in the middle . . *Erysimum*
   ww. Stems without such hairs
    xx. Siliques linear, terete . . . . *Descurainia*
    xx. Siliques flat, inflated . . . . *Smelowskia*

Alyssum L.

***Alyssum americanum** Greene
Perennial with slender, ascending or arching subligneous, leafy branches from a stout tap-root and branched caudex. Leaves alternate oblanceolate or ligulate, mostly less than 1 cm long and half as wide but occasionally longer; leaves and stems densely grey-canescent from appressed, flat, stellulate, glassy and scale-like hairs, each attached to a short stalk with its 4 to 5 branching radii interlocking with those of the surrounding scales. Inflorescence a terminal corymb or raceme; the petals pale yellow, 2 to 3 mm long. The fruit a 2-seeded, flat, elliptic, 5 to 7 mm long silique with a persisting, 2 mm long slender style.

Gravelly river terraces and benches.

General distribution: Endemic of unglaciated central Alaska and Y. T. A specimen in the National Herbarium of Canada labelled "Fort Simpson, 1853" most likely originated in Y. T.

Fig. 512 Map 591

Arabis L. Rock-cress

Biennial, or ours mostly short-lived perennials, mostly pubescent below and glabrous above, from a tap-root and branching base from which rise the leafy erect-ascending or somewhat spreading stems; basal leaves rosulate and very different from those of the stem (except in *A. alpina*). Flowers white, pale purple or pinkish, in terminal, elongating racemes. Fruit a linear, flat or flattened silique, from 1 to 3 mm broad, and from 2.5 to 8.0 cm long.

a. Stem leaves oblanceolate, not clasping, auriculate or sagittate; seeds wingless
 b. Young basal leaves lyrate-pinnatifid . . . . *A. lyrata* var. *kamchatica*
 b. Young basal leaves merely toothed, or entire
  c. Plant essentially glabrous . . . . *A. arenicola*

c. Plant soft pubescent on stem and on both sides of leaves; basal and stem leaves alike . . . . *A. alpina*

a. Stem leaves with auricled or sagittate base; seeds mostly winged
  d. Basal leaves grey, from a dense cover of minute, stellate hairs; mature siliques pendulous . . . . *A. Holboellii* var. *retrofracta*
  d. Basal leaves with simple, forked or variously branched hairs, or glabrous
    e. Mature siliques strictly erect
      f. Siliques about 3 mm broad, seeds in 2 rows . . . . *A. Drummondii*
      f. Siliques about 1 mm broad, seeds in one row . . . . *A. hirsuta* ssp. *pycnocarpa*
    e. Mature siliques arcuate, divaricate or horizontally spreading . . . . *A. divaricarpa*

Arabis alpina L.
Coarse, fetid and matted, with a slender and freely branched leafy caudex. Leaves all similar, 2 to 5 cm long and 1 to 2 cm wide, oblanceolate to obovate, sessile, fresh green and somewhat fleshy, coarsely dentate or sub-entire, usually densely stellate pubescent on both sides. Flowering stems erect-ascending, leafy, simple or branched, terminating in an elongating raceme. Petals white, 7 to 9 mm long and 2 to 4 mm broad. Siliques glabrous, 4 to 5 cm long and 1.5 to 2.5 mm broad, ascending and often somewhat curved, on pedicels one-third to one-half as long as the siliques.

Moist but well drained places, often near running water.

General distribution: Amphi-Atlantic, eastern subarctic west to Hudson Bay. From our area known from a single collection labelled Dubawnt L., District of Keewatin.

Fig. 513 Map 592

Arabis arenicola (Richards.) Gel.
Caespitose, essentially glabrous perennial from a much-branched base. Basal leaves 2 to 4 cm long, on slender petioles, the blade oblanceolate-cuneate, repand or sparsely toothed but never lyrate-pinnatifid, dark green, and somewhat fleshy. Flowering stems somewhat flexuous, ascending or decumbent, bearing 3 to 4 sessile leaves, terminating in a rapidly elongating 5- to 20-flowered raceme. Petals narrowly oblanceolate, 5 mm long, white or pale pink. Siliques glabrous, greenish-purple, 2.0 to 2.5 mm long, and 2 mm wide, on pedicels half as long as the siliques. The seeds mucilaginous when wetted.

Calcareous sand and gravel, often by lake shores or river banks.

General distribution: Endemic of subarctic and arctic Eastern N. American Arctic from E. Greenland westward to long. 110°W.

Fig. 514 Map 593

Arabis divaricarpa A. Nels.
Resembling *A. Holboellii* var. *retrofracta* but usually biennial, with a weak tap-root, and glabrous or sparsely pubescent basal leaves. The mature siliques are 5 to 7 cm long and about 2 mm broad, usually arcuate, erect-ascending at first, but never pendulous as in *A. Holboellii.*

Dry sandy or gravelly slopes mainly along the Mackenzie Valley north to its delta.

General distribution: Boreal N. American, Que. to Alaska and the Rocky Mts.

Fig. 515 Map 594

Arabis Drummondii Gray
Glabrous or nearly glabrous, short-lived glaucous perennial, from a simple or branching base; stems one to several, simple or branching above; basal leaves rosulate, usually entire-margined, glabrous or sparsely to moderately hirsute from variously branched hairs. Stem-leaves sessile, glabrous and glaucous. Flowers in a few- to many-flowered raceme, the pink petals narrowly-linear, 6 to 8 mm long. Siliques stiffly erect, 6 to 8 cm long and about 3 mm broad.

Rich herbmat slopes, north in the Mackenzie Valley to the Delta.

General distribution: N. America, from Lab. and Nfld. to Alaska and southwards in the Rocky Mts.

Fig. 516 Map 595

Arabis hirsuta (L.) Scop.
ssp. **pycnocarpa** (Hopkins) Hultén
Loosely tufted, short-lived perennial from a weak, fibrous root; the leafy, erect-ascending, single or branched stems 2 to 3 dm tall from a basal rosette in which at least the earliest leaves are distinctly lyrate-pinnatifid, whereas the later ones may be merely lobed or shallowly dentate, and oblanceolate in outline. The flowers in a short raceme, much elongated in fruit; petals white or pale pink. Siliques flat, erect-ascending, 2 to 3 cm long, about 1 mm broad. The style short with entire stigma.

Moist sandy lake shores or rocky ledges, with us north to Great Bear Lake and the Mackenzie Delta.

General distribution: N. America, from Que. to central Alaska.
Fig. 517 Map 596

Arabis Holboellii Hornem.
var. **retrofracta** (Grah.) Rydb.
Perennial; stems erect, simple or often branched from the base, 3 to 9 dm tall, short stellate-pubescent below, glabrous and glaucous above. Rosette leaves linear-oblanceolate, entire or sinuately dentate, minutely stellate pubescent, soon wilting. Cauline leaves linear, 1 to 2 cm long with a clasping base, finely stellate pubescent. Flowers in an elongating raceme, the petals white or pale purple about 8 mm long. Mature siliques 4 to 7 cm long and about 1.5 mm wide, first spreading but at length reflexed and pendulous. Stigma sessile.
The var. *retrofracta,* of central and western North America, differs by its narrower siliques from the var. *Holboellii* in which the siliques are from 2 to 2.5 mm broad.
Sunny, calcareous slopes.
General distribution: *A. Holboellii s. lat.* E. and W. Greenland, Que. to B. C. and Washington, north to Alaska.
Fig. 518 Map 597

Arabis lyrata L.
var. **kamchatica** Fisch.
Tufted, glabrous (ours), perennial, freely branching from the base, with erect-ascending or even declining or spreading flowering stems 1 to 3 dm tall. Basal leaves oblanceolate in outline, lyrate-pinnatifid; raceme lax, elongating in fruit. Petals white, 5 to 7 mm long, oblanceolate. Siliques slightly flattened, 2.0 to 2.4 cm long and about 1 mm broad. Stigma sessile.
Gravelly river flats and floodplains.
General distribution: Amphi-Beringian species whose range in the District of Mackenzie barely reaches the western limits of the somewhat similar but entirely eastern *A. arenicola.*
Fig. 519 Map 598

Barbarea R. Br. Winter-Cress

Barbarea orthoceras Ledeb.
Glabrous biennial or short-lived perennial from a weak, simple or branched tap-root; the 4 to 6 dm tall stems and often also the lower leaves dark purplish tinged; lower leaves and sometimes also those of the middle stem lyrate-pinnatifid. One to several racemes of small, pale-yellow flowers from the upper leaf-axils; the racemes first dense, but elongating in fruit. Siliques 2 to 3 cm long, linear and terete.
In damp places along streams and sheltered lake shores within the forested areas.
General distribution: Lab. to Alaska. Also E. Asia.
Fig. 520 Map 599

Brassica L. Mustard

Ephemeral weedy, introduced Old World annuals or biennials with coarsely toothed and often partly lyrate-pinnatifid leaves, and racemose yellow flowers; ours with slender siliques tapering into a stout beak.

a. Upper leaves not clasping . . . . *B. juncea*
a. Upper leaves clasping
  b. Root thickened, tuber-like; radicle leaves bristly . . . . *B. Napus*
  b. Root little thickened; leaves glabrous or remotely hispid when young . . . . *B. campestris*

Brassica campestris L.
B. rapa L.
Bird rape
Glaucous usually unbranched herb to 7 or more dm in height, glabrous or remotely hispid when young; lower leaves petioled, toothed or pinnatifid, the upper sessile and auriculate-clasping, subentire; petals pale yellow, 6 to 7 mm long.
Weed of cultivated fields collected once at Yellowknife.
General distribution: Cosmopolitan.

Brassica juncea (L.) Coss.
Indian mustard
Nearly glabrous, somewhat glaucous usually branched herb to 5 or more dm in height; lower leaves lyrate and petioled, the upper entire or nearly so, tapering to short petioles; petals pale yellow, about 8 mm long.
Weed of cultivated fields collected only once in our area.
General distribution: Cosmopolitan.

Brassica Napus L.
Similar to *B. rapa,* but greener, and with a thickened tuber-like base; radical leaves bristly; petals darker yellow and smaller.

Persistent after cultivation and often found in waste ground.

General distribution: Cosmopolitan.

Braya Sternb. & Hoppe

Perennial, calciphilous, green or purplish-tinged herbs of tufted habit, with simple, linear or narrowly spatulate, entire or shallowly toothed leaves. The flowers small, white or lilac, in capitate inflorescences that in some species become markedly elongated in fruit. Siliques cylindric and somewhat torulose, the valves nerveless. Seeds in two rows.

a. Flowering stems with one or more leaves, the siliques linear, about 1 mm wide
  b. Plant essentially glabrous, strongly purplish-tinged, leaves linear-spatulate
    c. Siliques erect at an acute angle; non-arctic . . . . *B. humilis* ssp. *humilis*
    c. Siliques ascending at an angle of about 45°; high-arctic . . . . . . *B. humilis* ssp. *arctica*
  b. Plant grey-canescent; leaves coarsely toothed . . . . *B. Richardsonii*
a. Flowering stems scapose, from branching caudex and usually stout tap-root, the siliques mostly over 1 mm wide
  d. Fruiting raceme not much elongated
    e. Dwarf species 4-10 cm tall in fruit; scapes slender; siliques usually less than 1 cm long
      f. Scapes stifly erect; alpine . . . . *B. Henryae*
      f. Scapes decumbent-ascending; arctic . . . . *B. Thorild-Wulffii*
    e. Taller in fruit 10-15 cm high, the scapes stouter, arching . . . . *B. purpurascens*
  d. Fruiting raceme much elongated; stems 10-20 cm high . . . . *B. glabella*

Braya glabella Richards.
B. ? pilosa Hook.
B. ? Bartlettiana Jord.
A slender green or purplish-tinged, glabrous or short-pubescent plant from a weak tap-root; basal rosettes few, the 3 to 5 cm long leaves linear in small fasicles, their petioles characteristically dilated from a geniculate and swollen base. Scapes stiffly erect, the siliques 1.2 to 1.4 cm long and about 1.5 mm in diameter. Style 1.2 to 1.5 mm long.

The type of *B. glabella* came from Copper Mountains, north of Dismal Lake, halfway between the north arm of Great Bear Lake and Coronation Gulf. Flowering and fruiting specimens matching the Richardson specimens were collected not far from the type locality, on 23-26 June, 1928 (Porsild Nos. 4698-9 CAN).

In mineral soil of damp tundra.

General distribution: Endemic of N. W. District of Mackenzie, N. W. Y. T. and N. Alaska.

Map 600

Braya Henryae Raup
Caespitose from a deep tap-root. Leaves narrowly spatulate; siliques lanceolate, 8 to 12 mm long and 1 to 2 mm in diameter at the base. Style 1.0 to 1.6 mm long, the stigma shallowly lobed.

Alpine, barren limestone flats.

Rather similar to *B. purpurascens* but smaller in all parts and, perhaps should be considered a geographical race of that species; thus far it is known only from the eastern slope of Richardson and Mackenzie Mountains.

General distribution: Endemic of the northern Rocky Mts.

Map 601

Braya humilis (C. A. Mey.) Robins.
Basal leaves oblanceolate, toothed or entire, glabrous or somewhat pubescent with simple or branching hairs. Flowering stems simple or branching, erect or ascending; inflorescence capitate, much elongated in fruit. Petals white or lilac; the mature siliques 1 to 2 cm long. With us represented by two rather distinct races of which one is high-arctic and the other subarctic.

ssp. *humilis:* Plant commonly purplish tinged; mature siliques erect-ascending; style slender and the stigma coronate. Non-arctic favouring moist, alluvial clay soils.

ssp. *arctica* (Bocher) Rollins, (*B. intermedia* Sør.): Siliques spreading at a more obtuse angle; styles short and stout with a distinctly bilobed stigma. High arctic, xerophytic.

General distribution: ssp. *humilis:* Nfld. to Alaska in widely disjunct populations restricted to areas underlain by calcareous rocks; ssp. *arctica:* From northernmost E. and W. Greenland across the northernmost Canadian arctic islands reaching the mainland coast mainly north of lat. 70°.

Fig. 521 (ssp. *arctica*) Map 602 (*B. humilis s. lat.)*

Braya purpurascens (R. Br.) Bunge
Caespitose, with a stout tap root and freely branching caudex. Leaves mostly basal, rosulate, somewhat fleshy, linear-oblanceolate, entire, obtuse, 2 to 3 cm long, 2 to 3 mm wide, glabrous, and mostly purplish tinged. Stems scapose, or occasionally with a single leaf supporting the lowermost flower, stiffly erect or ascending, often somewhat flexuous, glabrate or soft pubescent, dark purple, 5 to 15 or rarely 20 cm high in fruit. Flowering raceme at first capitate, soon somewhat elongated, 5- to 25-flowered. Sepals tardily deciduous; petals white or dilute purplish. Siliques plump, 1.0 to 1.4 cm long, 2 to 3 mm wide, glabrous, glabrate or soft pubescent.

Calcareous clay and gravel barrens.

General distribution: Circumpolar, high-arctic.

Fig. 522 Map 603

Braya Richardsonii (Rydb.) Fern.
Leaves, stems and siliques clothed by soft and white variously forked and branched hairs; basal leaves oblanceolate 1 to 3 cm long. Flowering stems leafy and in large specimens often greatly elongated, in fruit up to 3 dm long, much branched, ascending and decumbent at the base. Siliques 1 to 2 cm long, the style about 0.5 mm long, shallowly bi-lobed.

Calcareous gravelly riverflats.

General distribution: Endemic of the Canadian Rocky Mts. from southern Alta., northeastern B. C., western District of Mackenzie, Y.T., and central and N. W. Alaska.

Map 604

***Braya Thorild-Wulffii** Ostenf.
Low, caespitose with 5 to 10 cm long decumbent-ascending, crisply pubescent flowering and fruiting scapes. Basal leaves somewhat fleshy, oblong, entire, about 1 cm long and crisply ciliated. Flowering raceme at first capitate but soon elongating. Flowers small, the sepals hirsute near the tip, slightly shorter than the white or rose-pink petals. Siliques ovate and plump, 5 to 8 mm long and 2 to 3 mm wide, the valves convex, densely pilose; stigmas bilobed, sessile.

Calcareous, gravelly barrens.

General distribution: High-arctic endemic of northernmost Greenland and the Canadian Arctic Archipelago, but probably occurring also on the mainland west of Coronation Gulf.

Fig. 523 Map 605

Capsella Medic. Shepherd's-purse

Capsella Bursa-pastoris (L.) Medic.
Weedy, glabrous or somewhat hairy winter annual with a rosette of pinnately lobed or toothed basal leaves; flowering stem simple or branched, 2 to 6 dm high. The small, white flowers in an elongating raceme. The specific name is in allusion to the purse-shaped obcordate fruit.

A common weed in gardens and by roadsides.

General distribution: Cosmopolitan.

Cardamine L. Bitter Cress

Ours mainly glabrous, perennial herbs with white or purple, scentless flowers, all (except one) with pinnate leaves, and with linear, flattened or terete siliques that open elastically from the base; seeds in one row.

a. Dwarf species with ovate, entire leaves . . . . *C. bellidifolia*
a. Taller, with pinnate leaves
  b. Flowers large, with petals 5 mm long or longer, white or purple
    c. Leaflets of cauline leaves numerous, 9-17, linear or ovate, 5-8 mm long; flowers usually white, rarely pale purple . . . . *C. pratensis*
    c. Leaflets of cauline leaves fewer
      d. Leaflets 5-7, linear 1.0-2.5 cm long, flowers white . . . . *C. digitata*
      d. Leaflets 3-5, round or oval, the terminal usually 3-lobed, flowers normally purple, less often white

e. Plant totally glabrous . . . . *C. microphylla*
e. Plant always somewhat pubescent at least above . . . . *C. purpurea*
b. Flowers small and always white, the petals less than 4 mm long
f. Perennial, leaflets oval or cordate . . . . *C. umbellata*
f. Biennial or annual
g. Leaflets oval or oblong, decurrent on the rachis . . . . *C. pensylvanica*
g. Leaflets linear, not decurrent on rachis . . . . *C. parviflora* var. *arenaria*

Cardamine bellidifolia L.
Low and tufted with oval, entire, slender petioled leaves and erect-ascending mostly from 2 to 10 cm high flowering stems. Inflorescence 2- to 5-flowered, first capitate but soon elongating. Petals milky white, twice as long as the purple sepals. Siliques linear, short-peduncled, 15 to 25 mm long and 1.0 to 1.5 mm wide, stiffly erect with a short, thick style.

Wet mossy places, often by cold brooks or in shaded rock crevices.

General distribution: Circumpolar, arctic-alpine.

Fig. 524 Map 606

Cardamine digitata Richards.
C. hyperborea O.E. Schulz
C. Richardsonii Hult.
Stems simple and ascending 5 to 20 cm high from a slender and fragile creeping rhizome, which is deeply buried in the soil. Basal leaves slender-petioled, composed of from 5 to 7 linear-lanceolate, somewhat crowded, spreading leaflets 10 to 25 mm long and 2 to 5 mm wide. Inflorescence of from 10 to 20 flowers, at first subcapitate, soon elongating. Flowers about 6 mm high, the petals milky white, twice as long as the sepals. Mature siliques from 2 to 4 cm long and 2 mm wide, gradually tapering into a slender, 2 to 3 mm long beak.

Moist, turfy places, often growing on the peaty sides, or between large hummocks, in tundra. The species is strongly nitrophilous, responding to animal manure by luxuriant development of the vegetative parts.

General distribution: Endemic of arctic western N. America and eastern Asia.

Fig. 525 Map 607

Cardamine microphylla Adams
C. minuta Willd.
Plant totally glabrous with 10 to 15 cm tall stems from a horizontally creeping, slender rhizome; leaves few, 3- to 5-lobed. Inflorescence racemose, of from 5 to 20 long-petioled white or purplish flowers, much elongating in fruit, when the 2 to 3 cm long linear siliques rise stiffly erect like candles of a candelabrum.

Moist, alpine herbmats.

General distribution: Amphi-Beringian reaching northwestern District of Mackenzie.

Fig. 526 Map 608

Cardamine parviflora L.
var. **arenicola** (Britt.) O.E. Schulz
Weedy annual or biennial resembling *C. pensylvanica* but usually smaller, more slender and more freely branching from below.

Sandy, open places or rocky ledges.

General distribution: Eastern N. America. With us almost certainly a recent introduction.

Fig. 527 Map 609

Cardamine pensylvanica Muhl.
Biennial or short-lived, essentially glabrous perennial from a short, fibrous root. Stems 3 to 5 dm tall, simple but freely branching above, stiffly erect and uniformly leafy below the inflorescence. The first year's basal leaves soon withering. Cauline leaflets oblanceolate, 3 to 4 pairs, the terminal one largest. Inflorescence of small white flowers, much elongating in fruit; the siliques filiform, 1.5 to 2.0 cm long.

Wet, gravelly streambanks or lake shores.

General distribution: Eastern boreal forest species barely entering S. W. District of Mackenzie.

Fig. 528 Map 610

Cardamine pratensis L. *s. lat.*
Stems simple and ascending from a short rhizome. Leaflets numerous and small. Flowers large with white or pale purplish petals thrice the length of the sepals. A variable species, which in the arctic parts of our area reproduces mainly vegetatively, by leaf buds.

Occasional to common in wet meadows and on flood plains.

General distribution: Circumpolar, arctic and subarctic.

Fig. 529 (var. *angustifolia*) Map 611

***Cardamine purpurea** Cham. & Schlecht.
Resembling *C. microphylla* but of tufted habit, from a short, branching rhizome. Basal leaves

numerous; flowering stems ascending, rarely over 10 cm long, soft pubescent, at least in the upper part. Flowers smaller than in *C. microphylla*, purple, rarely white, in a short, not much elongating raceme; siliques 1 to 2 cm long, ascending or spreading.

Moist, alpine herbmats.

General distribution: Amphi-Beringian. A recent collection in the Y.T., from near the Continental Divide, suggests that the species may also occur east of the District of Mackenzie boundary.

Map 612

Cardamine umbellata Greene
Glabrous perennial with leafy and branching stems up to 25 cm tall, resembling *C. pratensis* but more slender, its flowers smaller and always white, and its leaflets fewer and much larger, rounded, oval and sometimes obscurely 3-dentate. The siliques are narrowly linear, 2.0 to 2.5 cm long.

Wet, seepy stream banks.

General distribution: Amphi-Beringian species. Barely entering our area along the central east slope of the Mackenzie Mts.

Fig. 530 Map 613

Cochlearia L. Scurvy-Grass

Cochlearia officinalis L. *s. lat.*
Glabrous somewhat fleshy halophytic biennial which in the first year forms a rosette of reniform, ovate to rounded-deltoid, entire or shallowly dentate, slender-petioled leaves. The small white flowers appear early the second season and are first hidden among the leaves, but the raceme soon elongates, and the fruiting stems become arched or even decumbent, with a few scattered, repand-dentate oblong leaves. The glabrous ellipsoid-oblong siliques are 5 to 8 mm long and are borne on slender erect-spreading peduncles.

Two weakly differentiated and perhaps overlapping races or varieties of the polymorphic *C. officinalis* occur in our area. One, with single or freely branched, mostly erect stems (ssp. *arctica* (Schlecht.) Hult.) appears to be predominantly western, whereas a lower and often prostrate or depressed-stemmed race (ssp. *groenlandica* (L.) Porsild) perhaps is restricted to the Eastern Arctic.

The scurvey-grass is common in not too dry places near the sea shore. It is strongly nitrophilous and for this reason is much favoured by the manure of nesting sea-birds. The succulent leaves are edible, either fresh or cooked, and are reputed to be rich in ascorbic acid.

General distribution: *C. officinalis s. lat.* Circumpolar.

Fig. 531 (ssp. *arctica*), 532 (ssp. *groenlandica*)
Map 614 (*C. officinalis s.l.*)

Descurainia Webb & Berth. Tansy-Mustard

Annual, biennial or short-lived perennials of weedy habit, with grey-green short pubescence of variously forked or stellate hairs; leaves pinnatifid or finely dissected, those of the basal rosette usually withered at flowering time. The flowers small and pale yellow, in elongating racemes, the siliques terete on slender ascending or spreading pedicels.

a. Leaves and stems with stellate and simple, non-glandular hairs; plant canescent
  b. Siliques about 1 cm long, or less . . . . *D. Richardsonii*
  b. Siliques 1.5-2.0 cm long, curved . . . . *D. Sophia*
a. Leaves and stems with gland-tipped hairs; plant green
  c. Siliques clavate, distant, on widely spreading pedicels . . . . . *D. pinnata* var. *brachycarpa*
  c. Siliques linear, crowded, curved and spreading . . . . *D. sophioides*

Descurainia pinnata (Walt.) Britt.
var. **brachycarpa** (Richards.) Fern.
Similar to *D. Richardsonii* but gland-tipped hairs mixed with the grey-green pubescence of the leaves and stems. Fruiting peduncles spreading nearly horizontally; siliques slightly clavate and usually less than 1 cm long with 8 to 10 seeds, arranged in two rows within each locule.

Sandy and rocky places, barely reaching the upper Mackenzie Valley.

General distribution: N. America: Que. to S. W. District of Mackenzie and south.

Map 615

Descurainia Richardsonii (Sweet) O.E.Schulz
Stems simple or branching above, 2 to 10 dm tall, the lower part leafy, the leaves oblong, bipinnately divided and, like the stem, canescent from variously branched and simple but non-glandular hairs. Silique slightly curved about 1 cm long with 6 to 8 seeds in one row within each locule.

Calcareous sandy and gravelly places north to the limit of trees.

General distribution: N. America, boreal.

Fig. 533 Map 616

Descurainia Sophia (L.) Webb
Similar to *D. Richardsonii* but the pubescence less dense and the plant, therefore, green rather than grey. Siliques 1.5 to 2.0 cm long, distinctly curved, with 10 to 20 seeds in one row within each locule.

Roadsides and waste places, but also readily spread to abandoned camp sites far beyond settled areas.

General distribution: A weed introduced from Europe, common in the southern parts of the Mackenzie Valley.

Descurainia sophioides (Fisch.) O.E.Schulz
Puberulent or slightly glandular biennial of weedy habit with simple or branching, sparsely-leaved stems 15 to 100 cm high, from a basal rosette of mostly bipinnatifid leaves, which remain green throughout the first winter and usually wither at the time of flowering. Terminal racemes usually much elongated toward maturity. Flowers small and pale yellow. Siliques from 10 to 30 mm long and about 1 mm wide, somewhat curved, on short, very slender, spreading pedicels 5 to 7 mm long.

An indigenous but weedy species, no doubt often spread by man and frequently growing on refuse heaps near human habitations. Not uncommon, however, on damp soil especially in places where landslips or soil movement has occurred.

General distribution: Amphi-Beringian, arctic and subarctic, western N. America.

Fig. 534 Map 617

Draba L.

Low, caespitose or tufted chiefly perennial and calciphilous arctic-alpine plants, most of them conspicuously pubescent, with entire or somewhat obscurely toothed leaves, small white or yellow flowers in a short, but often elongating raceme, and flat, lanceolate, ovate or oblong siliques from one to three (or more) times as long as broad.

a. Winter annuals or short-lived perennials
  b. Plant entirely glabrous or nearly so, 2-10 cm tall; petals pale yellow ...... *D. crassifolia*
  b. Plant with pubescent leaves and leafy stem 30 cm tall in fruit .... *D. nemorosa*
a. Perennials
  c. Petals yellow
    d. Stoloniferous .... *D. ogilviensis*
    d. Caespitose or matted
      e. Plant scapose or nearly so
        f. Siliques glabrous
          g. Leaves merely ciliate, dwarf species .... *D. pilosa*
          g. Leaves hirsute with variously branched as well as simple hairs
            h. Pubescence velvety of nearly sessile, stellate hairs ..... *D. Palanderiana*
            h. Pubescence sparse, of simple or branched but not stellate hairs .... *D. alpina*
        f. Siliques hairy
          i. Siliques spherical or not much longer than broad .... *D. oligosperma*
          i. Siliques more than twice as long as broad
            j. Style very short; high-arctic plants .... *D. corymbosa*
            j. Style prominent, 1 mm long, or longer; Cordilleran species .... *D. incerta*
      e. Plant leafy-stemmed; siliques prominently twisted .... *D. aurea*
  c. Petals white
    k. Biennial or short-lived perennial with glabrous siliques and obsolete styles .... *D. albertina*
    k. Perennial

l. Plant scapose or nearly so
  m. Pubescence of leaves closely pannose-stellate
    n. Scapes sparsely stellate-pubescent; mature siliques 6-8 mm long . . . . *D. nivalis*
    n. Scapes always glabrous; mature siliques 8-12 mm long and commonly twisted . . . . *D. lonchocarpa*
  m. Pubescence of leaves of simple or variously forked hairs
    o. Inflorescence capitate, but elongating in fruit; siliques lanceolate
      p. Leaves with simple hairs only . . . . *D. fladnizensis*
      p. Leaves with simple and also branched or forking hairs
        q. Stellate hairs less than ¼ mm broad, rays spreading . . . . *D. lactea*
        q. Stellate hairs more than ¼ mm broad, rays mostly parallel to leaf surface . . . . *D. Porsildii*
    o. Inflorescence capitate even in fruit
      r. Siliques ovate-orbicular, 4-5 mm in diam.; high-arctic . . . . *D. subcapitata*
      r. Siliques oblong-elliptic, 5-7 mm long; Cordilleran . . . . *D. Macounii*
l. Plant leafy-stemmed
  s. Siliques glabrous or nearly so; stem-leaves commonly 2-3
    t. Pedicel of mature silique longer than silique . . . . *D. longipes*
    t. Pedicel of mature silique shorter than silique
      u. Style prominent; pedicel of silique ¾ to 1 times as long as silique . . . . . . *D. glabella*
      u. Style nearly obsolete; pedicel of silique ½ as long as silique . . . . *D. norvegica*
  s. Siliques distinctly pubescent
    v. Stem leaves usually less than 4
      w. Leaves densely stellate-pannose; style distinct
        x. Mature siliques broadly lanceolate, not twisted
          y. Basal leaves with felt-like cover of mainly stellate hairs . . *D. cinerea*
          y. Basal leaves ragged-pilose with simple and variously branched hairs . . . . *D. oblongata*
        x. Mature siliques narrowly lanceolate, mostly twisted . . . . *D. cana*
      w. Leaves thinly stellate-pubescent; style very short or obsolete . . . . *D. praealta*
    v. Stem leaves usually 4-7 and, like the stem, commonly purplish . . . . *D. borealis*

Draba albertina Greene
D. stenoloba sensu Am. Auth. *pro parte*
Biennial or short-lived perennial, from a simple or branching base. Leaves mainly basal, oblanceolate-ovate, obscurely toothed, 0.5 to 2.0 cm long, the margins ciliate, otherwise thinly covered by variously forked hairs. Flowering stem 3 to 30 cm tall, glabrescent. Inflorescence 3- to 6-forked, sepals with few simple hairs; petals pale pink or purple, fading white, siliques about 1 cm long, glabrous, narrowly lanceolate, spreading, on slender pedicels about as long as the silique; style obsolete.

Pioneering on disturbed soil of moist alpine or subalpine slopes.

General distribution: S. Yukon, east to Mackenzie Mts. south through the mountains of B.C. and W. Alta. to Calif. and Colo.

Fig. 535 Map 618

Draba alpina L.
Densely tufted scapose plants with lanceolate, somewhat fleshy leaves with a prominent midrib, and long, rather stiff hairs along the margins; the upper and under surface glabrous or sparingly beset with simple or variously branched but not stellate hairs. Scapes stiffly erect from a few to 15 cm high, sparsely covered by simple or forked, soft, spreading hairs; inflorescence short and corymbose with from 3 to 10 rather large flowers; petals yellow, about twice as long as the hairy sepals; siliques glabrous, broadest below the middle, about 10 mm long or as long as their spreading pedicels; styles prominent, 1 mm long.

Common on snowbeds, in moist tundra, or in clay on wet gravelly barrens.

General distribution: Circumpolar, high-arctic, alpine.
Fig. 536 Map 619

Draba aurea M. Vahl
Tufted perennial from a branched, leafy caudex; basal leaves oblanceolate, densely stellate pubescent; stems 10 to 30 cm high, leafy. Petals yellow; siliques 10 to 15 mm long, hairy in the main species, often twisted in age, erect on 5 mm long pedicels; styles long.

A few collections from the southern Mackenzie may be separated as belonging to the var. *leiocarpa* (Payson & St. John) Hitchc. distinguished chiefly by its glabrous siliques.

Dry, calcareous slopes.

General distribution: Alaska to E. Greenland; subarctic-alpine.
Fig. 537 Map 620

Draba borealis DC.
D. luteola auctt. non Greene
Short-lived perennial from a simple or branched caudex; basal leaves rosulate, oblanceolate, 1.5 to 2.0 cm long; stem leaves 6 to 8, elliptic-oblong usually dentate in the upper third; the lower leaves and the stem usually dark purplish pigmented under the soft indument of simple and variously branched hairs. Flowering stems simple, stiffly erect, solitary or several to many, 15 to 40 cm tall and still elongating in fruit, the upper part with close pubescence of branched hairs, near the base mixed with much longer simple hairs. Inflorescence capitate, soon elongating. Petals white. Siliques sparsely pubescent, lanceolate, rarely over 10 mm long and about half as broad, often somewhat twisted, on pedicels not quite as long as body. Style slender about 1.0 mm long.

Moist, sandy or gravelly river banks and lake shores, often among willows.

General distribution: Amphi-Beringian, subarctic-alpine.
Fig. 538 Map 621

Draba cana Rydb.
D. lanceolata auctt. non Royle
Tufted, short-lived perennial from a simple or branching caudex. Leaves 1.5 to 2.0 cm long, narrowly oblanceolate, cinereous with dense and soft stellate tomentum, entire, or often with one or two teeth near the apex. Stems leafy, few to several, simple or somewhat branched, characteristically ascending-erect from the base. Inflorescence of small flowers, greatly elongating in fruit, the petals white. Siliques 6 to 14 mm long and 1.5 to 2.0 mm wide, often twisted, stellate pubescent; style short.

Open places on calcareous, gravelly slopes and river terraces, north to or near the limit of trees.

General distribution: N. American, boreal-subarctic, W. Greenland to Alaska.
Fig. 539 Map 622

Draba cinerea Adams
Densely caespitose with 0.5 to 1.5 cm long, entire, cuneate, oblanceolate or spatulate leaves which are ashy grey from a dense, felt-like covering of stellate hairs, mixed along the margins with long simple hairs. Flowering stems simple or forked, 10 to 25 cm high, stiffly erect or ascending, stellate pubescent throughout, and bearing 1 to 3 leaves. Inflorescence, 5- to 16-flowered, soon elongating into a lax raceme. Petals white. Siliques greenish-grey from dense stellate pubescence, narrowly lanceolate, usually 7 to 10 mm long and slightly shorter than their pedicels. Styles long.

Calcareous, rocky barrens and sunny cliffs.

General distribution: Circumpolar, arctic, alpine.
Fig. 540 Map 623

Draba corymbosa R. Br.
D. macrocarpa Adams
D. Bellii Holm
Similar to *D. alpina* but more densely caespitose, forming rather compact, hemispherical tussocks of tightly packed caudices from a many-headed tap-root. The marcescent, densely crowded rosette leaves conspicuously villous from simple as well as from variously forked hairs. Flowers pale yellow or creamy white; the conspicuously pubescent siliques elliptic-oblong, rarely twice as long as broad.

Common in dry, calcareous barrens and in rocky places.

General distribution: Circumpolar, arctic-alpine.
Fig. 541 Map 624

Draba crassifolia Grah.
Dwarf species from a weak tap-root, with tufted, lanceolate, entire, glabrous, or at most ciliate basal leaves. Flowering stems ascending or arching, 2 to 10 cm high, glabrous, and usually without leaves. Inflorescence with few and pale yellow flowers and glabrous, lanceolate siliques.

Rare in herbmats and on snowbeds.

General distribution: N. America, subarctic-alpine. N. W. Europe but not Asia.
Fig. 542 Map 625

Draba fladnizensis Wulfen
Very similar and often with difficulty separated from the more common *D. lactea*. However, the leaves of *D. fladnizensis* lack forked or branching hairs, the midvein is less prominent, and the petals are pure white.
Gravelly alpine slopes.
General distribution: Circumpolar, arctic.
Map 626

Draba glabella Pursh
D. daurica DC.
D. hirta auctt. non L.
Coarse, tufted or loosely matted with 1 to 5 cm long entire or few-toothed, cuneate, oblanceolate, or spatulate fresh green basal leaves, sparingly to densely covered by soft, stellate, forked or simple hairs. Flowering stems simple or forking, 10 to 25 cm high, soft stellate pubescent or glabrate above, bearing from 1 to 7 cauline leaves. Inflorescence 5- to 15-flowered, first capitate, soon elongating to a lax raceme. Petals white. Siliques glabrous, narrowly lanceolate, 6 to 15 mm long, usually as long as their pedicels, and with a very short style.
Common in not too dry rocky or grassy situations; strongly nitrophilous, favouring animal dung.
General distribution: Circumpolar, wide ranging, arctic-alpine.
Fig. 543 Map 627

Draba incerta Payson
Loosely tufted or matted, occasionally with up to 200 scapes from one root; leaves all in basal rosettes, linear, 8 to 12 mm long and about 2 mm wide with a prominent mid-rib, loosely covered on both sides by sessile variously stellate and pectinately branched hairs. Scapes 5 to 15 cm long, sparsely covered by short, forked hairs. Petals pale yellow, fading white. Siliques linear-lanceolate, 5 to 12 mm long and 2 to 3 mm wide, glabrate or with sparse pubescence of forked hairs. Style prominent, 1 mm long or longer.
Calcareous screes.
General distribution: North Cordilleran species.
Fig. 544 Map 628

Draba lactea Adams
D. fladnizensis Wulfen
var. *heterotricha* (Lindbl.) Ball
Dwarf species forming small loose mats. Leaves linear-oblong, ciliate with long simple, but usually also with some forked or even stellate hairs, especially near the apex, and keeled by the short but prominent mid-rib that is particularly noticeable in the old, marcescent leaves. Scapes 5 to 10 cm high, glabrous. Inflorescence short-racemose, usually from 3- to 7-flowered, much elongated in fruit. Petals creamy-white. Siliques ovate-lanceolate, 6 to 8 mm long and 3 to 4 mm wide, dark green and glabrous, equalling or slightly longer than their spreading pedicels.
Common in not too dry turfy places in tundra and on snowbeds.
General distribution: Circumpolar, high-arctic.
Fig. 545 Map 629

Draba lonchocarpa Rydb.
Scapose dwarf species in habit similar to *D. nivalis* from which it differs by its always glabrous scapes and pedicels, and by its much longer siliques that are usually somewhat twisted, and from seven to ten times longer than broad.
Moist alpine slopes; barely enters our area in southern Mackenzie Mountains.
General distribution: Cordilleran.
Map 630

Draba longipes Raup
Essentially glabrous perennial of loosely matted habit. Leaves mainly basal but not rosulate, oblanceolate, up to 2.5 cm long and 8 mm broad, glabrate and fresh green, usually with ciliate margins, entire or obscurely dentate towards the obtuse apex. Flowering stems slender, 10 to 15 cm tall usually with one or two leaves near its base. Flowers white, corymbose, but the axis elongating in fruit. The siliques usually glabrous, narrowly lanceolate, 10 to 12 mm long, with a short but distinct style, the lowermost on spreading pedicels much longer than the silique.
Common in alpine herbmats.
General distribution: Alaska, Y.T., Mackenzie Mts., S.E. Victoria Is., northern B.C. and S.W. Alta.
Fig. 546 Map 631

Draba Macounii O.E. Schulz
Dwarf, caespitose, alpine species with a branching caudex, each branch terminating in a tiny rosette of oblanceolate leaves 5 to 7 mm long and half as wide, their margins ciliate, the upper, but especially the underside, with tangled, simple and variously branched hairs.

Flowering scape 2 to 3 cm long, sparsely floccose-stellate, elongating in fruit. Inflorescence short racemose or corymbose, of 4 to 8 flowers. Petals pale yellow in life, drying white. Siliques glabrous, dark green, oblong-elliptic, 5 to 7 mm long and about half as wide, on peduncles slightly shorter than the silique. Stigma coronate, almost sessile.

Moist, alpine screes.

General distribution: Cordilleran endemic, north to S.W. Mackenzie District and central Y.T.

Map 632

Draba nemorosa L.
var. **leiocarpa** Lindbl.
Winter annual, with a leafy, simple or much branched stem, 30 to 45 cm tall in fruit. Leaves broadly lanceolate, 2 to 3 cm long, soft pubescent on both sides, toothed. Flowering raceme elongated and lax in fruit; flowers small with pale yellow petals. Siliques glabrous, narrowly clavate, about 10 mm long, on slender spreading pedicels twice as long as the body.

Occasional, mainly in settlements along the Mackenzie River north to Good Hope.

General distribution: Nearly circumpolar.

Fig. 547 Map 633

Draba nivalis Liljebl.
Low and caespitose, forming small, loose mats that may be recognized even from a distance by their characteristic pale glaucous colour which is due to the dense stellate pubescence of the crowded, narrowly oblong or ligulate leaves. Flowering stems scapose, or at most with a single cauline leaf near its base, 5 to 10 cm high, sparsely stellate pubescent. Flowers small, white; the inflorescence much elongated in fruit. Siliques linear-elliptic, lanceolate, glabrous, somewhat shiny, usually 6 to 8 mm long, or as long as their pedicels. Style long.

Common in dry, rocky, or gravelly situations.

General distribution: Circumpolar, arctic-alpine.

Fig. 548 Map 634

Draba norvegica Gunn.
D. rupestris R. Br.
Loosely tufted dwarf species. Basal leaves oblanceolate, entire or often with a few coarse teeth near the apex, hispid-pilose from simple and forked hairs. Stems 5 to 10 cm high, with one to several leaves near its base, in tall specimens occasionally forked below. Flowers large and white, the inflorescence soon elongating. Siliques elliptic, 5 to 8 mm long, sparingly hispid along the margins, the valves mostly hirtellous. Style short and thick.

Stony and gravelly, not too dry places.

General distribution: Amphi-Atlantic subarctic, in our area known from a single collection from north of the east arm of Great Slave L.

Fig. 549 Map 635

***Draba oblongata** R. Br.
D. groenlandica El. Ekm.
Resembling *D. cinerea* but flowering stems simple and rarely more than 10 cm tall with a single cauline leaf. Basal leaves entire, hispid-pilose, with a mixture of simple and variously forked hairs. Siliques short stellate-pubescent; style short and thick.

Barren, gravelly places.

General distribution: Endemic of Greenland and the Canadian high-arctic; to be expected along the mainland coast.

Fig. 550 Map 636

Draba ogilviensis Hult.
D. sibirica sensu Porsild
Loosely matted with long, trailing and freely branching leafy shoots from the forks of which rise the long and mostly naked flowering peduncles. Leaves deep green, glabrous, or with sessile or stalked 2-forked hairs, mainly along the margins. Inflorescence greatly elongating in fruit. Petals deep yellow. Siliques linear-oblong, 6 to 8 mm long, with a short but prominent style. Pedicels slender, twice as long as the silique.

Alpine herbmats.

General distribution: Endemic to the Ogilvie Mts. of central Y. T. and the central Mackenzie Mts.

Map 637

Draba oligosperma Hook.
Caespitose to matted. Leaves linear or narrowly oblanceolate, prominently keeled, all basal and rather densely imbricated, 3 to 8 mm long and 1.0 to 1.5 mm wide, the upper surface covered by pectinately or double-pectinately branched hairs, the underside glabrous. Scapes glabrous, 5 to 15 cm long, greatly elongating in fruit. Petals pale yellow, sepals glabrous and the siliques ovate, 2 to 5 mm long, sparsely pubescent, on pedicels twice the length of the silique.

A variable species somewhat similar to *D. in-*

certa from which, however, it is easily distinguished by its short siliques.

Unconsolidated calcareous screes.

General distribution: Cordilleran.

Fig. 551 Map 638

Draba Palanderiana Kjellm.
D. caesia auctt. non Adams
Loosely caespitose, scapose dwarf species from a branching multicipital caudex. Leaves 6 to 10 mm long and about 2.0 to 2.5 mm wide, velvety grey on both sides from minute, stalked, stellate and pectinately branched hairs. Flowering stems 6 to 10 cm tall, hirtellous or glabrate. Petals 4 to 5 mm long, white or pale yellow commonly drying darker yellow. Sliques linear-lanceolate, 5 to 12 mm long and 1.5 to 2.5 mm wide, often somewhat oblique, glabrous and with a short but prominent style, on pedicels as long or slightly longer than the siliques.

In habit rather similar to *D. nivalis* but rather stouter and with much larger and rather showy flowers, never milky-white as in *D. nivalis.* When past flowering the prominent midrib on the underside of the leaves at once distinguishes *D. Palanderiana.*

Moist, calcareous screes and slopes.

General distribution: Wide ranging in mountains of northern and central Alaska and Y. T., eastward to the Richardson and Mackenzie Mts.

Fig. 552 Map 639

Draba pilosa Adams
Scapose dwarf species from a simple or not much branched caudex, each branch covered by the remains of former season's leaves, terminating in a tiny tuft of very stiff and fresh green leaves in shape linear-oblong, 6 to 8 mm long and 1.5 to 2 mm broad with a very prominent keel. Leaves prominently ciliate along the margins, but otherwise glabrous, or occasionally the underside sparsely covered by variously branched hairs. Scape very slender, 3 to 6 cm long, glabrous. Inflorescence corymbose, 3- to 5-flowered, petals yellow or creamy white, drying yellow. Siliques lanceolate, 5 to 8 mm long. Style short.

Peaty tundra hummocks.

General distribution: Amphi-Beringian, arctic-alpine: From Bering Strait east to northern Hudson Bay and south in the Mackenzie Mts.

Map 640

Draba Porsildii G. A. Mulligan
Low, and loosely tufted. The marcescent basal leaves oblanceolate, ciliate, the surfaces glabrate or with simple, forked, cruciform and stellate hairs. Scapes 2.0 to 6.5 cm tall, glabrous to sparingly pubescent, occasionally with 1 oblanceolate leaf. Inflorescence capitate, elongating in fruit, 3- to 6-flowered; petals white; siliques ovate, 4.0 to 8.0 mm long, 2.0 to 3.0 mm broad, glabrous, on pedicels shorter than the siliques; style about 0.25 mm long.

Limestone talus slopes.

General distribution: Cordilleran, S. W. Alta and adjacent B. C. to S. Yukon and S. W. District of Mackenzie.

Map 641

Draba praealta Greene
Biennial or short-lived perennial with a simple or weakly branched caudex. Leaves mainly basal, lanceolate, entire or sometimes few-toothed, 1 to 3 cm long, including the slender petiole, pubescent on both sides with simple or variously branched, stalked hairs. Flowering stems up to 30 cm tall, simple or in strong plants branching from the base, with 2 to 3 reduced leaves. Inflorescence racemose 3- to 30-flowered, much elongated in fruit. Petals white, siliques linear, 5 to 10 mm long, thinly stellate pubescent, on arching pedicels half as long as the silique. Style short and thick.

Dry slopes and rocky ledges.

General distribution: Cordilleran species, with us north to Great Bear Lake.

Fig. 553 Map 642

Draba subcapitata Simm.
Low, and densely tufted. The marcescent basal leaves linear-oblong or spatulate with a prominent midrib, and densely ciliate. Scapes usually 3 to 5 cm high, softly pubescent with variously forked hairs. Inflorescence few-flowered, capitatex scarcely if at all elongating in fruit. Siliques 4 to 5 mm long and 2 to 3 mm wide, elliptic-lanceolate, glabrous, dark green or purplish, sub-sessile or on pedicels shorter or at most equalling the length of the spreading siliques. Styles short.

A pronounced calciphile of dry, gravelly tundra.

General distribution: Amphi-Atlantic, high-arctic; from the N. American mainland thus far known only from Boothia and Melville Pen.

Fig. 554 Map 643

Erucastrum gallicum (Willd.) O. E. Schulz
Branched annual or winter annual to 70 cm in height; stems with downward-directed hairs; leaves oblong, deeply pinnatifid; flowers leafy-bracted, the petals pale yellow, about 7 mm

long; siliques 3 to 5 cm long tipped with a short slender beak.

Weed of waste places in southwestern District of Mackenzie.

General distribution: A relatively recent introduction to North America which is apparently spreading quickly.

Erysimum L. Wallflower

Winter annuals or short-lived perennials flowering only once. Leaves and stems with appressed, simple or 3- to 4-forked, sessile or very short-stalked hairs, the first kind are always attached in the middle and their branches always oriented in the direction of the organ on which they are attached. Flowers yellow or purple. Seeds oblong, wingless.

a. Flowers yellow; tall, non-arctic plants
  b. Siliques less than 2 cm long . . . . *E. cheiranthoides*
  b. Siliques 3-5 cm long . . . . *E. inconspicuum*
a. Flowers purple; arctic plants, siliques usually somewhat curved, 4-11 cm long . . . *E. Pallasii*

Erysimum cheiranthoides L.
Tall, weedy annual or winter-annual with leafy and much branched stems up to 4 to 5 dm tall. Leaves and stems mainly with short-stalked 3 to 4-forked hairs. Flowers small, the petals pale yellow, 3 to 4 mm long, the siliques on slender, ascending peduncles, about one half as long as the straight silique.

In moist, turfy places commonly near human settlements or camps, but sometimes also by lake shores and on creek-banks. In the Mackenzie Valley north nearly to the arctic coast.

General distribution: Circumpolar.

Fig. 555 Map 644

Erysimum inconspicuum (S.Wats.) MacMill.
Similar to *E. cheiranthoides* but not an annual, frequently requiring two or more seasons to flower and fruit. Stems 2 to 5 dm tall, simple, or branched mainly above. Petals about 10 mm long, the siliques on stout, curved peduncles, about one fifth as long as the silique.

Grassy hillsides and riverside terraces, north to the limit of trees.

General distribution: N. America, from Nfld. to Alaska.

Fig. 556 Map 645

Erysimum Pallasii (Pursh) Fern.
Biennial or short-lived perennial which, after two to several years in the rosette stage, flowers, matures seeds and dies. Basal leaves linear-lanceolate, 5 to 7 cm long, the entire or repand-dentate blade 2 to 4 mm wide, gradually tapering into a narrow petiole. Inflorescence, racemose, 50- or more-flowered; when the first flowers expand, the flowering axis is so short that the flowers appear as if developed at the base of the leaves, among which they are hidden. The axis soon elongates, becoming 15 to 35 cm long. In plants coming into flower late in the season, this elongation does not take place, and the siliques, therefore, form among the withering leaves. This pseudo-acaulescent form looks very odd but apparently is of no taxonomic significance (f. *humilum* (Tolm.) Polunin). The large, purple flowers are very fragrant. Siliques linear, somewhat curved, 4 to 11 cm long and 2 to 3 mm wide. Seeds 2 mm long.

A pronounced dung-loving calciphile inhabiting sunny, grassy places, often near animal burrows or human habitation, or below bird cliffs.

General distribution: Circumpolar, high-arctic.

Fig. 557 Map 646

Eutrema R. Br.

Eutrema Edwardsii R. Br.
Glabrous and slightly fleshy perennial with a slender tap-root and from 1 to 20 simple, erect, few-leaved stems 10 to 30 cm high. Basal leaves slender and long-petioled, with entire, lanceolate-ovate blades 1.5 to 2.5 cm long; the 3 to 6 cauline leaves sessile. Inflorescence terminal, 7- to 20-flowered, at first capitate but soon elongating into a 5 to 10 cm long raceme. Flowers small, the petals white, twice as long as the greenish-purple sepals. Mature siliques dark purple, linear-lanceolate, 10 to 15 mm long, erect or ascending on slender pedicels of approximately the same length. The valves prominently keeled and the septum always perforated.

In not too dry, turfy places in tundra; the species is strongly nitrophilous and mostly restricted to calcareous soils.

General distribution: Circumpolar, arctic or high-arctic.

Fig. 558 Map 647

Halimolobos Tausch

Halimolobos mollis (Hook.) Rollins
Arabis Hookeri Lange
Coarse biennial of weedy habit, usually with several erect-ascending, 10 to 50 cm high, simple or branched stems, from a simple base. Stems hirsute below, glabrous, or with a few simple hairs above. Basal leaves 3 to 6 cm long, oblanceolate, sinuate-dentate, densely hirsute with forked hairs; cauline leaves 1 to 3 cm long, sessile, auricled or sagittate, with dentate to entire margins. Inflorescence at first capitate, soon elongating. Flowers small, white. Siliques 3 to 5 cm long and about 2 mm wide, glabrous and with strongly nerved valves; seeds sticky when wet.

A dung-loving calciphilous species of peculiar, disrupted range and undoubtedly often dispersed by man and by animals.

General distribution: Endemic of arctic N. America, from Alaska to W. Greenland.

Fig. 559 Map 648

Lepidium L. Pepperwort

Annual or biennial (ours) of weedy habit, glabrous or sparingly short-hirsute. Leaves pinnate, bipinnate or merely toothed, but commonly wilted and missing in fruiting specimens. Flowers small, white (ours) in dense terminal racemes. Fruit a dehiscent silique, strongly keeled or winged, usually containing a single seed.

a. Sililiques 5-6 mm long; flower with 6 stamens (not native) . . . . *L. sativum*
a. Siliques 2-3 mm long; flower with 2 stamens (native species although perhaps recent introductions within our area)
  b. Siliques puberulent, at least on the margin
    c. Inflorescence congested into numerous axillary as well as terminal racemes; siliques elliptic . . . . *L. ramosissimum*
    c. Inflorescence in simple, or branched racemes; siliques round-obcordate . . . . *L. densiflorum*
  b. Siliques glabrous, ovate . . . . *L. Bourgeauanum*

Lepidium Bourgeauanum Thell.
Biennial, 1.5 to 6 dm tall, sparsely to densely puberulent; lower and middle leaves incised, the upper linear, entire or slightly toothed.

Prairie or grassland species fairly common in the upper Mackenzie Valley where it was first reported (as *L. ruderale* L.) 140 years ago, and where it is thought to be native.

General distribution: N. America.

Map 649

Lepidium densiflorum Schrad.
Annual or winter annual, puberulent to pubescent, 1 to 5 dm tall and usually somewhat branched above the middle.

On disturbed soil near settlements of the upper Mackenzie Valley.

General distribution: N. America.

Lepidium ramosissimum A. Nels.
Biennial, 1 to 4 dm tall, sparsely to densely puberulent. Stem erect, usually profusely branched; lower and middle leaves sessile, pinnate or bipinnate, the upper usually with at least one pair of linear lobes near the apex.

Obviously a recent introduction thus far collected but twice, on recently disturbed soil within town sites.

General distribution: N. America, prairies and plains.

Lepidium sativum L.
Annual; lower leaves long-stalked, lyrate, with toothed ovobate lobes, the upper bipinnate or sometimes entire. At once distinguished from the other three North American species known from the Northwest Territories by its larger siliques and by its flowers having 6 rather than 2 stamens.

General distribution: An introduced European weed, thus far collected but once, at Norman Wells.

Lesquerella S. Wats.

Silvery-stellate pubescent, caespitose perennials with a stout taproot and simple or branched caudex. Basal leaves with a spatulate or oblanceolate blade tapering into a short petiole. Flowering stems decumbent, ascending or erect, 5 to 20 cm long, with a few reduced leaves. Flowers 2 to 14, pale yellow, on stout pedicels. Siliques globular or slightly pear-shaped on ascending, 1 to 2 cm long pedicels.

a. Petals 5-6 (7) mm long and about half as broad above the middle
  b. Siliques glabrous . . . . *L. arctica* ssp. *arctica*
  b. Siliques "frosted" from minute stellate trichomes . . . . *L. arctica* ssp. *Purshii*
a. Petals 7-9 (10) mm long and almost as broad above the middle . . . . *L. Calderi*

Lesquerella arctica (Wormskj.) S.Wats.
The ssp. *arctica* is wide ranging from North East Greenland across the Canadian Arctic to Alaska.

The Cordilleran ssp. *Purshii* (S. Wats.) Porsild, is distinguished by its siliques being "frosted" by minute stellate trichomes and with us restricted to the western District of Mackenzie.

Calcareous cliffs and stony barrens.

General distribution: *L. arctica s. lat:* Eastern Greenland to Eastern Siberia.

Fig. 560 Map 650 (*L. arctica s. lat.*)

Lesquerella Calderi Mulligan & Porsild
Similar to *L. arctica* ssp. *arctica* from which it differs by its larger petals and siliques, and by its longer styles.

General distribution: Thus far known only from a few collections on the east and west slopes of Richardson Mts., and from the Ogilvie Mts. of central Y.T.

Map 651

Neslia Desv. Ball-Mustard

Neslia paniculata (L.) Desv.
Erect annual or biennial with leafy 3 to 6 dm tall, stellate-pubescent stems, usually freely branched above. Leaves lanceolate with sagittate or clasping bases and, like the stems, rough-pubescent. The small pale-yellow flowers on slender peduncles, in a raceme, much elongated in fruit. Siliques globular, about 2 mm in diameter.

Occasional introduced weed around settlements along the Mackenzie drainage.

General distribution: Circumpolar.

Parrya R. Br.

Caespitose, essentially glabrous perennials with thick-branching caudices, entire or repand-dentate, somewhat fleshy basal leaves, and scapose stems. Inflorescence racemose, with large lavender or white flowers. Siliques linear-lanceolate, keeled, 3 to 6 mm wide, torulose and somewhat sinuate. Seeds large, with a broad wing formed by the honeycombed, loosely fitting epidermis of the seed-coat.

a. Flowers about 1 cm in diameter . . . . *P. arctica*
a. Flowers 1.5 to 3.3 cm in diameter . . . . *P. nudicaulis*

Parrya arctica R. Br.
Leaves 2 to 3 cm long, narrowly oblanceolate, short-petiolate, entire or rarely obscurely toothed. Scapes 5 to 10 cm high, or up to 20 cm in fruit. Inflorescence short racemose, 7- to 12-flowered. Flowers scentless, petals purple or creamy white, about 8 mm long; anthers short and ovate. Siliques 2 to 3 cm long and 3 to 4 mm wide, straight, strongly keeled when fresh. Seeds 2 to 3 in each locule, large, 4.0 x 5.5 mm, including the broad wing.

In Banks and Victoria islands purple- and white-flowered forms are equally common, and often grow side by side.

Wet calcareous clay and gravel barrens.

General distribution: High-arctic endemic of the palaeozoic parts of the Canadian Arctic Archipelago, south to Great Bear L.

Fig. 561 Map 652

Parrya nudicaulis (L.) Regel
Leaves 5 to 10 cm long, narrowly oblanceolate,

acute, long-petioled, entire or toothed, mostly glabrous and somewhat fleshy. Scapes 10 to 20 cm or even 30 cm high in fruit, glabrous or glandular-hispid. Flowers large and very fragrant; petals rose-purple or white, 1.0 to 1.5 cm long, prominently clawed; anthers linear-oblong 1.5 to 2.0 mm long. Siliques 3 to 5 cm long and 4 to 6 mm wide, strongly sinuate, rarely straight, glabrous or hispidulous, with 2 to 3 large seeds in each locule.

Calcareous, alpine meadows.

General distribution: Amphi-Beringian, arctic-alpine.

Fig. 562 Map 653

Rorippa Scop. Yellow Cress

Ours perennial or biennial terrestrial plants with pinnate, pinnatifid or entire leaves; the fruit a short-cylindrical or globular silique.

a. Perennial
  b. Rhizome slender; petals yellow . . . . *R. calycina*
  b. Rhizome thick; petals whitish . . . . *R. crystallina*
a. Biennial (rarely annual) with a weak tap-root
  c. Silique oblong, elliptic, about as long as its pedicel . . . . *R. islandica*
  c. Silique nearly globular, much shorter than its pedicel . . . . *R. barbareaefolia*

Rorippa barbareaefolia (DC.) Kitagawa
Biennial with a simple or somewhat branching stem, hispid-pubescent, 30 to 50 cm high; stem and radical leaves lanceolate, pinnate, sparingly hispid; the siliques 4-valved, dehiscent at the apex; the style about 0.75 mm long.

Non-arctic species of weedy habit, common locally and mainly in disturbed soil through Alaska and Yukon Territory; it is known from a single station in the Mackenzie Delta where it may be a recent introduction.

General distribution: Amphi-Beringian.

Map 654

Rorippa calycina (Engelm.) Rydb.
Rhizomatose perennial about 1 to 2 dm high with oblong, regularly pinnately lobed, sessile and auriculate leaves. Flowers small in an elongating raceme. The petals pale yellow and the siliques short and ovoid, about 3 mm long, densely covered with short, stiff hairs. The style about 2 mm long.

In Canada known from a single and recent collection from low deltaic flats of the Anderson River on the Arctic Coast, nearly 3200 km from its nearest known station, in Montana.

General distribution: N. America, non-arctic.

Fig. 563 Map 655

Rorippa cystallina Rollins
Glabrous perennial from a thick branching rhizome; stems one to several, unbranched, arising below a fascicle of leaves; basal leaves petiolate, ovate to lanceolate, dentate or deeply pinnatifid, the cauline narrower, dentate to nearly entire, the lower petioled, the upper sessile; petals whitish, 6 to 8 mm long; siliques terete, 1.5 to 2.5 cm long.

Sedge-grass meadows.

General distribution: endemic of the lowland area northwest of Great Slave Lake.

Like Rollins we are not completely happy with the disposition of this species in the genus *Rorripa*. The crystalline pustules of calcium oxalate on the leaves and stems of the type material were induced by the formaldehyde method used by the collector. These pustules are not present on fresh material, but have been induced by treating plants grown at Ottawa with formaldehyde, prior to drying.

Fig. 564 Map 656

Rorippa islandica (Oeder) Borbas
R. obtusa sensu Porsild, (1943) and Raup, (1947) *non* (Nutt.) Britt.
Essentially glabrous annual or biennial with a simple or branched stem up to 1 m tall. The leaves mainly cauline, oblanceolate in outline, pinnately divided or toothed. Flowers small, in an elongating raceme; petals pale yellow, the siliques cylindric, rarely more than 6 mm long and about equalling their slender, ascending peduncles. Style short but distinct.

Common north to the limit of trees, in wet lake shore meadows; readily invading disturbed soil.

General distribution: Circumpolar, wide ranging, non-arctic.

Fig. 565 Map 657

Sinapis L.

Sinapis arvensis L.
Brassica Kaber (DC.) L.C. Wheeler
var. *pinnatifida* (Stokes) L.C. Wheeler
Charlock
Simple or branching annual, 3 to 9 dm tall, glabrous or hirsute-hispid especially in the lower parts; leaves oblanceolate, sinuate-dentate or lyrate-pinnatifid. The pale yellow flowers in elongating racemes from the upper leaf-axils. Siliques cylindric, somewhat torulose, stout, 2.5 to 4.5 cm long, including the slender tapering beak.

Introduced weed mainly in fields and gardens.

General distribution: Cosmopolitan

Sisymbrium L.

Sisymbrium altissimum L.
Erect, freely branching annual 3 to 10 dm tall, sparsely to densely hirsute but mainly in the lower parts. Leaves oblanceolate, up to 20 cm long, pinnatifid, the segments usually coarsely toothed. Flowers in elongating racemes; petals pale yellow or white, 6 to 8 mm long, siliques linear, terete, 5 to 10 cm long.

Introduced weed occasional in gardens and waste places.

Smelowskia C.A. Mey.

Low, tufted, cinereous-pubescent perennials from a sub-ligneous taproot. Inflorescence racemose and elongating, of small creamy white or purplish flowers; siliques narrowly lanceolate or pear-shaped.

a. Flowers purple; stems simple . . . . *S. borealis*
a. Flowers creamy white; stems mostly branched from near the base . . . . *S. calycina*

Smelowskia borealis (Greene) Drury & Rollins
Melanidion boreale Greene
Hoary or densely grey-felted perennial with a long, sub-ligneous tap-root, terminating in a simple or branching caudex bearing rosulate leaves 3 to 8 cm long, the blade cuneate, narrowing to a slender petiole. Flowering takes place very early when head-like rapidly elongating racemes appear among the basal leaves. The woolly rachis may be simple, erect, or it may branch from the base into half a dozen or more decumbent branches, each bearing from 20 to 40 small flowers on ascending or curved 1 cm long hoary peduncles; sepals hoary, persisting, the petals about 4 mm long, purple. Fertility apparently is low and uncertain for commonly only a few flowers in each raceme may mature seeds. Siliques sub-clavate, 8 to 10 mm long, with from 1 to 3 seeds in each locule. Seeds 2 to 3 mm long, light brown. Style filiform about 1 mm long.

A second flowering may occur later in the summer when slender, leafy, whip-like shoots issue from the rosettes; the leaves of these shoots are linear and quite different from those of the rosettes. Single or paired flowers occur scattered along the shoot.

The often striking variations in the habit of individual plants, even within local or regional populations is, at least in part, due to the instability of habitat where individual plants or colonies of plants may frequently be disturbed by rock slides.

Calcareous screes and slide-rock.

General distribution: Endemic of unglaciated mountains of central and northern Alaska, Y.T. and western District of Mackenzie.

Fig. 566 Map 658

Smelowskia calycina (Stephan) C.A.Mey.
var. **media** Drury & Rollins
Densely caespitose, thinly canescent perennial with rather firm, branched caudices that are often greatly elongated beyond the stout taproot, and covered by persistent leaf-bases. Stems few to several, simple, 5 to 15 cm tall; basal leaves 1 to 3 cm long, petioled, the blade entire, lobed, or pinnately divided; the cauline leaves few, linear or pinnate. Inflorescence first capitate, elongated in fruit, the pedicels widely divaricate. Siliques oblong, terete, 5 to 8 mm long, broadest above the middle. Style about 0.5 mm long. Seeds about 2 mm long, 1 or 2 in each locule.

Alpine ridges or scree slopes.

General distribution: Endemic to N.E. Alaska, Yukon, and the Richardson and Mackenzie Mts.

Fig. 567 Map 659

Subularia

Subularia aquatica L.
ssp. **americana** Mulligan & Calder
Dwarf, stemless, pale green aquatic rarely more than 5 to 10 cm tall. Leaves tufted, awl-shaped, ultimately overtopped by the elongating, few-flowered scape. The tiny, non-emergent flowers white; siliques ovoid, 2 to 3 mm in diameter.

The North American race ssp. *americana* is distinguished from the largely Eurasian ssp. *aquatica* by its oblong ovoid rather than elliptic siliques and by its persisting sepals.

Silty bottom of shallow, clear ponds.

General distribution: *S. aquatica s. lat.* Circumpolar (with large gaps).

Fig. 568 Map 660

Thellungiella O.E.Schulz

Thellungiella salsuginea (Pall.) O.E.Schulz
Arabidopsis glauca (Nutt.) Rydb.
Glabrous and glaucous annual with spreading or erect stems from 3 to 40 cm, but commonly about 10 cm tall, usually simple but in tall specimens freely branched above. Basal leaves small, round or oblong on slender petioles, soon wilting; the cauline leaves oblong and sessile, clasping. Racemes few to many-flowered, in strong specimens up to 20 cm long in fruit. Petals milky-white, about 2 mm long and longer than the dark purplish sepals. Siliques 1.0 to 1.5 cm long, somewhat arched, on ascending pedicels 3 to 4 mm long. The style minute.

A very rare plant; in America first collected by members of the Franklin Expedition, on the Arctic Coast near the Mackenzie River Delta, where since, it has been taken repeatedly; now known also from a few widely disjunct stations in the upper Mackenzie Valley and the Prairie Provinces.

General distribution: Amphi-Beringian. Widely disjunct stations in N. America and Asia.

Fig. 569 Map 661

Thlaspi L. Penny-cress

Glabrous (ours) with entire, clasping or sessile leaves and small white flowers in an elongating raceme; siliques clavate or, if oblong, winged.

a. Annual; siliques flat, oblong about 1 cm long . . . . *T. arvense*
a. Perennial; siliques clavate 6 to 7 mm long . . . . *T. arcticum*

***Thlaspi arcticum** Porsild
Glabrous perennial from a short branching caudex; basal leaves 1.0 to 2.5 cm long, somewhat fleshy, the blade oblanceolate or oval with entire, or the earliest obscurely 2-toothed. Flowering stems one to several, each bearing 2 or 3 smaller and distinctly sagittate leaves. Inflorescence short-racemose but greatly enlongated in fruit. The mature silique clavate, 6 to 7 mm long and 2.0 to 2.5 mm broad with a filiform style about 1 mm long.

Rare, in turfy places in tundra.

General distribution: Thus far known from only two rather restricted areas, one on the arctic slope of Alaska—Y.T., between the Mackenzie Delta and Sadlerochit R., and the other in St. Elias Mts. in S.W.Y.T. Its closest relative is probably *T. cochleariforme* DC. of the Altai and Ural Mts., USSR.

Fig. 570 Map 662

Thlaspi arvense L.
Penny-Cress
Stems smooth and erect, up to 3 to 4 dm high; lower leaves petioled, soon dropping; the upper sessile and sagittate.

Introduced weed by roadsides and in waste places.

512. *Alyssum americanum*

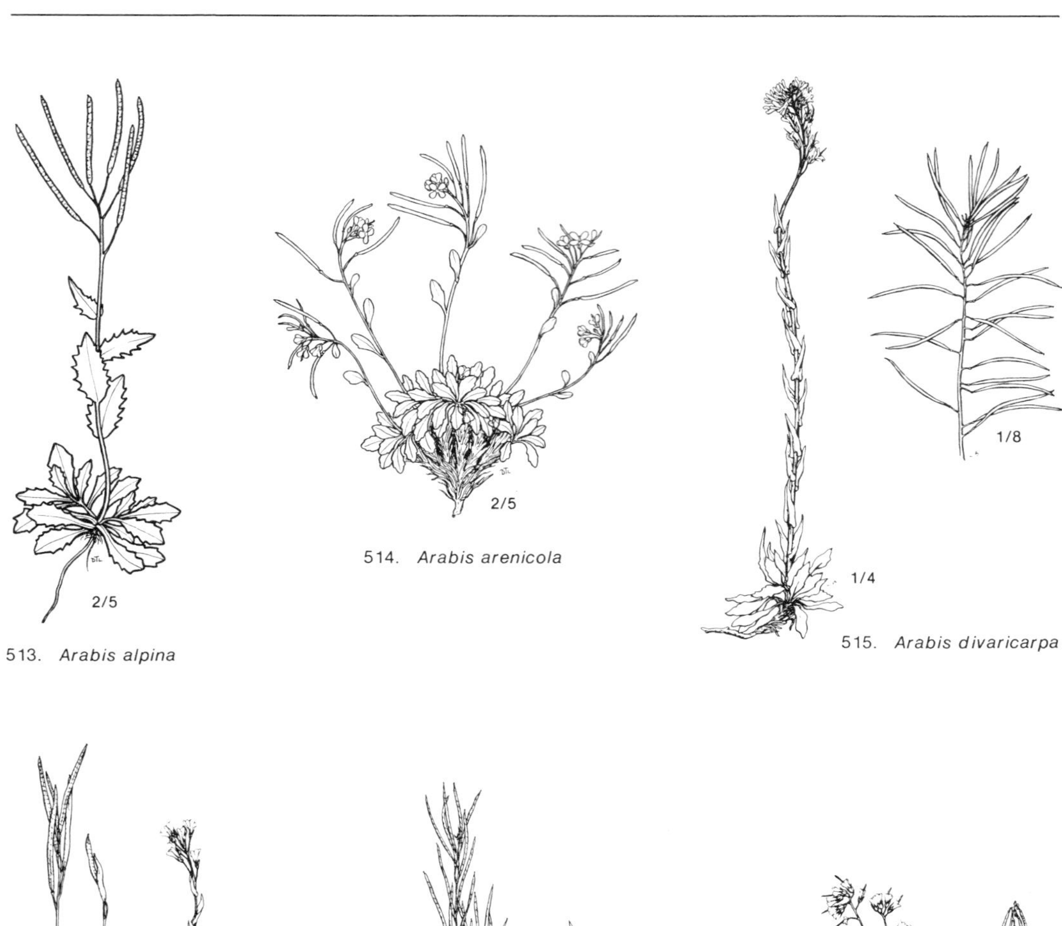

513. *Arabis alpina*

514. *Arabis arenicola*

515. *Arabis divaricarpa*

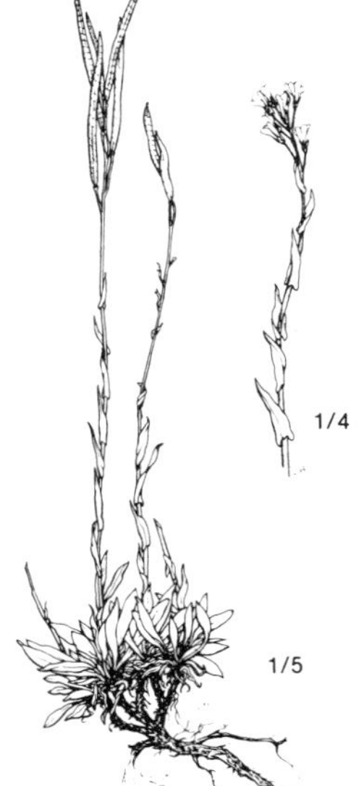

516. *Arabis Drummondii*

517. *Arabis hirsuta* ssp. *pycnocarpa*

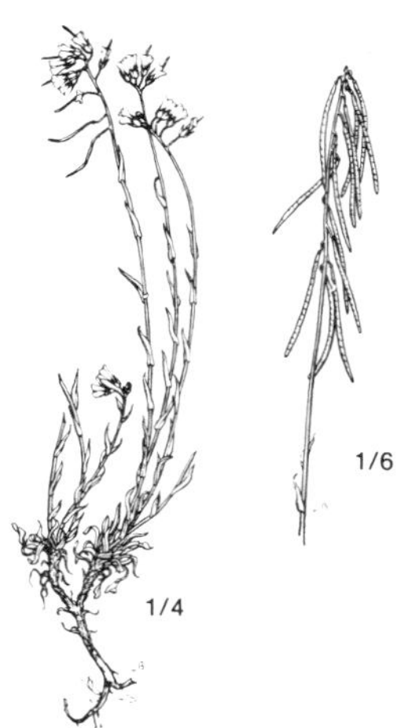

518. *Arabis Holboellii* var. *retrofracta*

519. *Arabis lyrata* var. *kamchatica*

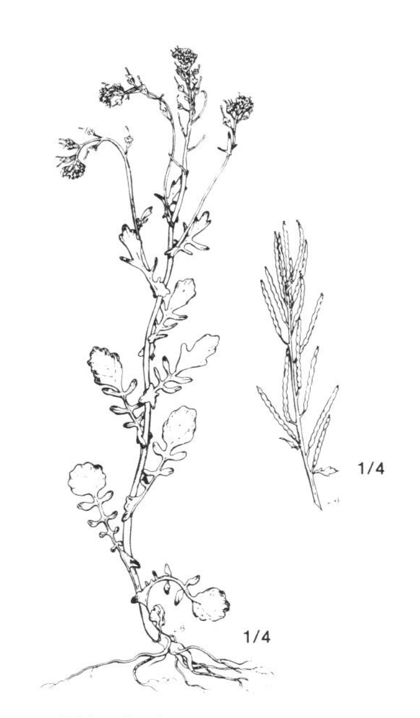

520. *Barbarea orthoceras*

521. *Braya humilis* ssp. *arctica*

522. *Braya purpurascens*

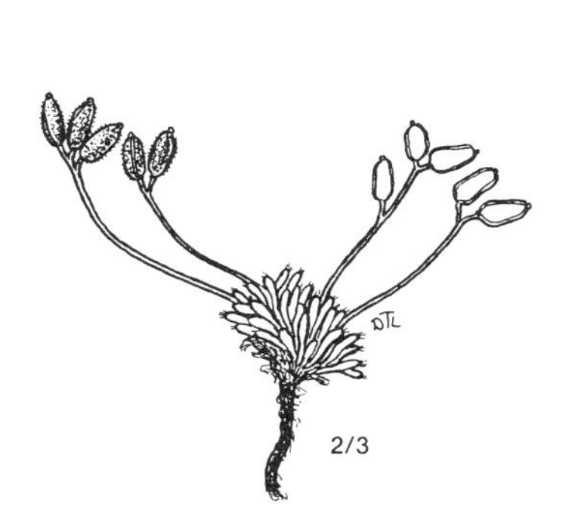

523. *Braya Thorild-Wulffii*

524. *Cardamine bellidifolia*

525. *Cardamine digitata*

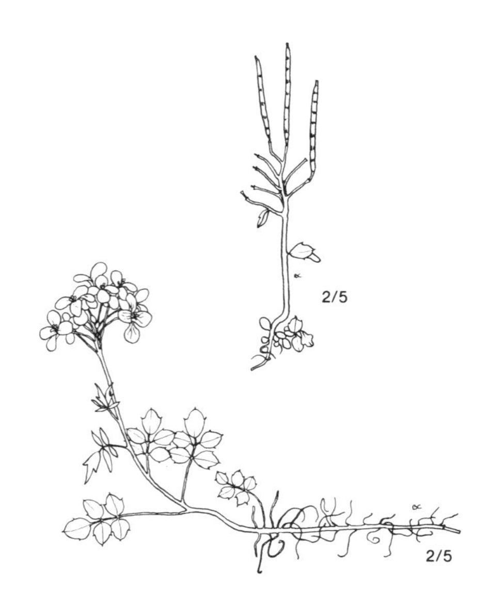

526. *Cardamine microphylla*

527. *Cardamine parviflora* var. *arenicola*

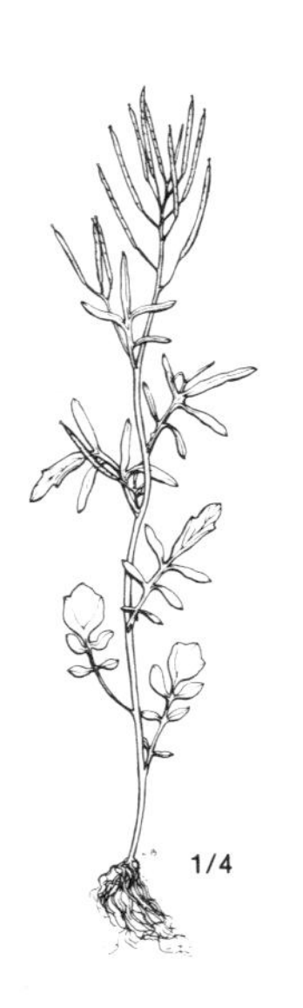

528. *Cardamine pensylvanica*

529. *Cardamine pratensis* var. *angustifolia*

530. *Cardamine umbellata*

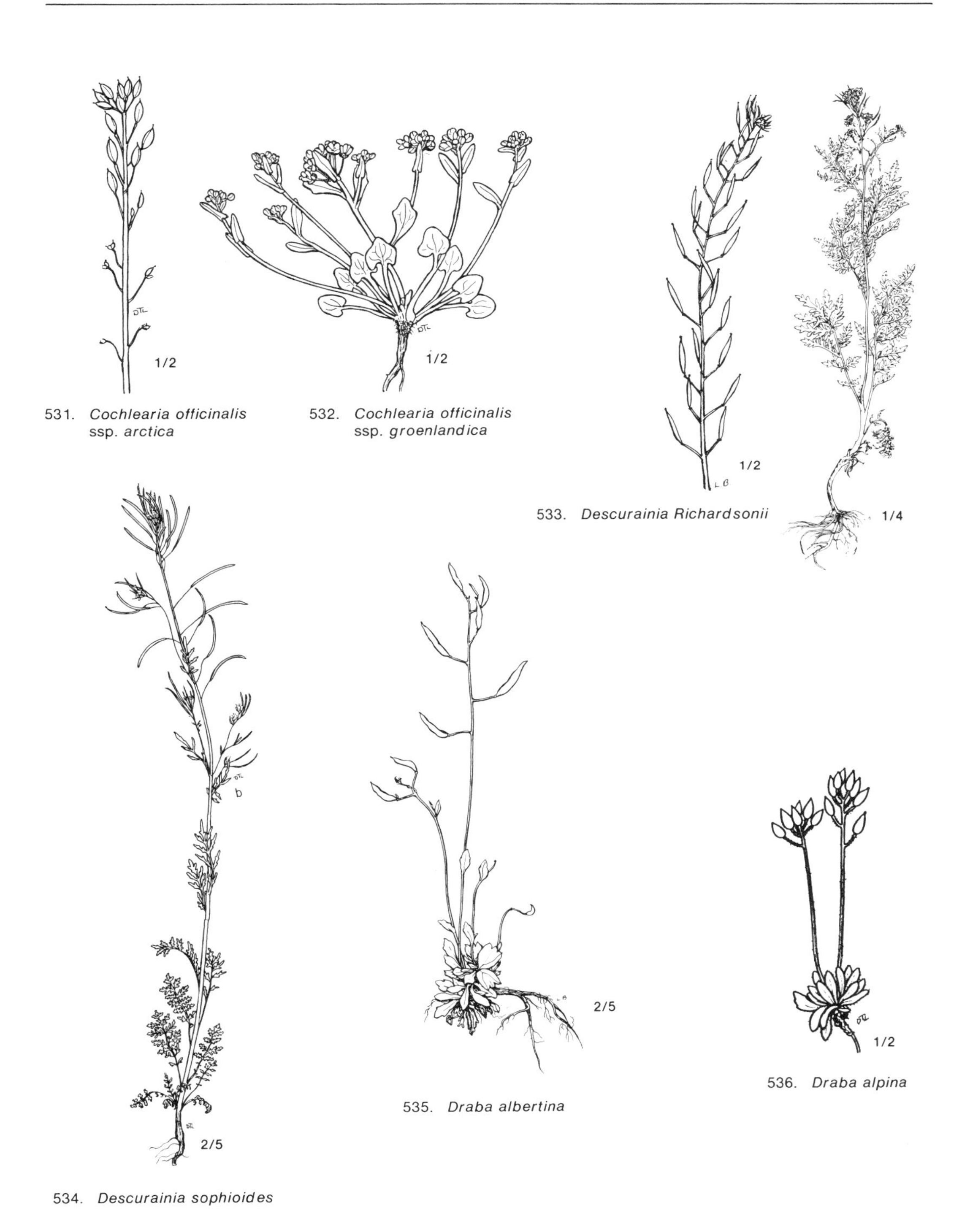

531. *Cochlearia officinalis* ssp. *arctica*

532. *Cochlearia officinalis* ssp. *groenlandica*

533. *Descurainia Richardsonii*

534. *Descurainia sophioides*

535. *Draba albertina*

536. *Draba alpina*

537. *Draba aurea*

538. *Draba borealis*

539. *Draba cana*

540. *Draba cinerea*

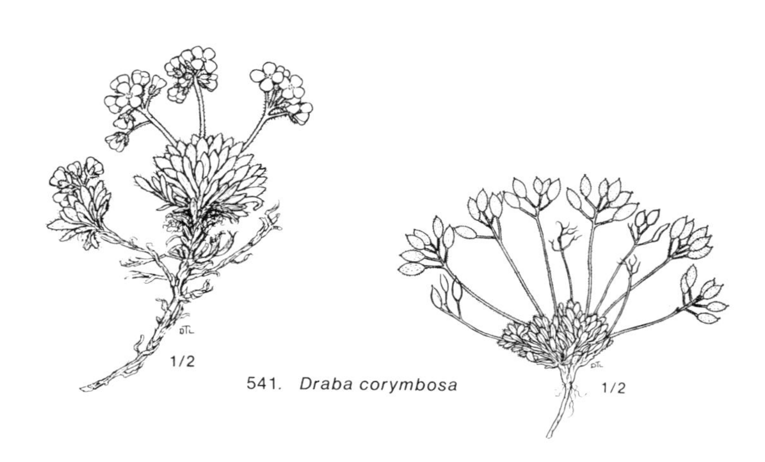

541. *Draba corymbosa*

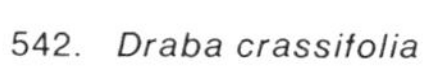
542. *Draba crassifolia*

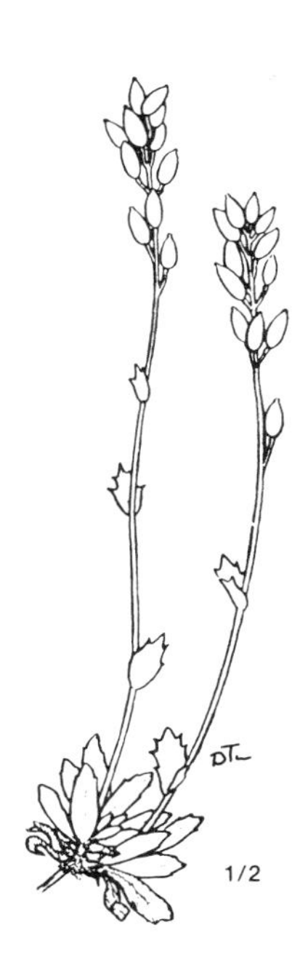

543. *Draba glabella*

544. *Draba incerta*

545. *Draba lactea*

546. *Draba longipes*

547. *Draba nemorosa* var. *leiocarpa*

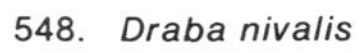
548. *Draba nivalis*

549. *Draba norvegica*

550. *Draba oblongata*

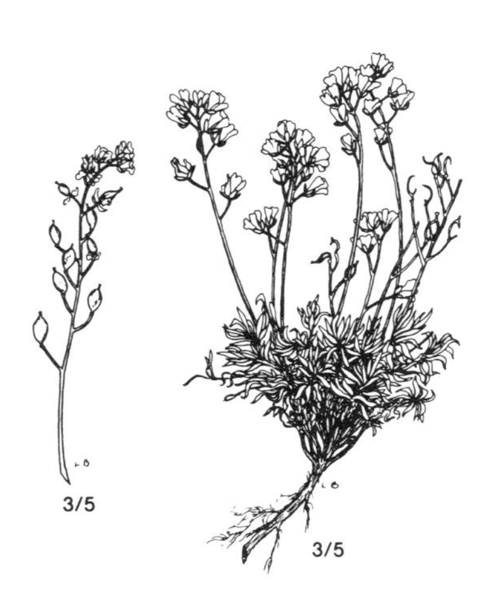

551. *Draba oligosperma*

552. *Draba Palanderiana*

553. *Draba praealta*

554. *Draba subcapitata*

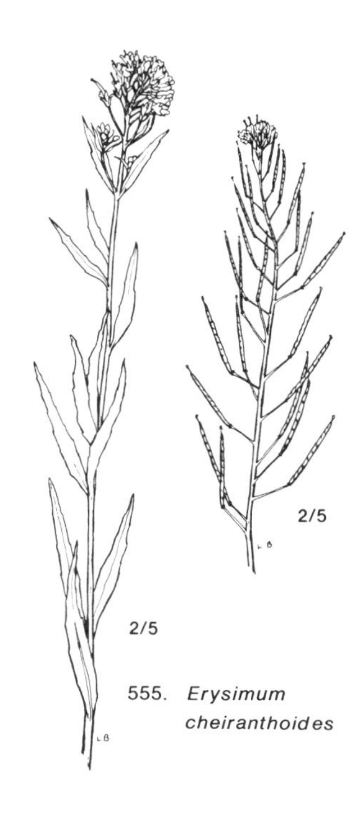

555. *Erysimum cheiranthoides*

556. *Erysimum inconspicuum*

557. *Erysimum Pallasii*

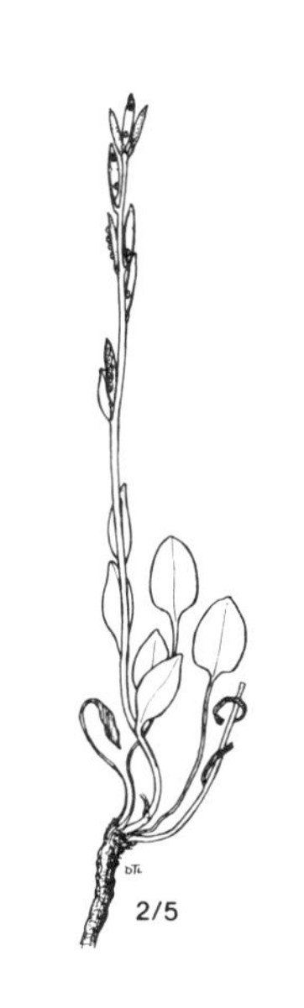

558. *Eutrema Edwardsii*

559. *Halimolobos mollis*

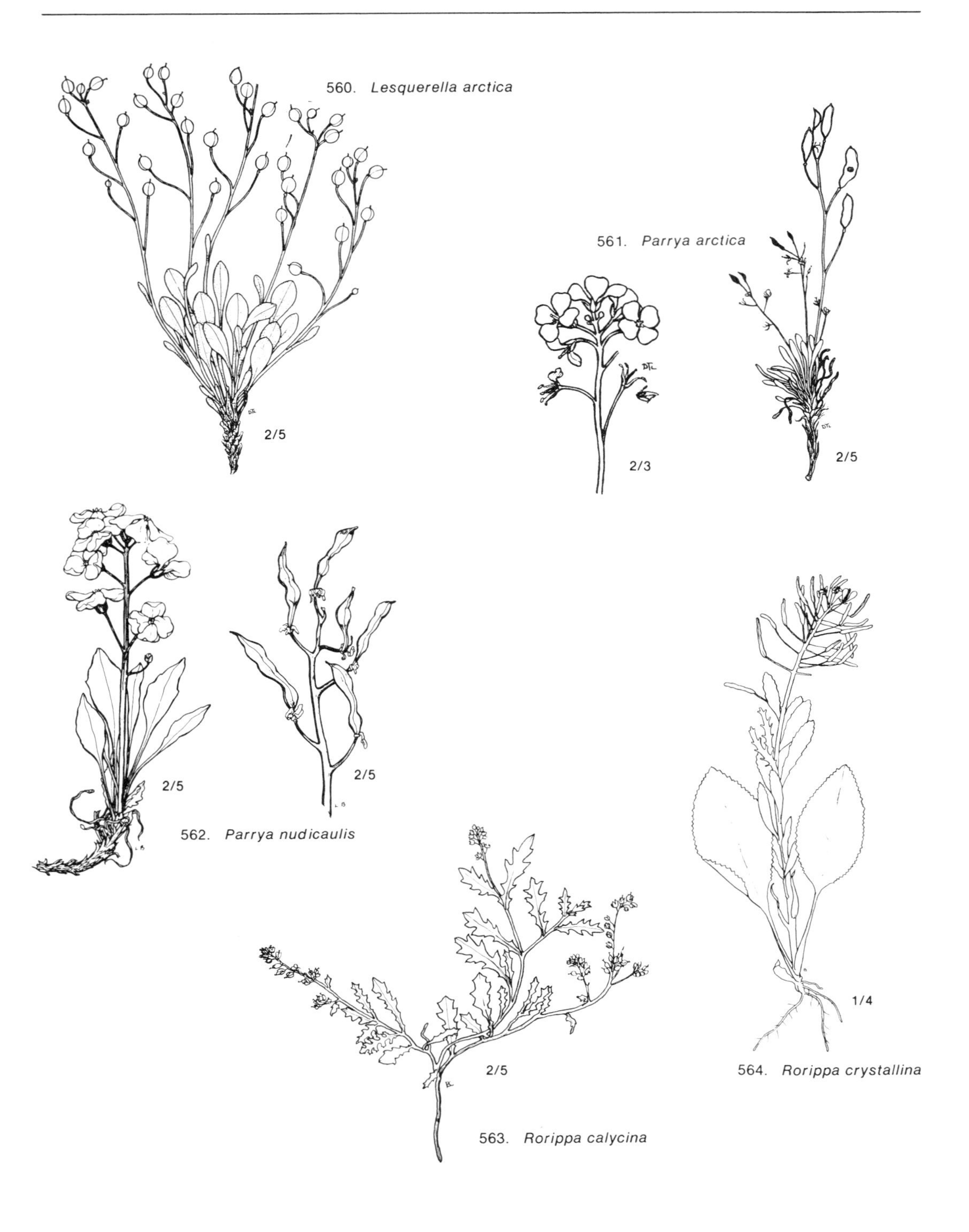

560. *Lesquerella arctica*

561. *Parrya arctica*

562. *Parrya nudicaulis*

563. *Rorippa calycina*

564. *Rorippa crystallina*

565. *Rorippa islandica*

566. *Smelowskia borealis*

567. *Smelowskia calycina* var. *media*

568. *Subularia aquatica* ssp. *americana*

569. *Thellungiella salsuginea*

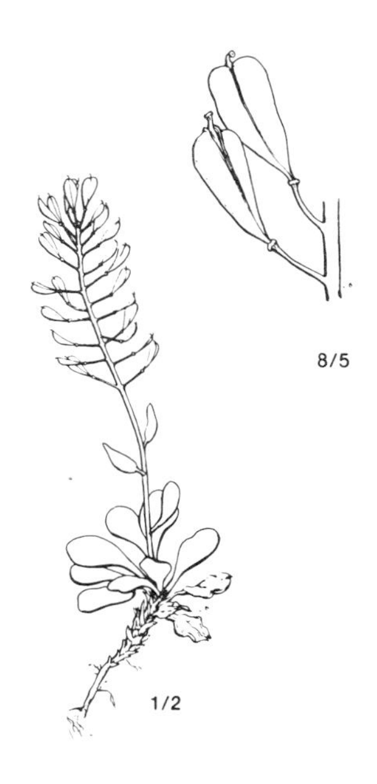

570. *Thlaspi arcticum*

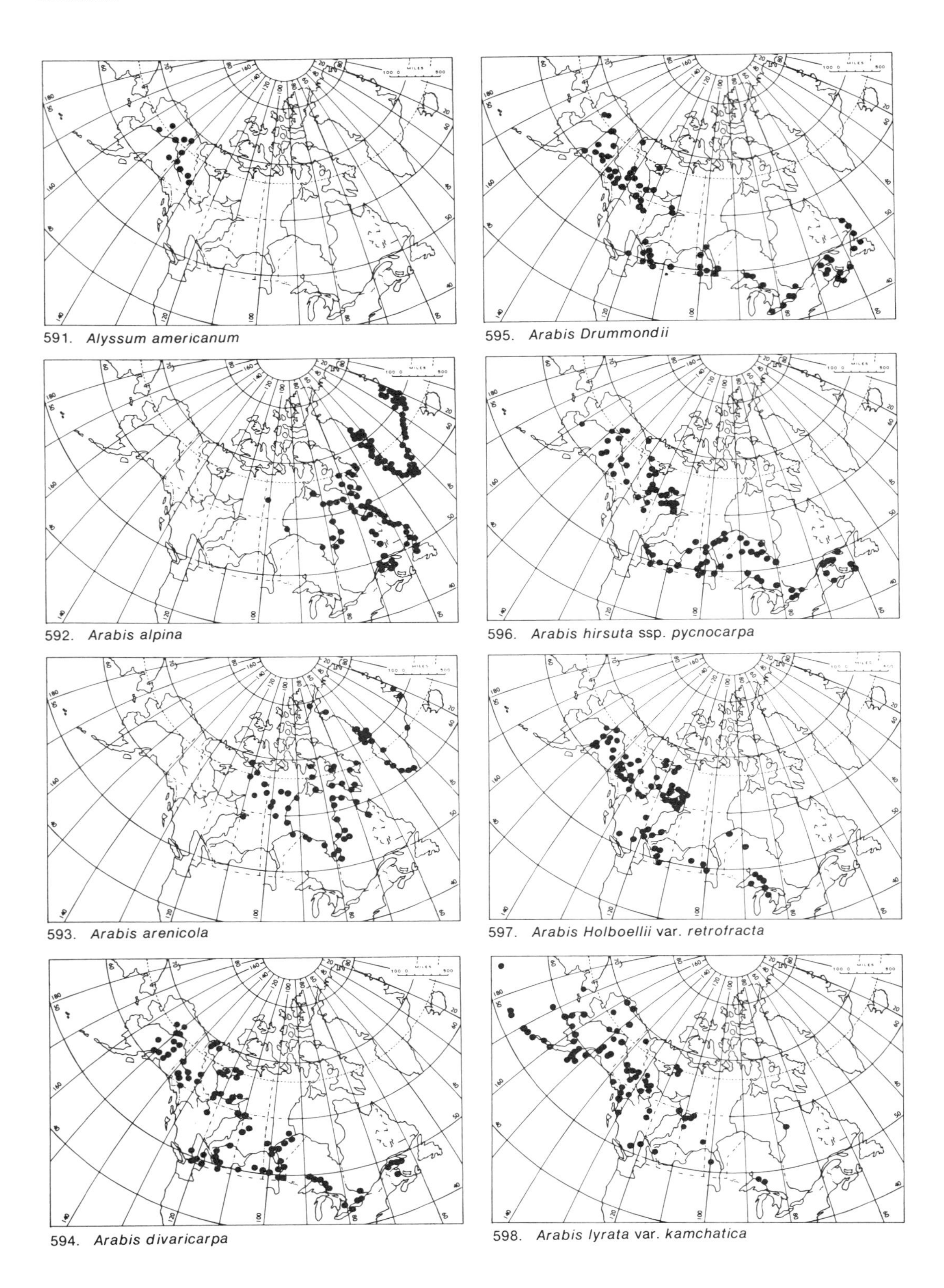

591. *Alyssum americanum*

595. *Arabis Drummondii*

592. *Arabis alpina*

596. *Arabis hirsuta* ssp. *pycnocarpa*

593. *Arabis arenicola*

597. *Arabis Holboellii* var. *retrofracta*

594. *Arabis divaricarpa*

598. *Arabis lyrata* var. *kamchatica*

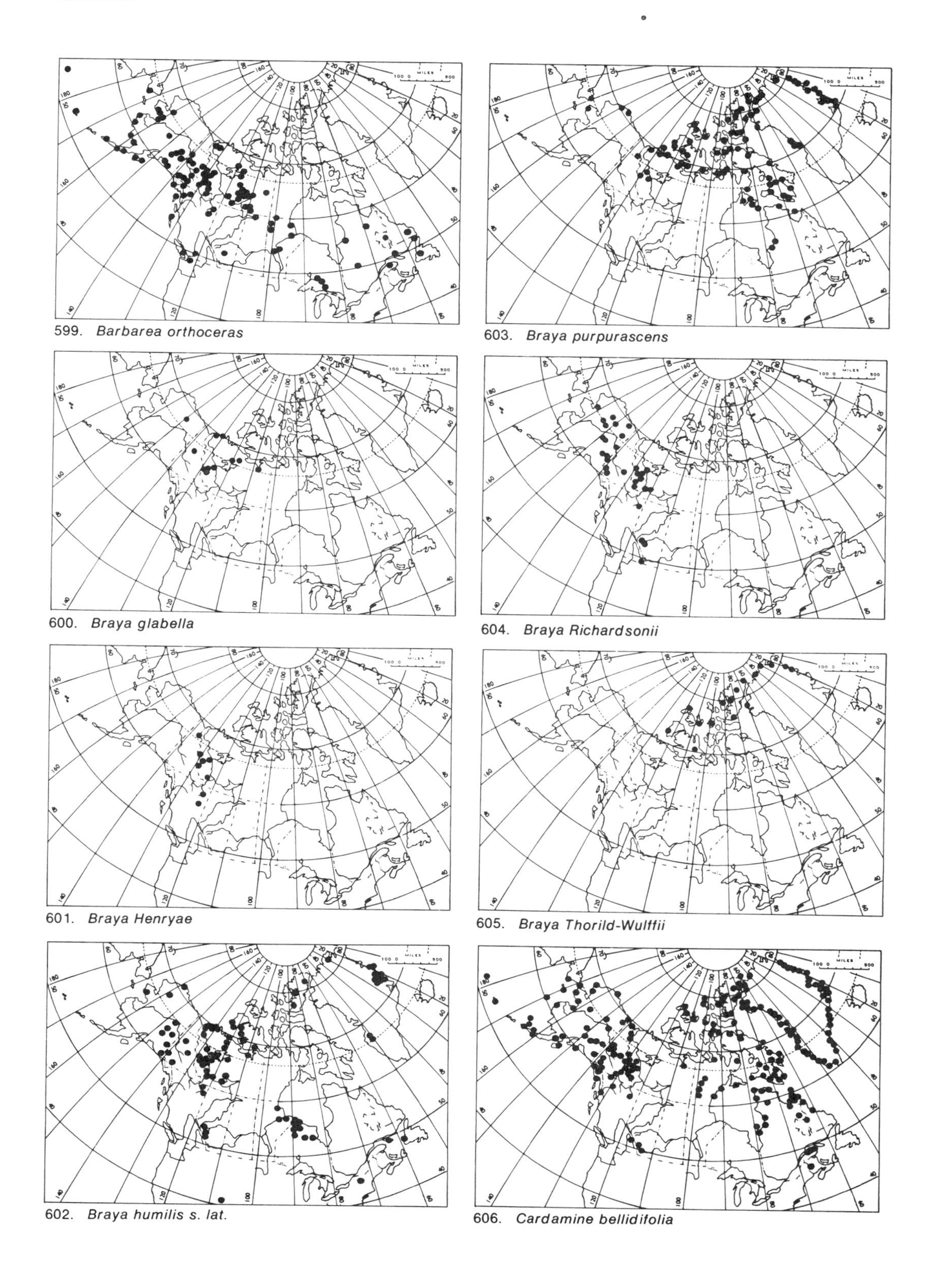

599. *Barbarea orthoceras*

600. *Braya glabella*

601. *Braya Henryae*

602. *Braya humilis s. lat.*

603. *Braya purpurascens*

604. *Braya Richardsonii*

605. *Braya Thorild-Wulffii*

606. *Cardamine bellidifolia*

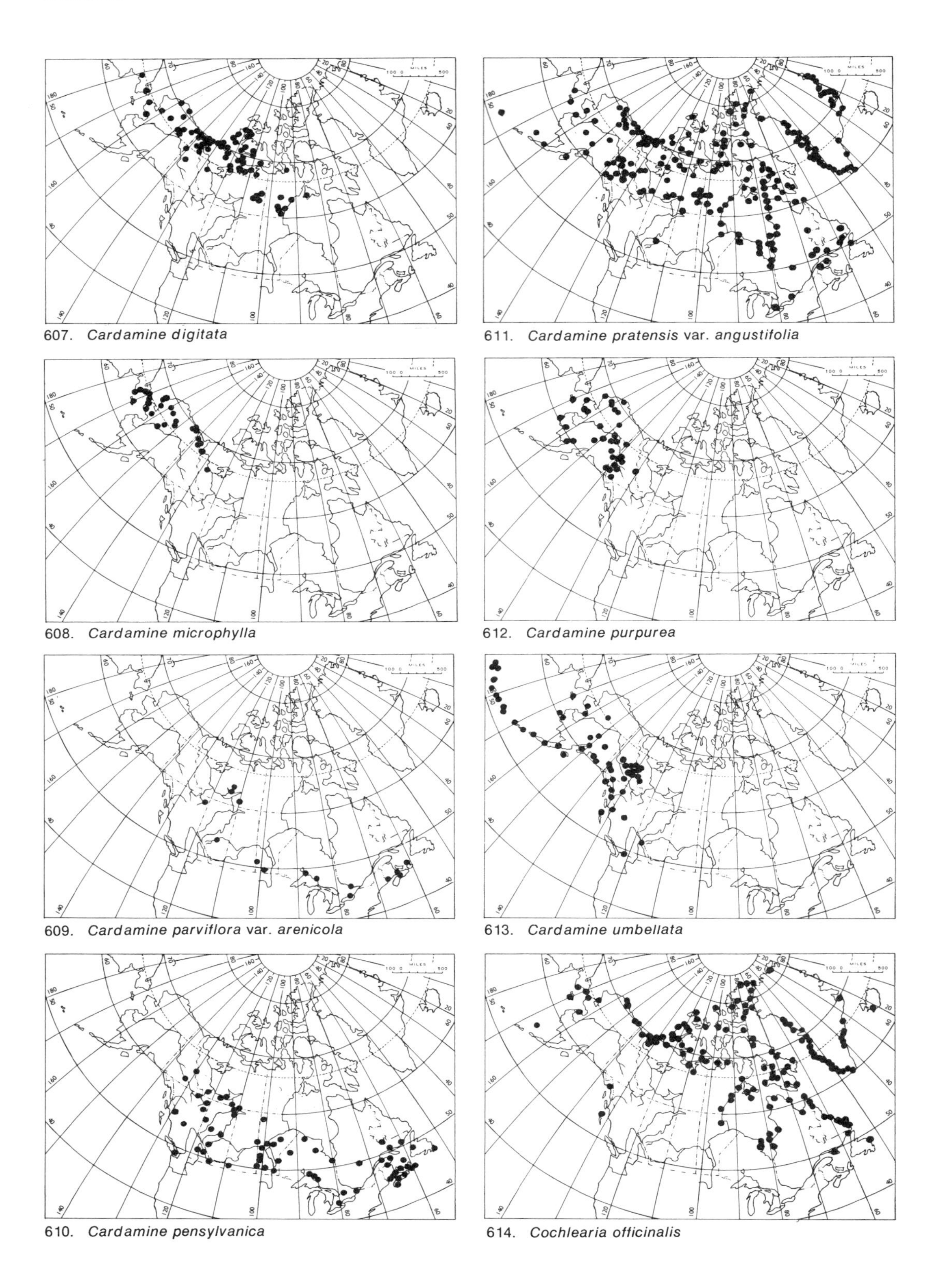

607. *Cardamine digitata*

611. *Cardamine pratensis* var. *angustifolia*

608. *Cardamine microphylla*

612. *Cardamine purpurea*

609. *Cardamine parviflora* var. *arenicola*

613. *Cardamine umbellata*

610. *Cardamine pensylvanica*

614. *Cochlearia officinalis*

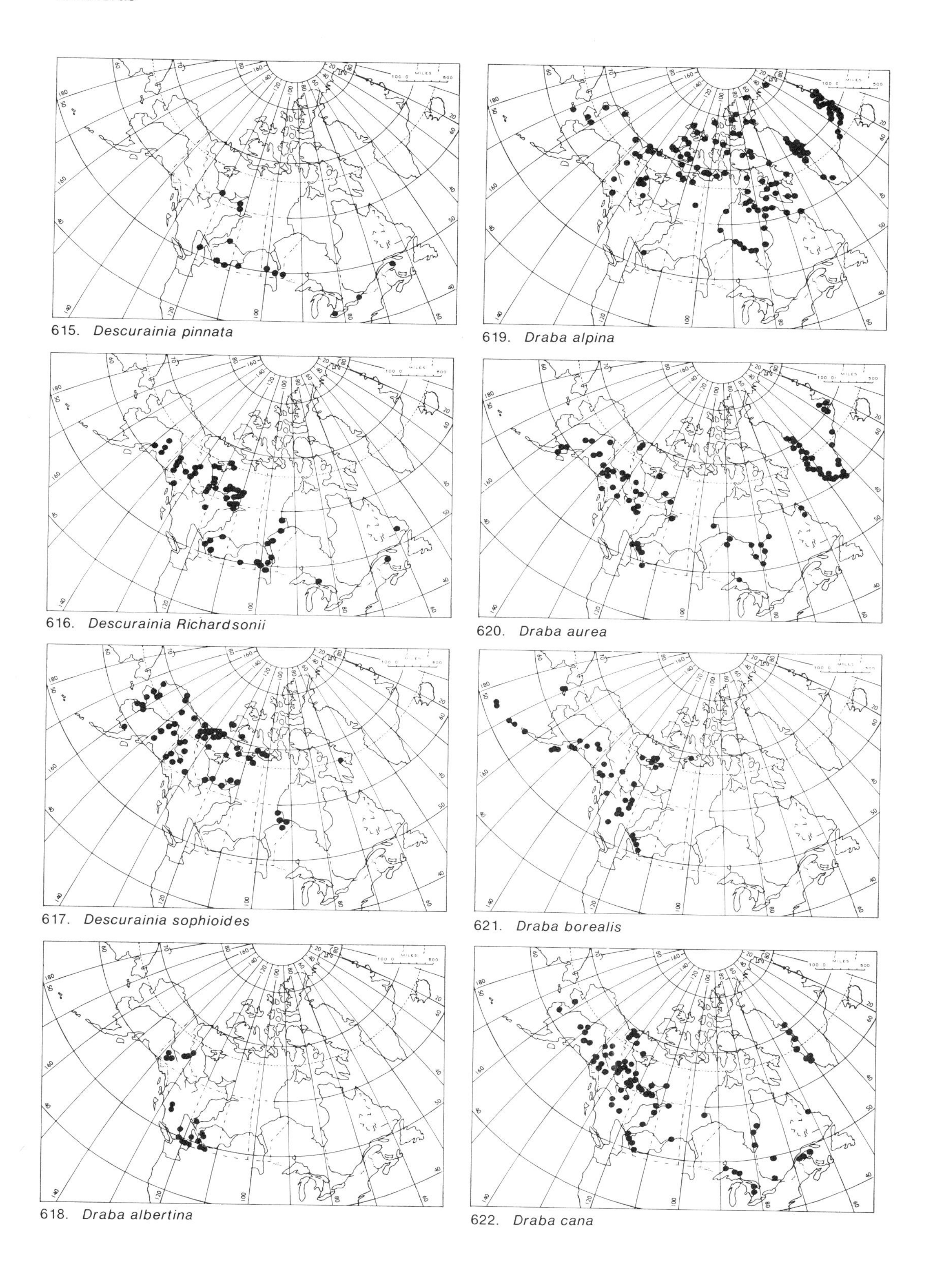

615. *Descurainia pinnata*

616. *Descurainia Richardsonii*

617. *Descurainia sophioides*

618. *Draba albertina*

619. *Draba alpina*

620. *Draba aurea*

621. *Draba borealis*

622. *Draba cana*

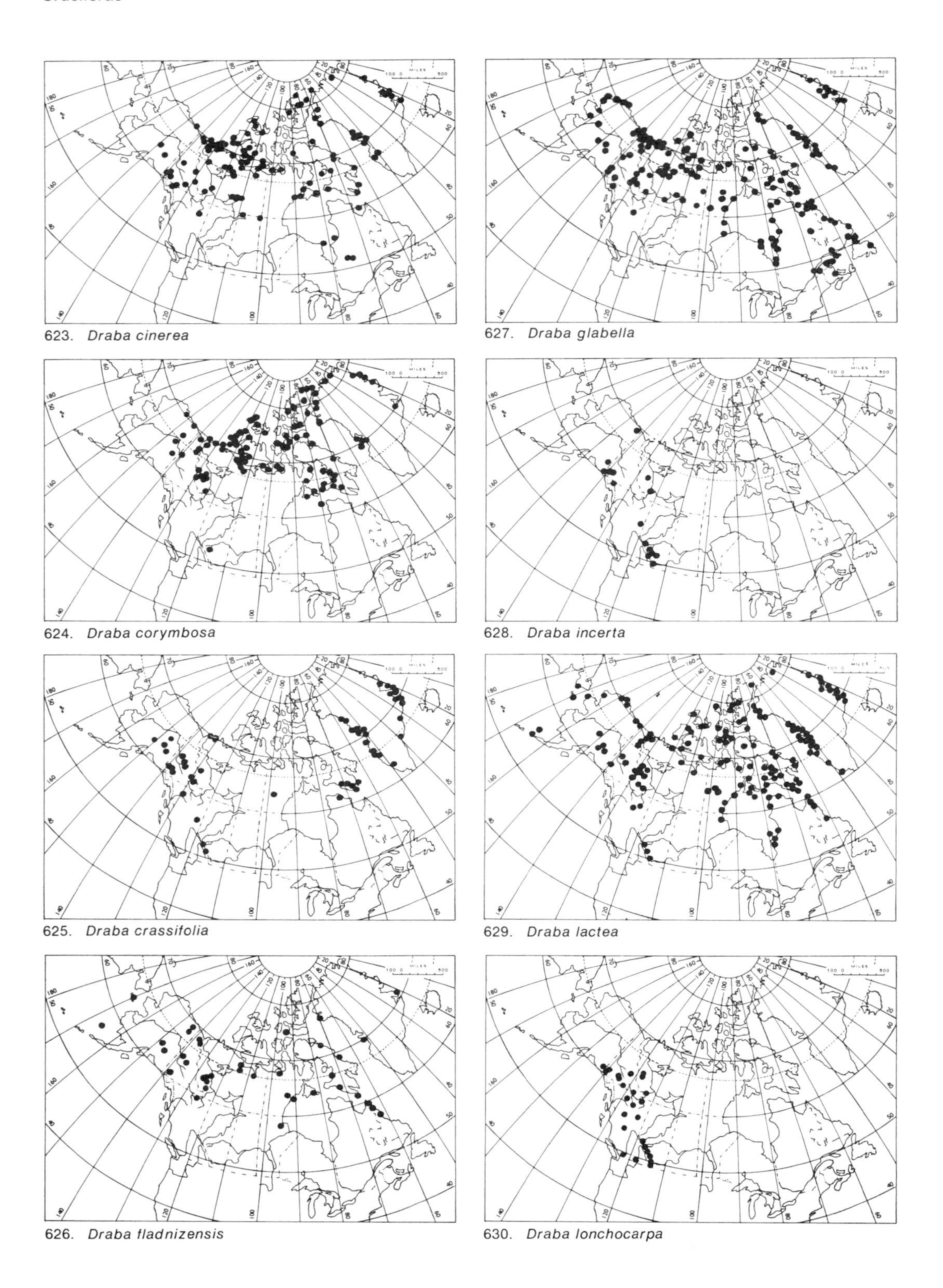

623. *Draba cinerea*

627. *Draba glabella*

624. *Draba corymbosa*

628. *Draba incerta*

625. *Draba crassifolia*

629. *Draba lactea*

626. *Draba fladnizensis*

630. *Draba lonchocarpa*

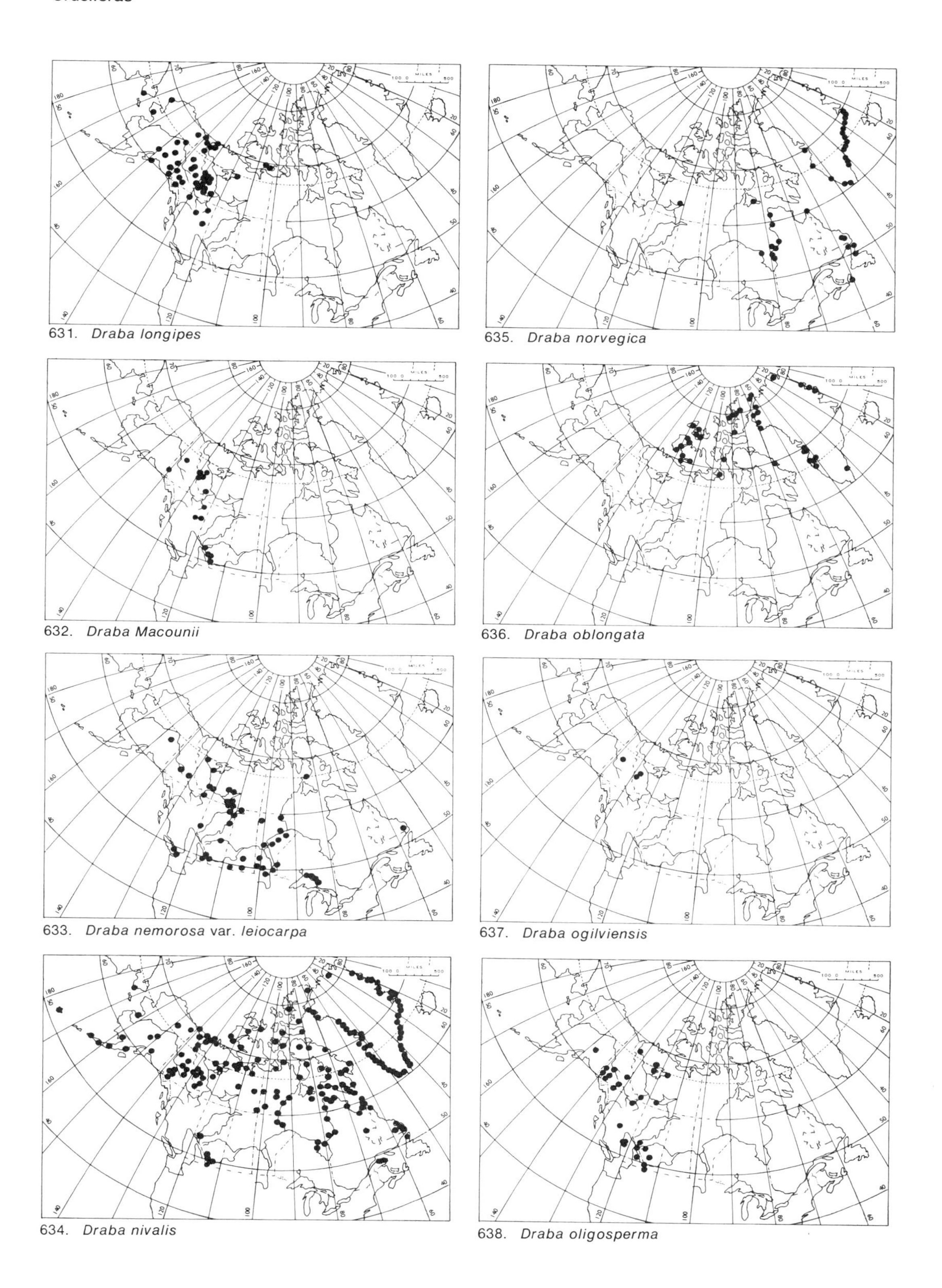

631. *Draba longipes*

632. *Draba Macounii*

633. *Draba nemorosa* var. *leiocarpa*

634. *Draba nivalis*

635. *Draba norvegica*

636. *Draba oblongata*

637. *Draba ogilviensis*

638. *Draba oligosperma*

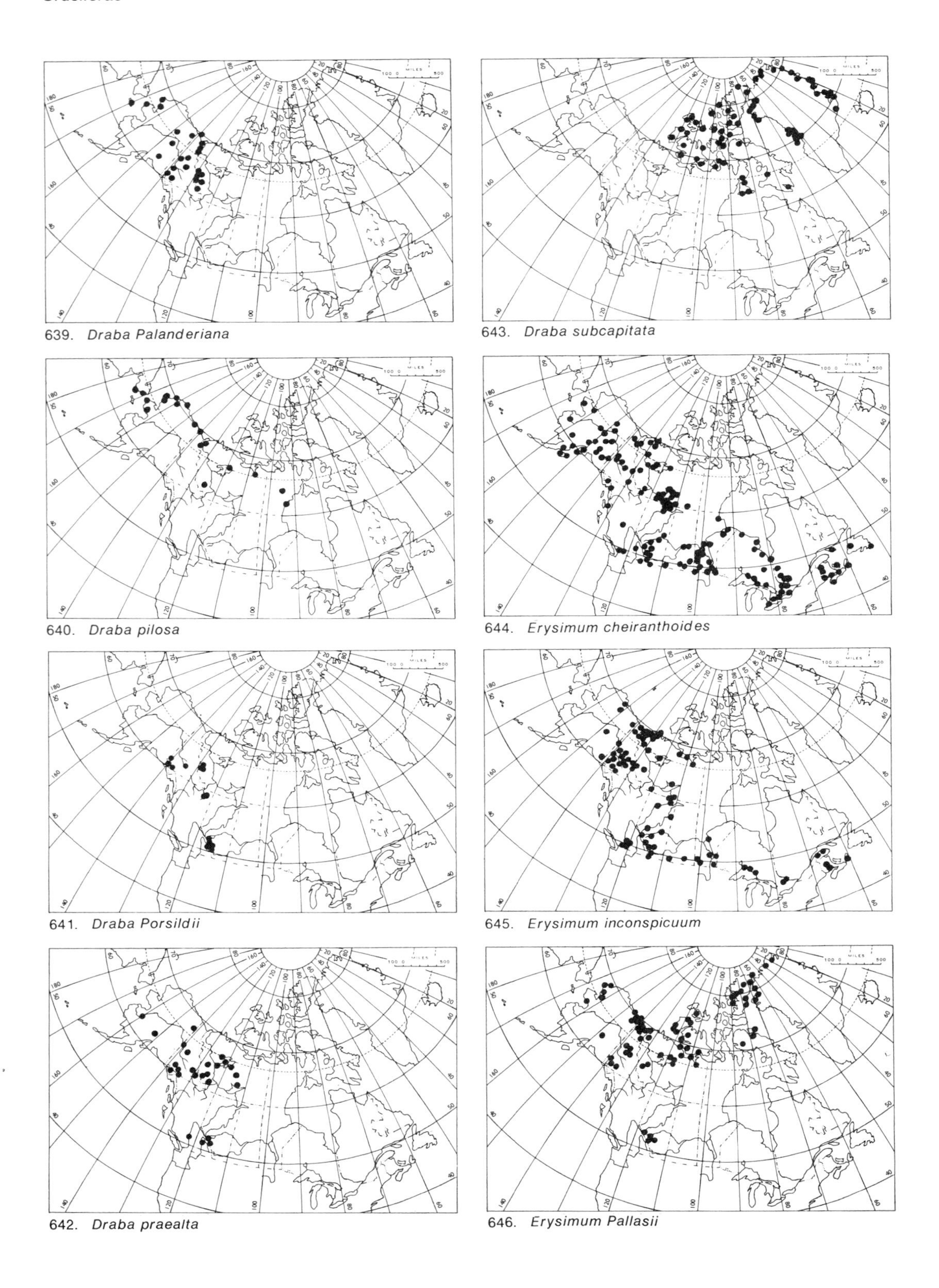

639. *Draba Palanderiana*

640. *Draba pilosa*

641. *Draba Porsildii*

642. *Draba praealta*

643. *Draba subcapitata*

644. *Erysimum cheiranthoides*

645. *Erysimum inconspicuum*

646. *Erysimum Pallasii*

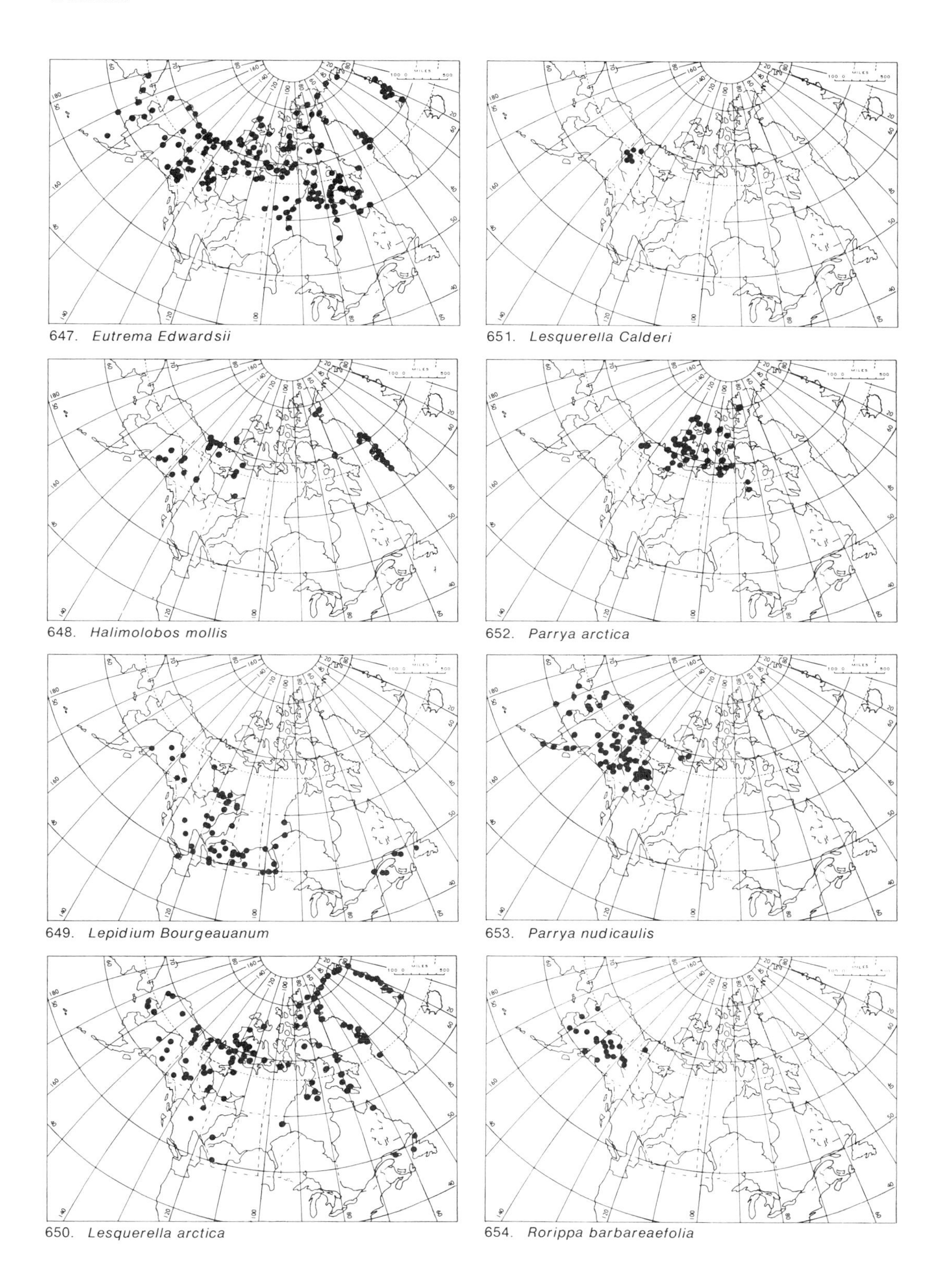

647. *Eutrema Edwardsii*

648. *Halimolobos mollis*

649. *Lepidium Bourgeauanum*

650. *Lesquerella arctica*

651. *Lesquerella Calderi*

652. *Parrya arctica*

653. *Parrya nudicaulis*

654. *Rorippa barbareaefolia*

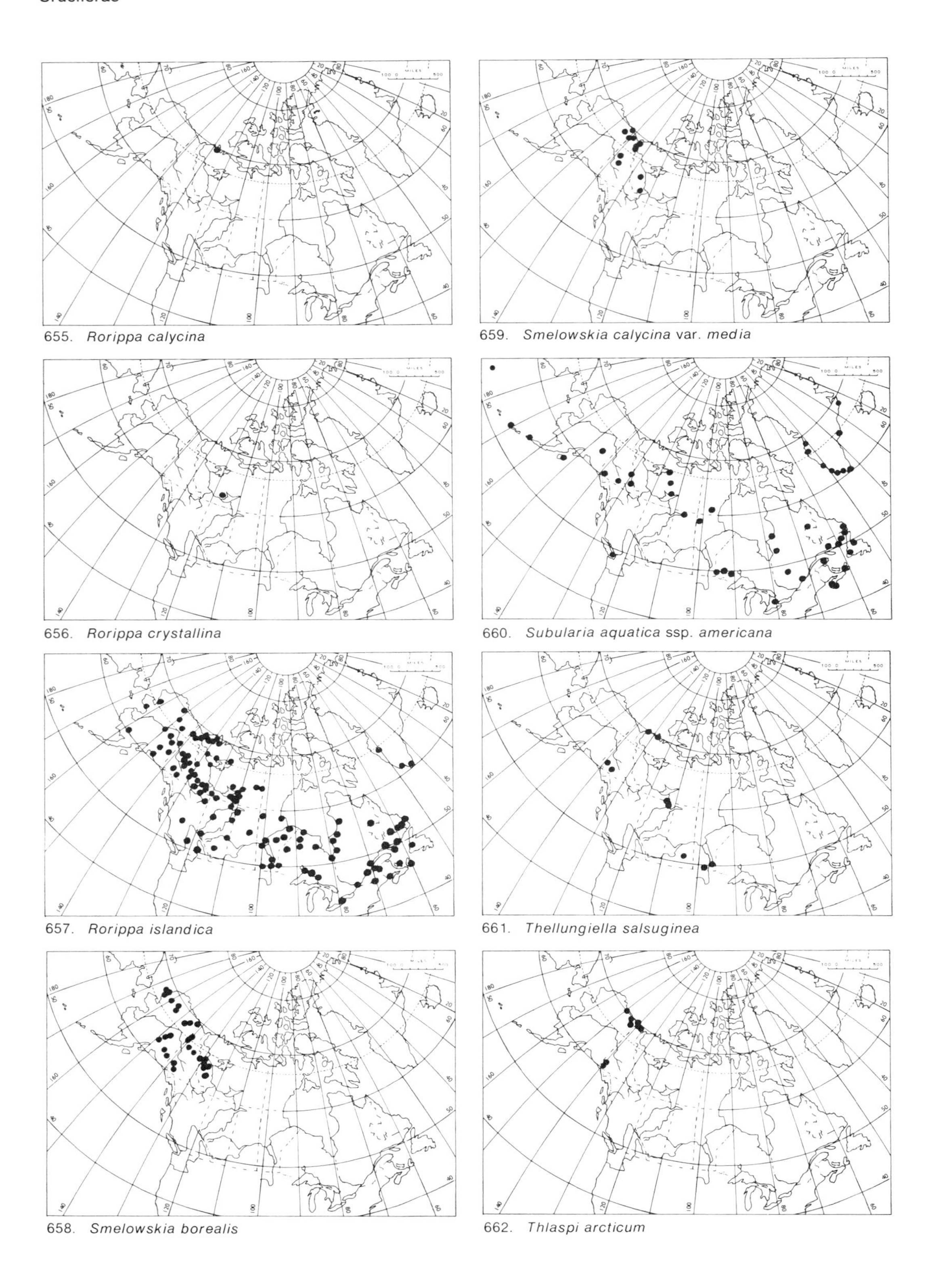

655. *Rorippa calycina*

659. *Smelowskia calycina* var. *media*

656. *Rorippa crystallina*

660. *Subularia aquatica* ssp. *americana*

657. *Rorippa islandica*

661. *Thellungiella salsuginea*

658. *Smelowskia borealis*

662. *Thlaspi arcticum*

SARRACENIACEAE Pitcher-plant Family
Sarracenia L. Pitcher-plant

Sarracenia purpurea L.
Pitcher-plant
Perennial, yellowish-green or purplish bog plants with hollow, pitcher-like leaves and solitary regular, pentamerous nodding flowers on slender, 2 to 4 dm long scapes.

The hood-like and purplish coloured lip of the "pitcher", as well as its inside surface, is covered by minute, down-pointed, smooth hairs that cause insects landing there to fall into the partly water-filled "pitcher". Unable to climb out against the down-pointing hairs, the insects eventually drown and decompose, thereby providing nitrogen and other plant nutrients that are lacking in the peat-bog environment in which the pitcher plant is at home.

General distribution: N. America from Nfld. to northern B.C. and southwestern District of Mackenzie.

Fig. 571 Map 663

571. *Sarracenia purpurea*

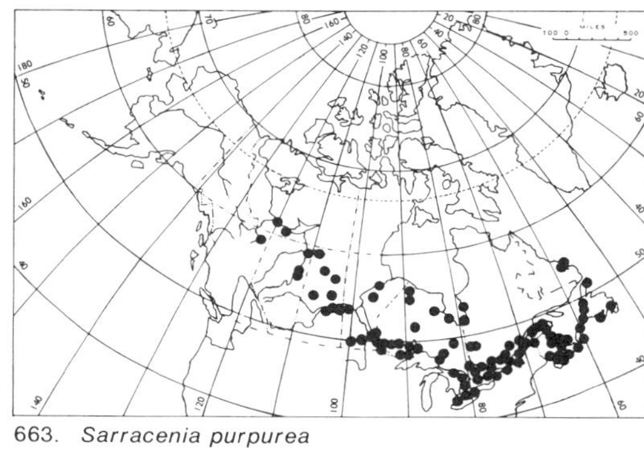

663. *Sarracenia purpurea*

DROSERACEAE Sundew Family
Drosera L. Sundew

Small, scapose, perennial bog plants with sticky, gland-tipped hairs on both sides of the leaves that trap and "digest" small insects. Flowers small, regular and pentamerous, in a few- to several-flowered spike-like raceme. Fruit a 1-locular, 3-valved capsule containing numerous seeds.

a. Leaf-blade linear or linear-spatulate to narrowly cuneate-obovate, tapering into a linear petiole much longer than the blade
  b. Leaf-blade linear-spatulate to narrowly cuneate-obovate . . . . *D. anglica*
  b. Leaf-blade linear . . . . *D. linearis*
a. Leaf-blade sub-orbicular to ovate, abruptly tapered into a short petiole . . . . *D. rotundifolia*

Drosera anglica Huds.
Leaves erect-ascending, in a basal rosette shorter than the stiffly erect scape that with us rarely exceeds 10 cm in length. Flowers white or pale pink; seeds black, spindle-shaped, the loose coat longitudinally striate-areolate.

On damp and usually marly and wet calcareous bog- and pond-margins. With us always rare and local; north to the limit of forest.

General distribution: Circumpolar with several gaps.

Fig. 572 Map 664

Drosera linearis Goldie
Similar to the preceding, but the leaf-blades linear, 1 to 4 cm long; seeds black, oblong-ovate, the close seed-coat with irregular shallow pits.

In similar habitats and often growing with *D. anglica* in southwestern District of Mackenzie.

General distribution: Nfld. to B.C. and S.W. District of Mackenzie and southward.

Map 665

Drosera rotundifolia L.
Round-leaved Sundew
Similar to preceding but the leaves in a spreading rosette on, or near, the ground; seeds spindle-shaped, with the loose seed-coat prolonged at the tips, longitudinally striate.

Chiefly confined to acid peaty bogs.

General distribution: Circumpolar, in our area north to, or slightly beyond the limit of trees.

Fig. 573 Map 666

572. *Drosera anglica*

573. *Drosera rotundifolia*

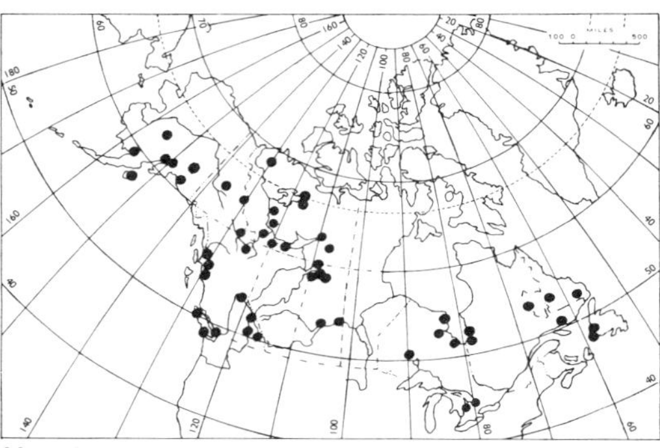

664. *Drosera anglica*

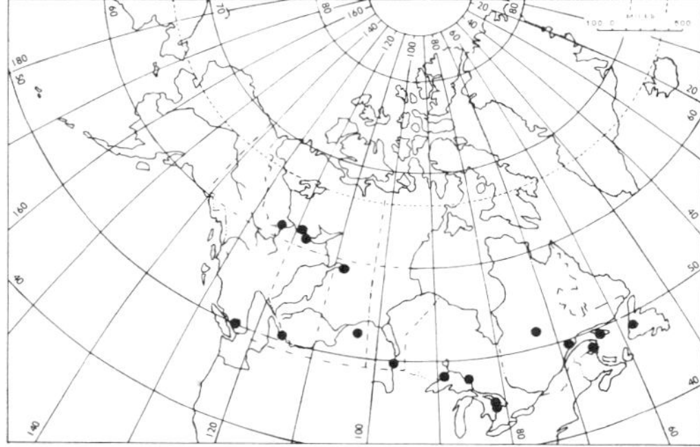

665. *Drosera linearis*

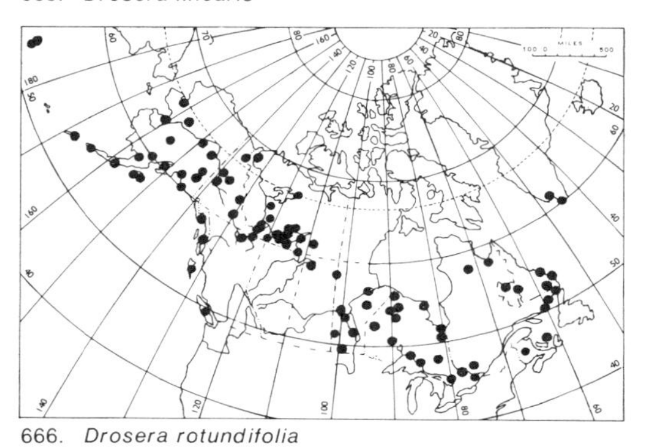

666. *Drosera rotundifolia*

CRASSULACEAE Stonecrop Family

Glabrous or somewhat succulent herbs with leafy stems; flowers small and regular, in terminal cymes, or axillary in *Tillaea*. Fruit a dry, dehiscent follicle.

a. Perennial, terrestrial
 b. Flowers dioecious, 4-merous . . . . *Rhodiola*
 b. Flowers perfect, 5-merous . . . . *Sedum*
a. Annual, tiny aquatic with small, axillary 3-4-merous flowers . . . . *Tillaea*

Rhodiola L.

Rhodiola integrifolia Raf.
Roseroot
Succulent, dioecious herb with a thick and much-branched, scaly rhizome bearing numerous, leafy stems 5 to 20 cm high. Leaves alternate, pale, glaucous, and somewhat spoon-shaped, oblong-lanceolate, entire or dentate, 1 to 4 cm long. Flowers in a dense terminal cluster, those of the male plant bright yellow, the female ones usually purple; follicles reddish, plump, and erect.

The succulent young stems and leaves of the roseroot are edible and may be used as salad or pot-herb.

Moist rocky alpine ledges but also on gravelly beaches and lagoon shores. The species is strongly nitrophilous and for this reason favours bird cliffs and the proximity of human habitations.

General distribution: The Amphi-Beringian *R. integrifolia* differs from the closely related, Amphi-Atlantic *R. rosea* by narrower and less succulent leaves. The two species are separated by a nearly 3000 km broad gap.

Fig. 574 Map 667

Sedum L. Stonecrop

***Sedum lanceolatum** Torr.
Plant slender, erect-ascending, 15 to 20 cm tall, from a branching, sub-ligneous rhizome. Leaves linear-oblong, terete or flattish, 7 to 20 mm long, densely crowded on the sterile branches, those of the flowering stems lanceolate and about half as wide as long, soon deciduous. Inflorescence a dense cyme of small, yellow flowers.

General distribution: Cordilleran species reaching mountains of southern Y.T., and to be expected in adjacent parts of the District of Mackenzie.

Fig. 575 Map 668

Tillaea L. Pigmyweed

Tillaea aquatica L.
Mat-forming with slender, much-branching stems with entire, linear-oblong, scale-like leaves, and tiny, solitary near-sessile, axillary greenish-white flowers.

In shallow fresh or saline pools, with us known only from a single station on the north shore of Great Slave Lake, disjunct from its nearest known station in southern British Columbia by more than 1200 km.

General distribution: Circumpolar, with large gaps; non-arctic.

Fig. 576 Map 669

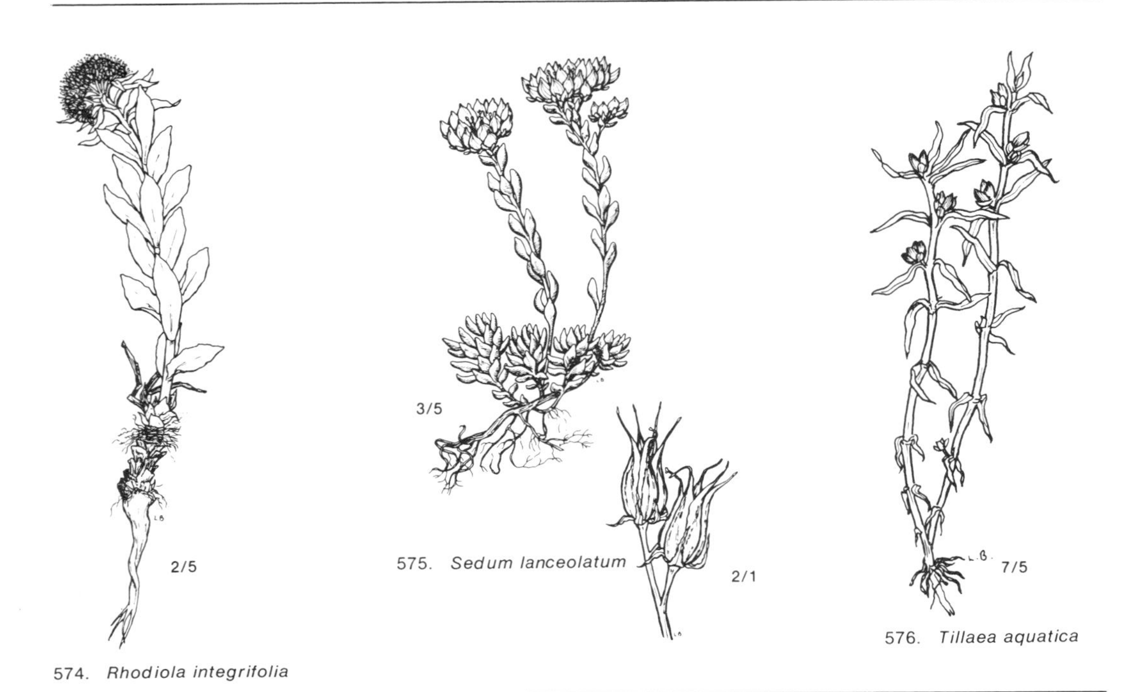

574. *Rhodiola integrifolia*

575. *Sedum lanceolatum*

576. *Tillaea aquatica*

667. *Rhodiola integrifolia*

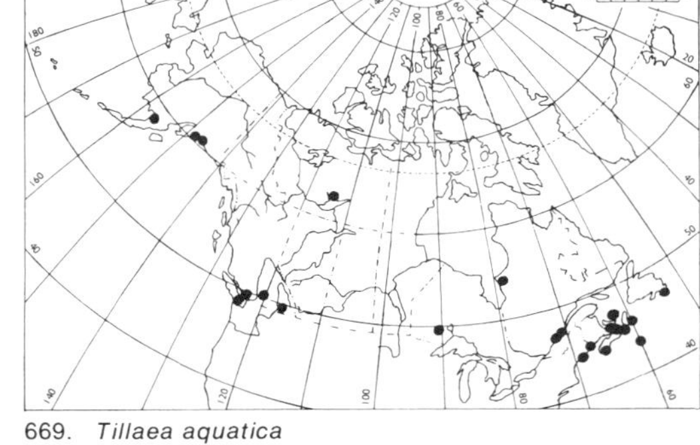

669. *Tillaea aquatica*

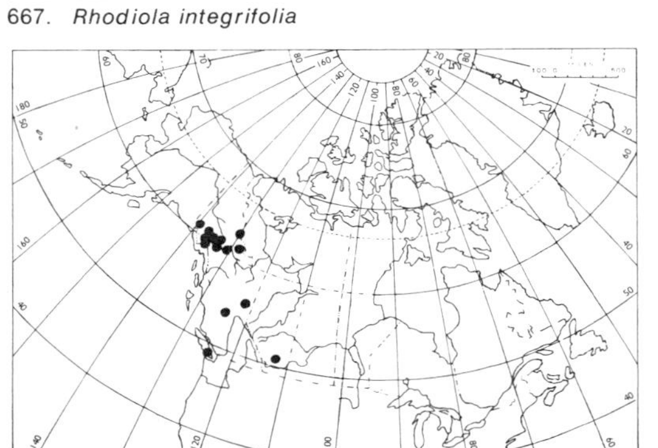

668. *Sedum lanceolatum*

SAXIFRAGACEAE Saxifrage Family

Perennial herbs or shrubs with alternate or rarely opposite, simple or lobed leaves, usually lacking stipules. Flowers perfect and regular, solitary or more often in a racemose or cymose inflorescence; calyx free or adnate to the ovary; stamens 5 to 10, usually twice as many as the petals, and inserted with them on the calyx; styles usually two, often partly united; fruit a capsule, follicle, or berry (in *Ribes*).

a. Shrubs; fruit a berry . . . . *Ribes*
a. Herbs; fruit a dry capsule
  b. Petals lacking; low delicate herbs; with 4-merous flowers . . . . *Chrysosplenium*
  b. Petals present, or if lacking, the flowers 5-merous
    c. Sterile filaments (staminodia) present and alternating with the 5 regular stamens . . . . *Parnassia*
    c. Sterile filaments none
      d. Petals laciniately lobed . . . . *Mitella*
      d. Petals not lobed or laciniate
        e. Stamens 5
          f. Capsule 1-locular; leaves mostly basal . . . . *Heuchera*
          f. Capsule 2-locular; stem leafy . . . . *Boykinia*
        e. Stamens 10
          g. Leaves evergreen
            h. Leaves thick and broad . . . . *Leptarrhenia*
            h. Leaves narrow . . . . *Saxifraga*
          g. Leaves deciduous . . . . *Saxifraga*

Boykinia Nutt.

***Boykinia Richardsonii** (Hook.) A. Gray
Therofon Richardsonii (Hook.) Ktze.
Rhizome stout and ascending, densely covered by persisting leaf-sheaths; basal leaves petioled, the blade reniform, 5 to 10 cm in diameter, shallowly double-crenate; the leaf-margins, the nerves of the lower surface of the blade, and also its petiole as well as the 3 to 5 dm tall and leafy flowering stem, prominently covered by dark brown, non-mucilaginous, capitate glands. Inflorescence a congested, spike-like panicle; the flowers showy, about 1.5 cm long, with oblanceolate, white or pale pink petals.

Moist alpine herbmats with ample snow cover in winter.

General distribution: A very showy and ornamental endemic of mountains of northern and central Alaska and northwestern Y.T.; to be looked for on the east slope of the Richardson Mts., west of the Mackenzie Delta.

Fig. 577 Map 670

Chrysosplenium L. Golden Saxifrage

Small, delicate, yellowish-green, essentially glabrous perennials, ours usually 3 to 10 cm tall, from a slender, erect-ascending or creeping rhizome; leaves alternate, the blade reniform, shallowly 3- to 7-lobed, the lower slender-petioled, the upper short-petioled or sessile. Flowers small and inconspicuous, solitary or a few together in the axils of the upper leaves; sepals yellowish-green, petals none, stamens 4 to 8. Seeds small, numerous and shiny.

a. Stamens 4, calyx green, petioles glabrous . . . . *C. tetrandrum*
a. Stamens 8, calyx minutely purple-dotted; petioles usually with long, crinkly and pale rusty hairs . . . . *C. Wrightii*

Chrysosplenium tetrandrum (Lund) Fries
Moist, somewhat shady places and commonly in soil enriched by animal manure, hence frequently growing among ruins of former Eskimo dwellings, inside collapsed stone burials and meat caches, or near animal dens.

General distribution: Circumpolar, wide spread, arctic-alpine.

Fig. 578 Map 671

Chrysosplenium Wrightii Franch. & Sav.
Similar to but slightly more robust than *C. tetrandrum* from which it is easily distinguished by the presence of purple spots on the calyx and pale, rusty hairs on the petioles.

Moist alpine tundra and by seepages.

General distribution: Amphi-Beringian ranging eastward to the Richardson and Mackenzie Mts.

Map 672

Heuchera L. Alumroot

Heuchera Richardsonii R. Br.
Stems 3 to 4 dm tall, hispid and somewhat glandular from a scaly, ascending rhizome; basal leaves petiolate with a reniform, shallowly 5- to 9-lobed blade. Inflorescence an elongated panicle. Flowers about 1 cm long, the hypanthium markedly oblique, finely glandular-dotted, the petals spatulate, pale purplish, slightly longer than the sepals.

Open woodland meadows.

General distribution: N. America; W. Ont. to B.C., reaching S.W. District of Mackenzie.

Fig. 579 Map 673

Leptarrhena R. Br.

Leptarrhena pyrolifolia (D. Don) Ser.
Leather-leaved Saxifrage
Flowering stems 1 to 4 dm tall, simple, erect, from a freely branching, horizontally spreading rhizome; leaves mainly basal, the blade crenate-serrate, oval oblong, leathery, dark green and glabrous above, paler and yellowish-brown beneath. Inflorescence of small, white flowers in 3 to 7 stalked cymules at first congested, elongating in fruit.

Moist alpine slopes.

General distribution: Cordilleran—Pacific coast species reaching S.W. Y.T. and the Mackenzie Mts.

Fig. 580 Map 674

Mitella L. Miterwort

Mitella nuda L.
Dwarf species with a creeping, slender rhizome from which issue elongated stolons. Leaves basal, 3 to 5 cm long, the blade shallowly crenate, cordate, sparsely hirsute above; scapes erect, up to 20 cm tall but mostly less, finely glandular. Inflorescence racemose, of from few to 10 short-pedicelled flowers; petals 5, greenish. Stamens 10.

Wide ranging boreal forest species of cold spruce woods northward to the upper Mackenzie Valley. The Cordilleran *M. pentandra* which lacks stolons and has larger, cordate leaves and 5 stamens has been reported from subalpine meadows north to southern Yukon Territory and should be looked for also on the east slope of Mackenzie Mountains.

General distribution: Lab. to S.W. Y.T., E. Asia.

Fig. 581 Map 675

Parnassia L. Grass-of-Parnassus

Scapose, glabrous perennials with ascending rhizome and basal, petiolate leaves. Flowers solitary, perfect and regular; the petals white, conspicuously veined; stamens 5, alternating with the petals and the 5 gland-tipped staminodia; fruit a 1-locular capsule opening at the apex.

a. Petals fimbriate along the sides . . . . *P. fimbriata*
a. Petals entire
  b. Petals 5-9 veined
    c. Petals nearly twice as long as sepals; staminodia 9-15 in each fascicle . . . *P. palustris*
    c. Petals not much longer than sepals; staminodia 7-9 in each fascicle . . . . *P. montanensis*
  b. Petals 3-veined, equalling the sepals or shorter . . . . *P. Kotzebuei*

Parnassia fimbriata Koenig
Rhizome stout; basal leaves numerous, their petiole slender, usually 4 to 5 times as long as the cordate, 2 to 3 cm broad blades. Scapes 1 to 3 dm tall, with a single clasping and reniform leaf inserted near or above the middle.

Moist alpine ravines and in wet moss by brooks.
General distribution: Cordilleran species reaching the east slope of the Mackenzie Mts.
Fig. 582 Map 676

Parnassia Kotzebuei Cham. & Schlect.
Leaves broadly ovate-cordate, petioled to subsessile. Scapes slender, 6 to 20 cm high, leafless or with a single sessile leaf inserted near the base. Sepals oblong, 5 to 7 mm long, equalling or exceeding the elliptic-oval, white petals. Staminodia short, with 4 to 5 filaments. Capsule ovoid, twice as long as the calyx.
Common or occasional in wet sand by lake shores north to the Arctic Coast.
General distribution: Wide ranging N. American species reaching W. Greenland and eastern Asia; low arctic-alpine.
Fig. 583 Map 677

Parnassia montanensis Fern. & Rydb.
Leaf-blades ovate with subcordate base, 2 to 3 times as long as their petiole. Flowering scapes 2 to 3 dm tall, their cauline leaf narrowly triangular, inserted at or well below the middle. Flowers 1.0 to 1.5 cm in diameter.
Moist, sandy river banks and lake shores.
General distribution: Cordilleran species, occasional to rare along the upper Mackenzie Valley to Great Bear L.
Fig. 584 Map 678

Parnassia palustris L.
var. **neogaea** Fern.
Similar to *P. montanensis* from which it differs by its much larger flower, distinctly cordate basal leaves and more numerous staminodia.
Wet calcareous soil; often also in herbmats by brooks or ponds.
General distribution: Wide ranging from Lab. to Alaska, north beyond the limit of trees.
Fig. 585 Map 679

Ribes L. Currant or Gooseberry

Shrubs with erect-ascending smooth or prickly branches, and palmately lobed cordate leaves. Flowers 5-merous in few- to many-flowered racemes, the petals shorter than the coloured calyx; stamens alternating with the petals on the calyx tube. Fruit a berry.

a. Stems prickly or spiny
  b. Flowers mostly single . . . . *R. oxyacanthoides*
  b. Flowers in racemes . . . . *R. lacustre*
a. Stems without prickles or spines
  c. Fruit red; petals pink or purplish
    d. Fruit smooth and shiny . . . . *R. triste*
    d. Fruit glandular bristly . . . . *R. glandulosum*
  c. Fruit black; petals white . . . . *R. hudsonianum*

Ribes glandulosum Grauer
Skunk Currant
Branches ascending or often trailing. The leaves deeply 5- to 7-lobed, fresh green, glabrous, sparsely glandular on the veins. Flowering racemes erect 5- to 8-flowered; berries red, glandular-bristly. Plant ill-smelling when bruised.
Damp woods and thickets.
General distribution: N. America, from Lab. to Alaska and southward.
Fig. 586 Map 680

Ribes hudsonianum Richards.
Black Currant
Branches smooth, erect-ascending, 1 to 2 m tall. Leaves 3-lobed, 5 to 7 cm in diameter, the underside pale, resinous-dotted and villous on the veins. Flowers in ascending, 8- to 10-flowered racemes; the petals whitish, 4 mm long. Berries black and smooth, edible.
Moist woods, often in shade amongst fallen rocks. Boreal forest species extending north nearly to the limit of trees.
General distribution: N. America from W. Que. to Alaska and southward.
Fig. 587 Map 681

Ribes lacustre (Pers.) Poir.
Branches erect-ascending, rarely up to 1 m tall, densely prickly. Leaves 3- to 5-lobed, the blade 3- to 4 cm broad, glabrous or nearly so, the petioles sparsely bristly. Calyx saucer-shaped, pink; berries about 5 to 6 mm in diameter, purplish-black and glandular-bristly.
Boreal forest species of sunny, calcareous

slopes and open woods, reaching the upper Mackenzie Valley.

General distribution: N. America from Lab. to Alaska and southward.

Fig. 588 Map 682

Ribes oxyacanthoides L.
Wild Gooseberry
Low, with prickly ascending or sometimes prostrate branches rarely more than 5 dm tall. Leaves 3- to 5-lobed, the blade 3 to 4 cm broad, pubescent especially below. Calyx campanulate, yellowish-green. Berries purplish-black and smooth, edible, about 1 cm in diameter.

Calcareous rocky hillsides or stony prairies, north to Great Bear Lake.

General distribution: N. America; Hudson Bay to S.E. Y.T., and southward.

Fig. 589 Map 683

Ribes triste Pall.
Wild Red Currant
Of somewhat similar habit as *R. hudsonianum* but its branches often declining and rooting. Leaves thin, glabrous, or in youth thinly pubescent beneath. Racemes drooping, 8- to 15-flowered. Petals reddish, and the smooth berries bright red and edible, in flavour similar to those of the garden red currant.

Boreal forest species ranging north to the limit of trees.

General distribution: Lab. to Alaska and southward; E. Asia.

Fig. 590 Map 684

Saxifraga L. Saxifrage

Ours all perennial, mostly with alternate, rarely opposite leaves and white, yellow, purple or pink flowers. Capsule 2-beaked and 2-locular, or sometimes almost separate follicles; stamens 10. In *S. cernua* and *S. foliolosa* at least some of the flowers normally are replaced by axillary bulblets, or by small tufts of leaves that root when detached from the mother plant. *S. radiata* and *S. rivularis,* besides regular and normal functional flowers, produce bulbils from the base of the flowering stems.

a. Leaves opposite and imbricated; low plant with trailing branches and solitary purple flowers on a short, leafy peduncle . . . . *S. oppositifolia*
a. Leaves alternate
  b. Flowering stems naked
    c. Basal leaves obricular or oblong, entire or merely serrate
      d. Plant pulvinate or spreading; flowers small, yellow . . . . *S. Eschscholtzii*
      d. Plant not pulvinate; flowers white or pale pink
        e. Leaf-blade orbicular, toothed, on a slender petiole . . . . *S. punctata*
        e. Leaf-blade oblong, apically toothed, lacking a distinct petiole
          f. Leaves cuneate, gradually tapering into the petiole
            g. Inflorescence much branched, the flowers mostly normal and functional although bulbils sometimes present . . . . *S. ferruginea*
            g. Inflorescence with short, spreading branches, the terminal with a single functional flower, the remainder with small, green bulbils . . . . *S. foliolosa*
          f. Leaves fan-shaped
            h. Filaments clavate or spindle-shaped . . . . *S. Lyallii*
            h. Filaments subulate . . . . *S. davurica* ssp. *grandipetala*
    c. Basal leaves oval-oblong, with dentate or crenate margins
      i. Inflorescence a spike-like panicle . . . . *S. hieracifolia*
      i. Inflorescence paniculate, or flowers in a terminal cluster
        j. Sepals reflexed; leaves pubescent on both sides . . . . *S. reflexa*
        j. Sepals not reflexed; leaves glabrous above
          k. Underside of leaves with coarse, rust-coloured hairs . . . . *S. tenuis*
          k. Underside of leaves lacking such hairs . . . . *S. nivalis*
  b. Flowering stem leafy
    l. Basal leaves not petiolate
      m. Petals yellow; leaves not marcescent

n. Plant with long, flagellate and naked stolons terminating in a rooting off-set . . . . *S. flagellaris*
n. Plants without such stolons
o. Leaves small, linear and fleshy
p. Plant small and delicate, loosely tufted or matted, with small, solitary flowers on a slender, bracted peduncle . . . . *S. serphyllifolia*
p. Plant coarser and matted; flowers 3 or more . . . . *S. aizoides*
o. Leaves, or at least the basal ones oblanceolate and petioled; flowering stem leafy, usually 1-flowered . . . . *S. Hirculus*
m. Petals white or cream-coloured; leaves leathery, marcescent; inflorescence corymbose
q. Leaves oblanceolate, green . . . . *S. Aizoon*
q. Leaves narrow
r. Leaves linear-lanceolate, mucronate, margins ciliate . . . . *S. bronchialis*
r. Leaves narrowly cuneate, reddish and shiny, entire-margined, sharply 3-toothed . . . . *S. tricuspidata*
l. Basal leaves petiolate, with reniform, cordate or cuneate, 3- to 5-lobed blade; flowers white or pale pinkish
s. Bulbils present in the inflorescence or at the base of the flowering stem
t. Inflorescence of one terminal and functional flower, and reddish or black bulbils in the axils of upper stem-leaves . . . . *S. cernua*
t. Inflorescence of normal flowers
u. Petals about 1 cm long; white bulbils nearly always present at the base of flowering stem . . . . *S. radiata*
u. Petals usually less than 5 mm long; bulbils but rarely present at the base of flowering stem where commonly replaced by short, rooting stolons . . . . *S. rivularis*
s. Bulbils lacking
v. Plant densely matted or tufted; leaves green and somewhat clammy . . . . *S. caespitosa*
v. Plant from tiny, rooted reddish-green rosette . . . . *S. adscendens*

Saxifraga adscendens L.
ssp. **oregonensis** (Raf.) Bacigalupi
Small, tufted short-lived perennial, 2 to 8 cm tall with a rosette of usually reddish-purple, oblanceolate, entire or 3-toothed leaves. Flowering stem simple or branched, leafy, glandular-pubescent, terminating in a few- to several-flowered cyme of small flowers with purple sepals and whitsh petals twice as long as the sepals.
Moist gravelly alpine situations.
General distribution: Cordilleran, with us reaching the southern Mackenzie Mts.
Fig. 591 Map 685

Saxifraga aizoides L.
Yellow Mountain Saxifrage
Low and matted with leafy decumbent and trailing branches terminating in a few-flowered raceme; leaves somewhat imbricated, linear and fleshy, 0.5 to 1.5 cm long; petals yellow, and mostly orange-dotted.
Moist calcareous clay and gravel.
General distribution: Amphi-Atlantic, arctic-alpine.
Fig. 592 Map 686

Saxifraga Aizoon Jacq.
var. **neogaea** Butters
White Mountain Saxifrage
Basal leaves oblanceolate or spatulate, thick and leathery, serrulate with a conspicuous lime-encrusted pore at the base of each tooth. Flowering stems erect, 10 to 25 cm high, with alternate, much-reduced leaves. Inflorescence paniculate-corymbiform. Petals creamy white, usually with orange dots.
Rocky ledges.
General distribution: Amphi-Atlantic and restricted to the Precambrian Shield area. With us known only from the extreme east end of Great Slave L., disjunct by 1600 km from its nearest station in S. Baffin Island.
Fig. 593 Map 687

Saxifraga bronchialis L.
ssp. **Funstonii** (Small) Hult.
Densely matted, forming flat cushions. The leaves coriaceous, linear-lanceolate, 6 to 7 mm long and 1.5 mm broad, mucronate and white-ciliate along the margins, densely crowded on the branches. Flowering stems 5 to 10 cm tall with 5 to 7 linear leaves; sepals obtuse about

one third as long as the orange-spotted whitish petals.

In rocky or gravelly alpine situations. In general habit resembling *S. tricuspidata* which, however, is readily distinguished by its prickly 3-toothed leaves.

General distribution: Amphi-Beringian, barely entering the District of Mackenzie in the northern Richardson Mts.

Fig. 594 Map 688

Saxifraga caespitosa L. *s. lat.*
Tufted Saxifrage
Loosely caespitose or matted, with numerous short, crowded sterile branches; basal leaves cuneate-flabellate 3- or less commonly 5-lobed, the lobes linear-oblong, obtuse, with more or less glandular-ciliate margins. Flowering stems slender, erect, 5 to 20 cm tall, glandular, especially above, with one to several reduced leaves; flowers 3 or more together in a corymb (or solitary in ssp. *uniflora*); petals creamy white, broader and longer than the triangular and glandular calyx lobes.

A polymorphous arctic-alpine species of rocky or gravelly situations, in North America represented by a number of more or less well-marked geographical races of which four enter our area:

a. Flowers large, 8-10 mm long, usually 3, rarely 1 or 2 (Amphi-Atlantic, arctic) . . . . ssp. *caespitosa* Engl. & Irmsch.
a. Flowers smaller, 4-7 mm long
  b. Flowers mostly solitary, 6-7 mm long; calyx purplish black (Circumpolar with large gaps, high-arctic) . . . . ssp. *uniflora* (R. Br.) Porsild
  b. Flowers mostly 3-4 together, 4-6 mm long
    c. Calyx greenish-purple, the lobes rounded (Endemic of Palaeozoic rocks from Hudson Bay area to Ellesmere Island) . . . . ssp. *exaratoides* (Simm.) Engl. & Irmsch. *emend* Porsild
    c. Calyx purplish-black, lobes narrowly triangular (Cordilleran, alpine) . . . . . . ssp. *monticola* (Small) Porsild

General distribution: *S. caespitosa s. lat.:* Circumpolar.

Fig. 595 (ssp. *caespitosa*), 596 (ssp. *exaratoides*), 597 (ssp. *monticola*), 598 (ssp *uniflora*) Map 689 (ssp. *caespitosa*), 690 (ssp exaratoides), 691 (ssp. *monticola*), 692 (ssp. *uniflora)*

Saxifraga cernua L.
Nodding Saxifrage
Stems slender, erect, simple, and leafy, 10 to 25 cm high, with a cluster of white bulblets at their bases. Basal and lower cauline leaves reniform, 3- to 5-lobed, petioled, upper stem leaves reduced, bractlike, subtending clusters of small purple bulblets. Flower single and terminal or sometimes lacking, nodding in youth; petals white, 5 to 12 mm long. Although stamens and pistil are present, viable seeds are not formed.

Moist ledges and gravelly places, sometimes in moss or wet sand by brooks and lake shores.

General distribution: Circumpolar, wide spread arctic-alpine.

Fig. 599 Map 693

Saxifraga davurica Willd.
ssp. **grandipetala** (Engl. & Irmsch.) Hult.
Tufted, from a simple or branching erect-ascending rhizome. Leaves all basal, 3 to 5 cm long, the oval-oblanceolate, coarsely toothed blade gradually tapering to a slender petiole; flowering scapes 8 to 12 cm tall, usually dark purple. The inflorescence paniculate, elongating in age; petals white or purplish, narrowly elliptic-oblanceolate, slightly longer than the reflexed, dark purple calyx lobes. The purplish black 6 to 10 mm long capsules stiffly erect.

Moist alpine herbmats.

General distribution: Amphi-Beringian, reaching the eastern slope of Richardson and Mackenzie Mts.

Map 694

***Saxifraga Eschscholtzii** Sternb.
Densely caespitose, forming small, compact and hemispherical cushions; the branches densely compacted, from a slender, multicipital tap-root; the scale-like leaves marcescent, bowl-shaped, ciliate and densely imbricated. The tiny, solitary, yellowish-golden brown flowers first subsessile, later, as the fruit ripens, the 2 mm long capsules with their reflexed sepals are raised above the plant cushion, on filiform, 1 cm long penduncles.

Calcareous gravelly slopes.

General distribution: Amphi-Beringian, alpine species ranging eastward along the Brooks Range. In Canada thus far known from the arctic coast of Y.T., and from a single station in

Prince Patrick Island, but to be looked for in the Richardson Mts. west of the Mackenzie Delta.
Fig. 600 Map 695

Saxifraga ferruginea Grah.
Leaves tufted from a fibrous, ascending rhizome, oblanceolate-spatulate, sharply toothed in the upper half, fresh green and glabrous above, paler beneath, with scattered simple, white hairs. Flowering stems mostly single but usually freely branching from the base, 15 to 30 cm tall, and sparsely glandular-pubescent. Inflorescence paniculate, each branch supported by a small leafy bract; sepals reflexed, purplish, about one fourth as long as the narrowly lanceolate, white and sometimes yellow-spotted petals; filaments filiform.

Moist rocky ledges.

General distribution: Pacific Coast species with us known from a single and widely disjunct station on the lower Peel R. west of the Mackenzie Delta.

Fig. 601 Map 696

Saxifraga flagellaris Willd.
ssp. **flagellaris**
Spider Plant
Stems single, erect, leafy, and glandular, 3 to 15 cm high, from a basal rosette, from which issue long, filiform and naked stolons, each terminating in a tiny, rooting rosette. Leaves entire, oblanceolate, acute, the basal ones with margins beset with a single row of spine-like, pale bristles; those of the stem narrower and glandular-ciliate. Flowers 1 to 3, large, with pale yellow petals twice as long as the linear-oblong and free sepals.

The ssp. *flagellaris* in which the hypanthium is broadly campanulate and green, is of Amphi-Beringian range, and with us is known only from calcareous rocks along the east slopes of the Mackenzie and Richardson Mountains. The high-arctic ssp. *platysepala* (Trautv.) Porsild is nearly circumpolar and mostly 1-flowered, its hypanthium is turbinate rather than campanulate and dark purple; its sepals are obovate-oblong, united in the lower third, and covered by stalked, purple-headed glands. It is wide ranging across the Canadian Arctic Archipelago and northern Greenland but, as yet has not been reported from the Canadian mainland.

Wet calcareous gravels.

General distribution: *S. flagellaris s. lat.:* Circumpolar, arctic-alpine.

Fig. 602 Map 697

Saxifraga foliolosa R. Br.
S. stellaris L.
var. *comosa* Retz.
Tufted, with a short and often branched caudex, from which often issue short, slender, leafy stolons terminating in leafy rosettes. Leaves all basal and rosulate, cuneate-oblanceolate to spatulate, thin and fresh green, serrate or toothed above the middle. Scapes slender, erect, simple or branched above, 5 to 25 cm high. Inflorescence racemose or paniculate, the flowers largely replaced by bulblets or small leafy buds, but in vigorous specimens usually with a single normal flower at the end of each branch. Petals white, lanceolate, with a short claw.

Mossy and springy places, often by the edge of brooks, by lake shores or in wet tundra.

General distribution: Circumpolar, wide ranging, high-arctic.

Fig. 603 Map 698

Saxifraga hieracifolia Waldst. & Kit.
Tufted, from a short, thick caudex. Leaves somewhat leathery, all basal and rosulate, essentially glabrous, oblong-lanceolate, short petiolate, 3 to 5 cm long, entire with serrate to coarsely toothed margins. Scapes 10 to 50 cm high; flowers in small dense clusters in an interrupted, spike-like raceme the lowermost subtended by leafy bracts. Flowers small and inconspicuous, with greenish-purple petals about as long as the sepals; follicles widely divergent, plump, purplish black, 5 to 7 mm long.

Uncommon and sporadic, in rather moist, turfy, and chiefly calcareous habitats.

General distribution: Circumpolar, arctic-alpine.

Fig. 604 Map 699

Saxifraga Hirculus L.
var. **propinqua** (R.Br.) Simm.
Yellow Marsh Saxifrage
Tufted with freely branching caudices bearing numerous glabrous and entire, petiolate, lanceolate-ligulate, obtuse leaves 1 to 4 cm long and about 2 cm wide. Flowering stems erect, bearing 3 to 4 linear, sessile leaves, 5 to 25 cm high, glabrous below, rufous pubescent below the hypanthium. Flowers mostly solitary, nodding at first; petals pale yellow, elliptic, clawless, about 1.0 cm long, about twice as long as the oblong elliptic, rufous-ciliated sepals; follicles 8 to 12 mm long, joined almost to the slightly spreading tips.

Wet and mossy tundra.

General distribution: In the Mackenzie and Richardson Mts. and in Y.T. and Alaska largely replaced by ssp. *Hirculus* which is a somewhat more robust plant, occasionally 2-flowered, and with petals 1.5 cm long. *S. Hirculus s. lat.*: Circumpolar, arctic-alpine.

Fig. 605 Map 700

Saxifraga Lyallii Engler

Scapes 20 to 40 cm tall from a black and fibrous ascending rhizome. Rather similar to *S. davurica* ssp. *grandipetala* from which, however, it is easily distinguished by its leaves in which the blade is flabellate, tapering abruptly into the petiole; also the inflorescence in *S. Lyallii* is more open, the flowering penducles longer and more slender and the flowers 5 to 6 mm in diameter, and twice as large as those of *S. davurica;* also, the filaments in the latter are filiform and thread-like but distinctly spindle-shaped in *S. Lyallii.*

Moist places by alpine brooks.

General distribution: Cordilleran, reaching the east slope of Mackenzie Mts.

Fig. 606 Map 701

Saxifraga nivalis L.

Alpine Saxifrage

Scapose, tufted with black, fibrous roots and spreading somewhat leathery basal leaves 2 to 4 cm long; the blade oval, obtuse, crenate-serrate, abruptly tapering into a broad petiole, dark green above, reddish-purple beneath, glabrous except for the pubescent margins. Scapes erect and slender, solitary or a few together, 5 to 20 cm high, purple and glandular-viscid. Inflorescence simple or branched, the small flowers in dense clusters, each subtended by a small bract. Petals white, 2 to 3 mm long, somewhat clawed, only slightly longer than the deltoid, purple sepals. Follicles 3 to 4 mm long, plump, with widely spreading tips.

Dry, rocky slopes and ledges, chiefly on Precambrian or acid rocks.

General distribution: Circumpolar, arctic.

Fig. 607 Map 702

Saxifraga oppositifolia L.

Purple Saxifrage

Densely or loosely matted with condensed, crowded, or trailing branches, with imbricated, scale-like, 4-ranked, leathery, marcescent and bristly-ciliated leaves. Flowering peduncles 1 to 3 cm long, dark purple, usually with a few alternate, linear leaves, terminating in a single flower. Petals lilac or purple, very rarely white, elliptic to oval 6 to 9 mm long, much longer than the ovate, bristly-ciliate sepals. Follicles 6 to 8 mm long, with slender, widely spreading tips.

Common in moist, calcareous gravels and on wet cliffs.

General distribution: Circumpolar, wide ranging arctic-alpine.

Fig. 608 Map 703

Saxifraga punctata L.

Acaulescent from a cord-like rhizome; leaves petioled, the petiole commonly twice as long as the diameter of the sub-orbicular, 7- to 14-toothed blade. Scapes slender, 15 to 25 cm long, terminating in a short raceme of white or sometimes slightly purplish flowers. Calyx lobes dark purple, reflexed, and the filaments spindle-shaped.

With us represented by two rather distinct geographical races: ssp. *Nelsoniana* (D. Don) Hult. in which the leaf-blade is 2 to 3 cm broad, dark green above and paler beneath, sparsely pubescent on both sides, with short, flat and more or less appressed hairs; the leaf-margin fringed by crinkled and septate hairs, and the branches of the inflorescence covered by gland-tipped hairs. Its range is distinctly Amphi-Beringian and arctic. With us it barely enters the District of Mackenzie west of the Mackenzie Delta. Ssp. *Porsildiana* Calder & Savile in which the leaf-blades are light green, glabrous, smaller and thinner, their margins sparingly if at all fringed by hairs. The range of *S. punctata* ssp. *Porsildiana* is an unusual one in that it extends from mountains of central and unglaciated Alaska and Yukon Territory, where it overlaps that of ssp. *Nelsoniana,* eastward over Great Bear Lake to eastern District of Keewatin, reaching the Arctic Coast only in Coronation Gulf, southward along the eastern continental divide to mountains of Jasper Park, Alberta.

Moist, open hillsides.

General distribution: *S. punctata s. lat.:* Amphi-Beringian.

Fig. 609 Map 704 (ssp. *Nelsoniana*), 705 (ssp. *Porsildiana*)

Saxifraga radiata Small

S. exilis Steph. *non* Pall.

Leaves mainly basal, long-petioled with a reniform, shallowly 5- to 7-lobed blade. Flowering stems mostly several, 15 to 20 cm tall, with egg-shaped, white bulbils at their base. Flowers 2 to 4, the terminal always largest; the petals white with pale, purple veins, those of the terminal flower 1.5 cm long. In habit somewhat similar to

S. cernua in which all but the terminal flowers are replaced by tiny reddish-brown bulbils.
Moist alpine slopes.
General distribution: Amphi-Beringian, with us known only from the Richardson and Mackenzie Mts.
Fig. 610 Map 706

Saxifraga reflexa Hook.
Similar to *S. nivalis* but at once distinguished by its more open inflorescence, pubescent leaves, reflexed sepals, and clavate filaments.
Dry, calcareous slopes and ledges.
General distribution: Endemic of arctic-alpine N. W. America.
Fig. 611 Map 707

Saxifraga rivularis L. *s. lat.*
Brook Saxifrage
Tufted dwarf species with reniform, palmately 3- to 5-lobed, essentially glabrous and slender-petioled basal leaves. Flowering stems 3 to 6 cm high, capillary, 1- to 3-leaved, soft pubescent, bearing 1 to 5 small flowers. Petals oblong-ovate, 3 to 5 mm long, white or pale pink, twice as long as the oblong-ovate sepals. Follicles 3 to 6 mm long, joined almost to the short, spreading tips. Basal bulbils occur, but are rarely seen, and in damp places may be replaced by slender, thread-like, rooting stolons.
The twice as tall Amphi-Beringian var. *flexuosa* (Sternb.) E. & I. with slender but strict and elongating penduncles barely enters our area in the Mackenzie and Richardson Mountains.
Wet, gravelly, and mossy places on wet cliffs and by brooks; a pronounced nitrophile frequently growing on wet slopes below bird cliffs and near human habitations.
General distribution: *S. rivularis s. lat.*, circumpolar wide ranging, arctic-alpine; var. *flexuosa,* Amphi-Beringian.
Fig. 612 Map 708

Saxifraga serpyllifolia Pursh
Low, delicate of loosely matted habit, with trailing sterile or fertile shoots rising from leafy, overwintering rosettes. Leaves linear-oblong, glabrous and somewhat fleshy, 5 to 10 mm long; flowers solitary, their peduncles 2 to 5 cm long, very slender, 2- to 3-bracted; petals bright yellow, about 5 mm long.
Moist gravelly scree-slopes.
General distribution: Amphi-Beringian arctic-alpine, reaching the east slopes of the Richardson and Mackenzie Mts., with two widely disjunct stations on Melville Island in the Canadian Arctic Archipelago.
Fig. 613 Map 709

***Saxifraga tenuis** (Wahlenb.) H. Sm.
Resembling *S. nivalis* but at once distinguished by its lower stature and more delicate habit, its fewer flowers, and more open inflorescence, and by the presence of coarse, rust-coloured hairs on the underside of the leaves and on the petioles.
Wet, gravelly places or in wet moss on cliffs, by brooks, or by lake shores; to be looked for in eastern District of Keewatin.
General distribution: Amphi-Atlantic, arctic.
Fig. 614 Map 710

Saxifraga tricuspidata Rottb.
Prickly Saxifrage
Densely matted, often forming large flat cushions; the leaves of the densely crowded branches often reddish tinged, rigid, leathery, and long marcescent, cuneate-spatulate with 3 prickly, tooth-like lobes at the apex. Flowering stems 6 to 15 cm tall, leafy. Inflorescence corymbose, nodding in youth; petals ellpitic-oblong, clawless, 6 to 7 mm long, creamy white and usually orange-dotted, much longer than the obtuse, somewhat ciliate sepals.
Dry, rocky and gravelly places.
General distribution: N. American wide ranging arctic-alpine.
Fig. 615 Map 711

577. *Boykinia Richardsonii*

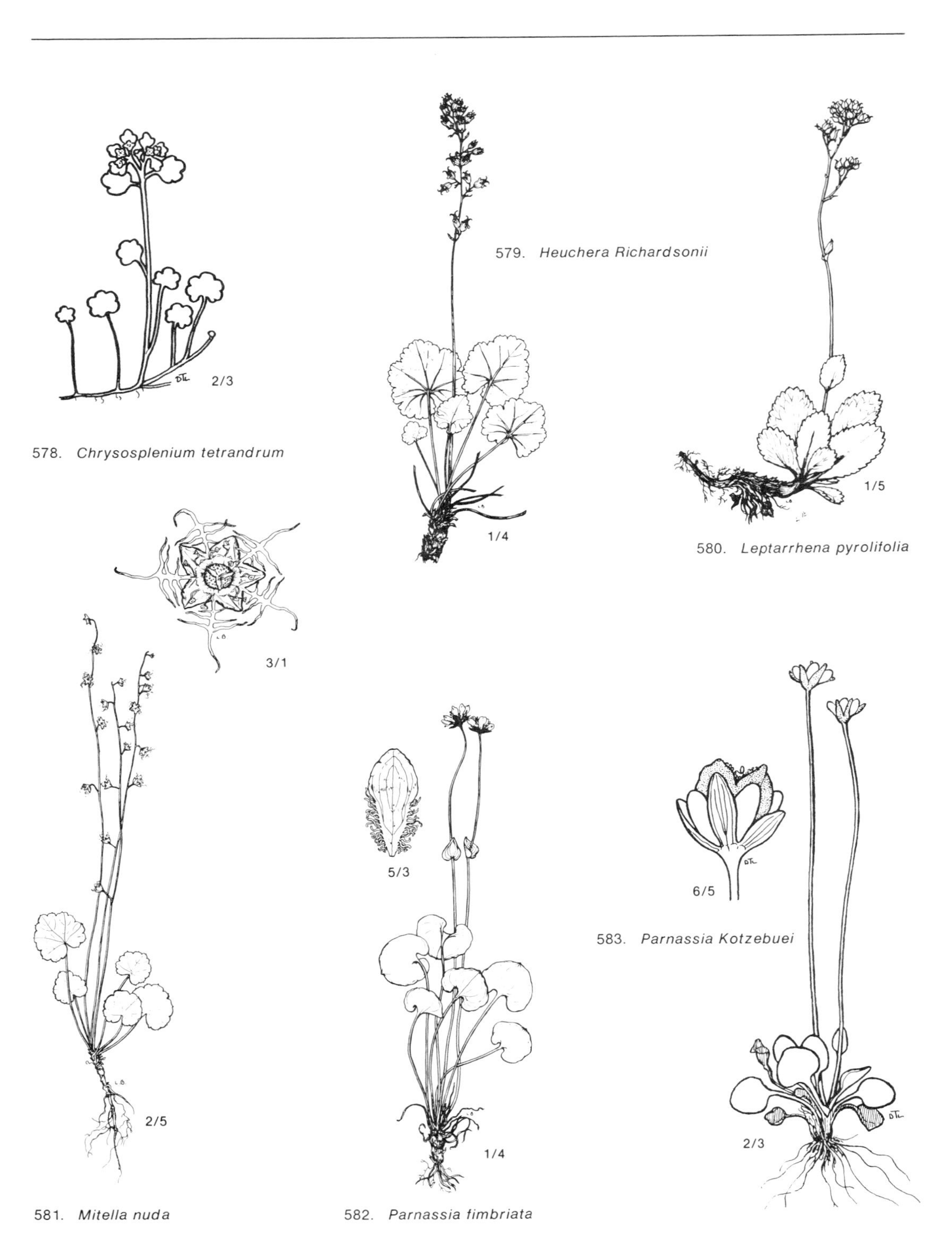

578. *Chrysosplenium tetrandrum*

579. *Heuchera Richardsonii*

580. *Leptarrhena pyrolifolia*

581. *Mitella nuda*

582. *Parnassia fimbriata*

583. *Parnassia Kotzebuei*

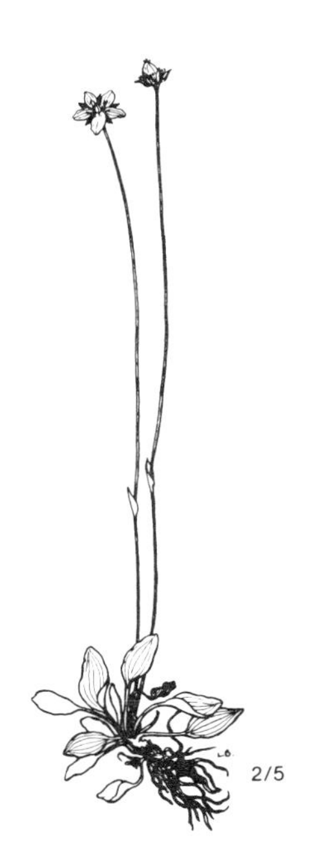

584. *Parnassia montanensis*

585. *Parnassia palustris* var. *neogaea*

586. *Ribes glandulosum*

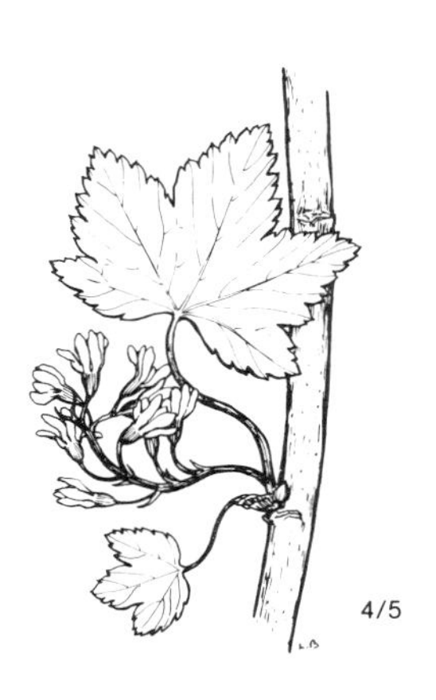

587. *Ribes hudsonianum*

588. *Ribes lacustre*

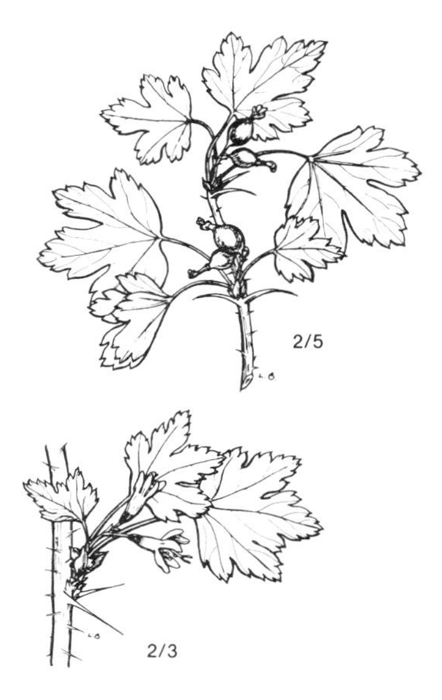

589. *Ribes oxyacanthoides*

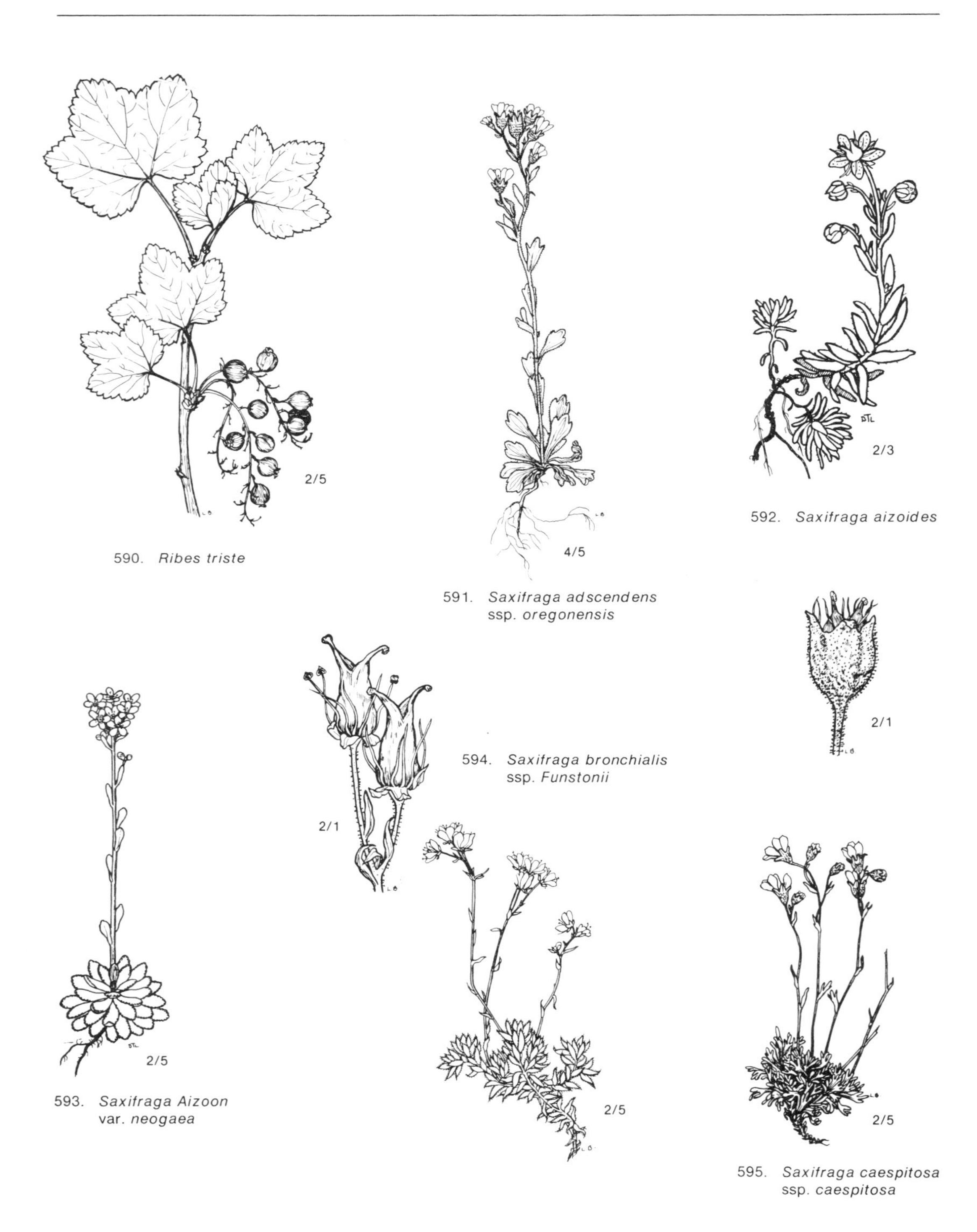

590. *Ribes triste*

591. *Saxifraga adscendens* ssp. *oregonensis*

592. *Saxifraga aizoides*

593. *Saxifraga Aizoon* var. *neogaea*

594. *Saxifraga bronchialis* ssp. *Funstonii*

595. *Saxifraga caespitosa* ssp. *caespitosa*

596. *Saxifraga caespitosa* ssp. *exaratoides*

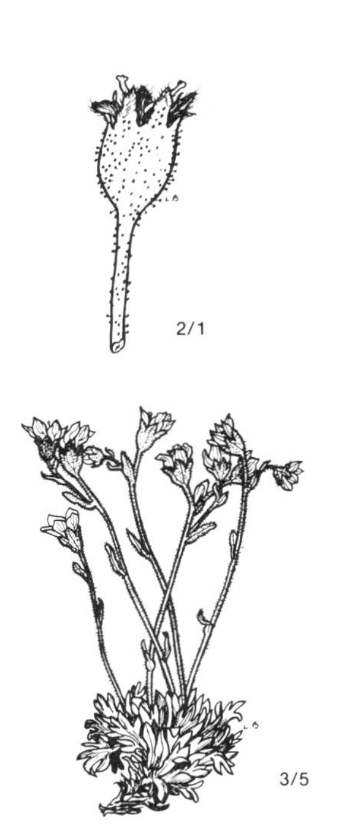

597. *Saxifraga caespitosa* ssp. *monticola*

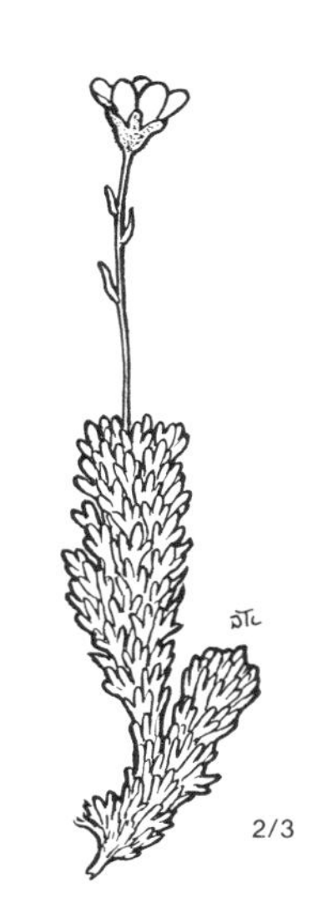

598. *Saxifraga caespitosa* ssp. *uniflora*

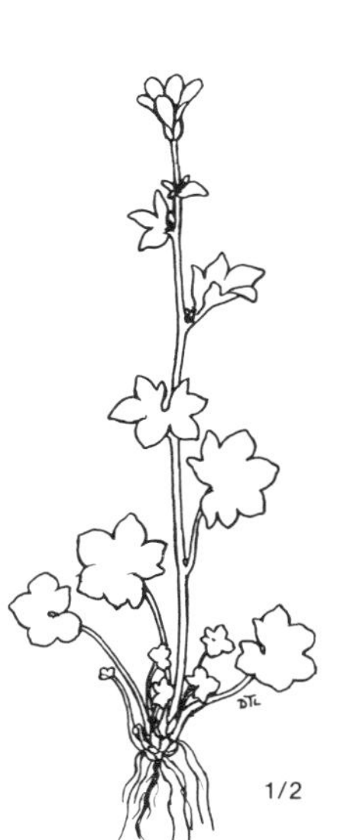

599. *Saxifraga cernua*

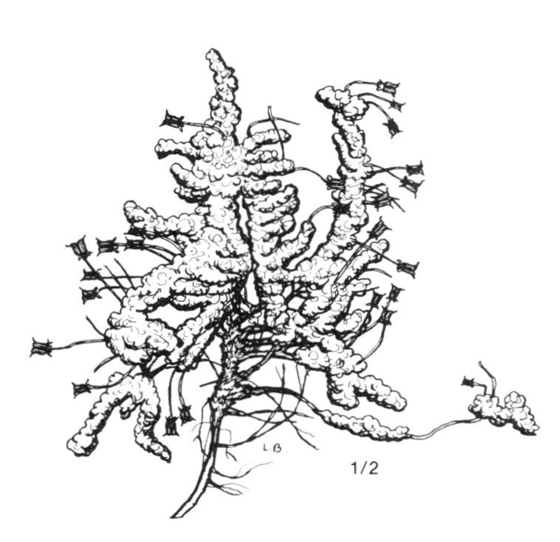

600. *Saxifraga Eschscholtzii*

601. *Saxifraga ferruginea*

602. *Saxifraga flagellaris* ssp. *flagellaris*

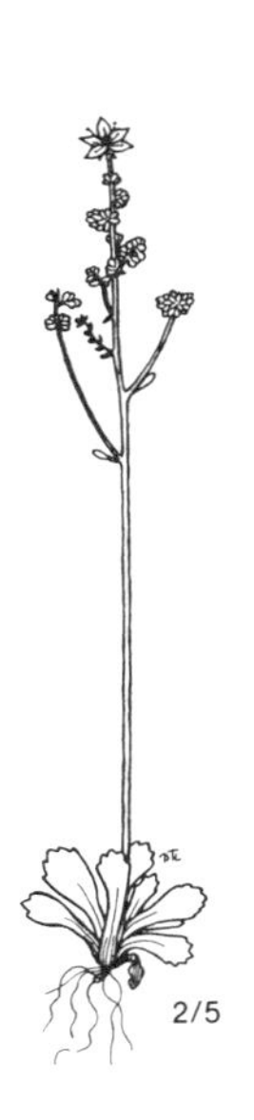

603. *Saxifraga foliolosa*

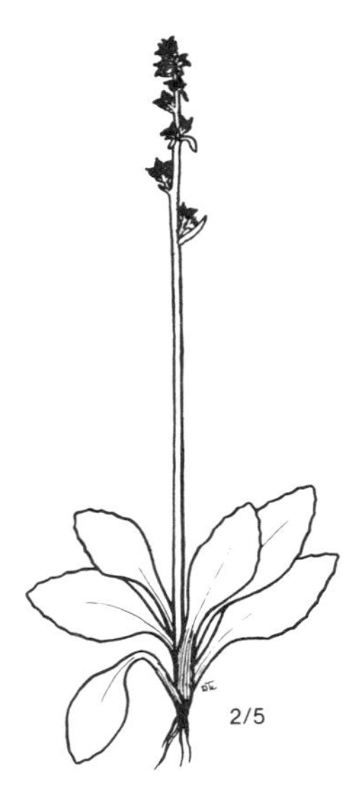

604. *Saxifraga hieracifolia*

605. *Saxifraga Hirculus* var. *propinqua*

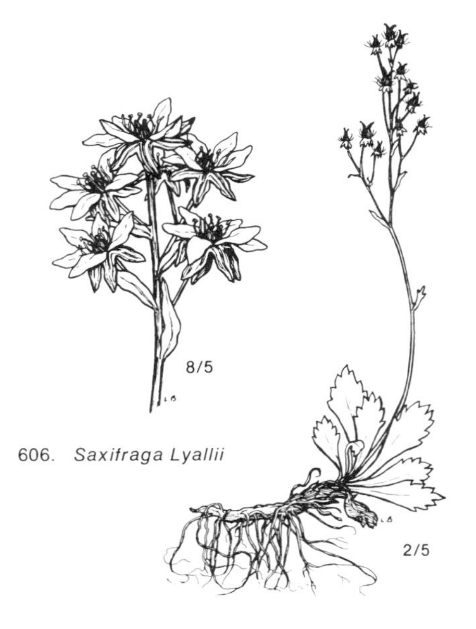

606. *Saxifraga Lyallii*

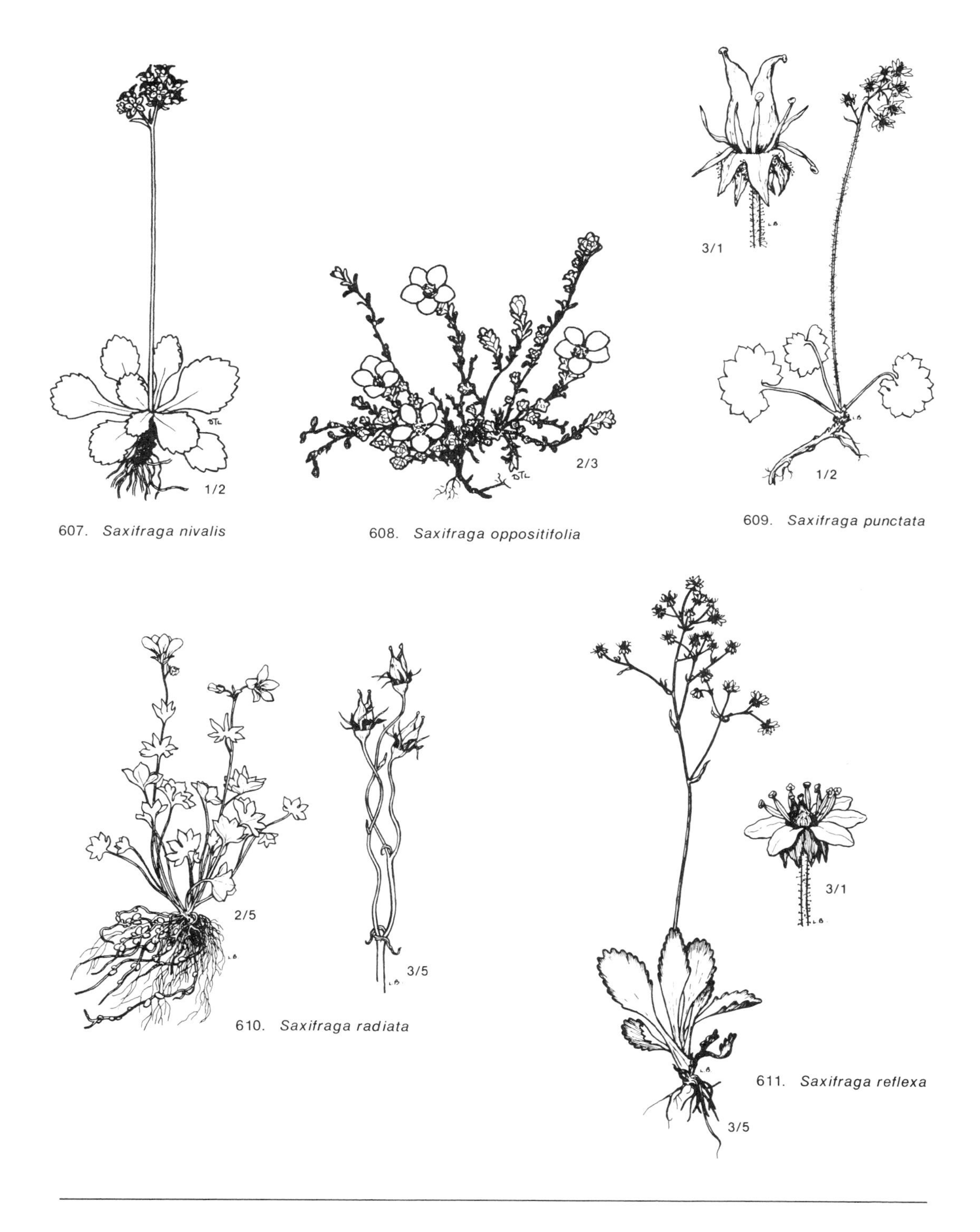

607. *Saxifraga nivalis*

608. *Saxifraga oppositifolia*

609. *Saxifraga punctata*

610. *Saxifraga radiata*

611. *Saxifraga reflexa*

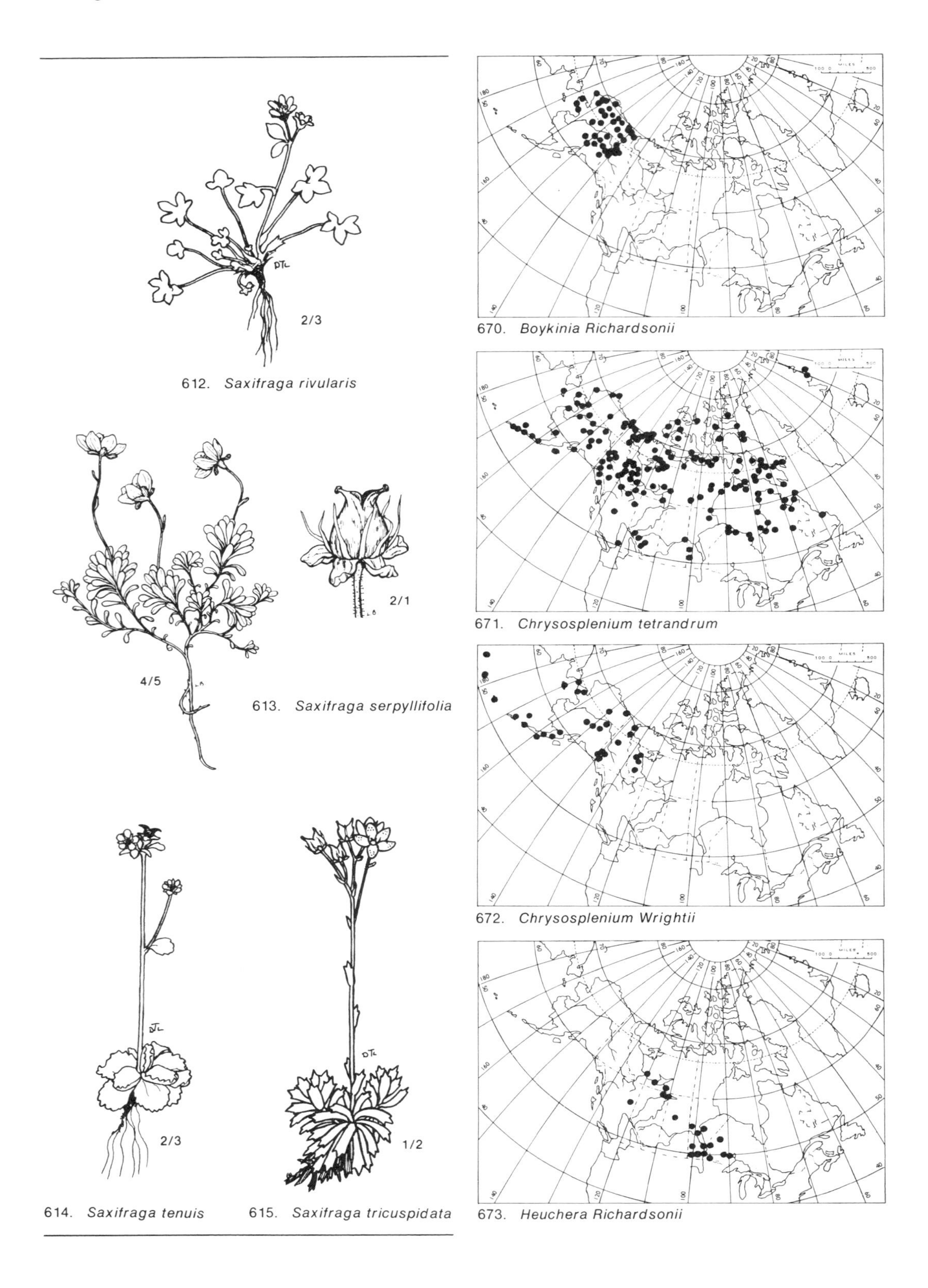

612. *Saxifraga rivularis*

613. *Saxifraga serpyllifolia*

614. *Saxifraga tenuis*

615. *Saxifraga tricuspidata*

670. *Boykinia Richardsonii*

671. *Chrysosplenium tetrandrum*

672. *Chrysosplenium Wrightii*

673. *Heuchera Richardsonii*

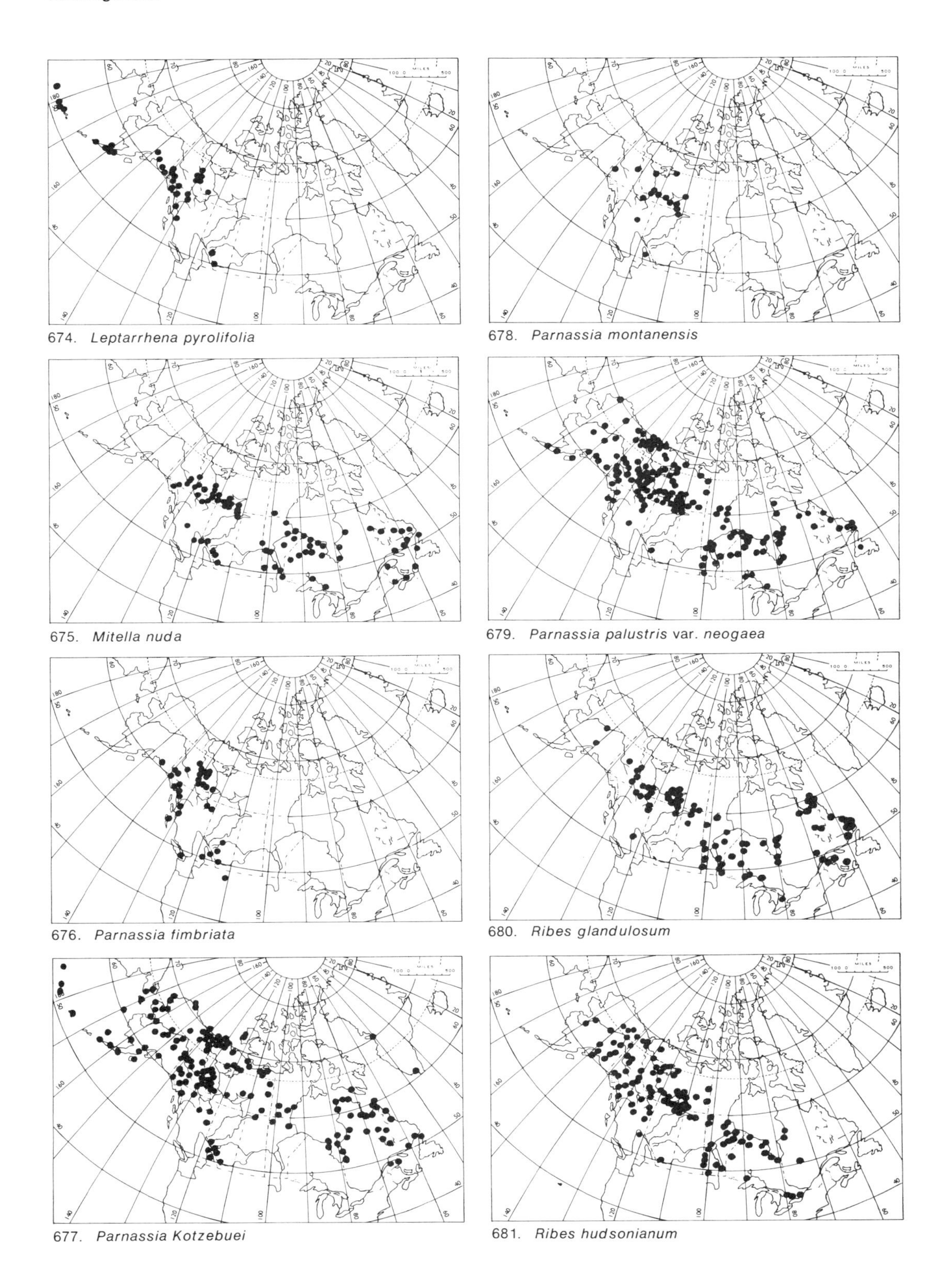

674. *Leptarrhena pyrolifolia*

675. *Mitella nuda*

676. *Parnassia fimbriata*

677. *Parnassia Kotzebuei*

678. *Parnassia montanensis*

679. *Parnassia palustris* var. *neogaea*

680. *Ribes glandulosum*

681. *Ribes hudsonianum*

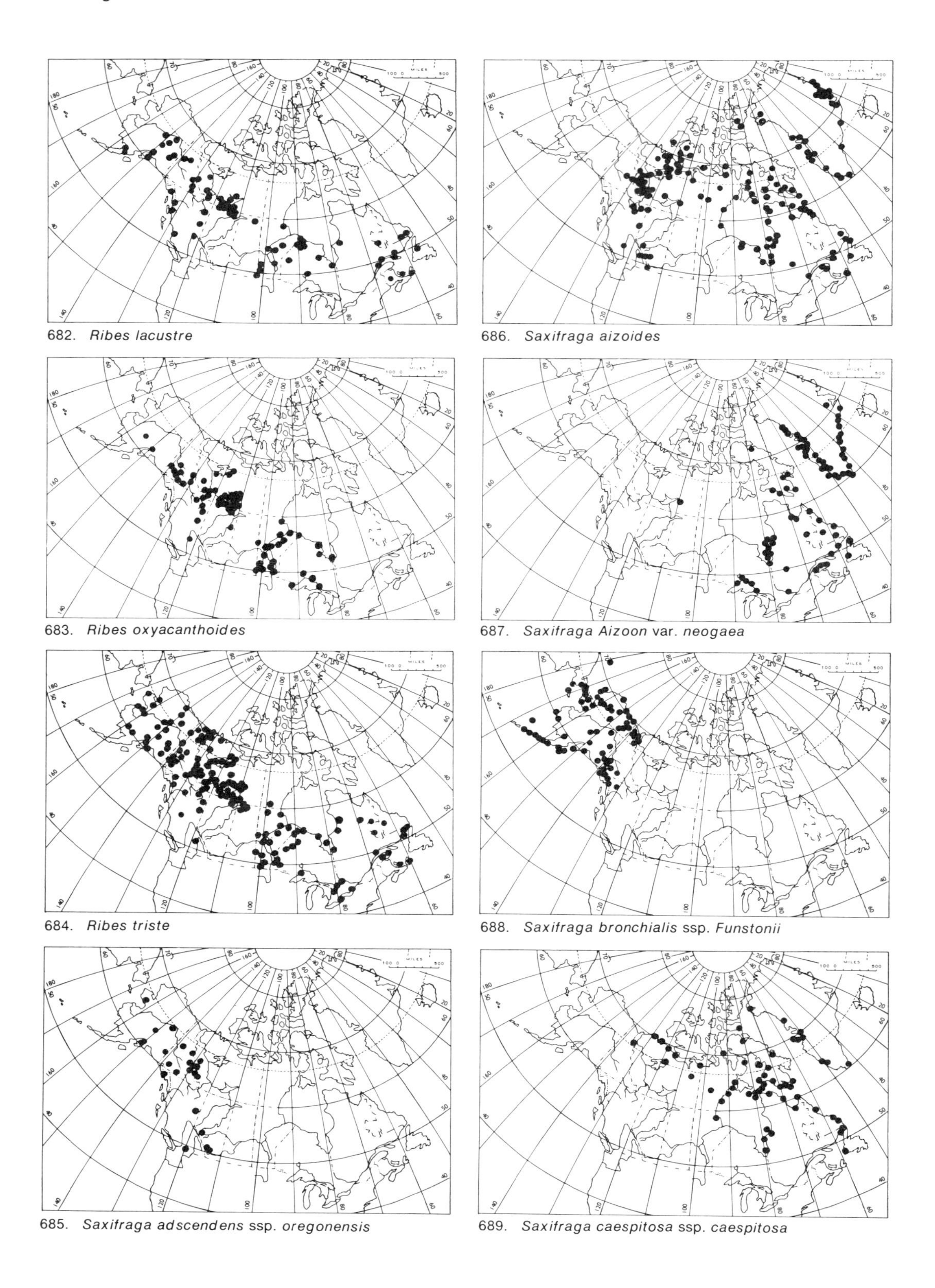

682. *Ribes lacustre*

683. *Ribes oxyacanthoides*

684. *Ribes triste*

685. *Saxifraga adscendens* ssp. *oregonensis*

686. *Saxifraga aizoides*

687. *Saxifraga Aizoon* var. *neogaea*

688. *Saxifraga bronchialis* ssp. *Funstonii*

689. *Saxifraga caespitosa* ssp. *caespitosa*

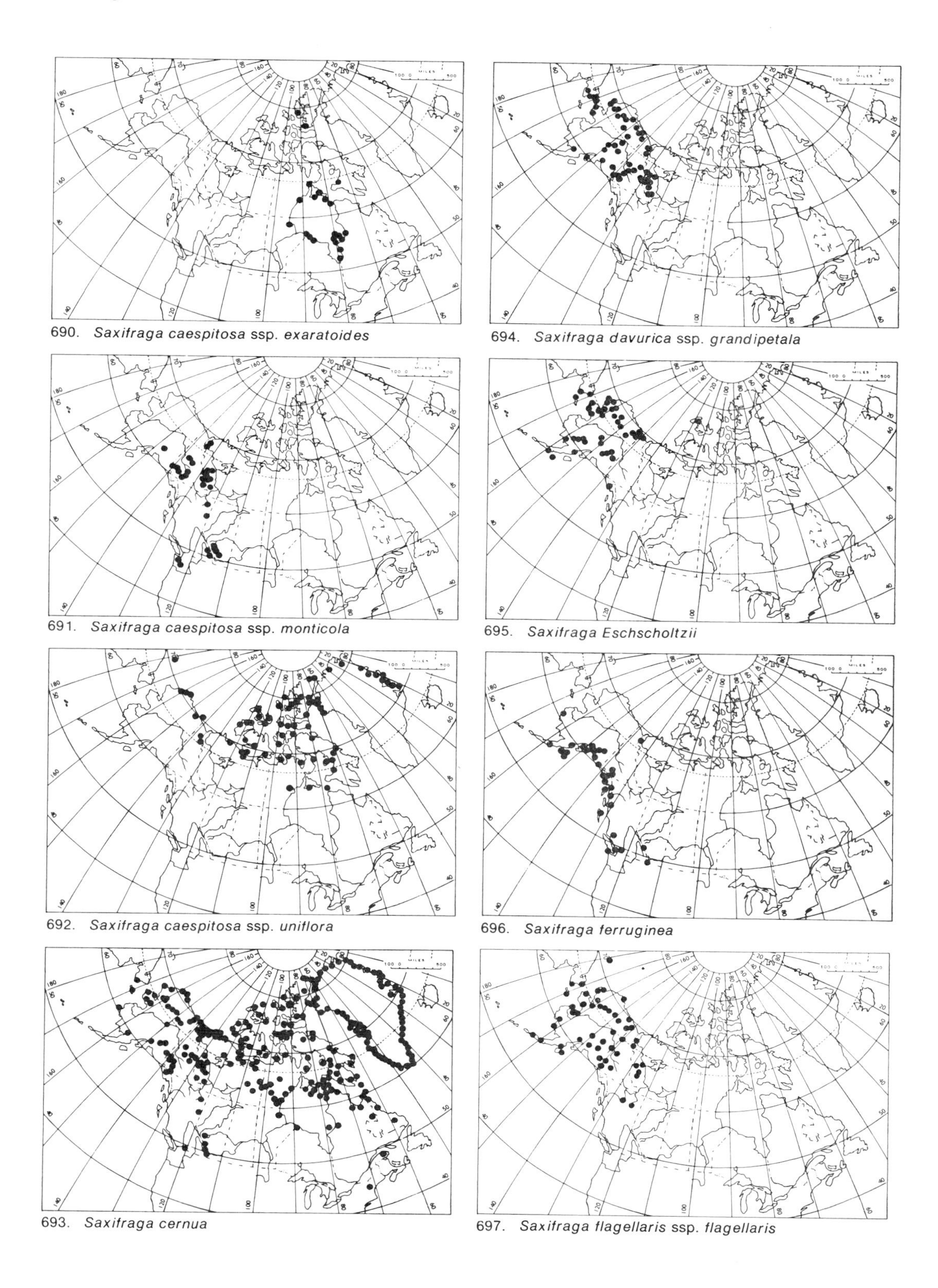

690. *Saxifraga caespitosa* ssp. *exaratoides*

691. *Saxifraga caespitosa* ssp. *monticola*

692. *Saxifraga caespitosa* ssp. *uniflora*

693. *Saxifraga cernua*

694. *Saxifraga davurica* ssp. *grandipetala*

695. *Saxifraga Eschscholtzii*

696. *Saxifraga ferruginea*

697. *Saxifraga flagellaris* ssp. *flagellaris*

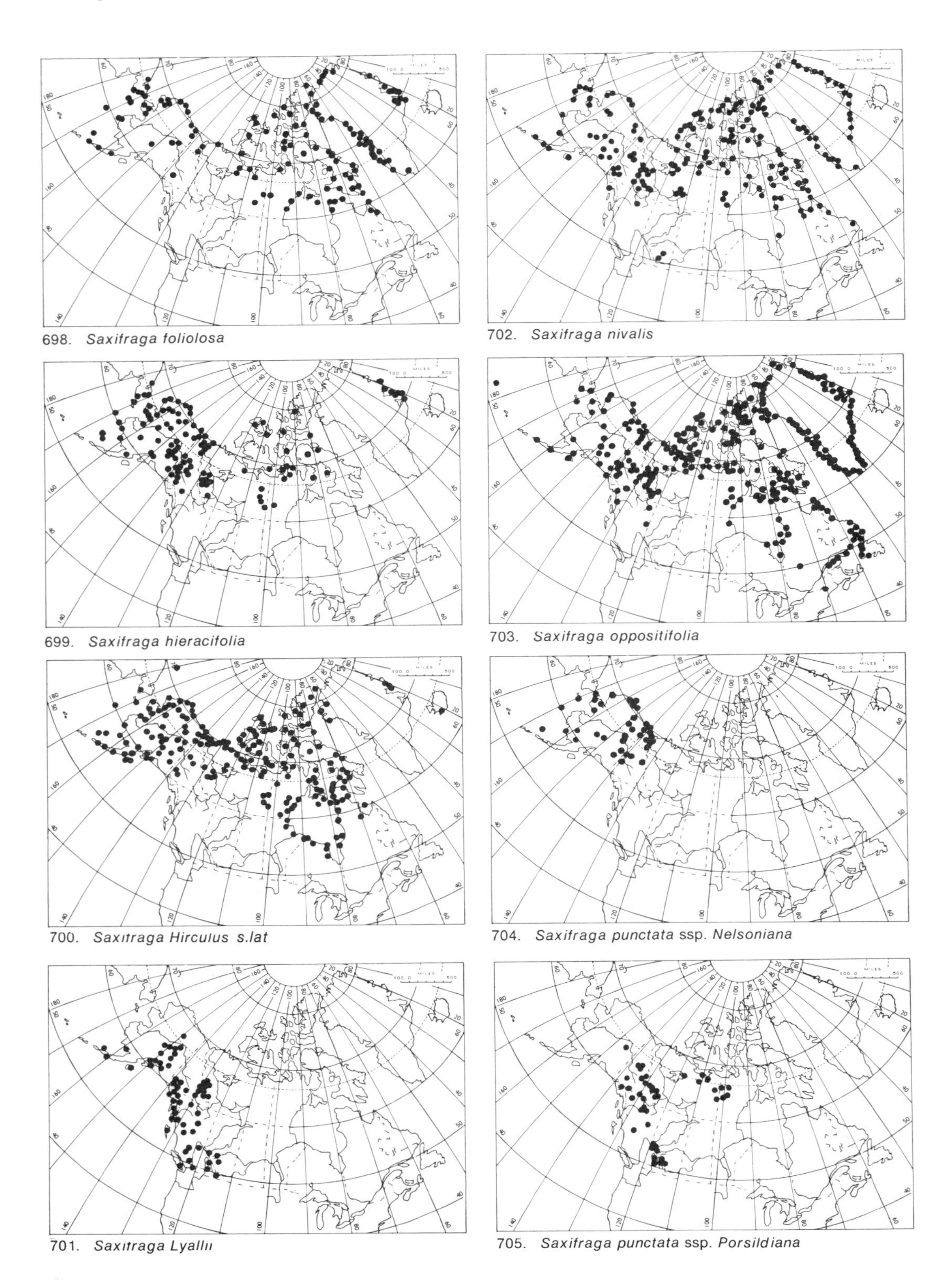

698. *Saxifraga foliolosa*

699. *Saxifraga hieracifolia*

700. *Saxifraga Hirculus* s.lat

701. *Saxifraga Lyallii*

702. *Saxifraga nivalis*

703. *Saxifraga oppositifolia*

704. *Saxifraga punctata* ssp. *Nelsoniana*

705. *Saxifraga punctata* ssp. *Porsildiana*

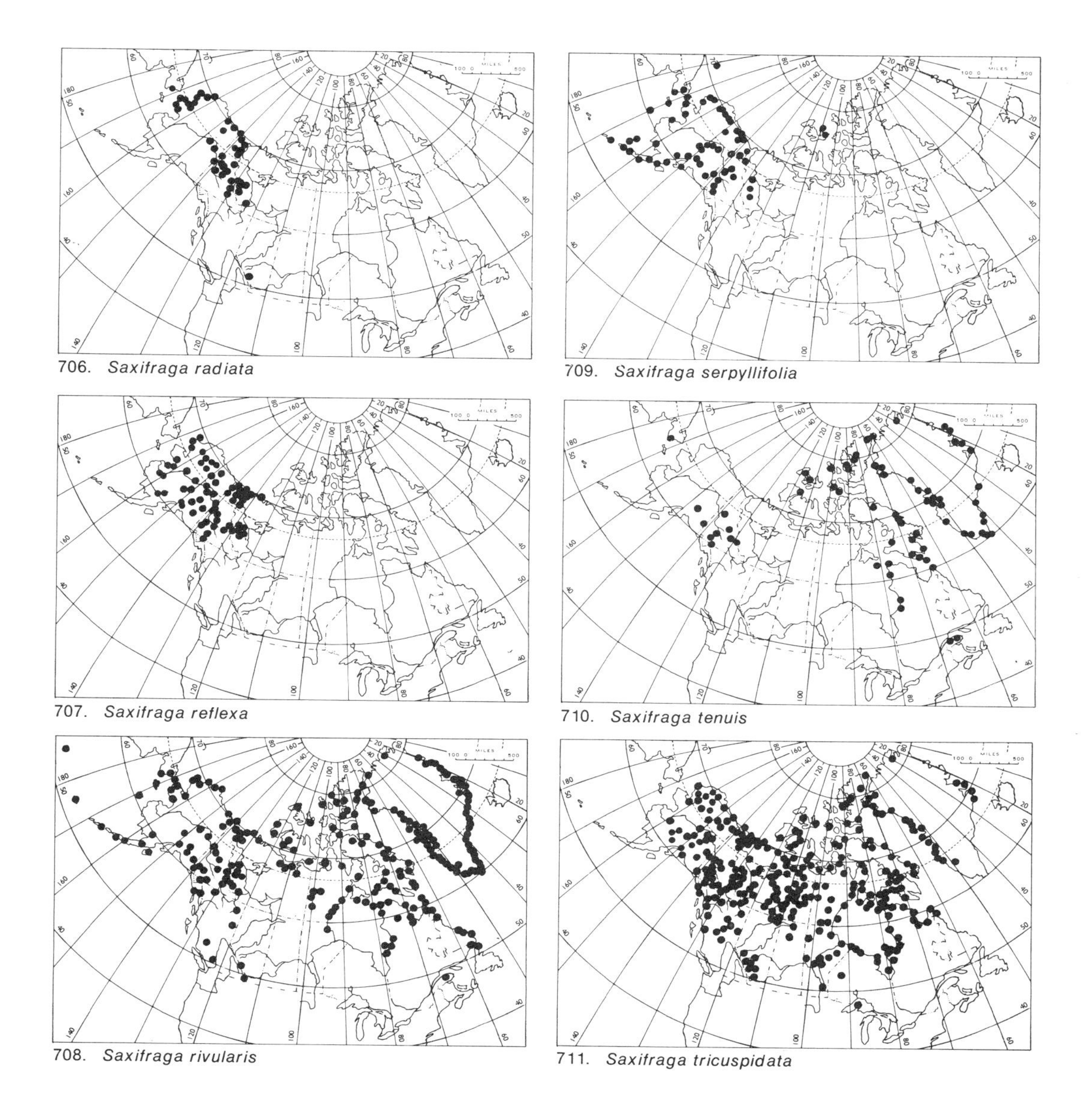

706. *Saxifraga radiata*

709. *Saxifraga serpyllifolia*

707. *Saxifraga reflexa*

710. *Saxifraga tenuis*

708. *Saxifraga rivularis*

711. *Saxifraga tricuspidata*

ROSACEAE Rose Family

Herbs, mostly with woody base, or shrubs, with alternate, simple or compound and usually stipulate leaves; flowers in racemes, cymes, or solitary, perfect and regular, the hypanthium saucer-shaped; sepals and petals 5; stamens numerous, rarely less than 10; carpels 1 to many, and the ovary superior or inferior, separate or united; fruit dry achenes, follicles, drupes, or berry-like pomes in *Amelanchier* and *Sorbus*.

a. Shrubs or dwarf trees
  b. Fruit a drupe or berry-like
    c. Fruit a drupe . . . . *Prunus*
    c. Fruit berry-like
      d. Leaves pinnate . . . . *Sorbus*
      d. Leaves simple . . . . *Amelanchier*
  b. Fruit dry
    e. Leaves compound
      f. Carpels few; dwarf shrub with biternately dissected leaves . . . . . . *Luetkea*
      f. Carpels numerous
        g. Fruit dry; bractlets alternating with the sepals . . . . . . *Potentilla*
        g. Fruit fleshy; bractlets none
          h. Carpels becoming drupelets (raspberry-like) . . . . . . *Rubus*
          h. Carpels becoming achenes enclosed in the fleshy receptacle (hip) . . . . *Rosa*
    e. Leaves simple
      i. Flowers normally solitary; petals 8; fruit a dry achene with a plumose persistent style . . . . *Dryas*
      i. Flowers in umbels, corymbs or racemes
        j. Carpels numerous, becoming fleshy aggregates of drupelets . . . . *Rubus Chamaemorus*
        j. Carpels 2-5; fruits a dry follicle . . . . *Spiraea*
a. Herbs with a woody base, but otherwise herbaceous; leaves compound
  k. Inflorescence a dense, spike-like head; petals lacking but sepals petaloid . . . . *Sanguisorba*
  k. Inflorescence not spike-like; petals and sepals present
    l. Calyx lobes 5, bractlets lacking; carpels numerous, becoming drupelets . . . . . *Rubus*
    l. Calyx lobes 5, alternating with the bractlets; carpels 1-several, becoming dry achenes
      m. Stamens 5; carpels 5-20; petals equalling the sepals
        n. Leaves 2-4-ternately divided into linear lobes . . . . . . *Chamaerhodos*
        n. Leaves 3-foliolate, the leaflets wedge-shaped, toothed at the apex; petals minute, yellow . . . . *Sibbaldia*
      m. Stamens 10 or more; carpels 10 to many; petals commonly larger than the sepals
        o. Basal leaves ternate
          p. Petals white; plants with slender, trailing stolons . . . . . . *Fragaria*
          p. Petals yellow; fruit dry (achenes) . . . . *Potentilla*
        o. Basal leaves with more than 3 leaflets
          q. Style articulate, deciduous and not plumose . . . . . . *Potentilla*
          q. Style persistent and plumose . . . . *Geum*

Amelanchier Medic. Serviceberry

Amelanchier alnifolia Nutt.
Arborescent, but with us never tree-like, and rarely over 6 feet tall, stoloniferous and forming small colonies or thickets. Bark grey or dark brown, smooth. Leaves simple, alternate, elliptic-ovate, rounded at the apex, the margin sharply serrate at least towards the apex, glabrous at least when mature. The white flowers in dense axillary racemes. Fruit berry-like, juicy, purplish-black when ripe, edible and sweet and

may be eaten raw or cooked. By the Indians formerly used to "sweeten" their pemmican.
Open woodland.
General distribution: N. America, from central Alaska to western Ont., south to Colorado and Oregon.
Fig. 616 Map 712

Chamaerhodos Bunge

Chamaerhodos erecta (L.) Bunge
ssp. **Nuttallii** (T. & G.) Hult.
Glandular-pubescent biennial or short-lived hapaxanthic perennial with a strong, sub-ligneous tap-root terminating in a small, compact rosette of bi- or tri-ternately dissected leaves. Stems 1 to several, simple or branching above, up to 30 cm tall. Flowers small with white petals, in small clusters at the end of the branches. Fruit a pear-shaped, smooth and dry achene, 1.5 mm long.
A lowland species of dry, calcareous slopes.
General distribution: N. America. Western foothills and plains. The sole N. American representative of a small, otherwise central Asiatic genus.
Fig. 617 Map 713

Dryas L. Mountain Avens

Low, mat-forming undershrubs, with freely rooting branches. Leaves leathery, tardily deciduous, petioled, simple, crenate or entire, dark green and mostly glabrous above, white-tomentose beneath. Stipules linear-lanceolate, rising from the leaf-axils, tomentose and usually with gland-tipped hairs. Flowers scapose, normally solitary. Hypanthium saucer-shaped, flattened or convex. Sepals and petals 8 to 10, the petals white or pale yellow; stamens and pistils numerous, the style terminal, persisting, much elongated and plumose in fruit.
All members of the genus, but especially *D. Drummondii*, are distinctly calciphilous. Therefore, the presence of *Dryas* on non-calcareous bed rock or on soils derived from such rocks, indicates the presence of local calcicolous mineral veins or of glacier or water transported calcareous soil particles.
Fig. 618

a. Petals yellow; filaments hairy, the flowers nodding in anthesis .... *D. Drummondii*
a. Petals white; filaments glabrous, the flowers erect in anthesis
  b. Median vein on the underside of the leaves prominent, with sessile or stalked glands, these naked or bearing tufts of hairs
    c. Upper surface of leaves dark green, dull, without punctiform glands ... *D. octopetala*
    c. Upper surface of leaves glandular-viscid along the leaf-folds
      d. Gland-tipped hairs on the underside of the leaves bearing tufts of white or brown hairs .... *D. punctata*
      d. Gland-tipped hairs on the underside of the leaves always naked
        e. Leaves elliptic, 1.5 to 2.5 times longer than broad, underside of the leaves densely tomentose between the nerves .... *D. Hookeriana*
        e. Leaves oblong, 2.5 to 3.5 times longer than broad, glabrate or thinly tomentose beneath .... *D. alaskensis*
  b. Median vein of the underside of the leaves often submerged under the tomentum and lacking stalked or hairy glands
    f. Leaf-base cordate, the margins revolute
      g. Leaf-margins entire or merely with a few teeth near the base
        h. Leaves lanceolate, 3 times as long as broad .... *D. integrifolia*
        h. Leaves short, ovate-elliptic, 1.5 times longer than broad ..... *D. Chamissonis*
      g. Leaf margins crenate to the tip .... *D. crenulata*
    f. Leaf-base truncate, the blade linear-oblong, flat, the margin entire or nearly so .... *D. sylvatica*

Dryas alaskensis Porsild
Resembling and undoubtedly most closely related to the Cordilleran *D. Hookeriana* from which, however, it is easily distinguished by its oblong, parallel-margined leaves, much longer

than those of the characteristically elliptic leaves of *D. Hookeriana*.

Open, well watered gravelly flood-plains and erosion fans, rarely in tundra or rocky places.

General distribution: Endemic of N.W. America. The main range of *D. alaskensis* is from Bering Strait eastward along the unglaciated central plateaux of Alaska and Y.T. A few isolated and disjunct stations are known in the Richardson and Mackenzie Mts.

Figs. 618, 619 Map 714

***Dryas Chamissonis** Juz.
Leaves ovate to ovate-elliptic, broadest below the middle, deeply crenately incised above the middle and less commonly along the entire length of the leaf; the upper leaf surface rugose, deep-green, and mostly glabrous and shiny, canescent-tomentose beneath. Flowering peduncles 5 to 10 cm long, white-tomentose, with a few gland-tipped black hairs below the flower. Petals often somewhat pubescent on the outside.

Gravelly places commonly by the seashore.

General distribution: Amphi-Beringian: Both shores of Bering Strait and disjunct on Banks Island by 1200 km from its nearest known station in Alaska. To be looked for along the Arctic Coast east and west of the Mackenzie R. Delta.

Map 715

Dryas crenulata Juz.
In habit and general appearance similar to *D. octopetala* or *D. punctata* but the veins on the underside of the leaves as well as the petioles and stipules are entirely lacking the stipitate glands or white-tufted hairs so distinctive of those species.

Alpine tundra.

General distribution: General range uncertain but probably of the Amphi-Beringian type. In N. America thus far known chiefly from mountains of central Alaska, Y.T., eastward to the Mackenzie R.

Fig. 618 Map 716

Dryas Drummondii Richards.
At once distinguished from all other North American members of the genus by its obovate, dark-green and rugose leaves and pale yellow, commonly nodding, but never fully expanded flowers, on up to 25 cm tall peduncles; these commonly with 1 to 4 vestigial bracts, the uppermost sometimes supporting adventitious but abortive flower-buds.

Non-arctic, sub-alpine, calcareous and gravelly flood plains and scree-slopes.

General distribution: N. America; Alaska south through the Cordillera to Oregon, east to Great Bear L. and the east end of Great Slave L., with widely disjunct stations on the north shore of L. Superior and in the Gulf of St. Lawrence, the first more than 1600 km from its nearest known western station.

Figs. 618, 620 Map 717

Dryas Hookeriana Juz.
For differences between this and its nearest relative, *D. alaskensis* see key and comments under the latter.

The preferred habitat of *D. Hookeriana* is well watered snow-bed slopes well above timberline; in the Canadian Rockies it ascends at least to 3000 m.

General distribution: The main range of *D. Hookeriana* is in the central Rocky Mts. between lat. 40° and 50° N. with disjunct populations in the Mackenzie Mts., S.W.Y.T. and Kenai Pen., Alaska, the nearest separated from the main range of the species by a gap of 800 km.

Map 718

Dryas integrifolia M. Vahl
Leaves lanceolate-oblong, about 3 times longer than broad, distinctly broadest below the middle, with a cordate or truncate base; margins entire or with a few teeth in the lower half, usually more or less revolute, dark green, shiny and glabrous above (or canescent-tomentose in var. *canescens* Simm.), the nerves not deeply impressed; white-tomentose beneath. The nerves lacking glands or branching hairs.

An ubiquitous pioneer species in rocky or gravelly places such as river flats; less common in tundra or heath where it cannot long survive competition from more aggressive tundra plants.

General distribution: N. America. *D. integrifolia s. str.* arctic-alpine and by far the most common and wide-ranging member of the genus. The var. *canescens* has a similar range and habit, but is mainly confined to high-arctic regions.

Figs. 618, 621 Map 719

Dryas octopetala L.
Leaves ovate-oblong, 1.5 to 2.5 times longer than broad, crenate-dentate, with more or less revolute margins and cordate or truncate base, strongly rugose, dull, dark green, glabrous or sparingly hirsute above; the underside white tomentose with prominent veins bearing gland-

tipped brown hairs and much branched white hairs.

Alpine tundra.

General distribution: Amphi-Beringian but apparently rather rare and local in N. America where it is restricted to the Bering Sea Coast and to mountains of central and northern Alaska and Y.T., barely reaching the District of Mackenzie along the east slopes of Richardson and Mackenzie Mts.

Fig. 618 Map 720

Dryas punctata Juz.

Leaves linear-oblong 2 to 3.5 times longer than broad, broadest in the middle with a blunt tip and narrowly cordate or truncate base, coarsely and deeply incised-crenate, the margins commonly strongly revolute; the upper surface glabrous and usually glandular-punctate, in life somewhat viscid, the median and lateral nerves deeply impressed; the underside densely canescent-tomentose between the prominent nerves that, with the petioles, are covered with short-stalked yellowish-brown glands mixed with much branched tufted brown- or white-plumose hairs similar to those of *D. octopetala.*

Rocky slopes and summits above timberline.

General distribution: Circumpolar, arctic-alpine, with a large gap between E. Greenland and the western islands of the Canadian arctic archipelago.

Fig. 618 Map 721

Dryas sylvatica (Hult.) Porsild

In general habit similar to *D. integrifolia* from which it is, as a rule, easily distinguished by its linear, entire margined, flat or slightly revolute-margined leaves, 4 to 6 times longer than broad, with cuneate or rounded rather than cordate or truncate bases.

A lowland species usually growing in open spruce forest and on gravelly river terraces.

General distribution: Endemic of wooded parts of central Alaska through Y.T. and western District of Mackenzie.

Fig. 618 Map 722

Fragaria L. Strawberry

Perennial herbs with scaly rhizomes, producing long, slender, nodally rooting stolons. Leaves basal, long-petioled, stipulate, 3-foliolate, the leaflets coarsely serrate and silky beneath. Flowers showy, in a bracteate, few-flowered cyme, on a slender scape; petals white. Receptacle much enlarged, red and juicy in fruit, the tiny achenes shallowly embedded in its surface.

a. Leaves fresh green, thin, not glaucous; the terminal tooth of each leaflet distinctly larger and usually projecting beyond those next to it . . . . *F. vesca* ssp. *americana*
a. Leaves glaucous, of thicker texture; the terminal tooth of each leaflet distinctly smaller and shorter than those next to it . . . . *F. virginiana* ssp. *glauca*

Fragaria vesca L.
ssp. **americana** (Porter) Staudt

Of more southern range than *F. virginiana* ssp. *glauca*, and with us reported only a few times along the upper Mackenzie drainage north to Fort Simpson.

General distribution: Boreal N. America, Gaspé, Que. to N. B.C. and S.W. District of Mackenzie.

Map 723

Fragaria virginiana Duchesne
ssp. **glauca** (Wats.) Staudt

This is the common wild strawberry of the interior Northwest, where it is confined mainly to the major drainage systems, but in our area not extending east much beyond the Mackenzie Valley, and north nearly to its delta.

General distribution: Endemic of Western N. America

Fig. 622 Map 724

Geum L. Avens

Perennial herbs with stout, sub-ligneous rhizomes and pinnate basal leaves; calyx with persistent bractlets in the sinuses between the lobes; 5 petals and numerous stamens and carpels, the latter on a dry conical receptacle, and with a long, persistent, plumose or sharply curved or hooked jointed style, of which the upper part is often deciduous.

a. Leaflets more or less alike
  b. Flowers pale purple (prairie grassland species) . . . . *G. triflorum*

- b. Flowers yellow (alpine or Arctic species)
  - c. Leaves and stems conspicuously villous; flowers 4-5 cm in diameter ..... *G. glaciale*
  - c. Leaves and stems glabrous or glabrate; flowers 2-3 cm in diameter ...... *G. Rossii*
- a. Terminal leaflet the largest and of different shape than the rest
  - d. Upper cauline leaves divided into 3 distinct leaflets .... *G. aleppicum* var. *strictum*
  - d. Upper cauline leaves merely 3-lobed .... *G. macrophyllum* var. *perincisum*

Geum aleppicum Jacq.
var. **strictum** (Ait.) Fern.
Up to 1 m tall, sparsely hirsute; the basal leaves 2 to 3 dm long, interruptedly pinnate, the terminal and lateral leaflets cuneate, 3-lobed and coarsely toothed, those of the flowering stem short-petioled or sub-sessile. Flowers 1 to 2 cm in diameter, in an open 3-flowered cyme; petals pale yellow, the sepals reflexed; fruiting head globose-ovoid about 1.5 cm long, receptacle long-hirsute, the achenes with a long, hooked and downward bent hairy beak.

Damp thickets and grassy clearings.

General distribution: Boreal N. American forest species, with us confined to the upper Mackenzie Basin.

Fig. 623 Map 725

Geum glaciale Adams
Densely tufted from a stout, sub-ligneous base. Leaves mainly basal, short-petioled, pinnate, with 5 to 7 pairs of entire or sub-entire leaflets; these glabrous above, their underside, as well as the petiole and flowering stem, densely villous from soft yellowish-white hairs. Flowers solitary, 3 to 4 cm in diameter, the bractlets linear, shorter than the calyx lobes; petals 5 to 8, yellow, oval; achenes hairy with 2 to 3 cm long plumose styles.

Stony alpine slopes or dry heath where its showy flowers appear soon after the snow melts.

General distribution: Amphi-Beringian, arctic-alpine barely entering the District of Mackenzie in the Richardson and Mackenzie Mts.

Fig. 624 Map 726

Geum macrophyllum Willd.
var. **perincisum** (Rydb.) Raup
Somewhat similar to *G. aleppicum* but usually lower, and the basal leaves distinctly lyrate, the terminal leaflet being cordate and merely crenate-dentate, much larger than the progressively smaller lateral leaflets. Inflorescence of smaller but more numerous flowers, their petals pale yellow, and the hooked beak of the achenes glabrous in the upper half; receptacle glabrous or short-hispid.

Damp thickets. Boreal forest species, with us ranging north in the Mackenzie Basin as far as Good Hope.

General distribution: Western N. America, east to Hudson Bay and the Great Lakes.

Fig. 625 Map 727

Geum Rossii (R.Br.) Ser.
Essentially glabrous herb from a very stout ascending, sub-ligneous base, densely covered by the marcescent remains of old leaf-bases, and terminating in a rosette of 6 to 10 cm long, erect, interruptedly pinnate leaves bearing about 7 pairs of variously toothed, cuneate leaflets. Flowering stems, one- or rarely 2-flowered, erect with a few reduced leaves, at first low but elongating in fruit, up to 10 to 28 cm high. Flowers large and showy, 2 to 3 cm in diameter; petals pale yellow, orbicular, longer than the sepals; achenes hirsute; styles not much elongated in fruit, and not plumose.

Calcareous clay in moist, alpine tundra; often growing on solifluction lobes.

General distribution: Amphi-Beringian, bilateral, from E. Asia across unglaciated Alaska and Y.T., with widely disjunct stations in Melville and Ellesmere Islands. Arctic-alpine.

Fig. 626 Map 728

Geum triflorum Pursh
Densely tufted, 2 to 4 dm tall from a stout, horizontal rhizome of which the upper half is exposed and covered by shaggy brown hairs. Leaves soft-hirsute, mainly basal, with 6 to 8 pairs of crowded leaflets, these cuneate-obovate, deeply cleft into linear divisions; the cauline leaves few and much reduced. Inflorescence 1- to several-flowered, the flowers nodding in anthesis; hypanthium and the long linear bractlets purplish; petals pale yellow or often pinkish. Styles not jointed, elongating up to 4 cm long and plumose except for the early deciduous tips.

Dry prairie and grassland species which barely enters the District of Mackenzie.

General distribution: N. America from B.C. east to Ont. and southward.

Fig. 627 Map 729

Luetkea Bong.

Luetkea pectinata (Pursh) Ktze.
Evergreen dwarf shrub with slender, prostrate branches terminating in clusters or fascicles of glabrous, bi-ternate leaves around the base of the 8 to 15 cm tall and erect flowering stems. Flowers small and white, in short racemes that elongate in fruit. Carpels 5, ripening into follicles, each containing a few dry, spindle-shaped, 3 mm long seeds.

Alpine herbmats or snowbed-slopes.

General distribution: Alaska and Y.T. southwards mainly through the western ranges of the Rocky Mts., barely entering the District of Mackenzie along the Mackenzie—Y.T. Divide.

Fig. 628 Map 730

Potentilla L. Cinquefoil

Tufted or caespitose and mainly perennial herbs (or shrub-like in *P. fruticosa*) with freely branched and mostly sub-ligneous caudices that are commonly covered by the marcescent remains of former year's leaves. Radical leaves compound; flowering stems erect, or ascending, simple or often branched above. Flowers 5-merous, solitary or in cymes; in ours the petals are mostly yellow, but one white and one purplish-red. Fruits of small, dry achenes inserted on the hemispherical dry, often hairy receptacle.

a. Basal leaves digitate, approximately as long as broad
- b. Basal leaves mainly 3-parted, although some occasionally with a second pair of much reduced leaflets
  - c. Flowers always white . . . . *P. tridentata*
  - c. Flowers yellow
    - d. Biennial or short-lived perennial . . . . *P. norvegica*
    - d. Perennial and caespitose
      - e. Lower surface of leaves densely white or yellowish hirsute-tomentose
        - f. Flowers in cymes; plant loosely tufted
          - g. Lower surface of leaves merely white-tomentose . . . . *P. nivea*
          - g. Lower surface of leaves shaggy from long, silky guard-hairs above the white tomentum . . . . *P. rubricaulis*
        - f. Flowers solitary, or rarely 2 or 3; plant densely tufted or matted and very hirsute or shaggy from long, silky guard-hairs
          - h. Guard-hairs yellow, imparting a yellowish sheen to entire plant (Eastern species) . . . . *P. Vahliana*
          - h. Guard-hairs dull white (Western species) . . . . *P. uniflora*
      - e. Leaves merely hirsute beneath, lacking a true tomentum; petals pale yellow, often drying white or even pinkish
        - i. Compact dwarf species with tiny, solitary flowers barely raised above the flat cushion; petals often drying pink . . . . *P. elegans*
        - i. Densely tufted with erect-ascending flowering stems well raised above the sub-ligneous base
          - j. Leaflets linear, cleft to the base . . . . *P. biflora*
          - j. Leaflets lanceolate, their margins merely dentate . . . . *P. hyparctica*
- b. Basal leaves 5 (7)-parted
  - k. Petals purple; marsh plants . . . . *P. palustris*
  - k. Petals yellow; plants of non-marshy habitats
    - l. Plant shrubby with reddish-brown shreddy bark . . . . *P. fruticosa*
    - l. Plant herbaceous
      - m. Basal leaves mainly 5-parted
        - n. Upper surface of leaves dark green, the lower surface sparsely grey-tomentose and lacking long silky guard-hairs . . . . *P. furcata*
        - n. Leaves and petioles dense shaggy from white-villous pubescence, the upper surface glabrate in age . . . . *P. rubricaulis*

m. Basal leaves mainly 7-parted, fresh green, silky pubescent when young, almost glabrous in age . . . . *P. diversifolia*

a. Basal leaves, and often also the lower stem-leaves pinnately divided; petals yellow
  o. Flowers solitary; plants with long and slender freely rooting stolons
    p. Leaflets usually 10 to 12 pairs, commonly silvery pubescent; stolons always pubescent . . . . *P. anserina*
    p. Leaflets 5 to 7 not silvery pubescent, stolons glabrous . . . . *P. Egedii*
  o. Flowers several, in cymes
    q. Plant not glandular, individual leaflets deeply incised
      r. Leaflets 3 to 5 pairs, their lower surface soft-pubescent
        s. Leaflets merely toothed . . . . *P. pensylvanica*
        s. Leaflets pinnately divided . . . . *P. multifida*
      r. Leaflets 2, or rarely 3 pairs, their lower surface merely silky . . . . *P. pulchella*
    q. Plant glandular throughout, leaflets oval, merely dentate; petals pale yellow, drying white . . . . *P. arguta*

Potentilla anserina L. *s. lat.*
Silverweed
Herbs of sea-shores, damp river banks or lake shores, with tufted, interruptedly pinnate basal leaves 10 to 20 cm long, and pubescent, slender, prostrate and freely rooting stolons bearing solitary, yellow flowers from the axils of leaves issuing from the rooting nodes.

In 'typical' *P. anserina* the leaves are densely silvery white-silky beneath, and sometimes also on the upper surface. It is a common and well established weed near human settlements along the upper Mackenzie and some of its tributaries. It is equally common in similar places in Yukon and Alaska where it has been separated as *P. yukonensis* Hult. or as *P. Egedii* ssp. *yukonensis* (Hult.) Hult., said to differ from the widely introduced and weedy *P. anserina* by having entire or non-divided bractlets below the calyx, and the achenes 'not, or indistinctly grooved'. However, an examination of the large material now available of 'Silverweed' from the Alaska, Yukon and the Mackenzie River basins reveals that toothed or divided bractlets are as common as are entire ones. The achenes are in no way distinguishable from those of *P. anserina* from Eastern Canada and the United States where this weedy species is as variable according to habitat as it is in Europe.

Gravelly or sandy lake-shores or river banks but commonly weedy also near human settlements along the Mackenzie R. and its tributaries.

General distribution: Circumpolar, non-arctic.

Fig. 629 Map 731

Potentilla arguta Pursh *s. lat.*
Coarse, weedy species with stout, erect stems up to 1 m tall, generally glandular or viscid; basal leaves tufted, viscid, short-pinnate; the leaflets obovate, serrate, the terminal largest, the lower progressively smaller. Flowers crowded in one large, terminal, or in several smaller, lateral cymes; petals pale yellow, drying white.

Prairies or grassy river-banks or road-sides, barely entering the District of Mackenzie along Slave River and the upper Mackenzie River.

General distribution: N. America.

Fig. 630 Map 732

Potentilla biflora Willd.
Densely caespitose dwarf shrub from a branching caudex densely covered by marcescent, shiny, oblong, 3 to 4 cm long and prominently veined stipules; leaf-blades fan-shaped, deeply divided, divisions linear, with revolute margins. Flowering peduncles scapose, 8 to 12 cm long; flowers commonly 2 together, about 1.5 cm in diameter; bracts and sepals triangular, of equal length; petals longer than sepals, pale yellow, drying white. Receptacle densely white-woolly.

Alpine gravelly slopes.

General distribution: Amphi-Beringian; in N. America across alpine parts of Alaska—Y.T. to the east slope of Mackenzie Mts., south to the northeast corner of B.C.

Fig. 631 Map 733

Potentilla diversifolia Lehm.
ssp. **glaucophylla** Lehm.
Loosely tufted with a stout, subligneous, branching base thickly covered by the marcescent remains of stipules and petioles. Basal leaves long-petioled, the up to 5 cm broad blade digitately 5-parted, glaucous, and almost glabrous, at least in age. Flowering stems erect-ascending, 15 to 45 cm tall, often repeatedly branched above, few- to many-flowered. Petals pale yellow.

Moist alpine or gravelly places.

General distribution: North Cordilleran species reaching central Y.T., S.E. Alaska and the southern Mackenzie Mts., N.W.T.

Fig. 632 Map 734

Potentilla Egedii Wormskj.
Tufted, with a short, thick and branched caudex ending in a rosette of interruptedly pinnate leaves from which issue long, slender freely rooting, flowering stolons. Generally similar to *P. anserina* but differs by its smaller size, glabrous and glabrate stolons, and fewer, and not silvery-pubescent leaflets.

Halophyte; rarely found beyond the tidal zone.

General distribution: circumpolar: sea coasts of Greenland, Labrador, Hudson Bay, Arctic Coast and Alaska.

Fig. 633 Map 735

Potentilla elegans Cham. & Schl.
Loosely caespitose dwarf species; stems erect-ascending, the single terminal flower barely over-topping the foliage, 6 to 8 mm in diameter; the petals pale yellow or almost white. Leaves all basal, the blade about 1 cm in diameter, its 3 cuneate segments glabrous or soft-ciliate and deeply incised into linear segments.

High-alpine species of rocky or gravelly places.

General distribution: Amphi-Beringian: With us known from a few alpine stations in the Mackenzie Mts.

Fig. 634 Map 736

Potentilla fruticosa L.
Much branched dwarf shrubs up to 1.5 m tall but commonly lower, with shreddy, reddish-brown outer bark and sheathing membranaceous, pinkish stipules. Leaves sessile or short-petioled, 3- to 5-foliolate, the leaflets narrowly oblanceolate, about 1 cm long, sparsely silky on both sides, in small fascicles at the end of the branchlets. Flowers axillary, solitary or in few-flowered cymes, 2 to 3 cm in diameter; achenes hirsute.

Common in open and partly wooded muskeg or tundra north beyond the limit of trees.

General distribution: Circumpolar with large gaps.

Fig. 635 Map 737

***Potentilla furcata** Porsild
Stems slender, erect-ascending, 20 to 30 cm tall from a sub-ligneous base. Basal leaves mostly 5-parted, the leaflets lanceolate-oblong, deeply incised into 7 to 9 narrow segments, appressed-pilose or glabrate above, the underside prominently nerved and short tomentose. The inflorescence of relatively small flowers, repeatedly forked (hence the specific name).

Dry, calcareous sub-alpine slopes.

General distribution: Cordilleran species described from the Rose-Lapie R. Pass in S.E. Y.T., but has subsequently become known as common also south to Jasper and Banff National Parks, Alta., and to be looked for in the Mackenzie Mts.

Fig. 636 Map 738

Potentilla hyparctica Malte
var. **elatior** (Abrom.) Fern.
P. emarginata Pursh, *non* Desf.
Densely tufted, with the branches of the caudex covered by marcescent brown stipules. Leaves all ternate, green, sparingly villous above, and pale and thinly soft pubescent beneath; petioles 3 to 4 times as long as the blade, and long-villous. Flowering stems erect-ascending and often somewhat flexuous, 5 to 12 cm high, usually 1- to 3-flowered. Flowers about 1.5 cm in diameter, the petals very pale yellow, obcordate, and prominently emarginate.

Rocky places such as ravines and talus slopes.

General distribution: Circumpolar, wide-ranging low-arctic. Typical *P. hyparctica* Malte is high-arctic and does not reach the Canadian mainland.

Fig. 637 Map 739

Potentilla multifida L.
Stems often greenish-purple, erect-ascending or spreading from a stout, subligneous and many-headed base. Leaves deep green and glabrous above, thinly tomentose beneath, nearly as long as the flowering stems, the 3 to 5 leaflets pinnately divided, their margins commonly involute. Flowers small and numerous, in leafy cymes.

Common on sandy river banks, lake shores and on stable dunes north to lat. 69°, but in the Northwest Territories narrowly confined to the Mackenzie drainage where it is often a common and well established weed near human habitations.

General distribution: Circumpolar with some large gaps.

Fig. 638 Map 740

Potentilla nivea L. *s. lat.*
Loosely tufted with a freely branched caudex. Leaves ternate, dark green and glabrous above, dull white-tomentose to glabrate below. Flowering stems erect-ascending, freely branched, 15 to 25 cm high, 3- to 15-flowered.

Potentilla nivea is represented in North America by several well-marked geographical races, of which, in addition to the typical plant, an eastern and western race enter our area.

a. Leaf petioles merely tomentose, lacking long, spreading or appressed guard-hairs . . . . *P. nivea* ssp. *nivea*
a. Leaf petioles with long and spreading hairs several times longer than the diameter of the petiole
  b. Petioles 2-3 times as long as the blade, with a short, dense pubescence beneath the long guard-hairs; flowers small, in rather dense cymes . . . . *P. nivea* ssp. *Hookeriana* (Lehm.) Hiit.
  b. Petioles longer, without a dense pubescence of short hairs beneath the long guard-hairs . . . . *P. nivea* ssp. *Chamissonis* (Hultén) Hiit.

Common and widespread in rocky and sunny places; often near animal burrows and below bird cliffs. The leaves and stems of *Potentilla nivea* are said to be a particularly rich source of vitamin C.

General distribution: Arctic-subarctic, *P. nivea* ssp. *nivea*: Circumpolar, arctic-alpine; ssp. *Chamissonis*: Amphi-Atlantic, arctic; ssp. *Hookeriana*: Asia—Northwest America, arctic-alpine.

Fig. 639 (ssp. *nivea*), 640 (ssp. *Chamissonis*), 641 (ssp. *Hookeriana*)

Map 741 (ssp. *nivea*), 742 (ssp. *Chamissonis*), 743 (ssp. *Hookeriana)*

Potentilla norvegica L.
Stems mostly simple, or branched above, leafy and hirsute, or glabrous in var. *labradorica* (Lehm.) Fern; the leaves 3-parted and the lowermost with long petioles. The inflorescence dense and leafy and the flowers small, with pale yellow petals.

In the southern part of its range *P. norvegica* may be biennial or even annual, but near its northern limit it is biennial or even a short-lived perennial and, as elsewhere, of weedy habit, readily invading areas of disturbed soil, clearings or burns.

General distribution: Circumpolar, non-arctic.

Fig. 642 Map 744

Potentilla palustris (L.) Scop.
Emergent bog species with long, freely branching, prostrate and submerged sub-woody rhizomes from which rise the erect aerial shoots; these often dark-reddish tinged, as are the 5- or less commonly 7-foliolate leaves. Flowers dark reddish purple, about 2 cm broad, solitary or few together in the axils of the uppermost leaves; the petals shorter than the sepals. Anthers deep purple.

Common in wet marshes and bogs, north to the Arctic Coast.

General distribution: Circumpolar, wide-ranging.

Fig. 643 Map 745

Potentilla pensylvanica L.
P. pectinata auctt. non Raf.
Loosely tufted from a simple or branched, erect-ascending base; stems few to several, 25 to 75 cm tall, simple or branching above, thinly tomentose, usually with 3 to 4 evenly spaced cauline leaves. Cauline and basal leaves similar, 5- to 9-foliolate and commonly, but not always, white tomentose beneath. Inflorescence of small flowers, in one or several dense clusters; the petals bright yellow, scarcely longer than the sepals.

Common in grassland and in rocky openings in the forest along the Mackenzie R. drainage, but not quite reaching the northern limit of trees.

General distribution: N. America. Gaspé, Que. to S. E. Alaska and southward.

Fig. 644 Map 746

Potentilla pulchella R. Br.
Caespitose or tufted with decumbent or arching 1- to 3-flowered stems. Basal leaves pinnate, with from 3 to 5 deeply incised linear-lanceolate leaflets; the terminal and the first pair largest, the second pair smaller or sometimes much reduced or absent, glabrate above, silky-tomentose beneath; petioles silky-villous, stipules large and prominent, essentially glabrous, rusty-brown. Flowers small, the petals pale yellow, not much longer than the calyx. In the western arctic islands and along the Arctic Coast often represented by var. *gracilicaulis* Porsild, which differs from typical *P. pulchella* by its erect-ascending flowering stems, 25 to 30 cm high, and by its less hirsute leaves.

Common in sandy and gravelly places, and in dry tundra; although perhaps not truly halophilous, the species usually grows on strand-flats, associated with seashore species.

General distribution: Circumpolar (with large gaps), high-arctic.

Fig. 645 (var. *gracilicaulis*) Map 747 *P. pulchella s.lat.*

Potentilla rubricaulis Lehm.

Caespitose with a very stout, sub-ligneous and freely branched caudex covered by dark brown marcescent stipules; flowering stems stout, reddish-brown under the long, white pubescence, erect-ascending, 10 to 30 cm high, freely forked, and few- to many-flowered. Basal leaves 5 to 8 cm long, the blade about half as long as the densely white-villous petiole, cordate-obovate, pinnate, usually with 5 approximate leaflets, these oblanceolate, deeply cleft into linear-lanceolate teeth, in youth sparingly white-villous above, glabrate in age, the underside densely covered by a long, soft, silky-white tomentum, which is drawn out at the tip of each tooth into a conspicuous white, brush-like tuft; the stipules glabrous, large and rusty brown. Flowers 3 to 7, large, 1.5 to 2.5 cm in diameter, with dark yellow, oblanceolate petals, overlapping each other, much longer than the silky-villous sepals.

A very showy species which forms large cushion-like mats, commonly in rocky or gravelly places. In the Far North and in exposed situations sometimes represented by a lower and more compact, 1- to few-flowered form which may appear to be 3-foliate. A close examination will, however, nearly always reveal a second, if much reduced, pair of leaflets.

The type locality of *P. rubricaulis* is "About Bear Lake, in lat. 66°" which is approximately the site of Fort Franklin at the head of Great Bear River.

General distribution: Endemic of the Arctic Archipelago, the adjacent mainland, and N. E. Greenland to S. E. Alaska, with disjunct, alpine stations in southern Alta.

Fig. 646 Map 748

Potentilla tridentata Sol.

A dwarf species with a sub-ligneous creeping stem, leathery, evergreen 3-foliate leaves, and small white flowers in open cymes.

Dry, rocky, gravelly or sandy places on Precambrian rock barely entering the Mackenzie district south of Great Slave Lake.

General distribution: Southern E. and W. Greenland, Lab. to L. Athabasca, south to Nfld. and mountains in eastern N. S., Great Lakes region, Alta. to North Dakota.

Fig. 647 Map 749

Potentilla uniflora Ledeb.

P. Ledebouriana Porsild

Densely to loosely caespitose with firm sub-ligneous and branching caudices covered by the long-marcescent brown or blackened remains of leaves; petioles of basal leaves long silky-pubescent, the blade ternately divided, the leaflets equal in size, broadly cuneate or obovate, each with 7 large and regular teeth; the green upper surface of the leaflet distinctly visible through the silky pubescence, the lower surface densely white-tomentose beneath a more open cover of white, silky hairs. Flowering stems erect-ascending, from 5 to 25 cm tall, simple or branched, with 1 to 3 flowers, these rarely over 1.5 cm in diameter; the petals deep yellow, drying pale yellow, the base of the style glandular-papillose, and the bractlets linear-oblong, distinctly longer than the sepals.

In alpine and exposed situations *P. uniflora* is usually dwarfed and may form dense cushions simulating the arctic or high-arctic *P. Vahliana* which is endemic to West and North Greenland, the Canadian Arctic Archipelago and the northernmost parts of the continent westward to the 130th meridian. *P. Vahliana* is readily distinguished from *P. uniflora* by the yellowish appearance of its foliage due to the dense, yellow guard hairs, by its broadly elliptic bracts and sepals that are nearly of equal length and by its larger and nearly always solitary flower.

General distribution: Cordilleran-Amphi-Beringian species ranging from eastern Siberia over Alaska, Y. T. and western District of Mackenzie south to Colorado.

Fig. 648 Map 750

Potentilla Vahliana Lehm.

Densely caespitose with stout, compact, sub-ligneous, branching caudices covered with the brown marcescent stipules. Flowering stems scapose, 5 to 10 cm high, mostly 1-flowered, densely covered with long, yellowish hairs; basal leaves short-petioled, ternate, densely silky-villous above, and densely yellowish tomentose beneath; leaflets cuneate, deeply cleft into linear-lanceolate teeth. Flowers 1.5 to 2.0 cm in diameter; calyx yellowish silky-villous; petals broadly obcordate, overlapping each other, deep yellow with a dark orange base, and twice as long as the sepals; style barely thickened at the base, and not glandular-papillose.

A pronounced calciphile and common mainly on limestone barrens where it forms large cushions often one meter in diameter, that may be completely covered by the large, showy flowers.

General distribution: High-arctic, endemic of the Arctic Archipelago and adjacent mainland; central W. Greenland north to Peary Land.

Fig. 649 Map 751

Prunus L. Cherry

Shrubs or dwarf trees (with us rarely over 1.5 m tall) with alternate, simple and finely serrate leaves, small white flowers, each producing a 1-seeded, fleshy fruit (drupe).

a. Flowers in umbellate clusters . . . . *P. pensylvanica*
a. Flowers racemose . . . . *P. virginiana*

Prunus pensylvanica L.
Leaves lanceolate-acuminate. Flowers few, in terminal or axillary umbellate clusters. Fruit red when mature.

Local along river banks of the upper Mackenzie River and tributaries north to Fort Simpson.

General distribution: N. America, from Lab. west to B. C. and southward.

Fig. 650 Map 752

Prunus virginiana L.
Leaves elliptic-obovate. Flowers numerous in terminal racemes. The mature fruit astringent, purplish-black when mature.

North of lat. 60° known only from south of Great Slave Lake where it is common locally along the Enterprise—Mackenzie River Highway and also known from Nahanni National Park.

General distribution: N. America, from Nfld. to B. C. and southward.

Fig. 651 Map 753

Rosa L. Rose

Shrubs mostly with prickly or thorny stems and odd-pinnate leaves with prominent stipules adnate to the petiole. Flowers pink and showy, up to 5 cm in diameter; petals and sepals five; calyx tube (hip) globose or oblong, first green but later reddish-orange and pulpy, enclosing the ripe hairy achenes. The ripe hips are rich in vitamin C and may be candied or made into jelly.

a. Stems bristly or prickly . . . . *R. acicularis*
a. Stems smooth or at most with scattered thorns
  b. Leaflets rarely over 2 cm long; stems with occasional thorns . . . . *R. Woodsii*
  b. Leaflets commonly 3-5 cm long, stems lacking thorns . . . . *R. blanda*

Rosa acicularis Lindl. *s. lat.*
Straggling or sometimes erect stems up to 1.5 m tall.

Common especially on riverbanks and in clearings or burns throughout the wooded parts of the District of Mackenzie where it is the only common member of the genus.

General distribution: Nearly circumpolar.

Fig. 652 Map 754

Rosa blanda Ait.
In the District of Mackenzie known from only two collections, both from gravelly river terraces along the Mackenzie River in approximate lat. 63° N. where it was first collected in 1922 and again 22 years later. The station is nearly 1600 km northwest of the nearest known station for *R. blanda* in Alberta.

General distribution: N. America; Que. to B.C. and southward.

Fig. 653 Map 755

Rosa Woodsii Lindl.
Not uncommon on open river banks and lake shores of the upper Mackenzie Valley. At once distinguished by its numerous small and deep green leaflets and small, clustered and fragrant flowers.

General distribution: Western N. America; W. Ont. to B. C., north to central Alaska.

Fig. 654 Map 756

Rubus L. Raspberry

Somewhat shrubby, tufted or trailing perennials with 3-lobed or 3- to 5-parted, stipulate leaves. Flowers regular and perfect, except the dioecious *R. Chamaemorus,* with numerous stamens, 5-lobed persistent calyx and 5 deciduous petals. The carpels numerous, becoming fleshy druplets forming an aggregate fruit around the enlarged and succulent receptacle.

a. Erect shrubs with freely branching prickly or bristly canes
  b. Petals white . . . . *R. strigosus*
  b. Petals pink . . . . *R. alaskanus*
a. Low, mainly herbaceous sub-shrubs lacking prickles
  c. Leaves rounded and shallowly lobed
    d. Petals white; fruit large and yellow . . . . *R. Chamaemorus*
    d. Petals pink; flowers mostly sterile . . . . *R. stellatus*
  c. Leaves 3-parted
    e. Plant tufted; stolons lacking or weakly developed
      f. Plant low, 5-10 cm tall, fruits red . . . . *R. acaulis*
      f. Plant 10-20 cm tall; flowers mostly sterile . . . . *R. arcticus*
    e. Plant thicket-forming, with long, trailing and rooting stolons
      g. Petals pink; leaflets blunt . . . . *R. paracaulis*
      g. Petals white; leaflets with acute tips . . . . *R. pubescens*

Rubus acaulis Michx.
Tufted, herbaceous dwarf species with large, pink and fragrant flowers. The fruits are red, relatively small, but sweet and aromatic.
Common locally in not too dry, turfy places.
General distribution: N. America from Lab. to Alaska, south in the mountains of Alta. and B.C., northward to or slightly beyond the limit of trees.
Fig. 655 Map 757

***Rubus alaskensis** Bailey
A very handsome, thicket-forming somewhat bristly shrub with erect and leafy flowering canes up to 40 cm tall, from stout, woody and widely trailing stolons. The pink flowers are scentless; thus far fruiting specimens have not been reported.
General distribution: A rare and little known species, probably of hybrid origin and thus far collected only a few times in the Susitna Valley south of Mt. McKinley Park, Alaska, and in alpine meadows on Mt. Sheldon, and in the Selwyn Mts. on the Yukon—Mackenzie divide. To be looked for on the east slope of the Mackenzie Mts.
Fig. 656 Map 758

Rubus arcticus L.
Tufted, from a freely branching, sub-ligneous base from which rise the 10 to 20 cm tall, 1- to 2-leaved, flowering canes; stolons lacking. Flowers fragrant, solitary, deep pink or purple and very showy. Fruit red, sweet and juicy.
In subalpine meadows often among willows.
General distribution: Amphi-Beringian; east to beyond the Mackenzie Delta.
Fig. 657 Map 759

Rubus Chamaemorus L.
Cloudberry or Baked-Apple
Dioecious, low, creeping, glabrous herb with large, round-reniform, mostly 5-lobed, serrate, and somewhat leathery leaves; stems simple, erect, bearing 1 to 3 leaves, and terminating in a solitary, large, white flower. The immature fruit first bright red, later amber-coloured and closely embraced in the calyx. When ripe, the calyx lobes become reflexed and the large fruit turns soft and pale yellow.
Common in moist peaty and turfy places well beyond the limit of trees. The Cloudberry is the favourite native fruit of the Eskimo, who preserve it in seal oil.
General distribution: Circumpolar, low-arctic.
Fig. 658 Map 760

Rubus paracaulis Bailey
Similar to *R. pubescens* but with blunter leaflets and pink petals. Possibly a hybrid between *R. pubescens* and *R. acaulis.*
In the Northwest Territories thus far reported from several collections south of Great Slave Lake. In its main range from across Canada to the Rocky Mountains it fruits well.
Open woodlands.
General distribution: Lab. to Alta.
Map 761

Rubus pubescens Raf.
Tufted, with erect-ascending leafy and non-prickly canes, from a woody base; stolons long, trailing and freely rooting. Flowers 1 to several, rarely more than 1 cm in diameter, petals white; fruit red, juicy and sweet.
Gravelly river banks and openings in woods.
General distribution: N. America, Nfld. to S. W. District of Mackenzie.
Map 762

***Rubus stellatus** Sm.
R. arcticus L.
ssp. *stellatus* (Sm.) Boiv. *emend.* Hult.
In habit and general appearance similar to *R. arcticus* from which it differs mainly by its 3-lobed rather than 3-parted leaves. As in *R. arcticus* the flowers are solitary, large and very fragrant.
Alpine herbmats.
General distribution: Amphi-Beringian, but thus far not reported beyond the Y.T.—Mackenzie Divide.
Fig. 659 Map 763

Rubus strigosus Michx.
Wild Raspberry
Erect, up to two meters tall shrub with freely branching biennial and prickly canes. Leaves 3-parted, their petiole, and often also the mid-rib glandular and prickly. Flowers white; the abundant red fruits make an excellent jam.
Woodland species common in openings in the forest, in the District of Mackenzie not quite to the limit of trees.
General distribution: Lab.—Alaska.
Fig. 660 Map 764

Sanguisorba L. Burnet

Glabrous perennial herbs with pinnate leaves. The small flowers in a dense, long-peduncled spike. Petals lacking but the 4 calyx lobes petal like.

a. Spikes oblong 1 to 1.5 cm long, purple; stamens not or barely exserted ..... *S. officinalis*
a. Spikes cylindrical, 3 to 10 cm long, greenish white; stamens exserted .... *S. sitchensis*

Sanguisorba officinalis L.
S. microcephala Presl
An Old World species of weedy habit, not uncommon in damp meadows in eastern Canada and the United States where it is considered non-native. In Alaska, Yukon and northern District of Mackenzie, *S. officinalis,* likewise is known mainly from places near human habitation, but also from stations where its presence could not be ascribed to human agencies.
General distribution: Circumpolar.
Fig. 661 Map 765

***Sanguisorba sitchensis** C.A. Mey.
A Pacific Coast species very similar to *S. canadensis* of eastern North America. It reaches the south and central parts of Alaska and southern Yukon where it has been collected mainly in sub-alpine or flood-plain meadows. *S. sitchensis* has not been reported from the District of Mackenzie, but should be looked for along the lower Nahanni and Liard rivers.
General distribution: Amphi-Beringian.
Fig. 662 Map 766

Sibbaldia L.

Sibbaldia procumbens L.
Low, depressed, and matted, from freely branched and creeping rhizomes; flowering stems 10 cm high or less, villous, few-leaved; basal leaves ternate, the leaflets wedge-shaped and 3-toothed and sparsely appressed pilose, on short slender petioles. Flowers small and inconspicuous in few-flowered, dense cymes. Petals 5, yellowish-green, linear-oblong, minute, shorter than the sepals, stamens 5.
In moist, gravelly places, in herbmats and places where the snow remains late.
General distribution: Circumpolar (with large gaps); subarctic-alpine.
Fig. 663 Map 767

Sorbus L. Mountain Ash

Sorbus scopulina Greene
A much branched, up to 3 m tall thicket-forming shrub with odd-pinnate, 10 to 15 cm long, glabrous and somewhat shiny leaves. The small,

white flowers in dense paniculate terminal or axillary clusters. Fruit fleshy, berry-like, and bright red.

Moist alpine slopes.

General distribution: Cordilleran, reaching southern Y.T. and southern and central Alaska. It barely enters S. W. District of Mackenzie.

Fig. 664 Map 768

Spiraea L. Spiraea

Spiraea Beauverdiana Schneid.
Erect-ascending shrub with slender and much branched stems occasionally 1 m tall, but mostly lower; bark dark brown and shreddy. Leaves alternate, oval, 1 to 2 cm long, glabrous, with shallowly crenate margins. The small, white flowers in 2 to 3 cm broad compound, terminal or axillary panicles. Fruit a dry capsule.

Open muskegs and alpine meadows north to the tree line.

General distribution: An Amphi-Beringian species common in Alaska and Y.T., reaching the Arctic Coast east of the Mackenzie Delta, but thus far not otherwise reported from east of the Mackenzie Valley.

Fig. 665 Map 769

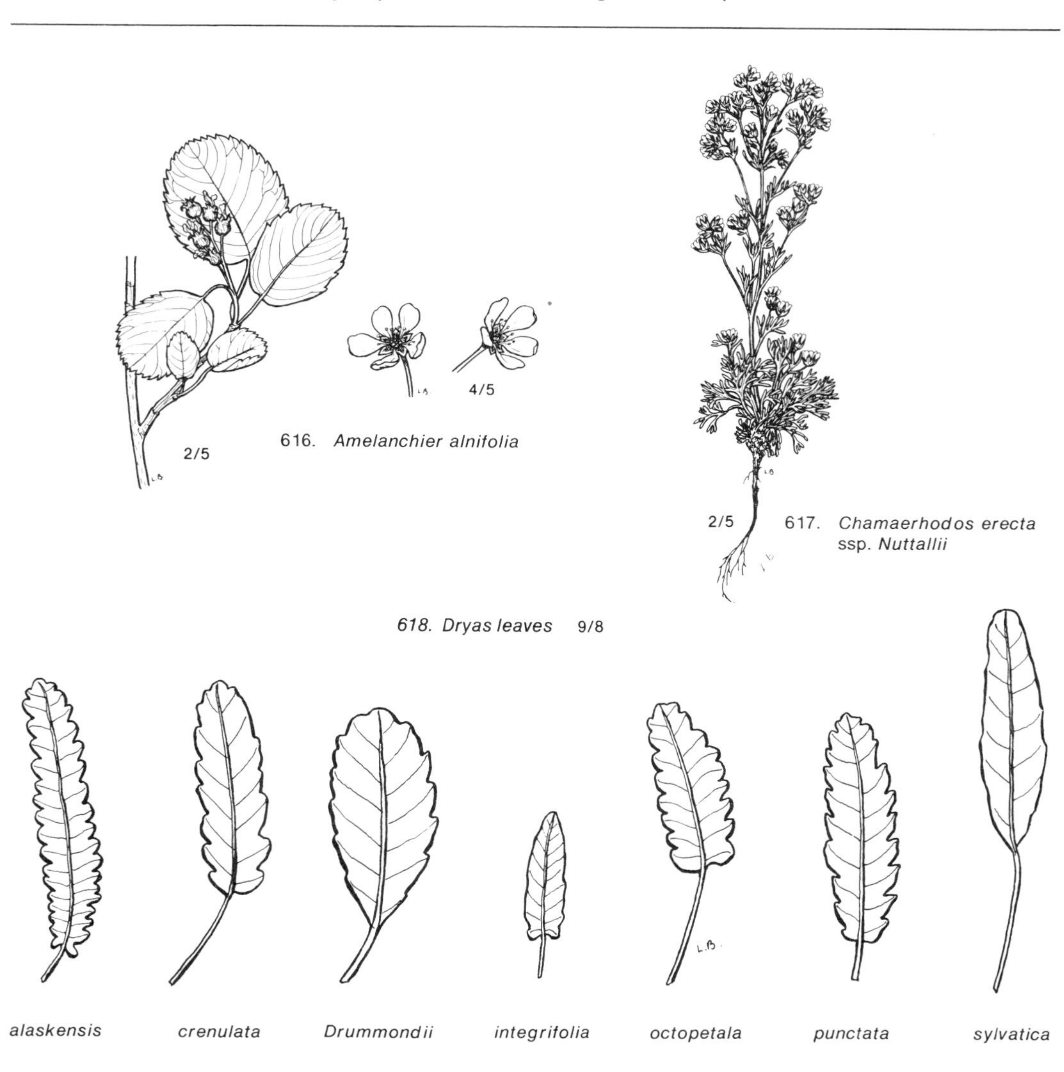

616. *Amelanchier alnifolia*

617. *Chamaerhodos erecta* ssp. *Nuttallii*

618. *Dryas leaves*

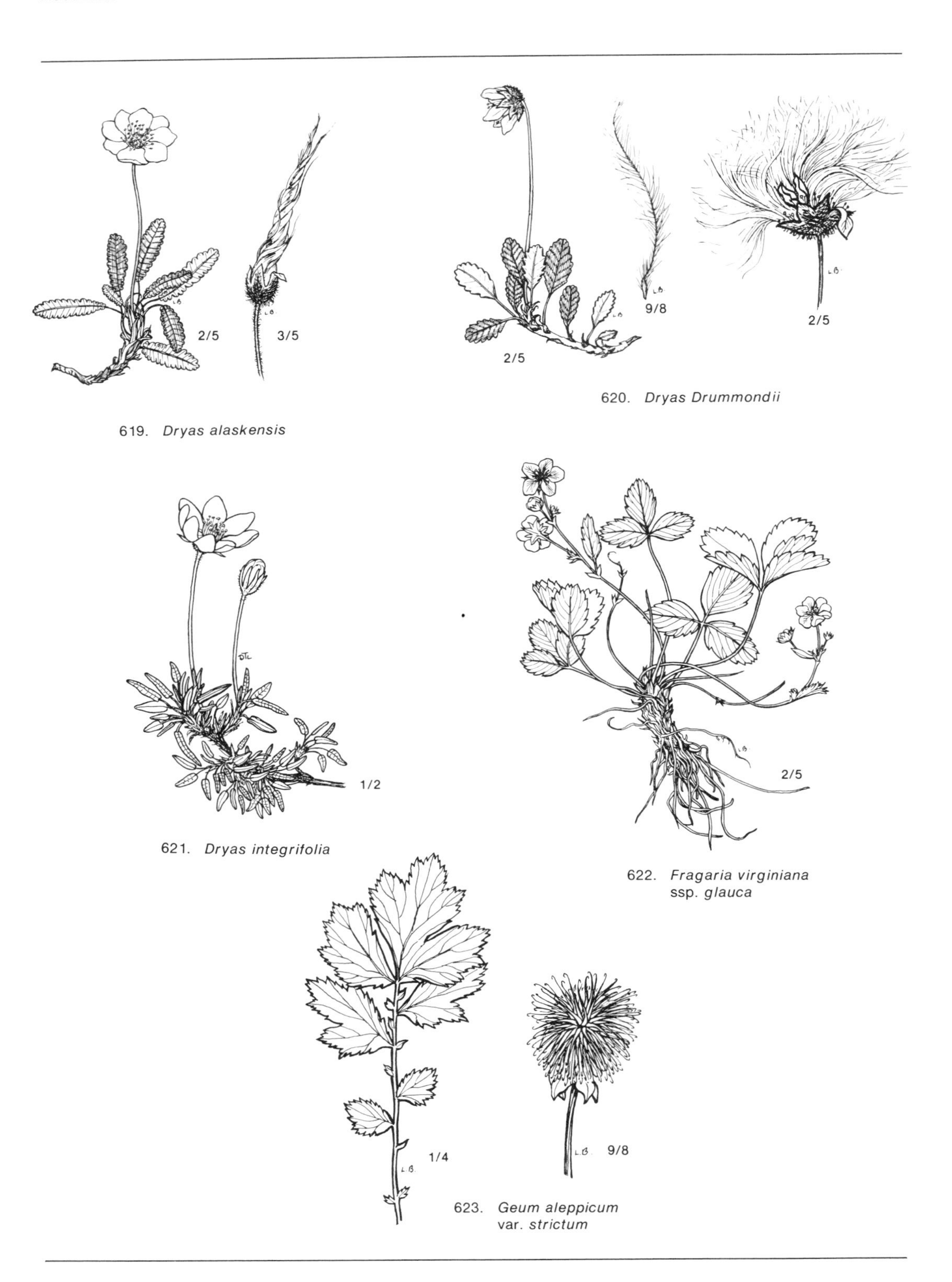

619. *Dryas alaskensis*

620. *Dryas Drummondii*

621. *Dryas integrifolia*

622. *Fragaria virginiana* ssp. *glauca*

623. *Geum aleppicum* var. *strictum*

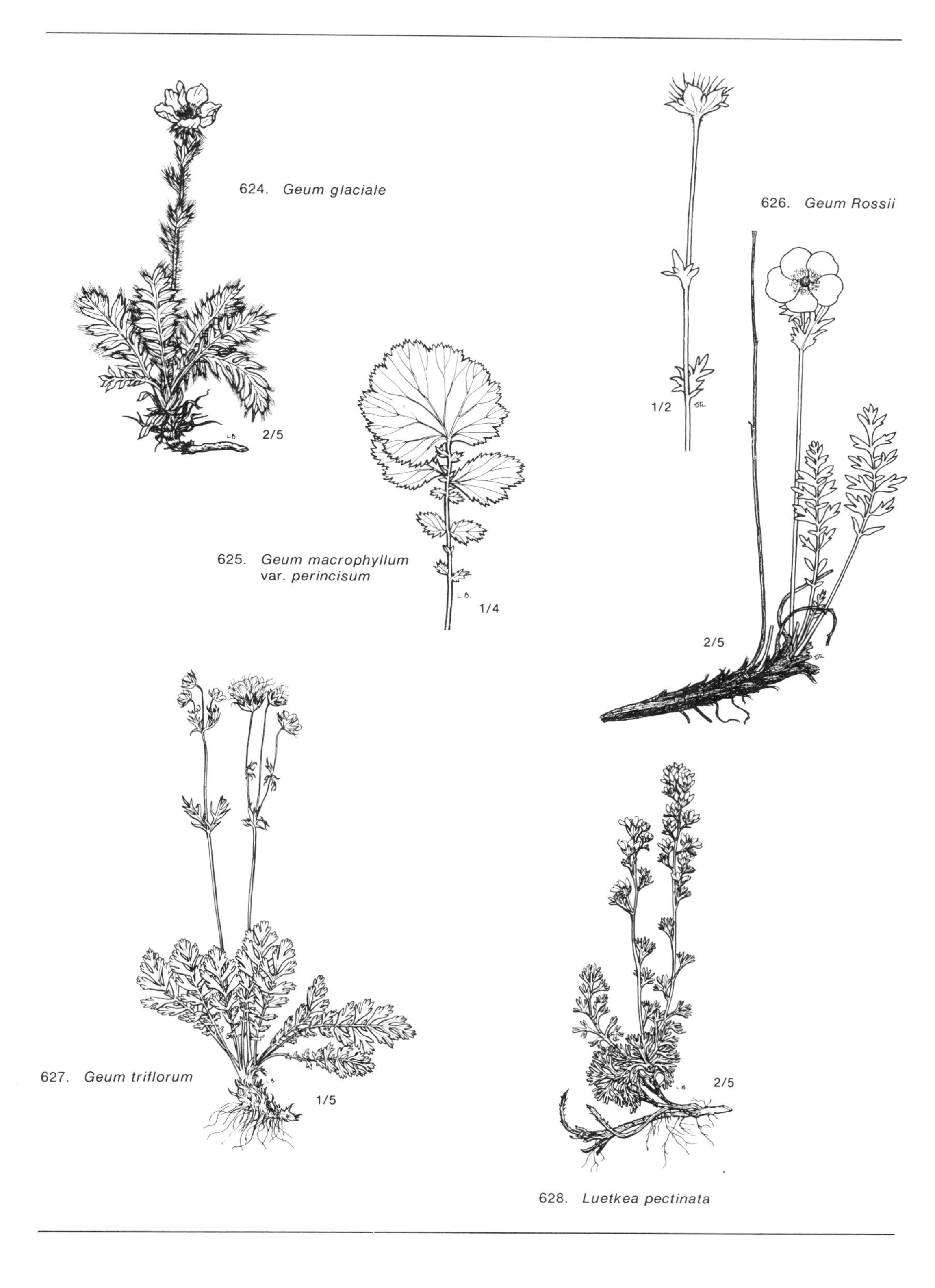

624. *Geum glaciale*

625. *Geum macrophyllum* var. *perincisum*

626. *Geum Rossii*

627. *Geum triflorum*

628. *Luetkea pectinata*

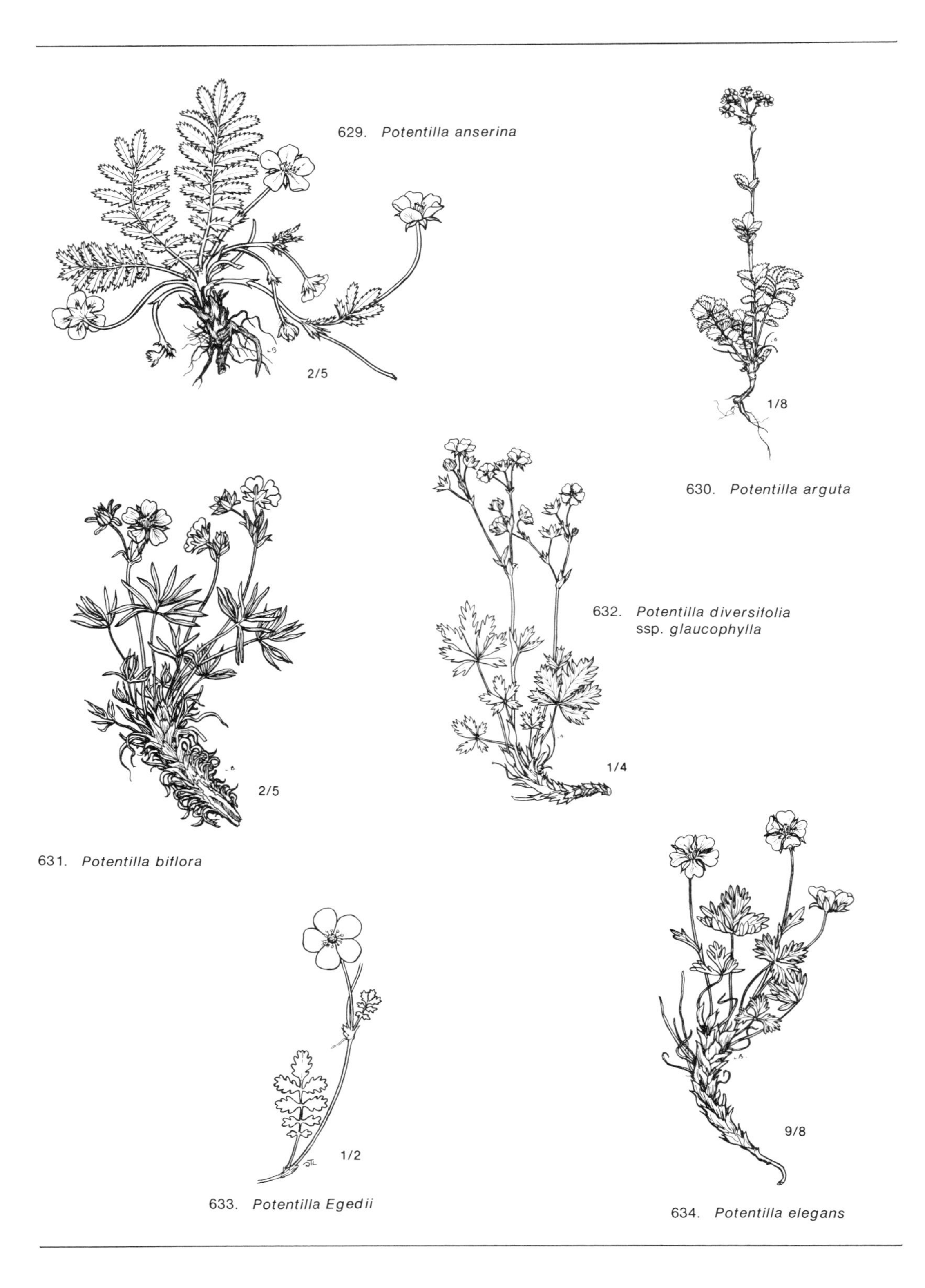

629. *Potentilla anserina*

630. *Potentilla arguta*

631. *Potentilla biflora*

632. *Potentilla diversifolia* ssp. *glaucophylla*

633. *Potentilla Egedii*

634. *Potentilla elegans*

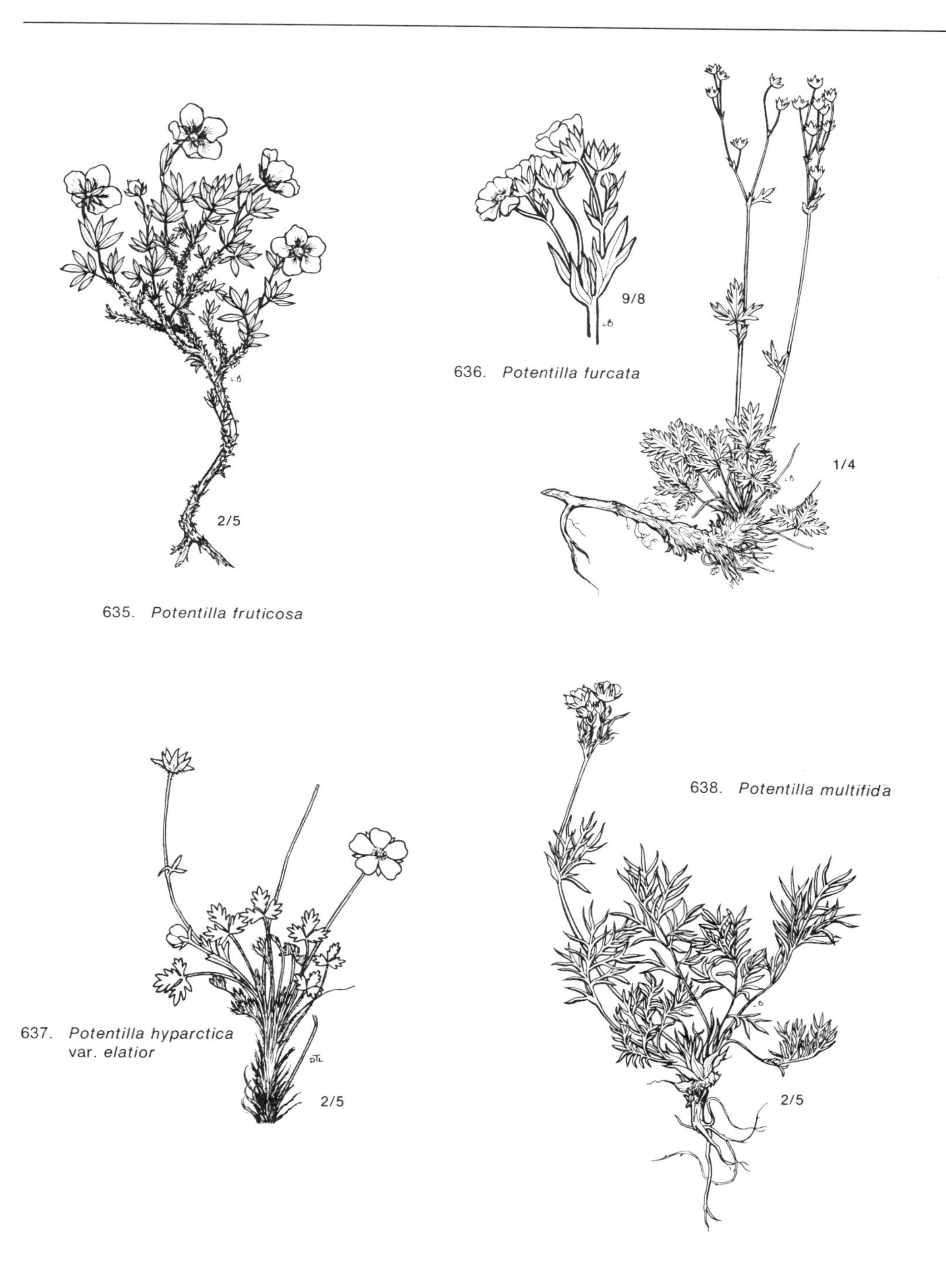

635. *Potentilla fruticosa*

636. *Potentilla furcata*

637. *Potentilla hyparctica* var. *elatior*

638. *Potentilla multifida*

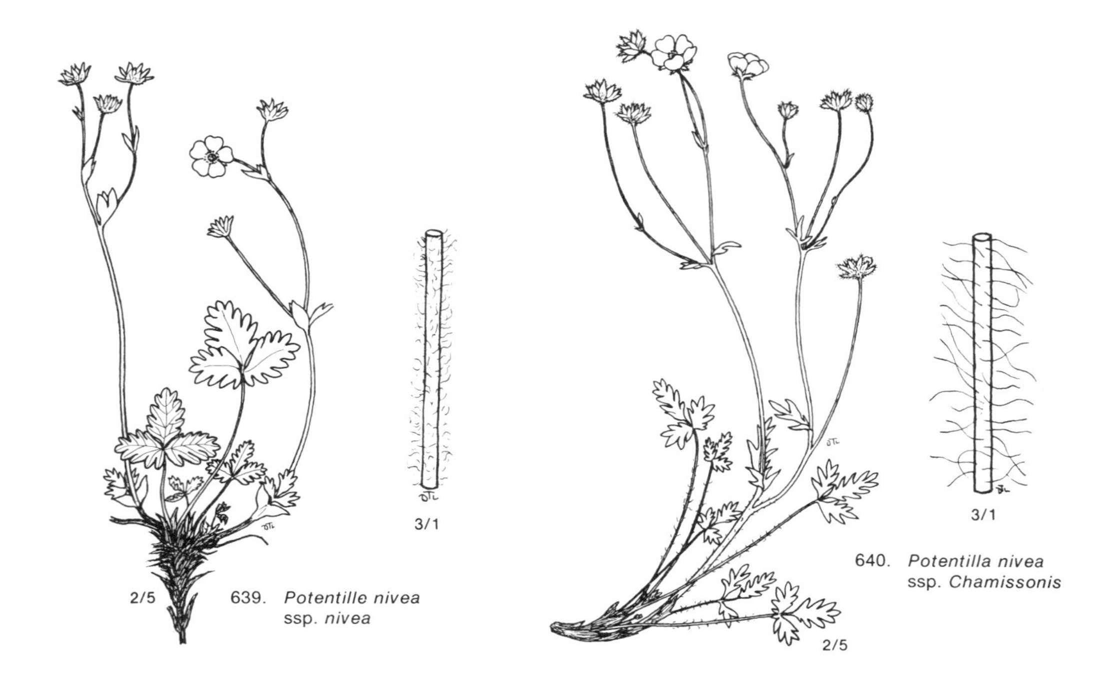

639. *Potentille nivea* ssp. *nivea*

640. *Potentilla nivea* ssp. *Chamissonis*

641. *Potentilla nivea* ssp. *Hookeriana*

642. *Potentilla norvegica*

643. *Potentilla palustris*

644. *Potentilla pensylvanica*

645. *Potentilla pulchella* var. *gracilicaulis*

646. *Potentilla rubricaulis*

647. *Potentilla tridentata*

648. *Potentilla uniflora*

649. *Potentilla Vahliana*

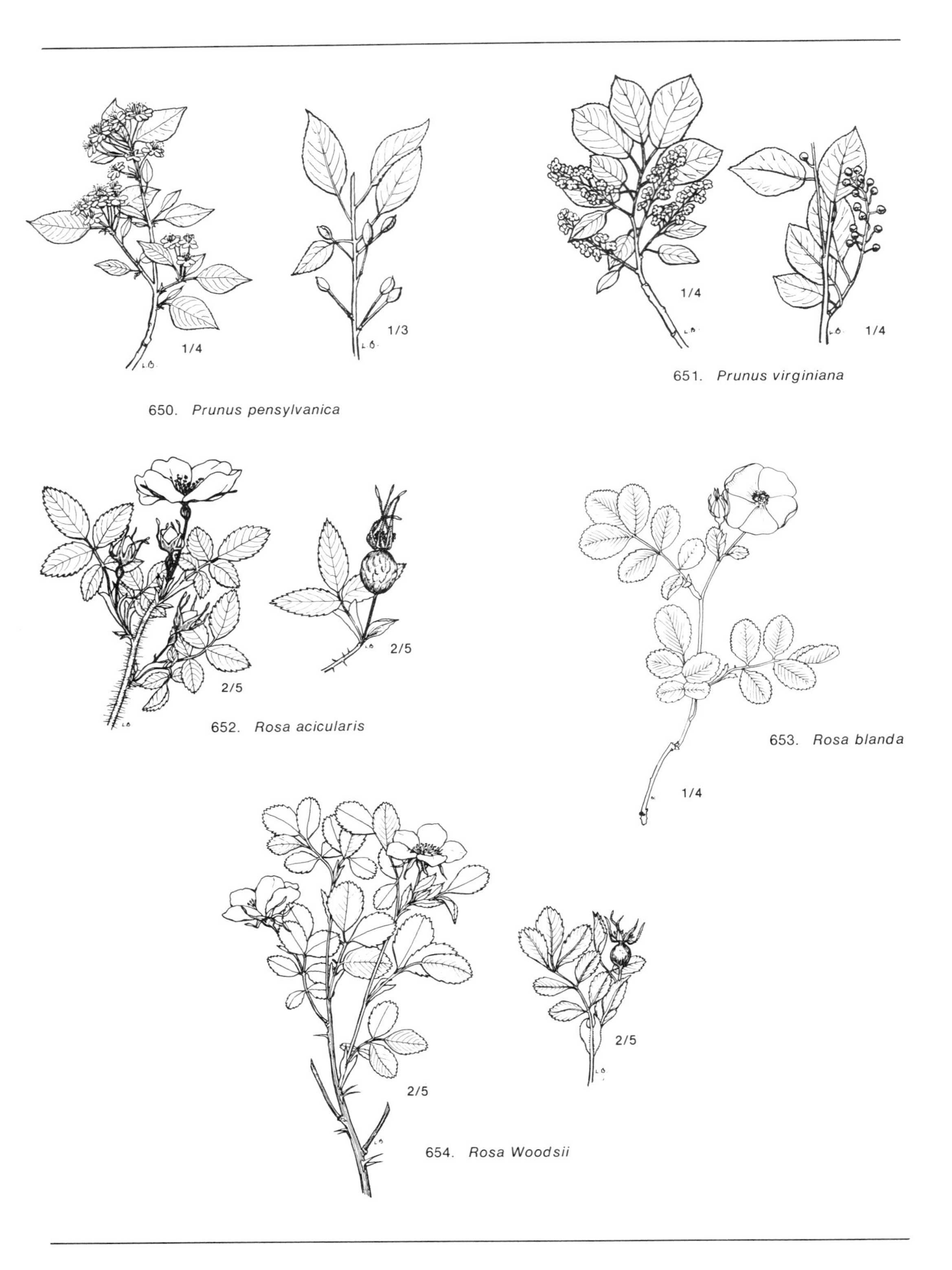

650. *Prunus pensylvanica*

651. *Prunus virginiana*

652. *Rosa acicularis*

653. *Rosa blanda*

654. *Rosa Woodsii*

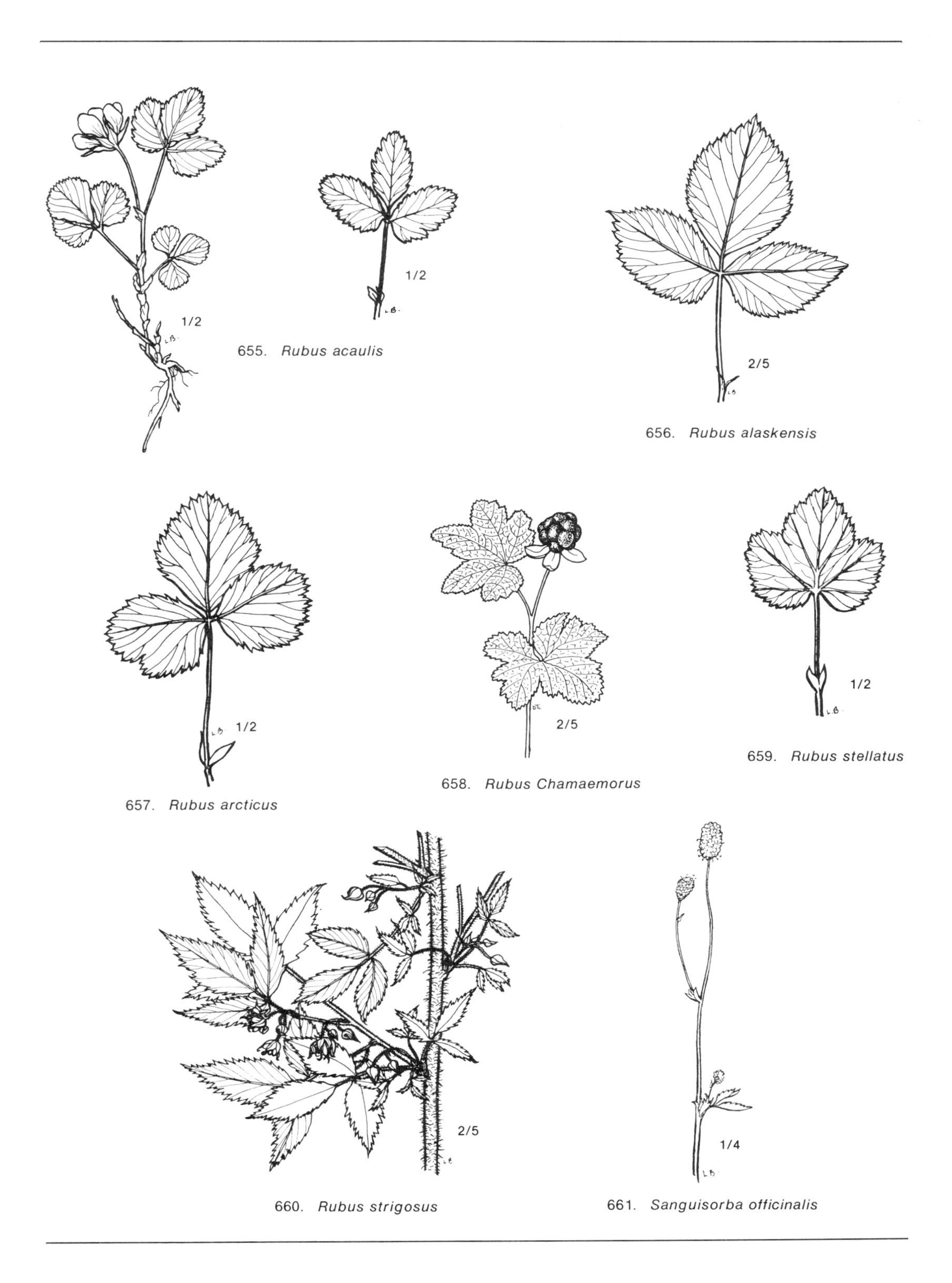

655. *Rubus acaulis*

656. *Rubus alaskensis*

657. *Rubus arcticus*

658. *Rubus Chamaemorus*

659. *Rubus stellatus*

660. *Rubus strigosus*

661. *Sanguisorba officinalis*

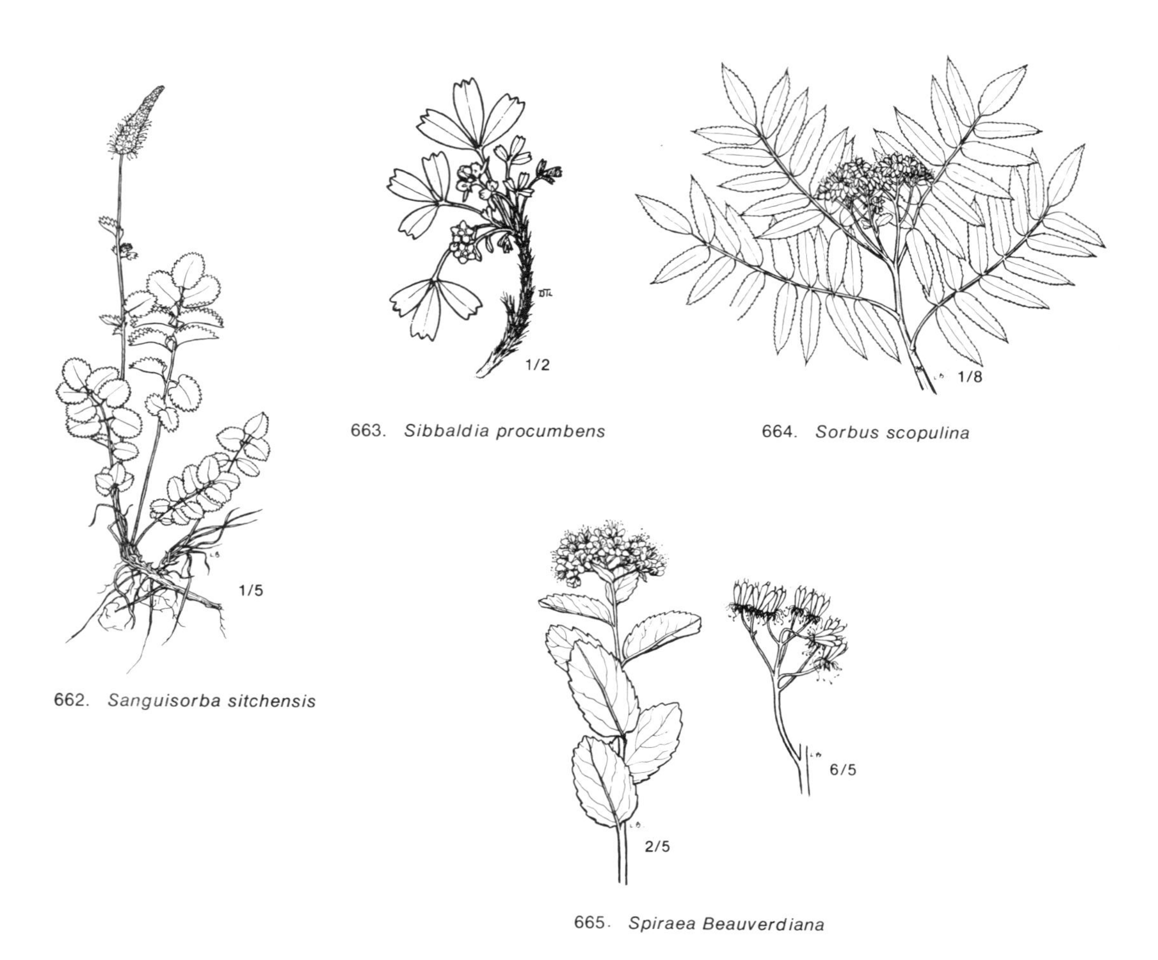

662. *Sanguisorba sitchensis*

663. *Sibbaldia procumbens*

664. *Sorbus scopulina*

665. *Spiraea Beauverdiana*

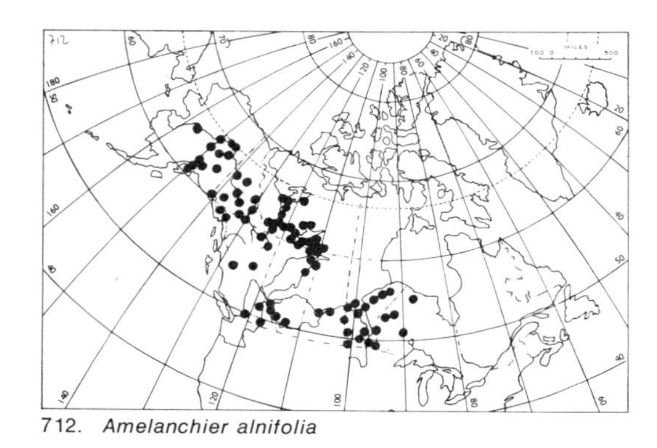

712. *Amelanchier alnifolia*

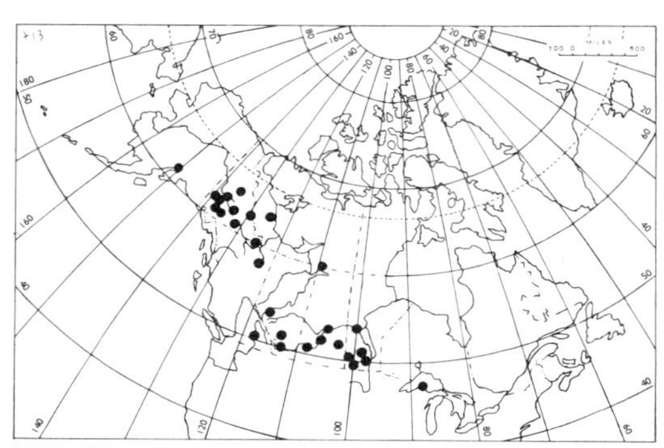

713. *Chamaerhodos erecta* ssp. *Nuttallii*

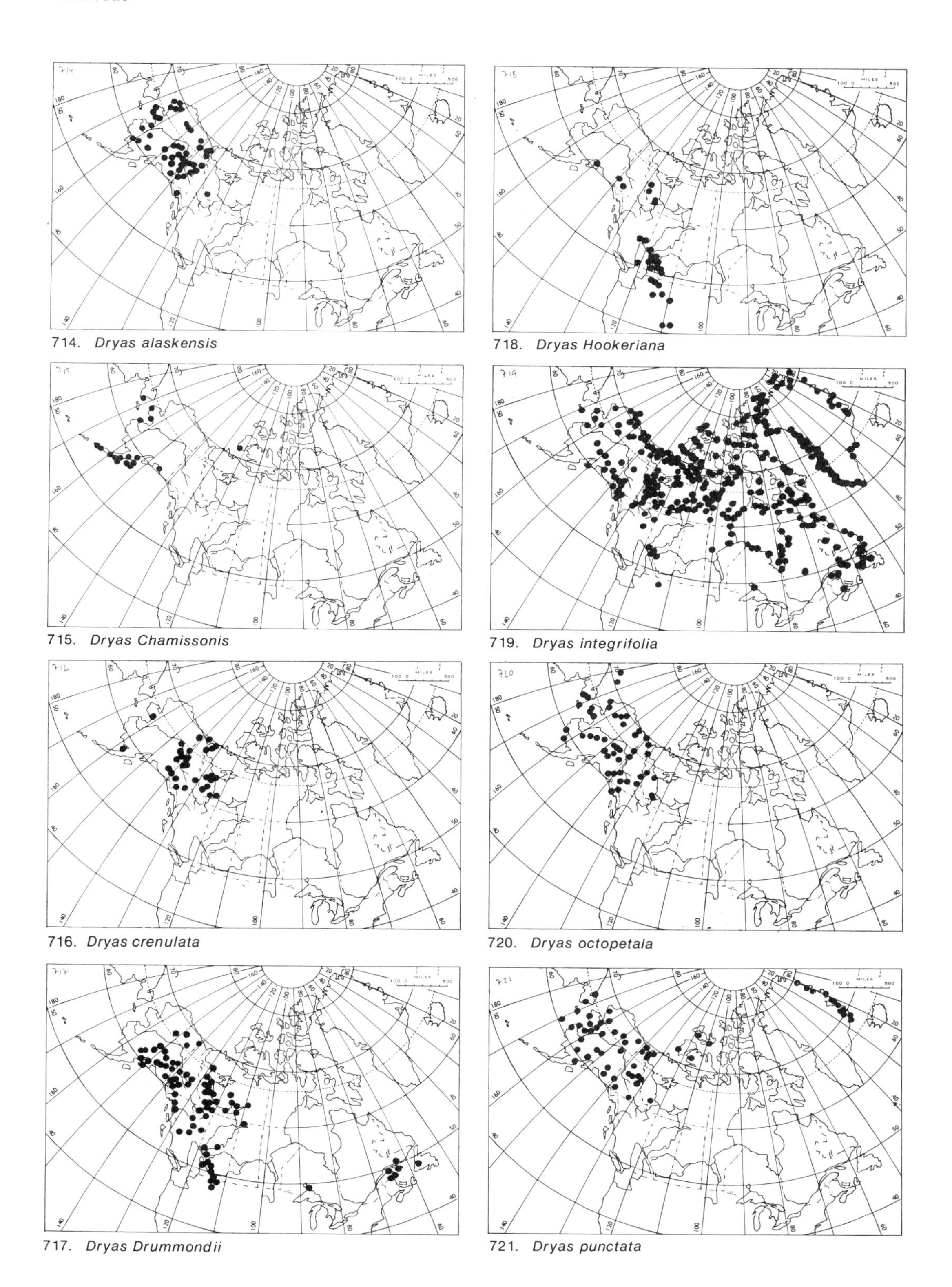

714. *Dryas alaskensis*

718. *Dryas Hookeriana*

715. *Dryas Chamissonis*

719. *Dryas integrifolia*

716. *Dryas crenulata*

720. *Dryas octopetala*

717. *Dryas Drummondii*

721. *Dryas punctata*

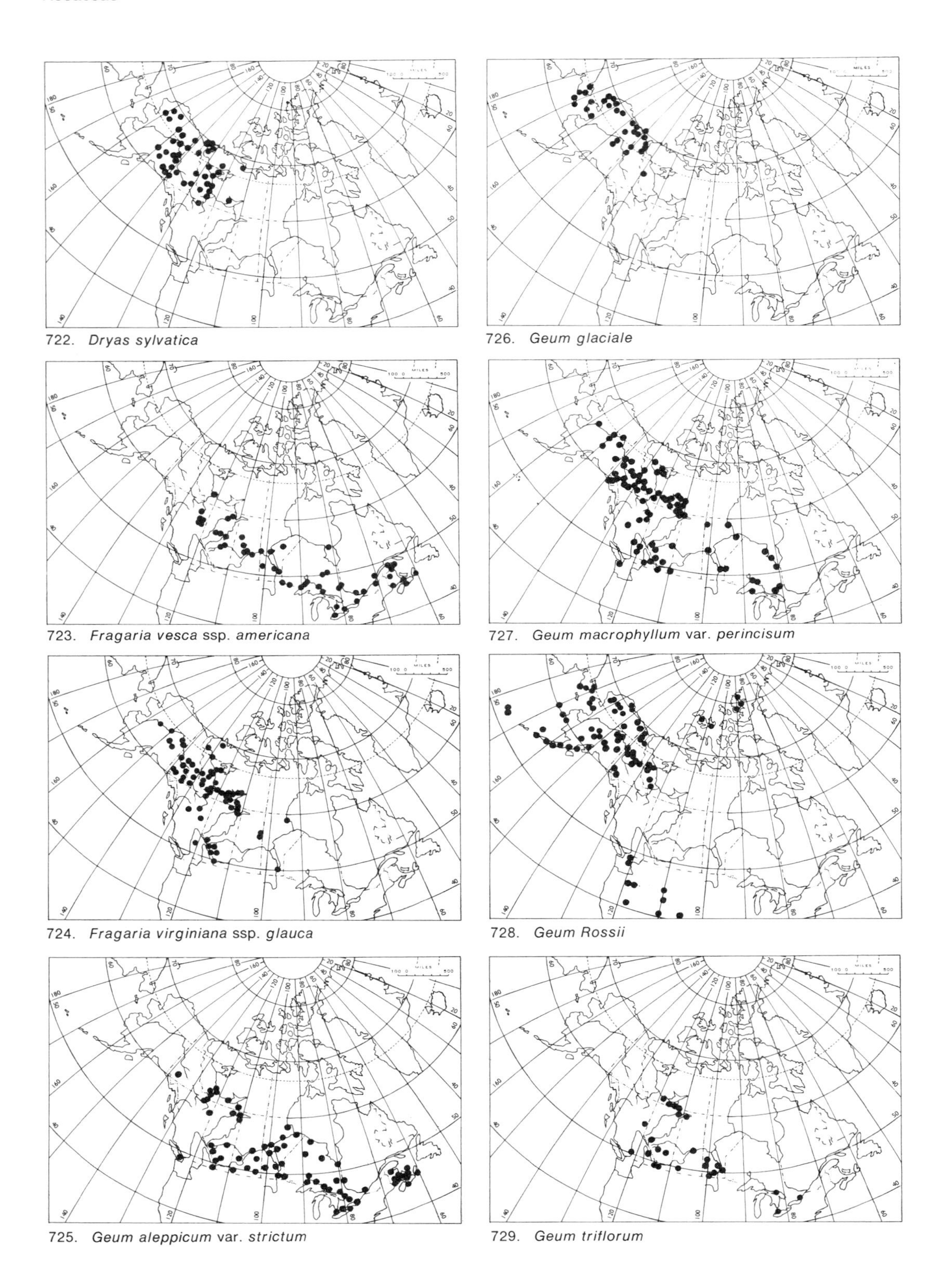

722. *Dryas sylvatica*

726. *Geum glaciale*

723. *Fragaria vesca* ssp. *americana*

727. *Geum macrophyllum* var. *perincisum*

724. *Fragaria virginiana* ssp. *glauca*

728. *Geum Rossii*

725. *Geum aleppicum* var. *strictum*

729. *Geum triflorum*

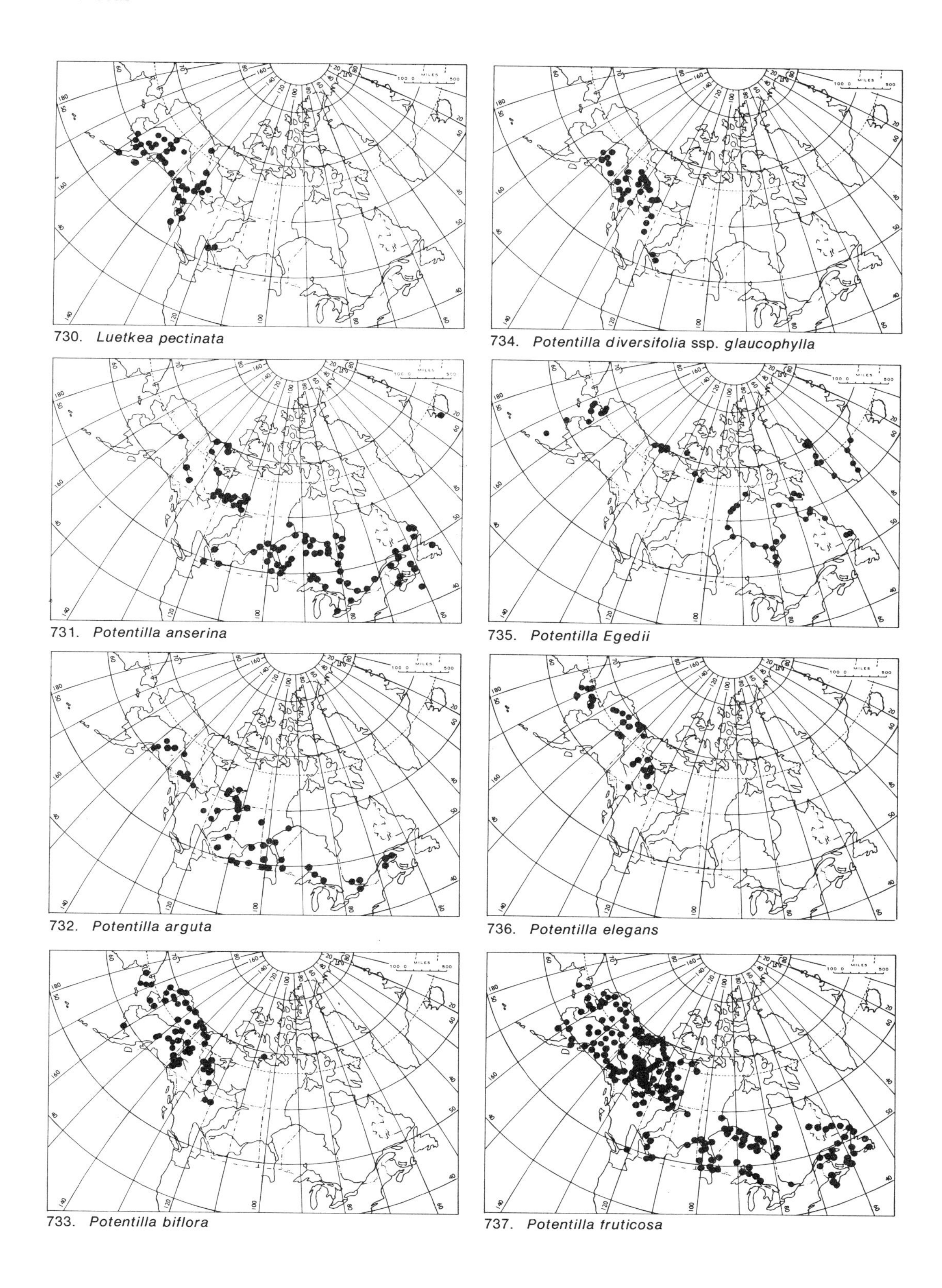

730. *Luetkea pectinata*

731. *Potentilla anserina*

732. *Potentilla arguta*

733. *Potentilla biflora*

734. *Potentilla diversifolia* ssp. *glaucophylla*

735. *Potentilla Egedii*

736. *Potentilla elegans*

737. *Potentilla fruticosa*

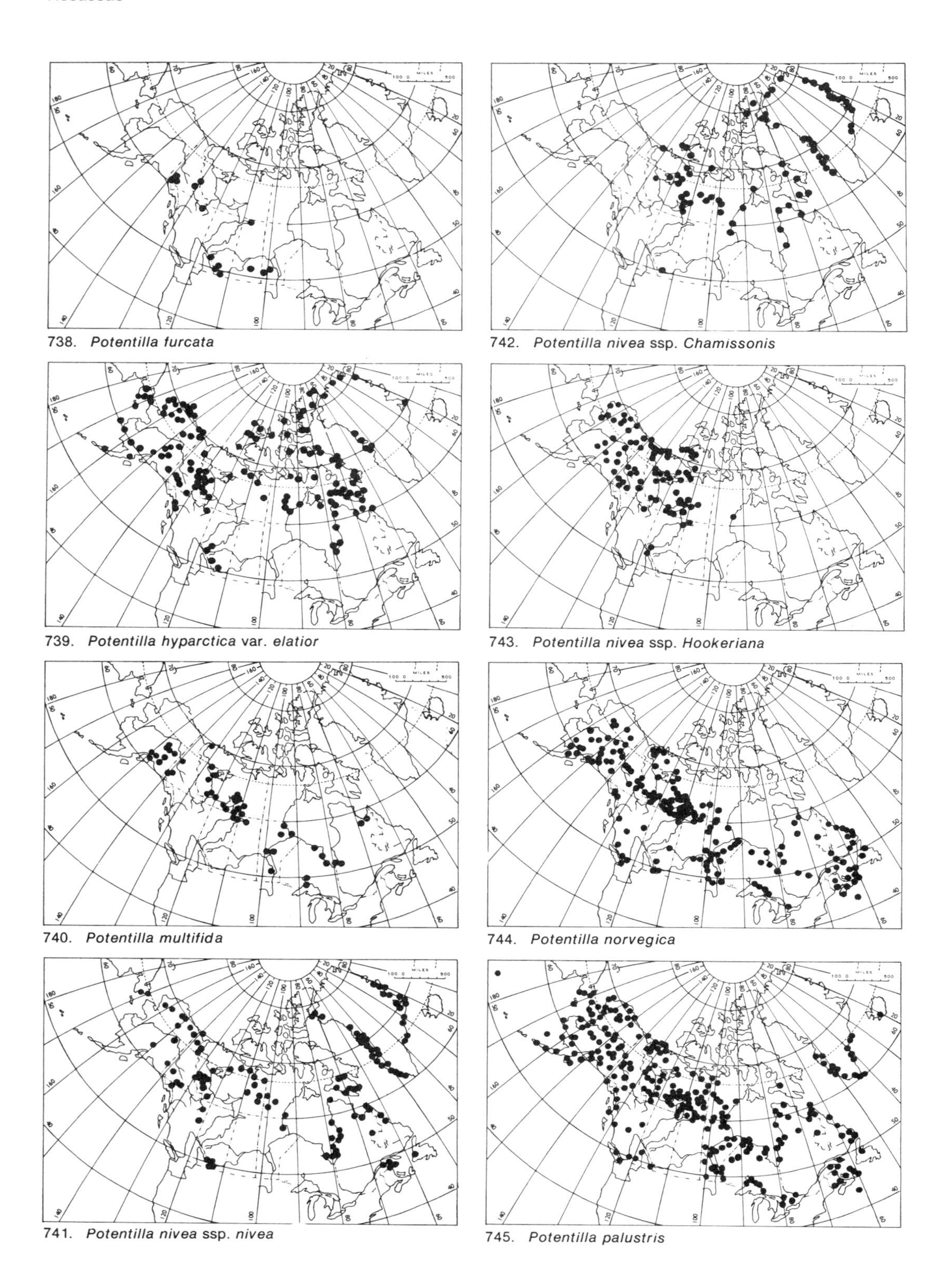

738. *Potentilla furcata*

739. *Potentilla hyparctica* var. *elatior*

740. *Potentilla multifida*

741. *Potentilla nivea* ssp. *nivea*

742. *Potentilla nivea* ssp. *Chamissonis*

743. *Potentilla nivea* ssp. *Hookeriana*

744. *Potentilla norvegica*

745. *Potentilla palustris*

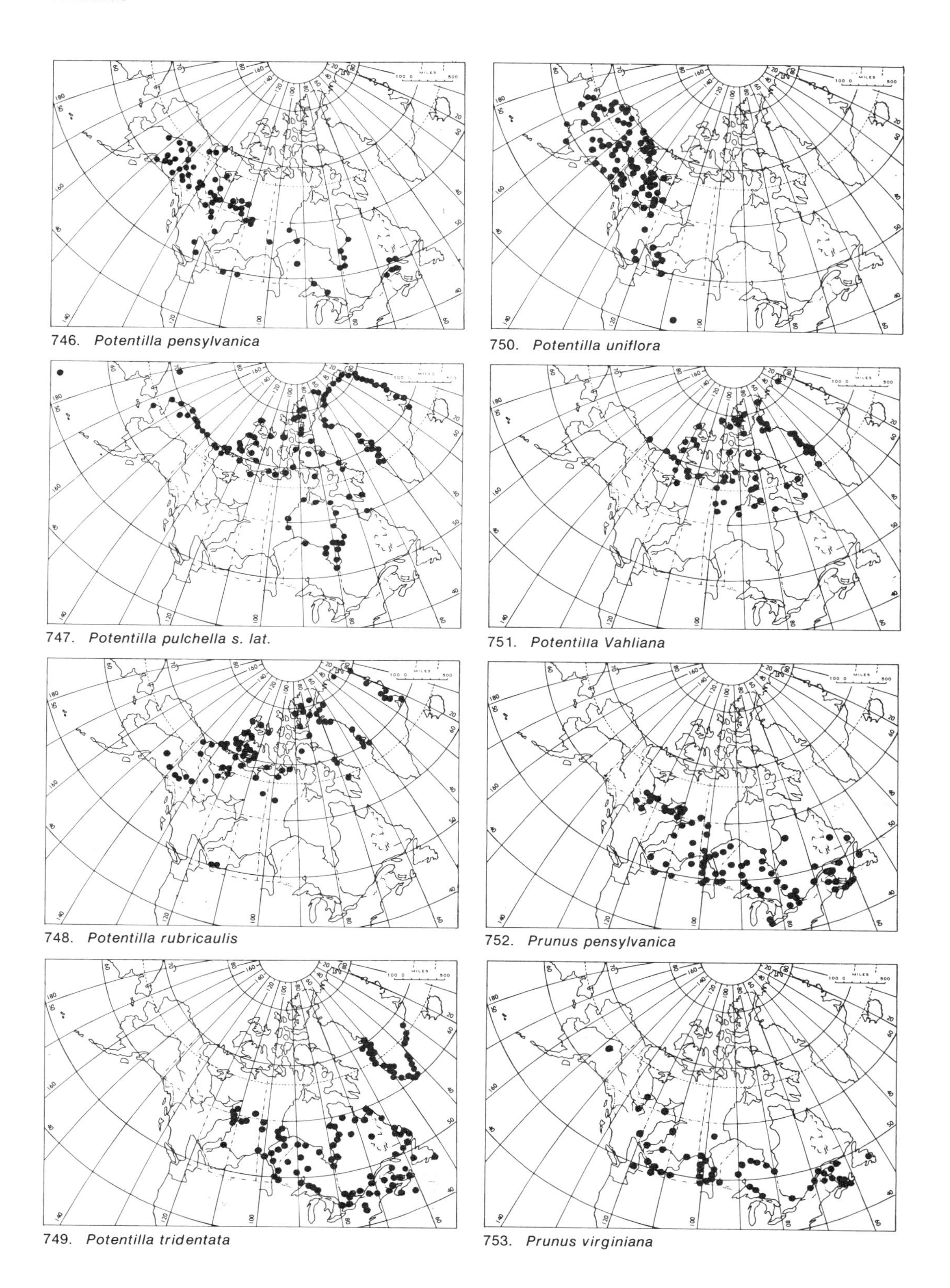

746. *Potentilla pensylvanica*

747. *Potentilla pulchella s. lat.*

748. *Potentilla rubricaulis*

749. *Potentilla tridentata*

750. *Potentilla uniflora*

751. *Potentilla Vahliana*

752. *Prunus pensylvanica*

753. *Prunus virginiana*

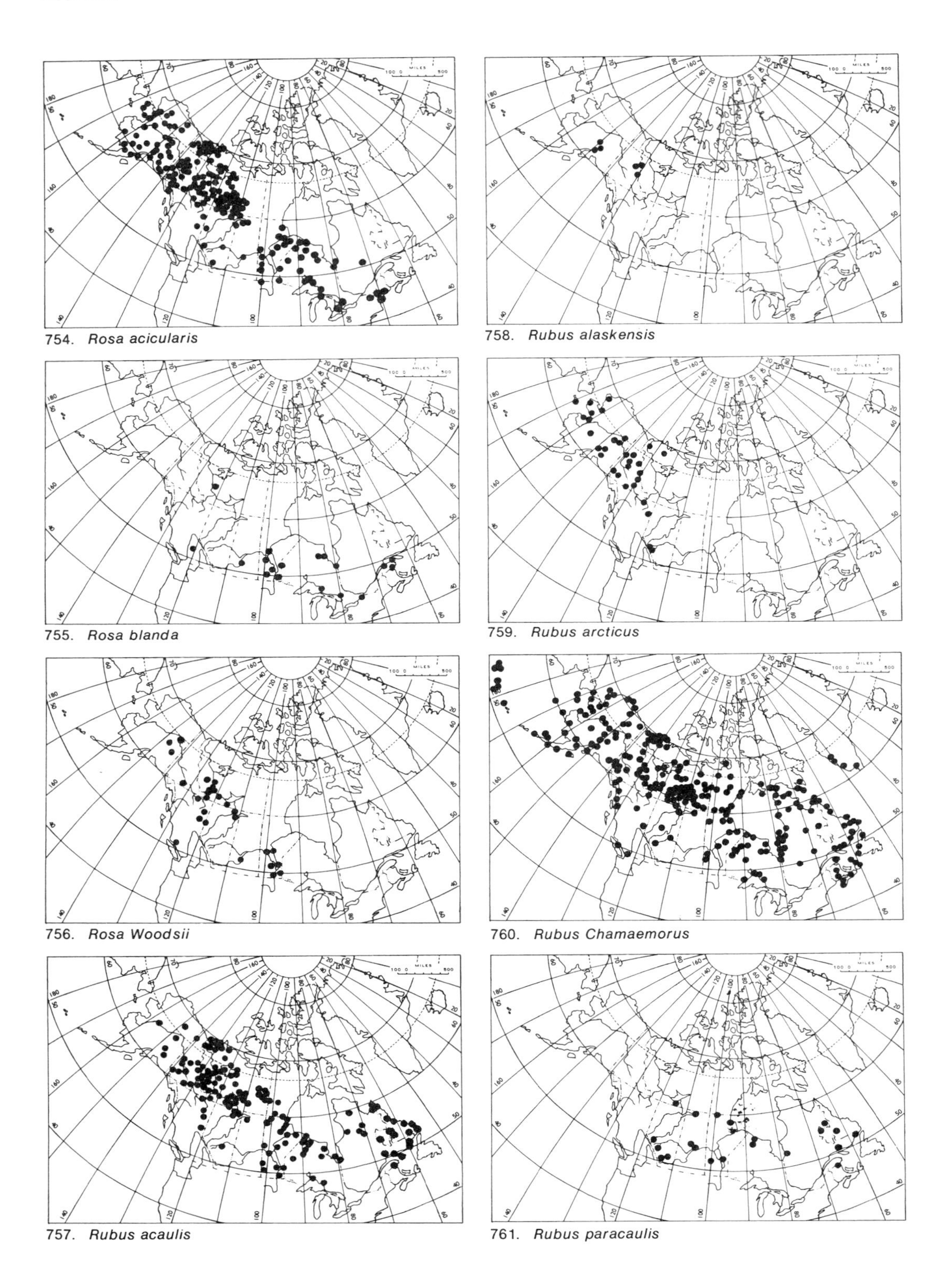

754. *Rosa acicularis*

755. *Rosa blanda*

756. *Rosa Woodsii*

757. *Rubus acaulis*

758. *Rubus alaskensis*

759. *Rubus arcticus*

760. *Rubus Chamaemorus*

761. *Rubus paracaulis*

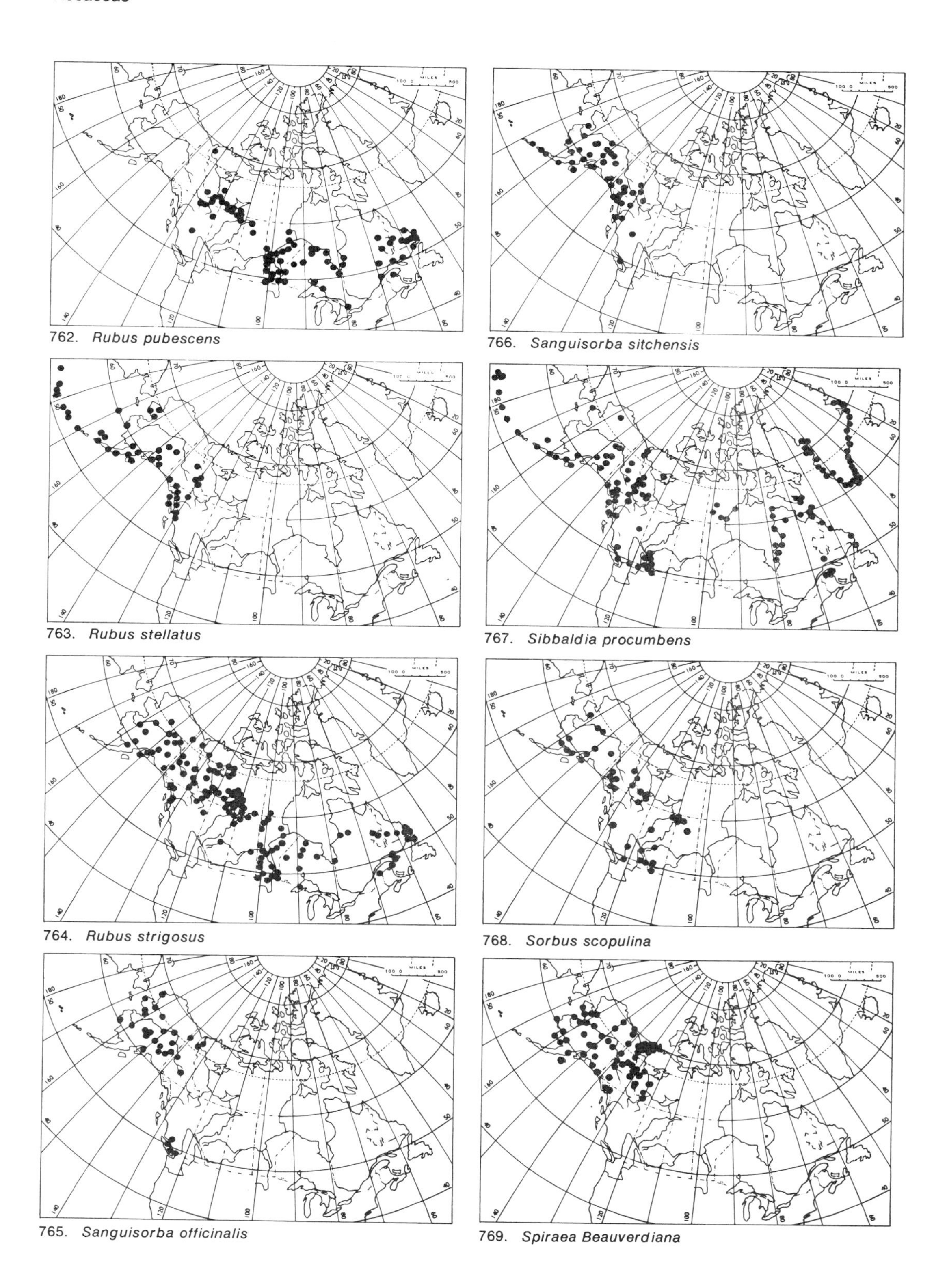

762. *Rubus pubescens*

766. *Sanguisorba sitchensis*

763. *Rubus stellatus*

767. *Sibbaldia procumbens*

764. *Rubus strigosus*

768. *Sorbus scopulina*

765. *Sanguisorba officinalis*

769. *Spiraea Beauverdiana*

LEGUMINOSAE Pea Family

Perennial herbs (ours) with alternate, compound and stipulate leaves; flowers perfect, irregular, of 5 more or less united sepals and 5 unlike petals, of which the upper and largest is called the standard or vexillum, the two lateral and similar are called the wings, and below and partly enclosed by them is the keel, which is formed by the two lowermost and united petals. Stamens 10 of which 9 are united, and one free; the fruit is a variously-shaped legume.

a. Leaflets more than 3 (native species)
  b. Leaves palmately lobed . . . . *Lupinus*
  b. Leaves pinnate
    c. Leaves without a terminal tendril
      d. Legumes sub-cylindric, dehiscent
        e. Plant caulescent; keel of the corolla blunt, without appendage at the tip . . . . *Astragalus*
        e. Plant caespitose; keel of the corolla tipped into an erect point . . . . . . *Oxytropis*
      d. Legumes flat, indehiscent, composed of articulate sections (loments) . . . . *Hedysarum*
    c. Leaves with a slender, terminal tendril
      f. Style filiform, bearded near the apex . . . . *Vicia*
      f. Style flattened, bearded down the inner side . . . . *Lathyrus*
a. Leaflets 3 (introduced, weedy species)
  g. Inflorescence globose . . . . *Trifolium*
  g. Inflorescence spike-like or racemose
    h. Legumes ovoid . . . . *Melilotus*
    h. Legumes twisted or coiled . . . . *Medicago*

Astragalus L. Milk-Vetch

Ours perennial caulescent and mostly calciphilous herbs with odd-pinnate leaves and lavender, blue-purple, or yellow flowers in short or elongating racemes. Legumes at least partly 2-locular by the sutures projecting inward into the locule, thus dividing the cavity lengthwise.

a. Caulescent, stems erect-ascending from subterranean horizontally spreading branches; flowers yellowish-white
  b. Stems up to 1 m tall, legumes glabrous
    c. Leaflets 11-13; legumes pendulous . . . . *A. americanus*
    c. Leaflets 13-25; legumes erect . . . . *A. canadensis*
  b. Stems 20-30 cm tall, legumes black-hirsute . . . . *A. umbellatus*
a. Caulescent from a central tap-root terminating in a many-headed crown
  d. Stems prostrate or ascending, forming tufts or low mats, legumes pubescent
    e. Stems low and trailing, flowers purple . . . . *A. Bodinii*
    e. Stems erect-ascending, forming low tufts or mats
      f. Flowers loose; legumes black-hirsute, pendulous . . . . *A. alpinus*
      f. Flowers dense; legumes grey-hirsute, erect and sessile
        g. Plants low, 5-20 cm tall, often mat-forming . . . . *A. agrestis*
        g. Plants 10-30 cm tall, tufted from a stout sub-ligneous root . . . . *A. striatus*
  d. Stems erect-ascending, mostly over 25 cm tall
    h. Legumes hirsute, less than 1 cm long, pendulous, in an elongated raceme . . . . *A. eucosmus*
    h. Legumes glabrous, 1.5-2.0 cm long
      i. Flowers pale yellow, plant essentially glabrous, drying black . . . . *A. tenellus*
      i. Flowers purple; plant grey-hirsute
        j. Stipules grey-brown pubescent; stipe of mature legume much longer than calyx lobes; non-arctic . . . . *A. aboriginum*
        j. Stipules black-pubescent; stipe of mature legume shorter than calyx lobes; arctic . . . . *A. Richardsonii*

Astragalus aboriginum Richards.
A. linearis (Rydb.) Porsild
Tufted with erect-ascending 15 to 30 cm tall stems from a yellowish, fleshy tap-root and a much branched crown. Leaves sessile, the 9 to 11 linear leaflets silvery grey from appressed hairs. Stipules grey-hirsute, connate, ovate-acute, 3 to 5 mm long. Inflorescence a dense but soon elongating 10- to 20-flowered raceme of yellowish-white or pale purplish flowers. Legumes sickle-shaped, turgid when ripe, dark grey, glabrous and about 1.5 cm long, prominently stipitate within the calyx. The roots are edible and by Richardson were compared to "those of *Glycyrrhiza* gathered in the spring by Cree and Stony Indians, as an article of food".

Stream banks, meadows and woodland clearings.

General distribution: Cordilleran prairie and foothill species east to central Sask., north to southeastern Alaska and the upper Mackenzie Basin. In northern Alaska the closely related or, perhaps, doubtfully distinct *A. Lepagei* Hult.

Fig. 666 Map 770

Astragalus agrestis Dougl.
A. goniatus Nutt.
Slender, loosely tufted and diffuse, with us rarely over 15 cm tall and often less. Leaflets 9 to 23, mostly less than 10 mm long, glabrous above, strigose beneath, commonly retuse; racemes dense, not elongating, becoming spherical in fruit. Corollas purple, the legumes about 1 cm long, ovoid and densely villous.

Damp prairies and flood-plain meadows.

General distribution: N. American prairie and foothill species; upper Mackenzie R. valley, north to Good Hope.

Map 771

Astragalus alpinus L.
Alpine Milk-Vetch
Low and matted with creeping and freely branched weak stems. Leaflets 15 to 23, oblong-elliptic or broadly lanceolate, white-strigose beneath, glabrous or sparingly appressed hirsute above. Peduncles axillary, longer than the leaves, terminating in short, non-elongating racemes of spreading pale bluish-violet flowers. Legumes pendulous, stipitate, and conspicuously black or brownish hirsute.

In well-watered calcareous sandy or gravelly places.

General distribution: Circumpolar, wide spread arctic-subarctic, alpine.

Fig. 667 Map 772

Astragalus americanus (Hook.) M.E. Jones
Erect from a somewhat woody and branching base. Stems up to 1 m tall; leaves pinnate, leaflets 11 to 13, oblong, 3 to 4 cm long, fresh green and glabrous above, paler and sparingly hairy beneath; stipules oblong, 1 cm long. Racemes 12- to 20-flowered, short and not elongating, in a terminal cluster. Flowers yellowish-white turning brownish in age. Legumes lanceolate, about 2 cm long, the valves glabrous and membranaceous.

Openings in thickets on alluvial plains and lake shores; with us restricted to the Upper Mackenzie drainage.

General distribution: N. America; N. W. Ont. to B.C. and S. E. Alaska.

Map 773

Astragalus Bodinii Sheld.
A. yukonis M.E. Jones.
Mat-forming and spreading from a central taproot. Stems very numerous, reclining, 30 to 50 cm long. Leaflets 11 to 19, lanceolate 5 to 10 mm long, glabrous above, sparingly short appressed-hirsute beneath, the tiny bluish-purple flowers in an elongating raceme on a slender up to 25 cm long peduncle. Legumes 6 to 8 mm long, ovoid, apiculate, black-strigose.

Damp, clayey river bank meadows northward along the Mackenzie River to the Arctic Coast.

General distribution: Endemic of N.W. America. In the East a closely related or perhaps identical species, *A. stragulus* Fern., thus far known only from Pistolet Bay in Newfoundland.

Fig. 668 Map 774

Astragalus canadensis L.
Stems stout, sub-ligneous, branching above and up to 1 m tall, from a creeping rhizome. Leaflets 13 to 25, oval, 1.5 to 2.0 cm long, glabrous above and appressed pilose beneath. Flowers greenish-yellow, very numerous, in dense racemes; legumes glabrous, 1.5 cm long, cylindrical, very firm and long-persisting, on stout peduncles.

Moist open woods and river banks, barely entering the District of Mackenzie along the lower Liard River.

General distribution: N. America; W. Que. to B.C.

Fig. 669 Map 775

Astragalus eucosmus Robins.
Stems slender, erect, 30 to 60 cm high from ascending, slender, branched caudices. Leaflets 9 to 17, linear-oblong to oblanceolate and

obtuse, usually glabrous and green above, white-strigose beneath. Peduncles axillary, longer than the leaves, terminating in a dense, but soon elongating, 6- to 16-flowered raceme of small, deep purple flowers. Legumes sessile, reflexed, densely black or greyish hirsute.

Calcareous gravels, often among willows on sand and gravel bars in rivers and by sheltered lake shores.

General distribution: N. America, wide-ranging, subarctic.

Fig. 670 Map 776

Astragalus Richardsonii Sheldon
A. aboriginum aucct. pro pte. non Richardson
Richardson's Milk-Vetch
Coarse plant with a stout, sub-ligneous tap-root terminating in a many-headed crown and ascending leafy stems 10 to 25 cm high. Leaves short and densely ashy-grey tomentose, with 9 to 15 linear-lanceolate leaflets about 1 cm long. Inflorescence axillary, on peduncles 5 to 8 cm long, short and capitate during anthesis, elongated in fruit. Flowers 10 to 20, about 1.5 cm long, scentless; calyx densely black villous, half as long as the corolla in which the standard and wings are white or pale pink, with prominent green veins, and only the keel prominently purplish tipped. Legumes short stipitate within the calyx, glabrous even in youth, 2 cm long when full-grown, strongly inflated and bladder-like, dark wine red and somewhat translucent.

Astragalus Richardsonii is closely related to the prairie species *A. aboriginum,* from which it differs by larger flowers, glabrous legumes, more densely tomentose leaves and stems, and by its larger, 5 to 10 mm long, black-hirsute stipules, and woody tap-root.

Calcareous sandy and gravelly places.

General distribution: Endemic of northern District of Mackenzie and the western islands of the Arctic Archipelago.

Fig. 671 Map 777

Astragalus striatus Nutt.
A. adsurgens Pall.
var. *robustior* Hook.
Stems 10 to 30 cm tall, erect-ascending and freely branching from a stout, sub-ligneous root terminating in a many-headed crown. Leaflets 19 to 23, oblong, 15 to 20 mm long. Flowers purplish, in a dense, rarely over 2 cm long head, at the end of a 10 to 15 cm long peduncle; the legumes oblong, sessile, 8 to 10 mm long.

Grasslands and open woods, in the District of Mackenzie collected only a few times north to Ft. Simpson.

General distribution: N. America: W. Ont. to S. Y.T.

Fig. 672 Map 778

Astragalus tenellus Pursh
Tufted, with few to numerous dark-purplish 5 to 7 dm tall stems, from a stout, sub-ligneous root terminating in a branching crown. Leaflets linear or narrowly oblanceolate, 11 to 15, glabrous or sparsely hairy on the upper surface, drying black. Flowers small, white or pale yellow, in few-flowered racemes. Legumes 1.2 to 2.0 cm long, glabrous and spindle-shaped, stipitate within the calyx, black and drooping when mature.

Prairie and grassland species.

General distribution: Western N. America: In the District of Mackenzie confined mainly to the Mackenzie River valley, north to lat. 67°30′.

Fig. 673 Map 779

Astragalus umbellatus Bunge
A. frigidus (L.) Bunge
var *littoralis* (Hook.) Wats.
Stems 20 to 30 cm tall, solitary or few together from a horizontally spreading rhizome. Leaflets 7 to 9, oblong, 2.0 to 2.5 cm long and up to 1.0 cm wide, dark green and glabrous above, sparsely white-hirsute beneath. Flowers in a short raceme, pendulous in age. Calyx cylindrical, greenish-yellow in life, drying brown, corolla deep yellow with white fringes; legumes black-hirsute, pendulous.

A very handsome species often forming small and pure colonies in rich herbmat meadows.

General distribution: Amphi-Beringian, arctic-alpine reaching the District of Mackenzie along the east slope of Richardson—Mackenzie Mts.

Fig. 674 Map 780

Hedysarum L. Liquorice-Root

Perennial herbs with odd-pinnate leaves, and showy flowers in axillary, long-peduncled racemes. Legumes linear, thin, flattened, composed of several readily separable, roundish 1-seeded sections or loments.

a. Calyx teeth deltoid . . . . *H. alpinum*
a. Calyx teeth lance- or linear-subulate . . . . *H. Mackenzii*

Hedysarum alpinum L.
var. **americanum** Michx.
Liquorice-Root
Stems few, erect, up to 60 cm tall, from brown, scaly caudices and large, fleshy rhizomes. Leaves fresh green and glabrous, with 9 to 13 lanceolate leaflets, and conspicuous, brown stipules. Flowering racemes 3 to 12 cm long on straight peduncles 5 to 10 cm long; flowers 10 to 20, pink or pale purple, 1.75 to 2.0 cm long. Loments of legume 2 to 3, oval, glabrous, conspicuously net-veined.

Beyond the tree line, in Alaska, Yukon and North West District of Mackenzie a lower and fewer-flowered plant has somewhat tentatively been referred to the Siberian *H. alpinum* L. var. *alpinum* (*H. arcticum* Fedtsch. or *H. hedysaroides* (L.) Schinz. & Thell.). However, the transitional differences between the arctic and more temperate plant appear to be quite gradual and to be merely an expression of climatic differences.

Calcareous sands and gravels, often on riverbanks and by sheltered lake shores. The fleshy roots, which in mature plants may be half an inch thick, are edible, and when cooked taste somewhat like young carrots.

General distribution: N. America; Nfld. to Alaska.

Fig. 675 Map 781

Hedysarum Mackenzii Richards.
Stems numerous, ascending or arching, 15 to 35 cm high, from a much-branched, sub-ligneous caudex, and thick, fibrous tap-root. Leaves with 5 to 13 linear-lanceolate, 1.5 cm long leaflets, glabrous and green above, the lower surface appressed, silvery pubescent; stipules inconspicuous, grey. Flowering racemes at first sub-capitate, but soon elongating, 2 to 8 cm long, on curved, arching peduncles; flowers 5 to 25, deep purple, showy and sweet scented, 2.5 to 3.0 cm long. Loments of the legume 3 to 6, oval or circular in outline, minutely pubescent and conspicuously transversely veined.

In calcareous clays and gravels, often along river banks and lake shores.

General distribution: Arctic-subarctic. N. America from Nfld. to Alaska.

Fig. 676 Map 782

Lathyrus L. Wild Pea

Perennial herbs (ours), climbing by means of tendrils at tip of the pinnate leaves. Flowers showy, in axillary racemes.

a. Flowers bluish-purple; plants of gravelly sea-shores .... *L. japonicus* var. *aleuticus*
a. Flowers yellowish-white; inland species .... *L. ochroleucus*

Lathyrus japonicus Willd.
var. **aleuticus** (Greene) Fern.
Beach Pea
Matted from widely creeping and freely forking, cord-like rhizome. Leaves glaucous-green, somewhat fleshy.

The var. *aleuticus* is distinguished from the essentially glabrous typical plant by more consistent pilosity of the stem and lower surface of the leaves. However, in the Mackenzie Delta region the sea-shore plant is more pilose than that of lakeshores and upstream.

Sheltered beaches and river banks.

General distribution: Circumpolar; in the District of Mackenzie confined to the Mackenzie Delta and the Coronation Gulf area.

Fig. 677 Map 783

Lathyrus ochroleucus Hook.
Stems up to 1 m tall, freely climbing among low shrubs along river banks and in openings in the forest.

General distribution: N. America; Que. to B.C. and southward. With us restricted to the upper Mackenzie R. drainage north to lat. 63° N.

Fig. 678 Map 784

Lupinus L. Lupine

Lupinus arcticus Wats.
Tufted perennial with numerous palmately divided, sparsely pubescent, dark green leaves, lowish-villous, the dehiscent valves strongly twisted.

Moist tundra and grassy alpine slopes.

and ascending-erect, 30 to 50 cm high stems, terminating in showy racemes of large, bluish-purple flowers. Legumes oblong, flattened, yel-

General distribution: N. W. America, arctic-alpine endemic.

Fig. 679 Map 785

Medicago L. Medick

Annual or perennial herbs with pinnately 3-foliolate leaves. Inflorescence racemose or spike-like. Legumes coiled.

With us as casually introduced and likely ephemeral roadside weeds.

a. Prostrate or decumbent annual with small, yellow flowers in a dense, spike-like raceme . . . . *M. lupulina*
a. Erect-ascending perennial with blue-violet flowers in sub-capitate racemes . . . . . *M. sativa*

Medicago lupulina L.
Black Medick
Annual weed with slender, soft-hairy, procumbent or reclining stems freely branching from the base. Flowers small, bright yellow in a short and dense spike. Legume 1-seeded.

A single specimen reported from garden at Fort Smith.

General distribution: Cosmopolitan weed.

Medicago sativa L.
Alfalfa
Stems freely branched, erect or ascending from a stout perennial root. Flowers purple, in oblong racemes. Legumes 10- to 20-seeded.

An occasional escape, reported in the Mackenzie River region, from Alexandra Falls on the Hay River, and at Fort Simpson.

General distribution: Cosmopolitan

Melilotus Mill. Sweet Clover

Similar to *Medicago* but legumes ovoid. With us as casually introduced roadside weeds.

a. Corolla 4-6 mm long, petals yellow . . . . *M. officinalis*
a. Corolla 2.0-2.5 mm long, petals white . . . . *M. alba*

Melilotus alba Desr.
Stems upright, branching from the base up to 1 m tall. Flowers white, the legumes 3 to 4 mm long.

Introduced roadside weed increasingly common in the upper Mackenzie Valley.

General distribution: Cosmopolitan.

Melilotus officinalis (L.) Lam.
Similar to preceding, but flowers yellow.

Recently reported from waste places in southernmost Mackenzie Valley.

General distribution: Cosmopolitan.

Oxytropis DC.

Ours caespitose or tufted, distinctly calciphilous perennials with stout tap-roots terminating in thick, many-headed caudices, odd-pinnate leaves, scape-like peduncles, and short spike-like inflorescences. The flowers are similar to those of the genus *Astragalus* but differ by the abruptly pointed keel and by the 1-locular legumes.

a. Caulescent; stems with one or more internodes from a branching rhizome
  b. Inflorescence not elongating in fruit; legume black-hirsute . . . . . *O. deflexa* var. *foliolosa*
  b. Inflorescence elongating in fruit; legume grey-hirsute . . . . *O. deflexa* var. *sericea*
a. Caespitose; stems from a central tap-root terminating in a many-headed crown
  c. Plant glandular-viscid; calyx lobes strongly glandular-verrucose
    d. Calyx lobes deltoid, one-quarter to one-third as long as the tube . . . . . *O. hudsonica*
    d. Calyx lobes subulate, one-half to two-thirds as long as the tube
      e. Stipules glabrous, their margins ciliate . . . . *O. sheldonensis*
      e. Stipules densely hirsute on the back
        f. Calyx and legume densely black-hirsute . . . . *O. glutinosa*
        f. Calyx and legume sparsely yellowish-grey hirsute . . . . *O. viscida*
  c. Plant not glandular-viscid
    g. Stipules chestnut-brown; flowers yellow

h. Calyx, bracts and also the legumes shaggy from mixed, long black and white hairs . . . . *O. Maydelliana* ssp. *Maydelliana*
h. Calyx, bracts and legumes uniformly black-hirsute . . . . *O. Maydelliana* ssp. *melanocephala*
g. Stipules papery white or yellow, glabrous or hirsute
i. Leaflets, or at least some of them in whorls of 3 or 4
j. Stipules densely pilose; flowers purple
k. Inflorescence of from 3-6 flowers, not elongating in fruit . . . . *O. Bellii*
k. Inflorescence of many but small flowers, not much elongated in fruit . . . . *O. splendens*
j. Stipules sparingly villous; flowers yellow . . . . *O. hyperborea*
i. Leaflets opposite or alternate
l. Dwarf species with loosely tufted or compact habit; flowers blue-purple
m. Loosely tufted; flowers 1-3
n. Leaflets fresh green; stipules papery and glabrous . . . . . *O. Scammaniana*
n. Leaflets glabrate above and white-villous beneath; stipules densely pilose . . . . *O. arctica*
m. Densely caespitose or pulvinate
o. Entire plants cushioned, permanently white-velutinous-silky; flowers solitary . . . . *O. arctobia*
o. Leaves appressed silky; flowers 2-3 . . . . *O. nigrescens*
l. Taller loosely caespitose, not pulvinate from a central tap-root; flowers yellow
p. Leaflets 4-6 pairs, thinly appressed pubescent or glabrate in age . . . . *O. Jordalii*
p. Leaflets 5-15 pairs, appressed silky on both surfaces
q. Stipules with clavate processes . . . . *O. varians*
q. Stipules lacking clavate processes . . . . *O. spicata*

Oxytropis arctica R. Br.
O. coronaminis Fern.
Densely caespitose, with a stout tap-root and freely branched caudex, often forming soft, cinereous-grey cushions or hassocks 20 to 30 cm in diameter. Leaves 4 to 6 cm long with 9 to 13 opposite, oblong-lanceolate leaflets, 9 to 10 mm long and 3 mm wide, which are densely white-villose beneath and glabrate above; the rachis and petiole, but especially the base of the stipules, shaggy with long snowy-white or slightly yellowish hairs, the free part papery white, glabrous, with long white hairs and tiny clavate processes along the margins. Scapes 8 to 10 cm high, erect; spike 3- to 5-flowered, sub-capitate, not elongating in fruit. Flowers sweetly perfumed, very large and showy; calyx about 1.5 cm long, the teeth subulate, one-half as long as the tube; corolla 2.5 cm long, dark purple, turning blue in drying. Mature legumes dark olive-green and pubescent, somewhat sickle-shaped, erect or spreading, 3 cm long, including the long beak.

Ubiquitous on dry, open tundra on the flat, windswept calcareous barrens of the central islands of the Archipelago north to Prince Patrick Island, south to the tree line north of Great Slave Lake; less common westward along the north and northwest coast of Alaska.

General distribution: Endemic of arctic N. America.

Fig. 680 Map 786

Oxytropis arctobia Bunge
Low and densely caespitose or pulvinate, forming flat or hemispherical, velvety-grey cushions that may attain a diameter of 25 to 30 cm. Leaves 1.5 to 2.0 cm long with 5 to 9 minute, often conduplicate, densely grey-villous leaflets. Scapes barely exceeding the leaves, 1- to 2-flowered, white- and black-villous; calyx densely black-villous; corolla purple, about 1.0 cm long; legumes sessile within the calyx, subcylindric, 2.5 to 3.0 cm long, black and grey-villous.

In anthesis *O. arctobia* is a strikingly handsome species when the grey cushions of large individuals may be covered by several hundreds of flowering scapes, whereas in fruit only the large, black legumes show up against the ashy-grey foliage, which, otherwise, completely blends with the limestone gravel on which it grows.

General distribution: Endemic of the islands

of the Arctic Archipelago and the arctic coast east to Baffin Island and northern Hudson Bay but very rare inland, south to the north shore of Great Bear L. and northern District of Keewatin.
Fig. 681 Map 787

Oxytropis Bellii (Britt.) Palibine
Loosely caespitose with freely branched caudices, forming large, flat cushions up to 30 cm in diameter. Leaves 3 to 10 cm long, with 15 to 31 opposite, sub-opposite, or verticillate white villous, somewhat crowded, 3 to 9 mm long leaflets, and rather small densely villous, pale stipules. Scape reddish-brown, 5 to 10 cm long; spike sub-capitate, 3- to 6-flowered, not elongating in fruit. Calyx wine red in life under the mixed black and white pubescence; the black, subulate teeth, one-third as long as the tube; corolla about 2.0 cm long, reddish-violet in life, drying violet, with a pale spot on the standard; legumes narrowly ovoid, 2.5 cm long, including the slender beak.
On dry, gravelly slopes and in rocky tundra. The flowers are faintly fragrant.
General distribution: Endemic of the northern Hudson Bay region.
Fig. 682 Map 788

Oxytropis deflexa (Pall.) DC.
var. **foliolosa** (Hook.) Barneby
Loosely tufted, delicate plant with slender, erect-ascending stems. Leaflets 15 to 23, in approximate pairs, lanceolate-elliptic, fresh green and glabrous above, sparingly white-hirsute beneath. Peduncles slender, slightly longer than the leaves; inflorescence short, non-elongating, 5- to 9-flowered. Flowers blue-violet, less than 1 cm long, spreading, reflexed or secund in anthesis. Legumes short, black-hairy, pendulous.
On well-watered calcareous sand and gravel bars, often under willows by rivers and lake shores.
General distribution: Endemic of subarctic-alpine N. America.
Fig. 683 Map 789

Oxytropis deflexa (Pall.) DC.
var. **sericea** T. & G.
Similar to var. *foliolosa* but generally more pilose, taller and more robust, often caulescent and with one or more internodes. Inflorescence of smaller, white or pinkish flowers, much elongated in fruit; legumes grey-pubescent.
Open woods and grassland.
General distribution: N. America; western foothills and plains; in the Mackenzie Valley north to Norman Wells, south central Y.T. and Alaska.
Fig. 684 Map 790

Oxytropis glutinosa Porsild
O. borealis sensu Hultén *non* DC.
Similar and probably closely related to *O. hudsonica* from which it differs by its much longer, subulate teeth of the calyx and by its inflorescence, which, unlike that of *O. hudsonica* is considerably elongated in fruit. The foliage and also the peduncles of *O. glutinosa* are distinctly clammy because of the prescence of numerous glands clearly visible under a lens in fresh specimens. When fresh, the leaves are fragrant.
Calcareous screes and mountain slopes.
General distribution: Endemic of arctic-alpine N. W. America.
Fig. 685 Map 791

Oxytropis hudsonica (Greene) Fern.
Loosely caespitose with freely branched caudices; viscid-glandular throughout. Leaves 5 to 10 cm long, with 15 to 33 oblong leaflets, these involute margined, glabrous or sparingly pilose above, sparingly villous beneath. Scapes 10 to 15 cm long, spike short, 7- to 13-flowered, not elongated in fruit. Flowers fragrant; calyx campanulate, greyish-black hirsute, densely glandular-verrucose, the teeth deltoid, one-quarter to one-third as long as the tube. Corolla 1.5 to 1.7 cm long, purple; legumes thinly grey-pilose, about 2 cm long, narrowly ovoid, tapering to a long, straight beak.
In not too dry, calcareous sandy and gravelly places.
General distribution: Endemic of the Hudson Bay region west to long. 120°W. Smaller but perhaps closely related to the western *O. viscida* Nutt.
Fig. 686 Map 792

Oxytropis hyperborea Porsild
Caespitose with stout, freely branched caudices. Leaves spreading or ascending, 3 to 12 cm long, short petioled, and with pale, papery, sparingly villous stipules; leaflets 9 to 19, linear-lanceolate, opposite, sub-opposite, or in fascicles of 3, silky-grey on both sides, or sometimes glabrate above. Scapes 5 to 20 cm long, erect-ascending, or in the Far North often prostrate. Spikes 6- to 10-flowered, sub-capitate or lax-flowered, not greatly elongated in fruit. Calyx densely hirsute with mixed short black, and long, white hairs; the teeth deltoid, one-quarter to one-third as long as the tube. Corolla about 1.5 cm long, pale yellow or buff

in life; legumes ovoid, 1.5 cm long, including the long, divergent beak, villous, with mixed black and white hairs.

Calcareous sandy and gravelly places, and in dry tundra.

General distribution: *O. hyperborea* is a truly arctic-alpine tundra species, endemic of arctic Alaska, Y.T., N.W. District of Mackenzie, and the westernmost islands of the Arctic Archipelago.

Fig. 687 Map 793

Oxytropis Jordalii Porsild
non O. campestris (L.) DC.
ssp. *Jordalii sensu* Hultén
Tufted from a few-headed tap-root, each branch densely covered by blackened, marcescent petioles and stipules. Leaves 4 to 6 cm long, with 9 to 11 narrowly lanceolate leaflets, glabrous above, appressed, silvery pubescent beneath. Stipules papery white and translucent, strigose. Scapes 2 to 14 cm long, dark purplish, sparsely sericeous. Inflorescence oblong 3- to 7-flowered, scarcely elongating in fruit. Calyx grey-pubescent, the teeth subulate, one-third as long as the tube, corolla 10 to 12 mm long, pale greenish-yellow, drying ivory-white; legumes 12 to 18 mm long including the 5 mm long beak, sessile within the calyx, greenish under a dense but short and black indument.

Turfy limestone barrens near timberline.

General distribution: Endemic to mountains of northern Alaska and Y.T. to the Mackenzie and Richardson Mts.

Fig. 688 Map 794

Oxytropis Maydelliana Trautv.
Tufted, with a stout, freely branching caudex, densely covered by marcescent, chestnut-brown stipules. Leaves 3 to 6 cm long, usually with about 13 lanceolate-oblong, acute, sparingly white-villous to glabrate leaflets. Scapes stiffly erect, 5 to 15 cm long; spike short, 5- to 7-flowered, not elongating in fruit. Corolla 1.0 to 1.5 cm long, pale yellow. Legumes ovoid, about 1.3 cm long and abruptly beaked.

Common in not too dry, turfy places in tundra.

O. Maydelliana may be separated into two easily distinguished races (see key).

General distribution: *O. Maydelliana* ssp. *Maydelliana* is Amphi-Beringian and from East Asia extends across arctic and alpine parts of Alaska and Y.T. to the east slopes of the Mackenzie—Richardson Mts.; *O. Maydelliana* ssp. *melanocephala* (Hook.) Porsild is endemic to eastern arctic Canada where it extends east from the Mackenzie Valley to northern Lab., south to the tree line and north across the Arctic Archipelago to Melville Island.

Fig. 689 (ssp. *melanocephala*) Map 795 (*O. Maydelliana s. lat.*)

Oxytropis nigrescens (Pall.) Fisch.
Caespitose with few to numerous branches from a many-headed, stout tap-root. Stipules papery, white, glabrous or with a few long white hairs. Leaves 5 to 8 cm long, stipe slender, and purplish, sparsely white-pilose; leaflets 9 to 13, narrowly lanceolate, 5 to 8 mm long and 2 mm wide, boat-shaped, densely silky-canescent when young, in age sparsely grey-hirsute on both sides. Racemes mostly 2-flowered, on scapes barely over-topping the leaves; calyx black-hirsute, the teeth about half as long as tube; corolla about 15 mm long, purple; mature legumes 3.0 to 3.5 cm long, narrowly lanceolate, sessile, or very short-stipitate, the valves dark-grey pilose.

Stony alpine slopes.

General distribution: Amphi-Beringian: In an arctic environment *O. nigrescens* tends to become pulvinate (ssp. *pygmaea* (Fern.) Hult.), whereas in interior Alaska, Y.T. and the Richardson and Mackenzie Mts. it may grow taller and less compact (*O. nigrescens* ssp. *bryophila* (Greene) Hult.).

Fig. 690 (ssp. *pygmaea*) Map 796 (ssp. *byophylla*), 797 (ssp. *pygmaea*)

Oxytropis Scammaniana Hult.
Loosely tufted dwarf species with numerous cylindrical branches from a many-headed crown. Stipules papery-white and glabrous dorsally, only the blunt lobes black-ciliate. Leaflets 9 to 13, mostly glabrous above and white-pilose beneath. Flowering peduncles pale purple under a cover of short, white hairs; racemes 2- to 3-flowered, the calyx densely black pilose, with linear teeth half as long as the tube; petals pale purplish, and the legumes black hirsute.

Alpine lichen or herbmats.

General distribution: Endemic of central Alaska and Y.T., with disjunct stations in the Mackenzie Mts., N.W.T.

Fig. 691 Map 798

Oxytropis sheldonensis Porsild
Caespitose from a stout, many-headed fibrous tap-root; stipules glabrous, papery white, the lobes long-acuminate with long-ciliate margins. Leaves 5 to 10 cm long, the 25 to 31 leaflets lanceolate, 8 to 10 mm long and 2 to 3 mm

broad, glabrous above, their margins and underside sparsely white-villous. Racemes capitate, 5- to 12-flowered, elongating in age. Calyx turbinate, black-hirsute, the teeth subulate. Corolla roseate, drying bluish; the legumes dark-grey, about 2 cm long, including the long, curving beak.

Although in life the plant feels clammy or glutinous and, in drying, stains the drying paper, no glands can be seen on dried specimens, except along the margins of the stipules, and on the calyx lobes.

Alpine ledges.

General distribution: Endemic of N. America; thus far known only from widely disjunct stations in northern Alaska, S.W. Y.T., and the east slopes of Mackenzie Mts.

Map 799

Oxytropis spicata (Hook.) Standl.
O. sericea Nutt.
var. *spicata* (Hook.) Barneby
Caespitose from a very stout, branching tap-root. Leaves 12 to 15 cm long, leaflets 11 to 31, oblong, appressed silky on both surfaces. Scapes stout, much longer than the leaves. Racemes oblong, many-flowered, elongating in age; calyx 10 to 12 mm long, including the subulate teeth, the tube densely grey-pubescent. Corolla pale yellow; legumes oblong, grey-strigose, about 2 cm long, including the long, slender beak.

Grassland and river terraces.

General distribution: Cordilleran foothill species barely reaching the upper and central Mackenzie R. drainage.

Fig. 692 Map 800

Oxytropis splendens Dougl.
Tufted, copiously silky-villous throughout, from a strong, fibrous several-headed tap-root. Leaves 10 to 20 cm long, the numerous linear-lanceolate leaflets in whorls of 4. Scapes 20 to 40 cm tall, the racemes spike-like, many flowered; calyx densely silky-villous, the flowers 10 to 14 mm long, reddish purple, drying blue. Legumes ovoid about 1 cm long, densely villous.

River banks and clearings.

General distribution: N. W. America; prairies and foothills north to Great Bear L.

Fig. 693 Map 801

Oxytropis varians (Rydb.) K. Schum.
O. campestris (L.) DC.
var. *varians* (Rydb.) Barneby
Caespitose from a stout many-headed tap-root; stipules pilose, the margins with bristly ciliae and clavate processes. Leaves 3 to 13 cm long with 11 to 25 scattered or subopposite silky-pilose or glabrescent leaflets. Racemes capitate, 6- to 25-flowered, elongating in age. Calyx black and white hirsute, the teeth subulate. Corolla yellow; the legumes yellow-green with mixed black and white hairs, 1.5 to 2.0 cm long including the fine recurved beak.

Gravel river banks and terraces and open slopes.

General distribution: Endemic of northwestern North America.

Fig. 694 Map 802

Oxytropis viscida Nutt.
O. viscidula (Rydb.) Tidestr.
Caespitose with a strong tap-root and a much branched crown. Most green parts of the plant pilose and viscid due to wart-like glands. Leaves 10 to 15 cm long with 25 to 35 opposite or alternate leaflets; raceme many-flowered, sub-capitate, elongating in fruit. The calyx grey-hairy, more than half as long as the purplish-blue corolla, its subulate teeth distinctly warty; legumes grey-hirsute, 10 to 15 mm long, including the slender beak.

Western foothill prairies.

General distribution: Cordilleran endemic reaching the southern parts of the Mackenzie R. drainage and S. E. Alaska.

Fig. 695 Map 803

Trifolium L. Clover

Perennial herbs (ours) with 3-foliolate leaves; the small white, pink, or red flowers in the oval head and the small, few-seeded legumes membranaceous, often included in the calyx.

a. Flowers small, about 1.0 cm long
  b. Stems erect-ascending . . . . *T. hybridum*
  b. Stems creeping, rooting at the nodes . . . . *T. repens*
a. Flowers 1.0-2.0 cm long . . . . *T. pratense*

Trifolium hybridum L.
Alsike Clover
Stems branched, erect or ascending, lacking basal runners. Flowers pink or white.
Occasional escape from cultivation; in the District of Mackenzie near towns and settlements north to Liard River.
General distribution: Naturalized from Europe.

Trifolium pratense L.
Red Clover
Biennial or short-lived perennial with erect or spreading stems up to 8 dm tall; the lower leaves long-petioled; leaflets oval-elliptic, 2 to 5 cm long; heads nearly sessile; flowers red or pink.
Occasional escape from cultivation.
General distribution: Naturalized from Europe.

Trifolium repens L.
White Clover
Stems creeping and freely rooting. Flowers white or pale pink.
Weed of lawns, in the District of Mackenzie reported but once at Fort Simpson.
General distribution: Introduced and naturalized from Europe.

Vicia L. Vetch

Vicia americana Muhl.
Herbs climbing by means of a tendril at the tip of the pinnate leaves. Flowers in axillary racemes. In these characters *Vicia* resembles members of the genus *Lathyrus,* but in the latter the style is bearded down the inner surface, and the free stamens of equal length whereas in *Vicia* the style is bearded only at the summit, and the free stamen of unequal length.
Open thickets and meadows. Occasional to common along the Mackenzie River and its tributaries, north to the upper Mackenzie Delta.
General distribution: N. America.
Fig. 696 Map 804

666. *Astragalus aboriginum*

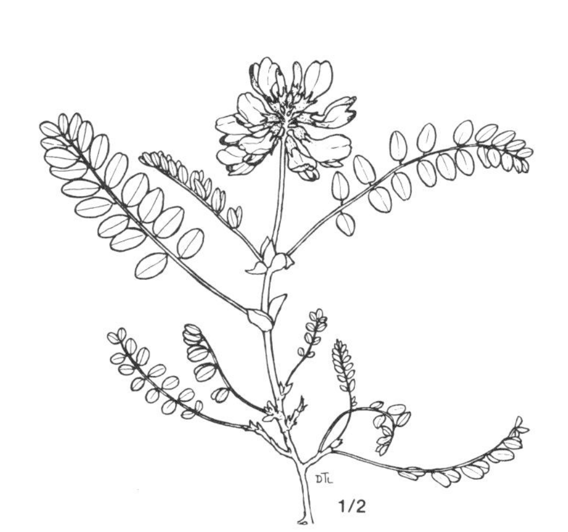

667. *Astragalus alpinus*

668. *Astragalus Bodinii*

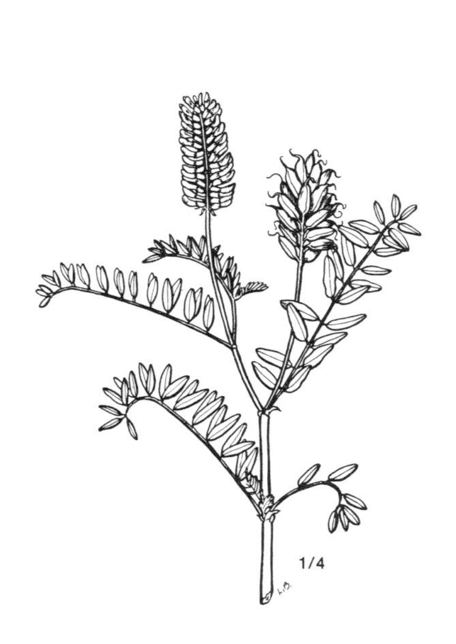

669. *Astragalus canadensis*

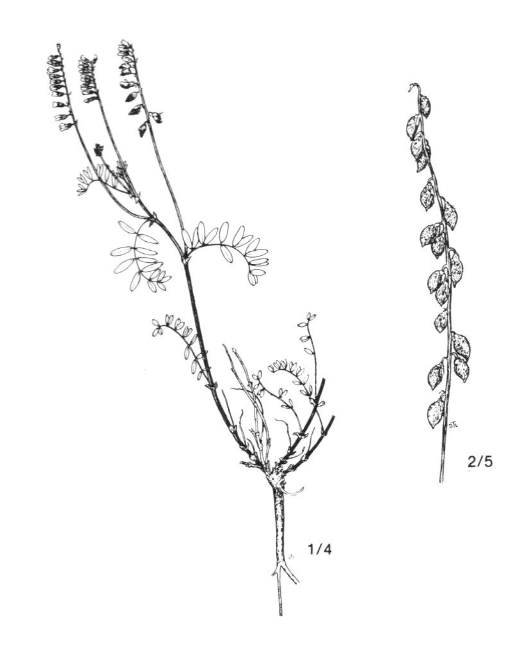

670. *Astragalus eucosmus*

671. *Astragalus Richardsonii*

672. *Astragalus striatus*

673. *Astragalus tenellus*

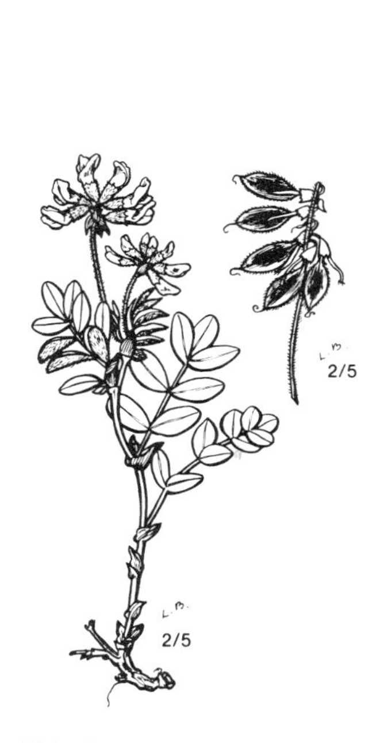

674. *Astralgalus umbellatus*

675. *Hedysarum alpinum* var. *americanum*

676. *Hedysarum Mackenzii*

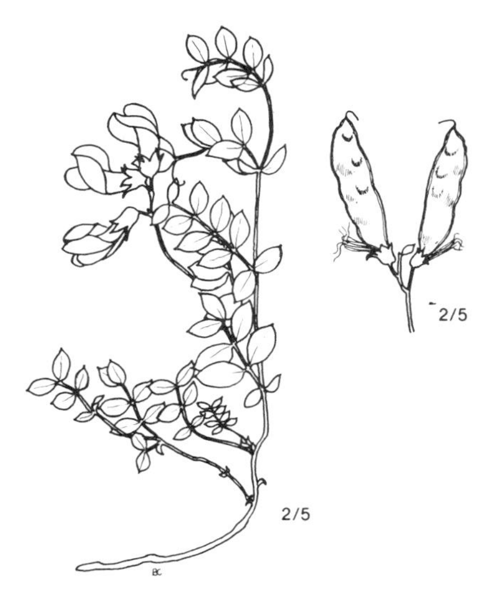

677. *Lathyrus japonicus* var. *aleuticus*

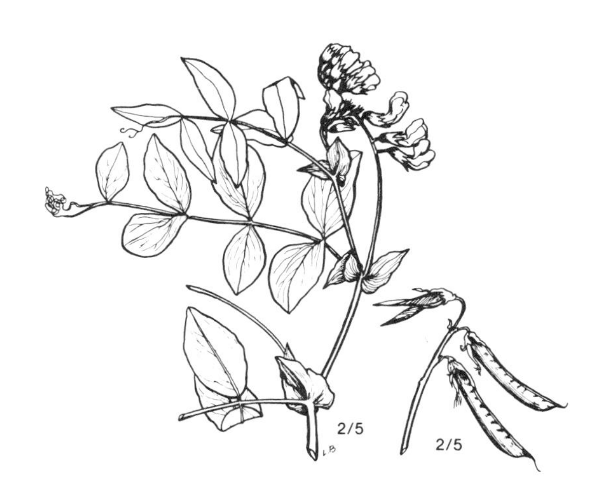

678. *Lathyrus ochroleucus*

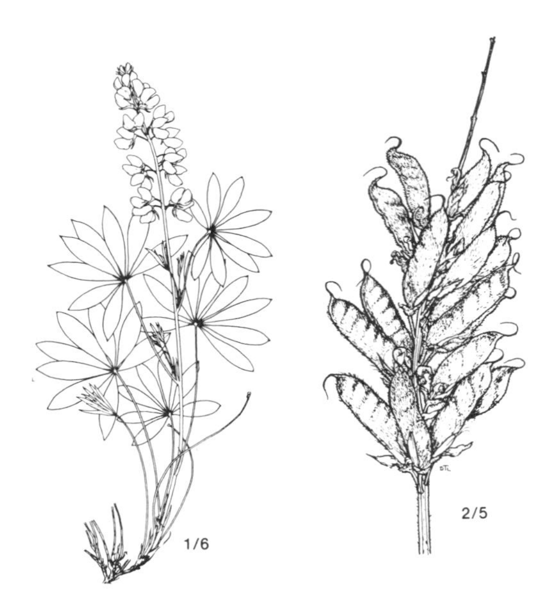

679. *Lupinus arcticus*

680. *Oxytropis arctica*

681. *Oxytropis arctobia*

682. *Oxytropis Bellii*

683. *Oxytropis deflexa* var. *foliolosa*

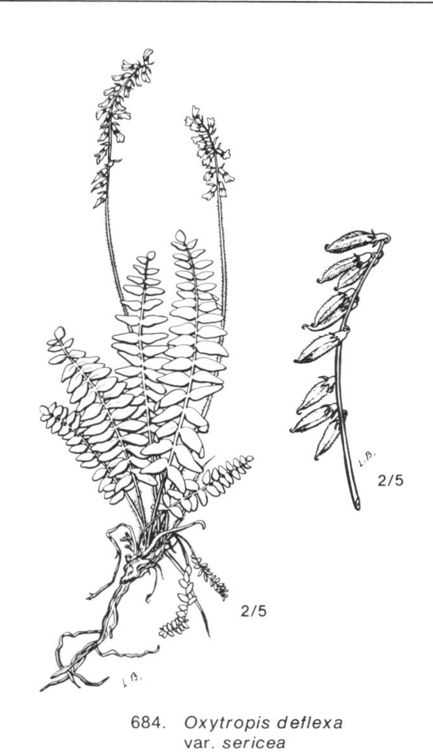

684. *Oxytropis deflexa* var. *sericea*

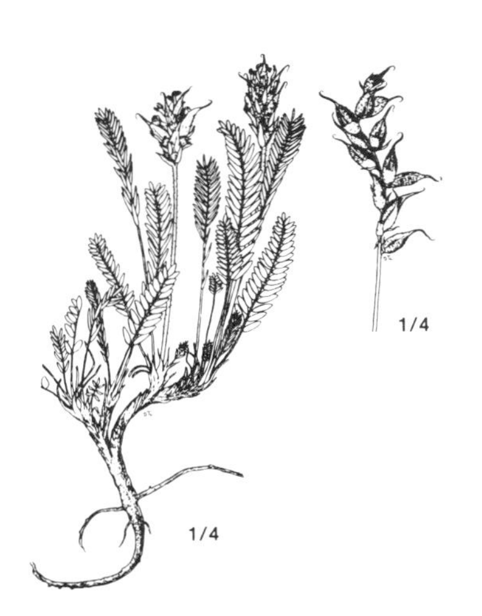

685. *Oxytropis glutinosa*

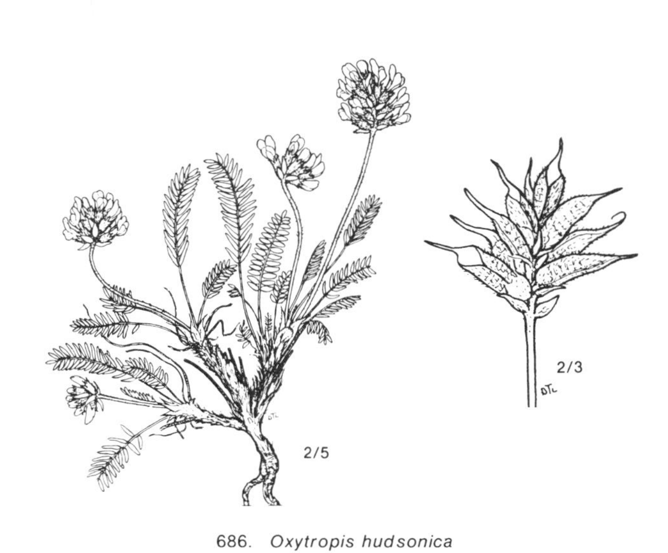

686. *Oxytropis hudsonica*

687. *Oxytropis hyperborea*

688. *Oxytropis Jordalii*

689. *Oxytropis Maydelliana* ssp. *melanocephala*

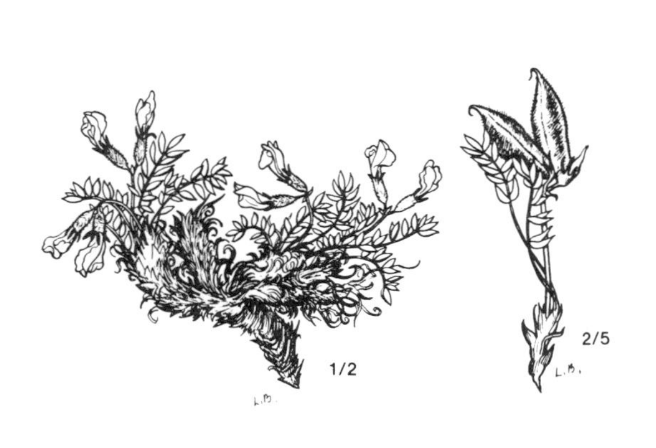

690. *Oxytropis nigrescens* ssp. *pygmaea*

691. *Oxytropis Scammaniana*

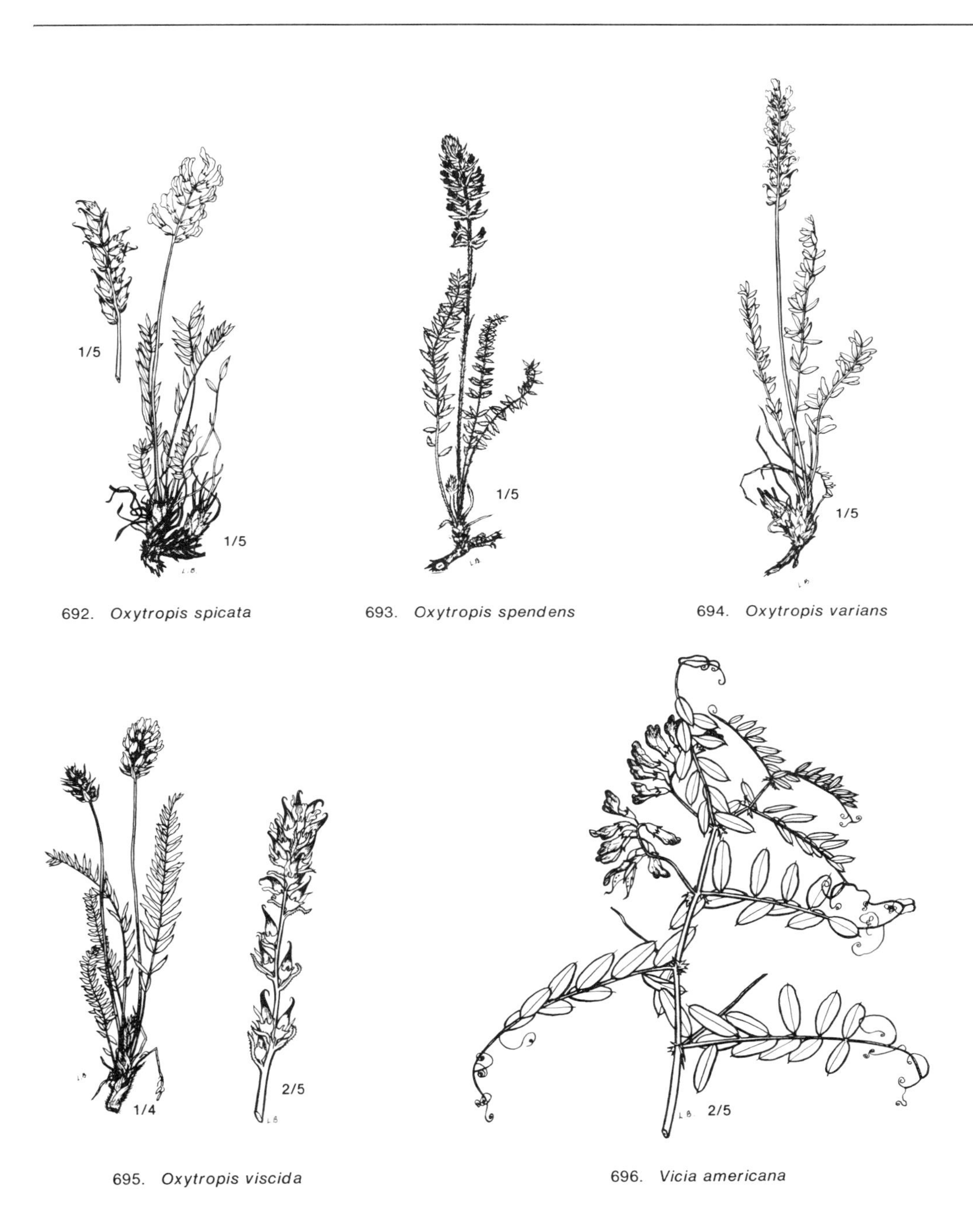

692. *Oxytropis spicata*

693. *Oxytropis spendens*

694. *Oxytropis varians*

695. *Oxytropis viscida*

696. *Vicia americana*

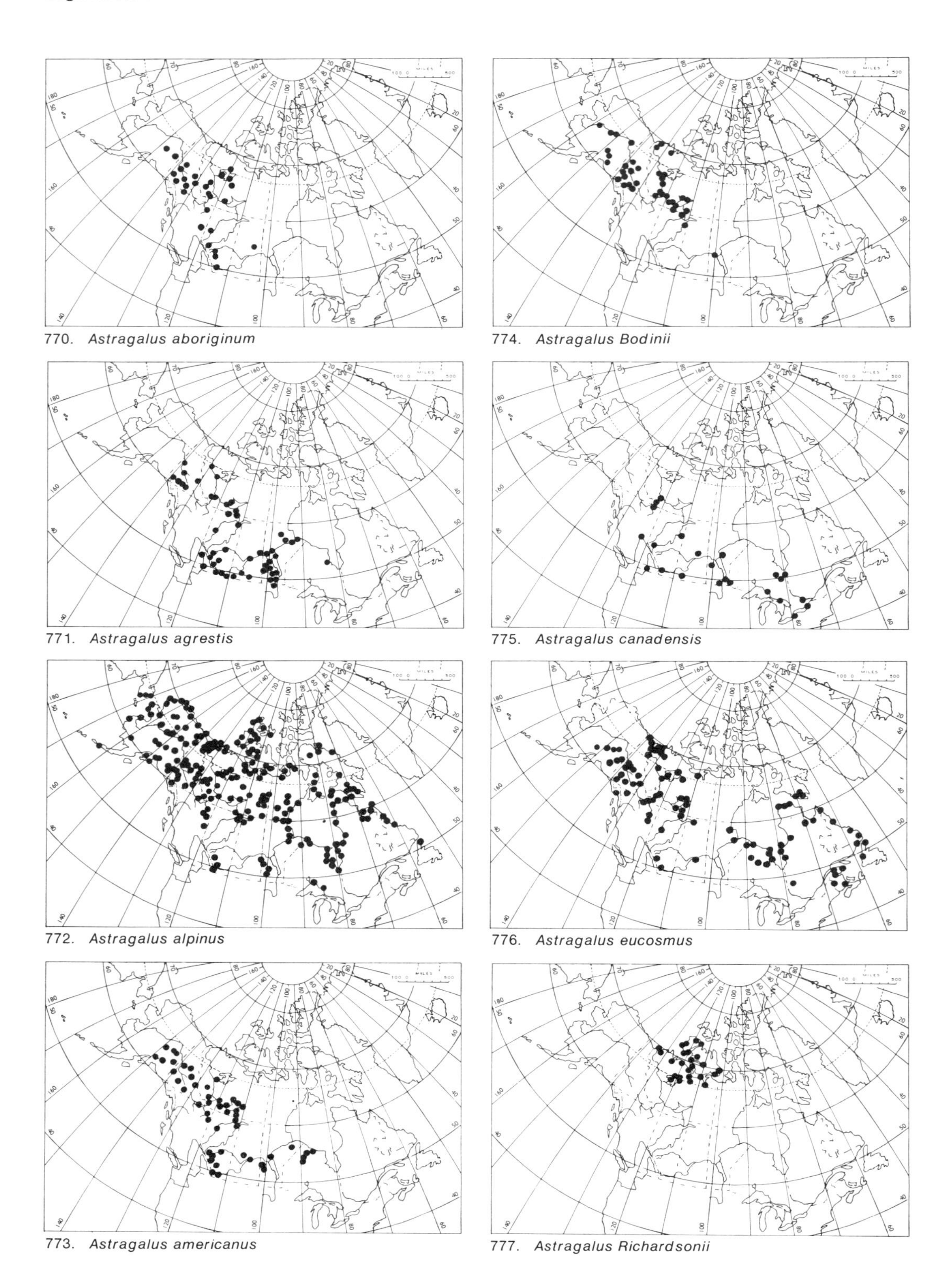

770. *Astragalus aboriginum*

771. *Astragalus agrestis*

772. *Astragalus alpinus*

773. *Astragalus americanus*

774. *Astragalus Bodinii*

775. *Astragalus canadensis*

776. *Astragalus eucosmus*

777. *Astragalus Richardsonii*

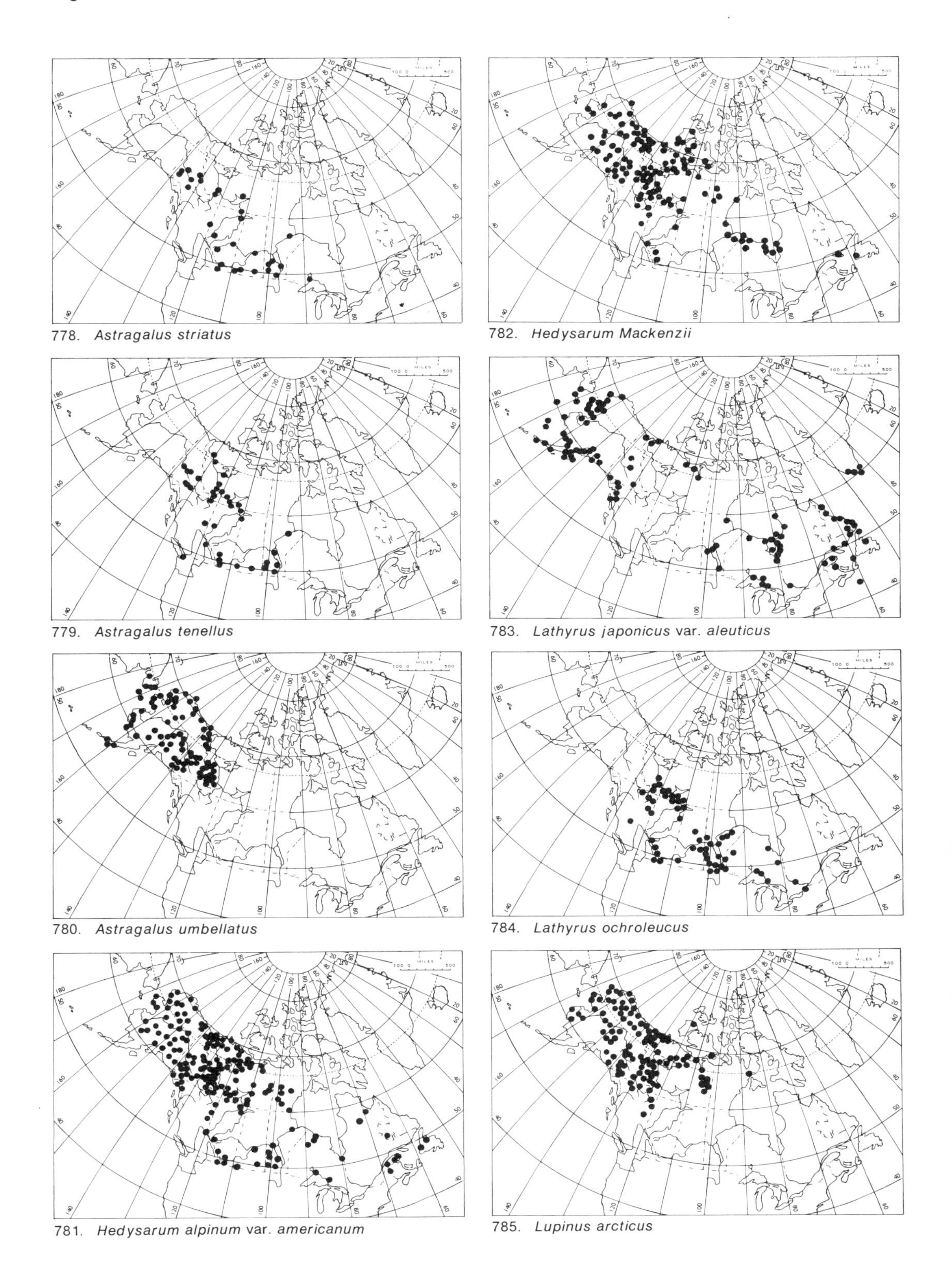

778. *Astragalus striatus*

779. *Astragalus tenellus*

780. *Astragalus umbellatus*

781. *Hedysarum alpinum* var. *americanum*

782. *Hedysarum Mackenzii*

783. *Lathyrus japonicus* var. *aleuticus*

784. *Lathyrus ochroleucus*

785. *Lupinus arcticus*

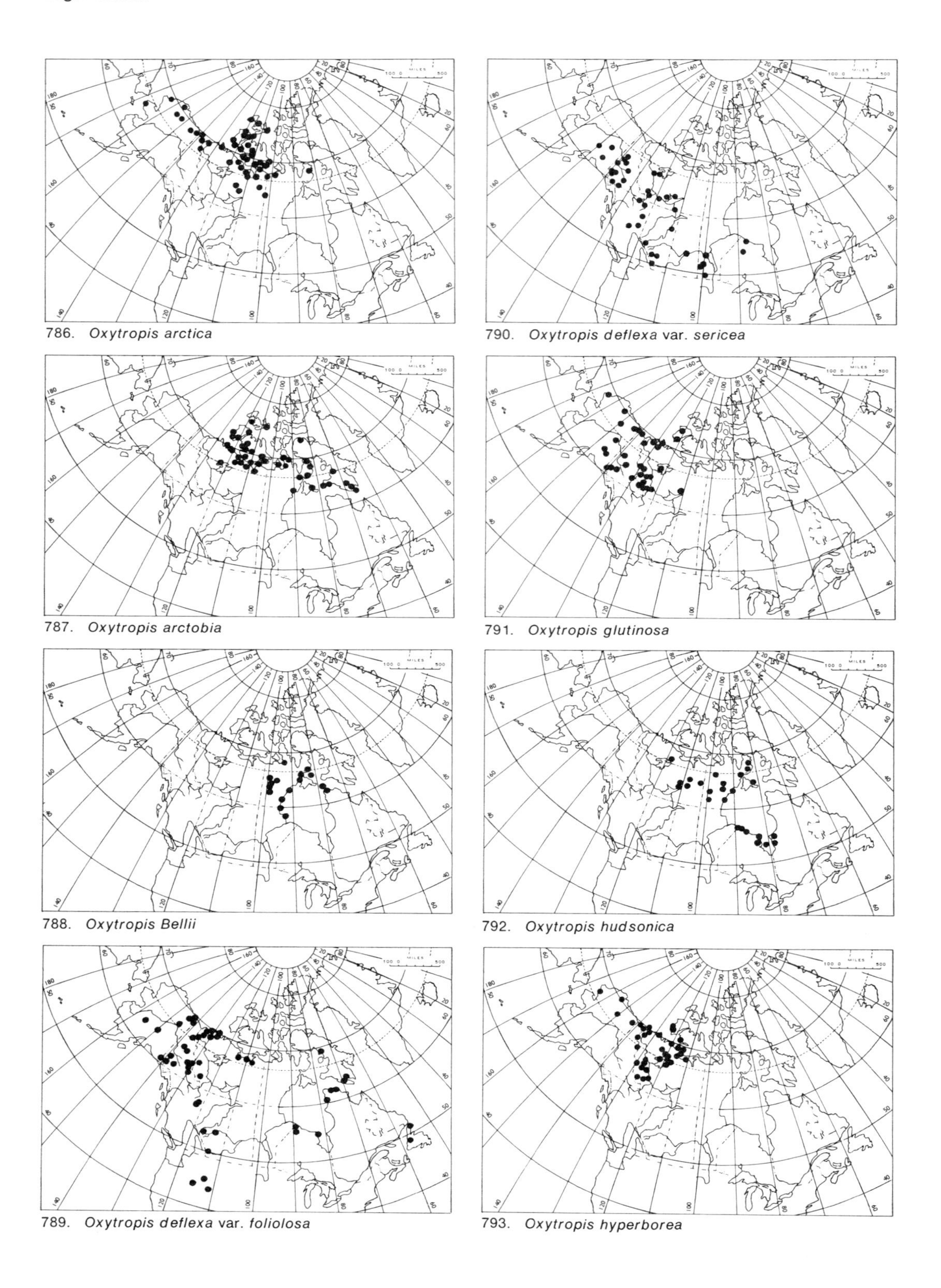

786. *Oxytropis arctica*

787. *Oxytropis arctobia*

788. *Oxytropis Bellii*

789. *Oxytropis deflexa* var. *foliolosa*

790. *Oxytropis deflexa* var. *sericea*

791. *Oxytropis glutinosa*

792. *Oxytropis hudsonica*

793. *Oxytropis hyperborea*

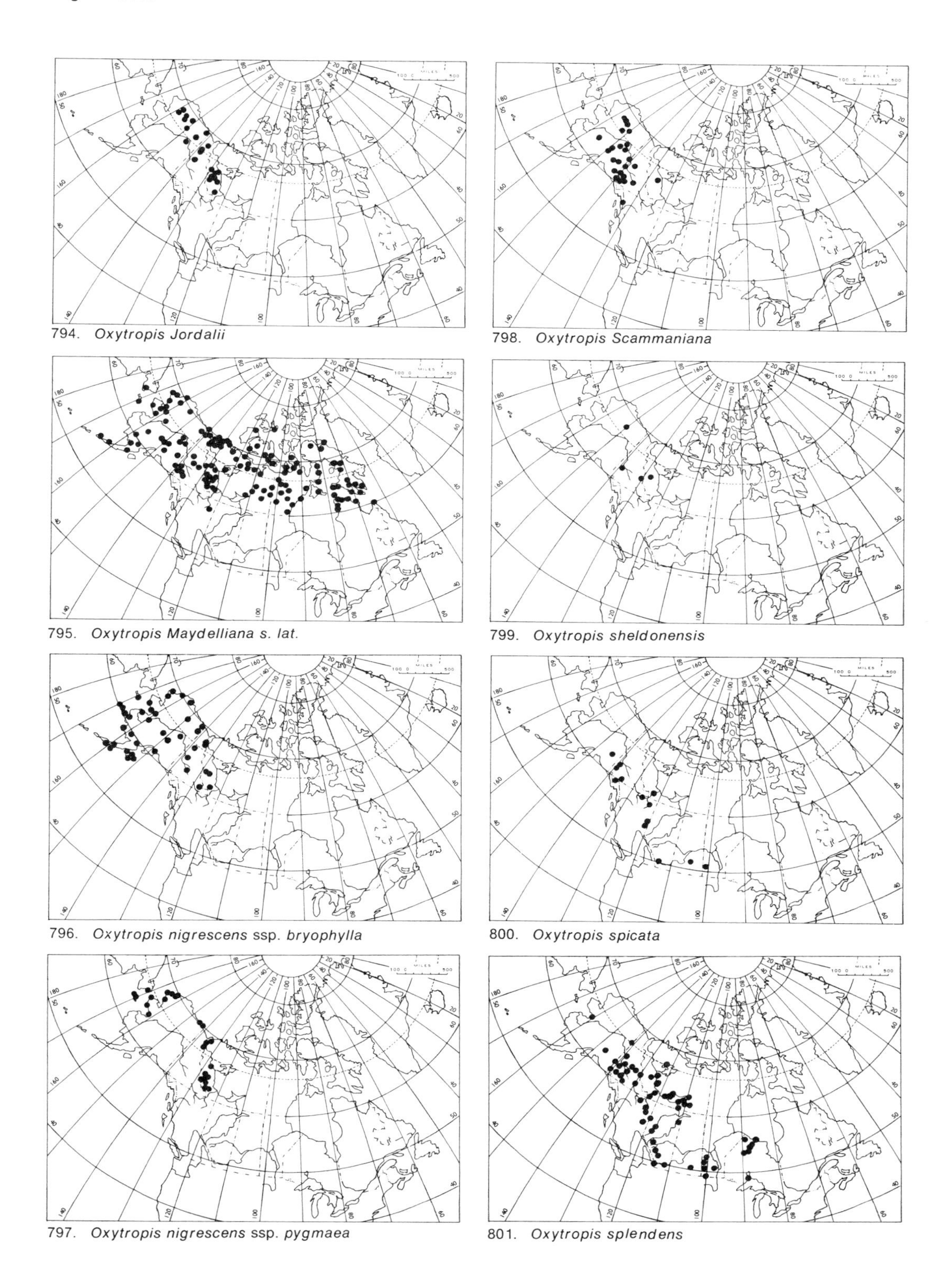

794. *Oxytropis Jordalii*

795. *Oxytropis Maydelliana s. lat.*

796. *Oxytropis nigrescens* ssp. *bryophylla*

797. *Oxytropis nigrescens* ssp. *pygmaea*

798. *Oxytropis Scammaniana*

799. *Oxytropis sheldonensis*

800. *Oxytropis spicata*

801. *Oxytropis splendens*

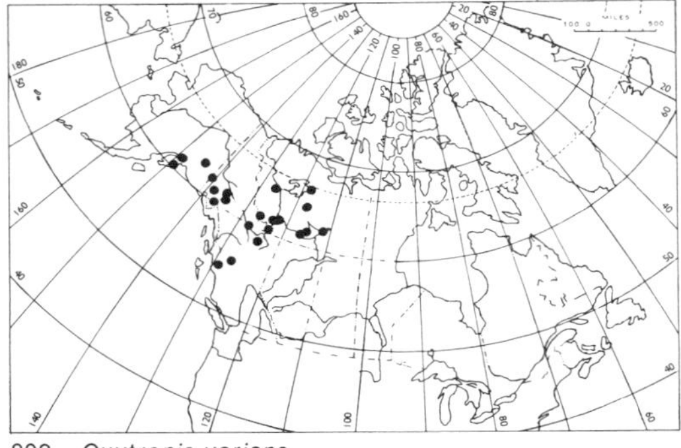

802. *Oxytropis varians*

804. *Vicia americana*

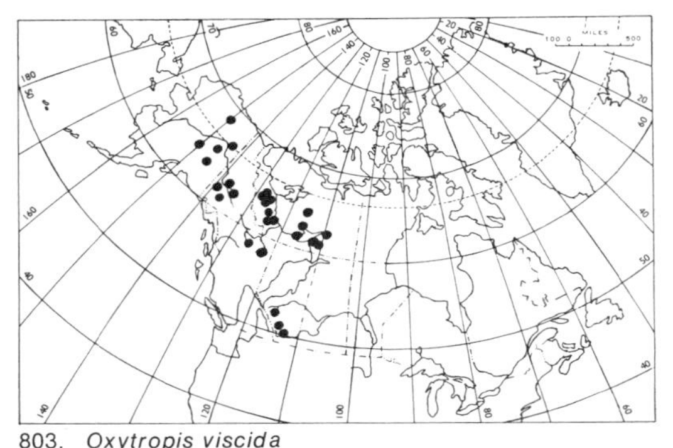

803. *Oxytropis viscida*

GERANIACEAE Geranium Family

Geranium L. Cranesbill

Ours erect herbs with opposite palmately divided leaves, the lower long-petioled, the upper sessile, or nearly so. Flowers cymose, 5-merous and regular. Ripe carpels dehiscing along the elongate central column, separating and curling elastically when freed from the axis.

a. Annual or biennial; petals dark pink, about as long as the sepals . . . . *G. Bicknellii*
a. Perennial; petals white or pale violet with darker veins, about twice as long as the sepals . . . . *G. Richardsonii*

Geranium Bicknellii Britt.
Low diffusely branched many-flowered annual or biennial; glands of hairs light-coloured; leaves deeply 5-parted, the segments cuneate and deeply oblong lobed, the lobes bluntish; flowers about 1 cm across.

Dry sandy situations, eroding river banks and shallow soil over igneous rocks in the south-western section; frequently invading disturbed roadsides.

General distribution: Wide spread boreal American, north to the upper Mackenzie valley.

Fig. 697 Map 805

Geranium Richardsonii Fisch. & Trautv.
Erect perennial with few ascending branches to

5 dm or more in height, few-flowered; glands of hairs blackish; leaves 5- to 7-parted, the segments rhombic and acutely lobed; flowers 2 to 3 cm across.

In our area known only from meadows near hot springs in the southern Mackenzie Mountains.

General distribution: Cordilleran.

Fig. 698 Map 806

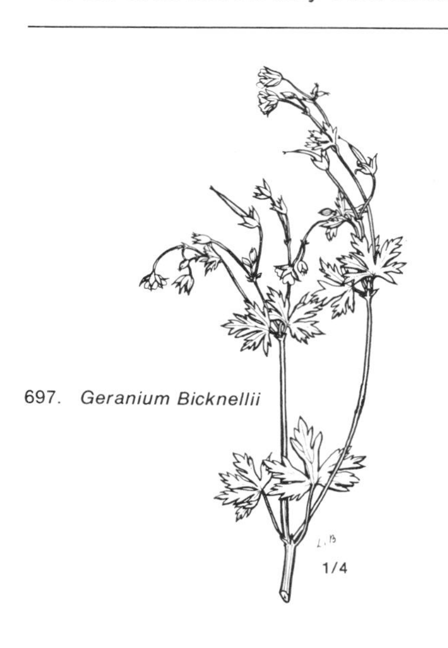

697. *Geranium Bicknellii*

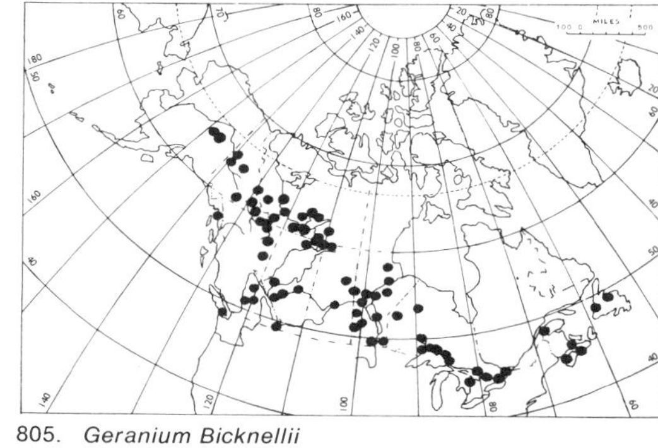

805. *Geranium Bicknellii*

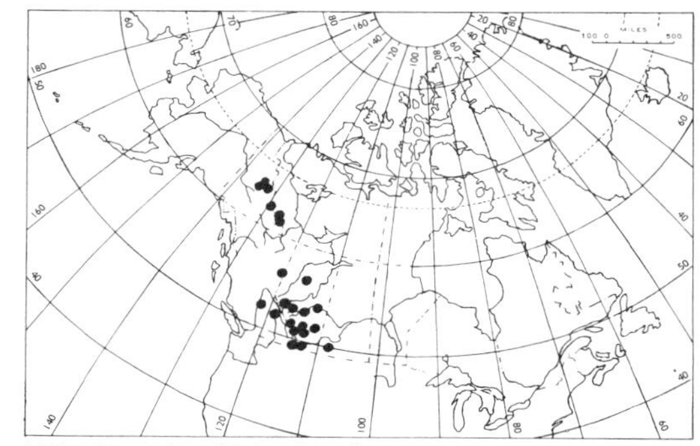

806. *Geranium Richardsonii*

698. *Geranium Richardsonii*

LINACEAE Flax Family
Linum L. Flax

Annual or perennial herbs, ours with linear or narrowly lanceolate, alternate leaves; flowers blue or rarely white, regular, 5-parted, the capsule of 5 united carpels, globose; seeds flattened, 2 to each carpel, separated by a false partition.

a. Perennial; pedicels arching and spreading; sepals not ciliate, short acuminate . . . *L. Lewisii*
a. Annual; pedicels erect; sepals ciliate, long-acuminate . . . . *L. usitatissimum*

Linum Lewisii Pursh
Glabrous native perennial herb with several densely leafy stems 15 to 60 cm tall, from a sub-ligneous base; leaves linear, 1 to 2 cm long, alternate; the large blue, or rarely white, flowers in leafy, often 1-sided racemes, the petals soon falling; capsules globose with two large seeds in each locule.

Calcareous, dry, rocky, or gravelly slopes.

General distribution: N. American endemic of prairies and foothills from Ont. to Alaska; in western District of Mackenzie north to Great Bear L. and the Arctic Coast and western Victoria Island.

Fig. 699 Map 807

Linum usitatissimum L.
Erect cultivated annual, readily distinguished from the native *L. Lewisii* by its usually taller stature, fewer more widely spaced and broader leaves.

A casual weed of waste places.

General distribution: Native of Europe.

699. *Linum Lewisii*

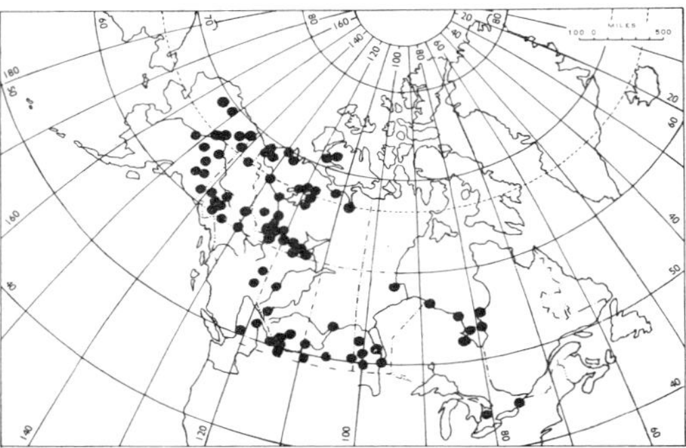

807. *Linum Lewisii*

CALLITRICHACEAE Water-Starwort Family

Callitriche L. Water-Starwort

Small, perennial (ours) rooted, tufted and normally submersed aquatics with weak, leafy stems and opposite, linear and in *C. anceps* and also *C. verna* floating, oblanceolate leaves. Flowers monoecious, usually solitary, in the leaf-axils; the tiny fruit nut-like, compressed and 4-lobed, separating at maturity into four 1-seeded carpels.

a. Plant dark green; fruit conspicuously winged . . . . *C. hermaphroditica*
a. Plant light green; fruit wingless or barely wing-margined above
  b. Submersed leaves linear oblanceolate . . . . *C. verna*
  b. Submersed leaves linear . . . . *C. anceps*

Callitriche anceps *Fern.*
By its distinctly winged stems and smaller fruit at once distinguished from the otherwise similar *C. verna.*

Rooted in silty bottoms of shallow ponds; in our area as yet known from only one locality in southwestern District of Mackenzie.

General distribution: Greenland, Lab.— Ungava to Nfld. and the Maritimes, with widely disjunct stations westward to Alaska.

Map 808

Callitriche hermaphroditica L.
C. autumnalis L.
Fully submersed, lacking floating leaves; stems much branched, 1 to 4 dm long. Fruit 1.5 to 2.5 mm in diameter, with broadly winged carpels.

Ponds and quiet streams in the lowlands of the Mackenzie River drainage north beyond the tree line east of the Mackenzie River Delta.

General distribution: Circumpolar with gaps.

Fig. 700 Map 809

Callitriche verna L.
Similar to *C. anceps* but with a narrower, angled stem; lowermost leaves linear but middle and floating leaves broadly oblanceolate, the latter often in crowded rosettes; fruit obovoid, 1.0 to 1.4 mm long, the carpels angled.

Mat forming on the silty bottom of shallow ponds or small, quiet streams, or occasionally stranded on wet mud, northward to the limit of trees.

General distribution: Circumpolar.

Fig. 701 Map 810

700. *Callitriche hermaphroditica*

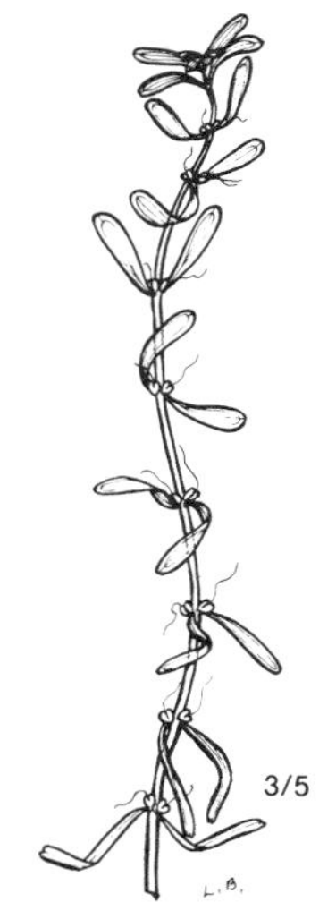

701. *Callitriche verna*

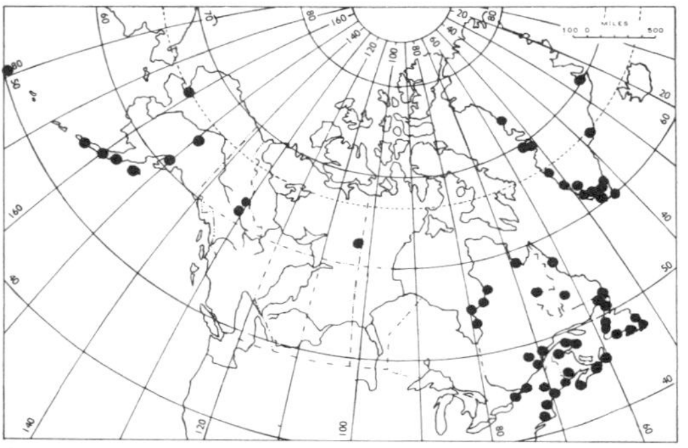

808. *Callitriche anceps*

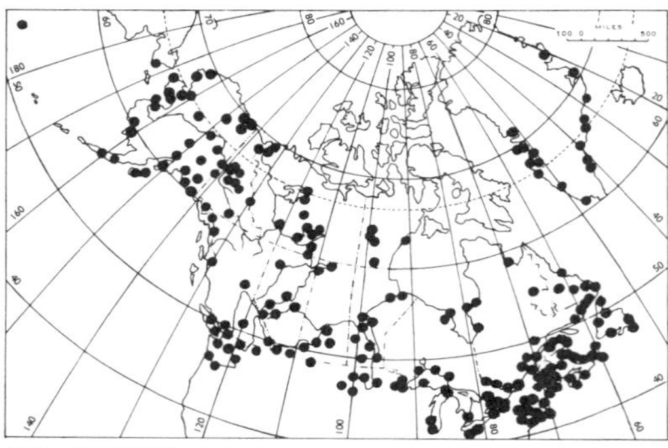

810. *Callitriche verna*

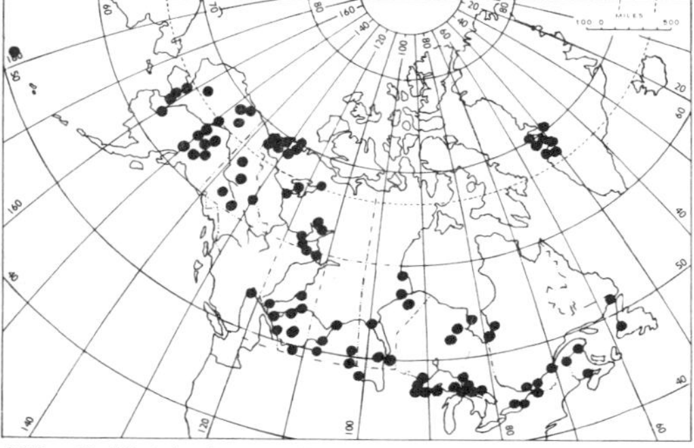

809. *Callitriche hermaphroditica*

EMPETRACEAE Crowberry Family

Empetrum L. Crow- or Curlewberry

Empetrum nigrum L.
ssp. **hermaphroditum** (Lge.) Böcher

Depressed and matted, freely branching, evergreen shrub. Leaves linear, about 5 mm long, spreading. The inconspicuous, dark purple flowers solitary in the leaf axils, monoecious, dioecious, or in the Arctic generally polygamous (ssp. *hermaphroditum*). The purplish-black and shiny fruits very juicy and sweet.

In sandy rocky, and always acid soils; reaching its perfection in a rather moist, maritime climate. The crowberry is the favoured fruit of the Eskimos who regularly harvest the berries.

General distribution: Circumpolar, wide ranging.

Fig. 702 Map 811

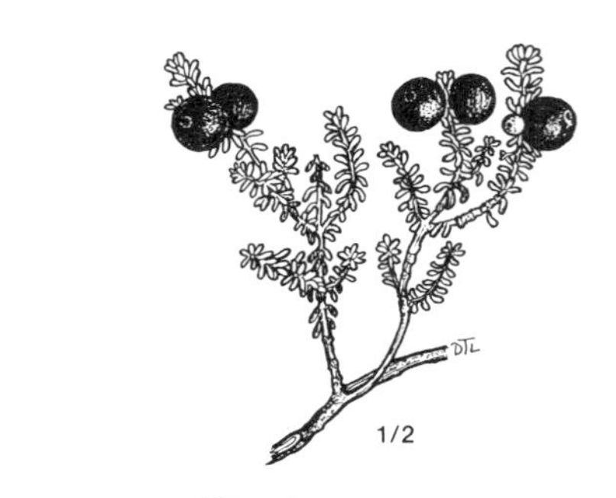

702. *Empetrum nigrum* ssp. *hermaphroditum*

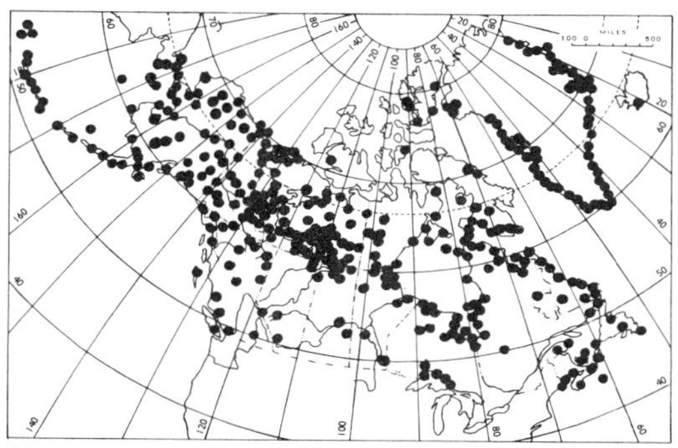

811. *Empetrum nigrum* ssp. *hermaphroditum*

ACERACEAE Maple Family

Acer L.

Acer Negundo L.
Manitoba Maple
Shrubby dioecious trees with opposite three to seven pinnate leaves; fruit, two one-seeded at length separable samaras or keys.

Escaped from cultivation about some townsites in S. W. District of Mackenzie.

General Distribution: W. North America but widely cultivated and naturalized across Canada.

BALASAMINACEAE Touch-me-not Family

Impatiens L. Touch-me-not

Impatiens capensis Meerb.
I. biflora Walt.
Delicate herbs with succulent stems to 1 m or more in height; leaves alternate, petioled, ovate to lanceolate, coarsely toothed and somewhat glaucous; flowers showy, the petals usually orange-yellow spotted with reddish-brown, the upper two small, and the lower saccate, abruptly constricting to a sharply recurved spur; fruit a 5-valved capsule which bursts open when touched, hence the vernacular name.

Low wet woodlands and moist banks; in our area known only from the Liard River Valley near the British Columbia border.

General distribution: Boreal America, common in the east, but of only sporadic occurrance westward, barely reaching the District of Mackenzie on the upper Liard R. An early report from Great Bear L., lat. 66°N., has not been verified.

Fig. 703 Map 812

703. *Impatiens capensis*

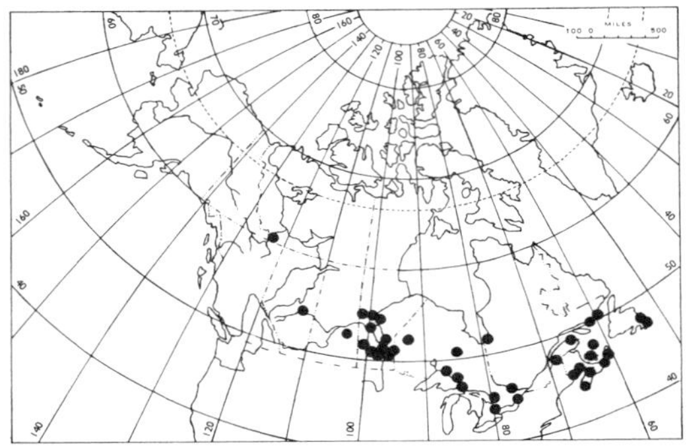

812. *Impatiens capensis*

ELATINACEAE Waterwort Family

Elatine L. Waterwort

Elatine triandra Schk. *s. lat.*
Tiny, matted semi-aquatic plant with slender stems rooting at the nodes; leaves opposite, linear, glabrous, 3 to 8 mm long. Flowers minute, perfect, 3-merous, sessile and axillary; fruit a septicidal, membranaceous, many-seed capsule; seeds straight or curved, areolate, with longitudinal rows of tiny pits.

Rooted in mud in shallow water by pond margins; known in our area from a single locality northwest of Yellowknife.

General distribution: Western America; Eurasia.

Fig. 704 Map 813

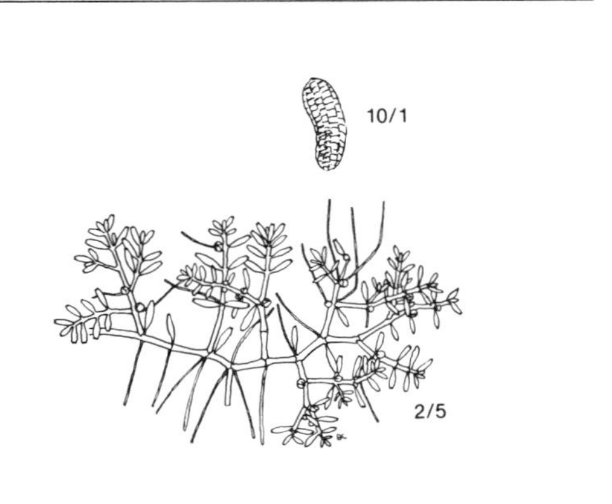

704. *Elatine triandra*

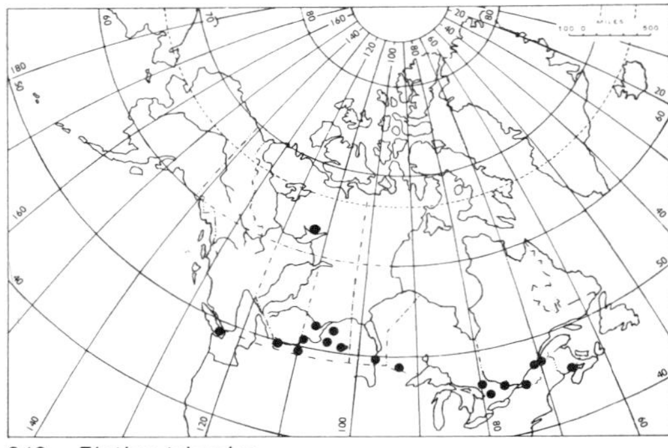

813. *Elatine triandra*

CISTACEAE Rockrose Family
Hudsonia L.

Hudsonia tomentosa Nutt.
Grey-tomentose, prostrate or bushy dwarf shrubs with persistent, closely appressed scale-like and ovate-lanceolate leaves; flowers small but numerous, at the end of short, leafy and lateral branches; petals 5, sulphur yellow; sepals often roseate-tinged, completely enclosing the glabrous few-seeded capsule.

Sand blow-outs, sandy beaches and open jack pine woods.

General distribution: Boreal America from the Atlantic Coast to Alta., but always restricted to sandy habitats; in our area known only from north of Great Slave L.

Fig. 705 Map 814

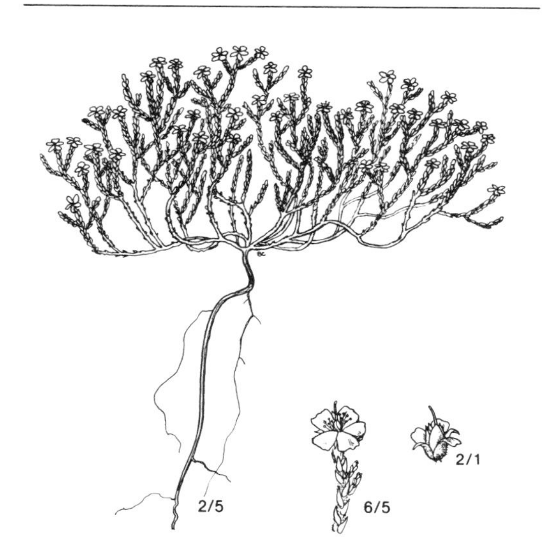

705. *Hudsonia tomentosa*

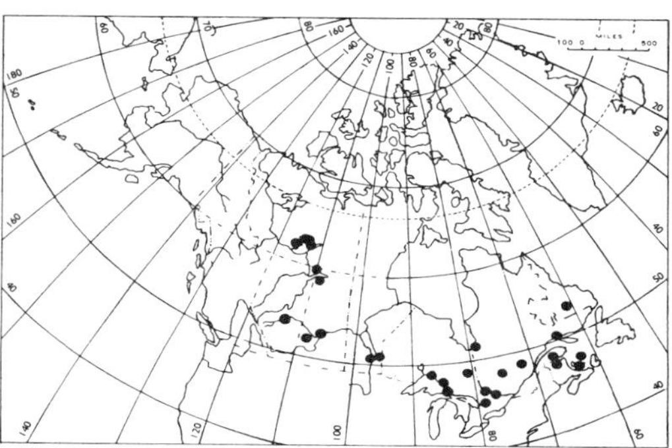
814. *Hudsonia tomentosa*

VIOLACEAE Violet Family
Viola L. Violet

Low perennial herbs with alternate or basal, simple, stipulate leaves; flowers 5-merous, irregular and solitary on slender bracted peduncles rising from the crown of the rhizome, or from leaf axils; the flowers usually of two kinds — the early showy, the latter smaller and cleistogamous; in the former the sepals are auricled at the base, and the lateral petals often bearded, while the lowermost and largest is spurred and often bearded in the throat; fruit a 3-valved capsule containing numerous seeds.

a. Plants with leafy stems
   b. Stipules large and laciniate . . . . *V. tricolor*

b. Stipules bract-like
  c. Flowers white to violet . . . . *V. rugulosa*
  c. Flowers blue or violet
    d. Stem and leaves ± puberulent . . . . *V. adunca*
    d. Stem and leaves glabrous or nearly so . . . . *V. adunca* var. *minor*

a. Stemless
  e. Flowers white
    f. Plants with thickened ascending rhizomes . . . . *V. renifolia* var. *Brainerdii*
    f. Plants arising directly from a threadlike rhizome . . . . *V. pallens*
  e. Flowers blue or lilac
    g. Plants with thin creeping rhizomes
      h. Leaves usually 1 or 2 at flowering time; bracts of pedicels usually above the middle . . . . *V. epipsila* ssp. *repens*
      h. Leaves usually more than 2 at flowering time; bracts of pedicels about the middle . . . . *V. palustris*
    g. Plants tufted from stout ascending rhizomes . . . . *V. nephrophylla*

Viola adunca J. E. Smith
Tufted, more or less densely puberulent; stems leafy; leaves subcoriaceous, orbicular or the upper ones gradually tapering to the apex, subcordate, crenulate, often spotted with brown; petals violet.

The var. *minor* (Hook.) Fern. [*V. labradorica* Schrank] is similar but glabrous or at most with a few hairs on the upper leaf surface and the leaves smaller and more rounded. With us of similar range as the species.

Open woods and rocky or sandy situations in the southwestern parts of our area.

General distribution: Boreal America, Que. to Alaska.

Fig. 706 (var. *adunca*), 707 (var. *minor*)
Map 815 (var. *adunca*), 816 (var. *minor*)

Viola epipsila Ledeb.
ssp. **repens** (Turcz.) Becker
Small stemless plants from a thin creeping rhizome and filiform leafy stolons that appear later in the season; leaves usually two, ovate-cordate, crenulate, delicate, glabrous; flowers 1 or 2 on long peduncles, the bracts usually above the middle; flowers violet to lilac; capsules elliptical.

Bogs and wet mossy thickets.

General distribution: Amphi-Beringian., reaching the Mackenzie R. valley.

Fig. 708 Map 817

Viola nephrophylla Greene
Tufted from a stout ascending rhizome; leaves subcoriaceous, cordate-ovate; the petals large, violet-purple, the lateral and lower ones bearded at the base.

Gravelly river banks, shores and open places; in the Mackenzie River valley north to Great Bear Lake.

General distribution: Boreal N. America; wide ranging from Nfld. west to S. E. Y.T.

Fig. 709 Map 818

Viola pallens (Banks) Brainerd
Stemless, the leaves and peduncles arising directly from the slender, thread-like rhizome; leaves cordate-ovate, glabrous, membranaceous, crenulate; flowers very fragrant, 7 to 10 mm long; petals white with purple veins; sepals glabrous; capsules green; seeds black 1.0 to 1.4 mm long.

Wet woods and thickets northward in our area to Great Bear Lake; infrequent.

General distribution: Boreal N. America, from Nfld. to B.C. and the District of Mackenzie.

Fig. 710 Map 819

Viola palustris L.
Similar to *V. epipsila* ssp. *repens* from which it may be separated by its rounder leaves, flowers often several, bracts of the pedicels usually at or below the middle, usually several leaves at flowering time, and by the absence of glands on the bracts and stipules.

Borders of swamps and moist banks; in our area known only from the southeastern part.

General distribution: Circumpolar; wide ranging.

Map 820

Viola renifolia Gray
var. **Brainerdii** (Greene) Fern.
Stemless, with thickened ascending rhizomes; leaves cordate-ovate, crenulate; flowers 1.0 to 1.5 cm long; petals white with purple veins; capsules usually purple; seeds brown 1.9 to 2.4 mm long.

Moist humus in woodlands, and sometimes in subalpine barrens, north to Great Bear Lake.

General distribution: Boreal N. America but not extending as far north in the eastern part of its range as the somewhat similar *V. pallens*.
Fig. 711 Map 821

Viola rugulosa Greene
Leafy-stemmed 2 dm or more tall often forming large colonies from long, much branched rhizomes; leaves thin, minutely pubescent to glabrate, somewhat rugulose, serrate, cordate-ovate, the upper obovate and abruptly tipped; petals white, sometimes tinged with mauve on the outside.
Moist woodland.
General distribution: Cordilleran, east to northwestern Ont.
Fig. 712 Map 822

Viola tricolor L.
Stems angled, glabrous or pubescent, freely branched; stipules foliaceous and laciniate or lyrate-pinnatifid; petals longer than the sepals, marked with purple, white or yellow.
Waste places.
General distribution: Native of Europe.

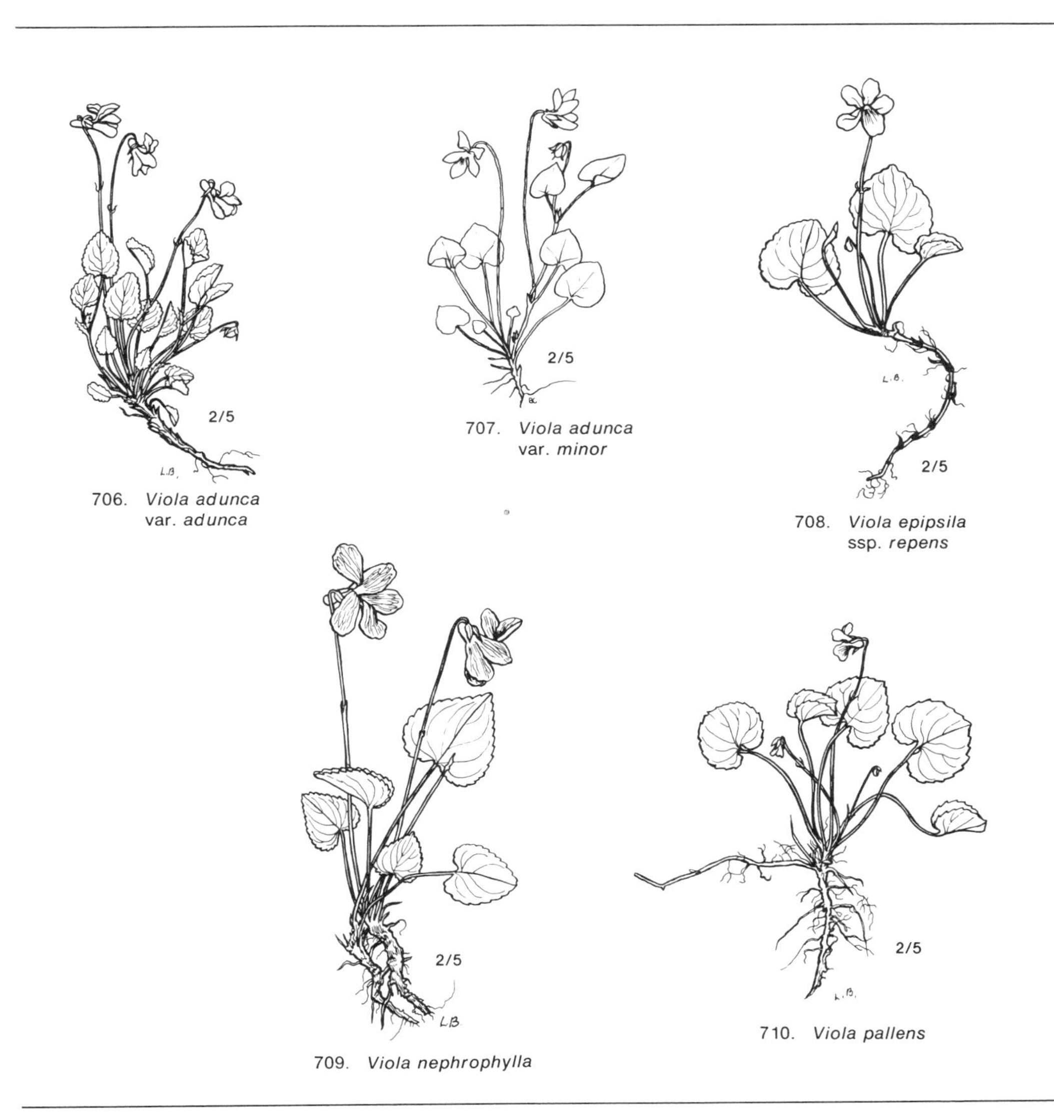

706. *Viola adunca* var. *adunca*

707. *Viola adunca* var. *minor*

708. *Viola epipsila* ssp. *repens*

709. *Viola nephrophylla*

710. *Viola pallens*

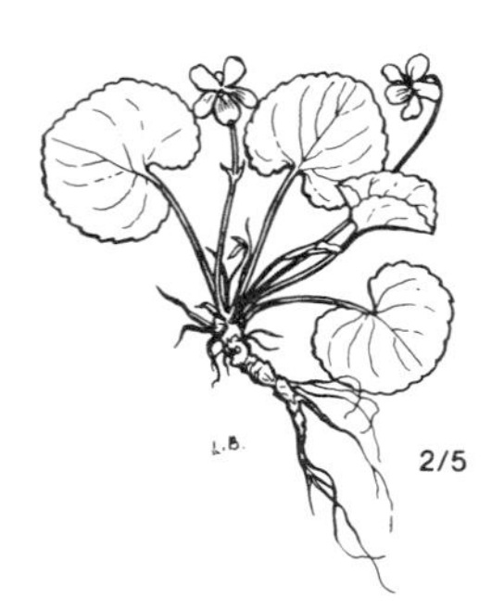

711. *Viola renifolia* var. *Brainerdii*

712. *Viola rugulosa*

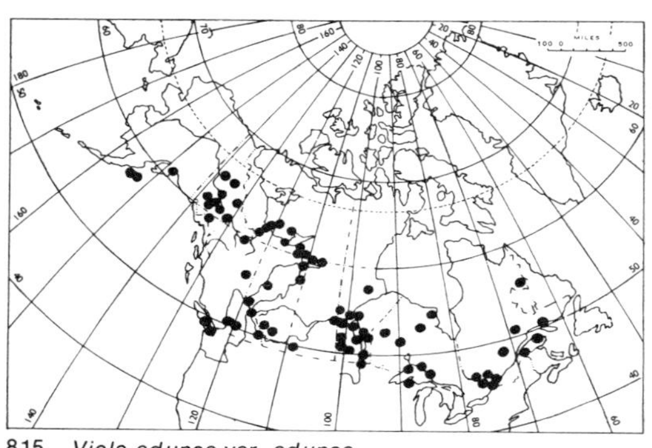

815. *Viola adunca* var. *adunca*

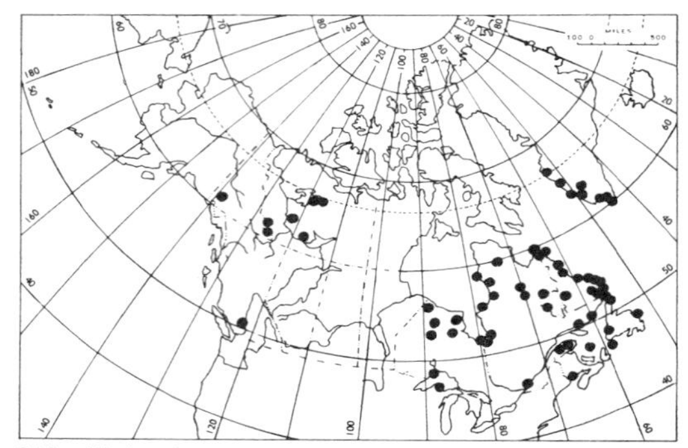

816. *Viola adunca* var. *minor*

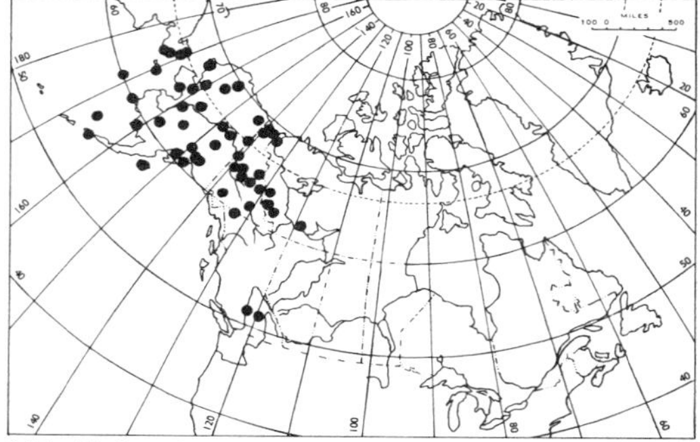

817. *Viola epipsila* ssp. *repens*

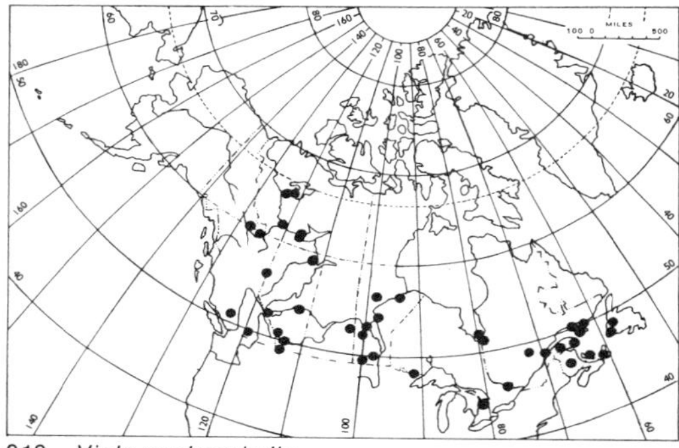

818. *Viola nephrophylla*

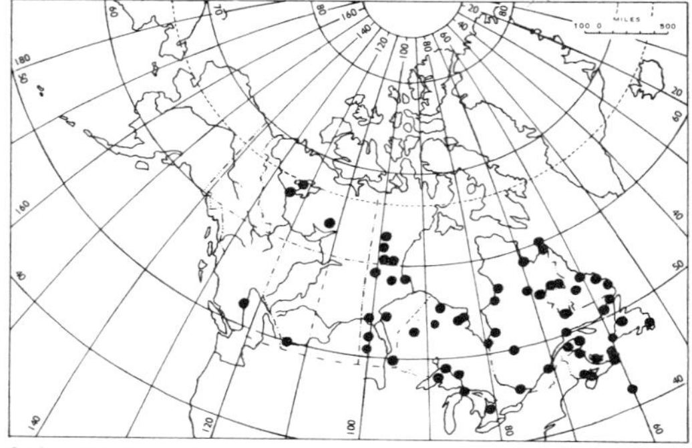

819. *Viola pallens*

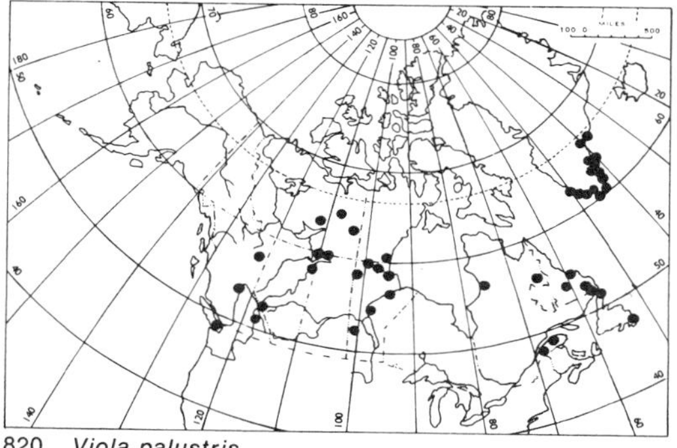

820. *Viola palustris*

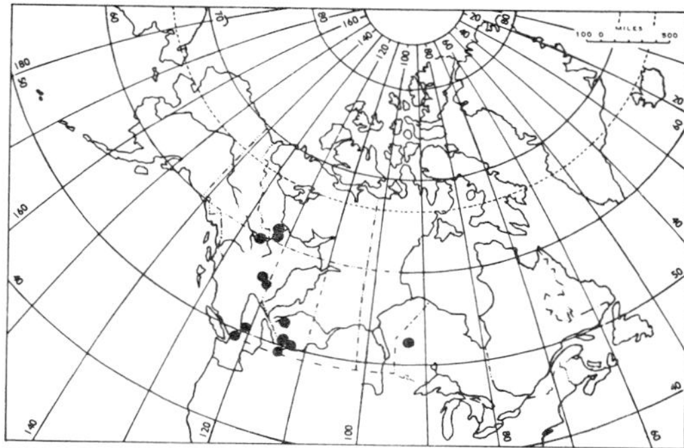

822. *Viola rugulosa*

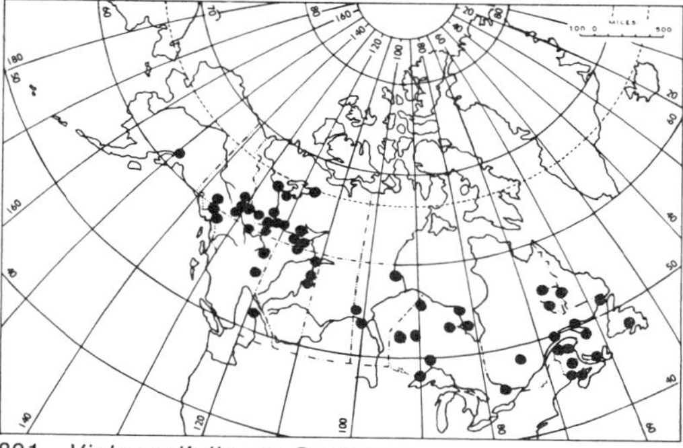

821. *Viola renifolia* var. *Brainerdii*

ELAEAGNACEAE Oleaster Family

Unarmed shrubs (with us) with silvery or reddish-scurfy leaves and stems; flowers with 4-parted calyx but no petals; stamens 4 or 8; fruit drupe-like, formed by the enlarged hypanthium enclosing the dry ovary.

a. Leaves alternate, silvery on both sides; stamens 4 . . . . *Elaeagnus*
a. Leaves opposite, dark green above, silvery-green below; stamens 8 . . . . *Shepherdia*

Elaeagnus L. Oleaster

Elaeagnus commutata Bernh.
Silverberry
Stoloniferous shrubs to 2 m tall (higher southwards); leaves elliptic to lanceolate, silvery scurfy on both sides, and somewhat ferruginous below; flowers sweet-scented, numerous, perfect, almost sessile in the axils of the leaves, the calyx yellow on the inside, silvery yellow on the outside; fruit about 1 cm long, on a short pedicel, silvery scurfy.

Occasional on slopes in open spruce and aspen woods in the south, but often forming patches several meters in diameter on the steep open banks of the Slave, Liard and Mackenzie Rivers north to lat. 67°N; in the Mackenzie Mountains ascending to 800 m in the Keele River drainage.

General distribution: N. America, Gaspé, Que. to Alaska and southward.

Fig. 713 Map 823

Shepherdia Nutt.

Shepherdia canadensis (L.) Nutt.
Soapberry
Erect-ascending shrubs; leaves elliptical to ovate with silvery stellate hairs and brown scurfy scales on the lower surface, and stellate hairs above; flowers, small, dioecious, in clusters or the pistillate single; fruit reddish-yellow, berry-like.

In season, the berry-like, drupaceous fruits of the soapberry are eagerly harvested by the northern chipmunk (*Eutamias minimus*) who climb the soapberry bushes and dexterously extract and store the large seeds in its cheek-pouches, but discard the pulp.

Usually dry calcareous soils in open woods and banks northward to, or slightly beyond, the limit of trees.

General distribution: N. America, Nfld. to Alaska and southward; wide ranging.

Fig. 714 Map 824

713. *Elaeagnus commutata*

714. *Shepherdia canadensis*

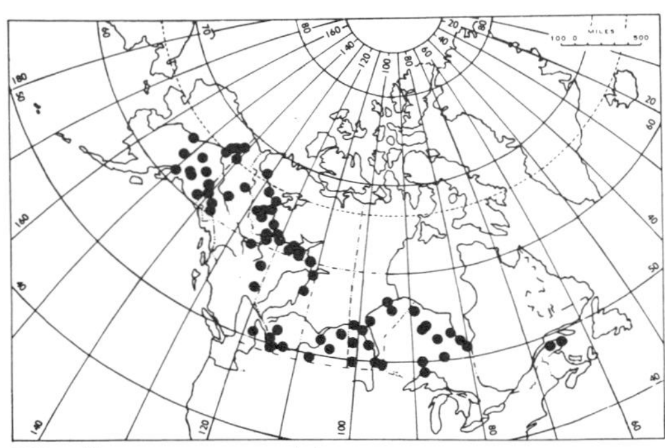

823. *Elaeagnus commutata*

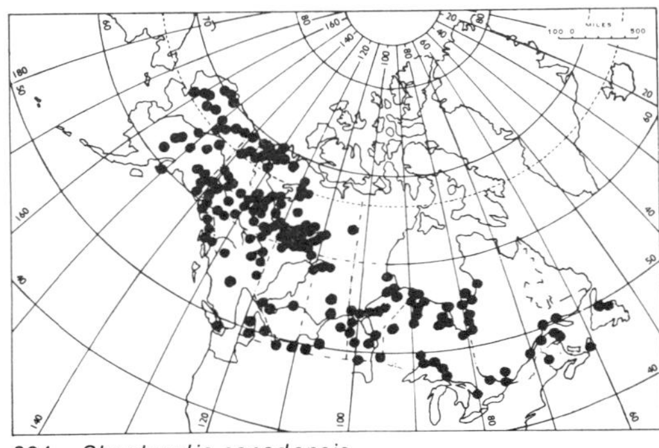

824. *Shepherdia canadensis*

ONAGRACEAE Evening Primrose Family

Perennial herbs (ours) with simple, alternate or opposite leaves and 2- or 4-parted, perfect and epigynous flowers; style 1, the stigma capitate, discoid or 4-lobed. Fruit a many seeded capsule or nut-like.

a. Sepals and petals 2, about 1.5 mm long; stamens 2; fruit bristly, indehiscent ..... *Circaea*
a. Sepals and petals 4, much longer; stamens 8; fruit a long, narrow and promptly dehiscent capsule, containing seeds bearing a tuft of white, silky hairs .... *Epilobium*

Circaea L. Enchanter's Nightshade

Circaea alpina L.
Delicate plants to 1.5 dm tall from a tuberous rhizome; leaves opposite, long-petioled, ovate, subcordate, shallowly undulate-dentate, pale green, flaccid; flowers tiny in terminal racemes, the calyx lobes and petals 2; the 2 mm long fruit covered with soft, hooked bristles.

In moist moss in cool woods and thickets; in our area known only from the upper Liard River valley and by a hot spring in the southwestern Mackenzie Mountains.

General distribution: Circumpolar

Fig. 715 Map 825

Epilobium L. Willow-herb

Perennial (ours) herbs with alternate or opposite, nearly sessile, ovate, lanceolate or linear leaves; flowers axillary or terminal in leafy racemes; petals pink, purple or white; capsule linear, splitting into valves; seeds numerous, bearing long silky tufts of white hairs at their summits.

a. Coarse plants with alternate leaves and large showy, purplish flowers
  b. Stems simple, erect, solitary or few together .... *E. angustifolium*
  b. Stems several, tufted, erect-ascending, often branched .... *E. latifolium*
a. Medium to low plants with mostly opposite leaves and few, small flowers
  c. Stem leaves linear, narrowly lanceolate or oblong
    d. Margins of leaves inrolled; upper surface thickly covered with incurved hairs .... *E. leptophyllum*
    d. Leaves flat, glabrous or nearly so
      e. Leaves linear-lanceolate, entire; plants with filiform stolons .... *E. palustre*
      e. Leaves oblong, remotely toothed; plants lacking filiform stolons
        f. Plants dark green to purplish, 5-15 cm tall; petals pink .... *E. arcticum*
        f. Plants green, to 35 cm tall; petals white .... *E. davuricum*
  c. Stem leaves lanceolate, ovate or elliptic
    g. Leaves lanceolate; tall plants often much branched above
      h. Plant with short crisped glandless hairs in the inflorescence .... *E. ciliatum*
      h. Glandular-pubescent with multicellular hairs in the inflorescence .... *E. glandulosum* var. *adenocaulon*
    g. Leaves ovate to elliptic; low plants, usually unbranched above
      i. Stems 10-15 cm tall; stem leaves small, 1.0-1.5 cm long ...... *E. anagallidifolium*
      i. Stems usually taller; leaves larger
        j. Leaves elliptic to oblong; petals pink; seed papillate .... *E. Hornemannii*
        j. Leaves ovate-oblong; petals white; seed smooth .... *E. lactiflorum*

Epilobium anagallidifolium Lam.
Dark green or reddish-tinged plant with weak, curved or S-shaped, 10 to 20 cm tall stems frequently with basal offsets; leaves opposite, ovate or oblanceolate, entire, or repand-dentate; petals reddish to pink; capsules more or less nodding, even at maturity; seeds smooth, about 0.8 mm long.

In wet moss by alpine brooks; in our area known only from the Mackenzie and Richardson Mountains and Great Bear Lake.

General distribution: Circumpolar (with several large gaps), subarctic-alpine.

Fig. 716 Map 826

Epilobium angustifolium L., *s.lat.*
Fireweed
Stems leafy up to 10 dm or more tall; the leaves membranaceous, green, veiny beneath. Raceme many-flowered, elongating; flowers rose-pink, rarely white, 2 to 3 cm in diameter.

A pioneer species on disturbed soil and in recently burnt over areas; northward and in the mountains occasional in willow thickets and on south-facing screes. *Epilobium angustifolium* is the territorial flower of Yukon Territory.

General distribution: Circumpolar-alpine, wide ranging beyond the limit of trees but not truly arctic.

Fig. 717 Map 827

Epilobium arcticum Samuelss.
E. davuricum Fisch.
var. *arcticum* (Samuelss.) Polunin
Stems dark green or purple, tufted or few together, 6 to 12 cm tall, stiff and mostly simple, bearing 2 to 3 pairs of narrowly oblong, repand-dentate leaves; basal offsets usually present; flowers pink or white, capsules stiffly erect; seeds papillose (when seen under a strong lens), about 1.5 mm long.

Rare, in wet clay in tundra barrens.

General distribution: Amphi-Atlantic, arctic.

Fig. 718 Map 828

Epilobium ciliatum Raf.
E. glandulosum
var. *perplexans* (Trel.) Fern.
Stems simple, or in tall specimens branched above, 15 to 60 cm tall, glabrous below, short-pubescent above; stem leaves 10 to 14 pairs, sessile or short-petioled, never clasping, lanceolate. Innovations short and sessile. Flowers 4 to 5 mm long, rose-purple; capsules 4 to 5 cm long, crisp-pubescent; seeds distinctly papillose.

Wet meadows and disturbed peaty soil; in our area known only from the vicinity of two hot springs in the southwestern Mackenzie Mountains.

General distribution: Boreal America; apparently sporadic, rare and local or perhaps often overlooked. Nfld. to B.C. and S. Y.T.

Map 829

Epilobium davuricum Fisch.
Stems green, few or solitary, 5 to 30 cm tall from a compact basal rosette; stem leaves linear, remotely denticulate, the lower opposite or sub-opposite, the upper alternate; flowers few, white; mature capsules erect; seeds fusiform, about 1.5 mm long.

A lowland species of wet places not often collected in our area.

General distribution: Circumpolar, subarctic.

Fig. 719 Map 830

Epilobium glandulosum Lehm.
var. **adenocaulon** (Haussk.) Fern.
Similar to *E. ciliatum* but often taller and more branched, the leaves narrowly lanceolate and the stems glandular-pubescent above.

Wet meadows and disturbed peaty soils, northward in the wooded zone to Norman Wells.

General distribution: Boreal America; Nfld. to Alaska; wide-spread.

Fig. 720 Map 831

***Epilobium Hornemannii** Rchb.
Stems 1.0 to 3.5 dm or more, usually simple, often somewhat curved; leaves thin, glabrous, elliptic to oblong, denticulate, mostly petiolate; petals about 6 mm long, pink; capsules upright at maturity; seeds about 1.0 mm long, papillose.

In wet moss by alpine brooks; known from a number of stations in the Yukon Territory and to be expected also along the east slope of the Mackenzie Mountains.

General distribution: Circumpolar (with large gaps); sub-arctic-alpine.

Fig. 721 Map 832

Epilobium lactiflorum Haussk.
Similar to *E. Hornemannii* from which it may be distinguished by its ovate-oblong rather than elliptic to oblong and mostly entire margined leaves and by its smaller and white petals, and smooth seed.

Wet meadows and moss by alpine brooks; known in our area only from the Mackenzie Mountains.

General distribution: Amphi-Atlantic arctic-alpine with a large gap from Que. — Lab. to B.C., north central Y.T. and Alaska.

Fig. 722 Map 833

Epilobium latifolium L.
Broad-Leaved Willow-Herb
Stems 1.5 to 4.0 dm tall, with dark green, glaucous or purplish and somewhat fleshy leaves; raceme of few flowers, showy, purple or rarely white, 3 to 5 cm in diameter.

The flowers of the willow-herb may be eaten raw as a salad, and the fleshy leaves are edible when cooked.

Common or even abundant throughout the Arctic in sandy and gravelly well-watered soils, such as gravel bars or flood plains.

General distribution: Circumpolar, arctic-alpine.
Fig. 723 Map 834

Epilobium leptophyllum Raf.
Stems 2.5 to 6.0 dm tall, simple or branched; leaves linear, their margins inrolled, often with axillary fascicles of shorter leaves, the upper surface and midrib, with short incurved hairs; capsules canescent; seeds smooth to minutely papillose.
Sedge meadows, marshes and bogs; in our area known only from the lowlands of the Liard River valley.
General distribution: Boreal N. America, Nfld. to northern B.C., S. W. Y.T. and S. W. District of Mackenzie.
Fig. 724 Map 835

Epilobium palustre L.
Stems 10 to 40 cm tall, stiffly erect, with filiform stolons issuing from their base, and terminating in a tiny and scaly winter bud; leaves linear-oblong, essentially glabrous.
Mossy edges of ponds and bogs.
General distribution: Circumpolar.
Fig. 725 Map 836

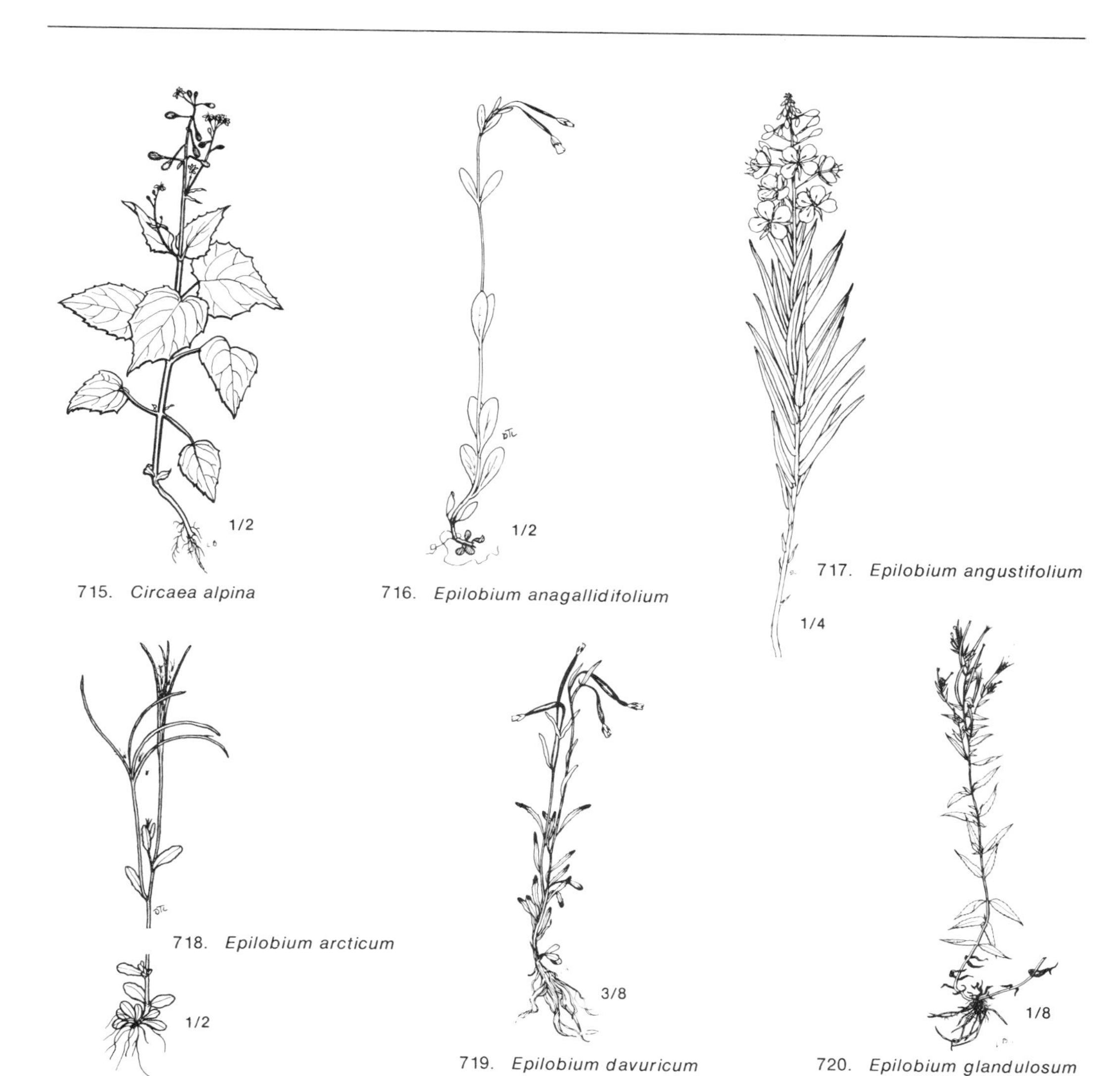

715. *Circaea alpina*

716. *Epilobium anagallidifolium*

717. *Epilobium angustifolium*

718. *Epilobium arcticum*

719. *Epilobium davuricum*

720. *Epilobium glandulosum* var. *adenocaulon*

721. *Epilobium Hornemannii*

722. *Epilobium lactiflorum*

723. *Epilobium latifolium*

724. *Epilobium leptophyllum*

1/4

725. *Epilobium palustre*

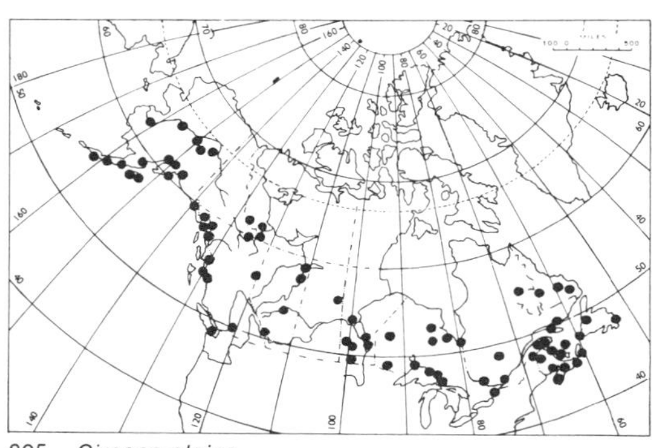

825. *Circaea alpina*

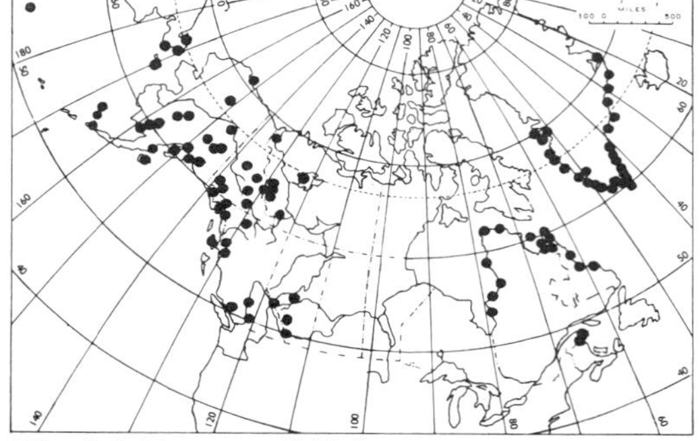

826. *Epilobium anagallidifolium*

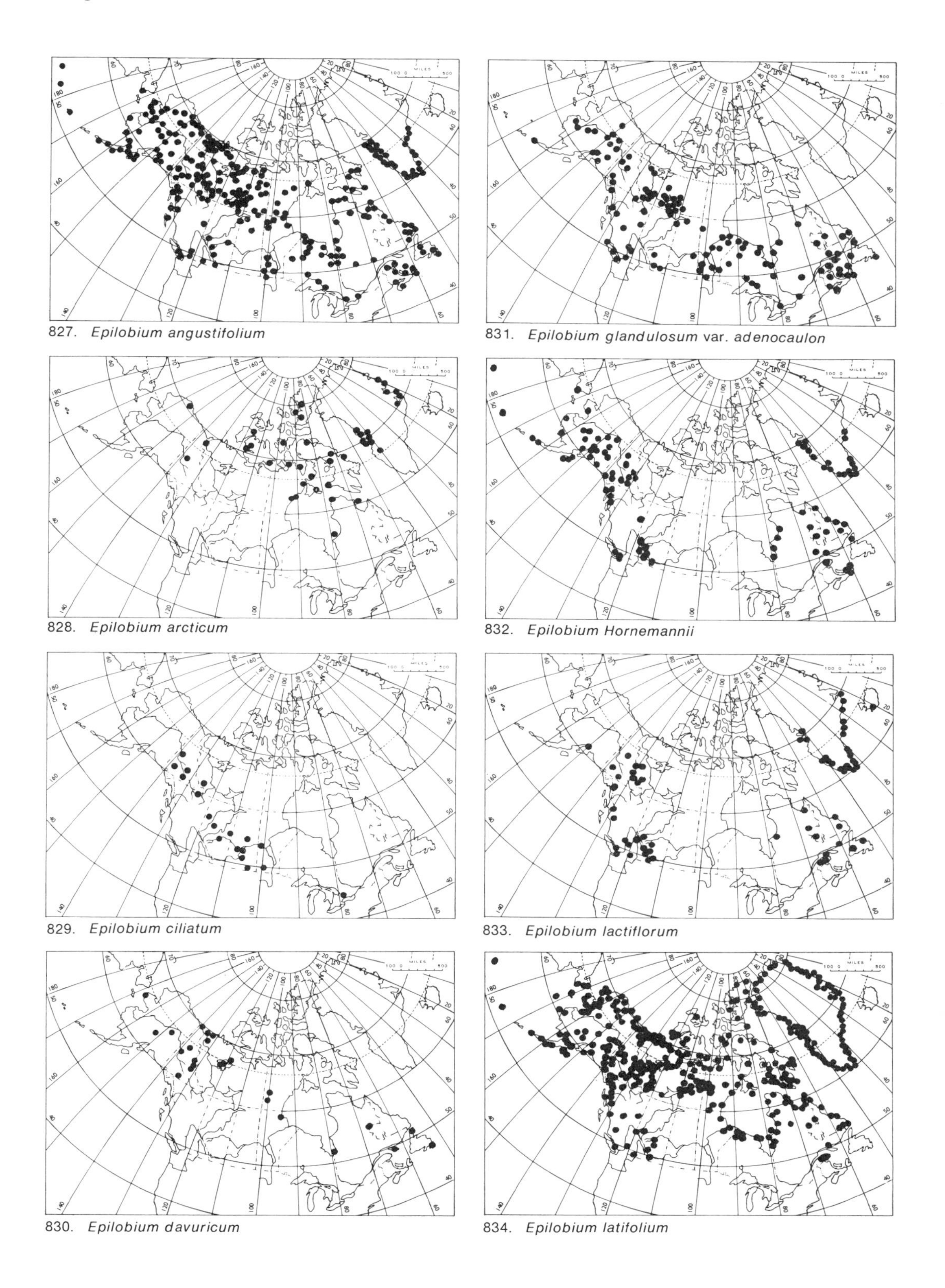

827. *Epilobium angustifolium*

828. *Epilobium arcticum*

829. *Epilobium ciliatum*

830. *Epilobium davuricum*

831. *Epilobium glandulosum* var. *adenocaulon*

832. *Epilobium Hornemannii*

833. *Epilobium lactiflorum*

834. *Epilobium latifolium*

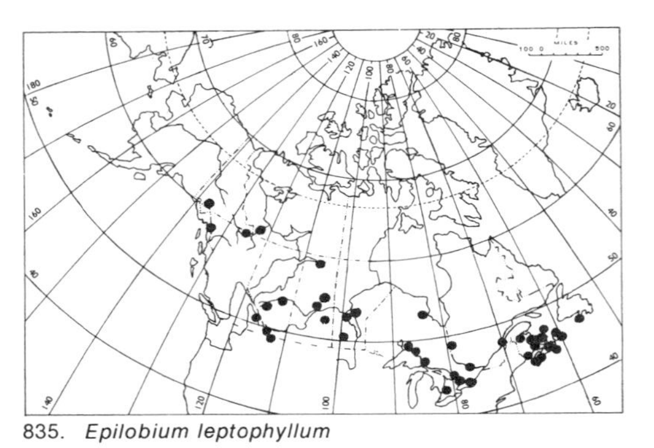

835. *Epilobium leptophyllum*

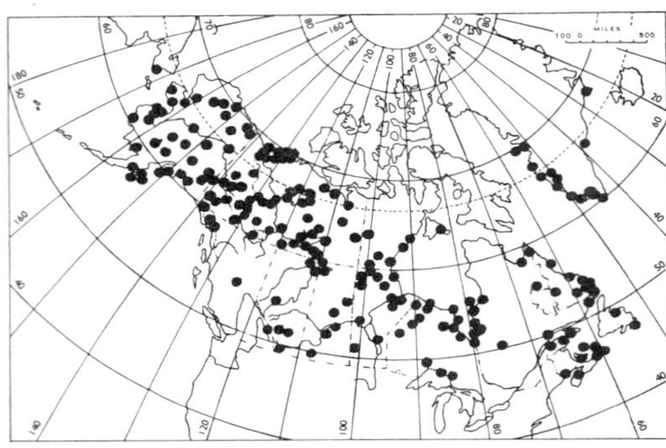

836. *Epilobium palustre*

HALORAGACEAE Water-Milfoil Family

Aquatic or emergent perennials with whorled leaves; flowers minute, sessile, perfect or monoecious; fruit nut-like.

a. Leaves entire, flowers axillary . . . . *Hippuris*
a. Leaves divided pinnately into capillary segments; flowers in terminal, emergent spikes . . . . *Myriophyllum*

Hippuris L. Mare's-Tail

Fleshy-stemmed, unbranched, glabrous emergent perennial aquatics with entire whorled leaves, arising from a creeping rhizome; flowers minute, in the axils of the leaves.

a. Stems 10 cm tall or less . . . . *H. montana*
a. Stems much longer
 b. Leaves 4-6 in a whorl, oblanceolate or elliptic, shorter than the internodes . . . . *H. tetraphylla*
 b. Leaves 6-12 in a whorl, linear-attenuate, equalling the internodes . . . . *H. vulgaris*

Hippuris montana Ledeb.
Leaves oblong-linear, 5 to 7 in a whorl.

In the District of Mackenzie known only from sterile specimens floating or emergent in pools formed by beaver dams, near Brintnell Lake in southwestern Mackenzie Mountains.

General distribution: N. W. N. America; from the Aleutian Islands and Alaska Pen. southward to Washington and the Selkirk Mts. of B.C.; disjunct in the Mackenzie Mts.

Map 837

Hippuris tetraphylla L. f.
Rhizomes about 5 mm thick; stems emergent, arising 3 to 6 cm or more above the water.

Strictly littoral, in shallow saline or brackish water.

General distribution: Circumpolar, but with large gaps.

Fig. 726 Map 838

Hippuris vulgaris L.
Similar to *H. tetraphylla* but a coarser plant with

much narrower leaves in whorls of 6 to 12; stems may be considerably elongated in deeper water when the submersed leaves become flaccid and up to 6 cm long.

In shallow ponds and lakes throughout our area.

General distribution: Circumpolar; wide-ranging.

Fig. 727 Map 839

Myriophyllum L. Water-Milfoil

Submersed aquatics with slender, simple or branched stems up to 1 m long; leaves pectinately divided, in whorls of 3 or 5, the divisions filiform. The flowers very small, 4-merous, sessile, bracted in a terminal, emergent spike; fruit nut-like, 4 lobed.

a. Upper flowers staminate, alternate . . . . *M. alterniflorum*
a. Flowers whorled
  b. Bracts of upper inflorescence entire or nearly so . . . . *M. exalbescens*
  b. Bracts of upper inflorescence pectinate . . . . *M. verticillatum* var. *pectinatum*

Myriophyllum alterniflorum DC.
Stems 1 to 2 mm thick, whitened in drying; leaves in whorls of 3 to 5, 3 to 12 mm long with 3 to 7 pairs of very fine stiffish capillary segments; lower flowers in whorls, the upper staminate and alternate; upper bracts entire or slightly toothed, soon deciduous, the lower pinnate.

Shallow lakes and ponds and slow moving streams; in our area known only from Great Bear Lake and the Eskimo Lake Basin.

General distribution: Amphi-Atlantic; in N. America, southern Greenland and in the boreal region from N.S. to Alaska, but much less common in the western part of its range.

Map 840

Myriophyllum exalbescens Fern.
Stems more robust than in *M. alterniflorum* and also whitened in drying; the leaves in whorls of 3 or 4, larger, 1.2 to 3.0 cm long and the divisions in 6 to 11 pairs, flaccid to slightly stiffish, about as long as the internodes; flowers all whorled; bracts entire to but slightly toothed.

Shallow ponds or sluggish streams.

General distribution: N. America; from central western Greenland to Alaska and southward; north to or slightly beyond the limit of trees.

Fig. 728 Map 841

Myriophyllum verticillatum L.
var. **pectinatum** Wallr.
Similar to *M. exalbescens* but the leaves not flaccid and much longer than the internodes; the stems not whitened in drying, and the bracts all deeply incised.

Ponds and quiet streams.

General distribution: The species circumpolar; var. *pectinatum* N. American from Nfld. to S. W. Alaska and northward along the Mackenzie R. valley to its delta.

Map 842

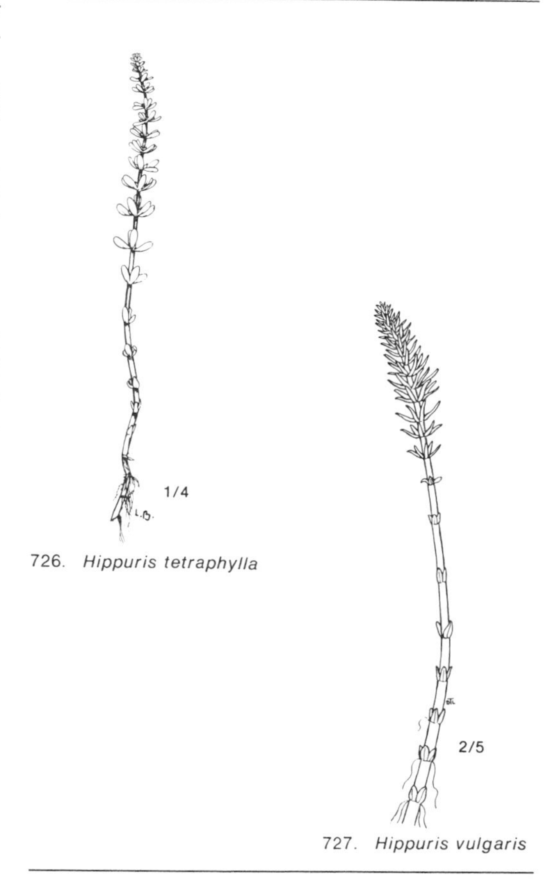

726. *Hippuris tetraphylla*

727. *Hippuris vulgaris*

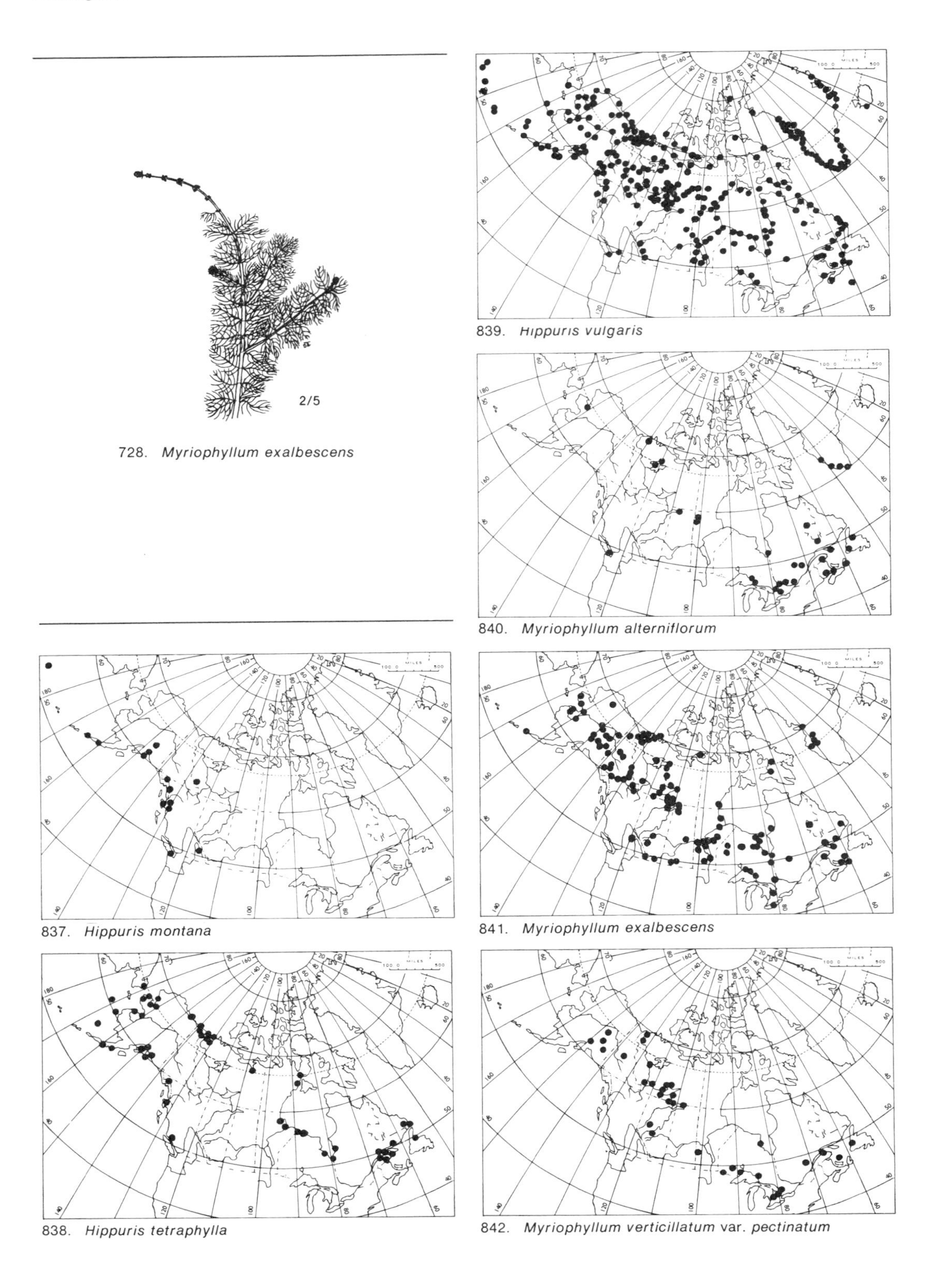

728. *Myriophyllum exalbescens*

839. *Hippuris vulgaris*

840. *Myriophyllum alterniflorum*

837. *Hippuris montana*

841. *Myriophyllum exalbescens*

838. *Hippuris tetraphylla*

842. *Myriophyllum verticillatum* var. *pectinatum*

ARALIACEAE Ginseng Family

Aralia L.

Aralia nudicaulis L.
Wild Sarsaparilla
Acaulescent perennial herb from a long, creeping rhizome; leaves ternate, long-petioled, the divisions pinnately 3- to 5-foliolate; peduncles 2 to 3 dm long, bearing one to several umbels of small, greenish flowers; fruit globose, purplish-black usually with 5 carpels, each containing one seed.

Moist to dry woodlands.

General distribution: Boreal N. America, Nfld. to B.C., S. E. Y.T., and southward; in the District of Mackenzie north to Ft. Simpson.

Fig. 729 Map 843

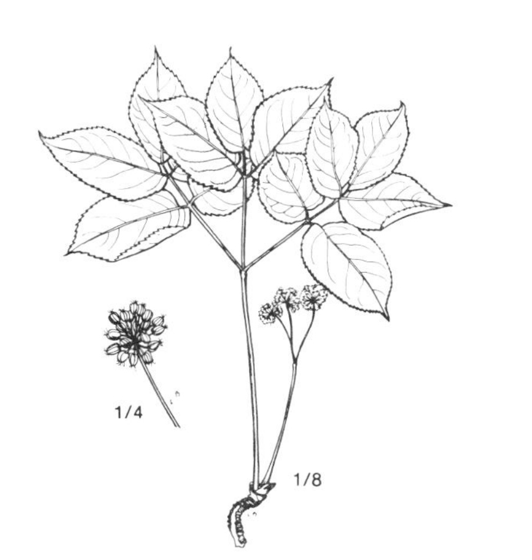

729. *Aralia nudicaulis*

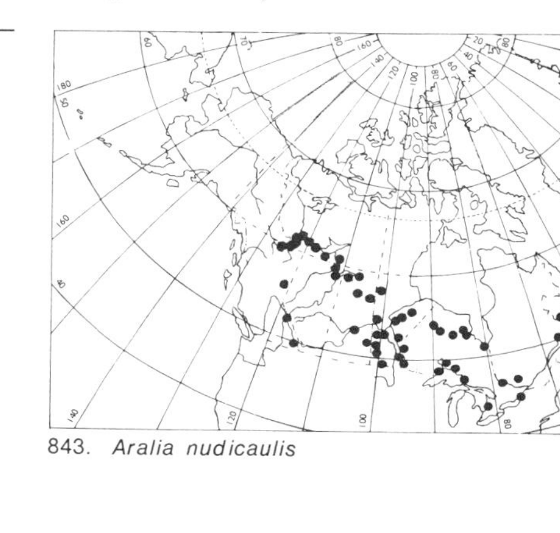

843. *Aralia nudicaulis*

UMBELLIFERAE Parsley Family

Ours perennial, small to robust herbaceous plants, usually with hollow stems and small flowers in simple or compound umbels; fruit dry, separating into 2 carpels, each with 5 primary ribs; leaves alternate or basal, compound or simple. Several species are aromatic, and members of the genus *Cicuta,* very poisonous. Mature fruits are generally required for ready determination.

In the District of Mackenzie the Parsley Family is represented by 8 genera, but all, except *Cicuta,* with only one species each.

a. Leaves entire . . . . *Bupleurum*
a. Leaves compound
  b. Leaves palmate; fruit with hooked spines . . . . *Sanicula*
  b. Leaves pinnate or 3-foliate; fruit not spiny
    c. Fruit linear, bristly hairy . . . . *Osmorhiza*
    c. Fruit oblong, less than 3 times as long as broad
      d. Fruit round in cross-section, or flattened

e. Leaves once-pinnate . . . . *Sium*
e. Leaves twice-pinnate . . . . *Cicuta*
d. Fruit flattened dorsally, the lateral ribs winged
f. Leaflets broad and few; stout and hairy plant . . . . *Heracleum*
f. Leaflets small, less than 5 cm wide; plant glabrous or merely puberulent
g. Leaves once-pinnate . . . . *Pastinaca*
g. Leaves twice or more times pinnate or ternate
h. Leaflets ovate, sharply toothed; fruit terete . . . . *Coelopleurum*
h. Leaflets finely pinnate; lateral ribs of fruit broadly winged . . *Conioselinum*

Bupleurum L.

Bupleurum americanum Coult. & Rose
Glabrous, with stems 1 to 3 dm tall, from a short, branching caudex; basal leaves linear-lanceolate, those of the stem usually broader and somewhat clasping; umbels compact, the involucels bright yellow and foliaceous; petals yellow or purplish; fruit oblong, flattened, 3.5 mm long with slender ribs.

Alpine meadows, moist sand and gravel banks and talus slopes.

General distribution: Amphi-Beringian; Alaska, Y.T. and N. W. District of Mackenzie, with apparently disjunct alpine areas in the Rocky Mts.

Fig. 730 Map 844

Cicuta L. Water-Hemlock

Stout, glabrous marsh plants from a stout, branched rhizome; leaves pinnate or pinnately compound; umbels large and compound; involucre usually lacking. Flowers white. The root especially, but also other parts, very poisonous.

a. Leaflets linear; axils of upper leaves with fascicles of bulblets . . . . *C. bulbifera*
a. Leaflets lanceolate; axils without bulblets
b. Leaflets lanceolate . . . . *C. maculata* var. *angustifolia*
b. Leaflets narrowly or linear-lanceolate . . . . *C. mackenzieana*

Cicuta bulbifera L.
Stems slender, up to 8 dm tall, from a tuberous root; leaves mainly cauline, 2- to 3-pinnate, the leaflets linear and the upper much reduced and bearing bulblets in the axils; umbels small, the fruits orbicular, but mostly abortive.

Rare and local in marshy places in the Mackenzie River valley north to Norman Wells.

General distribution: Boreal N. America, Nfld. to central Y.T.

Fig. 731 Map 845

Cicuta mackenzieana Raup
Stems 10 dm or more tall; similar to *C. maculata* var. *angustifolia*, but with narrower and longer leaf segments; usually fewer rays in the leading umbel, 9 to 21, and crowded umbellets with 50 or more rays; mature fruit 1.5 to 2.2 mm long, 2 to 3 mm wide.

In water up to 5 dm deep.

General distribution: Northern N. America. from James Bay to Alaska.

Fig. 732 Map 846

Cicuta maculata L.
var. **angustifolia** Hook.
C. Douglasii auctt. pro parte
Stems stout to 8 dm or more tall; leaves pinnately divided, the pinnae frequently ternately divided, the segments more or less lanceolate, their margins sharply serrate; umbels compound, 1 to several, the leading one with 18 to 28 rays, its umbellets each with 12 to 25 rays; mature fruit about 3 mm long and 2.2 mm wide.

Marshy lake-shores and stream-banks; rare, in our area known only in the upper Mackenzie drainage north to Fort Simpson.

General distribution: N. America, N.S. to B.C., Alaska and southward.

Fig. 733 Map 847

Coelopleurum Ledeb.

Coelopleurum Gmelinii (DC.) Ledeb.
Angelica lucida sensu Hultén
Stems stout up to 1 m tall; leaves essentially glabrous, twice or three times ternate, the irregularly sharply toothed segments ovate, the petioles much inflated; inflorescence and upper

stems scabrous-puberulent; umbels compound, the rays usually thick, the involucre dedicuous; fruit terete, 4 to 7 mm long, with prominent, thick corky ribs.

In our area known from a single locality in shrubby alpine tundra at 1150 m on Pointed Mountain in the Fisherman Lake area of the southern Mackenzie Mts.

General distribution: Amphi-Beringian and although predominantly littoral, with alpine stations in Alaska as well as in Y.T.; and extreme S. W. District of Mackenzie. The closely related *C. lucidum* (L.) Fern. is strictly coastal, from Lab. to Long Island, N.Y.

Map 848

Conioselinum Hoffm. Hemlock-parsley

Conioselinum cnidiifolium (Turcz.) Porsild
Glabrous, with simple or branched stems 6 dm or more tall, and doubly pinnate-pinnatifid leaves. Umbels compound with decisuous involucral leaflets; flowers and involucels purplish when young, the petals turning yellow; fruits ovate, about 5 mm long, with 5 winged ribs on each carpel.

Sandy river banks in the Mackenzie River Delta, the Eskimo Lakes and eastward to the Anderson River.

General distribution: Amphi-Beringian.

Fig. 734 Map 849

Heracleum L. Cow-parsnip

Heracleum lanatum Michx.
Very robust with stems up to 1.5 m tall from a stout tap-root. Leaves large, ternate, tomentose when young, the petioles conspicuously inflated, the petioled leaflets irregularly toothed; umbels compound, to 2 dm or more broad, the involucre deciduous; flowers white; fruits obcordate, strongly flattened, to 8 mm broad.

Common and often locally dominant in well drained subalpine meadows whereas specimens reported from near towns along the Mackenzie River and its tributaries probably are not indiginous. The fresh peeled stalks and the boiled roots of the Cow-parsnip are said to be edible.

General distribution: N. America; wide-ranging from Nfld. to B.C. and Alaska and E. Asia.

Fig. 735 Map 850

Osmorhiza Raf. Sweet Cicely

Osmorhiza depauperata Phil.
O. obtusa (Coult. & Rose) Fern.
Stems slender, glabrous or somewhat pubescent up to 8 dm tall, with 2- to 3-ternate leaves, the leaflets broadly lanceolate to ovate, coarsely serrate, incised or lobed, thin and delicate, more or less crisp-pubescent. Umbels compound, the 3 to 5 rays strongly ascending at flowering time, without involucre, the umbellets few rayed, widely divergent in fruit. Flowers greenish white and the fruit clavate, 8 to 12 mm long.

Rich woods; in our area known only from the slopes of Mount Coty adjacent to the Liard River.

General distribution: Boreal N. America, from Nfld. to south central Alaska, with several large gaps.

Map 851

Pastinaca L. Parsnip

Pastinaca sativa L.
Glabrous biennial to 6 dm or more tall, with grooved stems; leaves pinnate, the leaflets sessile or the lowest short-petioled, ovate to oblong, toothed. Flowers yellow in compound umbels, lacking involucres; fruit oval, flattened dorsally, winged; contact with the leaves may produce a severe skin irritation.

In the Northwest Territories thus far known only from a cabin clearing along the South Nahanni River.

General distribution: Introduced from Europe and growing as a weed, particularly in Ont. and Que.

Sanicula L. Black Snakeroot

***Sanicula marilandica** L.
Glabrous perennial to 7 dm or more tall with few, palmately lobed leaves, the basal ones long-petioled; leaflets obovate, cuneate, unequally serrate, the teeth spinulose-tipped; umbels paniculate-compound; involucre leaf-like. Umbellets dense with short-pedicelled yellowish flowers. Fruit bur-like, thickly covered with hooked prickles.

Moist meadows, thickets and open woods; to be looked for in the lowlands of the Liard River and especially near hot springs.

General distribution: Boreal N. America, Nfld. to B.C. and southward.

Map 852

Sium L. Water-parsnip

Sium sauve Walt.
Stout glabrous usually branched emergent marsh plants 10 dm or more tall; leaves pinnate, the segments sessile, linear or broadly linear, toothed; submersed leaves, if present, twice pinnate-pinnatifid; umbels compound, involucre of narrow segments; flowers white, mature fruit oval, about 2.5 mm long. Rather similar to *Cicuta maculata* var. *angustifolia* and *C. mackenzieana*, but readily distinguished by its once-pinnate leaves and presence of an involucre.

Wet meadows, muddy stream and lake shores of the upper Mackenzie drainage.

General distribution: Amphi-Beringian; widespread in boreal N. America from Nfld. to B.C. with disjunct stations in S. W. Y.T. and central Alaska.

Fig. 736 Map 853

730. *Bupleurum americanum*

731. *Cicuta bulbifera*

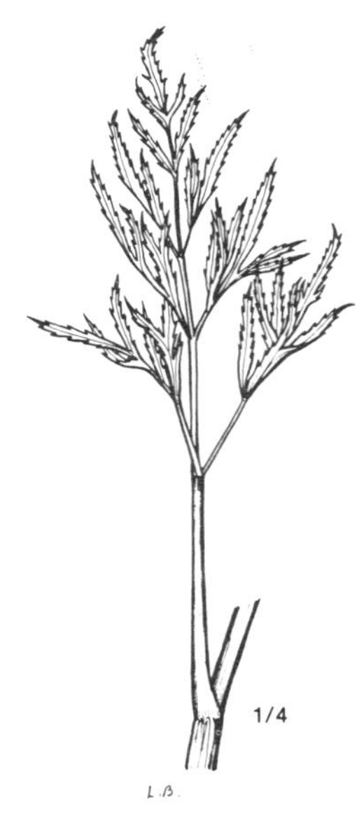

732. *Cicuta mackenzieana*

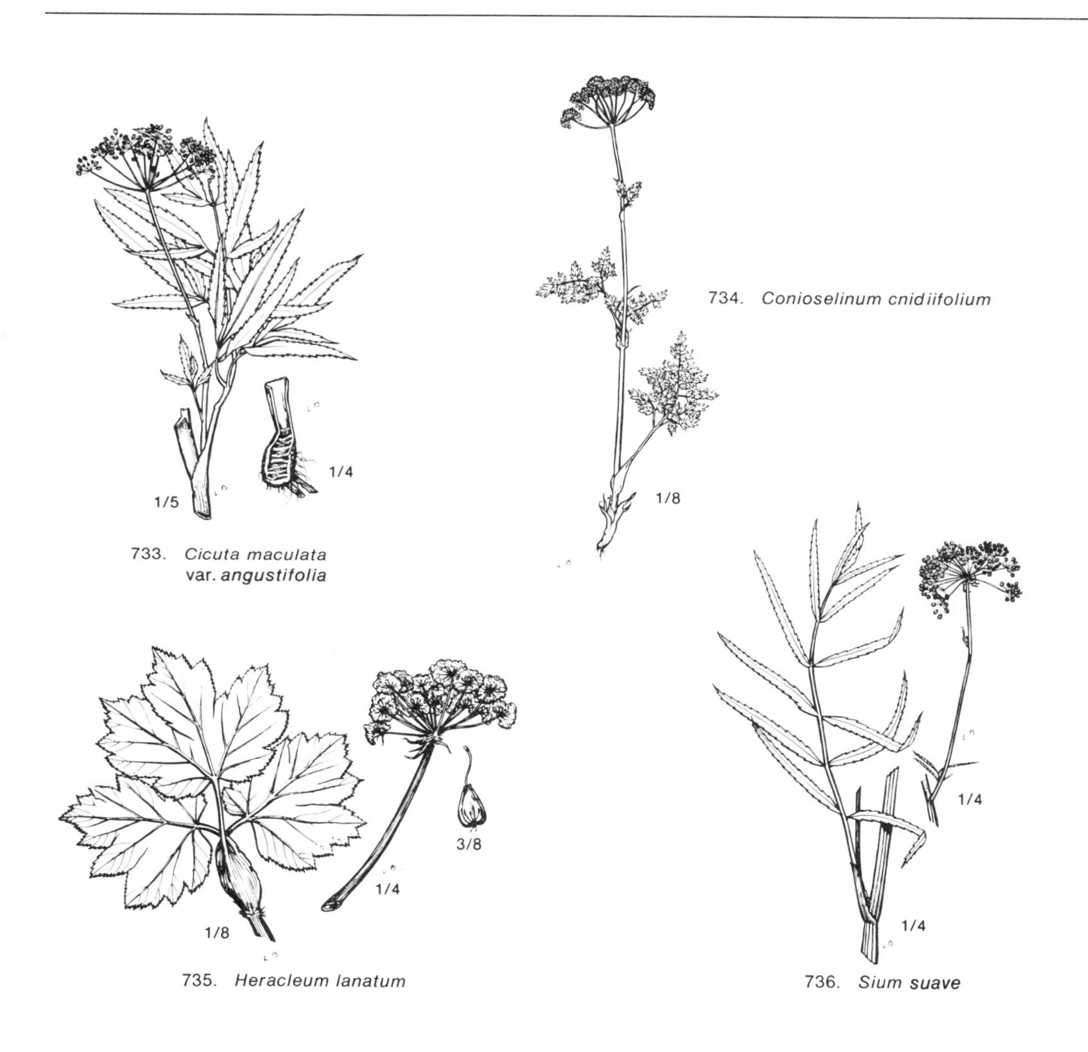

733. *Cicuta maculata* var. *angustifolia*

734. *Conioselinum cnidiifolium*

735. *Heracleum lanatum*

736. *Sium suave*

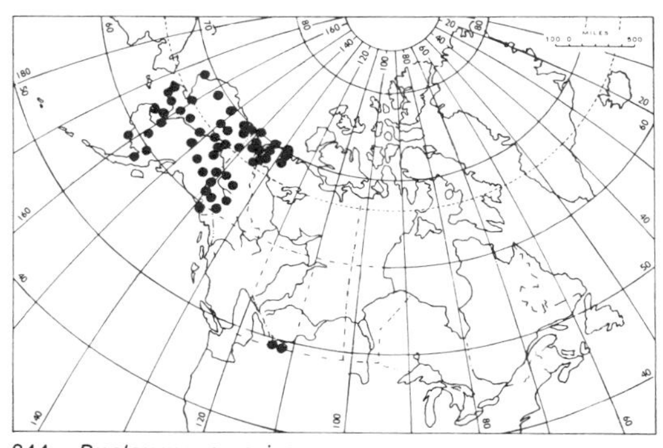

844. *Bupleurum americanum*

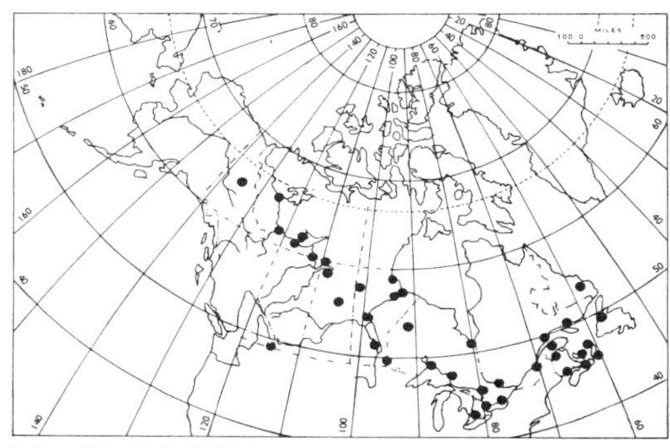

845. *Cicuta bulbifera*

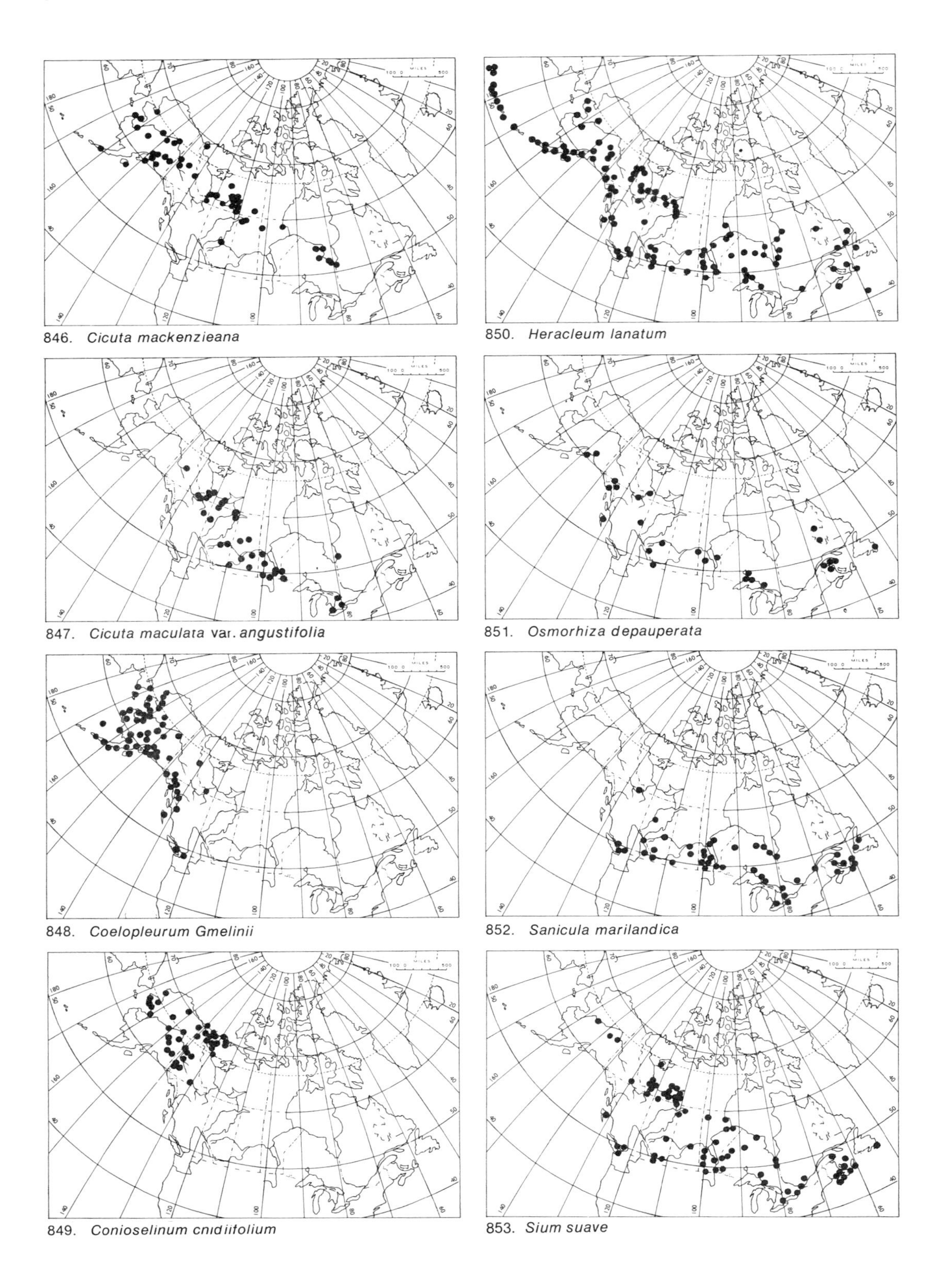

846. *Cicuta mackenzieana*

847. *Cicuta maculata* var. *angustifolia*

848. *Coelopleurum Gmelinii*

849. *Conioselinum cnidiifolium*

850. *Heracleum lanatum*

851. *Osmorhiza depauperata*

852. *Sanicula marilandica*

853. *Sium suave*

CORNACEAE Dogwood Family

Cornus L. Dogwood

Shrubs or herbs with broadly ovate-lanceolate, entire and opposite leaves; flowers 4-parted, in a small, dense cluster or surrounded by a corolla-like involucre of white, petal-like bracts; fruit a 1- to 2-seeded drupe.

a. Herbs; flowers subtended by 4 petal-like whitish bracts surrounding a tight cluster of small flowers
  b. Flowers white . . . . *C. canadensis*
  b. Flowers dark purple . . . . *C. suecica*
a. Shrubs with open cymes of small white flowers; bracts lacking . . . . *C. stolonifera*

Cornus canadensis L.
Bunchberry
Stems 8 to 15 cm tall from slender creeping rhizomes; leaves sub-sessile, the lower small, the upper 3 to 6 cm long, whorled; flowers yellowish or greenish-white, subtended by 4 ovate, white, petal-like bracts; fruits orange-red to red.

Moist woodland and clearings, in the Mackenzie drainage north to lat. 64°.

General distribution: Boreal America, from S. Greenland and Nfld. to Alaska and eastern Asia.

Fig. 737 Map 854

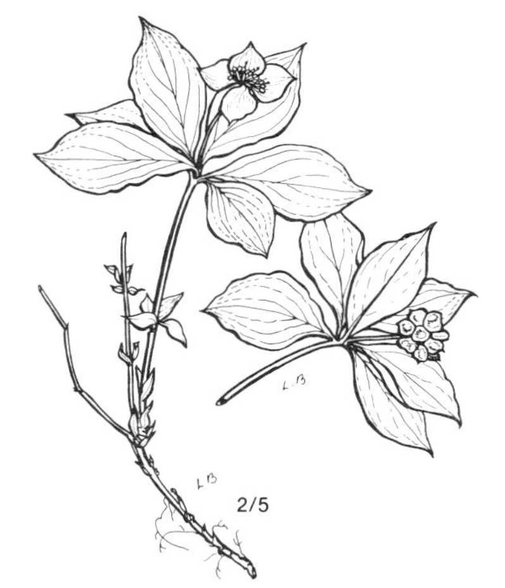

737. *Cornus canadensis*

Cornus stolonifera Michx.
Red Osier Dogwood
Stoloniferous shrub with 1 to 3 m tall, reddish-brown branches, the lower often prostrate and freely rooting; leaves ovate to oblong-lanceolate, acute, dark green above, glaucous beneath; flowers in an open, flat-topped cyme, the tiny petals white; fruit white.

Moist woods, thickets and clearings; along the Mackenzie River Valley to lat. 65°, with disjunct stations in the southern part of the delta.

General distribution: Boreal N. America, Nfld. to B.C. and Alaska and southward; widespread.

Fig. 738 Map 855

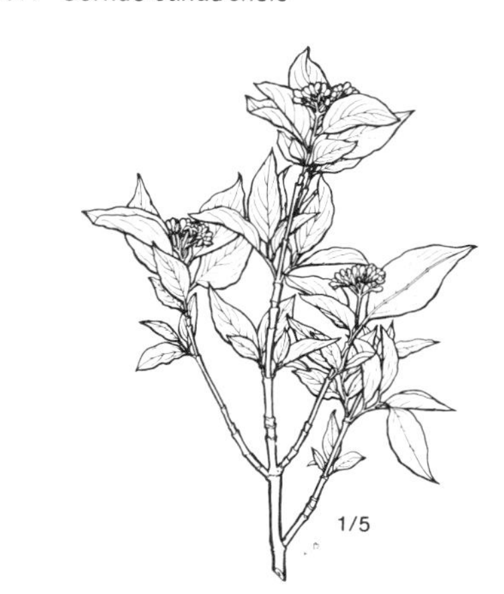

738. *Cornus stolonifera*

Cornus suecica L.
Similar to *C. canadensis* from which it is distinguished by its purple flowers and more numerous and smaller leaves, the uppermost 1.5 to 3.0 cm long.

Known in our area from a single station north of the east arm of Great Slave Lake.

General distribution: S. W. Greenland and eastern Canada with large gaps across boreal Canada to S. W. and Central Alaska.

Map 856

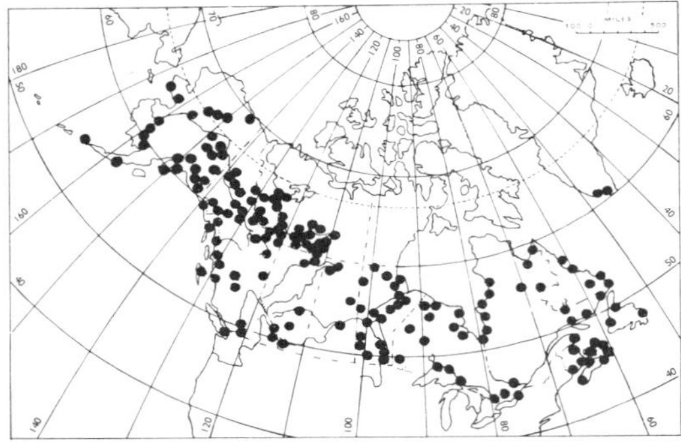
854. *Cornus canadensis*

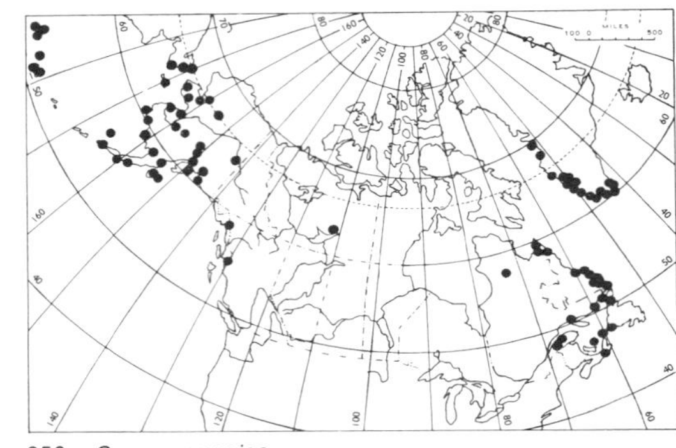
856. *Cornus suecica*

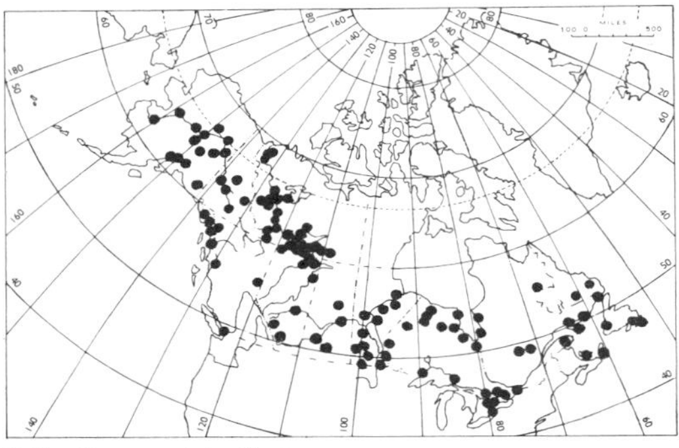
855. *Cornus stolonifera*

PYROLACEAE Wintergreen Family

Perennial herbs or low sub-shrubs with simple evergreen leaves, or saprophytes lacking chlorophyll; flowers with 5 sepals and petals, and 10 stamens; capsule 10-locular, containing numerous tiny seeds.

a. Plants with green leaves
  b. Stems leafy . . . . *Chimaphila*
  b. Stems scape-like, the leaves basal
    c. Flower solitary . . . . *Moneses*
    c. Flowers racemose . . . . *Pyrola*
a. Plants waxy-white; saprophytic . . . . *Monotropa*

Chimaphila Pursh Pipsissewa

Chimaphila umbellata (L.) Bart.
var. **occidentalis** (Rydb.) Blake
Evergreen dwarf woodland shrub from a creeping rhizome, with erect and leafy stems up to 2 dm tall; the leaves oblanceolate, sharply serrate, in several whorls or clusters; flowers white or pink, in a few-flowered terminal raceme.

Known in our area from a single sterile collection from Fort Smith where it was found at the outlook over the Rapids of the Drowned of the Slave River.

General distribution: *C. umbellata s. lat.*, circumpolar.

Map 857

Moneses Salisb. One-flowered Pyrola

Moneses uniflora (L.) Gray
Stems scapose, 1 dm tall, from a slender, creeping rhizome, the small, ovate and leathery leaves crowded at its base; the flower solitary, white, first nodding; capsule erect.

Frequent in deep, moist moss on wooded slopes northward to the Eskimo Lake Basin.

General distribution: Circumpolar; wide spread.

Fig. 739 Map 858

Monotropa L. Indian Pipe

***Monotropa uniflora** L.
Fleshy and smooth saprophyte lacking chlorophyll, with one or several scaly, white stems, from a fleshy ball of scaly rootlets. Flower solitary, waxy, white or pinkish, nodding; capsule erect. The entire plant turns black in drying.

To be looked for in rich woodland in the southern parts of our area.

General distribution: Circumpolar with large gaps.

Map 859

Pyrola L. Wintergreen

Herbs with creeping, slender rhizomes and basal, petioled, wintergreen leaves; the flowers racemose with 5 sepals, 5 petals, and 8 to 10 stamens; capsule with numerous small seeds.

a. Inflorescence 1-sided; petals greenish-white . . . . *P. secunda*
a. Inflorescence cylindrical
  b. Petals and anthers crimson to pale pink . . . . *P. asarifolia*
  b. Petals white, greenish-white or pinkish; anthers yellow
    c. Style short, included; petals white or pinkish . . . . *P. minor*
    c. Style longer, exserted
      d. Flowers small, greenish-white, the leaves small and rounded . . . . *P. chlorantha*
      d. Flowers large, creamy-white or pinkish; leaves larger, the blade 3-5 cm long . . . . *P. grandiflora*

Pyrola asarifolia Michx.
Pink-flowered Wintergreen
Stems 1 to 2 or more dm tall; leaves basal, long-petioled, the blade leathery, cordate-orbicular; flowers with crimson to pale pink petals, the calyx lobes lanceolate to deltoid; mature styles 5 to 10 mm long, deflexed and arched.

Frequent in damp, rich woods and thickets; north in the Mackenzie Valley to slightly beyond Fort Norman.

General distribution: Boreal N. America and E. Asia.

Fig. 740 Map 860

Pyrola chlorantha Sw.
P. virens Schweigg.
Stems 1 to 2 dm tall; leaves usually few, basal, long-petioled, the blade small, yellowish-green and somewhat crenate; raceme usually open, the flowers small with greenish-white converging petals; calyx lobes ovate-oblong to deltoid-ovate, and the anthers yellow.

Woods and thickets, in the western part of our area north to the upper delta of the Mackenzie River.

General distribution: Circumpolar.

Fig. 741 Map 861

Pyrola grandiflora Radius
Large-flowered Wintergreen
Similar to *P. asarifolia* but the flowers scented and larger, with creamy-white petals and yellow anthers.

P. grandiflora s. str. is the circumpolar, arctic-subarctic taxon described above; the western and predominantly woodland plant in which the flowers are smaller and scentless has been distinguished as var. *canadensis* (Andres) Porsild, whereas var. *Gormanii* (Rydb.) Porsild with ovate-acute and distinctly denticulate leaves and perfumed flowers is endemic to Alaska and Yukon.

Frequent to common in open boreal woodland and on sheltered, sunny tundra slopes.

General distribution: Circumpolar, arctic-

alpine, wide ranging from high-arctic or alpine tundra, southward into the boreal forest.
Fig. 742 Map 862

Pyrola minor L.
Similar to *P. chlorantha* but with oblong, dark green, dull leaves with flat, crenulated margins; inflorescence of small, scentless, globular, pink or white flowers in a slender, spike-like raceme; style straight, included to barely exserted; anthers yellow.

In sheltered, mossy places. Rare or local in the boreal woodland, north to Great Bear Lake and the upper Mackenzie Delta.

General distribution: Circumpolar, subarctic-alpine.

Fig. 743 Map 863

Pyrola secunda L.
var. **secunda**
One-sided Wintergreen
Stems 0.5 to 2.0 or more dm tall, with scattered leaves in the lower half; leaves short-petioled, the blade oblong-ovate, acute, fresh green and somewhat leathery, with crenulate-serrate margins; inflorescence 1-sided, of small whitish-green flowers.

Typical *P. secunda* is a woodland plant that, beyond the treeline and in alpine situations, is replaced by the var. *obtusata* Turcz. in which the leaves are elliptic-orbicular and rarely over 1 cm long (as opposed to 3 cm in var. *secunda*) and the inflorescence rarely more than 4-flowered.

Frequent in moist thickets and woodland, north beyond the limit of trees.

General distribution: Circumpolar.

Fig. 744 (var. *obtusata*) Map 864 (*P. secunda s. lat.*)

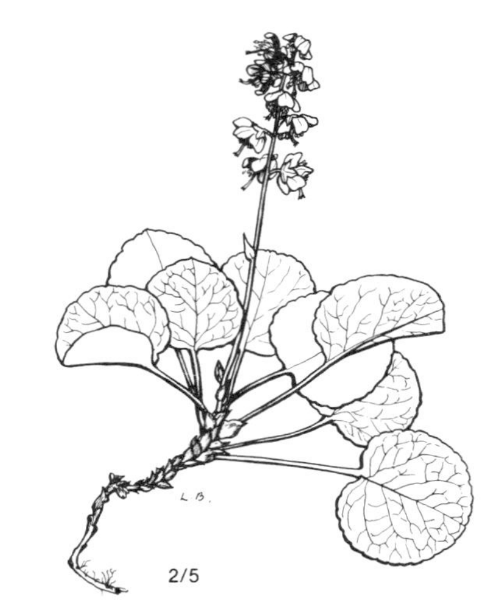

740. *Pyrola asarifolia*

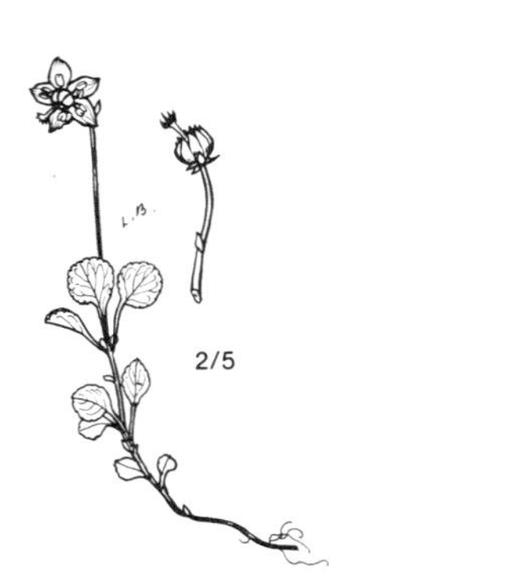

739. *Moneses uniflora*

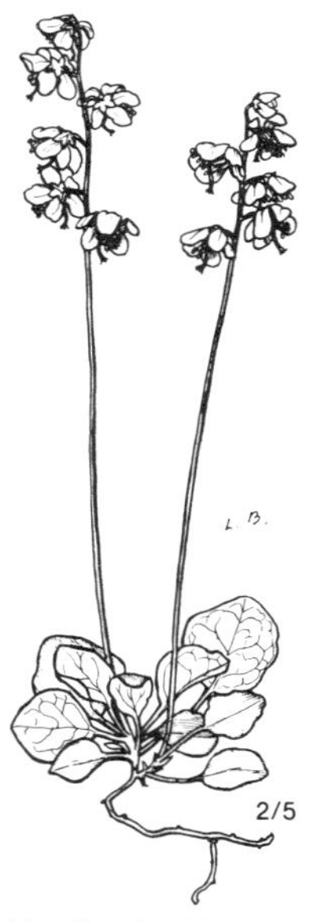

741. *Pyrola chlorantha*

742. *Pyrola grandiflora*

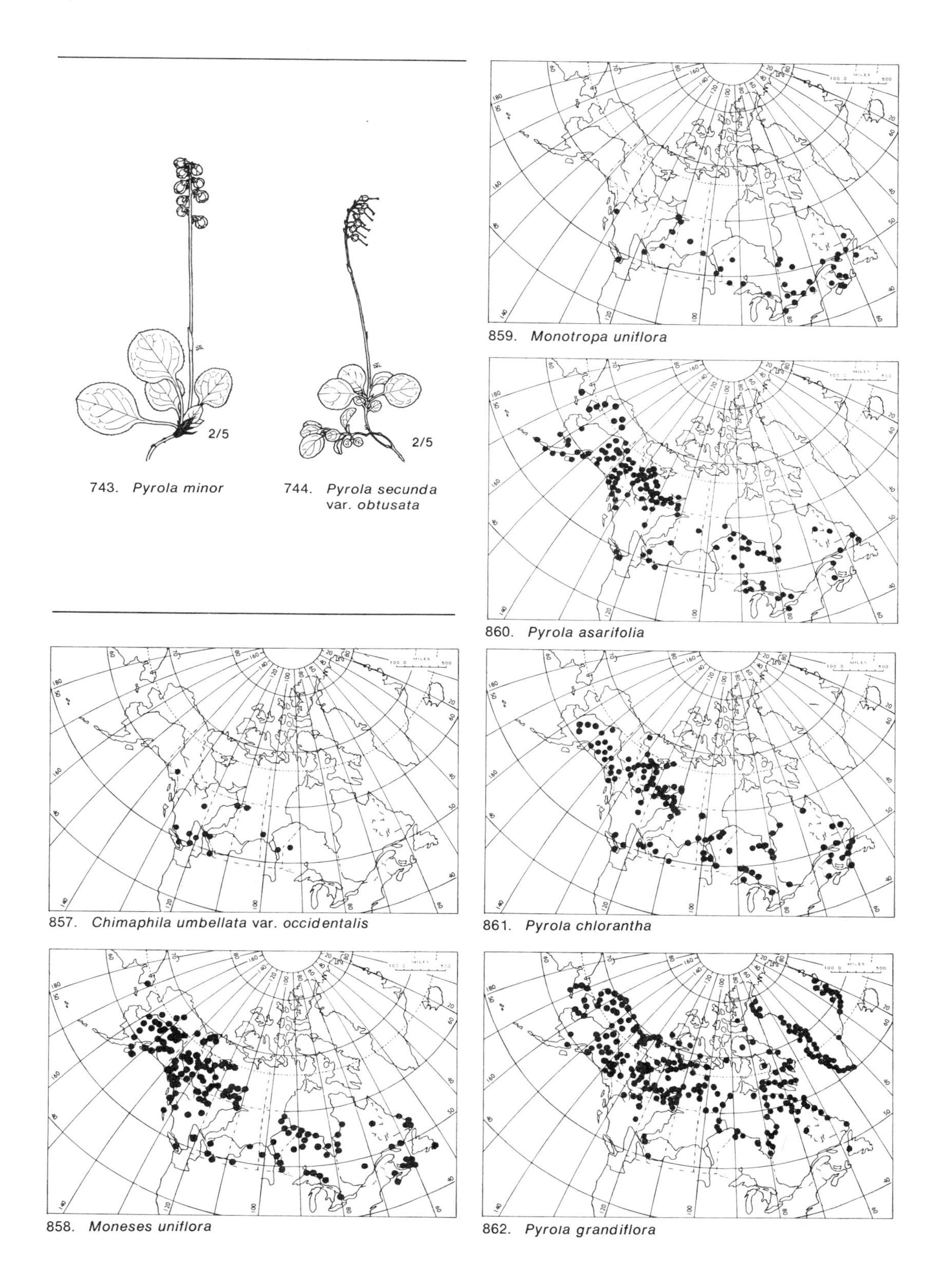

743. *Pyrola minor*

744. *Pyrola secunda* var. *obtusata*

859. *Monotropa uniflora*

860. *Pyrola asarifolia*

857. *Chimaphila umbellata* var. *occidentalis*

861. *Pyrola chlorantha*

858. *Moneses uniflora*

862. *Pyrola grandiflora*

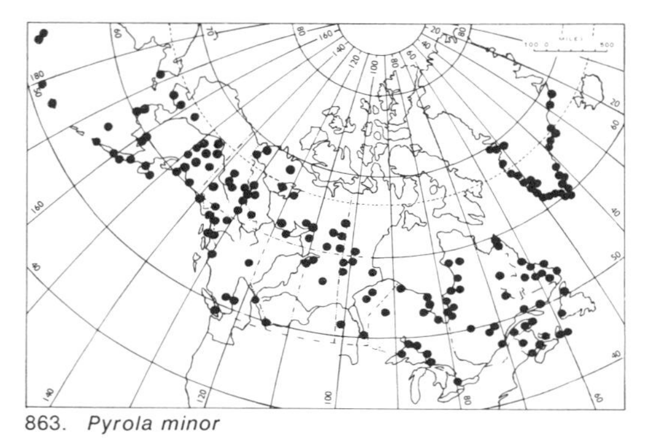

863. *Pyrola minor*

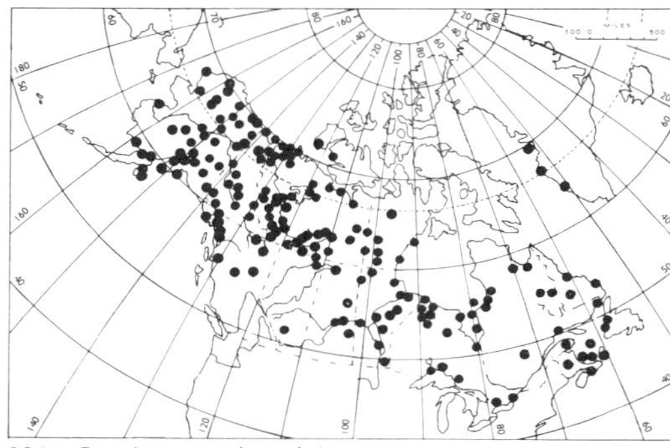

864. *Pyrola secunda s. lat.*

ERICACEAE

Ours mostly low shrubs with simple, and mostly leathery and persisting leaves; flowers regular, corolla 4- to 5-parted, gamopetalous (except *Ledum*) with stamens mostly twice as many as the corolla lobes; fruit a dry capsule, or a berry-like drupe.

a. Ovary superior
  b. Corolla polypetalous; flowers white; fruit a dry capsule .... *Ledum*
  b. Corolla gamopetalous; corolla pink, purple or white
    c. Fruit a dry capsule
      d. Anther cells awned
        e. Leaves minute, 2-4 mm long; flowers white .... *Cassiope*
        e. Leaves 1 cm long or more; flowers pink .... *Andromeda*
      d. Anther cells awnless
        f. Leaves linear, crowded on the branches .... *Phyllodoce*
        f. Leaves lanceolate or narrowly elliptic, not crowded on the branches
          g. Stamens 5; matted dwarf shrub .... *Loiseleuria*
          g. Stamens 10; erect, branched dwarf shrub
            h. Capsule globose .... *Kalmia*
            h. Capsule longer than broad
              i. Corolla broadly cup-shaped, purple .... *Rhododendron*
              i. Corolla urn-shaped or cylindric, white .... *Chamaedaphne*
    c. Fruit berry-like .... *Arctostaphylos*
a. Ovary inferior; fruit a berry
  j. Corolla divided nearly to the base .... *Oxycoccus*
  j. Corolla urn-shaped .... *Vaccinium*

Andromeda L. Andromeda

Andromeda Polifolia L.
Low shrub with prostrate base and ascending branches; leaves evergreen, alternate, thick, linear-lanceolate to oblong, strongly involute, deeply grooved on the upper surface, the lower surface glaucous and the midvein prominent; the small urceolate flowers in small terminal clusters.

Frequent in muskeg and damp tundra, north beyond the limit of trees, but not reaching the arctic islands.

General distribution: Circumpolar, wide ranging.

Fig. 745 Map 865

Arctostaphylos Adans. Bear Berry

Prostrate or trailing shrubs; leaves deciduous or evergreen; flowers urceolate in terminal racemes or clusters.

a. Leaves entire, evergreen, leathery . . . . *A. uva-ursi*
a. Leaves crenate-margined
  b. Leaves persisting for several years; fruit black . . . . *A. alpina*
  b. Leaves soon falling; fruit red . . . . *A. rubra*

Arctostaphylos alpina (L.) Spreng.
Alpine Bear Berry
Matted, trailing shrub with shreddy bark and subcoriaceous, marcescent, obovate or oblanceolate, rugose, crenulate leaves that turn a deep red in autumn; flowers small yellowish-green, urceolate, appearing in early spring before the leaves unfold; fruit edible but rather insipid, black and shiny when ripe.

Occasional in acid rocky and gravelly places and in rocky tundra.

General distribution: Circumpolar, arctic.

Fig. 746 Map 866

Arctostaphylos rubra (Rehd. & Wils.) Fern.
Similar to *A. alpina* but leaves readily deciduous and not deeply wrinkled; fruits scarlet and very juicy, but rather insipid.

Mossy places in open coniferous woodland, on peaty soils and rocky tundra; common in the southwestern part of our area, less frequent northward, but extending beyond treeline to the southern parts of the Arctic Archipelago.

General distribution: E. Asia and N. America, arctic-subarctic, alpine.

Fig. 747 Map 867

Arctostaphylos uva-ursi (L.) Spreng.
Common Bearberry; Kinnikinick
Prostrate much-branched shrub with reddish to dark grey, exfoliating bark; leaves obovate, entire, leathery; flowers in dense terminal racemes; corolla pink to white; fruit dull red, mealy.

Exposed rocks, river banks, eskers and sand plains northward to near the limit of trees.

General distribution: Circumpolar; wide ranging.

Fig. 748 Map 868

Cassiope D.Don White Heather

Dwarf shrubs with imbricated, scale-like evergreen leaves and white, nodding flowers on slender peduncles, appearing singly or few together near the tip of the branches.

a. Leaves needle-like; peduncles capillary . . . . *C. hypnoides*
a. Leaves lanceolate, deeply grooved dorsally; peduncles coarser
  b. Peduncles 1-2 cm long . . . . *C. tetragona* ssp. *tetragona*
  b. Peduncles much shorter . . . . *C. tetragona* ssp. *saximontana*

Cassiope hypnoides (L.) D. Don
Moss Heather
Moss-like, with tufted, procumbent stems; flowers small, bell-shaped, on soft-pubescent, capillary peduncles; corolla white, the calyx wine-red.

In sheltered, rocky places, especially along brooks and lake shores, where the snow remains until late into the summer; entering our area in the east adjacent to Hudson Bay, and reaching the District of Mackenzie in the Thelon Game Sanctuary.

General distribution: Amphi-Atlantic, arctic-alpine.

Fig. 749 Map 869

Cassiope tetragona (L.) D.Don
ssp. **tetragona**
Arctic White Heather
A coarse, heath-forming dwarf shrub with imbricated, dark green, deeply grooved, leathery leaves arranged in four rows, on the stiff, quadrangular branches; flowers faintly perfumed, larger than in the preceding; the calyx yel-

lowish-green and the long peduncles glabrous.

A dominant or even ubiquitous arctic species in not too dry tundra. Because of its local abundance, it was formerly important to the Eskimos, and to travellers in the Arctic as a source of fuel for outdoor cooking. Owing to its content of resin, the plant burns with a hot flame when dry.

General distribution: Circumpolar, arctic.

Fig. 750 Map 870

Cassiope tetragona (L.) D.Don ssp. **saximontana** (Small) Porsild

Similar to ssp. *tetragona*, from which it differs by its consistently shorter pedicels and somewhat smaller flowers.

Barely entering our area on subalpine slopes and tundra of the southern Mackenzie Mts.

General distribution: North Cordilleran.

Fig. 751 Map 871

Chamaedaphne Moench Leather-leaf

Chamaedaphne calyculata (L.) Moench

Much branched shrub 2 to 6 dm tall; leaves alternate, oblong-lanceolate, evergreen, coriaceous and scurfy, especially on the underside; flowers in small one-sided leafy racemes; corolla white; sepals 5, distinct.

Peaty swales, bogs, muskeg and lake margins in the wooded parts of our area north to the Mackenzie River delta.

General distribution: Circumpolar; wide ranging.

Fig. 752 Map 872

Kalmia L. Bog-Laurel

Kalmia polifolia Wang.

Low much branched shrub to 4 dm tall; leaves evergreen, opposite, lanceolate, oblong to obovate, the margins revolute, the lower surface whitened by dense, very short hairs; inflorescence umbelliform, the flowers long-pedicelled; corolla saucer-shaped, 5-lobed about 1.5 cm across, deep pink to crimson; calyx 5-parted, glabrous, the lobes papery, finely ciliate-margined.

Occasional in peaty soils, muskeg, bogs and lake margins north to the Arctic Circle and near to or just beyond the limit of trees.

The ssp. *microphylla* (Hook.) Calder & Taylor is similar but consistently of smaller stature and with smaller leaves and flowers. It is an alpine Cordilleran race reaching Yukon Territory and S.W. District of Mackenzie.

General distribution: Boreal N. America, Nfld. to B.C. and Y.T.

Fig. 753 Map 873

Ledum L. Labrador-tea

Low, much branched and strongly aromatic shrubs; leaves evergreen, alternate, the margins inrolled, glabrous above, densely rusty-hairy below; flowers in terminal umbel-like clusters; pedicels rusty-puberulent; petals white, 5, distinct; calyx small, 5-toothed.

a. Leaves linear, acutish . . . . *L. decumbens*
a. Leaves oblong to linear-oblong, blunt . . . . *L. groenlandicum*

Ledum decumbens (Ait.) Lodd.

Prostrate, decumbent or ascending dwarf shrub, with dark green, linear, strongly revolute leathery leaves; flowers white, small and spicy; fruiting pedicels abruptly bent below the capsule.

Common in not too wet dwarf shrub- or moss-lichen heath or on sunny cliffs and ledges.

General distribution: Eastern Asia, Alaska to W. Greenland; arctic.

Fig. 754 Map 874

Ledum groenlandicum Oeder

Erect shrub, 3 to 8 dm tall; leaves oblong to linear-oblong, blunt, usually much larger than in *L. decumbens*; pedicels arcuate in fruit.

Common in peaty soils northward to the limit of trees.

General distribution: N. American: W. Greenland, Lab. and Nfld. to B.C. and Alaska.

Fig. 755 Map 875

Loiseleuria Desv. Alpine Azalea

Loiseleuria procumbens (L.) Desv.
Depressed, much branched, tufted dwarf shrub with minute, opposite, leathery, elliptic, and glabrous leaves, and few-flowered clusters of small pink or white flowers.

Occasional to common in dry, stony heath mainly on acid soil.

General distribution: Circumpolar, arctic-alpine.

Fig. 756 Map 876

Oxycoccus Tourn. ex Adans. Cranberry

Small, trailing evergreen shrubs; stems creeping and rooting, the tips and flowering branches usually ascending; leaves entire, leathery, the margins somewhat revolute; flowers terminal, 1 to several, long-pedicelled, nodding; sepals 4; corolla dark pink, the 4 lobes reflexed in anthesis; stamens 8; fruit a reddish berry.

a. Stems short and slender; leaves 2-4 mm long, elliptic-ovate . . . . *O. microcarpus*
a. Stems thicker, elongate and much branched; leaves 4-9 mm long, narrowly ovate to lanceolate . . . . *O. quadripetalus*

Oxycoccus microcarpus Turcz.
Tiny shrub, stems filiform, usually buried in *Sphagnum*; leaves 2 to 4 mm long, elliptic ovate, the margins strongly revolute; flower pedicels glabrous; fruits 5 to 10 mm in diameter.

Frequent in muskegs, northward slightly beyond the limit of trees.

General distribution: Circumpolar.

Fig. 757 Map 877

Oxycoccus quadripetalus Gil.
Similar to *O. microcarpus*, but the stems coarser, longer, and more branched, the leaves larger and less revolute, the flower pedicels puberulent, and the fruit 8 to 14 mm in diameter.

Rare in our area; thus far known only from muskegs adjacent to the Liard and upper Mackenzie Rivers.

General distribution: Circumpolar; non-arctic.

Map 878

Phyllodoce Salisb. Mountain Heather

Low, heath-like shrubs; leaves linear, evergreen, densely crowded on the branches; corolla united, 5-lobed; stamens 10.

a. Corolla purple; calyx dark
  b. Calyx lobes glandular hairy on the back . . . . *P. coerulea*
  b. Calyx glabrous, the lobes finely ciliolate . . . . *P. empetriformis*
a. Corolla and calyx yellow . . . . *P. glanduliflora*

Phyllodoce coerulea (L.) Bab.
Leaves with serrulate margins, dark green and shiny above, grooved above and below along the mid-vein, the lower groove white with crisped hairs; pedicels thickly glandular-hairy over a fine pubescence; flowers nodding, the corolla urceolate, purple, with scattered white appressed hairs, and the reddish-purple calyx glandular hairy.

In turfy, rocky places where the snow remains late; rare in our area where it is known only from barren ground between Great Bear Lake and Hudson Bay, north to Victoria Island.

General distribution: Circumpolar with large gaps; arctic-alpine.

Fig. 758 Map 879

Phyllodoce empetriformis (Sm.) D. Don
Leaves ciliate on the obscure serrulations, fine pubescence lacking in the lower groove; corolla campanulate, purple, lacking hairs on the surface; calyx reddish-pink, glabrous on the back but densely short-ciliate on the margins.

Alpine, timberline meadows.

General distribution: Cordilleran: In our area

known only from the Mackenzie Mts. adjacent to the Y.T. border.
Fig. 759 Map 880

Phyllodoce glanduliflora (Hook.) Cov.
Leaves similar to those of *P. coerulea* but shorter, yellowish green, the obscure marginal serrulations glandular hairy; pedicels densely glandular-hairy over a fine pubescence, flowers nodding, the corolla urceolate, bright yellow.
Alpine herbmats.
General distribution: Cordilleran—Pacific Coast; the Mackenzie Mts. population is apparently disjunct from that to the west and south. In our area known only from the southwestern Mackenzie Mts.
Fig. 760 Map 881

Rhododendron L. Rhododendron

Rhododendron lapponicum (L.) Wahlenb.
Lapland Rose-bay
Depressed, matted, or erect, much-branched, dwarf shrub with scurfy twigs and elliptic-oblanceolate leaves 1 to 2 cm long, that are scurfy beneath; flowers showy, deep purple and very fragrant, 1 to 2 cm broad, in few-flowered terminal clusters.
In dry, rocky tundra and on stony slopes prostrate or nearly so, but in protected situations occasionally upright and up to 8 dm tall.
General distribution: Circumpolar, in N. America wide ranging, arctic-alpine.
Fig. 761 Map 882

Vaccinium L.

Depressed, creeping or upright dwarf shrubs with deciduous or evergreen leaves; flowers axillary or in terminal racemes; corolla campanulate or globose-urceolate; fruit a many-seeded berry.

a. Leaves deciduous; fruit sweet, blue-black, with a bluish bloom
  b. Leaves and twigs pubescent; flowers in dense clusters . . . . *V. myrtilloides*
  b. Leaves and twigs glabrous or nearly so; flowers solitary or few
    c. Leaves ovate or oblanceolate, rarely more than 1 cm long
      d. Leaves with entire margins . . . . *V. uliginosum*
      d. Leaves with serrate margins . . . . *V. caespitosum*
    c. Leaves elliptic-oblong, 2-5 cm long . . . . *V. membranaceum*
a. Leaves evergreen, shiny; fruits acid, red and shiny . . . . *V. Vitis-idaea* var. *minus*

Vaccinium caespitosum Michx.
Dwarf Bilberry
Low, densely branched shrubs to 2 dm or more tall; twigs glabrous or sparingly puberulent; leaves oblanceolate to obovate, cuneate, serrulate, sessile or nearly so; flowers axillary; berry light blue, sweet.
Alpine herbmats; in our region thus far known from a single collection on the west slope of the Liard Range of the Mackenzie Mts. at 1000 m altitude.
General distribution: Boreal N. America from Nfld. to Alaska and B.C.
Fig. 762 Map 883

Vaccinium membranaceum Dougl.
Tall Blueberry
Shrub usually about 3 dm tall, but much taller southward; twigs 4-angled, glabrous or somewhat puberulent; leaves elliptic-oblong, 2 to 5 cm long, more or less pointed, thin, sharply serrulate, short-pedicelled; flowers axillary; berry purplish to black, sweet.
A woodland species, in the District of Mackenzie thus far known from a single collection on the west slope of the Liard Range of the Mackenzie Mts. at 1300 m altitude.
General distribution: Cordilleran: mountain forests of B.C., western Alta., and southward; reported also from the upper Great Lakes region.
Fig. 763 Map 884

Vaccinium myrtilloides Michx.
Blueberry
Shrub to 4 dm tall, twigs warty and more or less densely hairy; leaves oblong-lanceolate, entire, hairy below and sometimes above; flowers in close terminal racemes; berries blue, with a bloom.
Thickets, mainly in dry or acid soil, in the District of Mackenzie known only from the Pre-

cambrian Shield Area south of Great Slave Lake.

General distribution: Boreal N. America: N.S. to B.C. and southward.

Fig. 764 Map 885

Vaccinium uliginosum L. *s. lat.*
Bilberry
Low much-branched, depressed or erect shrubs; twigs finely puberulent; leaves small, oval, firm, entire, dull green above, glaucous beneath; flowers axillary; berries blue to blackish, with a bloom.

Common in acid soil, in dry as well as in moist places; in the southern part of its range usually producing an abundance of delicious berries.

In alpine and arctic regions, north even beyond lat. 80°, represented by the smaller-leaved and smaller-fruited var. *alpinum* Bigel.

General distribution: *V. uliginosum s. lat.*: Circumpolar.

Fig. 765 (var. *uliginosum*), 766 (var. *alpinum*)
Map 886 (*V. uliginosum s. lat.*)

Vaccinium Vitis-idaea L.
var. **minus** Lodd.
Mountain Cranberry
Low, depressed-ascending evergreen shrub. The berries are smaller but in flavour similar or even superior to those of the commercially harvested *V. Vitis-idaea* of Europe.

Dominant in open, acid, turfy and boggy places throughout the District of Mackenzie and north to or even beyond the Arctic Coast.

General distribution: Circumpolar, subarctic-arctic.

Fig. 767 Map 887

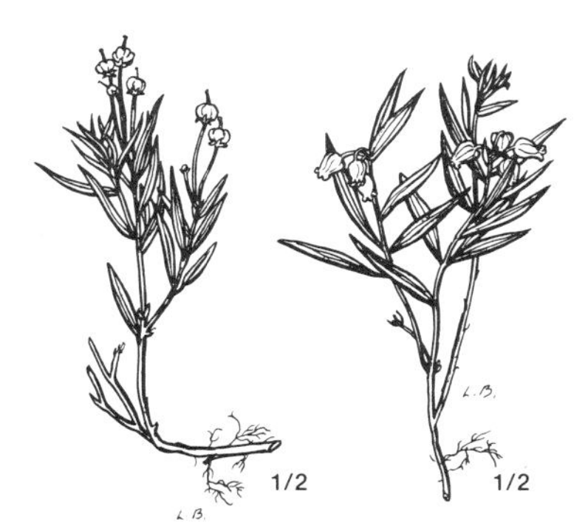

745. *Andromeda Polifolia*

746. *Arctostaphylos alpina*

747. *Arctostaphylos rubra*

748. *Arctostaphylos uva-ursi*

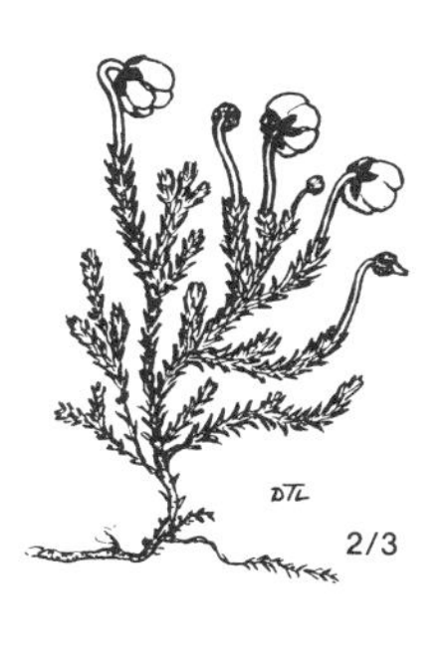

749. *Cassiope hypnoides*

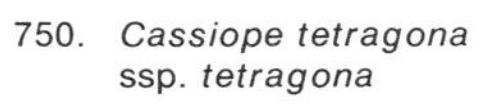

750. *Cassiope tetragona* ssp. *tetragona*

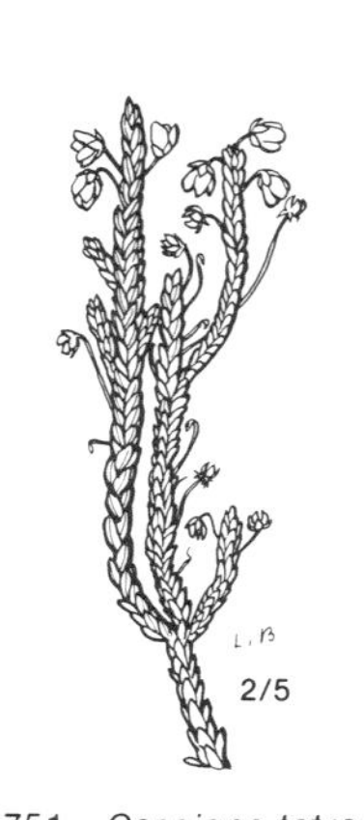

751. *Cassiope tetragona* ssp. *saximontana*

753. *Kalmia polifolia*

752. *Chamaedaphne calyculata*

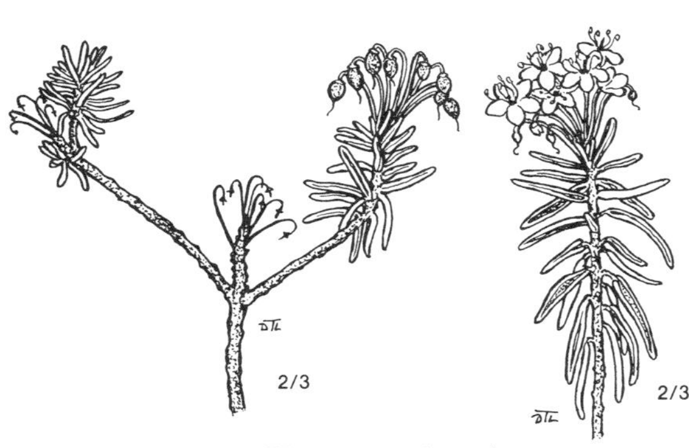

754. *Ledum decumbens*

755. *Ledum groenlandicum*

756. *Loiseleuria procumbens*

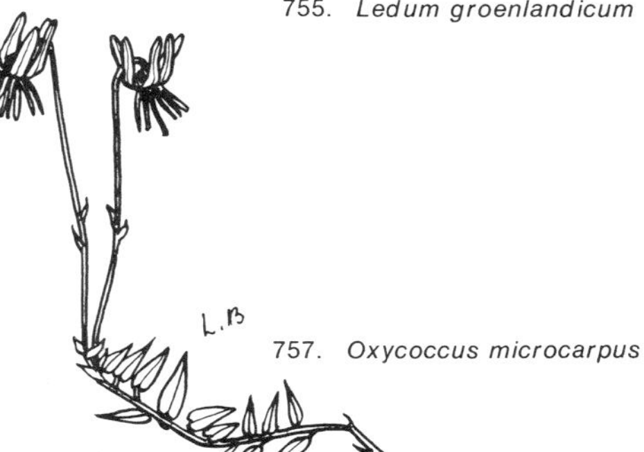

757. *Oxycoccus microcarpus*

758. *Phyllodoce coerulea*

759. *Phyllodoce empetriformis*

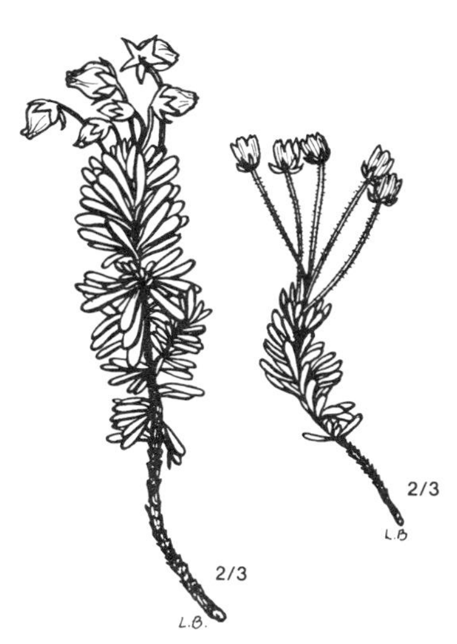

760. *Phyllodoce glanduliflora*

761. *Rhododendron lapponicum*

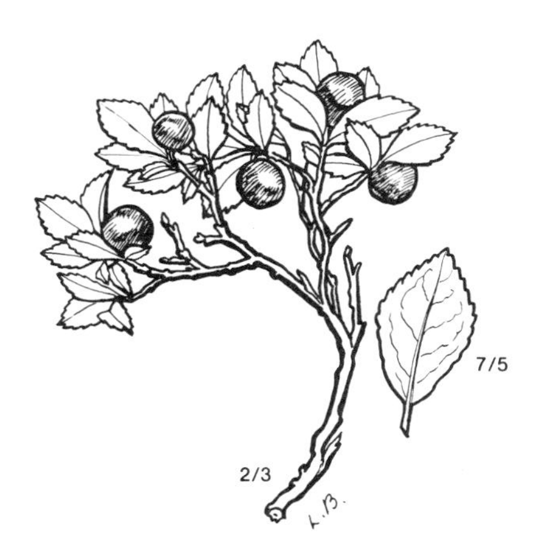

762. *Vaccinium caespitosum*

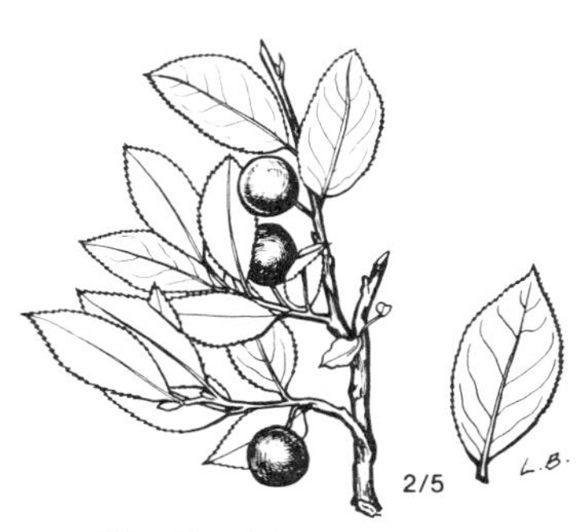

763. *Vaccinium membranaceum*

764. *Vaccinium myrtilloides*

2/5

765. *Vaccinium uliginosum* var. *uliginosum*

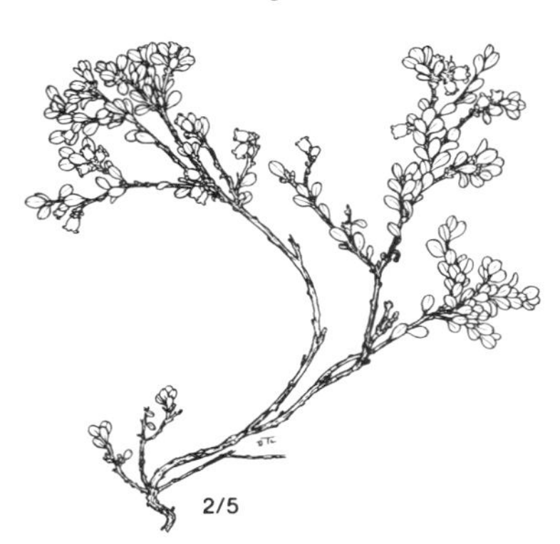

766. *Vaccinium uliginosum* var. *alpinum*

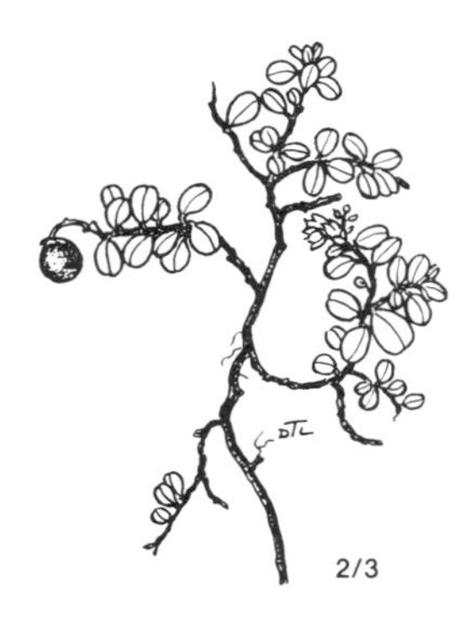

767. *Vaccinium Vitis-idaea* var. *minus*

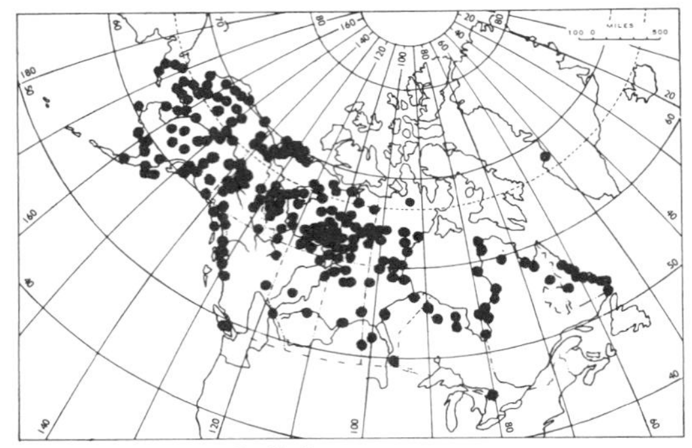

865. *Andromeda Polifolia*

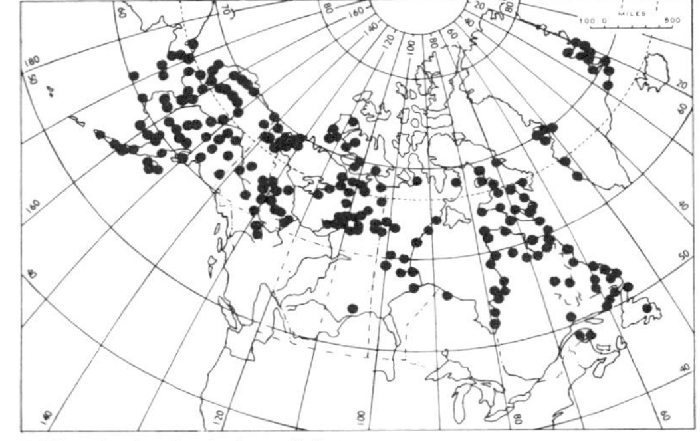

866. *Arctostaphylos alpina*

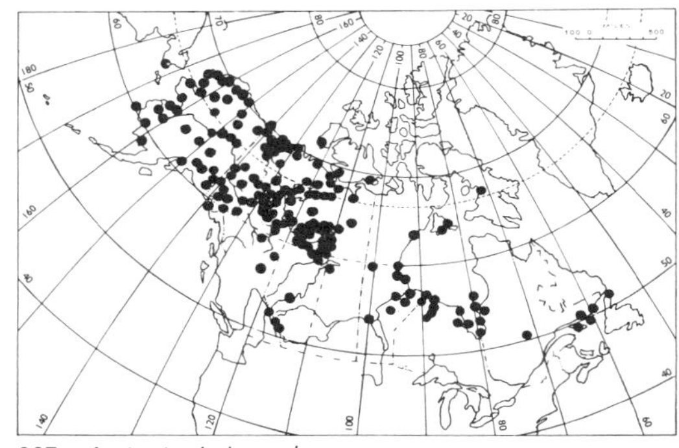

867. *Arctostaphylos rubra*

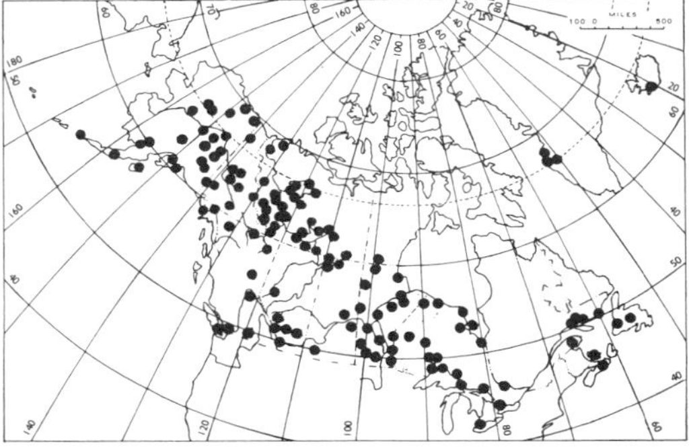

868. *Arctostaphylos uva-ursi*

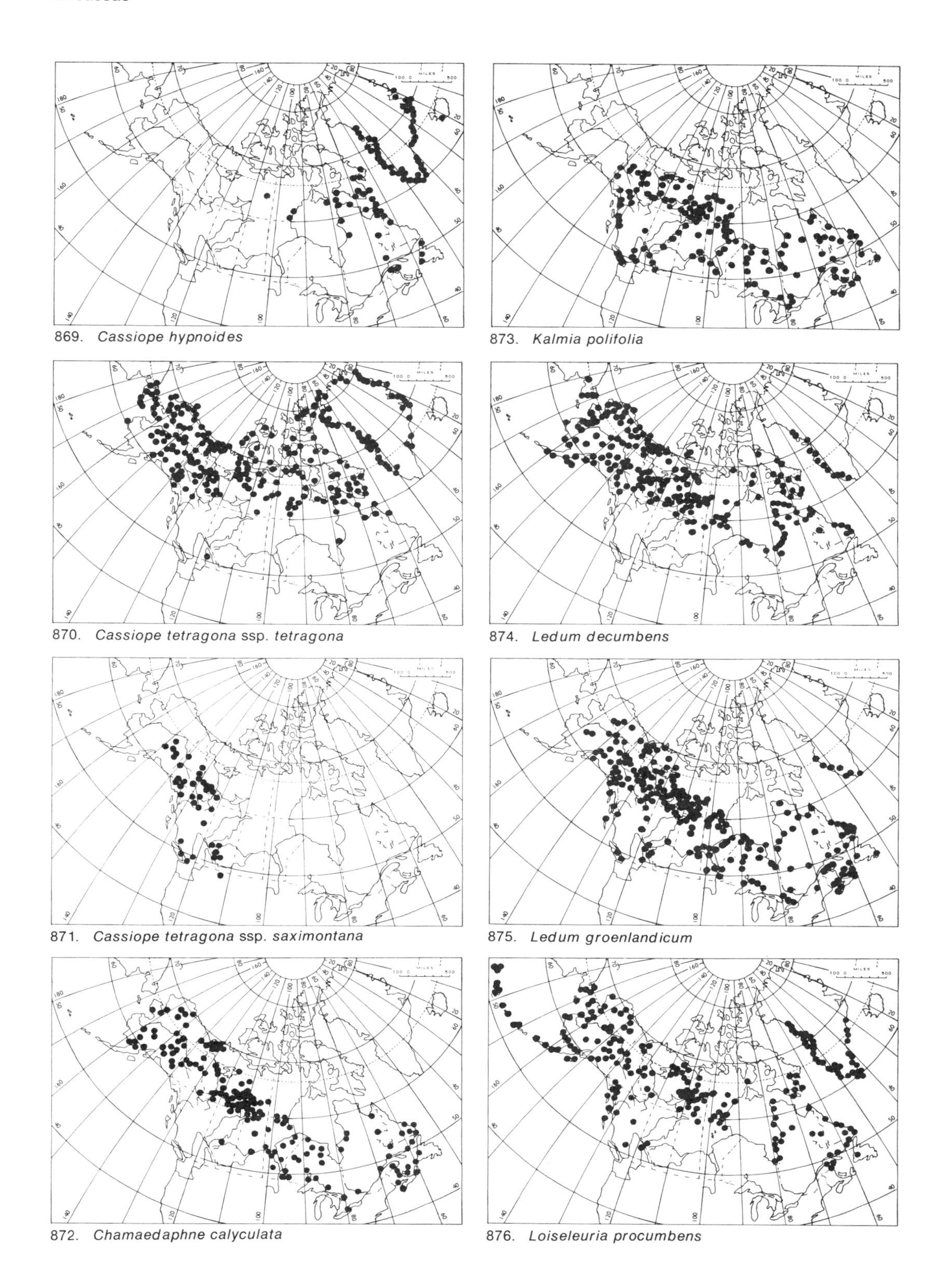

869. *Cassiope hypnoides*

873. *Kalmia polifolia*

870. *Cassiope tetragona* ssp. *tetragona*

874. *Ledum decumbens*

871. *Cassiope tetragona* ssp. *saximontana*

875. *Ledum groenlandicum*

872. *Chamaedaphne calyculata*

876. *Loiseleuria procumbens*

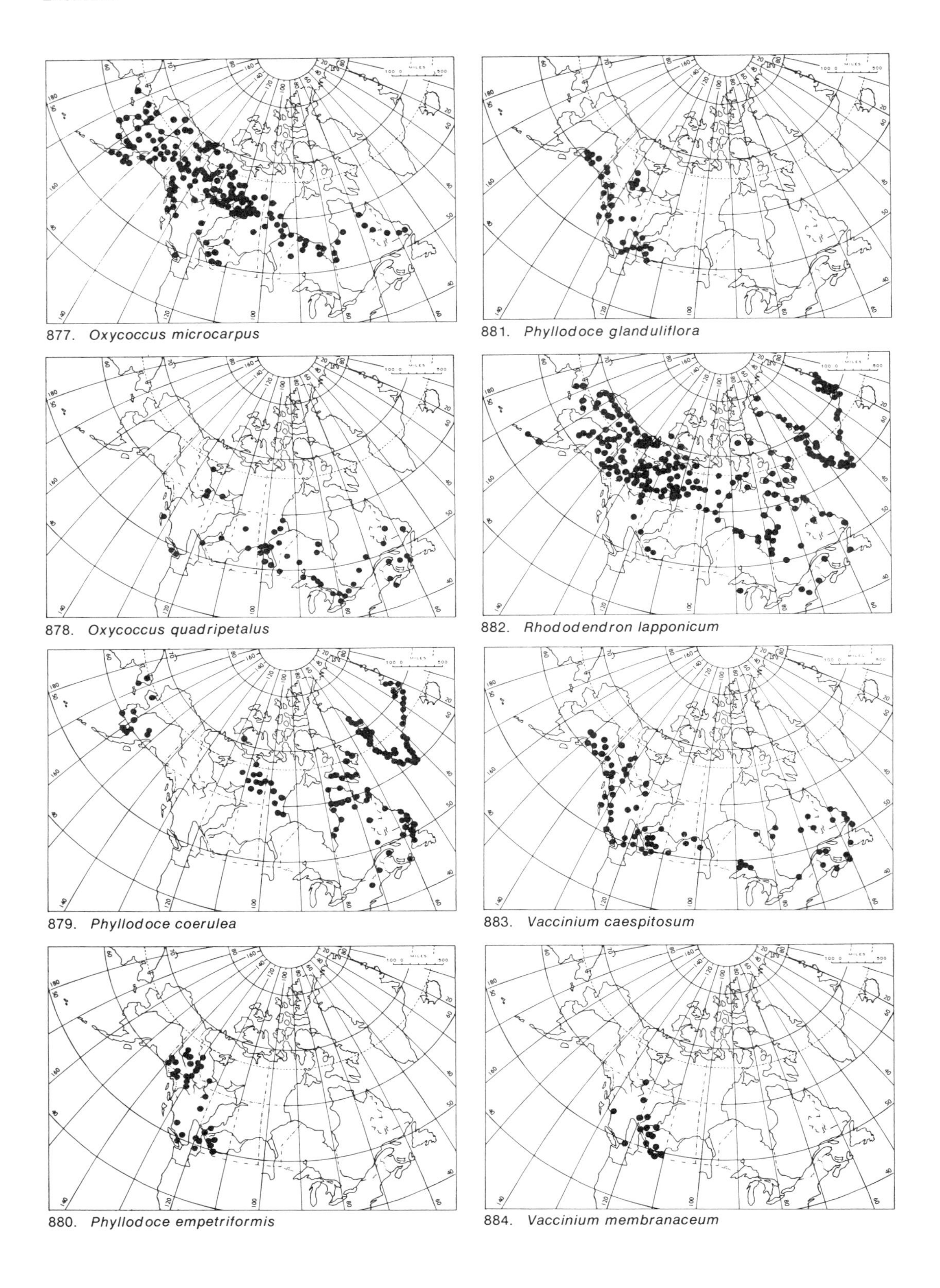

877. *Oxycoccus microcarpus*

878. *Oxycoccus quadripetalus*

879. *Phyllodoce coerulea*

880. *Phyllodoce empetriformis*

881. *Phyllodoce glanduliflora*

882. *Rhododendron lapponicum*

883. *Vaccinium caespitosum*

884. *Vaccinium membranaceum*

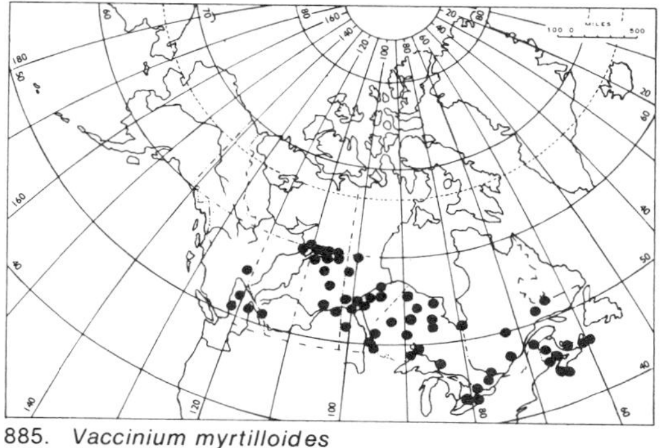

885. *Vaccinium myrtilloides*

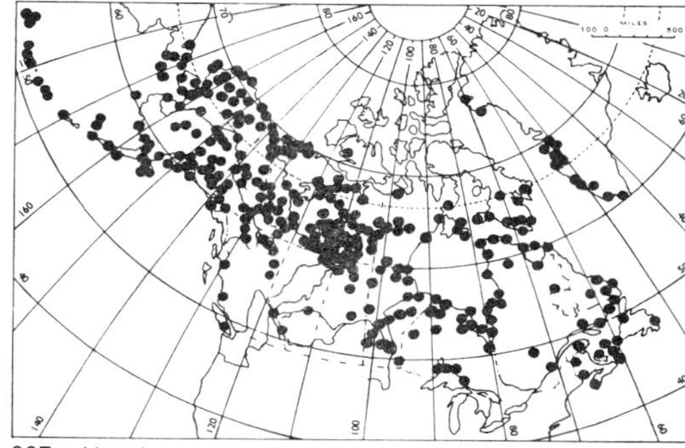

887. *Vaccinium Vitis-idaea* var. *minus*

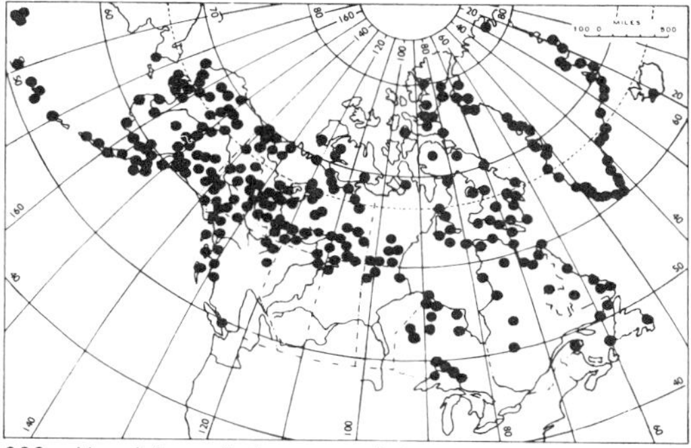

886. *Vaccinium uliginosum s. lat.*

DIAPENSIACEAE Diapensia Family

Diapensia L.

Perennial dwarf shrubs forming hemispherical cushions; stems with coriaceous, densely imbricated, evergreen, and usually curved leaves; flowers regular, 5-merous, solitary, on short, stiff peduncles; corolla white; capsule enclosed in the calyx, containing a few large seeds.

a. Plants densely tufted; leaves narrowly spatulate . . . . *D. lapponica*
a. Plants usually loosely tufted; leaves obovate to oblong-oblanceolate . . . . *D. obovata*

Diapensia lapponica L.
Forming dense hemispherical tussocks; leaves yellowish-green, narrowly spatulate, 1 to 2 cm long.

On Precambrian rocky ledges and in gravelly places of the District of Keewatin and the eastern parts of the District of Mackenzie as far west as Bathurst Inlet.

General distribution: Amphi-Atlantic, arctic-alpine.

Fig. 768 Map 888

Diapensia obovata (Fr.Schm.) Nakai
Similar to *D. lapponica*, but usually more spreading and less dense, the leaves obovate to oblong-oblanceolate and shorter.

Rare on acid rocks and gravels; in our area known only from the Richardson and Mackenzie Mts.

General distribution: Amphi-Beringian; arctic-alpine.

Fig. 769 Map 889

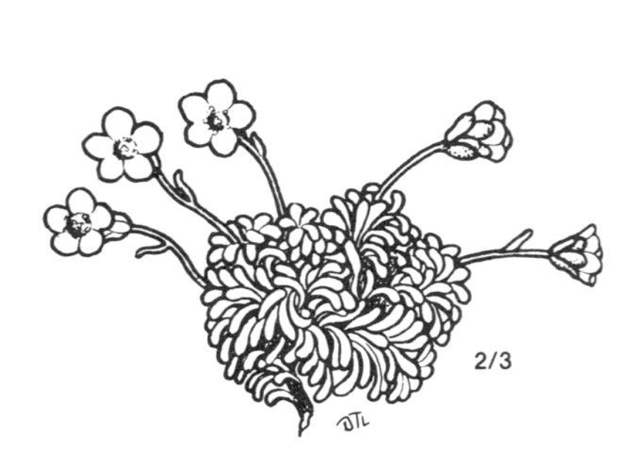

768. *Diapensia lapponica*

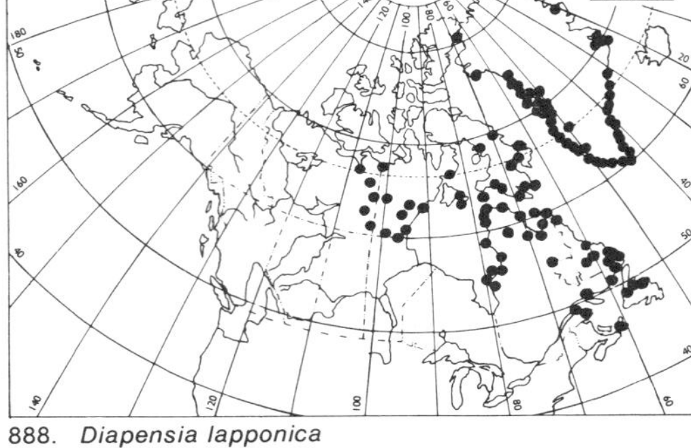

888. *Diapensia lapponica*

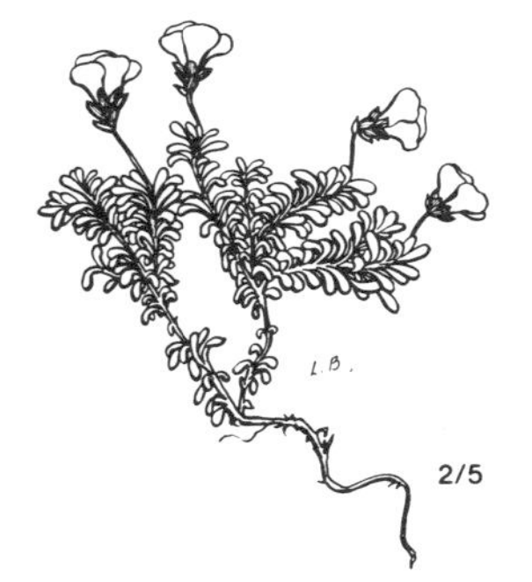

769. *Diapensia obovata*

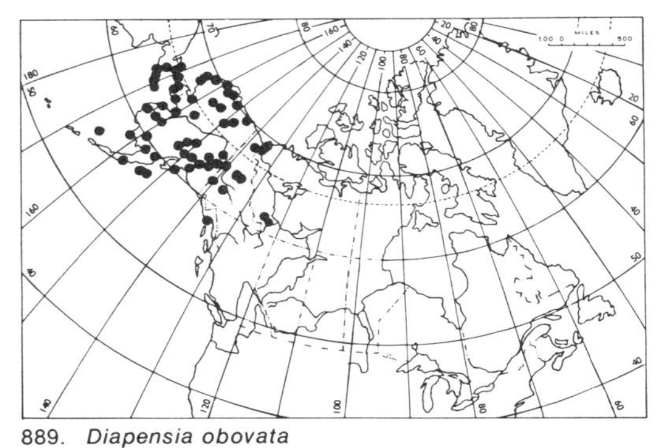

889. *Diapensia obovata*

PRIMULACEAE Primrose Family

Ours chiefly perennial herbs with simple, opposite or whorled leaves, and regular, gamopetalous flowers with the stamens inserted in the corolla tube; the ovary superior, 1-celled; fruit a capsule.

a. Plants scapose
  b. Plants densely caespitose, flowers solitary, short-pedicelled . . . . *Douglasia*
  b. Plants with basal rosettes; flowers in terminal umbels
    c. Calyx deeply cleft; corolla lobes reflexed . . . . *Dodecatheon*
    c. Calyx tubular; corolla lobes spreading
      d. Throat of corolla constricted . . . . *Androsace*
      d. Throat of corolla not constricted . . . . *Primula*
a. Plants with leafy stems
  e. Leaves opposite
    f. Tall plants with flowers in dense axillary racemes . . . . *Lysimachia*
    f. Low, succulent, grey-green plants with solitary and sessile flowers in the axils of the leaves . . . . *Glaux*
  e. Lower leaves small and alternate, the upper larger and whorled . . . . *Trientalis*

Androsace L.

Tufted or stoloniferous, scapose herbs with leaves in a basal rosette or at the tip of the branches; scapes few to several, each terminating in a few-flowered umbel; calyx 5-cleft; corolla 5-lobed, the tube constricted at the throat; capsule 5-valved.

a. Perennial; scapes villous with simple hairs . . . . *A. Chamaejasme*
a. Annual or biennial; scapes glabrous or with minute, branched hairs . . . . . *A. septentrionalis*

Androsace Chamaejasme Host.
var. **arctica** Knuth
Rock-Jasmine
Loosely caespitose with slender stolons terminating in rooting rosettes of small, densely imbricated leaves; scapes 2 to 15 cm high, villous, terminating in a few-flowered umbel of fragrant flowers; the corolla 8 to 10 mm across, creamy white with a yellow eye, turning pink in age.

Calcareous sand and gravel.

General distribution: Amphi-Beringian; extending eastward along the Arctic Coast and western arctic islands as far east as Coronation Gulf, south to Great Bear L., and the Richardson and Mackenzie Mts.

Fig. 770 Map 890

Androsace septentrionalis L.
Small, tufted annual or biennial from a basal rosette of entire or toothed, linear-lanceolate and often reddish-green leaves; scapes erect-ascending, few to several from each rosette, 5 to 30 cm high, and often of unequal length, each terminating in a few- to many-flowered umbel of small, white flowers.

A polymorphic species of not too dry, calcareous, sandy or gravelly places wide ranging northward across the Arctic Archipelago beyond lat. 80° N.

General distribution: Circumpolar.

Fig. 771 Map 891

Dodecatheon L. Shooting-star

Scapose, glabrous perennial herbs with a basal rosette of oblanceolate leaves from a short erect-ascending caudex; flowers umbellate, showy; corolla 5-parted, deeply cleft and the lobes strongly reflexed; the anthers unit to form a tube; fruit a capsule.

a. Filament tube not exposed; leaves oval to ovate, abruptly tapering into the petiole . . . . *D. frigidum*
a. Filament tube exposed; leaves elliptic, gradually tapering into a narrow petiole . . . . *D. pulchellum* ssp. *pauciflorum*

Dodecatheon frigidum C. & S.
Scapes 7 to 30 cm high; leaves oval to ovate, obtuse, with wavy margins; inflorescence and upper part of scape glandular puberulent; lobes of corolla magenta to lavender; anthers purplish-black on backs; capsule about 0.8 cm long.

Wet alpine meadows and creek banks; rare in the Mackenzie and Richardson Mts.

General distribution: Amphi-Beringian: Extreme eastern Siberia, and through Alaska and Y.T. to northern B.C. and the mountains of the District of Mackenzie.

Fig. 772 Map 892

Dodecatheon pulchellum (Raf.) Merr.
ssp. **pauciflorum** (Durand) Hult.
D. pauciflorum (Durand) Greene
Scapes 1.5 to 4.0 dm or more high; leaves elliptic, sometimes wavy margined; scape and inflorescence glabrous; lobes of corolla magenta to lavender; anthers usually dark maroon to black; capsule about 1.2 cm long.

Wet meadows and saline flats in the southwestern part of our area.

General distribution: Western U.S.A. and Canada extending northward into the District of Mackenzie, and central Y.T.—Alaska.

Fig. 773 Map 893

Douglasia Lindl.

Low tufted or matted perennials with short, linear leaves, imbricated and marcescent on the short and crowded branches; flowers singly, on a short pedicel; corollas pink or purple with spreading, obovate lobes; capsule turbinate, containing 2 or 3 brown and pitted seeds.

a. Leaves glabrous above . . . . *D. arctica*
a. Leaves short-pubescent above . . . . *D. ochotensis*

Douglasia arctica Hook.
Loosely caespitose; leaves linear, 4 to 10 mm long, flat and ciliate, glabrous above; corolla pink to purple; fruiting peduncle elongating in fruit; calyx about 6 mm long, the lobes long-triangular.

Rocky, mossy slopes; in our area known only from Richardson Mountains.

General distribution: Endemic of Alaska, Y.T. and the Richardson Mts. of N.W. District of Mackenzie.

Fig. 774 Map 894

***Douglasia ochotensis** (Willd.) Hultén
Androsace ochotensis Willd.
Densely caespitose; leaves shorter than in *D. arctica*, somewhat fleshy, recurved, ciliate and more or less pubescent on the upper surface; pedicels to 1.5 cm long, not elongating in fruit; corolla purple; calyx about 3 mm long, the lobes short triangular.

General distribution: Amphi-Beringian; in N. America known from rocky slopes of the Brooks Range in northern Alaska and the British Mts. in northern Y.T.; to be looked for in the Richardson Mts. in N.W. District of Mackenzie.

Map 895

Glaux L. Sea-Milkwort

Glaux maritima L.
var. **obtusifolia** Fern.
Low glaucous, glabrous, simple or diffusely branched and spreading perennial from a creeping rhizome; leaves opposite, sessile, oval to broadly oblong, 0.3 to 0.8 mm long; flowers about 3 mm long, solitary and sessile or nearly so, in the axils of the leaves; corolla lacking; calyx 5-lobed, pink; capsule orbicular, 5-valved.

In the District of Mackenzie known only from the Salt Plain west of Fort Smith where it is common about saline sloughs, along the south shore of Great Slave Lake, and the upper Mackenzie River.

General distribution: Circumpolar, with large gaps, and restricted to seashores and saline inland lake shores.

Fig. 775 Map 896

Lysimachia L. Loosestrife

Lysimachia thyrsiflora L.
Stems leafy, 2 to 8 dm tall, from creeping rhizomes; leaves opposite, sessile, black-glandular, the upper linear-lanceolate up to 10 cm long, the lower almost scale-like; flowers in short and dense spike-like racemes from the axils of the middle leaves; corolla light yellow, the 5 linear divisions as well as the lobes of the calyx purple-dotted; anthers 5, bright yellow, the filaments much longer than the corolla lobes.

Marshy lake shores, in mud and in sedge meadows, in our area thus far know from the upper Mackenzie River drainage and from a single station in the Mackenzie River Delta.

General distribution: Circumpolar; wide ranging.

Fig. 776 Map 897

Primula L. Primrose

Perennial, scapose herbs from a basal rosette of simple, green or somewhat farinose leaves; flowers in a terminal umbel, gamopetalous and regular, with the stamens inserted in the corolla tube; petals spreading, deeply notched or emarginate, mauve or purple (rarely white); capsule many-seeded, opening at the top.

a. Flowers large (about 2.0 cm long); lobes of corolla only slightly emarginate . . . . *P. tschuktschorum* ssp. *Cairnesiana*
a. Flowers much smaller; lobes of corolla deeply notched
  b. Robust plants, 15-45 cm tall; leaves strongly farinose beneath . . . . *P. incana*

b. Plants smaller, about 10 (-20) cm tall; leaves green or only slightly farinose
  c. Lobes of calyx glandular-ciliate . . . . *P. egaliksensis*
  c. Lobes of calyx without cilia
    d. Leaves usually long-petioled . . . . *P. borealis*
    d. Leaves with petioles about the length of the blade, or none
      e. Involucral bracts saccate . . . . *P. stricta*
      e. Involucral bracts not saccate . . . . *P. mistassinica*

Primula borealis Duby
Scapes one to several, 5 to 15 cm tall; leaves to 6 cm long, long-petioled, the blade broadly dilated, 0.5 to 1.5 cm long, entire or shallowly toothed. Involucral bracts more or less saccate at the base; flowers 2 to 7, the corolla about 1.5 cm across, lilac (or rarely white) with a yellow centre and tube.

In moist, usually saline meadows.

General distribution: Amphi-Beringian, eastward along the Arctic Coast to long. 130° W. Widely disjunct in southwestern Y.T. at 750 m above sea level.

Fig. 777 Map 898

Primula egaliksensis Wormsk.
Scape 8 to 15 cm tall, usually solitary; leaves to 6 cm long, the blade broadly dilated, similar to that of *P. borealis*; bracts decidedly saccate at the base; margins of calyx lobes and bracts glandular-ciliate; corolla about 0.6 cm across, lilac with a yellow centre and tube.

Meadows, and wet calcareous lake shores and river banks northward beyond the limit of trees.

General distribution: N. America: wide ranging from S.W. Greenland and northern Nfld. and Lab. to Alaska and south in the Rocky Mts.

Fig. 778 Map 899

Primula incana M. E. Jones
With us the tallest and most robust member of the genus. The underside of leaves, upper part of the scape and the inflorescence usually farinose; leaves 1.5 to 6 cm long, elliptic to oblong-obovate, shallowly denticulate, sessile or with winged petioles; corolla about 1.0 cm long and 0.6 cm across, lobes mauve, deeply notched, tube yellow.

Meadows, wet clearings and lake shores.

General distribution: N. America, from Hudson and James Bay to central Alaska, north along the Mackenzie R. to lat. 66°.

Fig. 779 Map 900

Primula mistassinica Michx.
Scapes 6 to 15 cm tall, often solitary; leaves oblanceolate to cuneate-obovate, sessile or tapering to the base, dentate; flowers few; bracts not saccate at the base; corolla about 1.5 cm across, lilac or bluish-purple with a yellow centre and tube.

Moist calcareous meadows and beaches north to Great Bear Lake.

General distribution: Boreal N. America, Nfld. to the Alaska—Y.T. border.

Fig. 780 Map 901

Primula stricta Hornem.
Scapes 5 to 30 cm tall, usually solitary; leaves oblanceolate to narrowly obovate, entire or obscurely dentate, sessile or nearly so, occasionally farinose below; inflorescence frequently somewhat farinose; bracts saccate at the base; corolla 0.8 cm or less across, lilac. Similar to *P. incana* but rarely as tall and usually less farinose and with somewhat smaller flowers.

Wet saline seepages on moist banks and along seashores.

General distribution: Amphi-Atlantic: In N. America from central E. and W. Greenland, and from Lab. to S.E. Alaska, north to the Arctic Coast of the District of Mackenzie, Banks and Victoria Islands.

Fig. 781 Map 902

Primula tschuktschorum Kjellm.
ssp. **Cairnesiana** Porsild
Scapes stout, 13 to 20 cm tall, not much elongated in fruit; leaves thin, narrowly lanceolate, 5 to 7 cm long and 0.6 to 1.0 cm wide; underside of leaves, the bracts and peduncles farinose in youth, glabrate in age. Umbels 3- to 5-flowered; corolla about 2.0 cm long, purple, the base of the tube white; calyx blackish-purple, cleft to the base. Seeds angular, 1.8 $\times$0.7 mm.

Rare and local in wet meadows and on stream banks.

General distribution: Amphi-Beringian east to the Richardson Mts. adjacent to the Yukon—District of Mackenzie border.

Fig. 782 Map 903

Trientalis L. Star-flower

Slender perennial herbs from a thin rhizome; stems simple, 1 to 2 dm tall; leaves broadly elliptic-lanceolate, whorled at the top of the stem, usually with a few and scale-like leaves below. Flowers 1 to a few, on slender pedicels from the leaf-axils; corolla white, rotate; capsule globular, few-seeded, the seeds white-reticulate.

a. Stem leaves scale-like . . . . *T. borealis*
a. Stem leaves larger and obovate . . . . *T. europaea* ssp. *arctica*

***Trientalis borealis** Raf.
Stems 0.5 to 2.0 dm tall; leaves broadly lanceolate, acuminate, in a single whorl; rhizome thickened and tuber-like at the apex; corolla pinkish.

General distribution: Boreal N. America, from Nfld. to northern Sask. and Alta.; to be looked for in the southern parts of our area.

Map 904

Trientalis europaea L.
ssp. **arctica** (Fisch.) Hult.
Similar to *T. borealis*, but leaves obovate; rhizome slender, somewhat thickened at the apex, but not tuber-like; corolla white.

Rare in spruce-poplar woods; in our area thus far known only from near the west end of Great Slave Lake, and from northern Wood Buffalo Park.

General distribution: Amphi-Beringian.

Fig. 783 Map 905

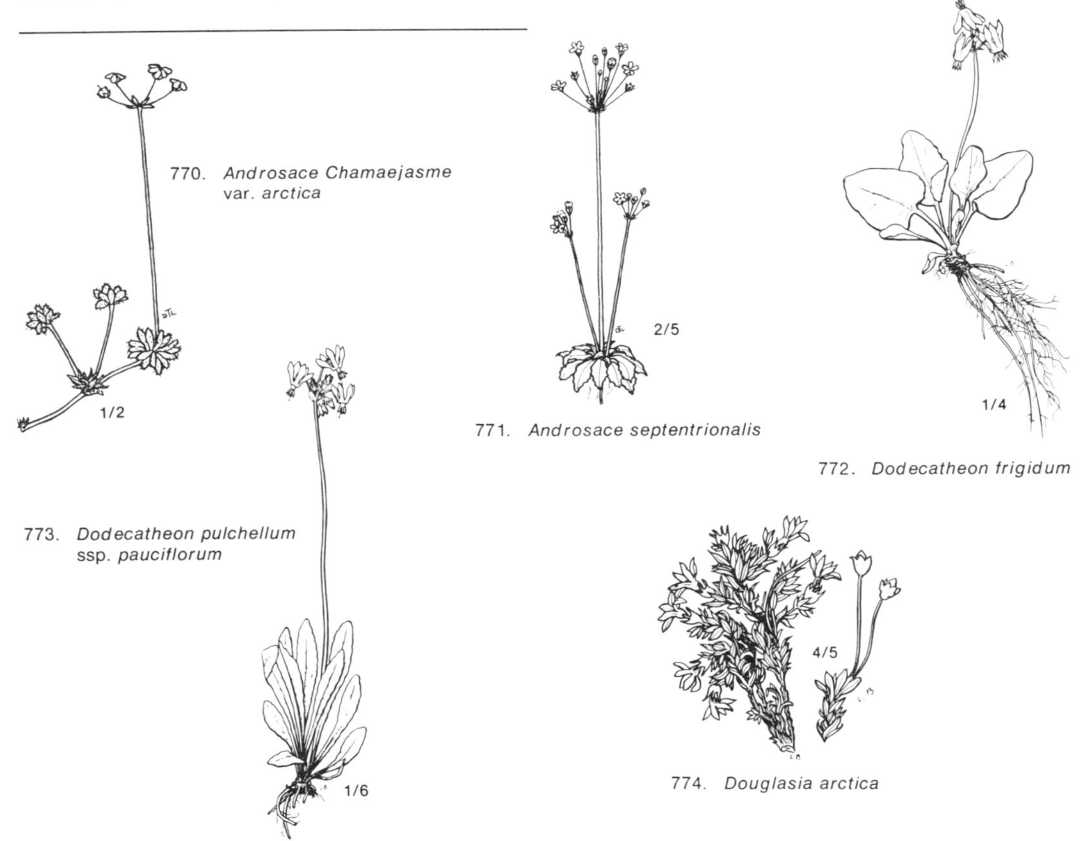

770. *Androsace Chamaejasme* var. *arctica*

771. *Androsace septentrionalis*

772. *Dodecatheon frigidum*

773. *Dodecatheon pulchellum* ssp. *pauciflorum*

774. *Douglasia arctica*

775. *Glaux maritima* var. *obtusifolia*

776. *Lysimachia thyrsiflora*

777. *Primula borealis*

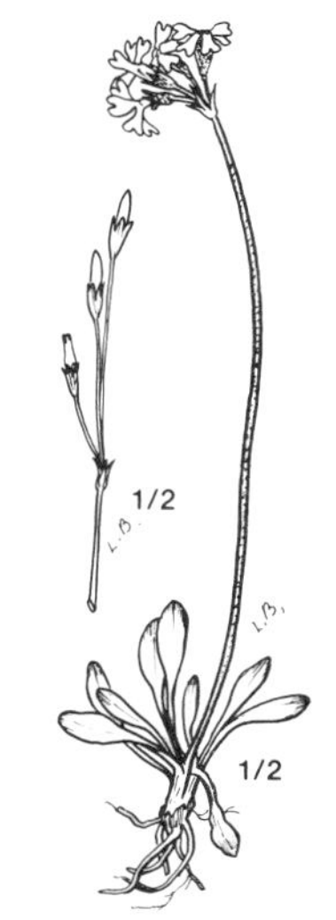

778. *Primula egaliksensis*

779. *Primula incana*

780. *Primula mistassinca*

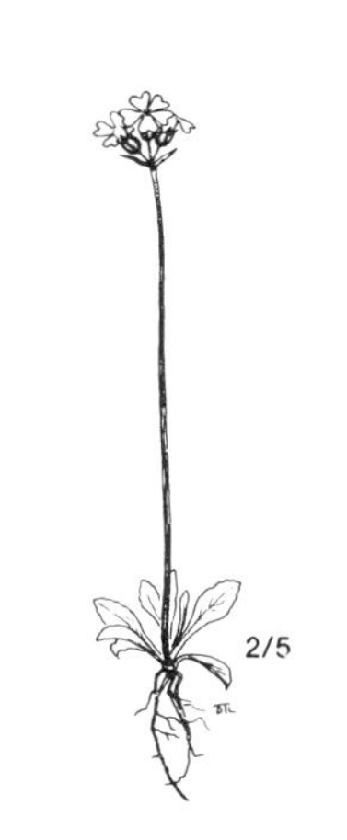

781. *Primula stricta*

782. *Primula tschuktschorum* ssp. *Cairnesiana*

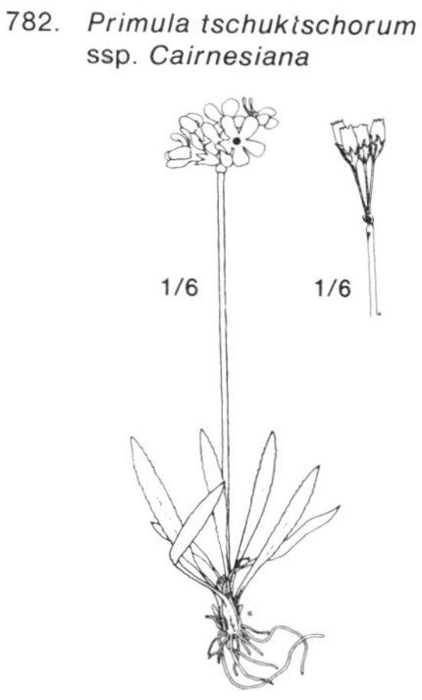

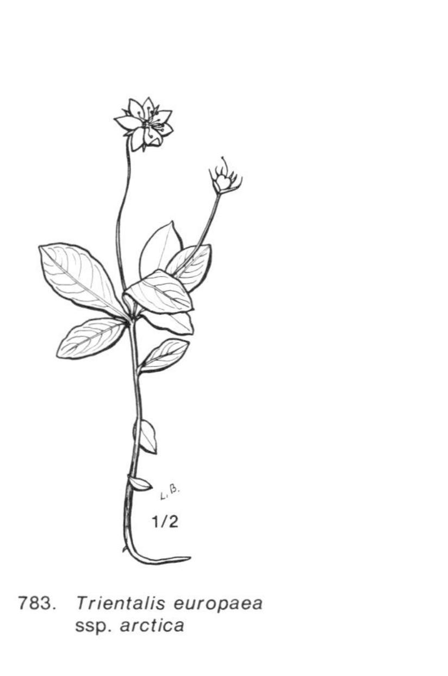

783. *Trientalis europaea* ssp. *arctica*

892. *Dodecatheon frigidum*

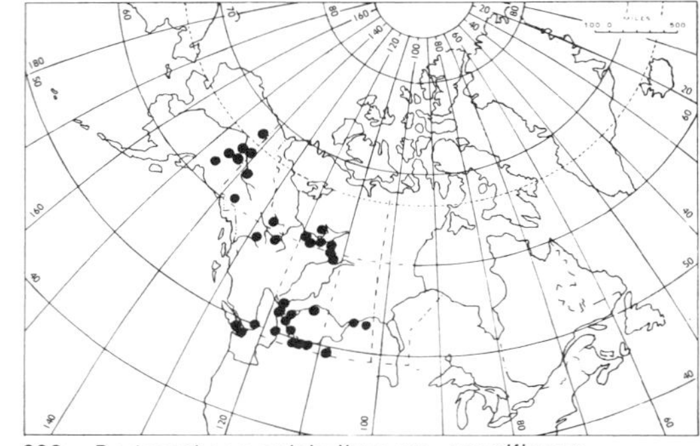

893. *Dodecatheon pulchellum* ssp. *pauciflorum*

890. *Androsace Chamaejasme* var. *arctica*

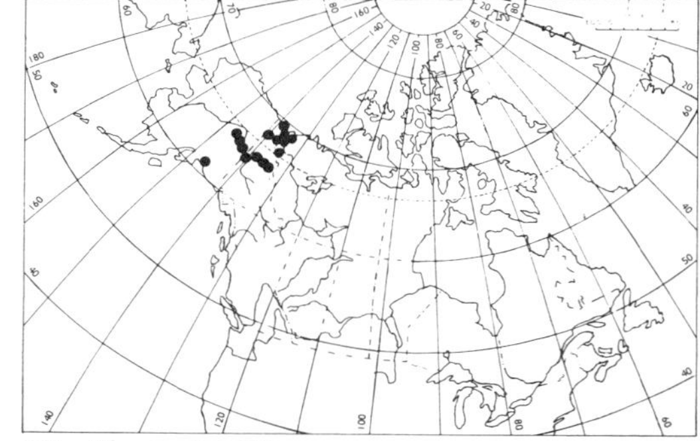

894. *Douglasia arctica*

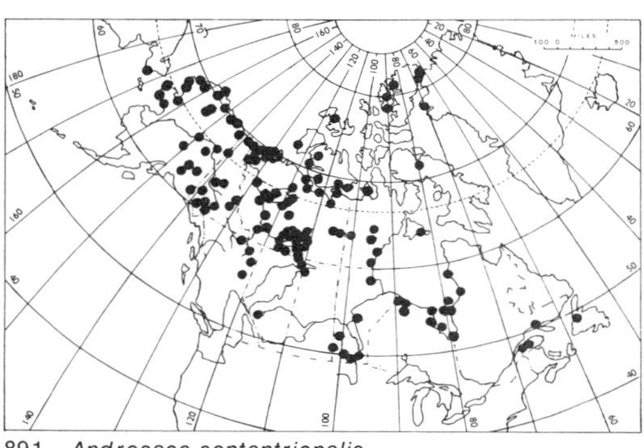

891. *Androsace septentrionalis*

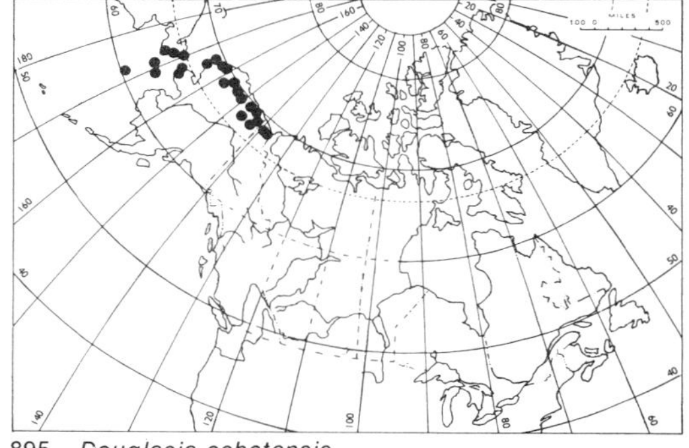

895. *Douglasia ochotensis*

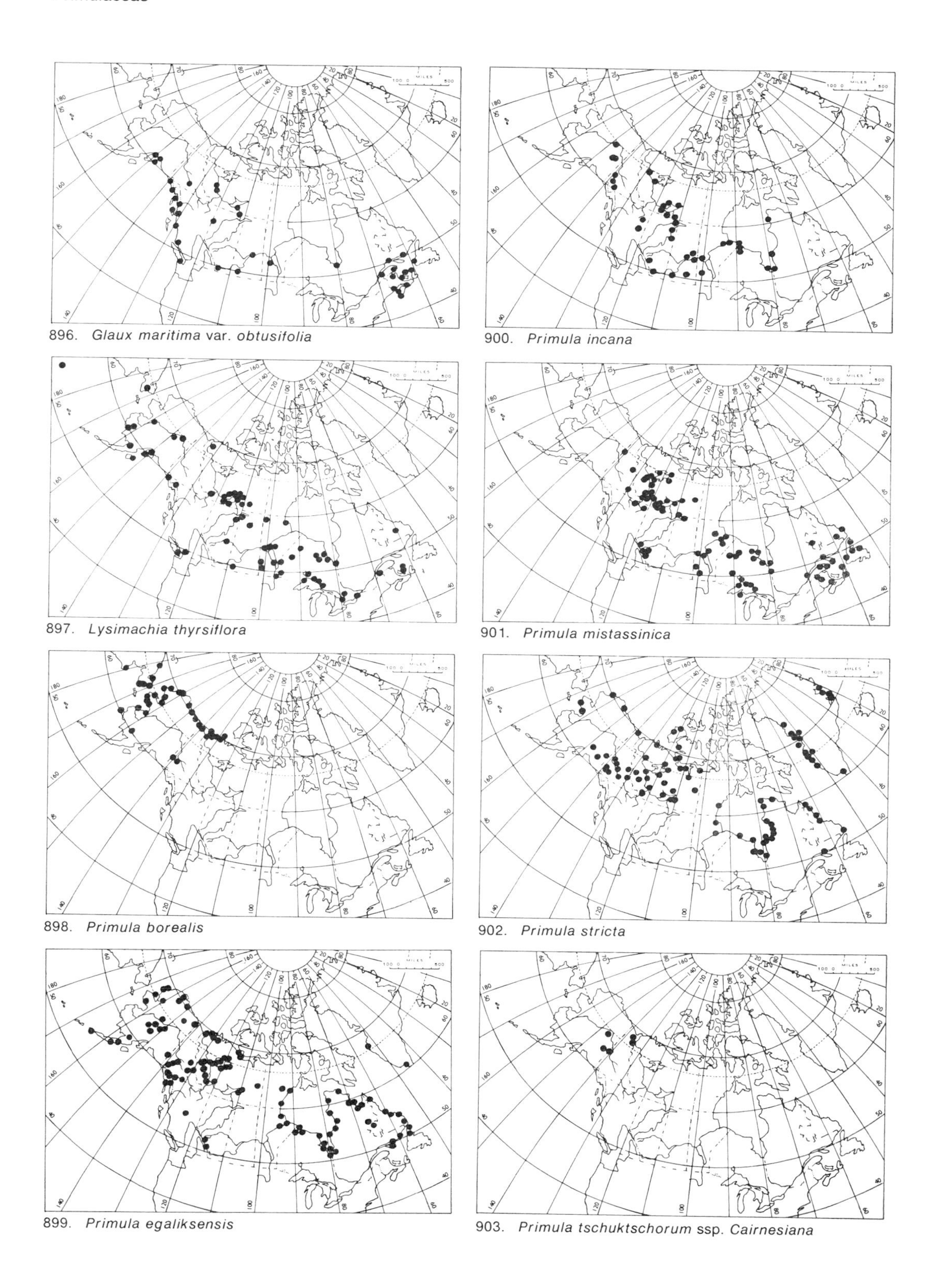

896. *Glaux maritima* var. *obtusifolia*

900. *Primula incana*

897. *Lysimachia thyrsiflora*

901. *Primula mistassinica*

898. *Primula borealis*

902. *Primula stricta*

899. *Primula egaliksensis*

903. *Primula tschuktschorum* ssp. *Cairnesiana*

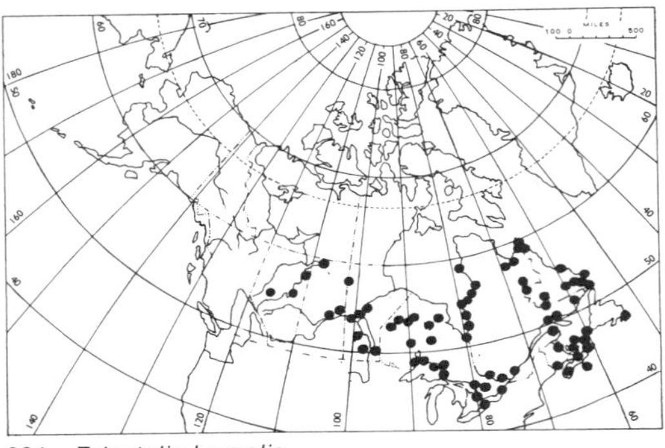

904. *Trientalis borealis*

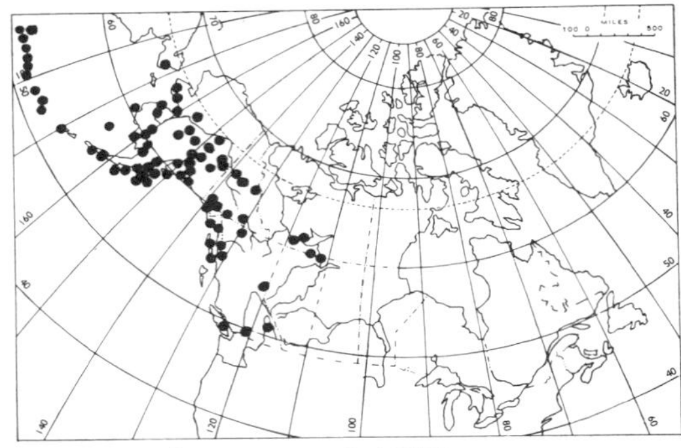

905. *Trientalis europaea* ssp. *arctica*

PLUMBAGINACEAE Leadwort Family

Armeria Willd. Thrift

Armeria maritima (Mill.) Willd.
Densely tufted herbs with linear, somewhat fleshy, greyish-green, marcescent leaves, and one to several, 5 to 30 cm tall scapes terminating in a hemispherical head of small pink flowers; calyx 6 to 7 mm long, dry and scarious with funnel-formed limb and pointed lobes, the tube densely pilose on the 10 ribs; petals rose pink; fruit indehiscent within the persistent calyx. Two subspecies occur in our area: ssp. *arctica* (Cham.) Hult., in which the grooves between the ribs of the calyx are glabrous, and ssp. *labradorica* (Wallr.) Hult., with pilose grooves between the ribs.

Gravelly tundra, flood plains and lake shores.

General distribution: The species circumpolar, arctic: The essentially littoral ssp. *arctica* is Amphi-Beringian, entering N. W. District of Mackenzie along the arctic coast east to Bernard Harbour, whereas ssp. *labradorica* is Amphi-Atlantic, wide ranging north of the treeline west to the Mackenzie Delta and Mackenzie Mts.

Fig. 784 (ssp. *arctica*), 785 (ssp. *labradorica*)
Map 906 (ssp. *arctica*), 907 (ssp. *labradorica*)

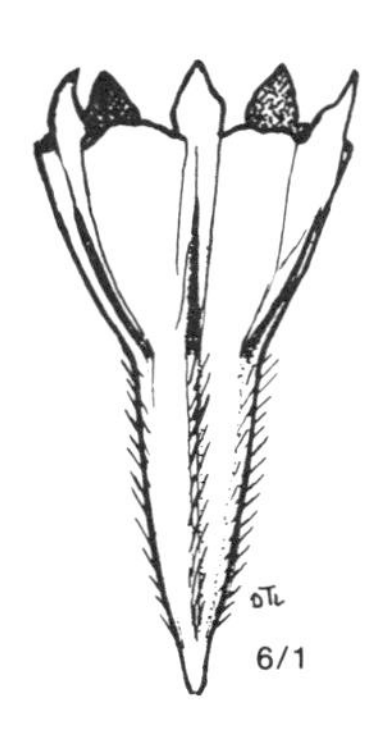

784. *Armeria maritima* ssp. *arctica*

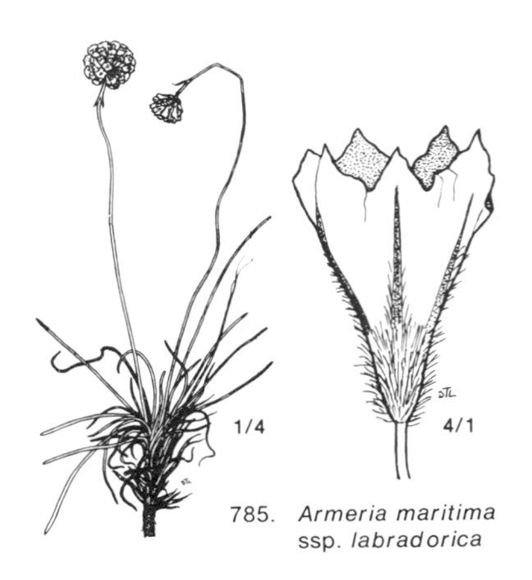

785. *Armeria maritima* ssp. *labradorica*

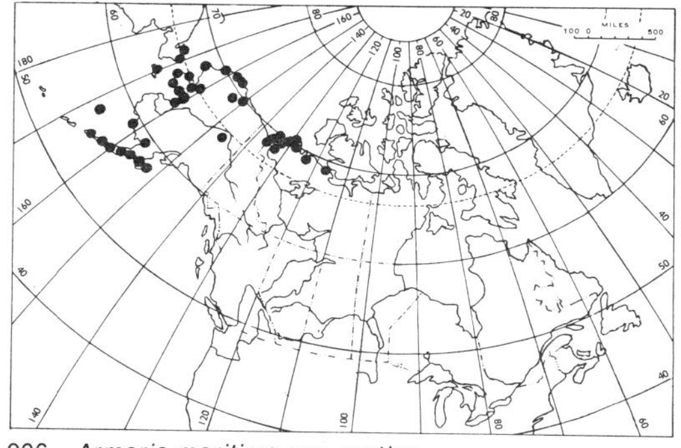

906. *Armeria maritima* ssp. *arctica*

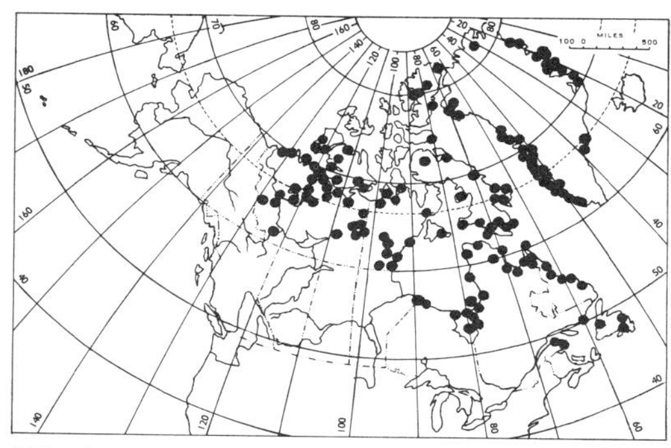

907. *Armeria maritima* ssp. *labradorica*

GENTIANACEAE Gentian Family

Small to medium-sized, terrestrial, erect and glabrous annual, biennial or perennial herbs, mostly with angular stems, opposite and entire leaves; flowers scentless, blue, purplish or white, solitary or in terminal or axillary clusters; corolla 4- or 5-lobed, the lobes entire or fimbriate; stamens as many as the corolla lobes, inserted on the corolla tube; fruit a capsule.

a. Corolla tubular or funnelform . . . . *Gentiana*
a. Corolla rotate . . . . *Lomatogonium*

Gentiana L. Gentian

Annual biennial or perennial glabrous herbs with opposite leaves; flowers blue or purple (or occasionally white), perfect, terminal or axillary; corolla 4-5 lobed, the stamens inserted on the tube; capsule ellipsoid, 2-valved, many-seeded.

a. Corolla tube with plaits or folds between the lobes
  b. Tiny annual or biennial; stems leafy, prostrate to ascending . . . . *G. prostrata*
  b. Perennial, larger, upright or decumbent
    c. Stems solitary, upright, with over-wintering rosettes from its base . . . . *G. glauca*
    c. Stems several, tufted, decumbent at the base . . . . *G. affinis*
a. Corolla tube without plaits or folds between the lobes
  d. Flowers large, 2.5-5.0 cm long, the lobes toothed, erose or somewhat lacerate
    e. Calyx keels papillose . . . . *G. Macounii*
    e. Calyx keels not papillose
      f. Stems 0.5-2.0 dm tall, simple or with weak branches from the base . . . . *G. Richardsonii*
      f. Stems usually taller and freely branched above . . . . *G. Raupii*
  d. Flowers smaller, less than 2.5 cm long, the lobes entire
    g. Corolla fimbriate in the throat
      h. Plants usually branched from the base; flowers long pedunculate . . . . . . *G. tenella*
      h. Plants usually branched above; flowers sessile or nearly so . . . . *G. acuta*
    g. Corolla not fimbriate in the throat
      i. Corolla lobes bristle-tipped . . . . *G. propinqua*
      i. Corolla lobes acute or acuminate, not bristle-tipped . . . . *G. arctophila*

Gentiana acuta Michx.
Gentianella Amarella (L.) Börner
ssp. *acuta* (Michx.) Gillett
Felwort
Annual or biennal, the stems 0.5 to 6.0 dm tall, simple or with strongly ascending branches; leaves sessile, the basal oblanceolate and obtuse, the upper oblong-lanceolate to linear; flowers often numerous, in small, stalked clusters in the axils of the upper leaves; calyx lobes narrowly lanceolate; corolla lobes mauve to purple (often turning blue in drying) or occasionally white; seeds ovoid.
Moist streambanks, meadows and clearings in the Mackenzie R. drainage north to the Arctic Circle.
General distribution: Boreal America from Nfld. to Alaska and southward in the Rocky Mts.
Fig. 786 Map 908

Gentiana affinis Griseb.
Prairie Gentian
Perennial; stems 2 dm or more tall, puberulent, tufted, decumbent at the base, bearing 6 to 12 pairs of leaves; the leaves ovate to oblong-lanceolate 1.5 to 3.0 cm long, minutely scabrous on the margin; flowers 5-parted, 1 to 3 in the axils of the upper leaves; calyx lobes oblong-lanceolate, equal or irregular; corolla 2.0 to 3.0 cm long, funnelform to tubular, bluish-green; seeds flattened, broadly winged.
Rare; in the District of Mackenzie long known only from a collection by McTavish in 1856 labelled "immediate vicinity of Fort Good Hope" but has recently been found on gravel bars of the Keele River in the Mackenzie Mts. and near Heart L. at Mile 180 Mackenzie Highway.
General distribution: Widely disjunct from its main range in the foothills of southern Alta., Sask. and B.C.
Fig. 787 Map 909

Gentiana arctophila Griseb.
G. propinqua Richards.
ssp. *arctophila* (Griseb.) Hultén
Similar to *G. propinqua* ssp. *propinqua*, but smaller; basal branches short or lacking and the corolla lobes acute or acuminate, but not bristle-tipped.
Arctic and alpine tundra herbmats and dry and sunny slopes.
General distribution: N. W. N. America.
Fig. 788 Map 910

Gentiana glauca Pall.
Perennial with a simple and solitary stem 5 to 12 cm high from a basal rosette of bluish-green, shiny and somewhat fleshy leaves. Flowers few, in a terminal cyme; corolla 1.2 to 1.8 cm long, bluish-green, with oblong and rounded lobes.
Moist alpine meadows and slopes in the Mackenzie and Richardson Mts.
General distribution: Amphi-Beringian, south to Alta. and B.C.
Fig. 789 Map 911

Gentiana Macounii Holm
Gentianella crinita (Froel.) G. Don
ssp. *Macounii* (Th. Holm) Gillett
Fringed Gentian
Erect annual or biennial with leafy, 4-angled stems up to 3 dm tall, simple, or often branched above; flowers 4 to 5 cm long, 1 or 2 at the end of each branch; corolla deep blue, the spreading lobes conspicuously fringed; the capsule spindle-shaped.
Rare and local on gravelly beaches of the upper Mackenzie Valley.
General distribution: N. America; James Bay to the foothills of the Rocky Mts. and southward. Widely disjunct in S. W. District of Mackenzie.
Fig. 790 Map 912

Gentiana propinqua Richards.
Gentianella propinqua (Richards.) Gillett
Glabrous, erect, dwarf annual or biennial from a rosette of obovate, sessile leaves; stems rarely over 20 cm tall, usually with lateral and weaker branches from the base; flowers mostly 4-parted, solitary at the end of the branches; corolla deep blue, that of the terminal and largest flowers 1.5 to 2.0 cm long, the lobes bristle-tipped; capsules slender, as long as the corolla.
Sunny slopes and open herbmats.
General distribution: Arctic-alpine, wide ranging from Nfld., Hudson Bay, District of Mackenzie to Northeast Asia and south in the Rocky Mts. through Alta. and B.C.
Fig. 791 Map 913

Gentiana prostrata Haenke
Dwarf biennial usually with one or several leafy, prostrate to erect-ascending stems, from a spindly root; the flower terminal and solitary; corolla 10 to 20 mm long, sky-blue, opening only during sunny periods; capsule long-stipitate.
Alpine meadows, lake and stream banks.
General distribution: Amphi-Beringian, alpine-arctic, with widely disjunct populations through its mainly Eurasian range; in N. Amer-

ica from Alaska and Y.T. and western District of Mackenzie, south to Colorado.
Fig. 792 Map 914

Gentiana Raupii Porsild
Gentianella detonsa (Rottb.) G. Don ssp. *Raupii* (Porsild) Gillett
Glabrous annual or biennial from a small, basal rosette; stems 2 to 6 dm tall, angular, with 2 or 3 pairs of linear-oblong leaves, simple or more often branched from the base; flowers solitary, on long, slender peduncles, the corolla bluish-green 1.5 to 3.0 cm long, its 4 lobes oblong-truncate, fimbriate; calyx purplish-green, deeply divided, the lobes sub-equal, but not prominently keeled as in the somewhat similar *G. Macounii.* Seeds oblong, densely short-papillose.

Alluvial meadows and clay banks of the Mackenzie River Basin, northward at least to the Delta.

General distribution: Endemic of the Mackenzie R. Basin.

Fig. 793 Map 915

Gentiana Richardsonii Porsild,
G. detonsa sensu Hultén (1947 and 1968) *non* Rottb.
Glabrous biennial usually with one simple, slender scape up to 20 cm long or, in stronger individuals with one or two shorter, lateral peduncles issuing from among the leaves of the basal rosette. The terminal and largest flower up to 3.5 cm long, the corolla lobes bluish-grey, narrowly oblong, truncate, not fimbriate or erose-dentate towards the tip. Seeds 0.5 ×0.3 mm, papillose. Similar to *G. detonsa* Rottb. of arctic Eurasia and Greenland, but differs by its broader and obtuse basal leaves, shorter and less attenuate calyx lobes, much narrower corolla lobes, and more papillose seeds.

Dry, sandy slopes, but not littoral.

General Distribution: Endemic of the arctic coast of the District of Mackenzie and Kotzebue Sound, Alaska.

Fig. 794 Map 916

Gentiana tenella Rottb.
Gentianella tenella (Rottb.) Börner
Glabrous annual, 10 to 15 cm tall, freely branched from base, each branch terminating in a long and slender peduncle bearing a single 4-parted flower; calyx deeply divided, the corolla light blue.

Apparently rare or local, on sandy beaches and gravelly mud flats along the Arctic Coast and on the shores of Hudson Bay.

General distribution: Circumpolar, arctic-alpine, with many and widely disjunct populations.

Fig. 795 Map 917

Lomatogonium A. Br.

Lomatogonium rotatum (L.) Fries
Small, purplish-green annual or biennal with simple or branched stems 5 to 20 cm high; basal leaves spatulate-lanceolate, soon withering, stem leaves linear-lanceolate, or oblong; inflorescence a few-flowered raceme; sepals like the upper stem-leaves; the corolla rotate, purplish-blue or rarely white.

Moist turfy or sandy sea shores or very rarely by inland saline lakes or springs.

General distribution: Circumpolar, subarctic-arctic.

The ssp. *tenuifolium* (Griseb.) Porsild differs from the circumpolar and always littoral *L. rotatum s. str.* by its taller, up to 4 dm high and freely branched stems, narrower leaves and calyx lobes, and smaller and more numerous flowers. It is a non-halophytic inland race apparently restricted to interior plains and foothills.

General distribution: Alaska and Y.T., the upper Mackenzie Basin, Canadian Rocky Mts. and eastward to western Sask.

Fig. 796 (ssp. *rotatum*) Map 918 (ssp. *rotatum*), 919 (ssp. *tenuifolium*)

786. *Gentiana acuta*

787. *Gentiana affinis*

788. *Gentiana arctophila*

789. *Gentiana glauca*

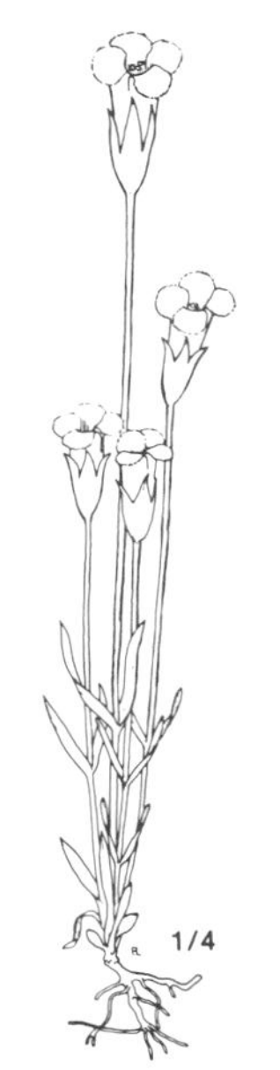

790. *Gentiana Macounii*

791. *Gentiana propinqua*

792. *Gentiana prostrata*

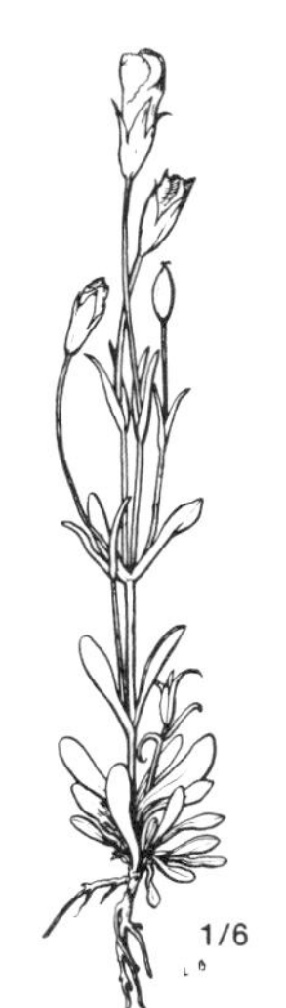

793. *Gentiana Raupii*

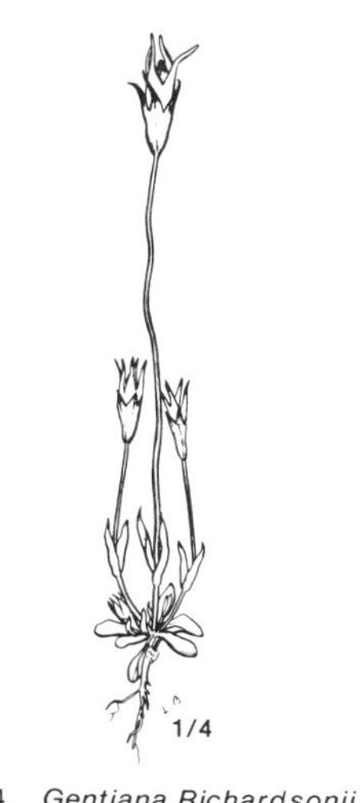

794. *Gentiana Richardsonii*

795. *Gentiana tenella*

796. *Lomatogonium rotatum* ssp. *rotatum*

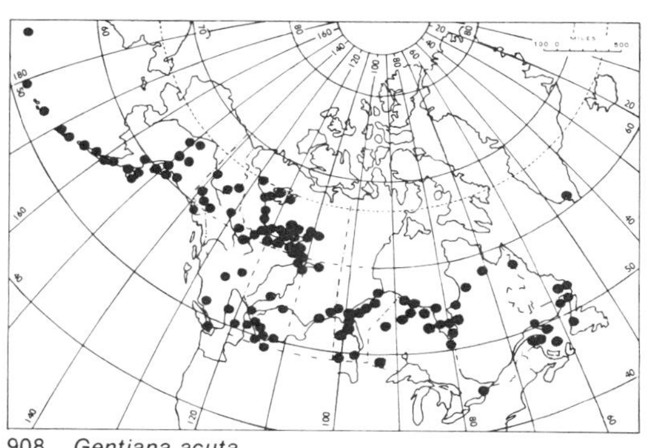

908. *Gentiana acuta*

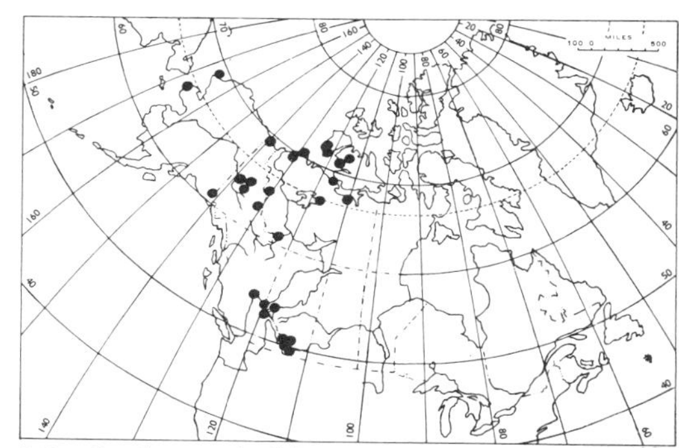

910. *Gentiana arctophila*

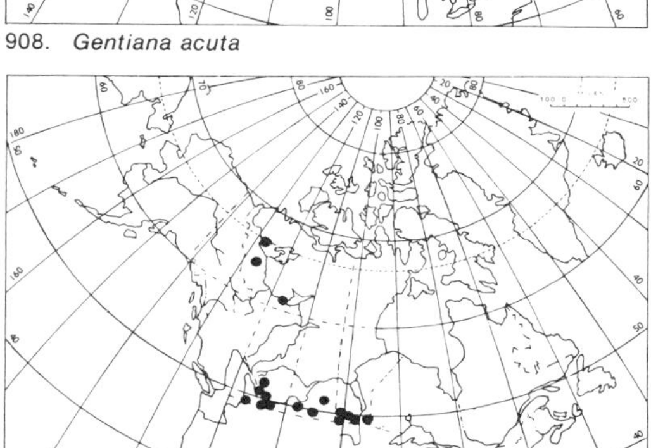

909. *Gentiana affinis*

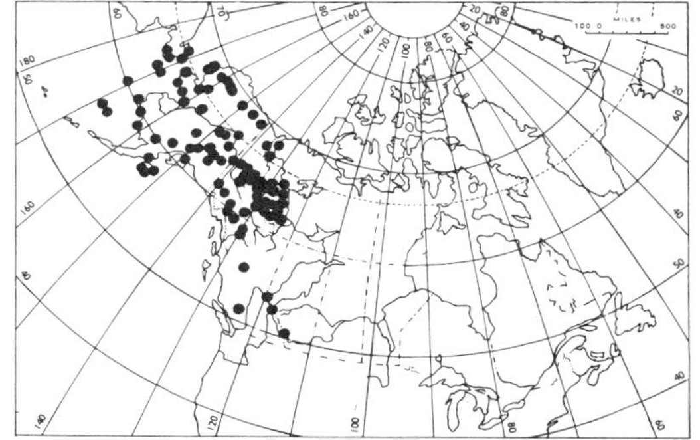

911. *Gentiana glauca*

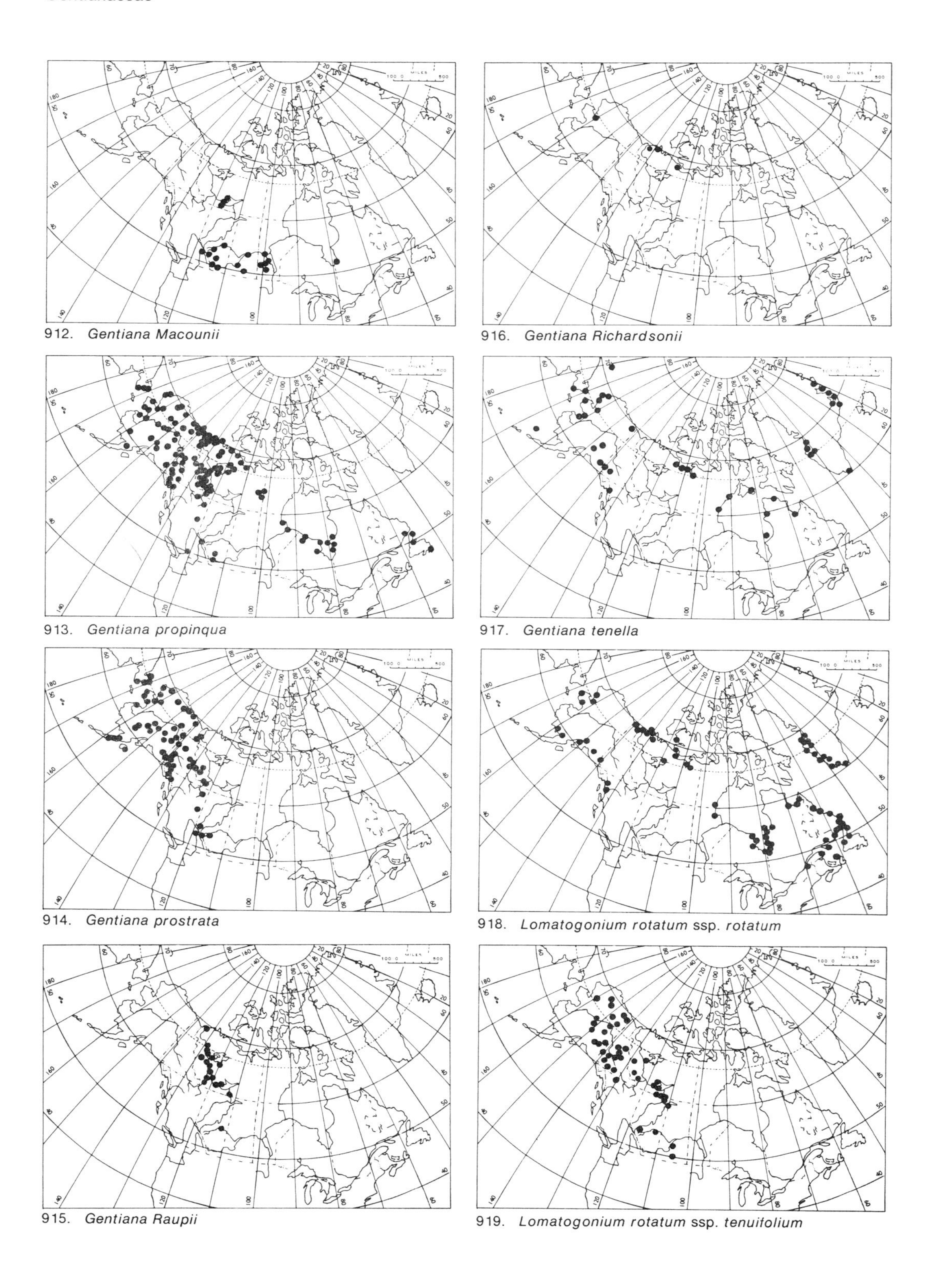

912. *Gentiana Macounii*

916. *Gentiana Richardsonii*

913. *Gentiana propinqua*

917. *Gentiana tenella*

914. *Gentiana prostrata*

918. *Lomatogonium rotatum* ssp. *rotatum*

915. *Gentiana Raupii*

919. *Lomatogonium rotatum* ssp. *tenuifolium*

MENYANTHACEAE Buckbean Family

Menyanthes L. Buckbean

Menyanthes trifoliata L.
Coarse perennial bog plant from a submerged, stout rhizome covered by persisting leaf bases, and fleshy rootlets; leaves trifoliate, long petioled with a large basal sheath; leaflets oblong to elliptic, sessile, wavy-margined; the white or pinkish flowers in a short raceme on a 1 to 3 dm long scape-like peduncle rising from the rhizome; capsule ovate containing large shiny seeds.

Floating bogs and margins of quiet lakes or streams, north and slightly beyond the limit of trees.

General distribution: Circumpolar; wide ranging.

Fig. 797 Map 920

797. *Menyanthes trifoliata*

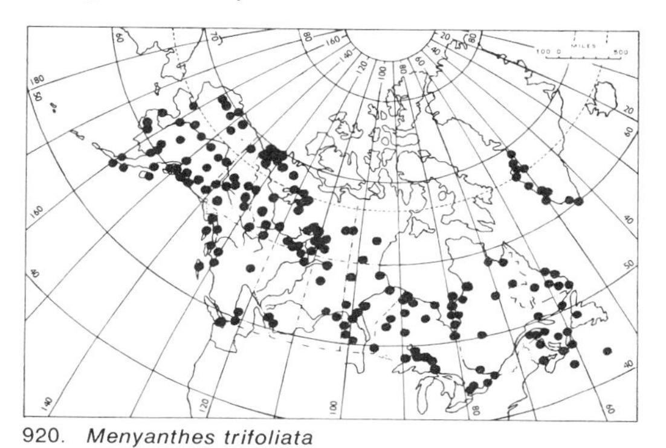

920. *Menyanthes trifoliata*

APOCYNACEAE Dogbane Family

Apocynum Dogbane

Upright branching perennial herbs with milky juice; leaves opposite, entire; flowers cymose, regular, 5-parted, the stamens inserted on the corolla; ovaries 2, with a large ovoid slightly 2-lobed stigma; fruit 2 long slender follicles containing numerous seeds, each seed with a tuft of silky hairs at the apex.

a. Leaves usually reflexed, the under surface more or less pubescent . . . *A. androsaemifolium*
a. Leaves usually ascending, glabrous . . . . *A. sibiricum*

Apocynum androsaemifolium L.
Spreading Dogbane
Stems branched to 5.0 dm or more tall; leaves short-petioled, ovate to ovate-oblong, mucronate, the under surface paler and more or less pubescent; flowers in cymes, both terminal and axillary, the corolla pink with darker stripes in the tube.

Dry sandy pine and poplar woods and dry river banks of the upper Mackenzie drainage.

General distribution: N. America; Nfld. to Alaska and southward.

Fig. 798 Map 921

Apocynum sibiricum Jacq.
Stems branched to 7.0 dm or more tall; leaves sessile, oblong-lanceolate to narrowly ovate, more or less cordate at the base, glabrous; cymes terminal and axillary; corolla milk-white.

Exposed river banks; in our area known only at Hay River and from the Mackenzie River opposite Fort Simpson.

General distribution: N. America from Nfld. to B.C.

Fig. 799 Map 922

798. *Apocynum adrosaemifolium*

799. *Apocynum sibiricum*

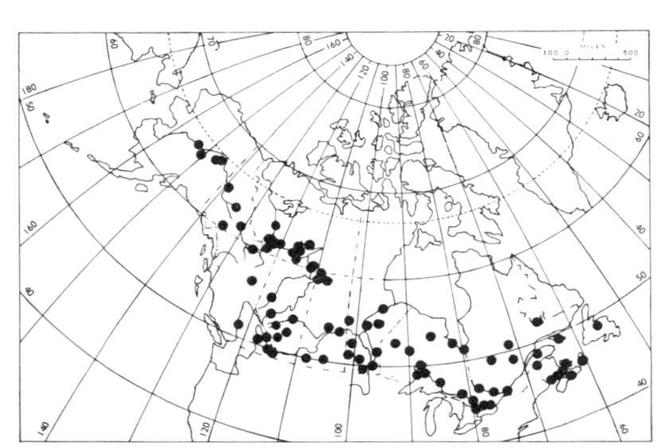
921. *Apocynum androsaemifolium*

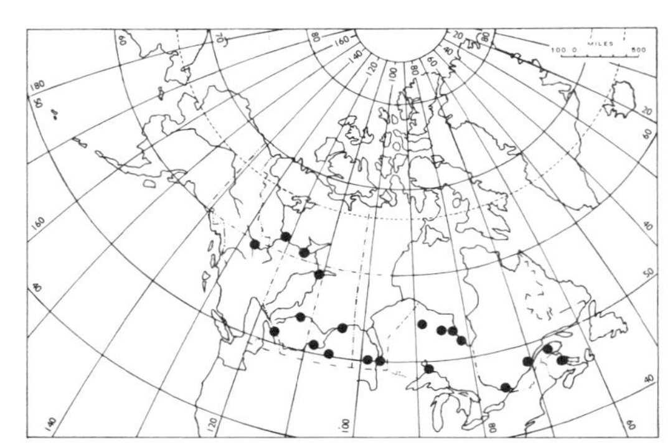
922. *Apocynum sibiricum*

POLEMONIACEAE Phlox Family

Annual or perennial herbs with alternate or opposite, simple or pinnate leaves, regular 5-merous flowers and 3-valved, 3-locular, and few-seeded capsules.

a. Leaves alternate
  b. Leaves entire . . . . *Collomia*
  b. Leaves pinnate . . . . *Polemonium*
a. Leaves opposite, entire . . . . *Phlox*

Collomia Nutt.

Collomia linearis Nutt.
Annual, finely pubescent herb with erect, leafy, simple or branched stems to 4.0 dm tall; leaves entire, narrowly lanceolate, sessile or clasping; flowers small and inconspicuous in terminal leafy-bracted clusters; corolla tubular, about 1.0 cm long, its ovate lobes about 2 mm long, pinkish purple to whitish; seeds numerous, mucilagenous and producing spiraling threads when wetted.

Introduced in townsites of the southern parts of our area.

General distribution: N. America; N.B. to B.C. but spreading as a weed in disturbed situations far beyond its natural range.

Fig. 800 Map 923

Phlox L. Phlox

Low tufted or matted perennials with opposite, more or less imbricated leaves; corolla with a long tube, the limb flat and showy; stamens unequally inserted on the tube; capsule ovoid, 3-valved, few seeded.

a. Leaves flat; flowers about 2 cm across, pink to whitish (drying blue) . . . . *P. alaskensis*
a. Leaves awl- or boat-shaped; flowers smaller
  b. Leaves awl-shaped . . . . *P. Hoodii*
  b. Leaves boat-shaped, with a groove along the midvein on the upper surface . . . .
  . . . . *P. Richardsonii*

Phlox alaskensis Jordal
P. borealis Wherry
P. sibirica quoad pl. Alaska, Yukon Territory and Mackenzie, *non* L.
Stems much branched from a sub-ligneous base, their lower part covered with marcescent leaves; leaves flattened, linear-lanceolate, acuminate, viscid with gland-tipped, septate hairs; flowers usually solitary at the ends of the branches; tube as long as the viscid-pubescent calyx; limb rotate, to 2.2 cm across, pink to whitish, drying mauve or blue.

Open stony slopes.

General distribution: Endemic of Alaska and Y.T. eastward to the Richardson Mts. west of the Mackenzie Delta.

Fig. 801 Map 924

***Phlox Hoodii** Richards.
Pulvinate dwarf species with decumbent or ascending simple or much-branched stems from a sub-ligneous base; leaves imbricated, awl-shaped, apiculate, arachnoid hairy towards the base; flowers mostly solitary at the ends of the branches; tube of corolla long, the limb 1 cm or more across, usually white.

Dry prairies and foothills.

General distribution: In the prairie provinces and south into the U.S.A., disjunct to the Upper Yukon R. region of Y.T. and Alaska; to be looked for in western Mackenzie Mts.

Map 925

Phlox Richardsonii Hook.
Richardson's Phlox
Loosely tufted from a much branched, sub-ligneous base; leaves imbricated, awl-shaped and hairy towards the base. Flowers solitary, sessile, pale lavender or white and very fragrant; corolla with a long tube and flat, rotate limb, about 1 cm in diameter, pale lavender, pale blue to almost pure white.

Sandy or gravelly hilltops or barrens.

General distribution: A rare endemic thus far known only from a few stations in northwest Alaska, Y.T., the Arctic Coast of the District of Mackenzie and from Banks Island.

Fig. 802 Map 926

Polemonium L. Jacob's Ladder

Erect (or decumbent at base) herbaceous or suffrutescent perennials from a horizontal or ascending rhizome; leaves alternate, pinnately divided; flowers regular, 5-merous, blue, in corymbs; stamens inserted at the summit of the short tube; fruit a few to several-seeded capsule.

a. Leaves glabrous; corolla lobes ciliate . . . . *P. acutiflorum*
a. Leaves glandular pubescent; corolla lobes not ciliate
  b. Flowers 15-20 mm long . . . . *P. boreale*
  b. Flowers 10-12 mm long . . . . *P. pulcherrimum*

Polemonium acutiflorum Willd.
Stems 1.5 to 4.0 dm tall, decumbent at base, usually singly from an ascending rhizome; leaves mostly basal, the leaflets lanceolate to elliptic, acute; flowers showy, the corolla blue to violet, 1.5 to 2.0 cm long, its lobes acutish and ciliate.

Moist places rich in peaty humus and particularly abundant in alpine meadows.

General distribution: Amphi-Beringian, arctic alpine; in the Mackenzie and Richardson Mts., and eastward through the Eskimo Lakes region.

Fig. 803 Map 927

Polemonium boreale Adams
Stems somewhat viscid-pubescent, loosely tufted, 10 to 15 cm tall, from a slender, branching caudex; leaves mostly basal, the elliptic leaflets about 0.5 cm long, commonly in 9 to 12 pairs; flowers in a dense corymb, showy, 1.5 to 2.0 cm long, purplish-blue with a yellow tube.

Calcareous, sandy or gravelly places, often near animal burrows.

General distribution: Amphi-Beringian; Alaska and Y.T. eastward in the Mackenzie and Richardson Mts. and along the Arctic Coast to about long. 117°W.

Fig. 804 Map 928

Polemonium pulcherrimum Hook.
Similar to *P. boreale* but somewhat taller, the leaves usually longer with up to 14 pairs of usually longer, ovate to ovate-lanceolate leaflets; flowers smaller, 1.0 to 1.2 cm long, blue or purplish, with a yellow tube.

Rare on rather dry sandy and rocky slopes.

General distribution: Wide ranging from Alaska, Y.T. and N.W. District of Mackenzie south in the foothills of B.C. and Alta.

Fig. 805 Map 929

800. *Collomia linearis*

801. *Phlox alaskensis*

802. *Phlox Richardsonii*

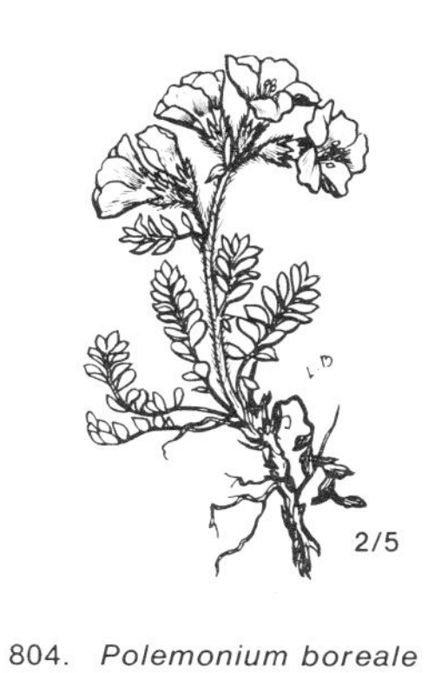

804. *Polemonium boreale*

803. *Polemonium acutiflorum*

805. *Polemonium pulcherrimum*

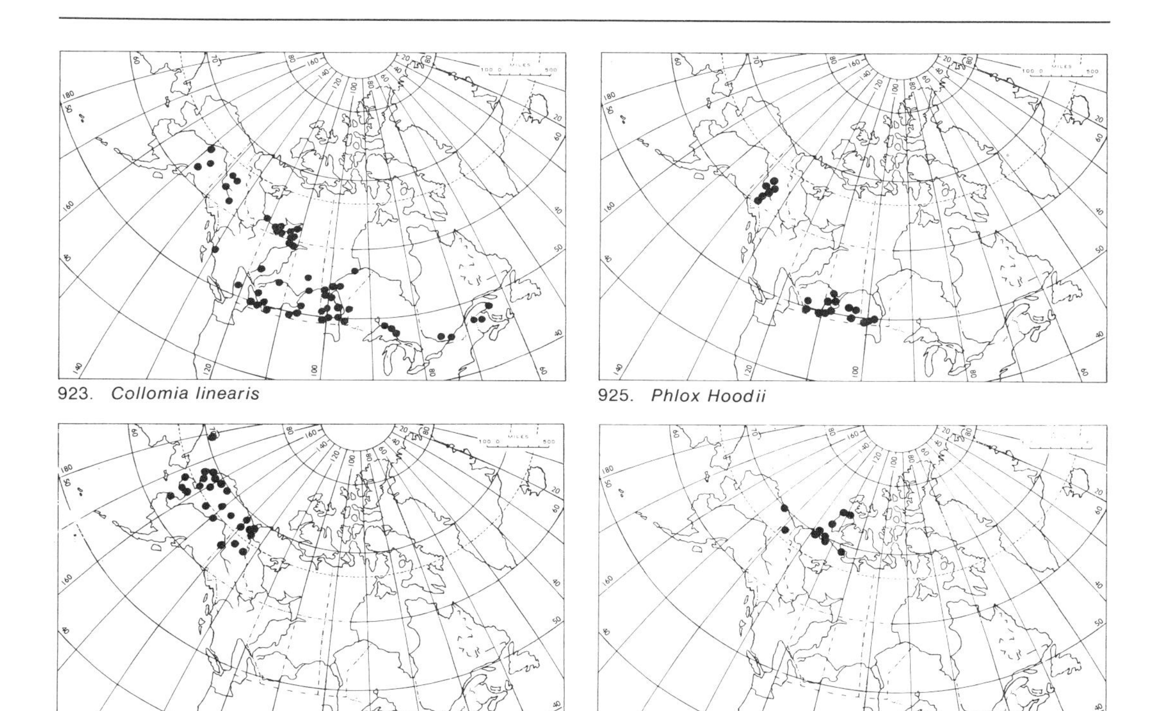

923. *Collomia linearis*

925. *Phlox Hoodii*

924. *Phlox alaskensis*

926. *Phlox Richardsonii*

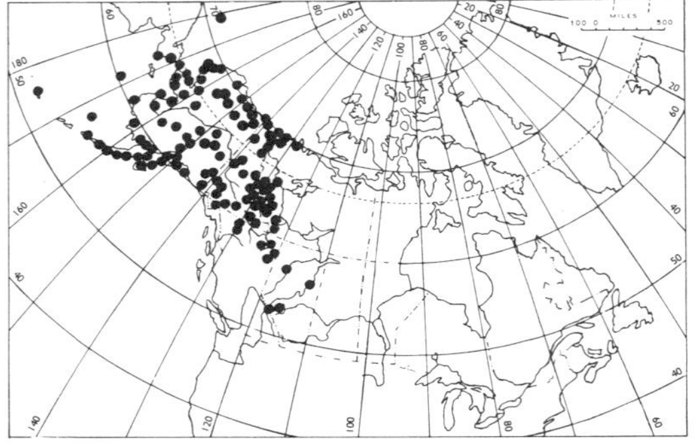

927. *Polemonium acutiflorum*

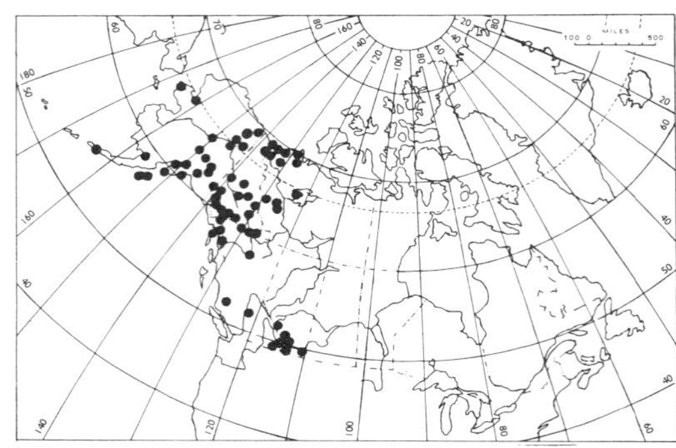

929. *Polemonium pulcherrimum*

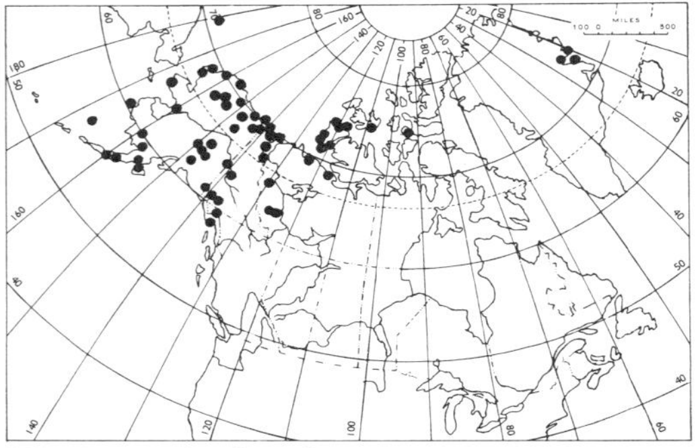

928. *Polemonium boreale*

HYDROPHYLLACEAE Waterleaf Family

Phacelia Juss. Scorpion-weed

Phacelia Franklinii (R.Br.) Gray
Biennial or winter annual with erect hairy and leafy stems to 5 dm in height; leaves alternate, hirsute, pinnatifid, the divisions linear-oblong to triangular and often toothed; flowers in raceme-like cymes, the calyx lobes linear, hispid, the bluish to whitish corolla rotate-campanulate; stamens and styles exserted; fruit a sessile, ovoid, many-seeded capsule; seeds dark brown, angular, reticulate-pitted.

Dry sandy or disturbed situations and burnt-over areas.

General distribution: N. America; from western Ont. to B.C. and central Alaska south into the U.S.A. In S.W. District of Mackenzie north to Great Bear L.

Fig. 806 Map 930

806. *Phacelia Franklinii*

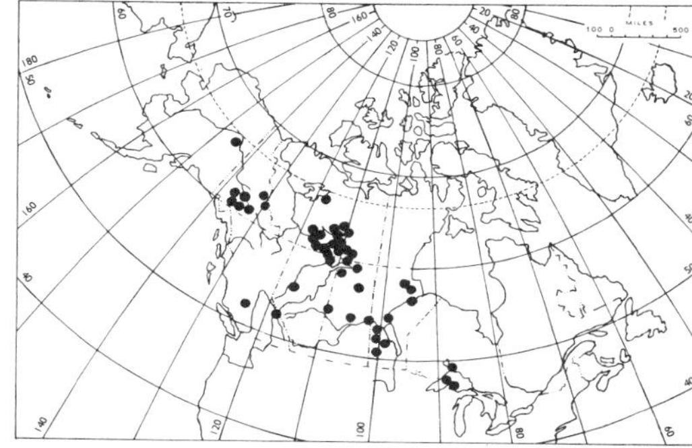

930. *Phacelia Franklinii*

BORAGINACEAE Borage Family

Herbaceous perennials with entire, usually rough-hairy leaves; flowers in one-sided cymes, the calyx 5-parted, the corolla regular, blue and 5-lobed; stamens 5, inserted on the tube; ovary usually 4-lobed, forming 4 seed-like nutlets.

a. Flowers minute, 4 mm or less long
  b. Stem leaves linear-lanceolate . . . . *Lappula*
  b. Stem leaves lanceolate to elliptic-oblong . . . . *Hackelia*
a. Flowers larger
  c. Plants densely caespitose; leaves narrowly oblanceolate to linear . . . . *Eritrichium*
  c. Plants loosely tufted
    d. Flowers with leafy bracts . . . . *Mertensia*
    d. Flowers without leafy bracts . . . . *Myosotis*

Eritrichium Schrad.

Low densely caespitose perennials with crowded marcescent leaves on short branches; corolla blue, funnelform; nutlets attached to a conic receptacle, smooth, with a toothed crown.

a. Flowers about 6 mm in diameter; leaves whitened by straight more or less spreading hairs . . . . *E. aretioides*
a. Flowers about 10 mm in diameter; leaves densely strigose with mostly appressed hairs . . . . *E. splendens*

***Eritrichium aretioides** (Cham. & Schlecht.) DC.
Leaves ovate-lanceolate to oblanceolate, acutish, villous, the hairs sometimes pustulate at the base; flowering stems 5 to 10 cm tall with a short and dense raceme of blue flowers; limb of corolla about 6 mm in diameter; nutlets with a crown of nearly smooth teeth.
General distribution: Amphi-Beringian; in Canada known from dry alpine slopes in the Y.T.; to be looked for in the Mackenzie and Richardson Mts.
Fig. 807 Map 931

Eritrichium splendens Kearney
Leaves linear to narrowly oblanceolate, densely appressed strigose; flowering stems to about 10 cm high with an open, few-flowered raceme of blue flowers; limb of corolla about 10 mm in diameter; nutlets with a crown of jagged and hispid teeth.
Occasional to rare on alpine slopes in the Richardson Mts.
General distribution: Endemic of Alaska, northern Y.T. and the Richardson Mts. of N.W. District of Mackenzie.
Map 932

Hackelia Opiz
Stickseed or Beggar's-lice

Hackelia americana (A. Gray) Fern.
Biennial or perennial herb with a rosette of ovate-lanceolate basal leaves which are usually shrivelled at the flowering time, and alternate mostly lanceolate stem leaves; stems 4 to 9 dm tall, branched in the upper part; leaves and stem strigose; flowers minute, blue, in terminal racemes; nutlets with a crown of barbed prickles.
Thickets, woods, clearings and banks.
General distribution: Que. to S.W. District of Mackenzie, B.C. and southwards.
Map 933

Lappula Moench Stickseed

Annual or winter annual hispid herbs; flowers minute, blue or white, in bracted racemes; nutlets with a crown of barbed prickles, muricate on the back.

a. Nutlets with 2 rows of marginal prickles . . . . *L. echinata*
a. Nutlets with 1 row of marginal prickles . . . . *L. Redowskii* var. *occidentalis*

Lappula echinata Gilib.
Stems to 5 dm or more tall, usually branched above; leaves linear-lanceolate to lanceolate, obtuse and sessile, or the lower petioled; corolla salverform; nutlets with two rows of marginal prickles.
Fields, roadsides and about buildings.
General distribution: An introduced cosmopolitan weed, occasional in the townsites in S.W. District of Mackenzie.

Lappula Redowskii (Hornem.) Greene var. **occidentalis** (Wats.) Rydb.
Similar to *L. echinata* but nutlets with only one row of prickles. Roadsides, clearings and fields.
General distribution: Western Canada and U.S.A., in the Mackenzie R. drainage where likely introduced, and in eastern N. America.
Fig. 808 Map 934

Mertensia Roth Lungwort

Perennial herbs with one- to several erect, ascending or decumbent leafy stems; inflorescence cymose, with campanulate, blue (rarely white) flowers; nutlets smooth or muricate.

a. Decumbent sea-shore plant with ± fleshy, glabrous, glaucous leaves . . . . *M. maritima*
a. Erect non littoral species, with more or less pubescent non-glaucous leaves
  b. Leaves with short, stiff hairs on the margins and on the upper surface . . . . *M. Drummondii*
  b. Leaves more or less rough-hairy on both sides
    c. Calyx lobes green, hairy . . . . *M. paniculata* var. *paniculata*
    c. Calyx lobes glaucous, glabrous on the back, ciliate on the margin . . . . *M. paniculata* var. *alaskana*

Mertensia Drummondii (Lehm.) G. Don
Drummond's Lungwort
Erect-ascending stems, 7 to 15 cm high, from a slender tap-root; basal leaves long-petioled, oblong-lanceolate, the cauline leaves sessile, oblong to ovate-oblong; cymes few-flowered; styles about 8 mm long in fruiting specimens; nutlets smooth to rugose.

Sandy banks and eskers.

General distribution: A very rare plant, endemic to the Arctic Coast of the District of Mackenzie, southern Victoria Island and northwestern Alaska, but not a sea shore species.

Fig. 809 Map 935

Mertensia maritima (L.) S.F. Gray
Sea-Lungwort
Spreading, decumbent sea-shore plant with smooth, somewhat fleshy, glaucous ovate or spatulate leaves; and small, purplish-pink, pale blue or rarely white flowers in a terminal cyme; style about 2 mm long in fruiting specimens; nutlets smooth under the spongy outer coat which becomes inflated thereby making them adaptable to dispersal by sea.

Occasional to common on gravelly or shingly sea-beaches.

General distribution: Wide ranging from Alaska to northwestern Europe.

Fig. 810 Map 936

Mertensia paniculata (Ait.) G. Don
var. **paniculata**
Bluebell
Stems erect, 2 to 7 dm tall; leaves more or less rough hairy, the lower long-petioled with cordate-ovate to lance-elliptic blade up to 15 cm long, and the upper sessile, lance- to ovate-acuminate, to 12 cm long; inflorescence dense, later paniculate; corolla pink in bud, becoming blue; calyx lobes pilose beneath; style about 12 mm long; nutlets wrinkled.

A common lowland species of river banks, open woods and clearings occasionally ascending to and above timberline.

General distribution: Boreal N. America: James Bay, and northern Ont. to Alaska, north along the Mackenzie R. Valley to the Arctic Coast west of the Delta.

The var. *alaskana* (Britt.) Williams differs in having the calyx lobes glaucous, glabrous on the back and ciliate on the margin. It is endemic of Central Alaska, western Y.T. and the central Mackenzie Mts.

Fig. 811 Map 937 (var. *paniculata*), 938 (var. *alaskana*).

Myosotis L. Forget-me-not

Myosotis alpestris Schm.
ssp. **asiatica** Vestergr.
Stems 1 to several to 2 dm or more tall; basal leaves long-petioled, lanceolate, the cauline sessile, oblong to lanceolate, pubescent with appressed and spreading hairs; flowers small, blue (or occasionally white) in more or less compact racemes, elongating in fruit; nutlets smooth.

Moist tundra, late snow patches, sandy or gravelly banks and moist open thickets.

General distribution: Amphi-Beringian: Western N. America across Asia; reaching the Mackenzie Mts., Richardson Mts., and Caribou Hills east of the Mackenzie R. Delta.

Fig. 812 Map 939

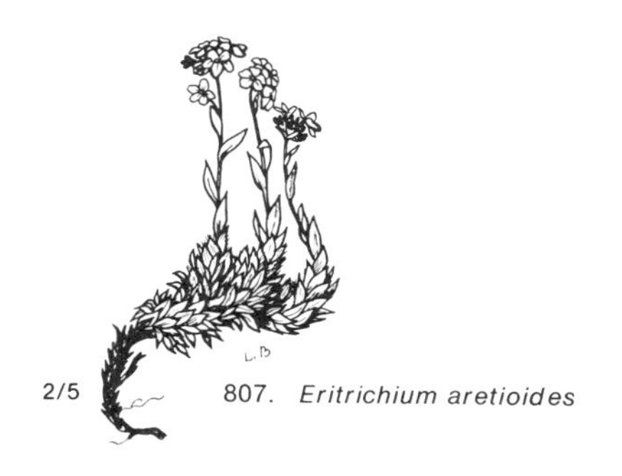

2/5 807. *Eritrichium aretioides*

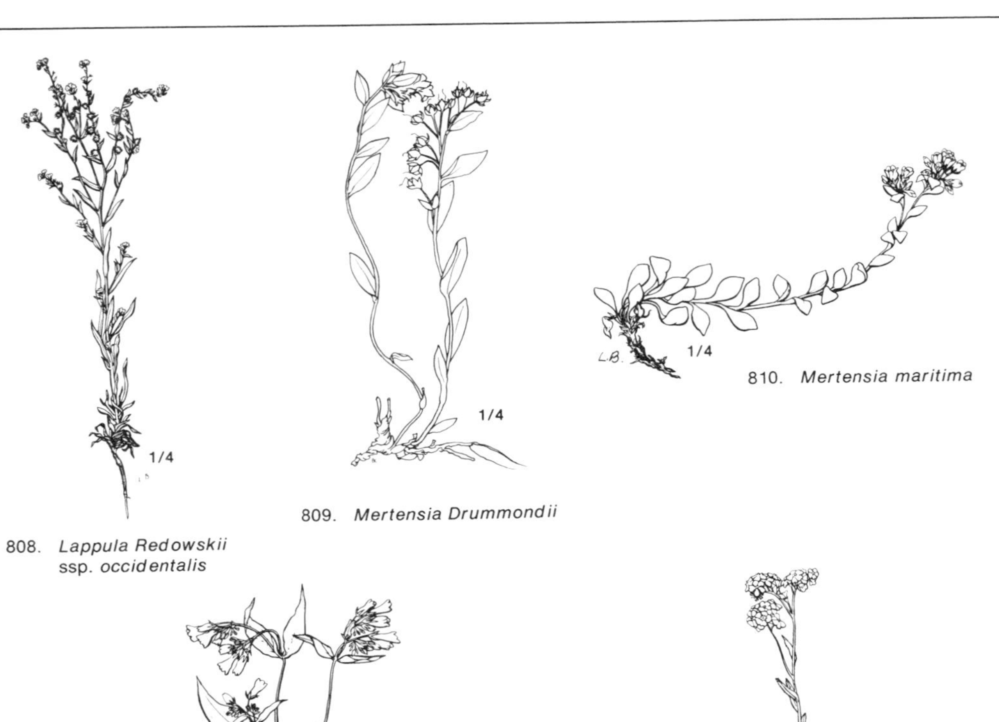

808. *Lappula Redowskii* ssp. *occidentalis*

809. *Mertensia Drummondii*

810. *Mertensia maritima*

811. *Mertensia paniculata* var. *paniculata*

812. *Myosotis alpestris* ssp. *asiatica*

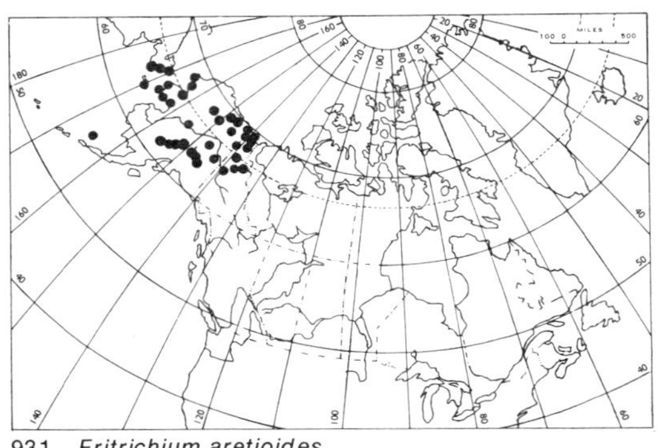

931. *Eritrichium aretioides*

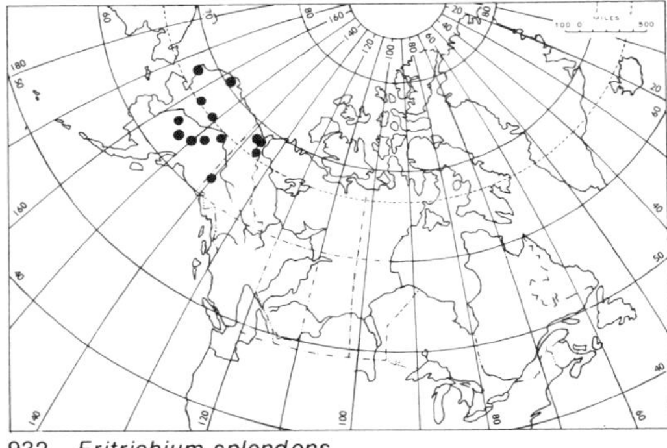

932. *Eritrichium splendens*

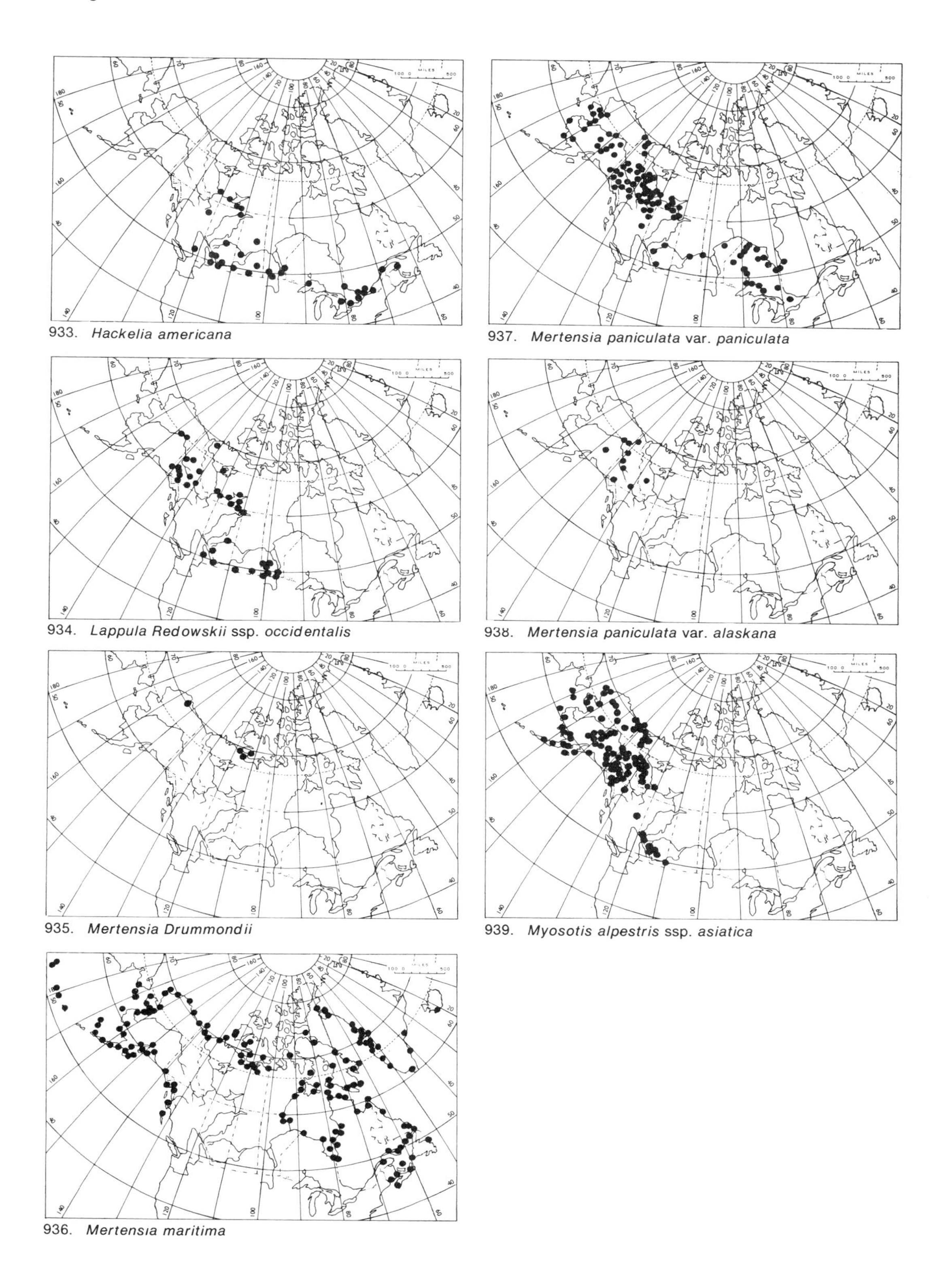

933. *Hackelia americana*

934. *Lappula Redowskii* ssp. *occidentalis*

935. *Mertensia Drummondii*

936. *Mertensia maritima*

937. *Mertensia paniculata* var. *paniculata*

938. *Mertensia paniculata* var. *alaskana*

939. *Myosotis alpestris* ssp. *asiatica*

LABIATAE Mint Family

Herbs with square stems and opposite leaves; usually aromatic; flowers axillary or in bracted terminal spikes; corolla more or less 2-lipped, the upper 2-lobed, the lower 3-lobed; stamens 2 or 4, frequently of different length; ovary deeply 4-lobed, dividing into four 1-seeded nutlets.

a. Leaves round-ovate to reniform . . . . *Lamium*
a. Leaves linear, oblong, ovate or lanceolate
  b. Flowers in the axils of leaves
    c. Corolla 12-18 mm long
      d. Flowers singly, corolla blue; calyx 2-lipped . . . . *Scutellaria*
      d. Flowers several, whorled, corolla purplish or white; calyx lobes bristle-tipped . . . . *Galeopsis*
    c. Corolla smaller
      e. Flowers about 4 mm long, corolla mauve; stolons without tubers . . . . *Mentha*
      e. Flowers about 2.5 mm long, corolla white; stolons with tubers . . . . *Lycopus*
  b. Flowers in terminal leafy or bracted spikes
    f. Corolla barely longer than the calyx . . . . *Dracocephalum*
    f. Corolla longer than the calyx
      g. Flowers large and showy; corolla rose-pink to purplish; bracts small . . . . *Physostegia*
      g. Flowers small, leafy-bracted
        h. Corolla mottled rose-purple; calyx lobes lance-subulate . . . . *Stachys*
        h. Corolla blue; calyx violet or bluish . . . . *Agastache*

Agastache Clayton Giant Hyssop

Agastache Foeniculum (Pursh) Ktze.
Perennial herbs from a creeping rhizome; stems usually simple, to 8 dm or more tall; leaves petioled, the blades ovate, acute, coarsely toothed and whitened beneath with a dense fine pubescence; flowers small, leafy-bracted in an interrupted spike; corolla blue; calyx violet or bluish.

Thickets and grassy clearings.

General distribution: N. American plains, and occasionally disjunct or adventive in the east. Rare in S.W. District of Mackenzie where it is known only from two early collections (*Anderson*, Mackenzie R., and *Kennicott*, Resolution).

Fig. 813 Map 940

Dracocephalum L. Dragonhead

Dracocephalum parviflorum Nutt.
Moldavica parviflora (Nutt.) Britt.
Coarse upright annual or biennial herb with stems 5 dm or more tall, solitary or branched from the base; leaves petioled, the blade lanceolate to lance-ovate, sharply toothed; spike short and dense; bracts shorter than the calyx, their sharp teeth usually spine-tipped; corolla blue, mauve or pink, barely longer than the calyx.

Woodland clearings.

General distribution: N. America: Que. to Alaska and southward; in the Mackenzie R. drainage north to the Arctic Circle.

Fig. 814 Map 941

Galeopsis L. Hemp-Nettle

Galeopsis tetrahit L.
var. **bifida** (Boenn.) Lej. & Court.
Coarse, weedy annual with simple or branched stems, bristly hirsute and swollen below the nodes; leaves petioled, ovate, cuneate to the base, acuminate and coarsely toothed; flowers axillary and whorled; corolla purplish or white, about 15 mm long; calyx lobes bristle-tipped.

Waste places and near settlements.

General distribution: Nfld. to B.C. In the District of Mackenzie north to Ft. Simpson. Naturalized from Europe.

Lamium L. Dead-Nettle

Lamium amplexicaule L.
Annual or biennial with stems branching from the base; leaves long-petioled at the base to clasping-sessile on the upper stem, the blades crenate, round-ovate to reniform; flowers whorled in the axils of the upper leaves; corolla about 1.4 mm long with a slender tube and pink or purple tips; calyx lobes equal, awl-shaped.
Waste ground in gardens, Fort Providence.
General distribution: Naturalized from Europe.

Lycopus L. Bugleweed

Lycopus uniflorus Michx.
Stoloniferous and tuber-bearing perennial herb; stems single, erect, to 2 dm or more tall; leaves ovate or ovate-oblong, tapering at both ends, coarsely toothed, finely glandular-punctuate, glabrate; flowers minute, clustered, in the axils of the upper leaves; corolla white; calyx lobes triangular.
In wet sand of lake shores and along streams.
General distribution: N. America: Nfld. to central Alaska and southward, disjunct in eastern Asia. Rare in the Precambrian Shield area south of Great Slave L.
Fig. 815 Map 942

Mentha L. Mint

Mentha arvensis L.
var. **villosa** (Benth.) Stewart
Strongly aromatic perennial herb from a slender, branching rhizome; stems single or two or more together, to 5 dm tall; leaves short-petioled, ovate, oblong or lanceolate, serrate, glandular-punctate, tapered at both ends; flowers small, whorled in the axils of the middle and upper leaves; corolla mauve; calyx lobes almost equal, triangular to subulate.
Occasional in grassy swales, meadows, moist ditches, river banks and lake shores.
General distribution: Circumpolar; in S.W. District of Mackenzie and north along the Mackenzie R. to about lat. 64°30′.
Fig. 816 Map 943

Physostegia Benth. False Dragonhead

Physostegia parviflora Nutt.
Dracocephalum nuttallii Britt.
Smooth, tall perennial herb with sharply serrate, sessile leaves and pink or purplish flowers in a bracted terminal spike.
Clay banks and depressions.
General distribution: Man. to B.C. and southward. In the District of Mackenzie known only from the Salt R. in the Wood Buffalo Park.
Map 944

Scutellaria L. Skullcap

Scutellaria galericulata L.
var. **pubescens** Benth.
S. epilobiifolia A. Ham.
Stems simple or branched up to 4 dm tall from a slender, creeping rhizome; leaves ovate or lanceolate, serrate, pubescent beneath; flowers solitary in the axils of the middle and upper leaves; corolla blue, 12 to 18 mm long, arched; calyx 2-lipped.
Moist swales, thickets and river banks in the Mackenzie lowlands as far downstream as Norman Wells.
General distribution: Boreal; Nfld. to Alaska and southward; the species circumpolar.
Fig. 817 Map 945

Stachys L. Hedge-nettle

Stachys palustris L. *s. lat.*
Stems usually unbranched up to 6 dm tall from a slender, creeping rhizome; leaves oblong-ovate to lanceolate, crenate, and pubescent on both surfaces; flowers in leafy-bracted whorls forming a short, interrupted spike; corolla rose-

purple, about 14 mm long; the calyx lobes lance-subulate, about as long as the corolla.

Moist meadows, river banks and lake shores.

General distribution: Circumpolar, wide ranging and very variable. In the Mackenzie R. lowlands north to the Delta.

Fig. 818 Map 946

813. *Agastache Foeniculum*

814. *Dracocephalum parviflorum*

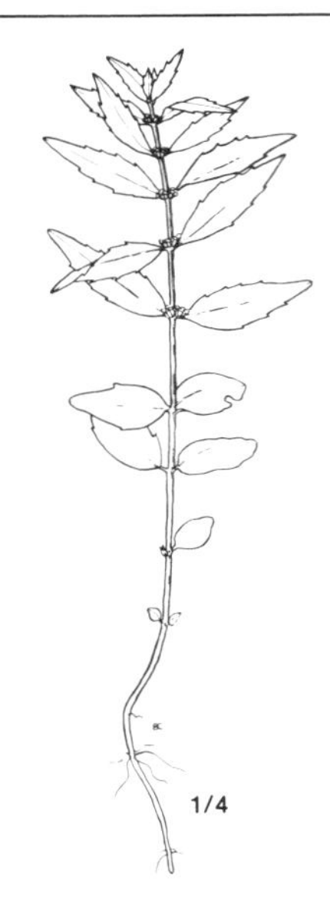

815. *Lycopus uniflorus*

816. *Mentha arvensis* var. *villosa*

817. *Scutellaria galericulata* var. *pubescens*

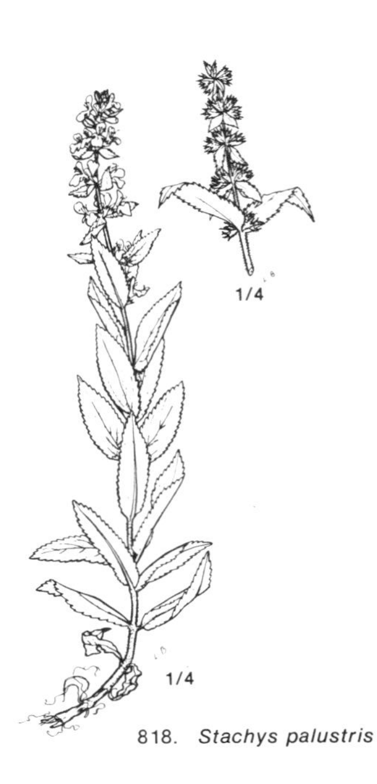

818. *Stachys palustris*

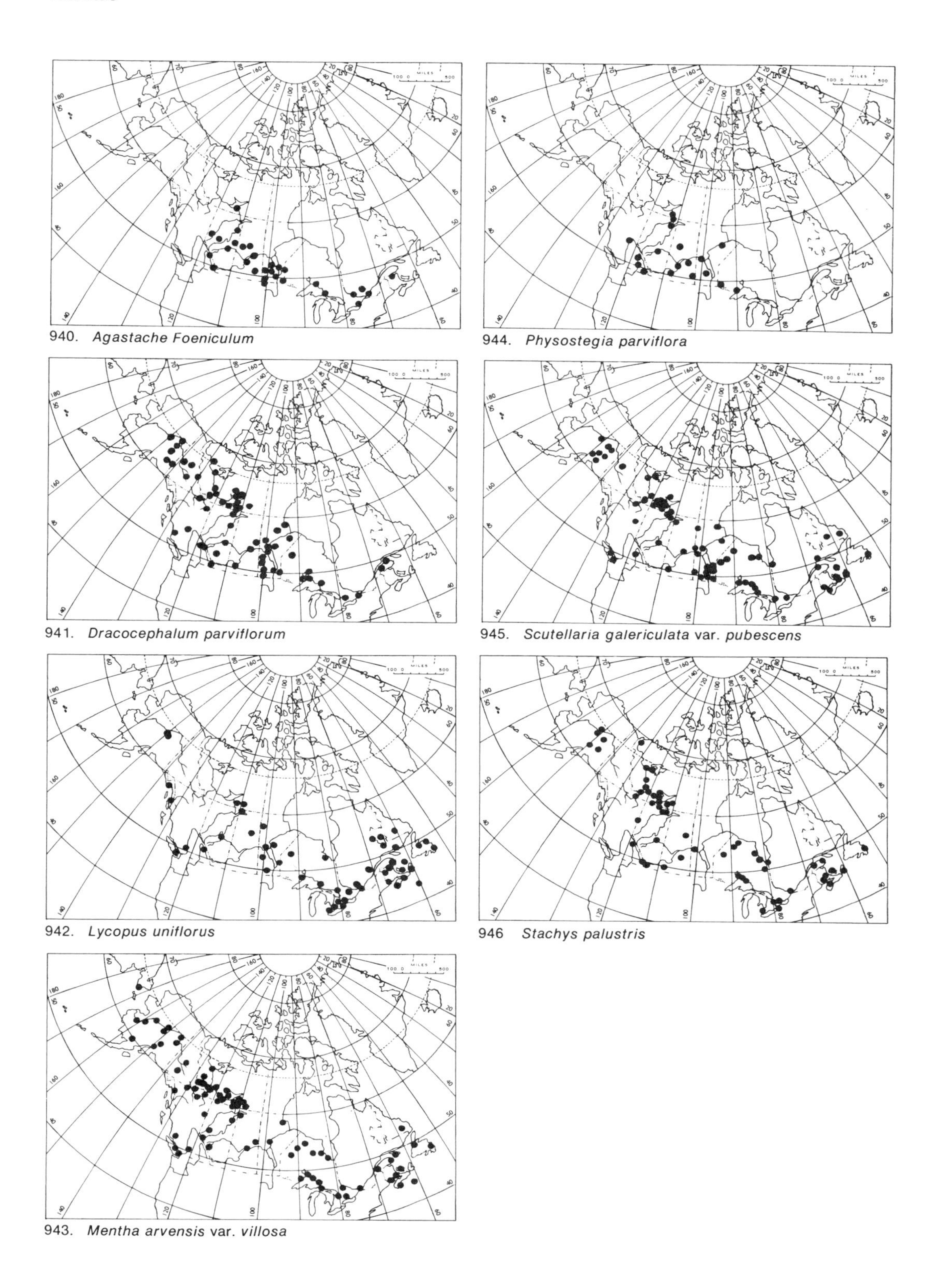

940. *Agastache Foeniculum*

941. *Dracocephalum parviflorum*

942. *Lycopus uniflorus*

943. *Mentha arvensis* var. *villosa*

944. *Physostegia parviflora*

945. *Scutellaria galericulata* var. *pubescens*

946 *Stachys palustris*

SCROPHULARIACEAE Figwort Family

Ours chiefly perennial herbs with opposite, alternate or verticillate leaves and axillary, racemose, or spicate inflorescence of mostly irregular flowers, and 2-locular, many-seeded capsules.

a. Aquatic annuals with linear basal leaves; flowers solitary from leaf-axils . . . . *Limosella*
a. Terrestrial
  b. Corolla spurred, bright yellow (not native) . . . . *Linaria*
  b. Corolla not spurred
    c. Stamens 2
      d. Corolla rotate . . . . *Veronica*
      d. Corolla 2-lipped
        e. Plant glabrous, somewhat fleshy . . . . *Lagotis*
        e. Plant villous, especially in the inflorescence . . . . *Synthyris*
    c. Stamens 4
      f. Leaves opposite
        g. Annuals
          h. Calyx inflated; corollas yellow . . . . *Rhinanthus*
          h. Calyx not inflated
            i. Corollas small, white and purple . . . . *Euphrasia*
            i. Corollas yellow . . . . *Orthocarpus*
        g. Perennials
          j. Entire plant viscid-villous, dark-purplish; flowers dark-purplish . . . . . . *Bartsia*
          j. Plant essentially glabrous, green; flowers large and yellow . . . . *Mimulus*
      f. Leaves alternate or verticillate
        k. Leaves pinnately lobed, mainly in a basal rosette; floral bracts not highly coloured . . . . *Pedicularis*
        k. Leaves irregularly cut or lobed, not rosulate; floral bracts petaloid and strongly coloured . . . . *Castilleja*

Bartsia L.

***Bartsia alpina** L.
Tufted, dark-coloured perennial with a stout, scaly rhizome and viscid-villous, ascending-erect stems, 5 to 20 cm high; leaves opposite, sessile, oval, crenate-serrate, the lowermost scaly, the upper 1 to 2 cm long, dark purple, drying black. Flowers 2 cm long, dark purple, solitary, in the uppermost leaf-axils.

Moist, turfy, and chiefly calcareous south-facing slopes.

General distribution: Amphi-Atlantic, subarctic-alpine. As yet not reported from continental N.W.T. where it should be looked for in eastern District of Keewatin.

Fig. 819 Map 947

Castilleja Mutis Indian Paintbrush or Painted Cup

Ours tufted perennials from a short, weak and branching tap-root, partly parasitic on the roots of grass-like plants; stems ascending-erect; leaves alternate, linear-caudate, entire or with a few linear, spreading lobes near the tip, the uppermost bract-like and usually highly coloured and partly concealing the lowermost flowers of the spike-like inflorescence; corolla 2-lipped, yellow or reddish; stamens 4. Capsule 2-roomed, splitting longitudinally; seeds numerous, wedge- or top-shaped, about 1.5 to 2.0 mm long, the seed-coat of large, reticulate and translucent cells.

a. Lower lip of corolla 3-8 mm below the tip of the upper
  b. Inflorescence merely hirsute; bracts yellow or yellowish; stems and leaves glabrate; leaves caudate . . . . *C. caudata*
  b. Inflorescence villose; bracts violet-purple . . . . *C. elegans*
a. Lower lip of corolla 3 mm or less below tip of the upper
  c. Bracts violet; stem-leaves entire . . . . *C. Raupii*

c. Bracts yellow or yellowish-green; at least the upper stem-leaves with divaricate lobes
  d. Upper and middle stem-leaves with long, divaricate lobes .... *C. hyperborea*
  d. Upper stem-leaves only with lobes .... *C. yukonis*

Castilleja caudata (Pennell) Rebr.
C. pallida (L.) Spreng.
ssp. *caudata* Pennell
Stems few to several, 20 to 40 cm tall, glabrous, or glabrate in the lower half, finely pubescent or puberulent upwards, and somewhat hirsute in the inflorescence. Lowermost leaves narrowly lanceolate, the upper distinctly caudate, in tall specimens up to 1 cm broad at the point of attachment, glabrous above and finely puberulent below. Inflorescence 5- to 12-flowered, elongating in age, the bracts somewhat hirsute, and the uppermost greenish-yellow; corolla 2 to 3 cm long, yellow and usually cobwebby.

In alpine-subarctic, moist upland tundra and stony lake shores.

General distribution: Amphi-Beringian; eastward to Coronation Gulf and Great Bear L. and southward to Great Slave L.

Fig. 820 Map 948

Castilleja elegans (Ostenf.) Malte
Stems several, 5 to 25 cm high, appressed pubescent or glabrate below, loosely pubescent or long-villous through the inflorescence. Leaves pubescent, linear, caudate, entire or frequently with a pair of lateral and linear lobes near the tip; bracts purplish, sometimes with yellowish tips.

Moist, calcareous tundra or rocky or sandy lake shores.

General distribution: Arctic; from northern Alaska eastward to northern District of Keewatin and the southern islands of the Arctic Archipelago, southward to the north shore of Great Bear L. Disjunct in alpine tundra of Rocky Mt. Nat. Parks, Alta.

Fig. 821 Map 949

Castilleja hyperborea Pennell
Stems two to several, 8 to 15 cm tall, erect-ascending, simple or somewhat branching above. Inflorescence elongating, usually one third the length of the stem; floral bracts large, yellowish-green, often drying much darker. Leaves glabrate, short and caudate, each with a pair of linear and usually curling lobes below the tip.

Alpine herbmats and stony slopes.

General distribution: Amphi-Beringian. Eastward to the Richardson and Mackenzie Mts., and along the Arctic Coast near Cape Parry, 125°W. long.

Fig. 822 Map 950

Castilleja Raupii Pennell *s. lat.*
Stems one to several, 20 to 40 cm tall, simple or branched above, green or purplish, cobwebby-villose. Leaves linear or narrowly attenuate, 4 to 5 cm long, finely pubescent on both surfaces. Inflorescence elongating at maturity, the bracts purplish and the lower lip of the 2 cm long corolla about half as long as the upper.

The ssp. *ursina* Pennell was based on three collections from the middle Mackenzie River and, perhaps, is nothing more than a stouter and more branched edaphic variation of *C. Raupii.* Whatever its taxonomic status, ssp. *ursina* is now known along the entire Mackenzie Valley where it is commonly associated with typical *C. Raupii.*

Common along river banks and lake shores.

General distribution: Mackenzie R. and tributaries reaching north to its delta and eastward from Great Slave L. to Hudson and James B.

Fig. 823 Map 951

Castilleja yukonis Pennell
Stems slender, 20 to 25 cm tall, purplish short-pubescent, when one or few together commonly branched above, but simple when, as often, numerous stems are clustered into one compact tussock. Leaves linear, entire, 3-nerved, finely crisp-pubescent. Bracts yellowish-green, lanceolate, hirsute. Corolla dark purple, about 20 mm long, the upper lip very narrow and much longer than the 3-lobed lower lip.

Stony riverbanks and hillsides.

General distribution: Long known only from a few stations in S. W. Y. T., and central interior Alaska but now identified also from the Mackenzie Delta.

Map 952

Euphrasia L. Eyebright

Simple or branched, 2 to 25 cm tall, slender annuals with small, mostly opposite, short-pubescent, somewhat glandular, oval, coarsely crenate and sessile leaves; the small flowers in a rather dense, leafy spike.

a. Corolla 5-8 mm long . . . . *E. arctica*
a. Corolla 3.0-3.5 mm long . . . . *E. subarctica*

Euphrasia arctica Lge.
Simple or branched, 2 to 25 cm high, with small, oval, coarsely crenate sessile leaves; the small flowers in a rather dense, leafy spike. Corollas 5 to 8 mm long, white with lavender or pale blue veins, the upper lip bluish tinged.

Sunny, grassy slopes.

General distribution: E. and W. Greenland over southern Baffin Island and Ungava Pen. to the west shore of Hudson B., north to the mouth of the McConnell River, District of Keewatin.

Fig. 824 Map 953

Euphrasia subarctica Raup
Similar to, and in similar habitats as, *E. arctica* from which it may be distinguished by its smaller, 3.0 to 3.5 mm long corollas.

General distribution: S. W. District of Mackenzie over southern Y. T. to central and southern Alaska, disjunct in alpine parts of Alta.

Fig. 825 Map 954

Lagotis Gaertn.

Lagotis Stelleri (Cham. & Schlecht.) Rupr.
L. glauca Gaertn. var. *Stelleri* (Cham. & Schlecht.) Trautv.
Glabrous, somewhat fleshy perennial from a stout, ascending rhizome; stems 1 to several, 15 to 20 cm tall. Basal leaves petiolate, their blades lanceolate-oblanceolate, crenate-dentate; stem leaves much smaller, sessile and entire. Inflorescence spike-like, the flowers pale blue, each supported by a small, leafy bract. The stem and leaves tend to blacken in drying.

Moist alpine herbmats.

General distribution: Amphi-Beringian; barely reaching the District of Mackenzie on the east slope of the Mackenzie and Richardson Mts.

Fig. 826 Map 955

Limosella L. Mudwort

Limosella aquatica L.
Glabrous dwarf annual with slender decumbent runners terminating in a small cluster of long-petioled, elliptic and somewhat fleshy leaves. The small pink flowers on naked peduncles appearing from the base of the leaves.

Wet, muddy or sandy pond margins.

General distribution: Circumpolar; rare, or no doubt often overlooked. In the N.W.T. thus far reported only from two stations in McTavish B., Great Bear L., and from one station on the west coast of Hudson Bay.

Fig. 827 Map 956

Linaria Mill. Toadflax

Linaria vulgaris Hill
Perennial, glabrous herb 2 to 8 dm tall, with simple or branching stems, numerous linear leaves, and yellow, spurred flowers in an open raceme.

Roadsides or waste places.

General distribution: Introduced cosmopolitan weed reported a few times from townsites of the upper Mackenzie R. drainage.

Mimulus L. Monkey-flower

Mimulus guttatus DC.
M. Langsdorffii Donn
Yellow Monkey-flower
Glabrous perennial with erect-ascending simple stems, 5 dm tall or more, from a scaly rhizome; leaves cauline, opposite, oval and finely crenulate, the lower short-petioled, the upper sessile or somewhat clasping. Inflorescence of

few and large flowers, in an open raceme, on glandular-pubescent peduncles; corolla 3 to 4 cm long, yellow, sometimes spotted in the throat; calyx half as long as the tube, with broadly triangular teeth.

In wet moss by brooks or warm springs.

General distribution: Alaska northwest Pacific, with alpine inland stations on the S. Nahanni R. in the southern District of Mackenzie and in the Rocky Mts.

Map 957

Orthocarpus Nutt. Owl's Clover

Orthocarpus luteus Nutt.
Pubescent annual with simple, leafy stems 1 to 2 dm tall, the leaves sessile, linear-lanceolate, entire or pectinately divided. The small, yellow flowers in a narrow, leafy-bracted spike.

Sandy riverbanks and lakeshores.

General distribution: Man. to B.C., north to the Slave R. in lat. 60°30′N.

Map 958

Pericularis L. Lousewort or Fernweed

Perennial or biennial herbs with alternate, pinnatifid or verticillate leaves, and irregular flowers in a racemose or spike-like inflorescence. Calyx tubular, somewhat inflated; the corolla 2-lipped, the upper lip laterally compressed, helmet-like, enclosing the anthers, the lower 3-lobed; stamens 4; capsule compressed, containing relatively few, large seeds.

The roots and young flowering stems of *P. lanata, P. arctica, P. hirsuta,* and *P. sudetica* may be eaten raw, or cooked as a pot-herb. Eskimo children suck the sweet nectar from the base of the long corolla-tube of *P. lanata* and *P. arctica.*

a. Stems branched; biennial or short-lived perennials
  b. Flowers yellow in few-flowered elongating spikes . . . . *P. labradorica*
  b. Flowers purple, in short spikes, or solitary in the leaf axils . . . . *P. parviflora*
a. Stems simple; evidently perennials
  c. Corollas predominantly yellow
    d. Helmet prolonged into a distinct beak . . . . *P. lapponica*
    d. Helmet beakless or nearly so
      e. Inflorescence, few-flowered, capitate, flowers creamy yellow, sometimes with a reddish tinge . . . . *P. capitata*
      e. Inflorescence spicate
        f. Corolla bright yellow; tip of helmet darker . . . . *P. Oederi*
        f. Corolla orange-yellow, the upper half of helmet deep purple . . . . *P. flammea*
  c. Corollas pink or purple
    g. Cauline leaves verticillate . . . . *P. verticillata*
    g. Cauline leaves alternate, or lacking
      h. Helmet with long and slender, curved beak . . . . *P. groenlandica*
      h. Helmet not beaked
        i. Plants with distinct tap-root
          j. Spike dense, stem white-woolly
            k. Corolla deep-pink, spike long . . . . *P. lanata*
            k. Corolla pale pink, spike much shorter . . . . *P. hirsuta*
          j. Spike and stem glabrous or nearly so . . . . *P. arctica*
        i. Plants with branching rhizome and scapose stem . . . . *P. sudetica*

Pedicularis arctica R. Br.
P. Langsdorffii var. *arctica* (R. Br.) Polunin
Tufted, with a pale yellow tap-root; stems one to several, 10 to 25 cm high, essentially glabrous, leafy, terminating in a showy, loose-flowered, leafy spike of bright pink, spreading flowers. Corolla 20 to 25 mm long with a prominently arched helmet which has two filiform teeth, 1 to 2 mm long, near the tip. Seeds smooth, brown.

Moist tundra barrens.

General distribution: Endemic of arctic-alpine N. America.

Fig. 828 Map 959

Pedicularis capitata Adams
Dwarf species with a slender, branched rhizome, and simple, scapose, glabrous or short, pubescent stems, 5 to 15 cm high, terminating in a capitate 2- to 4 (rarely 6)-flowered head. Flowers very large and scentless; corollas 3 to 4 cm long, creamy yellow, often with a reddish tinge.

Gravelly, calcareous tundra or heath.

General distribution: Circumpolar (with large gaps in arctic Europe, E. and W. Greenland); arctic-alpine.

Fig. 829 Map 960

Pedicularis flammea L.
Roots spindly; stems one to several, simple, reddish-purple, essentially glabrous, sparingly leafy, 8 to 15 cm high, terminating in a spike-like raceme of small, scentless flowers. Corolla 6 to 7 mm long, the lip bright yellow, the helmet with a reddish-purple tip.

In rather moist calcareous tundra, on snowbeds and by lake shores.

General distribution: Amphi-Atlantic, arctic-alpine.

Fig. 830 Map 961

***Pedicularis groenlandica** Retz.
Elephant Head
Stems one to several, glabrous and dark reddish-purple, 3 to 6 dm tall; basal leaves petioled, the upper sessile, the blade deeply pinnatifid. Spike dense, elongating in fruit, 6 to 15 cm long; corolla reddish-purple, its upper lip or helmet with a long and slender up-curved beak resembling the head and trunk of an elephant.

Wet calcareous meadows.

General distribution: A very striking species endemic to S. W. Greenland and boreal N. America. Except for a single station in S. E. Y.T., not yet reported from north of lat. 60°, but to be expected in southern Districts of Mackenzie and Keewatin.

Fig. 831 Map 962

Pedicularis hirsuta L.
Hairy Lousewort
Similar to *P. lanata* but much less woolly, and with a pale, spindly tap-root and slender stems, 5 to 20 cm high, terminating in a sub-capitate, soon elongating spike of small, pale pink flowers. Helmet short and stubby. Seeds smooth, brown.

Moist, stony, and sandy places in tundra and by lake and river banks.

General distribution: Amphi-Atlantic, eastern arctic, in continental N.W.T., barely entering N. E. District of Mackenzie.

Fig. 832 Map 963

Pedicularis labradorica Wirsing
Labrador Lousewort
Biennial with a weak, spindly tap-root and simple or freely branched leafy, glabrous or short-pubescent stems, 15 to 30 cm high. Inflorescence short-spicate, 5- to 10-flowered, first subcapitate, soon elongating. Flowers yellowish, scentless; the helmet often somewhat purple-tinged with two slender teeth near the tip.

In muskegs and open, mossy, not too dry heath.

General distribution: Circumpolar (with large gaps), subarctic-alpine.

Fig. 833 Map 964

Pedicularis lanata Cham. & Schlecht.
Woolly Lousewort
Similar to *P. hirsuta* but with a bright lemon-yellow tap-root, stouter, densely white-woolly stem, and a dense-flowered, soon elongating, copiously woolly spike of faintly scented, showy flowers. Corolla deeper pink, or rarely white, the helmet not prominently arched, as in *P. arctica,* and lacking teeth near the apex. Seeds large, with a loose-fitting, ashy-grey, honeycombed, seed-coat; by this character alone *P. lanata* may be distinguished from fruiting (or even over-wintered) specimens of *P. hirsuta* or *P. arctica.*

Common in moist, stony tundra.

General distribution: N. America, arctic-alpine.

Fig. 834 Map 965

Pedicularis lapponica L.
Lapland Lousewort
Essentially glabrous dwarf species with a slender, branching rhizome; stems purplish brown, simple, one or several together, 10 to 20 cm high, leafy. Inflorescnece short-spicate, few-flowered; flowers very fragrant, with pale yellow corollas.

In rather dry, turfy places.

General distribution: Circumpolar, subarctic.

Fig. 835 Map 966

Pedicularis Oederi M. Vahl
Stems glabrous, mostly solitary, 15-20 cm tall, from a cluster of slender, spindle-shaped roots; leaves pinnatifid, the segments triangular-retrorse. Inflorescence a few- to many-

flowered spike; corollas bright yellow, except for the darker tip of the helmet. In habit similar to *P. flammea*, but taller and with larger corollas.

Damp, alpine tundra.

General distribution: Amphi-Beringian species barely entering the District of Mackenzie in the Richardson Mts. but with disjunct alpine stations in the Rocky Mts.

Fig. 836 Map 967

Pedicularis parviflora J. E. Smith
P. macrodonta Richards.
Biennial or (with us) a short-lived perennial from a weak, simple or branching root; stems simple or when branched, spreading from the base. Flowers in a short spike, or solitary in the axils of stem-leaves; corollas purple, 10 to 12 mm long. Wet bogs and meadows.

General distribution: Hudson Bay to S. W. Alaska, north of lat. 60° N., thus far known only from the mouth of McConnell R. on the west coast of Hudson Bay, lat 60° 50′N., and from S. W. District of Mackenzie.

Fig. 837 Map 968

Pedicularis sudetica Willd.
Stems scapose, glabrous, dark purple, singly or several together, from a stout, branched rhizome. The white-woolly inflorescence first capitate but soon elongating. Corolla dark reddish purple, the helmet with two prominent teeth near the apex.

Rather wet, calcareous tundra and lake shores.

General distribution: Circumpolar (but lacking in Greenland) wide ranging, arctic-alpine.

Fig. 838 Map 969

Pedicularis verticillata L.
Short-lived perennial from a weak, branching tap-root. Stems few to 25 or even more, simple, erect-ascending, 10 to 15 cm tall. Leaves mainly basal, those of the stem and inflorescence verticillate. Flowers purple or rarely white, in dense but soon elongating spikes.

A very striking and handsome species of damp meadows, often near beaches or lake shores.

General distribution: Amphi-Beringian species barely entering the District of Mackenzie in mountains west of the Mackenzie Delta.

Fig. 839 Map 970

Penstemon Mitchell Beard-tongue

Perennial with opposite leaves, 5-parted calyx, 4 stamens and tubular 2-lipped corollas (ours blue).

a. Inflorescence hirsute; corollas about 2 cm long . . . . *P. Gormanii*
a. Inflorescence glabrous; corollas smaller . . . . *P. procerus*

Penstemon Gormanii Greene
Stems tufted, 15 to 40 cm tall, one to several from a simple or branching tap-root; leaves glabrous, narrowly lanceolate.

Gravelly riverbanks and terraces.

General distribution: A very handsome species endemic to central Alaska and Y. T., barely reaching interior northern B.C. and the east slope of the Mackenzie Mts., N.W.T.

Fig. 840 Map 971

***Penstemon procerus** Dougl.
Stems slender, rarely over 25 cm tall, singly or several together from an ascending and branching rhizome. Leaves oblanceolate, mainly cauline. Flowers small, deep blue, verticillate, in an interrupted spike.

Gravelly or sandy river terraces or sub-alpine meadows.

General distribution: North Cordilleran species reaching southern Y.T. and to be looked for on the east slope of southern Mackenzie Mts., N.W.T.

Map 972

Rhinanthus L. Yellow-rattle

Rhinanthus borealis (Sterneck) Chab.
Annual; stems simple or often freely branched above, 3 to 5 dm tall; leaves opposite, lanceolate and sessile, the margins serrate. Calyx ter including the broad, membranaceous wing. membranaceous and much inflated in fruit, hence the vernacular name; the corolla pale yellow and barely exserted. Seeds 4 mm in diame-

Dry gravelly slopes often pioneering on disturbed soil.

General distribution: N. American boreal forest species; in the Mackenzie Valley north to about lat. 65°.

Fig. 841 Map 973

Synthyris Benth.

Synthyris borealis Pennell
Tufted perennial from an ascending and often freely branched rhizome. Basal leaves petioled, the kidney-shaped blade glabrous except for the ciliated margins. Flowering stems erect-ascending, 1 to 2 dm tall, usually with 2 to 3 pairs of small, sessile leaves. Inflorescence an elongating spike, the small deep-blue flowers barely emerging behind white-hirsute bracts. Capsule heart-shaped, containing few and pale brown, smooth seeds.

Alpine herbmats.

General distribution: Endemic to central Alaska and Y.T., barely entering the District of Mackenzie in the Richardson Mts.

Map 974

Veronica L. Speedwell

Ours low and mainly perennial herbs with opposite leaves and small, blue or white flowers solitary in leaf axils or in terminal or axillary racemes. Calyx and corolla 4-lobed; stamens 2, and the capsule obcordate, flattened, few- to many-seeded.

a. Annual; flowers in the axils of alternate leaves .... *V. peregrina*
a. Perennial
  b. Racemes terminal .... *V. Wormskjoldii*
  b. Racemes lateral
    c. Leaves linear, entire and sessile .... *V. scutellata*
    c. Leaves lanceolate or narrowly ovate and serrate, short-petioled ...... *V. americana*

Veronica americana Schwein.
Stems decumbent, rooting at the base; racemes few to several from the axils of the uppermost leaves; flowers small, corolla bluish-violet.

In wet places by streams or by springs.

General distribution: N. America: Nfld. to Alaska. In the N.W.T. known only from hot spring meadows in the southern Mackenzie Mts.

Fig. 842 Map 975

Veronica peregrina L.
var. **xalapensis** (HBK.) St. John & Warren
Glandular pubescent weedy annual with simple or branched 1 to 3 dm tall stems from a spindly, rooting base. Leaves sessile or nearly so, linear-lanceolate, with entire or somewhat dentate margins. The small whitish flowers solitary in the axils of the uppermost and bract-like leaves.

Moist places of settled areas.

General distribution: N. America from Que. to Alaska; in the N.W.T. known from a few collections along the upper Mackenzie R. north to Ft. Simpson.

Fig. 843 Map 976

Veronica scutellata L.
Glabrous or somewhat pubescent perennial with weak, decumbent or ascending and leafy stems rooting at the nodes; leaves linear; flowers small, pale blue or white, in axillary racemes much elongated in fruit; capsule flat, deeply notched.

Wet thickets, often near springs.

General distribution: Circumpolar, non-arctic.

Fig. 844 Map 977

Veronica Wormskjoldii Roem. & Schult.
Small dark green herbs with slender, creeping rhizomes and ascending, leafy stems; leaves mostly opposite, long-ciliate, shallowly toothed or entire, blackened in drying. Inflorescence racemose, the flowers small with dark blue corollas. Capsules bluish-black with a greenish sheen, and long gland-tipped hairs. Seeds small, bright yellow.

In wet mossy places by brooks and on snowbeds.

Two geographical races have been recognized within our area:

a. Leaves oval-elliptic, shallowly toothed; raceme dense, not elongating in fruit. . . . . . ssp. *Wormskjoldii* (*V. alpina* var. *unalaschcensis* Cham. & Schlect.)

a. Leaves narrowly lanceolate, entire; raceme interrupted, elongating in fruit . . . . . ssp. *alterniflora* (Fern.) Pennell (*V. alpina* var. *alterniflora* Fern.)

General distribution: ssp. *Wormskjoldii*, arctic-alpine, Alaska to Lab.; ssp. *alterniflora*, Cordilleran alpine, barely entering the District of Mackenzie in the central Mackenzie Mts.

Fig. 845 (ssp. *Wormskjoldii*) Map 978 (ssp. *Wormskjoldii*), 979 (ssp. *alterniflora*)

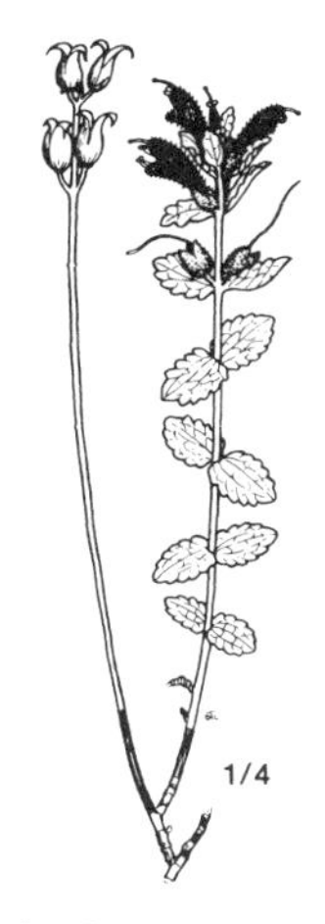

819. *Bartsia alpina*

820. *Castilleja caudata*

821. *Castilleja elegans*

822. *Castilleja hyperborea*

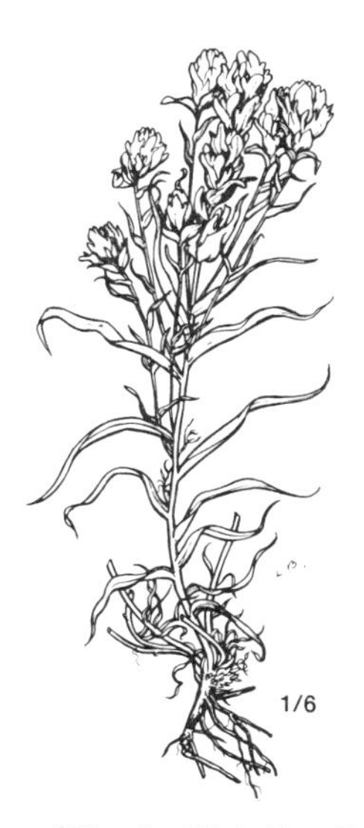

823. *Castilleja Raupii*

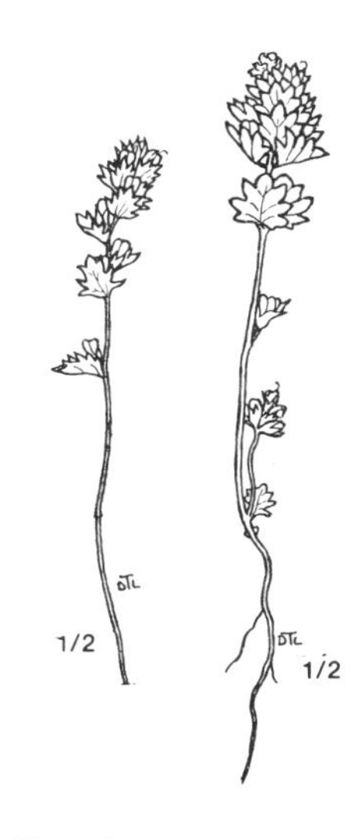

824. *Euphrasia arctica*

825. *Euphrasia subarctica*

826. *Lagotis Stelleri*

827. *Limosella aquatica*

828. *Pedicularis arctica*

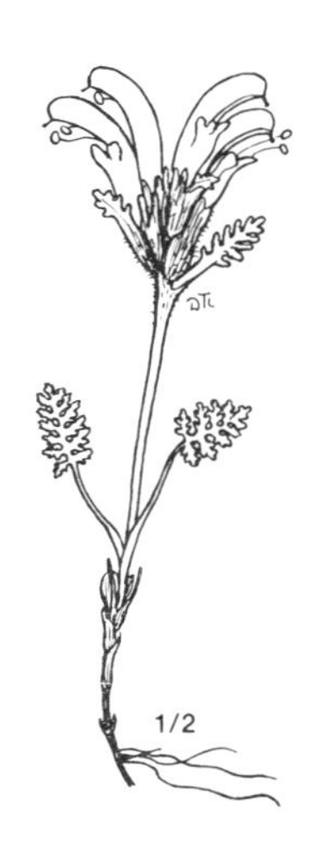

829. *Pedicularis capitata*

830. *Pedicularis flammea*

831. *Pedicularis groenlandica*

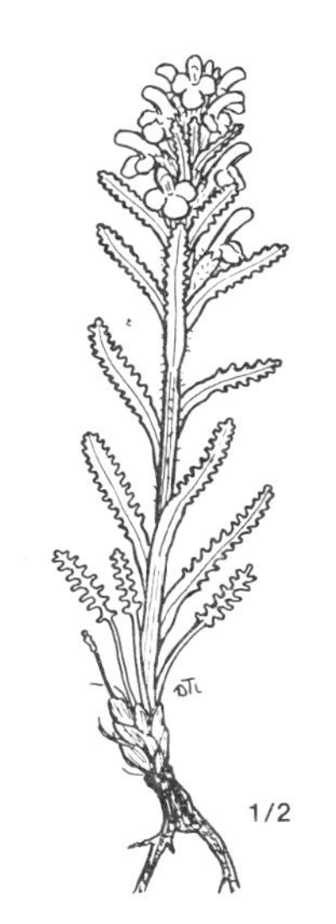

832. *Pedicularis hirsuta*

833. *Pedicularis labradorica*

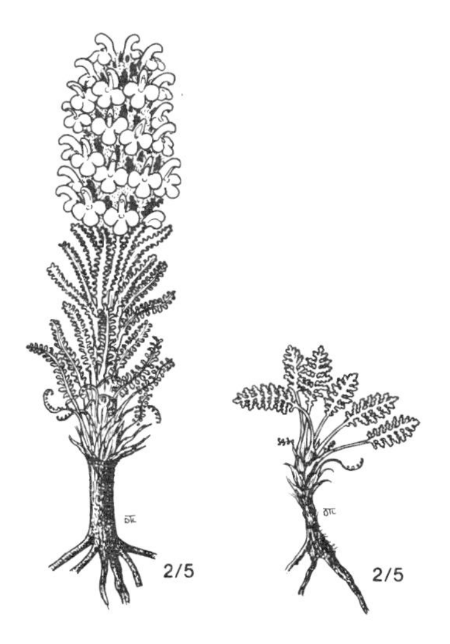

834. *Pedicularis lanata*

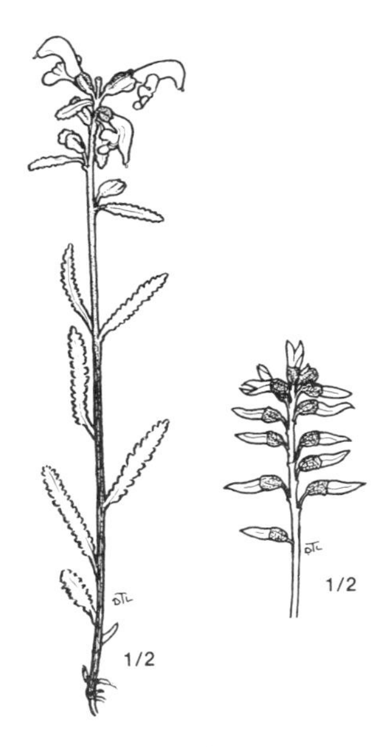

835. *Pedicularis lapponica*

836. *Pedicularis Oederi*

837. *Pedicularis parviflora*

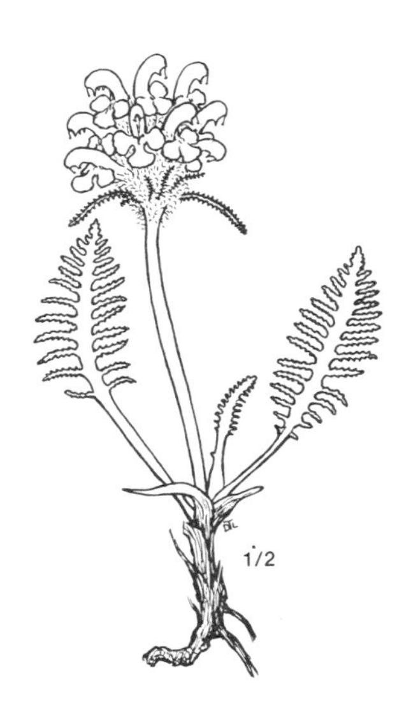

838. *Pedicularis sudetica*

839. *Pedicularis verticillata*

840. *Penstemon Gormanii*

841. *Rhinanthus borealis*

842. *Veronica americana*

843. *Veronica peregrina* var. *xalapensis*

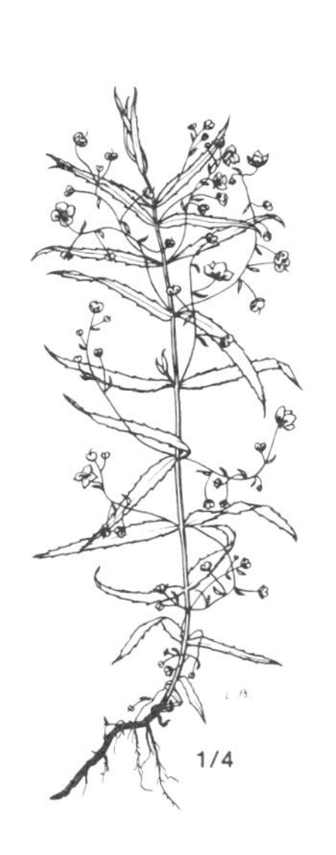

844. *Veronica scutellata*

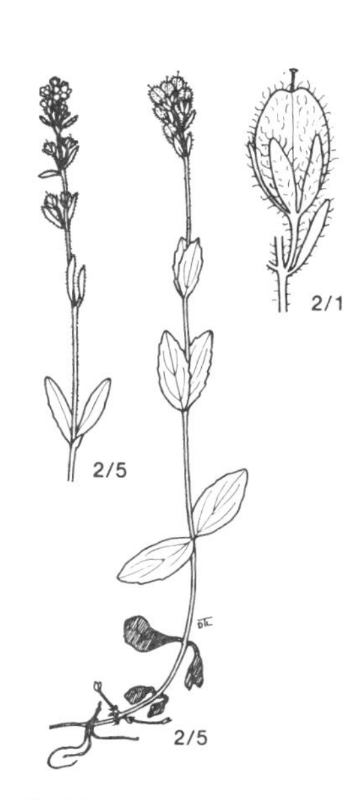

845. *Veronica Wormskjoldii* ssp. *Wormskjoldii*

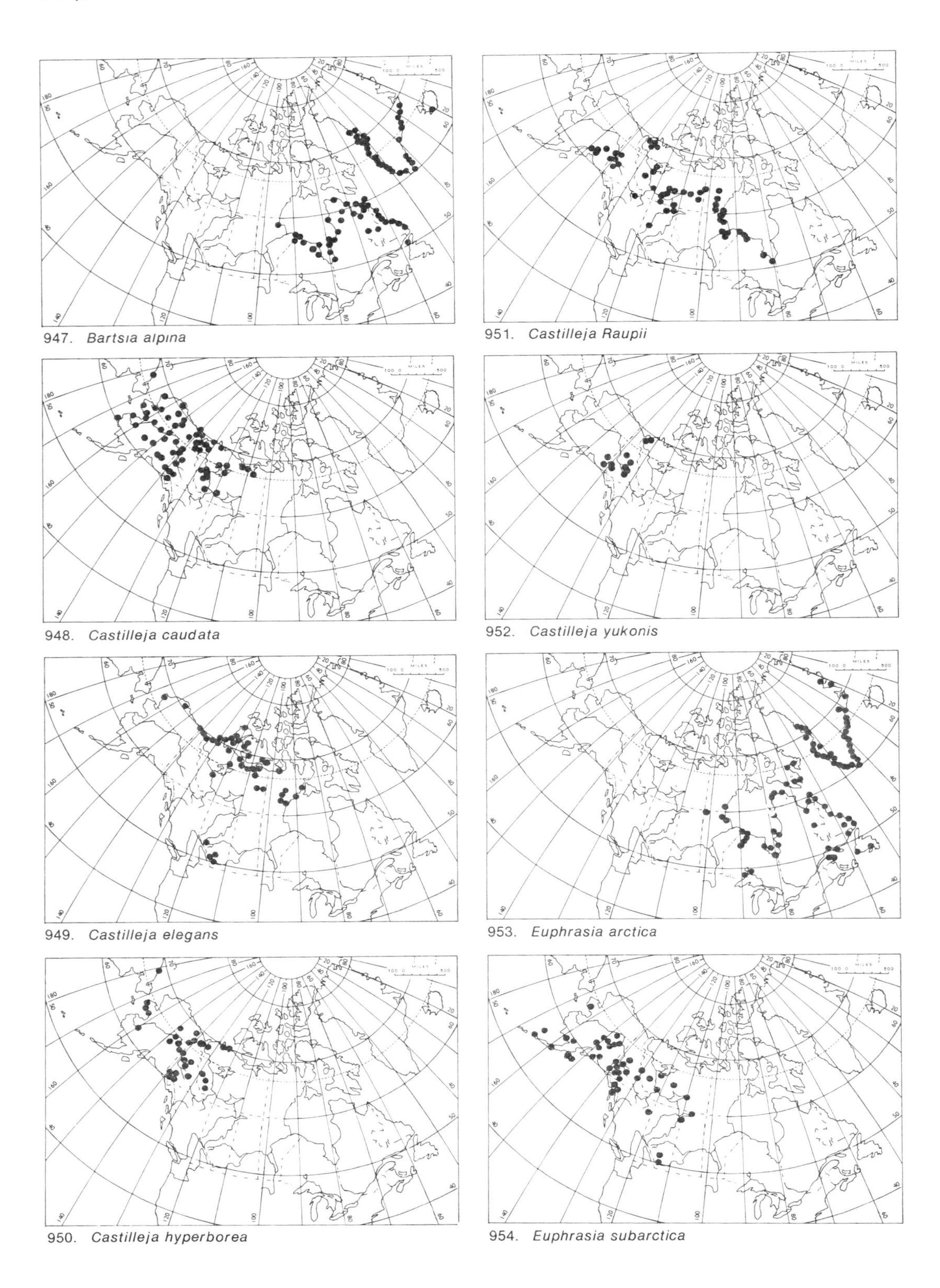

947. *Bartsia alpina*

948. *Castilleja caudata*

949. *Castilleja elegans*

950. *Castilleja hyperborea*

951. *Castilleja Raupii*

952. *Castilleja yukonis*

953. *Euphrasia arctica*

954. *Euphrasia subarctica*

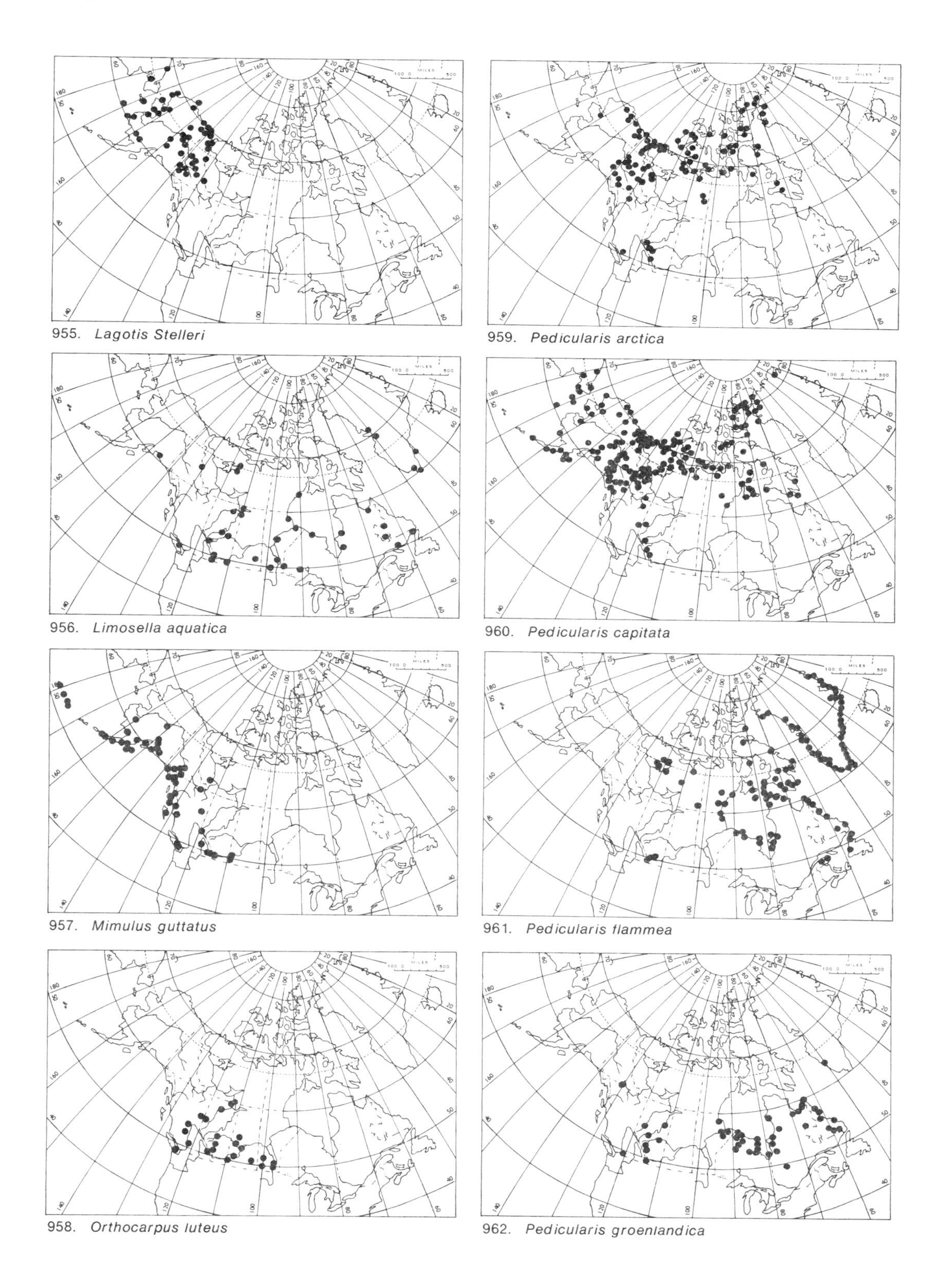

955. *Lagotis Stelleri*

956. *Limosella aquatica*

957. *Mimulus guttatus*

958. *Orthocarpus luteus*

959. *Pedicularis arctica*

960. *Pedicularis capitata*

961. *Pedicularis flammea*

962. *Pedicularis groenlandica*

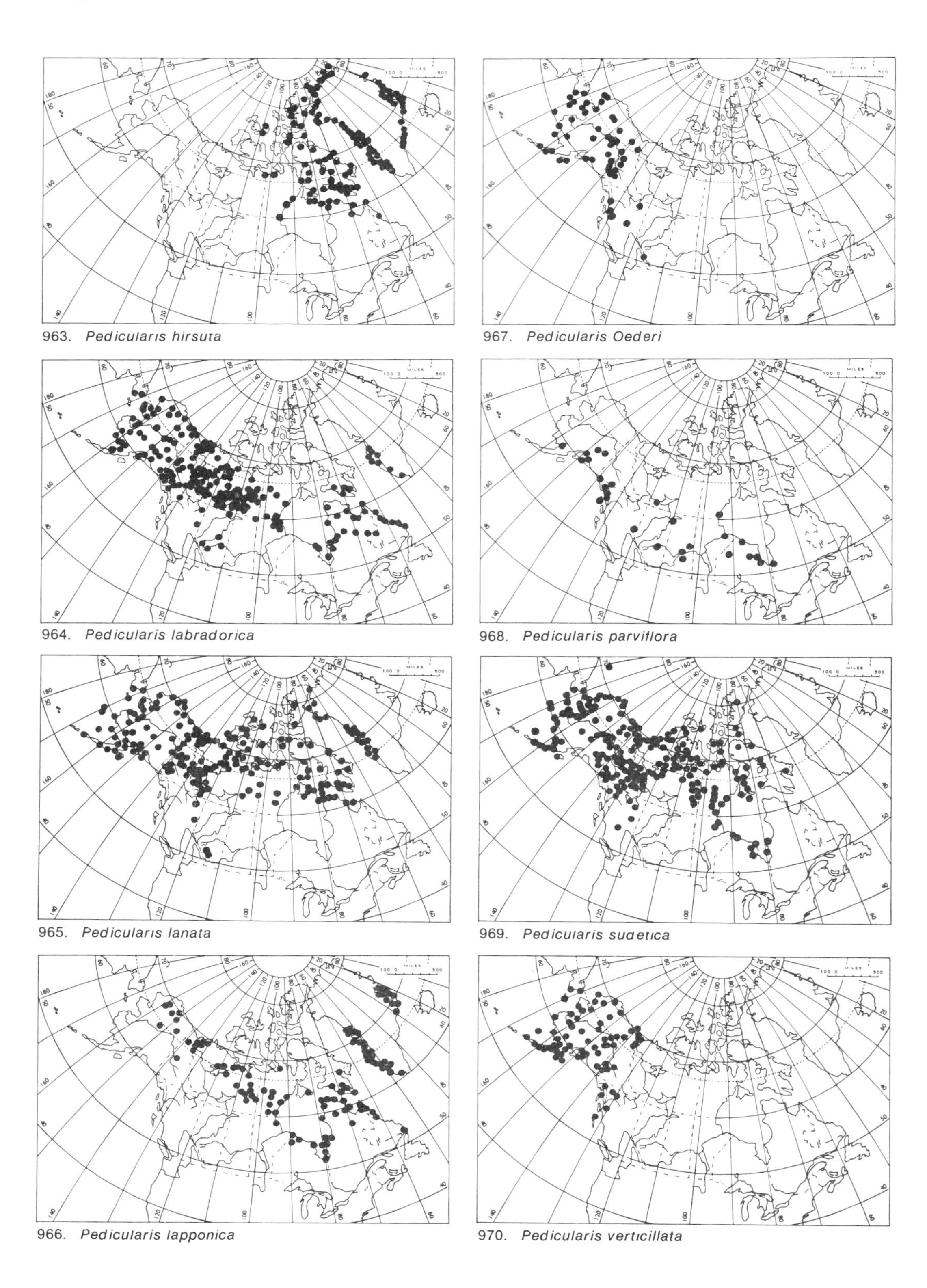

963. *Pedicularis hirsuta*

964. *Pedicularis labradorica*

965. *Pedicularis lanata*

966. *Pedicularis lapponica*

967. *Pedicularis Oederi*

968. *Pedicularis parviflora*

969. *Pedicularis sudetica*

970. *Pedicularis verticillata*

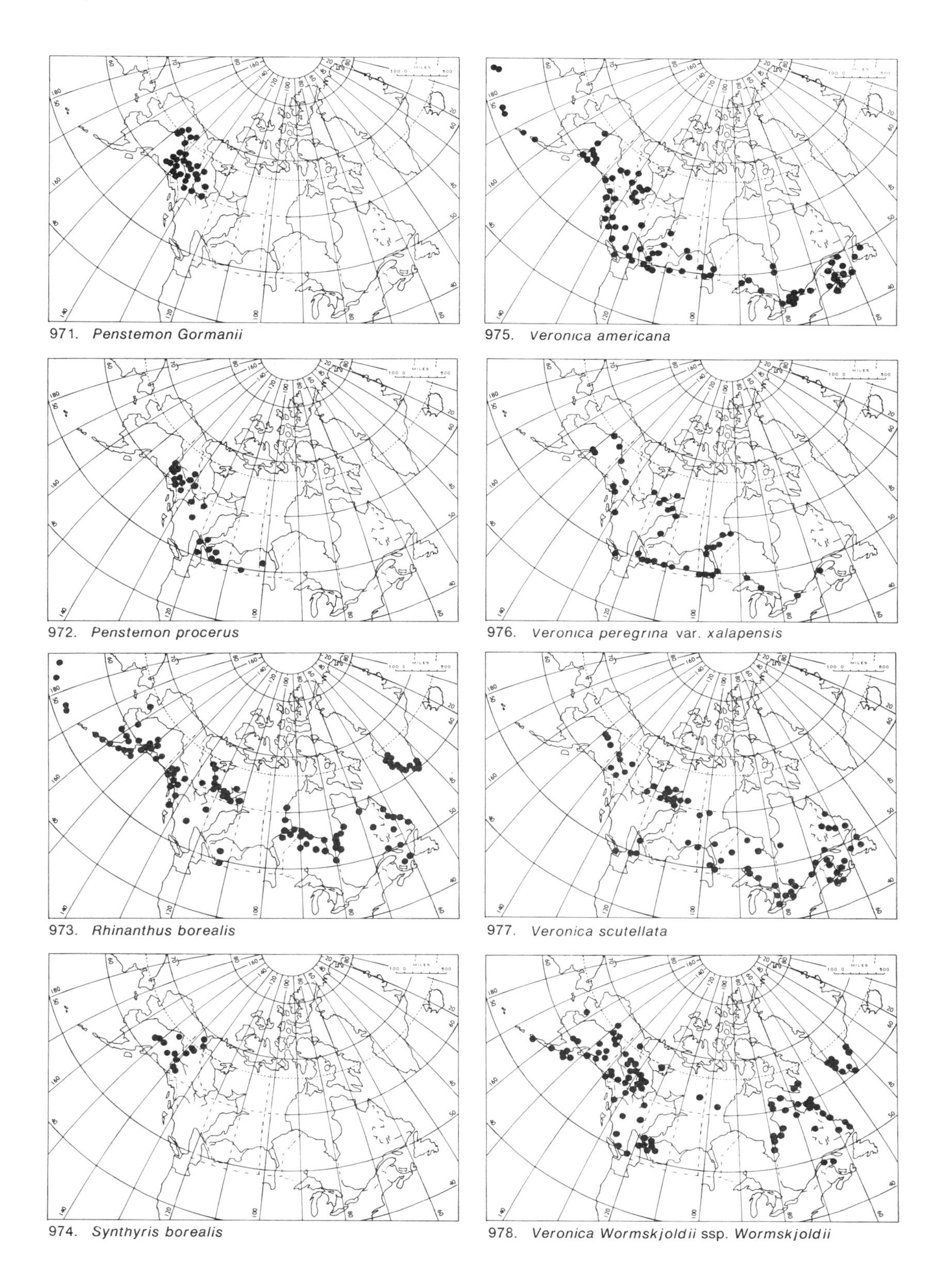

971. *Penstemon Gormanii*

972. *Penstemon procerus*

973. *Rhinanthus borealis*

974. *Synthyris borealis*

975. *Veronica americana*

976. *Veronica peregrina* var. *xalapensis*

977. *Veronica scutellata*

978. *Veronica Wormskjoldii* ssp. *Wormskjoldii*

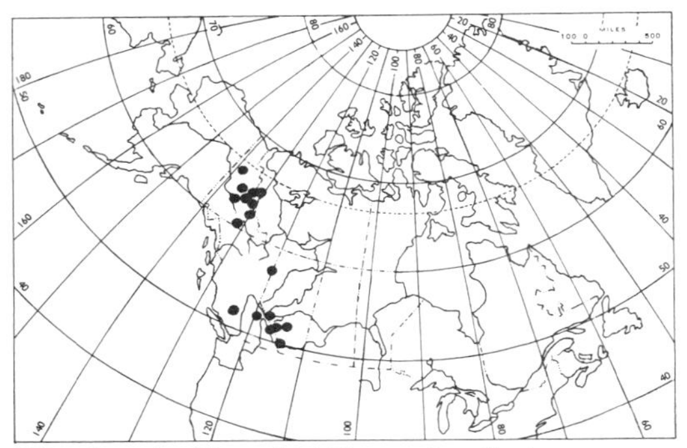
979. *Veronica Wormskjoldii* ssp. *alterniflora*

OROBANCHACEAE Broom-Rape Family

Boschniakia C. A. Mey.

Boschniakia rossica (Cham. & Schlecht.) Fedtsch.
Root parasite lacking freen foliage. The stems glabrous, club-like, singly, or a few together from a short and thick base by which they are attached to the host plant. The brownish flowers in a dense, terminal spike, each in the axil of a brown, scale-like leaf; seeds very small and numerous.

Parasitic on the roots of *Alnus crispa, Picea glauca* but occasionally on some other woody plants.

General distribution: Amphi-Beringian; boreal forest species ranging eastward over Alaska, Y.T. and the District of Mackenzie, northward to slightly beyond the limit of coniferous forest.

Fig. 846 Map 980

846. *Boschniakia rossica*

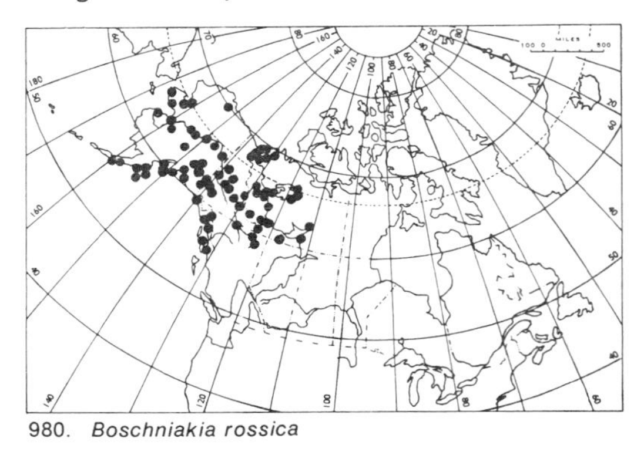
980. *Boschniakia rossica*

LENTIBULARIACEAE Bladderwort Family

Small green aquatic or terrestrial herbs equipped to trap and digest small insects and other tiny animals by means of small, submersed, bladder-like traps, or by sticky leaves. Flowers irregular, the corolla deeply 2-lipped, with a spur from its base.

a. Terrestrial plants with fleshy basal leaves and purplish flowers . . . . *Pinguicula*
a. Aquatic submerged (ours) plants with dissected leaves and yellow emergent flowers . . . . *Utricularia*

Pinguicula L. Butterwort

Dwarf, scapose perennial herbs with a rosette of entire, elliptic-spatulate, yellowish-green and succulent leaves that are glandular-viscid on the upper surface, and act as traps for small insects held and digested by the mucilaginous secretions. Scapes one to several, the corolla bright violet, 2-lipped, the lower lip terminating in a straight spur.

a. Scape glandular-villous; corolla less than 10 mm long . . . . *P. villosa*
a. Scape essentially glabrous; corolla 15 mm or more long . . . . *P. vulgaris*

Pinguicula villosa L.
Basal leaves less than 10 mm long. Scape slender, solitary, commonly less than 10 cm tall, distinctly villous in the lower half. Flowers 10 mm long or less.

Hummocks in sphagnum bogs.

General distribution: Circumpolar, sub-arctic with large gaps; north to or slightly beyond the limit of trees.

Fig. 847 Map 981

Pinguicula vulgaris L.
Similar to *P. villosa* but larger and more robust. Basal leaves up to 2.5 cm long, the scapes stouter and essentially glabrous, and the corollas usually 15 mm long.

In damp calcareous soil by the edge of small brooks or ponds or by seepages.

General distribution: Circumpolar; sub-arctic alpine, north beyond the limit of trees.

Fig. 848 Map 982

Utricularia L. Bladderwort

Non-rooting, immersed or floating (ours) aquatic plants with finely dissected pale green leaves and small, colourless bladder-like "traps" in which tiny aquatic animals are effectively trapped and, in time, decomposed and "digested". Flowers 2-lipped and spurred, on naked scapes raised above the water.

In Canada four species of this otherwise subtropical or temperate genus are wide-ranging north to, or even beyond, the limit of trees and always restricted to small, eutrophic and shallow ponds.

Under present climatic conditions *U. minor* and *U. ochroleuca* rarely flower in Canada north of Lat. 60°; *U. intermedia* and *U. vulgaris* flower regularly but so late in the season that seeds rarely, if ever, mature. In all four, however, turions or winterbuds are formed at the tip of the branches and, when the rest of the plant dies at the end of the season, sink to the bottom to sprout the following spring.

a. Tip and margin of leaf-segments lacking teeth or bristles . . . . *U. minor*
a. Tip and margins of leaf-segments with teeth or bristles
  b. Bladders mainly on separate and pale branches lacking leaves; scapes 2- 5-flowered . . . . *U. intermedia*
  b. Bladders on leafy branches
    c. Branches slender, mostly simple; leaf-segments flat . . . . *U. ochroleuca*
    c. Branches coarse, up to 1 m long, freely branching scapes; leaf-segments terete . . . . *U. vulgaris*

Utricularia intermedia Hayne
At once distinguished by having bladders on separate and leafless branches. Flowers 1.0 to 1.5 cm long.

Often in boggy or springy places.

General distribution: Circumpolar; Nfld. to Alaska, north to the limit of trees.

Fig. 849 Map 983

Utricularia minor L.
Tiny, delicate plant, submerged in shallow water or sometimes emergent on the wet margins of pools. Owing to its small size is very easily overlooked; within the northern parts of its range nearly always sterile.

General distribution: Circumpolar. Range similar to that of *U. intermedia* although less common; in the District of Mackenzie not reported from north of Great Bear L.

Fig. 850 Map 984

Utricularia ochroleuca Hartm.
Similar in size to *U. intermedia* from which it differs by having bladders also on the leafy branches.

The few collections from our area, all from the Mackenzie Basin, north to the Arctic Coast, are sterile.

General distribution: Circumpolar.

Fig. 851 Map 985

Utricularia vulgaris L.
Coarse plant with freely floating and flowering branches up to 1 m long and, perhaps the most common member of the genus in Continental Northwest Territories where, in common with the other three, it is almost entirely restricted to the Mackenzie Basin.

General distribution: Circumpolar.

Fig. 852 Map 986

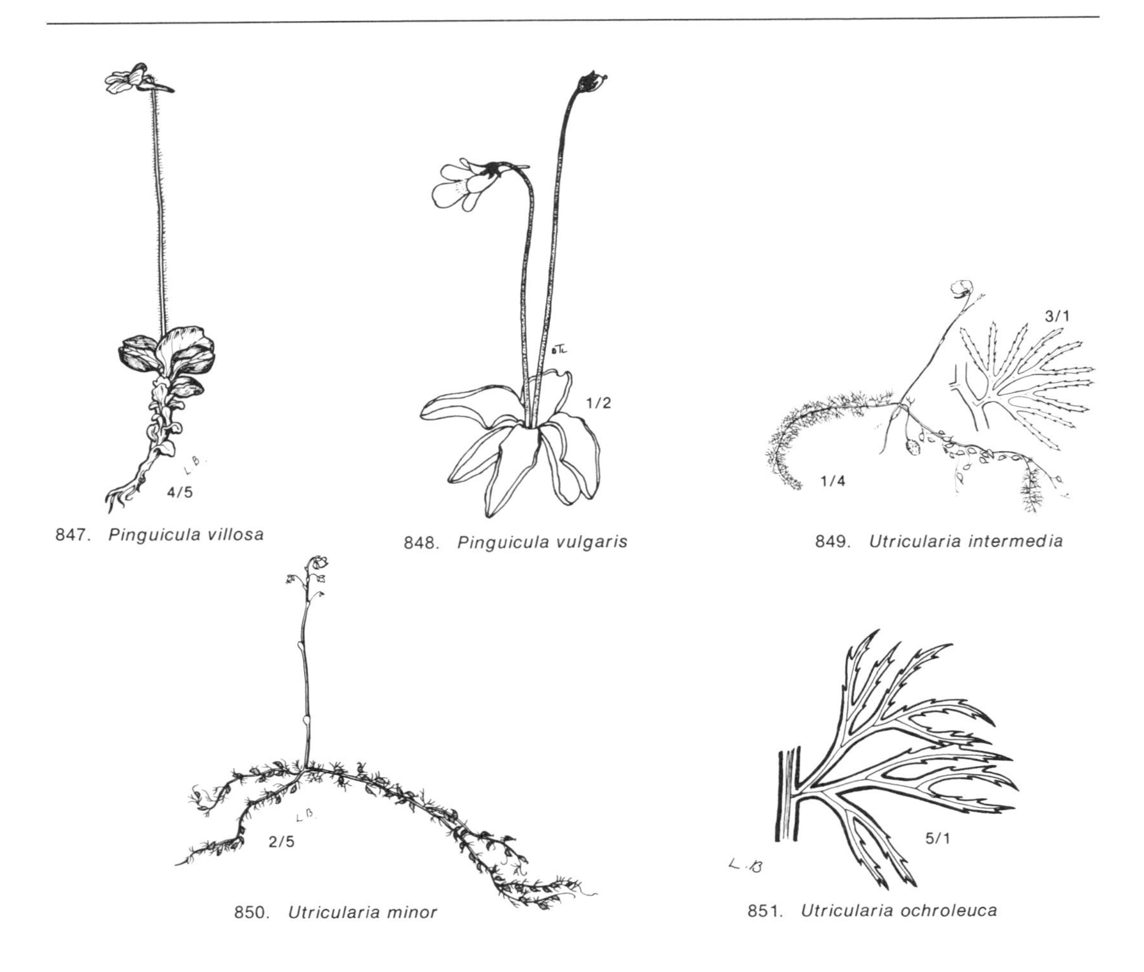

847. *Pinguicula villosa*

848. *Pinguicula vulgaris*

849. *Utricularia intermedia*

850. *Utricularia minor*

851. *Utricularia ochroleuca*

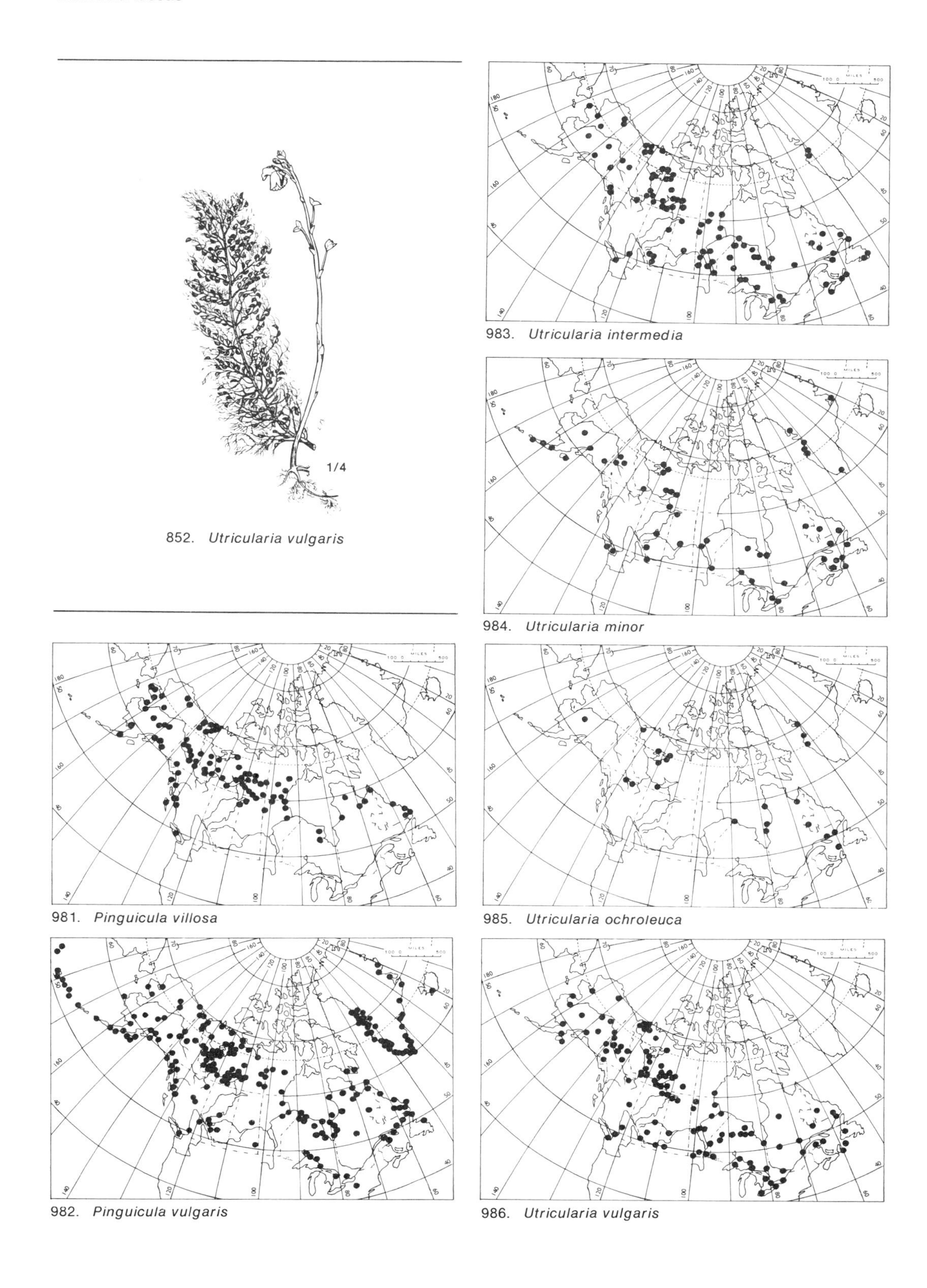

852. *Utricularia vulgaris*

983. *Utricularia intermedia*

984. *Utricularia minor*

981. *Pinguicula villosa*

985. *Utricularia ochroleuca*

982. *Pinguicula vulgaris*

986. *Utricularia vulgaris*

PLANTAGINACEAE Plantain Family
Plantago L. Plantain

Tufted, scapose, perennial herbs with linear or linear-lanceolate or oval leaves in radical rosettes, and small, regular 4-merous flowers in bracted heads or spikes. Capsule 2-locular, opening by a cup-shaped lid.

a. Leaves linear and somewhat fleshy; sea-shore plants; occasionally inland by saline springs . . . . *P. juncoides*
a. Leaves lanceolate or oval
  b. Leaf-blade lanceolate, gradually tapering into petiole
    c. Plant thinly pubescent; base of leaves brownish-woolly . . . . *P. eriopoda*
    c. Plant villous; leaf bases not woolly . . . . *P. canescens*
  b. Leaves oval, the blade abruptly tapering into a narrow petiole . . . . *P. major*

Plantago canescens Adams
P. septata Morris
Leaves linear-lanceolate, the blade 3 to 15 mm wide, distinctly rough from short, stiff pubescence, especially on the slender petioles. Scapes stiffly erect, 15 to 30 cm long, much longer than the leaves; spikes oblong-ovate, 2 to 4 cm long and 8 to 10 mm wide.

Riverbanks and calcareous cliffs and screes.

General distribution: Amphi-Beringian, extending to the northern Cordillera and the District of Mackenzie and north to the western islands of the Arctic Archipelago.

Fig. 853 Map 987

Plantago eriopoda Torr.
Leaves lanceolate, coriaceous, pubescent to nearly glabrous, 5 to 25 cm long from a thick crown more or less covered with brown woolly hairs; scapes one to several, exceeding the leaves; spikes loosely flowered, 3.5 to 20.0 cm long.

Saline seepages and by sea-shores.

General distribution: N. America: Que. to S. E. Alaska, north along the Mackenzie R. Valley to the Arctic Coast.

Fig. 854 Map 988

Plantago juncoides Lam.
var. **glauca** (Hornem.) Fern.
Seaside Plantain
Leaves linear, 2 to 3 mm wide, fleshy and glabrous, equalling or slightly shorter than the erect, ascending or spreading, 5 to 10 cm high, dark purple, sparingly hirsute scapes. Spikes short, cylindrical, 1 to 2 cm long and 5 to 6 mm wide.

Cliffs and sea-beaches, or occasionally by saline springs inland.

General distribution: Amphi-Atlantic, subarctic. With us thus far reported only from the west coast of Hudson Bay and from the north shore of McTavish Bay, Great Bear L.

Fig. 855 Map 989

Plantago major L.
Common as a road-side weed now well established in and near towns and settlements along the Mackenzie River and tributaries, northward nearly to the Mackenzie Delta. However, in the District of Mackenzie and also in Yukon Territory and Alaska a plant reported as *P. major* var. *asiatica* has been observed as "decidedly a component of the aboriginal flora".

General distribution: *P. major s. lat.,* circumpolar road-side weed.

Fig. 856 Map 990

853. *Plantago canescens*

854. *Plantago eriopoda*

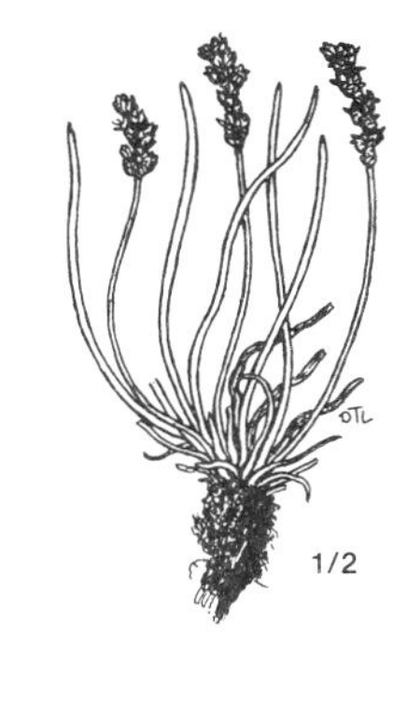

855. *Plantago juncoides* var. *glauca*

856. *Plantago major*

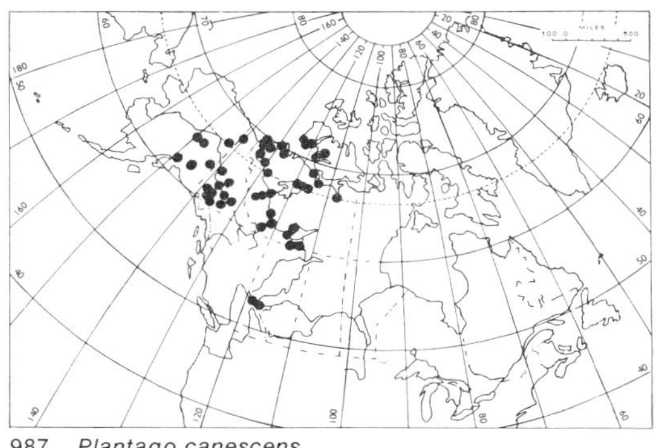

987. *Plantago canescens*

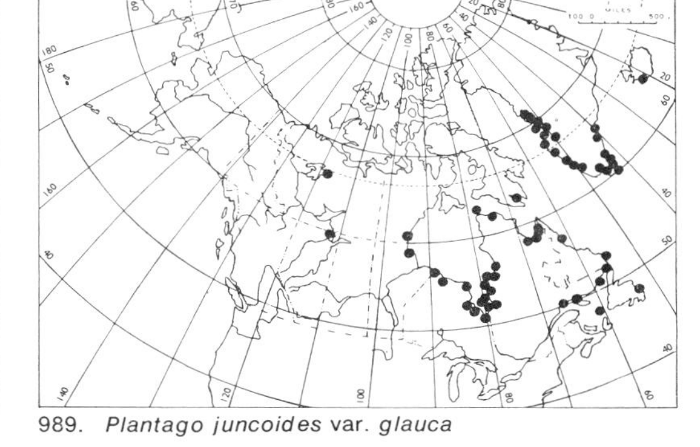

989. *Plantago juncoides* var. *glauca*

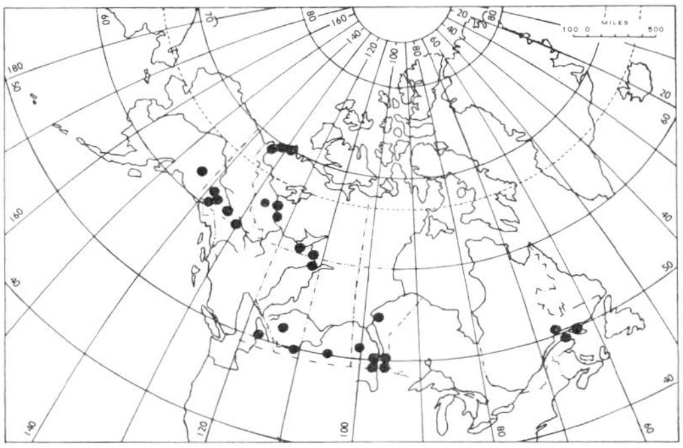

988. *Plantago eriopoda*

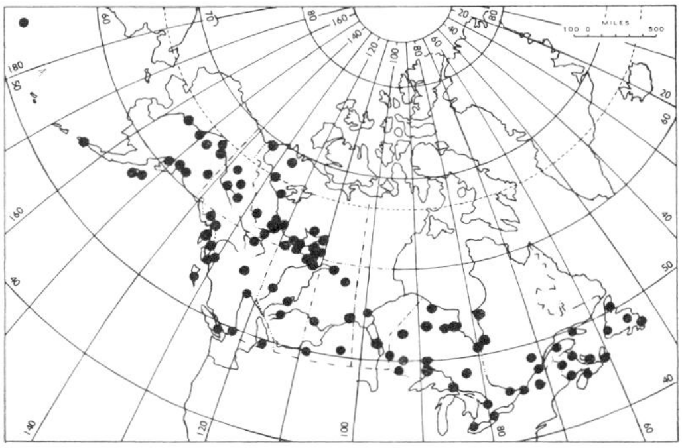

990. *Plantago major*

RUBIACEAE Madder Family

Galium L. Bedstraw

Slender perennial (ours) herbs with 4-angled stems, whorled leaves and tiny flowers in few-flowered lateral or terminal cymes. Corolla 3- to 4-lobed, greenish or white; the fruit globular, bristly or smooth, when ripe dividing into two indehiscent 1-seeded carpels.

a. Leaves in whorls of 6, broadly lanceolate, mucronate, 7 to 8 mm broad and 15 mm long . . . . . *G. triflorum*
a. Leaves in whorls of 4
  b. Plant stout, stiffly erect and tufted from a slender rhizome . . . . . *G. boreale*
  b. Plant weak and slender, of matted growth
    c. Corollas 3-lobed
      d. Leaves broadly oval, 10 mm broad and 20 mm long . . . . . *G. kamtschaticum*
      d. Leaves 1 to 2 mm broad and rarely 10 mm long
        e. Stems and leaf-margins retrorsely scabrous
          f. Pedicels glabrous . . . . . *G. tinctorium*
          f. Pedicels retrorse scabrous . . . . . *G. trifidum*
        e. Stems and leaf-margins smooth . . . . . *G. Brandegei*
    c. Corollas 4-lobed, leaves reflexed . . . . . *G. labradoricum*

Galium boreale L.
Stems stout, 3 to 7 dm tall, simple or branched. Leaves 3 to 4 cm long, 3-nerved, broadest near the base. The flowers pure white, drying yellow, in richly branched cymes. Fruit hairy.

Dry open or gravelly places, often pioneering in disturbed soil.

General distribution: Circumpolar, north along the Mackenzie Valley to the Arctic Coast.

Fig. 857 Map 991

Galium Brandegei Gray
Totally glabrous dwarf species with matted, 5 to 10 cm long, simple or branching stems. Leaves oblanceolate, 5 to 7 mm long. The small white or greenish flowers in 2- to 3-flowered cymes. Fruit black, glabrous.

Low and matted, in damp sphagnum bogs.

General distribution: Eastern sub-arctic alpine, reaching the west coast of Hudson Bay.

Map 992

Galium kamtschaticum Steller
Stems 10 to 20 cm tall, solitary or few together from slender, filiform rhizome and stolons. Leaves 3-nerved, broadly oval, usually in 3 whorls, the uppermost the largest, 2 to 3 cm long. Flowering peduncles 1 to 3, terminal, each with 1 to 3 flowers. The fruit densely covered by shiny hooked bristles.

General distribution: In Canada north of lat. 50° known from a single station east of the Mackenzie Delta, geographically and climatically remote from its nearest stations, in damp cool woods, on the Pacific Coast of Alaska, and from its even more remote and isolated stations in Nfld., northwest Que., N.B., eastern Ont. and northern N. Y.

Fig. 858 Map 993

Galium labradoricum Wieg.
Stems slender, smooth, freely branching, erect-ascending, 2 to 3 dm tall; the linear to narrowly oblanceolate leaves characteristically reflexed in age. Cymes terminal, few flowered, the corollas white, 4-lobed, about 2 mm in diameter.

Moist, mossy thickets and open woods.

General distribution: Nfld. to Alta., northward barely beyond lat. 60° N.

Map 994

Galium tinctorium L.
var. **subbiflorum** (Wieg.) Fern.
Stems slender, scabrous, ascending and commonly matted, freely branched. Leaves narrowly oblanceolate, rarely over 10 mm long, not reflexed, their margins scabrous. At once distinguished from *G. labradoricum* by its 3-lobed corolla, and from *G. trifidum* by its smooth peduncles.

Wet places in thickets and open woods.

General distribution: Nfld. to Alaska, barely entering the District of Mackenzie, south of Great Slave L.

Fig. 859 Map 995

Galium trifidum L.
Similar to *G. tinctorium* but its stems weaker and usually densely matted, the leaves somewhat broader and the flowering peduncles at least somewhat scabrous.

Common, although no doubt often overlooked, among tall sedges in wet woodland bogs.

General distribution: Circumpolar; wide ranging from Nfld. to Alaska, north to the limit of trees.

Fig. 860 Map 996

Galium triflorum Michx.
Sweet-scented Bedstraw
Plants with weak, mainly simple, 5 to 10 dm tall stems. With us at once distinguished from other members of the genus by its broad, distinctly cuspidate leaves in whorls of six. In drying, the entire plant is sweet-scented.

Openings in rich woods.

General distribution: Circumpolar; Nfld. to Alaska; in the District of Mackenzie north to lat. 62°N.

Fig. 861 Map 997

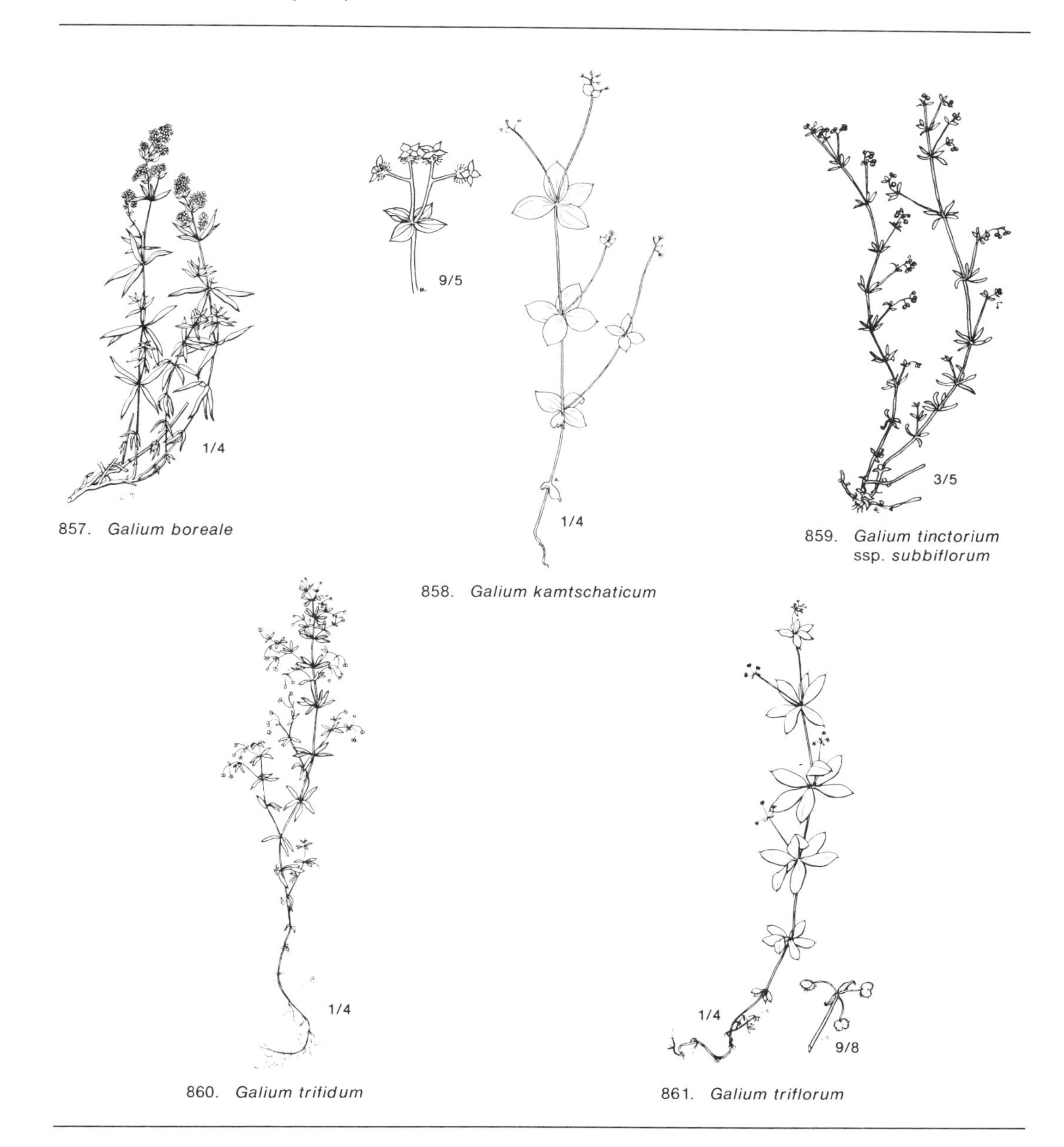

857. *Galium boreale*

858. *Galium kamtschaticum*

859. *Galium tinctorium* ssp. *subbiflorum*

860. *Galium trifidum*

861. *Galium triflorum*

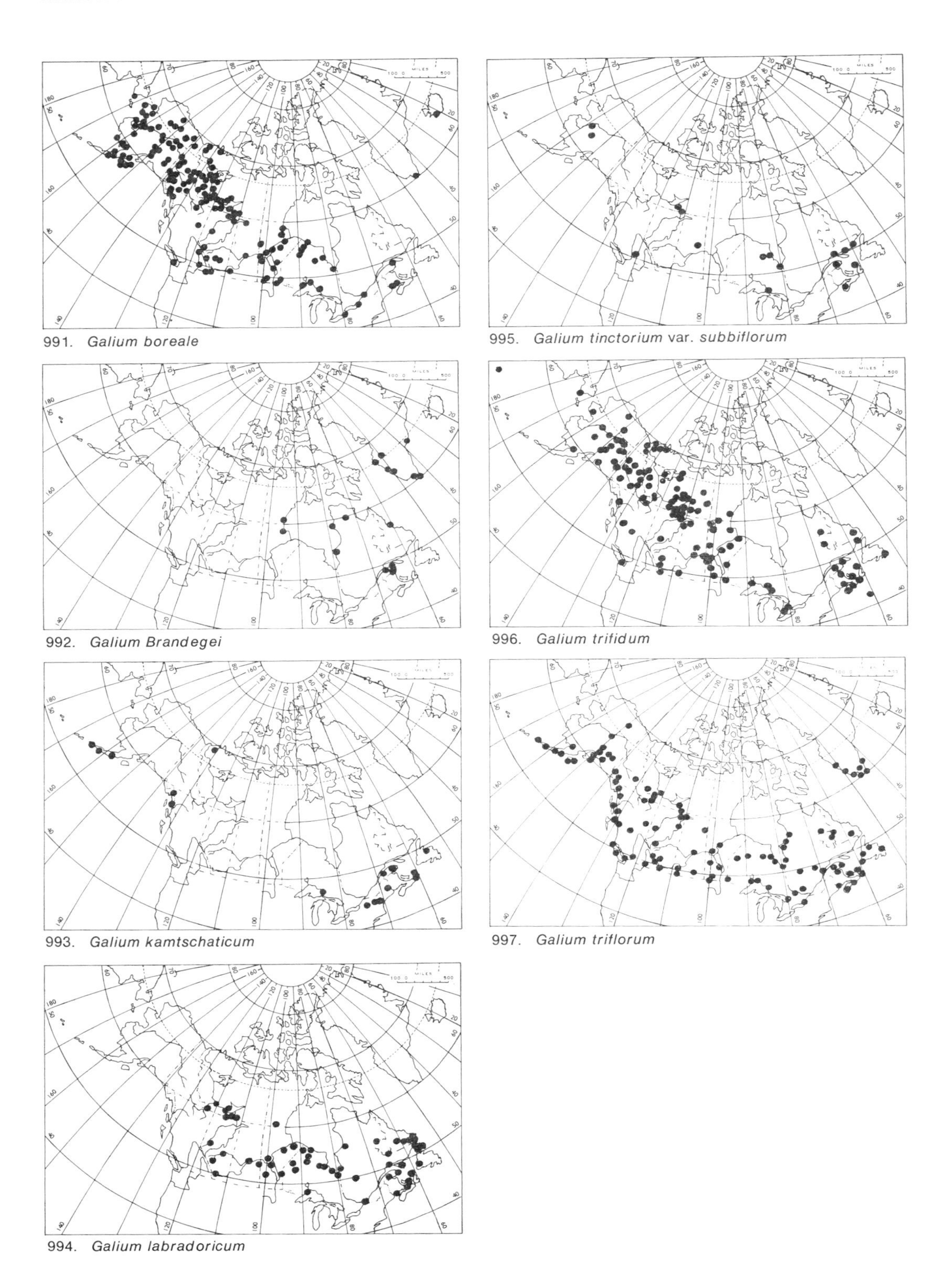

991. *Galium boreale*

992. *Galium Brandegei*

993. *Galium kamtschaticum*

994. *Galium labradoricum*

995. *Galium tinctorium* var. *subbiflorum*

996. *Galium trifidum*

997. *Galium triflorum*

CAPRIFOLIACEAE Honeysuckle Family

Erect or creeping shrubs with opposite leaves. Corolla tubular or rotate. Fruit a berry, or a dry, 1-seeded capsule.

a. Erect shrubs; fruit berry-like
  b. Corolla rotate, flowers numerous in compound cymes . . . . *Viburnum*
  b. Corolla funnel-form or campanulate
    c. Corolla tubular, fruit red, several seeded . . . . *Lonicera*
    c. Corolla short-campanulate; fruit white, 2-seeded . . . . *Symphoricarpus*
a. Low, trailing dwarf shrub; fruit dry . . . . *Linnaea*

Linnaea Gronov. Twinflower

Linnaea borealis L.
var. **americana** (Forbes) Rehd.
Dwarf shrub with widely trailing and freely branching stems and winter-green, sub-sessile, oval, crenate leaves among which rise the slender, erect scapes forked above into two short peduncles, each bearing a sweet-scented and nodding flower; corolla funnel-shaped, creamy white or faintly rose-purple, 8 to 15 mm long. Fruit a tiny dry, hooked-bristly 1-seeded capsule.

The genus is monotypic and circumpolar-alpine; its name commemorates the Swedish botanist, Linnaeus, father of modern nomenclature, with whom it was a favorite.

Common in mossy and turfy openings in thickets or open woods northward to or slightly beyond the limit of trees.

General distribution: *L. borealis s. lat.* circumpolar; var. *americana* N. America.

Fig. 862 Map 998

Lonicera L. Honeysuckle

Lonicera dioica L.
var. **glaucescens** (Rydb.) Butters
Shrubs with freely branching, twining or ascending stems with white or yellowish, shreddy bark. Leaves entire, oblong-ovate, short-petioled or almost sessile, the uppermost perfoliate-connate. Flowers 2 to 4, yellowish-red, in short, terminal clusters. Fruit a several-seeded edible red berry.

Open woods.

General distribution: N. America; western Que. to B.C., north to lat. 62° in the District of Mackenzie.

Fig. 863 Map 999

Symphoricarpos Duham. Snowberry

Low, freely branching shrubs with entire, short-petioled, oval-elliptic, deciduous leaves. Flowers white or pinkish, in small, axillary or sub-terminal clusters. Fruit a 2-seeded, berry-like drupe.

a. Leaves thin; corolla lobes shorter than tube . . . . *S. albus*
a. Leaves thick, corolla lobes as long as tube . . . . *S. occidentalis*

Symphoricarpos albus (L.) Blake
Low, ascending, thicket-forming shrub up to 1 m tall, with reddish-brown, shreddy bark. Fruits depressed-globose, white.

Abundant in prairies and open aspen woods.

General distribution: N. America; Que. to B.C., and north slightly beyond lat. 60° on the Slave R.

Fig. 864 Map 1000

Symphoricarpos occidentalis Hook.
Similar to *S. albus* but a coarser and more robust shrub from a stout, stoloniferous base. The bark is reddish-brown, turning black in age, not shreddy. Fruit greenish-white.

Stony or gravelly places.

General distribution: N. America; Ont. to B.C., north along the Mackenzie R. to Ft. Simpson.

Map 1001

Viburnum L.

Viburnum edule (Michx.) Raf.
Erect, branching up to 2 m tall shrub with smooth, dark-grey bark. Leaves petioled, the blade oval, 4 to 7 cm long, the lowermost shallowly 3-lobed, the upper merely serrate. Flowers milky-white. The fruit is edible, bright red, juicy and acid when ripe.

Woodland thickets.

General distribution: N. America; wide ranging from Nfld. to Alaska, northward almost to the limit of trees.

Fig. 865 Map 1002

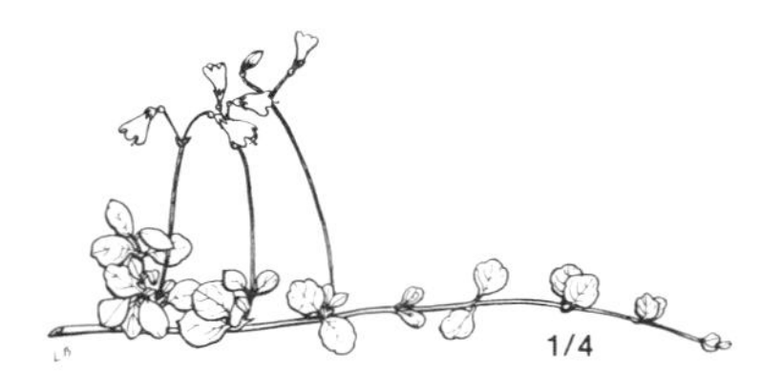

862. *Linnaea borealis* var. *americana*

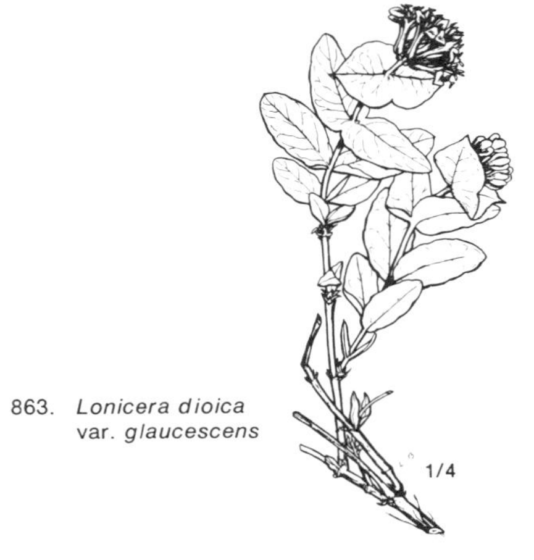

863. *Lonicera dioica* var. *glaucescens*

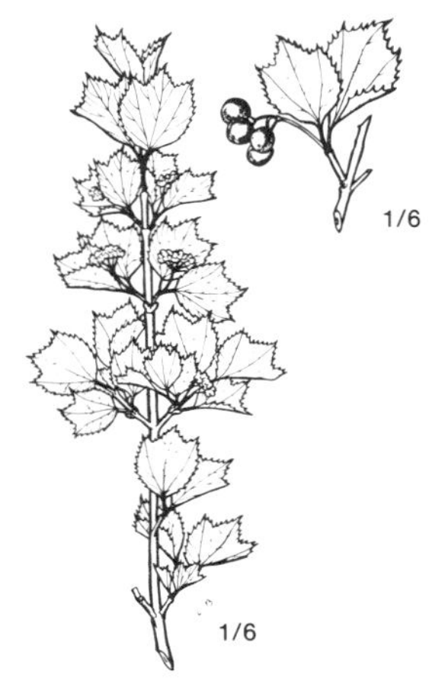

865. *Viburnum edule*

864. *Symphoricarpos albus*

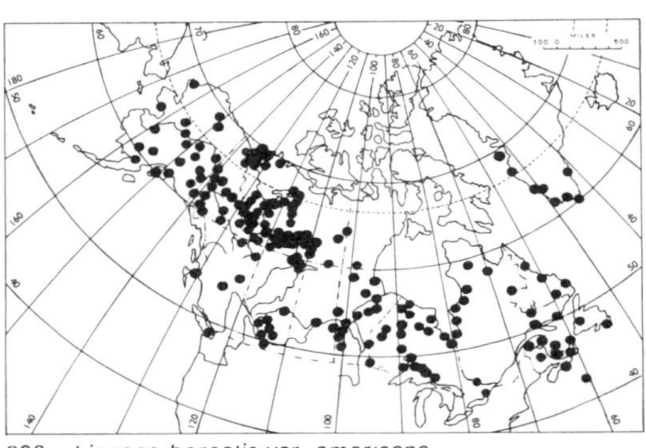

998. *Linnaea borealis* var. *americana*

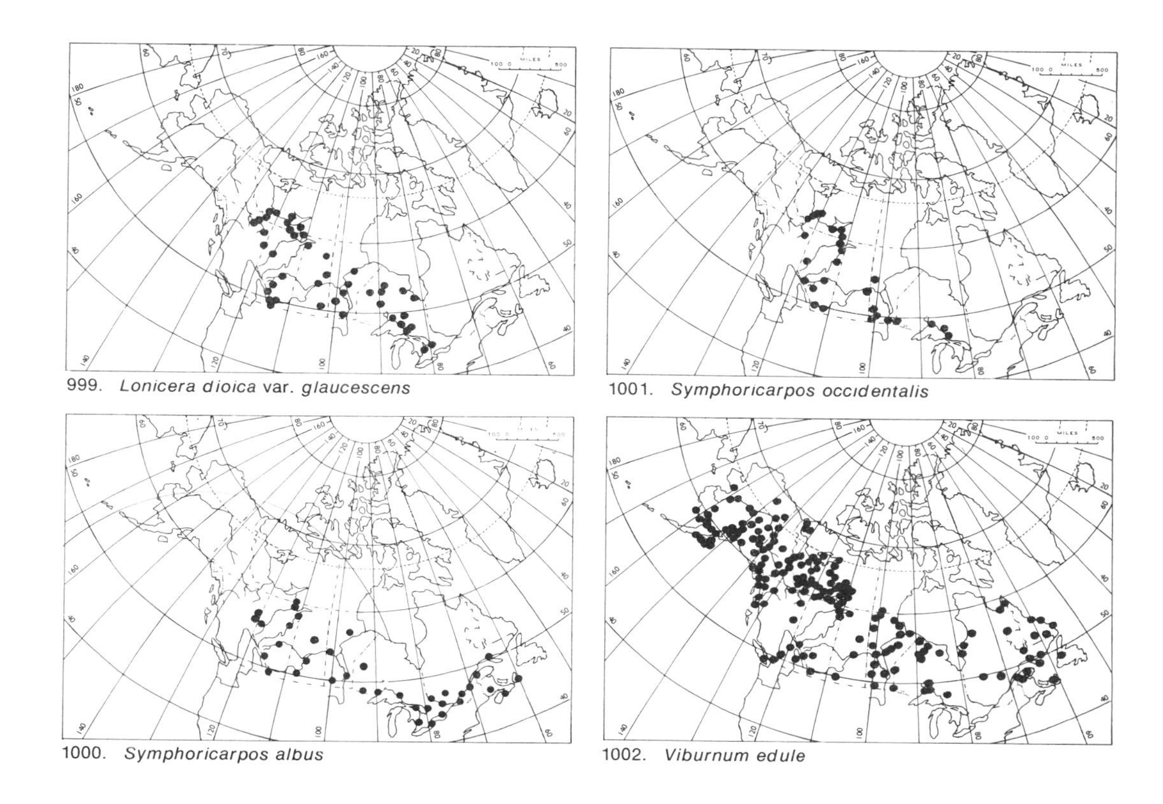

999. *Lonicera dioica* var. *glaucescens*

1001. *Symphoricarpos occidentalis*

1000. *Symphoricarpos albus*

1002. *Viburnum edule*

ADOXACEAE Moschatel Family

Adoxa L. Moschatel

Adoxa Moschatellina L.
Delicate, 10 to 15 cm tall perennial from a white scaly rhizome. Leaves ternately compound, mainly basal, the cauline opposite. Flowers small, yellowish-green, 3 or more in a dense cluster less than 1 cm broad, on a slender peduncle. The fruit a small, greenish, dry drupe.

Adoxa Moschatellina is monotypic in the genus *Adoxa* which, again, is the only genus of the family *Adoxaceae*.

Throughout its entire range its local distribution is peculiar and spotty, perhaps due to special light, soil and moisture requirements. In the Canadian northwest *Adoxa* may be common locally in rich leaf-mould in moist partly shaded alder and poplar woods.

General distribution: Circumpolar, with some large gaps.

Fig. 866 Map 1003

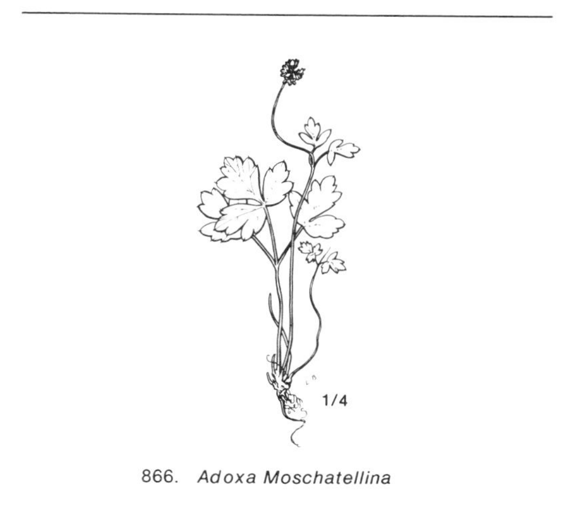

866. *Adoxa Moschatellina*

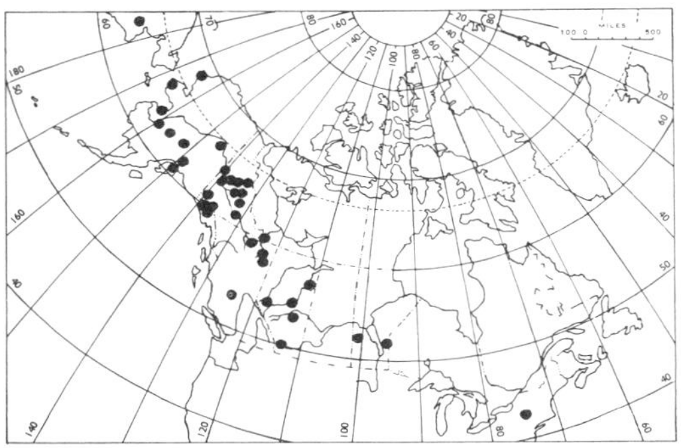

1003. *Adoxa Moschatellina*

VALERIANACEAE Valerian Family
Valeriana L. Valerian

Perennial, essentially glabrous, herbaceous plants from an ascending rhizome, with opposite, simple or pinnately divided leaves, and small white or pinkish flowers in a terminal head that, in fruit, elongates into an open and branching raceme or thyrsoid panicle. The rhizome, but also other parts of the plant, with a strong, musky odour that persists long after the plant is dried.

a. Stem leaves all pinnate; corolla 2 to 3 mm long .... *V. septentrionalis*
a. Stem leaves variously divided; corollas 5 to 7 mm long
  b. Bractlets of inflorescence ciliate .... *V. sitchensis*
  b. Bractlets of inflorescence glabrous .... *V. capitata*

Valeriana capitata Pall.
Stems 3 to 10 dm tall, simple, leafy, from an ascending simple or branching rhizome. The lowermost stem-leaves, and those of sterile shoots, slender petioled with an oval, or cordate and entire blade; middle or upper stem-leaves sessile and small, their blade narrowly laciniate, but not divided.

Moist, turfy wood-land bogs, tundra and river flats.

General distribution: Amphi-Beringian, barely reaching the District of Mackenzie, mainly along the east slope of Mackenzie and Richardson Mts.

Fig. 867 Map 1004

Valeriana septentrionalis Rydb.
Stems simple, erect, 5 to 10 dm tall; basal leaves petiolate, with oval or oblong, entire blade; stem-leaves 2 to 4 pairs, pinnate with 5 to 7 pairs of lanceolate leaflets. Flowering head 2 to 3 cm in diameter in anthesis, but soon elongating into an interrupted fruiting panicle, 16 to 20 cm long and half as broad.

Damp calcareous bogs and lake shore meadows.

General distribution: N. America; Nfld. to S. E. Y.T., rare and local in the upper Mackenzie drainage, north to Great Bear L.

Fig. 868 Map 1005

Valeriana sitchensis Bong.
Similar to *V. capitata* but more leafy, the lower and middle cauline leaves long-petioled, 3 to 5 lobed, the terminal leaflet, largest, ovate, and up to 5 cm long.

Moist, alpine and non-calcareous alpine meadows.

General distribution: A Cordilleran species reaching southern and central Y.T. and Alaska, but barely entering the District of Mackenzie along the east slope of the Mackenzie Mts.

Fig. 869 Map 1006

867. *Valeriana capitata*

868. *Valeriana septentrionalis*

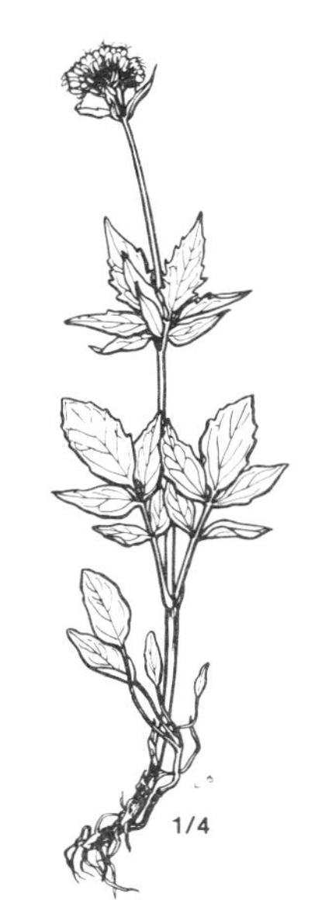

869. *Valeriana sitchensis*

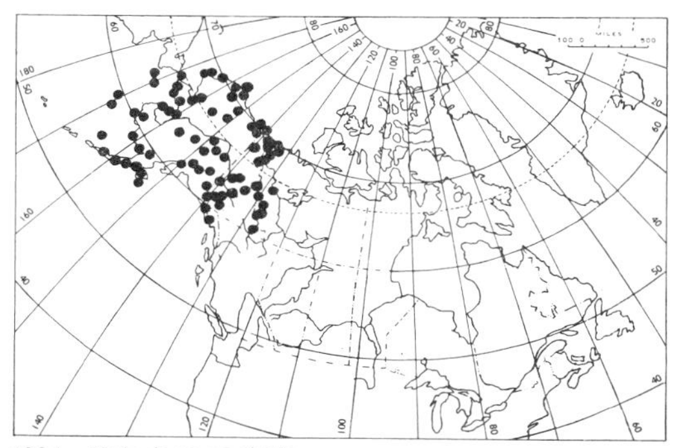

1004. *Valeriana capitata*

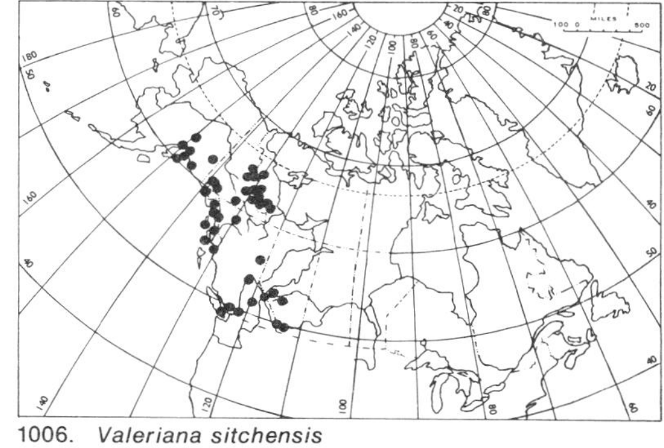

1006. *Valeriana sitchensis*

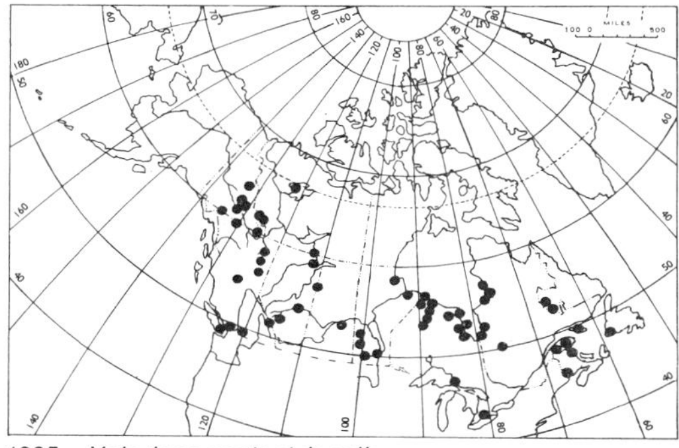

1005. *Valeriana septentrionalis*

CAMPANULACEAE Bluebell Family

Campanula L. Bellflower

Perennial herbs with milky juice and alternate leaves. Flowers bell-shaped, regular, with 5-parted, united corolla, 5 stamens, one pistil, and inferior ovary; capsule many-seeded, opening by slits or pores.

a. Calyx pubescent
  b. Leaves entire; corolla funnel-shaped, the lobes as long as the tube . . . . *C. uniflora*
  b. Leaves dentate; corolla bell-shaped, shallowly lobed . . . . *C. lasiocarpa*
a. Calyx glabrous
  c. Corolla rotate, deeply lobed . . . . *C. aurita*
  c. Corolla bell-shaped, shallowly lobed . . . . *C. rotundifolia*

Campanula aurita Greene
Stems slender, 10 to 30 cm tall, one to several, from a freely branching rhizome; leaves glabrous, all cauline, oblanceolate to narrowly lanceolate. Flowers solitary or in few-flowered racemes.

Common locally in turfy or gravelly places and perhaps always on calcareous soil.

General distribution: Endemic of interior and mainly alpine parts of Alaska, Y.T. and central western District of Mackenzie and northern B.C.

Fig. 870 Map 1007

Campanula lasiocarpa Cham.
Flowering stems mostly solitary, 5 to 12 cm tall, from a basal rosette of slender-petioled, lanceolate leaves. Flowers solitary, or very rarely with a second and smaller lateral flower; corolla 2.0 to 2.5 cm long, purplish blue, often drying white; calyx lobes linear, with one or two pairs of filiform teeth.

Alpine, gravelly tundra and mainly on non-calcareous soil.

General distribution: N. America; high mountains of Alta. and B.C., westernmost District of Mackenzie, Y.T. and Alaska, barely reaching E. Asia.

Fig. 871 Map 1008

Campanula rotundifolia L.
Stems usually numerous, from a freely branching rhizome, glabrous, slender, erect, or ascending, 10 to 25 cm high, leafy mainly in the lower half. Leaves thin, fresh green, the basal ones oval or round-cordate, long-petioled, soon wilting, those of the stem linear. The large, blue, or rarely white flowers broadly campanulate, 10 to 20 mm in diameter, solitary or several together, the corolla much longer than the calyx lobes.

Rocky, gravelly, or turfy places.

General distribution: *C. rotundifolia s. lat.*: Circumpolar (with large gaps) arctic-alpine.

Fig. 872 Map 1009

Campanula uniflora L.
Stems solitary or few together, glabrous, rather stout, 5 to 30 cm high, leafy. Leaves dark green, subcoriaceous, linear-lanceolate. Flower always solitary, slightly nodding during anthesis. Corolla narrowly campanulate, pale blue, 4 to 8 mm in diameter, slightly longer than the pubescent calyx lobes.

Calcareous cliffs and gravelly screes.

General distribution: Circumpolar; wide ranging high-arctic-alpine with a large gap between Bering Strait and the Ural Mts.

Fig. 873 Map 1010

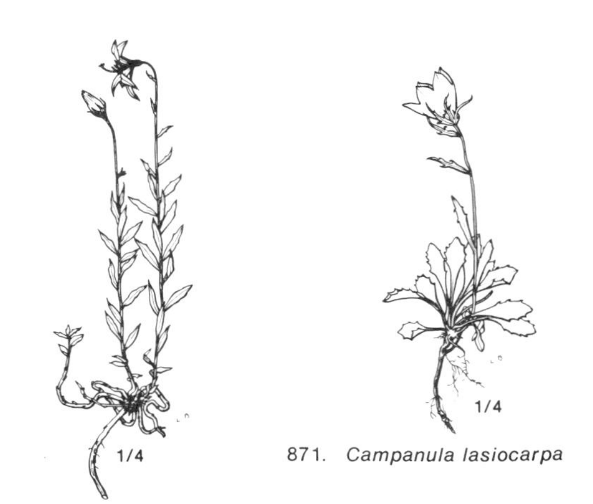

870. *Campanula aurita*

871. *Campanula lasiocarpa*

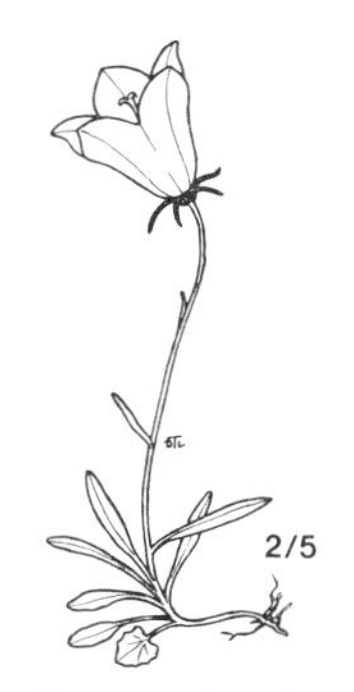

872. *Campanula rotundifolia*

873. *Campanula uniflora*

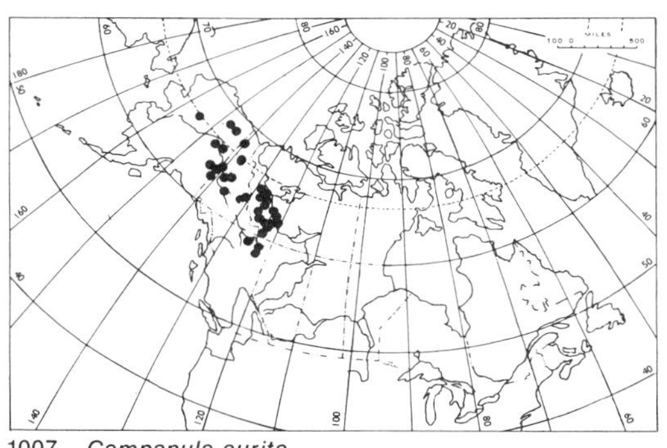

1007. *Campanula aurita*

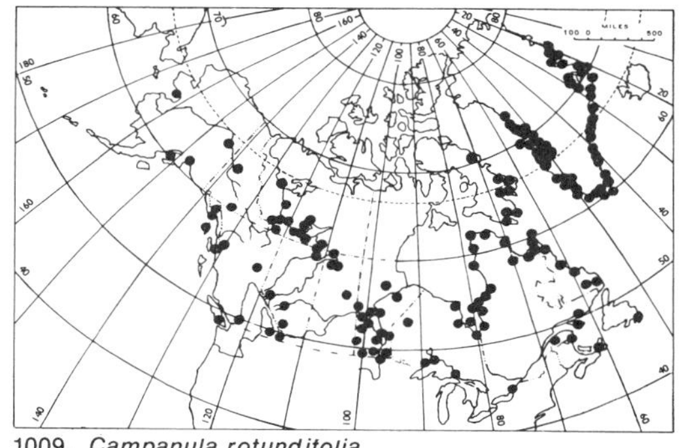

1009. *Campanula rotundifolia*

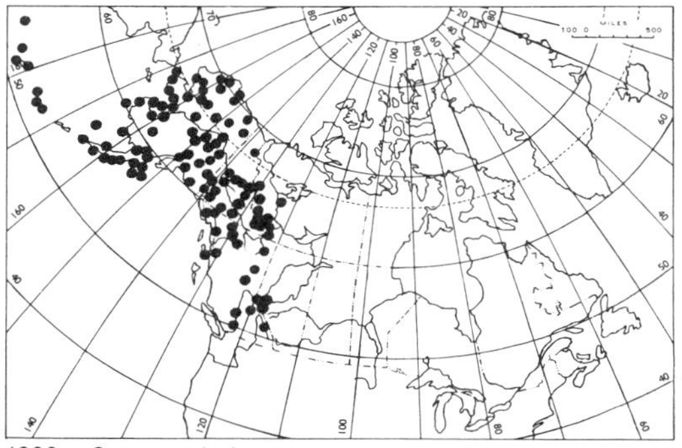

1008. *Campanula lasiocarpa*

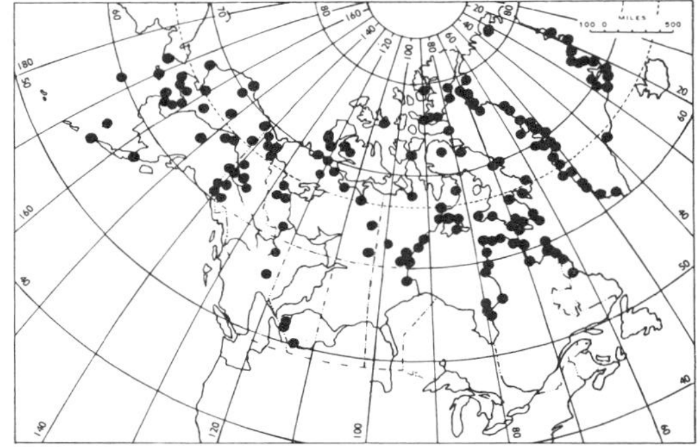

1010. *Campanula uniflora*

LOBELIACEAE Lobelia Family
Lobelia L. Lobelia

Herbs with alternate or basal leaves, and a raceme of bracted, blue flowers; calyx 5-parted; corolla tubular, 2-lipped, the upper lip with 2 erect lobes, the lower lip 3-cleft; fruit a 2-locular, many-seeded capsule.

a. Scapose with a rosette of hollow curved leaves . . . . *L. Dortmanna*
a. Leafy-stemmed with flat leaves . . . . *L. Kalmii*

Lobelia Dortmanna L.
Water Lobelia
Glabrous aquatic or emergent perennial; leaves numerous in a rosette, terete, blunt, hollowed by two tubes separated by the midvein. Stems scapose, simple, or rarely branched, hollow, bearing a few bracts and holding the few-flowered raceme above the water. Corolla light blue.

Rooted in sand or muck in shallow water of lakes and ponds.

General distribution: Amphi-Atlantic; in N. America from Nfld. to Sask. and southern District of Mackenzie near Abitau Lake, and disjunct to western B.C. and Wash. and Oregon.

Map 1011a

Lobelia Kalmii L.
Biennial; stems simple or with lateral branches, from a basal rosette of spatulate-obovate, pubescent, purplish leaves; cauline leaves alternate, narrowly oblanceolate. Flowers in a few-flowered elongating raceme. Corolla blue, with a white throat.

Wet calcareous meadows or pond-margins.

General distribution: N. America; Nfld. to interior B.C., north to southwestern District of Mackenzie.

Fig. 874 Map 1011b

874. *Lobelia Kalmii*

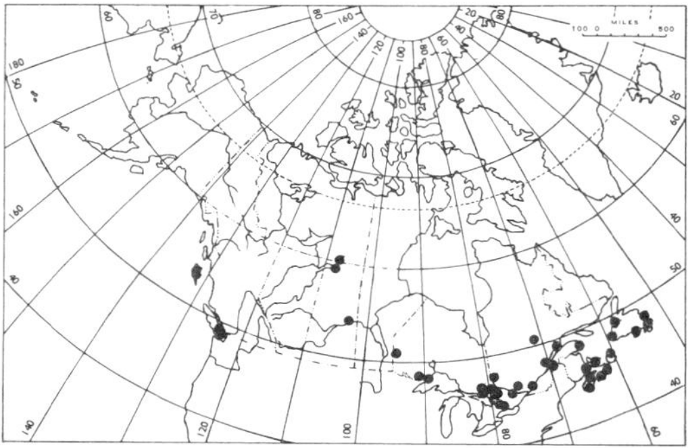

1011a. *Lobelia Dortmanna*

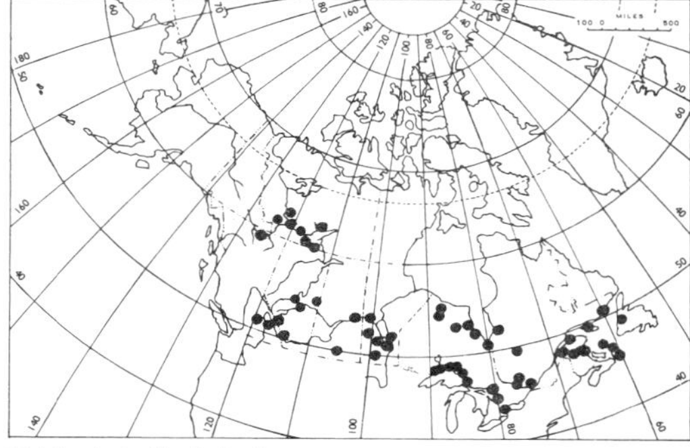

1011b. *Lobelia Kalmii*

COMPOSITAE Composite Family

Ours perennial (or biennial) herbs (*Artemisia* subligneous) with simple or compound leaves lacking stipules. Flowers small, aggregated on a flat, concave, or convex common receptacle into a close head (resembling a single large flower) surrounded by an involucrum of one to several rows of bracts (phyllaries). The flowers, of which the head is composed, may all be alike, or those of the centre may be different from those of the margin. The head is said to be discoid when all flowers have regular, tubular corollas (as in a thistle); ligulate when the corollas are irregular, one-sided, and strap-like (as in a dandelion); and radiate when the central flowers are discoid or tubular, and the radial, ligulate (as in a daisy). Calyx none, or its tube united with the ovary, the limb or pappus developed into persistent scales, teeth, capillary bristles or hairs that serve in the dispersal of the fruits by wind or by animals. Corollas united; stamens 5, their anthers usually united into a tube through which the simple, usually 2-cleft, style protrudes. Fruit an achene.

The largest, and taxonomically the most complex family in the flora of continental Northwest Territories, where it is represented by 28 genera with a total of about 160 species.

Key to genera adapted in part from E. H. Moss, Flora of Alberta.

Key to Groups of Genera

a. Flowers all ligulate and perfect; juice milky .... Group I
a. Flowers not all ligulate; the ray flowers when present, either pistillate or neutral; juice watery
  b. Heads radiate
    c. Rays yellow or orange
      d. Pappus chaffy, or of firm awns, or none; receptacle chaffy, bristly or naked .... Group II
      d. Pappus partly or wholly of capillary bristles, sometimes plumose; receptacle naked .... Group III
    c. Rays white, pink or purple, never yellow or orange .... Group IV
  b. Heads discoid (without rays)
    e. Pappus of numerous capillary bristles, sometimes plumose .... Group V
    e. Pappus of scales, awns, very short chaffy bristles, or a mere crown, or none .... Group VI

Group I

(Flowers all ligulate and perfect; juice milky)

a. Pappus bristles feathery; involucre 1-seriate .... *Tragopogon*
a. Pappus bristles simple; involucre in more than one series
  b. Plants scapose; the leaves all basal, and the heads solitary on erect scapes; achenes beaked
    c. Achenes spinulose, at least toward the summit .... *Taraxacum*
    c. Achenes not spinulose .... *Agoseris*
  b. Plants more or less leafy-stemmed although upper leaves often mere bracts; heads mostly several; achenes beaked or beakless
    d. Achenes flattened; flowers yellow or bluish-purple
      e. Achenes beakless; leaves prickly-margined; flowers yellow .... *Sonchus*
      e. Achenes beaked; flowers bluish-purple .... *Lactuca*
    d. Achenes not flattened; flowers yellow (except *Hieracium albiflorum*)
      f. Pappus sordid or brownish; plants with fibrous roots .... *Hieracium*
      f. Pappus white; plants with tap-root .... *Crepis*

Group II

(Rays yellow or orange; pappus chaffy, or of firm awns, or none)

a. Involucral bracts of two distinct and dissimilar series; receptacle chaffy; weedy annual . . . . *Bidens*
a. Involucral bracts similar, in one or more series
  b. Receptacle naked (Native species)
    c. Pappus of few, deciduous awns; involucre resinous . . . . *Grindelia*
    c. Pappus of chaffy or scarious scales . . . . *Helenium*
  b. Receptacle chaffy . . . . *Gaillardia*

Group III

(Rays yellow or orange; pappus of capillary bristles; receptacle naked)

a. Leaves opposite . . . . *Arnica*
a. Leaves alternate, or all basal
  b. Involucral bracts in one series . . . . *Senecio*
  b. Involucral bracts imbricated, in several series
    c. Heads mostly solitary on each stem (ours) . . . . *Haplopappus*
    c. Heads several on each stem . . . . *Solidago*

Group IV

(Rays present, not yellow or orange)

a. Pappus, or at least in part, of capillary bristles; receptacle naked
  b. Basal leaves cordate or sagittate . . . . *Petasites*
  b. Basal leaves not cordate or sagittate
    c. Involucral bracts subequal or more or less imbricated, often green in part, but not definitely leafy, nor with chartaceous base and herbaceous green tip . . . *Erigeron*
    c. Involucral bracts either subequal and the outer leafy, or more commonly imbricated, with chartaceous or distinctly green tips, or sometimes chartaceous throughout . . . . *Aster*
a. Pappus of scales or awns, or flattened, chaffy bristly, or a mere crown, or none; leaves inconspicuously toothed to pinnately dissected (except *Chrysanthemum integrifolium*); rays white (or rarely pinkish in *Achillea*)
  d. Leaves toothed or entire . . . . *Chrysanthemum*
  d. Leaves all pinnately dissected
    e. Heads small and numerous, the rays about 5 in number, and 2-3 mm long . . . . *Achillea*
    e. Heads few, larger; rays numerous and the receptacle naked . . . . *Matricaria*

Group V

(Heads discoid; pappus capillary)

a. Receptacle densely bristly
  b. Leaves spiny-margined . . . . *Cirsium*
  b. Leaves not spiny-margined . . . . *Saussurea*
a. Receptacle naked
  c. Flowers perfect, yellow or orange
    d. Plants with a stout tap-root . . . . *Haplopappus*
    d. Plants with fibrous-rooted and branching base
      e. Leaves mainly opposite . . . . *Arnica*
      e. Leaves alternate

f. Involucral bracts essentially in a single series . . . . *Senecio*
f. Involucral bracts in more than one series . . . . *Erigeron*
c. Radical, or all flowers, at least of some heads pistillate
g. Basal leaves cordate or sagittate . . . . *Petasites*
g. None of leaves cordate or sagittate
h. Plants more or less white-woolly, involucral bracts mostly with dry, scarious, thin, white or yellowish tips
i. Basal leaves soon deciduous but otherwise similar to the cauline . . . . *Anaphalis*
i. Basal leaves persistent; plants strictly dioecious (but male plants rare in *Antennaria*)
j. Annual, low and branched from the base (ours) . . . . . *Gnaphalium uliginosum*
j. Perennials of low and matted habit . . . . *Antennaria*
h. Plants not white-woolly (see also Group IV)
k. Involucral bracts in a single row . . . . *Erigeron*
k. Involucral bracts imbricated . . . . *Aster*

Group VI

(Heads discoid; pappus not capillary)

a. Sub-woody, usually aromatic plants from branching base . . . . *Artemisia*
a. Herbaceous plants
b. Involucre strongly resinous; pappus of deciduous awns . . . . *Grindelia*
b. Involucre not resinous
c. Annual or biennial, with glabrous and finely dissected leaves . . . . *Matricaria*
c. Perennial, leaves 2-3-pinnate, glabrous or villous . . . . *Tanacetum*

Achillea L. Yarrow

Perennial, somewhat aromatic herbs from a branching rhizomatose base. Leaves alternate, pinnate or bi-pinnately dissected. Inflorescence corymbiform, of several small, radiate heads. Achenes oblong, flattened and glabrous, lacking pappus.

a. Leaves linear, pinnate or merely toothed . . . . *A. sibirica*
a. Leaves bi-pinnately dissected
b. Involucral bracts with dark brown or black margins . . . . *A. nigrescens*
b. Involucral bracts with pale green or yellowish margins
c. Herbage mostly green and strongly aromatic, the leaves ample; non-native . . . . *A. Millefolium*
c. Herbage and stems mostly yellowish-woolly, leaves linear; native . . . . *A. lanulosa*

Achillea lanulosa Nutt.
Stems simple or branched above, 2 to 6 dm tall, singly or several, from a branching horizontally creeping rhizome. Stems, leaves and even flowering peduncles and the involucral bracts densely covered by shaggy yellowish hairs.

Prairie and foothills.

General distribution: N. America; common along the upper Mackenzie R. north to lat. 65°.

Map 1012

Achillea Millefolium L.
Tufted, from a slender, freely forking rhizome. Stems 3 to 6 dm tall, simple or freely forking above. Leaves mostly fresh green; the sub-divisions of the leaves 1.0 to 1.5 cm long. Ray flowers white or pale pink, 3 to 4 mm long.

An aromatic, medicinal oil was formerly extracted from this plant.

General distribution: Circumpolar. A weedy Old World species widespread and fully naturalized across the continent, north to the limit of agriculture. In the District of Mackenzie collected a few times near settlements north to Ft. Simpson.

Achillea nigrescens (E.Mey.) Rydb.
A. borealis Bong.
Stems 3 to 5 dm tall, singly or more often several together. Similar to *A. Millefolium* in habit,

but at once distinguished from that, as well as from other species of *Achillea*, by the dark brown or black margins of its involucral bracts, and by the 5 mm long and almost as broad corolla of its ray flowers.

Damp, sandy river banks and lake shores.

General distribution: N. American boreal-arctic species, wide ranging from Nfld. to Alaska, northward to the Arctic Coast and Banks Island.

Map 1013

Achillea sibirica Ledeb.
Stems simple or freely branching above, 5 to 8 dm tall, singly or several together from short, branching rhizome. Leaves 5 to 10 cm long, fresh green, glabrous or sparingly hirsute. Flowering heads 5 to 8 mm in diameter; ray flowers white, about 2 mm long.

Rocky or gravelly river banks or lake shores.

General distribution: Man. to Alaska and E. Asia, north along the Mackenzie R. to near its delta.

Map 1014

Agoseris Raf.

Perennial, scapose, tap-rooted herbs with milky latex and tufted basal leaves; heads solitary on a long, slender scape. Flowers, all ligulate, yellow or pink; the achenes smooth, more or less beaked, with a well-developed pappus of capillary bristles.

A North American genus resembling *Taraxacum* but with smooth rather than spinulose achenes.

a. Leaves linear-lanceolate, entire or less often remotely pinnate-incised; flowers orange or pink; beak of achene slender . . . . *A. aurantiaca*
a. Leaves glaucous, entire, the oblanceolate, entire or somewhat toothed blade tapering into a slender petiole; flowers yellow; beak of achene stout . . . . *A. glauca*

Agoseris aurantiaca (Hook.) Greene
Cordilleran foothill species ranging north into southern Yukon Territory and southwestern Mackenzie District where, thus far, it is known only from the vicinity of hot springs along a tributary to the South Nahanni River.

Moist, open woodland.

General distribution: N. America, Cordilleran.

Fig. 875 Map 1015

Agoseris glauca (Pursh) Raf.
ssp. **scorzoneraefolia** (Schrad.) Piper
Western foothill species, which, north of lat. 60° is known only from a few collections on south facing gravelly slopes in the Mackenzie Delta region eastward to the Anderson River.

General distribution: N. America, B.C. to Hudson Bay.

Fig. 876 Map 1016

Anaphalis DC. Everlasting

Anaphalis margaritacea (L.) Benth. & Hook.
var. **subalpina** Gray
Perennial, soft velvety, grey-green herbs with simple, 2 to 5 dm tall, solitary, or clustered stems, from a branching, horizontal rhizome. Leaves all cauline, alternate, narrowly lanceolate, 3 to 6 cm long, green above. Inflorescence a rather open, many-headed corymb; heads discoid, 6 to 10 mm in diameter, the phyllaries pearly white, in several series.

Subalpine wooded slopes.

General distribution: *A. margaritacea s. lat.* Nfld. to southern Alaska; var. *subalpina*, western, in the District of Mackenzie thus far known from a single collection in the Mackenzie Mts., disjunct by 800 km from the nearest population in B.C.

Fig. 877 Map 1017

Antennaria Gaertn. Everlasting

Perennial, dioecious, woolly or silky herbs of tufted or matted habit, with alternate, entire leaves. Heads solitary or more often several together in a dense and head-like or open and corymbose inflorescence. Flowers all tubular; the involucre imbricated, of several rows of dry, scarious bracts; achenes terete or flattened, commonly from one to two mm long, glabrous or sparsely papillate; pappus bristles of the fertile flower capillary, that of the sterile barbellate.

Although the genus normally is dioecious, the male plant is known only from six of the 23 species thus far recognized in continental Northwest Territories where most, if not all, species produce viable seeds without fertilization.

a. Basal leaves prominently 3-nerved, erect, 4-16 cm long, similar to cauline leaves; tufted plants up to 5 dm (*Carpaticae*) . . . . *A. pulcherrima*
a. Basal leaves 1-nerved (or at most with obscure lateral nerves); cauline leaves reduced, usually with scarious tips
  b. Tall, dioecious and broad-leaved species; stolons of the year flagellate, ending in a leafy rosette; heads large, with involucre about 10 mm high (*Plantaginifoliae*) . . . . *A. campestris*
  b. Tall, medium to dwarf species; stolons of the year not flagellate, leafy throughout; heads smaller, with involucres usually less than 8 mm high
    c. Bracts of involucres (phyllaries) with pale, greenish-brown to olivaceous or dark brown and usually acuminate and erose tips (*Alpinae*)
      d. Male and female plant usually present (dioecious)
        e. Involucres of pistillate heads small, 4-5 mm high, strictly monocephalous; plant densely tufted and lacking stolons . . . . *A. monocephala*
        e. Involucre of pistillate heads about 6 mm high
          f. Plant monocephalous, stoloniferous and almost entirely glabrous (Northwestern arctic-apline) . . . . *A. philonipha*
          f. Plant pleiocephalous, densely caespitose, with short, rhizomatose base
            g. Basal leaves spatulate, 10-15 mm long, not imbricated, thinly appressed tomentose on both surfaces; achenes papillose . . . . *A. neoalaskana*
            g. Basal leaves densely imbricated, about 6 mm long, achenes glabrous . . . . *A. densifolia*
      d. Only the female plant known
        h. Plant normally monocephalous, densely tufted and lacking stolons
          i. Inner phyllaries normally with blunt, pale-coloured tips (*Dioicae*) . . . . *A. pygmaea*
          i. Inner phyllaries with attenuate, olive-brown tips
            j. Plant fresh green, almost totally glabrous . . . . *A. glabrata*
            j. Plant grey-green; basal leaves appressed tomentose beneath, glabrous above; stems floccose-tomentose . . . . *A. angustata*
        h. Plant normally pleiocephalous
          k. Caespitose, with stout, rhizomatose caudices and numerous, sessile, sterile rosettes
            l. Basal leaves narrowly lanceolate, gradually tapered towards the tip; pappus rufidulous; achenes glabrous or minutely papillose . . . . *A. Ekmaniana*
            l. Basal leaves spatulate-ovate, rounded at the tip; pappus white; achenes glabrous
              m. Basal leaves short-obovate, white-lanate tomentose; stems floccose; plant pulvinate . . . . *A. compacta*
              m. Basal leaves longer, grey, thinly appressed-tomentose; stems purplish under the thin, non-floccose tomentum; plant caespitose but not pulvinate . . . . *A. subcanescens*
          k. Caespitose or matted, with basal rosettes borne on well-developed prostrate or ascending stolons
            n. Achenes glabrous (eastern)
              o. Stems rarely over 15 cm high; inflorescence compact or heads at most with peduncles less than 1 cm long . . . . *A. canescens*
              o. Stems slender up to 25 cm high, heads few, on 2-3 cm long peduncles . . . . *A. ungavense*
            n. Achenes strongly papillose, even when immature; heads 4-10, with involucres 6-7 mm high (western)

p. Inflorescence rather compact; inner phyllaries dark brown . . . . *A. stolonifera*

p. Inflorescence open, the lower heads remote, on long peduncles; inner phyllaries pale brown . . . . *A. pedunculata*

c. At least the inner phyllaries with papery, white, straw-coloured or pink, and usually spatulate-ligulate tips; in our area only the pistillate plants known (*Dioicae*)

q. Inner phyllaries pink (in *A. elegans* soon turning pale grey or straw-coloured)

r. Tall plants with fruiting stems usually 20 cm tall or over

s. Plants with well-developed stolons, and forming colonies

t. Basal leaves fresh green and glabrate above, cauline leaves ample; fruiting stem 20-30 cm tall; heads showy, nodding in youth . . . . *A. alborosea*

t. Basal leaves neither fresh-green, nor glabrate above, cauline leaves small

u. Basal leaves narrowly spatulate, appressed-sericeous-tomentose; phyllaries pale pink only in youth, soon turning pale grey or straw-coloured . . . . *A. elegans*

u. Basal leaves oblanceolate, dull greyish-tomentose; phyllaries pink even in age . . . . *A rosea*

s. Plants with short, sessile offsets forming small tufts; basal leaves, spatulate-obovate, with appressed grey tomentum . . . . *A. oxyphylla*

r. Dwarf plants with fruiting stems less than 20 cm tall; basal leaves erect, oblanceolate, sericéous-tomentose . . . . *A. incarnata*

q. Inner phyllaries papery white or straw-coloured, never pink; dwarf to medium sized plants with fruiting stems less than 20 cm tall

v. Monocephalous; densely tufted, stems rarely over 10 cm tall; basal leaves small, rarely over 1 cm long . . . . *A. pygmaea*

v. Pleiocephalous, usually with from 4-10 heads

w. Inflorescence nodding in youth, glomerulate; phyllaries firm; the upper cauline leaves with slender, scarious tips; basal leaves appressed sericeous-tomentose, almost silvery . . . . *A. nitida*

w. Inflorescence not nodding in youth, open, with the lateral heads pedunculate; phyllaries thin and soft, their tips often spreading in age

x. Pappus pale rufidulous; plant pulvinate with short, sessile off-sets . . . . *A. crymophila*

x. Pappus white; plant not pulvinate

y. Upper 1-3 cauline leaves with slender, scarious tips . . . . *A. subviscosa*

y. Upper 5-8 cauline leaves with brown, flat and very prominent scarious appendages . . . . *A. isolepis*

Antennaria alborosea Porsild
Stolons creeping, freely branching, 5 to 10 cm long. Basal leaves oblanceolate-cuneate, tapering into a narrow petiole, mucronate, 2.0 to 3.0 cm long, 0.5 to 0.6 cm wide, obscurely 3-nerved, glabrate or glabrous and deep green above, sericeous-appressed tomentose beneath. Flowering stems 20 to 30 cm high, slender, stiff or sometimes arcuate, thinly arachnoid-tomentose, glabrate and greenish-purple. Stem leaves 15 to 20, linear-lanceolate, glabrate and fresh green, long acuminate but scarcely appendiculate, the lower 2.0 cm long, the upper much smaller. Inflorescence lax cymose, nodding in youth, with from 5 to 8 pistillate heads on 0.6 to 1.0 cm long peduncles. Involucres 6 to 7 mm high, the phyllaries 3- to 4-seriate, conspicuously imbricated, thin and oblanceolate; the outer green with thinly lanate bases, the inner first roseate, later straw-coloured. Corolla with greenish tips. Style scarcely exserted. Achenes dusky brown, glabrous or minutely and sparingly papillose. Pappus silky and snowy white. Staminate plant unknown.

Sandy or gravelly and not too dry places on acidic soils, where it forms small colonies, sometimes one metre across.

General distribution: Central Alaska through central and S. E. Y. T. east to Great Bear L., south to mountains of S. E. Alaska and northern B.C.

Fig. 878 Map 1018

Antennaria angustata Greene
Caespitose with sessile or subsessile, erect basal offsets. Basal leaves narrowly oblanceolate, about 10 mm long and 2 mm wide, mucronate, appressed-tomentose beneath, glabrate or even glabrous above. Flowering stems 4 to 10 (14) cm high, thinly floccose-tomentose, bearing from 7 to 11 linear leaves; these about 10 mm long with prominent, flat scarious tips. Heads large, solitary, or very rarely with one or two small lateral heads; involucres 8 to 10 mm high, thinly lanate at the base; the phyllaries obscurely 3-seriate, long-attenuate, olivaceous; style included or short-exserted; pappus white; achenes glabrous, about 1.4 mm long. The staminate plant not known.

Slopes and depressions where the snow remains late.

General distribution: Wide ranging N. American arctic, subarctic-alpine.

Fig. 879 Map 1019

Antennaria campestris Rydb.
Low and creeping; the old stolons slender, cord-like, subterranean, freely branching with numerous sterile rosettes; young stolons of the year above the surface, flagellate, up to 10 cm long, issuing from the base of the fertile stems. Basal leaves obovate-cuneate, 2 to 3 cm long and about 1 cm wide, short-petioled, with one prominent median nerve and two indistinct lateral ones; the upper leaf-surface dark green and glabrate in age, the lower covered by a dense, permanent and yellowish-white felt. Flowering stems slender, from 10 to 25 cm tall, bearing from 5 to 10 much reduced, scarious-tipped leaves. Inflorescence dense, glomerulate; pistillate heads large, from 2 to 5, their involucres about 10 mm high; phyllaries indistinctly 3-seriate, linear-lanceolate, greenish-brown and lanate at the base, with long-acuminate, papery, white tips; pappus tawny; mature achenes olivaceous, shiny and distinctly papillose. Staminate plant similar, the heads somewhat smaller, the phyllaries with obtuse-erose, more or less reflexed tips.

Dry sandy plains and prairies.

General distribution: Western N. America. The only species in the section *Plantaginifoliae* which reaches north into our area.

Fig. 880 Map 1020

Antennaria canescens (Lge.) Malte
Densely matted, with well-developed, creeping and spreading stolons. Basal leaves permanently and densely ashy grey, appressed-tomentose on both sides. Flowering stems 5 to 15 cm high, terminating in a dense corymb, usually of three heads. Staminate plant not known.

Non-calcareous, sunny slopes and in herb-mats.

General distribution: Endemic of subarctic, eastern N. America and Greenland.

Fig. 881 Map 1021

Antennaria compacta Malte
Densely caespitose almost pulvinate, with a stout ascending and often branched rhizome. Basal leaves broadly oblanceolate-obovate, less than 1 cm long, permanently ashy grey on both sides, from appressed, felt-like hairs. Flowering stems 5 to 10 cm high, slender, and usually somewhat arched, copiously floccose-tomentose. Heads rather small, usually 2 to 4 in a dense corymb, but occasionally with one or more long-peduncled supernumerary heads from the axils of the uppermost stem-leaves; corolla lobes deep purple, styles barely exserted; achenes glabrous. Staminate plant not known.

Dry gravelly and rocky places.

General distribution: N. America, high-arctic, from Alaska to E. Greenland.

Fig. 882 Map 1022

Antennaria crymophila Porsild
Plant densely caespitose or even pulvinate with numerous short, sessile and erect sterile rosettes. Basal leaves spatulate, obtuse, about 1.0 cm long and 0.3 cm broad, appressed white-tomentose on both sides, in age becoming dark grey. Flowering stems about 10 cm high, covered by a floccose tomentum. Stem leaves 6 to 8, rather distant, linear, the lower about 1.0 cm long, the upper much reduced and bearing long, scarious tips. Inflorescence open, mostly with 4 heads; the lateral on 1 to 2 cm long peduncles. Involucres about 0.6 cm high; phyllaries 4- to 5-seriate, brownish black at the base, with obtuse, straw-coloured, papery tips. Pappus rufidulous; style barely exserted. Mature achenes olivaceous, minutely but distinctly papillose. Staminate plant unknown.

Antennaria crymophila is perhaps most closely related to *A. pulvinata* Greene of mountains of Alberta and British Columbia, but differs from it by its open inflorescence, taller flowering stem and strongly developed thick stolon.

Gravelly river banks.

General distribution: Long known only from the type locality: Caribou Hills, east of the Mackenzie Delta, *A. crymophila* is now known

also from mountains in S. E. Alaska and in S. W. Y.T.
Fig. 883 Map 1023

Antennaria densifolia Porsild
Densely caespitose, with short crowded sessile or subsessile basal offsets. Basal leaves densely congested or imbricated, cuneate-obovate or broadly oblanceolate, obtuse, not mucronate, 5 to 6 mm long, 3 mm broad, densely yellowish tomentose on both surfaces. Flowering stems 6 to 9 cm high, slender, floccose-tomentose; stem leaves 5 to 7 close together, linear, the lowermost 6 to 10 mm long and 2 mm broad, acute, with subulate tips; the upper smaller and bearing scarious, lanceolate appendices. Pistillate heads 2 to 4, campanulate, cymose, on 0.5 cm long pedicels; involucre 0.5 cm high, densely lanate at the base; phyllaries 2- to 3-seriate, subequal, lanceolate, subacute, the exterior chestnut-brown, greenish at the base, the interior brownish, erose; corolla lobes purplish; style exserted, bifid; pappus white; achenes glabrous. Staminate heads smaller with 4 mm high involucres; phyllaries broadly oblanceolate, subequal and spreading; pappus bristles plumose somewhat clavellate toward the tip. The staminate plant about as common as the pistillate.

Antennaria densifolia superficially may resemble depauperate specimens of *A. compacta* from which it differs by its always shorter and more densely crowded leaves, and by being dioecious.

Calcareous alpine slopes.

General distribution: N. America. Endemic of S. W. District of Mackenzie and central Y.T.
Fig. 884 Map 1024

Antennaria Ekmaniana Porsild
Densely caespitose with sessile or subsessile, erect basal rosettes; leaves 1.0 to 2.2 cm long, 2 to 3 mm wide, linear-lanceolate, distinctly mucronate, densely appressed-tomentose on both sides, and usually somewhat purplish tinged at the base. Flowering stems 10 to 20 cm high, distinctly purplish tinged under the thin, appressed tomentum, with from 6 to 8 linear and prominently apiculate lanate-tomentose leaves. Heads from 1 to 7, but usually 3 to 4; involucres about 7 mm high, thinly lanate at the base; phyllaries, subequal, the outer ones lanceolate, olive-green, the inner long-attenuate, light chestnut-brown; pappus sub-rufescent; style exserted, deeply cleft; achenes glabrous or minutely papillose. Staminate plant unknown.

Rocky, lichen-covered slopes, above timberline.

General distribution: N. America; wide ranging, arctic-alpine.
Fig. 885 Map 1025

Antennaria elegans Porsild
Humifuse, with well-developed, creeping and freely branching, subligneous stolons. Basal leaves spreading, linear-oblanceolate, acute, 10 to 20 mm long, 2 to 3 mm wide, thinly silky-tomentose on both sides. Flowering stems slender and thinly floccose, 12 to 18 cm high, bearing about 10 appressed-tomentose, linear-lanceolate leaves, 10 mm long and 2 mm broad; the lower ones more or less obtuse, the uppermost one or two with thin, scarious tips. Pistillate heads from 1 to 8, but usually 4, loosely glomerate at the time of florescence, but soon after the lowermost peduncles elongate so that the inflorescence becomes flat-topped, or the lateral heads may even overtop the central ones. Involucres 5 to 6 mm high, their phyllaries in 3 or 4 rows, at first densely imbricated, later somewhat spreading. The outer phyllaries ovate, with thinly lanate, brownish bases and stramineous tips, the inner oblong-lanceolate, acuminate and erose, at first pale pink, later becoming straw-coloured. Corolla lobes greenish. Styles barely exserted. Achenes brown and smooth. Staminate plant unknown.

Antennaria elegans is well marked by its matted growth, narrow, appressed-sericeous tomentose and spreading basal leaves; its slender, up to 20 cm tall, and often somewhat flexuous stems; and by its numerous, rather ample, oblanceolate, obtuse stem leaves of which all but the topmost lack scarious appendages. In life, the young inner phyllaries are pale rose but soon turn pale grey or straw-coloured.

Rocky alpine slopes.

General distribution: Mountains of southeast and central Y.T., northern B.C., east to Great Bear and Great Slave Lakes, N.W.T.
Fig. 886 Map 1026

Antennaria glabrata (J.Vahl) Greene
Caespitose, with numerous sterile offsets, forming small, flat cushions; basal leaves lanceolate-spatulate, prominently apiculate, 1.0 to 2.5 cm long and 2 to 4 mm wide, fresh green and totally glabrous on both surfaces, or very rarely thinly tomentose in youth. Flowering stems 8 to 12 cm high, slender, often somewhat flexuous, glabrous, bearing from 3 to 8 linear oblong, 10 to 15 mm long and 1.5 to 2.5 mm wide leaves,

the uppermost terminating in a conspicuous, scarious tip. Head normally solitary and large, less commonly 2, and very rarely 3; involucral bracts glabrous, olive-green with slightly paler, erose and hyaline tips. Achenes glabrous. Staminate plant not known.

By its firm caespitose habit and totally glabrous fresh green, or in age yellowish-green leaves, and large, mostly solitary heads *A. glabrata* is a well marked species until recently considered endemic of Greenland. By its green and glabrous leaves it resembles *A. Porsildii* El. Ekm. of arctic Scandinavia and central East and West Greenland which, however, is easily distinguished by its well developed stolons, and shorter and attenuate basal leaves, and several and smaller heads.

Fig. 887 Map 1027

***Antennaria incarnata** Porsild

Humifuse, with subligneous, freely branching 3 to 5 cm long, suberect-ascending stolons. Basal leaves oblanceolate or spatulate, acute, 10 mm long and 3 mm wide, on both sides covered by a grey, silky tomentum. Flowering stems 10 to 12 cm high, appressed silky-tomentose, bearing 7 to 8 linear, 10 to 15 mm long and 2 mm broad leaves, their tips acute or attenuate, lacking scarious appendages. Inflorescence subglobose with from 5 to 8 heads on non-glandular peduncles of which the lowermost are sometimes branched. Involucres 5 mm high, their phyllaries almost of equal length in 2 or 3 series, the exterior greenish and densely lanate below and with pink tips, the interior oblong-lanceolate, pale roseate, in life purplish-puncticulate. Corolla lobes pale purplish. Styles exserted. Achenes smooth. Staminate plant unknown.

By its much smaller size alone *Antennaria incarnata* is easily distinguished from *A. alborosea* and from the somewhat polymorphous *A. rosea*.

Forms small but rather dense colonies in turfy places on river terraces, outwash plains, and in open birch thickets.

General distribution: Thus far known only from mountains of central Y.T. and Alaska but likely to occur also in the east slopes of Mackenzie Mts.

Fig. 888 Map 1028

Antennaria isolepis Greene

A. leontopodioides Cody

Humifuse, with leafy stolons up to 5 cm long. Basal leaves oblanceolate, 1.5 to 2.0 cm long and 3 to 4 mm broad, subacute or short mucronate, appressed white-tomentose on both surfaces but more densely so underneath. Flowering stems 10 to 15 cm high, somewhat flexuous, floccose-tomentose; cauline leaves numbering about 10, linear-oblanceolate, floccose-tomentose, the lower about 1.5 cm long, with subulate tips, the upper not much reduced and with broad, flat, scarious tips; heads 3 to 6, usually 4, the lateral on 1 cm long peduncles; involucres 6 to 7 mm high; phyllaries 3-seriate, oblanceolate, with oblong, erose, papery white and spreading tips. Pappus bristles white. Achenes glabrous. Staminate plant unknown.

Stony slopes.

General distribution: Wide ranging arctic-alpine, ranging from Ungava to central Alaska south to mountains of B.C.

Fig. 889 Map 1029

Antennaria monocephala DC.

Dwarf, often forming small mats although the offsets are rarely more than a few centimetres long, erect or suberect; their leaves spatulate-obcuneate toward the base, about 1 cm long, mucronate, thinly floccose-tomentose beneath, green and glabrous above. Flowering stems usually about 5 cm high but often only 2, but occasionally up to 15 cm high, bearing from 5 to 10 thinly lanate, linear leaves with prominent, 2 mm long, scarious tips. Heads always solitary; involucres about 4 mm high; phyllaries of the pistillate heads, green at the base, dark brown to almost black in the middle and olivaceous to golden brown towards the attenuate tips. Pappus bristles white; style long-exserted; achenes glabrous. Staminate heads slightly smaller, their phyllaries broadly lanceolate with papery, obtuse and often spreading tips; pappus bristles clavellate.

A common species above timberline chiefly on Precambrian rocks in snow patch vegetation, ravines and avalanche paths.

General distribution: Amphi-Beringian arctic-alpine from Kamtchatka and Eastern Chukotsk Pen., and Alaska eastward to Mackenzie and Richardson Mts., N.W.T., south in the Rocky Mts. to Jasper Park, Alta.

Fig. 890 Map 1030

Antennaria neoalaskana Porsild

Dioecious, with a short, rather stout, freely branched, subligneous and rhizomatose base; the offsets sessile, their leaves spatulate 1.0 to 1.5 cm long, 2 to 3 mm wide, mucronate, thinly appressed tomentose on both surfaces, in age becoming glabrate. Flowering stems 5 to 12 cm high, thinly floccose, stout, stiff and usually somewhat arching, bearing from 4 to 7 linear-spatulate, scarious tipped leaves. Inflorescence glomerate, the lateral heads on 5 mm long peduncles; pistillate heads mostly 3. Involucre about 6 mm high; phyllaries 2-seriate, long-attenuate, the outer somewhat lanate below, in youth with olive-green tips which in age become pale brown. Pappus white; tips of corolla pale yellow; styles much exserted. Achenes minutely papillose. The staminate plant similar, but heads smaller, their phyllaries ovate-lanceolate, with obtuse, erose, pale stramineous tips, and clavellate pappus bristles.

Perhaps most closely related to *A. alaskana* which differs by its smaller heads and characteristic purplish leaf petioles. Superficially *A. neoalaskana* resembles *A. subcanescens* from which it differs by the shape and vesture of its leaves, and by being dioecious.

Alpine gravelly ridges and snowbed slopes.

General distribution: From eastern Brooks Range, Alaska south over Richardson Mts. to central Mackenzie Mts., N.W.T.

Fig. 891 Map 1031

Antennaria nitida Greene
Humifuse and densely matted with well developed, subligneous, freely branching and leafy stolons. Basal leaves small and spreading, obovate-oblanceolate, mucronate, narrowly petiolate, 5 to 15 mm long, 4 to 5 mm wide, somewhat involute and almost keeled by the prominent median nerve, silvery white and almost shiny on both surfaces due to the finely appressed tomentum. Flowering stems of the pistillate plant from 5 to 25 cm high, slender and somewhat flexuous, pale green and thinly arachnoid-tomentose, bearing from 8 to 20 linear leaves, the lowermost merely mucronate, the upper with attenuate brownish and scarious tips. Inflorescence cymose, densely glomerulate and nodding in youth, in age open and branched, the lower branches often distant and bearing up to 4 heads. Involucres 6 to 7 mm high, viscid-hirsute; phyllaries in 4 rows, pale green below with oblong, obtuse and entire, papery white tips. Pappus white, scarcely exceeding the mature corollas. Styles barely exserted; achenes glabrous. Staminate plant less common, similar in stature but with broader and often reflexed tips of the phyllaries and with conspicuously barbellate pappus bristles.

Prairies and plains, mainly on calcareous soil.

General distribution: Central Alaska through Y.T. and the District of Mackenzie north to Bear L., east to James Bay, south to Man., Alta. and interior B. C., and in the Rocky Mts. south to New Mexico.

Fig. 892 Map 1032

Antennaria oxyphylla Greene
Humifuse, with short, leafy, ascending-erect stolons, forming small, rather compact colonies. Basal leaves 1.5 to 2.0 cm long and 6 to 7 mm broad, obovate-oblanceolate, distinctly mucronate, silvery grey, appressed tomentose on both surfaces but more densely so on the underside. Flowering stems 20 to 25 cm high, slender and somewhat arched, the cortex greenish to purplish brown under the thin, caducous, floccose tomentum; stem leaves about 9 to 10, evenly spaced, linear-lanceolate, the lower about 1.5 cm long and 3 mm wide, merely acuminate, the upper rapidly diminishing in size and terminating in thin, scarious tips. Inflorescence glomerulate, with from 3 to 5 heads. Involucres 6 to 7 mm high, lanate below; phyllaries 4-seriate, the outer pale grey-green, the inner white with oblong, attenuate somewhat erose, non-spreading tips. Pappus bristles white, much longer than the reddish-purple corollas. Styles barely exserted; achenes 1.4 mm long, glabrous. Staminate plant not known.

Prairies and dry meadows.

General distribution: Central Alaska, Y.T., and the Mackenzie drainage east to northern Sask. and south through the Rocky Mts. to Nebraska and S. Dakota.

Fig. 893 Map 1033

***Antennaria pedunculata** Porsild
Humifuse, with freely branching stems 5 to 8 cm long. The well-developed stolons, 5 to 10 cm long, bearing oblanceolate and mucronate, 2 cm long and 0.5 cm broad leaves that in youth are thinly silky-tomentose, in age becoming glabrescent and showing a single, prominent nerve on the underside. Flowering stems 15 to 22 cm high, glabrescent, bearing 8 to 10 linear-lanceolate leaves; the lowermost obtuse, the upper with scarious tips. The pistillate heads cymose, 1 to 5, the lateral ones rather remote on 3 to 6 cm long peduncles. Involucre 7 to 10 mm high, lanate at base. Phyllaries of unequal length, 3-seriate, the outer oblanceolate, dusky brown, the inner oblanceolate, subacute, with whitish tips. Corolla purple. Styles much exserted, bifid. Achenes 1.0 mm long, olivaceous, strongly papillose. Staminate plant unknown.

By its open inflorescence of long-peduncled and remote lower heads, and pale, chartaceous-tipped inner phyllaries, *A. pedunculata* differs strikingly from other northwestern members of the genus.

Alpine slopes.

General distribution: Thus far known from the upper Rose R. Valley in the Pelly Mts. of east-central Y.T., and from the British Mts. of northern Y.T. It may, therefore, be expected to occur also in the District of Mackenzie along the east slope of the Richardson and Mackenzie Mts.

Map 1034

Antennaria philonipha Porsild
Dioecious, matted, forming small colonies by stolon-like offsets 5 to 10 cm long; basal leaves spatulate-obovate about 1.5 cm long and 4 mm wide, mucronate, glabrous above, appressed silky-tomentose beneath, becoming glabrate in age. Stems very slender and weak, thinly floccose-tomentose, especially above, becoming glabrate in age, with 5 to 10 very reduced stem leaves tipped with very conspicuous brown, flat, obtuse and scarious 2 mm long appendages. Stem of pistillate plant 10 to 14 cm tall, that of the staminate as a rule less than 10 cm. Heads solitary or very rarely 2 to 3 together, the staminate small, nodding when young, their corolla lobes pink, involucres about 5 mm high with obovate, lead-coloured bracts and with pappus scarcely barbellate, slightly thickened at the apex. Pistillate heads somewhat larger, their involucres 6 to 7 mm high, the bracts of equal length, thin and hyaline, acuminate, dark brown below the tips, olivaceous when young, becoming pale straw-coloured in age. Pappus rufidulous when mature, the rays merely scabrous, not at all plumose; style long, exserted, bilobed, pale rose, turning reddish purple. Achenes glabrous.

Antennaria philonipha is a very well marked species, perhaps nearest related to *A. monocephala* with which it, apparently, hybridizes when the two species meet. From *A. monocephala* it is at once distinguished by its tall and slender flowering stems and larger pistillate heads, by its long stolon-like offsets and by the much thinner indument of the leaves and stem. It is an arctic-alpine species which, as the name implies, is ecologically bound to moist, grassy places near alpine snow banks.

General distribution: Arctic-alpine: From Seward Pen. east through high mountains of central Alaska and Y.T., east along the east slope of the Mackenzie and Richardson Mts. to the Mackenzie Delta and Anderson R., south to mountains of northern B.C.

Fig. 894 Map 1035

Antennaria pulcherrima (Hook.) Greene
Stems solitary, 20 to 50 cm tall, leafy, from a branching, slender subligneous rhizome; basal leaves 8 to 12 cm long and 8 to 14 mm broad, oblanceolate, acute, long-petioled, prominently 3- to 5-nerved and sericeous-tomentose on both surfaces, the lower cauline leaves similar to the basal. Inflorescence glomerulate with from 4 to 12 heads, commonly 2 or 3 on each branch; the pistillate large, 7 to 8 mm high. The staminate plant similar but usually smaller.

Well watered alluvial soils on river flats and meadows, or on alpine slopes, and mainly on calcareous soil.

General distribution: N. America. Boreal-alpine but not arctic. Wide ranging from central Alaska to Nfld., north to the Arctic Coast of Y.T.

Fig. 895 Map 1036

Antennaria pygmaea Fern.
Humifuse, with very short erect-ascending stolons, forming small, compact tufts. Basal leaves oblanceolate, prominently mucronate, 8 to 14 mm long and 3 to 4 mm broad, glabrous or glabrate above, thinly appressed tomentose beneath. Flowering stems from 4 to 14 cm high, stiff, dark purplish brown and finely papillose under the thin arachnoid indument, bearing about 9 somewhat crowded linear-oblanceolate, 6 to 14 mm long, glabrate leaves, all bearing prominent up to 4 mm long narrowly deltoid to subulate, scarious brown tips. Heads large, solitary or occasionally two side by side or occasionally with two smaller lateral heads below the terminal one; involucre hemispherical, about 7 mm high, lanate at the base. Phyllaries densely imbricated, in 3 or 4 rows, oblong and obtuse, with straw-coloured tips. Pappus bristles silky, white, longer than the corollas; styles barely exserted. Achenes glabrous. Staminate plant not known.

Superficially *A. pygmaea* resembles the pistillate plant of *A. monocephala* from which it may at once be distinguished by its larger heads and pale, straw-coloured phyllaries.

Open stony slopes.

General distribution: Arctic-alpine: Lab. to high mountains of S.E. Y.T.

Fig. 896 Map 1037

Antennaria rosea (Eat.) Greene
Humifuse, with strongly developed, long, subligneous and freely branching stolons, each ending in an erect or ascending leafy rosette which the following year produces a flowering stem. Basal leaves oblanceolate, acute, about

2 cm long and 4 mm wide, often involute in age, with a densely appressed pale yellowish-green tomentum. Flowering stems rather weak, in our area usually 10 to 15 cm high, but farther south frequently twice as high, with 8 to 10 linear, 2 cm long, acuminate leaves, all without scarious appendages. The inflorescence rather compact with 3 to 5 rather small heads in a hemispherical cyme. Involucres 4 to 5 mm high; phyllaries 3-seriate, the outer ones pale green and lanate, the inner lanceolate, acuminate, dark rose or pink, fading in age. Pappus dirty white, not lustrous; styles not exserted, and the achenes glabrous. The staminate plant unknown.

Sandy places, often in open pine woods.

General distribution: South central Alaska east to James Bay.

Fig. 897 Map 1038

Antennaria stolonifera Porsild
Humifuse with well-developed, subligneous, freely creeping and spreading 5 to 10 cm long stolons; old individuals frequently attaining a diameter of one metre or more. Basal leaves spatulate, obtuse and flat, 20 to 25 mm long and 4 to 5 mm broad, thinly silky-tomentose on both surfaces, rarely glabrate, distinctly 1-nerved. Flowering stems about 10 cm high at the time of florescence, soon elongating, becoming 14 to 18 cm high toward maturity, robust and stiffly erect; the cortex purplish under the white, floccose tomentum. Stem leaves 8 to 10, the lowermost well developed, linear-lanceolate, 15 to 20 mm long and lacking scarious tips, the upper smaller and bearing long scarious appendages. Pistillate heads 3 to 5, densely glomerate, at the time of florescence; later the lower peduncles elongate so that the lateral heads in time overtop the central ones. Phyllaries 3-seriate, the exterior densely lanate, with obtuse greyish tips; the interior linear-oblong, obtuse, olivaceous. Corolla purple; styles barely exserted. Achenes small, 0.8 mm long, with numerous long, hispid papillae. Pappus dirty white, finely barbellate. Staminate plant unknown.

Alpine slopes.

General distribution: Common locally in central Alaska, central and S.E. Y.T., and on the east slope of the Mackenzie Mts.

Fig. 898 Map 1039

Antennaria subcanescens Ostenf.
Densely caespitose, forming flat cushions 10 to 15 cm in diameter, with short, ascending-erect, leafy offsets. Basal leaves oblanceolate, obscurely mucronate, 1.5 to 2.5 cm long, including the well-developed petiole, and 4 to 5 mm wide, dark cinereous and thinly appressed tomentose on both surfaces. Flowering stems 5 to 10 cm high, dark purplish, distinctly glandular- papillose under the sparse, appressed-woolly indument. Cauline leaves 5 to 7, linear, thinly lanate, 10 to 12 mm long, the upper with 2 mm long, scarious appendages. Inflorescence glomerulate; heads mostly 3. Involucres about 7 mm high; phyllaries slightly lanate and dark brown at the base, with long, narrow and greenish-brown tips. Styles exserted, bifid; pappus bristles dirty white. Achenes glabrous. Staminate plant unknown.

Antennaria subcanescens, like *A. compacta*, is densely caespitose but differs from that species by the very characteristic dark ashy-grey basal leaves, dark purplish flowering stems with long gland-tipped papillae beneath the thin, non-floccose indument.

Stony flats and ridges.

General distribution: Arctic: Mountains of Alaska and Y.T., east along the Arctic Coast to southern Baffin Island and central W. & E. Greenland.

Fig. 899 Map 1040

Antennaria subviscosa Fern.
Humifuse, with freely branching, subligneous stolons ending in leafy, erect or ascending rosettes. Basal leaves oblanceolate-spatulate, obtuse and scarcely if at all mucronate, 1.0 to 1.5 cm long and 3 to 5 mm broad, densely white-tomentose on both surfaces. Flowering stems 8 to 11 cm high, floccose, slender and more or less flexuous, with about 10 linear-oblanceolate leaves, the lower about 1.0 cm long, merely acuminate, the upper but slightly reduced in size, with thin, subulate, brown and scarious tips. Inflorescence glomerulate, hemispherical, with from 3 to 6 heads. Involucre 5 to 6 mm high, densely white-lanate below; phyllaries 3-seriate; the outer ones green below, with a brownish centre and stramineous blunt, papery tips. Pappus white; styles barely exserted. Achenes glabrous. Staminate plant unknown.

General distribution: Widely disjunct stations from Gaspé, Que., northern Ungava, L. Superior and alpine situations in S. W. District of Mackenzie and southern and central Y.T.

Fig. 900 Map 1041

Antennaria ungavensis (Fern.) Malte
Tufted, from a subligneous, rhizomatose base from which issue short stolons; basal leaves oblanceolate, short but distinctly mucronate,

tapering to a narrow base, 1.3 to 2.0 cm long and 2 to 4 mm wide, at maturity fresh green and glabrous on the upper surface, appressed silky-tomentose beneath. Flowering stems slender and somewhat flexuous, thinly floccose-tomentose, 22 to 25 cm tall (in the type) but more commonly 12 to 18 cm tall, bearing 7 to 8 well-developed leaves, the uppermost with thin, scarious appendages, the lower merely apiculate; heads 1 to 3 (rarely 4 to 5) the lateral often on long peduncles, rising from the axils of the upper cauline leaves; heads 7 to 9 mm high, floccose at the base. Involucral bracts unequal, in 2 or 3 series, dusky brown to light greenish-brown. Styles barely exserted; achenes glabrous. Staminate plant not known.

Gravelly open tundra.

General distribution: Endemic of northern Ungava, the District of Keewatin, and the mountains of S. W. Alberta.

Fig. 901 Map 1042

Arnica L.

Perennial rhizomatose, caulescent herbs with opposite leaves. Heads radiate and showy, solitary or in two or more pairs below the terminal head; involucre campanulate, the bracts in one series. Ray-flowers bright yellow, the disc-flowers slightly darker. Achenes elongate, usually about 4 mm long, 5- to 10-ribbed, hirsute or glabrous. Pappus milky white or straw-coloured, in a single series of capillary and barbellate bristles.

a. Pappus straw-coloured
  b. Head solitary, nodding; anthers dark purple . . . . *A. Lessingii*
  b. Heads usually several, not nodding; anthers yellow
    c. Stem-leaves in five or more pairs
      d. Tip of involucral bracts with tuft of long white hairs
        e. Leaves fresh green . . . . *A. Chamissonis* ssp. *foliosa*
        e. Leaves silvery cancescent-tomentose . . . . *A. Chamissonis* ssp. *incana*
      d. Tip of involucral bract lacking tuft of white hairs . . . . *A. amplexicaulis*
    c. Stem-leaves mostly 2-4 pairs, narrowly lanceolate, the lowermost long-petioled . . . . *A. mollis*
a. Pappus milky white
  f. Heads solitary, nodding . . . . *A. louiseana* ssp. *frigida*
  f. Heads erect, 1 to several
    g. Basal and lower stem-leaves elliptic or cordate, usually long-stalked
      h. Achenes hirsute . . . . *A. cordifolia*
      h. Achenes glabrous . . . . *A. latifolia*
    g. Basal leaves narrowly lanceolate
      i. Basal leaves long-stalked, the blade denticulate . . . . *A. lonchophylla*
      i. Basal leaves sessile or short-stalked, the blade entire-margined
        j. Flowering stems 20-30 cm tall; heads mostly solitary, or sometimes with a pair of somewhat smaller, lateral heads
          k. Involucral bracts and the peduncle densely white woolly . . . . *A. alpina* ssp. *tomentosa*
          k. Involucral bracts and peduncle not woolly . . . . *A. alpina* ssp. *angustifolia*
        j. Flowering stem commonly 30-50 cm tall, heads 3-7, long-peduncled. . . . . *A. alpina* ssp. *attenuata*

Arnica alpina (L.) Olin
ssp. **angustifolia** (J. Vahl) Maguire

Tufted, with a short, stout rhizome, terminating in a rosette of lanceolate, petioled leaves, and a simple stem, 20 to 30 cm high, with one to two pairs of linear, sessile leaves, and one large, or less often two smaller lateral heads.

Dry, sandy or gravelly slopes, but often also below bird cliffs and near animal dens.

General distribution: N. America, wide ranging, arctic-alpine .

Fig. 902 Map 1043

Arnica alpina (L.) Olin
ssp. **attenuata** (Greene) Maguire

Differs from *A. alpina* ssp. *angustifolia* by its up to 50 cm tall flowering stems, longer, narrower and sub-sessile basal leaves, and by having one large and often 4 smaller lateral heads.

River banks and flood-plains, sandy or gravelly hills or grassy open woods.

General distribution: N. America; from James Bay to central eastern Alaska north to or slightly beyond the limit of trees.

Map 1044

Arnica alpina (L.) Olin
ssp. **tomentosa** (J.M. Macoun) Maguire
Similar in habit to *A. alpina* ssp. *angustifolia* from which it differs by its somewhat smaller size and by its copious villose pubescence of the involucre, the flowering peduncle and both surfaces of the leaves; also the ligules are of paler yellow than in ssp. *angustifolia*.

Arnica tomentosa was described as being "generally monocephalous" but specimens with 2 or even 3 heads are by no means uncommon, within its western range.

Stony and gravelly slopes.

General distribution: Alpine-arctic from Montana north through the Canadian Rocky Mts., the Mackenzie and Richardson Mts., with disjunct stations in Great Bear L. and Nfld.

Fig. 903 Map 1045

Arnica amplexicaulis Nutt.
Stems 5 to 6 dm tall, finely rough-pubescent, few or several together from a short and stout rhizome; cauline leaves commonly 5 to 6 pairs, ovate or broadly lanceolate, those of the middle largest; the lowermost smallest and short-petioled, the rest sessile, their margins conspicuously and sharply dentate. Leaves of sterile basal offshoots with slender petioles as long as the blade. Heads 3 to 5, broadly campanulate, on slender peduncles. Achenes hirsute.

Subalpine, moist woods.

General distribution: N. America; Cordilleran—Pacific N.W. In the District of Mackenzie thus far known only from two stations on the east slope of Mackenzie Mts.

Fig. 904 Map 1046

Arnica Chamissonis Less.
ssp. **foliosa** (Nutt.) Maguire
Stems solitary, leafy, from a slender rhizome. Leaves fresh green, sparingly pubescent and slightly glutinous, their margins entire or finely denticulate, the upper lanceolate and sessile, the lower about 10 cm long and tapering into a narrow and purplish base. Heads 3 to 7, broadly campanulate, the lowermost generally long-peduncled; involucral bracts fresh green; ligules pale yellow and the ripe achenes coal-black, sparingly barbellate, about 4 mm long, their pappus pale straw-coloured.

Moist grassland.

General distribution: N. America; from James Bay to the western foot-hills north to the upper Mackenzie Valley and south-central Yukon.

The ssp. *incana* (Gray) Maguire of similar habit as ssp. *foliosa* but distinguished by its silvery canescent-tomentose stems and foliage, is known from damp prairies of southern Yukon Territory and from the lower Slave, and Upper Mackenzie Valleys south to Calif.

Fig. 905 Map 1047 (ssp. *foliosa*), 1048 (ssp. *incana*)

Arnica cordifolia Hook.
Stems 40 to 50 cm high, singly or few together from a slender rhizome. Stem-leaves 3 to 4 pairs, the lower, and also those of sterile offshoots, with cordate and shallowly serrate blades on slender petioles as long as the blades, the upper progressively smaller and nearly sessile. Flowering heads 1 to 3, the terminal 2 to 3 cm in diameter; achenes hirsute, and the pappus milky white.

Moist *Abies lasiocarpa* forest.

General distribution: Cordilleran species barely entering our area along the lower Liard R. and along the east slope of the Mackenzie Mts.

Fig. 906 Map 1049

Arnica latifolia Bong.
Similar to *A. cordifolia* but a more delicate and not so tall plant, from a stout, ascending rhizome. Basal leaves, and those of sterile offshoots, with thin, cordate or oval blade, on a slender petiole, usually twice as long as the blade; cauline leaves smaller, usually three pairs, oval in outline, the lowermost short-petioled. Heads 1 to 3, much smaller than in *A. cordifolia;* achenes glabrous, their pappus white.

Alpine meadows.

General distribution: Cordilleran—Pacific Coast species which has barely crossed the Mackenzie—Y.T. watershed in MacMillan Pass and on the upper Flat R.

Fig. 907 Map 1050

Arnica Lessingii Greene
Stems solitary, 10 to 25 cm tall from a slender, ascending rhizome. Leaves oblanceolate, with finely serrulate margins, and mainly basal. Head solitary, always nodding. Achenes pale brown, cylindrical, nearly glabrous. Pappus brownish-yellow; the anthers, unlike those of

the other *Arnicas* of our area, are dark purplish.

Alpine lichen heath.

General distribution: Amphi-Beringian, arctic-alpine, reaching the eastern slopes of Mackenzie and Richardson Mts.

Fig. 908 Map 1051

Arnica lonchophylla Greene

Stems slender, up to 40 cm tall, from a scaly, ascending and somewhat branching rhizome. Cauline leaves 2 to 3 pairs, narrowly lanceolate, sessile, about 3 cm long; those of the base and the innovations 10 to 15 cm long, the blade narrowly lanceolate, rather prominently 3-nerved, denticulate, gradually tapering into a slender petiole about as long as the blade. Heads small, 3 (5), campanulate, the involucrum somewhat glandular, the terminal and the largest head 1.0 to 1.4 cm in diameter. Achenes hirsute; pappus white.

A lowland species of similar range as *A. alpina* ssp. *attenuata* from which it is at once distinguished by its slender habit and much smaller heads.

River banks, sandy or gravelly places.

General distribution: N. America, from Hudson Bay to central Alaska, north to or slightly beyond the limit of trees.

Fig. 909 Map 1052

Arnica louiseana Farr.
ssp. **frigida** (Meyer ex Iljin) Maguire

Stems simple, 10 to 30 cm tall, from an ascending, branched and black-scaly rhizome. Leaves mainly basal, the blade oblanceolate or obovate, 3 to 6 cm long, narrowing to a short, winged petiole, glabrate, or sparingly pubescent especially along the denticulate margins. Heads solitary, nodding in youth, the involucrum campanulate, dark purple, yellowish woolly in the lower half; peduncles yellowish woolly above.

Alpine herbmats.

General distribution: Amphi-Beringian arctic-alpine.

Fig. 910 Map 1053

Arnica mollis Hook.

Stems simple, 20 to 40 cm tall, one to several from an ascending and branched rhizome. Cauline leaves sessile, 3 to 4 pairs, narrowly ovate or lanceolate, 3 to 6 cm long, their margins denticulate; basal leaves and those of sterile offshoots distinctly petioled. Heads solitary, or less often with one or two lateral and smaller heads. Achenes hirsute; pappus tawny.

Alpine meadows.

General distribution: N. American Cordilleran species north of lat. 53° known only from a few widely disjunct alpine stations in the southern Mackenzie Mts., N.W.T., and in mountains of southern Y.T. and Central Alaska.

Fig. 911 Map 1054

Artemisia L. Wormwood

Ours chiefly perennial herbs from a sub-ligneous and often much branched base, several strongly aromatic; leaves mostly alternate, entire, 3-lobed or more commonly once or twice pinnate; heads small, commonly nodding, in panicled, racemose or spicate inflorescences.

a. Biennial . . . . *A. biennis*
a. Perennials
  b. Plants with a stout rhizome
    c. Leaves entire, or the lowermost, 3-cleft
      d. Leaves glabrous . . . . *A. Dracunculus*
      d. Leaves strongly tomentose . . . . *A. ludoviciana* var. *gnaphalodes*
    c. Leaves pinnatifid, glabrate or glabrous above, thinly tomentose beneath . . . . *A. Tilesii*
  b. Plants distinctly caespitose, with a strong tap-root ending in a much branched crown
    e. Leaves green, glabrous or glabrate
      f. Involucral bracts glabrous or nearly so
        g. Heads numerous, 2-3 mm in diameter, in a simple or more often compound raceme . . . . *A. canadensis*
        g. Heads fewer, 6-8 mm in diameter, mostly solitary and nodding, on a slender peduncle . . . . *A. arctica*
      f. Involucral bracts villous . . . . *A. borealis*

e. Leaves distinctly grey from a close indument
  h. Plant distinctly hoary; the more or less sessile heads in a short, dense spike . . . . *A. Richardsoniana*
  h. Plant silvery-canescent
    i. Plant not mat-forming; heads 5-6 mm in diameter
      j. Involucral bracts yellowish, with a pale darker margin; heads erect, not nodding on their peduncles . . . . *A. hyperborea*
      j. Involucral bracts with prominent dark margins; heads long-peduncled, nodding . . . . *A. alaskana*
    i. Plant mat-forming; heads 3-4 mm in diameter, nodding, on short peduncles . . . . *A. frigida*

Artemisia alaskana Rydb.
A. Tyrrellii Rydb.
Caespitose, from a stout, sub-ligneous and many-headed tap-root. Stems simple, ascending, somewhat flexuous, 15 to 40 cm tall, and like the leaves, light-grey tomentose. Heads mostly solitary on slender peduncles, the uppermost sessile. Basal leaves fan-shaped, bi- or tri-pinnate, the ultimate divisions linear, flat, with a blunt tip.

Calcareous cliffs and screes.

General distribution: Endemic of alpine parts of Alaska—Y.T., and in the District of Mackenzie where, thus far, it is known from a few collections on the east slope of the Richardson Mts.

Map 1055

Artemisia arctica Less.
ssp. **arctica**
Caespitose, from a stout, erect-ascending, sub-ligneous and freely branching base. Stems singly or few together, 20 to 50 cm tall, glabrous and fresh green or purplish. Leaves mainly basal, those of the sterile off-sets 5 to 20 cm long, with slender petioles as long as the twice pinnately divided blade. Flowering heads mostly solitary and nodding, up to 10 mm in diameter, on slender up to 5 cm long peduncles. Involucral bracts 5 mm long and 4 mm broad, with scarious black margins strongly contrasting with the narrower green centre.

Moist alpine meadows.

General distribution: Amphi-Beringian species extending south in the Rocky Mts. beyond lat. 60°, but eastward only along the east slope of the Richardson—Mackenzie Mts.

The ssp. *comata* (Rydb.) Hult. differs from ssp. *arctica* by its denser, spike-like inflorescence of somewhat larger and nearly sessile heads, and is endemic to northern Alaska, northern and central Yukon Territory, east to the Richardson Mts., west of the Mackenzie Delta.

Map 1056

Artemisia biennis Willd.
Glabrous, weedy biennial from a simple and slender tap-root. Stems 3 to 10 dm tall. Leaves bi-pinnate, the secondary divisions oblong, irregularly toothed. Heads small and numerous, in dense, leafy and sub-sessile panicles.

General distribution: A cosmopolitan weedy species, in the District of Mackenzie confined to waste places near human habitations along the Mackenzie R. and tributaries, north to the Arctic Circle.

Artemisia borealis Pall.
Stems simple or branched, 1 to 3 dm tall, glabrate or somewhat pubescent; basal leaves on long, slender petioles, the blade bi-pinnate, with linear divisions; inflorescence elongate, of small globular heads, 3 to 4 mm in diameter. Plant not strongly aromatic.

Dry, sandy places, often by sea or lake shores, or on sandy river banks.

General distribution: Circumpolar, subarctic.

Fig. 912 Map 1057

Artemisia canadensis Michx.
Biennial or short-lived perennial with 50 to 70 cm tall and commonly dark-purplish stems, singly or several together from a stout tap-root. Leaves fresh green, bi-pinnately divided, the leaflets linear, about 1 mm wide, silky canescent. Flowering heads globular, 2 to 3 mm in diameter, sessile or slightly nodding, in narrow, many-headed, leafy panicles.

Open shallow soil over rocks, sandy banks and disturbed situations.

General distribution: Nfld. to B.C. and Alaska, north in the Mackenzie Valley to slightly beyond the Arctic Circle.

Fig. 913 Map 1058

***Artemisia Dracunculus** L.
Essentially glabrous and fresh green perennial with leafy stems up to 1 m high, from a stout

sub-woody base. Leaves linear, 2 to 5 cm long, entire or the lowermost ternately divided. Heads small, 2 to 4 mm high, sessile or on slender leafy branches.

General distribution: A wide ranging nearly circumpolar weedy prairie or grassland species to be expected, but as yet not reported from north of lat. 60° on the Slave R.

Map 1059

Artemisia frigida Willd.
Prairie-Sagewort
Mat-forming, strongly aromatic perennial from a stout, sub-woody and much branched central tap-root. Basal leaves slender-petioled, silvery grey-tomentose, each blade twice ternately divided into linear, about 1 mm wide segments, the petiole approximately as long as the blade. Flowering stems slender, erect-ascending, 3 to 5 dm tall.

Locally dominant on river terraces, foothills and subalpine plains.

General distribution: Central Asia, Western N. America from Alaska to Sask. and eastward, where naturalized.

Map 1060

***Artemisia globularia** Cham.
Tufted, canescent dwarf species from a stout, branching tap-root. Basal leaves crowded, their blades once or twice ternately divided and tapering into a short, flat petiole. Inflorescence globular, of 5 to 8 dark heads; involucral bracts brown with black, scarious margins; the flowers purplish-black.

Alpine, gravelly slopes.

General distribution: Amphi-Beringian with disjunct stations in the British Mts. of northern Y.T., and to be looked for along the east slope of Richardson Mts.

Map 1061

***Artemisia glomerata** Ledeb.
Tufted or matted from a stout branching base; basal leaves short, twice ternately divided. Flowering stems slender, 10 to 12 cm tall with 2 or 3 cauline leaves similar to the basal. Inflorescence a short, elongating raceme of small, yellow heads; the involucral bracts oblong, woolly, with narrow, black margins.

Sandy or gravelly slopes.

General distribution: Amphi-Beringian, arctic-alpine, barely entering northern Y.T., and to be looked for in the Richardson Mts. west of the Mackenzie Delta.

Map 1062

Artemisia hyperborea Rydb.
A. furcata aucct. non M. Bieb.
Caespitose with numerous sterile rosettes from a subligneous, multicipital caudex. Leaves strongly aromatic, about 3 cm long, twice pinnate, finely silvery tomentose, the blade about 1 to 1.5 cm long, the ultimate divisions flat and obtuse. Flowering stems 10 to 20 cm high, simple, thinly appressed tomentulose, bearing from 2 to 4 reduced leaves; inflorescence racemose, at first capitate but soon elongating, becoming 6 to 8 cm long, with from 15 to 25 erect heads, on 3 to 5 mm long, woolly peduncles, each supported by a leafy bract. Heads about 5 mm broad and 4 mm high; phyllaries broadly oval with a woolly, greyish-green centre and a broad, dark brown scarious margin; the disc flowers are fertile, the corollas bright yellow, and the receptacle naked.

Dry and sunny calcareous rocks.

A. hyperborea has erroneously been referred to the Amphi-Beringian *A. furcata* M. Bieb. which is a much smaller and more delicate plant.

General distribution: Endemic of N.W. District of Mackenzie and the southwestern Arctic Islands, with a somewhat disjunct population in the southern Mackenzie Mts., St. Elias Mts. of S. W. Y.T., and N. W. Alaska.

Fig. 914 Map 1063

Artemisia ludoviciana Nutt.
var. **gnaphalodes** (Nutt.) T. & G.
Perennial with often clustered, 3 to 9 dm tall stems from a slender, horizontally creeping rhizome; leaves entire or somewhat toothed, narrowly lanceolate, densely grey-velvety tomentose. Heads small and numerous in a leafy panicle.

Prairie and open woodland species barely reaching the District of Mackenzie south of Great Slave Lake.

General distribution: N. America: S. Ont. to B.C. and southward.

Map 1064

Artemisia Richardsoniana Bess.
Densely caespitose from a stout tap-root. Flowering stem 8 to 25 cm tall bearing 2 to 3 simple and linear leaves. Inflorescence first compact and almost capitate, tardily elongating, the lower heads each supported by a leafy bract. Basal leaves numerous, short-pinnate or 5 parted. Entire plant strikingly hoary-villous.

Dry calcareous, sandy or gravelly places.

General distribution: Endemic of the western

N. American Arctic from N.W. Alaska to Banks and Victoria Islands and the Arctic Coast east to Coronation Gulf.
Fig. 915 Map 1065

Artemisia Tilesii Ledeb.
Stems freely branched, 10 to 50 cm high, from a thin rhizome. The large, pinnately lobed leaves green and glabrous above, grey-canescent beneath. Heads small, somewhat nodding. The var. *elatior* T. & G., with broader leaf-segments is a taller plant of less northern range.
In sandy places, often near the beach, often of weedy habit.
General distribution: Asia, N.W. America, arctic-subarctic; the var. *Tilesii* is Amphi-Beringian and the var. *elatior* N. American, wide ranging east to James Bay.
Fig. 916 Map 1066

Aster L. Aster

Late-flowering, mostly leafy, perennial herbs (except the annual *A. brachyactis*) with alternate, entire or toothed leaves, and few (rarely solitary) to many flowering heads. The heads mainly radiate, the rays pistillate, white, blue or pink. In general appearance similar to some members of the genus *Erigeron* in which, however, nearly all members are early-flowering.

a. Stems leafy, usually over 6 dm tall
  b. Stems crisply pubescent; leaves retrorsely pubescent . . . . *A. puniceus*
  b. Stems and leaves essentially smooth
    c. Basal and lower stem leaves abruptly narrowed into a winged petiole longer than the broadly ovate blade . . . . *A. ciliolatus*
    c. Basal and lower stem leaves lanceolate, gradually narrowed into a clasping base
      d. Heads small, in a compound panicle . . . . *A. laevis* var. *Geyeri*
      d. Heads larger but fewer, in a leafy cluster
        e. Leaves lanceolate, fresh green, the lower 6-10 cm long and about 2 cm wide . . . . *A. modestus*
        e. Leaves linear-oblong, dark green, 6-8 cm long and 0.5-0.7 cm wide . . . . *A. Franklinianus*
a. Stems less than 6 dm tall
  f. Heads mostly solitary (low arctic-alpine species)
    g. Leaves lanceolate, serrate . . . . *A. sibiricus*
    g. Leaves oblanceolate, entire
      h. Leaves mainly basal, rough-pubescent; ray flowers white or lavender . . . . *A. alpinus* ssp. *Vierhapperi*
      h. Leaves mainly cauline, soft-pubescent; ray flowers purple
        i. Stems ascending . . . . *A. pygmaeus*
        i. Stems erect . . . . *A. yukonensis*
  f. Heads few to many (non-arctic lowland species)
    j. Annual; ray-flowers lacking . . . . *A. brachyactis*
    j. Perennial; ray-flowers present; leaves linear, linear-lanceolate or narrowly oblanceolate, 8-12 times longer than broad
      k. Heads terminal at the end of the stem or weak branches . . . . *A. nahanniensis*
      k. Heads in panicles or racemes
        l. Heads in open panicle
          m. Stems tufted, from a short and fibrous root . . . . *A. pauciflorus*
          m. Stems solitary from a slender, horizontal rhizome
            n. Stem and leaves commonly dark purple; heads small, the involucre about 10 mm broad; ray flowers white . . . . *A. junciformis*
            n. Stem and leaves green, heads larger with 15 mm broad involucre; ray flowers pink . . . . *A. spathulatus*
        l. Heads in dense racemes
          o. Stems clustered from a short, thick and ascending rhizome; heads small and numerous; leaves numerous 1-2 cm long and hairy . . . . *A. pansus*
          o. Stems solitary from a horizontal rhizome; heads few and larger; leaves 2-3 (-8) cm long, smooth . . . . *A. falcatus*

Aster alpinus L.
ssp. **Vierhapperi** Onno
Stems from an ascending, subligneous, stout and branched rhizome. Basal leaves numerous, oblanceolate, obtuse, 3 to 6 cm long and 0.6 to 1.0 cm wide, the cauline leaves 5 to 8, linear, the uppermost much reduced. The leaves and stems rough from stiff, spreading hairs. Flowering stems simple, stiffly erect, not ascending, 10 to 15 (25) cm tall; head solitary and showy, 3 to 4 cm in diameter; ligules papery, white or lavender. Entire plant viscid-mealy in life due to tiny resinous glands.

Dry, sunny slopes and terraces chiefly on calcareous rock.

General distribution: Cordilleran; north along the Mackenzie R. and its tributaries to Great Bear L. and to the east slope of the Richardson Mts. west of the Mackenzie Delta.

Fig. 917 Map 1067

Aster brachyactis Blake
A. angustus (Lindl.) T. & G.
Annual from a weak tap-root; stems simple or bushy-branched, 3 to 6 dm tall. Leaves linear or narrowly lanceolate with short, ciliate margins, all cauline and sessile and usually greenish-purple. Heads small but numerous, in an open panicle, solitary at the end of leafy branches; rays mostly lacking, and the disc flowers usually concealed by the pappus.

Damp calcareous or saline river banks and lake shore meadows.

General distribution: Boreal N. America from Que. to S. E. Y.T. and the upper Mackenzie drainage.

Fig. 918 Map 1068

Aster ciliolatus Lindl.
A. Lindleyanus T. & G.
Tufted, essentially glabrous perennial readily spreading by slender rhizomes. Stems simple or much branched above, up to 1 m tall; stem leaves broadly lanceolate, the blade 8 to 10 cm long and 3 to 4 cm wide, with shallow, antrorsely serrated margins; the lower as well as the basal leaves with a winged petiole equalling or longer than the blade. Heads few to many, in an open panicle; involucral bracts 7 to 8 mm long, narrowly lanceolate with a green centre, and prominent white-leathery margins; the rays pale violet or light blue.

Open woodland thickets or prairie.

General distribution: N. America: B.C. to N.S., north along the Mackenzie R. and its tributaries to Great Bear L.

Fig. 919 Map 1069

Aster falcatus Lindl.
Stems 3 to 6 dm tall, mostly solitary; ascending, from an elongated rhizome. Leaves linear-oblong, 2 to 3 cm long and 3 to 4 mm wide, with a spiny tip, blue-green and densely covered on both sides by short and stiff hairs. Heads about 8 to 10 mm in diameter, mostly solitary, at the end of short, lateral branches; the rays white, narrowly lanceolate.

General distribution: Western prairie species barely entering the District of Mackenzie west of Fort Smith.

Fig. 920 Map 1070

Aster Franklinianus Rydb.
A. salicifolius Richards. *non* Lam.
Essentially glabrous perennial from a slender, horizontal rhizome. Stems slender, up to 1 m tall, simple or branching above. Leaves entire-margined, 8 to 10 cm long and 6 to 10 mm wide, gradually tapering from a more or less auriculate-clasping base. Heads few to many, 1.5 to 2.0 cm in diameter, in an elongating, leafy raceme. Bracts not clearly 3-seriate, linear, of unequal length, with a narrow, green centre and broader, white margins; the rays pale blue.

River banks and sheltered lake shore meadows.

General distribution: A western prairie species, in the District of Mackenzie common locally along the Mackenzie R. and its tributaries, north to about lat. 62°.

Fig. 921 Map 1071

Aster junciformis Rydb.
Similar to *A. Franklinianus* but a lower and more delicate plant with stems rarely over 3 dm tall, fewer and smaller heads, and shorter and narrower leaves. Rays always white, and the involucral bracts distinctly 3-seriate, and each bract uniformly dark green. By the latter character alone *A. junciformis* is readily distinguished from small and weak specimens of *A. Franklinianus*.

Damp, mossy openings in muskeg forest.

General distribution: Wide ranging from N. S. to Alaska, north along the Mackenzie R. and its tributaries to the Arctic Circle.

Fig. 922 Map 1072

***Aster laevis** L.
var. **Geyeri** A.Gray
Essentially glabrous perennial from a short, ascending rhizome. Stems simple, 4 to 6 dm tall, purplish. Leaves entire, somewhat leathery, the basal 10 to 20 cm long, the blade lanceolate, as long as the narrowly winged petiole. Cauline

leaves 8 to 10 cm long, with a clasping or auriculate base. Heads small, in a compound panicle; involucres 6 to 8 mm high, the bracts imbricated, linear, each with firm white margins and narrow, green centre and tip; rays blue or deep purple; pappus tawny.

Foothill prairies.

General distribution: From Ont. west to northern B.C. and in Alta. north almost to lat. 60°. To be looked for in the upper Mackenzie drainage.

Map 1073

***Aster modestus** Lindl.

Stems mostly single, 6 to 8 dm tall from a slender horizontally creeping rhizome, slender and very leafy, reddish-brown, smooth or short pubescent, and stipitate glandular above. Leaves sessile, lanceolate, 6 to 10 cm long and 1 to 2 cm broad, soft pubescent beneath, their margins with a few sharp teeth near the apex. Heads few to a dozen or more, in an open leafy panicle. Involucral bracts linear, glabrous or somewhat glandular, about 10 mm long; rays linear and purplish.

Moist woodland thickets and stream banks.

General distribution: N. America from Man. to B.C. north to central Y.T. and the upper tributaries of the Mackenzie R., but as yet not reported from the District of Mackenzie.

Map 1074

Aster nahanniensis Cody

Stems weakly branched, to 3.5 dm in height; leaves glabrous, but upwardly scabrous on the margins, the basal short-spatulate, blunt, to 1.3 cm long, the stem leaves linear to linear-lanceolate, acute, to 6.5 cm long, sessile and partly clasping, the leaves of the branches 0.5 to 1.8 cm long, oblong to ovate-oblong, acutish; heads 2 cm or more in diameter, terminal on the stem or the weak branches; bracts in 3 or 4 ranks, the outer spatulate, the inner lanceolate, their margins papery and erose-ciliate, tinged with purple; rays white, the ligule about 1.5 cm long, disc florets yellow; achenes 2 to 3 mm long, pubescent, the pappus white.

By warm springs in the mountains.

General distribution: Endemic to the southern Mackenzie Mts., N.W.T.

Map 1075

Aster pansus (Blake) Cronq.

A. ericoides aucct. non L.

Stems one to several, clustered, from a short erect rhizome; the entire plant glaucous-green and rough-hairy from spreading, stiff hairs. Leaves small and narrow, very numerous. Inflorescence racemose, rather dense, with numerous spreading one- to several-headed branches; involucral bracts imbricated, of unequal length, prominently ciliate, their tips spreading. Rays white, turning yellowish in drying. By its clustered stems and erect rootstock alone, readily distinguished from the usually lower *A. falcatus* which has a slender, horizontal rhizome and fewer and larger heads.

Grassy clearings.

General distribution: A western grassland species in Canada ranging north from Alta. to the upper Mackenzie R., southern Y.T. and central Alaska.

Fig. 923 Map 1076

Aster pauciflorus Nutt.

Stems slender, 2 to 4 dm tall, simple or freely branched from a short base, glabrous below, short stipitate-glandular above. Leaves glabrous, narrowly lanceolate, 3 to 12 cm long and 2 to 4 mm wide. Heads few to many, singly on the branches of the simple to freely branched inflorescence. Heads very variable in size, their involucral bracts stipitate-glandular, 4 to 6 mm long, in a single row. Rays lilac or white.

Wet, marshy and usually saline meadows or lake shores.

General distribution: N. America: Man. to Alta. foothills, north just beyond lat. 60° on the Salt Plain west of Fort Smith.

Map 1077

***Aster puniceus** L.

Stems from a thick, ascending rhizome, stout, from 6 to 15 dm tall, reddish-purple and crisply pubescent, leafy and usually much branched above. Leaves 8 to 12 cm long, 1 to 2 cm broad, sessile, oblong or lanceolate with a clasping and usually auriculate base, finely scabrous on both surfaces, and the margins shallowly toothed towards the apex. Heads 5 to 8 on each branch, relatively large and showy; involucral bracts linear, about 10 mm long, in a single row, glabrous or their margins ciliate in the lower half. Rays about 2 cm long, pale blue or purplish.

In moist or swampy places.

General distribution: N. America; Nfld. to the Alta. foothills. To be looked for in the upper Mackenzie R. drainage.

Map 1078

Aster pygmaeus Lindl.

Dwarf species from a branching base; stems 4 to 11 cm tall, ascending, purple, somewhat woolly, especially in the lower part, and densely so below the involucrum; radical and stem leaves sessile, narrowly oblong, glabrous or woolly along the margins; flowering heads single, involucrum densely white-woolly, the bracts linear-oblong, dark purple; ray flowers lilac; disc flowers numerous, the tube yellow with purple lobes; pappus reddish-brown; achenes hirsute. The somewhat similar, and perhaps closely related *A. yukonensis* is taller, the stems simple, erect, and the leaves linear and longer than in *A. pygmaeus*.

Gravelly places.

General distribution: A rare and little collected species apparently endemic to the western arctic islands and adjacent mainland.

Fig. 924 Map 1079

Aster sibiricus L.

Freely branched leafy perennial from a slender, spreading and freely branched rhizome. Stems mostly several together, 2 to 3 dm tall, glabrate or copiously woolly. Leaves variable from lanceolate to oblanceolate, sessile or short-petioled, generally hairy beneath and glabrous or glabrate above, their margins generally serrate. Heads one to several; involucral bracts mostly in two series, those of the outer series green. Rays purple and the pappus reddish-brown.

Gravelly river flats, dry meadows or in open woods.

General distribution: Amphi-Beringian species extending over Alaska and Y.T. well into the District of Mackenzie, north to the Arctic Coast; in the Rocky Mts. represented by the var. *meritus* (A.Nels.) Raup which differs from *A. sibiricus* mainly in its smaller heads and by its scabrous rather than pubescent leaves and stems.

Fig. 925 Map 1080

Aster spathulatus Lindl.

Essentially glabrous perennial with slender 3 to 4 dm tall branching stems from a slender, horizontally spreading rhizome. Leaves oblanceolate, entire, 8 to 10 cm long and 0.4 to 0.5 cm broad, with a clasping base. Heads 3 to 4, the involucral bracts of equal length, in 2 to 3 series, the individual bracts linear, glabrous except for the ciliate margins, the lower half whitish, the upper green. Rays lilac, drying to yellow.

River banks and lake shores.

General distribution: Endemic of central Mackenzie drainage.

Fig. 926 Map 1081

Aster yukonensis Cronq.

Stems leafy, 1 to several, erect, sparsely white-villous, not glandular, 10 to 15 cm tall from a branching tap-root; leaves mainly cauline, clasping, linear, 2 to 3 mm broad, with a prominent midrib, sparsely white-villous. Heads 1.5 to 2.0 cm in diameter, solitary, or less often with a second and smaller head; involucral bracts linear, with a dark tip, villous but not glandular; ligules 20 to 25, pale blue, linear and about 1 mm broad; pappus pale brown; achenes sparsely hirsute. The somewhat similar *A. pygmaeus* differs by its always ascending stems, broader and shorter leaves, and distinctly arctic range.

Subalpine, in stony and silty and sometimes saline places.

General distribution: Described from Kluane L. in S.W. Y.T., *A. yukonensis* is now known also from Bettles in northern Alaska, and from the Mackenzie Mts. and middle Mackenzie R. valley.

Map 1082

Bidens L. Bur-Marigold

Bidens cernua L.

Glabrous, weedy annual with freely branching stems 15 to 40 cm high; leaves opposite and connate, the blades oblanceolate, sub-entire or often coarsely serrate. Flowering heads yellow, radiate, globular, and often nodding in fruit. Achenes cuneate, the pappus consisting of 4 retrorsely toothed awns.

Moist meadows, stream banks and lake shores.

General distribution: Circumpolar, often of weedy habit.

Fig. 927 Map 1083

Chrysanthemum L. Daisy

Perennial (ours) herbs from a branching base or short rhizome; leaves glabrous, entire or toothed; heads many-flowered, solitary, or few; ray-flowers white, pistillate, disc-flowers yellow, perfect; involucral bracts in 2 to 3 series.

a. Leaves linear, entire . . . . *C. integrifolium*
a. Leaves crenate or lobed
  b. Basal leaves cuneate, glabrous and fleshy . . . . *C. arcticum*
  b. Basal leaves spatulate . . . . *C. leucanthemum*

Chrysanthemum arcticum L.
Stems erect-ascending, 15 to 25 cm high from a branching base, with one (or very rarely 2 or 3) heads, 3 to 4 cm in diameter.
Moist, saline meadows, rocky crevices, or in moist gravel by the sea shore.
General distribution: Circumpolar, low-arctic; in N. America from Alaska and Y.T. east beyond the Mackenzie R. Delta, and reappearing again along the shores of Hudson and James Bays.
Map 1084

Chrysanthemum integrifolium Richards.
Tufted dwarf perennial from a branched base. Leaves basal, linear, 15 to 20 mm long. Flowering stems 5 to 10 cm long, woolly above, with one or more reduced leafy bracts, terminating in a single head from 1.0 to 1.5 cm in diameter.
Stony, calcareous barrens.
General distribution: N. America, wide ranging arctic-alpine.
Fig. 928 Map 1085

Chrysanthemum leucanthemum L.
Ox-eye Daisy
Glabrous perennial with leafy, erect, simple or forked stems up to 8 dm tall, from a short, subligneous rhizome; leaves in a basal rosette of spatulate, toothed or pinnatifid leaves; heads solitary, 3 to 4 cm in diameter, on bracted peduncles.
Meadows and roadsides.
General distribution: Cosmopolitan weed, with us reported only from near Ft. Smith.

Cirsium Mill. Thistle

Perennial herbs (ours) with alternate, prickly, pinnatifid leaves; heads many-flowered, hemispherical or globose; involucral bracts in several rows, spine-tipped; corollas all alike, tubular and deeply 5-cleft; pappus of slender, plumose bristles.

a. Native, monoecious, from a stout tap-root
  b. Leaves not tomentulose; head usually solitary, corollas reddish-purple . . . *C. Drummondii*
  b. Leaves more or less tomentose; heads usually several and clustered; corollas white or pale pink . . . . *C. foliosum*
a. Introduced, dioecious and weedy, from a deeply buried rhizome; stems and leaves prickly, but not tomentulose . . . . *C. arvense*

Cirsium arvense (L.) Scop.
Dioecious, very prickly but otherwise glabrous perennial from a deeply buried and freely branched rhizome. Heads small in a much branched panicle, those of the male plant with red, and the female with purple flowers.
General distribution: Cosmopolitan, roadside weed, reported once from Ft. Simpson.

Cirsium Drummondii T. & G.
Biennial herb, acaulescent or sometimes 15 to 30 cm high from a tap-root; stem unbranched, fleshly, with deeply lobed, marginally spiney, glabrous to pilose, spreading ascending leaves; heads 1 to several in a compact group, 5 to 6 cm high, the phyllaries in several appressed rows; flowers rose purple.
Prairies.
General distribution: N.W. Ont. to E. B.C. north in the Mackenzie drainage to Bear R.
Map 1086

Cirsium foliosum (Hook.) DC.
Biennial herb 25 to 70 cm high from a tap-root; stem unbranched, fleshy, hidden by the numerous ascending shallowly lobed marginally spiney lightly tomentose leaves; flower heads several, 3 to 4 cm high; the olive green phyllaries in 4 or 5 appressed rows; flowers white or pale pink.
Prairie and foothills.
General distribution: B.C. and Alta. north to the lower Slave R. and S.E. Y.T.
Map 1087

Crepis L. Hawk's-beard

Herbs with milky juice; leaves alternate or basal; heads 1 to several. Flowers yellow, ligulate and perfect; achenes cylindric, many-ribbed, beakless (ours), pappus of capillary white bristles.

a. Perennial native species
  b. Stemless; leaves spreading in a flat basal rosette; heads singly, or few together, barely emerging above the rosette . . . . *C. nana*
  b. Stems erect, branched . . . . *C. elegans*
a. Annual, weedy species with branching, leafy stems . . . . *C. tectorum*

Crepis elegans Hook.
Tufted dwarf perennial from a central many-headed tap-root. Basal leaves purplish-green, 3 to 7 cm long and usually entire, the oval blade tapered into a long and narrow petiole; in exceptionally vigorous specimens the blade of some, or all leaves may be deeply toothed or even runcinate. Flowering stems 12 to 14 cm tall, branched above the middle. The small, cylindrical heads in terminal clusters of 2 to 5. Ligules pale purple, the pappus short and pure white.

Sandy flood plains and river banks.

General distribution: Western N. America; central Alaska, Y.T., and western District of Mackenzie south to Wyoming.

Fig. 929 Map 1088

Crepis nana Richards.
Similar to *C. elegans* but of more dwarfed and condensed habit; the small yellow heads crowded and barely emerging above the leaves.

Dry, calcareous screes or gravelly places.

General distribution: Wide ranging arctic-alpine from central and eastern Asia, across N. America to Nfld. and south in the Rocky Mts.

Fig. 930 Map 1089

Crepis tectorum L.
Weedy annual. Stems simple, branching above; stem leaves linear, entire, the lower variously incised. Heads few to many in an open panicle. Ligules pale yellow.

General distribution: Cosmopolitan weed of roadsides and town sites in the southeastern part of our area.

Erigeron L. Fleabane

Early flowering, and mainly low to medium-sized perennial or less commonly annual or biennial herbs with leafy, or nearly scapose stems. Leaves alternate, in ours mostly entire, linear or oblanceolate, but ternately divided in *E. compositus*. Heads solitary, or in open, few-headed racemes or corymbs; involucrum hemispherical, mostly somewhat woolly; ray flowers pistillate, white, lilac, pink or purplish; disc-flowers perfect, with yellow corollas; achenes flattened, the pappus capillary, in ours light yellowish-grey or purplish.

a. Leaves not entire-margined; heads solitary on naked peduncles
  b. Leaves ternately lobed or dissected . . . . *E. compositus*
  b. Leaves, at least the early ones, shallowly lobed . . . . *E. pallens*
a. Leaves entire (sometimes toothed in *E. acris*), heads mostly several
  c. Ligules short (or sometimes lacking)
    d. Annual; heads 5-6 mm in diameter and very numerous . . . . *E. canadensis*
    d. Biennial, or short-lived perennial; heads 8-15 mm in diameter
      e. Heads on short, erect branches . . . . *E. lonchophyllus*
      e. Heads in an open raceme
        f. Peduncles and involucres glandless . . . . *E. elatus*
        f. Peduncles and involucres more or less glandular
          g. Plant tall, with numerous heads . . . . *E. acris* var. *asteroides*
          g. Plant mostly less than 3 dm tall with few heads . . . . . . *E. acris* var. *debilis*
  c. Ligules conspicuous, usually more than 4 mm long, white, pink or purplish

h. Biennial or short-lived perennial; heads mostly several, in an open panicle . . . . *E. philadelphicus*
h. Perennials, tufted, from a branching base
  i. Heads always solitary
    j. Involucral bracts glabrous, ligules pink; leaves narrowly oblanceolate, glabrous or thinly pubescent . . . . *E. hyssopifolius*
    j. Involucral bracts villous or woolly; ligules purple, leaves oblanceolate, more or less ciliate
      k. Pappus light brown . . . . *E. purpuratus*
      k. Pappus white or yellowish
        l. Heads about 10 mm in diameter, involucral hairs with purplish cross-walls . . . . *E. humilis*
        l. Heads about 2-3 cm in diameter, involucral hairs with clear, white or pink cross-walls
          m. Involucral bracts densely yellowish-white, woolly; ray flowers white . . . . *E. eriocephalus*
          m. Involucral bracts woolly only at the base; ray flowers pale lavender . . . .*E. grandiflorus* ssp. *arcticus*
  i. Heads several, rarely solitary; stems simple, or in tall specimens branched above
    n. Stems 3-6 dm tall
      o. Plant essentially glabrous . . . . *E. yukonensis*
      o. Plant hirsute . . . . *E. glabellus* ssp. *pubescens*
    n. Plant canescent, with erect ascending stems, 1.0-1.5 dm tall . . . *E. caespitosus*

Erigeron acris L.
var. **asteroides** (Andrz.) DC.
E. angulosus Gaud. var. *kamtschaticus* (DC.) Hara
Medium to tall biennial or short-lived perennial from a simple or short-branched base. Flowering stems smooth, leafy, from a few to 7 to 8 dm tall. Leaves thinly, soft-pubescent, the basal and lower stem-leaves oblanceolate, the blade tapered into a narrow, petiole-like base. Flowering heads few to many, in an open raceme; the mature and expanded heads up to 2 cm in diameter. Involucral bracts linear, somewhat glandular. Ligules pale pink; pappus ample, drying reddish-brown.

Boreal woodland species of damp sandy river banks and lakeshore meadows.

General distribution: Circumpolar.

Fig. 931 Map 1090

Erigeron acris L.
var. **debilis** Gray
E. jucundus Greene
Rather similar to *E. acris* var. *asteroides* from which it differs by lower stature, being rarely over 3 dm tall and by its fewer heads on slender and often arcuate or ascending peduncles.

Alpine, gravelly slopes.

General distribution: Cordilleran, barely reaching S.E. Alaska and S.W. District of Mackenzie.

Fig. 932 Map 1091

***Erigeron caespitosus** Nutt.
Tufted, canescent perennial from a stout taproot and branched base; stems few to many, erect-ascending, 1 to 1.5 dm long, leafy; basal leaves narrowly oblanceolate, those of the stem linear, reduced in size upwards. Heads solitary or few, 1.5 to 2.5 cm in diameter, the ligules creamy-white or pinkish, much longer than the involucral bracts.

Dry, rocky places.

General distribution: Prairies and foothills from S.W. Sask. to S.W. Alta., north to S.W. Y.T. and central Alaska. To be looked for along the east slope of the Mackenzie Mts.

Erigeron canadensis L.
Conyza canadensis (L.) Cronq.
Coarse hairy annual with from one to several very leafy stems from a weak, simple or branching base. Leaves linear-oblanceolate with ciliate margins. Flowering heads numerous, about 5 mm in diameter, in an elongating and often leafy panicle.

General distribution: Roadside weed barely entering the District of Mackenzie near Ft. Smith.

Erigeron compositus Pursh
Low, densely caespitose, somewhat glandular and hispid, from a much branched base. Leaves basal, densely crowded, once ternate in var. *discoideus* Gray but 2- to 3-ternate in var. *gla-*

bratus Macoun, the segments linear. Flowering stems scapiform, 5 to 15 cm tall. Ray flowers white or pale lilac, in some forms wanting.

Dry gravelly and mainly calcareous soils.

General distribution: N. American. Arctic-alpine.

Fig. 933 (var. *discoideus*), 934 (var. *glabratus*) Map 1093 (*E. compositus s. lat.*)

Erigeron elatus (Hook.) Greene
E. acris L.
var. *elatus* (Hook.) Cronq.
Similar to *E. acris* var. *asteroides* but a smaller and more slender plant, with narrower and fewer leaves, and solitary or at least fewer heads; these on erect rather ascending or spreading peduncles, the latter, as well as the involucres always distinctly hirsute but not glandular and sub-glabrate as in *E. acris* var. *asteroides*.

Flood-plain meadows and open, boggy woods.

General distribution: Boreal N. America: Wide ranging from central Alaska to Lab., north to the limit of forest.

Fig. 935 Map 1094

Erigeron eriocephalus J. Vahl
Stems leafy, 5 to 15 cm tall, one to several from a short and fibrous root; stems, leaves and involucrum grey-villous from densely crinkly grey hairs lacking purplish cross-walls. Heads solitary, ray flowers white, turning pale lilac; pappus light brown.

Stony, gravelly places, mainly on calcareous soil.

General distribution: Circumpolar, arctic-alpine.

Fig. 936 Map 1095

Erigeron glabellus Nutt.
ssp. **pubescens** (Hook.) Cronq.
Ours coarsely pubescent perennial with simple or somewhat branching leafy stems 3 to 6 dm tall; white hairs lacking cross-walls. Leaves oblanceolate, tapering into a winged petiole, 10 to 15 cm long and 1 to 2 cm broad, their margins entire or minutely toothed. Heads 1 to 3, the terminal 2 to 3 cm in diameter; ligules numerous, pink, pale purple, or white; achenes hairy, pappus ample, drying reddish-brown.

Stony river bank meadows.

General distribution: Western N. America, with us mainly along the Mackenzie R. north to the Arctic Coast.

Fig. 937 Map 1096

Erigeron grandiflorus Hook.
ssp. **arcticus** Porsild
Perennial herbs; stems 15 to 25 cm tall, the narrowly oblanceolate, 5 to 6 cm long densely grey-villous leaves tapering into a narrow petiole; heads 2.5 to 3.5 cm in diameter the involucral bracts and peduncles lacking glands.

Calcareous slopes and dry tundra.

General distribution: Alaska, Y.T., western District of Mackenzie and Victoria Island.

Fig. 938 Map 1097

Erigeron humilis Grah.
E. unalaschkensis (DC.) Vierh.
In habit and size similar to *E. eriocephalus* but heads smaller and the involucral bracts bluish-black rather than grey, from densely crinkly hairs in which the cross-walls are purplish.

In moist, grassy places, often growing in herbmats below perennating snow banks.

General distribution: Circumpolar, arctic-alpine.

Fig. 939 Map 1098

Erigeron hyssopifolius Michx.
Perennial; stems delicate and slender, 1.5 to 2.5 dm tall, thinly pubescent or nearly glabrous, singly or often several together from a slender and much branched rhizome. Leaves very numerous, linear-oblong, 2 to 3 cm long and 2 to 3 mm wide. Flowering heads commonly less than 15 mm in diameter, solitary or sometimes several, on slender peduncles. Ligules white or pale lilac.

Moist, non-arctic, calcareous river or lake shore meadows.

General distribution: N. America, Nfld. to N.W. District of Mackenzie and central Y.T., north to or slightly beyond the limit of trees.

Fig. 940 Map 1099

Erigeron lonchophyllus Hook.
Biennial or short-lived perennial from a weak and fibrous root. Basal leaves mostly oblanceolate, up to 15 cm long, the cauline narrower and commonly linear. In key-characters rather similar to the usually taller and more robust *E. acris* var. *asteroides* from which, as a rule, it is readily distinguished by its compact and narrowly tufted habit and by its smaller heads on erect and straight peduncles; involucral bracts hispid, first green, turning solid purplish in age, whereas in *E. acris* var. *asteroides* they are glandular, with deep purplish-brown centres and pale, narrow margins.

Rather wet meadows or wet places by muskeg ponds.
General distribution: N. America: From Gulf of St. Lawrence to E. central Alaska north along the Mackenzie Valley to Norman Wells.
Map 1100

Erigeron pallens Cronq.
Low, tufted perennial from a branching tap-root; leaves oblanceolate, entire, or the early somewhat lobed, their margins soft-ciliate. Heads solitary, about 1 cm in diameter on 3 to 5 cm long scapiform peduncles; involucral bracts sparsely villous, dark purplish, the ligules white or yellowish-pink, filiform or rarely 1 mm wide.
Shaly, alpine slopes.
General distribution: Long known only from a few alpine stations in Alta. and B.C., *E. pallens* has recently been discovered in the Mackenzie Mts., N.W.T.
Map 1101

Erigeron philadelphicus L.
Biennial or short-lived perennial from a weak, fibrous root. Stems leafy, 2 to 10 dm tall. Leaves oblanceolate, entire to crenate, the cauline with a clasping base. Heads showy, about 10 to 15 mm in diameter, in a simple or compound panicle; the ligules linear, pale purple.
Moist river and lake shores; readily established by roadsides and in clearings.
General distribution: Wide ranging boreal species; not uncommon along the upper Mackenzie north to lat. 63°, with a widely disjunct station in northern Y.T. where, probably, a recent introduction.
Fig. 941 Map 1102

Erigeron purpuratus Greene
Dwarf, loosely matted perennial from a slender, much branched tap-root. Basal leaves oblanceolate, entire or occasionally shallowly lobed, about 10 to 15 mm long and about 6 mm wide. Heads solitary, 10 to 15 mm in diameter, on slender villous, 3 to 6 cm long peduncles; ligules linear, purplish or less often white, and the pappus pale purplish.
Open gravelly places.
General distribution: Endemic of N.W. America, from N. W. Alaska to the east slope of Mackenzie Mts., N.W.T.
Fig. 942 Map 1103

Erigeron yukonensis Rydb.
Perennial from a short, ascending rhizome; stems leafy, 2 to 4 dm tall, simple, or in tall specimens branched above, sparsely villous-hirsute; leaves linear-oblanceolate, 3 to 10 cm long, the lower tapering into a narrow petiole; flowering heads 1 to 3, and 2 to 3 cm in diameter, the lateral on slender peduncles; the involucre white-villous with multicellular, crinkly hairs in which the cross-walls are white; involucral bracts linear, dark purple, their tips somewhat reflexed or spreading; ligules 1.2 to 1.5 cm long, and 2 to 3 mm wide, pink or purplish, turning yellow in drying; achenes hirsute; pappus yellowish.
Calcareous, stony slopes.
General distribution: Long known only from the type locality near Dawson, Y.T., *E. yukonensis* is now known from a dozen or so stations from central and southwestern Y.T., east to the Mackenzie Delta, and along the Arctic Coast to Coronation Gulf.
Map 1104

Gaillardia Foug.

Gaillardia aristata Pursh
Finely pubescent perennial with leafy, simple or branched stems 2 to 6 dm tall, from a slender tap-root. Leaves alternate, oblanceolate and deeply incised, the lower petioled. Heads radiate, mostly solitary, up to 6 to 8 cm in diameter, or in strong plants occasionally with a smaller, lateral head; disc-flowers deep purple, the ligules orange-yellow.
A western prairie and foothill species readily established in disturbed soil.
General distribution: In N.W.T. reported as a recent introduction on farm land near Ft. Simpson.

Gnaphalium L. Everlasting

Gnaphalium uliginosum L.
Low, floccose-tomentose, simple or diffusely branching annual frequently only 5 cm tall, with linear or narrowly oblanceolate leaves and tiny flowering heads in leafy glomerules.
Moist, open and gravelly places, often pioneering in disturbed soil by roadsides or in recent clearings.
General distribution: Circumpolar weedy species, in N.W.T. known from a single collection near Yellowknife.
Fig. 943

Grindelia Willd. Gumweed

Grindelia squarrosa (Pursh) Dunal
Glabrous biennial or short-lived perennial of weedy habit, with a slender and simple tap-root. Stems dark purple, 3 to 5 dm tall, freely branched above. Leaves oblanceolate, finely serrulate and glandular dotted, the upper clasping. Heads few to several, 2.5 to 3.0 cm in diameter, involucral bracts resinous and the ligules yellow.

Dry and often somewhat saline flats.

General distribution: Wide ranging western prairie or foothill species, barely entering the District of Mackenzie on Salt Plains west of Ft. Smith.

Map 1105

Haplopappus Endl.

Haplopappus uniflorus (Hook.) T. & G.
H. lanceolatus (Hook.) T. & G.
var. *sublanatus* Cody
Perennial from a stout tap-root with a simple or branching crown. Flowering stems 20 to 30 cm tall, characteristically curved or ascending, with a few reduced leaves near the base. Basal leaves 10 to 15 cm long, including the slender petiole, the blade narrowly lanceolate, glabrate above, thinly tomentose beneath, sub-entire to sharply toothed. Heads mostly solitary, 15 to 20 mm in diameter; involucral bracts hirsute, in one or two rows; ligules yellow. Achenes hirsute, their pappus light yellowish-grey.

Moist, saline meadows.

General distribution: A western prairie-foothill species barely entering the District of Mackenzie on Salt Plains west of Ft. Smith.

Map 1106

Helenium L. Sneezeweed

Helenium autumnale L.
var. **grandiflorum** (Nutt.) T. & G.
H. macranthum Rydb.
Fibrous-rooted, essentially glabrous perennial. Stems up to 1 m tall, freely branched, prominently winged or ribbed from the decurrent leaf-bases. Leaves broadly lanceolate with dentate or entire margins. Heads few to several, hemispheric or nearly globose, the disc-flowers brownish-yellow and very numerous. Ligules cuneate, usually 3-lobed, soon reflexed.

Moist stream banks and meadows.

General distribution: Wide ranging from Que. to B.C., north along the upper Mackenzie R. to Ft. Simpson where, perhaps, introduced.

Fig. 944 Map 1107

Hieracium L. Hawkweed

Ours tufted perennial herbs from a branched base. Stems erect, leafy, or in some species scapose. Heads many-flowered, few to several in an open paniculate-corymbiform inflorescence. Flowers all ligulate, yellow or white; the achenes smooth, their pappus of capillary straw-coloured bristles.

a. Heads few and large, 1-2 cm in diameter; stems leafy . . . . *H. scabriusculum*
a. Heads mostly less than 1 cm in diameter, few to several; stems with few and reduced leaves
  b. Ligules white; lower third of stem retrorse-pilose; achenes straw-coloured . . . . *H. albiflorum*
  b. Ligules yellow
    c. Upper part of stem, peduncles and the involucral bracts black-villous; achenes black . . . . *H. triste*
    c. Upper part of stem, peduncles and involucral bracts sparsely grey-hirsute; achenes red . . . . *H. gracile*

Hieracium albiflorum Hook.
White Hawkweed
Stems 4 to 8 dm tall, leafy, glabrous above, but the lower half conspicuously pilose from tawny, retrorse hairs. Basal and lower stem leaves 10 to 15 cm long, the blade oblanceolate, tapering

into the petiole, the upper much reduced. Heads few to numerous, in an open panicle. Achenes straw-coloured.

Floodplain meadows and openings in subalpine woods.

General distribution: Cordilleran subalpine species barely entering the District of Mackenzie where, thus far, it is known from a single collection on the east slope of the Mackenzie Mts.

Fig. 945 Map 1108

Hieracium gracile Hook.
Slender Hawkweed
Stems slender, 1 to 3 dm tall, the upper half puberulent; the leaves mainly basal, the blade oblanceolate and long-petioled. Heads small and few; the involucres with long, black hairs covering a shorter and dense grey pubescence. The ripe achenes bright red.

Alpine meadows.

General distribution: Cordilleran species reaching south and central Y.T., in the District of Mackenzie north to Great Bear L.

Fig. 946 Map 1109

Hieracium scabriusculum Schwein.
Narrow-leaved Hawkweed
Stems stiffly erect, leafy, essentially glabrous, up to 1 m tall. Leaves rather stiff and firm, lanceolate or oblong, entire or sparsely toothed, glabrous except for the distinctly scabrous margins. Heads large and few; ligules yellow and the achenes reddish-brown.

Woodland clearings.

General distribution: N. America from N.B. to B.C., north to Great Bear L.

Fig. 947 Map 1110

Hieracium triste Willd.
Mourning Hawkweed
Stems slender, mostly solitary, 2 to 3 dm tall, the upper third, and also the involucral bracts prominently black-villous; leaves mainly basal, with entire-margined, obovate and essentially glabrous blade, tapering into the slender petiole.

Alpine meadows.

General distribution: Beringian and mainly coastal species recurring in alpine situations in central Y.T. and in the Mackenzie Mts., N.W.T.

Fig. 948 Map 1111

Lactuca L. Lettuce

Lactuca pulchella (Pursh) DC.
Blue Lettuce
Perennial herbs with milky juice, from a deeply buried and freely branching rhizome. Stems leafy, simple, 3 to 6 dm tall. The leaves subentire or more often pinnatifid, with few large and retrorse lobes. Inflorescence corymbose-paniculate; heads about 15 mm long, the involucral bracts in 3 rows; ligules purplish-blue.

Sandy and loamy river banks and moist calcareous meadows.

General distribution: James Bay to Alaska, north to about lat. 68° along the Mackenzie and its tributaries.

Fig. 949 Map 1112

Matricaria L. Wild Chamomile

Annual or short-lived, glabrous herbs with finely divided leaves and simple or branched stems bearing one to several ligulate or discoid flowering heads. Ligules white (or sometimes lacking).

a. Heads with prominent white rays
  b. Involucral bracts with dark brown, broad scarious margin . . . . *M. ambigua*
  b. Involucral bracts with light brown, narrow, scarious margin . . . *M. maritima* var. *agrestis*
a. Heads lacking white rays; introduced weed . . . . *M. matricarioides*

Matricaria ambigua (Ledeb.) Kryl.
Sea-shore Chamomile
A short-lived, glabrous and scentless perennial, with erect or ascending, simple or branched stems, 10 to 30 cm high, bearing from one to several large, flowering heads, 3 to 5 cm in diameter when fully expanded. Ray flowers pure white.

In moist, sandy places by the sea shore, and

sometimes becoming weedy near human habitations.

General distribution: Low-arctic, nearly circumpolar.

Fig. 950 Map 1113

Matricaria maritima L.
var. **agrestis** (Knaf) Wilmott
Annual, erect or ascending simple or branched stems to 5 dm or more high bearing one to several long peduncled flowers; heads 1.5 to 3.5 cm in diameter. Rays white.

An occasional roadside introduced weed in towns along the Mackenzie River.

General distribution: Cosmopolitan weed.

Matricaria matricarioides (Less.) Porter
M. suaveolens (Pursh) Buchen.
Pineapple Weed
Much branched, leafy annual with numerous small, conical heads, lacking ray-flowers. Flowering heads and leaves fragrant when crushed.

Waste places and roadsides.

General distribution: Cosmopolitan weed.

Petasites Mill. Sweet Coltsfoot

Coarse perennial herbs from a cord-like creeping rhizome; leaves basal, petioled, somewhat fleshy, the blade large, in ours cordate or triangular and variously toothed or lobed. Heads many-flowered, mainly dioecious, terminal, on somewhat fleshy and scaly scapes, appearing early in spring before the leaves, and for this reason often poorly represented in herbarium specimens; flowers white and mostly fragrant.

The young leaves and stems may be eaten raw as salad, or cooked as a potherb.

a. Leaf-blade oblong-cordate; the margin shallowly dentate . . . . *P. sagittatus*
a. Leaf-blade triangular-cordate; the margin variously toothed or incised
  b. Leaf-margin toothed or shallowly lobed
    c. Blade distinctly triangular, the lobes numerous, about 1 cm long . . . . *P. frigidus*
    c. Blade cordate, palmately 5-7-lobed halfway to the centre . . . . *P. vitifolius*
  b. Leaf-blade deeply lobed
    d. Blade totally glabrous, except for the glandular margin . . . . *P. arcticus*
    d. Blade thinly white-tomentose beneath
      e. Blade deeply palmate 5-7-lobed, the divisions merely toothed . . . . . . *P. palmatus*
      e. Blade 3-5-lobed, each lobe again deeply incised . . . . *P. hyperboreus*

Petasites arcticus Porsild
Leaf-blade deep green, totally glabrous on both sides, up to 20 cm in diameter, deeply 5 to 7-lobed.

Common locally on moist clay or shaly slopes, but never in boggy places.

General distribution: Apparently endemic to the Mackenzie Delta region and foothills west of the delta, south along the east slope of Richardson and Mackenzie Mts.

Fig. 951 Map 1114

Petasites frigidus (L.) Fries
Leaves more or less triangular, rather coarsely toothed but never deeply lobed. Upper surface of leaf glabrous and deep green, the lower densely white-tomentose.

In moist, sandy places by sheltered lake shores, and by the edge of small streams.

General distribution: Nearly circumpolar, arctic but not alpine, and with large gaps in eastern arctic N. America.

Fig. 952 Map 1115

Petasites hyperboreus Rydb.
Similar to *P. arcticus*, but leaves smaller, rarely over 10 cm in diameter, and the underside always tomentose.

Wet, shaly lake shores or stream banks.

General distribution: A Cordilleran Pacific Coast species, barely entering the District of Mackenzie along the Mackenzie-Y.T. divide, but wide ranging through Y.T. to Bering Strait.

Fig. 953 Map 1116

Petasites palmatus (Ait.) Gray
Leaf-blade 8 to 20 cm in diameter, the divisions of the blade reaching more than half-way to the centre. A rather delicate plant with slender rhizomes and leaf-stalks. Leaves glabrous or thinly tomentose beneath.

Moist or swampy woods.

General distribution: N. America; Nfld.—Lab. to B.C., north in S. W. District of Mackenzie and S. E. Y.T.

Fig. 954 Map 1117

Petasites sagittatus (Banks) A.Gray
Coarse plant with leaf-blades up to 30 cm long and 25 cm broad, although commonly smaller, dark green above and soft velvety-tomentose beneath.

Wet places by lake shores and pond margins.

General distribution: N. America. Its general range is similar to that of *P. palmatus* but extends northward beyond the limit of trees and reaches the Arctic Coast.

Fig. 955 Map 1118

Petasites vitifolius Greene
A delicate woodland species in general appearance intermediate between *P. frigidus* and *P. palmatus*, but its leaf-blades cordate rather than triangular, and shallowly lobed rather than merely toothed, and thinly tomentose beneath.

Meadows, swales and boggy woods.

General distribution: N. America; non-arctic and barely reaching southern Districts of Mackenzie and Keewatin.

Map 1119

Saussurea DC.

Saussurea angustifolia (Willd.) DC.
Perennial herb from a creeping, cord-like rhizome. Stems dark purple, 10 to 30 cm tall. Leaves all cauline, glabrous or floccose beneath, the lower 5 to 10 cm long and 0.5 to 1.0 cm wide, sinuate-margined, tapering into a narrow petiole, the upper much reduced. Heads discoid, 3 to 5, in a terminal corymb; involucral bracts in 3 to 4 rows. Ligules linear, purple, and the pappus tawny.

The var. *yukonensis* Porsild (*S. viscida* Hult. var. *yukonensis* (Porsild) Hult.) differs from typical *S. angustifolia* by its dwarf habit, and by its leaves that are usually somewhat broader and often with sinuate-dentate margins, but never viscid; the underside may be glabrous, or floccose as in *S. angustifolia*.

Dry alpine or arctic tundra.

General distribution: var *angustifolia*: Amphi-Beringian, from E. Asia east to northern Hudson Bay and north to the Arctic Coast; var. *yukonensis* endemic of mountains of Alaska, Yukon and District of Mackenzie.

Fig. 956 (var. *angustifolia*), 957 (var. *yukonensis*)

Map 1120 (var. *angustifolia*), 1121 (var. *yukonensis*)

Senecio L. Groundsel

A very large and taxonomically complex genus variously estimated, within its world-wide range, to include some 1300 species, among them annual or perennial herbs, shrubs, and even tree-like species.

Ours all caulescent herbs with alternate, entire or variously cut or divided leaves. Heads radiate or discoid, solitary, or in the majority of ours in few- to many-headed cymes. Ray flowers (when present) yellow or orange, pistillate and fertile, the more numerous disc-flowers bi-sexual and fertile, their corollas tubular, yellow or orange. Achenes nearly cylindrical, prominently ribbed, glabrous, or in some species hispid, their pappus of soft white and hair-like bristles.

a. Leaves all cauline (no basal rosettes)
 b. Perennial
  c. Leaves triangular-hastate, their margins denticulate .... *S. triangularis*
  c. Leaves oblanceolate
   d. Leaves deeply incised .... *S. eremophilus*
   d. Leaves with merely repand-denticulate margins .... *S. sheldonensis*
 b. Annual or biennial
  e. Leaves deeply pinnatifid, heads discoid; introduced weed .... *S. vulgaris*
  e. Leaves entire or irregularly toothed, heads radiate; native species ..... *S. congestus*
a. Leaves mainly in basal rosettes and always broader and larger than those of the stem
 f. Heads solitary (or occasionally with one or two small, lateral heads)
  g. Involucral bracts glabrous
   h. Involucral bracts dark purple; achenes smooth .... *S. cymbalaria*
   h. Involucral bracts green; achenes hirsute .... *S. hyperborealis*
  g. Involucral bracts hirsute
   i. Heads always solitary
    j. Involucral bracts and upper third of stem densely woolly from brown, non-septate hairs .... *S. Kjellmanii*

j. Involucral bracts blackened by distinctly septate hairs (at times partly covered by white-floccose pubescence) . . . . *S. atropurpureus*
i. Head often with one or two smaller lateral heads; leaves, stem and involucrum prominently white-floccose-tomentose . . . . *S. Lindstroemii*
f. Heads several to many, all approximately of the same size
k. Plants with a well-developed ascending rhizome
l. Basal leaves sessile, oblanceolate, entire, or their margins finely denticulate; involucral bracts prominently black-tipped . . . . *S. lugens*
l. Basal leaves petioled; involucral bracts not black-tipped
m. Leaf-blade thin, entire-margined; heads few and relatively large; involucral bracts and upper half of stem usually densely lanate . . . . *S. yukonensis*
m. Leaf-blade somewhat fleshy, oblanceolate, dentate or merely notched; heads small, 6-12 in open corymb; involucral bracts glabrous . . . . *S. streptanthifolius*
k. Plants with fibrous roots; basal leaves petioled
n. Stems slender; basal leaves oval, subentire to serrate . . . . *S. pauperculus*
n. Stems stout; basal leaves with oblong-ovate mostly deeply toothed blade
o. Heads many, usually over six; ray flowers pale yellow . . . . *S. indecorus*
o. Heads few, usually less than six; ray flowers reddish-orange . . . . . *S. pauciflorus*

Senecio atropurpureus (Ledeb.) Fedtsch.
S. frigidus (Richards.) Less.
Plant glabrous or thinly white-floccose. Stems 5 to 20 cm high, mostly solitary from a slender creeping rhizome, bearing 4 to 6 sessile linear-lanceolate, slightly fleshy leaves. Basal leaves petiolate, the blade ovate 1 to 2 cm long, subentire. The solitary head normally with disc and ligulate flowers, but heads in which the latter are wanting are not uncommon.

Common in not too dry herbmats.

General distribution: Arctic and alpine Eurasia and N.W. America.

Fig. 958 (radiate), 959 (discoid) Map 1122

Senecio congestus (R.Br.) DC.
Stems solitary or clustered from a short and fibrous root, 10 to 100 cm high, thick, densely brownish-yellowish hirsute above, hollow and easily compressed, bearing numerous sessile, linear-oblong, sinuate-toothed leaves. The numerous small heads in one large, or several smaller, variously stalked glomerules.

A coarse, weedy plant, commonly found in moist but sheltered places, often near the sea shore and in disturbed places southward. The young leaves and flowering stems are edible and may be used as salad or potherb. Normally the plant is hapaxanthic or once-flowering, dying completely after having flowered and produced seeds, but in the high latitudes, specimens in which the flowers of the year were killed by early frost, may flower and fruit the following year.

General distribution: Circumpolar (except Greenland), arctic.

Fig. 960 Map 1123

Senecio cymbalaria Pursh
S. resedifolius Less.
Dwarf and usually somewhat tufted, glabrous and slightly fleshy-leaved perennial with stems 5 to 25 cm tall, from a rather stoutish, simple or branched base. Heads solitary or rarely with one or two long-peduncled lateral heads, up to 3 cm in diameter, commonly radiate, the rays yellow, but often purplish-tinged, or sometimes wanting.

Alpine tundra. In the Mackenzie Mts. ascending to 1800 m.

General distribution: Amphi-Beringian arctic-alpine, from the Mackenzie Mts. over Alaska and North and Central Asia to the Ural Mts. In eastern N. America disjunct in Gaspé and Nfld.

Fig. 961 Map 1124

Senecio eremophilus Richards.
Stout, glabrous perennial, 3 to 6 dm tall with simple or branched leafy stem from a short and fibrous base. Leaves deeply pinnatifid, the lower petioled, the upper sessile. Heads small, very numerous, the bracts black-tipped and the ligules pale yellow.

Damp woodland meadows and roadsides.

General distribution: North Dakota north to eastern B.C., Alta., Man. and Sask., barely entering the District of Mackenzie north of Ft. Smith.

Fig. 962 Map 1125

Senecio hyperborealis Greenm.
Perennial, with a short branched base. Leaves mainly basal, lyrate or oval, the blade and petiole glabrous but prominently white-floccose in the leaf-axils. Stems few to several, 5 to 25 cm

high. Heads solitary or with one or two, long-peduncled lateral heads. Rays pale yellow. Achenes hispid.

Calcareous screes and slopes.

General distribution: Endemic of alpine arctic and subarctic northwestern America.

Fig. 963 Map 1126

Senecio indecorus Greene

Glabrous perennial with simple 3 to 9 dm tall stems from a fibrous simple or somewhat branched base. Basal leaves 5 to 15 cm long, the blade somewhat fleshy or succulent, elliptic, finely toothed. Cauline leaves incised-pinnatifid. Heads pale yellow, mostly discoid, 6 to 12.

In habit rather similar to *S. pauciflorus* in which the heads are larger, orange-reddish and rarely exceed half a dozen.

Moist woodland meadows; rarely sub-alpine.

General distribution: N. America; Gaspé to S.E. Alaska, north to Great Bear L.

Fig. 964 Map 1127

Senecio Kjellmanii Porsild

S. atropurpureus (Ledeb.) Fedtsch.

spp. *tomentosus* (Kjellm.) Hult.

Dwarf but rather stout perennial with leafy stem rarely over 12 cm tall and commonly less, from a stout, ascending rhizome. Leaves oblanceolate, the upper sessile, the lower tapered into a slender petiole. Heads solitary, 1.5 to 2.5 cm in diameter. The involucral bracts as well as the upper half of the stem densely matted with brown or blackish hairs. Ligules strap-like, pale yellow.

Moist alpine herbmats or tundra.

General distribution: Endemic of arctic-alpine N.W. America and E. Asia, with a range similar to that of *S. Lindstroemii.*

Map 1128

Senecio Lindstroemii (Ostf.) Porsild

S. fuscatus sensu Hultén *non* Jord. & Fourr., nor *S. tundricula* Tolm. of West Siberia

Grey- or white-floccose-tomentose perennial mostly with a solitary and leafy stem from 5 to 25 cm tall from a short, fibrous base. Heads solitary, 3 to 4 cm in diameter, or sometimes with one or two smaller, lateral and pedunculate heads. Involucral bracts linear, dark purplish-tipped, thinly floccose. Ligules 1.5 to 2.0 cm long, linear, bright orange-yellow or sometimes purplish tinged. Disc-flowers yellow with reddish-purple lobes. Achenes sparingly strigose-hirsute.

Moist alpine tundra.

General distribution: Endemic of arctic and alpine N. W. America from the Arctic Coast west of Mackenzie Delta, the east slope of Richardson and Mackenzie Mts., mountains of Alaska, west to Seward Pen., disjunct by nearly 2100 km in mountains of Montana and Wyoming.

Fig. 965 Map 1129

Senecio lugens Richards.

Stems solitary or clustered 5 to 40 cm tall, from a stout, ascending rhizome; leaves mainly basal, finely denticulate and tapering to a narrow petiole. Heads few to a dozen or more, 10 to 15 mm in diameter, in a loose corymb; the inner involucral bracts prominently black-tipped.

The black tips of the involucral bracts inspired the specific name, from the Latin *lugeo* (to mourn), and refers to the massacre at Bloody Falls on the Coppermine River, of a group of unsuspecting Eskimo, by the Indian warriors who, in 1771, accompanied Samuel Hearne.

Moist herbmats, lake shores and river banks, often among willows.

General distribution: Alaska and Y.T. east to Coppermine R., south in the Rocky Mts. beyond the 50th parallel.

Fig. 966 Map 1130

Senecio pauciflorus Pursh

Essentially glabrous perennial, 3 to 6 dm tall; leaves somewhat fleshy, in general habit similar to *S. indecorus* from which it can usually be distinguished by its fewer but larger, reddish-orange heads. Although the general range of the two species is similar, *S. pauciflorus* tends to be mainly alpine.

Lake shores and herbmats.

General distribution: N. America: Lab. to S. E. Alaska, barely entering S. W. District of Mackenzie in the Mackenzie Mts.

Fig. 967 Map 1131

Senecio pauperculus Michx.

Fibrous-rooted short-lived perennial, 3 to 5 dm tall, glabrous but often somewhat floccose in youth. Heads rarely more than six, always radiate, their ligules pale yellow. In general habit and range similar to *S. indecorus* and *S. pauciflorus* from which it differs by its thinner and not at all fleshy or succulent, oblong leaves, and by its fewer and always radiate heads.

Woodland species of clayey river banks or lake shores.

General distribution: Lab. to E. Alaska, north to Great Bear L.
Fig. 968 Map 1132

Senecio sheldonensis Porsild
Slender, glabrous perennial from a short, many-headed rhizome. Stems singly to several, 3 to 6 dm tall, leafy; the leaves thin, broadly lanceolate, repand-denticulate, the lowermost conspicuously petioled, 6 to 10 cm long. Heads 3 (4), rays yellow. From *S. triangularis* and its var. *angustifolius*, *S. sheldonensis* differs by its more slender habit and always lanceolate, never triangular-hastate leaves, and by its dark-tipped involucral bracts.
Turfy places in subalpine meadows.
General distribution: Endemic of mountains of central Y.T., northern B.C., reaching the east slope of Mackenzie Mts., N.W.T.
Map 1133

Senecio streptanthifolius Greene
S. cymbalarioides Nutt.
var. *borealis* (T. & G.) Greenm.
Glabrous perennial 12 to 25 cm tall from a stout, simple or several-headed tap-root. Leaves somewhat fleshy, the basal 2 to 6 cm long, with a short, entire or few-notched blade tapered into a slender petiole. Heads 5 to 20, the ligules bright yellow, the bracts light-yellowish green.
Moist, stony river banks and lake shores.
General distribution: Cordilleran, north to southern Y.T. and the District of Mackenzie.
Fig. 969 Map 1134

Senecio triangularis Hook.
Glabrous perennial with leafy and often clustered stems up to 1.5 m tall, from a stout, ascending rhizome. Leaves triangular-hastate. Heads radiate, few to a dozen or more in a flat-topped cyme.
Moist, alpine herbmat slopes, alpine meadows or by alpine brooklets.
General distribution: Cordilleran, reaching the District of Mackenzie along the southern east slope of the Mackenzie Mts.
Fig. 970 Map 1135

Senecio vulgaris L.
Glabrous annual with erect, simple or much branched leafy stem from a fibrous root. Leaves pinnatifid, clasping, irregularly toothed. Heads small, lacking rays.
General distribution: Circumpolar weed, barely entering the N.W.T., near human settlements along the upper Mackenzie R.

Senecio yukonensis Porsild
S. alaskanus Hult.
Floccose-pubescent perennial from a short ascending rhizome producing numerous oblanceolate, sub-entire and somewhat fleshy basal leaves and one to several flowering stems 10 to 30 cm tall. Heads occasionally single but more often 2 to 5 in a dense cluster, their peduncles densly yellowish-woolly; involucral bracts dark green or black, and the ligules pale yellow.
Mossy, alpine heath.
General distribution: Endemic of Alaska—Y.T., east to the Richardson and Mackenzie Mts., north to the Arctic Coast.
Fig. 971 Map 1136

Solidago L. Goldenrod

Perennial herbs with alternate, sessile cauline leaves and erect-ascending stems from a creeping short-ascending rhizome. Heads relatively small, few to many, mostly radiate and pistillate. The corollas of disc and ray flowers yellow. Achenes ribbed, nearly terete, their pappus white, of simple, capillary bristles.
A large and taxonomically complex, non-arctic, chiefly North American genus.

a. Tall, plants from a creeping rhizome; leaves mainly cauline
  b. Heads few, in dense, cymose clusters . . . . *S. canadensis* var. *salebrosa*
  b. Heads numerous, in a flat-topped panicle . . . . *S. graminifolia*
a. Low, plants from a short, ascending rhizome; leaves mainly basal
  c. Margins of leaves and their petioles ciliate . . . . *S. multiradiata*
  c. Margins of leaves and their petioles not ciliate . . . . *S. decumbens* var. *oreophila*

Solidago canadensis L.
var. **salebrosa** (Piper) Jones
S. lepida DC.
var. *elongata* (Nutt.) Fern.
Stems leafy up to 1.3 m tall from a branched rhizome. The leaves numerous, lanceolate, sessile, glabrous and fresh green, their margins sharply serrate. Heads 5 to 6 mm long, very numerous and crowded along the spreading and often nodding branches of the panicle.
Open woodlands.
General distribution: Nfld., Lab. and the Hudson Bay region west to the Rocky Mts., Y.T. and central Alaska; in the N.W.T. north along the upper Mackenzie drainage to about lat. 66°.
Fig. 972 Map 1137

Solidago decumbens Greene
var. **oreophila** (Rydb.) Fern.
Stems solitary or few together, 2 to 4 dm tall, dark-purplish, from a short, ascending rhizome. Leaves mainly basal, oblanceolate, entire or shallowly toothed. Inflorescence racemose.
Damp river banks and lake shores or open woods.
General distribution: Western Man. to B.C., north through the upper Mackenzie drainage to near the Arctic Coast.
Fig. 973 Map 1138

Solidago graminifolia (L.) Salisb.
var. **major** (Michx.) Fern.
Stems 3 to 7 dm tall. Leaves broadly lanceolate, scabrous-margined, greyish-green, 3 to 8 cm long. Heads glomerulate, 5 to 8 mm long in small, compact clusters forming a flat-topped inflorescence.
Wet, clayey river banks and lake shores.
General distribution: Nfld. to Alta. In the District of Mackenzie thus far known from along the Mackenzie Valley near Ft. Simpson.
Map 1139

Solidago multiradiata Ait.
Stems 0.5 to 4.0 dm tall, few to several from a short, ascending sub-ligneous and divided rhizome. Leaves oblanceolate, the lower sparingly serrate, and conspicuously ciliate-margined at least on the petiole. Heads few to several in an open to compact corymbiform inflorescence.
River meadows and subalpine slopes below timberline.
General distribution: Arctic-alpine; wide ranging from Nfld., Lab., the Hudson Bay region, north to the arctic coast and western Arctic islands, west to Bering Strait. In the mountains of Alta., B.C., western District of Mackenzie, Y.T. and S. E. Alaska largely represented by the var. *scopulorum* Gray, distinguished by its acuminate and distinctly serrate basal leaves.
Fig. 974 Map 1140

Sonchus L. Sow-Thistle

Sonchus arvensis L.
Essentially glabrous perennial from a slender, horizontal rhizome. The stems leafy, up to 1 m tall, the leaves runcinate pinnatifid and prickly-margined, clasping. Heads several, in an open, terminal cluster. Flowers all ligulate, yellow.
General distribution: Cosmopolitan weed of roadsides and waste places; in the N.W.T. reported but once north of lat. 60°, along the Mackenzie Highway.

Tanacetum L. Tansy

Aromatic perennial, leafy-stemmed herbs with sessile, alternate and twice pinnatifid leaves. Heads discoid, few to many in a terminal corymb. Ray flowers pistillate, yellow or inconspicuous; disc-flowers perfect; pappus scale-like.

a. Heads few, about 2 cm in diameter; native . . . . *T. huronense*
a. Heads numerous, small; introduced and weedy . . . . *T. vulgare*

Tanacetum huronense Nutt.
Chrysanthemum bipinnatum L.
ssp. *huronense* (Nutt.) Hult.
Stems 2 to 6 dm tall from a creeping rhizome. The leaves finely dissected, villose. Heads few to several, about 2 cm in diameter.
Sandy river banks.
General distribution: N. America; Nfld. to Alaska, in the N.W.T. thus far known from a single station on the east branch of the Mackenzie R. delta.
Map 1141

Tanacetum vulgare L.
Stems 6 to 12 dm tall, from a short and stout rhizome. Leaves glabrous. Heads small and very numerous in a flat-topped corymb.

General distribution: Cosmopolitan, strongly aromatic roadside weed, in the N.W.T. reported from the southern District of Mackenzie.

Taraxacum Weber. Dandelion

Perennial herbs with milky juice, from a stout tap-root and a basal rosette of runcinate to sub-entire leaves. Flowering heads solitary, on a naked, hollow peduncle. Ligules yellow, or in some, and mainly arctic species, creamy white, or the outer row often pale purplish tinged. The involucral bracts in two or three rows, with or without corniculate tips or appendages. The lower or outer ones usually shorter and broader than the inner. The achenes fusiform, longitudinally ribbed, more or less spiny, and prolonged into a long, slender beak, bearing a whorl of slender, spreading pappus. Some species produce good pollen and may thus reproduce sexually, but in others fertile pollen is always lacking; in these the fruits are thus formed parthenogenetically, i.e. without fertilization. In this manner even minor variations persist as in horticulturally developed cultivars.

The leaves of the larger species, especially when young and tender, may be eaten as a salad or, when cooked, as a pot-herb.

a. Dwarf species with scapes less than 20 cm tall
  b. Leaves sub-entire
    c. Involucral bracts, at least those of the inner row, with prominently horned appendages; the achenes spiny in their upper half
      d. Outer ligules yellow . . . . *T. pellianum*
      d. Outer ligules white or purplish-tinged
        e. Outer ligules white . . . . *T. mackenziense*
        e. Outer ligules purplish tinged . . . . *T. integratum*
    c. Involucral bracts, lacking horned appendages; ligules yellow, and the achenes spiny throughout . . . . *T. phymatocarpum*
  b. Leaves runcinate-pinnatifid
    f. Outer ligules white . . . . *T. hyperboreum*
    f. Outer ligules yellow
      g. Achenes spiny throughout . . . . *T. pumilum*
      g. Achenes spiny only in their upper half
        h. The lobes of leaves all entire-margined . . . . *T. alaskanum*
        h. The lobes of leaves and the rachis with one or several minute but sharp teeth . . . . *T. sibiricum*
a. Medium to tall species with scapes mostly over 20 cm tall
  i. Outer ligules white . . . . *T. hyparcticum*
  i. Outer ligules yellow
    j. Achenes bright reddish-brown and spiny throughout . . . . *T. erythrospermum*
    j. Achenes mostly greenish- or olive-grey, spiny only in the upper half (except *T. dumetorum*)
      k. Outer involucral bracts strongly reflexed (weedy species) . . . . *T. officinale*
      k. Outer involucral bracts not reflexed (native species)
        l. Tips of involucral bracts lacking horned appendages . . . . *T. lapponicum*
        l. Tips at least of inner row of involucral bracts with horned appendages
          m. Leaves shallowly and unevenly incised
            n. Scapes very slender, usually 3-4 times as long as the narrowly oblanceolate leaves; achenes spiny in their upper half . . . *T. maurolepium*
            n. Scapes stout, about twice as long as the leaves, and hairy below the involucrum; achenes spiny throughout . . . . *T. dumetorum*
          m. Leaves deeply and irregularly incised; involucral bracts prominently white-margined
            o. Achene small, one third as long as its beak . . . . *T. lacerum*
            o. Achene half as long as its beak . . . . *T. pseudonorvegicum*

Taraxacum alaskanum Rydb.
Dwarf species with linear, oblanceolate, 3 to 5 cm long, runcinate-pinnatifid leaves; the lobes opposite, all entire-margined and the terminal spear-shaped and usually larger than the lateral. Scapes 5 to 12 cm long, commonly brownish-purple, rarely more than 2 to 3 to each rosette. Heads 2.0 to 3.5 cm in diameter; the involucral bracts dark green; ligules pale yellow, the outer often pink or pale purplish-tinged. Achenes 4 mm long, spiny above, half as long as the beak. Pollen scarce, but rarely lacking.

In the rather similar but usually much smaller and strictly Cordilleran *T. scopulorum* Rydb. the lateral lobes of the leaves are broadly triangular and the terminal lobe not larger than the rest.

Moist, arctic-alpine slopes, on fresh moraines or on snowflushes.

General distribution: From Bering Strait east along the Arctic Coast to long. 125°W., and southwards in the Richardson and Mackenzie Mts., and in Alaska and B.C. slightly beyond lat. 60°, with a disjunct alpine station in B.C. in lat. 49°04′N., and 120°12′W.

Map 1142

Taraxacum dumetorum Greene
T. ceratophorum auctt. non Ledeb.
Stout species from a simple or branching taproot. Leaves erect-spreading, oblanceolate, 10 to 25 cm long and up to 4 cm broad, their margins shallowly and unevenly laciniated. Scapes one or two from each rosette, up to 4 to 5 dm tall and usually reddish-brown, and prominently hairy below the involucrum. Involucral bracts in 3 to 4 rows, the outer somewhat reflexed, the inner linear, up to 2 cm long, green, their tips prominently bifid at the apex; ligules pale yellow; pollen lacking or very scarce. Achenes 3 mm long, pale greyish-green, spiny throughout, the slender beak 3 to 6 times as long as the body.

Meadows and open woodlands.

General distribution: From James Bay west to B.C., north to south-eastern Y.T. and the upper Mackenzie Valley, N.W.T., with an apparently disjunct population in the Mackenzie Delta area.

Map 1143

Taraxacum erythrospermum Andrz.
T. laevigatum (Willd.) DC.
T. ? Malteanum Dahlst.
Coarse, weedy species with scapes up to 3 to 4 dm tall. The leaves deeply runcinate-pinnatifid, often to near the mid-rib. Achenes 3 mm long, one third as long as beak, spiny throughout, and bright reddish-brown when mature.

General distribution: Road sides and waste places near towns of the Upper Mackenzie Valley, beside trading posts along the Arctic Coast.

***Taraxacum hyparcticum** Dahlst.
A strikingly handsome species with slender, oblanceolate leaves sparsely toothed or with subentire margins, the blade tapering into a narrow petiole as long or longer than the blade. Heads 2 to 3 cm in diameter when fully expanded, on slender scapes 10 to 25 cm long; involucral bracts dark green, drying almost black, the inner with paler margins, their tips somewhat callose but rarely with a small corniculate appendage. Ligules white or creamy yellow, the outer often purplish tinged; pollen lacking. Fully mature achenes about 3 mm long, dark slaty-grey, prominently spiny in the upper half, the beak slightly longer than the achene.

In grassy and well-manured places near animal or human habitations. Vigorous specimens, with upwards of 80 flowering scapes from one root, have been observed.

General distribution: Endemic of N.W. Greenland and the Canadian Arctic Archipelago, and to be looked for along the mainland coast.

Fig. 975 Map 1144

Taraxacum hyperboreum Dahlst.
Dwarf species with 10 to 15 cm high scapes from a simple to several-headed, stout taproot. Leaves 5 to 10 cm long and rarely more than 1 cm broad, irregularly toothed, the teeth short-triangular, often retrorse. Heads one or rarely more than two from each rosette, barely overtopping the leaves, relatively large, 2.0 to 2.5 cm high and 2.0 to 3.0 cm in diameter; ligules pale yellow or white. Involucral bracts medium to dark green, all prominently horned, the inner 3 mm broad, with pale margins, and small but conspicuous black and 2-pronged tips. Achenes straw-coloured, spinulose in the upper half; pollen lacking. In general habit *T. hyperboreum* may resemble a small specimen of *T. lacerum,* however, in the latter the leaves are narrower, more deeply lobed, and pollen is always present.

Somewhat moist places, often on mineral soil near the sea shore, but frequently also near animal or human habitations.

General distribution: Described from King William Island, *T. hyperboreum* appears to be

endemic to the central Canadian Arctic ranging from Baffin and Ellesmere Islands west to Banks Island, and to about long. 145° on the mainland south to Great Bear L.
Map 1145

Taraxacum integratum Hagl.
Low to medium sized species with linear-oblanceolate leaves 5 to 13 cm long and rarely more than 1.0 cm broad, with sub-entire to short sinuate-dentate margins, the blade narrowing into a slender petiole. Scapes mostly twice as long as the leaves, and usually strongly coloured. Heads about 2.0 cm high, the involucral bracts olive-green, all prominently horned; the outer ligules pale purplish tinged; pollen lacking or very sparse. Achenes light yellowish-brown, about 5 mm long, spinulose almost to the base; the beak about three times as long as the achene.
Alluvial river banks and lake shores.
General distribution: N.W. America; endemic.
Map 1146

Taraxacum lacerum Greene
T. groenlandicum Dahlst.
T. Carthamopsis M.P.Porsild
Usually robust with numerous leafy rosettes, and flowering scapes up to 5 to 6 dm long. Leaves narrow, mostly deeply runcinate, the midrib usually purplish. Heads commonly up to 5 cm in diameter in vigorous specimens. Involucral bracts dark green, prominently horned; ligules pale yellow. Pollen present and usually abundant. Achenes 3.5 mm long, yellowish-grey, the upper half spiny; beak 5 to 8 mm long.
In somewhat moist places, often on mineral soil near the sea shore, but frequent also near animal burrows, bird cliffs and human habitations.
General distribution: Common and widely distributed from S. W. Greenland across arctic and subarctic N. America south to the limit of trees.
Fig. 976 Map 1147

Taraxacum lapponicum Kihlm.
T. croceum Dahlst.
Robust species with scapes up to 30 cm long and broadly oblanceolate leaves, their margins entire but commonly with small, well-spaced, sharply pointed teeth, these less often interspaced by a few larger lobes. Flowering heads, when fully expanded, 5 to 6 cm in diameter; involucral bracts narrow, with entire tips, ligules bright yellow, the anthers lacking pollen.
Moist, grassy herbmat slopes.
General distribution: Amphi-Atlantic, subarctic-alpine; in N. America from E. Greenland to Y.T.
Fig. 977 Map 1148

Taraxacum mackenziense Porsild
Dwarf species with slender scapes 10 to 15 cm tall and narrowly oblanceolate, 0.5 to 0.7 cm broad, sub-entire or short and remotely denticulate leaves tapering into a narrow-winged, purplish or green petiole; heads small, one or rarely two from each root, 1.0 to 1.8 cm high; the involucral bracts light green with pale margins, all distinctly corniculate; ligules white or pale rose-purplish tinged, in some heads in a single row or entirely missing; anthers with sparse pollen, and the achenes pale olive, about 4 mm long, prominently spiny at the summit, the beak twice as long as the body.
General distribution: Thus far known only from dolomitic outcrops along the east bank of the lower Mackenzie R., and from the east branch of the Mackenzie Delta.
Map 1149

Taraxacum maurolepium Hagl.
Slender, delicate plant forming small tussocks from a thin tap-root. Leaves narrowly lanceolate or more often oblanceolate, 10 to 15 cm long and 1.0 to 1.5 cm broad, tapering into a slender petiole, their margins sub-entire or shallowly dentate. Scapes very slender, usually 3 to 4 times longer than the leaves, rarely more than two from each root. Heads 1.5 to 2.0 cm tall, the involucral bracts deep olive-green with narrow and distinctly paler margins, the inner, but often also those of the second row, with small but well-developed purplish appendages 0.5 to 2.5 mm long; the outer ligules white, with purple lines; pollen sparse or lacking. Mature achenes slaty grey, spiny in the upper half, about 4 to 5 mm long, the beak twice as long as the body.
General distribution: N. W. American endemic.
Map 1150

Taraxacum officinale Weber
Non-native, coarse and polymorphous species with variously pinnatifid to sinuate-dentate leaves. Heads 2 to 5 cm in diameter with orange-yellow ligules and green to olive

coloured, mostly unappendaged involucral bracts, the outer strongly reflexed in age. The achenes grey-olive, spiny above.
Roadsides and waste places.
General distribution: Cosmopolitan weed, in the N.W.T. known from near settlements along the upper Mackenzie R.

Taraxacum pellianum Porsild
A small species, with a stout, many-headed primary root, bearing 1 to several flowering scapes, besides sterile rosettes. Leaves linear-lanceolate, gradually tapering into a winged petiole; the early leaves sub-entire, the fully mature 6 to 8 cm long and 0.8 to 1.0 cm wide, their margins shallowly dentate, runcinate or even lobate. Scapes 8 to 10 cm high, glabrous except below the head. Involucre about 15 mm high, the bracts pale olive-green, the middle and inner prominently corniculate. Flowering heads tardily expanding; the ligules bright golden-yellow; anthers containing abundant pollen. Mature achenes about 4 mm long, dark straw-coloured, spiny in the upper half, and the beak but slightly longer than the achene.
Sunny, calcareous slopes.
General distribution: The type of *T. pellianum* came from Y.T. where it is now known from a number of stations in the southern half; it has been reported also from the lower Mackenzie R. and its delta, and has been collected at Little Carcajou L. in the Mackenzie Mts.
Map 1151

Taraxacum phymatocarpum J. Vahl
A slender, delicate dwarf species with narrow, subentire, dark green leaves. Scapes usually solitary, rarely 2 or 3 from each root, purple, slender and weak, often decumbent or reclining. The flowering heads narrow and turbinate, remaining closed during anthesis. Involucral bracts bluish-black, pruinose, broadly lanceolate with entire, obtuse tips. Ligules lemon-yellow; pollen present. Achenes dark olivaceous, 4.5 to 5.0 mm long, spiny to the very base; the beak rarely much longer than the achene.
Stony and gravelly places, usually on mineral soil.
General distribution: N. America, high-arctic and wide ranging from Bering Strait along the Arctic mainland and islands to E. Greenland, with a disjunct station in northern Nfld.
Fig. 978 Map 1152

Taraxacum pseudonorvegicum Dahlst.
Low to medium sized species with fresh green, narrowly lanceolate and spreading leaves, the inner with short and widely spaced teeth, the outer with much larger and backward curved lobes, the terminal lobe long and narrow. Heads 3.0 to 3.5 cm in diameter when fully expanded, one to few in each plant, rarely overtopping the leaves. Involucral bracts dark green, all prominently white-margined and distinctly horned, the outer ovate-lanceolate, the inner linear. Ligules pale yellow; the anthers without pollen. Achenes yellowish-brown, about 4 mm long, prominently spiny in the upper third, half as long as the beak.
Herbmats.
General distribution: Thus far known only from southwestern Baffin Island, Chesterfield Inlet, west to the upper Thelon R.
Map 1153

Taraxacum pumilum Dahlst.
A dwarf species with short, often spreading, dark green leaves, their margins with short but regular triangular lobes. Scapes usually 1 to 3, approximately 5 cm high, or as long as the leaves, and often decumbent. Flowering heads about 2 cm in diameter when fully expanded. Involucral bracts olive green, somewhat pruinose, their tips somewhat callose, but some, at least, with a small, corniculate appendage. Flowers dark yellow; pollen present. The ripe achenes pale, slaty-olive, spiny throughout.
In dry turfy places.
General distribution: A high-arctic, and rarely collected species thus far known from northernmost Greenland, the northern and northwestern Canadian Arctic Archipelago, and from a few stations along the mainland coast west to the Mackenzie Delta.
Map 1154

Taraxacum sibiricum Dahlst.
Dwarf species with erect-ascending slender scapes 5 to 10, or rarely 15 cm long, and deeply runcinate leaves, their lobes, and often also the rachis with one or more small but sharp teeth. Heads small, when fully expanded rarely more than 2 cm in diameter; involucral bracts dark green, tapered to a slender, entire tip, the outer row reflexed; ligules pale yellow, those of the outer row purplish-tinged; pollen abundant. The fully mature achenes 4.0 to 4.5 mm long, dark brown and spiny only at the summit, the beak rarely longer than the achene.
An arctic or high-alpine snowbed species.
General distribution: Described from near the mouth of the Lena R. in north central Siberia, *T. sibiricum* is now known to be wide ranging in N. America (although with large gaps) from

Seward Pen., Alaska, north to Herschel Island on the arctic coast of Y.T. and south to mountains of central Alaska, S. W. Y.T., east to Mackenzie Mts. and in the Canadian Rocky Mts. southward nearly to lat. 49°.

Map 1155

Tragopogon Goat's-beard

Tragopogon major Jacq.

Glabrous, weedy and short-lived perennial herb with milky juice. Stems simple or branching, up to 1 m tall, from a stout tap root. Leaves alternate, linear, from an enlarged, clasping base. Heads solitary, large. The flowers all ligulate, pale yellow, and the achenes spindle-shaped, tapering to a slender beak. Pappus white, in a single series, rather similar to that of a dandelion.

General distribution: Cosmopolitan roadside weed; in waste places north to Ft. Smith.

875. *Agoseris aurantiaca*

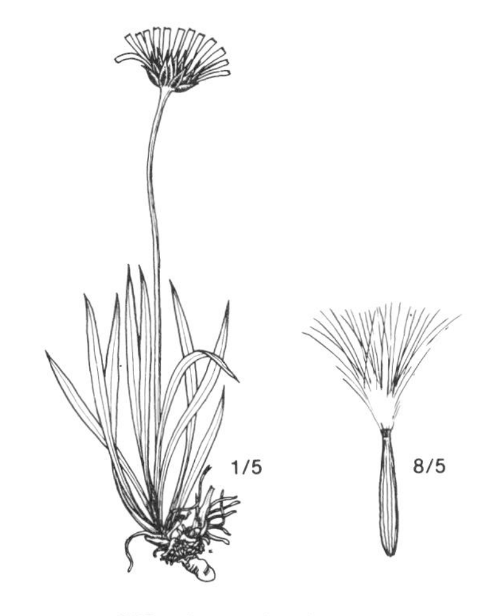

876. *Agoseris glauca* ssp. *scorzoneraefolia*

7. *Anaphalis margaritacea* var. *subalpina*

878. *Antennaria alborosea*

879. *Antennaria angustata*

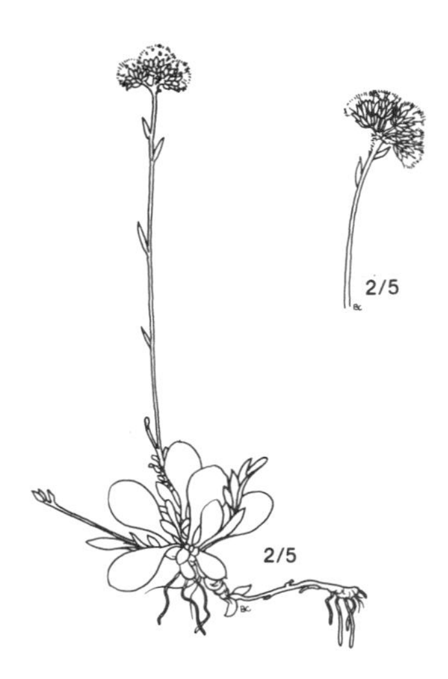

880. *Antennaria campestris*

881. *Antennaria canescens*

882. *Antennaria compacta*

883. *Antennaria crymophila*

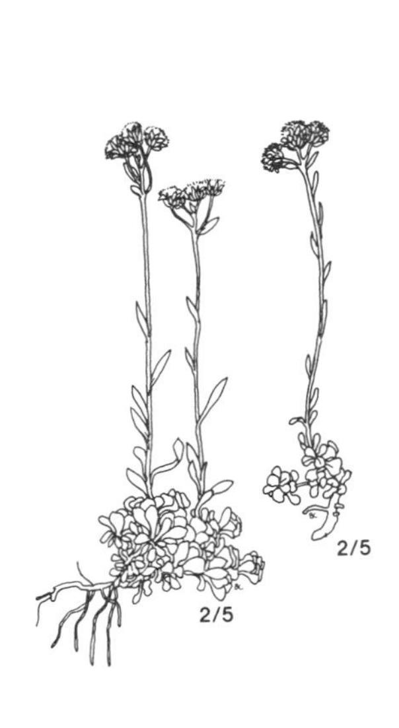

884. *Antennaria densifolia*

885. *Antennaria Ekmaniana*

886. *Antennaria elegans*

887. *Antennaria glabrata*

888. *Antennaria incarnata*

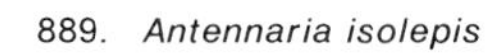

889. *Antennaria isolepis*

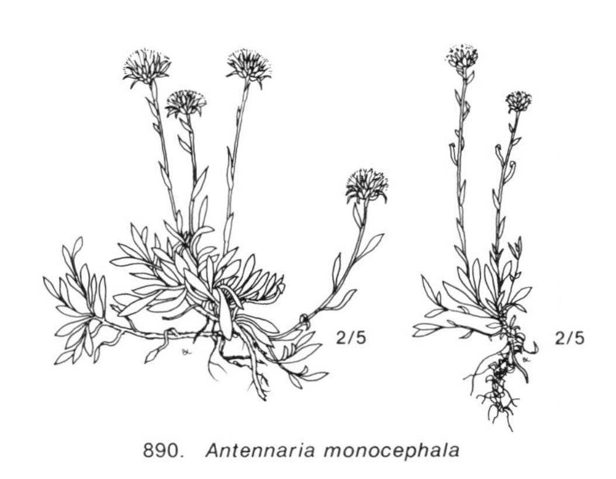

890. *Antennaria monocephala*

891. *Antennaria neoalaskana*

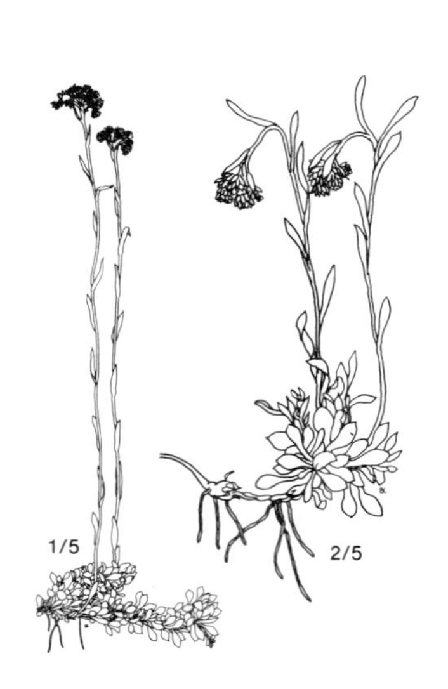

892. *Antennaria nitida*

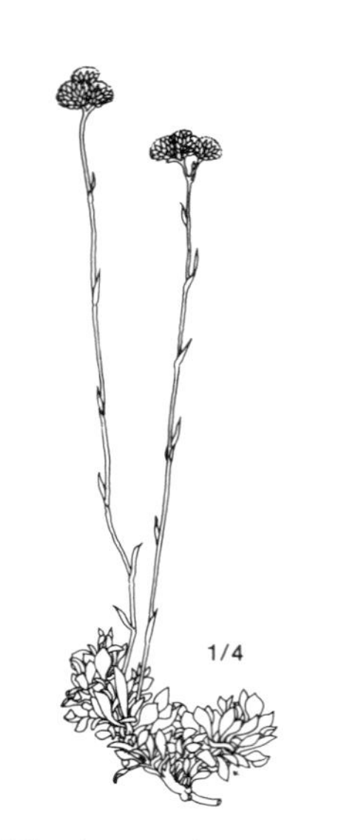

893. *Antennaria oxyphylla*

894. *Antennaria philonipha*

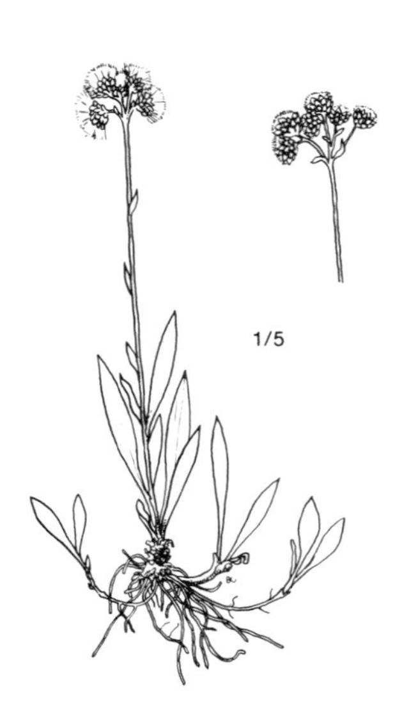

895. *Antennaria pulcherrima*

896. *Antennaria pygmaea*

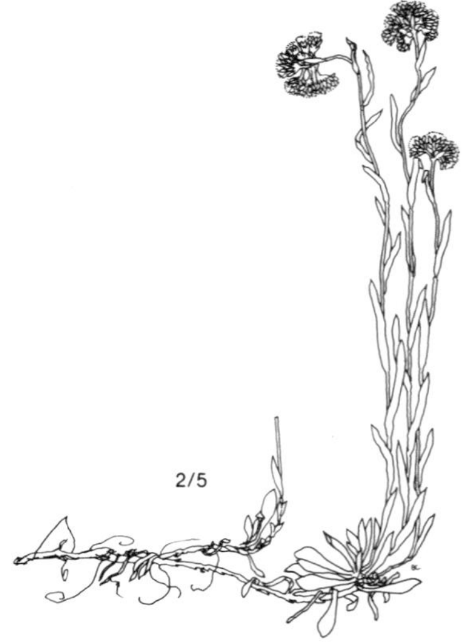

897. *Antennaria rosea*

898. *Antennaria stolonifera*

899. *Antennaria subcanescens*

900. *Antennaria subviscosa*

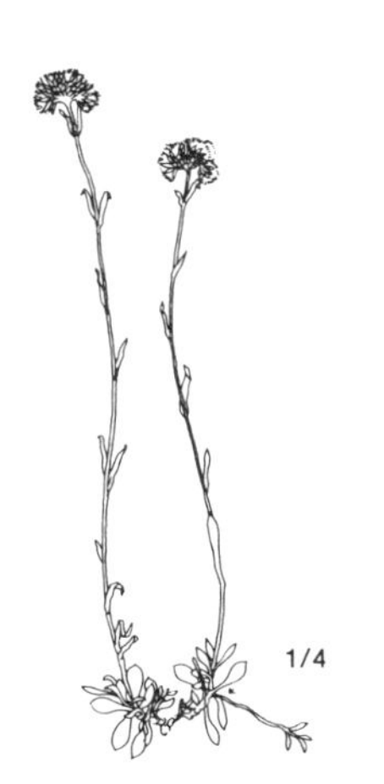

901. *Antennaria ungavensis*

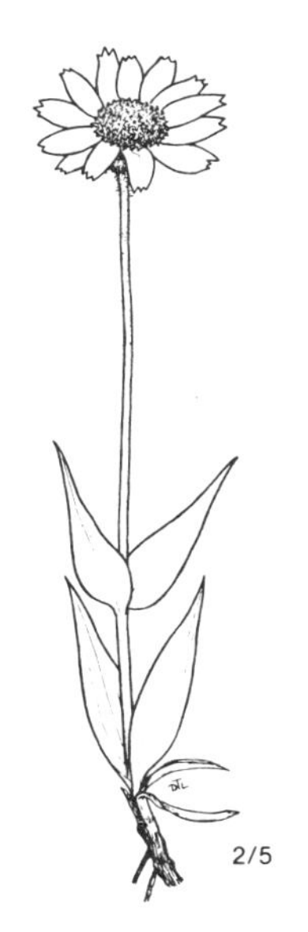

902. *Arnica alpina* ssp. *angustifolia*

903. *Arnica alpina* ssp. *tomentosa*

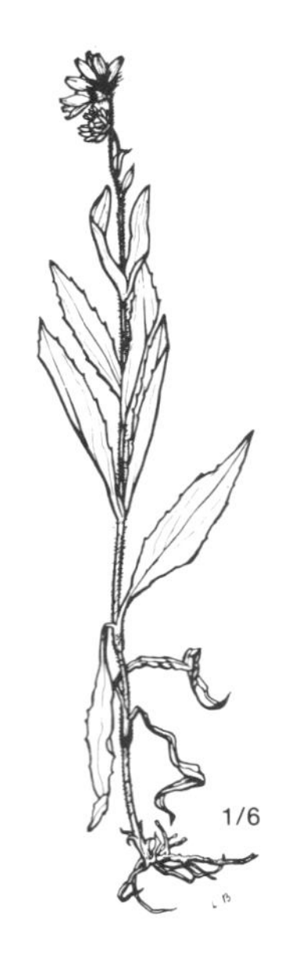

904. *Arnica amplexicaulis*

905. *Arnica Chamissonis* ssp. *foliosa*

906. *Arnica cordifolia*

907. *Arnica latifolia*

908. *Arnica Lessingii*

909. *Arnica lonchophylla*

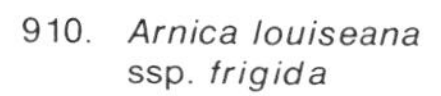
910. *Arnica louiseana* ssp. *frigida*

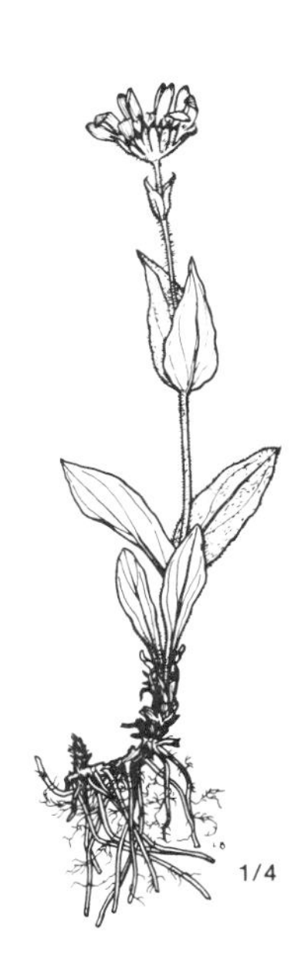

911. *Arnica mollis*

912. *Artemisia borealis*

914. *Artemisia hyperborea*

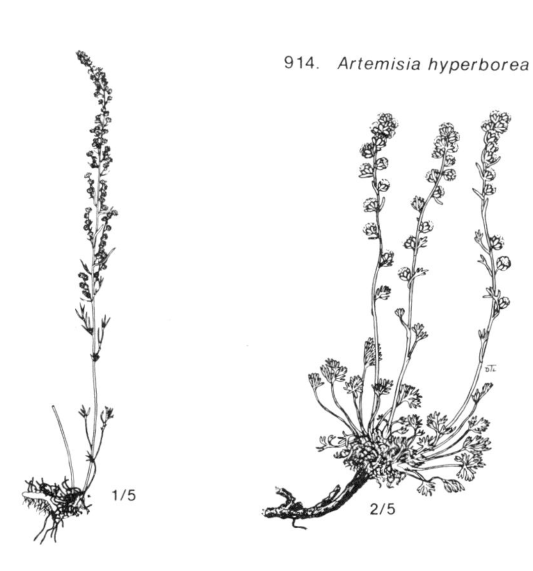

913. *Artemisia canadensis*

915. *Artemisia Richardsoniana*

916. *Artemisia Tilesii*

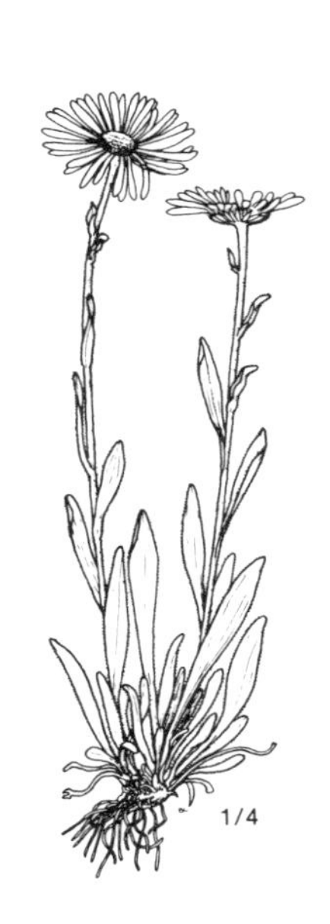

917. *Aster alpinus* ssp. *Vierhapperi*

918. *Aster brachyactis*

919. *Aster ciliolatus*

920. *Aster falcatus*

921. *Aster Franklinianus*

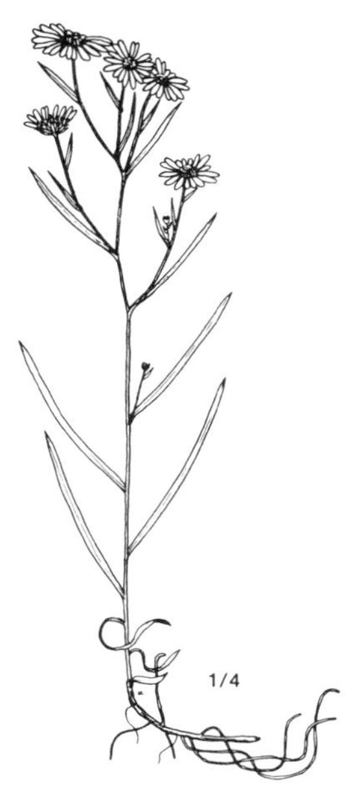

922. *Aster junciformis*

923. *Aster pansus*

924. *Aster pygmaeus*

925. *Aster sibiricus*

926. *Aster spathulatus*

927. *Bidens cernua*

928. *Chrysanthemum integrifolium*

929. *Crepis elegans*

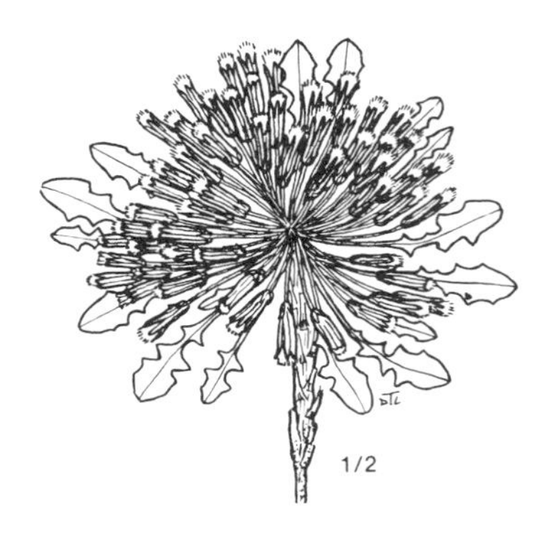

930. *Crepis nana*

931. *Erigeron acris* var. *asteroides*

932. *Erigeron acris* var. *debilis*

933. *Erigeron compositus* var. *discoideus*

934. *Erigeron compositus* var. *glabratus*

935. *Erigeron elatus*

936. *Erigeron eriocephalus*

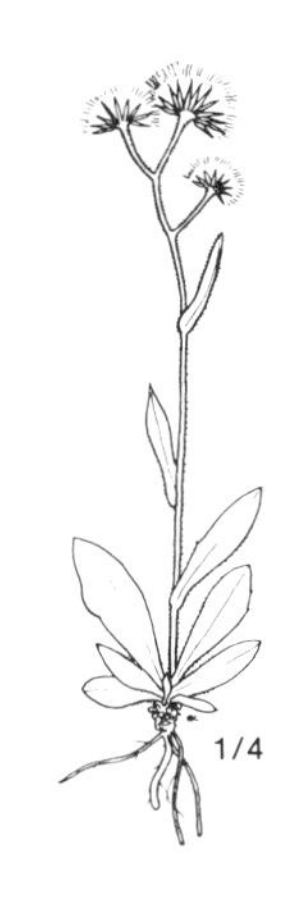

937. *Erigeron glabellus* ssp. *pubescens*

938. *Erigeron grandiflorus* ssp. *arcticus*

939. *Erigeron humilis*

940. *Erigeron hyssopifolius*

941. *Erigeron philadelphicus*

942. *Erigeron purpuratus*

943. *Gnaphalium uliginosum*

944. *Helenium autumnale* var. *grandiflorum*

945. *Hieracium albiflorum*

946. *Hieracium gracile*

947. *Hieracium scabriusculum*

948. *Hieracium triste*

949. *Lactuca pulchella*

950. *Matricaria ambigua*

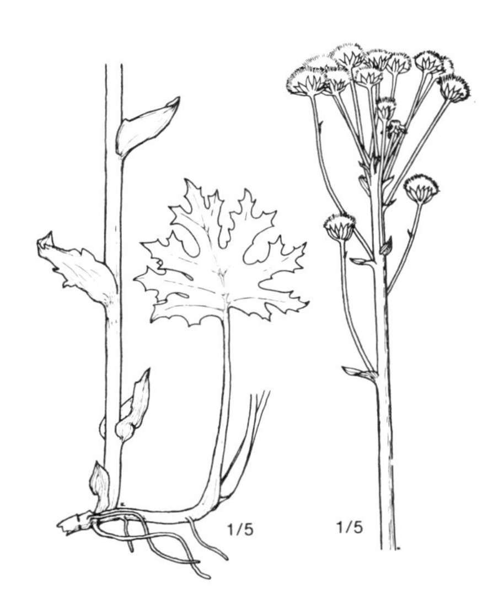

951. *Petasites arcticus*

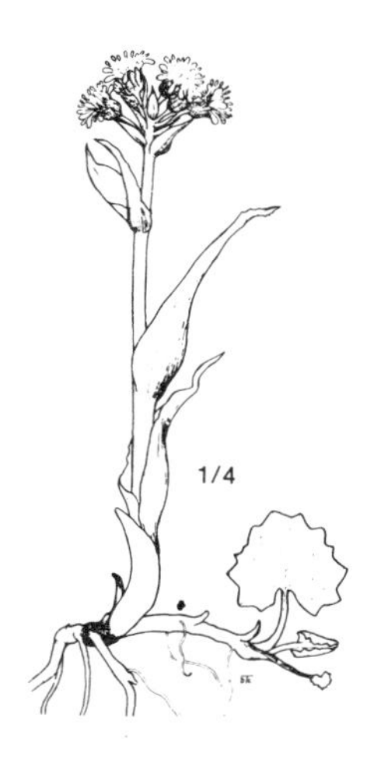

952. *Petasites frigidus*

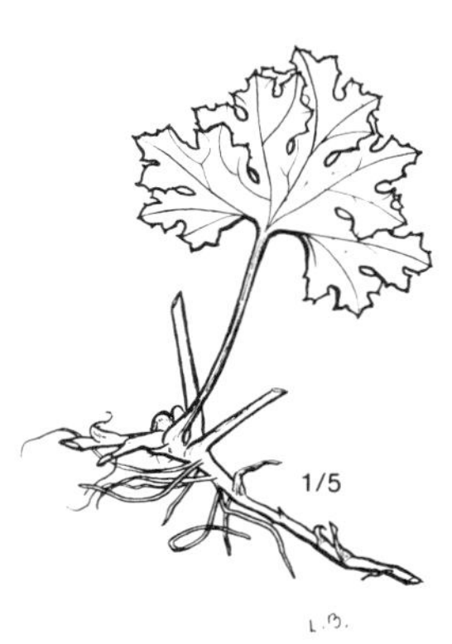

953. *Petasites hyperboreus*

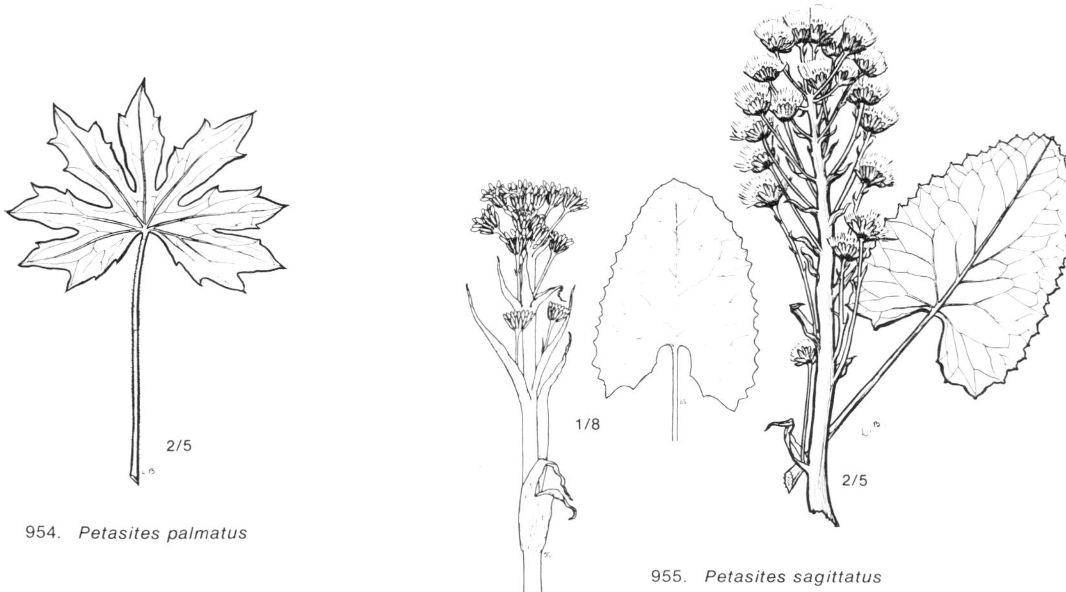

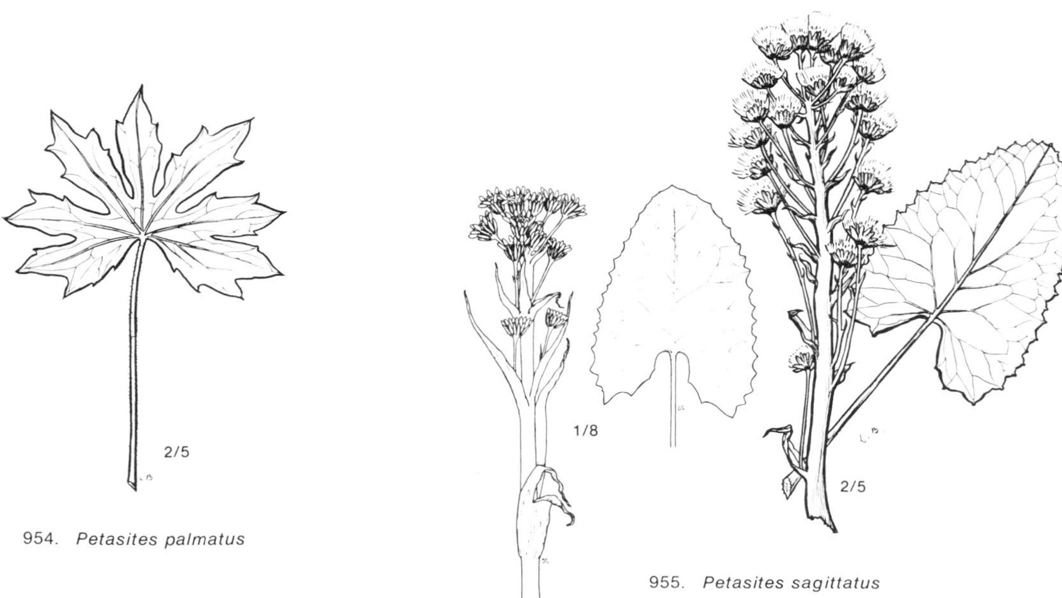

954. *Petasites palmatus*

955. *Petasites sagittatus*

956. *Saussurea angustifolia* var. *angustifolia*

957. *Saussurea angustifolia* var. *yukonensis*

958. *Senecio atropurpureus* (*radiate*)

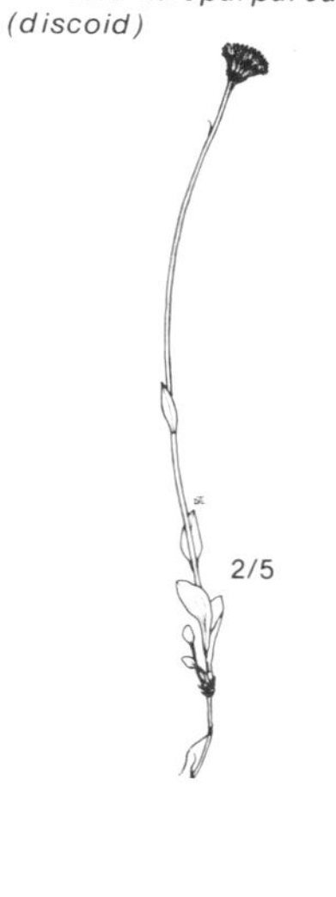

959. *Senecio atropurpureus* (*discoid*)

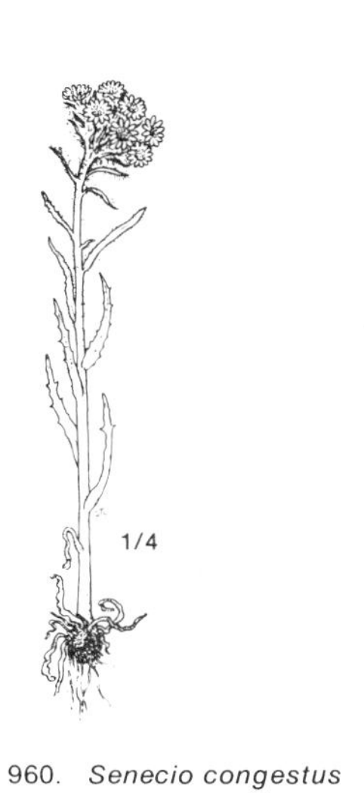

960. *Senecio congestus*

961. *Senecio cymbalaria*

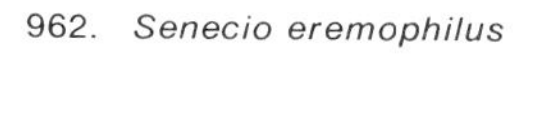
962. *Senecio eremophilus*

963. *Senecio hyperborealis*

964. *Senecio indecorus*

965. *Senecio Lindstroemii*

966. *Senecio lugens*

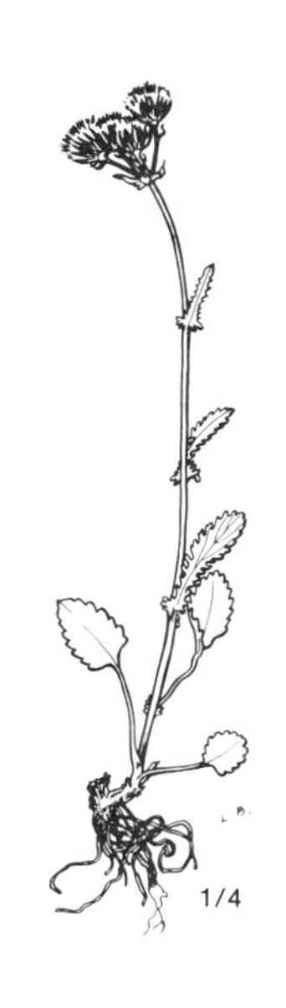

967. *Senecio pauciflorus*

968. *Senecio pauperculus*

969. *Senecio streptanthifolius*

970. *Senecio triangularis*

971. *Senecio yukonensis*

972. *Solidago canadensis* var. *salebrosa*

973. *Solidago decumbens* var. *oreophila*

974. *Solidago multiradiata*

975. *Taraxacum hyparcticum*

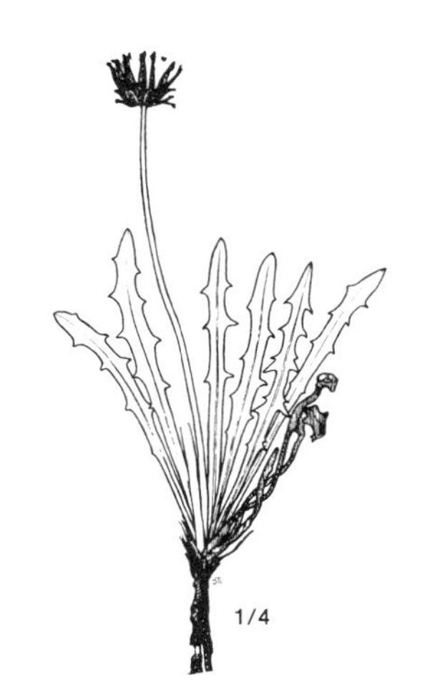

976. *Taraxacum lacerum*

977. *Taraxacum lapponicum*

978. *Taraxacum phymatocarpum*

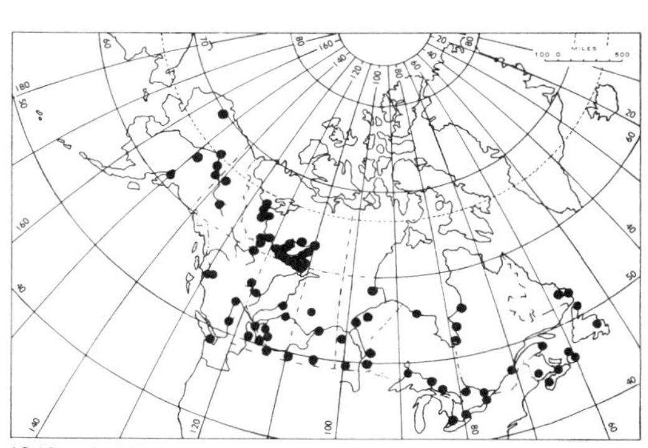

1012. *Achillea lanulosa*

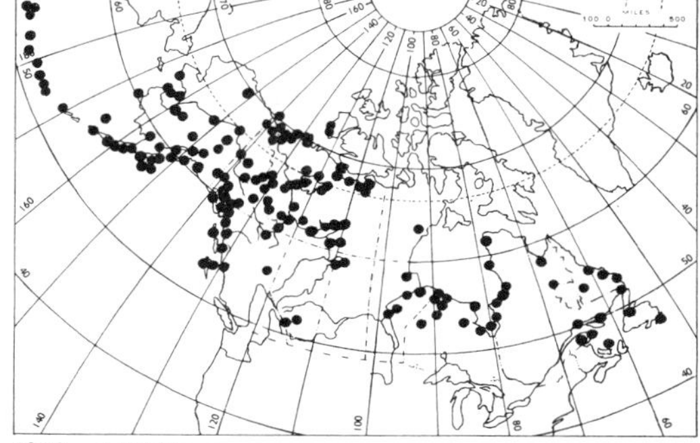

1013. *Achillea nigrescens*

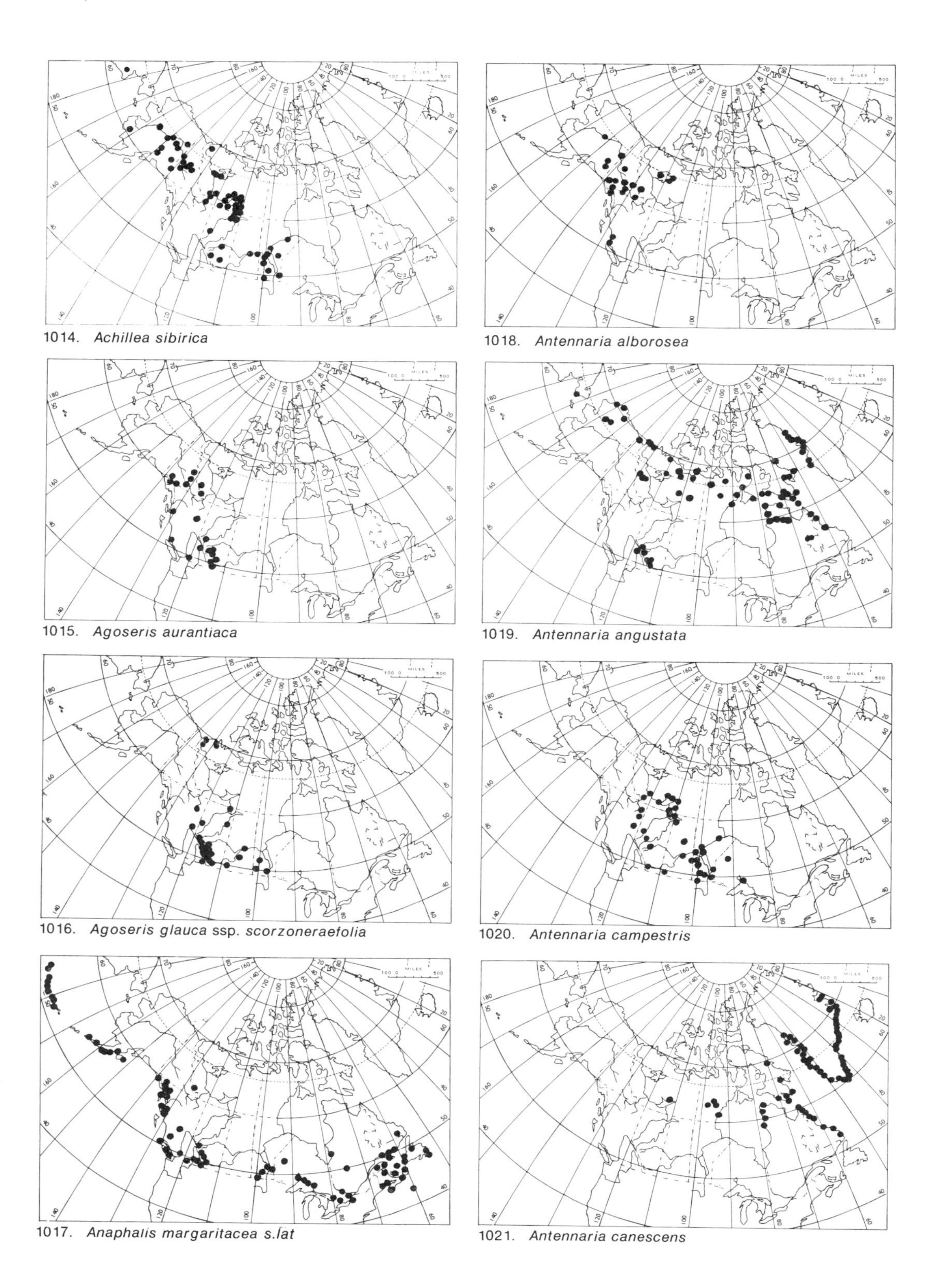

1014. *Achillea sibirica*

1018. *Antennaria alborosea*

1015. *Agoseris aurantiaca*

1019. *Antennaria angustata*

1016. *Agoseris glauca* ssp. *scorzoneraefolia*

1020. *Antennaria campestris*

1017. *Anaphalis margaritacea s.lat*

1021. *Antennaria canescens*

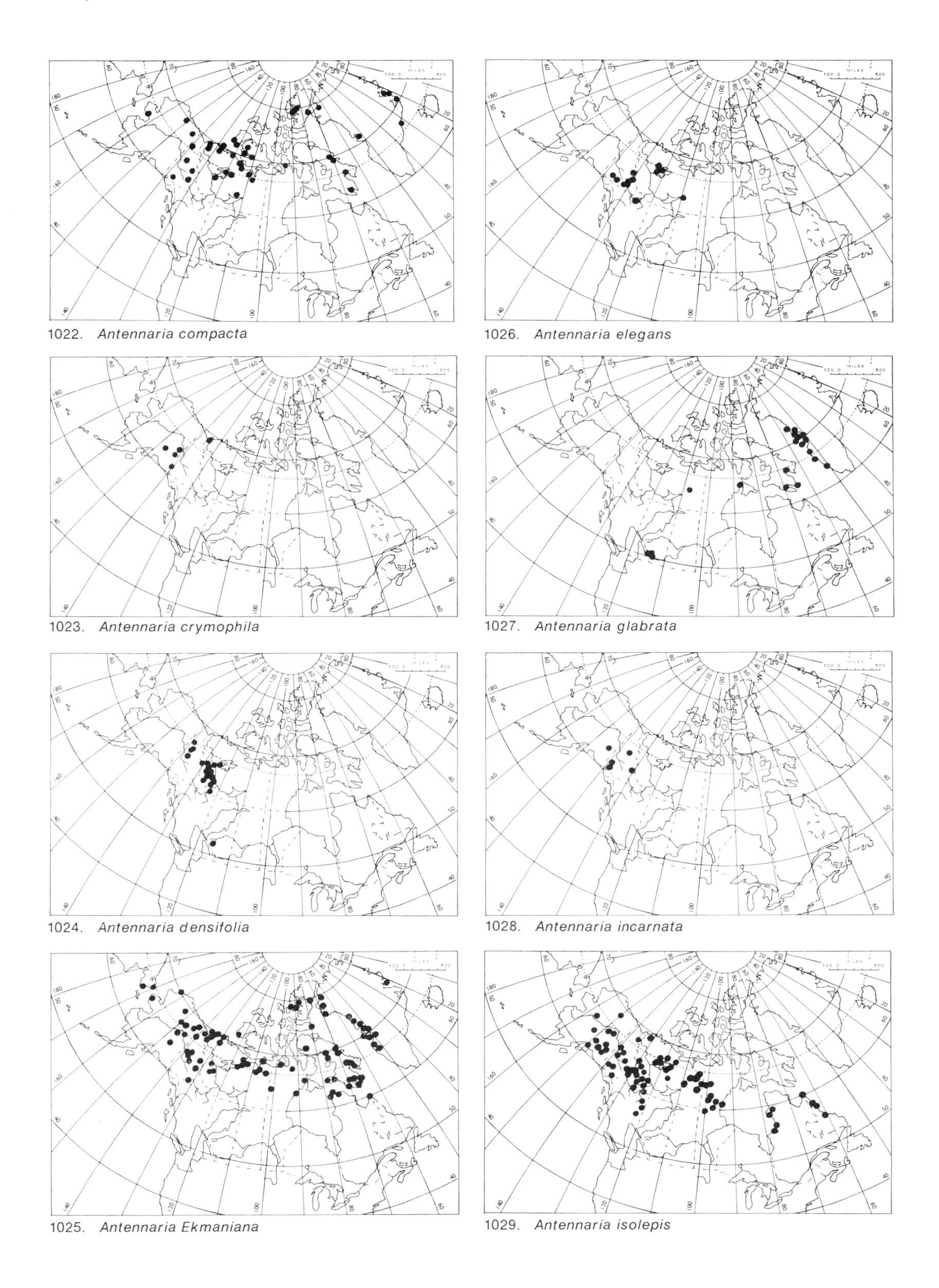

1022. *Antennaria compacta*

1023. *Antennaria crymophila*

1024. *Antennaria densifolia*

1025. *Antennaria Ekmaniana*

1026. *Antennaria elegans*

1027. *Antennaria glabrata*

1028. *Antennaria incarnata*

1029. *Antennaria isolepis*

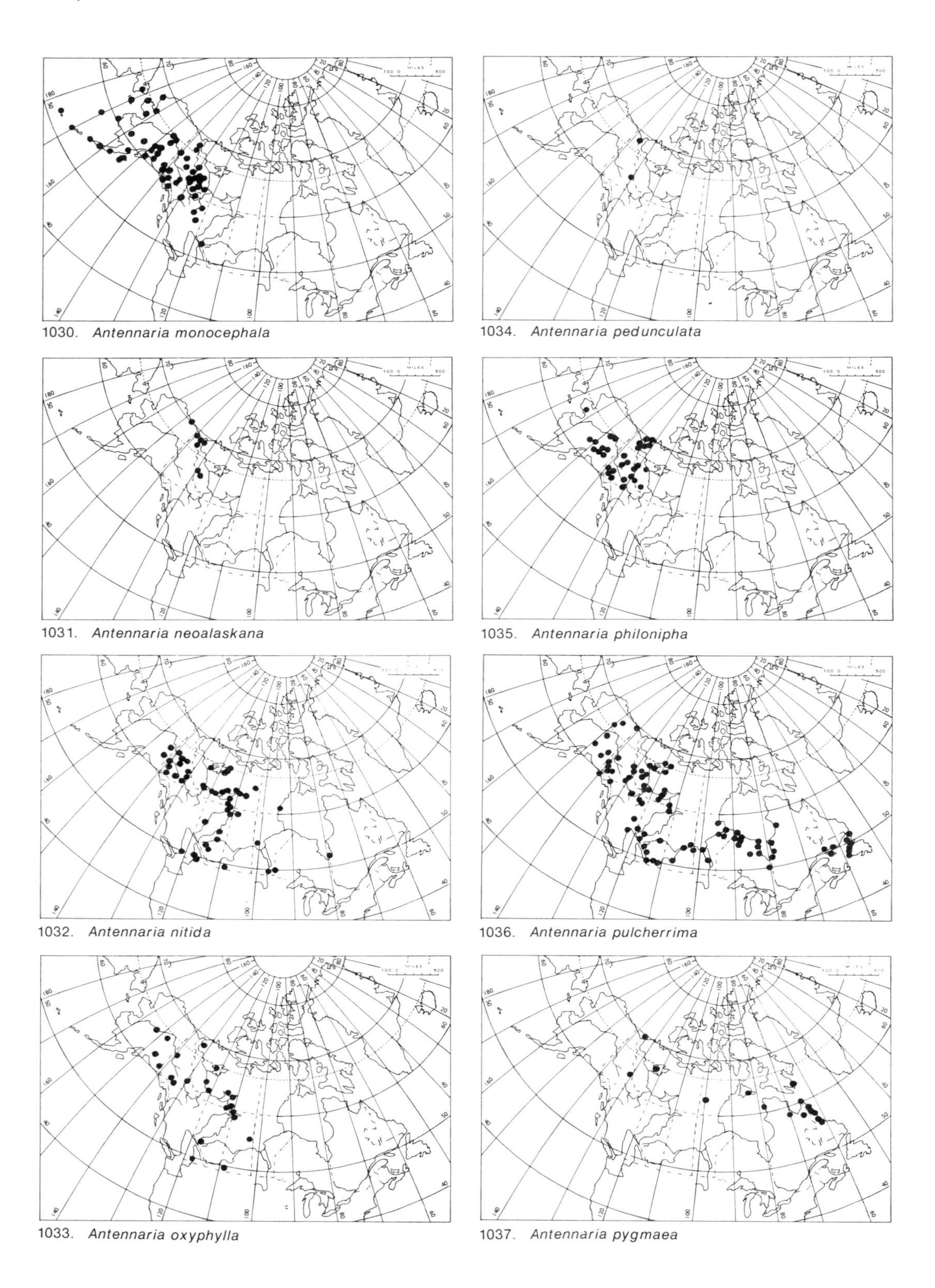

1030. *Antennaria monocephala*

1031. *Antennaria neoalaskana*

1032. *Antennaria nitida*

1033. *Antennaria oxyphylla*

1034. *Antennaria pedunculata*

1035. *Antennaria philonipha*

1036. *Antennaria pulcherrima*

1037. *Antennaria pygmaea*

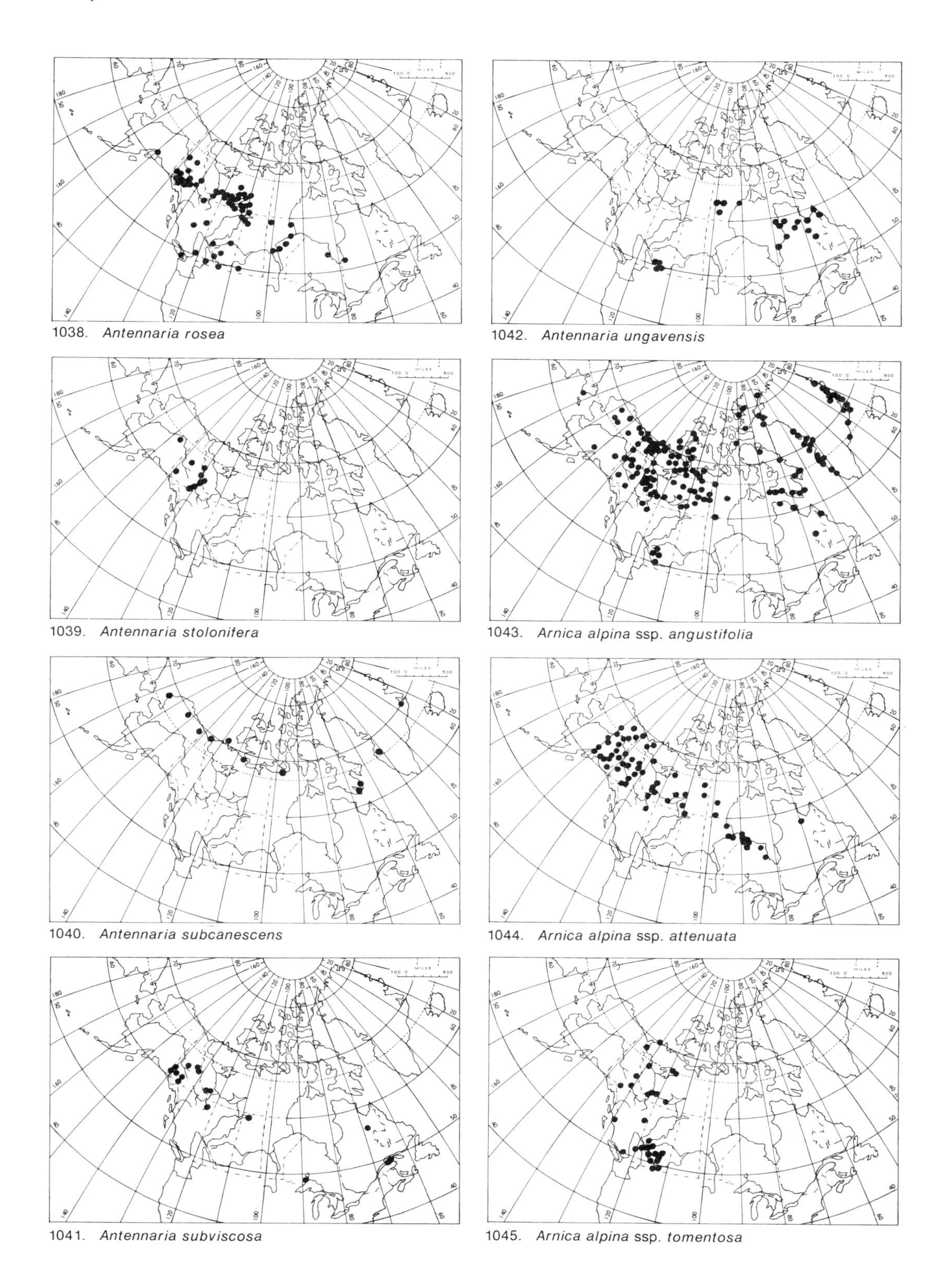

1038. *Antennaria rosea*

1039. *Antennaria stolonifera*

1040. *Antennaria subcanescens*

1041. *Antennaria subviscosa*

1042. *Antennaria ungavensis*

1043. *Arnica alpina* ssp. *angustifolia*

1044. *Arnica alpina* ssp. *attenuata*

1045. *Arnica alpina* ssp. *tomentosa*

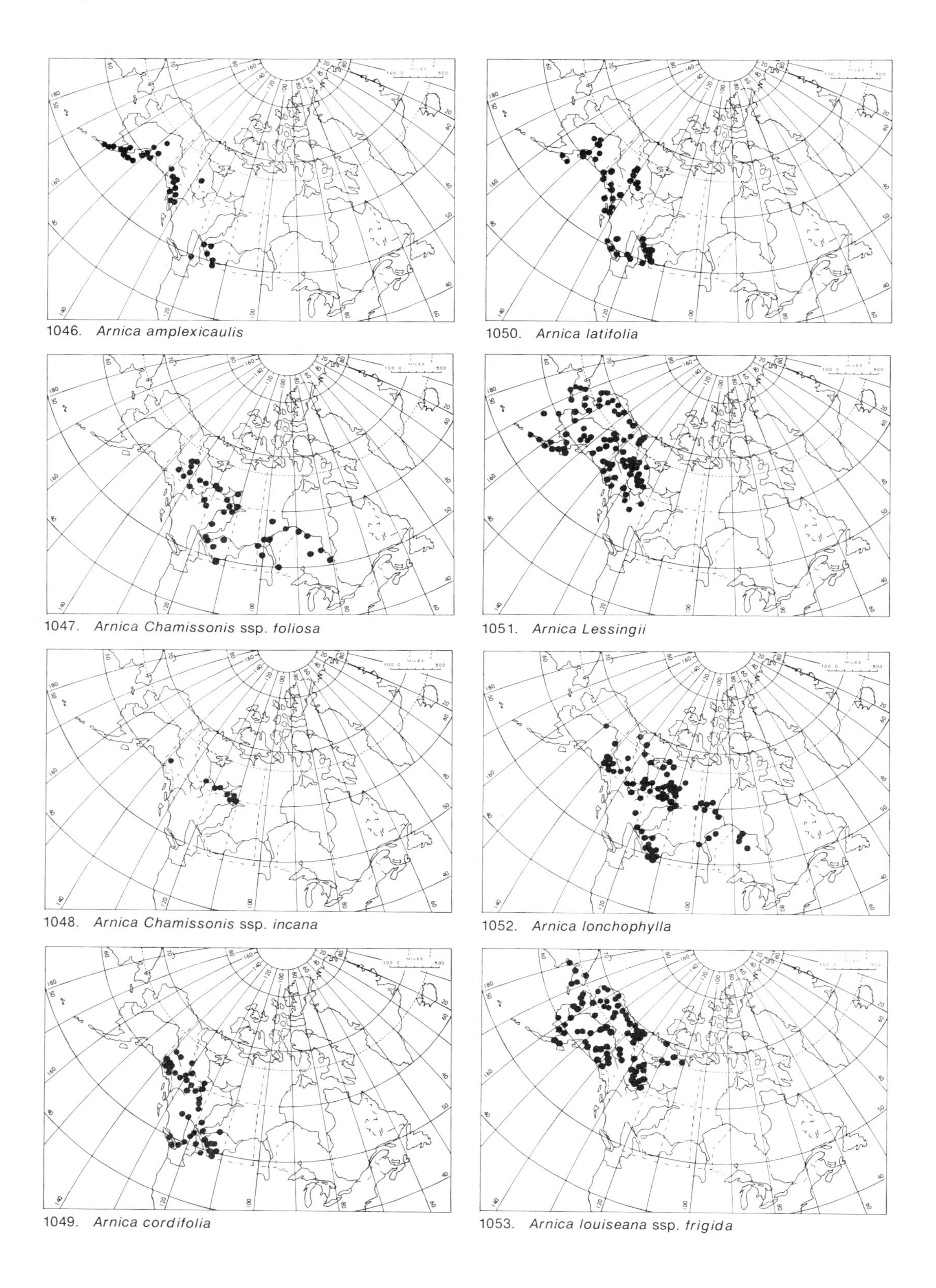

1046. *Arnica amplexicaulis*

1047. *Arnica Chamissonis* ssp. *foliosa*

1048. *Arnica Chamissonis* ssp. *incana*

1049. *Arnica cordifolia*

1050. *Arnica latifolia*

1051. *Arnica Lessingii*

1052. *Arnica lonchophylla*

1053. *Arnica louiseana* ssp. *frigida*

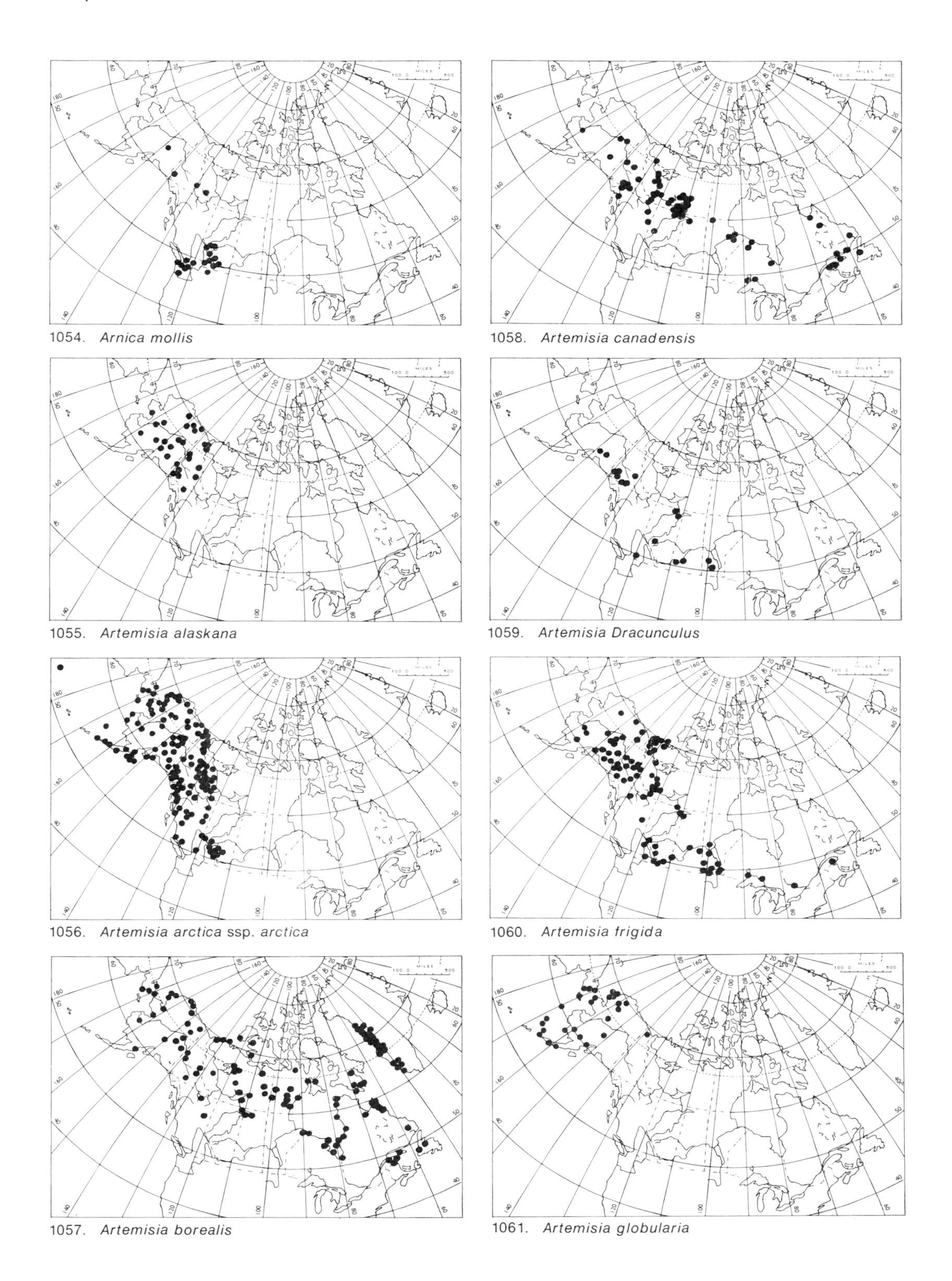

1054. *Arnica mollis*

1055. *Artemisia alaskana*

1056. *Artemisia arctica* ssp. *arctica*

1057. *Artemisia borealis*

1058. *Artemisia canadensis*

1059. *Artemisia Dracunculus*

1060. *Artemisia frigida*

1061. *Artemisia globularia*

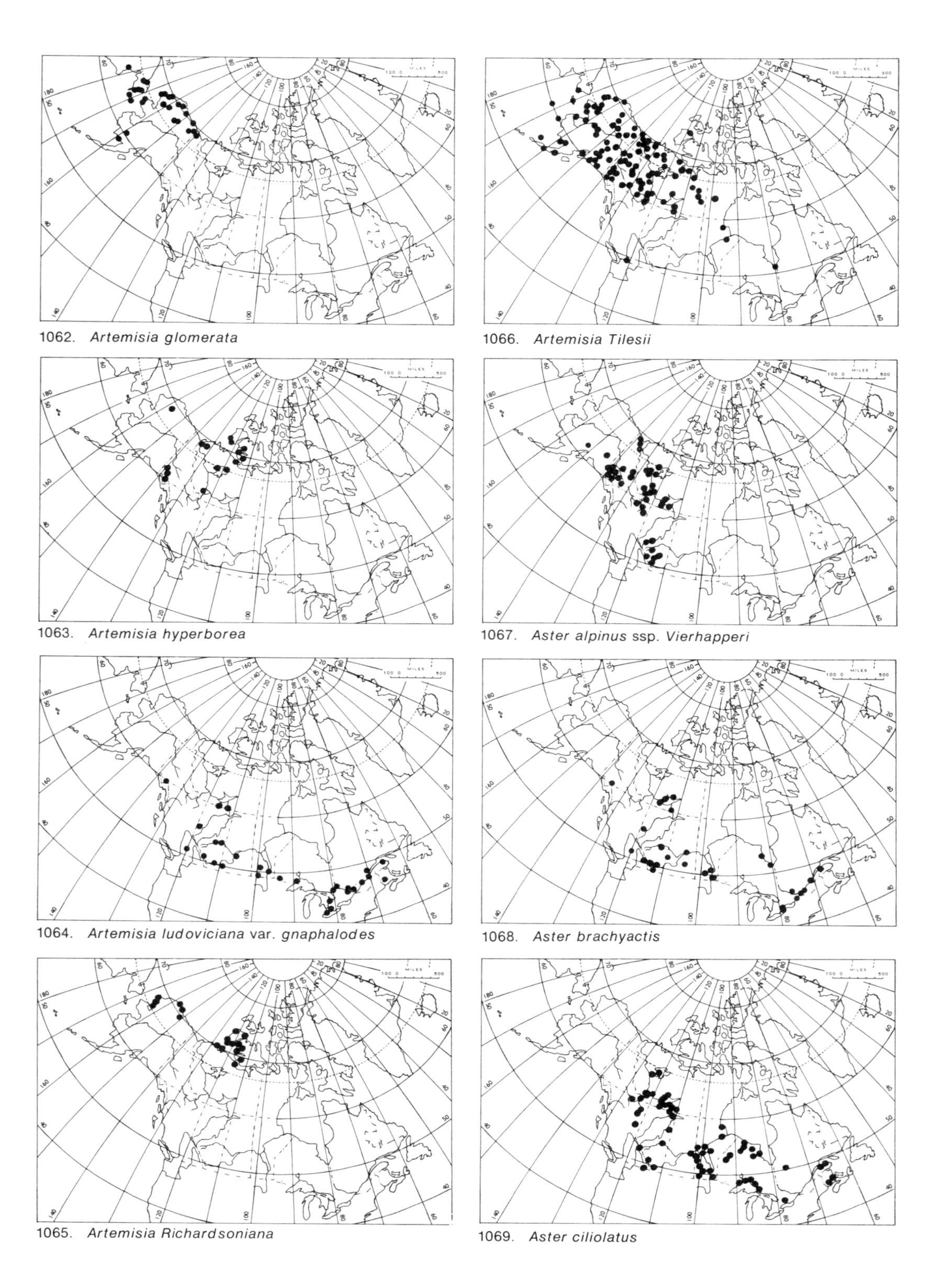

1062. *Artemisia glomerata*

1063. *Artemisia hyperborea*

1064. *Artemisia ludoviciana* var. *gnaphalodes*

1065. *Artemisia Richardsoniana*

1066. *Artemisia Tilesii*

1067. *Aster alpinus* ssp. *Vierhapperi*

1068. *Aster brachyactis*

1069. *Aster ciliolatus*

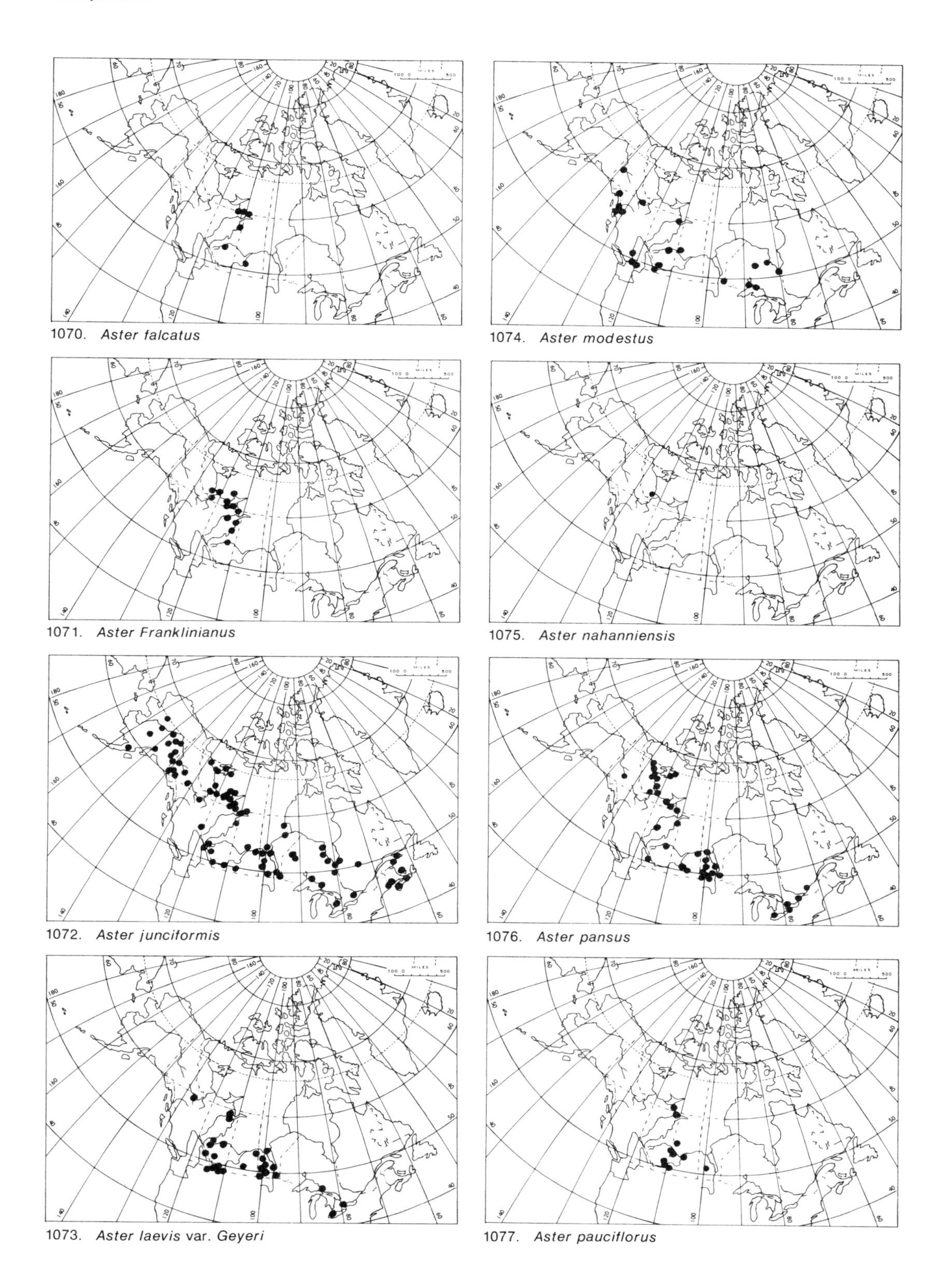

1070. *Aster falcatus*

1074. *Aster modestus*

1071. *Aster Franklinianus*

1075. *Aster nahanniensis*

1072. *Aster junciformis*

1076. *Aster pansus*

1073. *Aster laevis* var. *Geyeri*

1077. *Aster pauciflorus*

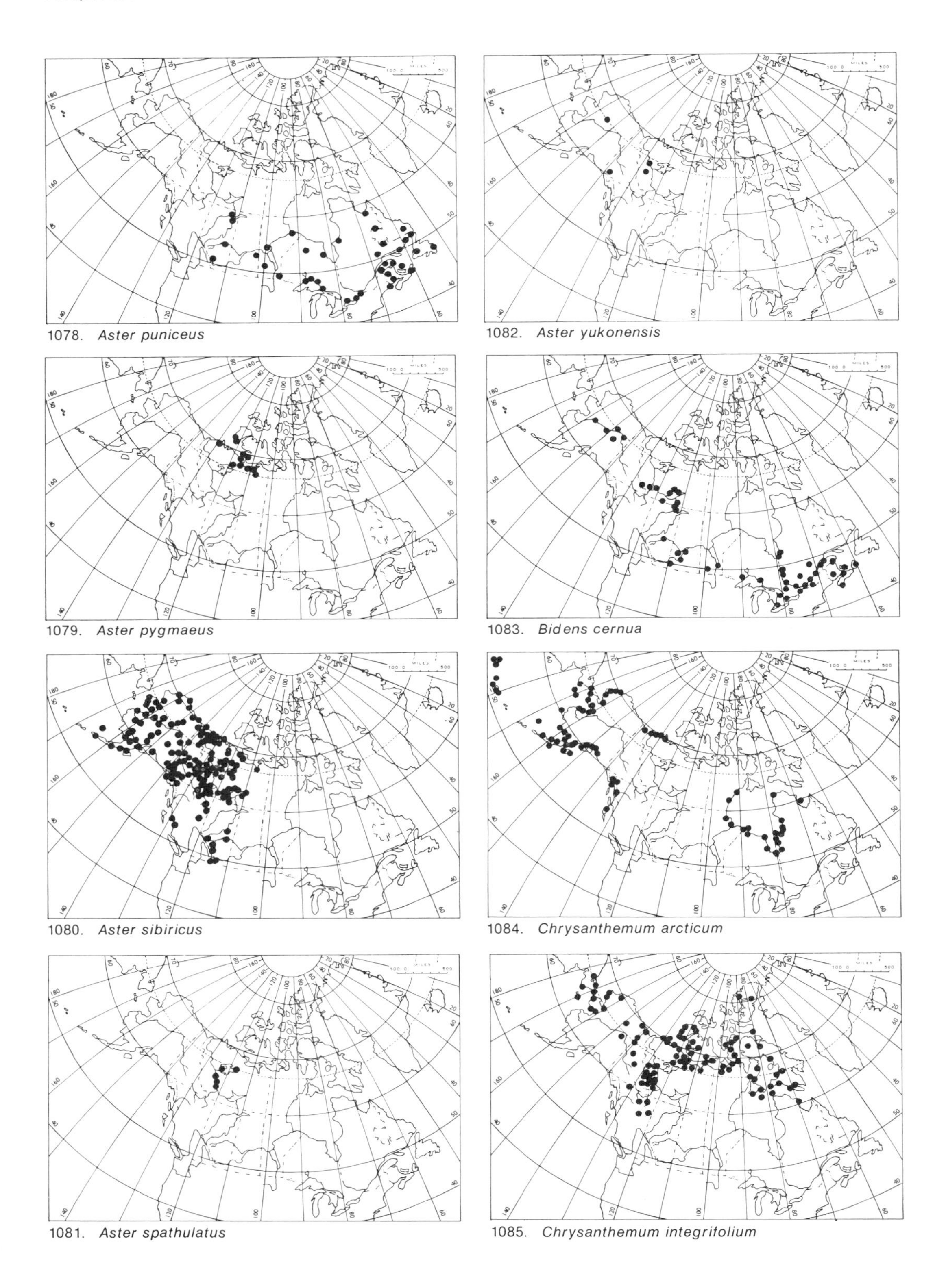

1078. *Aster puniceus*

1082. *Aster yukonensis*

1079. *Aster pygmaeus*

1083. *Bidens cernua*

1080. *Aster sibiricus*

1084. *Chrysanthemum arcticum*

1081. *Aster spathulatus*

1085. *Chrysanthemum integrifolium*

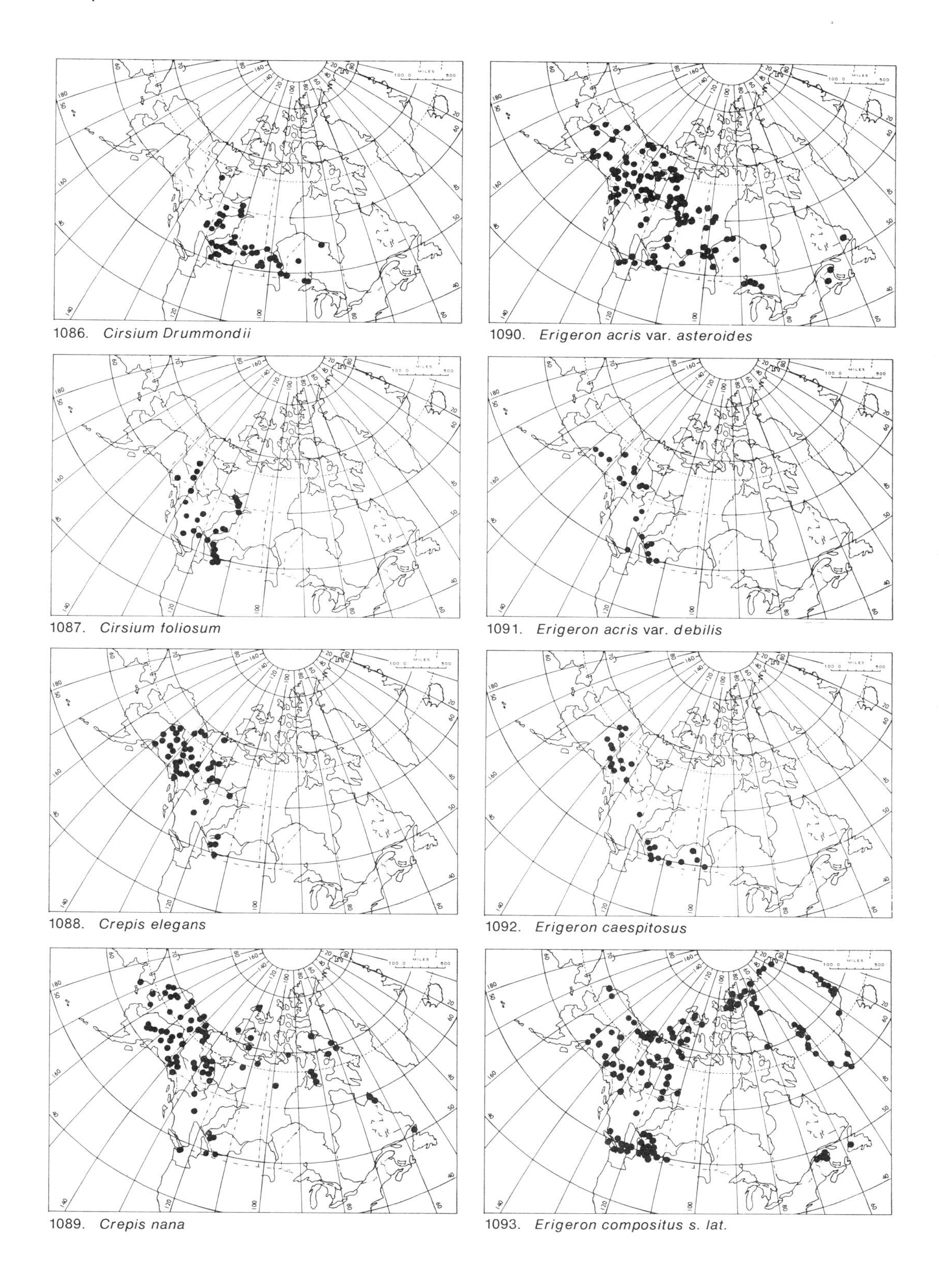

1086. *Cirsium Drummondii*

1087. *Cirsium foliosum*

1088. *Crepis elegans*

1089. *Crepis nana*

1090. *Erigeron acris* var. *asteroides*

1091. *Erigeron acris* var. *debilis*

1092. *Erigeron caespitosus*

1093. *Erigeron compositus s. lat.*

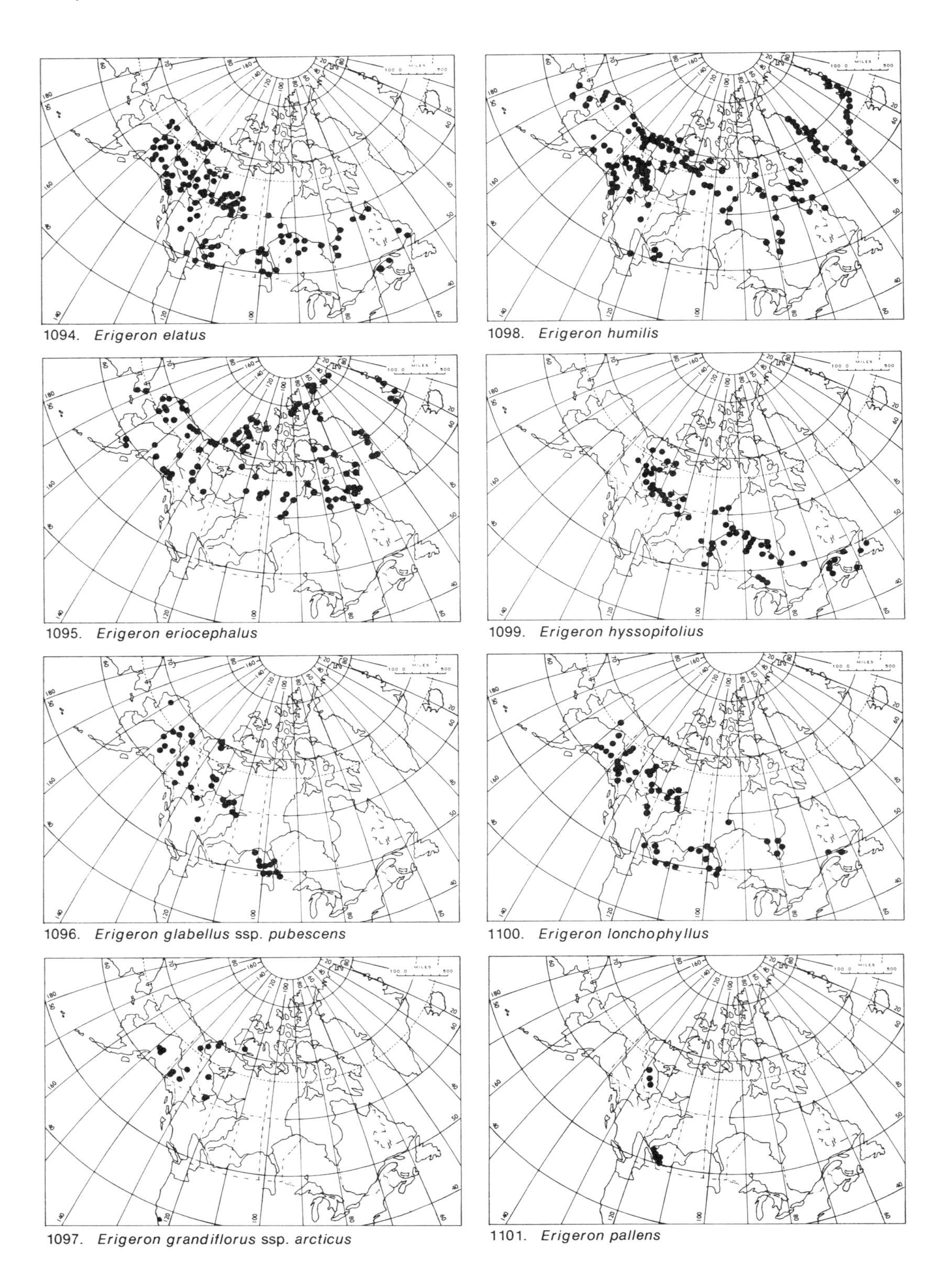

1094. *Erigeron elatus*

1098. *Erigeron humilis*

1095. *Erigeron eriocephalus*

1099. *Erigeron hyssopifolius*

1096. *Erigeron glabellus* ssp. *pubescens*

1100. *Erigeron lonchophyllus*

1097. *Erigeron grandiflorus* ssp. *arcticus*

1101. *Erigeron pallens*

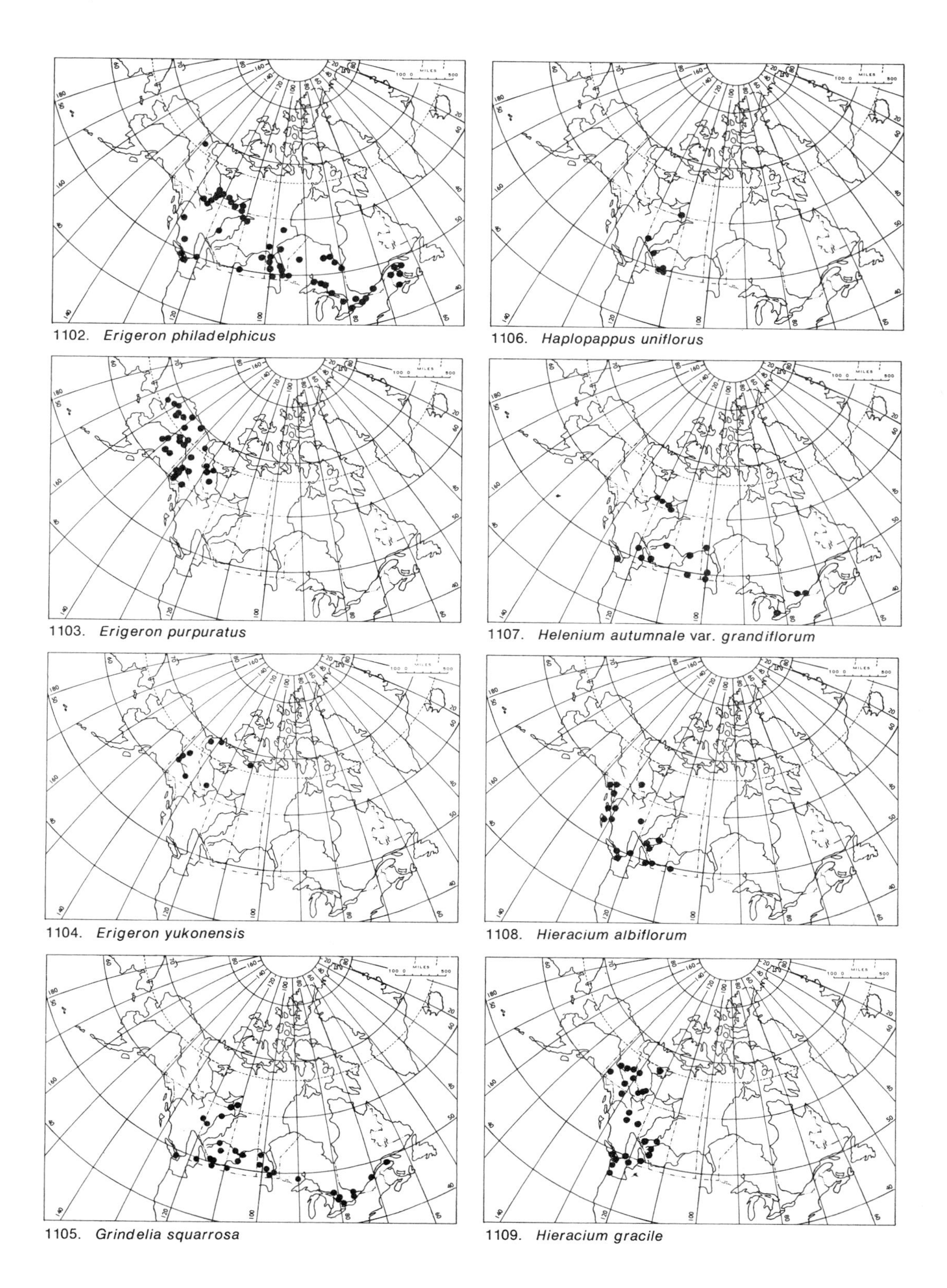

1102. *Erigeron philadelphicus*

1103. *Erigeron purpuratus*

1104. *Erigeron yukonensis*

1105. *Grindelia squarrosa*

1106. *Haplopappus uniflorus*

1107. *Helenium autumnale* var. *grandiflorum*

1108. *Hieracium albiflorum*

1109. *Hieracium gracile*

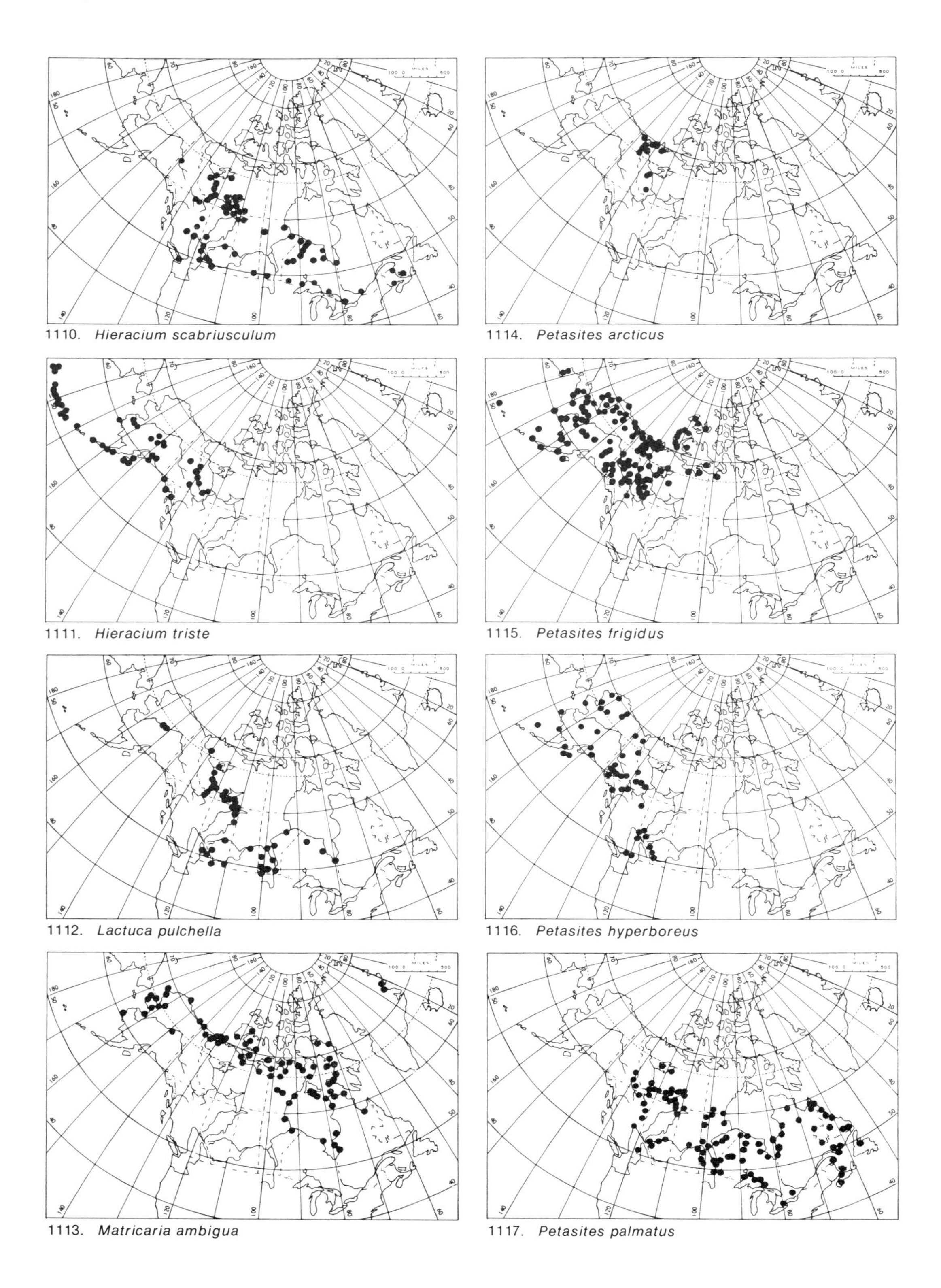

1110. *Hieracium scabriusculum*

1114. *Petasites arcticus*

1111. *Hieracium triste*

1115. *Petasites frigidus*

1112. *Lactuca pulchella*

1116. *Petasites hyperboreus*

1113. *Matricaria ambigua*

1117. *Petasites palmatus*

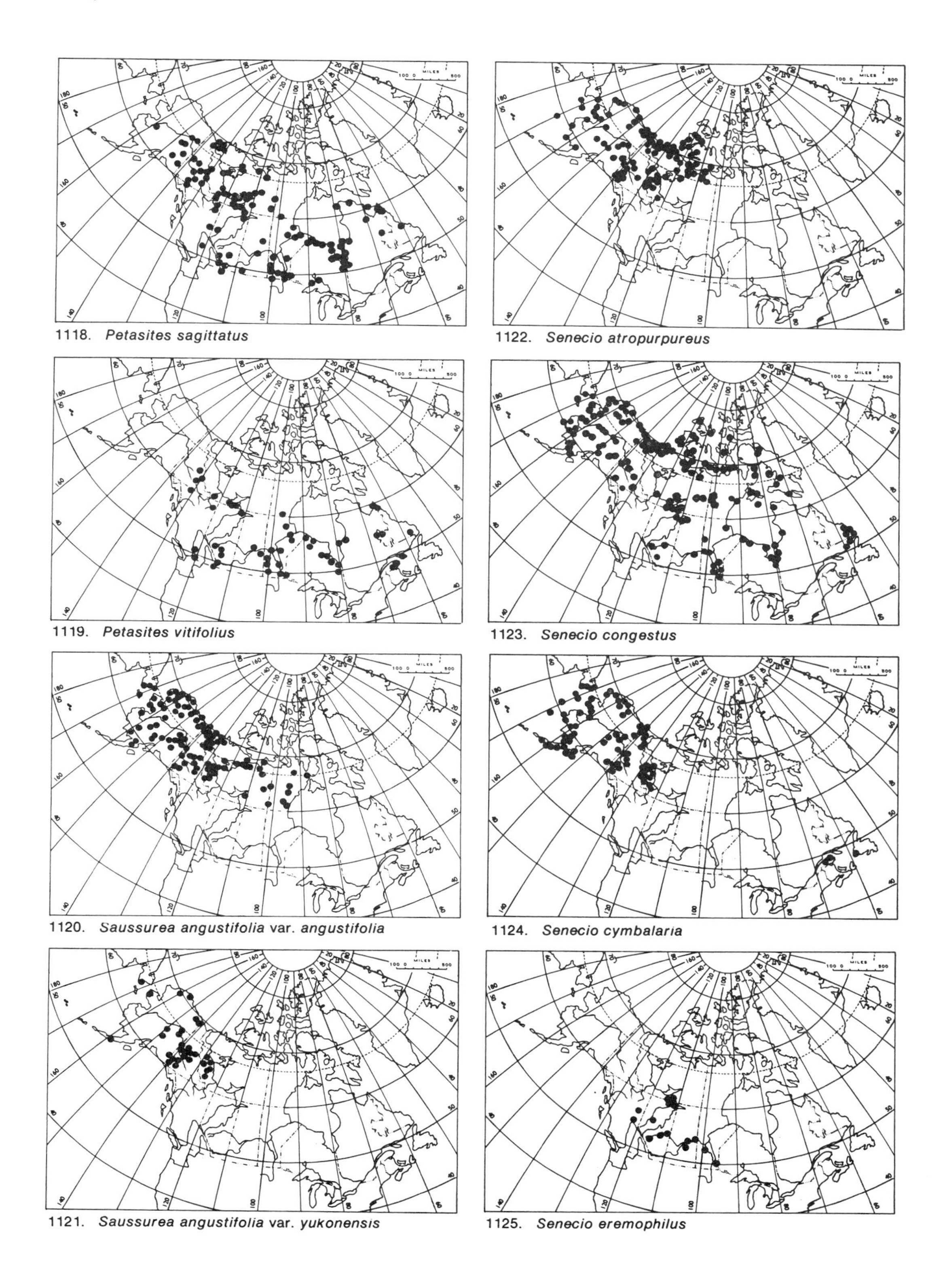

1118. *Petasites sagittatus*

1119. *Petasites vitifolius*

1120. *Saussurea angustifolia* var. *angustifolia*

1121. *Saussurea angustifolia* var. *yukonensis*

1122. *Senecio atropurpureus*

1123. *Senecio congestus*

1124. *Senecio cymbalaria*

1125. *Senecio eremophilus*

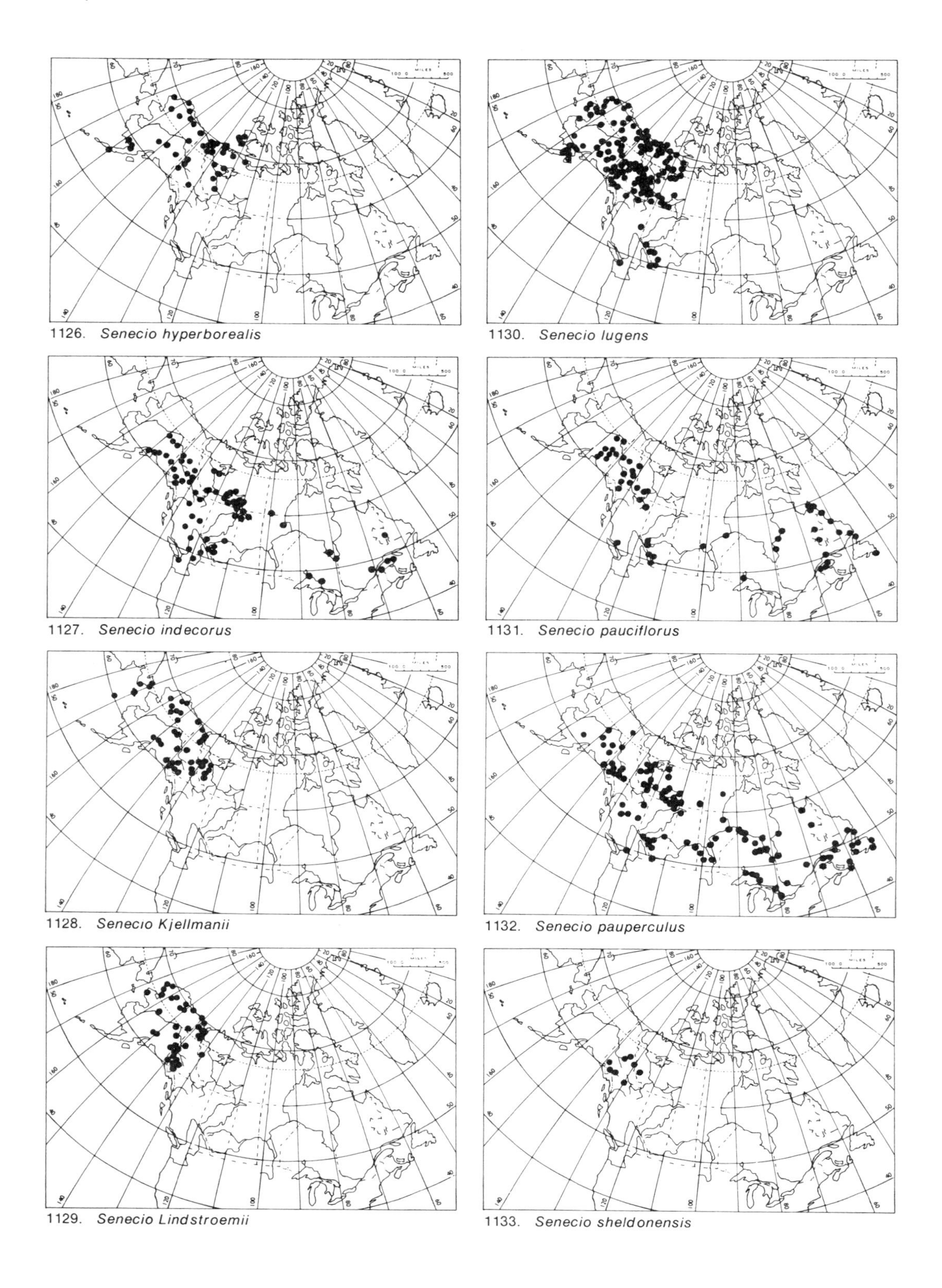

1126. *Senecio hyperborealis*

1127. *Senecio indecorus*

1128. *Senecio Kjellmanii*

1129. *Senecio Lindstroemii*

1130. *Senecio lugens*

1131. *Senecio pauciflorus*

1132. *Senecio pauperculus*

1133. *Senecio sheldonensis*

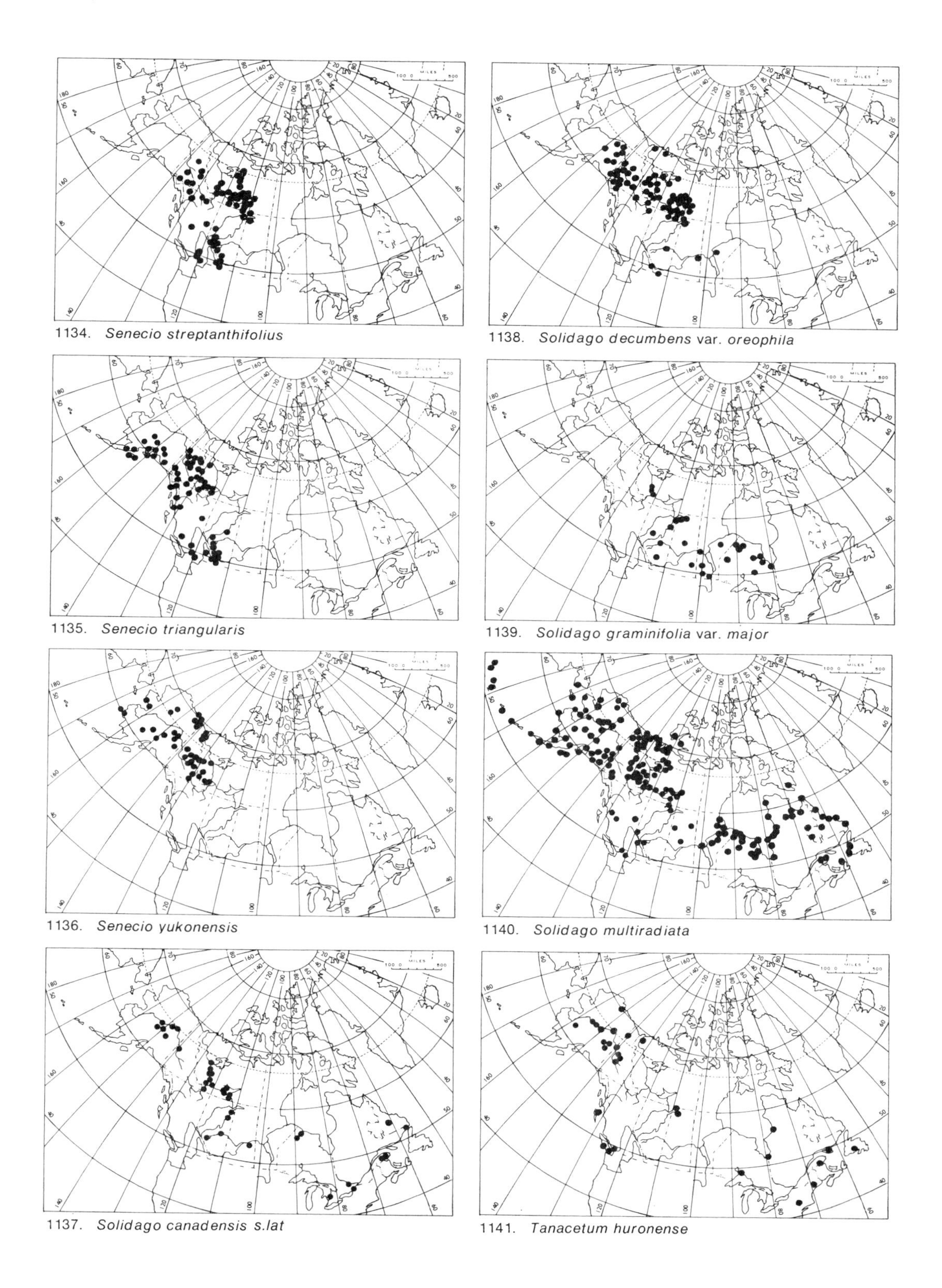

1134. *Senecio streptanthifolius*

1135. *Senecio triangularis*

1136. *Senecio yukonensis*

1137. *Solidago canadensis s.lat*

1138. *Solidago decumbens* var. *oreophila*

1139. *Solidago graminifolia* var. *major*

1140. *Solidago multiradiata*

1141. *Tanacetum huronense*

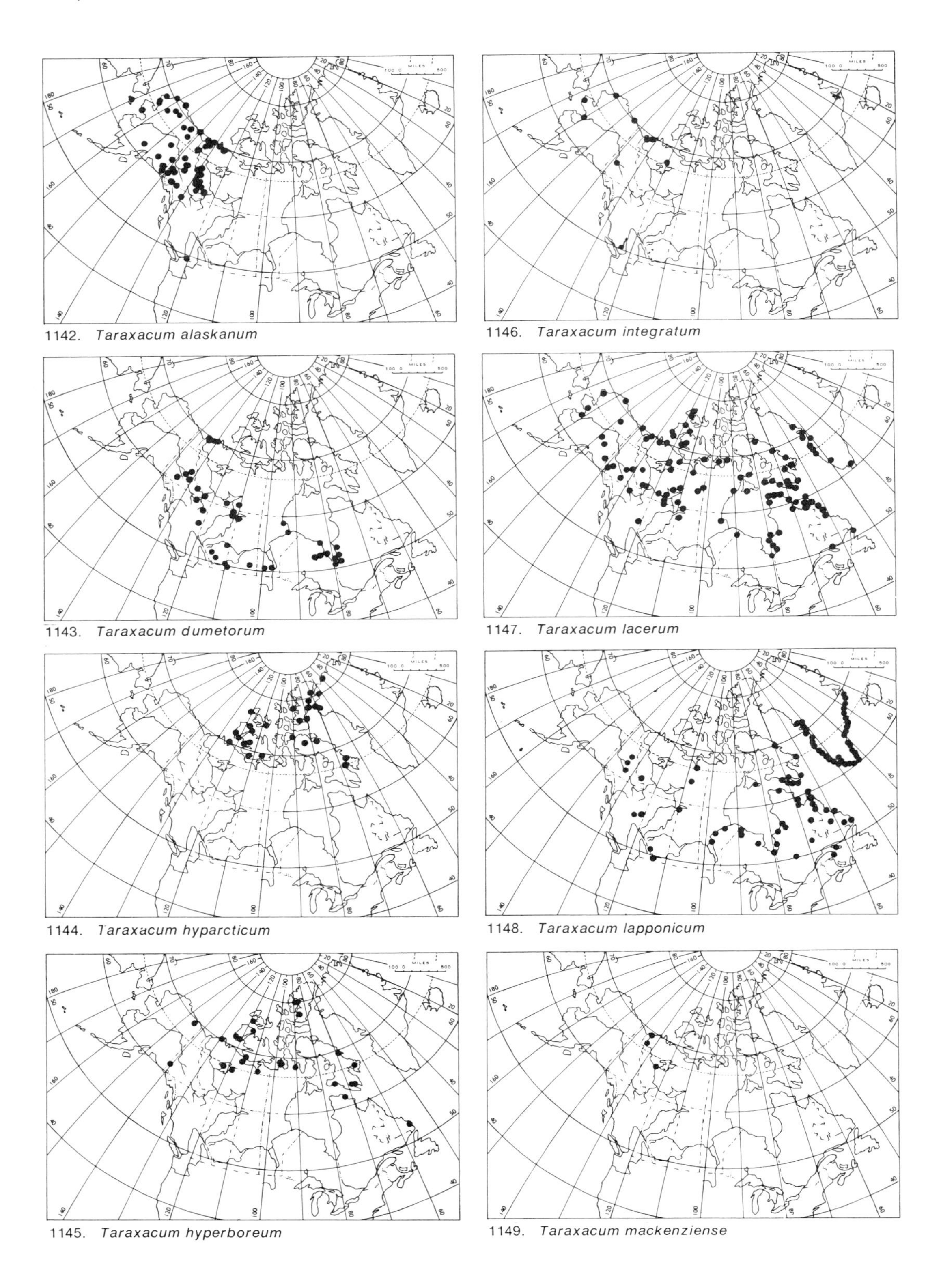

1142. *Taraxacum alaskanum*

1143. *Taraxacum dumetorum*

1144. *Taraxacum hyparcticum*

1145. *Taraxacum hyperboreum*

1146. *Taraxacum integratum*

1147. *Taraxacum lacerum*

1148. *Taraxacum lapponicum*

1149. *Taraxacum mackenziense*

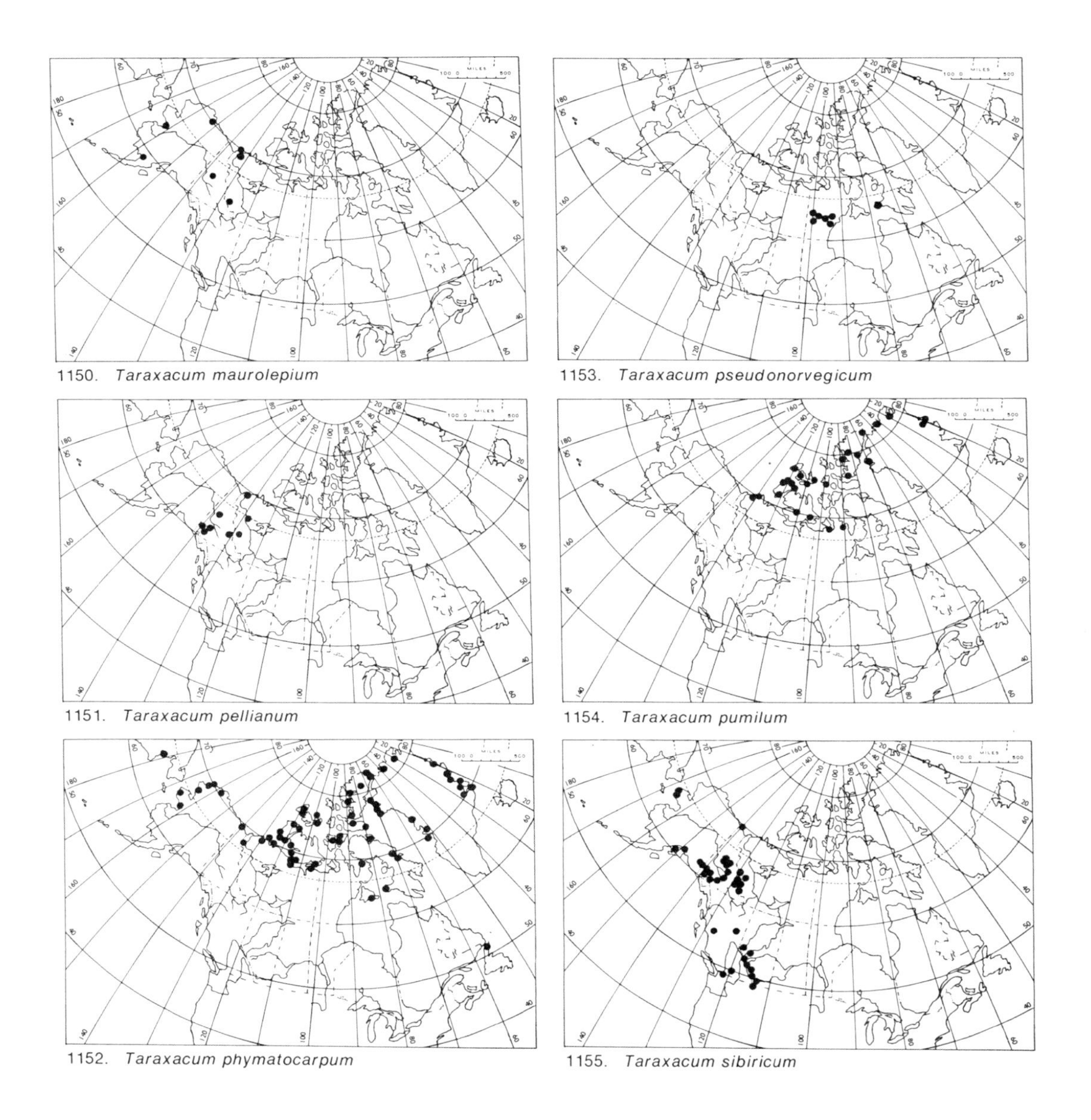

1150. *Taraxacum maurolepium*

1151. *Taraxacum pellianum*

1152. *Taraxacum phymatocarpum*

1153. *Taraxacum pseudonorvegicum*

1154. *Taraxacum pumilum*

1155. *Taraxacum sibiricum*

Abortive
Not perfectly developed
Acaulescent
Apparently without a stem, the leaves and inflorescence arising near the surface of the ground
Achene
A small dry indehiscent fruit, distinguished from a nutlet by its relatively thin wall
Acicular
Slenderly needle-shaped
Acidophyte
Of a plant growing by preference in a relatively acid environment.
Acrid
Unpleasantly or irritatingly sharp or unpleasant to the taste or smell
Actinomorphic
Exhibiting radial symmetry, as a regular flower
Acuminate
Tapering to a slender point
Acute
Forming an acute angle at base or apex
Adnate
Grown together or attached; applied only to unlike organs, as stipules adnate to the petiole
Adventitious
Not in the usual place
Adventive
Not fully established or only locally established
Aerial
Existing, growing or forming in the air rather than in the ground or water
Alternate
Situated singly at each node, as leaves on a stem or flowers along an axis; situated regularly between other organs, as stamens alternate with petals
Ament
See catkin
Amphi-Atlantic
Occurring on both sides of the Atlantic Ocean, but lacking on the Pacific side of the globe
Amphi-Beringian
Occurring on both sides of the Bering Strait, but lacking on the Atlantic side of the globe
Amphibious
Equally well adapted to life in water and on dry land
Androgynous (inflorescence in *Carex*)
Denoting a spike containing both staminate and pistillate flowers, the latter at the base
Annual
Of one year's duration
Anther
The distal part of a stamen in which pollen is produced, composed usually of two parts known as anther-sacs, pollen sacs or thecae
Anthesis
The period during which a flower is fully expanded and functional
Anthropochorus
Regularly distributed by man deliberately (as crop plants) or accidently (as weeds)
Antrorse
Directed more or less toward the summit of the plant or of an organ of a plant
Apex
Tip
Apiculate
Ending abruptly in a small, usually sharp tip
Apogamy
The development of fruits without fertilization, i.e. parthenogenesis
Apomictic
Capable of producing seed without fertilization
Appendiculate
Furnished with an appendage
Appressed
Lying close to or parallel to an organ, as hairs appressed to a leaf or leaves appressed to the stem
Approximate
Close together
Arachnoid (pubescence)
Cobwebby; thinly pubescent with relatively long, usually appressed and interlaced hairs
Arborescent
Treelike
Arcuate
Curved into an arc of a circle, without regard to direction
Areola, Areole
A small space; usually referring to the spaces bounded by veinlets on the surface of a leaf
Areolate
Marked with areolae
Aristate
Having an awn, usually terminal in position
Articulate
Jointed; provided with nodes or joints, or places where separation may naturally take place
Ascending
Growing obliquely upward (stems); directed obliquely forward in respect to the organ to which they are attached (parts of a plant)
Astringent
Tending to pucker the tissues of the mouth
Attenuate
Gradually tapering to a very slender point
Auricle
A small projecting lobe or appendage at the base of an organ
Auriculate
With an auricle
Awl-shaped
Gradually tapering from the base to a slender or stiff point
Awn
A slender terminal bristle, usually stiff in proportion to its size
Axil
The position (not a structure) between a lateral organ, especially a leaf, branch, or pedicel, and the axis
Axillary
Located in or arising from an axil
Axis
A portion of a plant from which a series of organs arises radially, as the axis of an inflorescence. (Compare rachis)

Barbed
Provided, usually laterally or marginally, with short reflexed points
Barbellate
Diminutive of barbed
Basal
Located at the base of a plant or of an organ of a plant
Beak
A comparatively short and stout terminal appendage on a thickened organ, as a seed or a fruit; not used for a flat organ, as a leaf
Bearded
Bearing or furnished with long or stiff hairs
Berry
In a strict sense, a fruit developed from a single ovary, fleshy or pulpy throughout, containing one to many seeds; loosely, any pulpy or juicy fruit. (Compare drupe)
Bi-
As a prefix, two or twice
Bidentate
Having two teeth
Biennial
Living two years only and blooming the second year
Bifid
Forked
Bipinnate
Doubly or twice pinnate
Bisexual
Having both sexes on the same individual; a hermaphrodite
Blade
The expanded terminal part of a flat organ, as leaf, petal, or sepal, in contrast to the narrowed basal portion
Bract
A specialized leaf from the axil of which a flower arises; differing from foliage leaves in size, shape or texture, but sometimes gradually modified from them. Sometimes applied to a specialized leaf subtending an inflorescence. (Compare spathe)
Bracteate, Bracted
With bracts
Bracteolate
Having bracteoles
Bracteole
Often denoting a small bract; better applied to small bract-like organs arising laterally on the pedicel
Branchlet
The ultimate division of a branch
Bud
An undeveloped stem, leaf, or flower. Buds are often enclosed by reduced or specialized leaves termed bud-scales
Bulb
A short, vertical, underground organ for food-storage or reproduction on which specialized leaves are prominently developed
Bulbil
A bulb arising from a mother bulb, or one produced on an aerial part of the plant and capable, when separated, of producing a new plant
Bulblet
A little bulb; usually applied to the bulb-like structures produced by some plants in the axils of the leaves or replacing the flowers
Bulbiform
Bulb-like
Bulbous
Resembling or suggesting a bulb

Caducous
Falling very early
Caespitose
Growing in dense tufts; usually applied only to plants of small size
Calcareous
Of soil rich in lime
Calcicole
Of a plant growing by preference on soil rich in lime
Calciphile
Plant addicted to calcareous soil
Calciphilous
Lime-loving
Callose
Hardened or thickened
Callus
As used in taxonomy, the swollen nodes of the rachilla in Gramineae
Calyx
The outer series of floral leaves forming the perianth of a flower, often green in colour, frequently enclosing the rest of the flower in bud, occasionally coloured or petal-like, or in some groups of plants greatly reduced or completely lacking
Calyx-lobe
In a gamosepalous calyx, the free projecting parts
Calyx-tube
In a gamosepalous calyx, the basal tubular portion
Cambium
A layer, usually one cell thick, of persistent meristematic tissue
Campanulate
Bell-shaped, usually descriptive of a corolla or calyx
Canaliculate
Channelled longitudinally
Canescent
Pale or grey, because of a fine close pubescence
Capillary
Hair-like
Capitate
Head-like; in a head
Capsule
A dry dehiscent fruit developed from a compound ovary and almost always containing two or more seeds. (Compare follicle)
Carnivorous
Feeding on animal tissue
Carpel
A sporophyll which bears ovules, often regarded as homologous with a single leaf. (Compare simple, compound)
Castaneous
Of chestnut colour

Catkin
A dense bracteate spike or raceme bearing many small, naked or apetalous flowers
Caudate
Having a tail-like terminal appendage
Caudex
The thickened base of a perennial plant
Caulescent
Having a well-developed stem above ground
Cauline
Situated on or pertaining to the stem
Chaff
The receptacular bracts of many Compositae
Channelled
Grooved longitudinally
Chartaceous
Papery in texture
Chlorophyll
The substance which gives the green colour to plants
Cilia
Marginal hairs
Ciliate
With marginal hairs
Ciliolate
Diminutive of ciliate
Cinereous
Ashy in appearance or colour
Circumscissile
Dehiscent by a transverse cleft usually extending completely around the organ
Clasping
Partly surrounding another organ at base
Clavate
Club-shaped, gradually increasing in diameter toward the summit
Claw
The narrow basal portion of some sepals and petals
Cleft
Deeply lobed, probably to below the middle. There is no sharp distinction between cleft, lobed, meaning less deeply cut, and parted, meaning more deeply cut
Cleistogamous
Descriptive of a flower which does not open
Coetaneous
In *Salix*, indicating that flowers and leaves appear at the same time
Coma
A tuft of soft hairs, usually terminal on a seed
Compound (leaf)
Composed of two or more separate leaflets
Conduplicate
Folded together lengthwise with the upper surface within, as in the blades of many grasses
Cone
A globose to cylindric arrangement of crowded bracts or scales subtending reproductive organs and usually hard or woody or long persistent; a structure of similar appearance although possibly of different morphological nature
Connate
Grown together or attached; applied only to like organs, as filaments connate into a tube or leaves connate around the stem. (Compare adnate)
Contiguous
Connected
Convex
Having a more or less rounded surface
Convolute
Rolled or twisted together when in an undeveloped stage
Cordate
Heart-shaped; applied sometimes to whole organs but more often to the base only
Cordilleran
Pertaining to the mountains of western North America
Coriaceous
Leathery in texture
Corm
A short, vertical, underground organ for food-storage or reproduction, consisting chiefly of a thickened portion of a stem with poorly developed scale-leaves or more commonly leafless
Corniculate
Horned or provided with horn-like appendages
Corolla
The second set of floral leaves of the perianth, often conspicuous by its size or colour, but in some plants small and inconspicuous, or reduced to nectaries, or lacking
Corona
A structure exhibited in some plants between the corolla and the stamens and often simulating an additional part of the perianth
Coronate
Crowned
Cortex
Bark
Corymb
A type of raceme in which the axis is relatively short and the lower pedicels relatively long, thereby producing a round-topped or flat-topped inflorescence; sometimes loosely applied to any type of flower-cluster of similar shape
Corymbiform
Shaped like a corymb
Corymbose
In a corymb
Costate
Longitudinally ribbed
Cotyledon
A leaf of the embryonic plant within the seed
Creeping
Growing along the surface of the ground and emitting roots at intervals, usually from the nodes
Crenate
Describing a toothed margin, usually of a leaf, the apex of each tooth blunt or rounded
Crenulate
Diminutive of crenate
Crisp
Curled
Crisped
Irregularly curled or crooked (hairs); irregularly curled along the margin (leaf)

Crisp-hairy
With curled hairs
Cruciform
Forming a cross
Culm
The stem of a grass or sedge
Cuneate
Wedge-shaped; narrowly triangular with the acute angle pointed downward
Cupuliform
Cup-shaped
Cusp
A sharp, abrupt, and often rigid point
Cuspidate
With a cusp
Cyme
A type of inflorescence in which each flower is strictly terminal either to the main axis or to a branch. (See raceme and racemose). Cymes assume many forms depending on the number and position of the branches. They are sometimes distinguished with difficulty from a racemose inflorescence, but may often be known by the position of the bracts opposite the base of the pedicel instead of below it.
Cymose
Having the flowers in a cyme

DBH
Diameter breast height
Deciduous
Falling after completion of the normal function. (See caducous, persistent)
Decumbent
Prostrate at base, erect or ascending elsewhere
Decurrent
Extending downward; applied usually to leaves in which the blade is apparently prolonged downward as two wings along the petiole or the stem
Deflexed
Bent abruptly downward
Dehiscence
The process or act of opening, usually in a fruit
Deltoid
Broadly triangular
Dentate
Toothed along the margin, the apex of each tooth sharp (compare crenate) and directed outward (compare serrate)
Denticulate
Diminutive of dentate
Depressed
Flattened
Diadelphous
Descriptive of stamens in which the filaments are united in two fascicles
Dichotomous
Forking more or less regularly into two branches of about equal size
Diffuse
Loosely spreading
Digitate
Having parts diverging from a common base, as the fingers of a hand, usually descriptive of leaflets or parts of an inflorescence

Dilate
Enlarge
Diminutive
Small
Dimorphic
Of two forms
Dioecious
Bearing staminate and pistillate flowers on separate plants; in the lower plants, bearing the male and female gametes on different plants
Discoid
Resembling a disc; in Compositae, denoting a head composed of tubular flowers only
Disc
An enlargement of or an outgrowth from the receptacle, appearing in the center of the flower of various plants. Also, in Compositae, the central part of the head, composed of tubular flowers
Distinct
Separate, not united
Divaricate
Widely spreading from the axis or rachis
Divided
Cut into distinct parts, usually describing a leaf cut to the midrib or to the base
Dolmitic
Containing dolomite
Dorsal
Located on or pertaining to the back of an organ
Downy
Pubescent with soft, fine hairs
Drupe
Fruit with fleshy or pulpy exocarp, the endocarp hard or bony and permanently enclosing a single seed, or with a portion of the endocarp separately enclosing each of several seeds
Drupaceous
Resembling a drupe
Drupelet
A small drupe

Ellipsoid
Shaped more or less like a football
Emarginate
Notched at the apex
Embryo
The rudimentary plant formed in a seed
Emergent
Of water plants only partly submersed
Endemic
Confined geographically to a single area
Entire
With continuous unbroken margin
Epidermis
The superficial layer of cells
Epigynous
Indicating a flower in which the hypanthium or the basal parts of the perianth are adnate to the ovary, the perianth and stamens then appearing to rise from the summit of an inferior ovary

Erect
Growing essentially in a vertical position (whole plant); describing the position of a structure which extends in the same direction as the organ which bears it (part of a plant)
Erose
Irregularly cut or toothed along the margin
Eutrophic
Rich in dissolved nutrients but frequently shallow and with seasonal oxygen deficiency
Evergreen
Remaining green through the winter
Exfoliate
To come off in scales or flakes
Exsert
Projecting out or beyond; often referring to stamens or styles which project beyond the perianth
Extravaginal
Beyond or outside the sheath

Falcate
Having a curved longitudinal axis, applied only to organs which in most plants have a straight axis
Farinose
Covered with a meal
Fascicle
A small bundle or cluster, without reference to the morphological details of arrangement
Fastigiate (branches)
Close together and nearly parallel
Fen
Low peaty land covered wholly or partly with water
Ferrugineous
Rust-coloured
Fertile
Capable of normal reproduction functions, as a fertile stamen produces pollen, a fertile pistil produces ovules, a fertile flower normally produces fruit although it may lack stamens
Fibrillose
Furnished or abounding with fine fibres
Fibrous
Resembling fibres
-fid
A suffix meaning deeply cut
Filament
The basal sterile portion of a stamen below the anther, usually slender, sometimes lacking; any thread-like structure
Filiform
Thread-like
Fimbriate
Fringed
Fistulose
Hollow; lacking pith
Flabellate
Fan-shaped
Flaccid
Flabby; lacking in stiffness
Flagellate
Bearing long slender shoots (as a stolon or runner)
Fleshy
Thick and juicy; succulent
Flexuous
Curved alternately in opposite directions
Floccose
Irregularly covered by tufts or flocks of soft hair or wool
Floret
A small flower, usually one of several in a cluster
Floriferous
Flower bearing
Fluted
Regularly marked by alternating ridges and groove-like depressions
Foliaceous
Leaf-like in flatness, colour, and texture
Foliose
Leafy
Follicle
A dry dehiscent fruit developed from a simple ovary and dehiscent usually along one suture only
Forked
Divided into equal branches
Free
Not adnate to other organs. (Compare distinct)
Frond
The leaf of a fern
Fruit
A ripened ovary, together with such other parts of the plant as are regularly associated with it
Fruticose
Shrub-like
Fulvous
Yellow, tawny
Funnelform
With the tube gradually widening and passing insensibly into the limb, as in *Campanula uniflora*
Fusiform
Descriptive of a solid body, as a fruit, thick near the middle and tapering to both ends

Gamopetalous
Having the petals wholly or partly united
Geniculate
Bent abruptly at the nodes
Glabrate
Essentially without pubescence
Glabrescent
Becoming glabrous
Glabrous
Lacking pubescence; smooth
Gland
A secreting organ, in plants usually producing nectar or volatile oil and either internal or external
Glandular
Containing or bearing glands
Glaucous
Grey, greyish green, bluish green with a thin coat of fine removable particles often waxy in nature
Globular
Spherical
Glomerate
In a compact cluster
Glomerule
A compact head-like cyme

Glume
A certain type of bract in the Gramineae
Glutinous
Covered with a sticky substance
Grain
A single small hard seed (usually of a grass)
Granulate
Giving the impression of roughness
Gynaecandrous
With staminate and pistillate flowers in the same spike, the pistillate at the apex

Habit
The general appearance of a plant
Habitat
The kind of place in which a plant grows, such as bogs, woods, etc.
Hair
An epidermal appendage, usually slender, simple, or variously branched
Halberd-shaped
Having two divergent basal lobes
Halophilous
Adapted to growth in saline soil
Halophyte
Plant adapted to growth in saline soil
Hapaxanthic
Flowering but once
Hastate
Having two divergent basal lobes
Head
A dense flower-cluster, composed of sessile or nearly sessile flowers crowded on a short axis
Heath
One of several shrubs belonging to the family Ericaceae
Helmet
The hood-shaped fused upper petals of some Scrophulariaceae
Herb
A plant, either annual, biennial, or perennial, with stems dying back to the ground at the end of the growing season
Herbaceous
Dying back to the ground at the end of the growing season (plant); leaf-like in colour or texture (part of a plant)
Hermaphrodite
With stamens and pistils in the same flower (a perfect flower)
Heterosporous
Producing microspores and megaspores
Hirsute
Pubescent with spreading hairs
Hirtellous
Softly or minutely hirsute or hairy
Hispid
Pubescent with stiff spreading hairs
Hispidulous
Diminutive of hispid
Hoary
Greyish-white, close pubescence
Homosporous
Producing asexual spores of only one kind
Humifuse
Spreading over the ground
Hyaline
Translucent or transparent
Hybrid
A cross between two species
Hypanthium
An expansion of the receptacle forming a saucer-shaped, cup-shaped, or tubular organ, often simulating a calyx-tube and bearing the sepals, the petals, and often the stamens at or near its margin
Hypogynous
Appearing on the receptacle beneath the ovary and free from it or from the calyx; term usually applied to petals and stamens

Imbricate
Literally, shingled; in botany, overlapping, either in width only, as the sepals or petals of various plants, or in both width and length, as the involucral bracts of many species of Compositae
Immersed
Growing completely under water
Incised
Deeply and irregularly cut
Incurved
Curved inward
Indehiscent
Not opening at maturity, usually applied to fruits
Indigenous
A native, not introduced
Indument
The epidermal appendages of a plant or an organ considered collectively, as its pubescence
Indurate
Hardened
Indusium
In ferns, an outgrowth of the frond wholly or partly covering the sorus
Inferior
Descriptive of an ovary adnate to the hypanthium or to the lower parts of the perianth and therefore appearing to be located below the flower at the summit of the pedicel
Inflorescence
A complete flower-cluster, including the axis and bracts
Innovation
An offshoot, usually from the base of a stem
Inserted
Attached to, referring to the point of origin of an organ
Insipid
Lacking taste or savour to such a degree as to be unpleasing to the palate
Internode
The portion of a stem between one node and the next
Involucel
Diminutive of involucre; an involucre subtending a definite part of an inflorescence only, in contrast with the involucre which subtends the whole
Involucral
Pertaining to an involucre

Involucrate
With an involucre
Involucre
A set of bracts closely associated with each other and subtending an inflorescence
Involucrum
Involucre
Involute
Rolled in, so that the lower side of the organ is exposed and the upper concealed. (Compare revolute)
Irregular
Descriptive of a flower in which the members of one or more sets of organs (usually the corolla) differ among themselves in size, shape, or structure; synonymous with zygomorphic

Keel
A sharp or conspicuous longitudinal ridge; also the two lower united petals in the Leguminosae

Lacerate
Torn; describing a flat organ, as a leaf or petal, with irregularly jagged margin
Laciniate
Deeply cut into narrow segments
Lacustrine
Living in lakes
Lanate
Woolly
Lance-attenuate
Lanceolate, with the tip slenderly tapering
Lance-oblong
Lanceolate and oblong
Lanceolate
Shaped like a lance-head, much longer than wide and widest below the middle
Lateral
Situated on or arising from the side of an organ, as a lateral inflorescence
Latex
The milky juice of some plants
Lax-flowered
Flowers loose or distant
Leaflet
A single segment of a compound leaf
Legume
A dry dehiscent fruit derived from a simple ovary and usually dehiscing along two sutures
Lemma
A certain type of bract in the Gramineae
Lenticular
Lens-shaped
Ligneous
Woody
Ligulate
Having a ligule; having the nature of a ligule
Ligule
A small, usually flat outgrowth from an organ, as seen at the junction of claw and blade in the petals of some Caryophyllaceae or at the junction of sheath and leaf-blade in Gramineae; the ligulate corolla of many species of Compositae

Limb
The expanded part of a gamopetalous corolla; the expanded part of any petal or leaf
Linear
Narrow and elongate with essentially parallel sides
Linear-capillary
Finely linear
Linear-subulate
Linear, with an awl-shaped tip
Lingulate
Tongue-shaped
Lip
Either portion of the limb of a bilabiate corolla or calyx, distinguished as upper lip and lower lip; the odd petal (usually the lowest) in Orchidaceae
Littoral
Growing on sea beaches
Lobe
A partial division of an organ such as a leaf. The term generally applies to a division less than half-way to the base of the midrib
Lobate
Divided into or bearing lobes
Locule
A cavity or one of the cavities within an ovary, a fruit, or an anther; used by many in preference to the older term cell
Loment
A legume composed of one-seeded articles
Lunate
Crescent-shaped
Lustrous
Glossy, shiny
Lyrate
Pinnately lobed with the terminal lobe the largest

Macrospore
The larger of the two kinds of spores in *Selaginella* and related plants
Marcescent
Withering and persistent, usually applied to petals or stamens after anthesis
Mealy
Covered with meal or with fine granules
Median
Pertaining to the middle
Megaspore
In some Pteridophytes, the spore which produces the female gametophyte. The term is also used for a similar spore in Spermatophytes
Megasporophyll
The organ upon which or within which megaspores are produced
Membranaceous, Membranous
Thin and flexible, as an ordinary leaf, in contrast to chartaceous, coriaceous, or succulent
Microspore
In some Pteridophytes, the spore from which the male gametophyte is developed
Microsporophyll
In some Pteridophytes, the organ upon which microspores are produced

Midrib
The median or central rib of a leaf
Minerotrophic
Rich in minerals, as applied to a bog
Monocephalus
Single headed
Monochrome
Having or consisting of one colour or hue
Moniliform
Literally, necklace-like; constricted at regular intervals
Monoecious
Bearing both staminate and pistillate flowers but not perfect ones
Moraine
An accumulation of earth and stones carried and finally deposited by a glacier
Mucilaginous
Slimy, composed of mucilage
Mucro
A short, sharp, slender point
Mucronate
Tipped with a short, sharp, slender point
Multicipital
Literally, many-headed; descriptive of a crown of roots or a caudex from which several stems arise
Muricate
Descriptive of a surface beset with small sharp projections
Mycorrhiza
The association of certain fungi with the roots of certain seed-plants

Naked
Lacking various organs or appendages, almost always referring to organs or appendages which are present in other similar plants, as a naked flower lacks perianth
Nectary
A gland which secretes nectar, usually on the corolla or disk or within the spur of a flower
Nerve
A prominent longitudinal vein of a leaf or other organ. The adjective nerved is often used as a suffix, as three-nerved
Nitrophilous
Preferring or thriving in a soil rich in nitrogen
Nodal
Located at or pertaining to a node
Node
A point on the stem from which leaves or branches arise, characterized internally by certain anatomical features
Nodose
Knotty or Knobby
Nodulose
Provided with little knots or knobs
Nut
A hard, dry, indehiscent, one-seeded fruit or part of a fruit
Nutlet
A small nut, loosely distinguished only by its size and scarcely separable from an achene except by the comparative thickness of its wall

Ob-
A prefix, signifying in a reverse direction, usually attached to an adjective indicating shape
Obcordate
Inverted heart-shaped, the notch being apical
Obcuneate
Inversely cuneate
Oblanceolate
Lanceolate with the broadest part above the middle
Oblique
Slanting, unequal-sided
Oblong
Descriptive of a flat organ broader than linear but maintaining its width with little change for a considerable part of its length. Also describing a solid object, such as a fruit or seed, which is essentially cylindric or prismatic and therefore appears oblong when viewed from the side
Obpyramidal
Inversely pyramidal
Obovate
Reversed ovate, the distal end broader
Obovoid
Appearing as an inverted egg
Obtuse
Blunt
Ocrea
A sheath around the stem just above the base of a leaf and derived from the stipules; used chiefly in the Polygonaceae
Offset
A short prostrate or ascending shoot, usually propagative in function, arising near the base of a plant
Olivaceous
Olive green; olive-coloured
Opposite
Situated diametrically opposite each other at the same node, as leaves, flowers, or branches; situated directly in front of another organ, as stamens opposite the petals
Orbicular
Essentially circular
Oval
Broadly elliptic
Ovary
The basal, usually expanded portion of a pistil within which the ovules are borne
Ovate
Descriptive of a flat organ widest below the middle and broader than lanceolate
Ovoid
Egg-shaped
Ovule
A reproductive organ within the ovary in which the female cell is produced and which after further development becomes a seed

Palea
A certain type of bract in the Gramineae
Palmate
With three or more lobes or nerves or leaflets or branches arising from one point; essentially synonymous with digitate

Paludal
Of or growing in marshes
Panicle
A compound or branched inflorescence of the racemose type; often applied to any compound inflorescence which is loosely branched and longer than thick
Panicled, Paniculate
Arranged in a panicle
Pannose
With the texture of felt
Papillae
Minute nipple-shaped projections
Papillate
Bearing minute nipple-shaped projections
Papilliform
Shaped like a papilla, which is a short, blunt, rounded, or cylindric projection of small size
Papillose
Descriptive of a surface beset with short, blunt, rounded, or cylindric projections; papillose hairs are slender above a papilliform base
Pappus
An outgrowth of hairs, scales, or bristles from the summit of the achene of many species of Compositae
Parasite
A plant which derives its food and water wholly or chiefly from another plant to which it is attached. (Compare epiphyte)
Parietal
Located on the inner side of the exterior wall of the ovary
Parthenogenetically
Developing without fertilization
Pectinate
Literally, comb-like; pinnatifid into narrow segments of uniform size and resembling a comb; closely ciliate with comparatively large or stiff and parallel hairs
Pedate
Palmately lobed with the lateral segments again divided
Pedicel
The stalk of a single flower in an inflorescence
Pedicellate
Born on a pedicel
Peduncle
The portion of a stem which bears an inflorescence or a solitary flower, either leafless or with bracts
Pedunculate
With a peduncle
Pellucid
Clear, transparent
Peltate
Attached by the surface instead of the margin
Pendulous
Hanging or drooping
Pentamerous
With parts in fives, as a corolla of five petals
Perennating
Living over from season to season
Perennial
Living several years
Perfect
Descriptive of a flower containing stamens and pistils
Perfoliate
Descriptive of a sessile leaf with its basal portion continuous around the stem
Perianth
The corolla and calyx considered together, or either of them if the other is lacking
Pericarp
The wall of the fruit
Perigynium
A special type of bract in *Carex*, as described in the text
Perigynous
Describing a flower with developed hypanthium or calyx-tube which is free from the ovary and bears the sepals or calyx-lobes, petals, and stamens at its margin or on its inner surface
Persistent
Remaining attached after the normal function has been completed
Petal
A separate segment of the corolla
Petaloid
Having the character or appearance of a petal
Petiolate
With a petiole
Petiole
The basal stalk-like portion of an ordinary leaf, in contrast with the expanded blade
Phyllaries
The involucral bracts in the Compositae
Phyllode
An expanded bladeless petiole
Phyllodial
Having the character of a phyllode
Phytogeographic
Related to the science of plant distribution
Pilose
Rather sparsely beset with straight spreading hairs
Pilosulous
Diminutive of pilose
Pingo
A dome-shaped mound of earth resulting from frost action, found in considerable numbers on the coastal plain east of the Mackenzie River Delta
Pinna
One member of a pinnatifid or pinnately compound organ
Pinnate
Having branches or lobes or leaflets or veins arranged on two sides of a rachis
Pinnatifid
With lobes, clefts, or divisions pinnately arranged
Pinnule
Diminutive of pinna, usually applied to a segment of a bipinnatifid or decompound leaf
Pistil
The innermost or central organ or organs of a flower, composed typically of ovary, style, and stigma
Pistillate
Having a pistil; usually applied to flowers which lack stamens

Placenta
The point or place of attachment of the ovules to the ovary or of the seeds to the fruit
Placentation
The disposition of placenta
Plano-convex
Flat on one side and convex on the other
Pleiocephalus
With several heads
Plumose
Feathery; applied to a slender organ or structure, such as a style, with dense pubescence
Pod
Strictly, a legume; loosely, often a synonym of capsule
Pollen
The spores borne within the anther which produce the male cells
Polymorphous
Variable as to form or habit
Pollen-tube
The slender tube which develops from a pollen-grain, penetrates the tissue of the ovary, and enters the ovule
Pollinate
To transfer pollen from a stamen to a stigma
Pollination
The act or process of pollinating
Pollinium
A mass of coherent pollen, as developed in Asclepiadaceae and Orchidaceae
Polygamous
Bearing partly perfect, partly unisexual flowers
Polypetalous
Composed of or possessing separate petals
Polysepalous
Composed of or possessing separate sepals
Precocious
Denoting flowers which appear in advance of the leaves
Prismatic
Of the shape of a prism
Procumbent
Prostrate or trailing, but not rooting at the nodes
Proliferous
Reproducing freely by offsets, bulbils, or other vegetative means
Prostrate
Flat on the ground
Pruinose
With a powdery, waxy secretion on the surface, i.e. a "bloom"
Puberulent, Puberulous
Minutely or sparsely pubescent
Pubescence
An indument of hairs, without reference to structure
Pubescent
Bearing hairs on the surface
Pulvinate
Cushion-shaped
Puncticulate
Minutely punctate
Punctiform
Marked or composed of points or dots
Punctate
Dotted; usually denoting the presence of glands either on the surface or within the tissues
Pyramidal
Pyramid-shaped
Pyriform
Pear-shaped
Pustule
Any small elevation or spot on a plant resembling a blister

Quagmire
Soft wet miry land that shakes or yields underfoot
Quinate
Five-parted

Raceme
A common type of inflorescence with an elongate unbranched axis and lateral flowers, the lowest opening first.
Racemose
A general type of inflorescence in which all flowers are axillary and lateral, the axis therefore theoretically capable of indefinite prolongation. (Compare cymose)
Rachilla
Literally, a little rachis; specifically, the rachis of a spikelet in the Gramineae and some Cyperaceae
Rachis
The central portion of a compound organ bearing its separate divisions laterally in one or two rows, as the rachis of a compound leaf; to be distinguished from axis, which bears the divisions radially. The two terms are often used loosely and interchangeably
Radiant, Radiate
Spreading from or arranged around a common centre
Radical
Belonging to the root
-ranked
With a numerical prefix, indicating the number of longitudinal rows in which leaves or other structures are arranged along an axis or rachis
Ray
The ligule or strap-like marginal flower in Compositae
Receptacle
The end of a pedicel or one-flowered peduncle which bears the floral organs, also known as torus. In Compositae, the apex of the pedicel upon which the flowers are inserted
Recurved
Curved backward
Reflexed
Bent backward
Regular
Describing a flower in which the members of each circle of parts are similar in size and shape
Remote
Scattered, not close together
Reniform
Kidney-shaped; wider than long, rounded in general outline, and with a wide basal sinus
Repand
With a shallowly sinuate or slightly wavy margin

Repent
Creeping or prostrate and rooting at the nodes
Resinous
Having resin
Reticulate
Netted; usually referring to the network of veins in a leaf
Retrorse
Directed backward
Retuse
With a small terminal notch in an otherwise rounded or blunt apex
Revolute
Rolled backward, so that the upper surface of the organ is exposed and the lower side more or less concealed
Rhizomatous
Having a rhizome
Rhizome
An underground stem, usually horizontal in direction, usually emitting roots from the lower side and leafy stems from the upper
Rhombic
With the outline of an equilateral oblique-angled figure
Rib
A primary and prominent vein of a leaf
Riparian
Growing by rivers or streams
Roseate
Rose-coloured
Rosette
A cluster of leaves crowded on very short internodes, often basal in position
Rostrate
Beaked
Rosulate
Arranged in a rosette
Rotate
Wheel-shaped; descriptive of a gamopetalous corolla or gamosepalous calyx widely spreading, without a contracted tube or with only a short and inconspicuous tube
Rotund
Essentially circular
Rufidulous
Somewhat reddish
Rugose
Describing a wrinkled surface; in leaves, usually depressed along the veins and veinlets and elevated between them
Rugulose
Diminutive of rugose
Runcinate
Sharply incised, with the segments pointing backward
Runner
An elongate, slender, prostrate branch taking root at the nodes or tip

Saccate
Sac-like or dilated
Sagittate
Arrow-shaped; lanceolate or triangular in outline with two retrorse basal lobes
Salverform
Descriptive of a gamopetalous corolla (or calyx) with well-developed slender tube and abruptly widely spreading limb
Saprophyte
A plant without green colour, deriving its food from organic material in the soil by mycorrhiza or otherwise
Scaberulous, Scabrellate, Scabrid
Diminutives of scabrous
Scabrous
Rough to the touch, due to the structure of the epidermis or to the presence of short stiff hairs
Scale
Any small thin or flat structure; also a single bract of the involucre in Compositae
Scape
A peduncle with one or more flowers arising directly from the ground or from a very short stem, and either leafless or with bracts only
Scapose
Arranged on or borne on a scape
Scarious
Thin and chaffy in texture and not green
Scorpioid
Describing an inflorescence usually cymose in structure; spiciform or racemiform in shape, and coiled at the tip before anthesis
Scree
Debris from a landslide
Scurfy
Covered with scale-like particles
Seed
A ripened ovule
Sepal
One separate segment of the calyx
Sepaloid
Sepal-like, usually green and thicker in texture than a petal
Septate
Provided with a septum
Septicidal
Descriptive of a capsule which dehisces along or through the septa which separate its locules or cells
Septum
A partition within an organ, as the septa of an ovary or of the leaf of a *Juncus*
Seriate
Disposed in a series of rows, either transverse or longitudinal
Sericeous
Silky, due to the presence of numerous soft appressed or ascending hairs
Serrate
Toothed along the margin, the apex of each tooth sharp (compare crenate) and directed forward (compare dentate)
Serrulate
Diminutive of serrate

Sessile
Without a petiole, petiolule, pedicel, peduncle, stipe, or other type of stalk
Seta
A bristle
Setaceous
Bristle-like or bristle-shaped
Setose
Beset with setae or bristles
Sheath
An organ which wholly or partly surrounds another organ at base, as the sheathing leaf of a grass
Silicle
A short silique
Silique
A special type of capsule in the Cruciferae, in which the two valves separate from a thin longitudinal partition known as the replum
Simple
Descriptive of a pistil organized from a single carpel and therefore one-celled with a single style and stigma; the term is also applied to the ovary alone; descriptive of a leaf with a single blade
Sinuate
Having a wavy margin
Sinus
The space or position (not a structure) between two lobes or other divisions
Slough
A wet or marshy depression
Solifluction
The slow creeping of wet soil and other saturated fragmented material down a slope
Sorus
A cluster of sporangia; used especially for flowerless plants
Spadix
A form of spike or head with a thick or fleshy axis
Spathe
A large, usually solitary bract subtending and often enclosing an inflorescence; the term is used only in the Monocotyledons
Spatulate
Shaped like a spatula, maintaining its width or somewhat broadened toward the rounded summit
Spicate
Arranged in a spike
Spiciform
Having the form of a spike but not necessarily the technical structure
Spike
An elongate inflorescence of the racemose type with sessile or subsessile flowers; the term is often loosely applied to an inflorescence of different morphological nature but of similar superficial appearance
Spikelet
A small or secondary spike subtended by a common pair of glumes or bracts as in grasses
Spine
A thorn
Spinule
A little thorn
Spinulose
Having little thorns
Spiral
Describing the arrangement of like organs, such as leaves, at regular angular intervals
Sporangium
An organ in which spores are produced. An anther and an ovule are sporangia, but the term is rarely used except in flowerless plants
Sporadic
Occurring here and there without continuous range
Spore
A one-celled asexual reproductive organ. The term is used almost exclusively in flowerless plants
Sporophyll
A specialized organ for the production of spores. Those of flowering plants (pistil, stamen) are often considered to be homologous with leaves
Spur
A hollow appendage projecting from the corolla or the calyx and usually nectarial in function
Squarrose
Spreading or recurved at the tip
Stamen
A member of the third set of floral organs, typically composed of anther and filament
Staminate
Bearing stamens; usually applied to a flower lacking pistils
Staminode, Staminodium
A sterile structure occupying the position of a stamen
Standard
The uppermost petal in a typical flower of the Leguminosae
Stellate
Star-shaped; usually applied to branched hairs
Stellulate
Finely stellate
Stem
A major division of the plant-body in contrast to root and leaf, distinguished from both by certain anatomical features and commonly also by general aspect
Sterigma
As used here, a very short persistent stipe (Plural, sterigmata)
Sterile
Unproductive
Stigma
The terminal (or by asymmetric growth occasionally lateral or even basal) portion of a pistil, adapted for the reception and germination of pollen
Stigmatic
Like or pertaining to a stigma
Stipe
The stalk of a structure or organ, applied only where the terms petiole, petiolule, pedicel, or peduncle cannot be used, as the stipe of an ovary
Stipitate
With a stipe or stalk
Stipular
Pertaining to or located on a stipule
Stipulate
With stipules

Stipules
A pair of small structures at the base of the petiole of certain leaves, varying from minute to foliaceous and from caducous to persistent
Stolon
A horizontal branch arising at or near the base of a plant and taking root and developing new plants at the nodes or apex
Stoloniferous
Producing stolons
Stramineous
Straw-coloured
Striate
Marked with fine and usually parallel lines
Strigose
Describing a type of pubescence in which the hairs are closely appressed to the surface and point in one direction (usually antrorse)
Strigulose
Diminutive of strigose
Strobile
An inflorescence resembling a spruce or fir cone, partly made up of imbricated scales
Style
The attenuated part of a pistil connecting the stigma to the ovary
Sub-
A prefix to many adjectives, meaning more or less, or somewhat
Subcoriaceous
Somewhat leathery in appearance or texture
Subligneous
Almost woody
Subopposite
Almost opposite
Subtend
To stand below and close to, as a bract below a flower or a leaf below a bud
Subulate
Awl-shaped
Succulent
Fleshy and juicy
Suffrutescent
Somewhat shrubby, or shrubby at base
Superior
Describing an ovary occupying a terminal or central position in the flower and not adnate to other floral organs

Talus
Rock debris at the base of a cliff or slope
Tendril
A portion of a stem or leaf modified to serve as a holdfast organ
Tentacle
A sensitive glandular hair, as those on the leaf of *Drosera*
Terete
Circular or essentially so in cross-section
Ternate
In threes
Terrestrial
Growing in the soil in distinction from growing in water or other habitats
Testa
The outer covering of a seed
Thallus
A plant body not clearly differentiated into stem and leaf and often also without roots or rhizoids
Thyrse
A compound inflorescence composed of cymes racemosely arranged; also commonly but loosely used to designate a compact panicle
Thyrsoid
With the appearance of a thyrse
Tomentose
Woolly, with an indument of crooked matted hairs
Tomentulose
Diminutive of tomentose
Tomentum
An indument of crooked matted hairs
Torulose
Cylindrical with contractions at regular intervals
Trailing
Prostrate but not rooting
Translucent
Partly transparent
Trichome
A hair-like outgrowth of the epidermis
Trichotomous
Forking regularly by threes
Trifid
Divided into three parts
Trifoliate
Having three leaflets
Trigonous
Three-angled
Truncate
With the base or apex transversely straight or nearly so, as if cut off
Tuber
A thickened portion of a rhizome or root, serving for food storage and often also for propagation
Tubercle
A minute swollen or tuber-like structure, usually distinct in color or texture from the organ on which it is borne, as the tubercle on the achene of *Eleocharis*; a nodule containing bacteria, as on the roots of Leguminosae
Tuberculate
Bearing tubercles
Tumid
Swollen
Turbinate
Top-shaped
Turgid
Swollen
Turion
A scaly, often thick and fleshy, shoot produced from a bud on an underground rootstock
Tussock
A tuft of grass or grass-like plants

Ubiquitous
Occurring everywhere

Umbel
A racemose type of inflorescence with greatly abbreviated axis and elongate pedicels. In a compound umbel the branches are again umbellately branched at the summit
Umbellate
Arranged in umbels
Umbellet
One of the small umbels collectively composing a compound umbel
Umbelliform
Resembling an umbel in appearance
Undulate
Wavy-margined
Unisexual
Bearing stamens or pistils but not both
Urceolate
Urn-shaped; descriptive of a gamopetalous corolla somewhat contracted at the throat and lacking a prominent limb
Utricle
A small bladdery sac enclosing an achene, as in *Carex*

Valvate
Opening by valves
Valve
One of the portions of the wall of a capsule into which it separates at dehiscence; in anthers opening by pores, the portion of the anther-wall covering the pore
Vascular bundle
A strand of wood fibers and associated tissues
Vein
Any of the vascular bundles externally visible, as in a leaf
Veinlet
A small vein
Velutinous
Velvety
Ventral
Situated on or pertaining to the lower side of a flat organ or to the adaxial side of a stamen or carpel
Vernal
Appearing in the spring
Verrucose
Warty; covered on the surface with low rounded protuberances
Verticil
A whorl of leaves or flowers
Verticillate
Arranged in whorls
Vexillum
The standard or large posterior petal in flowers of the Leguminosae
Villose
Covered with fine long hairs but not matted (compare tomentose) or only obscurely matted
Villosity
A villous indument
Villosulous
Diminutive of villous
Villous
Same as villose
Viscid
Sticky
Viviparous
Of adventitious buds sprouting or germinating on the parent plant

Whorl
A circle of three or more leaves or branches or pedicels arising from one node
Wing
Any flat structure emerging from the side or summit of an organ; also the lateral petals in Leguminosae
Winter bud
A shortened and crowded, hibernating vegetative shoot
Wintergreen
Remaining green throughout the winter

Zygomorphy
The type of symmetry exhibited in most irregular flowers, the upper half unlike the lower, the left half a mirror image of the right

Baldwin, W.K.W.
(1953). Botanical investigation in the Reindeer–Nueltin lakes area, Manitoba.
Nat. Mus. Canada Bull. 128: 110-142.

Barkley, T.M.
(1962). Revision of *Senecio aureus* L. and allied species. Trans. Kansas Acad. Sci. 65 (3&4): 318-408.

Barneby, R.C.
(1952). A revision of the North American species of *Oxytropis* DC. Proc. Calif. Acad. Sc. 27: 177-312.

Bassett, I.J., and C.W. Crompton
(1973). The genus *Atriplex* (Chenopodiaceae) in Canada and Alaska. III. Three hexaploid annuals: *A. subspicata, A. gmelinii* and *A. alaskensis*. Can. J. Bot. 51: 1715-1723.
(1978). The genus *Suaeda* (Chenopodiaceae) in Canada. Can. J. Bot. 56: 581-591.

Benson, Lyman
(1948). A treatise on the North American *Ranunculi*. Am. Midl. Nat. 40: 1-261.

Bliss, L.C., and R.W. Wein
(1972). Plant community responses to disturbances in the Western Canadian Arctic. Can. J. Bot. 50: 1097-1109.
(1972). Botanical studies of natural and man modified habitats in the eastern Mackenzie Delta Region and the Arctic Islands. ALUR 71-72-14, Dept. Indian & Northern Affairs, Ottawa.

Böcher, Tyge W.
(1950). The *Luzula multiflora* Complex. Medd. o. Grönl. 147: 9-23.

Boivin, B.
(1948). Centurie de Plantes Canadiennes. Nat. Can. 75: 202-227.
(1966). Enumeration des plantes du Canada II. Nat. Can. 93: 371-437.
(1972). Flora of the Prairie Provinces. Phytologia 23: 13 (Provancheria 4: 97).

Boivin, B., and D. Löve
(1960). *Poa agassizensis*, a new prairie bluegrass. Nat. Can. 87: 173-180.

Bowden, W.M.
(1960). Chromosome numbers and taxonomic notes on northern grasses II Tribe *Triticeae*. Can. J. Bot. 38: 117-131.
(1961). Chromosome numbers and taxonomic notes on northern grasses IV. Can J. Bot. 39: 123-138.
(1967). Taxonomy of intergeneric hybrids of the tribe *Triticeae* from North America. Can. J. Bot. 45: 711-724.

Bowden, W.M., and W.J. Cody.
(1961). Recognition of *Elymus sibiricus* L. from Alaska and the District of Mackenzie. Bull. Torrey Bot. Club 88: 153-155.

Calder, J.A.
(1952). Notes on the genus *Carex* I. Rhodora 54: 246-250.

Calder, J.A., and D.B.O. Savile
(1960). Studies in Saxifragaceae III. *Saxifraga odontoloma* and *lyallii* and North American subspecies of *S. punctata*. Can. J. Bot. 38: 409-435.

Clarke, C.H.D.
(1940). A biological Investigation of the Thelon Game Sanctuary. Nat. Mus. Canada Bull. 96: 1-135.

Cody, W.J.
(1953a). A plant collection from the west side of Boothia Isthmus, N.W.T. Canada. Can. Field-Nat. 67: 40-43.
(1953b). *Phyllodoce coerulea* in North America. Can. Field-Nat. 67: 131-134.
(1954a). New plant records from Bathurst Inlet, N.W.T. Can. Field-Nat. 68: 40.
(1954b). A history of *Tillaea aquatica (Crassulaceae)* in Canada and Alaska. Rhodora 56: 96-101.
(1954c). Plant Records from Coppermine, Mackenzie District, N.W.T. Can. Field-Nat. 68: 110-117.
(1956). New plant records for northern Alberta and southern Mackenzie District. Can. Field-Nat. 70: 101-130.
(1960). Plants of the vicinity of Norman Wells, Mackenzie District, N.W.T. Can. Field-Nat. 74: 71-100.
(1961). New plant records from the upper Mackenzie River Valley, Mackenzie District, N.W.T. Can. Field-Nat. 75: 55-69.
(1963a). A contribution to the knowledge of the flora of southwestern Mackenzie District, N.W.T. Can. Field-Nat. 77: 108-123.
(1963b). Some rare plants from the Mackenzie Mountains, Mackenzie District, N.W.T. Can. Field-Nat. 77: 226-228.
(1965a). New plant records from Northwestern Mackenzie District, N.W.T. Can. Field-Nat. 79: 96-106.
(1965b). Plants of the Mackenzie River Delta and Reindeer Grazing Preserve. Canada Department of Agriculture, Plant Research Institute, 56 pp. offset.
(1967). *Elymus sibiricus* (Gramineae) new to British Columbia. Can. Field-Nat. 81: 275-276.
(1971). A phytogeographic study of the floras of the continental Northwest Territories and Yukon. Naturaliste can. 98: 145-158.
(1975). *Scheuchzeria palustris* L. (Scheuchzeriaceae) in Northwestern North America. Can. Field-Nat. 89: 69-71.
(1978). Range extensions and comments on the vascular flora of the Continental Northwest Territories. Can. Field-Nat. 92: 144-150.

Cody, W.J., and J.G. Chilcott
(1955). Plant collections from Matthews and Muskox lakes, Mackenzie District, N.W.T. Can. Field-Nat. 69: 153-162.

Cody, W.J., G.W. Scotter and S.S. Talbot
(In press). Additions to the vascular plant flora of Nahanni National Park, Northwest Territories.

Cody, W.J., and A.E. Porsild
(1967). *Potamogeton illinoensis*, new to Mackenzie District. Blue Jay 25: 28-29.
(1968). Additions to the flora of Continental Northwest Territories, Canada. Can. Field-Nat. 82: 263-275.

Cody, W.J., and S.S. Talbot
(1973). The pitcher plant, *Sarracenia purpurea* L. in the northwestern part of its range. Can. Field-Nat. 87: 318-320.
(1978). Vascular plant range extensions to the Heart Lake area, District of Mackenzie, Northwest Territories. Can. Field-Nat. 92: 137-143.

Corns, I.G.W.
(1974). Arctic plant communities east of the Mackenzie Delta. Can. J. Bot. 52: 1731-1745.

Detling, L.E.
(1939). A revision of the North American species of *Descurainia*. Am. Midl. Nat. 22: 481-520.

Ekman, El.
(1936). In Grøntved, 1936.

Fernald, M.L.
(1928). The genus *Oxytropis* in Northeastern America. Rhodora 30: 137-155.
(1932). The linear-leaved North American species of *Potamogeton*, section *Axillares*. Mem. Gray Herb. Harvard Univ. 3: 1-183.

Flook, D.
(1959). An occurrence of Lodgepole Pine in the Mackenzie District, Can. Field-Nat. 73: 130-131.

Franklin, J.
(1823). Narrative of a journey to the shores of the Polar Sea in the years 1819-20-21-22. London.
(1828). Narrative of a second expedition to the shores of the Polar Sea in the years 1825, 1826 and 1827. London.

Gardner, G.
(1973). Analytic catalogue of plant species from the arctic and subarctic of Quebec and other regions of Canada. Montreal 235 pp. (mimeo).

Gill, D.
(1971). Vegetation and environment in the Mackenzie River Delta, Northwest Territories–a study in sub-arctic ecology. Ph.D. Dissertation, University of British Columbia, Vancouver, B.C.
(1973). Ecological modifications caused by removal of tree and shrub canopies in the Mackenzie Delta. Arctic 26: 95-111.

Grøntved, J.
(1936). Vascular Plants from Arctic North America. Report of the Fifth Thule Expedition, 1921-4. II, 1: 1-93.

Güssow, W.C.
(1933). Contribution to the knowledge of the Flora of Northern Manitoba and the North-Western Territories, Dominion of Canada. Can. Field-Nat. 47: 116-119.

Haglund, G.
(1943). *Taraxacum* in Arctic Canada. Rhodora 45: 337-343.
(1949). Supplementary notes on the *Taraxacum* flora of Alaska and Yukon. Sv. Bot. Tidsk. 43: 107-116.

Harper, F.
(1931). Some plants of the Athabaska and Great Slave Lakes region. Can. Field-Nat. 45: 97-107.

Hearne, S.
(1795). A journey from Prince of Wales Fort, in Hudson's Bay to the Northern Ocean, etc. London. See also a recent edition by J.B. Tyrrell, Champlain Soc. Toronto, 1911; also Tyrrell, J.B. 1934.

Hernandez, H.
(1973a). Natural plant recolonization of surficial disturbances, Tuktoyaktuk Peninsula Region, Northwest Territories. Can. J. Bot. 51: 2177-2196.
(1973b). Revegetation studies: Norman Wells, Inuvik and Tuktoyaktuk, N.W.T. and Prudhoe Bay, Alaska. Appendix V. *In* Interim Report No. 3, 1973. Towards an environmental impact assessment of the portion of the Mackenzie Gas Pipeline from Alaska to Alberta. Winnipeg: Environ. Prot. Board.

Holmgren, P.K., and W. Keuken
(1974). The Herbaria of the world, Part I. *In* Stafleu, F.A. Index Herbariorum. Regnum Veg. 92: 1-397.

Hooker, W.J.
(1829-1840). Flora Boreali-Americana. London.

Hultén, E.
(1937). Outline of the history of arctic and boreal biota during the Quaternary Period. Stockholm. 168 pp.
(1940). The history of botanical exploration in Alaska and Yukon Territories from their discovery to 1940. Bot. Not. 1940: 289-345.
(1940). Two new species of *Salix* from Alaska. Sv. Bot. Tidskr. 34: 73-76.
(1941-50). Flora of Alaska and Yukon. Lunds Univ. Årsskr. Lund.
(1968). Flora of Alaska and neighboring territories. Stanford University Press. Stanford. 1008 pp.

Jeffrey, W.W.
(1961). Notes on plant occurrence along lower Liard River, N.W.T. Nat. Mus. Canada Bull. 171: 32-115.

Kindle, E.M.
(1928). Canada north of fifty-six degrees. Can. Field-Nat. 42: 53-86.

Knaben, G.
(1959). On the evolution of the *Radicatum*-Group of the *Scapiflora* Papavers as studied in 70 and 56 chromosome species. Opera Botanica 3: 1-96.

Lambert, J.D.H.
(1968). Ecology and successional trends of tundra plant communities in the low arctic sub-alpine zone of the Richardson and British Mountains of the Canadian western Arctic. Ph.D. Thesis, University of British Columbia, Vancouver, B.C.
(1972a). Vegetation patterns in the Mackenzie Delta area. Northwest Territories *In* Mackenzie Delta monograph. *Edited by* D.E. Kerfoot. Brock University, St. Catherines, Ont. pp. 51-68.
(1972b). Plant succession on tundra mudflows: preliminary observations. Arctic 25: 99-106.

Larsen, J.A.
(1965). The Vegetation of the Ennadai Lake Area, N.W.T.: Studies in Arctic and Subarctic Bioclimatology. Ecological Monographs 35: 37-59.
(1967). Ecotonal Plant Communities of the Forest Border, Keewatin, N.W.T., Central Canada. ONR Contract No. 1202(07): Tech. Report No. 32. Dept. of Meteorology, University of Wisconsin, Madison, Wisconsin.
(1969). Vegetation of Fort Reliance, Northwest Territories. Can. Field-Nat. 85: 147-178.

Lindsey, A.A.
(1952). Vegetation of the ancient beaches above Great Bear and Great Slave Lakes. Ecology 33: 535-549.

Louis-Marie, Pere
(1961). Dutilliana I-IV, Contrib. de l'Inst. d'Oka. 14: 1-46.

Mackenzie, A.
(1801). Voyages from Montreal on the river St. Lawrence through the continent of North America to the Frozen and Pacific Oceans with a preliminary account of the rise, progress and present state of the Fur Trade of that country. London.

Macoun, J.M., and Th. Holm
(1921). The vascular plants of the Arctic Coast of America west of the 100th meridian. Report of the Canadian Arctic Expedition 1913-18. Vol. 5: Botany. Part A: Vascular Plants: 1A-50A.

Maguire, B.
(1943). A monograph of the genus *Arnica*. Brittonia 4: 386-510.

Malte, M.O.
(1934). *Antennaria* in Arctic America. Rhodora 36: 101-117.

Mulligan, G.A.
(1961). The genus *Lepidium* in Canada. Madrõno 16: 77-90.
(1974). Cytotaxonomic studies of *Draba nivalis* and its close allies in Canada and Alaska. Can. J. Bot. 52: 1793-1801.
(1975). *Draba crassifolia, D. albertina, D. nemorosa* and *D. stenoloba* in Canada and Alaska. Can. J. Bot. 53: 745-751.

Mulligan, G.A., and J.A. Calder
(1964). The genus *Subularia* (Cruciferae). Rhodora 66: 127-135.

Mulligan, G.A., and W.J. Cody
(1968). *Draba norvegica*, disjunct to the Mackenzie District, Northwest Territories, Canada. Can. J. Bot. 46: 1334-1335.

Mulligan, G.A., and A.E. Porsild
(1966). *Rorippa calycina* in the Northwest Territories, Canada. Can. J. Bot. 44: 1105.
(1969). A new species of *Lesquerella* (Cruciferae) in northwestern Canada. Can. J. Bot. 47: 215-216.

Nagy, J.A., A.M. Pearson, B.C. Goski and W.J. Cody
(In press). Range extensions of vascular plants in northern Yukon Territory and northwestern District of Mackenzie, Canada. Can. Field-Nat.

Ostenfeld, C.H.
(1910). Vascular plants collected in arctic North America by the Gjöa Expedition, 1904-06. Vidensk.-Selsk. Skr. 1. Math.-Naturv. Kl. 1908, no. 8.

Packer, J.
(1964). Chromosome numbers and taxonomic notes on Western Canadian and Arctic plants. Can. J. Bot. 42: 473-494.

Pennell, F.W.
(1934). *Castilleja* in Alaska and Northwestern Canada. Proc. Acad. Nat. Sc. Phil. 86: 517-540.

Petitot, E.
(1891). Autour du Grand Lac des Esclaves. Albert Savine. Paris. 369 pp.

Polunin, N.
(1940). Botany of the Canadian Eastern Arctic, Part I Pteridophyta and Spermatophyta. Nat. Mus. Can. Bull. 92: 1-408.

Porsild, A.E.
(1932). Notes on the occurrence of *Zostera* and *Zannichellia* in arctic North America. Rhodora 34: 90-94.
(1939a). *Nymphaea tetragona* Georgi in Canada. Can. Field-Nat. 53: 48-50.
(1939b). Contributions to the flora of Alaska. Rhodora 41: 141-183; 199-254; 262-301.
(1940a & 1940b). Miscellaneous contributions from the National Herbarium of Canada. I. Can. Field-Nat. 54: 54-55. II. ibid. 68-69.
(1942). Miscellaneous contributions from the National Herbarium of Canada, III. Can. Field-Nat. 56: 112-113.

(1943). Materials for a flora of the continental Northwest Territories of Canada. Sargentia 4: 1-79.
(1945a). The so-called *Woodsia alpina* in North America. Rhodora 47: 145-148.
(1945b). The alpine flora of the east slope of Mackenzie Mountains, N.W.T. Nat. Mus. Canada Bull. 101: 1-35.
(1947). The genus *Dryas* in North America. Can. Field-Nat. 61: 175-192.
(1950a). The genus *Antennaria* in Northwestern Canada. Can. Field-Nat. 64: 1-25.
(1950b). Vascular Plants of Nueltin Lake, Northwest Territories. Nat. Mus. Canada Bull. 118: 72-83.
(1951a). Botany of southeastern Yukon adjacent to the Canol Road. Nat. Mus. Canada Bull. 121: 1-400.
(1951b). Two new species of *Oxytropis* from arctic Alaska and Yukon. Can. Field-Nat. 65: 76-79.
(1954). The North American races of *Saxifraga flagellaris* Willd. Bot. Tidsskr. 51: 292-299.
(1955). The vascular plants of the western Canadian Arctic Archipelago. Nat. Mus. Canada Bull. 135: 1-226.
(1957a & 1946). Illustrated flora of the Canadian Arctic Archipelago. Nat. Mus. Canada Bull. 146: 1-209.
(1957b). The genus *Dryas* in North America. Can. Field-Nat. 61: 175-192.
(1961). The vascular flora of an alpine valley in the Mackenzie Mountains, N.W.T. Nat. Mus. Canada Bull. 171: 116-130.
(1963). *Stellaria longipes* Goldie and its Allies in North America. Nat. Mus. Canada Bull. 186: 1-35.
(1965). The genus *Antennaria* in Eastern Arctic and Subarctic America. Bot. Tidsskr. 61: 22-55.
(1966a). Some new or critical plants of Alaska and Yukon. Can. Field-Nat. 79: 79-90.
(1966b). Contributions to the flora of southwestern Yukon Territory. Nat. Mus. Canada Bull. 216: 1-86.
(1974). Materials for a flora of central Yukon Territory. Nat. Mus. Canada Pub. Bot. 4: 1-77.

Porsild, A.E., and W.J. Cody
(1968). Checklist of the vascular plants of Continental Northwest Territories. Plant Research Institute, Canada Agriculture, Ottawa. 102 pp.

Porsild, M.P.
(1939). A new *Taraxacum* from the Mackenzie Delta. Trans. Roy. Soc. Can. 3rd. Ser. Sec. V 33: 29-34.

Preble, E.A.
(1902). A biological investigation of the Hudson Bay region. N. Am. Fauna 22: 1-140.
(1908). A biological investigation of the Athabaska – Mackenzie Region. North American Fauna 27: 1-574.

Raup, H.M.
(1935). Botanical investigations in Wood Buffalo Park. Nat. Mus. Canada Bull. 74: 1-174.
(1936). Phytogeographic Studies in the Athabaska – Great Slave Lake Region I. Catalogue of the vascular plants. Jour. Arn. Arb. 17: 180-315.
(1943). The willows of the Hudson Bay Region and the Labrador Peninsula. Sargentia 4: 81-128.
(1946). Phytogeographic Studies in the Athabaska – Great Slave Lake Region II. Jour. Arn. Arb. 27: 1-85.
(1947). The Botany of Southwestern Mackenzie. Sargentia 6: 1-275.
(1959). The willows of boreal western America. Contr. Gray Herb. Harvard Univ. 185: 3-95.

Richardson, J.
(1851). Arctic Searching Expedition: A Journal of a Boat-Voyage Through Rupert's Land and the Arctic Sea, in Search of the Discovery Ships under Command of Sir John Franklin. London. 2 vols.

Ritchie, J.C.
(1974). Modern pollen assemblages near the arctic tree line, Mackenzie Delta region, Northwest Territories. Can. J. Bot. 52: 381-396.

Rollins, R.C.
(1962). A new crucifer from the Great Slave Lake area of Canada. Rhodora 64: 324-327.

Ross, B.R.
(1862). An account of the botanical and mineral products used by the Chippewayan Indians. Can. Nat. 7: 133.

Russell, F.
(1898). Explorations in the far north. Univ. of Iowa.

Rydberg, P.A.
(1916). North American Flora (*Artemisia*) 34(3): 261-262.

Savile, D.B.O., and J.A. Calder
(1952). Notes on the flora of Chesterfield Inlet, Keewatin District, N.W.T. Can. Field-Nat. 66: 103-107.

Scotter, G.W.
(1966). A contribution to the flora of the Eastern Arm of Great Slave Lake, Northwest Territories. Can. Field-Nat. 80: 1-18.

Scotter G.W., and W.J. Cody.
(1974). Vascular plants of Nahanni National Park and vicinity, Northwest Territories. Naturaliste Can. 101: 861-891.

Seeman, B.C.
(1852-7). The botany of the voyage of H.M.S. Herald under the command of Capt. Henry Kellett R.N.C.B., during the years 1845-51. London. 483 pp.

Sørensen, Th.
(1953). A revision of the Greenland species of *Puccinellia* Parl. Med. om Grønland 136(3): 1-179.

Thieret, J.W.
(1961a). A collection of plants from the Horn Plateau, District of Mackenzie. Can. Field-Nat. 75: 77-83.
(1961b). New plant records for southwestern District of Mackenzie. Can. Field-Nat. 75: 111-121.

(1962). New plant records from District of Mackenzie, Northwest Territories. Can. Field-Nat. 76: 206-208.
(1963a). Botanical survey along the Yellowknife Highway, Northwest Territories, Canada. I. Catalogue of the flora. Sida 1: 117-170.
(1963b). Additions to the flora of the Northwest Territories. Can. Field-Nat. 77: 126.

Tyrrell, J.B.
(1898). Plants collected [by J.W. Tyrrell] between Lake Athabasca and the West coast of Hudson Bay, north Shore of Hudson Straits and Fort Churchill. Ann. Rep. Geol. Surv. Can., IX, (new ser.), pp.205F-218F (1896).

Wein, R.W., and L.C. Bliss
(1973a). Changes in arctic *Eriophorum* tussock communities following fire. Ecology 54: 845-852.
(1973b). Experimental crude oil spills on arctic plant communities. Jour. Appl. Ecol. 10: 671-682.

Wein, R.W., L.R. Hettinger, A.J. Janz and W.J. Cody
(1974). Vascular plant range extensions in the northern Yukon Territory and northwestern Mackenzie District, Canada. Can. Field-Nat. 88: 57-66.

Younkin, W.E.
(1973). Autecological studies of native species potentially useful for revegetation, Tuktoyaktuk Region, N.W.T. *In* Botanical studies in the Mackenzie Valley, Eastern Mackenzie Delta Region, and the Arctic Islands. pp. 50-96.

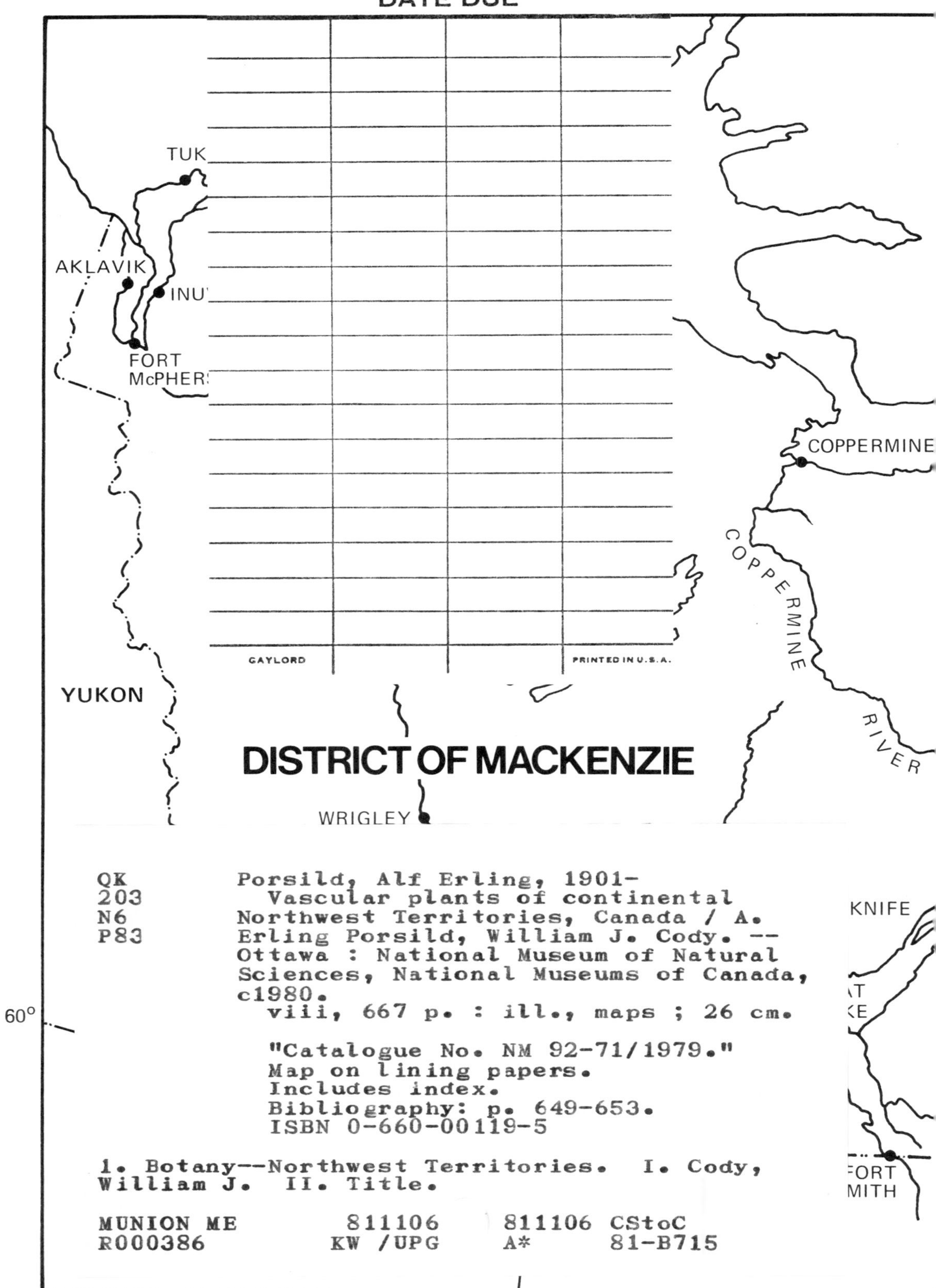
DATE DUE
GAYLORD
PRINTED IN U.S.A.
TUK
AKLAVIK
INU
FORT
McPHERS
COPPERMINE
COPPERMINE
RIVER
YUKON
DISTRICT OF MACKENZIE
WRIGLEY
KNIFE
FORT
MITH
60°